THE HPLC SOLVENT GUIDE
SECOND EDITION

THE HPLC
SOLVENT GUIDE
SECOND EDITION

PAUL C. SADEK

**WILEY-
INTERSCIENCE**

A John Wiley & Sons, Inc., Publication

For ordering and customer service, call 1-800-CALL-WILEY.

Library of Congress Cataloging-in-Publication Data:

CIP data is available

ISBN 0-471-41138-8

Printed in the United States of America.

10 9 8 7 6 5 4 3

CONTENTS

Preface   xiii

Introduction   xv

Abbreviations, Acronyms, and Symbols   xvii

Chapter 1   Physical and Chemical Solvent Properties   1

    1.1   UV Cutoff, 2

    1.2   Absorbance Versus Wavelength Curves, 3

    1.3   Reversed-Phase Solvents, 5

        1.3.1   Mobile Phases and Mobile Phase Modifiers, 9

    1.4   Normal-Phase Solvents, 11

    1.5   System Peaks, 14

    1.6   Lot-to-lot Solvent Variability, 16

    1.7   Viscosity, 19

    1.8   Miscibility and Solubility, 22

    1.9   Buffers and Other Mobile Phase Additives, 25

    1.10 Volatility,   29

    1.11 Solvents: Instability, Reactivity, and Denaturants,   31

        1.11.1   Ethers, 32

        1.11.2   Chlorinated Alkanes, 34

        1.11.3   Ethanol, 36

        1.11.4   Water, 36

        1.11.5   Acetone, 38

    1.12   Sample Stability in Solvents, 39

    1.13   Mobile Phase Interaction with Stationary Phase, 40

    1.14   Inherent Contaminants, 41

    1.15   Solvent Effects on Fluorescence, 42

    1.16   Particulates and Solvent Filtering, 43

    1.17   Manufacturers and Testing Protocols, 44

Chapter 2   Method Optimization   **45**

    2.1   Eluotropic Series and Solvent Strength Parameters, 46
          2.1.1   Eluotropic Series, 46
          2.1.2   Solvent Strength Parameters, 49
          2.1.3   Mathematical Representation of Solvent Strength, 52
    2.2   Triangulation and Other Algorithmic Methods, 53
    2.3   Scout Gradients, 58
    2.4   Window Diagrams, 60
    2.5   Brute-force Method, 60
    2.6   Other Optimization Techniques and Retention Parameters, 61

Chapter 3   Method Validation and Ongoing Performance Evaluation   **63**

    3.1   Qualification Process, 64
          3.1.1   Instrument Qualification, 64
          3.1.2   Design Qualification, 65
          3.1.3   Installation Qualification, 65
          3.1.4   Operational Qualification, 66
          3.1.5   Performance Qualification, 66
    3.2   Method Validation, 67
          3.2.1   Accuracy, 68
          3.2.2   Precision, 68
          3.2.3   Linearity, 69
          3.2.4   Linear and Working Ranges, 71
          3.2.5   Specificity, 72
          3.2.6   Detection and Quantitation Limit, 72
          3.2.7   Robustness, 73
          3.2.8   Ruggedness, 73
    3.3   System Suitability, 74

Chapter 4   Alcohols   **75**

    4.1   Impurities, 81
    4.2   General Analytes, 82
          4.2.1   Simple Substituted Benzene Analytes, 82
          4.2.2   Other Organic Compounds, 89
          4.2.3   Organometallic Compounds and Metal–Ligand Complexes, 93
          4.2.4   Summary, 96
    4.3   Environmentally Important Analytes, 96
          4.3.1   PAHs, Substituted PAHs, and Related Analytes, 96
          4.3.2   Nitrated and Chlorinated Nonpesticide/Herbicide Pollutant
                 Analytes, 100
          4.3.3   Pesticides, Herbicide, and Fungicides, 103
               4.3.3.1   Pesticides, 103
               4.3.3.2   Herbicides, 108
               4.3.3.3   Fungicides, 111
          4.3.4   Summary, 113

4.4   Industrial and Polymer Analytes, 113
    4.4.1   Surfactant and Additive Analytes, 113
    4.4.2   Polymers and Polymer Additives, 115
    4.4.3   Sunscreen Agents, 118
    4.4.4   Dyes, 120
    4.4.5   Other Industrial Analytes, 123
    4.4.6   Summary, 124
4.5   Biological Analytes, 124
    4.5.1   Carboxylic Acid Analytes, 124
    4.5.2   Basic Amine Analytes, 130
    4.5.3   Aflatoxins, Mycotoxins, and Other Toxic Analytes, 133
    4.5.4   Vitamins and Related Analytes, 139
        4.5.4.1   Water-soluble Vitamins and Related
                Compounds, 139
        4.5.4.2   Fat-soluble Vitamins and Related
                Compounds, 140
    4.5.5   Terpenoids, Flavonoids, and Related Compounds, 148
        4.5.5.1   Terpenoids, 148
        4.5.5.2   Flavanoids, 151
        4.5.5.3   Caffeine and Related Compounds, 153
        4.5.5.4   Other Compounds, 155
    4.5.6   Analytes Derived from Oils and Fats, 169
    4.5.7   Nucleotides, Nucleosides, and Related Analytes, 171
    4.5.8   Other Analytes, 174
    4.5.9   Summary, 178
4.6   Amino Acid, Peptide, and Protein Analytes, 178
    4.6.1   Amino Acid Analytes, 178
    4.6.2   Peptide Analytes, 180
    4.6.3   Protein Analytes, 181
    4.6.4   Summary, 184
4.7   Pharmaceutical Analytes, 184
    4.7.1   Drug Surveys and Screening Procedures, 184
    4.7.2   Retention Mechanisms for Drug Compounds, 186
    4.7.3   NSAIDs and Analgesic Drugs, 187
    4.7.4   Antibiotic Drugs, 189
    4.7.5   Anticancer Drugs, 192
    4.7.6   Antiepileptic Drugs, 195
    4.7.7   Steroidal Drugs, 196
    4.7.8   Anthelmintics, 198
    4.7.9   Illicit Drugs, 199
    4.7.10 Antihistamines, 201
    4.7.11 Anti-HIV Drugs, 202
    4.7.12 Antianxiety and Antipsychotic Drugs, 202
    4.7.13 Other Drug Analytes, 204
    4.7.14 Summary, 213

Chapter 5   Alkanes and Alkyl Aromatics   **214**

5.1   Impurities, 216
5.2   General Analytes, 221
       5.2.1   Simple Substituted Aromatic Analytes, 221
       5.2.2   Organometallics and Metal–Ligand Complexes, 226
       5.2.3   Summary, 227
5.3   Environmentally Important Analytes, 227
       5.3.1   PAHs, Substituted PAHs, and Related Analytes, 227
       5.3.2   Nitrated and Chlorinated Nonpesticide/herbicide Pollutant
                Analytes, 230
       5.3.3   Pesticide, Herbicide, and Fungicide Analytes, 231
       5.3.4   Summary, 233
5.4   Industrial and Polymer Analytes, 233
       5.4.1   Surfactant and Additive Analytes, 233
       5.4.2   Other Polymeric Analytes, 235
       5.4.3   Fullerenes, 236
       5.4.4   Summary, 237
5.5   Biological Analytes, 238
       5.5.1   Carboxylic Acid Analytes, 238
       5.5.2   Vitamins and Related Analytes, 239
       5.5.3   Terpenoids, Flavonoids, Steroids, and Related Compounds, 242
       5.5.4   Analytes Derived from Oils and Fats, 244
       5.5.5   Other Analytes, 248
       5.5.6   Summary, 250
5.6   Amino Acid and Peptide Analytes, 250
       5.6.1   Summary, 250
5.7   Pharmaceutical Analytes, 252
       5.7.1   Cardiac Glycosides, 252
       5.7.2   NSAIDs and Analgesic Drugs, 253
       5.7.3   Benzodiazepines, 253
       5.7.4   Other Analytes, 254
       5.7.5   Summary, 257

Chapter 6   Chlorinated Alkanes and Chlorinated Benzenes   **258**

6.1   Impurities, 260
6.2   General Analytes, 266
       6.2.1   Simple Substituted Benzene Analytes, 266
       6.2.2   Organometallics and Metal–Ligand Complexes, 269
       6.2.3   Summary, 269
6.3   Environmentally Important Analytes, 269
       6.3.1   PAHs, Substituted PAHs, and Related Analytes, 269
       6.3.2   Nitrated and Chlorinated Nonpesticide/herbicide Analytes, 271
       6.3.3   Pesticide and Herbicide Analytes, 271
       6.3.4   Summary, 272

6.4 Industrial and Polymer Analytes, 272
    6.4.1 Surfactant and Additive Analytes, 272
    6.4.2 Polymeric Analytes, 273
    6.4.3 Fullerenes, 274
    6.4.4 Summary, 275
6.5 Biological Analytes, 275
    6.5.1 Carboxylic Acid Analytes, 275
    6.5.2 Vitamins and Related Analytes, 275
    6.5.3 Analytes from Fats and Oils, 277
    6.5.4 Other Analytes, 278
    6.5.5 Summary, 280
6.6 Amino Acid and Peptide Analytes, 280
6.7 Pharmaceutical Analytes, 281

Chapter 7   Ethers   **285**

7.1 Impurities, 287
7.2 General Analytes, 288
    7.2.1 Simple Substituted Benzenes and Related Analytes, 288
7.3 Environmentally Important Analytes, 291
    7.3.1 PAHs, Substituted PAHs, and Related Analytes, 291
    7.3.2 Pesticide and Herbicide Residue Analytes, 292
7.4 Industrial and Polymer Analytes, 294
    7.4.1 Surfactant and Additive Analytes, 294
    7.4.2 Polymeric Analytes, 294
    7.4.3 Fullerenes and Other Industrial Analytes, 296
7.5 Biological Analytes, 297
    7.5.1 Carboxylic Acid Analytes, 297
    7.5.2 Vitamins and Related Analytes, 298
    7.5.3 Terpenoids, Flavonoids, Steroids, and Related Analytes, 300
    7.5.4 Analytes Derived from Oils and Fats, 302
    7.5.5 Other Analytes, 303
7.6 Amino Acid and Peptide Analytes, 305
7.7 Pharmaceutical Analytes, 305
    7.7.1 Summary, 311

Chapter 8   Ketones and Esters   **312**

8.1 Impurities, 314
8.2 General Analytes, 316
8.3 Environmentally Important Analytes, 318
8.4 Industrial and Polymer Analytes, 319
8.5 Biological Analytes, 320
8.6 Pharmaceutical Analytes, 325

Chapter 9   Nitriles and Nitrogenous Solvents   **327**

9.1   Impurities, 328

9.2   General Analytes, 329

   9.2.1   General Sample Solvent Considerations, 329

   9.2.2   Simple Substituted Hydrocarbons and Benzene Analyte Retention Studies, 332

   9.2.3   Other Compounds, 337

   9.2.4   Organometallic and Metal–Ligand Complexes, 341

   9.2.5   Summary, 343

9.3   Environmentally Important Analytes, 343

   9.3.1   Substitutes Benzenes and Related Analytes, 343

   9.3.2   PAHs, Substituted PAHs, and Related Analytes, 344

   9.3.3   Nitro-, Nitroso-, and Chlorinated Nonpesticide/herbicide Pollutant Analytes, 348

   9.3.4   Pesticide, Herbicide, and Fungicide Analytes, 350

      9.3.4.1   Pesticides, 350

      9.3.4.2   Herbicides, 363

      9.3.4.3   Fungicides, 368

   9.3.4   Summary, 369

9.4   Industrial and Polymer Analytes, 370

   9.4.1   Surfactant and Additive Analytes, 370

   9.4.2   Polymeric Analytes, 373

   9.4.3   Dyes and Related Analytes, 375

   9.4.4   Other Industrial Analytes, 378

   9.4.5   Personal Care and Cosmetic Analytes, 379

   9.4.6   Summary, 380

9.5   Biological Analytes, 381

   9.5.1   Carboxylic Acid Analytes, 381

   9.5.2   Basic Amine Analytes, 387

   9.5.3   Aflatoxins, Mycotoxins, and Other Toxic Analytes, 395

   9.5.4   Vitamins and Related Analytes, 401

      9.5.4.1   Water-soluble Vitamins and Related Compounds, 401

      9.5.4.2   Fat-Soluble Vitamins and Related Compounds, 402

   9.5.5   Terpenoids, Flavonoids, and Other Naturally Occurring Analytes, 405

      9.5.5.1   Terpenoids, 405

      9.5.5.2   Flavanoids and Related Compounds, 408

      9.5.5.3   Alkaloids and Related Compounds, 418

      9.5.5.4   Fats, Oils, and Related Analytes, 425

   9.5.6   Nucleotides, Nucleosides, and Related Analytes, 428

   9.5.7   Sugars and Related Analytes, 430

   9.5.8   Other Analytes, 433

   9.5.9   Summary, 450

9.6   Amino Acid, Peptide, and Protein Analytes, 450

9.6.1   Amino Acid Analytes, 450

9.6.2   Peptide Analytes, 452

9.6.3   Protein Analytes, 457

9.6.4   Summary, 459

9.7   Pharmaceutical Analytes, 459

9.7.1   Drug Surveys and Screening Procedures, 459

9.7.2   NSAIDs and Analgesic Drugs, 460

9.7.3   Antibiotic Drugs, 464

9.7.4   Anticancer Drugs, 476

9.7.5   Antiepileptic Drugs, 482

9.7.6   Anthelmintics, 485

9.7.7   Illicit and Related Drugs, 487

9.7.8   Antihistamines, 490

9.7.9   Antiretroviral Drugs, 492

9.7.10 Antidepressants, 496

9.7.11 Antibacterial Drugs, 499

9.7.12 Anesthetics, 503

9.7.13 Immunosuppressants, 504

9.7.14 Analgesics, 505

9.7.15 Antihypercholesterolemic and Antihyperlipidemic Drugs, 506

9.7.16 Antihypertension Drugs, 507

9.7.17 Antimalarials, 511

9.7.18 Steroids, 512

9.7.19 Antipsychotics, 515

9.7.20 Other Drug Analytes, 518

9.8   Summary, 527

Chapter 10   Water, Dimethyl Sulfoxide, and Common Acidic Modifiers   531

10.1   General Considerations and Impurities, 531

10.1.2   Solvent Preparation, 532

10.2   Industrial Analytes, 535

10.3   Biological Analytes, 538

10.3.1   Carboxylic Acid Analytes, 539

10.3.2   Basic Amine Analytes, 540

10.3.3   Toxins, 541

10.3.4   Vitamins, 542

10.3.5   Nucleotides, Nucleosides, Amino Acids, and Peptides, 543

10.3.6   Terpenoids, Flavonoids, and Related Analytes, 545

10.4   Pharmaceutical Analytes, 546

10.5   Summary, 547

10.6   Dimethyl Sulfoxide, 549

References   **552**

Index   **607**

PREFACE

HPLC separations are a fundamental analytical tool for the vast majority of testing laboratories. Over the last few years HPLC has expanded in two important ways. First, it become even more prevalent in the pharmaceutical, nutraceutical (natural product), and protein/peptide characterization areas. Second, along with this, mass spectrometry has become more widespread as a detector of choice.

The second edition has been updated to reflect these advances. In addition, the original 1123 references have been replaced, where possible, by newer citations and the reference base has been increased to over 1600. The original Chapter 1 has been reorganized into three chapters; the first still deals with solvent properties. The second expands the topics of method optimization somewhat, and the third deals briefly with method/system validation and operating parameters.

The most striking change is the addition of chemical structures. This was done to give the analyst a ready reference to the chemicals cited in a separation and their similarities/differences. Obviously, presenting the structures for all compounds was not feasible (and not necessary). For example, simple substituted benzenes, common acids like acetic, benzoic, etc., and base structures such as pentane (and 1-pentanol) are not included.

Great effort was made in trying to verify structures and to check and recheck their accuracy of depiction. However, with the size of the database and the paucity of available independent structure verification (i.e., second sources), there may be some errors.

As with the first edition, my hope is that the information contained in this volume will significantly decrease the time and effort for the analyst and in terms of both deriving final methods and avoiding the numerous potential problems associated with reaching that method.

Paul C. Sadek
Grand Rapids, MI

INTRODUCTION

This book serves two separate and important functions for the chromatographer: practical and operational. The first three chapters deal with the operational aspects of solvents. They contain information regarding solvents and solvent classes, method optimization techniques, and the definition and use of method validation protocols/system suitability parameters. These chapters describe solvents from a practical use-oriented point of view. Here the physical and chemical properties of numerous solvents are discussed with respect to their impact on the chromatographic system. A clear understanding of the implications presented by these properties will save the chromatographer considerable time and effort.

The method optimization chapter reviews five commonly used techniques for assisting bench chromatographers in method optimization or development: eluotropic series and solvent strength, triangulation, "scout gradient," window diagrams, and brute force. Detailed examples and explanations are presented so that these concepts can readily be put to use. Theory is *not* discussed.

Chapter 3 deals with validation issues (e.g., determination of whether or not a method produces the correct results [accuracy], reproducibly [precision and robustness], and from analyst to analyst and laboratory to laboratory [ruggedness]) as well as ongoing system suitability parameters (e.g., those variables that are monitored and recorded that ensure that the system is functioning properly during any given analysis). In total, these first three chapters embody the "use" aspect of solvents and chromatographic systems.

The last seven chapters (Chapters 4 to 10) present documented real-world chromatographic uses for each solvent class on an individual basis: alcohols, ketones and esters, nitriles and nitrogenous, ethers, and so on. These chapters contain details regarding individual separations so that they may be implemented directly or used in conjunction with the method optimization chapter to modify existing methods in order to meet the needs of new and specific separations.

ABBREVIATIONS, ACRONYMS, AND SYMBOLS

α	Selectivity factor [$\equiv (k_1'/k_2')$]
APCI	Atmospheric pressure chemical ionization
AU	Absorbance unit
BHT	2,6-di-t-butyl-4-methylphenol
C_4	Butyl-bonded phase
C_8	Octyl-bonded phase
C_{18}	Octadecyl-bonded phase
C_n	Alkyl chain n carbons in length
CN	Cyanopropyl-bonded phase
DMF	Dimethylformamide
DMSO	Dimethyl sulfoxide
ε°	Eluotropic strength
EC	Electrochemical
EDTA	Ethylenediaminetetraacetic acid
ELSD	Evaporative light scattering detector
em	Emission wavelength used in fluorescence detection
ex	Excitation wavelength used in fluorescence detection
FAB	Fast atom bombardment
fg	Femtogram (10^{-15} g)
fmol	Femtomole (10^{-15} mol)
FMOC	9-Fluoromethylchloroformate
FTIR	Fourier transform infrared
GPC	Gel permeation chromatography
HPLC	High-performance liquid chromatography
ICP	Inductively coupled plasma

IPA	Isopropyl alcohol, 2-propanol
k'	Capacity factor $[\equiv (t_R - t_0)/t_0)]$
λ	Wavelength
LC	Liquid chromatography
M	Molar, molarity (mol/L)
mg	Milligram (10^{-3} g)
mL	Milliliter (10^{-3} L)
mM	Millimolar (10^{-3} mol/L)
mmol	Millimole (10^{-3} mol)
μg	Microgram (10^{-6} g)
μL	Microliter (10^{-6} L)
μM	Micromolar (10^{-6} mol/L)
μmol	Micromole (10^{-6} mol)
MtBE	Methyl t-butyl ether
MW	Molecular weight
MS	Mass spectrometry
ng	Nanogram (10^{-9} g)
nmol	Nanomole (10^{-9} mol)
NP	Normal phase
OPA	o-Phthalaldehyde
P$'$	Solvent strength parameter
pg	Picogram (10^{-12} g)
pmol	Picomole (10^{-12} mol)
PAH	Polyaromatic hydrocarbon
RI	Refractive index
RP	Reversed phase
SDS	Sodium dodecyl sulfate
S/N	Signal-to-noise ratio
t_R	Solute retention time
t_0	Void volume of system
TEA	Triethylamine
TFA	Trifluoroacetic acid
THF	Tetrahydrofuran
UV	Ultraviolet

THE HPLC
SOLVENT GUIDE
SECOND EDITION

1

PHYSICAL AND CHEMICAL SOLVENT PROPERTIES

When a chromatographer develops a new high-performance liquid chromatography (HPLC) method, a significant amount of time is spent evaluating the choice of column with regard to its stated performance. This is not unexpected, since literally hundreds of columns are available on the market. Numerous documents indicate that the stationary phase plays a key role in the general retention process [1–4]. However, gross retention time, peak shape, functional group specificity, and important system operating parameters such as backpressure and detector background signal levels are also uniquely affected by the choice of an often overlooked variable—the solvent [5–7].

To effectively differentiate solvents in terms of the benefit that one offers over another, or the trade-off the chromatographer faces in choosing one solvent versus another, three fundamental factors need to be considered: (1) physical properties of the solvent, (2) the chemical properties of the solvent (especially with respect to system compatibility and safety aspects), and (3) the effects these properties have on the chromatographic process (i.e., system operation, chromatographic separation, detection limits, and analytical reproducibility). This chapter deals with the chemical and physical properties of HPLC solvent groups as well as important features, concerns, and limitations of individual solvents.

Before proceeding, however, it is important to define two HPLC terms: *solvent* and *mobile phase*. Quite clearly, a solvent is the liquid that is contained in the bottle obtained from the manufacturer. Also as obvious is the definition of the mobile phase: the liquid that is pumped through the column. In this book, the term solvent is always used when considering the chemical and physical properties of pure liquids and their mixtures, whereas the term mobile phase is used to indicate immanent use in the chromatographic system. The terms are used interchangeably in that gray area that occurs between each component being neat up to and including their preparation prior to use in the system.

1.1   UV CUTOFF

In and of itself, the UV cutoff is typically not a critical parameter on which to base solvent selection. Rather, the UV cutoff is a rapid way of assessing whether or not the gross characteristics of the solvents (1) make it an appropriate choice based on the system's operating wavelength, and (2) have changed from lot to lot. Most manufacturers provide this datum for each lot of solvent as part of the printed label affixed to each solvent bottle, since it is an easy test to perform.

To be sure that the UV cutoff value is properly interpreted, the working definition of the UV cutoff is: "The wavelength at which the absorbance of the solvent in a 1-cm cell (versus air as reference) is equal to unity." The mathematical relationship between the absorbance, the incident beam intensity, and the transmitted beam intensity at a specified wavelength is represented by Beer's law as

$$\log\left(\frac{1}{T}\right) = \log\left(\frac{I_0}{I}\right) = A = \varepsilon bc \tag{1.1}$$

where T is the transmittance, I_0 is the incident beam intensity, I is the transmitted beam intensity, A is the absorbance, ε is the molar absorptivity L/(mol·cm), b is the cell length (cm), and c is the concentration of the compound in solution (mol/L).

Close examination of Eq. 1.1 leads to the discovery of a misnomer: the UV incident radiation is not truly cut off when $A = 1$ (i.e., $T \neq 0$ at $A = 1$), but rather is markedly attenuated. As a result, the cutoff, as defined above, occurs when the transmitted beam intensity reaching the detector is attenuated to 10% of the incident radiation on the sample.

When impurities are present in any solvent, the overall absorbance for the solvent at a given wavelength, λ, is the sum of the absorption contributions of each component:

$$A(\lambda) = \varepsilon(\lambda)_{\text{solvent}} bc_{\text{solvent}} + \varepsilon(\lambda)_{\text{impurity a}} bc_{\text{impurity a}} + \cdots + \varepsilon(\lambda)_{\text{impurity n}} bc_{\text{impurity n}} \tag{1.2}$$

Consequently, for any given λ, low concentrations of impurities with large ε values or high concentrations of impurities with low ε values will create absorbance problems. Consistent removal of such impurities from solvents is therefore critical to reproducible chromatographic and analytical performance.

Table 1.1 lists the cutoff ranges by solvent class (e.g., the class identified as alkyl alcohols includes methanol, n-propyl alcohol, isopropyl alcohol, etc.). In general, a solvent with a UV cutoff higher than the working wavelength used for an analysis generates such a high background absorbance that it is excluded from further

TABLE 1.1   Approximate Cutoff Ranges for Solvent Classes

Solvent or Solvent Class [a]	Cutoff (nm)
Acetonitrile and water	<190
Alkanes (hexane, iso-octane, etc.)	190–205
Alkyl alcohols (methanol, isopropyl alcohol, etc.)	205–220
Alkyl ethers (diethyl ether, methyl *t*-butyl ether, etc.)	210–220
Alkyl chlorides (dichloromethane, chloroform, etc.)	220–270
Freons	225–245
Alkyl acetates (ethyl and butyl acetate, etc.)	250–260
Alkyl amides (dimethylformamide, dimethylacetamide, etc.)	260–270
Benzene and alkyl benzenes (toluene, xylene, etc.)	270–290
Chlorobenzenes (chlorobenzene, 1,2-dichlorobenzene, etc.)	280–310
Alkyl ketones (acetone, methyl propyl ketone, etc.)	320–340

[a] All solvents unpreserved.

consideration. For example, a high volume percent of an alcohol in the mobile phase is not recommended for use at $\lambda < 220$ nm and certainly is not compatible for use at $\lambda < 205$ nm.

One important exception occurs when the solvent is used as a low-level component of the mobile phase, for example, at 10% v/v (volume to volume) or less. In such cases, working at or near the cutoff for the 10% component leads to an equivalent background absorbance of only 0.1 AU (absorbance unit). This is often acceptable despite the concomitant increased noise level (which leads to higher detection limits), decreased linear working range, and lowered sensitivity.

Nevertheless, the use of the UV cutoff value as the only criterion for solvent selection leads to potentially missed opportunities for unique and powerful separations, as will be discussed in the following sections.

1.2   ABSORBANCE VERSUS WAVELENGTH CURVES

Of more practical importance is the absorbance versus wavelength (A vs. λ) spectrum that a solvent generates. Figure 1.1 shows spectra generated in a 1-cm cell for the common and representative reversed-phase (RP) solvents methanol, THF (tetrahydrofuran; unpreserved or UV-grade), and acetonitrile. Figure 1.2 shows representative spectra for common normal-phase (NP) solvents hexane, ethyl acetate, and dichloromethane (methylene chloride). It is important to recognize that not all solvent spectra are neatly represented by exponential or modified Gaussian curves. A detailed discussion of individual solvent spectra from the perspective of RP and NP separations will help illustrate the subtleties involved with and the limitations and advantages presented by various solvents.

By definition, a reversed-phase column has a nonpolar surface or bonded phase. Examples of RP columns therefore include octadecyl (C_{18}), octyl (C_8), and phenyl.

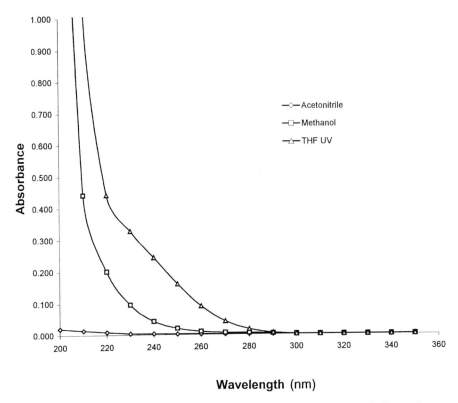

Wavelength (nm)

FIGURE 1.1 Typical absorbance–wavelength curves for common reversed-phase solvents: acetonitrile, methanol, and unpreserved tetrahydrofuran (THF). Note that even though the UV cutoffs for THF and methanol differ by only ~7 nm, the relative absorbance at the THF cutoff of 212 nm is only 0.37 AU for methanol.

(Note that the actual bonded moiety is typically a dimethylalkylsilane; the C_{18} column is really precisely defined as having a dimethyloctadecylsilane bonded phase.) Those solvents used in conjunction with RP columns are called reversed-phase solvents. The most common RP solvents are mixtures of water with water-soluble solvents such as acetonitrile, methanol, and tetrahydrofuran. Uncommon cases are the use of nonaqueous solvents in reversed-phase separations. These are classified as NARP (nonaqueous reversed-phase) separations. An example of an NARP mobile phase would be a 50/50 v/v methanol/acetonitrile mixture.

Conversely, normal-phase columns have polar surfaces or bonded phases such as silica aminopropyl and cyanopropyl (often as the ethoxy silane) and utilize nonpolar solvents such as hexane, iso-octane, and cyclohexane. These solvents, along with very low levels of polar solvents such as ethyl acetate, dichloromethane, or ethyl ether, are used as mobile phases. An example of a normal-phase mobile phase is 99.5/0.5 v/v hexane/ethyl acetate. Water in the solvent is typically avoided when

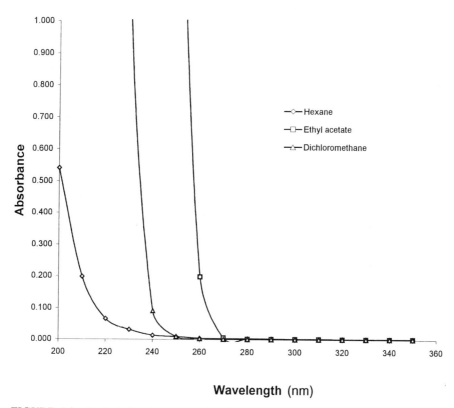

FIGURE 1.2 Typical absorbance–wavelength curves for common normal-phase solvents: hexane, ethyl acetate, and dichloromethane.

silica columns are used for normal-phase separations since water effectively deactivates the silica support.

Finally, there is a set of hybrid separations in which a normal-phase column (cyanopropyl, diol, aminopropyl, etc.) is used in conjunction with a reversed-phase solvent system. This is due to the fact that these phases are compatible with water and offer a range of polarities for use in conjunction with RP solvents.

1.3 REVERSED-PHASE SOLVENTS

Four of the most frequently used reversed-phase (RP) solvents will be discussed in detail: water, acetonitrile, methanol, and THF. From a UV absorbance point of view, the "perfect" UV solvent would not absorb from 195 nm (a typical lower operating limits for detectors) all the way through the entire UV range. From this definition alone, water is the only commonly used solvent that fits into this category. However, some solvents exhibit very low absorbances below 220 nm. Acetonitrile is one such and has a UV cutoff of <190 nm. Beyond this, acetonitrile has a very low background absorbance even at wavelengths as low as 200 nm (<0.05 AU). These nearly ideal spectroscopic qualities, coupled with excellent solubilizing capabilities and unique chromatographic properties, lead to acetonitrile being the most commonly used solvent in RP separations.

Methanol has a UV cutoff of 205 nm. If it assumed that the A vs. λ curve for methanol is similar to the curve for acetonitrile, then methanol should have little or no absorbance at 215 nm. The methanol spectrum shown in Figure 1.1 shows that this assumption is incorrect. In fact, the absorbance for methanol at 215 nm is greater than 0.3 AU. Closer examination of the methanol A vs. λ curve leads to the conclusion that achieving a background absorbance contribution from methanol of <0.05 AU requires either working at $\lambda > 235$ nm or limiting the methanol level in the solvent to <15% at $\lambda = 215$ nm. This clearly can present a problem for solutes with either small molar absorptivy (ε) values or chromophoric maxima of <235 nm.

Even more important is the low-level absorbance of methanol at longer wavelengths. The advent of extremely stable, reproducible, and noise-free pumps has somewhat alleviated pump-generated noise, but background mobile phase absorbance or baseline shifts (due to the change in solvent composition but not as a gradient) will always contribute to a decrease in the overall method accuracy and precision. When high-sensitivity work is done in conjunction with a mobile phase gradient, regardless of the pump system used, any low-level absorbance due to mobile phase constituents will amplify pump noise (seen as a detector sawtooth output synchronous with pump reciprocation) and cause significant baseline shifts as the mobile phase composition changes over the composition range. Examples of these effects are presented in Figure 1.3. Note the considerable difference in the baseline shift between acetonitrile/water and methanol/water at both wavelengths. Also note that the small baseline shift in the acetonitrile gradient monitored at 254 nm is not the consequence of absorbance but is a symptom of the changing refractive index of the mobile phase. This is clearly not the case for methanol at 254 nm; the typical absorbance is ~0.015 (in a 1-cm cell). This example illustrates the primary reason why many chromatographers prefer to use acetonitrile in gradients over methanol, especially at lower wavelengths (<230 nm).

Prior to the discussion of the spectroscopic characteristics of tetrahydrofuran (THF), it should be noted that THF and many other ether compounds are chemically unstable. Over time, THF breaks down to form peroxides. Peroxides are chemically reactive and unstable. In order to help control peroxide levels in a solvent such as THF, manufacturers typically add a chemical scavenger, called a *preservative* or *stabilizer* such as 2,6-di-*t*-butyl-4-methylphenol (2,6-di-*t*-butyl-*p*-cresol, BHT). This type of THF is referred to as *preserved* or *nonspectral grade* THF. BHT has large molar absorptivity values in the UV below 280 nm with the peak maximum at about 270 nm. As a consequence, preserved THF is rarely used in conjunction with UV detectors (see Fig. 1.4).

Preserved THF in conjunction with a refractive index (RI) detector finds common use in gel permeation chromatography (GPC). Here the absorbance of BHT in the

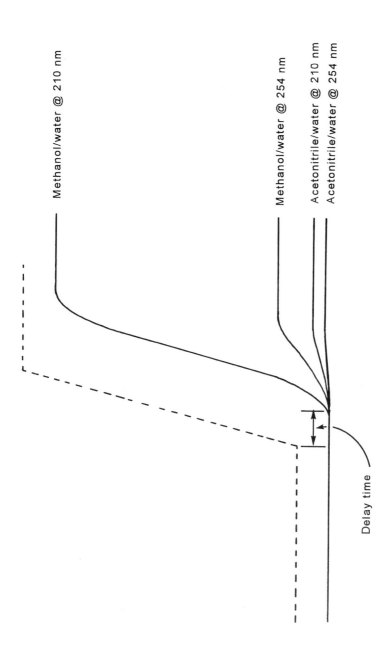

Methanol/water @ 210 nm

Methanol/water @ 254 nm

Acetonitrile/water @ 210 nm

Acetonitrile/water @ 254 nm

Delay time

FIGURE 1.3 Comparative baseline response for rapid water/methanol and water/acetonitrile (0 to 100% organic) gradients. The gradient profile itself is shown as the dashed line (– – –). The delay time indicates the time it takes for the mobile phase created at the pump head to reach the column and is a function of the volume between the pump and the column and the flow rate. From the gradient profiles it is easy to understand why acetonitrile is preferred for low-wavelength UV work.

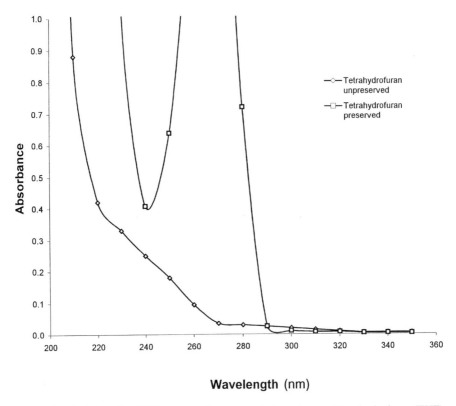

FIGURE 1.4   The effect BHT preservative on the UV spectrum of tetrahydrofuran (THF). For all practical purposes, the use of preserved THF is reserved for detectors that do not respond to its presence (e.g., refractive index) or in regions where it generates a low response (e.g., UV detection at >290 nm).

UV is unimportant, but the elimination of peroxides in the mobile phase, which can seriously degrade and irreversibly damage the polymer-based packing material, is critical.

Unpreserved THF, or THF UV-grade, has no chemical added to prevent or inhibit peroxide formation. The implications here are twofold. First, the spectroscopic characteristics are such that THF is effective in many techniques utilizing UV detectors. The cutoff for THF is slightly higher than for methanol, 212 nm versus 205 nm, but the critical difference between the spectra is that the THF spectrum is severely shouldered and skewed into the longer-wavelength UV region, making the lowest useful working wavelength for high volume percent THF mobile phases approximately 240 nm. Second, the absence of a peroxide scavenger leads to a more rapid buildup of peroxides in the THF UV-grade solvent.

Increases in the peroxide level are bad for two reasons. First, peroxides alter the absorbance vs. wavelength curve. Second, and potentially more important, peroxides are chemically active and can react with solutes, thereby producing a new chemical and a decrease in the original analyte concentration. Severe peak tailing, peak splitting, or the appearance of new peaks in the chromatogram often accompanies this change. Mobile phase components may also react with peroxides, resulting in potential changes in the overall chromatography seen as retention time shifts and/or the generation of spurious peaks.

To minimize the chances of using degraded THF (or any unstable material for that matter), manufacturers often print an expiration date on each bottle label so that the end user can easily ascertain when overall solvent integrity is a concern. For example, the peroxide-forming solvents THF and ethyl ether typically have expiration dates that are six months from the date of manufacture. To ensure the integrity of peroxide-forming solvents and minimize potentially deleterious effects due to their peroxide formation, one should pay close attention to these expiration dates. Judiciously rotate laboratory stocks of these solvents and never exceed the equivalent of three months' working reserve. To further minimize the rate of peroxide formation, store THF (as well as all other peroxide-forming solvents) in a cool, dry place and out of direct sunlight.

Solvents that form peroxides should also be routinely tested for peroxide prior to use to prove that peroxide levels are acceptable (see Section 1.11.1 for qualitative tests), because the presence of peroxides, even at low levels, presents an explosion hazard if the ethereal solution is subsequently concentrated, as is often the case in preparative chromatography or preconcentrations of sample extracts.

1.3.1 Mobile Phases and Mobile Phase Modifiers

As described earlier, a mobile phase is the liquid that is pumped through the column. Mobile phases are comprised of major and minor components. In RP separations, components are typically considered major if they are present at levels of >5% in the mobile phase. Minor components are present at <5% and are commonly referred to as *mobile phase modifiers*. Typical mobile phase modifiers (MPMs) are undiluted acids (e.g., phosphoric, trifluoroacetic, and acetic) and bases (e.g., triethylamine, triethanolamine, and diethylamine) as well as buffer systems (e.g., phosphate and acetate or mixed such as trifluoroacetic acid/triethylamine) and ion-pair reagents (e.g., sodium dodecyl sulfate and tetrabutylammonium phosphate).

Even though the MPMs are present at low concentrations, consideration of their absorbance versus wavelength curves is also very important. Examples of 3% aqueous solutions of MPMs are presented in Figures 1.5a–c. In these examples, phosphate has little or no absorbance down to 200 nm, whereas acetic and trifluoroacetic acids cut off between 230 and 240 nm. Notice also that aging can also affect the A vs. λ curve. Compare the fresh triethylamine curve to the one-year aged material (see Figure 1.5d). Both fresh and old have UV cutoff values between 240 and 250 nm. However, the old material has significant absorbance out to 290 nm (possibly due in part to the extraction of phenolic residues from the bottle cap),

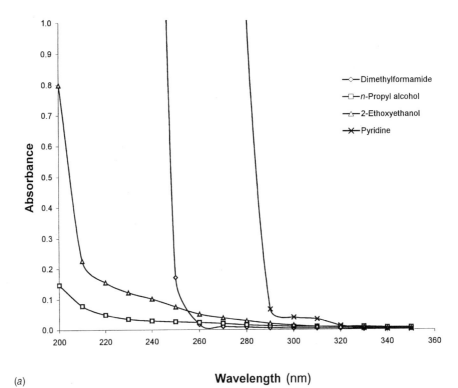

(a)   **Wavelength (nm)**

FIGURE 1.5   The absorbance–wavelength curves for 3% solutions of individual mobile phase modifiers (MPMs) in water. The MPMs are (*a*) dimethylformamide, *n*-propyl alcohol, 2-ethoxyethanol, and pyridine; (*b*) *N*-methylpyrrolidone, dimethyl sulfoxide, and ethyl ether. Similar curves for (*c*) MPM/buffer components in water: these are acetic acid, trifluoroacetic acid, and phosphoric acid. (*d*) The effect of fresh and aged 50 mM triehtylamine solutions on absorbance vs. wavelength spectra.

whereas the fresh material has negligible absorbance at 260 nm. Obviously, judicious choice of MPM identity, concentration, and age/storage is important.

Ion pair reagents present the same concerns. Figure 1.6 shows *A* vs. λ curves for 50 mM aqueous solutions of a number of typical ion-pair reagents. In this case tetrabutylammonium hydroxide has a cutoff above 220 nm, whereas tetramethylammonium chloride peaks sharply at 200–210 nm. Octanesulfonic acid and sodium dodecyl sulfate have moderate absorbance from 200 to 210 nm but are essentially transparent above this region.

Clearly, for each of the examples presented above, changes in solution concentration will yield an approximately proportional change in the absorbance and must be taken into account prior to use.

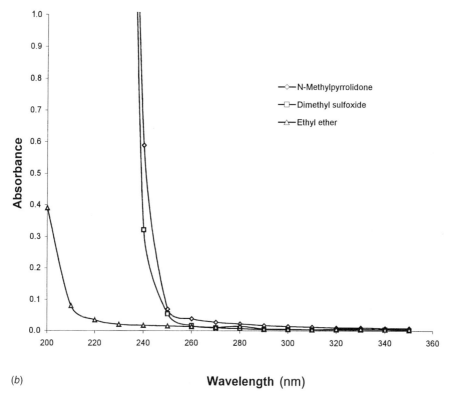

(b)   **Wavelength (nm)**

FIGURE 1.5   (*continued*)

1.4   NORMAL-PHASE SOLVENTS

Four frequently used normal-phase (NP) solvents will be discussed in detail: hexane, dichloromethane, isopropyl alcohol, and ethyl acetate. Hexane is one of the most frequently used NP solvents. As in the case of acetonitrile, the low UV cutoff wavelength of 195 nm and the low background absorbance at and above 210 nm make hexane an excellent choice for the majority of compounds analyzed in the NP mode (see Fig. 1.2). (Note that cyclopentane, cyclohexane, and heptane have similar UV characteristics.) In addition, under normal circumstances hexane has excellent long-term chemical stability. Because of the limited solubility of hexane, and of related hydrocarbons, in water and acetonitrile, their use in reversed-phase separations is rare.

The nonpolar character of hexane and the alkanes severely limits their ability to solubilize polar compounds. Since the NP support material typically has a polar surface (such as silica), some polar mobile phase component is needed to elute the solute. Therefore, a low level (typically ≪5% v/v) of a more polar component, such as dichloromethane or ethyl acetate, often needs to be included as part of the mobile

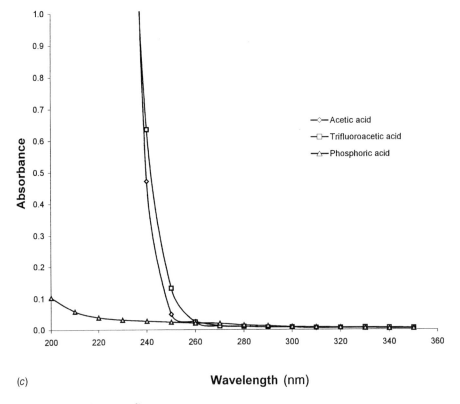

(c)

Wavelength (nm)

FIGURE 1.5 (*continued*)

phase to assist in the elution process. It should be noted that, due to the low levels of polar solvent needed to have a large effect in NP work, the term MPM is typically not used.

Dichloromethane has a UV cutoff of 233 nm and is effectively used for compounds with chromophores with $\lambda_{max} > 250$ nm. However, since it is used at such low levels in NP work, absorbance is rarely an issue. Dichloromethane (and chloroform) is unstable and degrades through a free-radical process. Amylene and cyclohexene are commonly used chloroalkane preservatives. Although commonly used as a low-volume mobile phase component in NP separations, dichloromethane has found only limited use in RP work because of its low water solubility. Conversely, solubility with alkanes and good sample-solubilizing characteristics make dichloromethane a very useful component in NP mobile phases.

Isopropyl alcohol, IPA (as well as methanol and ethanol), is a very strong NP solvent, due to the fact that the alcohols have strong hydrogen bond donating and accepting characteristics. As a consequence, the alcohols react strongly with the silica's silanol groups (i.e., residual surface functional groups, SiOH), causing solutes to be displaced from the surface and forced through the column. IPA is

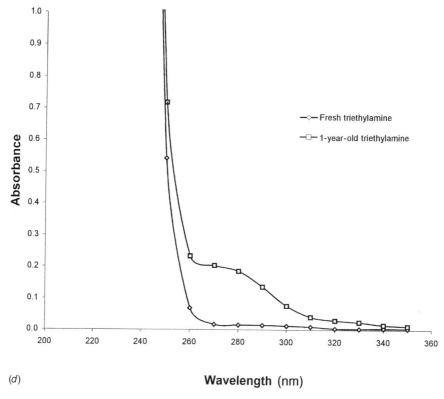

(d)   **Wavelength (nm)**

FIGURE 1.5 (*continued*)

also very useful because it is miscible with a wide range of solvents including nonpolar alkanes as well as the very polar solvent water. Therefore, IPA finds significant application in RP separations as well. IPA has a characteristic alcohol A vs. λ spectrum with a cutoff near 205 nm and a significant absorbance up to 235 nm.

Ethyl acetate is of intermediate solvent strength to dichloromethane and IPA. A major drawback is the very high UV cutoff of 256 nm. An advantage, though, is that it is stable and unreactive in normal HPLC use. It is immiscible with water and has limited use in RP methods.

Acetone is included in this discussion not because it is frequently used in either RP or NP separations but because (1) it has excellent solubilizing characteristics, evident through the fact that it is often referred to as a *universal solvent*, and (2) it exemplifies a spectral characteristic for ketones that may be of particular importance, albeit for a limited number of separations.

The published UV cutoff for acetone is 330 nm. If acetone use is judged solely on the basis of this datum, then the "window of opportunity" it gives cannot be exploited. Inspection of the 205–220 nm range for acetone in Figure 1.7 reveals that the absorbance for a 10% v/v aqueous solution drops to a reasonable level over this

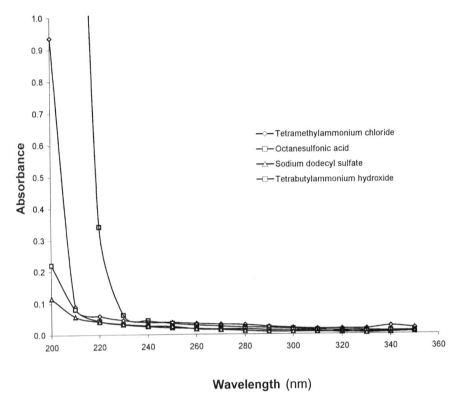

Wavelength (nm)

FIGURE 1.6   The absorbance–wavelength curves for 50 mM solutions of individual ion-pair reagents in water. The ion-pair reagents are tetramethyl ammonium chloride, octanesulfonic acid, sodium dodecyl sulfate, and tetrabutyl ammonium hydroxide.

range. Therefore, acetone in particular, and analogous ketones in general, may be utilized as low volume percent components for analyses monitored in the 205–220 nm range. Be aware, however, that since this region lies below the published UV cutoff, manufacturers do not monitor the absorbance in this region (i.e., there is no existing specification). As a consequence, exceedingly small changes in the finished product due to slight variation in raw-material feed stock and processing will have major effects on the absorbance in this 205–220 nm region. Prevalidation of individual lots of material may be necessary to assure acceptable performance.

1.5   SYSTEM PEAKS

An important result of working with mobile phases in spectral regions in which the background absorbance is significant is the potential appearance of large system peaks. System peaks represent depletion and enrichment zones of various mobile

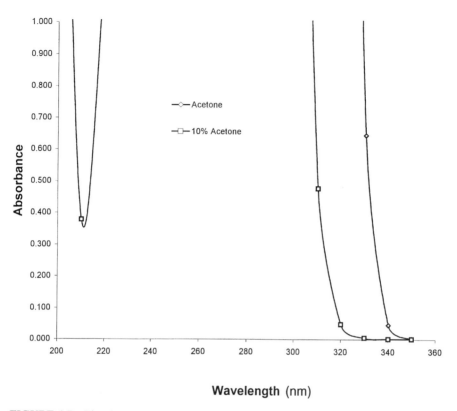

FIGURE 1.7   The absorbance–wavelength curves for neat acetone and a 90/10 v/v water/ acetone mixture.

phase components eluting from the column and are seen chromatographically as positive and negative peaks (see Fig. 1.8). The presence, size, and elution time of system peaks are dependent on the complex relationship between sample matrix, injection volume, detector wavelength, mobile phase composition, and type of packing material. Extensive discussions are found in papers describing the problems associated with system peaks (including negative peaks that co-elute with analyte peaks) in reversed-phase work [8,9] and in normal-phase work [10]. A complete theoretical discussion is also available [11].

The best ways to avoid system peaks are (1) to dissolve the sample in the mobile phase whenever possible, and (2) to work at a wavelength where the sample matrix and the mobile phase have little or no absorbance. It should be noted that even if the second condition is satisfied, the refractive index difference between the unmatched sample solution and the solvent system might generate refractive index-related "peaks" or other baseline shifts. These "peaks" may elute at any time during the elution process and could also be very disruptive, especially when working at sensitive detector settings.

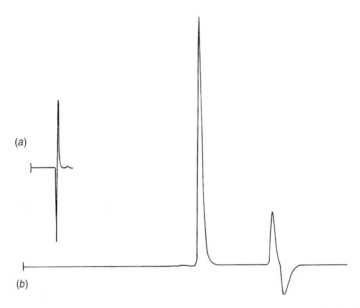

FIGURE 1.8   System peaks. (*a*) A commonly observed void volume refractive index pulse. It is indicative of a solvent mismatch. (*b*) A system peak that elutes after the analyte of interest.

1.6   LOT-TO-LOT SOLVENT VARIABILITY

The individual solvent examples presented above were chosen to illustrate the dramatic differences between the spectra of commonly used solvents and the limitations that these wide-ranging spectral characteristics impose on the chromatographer. However, in addition to the difference between solvents, a presentation covering the important spectral characteristics of HPLC solvents would be incomplete without a discussion of lot-to-lot solvent variability and its effect on chromatographic methods.

Lot-to-lot changes may manifest themselves in solvent spectra in one or both of the following ways: as spectral shifts and as spectral abnormalities. Over the course of the production of many lots of a specific solvent, the UV cutoff will vary. As an example, the UV cutoff for sequentially produced lots of methanol may vary from 202 to 205 nm. In the majority of analyses, this presents no problem. However, for a method monitoring analyte absorbance at 205 nm, the background absorbance may shift up a full 0.2 absorbance units. This is shown by the absorbance versus wavelength curve presented in Figure 1.9. A typical lot of methanol with a 202 nm cutoff shows an absorbance of approximately 0.84 at 205 nm. The methanol spectrum for the 205 nm cutoff by definition has an absorbance value of 1.0, so that the resulting ΔA at 205 nm is approximately 0.2 AU. This change alters both the sensitivity ($\Delta A/\Delta$concentration) and the linear working range of the method. Also, a

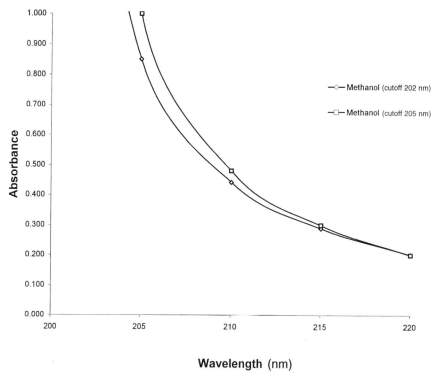

FIGURE 1.9 The absorbance–wavelength curves for two lots of methanol. One lot has a UV cutoff of 202 nm and the other 205 nm. Notice the sizeable different in total absorbance.

concomitant increase in background noise occurs, resulting in a larger standard deviation for the method. This is obviously a worst-case scenario, but chromatographers who are developing methods that will be mandated for use in perpetuity need to be particularly aware of this potential problem of working near the UV cutoff of a solvent.

The effects of spectral shifts on an analysis are also readily anticipated if both UV cutoff and complete spectral curve information are considered. In most cases, the analytical effects of a spectral shift can be virtually eliminated if the analysis is run at a wavelength more than 20 nm higher than the reported UV cutoff. In instances where such considerations cannot be met, specifically designated solvent specifications should be communicated to the manufacturer so that solvent lots are supplied that meet the requirements of the method.

Totally unanticipatable are the effects of spectral abnormalities. These are usually the result of slight changes in the raw material composition and/or the manufacturing process. These changes may manifest themselves as extra shoulders, plateaus, or in extreme instances, maxima in the spectrum. More aggravating still are aromatic contaminants in a solvent that create broad-wavelength, low-absorbance peaks in the solvent spectrum over the 250–280 nm range.

Spectral abnormalities are very troublesome in isocratic analyses since, depending upon the number and concentrations of the impurities, a significant amount of time is needed to fully equilibrate the system. The symptom of this problem is a baseline that constantly drifts until the column has been saturated with the impurities. Next, as injections are made, particularly from solvents that do not match the mobile phase, unexplained peaks (i.e., new system peaks) appear. These peaks will typically have a constant retention time but may increase in size with increasing time between injections, depending upon how long it takes for the system to re-equilibrate. They also tend to change in size if injection volumes are changed. Finally, when the contaminated lot of material is replaced, system re-equilibration must occur and so another long baseline drift process ensues.

In gradient work these contaminants lead to peaks with constant retention times as well. Once again, peak size is directly related to the volume of the weak solvent that is pumped onto the column between injections up to the point where the system is in equilibrium (see Fig. 1.10). These peaks are eliminated only through replacement of the contaminated solvent with uncontaminated solvent followed by elution of the column-absorbed contaminants.

It would be very advantageous for chromatographers if manufacturers could assure the reproducible production of contaminant-free solvents. Unfortunately, once a contaminant level in the solvent falls below 1 ppm (one part per million), simple UV spectra often cannot detect their presence. Therefore, in critical analyses, the individual laboratory must chromatographically prevalidate solvents in order to verify their suitability for use.

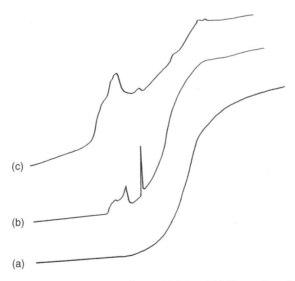

(c)

(b)

(a)

FIGURE 1.10   Three methanol/water gradients (20/80 to 100/0) monitored at 254 nm. (*a*) An excellent lot of methanol. (*b*) Low-level contamination of poorly to slightly retained materials. (*c*) Grossly contaminated material. Note that from these data alone it is not possible to determine the source of the contamination (i.e., water, methanol, or both).

Nevertheless, if a critical test is being performed or the analysis requires working near the UV cutoff of any solvent component, the routine availability of "select parameter" solvent should be discussed with the solvent manufacturer. After all, the only solvent performance standards guaranteed are the manufacturer's printed specifications.

Finally, a discussion of an often overlooked impurity is appropriate here. This impurity is dissolved oxygen. Molecular oxygen has a non-negligible absorbance in the UV region. To illustrate this, Brown et al. [12], sparged methanol with helium and obtained baseline absorbance readings at a wide range of wavelengths. The methanol was then saturated with air and the absorbance increased by 0.02 AU at 260 nm, 0.05 AU at 240 nm, 0.22 AU at 220 nm, and 0.38 AU at 210 nm. At 210 nm the absorbance is nearly double that of methanol alone (0.44 AU from Fig. 1.1). For the best analytical results it is therefore prudent to degas all solvents prior to use and, after the degassing is complete, then blanket the reservoir with a slow helium bleed above the solvent [13,14]. It should be noted that continuous sparging leads to preferential removal of the most volatile component of the solvent, thereby altering solvent composition and the chromatography. This may result in significant changes in retention time if the solvent has an initial low level of volatile component, and/or a degradation of peak shape if the low-level component is a mobile phase modifier such as trifluoroacetic acid or triethylamine.

1.7 VISCOSITY

Viscosity, as it affects the chromatographer, can be considered very simply as the resistance a fluid develops to forced flow through a constricted path. The mathematical relationship is [15]

$$\Delta P = \frac{\phi \eta L v}{d_{\mathrm{p}}^2} \tag{1.3}$$

where ΔP is the pressure drop across a column of length L (monitored as the backpressure of the system), ϕ is a flow resistance factor (related to the porosity of the packing material), η is the mobile phase viscosity, v is the mobile phase linear velocity, and d_{p} is the particle diameter. From this relationship it is easy to see that the pressure drop is directly proportional to column length, flow velocity, and viscosity and inversely proportional to the square of the particle diameter. Unfortunately, ϕ is not an easy parameter to work with, so a more practical relationship is given as [16]

$$\Delta P = 1000 \frac{\eta L F}{\pi r^2 d_{\mathrm{p}}^2} \tag{1.4}$$

where symbols are as above and F is the flow rate and r is the radius of the column. Now it is plainly evident that increases in flow rate or decreases in column radius will increase the pressure drop across the column.

Equation 1.3 explains one reason for the slow increase in backpressure associated with column aging. Silica particles fracture under the continual pressure swings from high pressure when in use to no pressure when not used (i.e., compression and decompression). The small fragments that are formed, called fines, collect in the interstitial column flow paths. This process blocks the available flow paths and, just as importantly, the presence of fines in the column decreases the average d_p; both processes lead to increased operating backpressure. Increased column backpressure also occurs when components precipitate on the packing material or in the pores in the inlet frit.

The Knox equation [17,18] predicts that increases in mobile phase viscosity will decrease overall chromatographic efficiency. However, the magnitude of the loss in efficiency due to viscosity effects in an HPLC system is typically small compared with the effects due to changes in mobile phase composition and temperature. Viscosity is of major importance only when very low-viscosity solvents are exchanged for considerably higher-viscosity ones, such as changing from hexane (0.31 cP [centipoise]) to iso-octane (0.50 cP).

Practically speaking, problems may arise in both low- and high-viscosity solvent systems. In low-viscosity situations such as 99/1 v/v hexane/ethyl acetate, little backpressure is generated across the column even at high flow rates. This may lead to improper seating of the inlet and outlet check valves and result in erratic flow rates. To rectify this problem, a flow restrictor may be placed after the pump and before the injector to create a higher and constant backpressure on the check valves.

With higher-viscosity solvents, column lifetime decreases (silica packing breaks down more rapidly) and pump maintenance requirements increase (e.g., piston seals wear faster). These costs are often minimal when compared with the savings realized through increased system efficiency and sample throughput. A good routine preventive maintenance program is recommended in these cases.

Gradient programs cause the most problems when viscous solvents are used. Figure 1.11 shows a plot of solvent viscosity versus composition for water/THF, water/methanol, and water/acetonitrile mixtures [19–21]. The resulting curves are nonlinear and always exhibit a maximum value higher than either pure component.

Acetonitrile, when used as water-based mobile phase, offers unique advantages over other solvents in that the maximum viscosity is only 20% greater than that of pure water. In comparison, the viscosity of a 45/55 v/v methanol/water is 1.8 times that of water. A similar mixture of n-propyl alcohol/water is over 2.5 times more viscous than water [22]. Such a wide range of viscosity range has the direct result of doubling or tripling the operating backpressure over the course of a 0–100% alcohol/water gradient. This becomes important in routine methods where the system is automated and unsupervised. System shutdown, malfunction, damage, and loss of data and/or time due to exceeding system backpressure limits is more likely to occur with high-viscosity solvents. Establishment of working parameters (e.g., flow rate, column length, and silica particle diameter) for which the highest

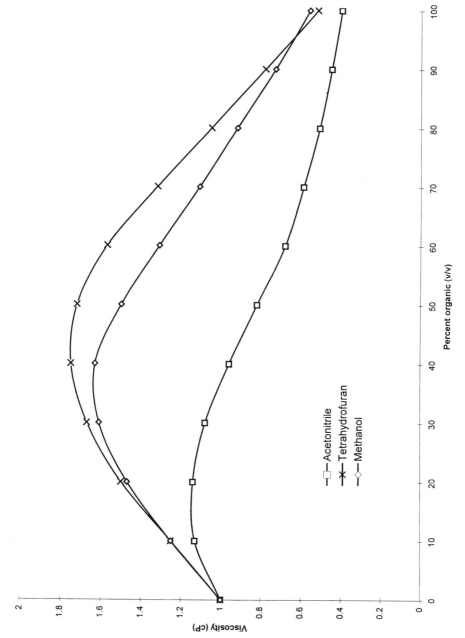

FIGURE 1.11 Viscosity *vs.* organic/water composition for acetonitrile, tetrahydrofuran, and methanol.

21

expected system pressure during a gradient is well below the system backpressure limit minimizes the chance for catastrophic results.

The highest expected system backpressure generated by a given mobile phase composition/column combination is readily estimated through multiplying the initial backpressure by the ratio of the highest viscosity encountered during the gradient to the initial mobile phase viscosity. For example, compare two gradients that run from 10/90 v/v organic/water to 90/10 v/v organic/water and an initial system back-pressure of 1000 psi. For a case where the organic component is acetonitrile, the backpressure goes from 1000 psi (this mobile phase has a viscosity of 1.1 cP and is near its maximum at the initial composition) to 360 psi at the end of the gradient (this mobile phase has a viscosity of 0.4 cP and the final operating viscosity is estimated as 1000 psi $\times$ [0.4 cP/1.1 cP] = 360 psi).

Conversely, an equivalent methanol/water gradient (1.2 cP initial viscosity), reaches a maximum operating backpressure at approximately 35% methanol (1000 psi $\times$ [1.6 cP/1.2 cP] = 1333 psi), and finishes at 667 psi (1000 psi $\times$ [0.8 cp/1.2 cP]).

Viscosity is a temperature-dependent property. For the most commonly used HPLC mobile phase mixtures, the viscosity decreases as the temperature increases, thereby lowering the system operating backpressure. Constant temperature for the solvent reservoir, the connecting tubing, the injector, and the column is necessary to ensure reproducible chromatographic results. The entire system, including the sample, must be heated to the temperature of the detector when a refractive index detector is used.

1.8   MISCIBILITY AND SOLUBILITY

The terms *miscible* and *soluble* are commonly used in a manner that may cause a great deal of confusion for the chromatographer. Unambiguous operational defini-tions for miscible and soluble are as follow:

Miscible: Two components that can be mixed together in all proportions without forming two separate phases are *miscible*. Conversely, two components that form separate layers when mixed are *immiscible*.

Soluble: A component that is present at any level in a solvent is *soluble* in that solvent. The maximum amount of solute A that will dissolve in a given amount of solvent B is the *solubility* (or *solubility limit*) of A in B.

Misnomers such as "partially miscible" (i.e., has a low solubility) or "totally soluble" (i.e., miscible) only lead to confusion and their use should be avoided.

Miscibility charts (Fig. 1.12) are handy references [23,24] but one should not read more into the chart than intended. When a solvent pair is designated "immiscible," it means only that the two components will form two separate phases when they are mixed at *some* proportion. This does not imply that the

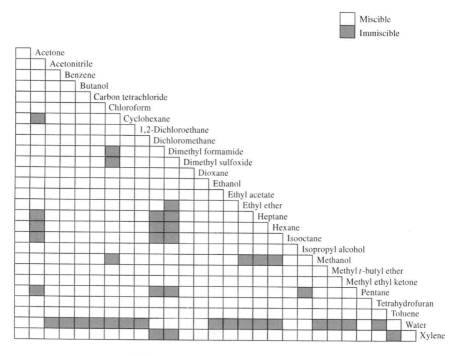

FIGURE 1.12   Solvent miscibility chart.

components are not soluble to an extent that is chromatographically useful (or harmful). For example, dichloromethane and water form two layers when combined at many proportions; hence they are immiscible in one another. Dichloromethane is, however, soluble in water at the 1.6% level and water is soluble in dichloromethane at the 0.2% level. The latter level (2000 ppm) is a nontrivial concentration for silica columns used in NP separations and causes surface deactivation on this type of column, resulting in concomitant loss of analyte retention.

In many gradient separations, the factor that limits the choice of solvent composition is typically the solubility of the organic solvent in water. On the one hand, Figure 1.12 shows that nearly 65% (17 of 26) of the solvents listed are immiscible with water. On the other hand, nearly 30% of those listed as immiscible with water have solubilities in water significant enough to be useful in HPLC work: methyl ethyl ketone 24%; ethyl acetate 9%; 1-butanol 8%; ethyl ether 7%; methyl t-butyl ether 5%. The use of a solvent that itself is miscible with an immiscible pair of solvents often generates a single-phase ternary mixture. For example, dichloromethane and water are immiscible, but the addition of isopropyl alcohol to a mixture will ultimately lead to a single-phase ternary mixture. This gives the chromatographer an important and additional degree of flexibility in choosing mobile phases.

Solubility is a strong function of temperature, and so care should be exercised when working with solvent compositions that are close to the solubility limit. This is even more critical when the process of mixing solvents is markedly exothermic (e.g., methanol/water) or endothermic (e.g., acetonitrile/water) and the final volume is obtained by diluting to volume. For example, if the mixing process is exothermic a mixture of solvents may at first present itself as miscible due to the increased temperature of the mix. However, upon cooling to room temperature the solution may become biphasic. This scenario is avoided and more consistent solvent compositions are obtained if exact volumes of each mobile phase component (all below the solubility limit) are measured separately, added together, and equilibrated, and the resulting solution is used.

Another reason for preparing solvent mixtures as described above is that many solvent combinations, when mixed, result in final volumes that are not the sum of the two individual parts. Table 1.2 shows a small sample of these effects. It should be noted that those mixtures with the largest changes in temperature upon mixing (dimethyl sulfoxide/water and acetonitrile/isopropyl alcohol) take over one hour for 100 mL to return to ambient temperature.

Temperature effects, i.e., changes in laboratory temperature, can be negated if the solvent/column system is placed in a controlled temperature environment. When uncontrolled ambient temperatures are used, one can be assured that phase separation will occur as soon as the experiment is half complete and the air conditioning kicks in! For these situations, in order to be confident that the solution will remain as a single phase, work at least 1% below the solubility limit if possible.

TABLE 1.2   Solvent Mixtures, 50/50 v/v

Solvent A	Solvent B	T_{init} (°C)	T_{final} (°C)	ΔT_{max} (°C)	V_{final} (mL)
Water	Methanol	21.4	30.2	+8.8	97.0
Water	Isopropyl alcohol	21.4	26.1	+4.7	97.1
Water	Acetonitrile	21.4	16.6	−4.8	99.2
Water	Acetone	21.4	27.4	+6.0	96.7
Water	Pyridine	21.4	31.2	+9.8	98.4
Water	Tetrahydrofuran	21.4	27.4	+3.0	97.8
Water	Dimethylformamide	21.4	37.7	+15.3	97.5
Water	Dimethyl sulfoxide	21.4	42.3	+20.9	98.2
Acetonitrile	Methanol	21.2	14.2	−7.0	99.7
Acetonitrile	Isopropyl alcohol	21.2	10.9	−10.3	100.8
Acetonitrile	Tetrahydrofuran	21.2	22.0	+0.8	99.9
Acetonitrile	Pyridine	21.2	20.5	−0.7	100.2
Methanol	Tetrahydrofuran	21.2	19.3	−1.9	100.0
Methanol	Pyridine	21.2	26.7	+5.5	99.1
Methanol	Dimethylformamide	21.2	22.7	+1.5	99.3
Tetrahydrofuran	Dimethylformamide	21.2	20.7	−0.5	100.2
Tetrahydrofuran	Pyridine	21.2	23.9	+2.7	100.0

Another way to avoid unwanted phase separation is to force the solvent system to be saturated in one component. For example, a distinct water layer is put in contact with water-saturated hexane and the solution is continuously stirred to ensure that the hexane will remain water-saturated. Unfortunately, the solubility is still temperature dependent, so that the solvent/column system must be temperature regulated to guarantee stable and reproducible chromatographic results.

1.9   BUFFERS AND OTHER MOBILE PHASE ADDITIVES

As described earlier, mobile phase modifiers (MPMs) are present at low levels— typically 2% or less. MPMs are typically considered as additives or buffers. To distinguish the role of additives and buffers, the following definitions are used. Additives are present at a defined level and the resulting equilibrium establishes the solvent characteristics. Buffers are MPMs that are added in order to hold a particular aspect of the mobile phase constant; a common case being pH. Examples of MPMs that are additives include 0.1% triethylamine (TEA), 0.2% trifluoroacetic acid (TFA), 1% phosphoric acid, and 50 mM sodium dodecyl sulfate.

Buffers are typically mixtures of solids and or a liquid and a solid. Examples of buffers include acetate buffer (50 mM at pH 4.0), phosphate buffer (100 mM at pH 7.0), and 0.1% TEA/0.1% TFA. Since the use of buffers in RP separations is commonplace, basic knowledge of the solubility of each buffer component in the mobile phase is critical. Consider the case of phosphoric acid and its series of conjugate acid/base pairs as the buffer for acetonitrile/water mobile phase. Phosphate has negligible UV absorbance down to 200 nm (see Fig. 1.5*d*), does not denature proteins, and has three effective buffer regions around the pH values of 2, 7, and 12. These properties make phosphate buffers very attractive for many HPLC separations.

The use of buffers is common and is critical enough to many separations to warrant a detailed discussion here. As mentioned above, HPLC buffers are commonly used to control solution hydronium ion, H^+. The H^+ concentration $[H^+]$ is represented by the pH scale where $pH \equiv -\log[H^+]$.

Buffers are chosen on the basis of their dissociation constants. For example, acetic acid has a dissociation constant, K_a, of 1.76×10^{-5} (also expressed as a pK_a value of 4.76). The chemical equilibrium for acetic acid is represented as

$$CH_3COOH \rightleftharpoons CH_3COO^- + H^+.$$

The corresponding equilibrium constant, K_a, is written as

$$K_a = \frac{[H^+][CH_3COO^-]}{[CH_3COOH]} \tag{1.5}$$

Solving Eq. 1.5 in terms of H^+ gives

$$[H^+] = \frac{K_a[CH_3COOH]}{[CH_3COO^-]} \tag{1.6}$$

From Eq. 1.6 it is easy to conclude that the most effective buffer, i.e., the one with the greatest capacity to resist changes in pH, is one that has $[CH_3COOH] = [CH_3COO^-]$. As a consequence, small changes in the acid or base content of a solution produce correspondingly small changes in $[CH_3COOH]/[CH_3COO^-]$ and so the $[H^+]$ (and the pH) remains constant. Also note that when $[CH_3COOH] = [CH_3COO^-]$, the ratio is 1 and $[H^+] = K_a$ (pH $= pK_a$). Therefore, buffers are constructed from acids and bases whose pK_a values are close to the desired pH. Table 1.3 lists some common components used for buffers.

When preparing a buffer solution, keep in mind that the buffer concentration should greatly exceed (at least by a factor of 10) any potential external source of acid or base. This is often not a problem in HPLC since very small volumes of low-level analyte are injected into the system. Also, the buffer must be compatible with the system (both soluble and nonreactive [other than as a buffer]). Representative buffer concentrations range from 5 to 200 mM. The use of straight acids (e.g., H_3PO_4, CF_3COOH) or bases ($[CH_3]_3N$, CH_3COONa) does not constitute a buffer since the conjugate base or acid is present in only negligible amounts.

A very important fact to remember when using buffers in HPLC mobile phases is that the pH system we commonly refer to is *based on an aqueous system*. In this system we are dealing with ostensibly "pure" water with a pH $\sim$ 7. Most HPLC mobile phases contain significant levels of organic components. Therefore, keep in mind that:

1. The *actual* pH of a water/organic buffer system is *not* that of a corresponding aqueous buffer.
2. The pH of a water/organic buffer obtained through the use of a pH electrode *does not* correspond to what is obtained for a corresponding aqueous buffer.

With respect to point (1), a 0.1 M aqueous acetate buffer of pH 4.2 does not have a pH equivalent to that of a 50/50 v/v methanol/water (0.2 M acetate buffer at pH 4.2). Considering point (2), a pH electrode responds to changes in $[H^+]$ through the changes it produces on the hydration layer of the glass electrode membrane. Obviously, a 50% solution of organic solvent will have a dramatic and variable effect on the functioning of the electrode and on the resulting "observed" pH. Examples of this are given in Table 1.4 where a series of 0.1 M aqueous pH values are compared against those of corresponding 0.1 M buffers in 50/50 v/v solution of water/acetonitrile or water/methanol (e.g., 50/50 v/v water/acetonitrile: 0.2 M acetate buffer at pH 2.21).

TABLE 1.3   Common Components Used for Buffers

Chemical Name	Reaction Equilibria	K_a	pK_a
Acetic acid	$CH_3COOH \overset{K_a}{\rightleftharpoons} CH_3COO^- + H^+$	1.74×10^{-4}	4.76
Ammonium	$NH_4^+ \overset{K_a}{\rightleftharpoons} NH_3 + H^+$	5.75×10^{-10}	9.24
Boric acid	$H_3BO_3 \overset{K_a}{\rightleftharpoons} H_2BO_3^- + H^+$	5.75×10^{-10}	9.24
Citric acid	$HOOCCH_2C(OH)CH_2COOH$ $\mid$ COOH $\overset{K_1}{\rightleftharpoons} HOOCCH_2C(OH)CH_2COO^- + H^+$ $\mid$ COOH	1.74×10^{-5}	3.13
	$HOOCCH_2C(OH)CH_2COO^-$ $\mid$ COOH $\overset{K_2}{\rightleftharpoons} HOOCCH_2C(OH)CH_2COO^- + H^+$ $\mid$ COO^-	1.74×10^{-5}	4.76
	$HOOCCH_2C(OH)CH_2COO^-$ $\mid$ COO^- $\overset{K_3}{\rightleftharpoons} {}^-OOCCH_2C(OH)CH_2COO^-$ $\mid$ COO^-	3.98×10^{-7}	6.40
Formic acid	$HCOOH \overset{K_a}{\rightleftharpoons} HCOO^- + H^+$	1.77×10^{-4}	3.75
Orthophosphoric acid	$H_3PO_4 \overset{K_1}{\rightleftharpoons} H_2PO_4^- + H^+$	7.08×10^{-3}	2.15
	$H_2PO_4^- \overset{K_2}{\rightleftharpoons} HPO_4^{2-} + H^+$	6.31×10^{-8}	7.20
	$HPO_4^{2-} \overset{K_3}{\rightleftharpoons} PO_4^{3-} + H^+$	4.17×10^{-13}	12.38
Triethanolammonium	$(HOCH_2(CH_2)_3NH^+ \overset{K_a}{\rightleftharpoons} (HOCH_2CH_2)_3N + H^+$	1.74×10^{-8}	7.76
Triethylammonium	$(CH_3CH_2)_3NH^+ \overset{K_a}{\rightleftharpoons} (CH_3CH_2)_3NH + H^+$	1.91×10^{-11}	10.72
Trifluoroacetic acid	$CF_3COOH \overset{K_a}{\rightleftharpoons} CF_3COO^- + H^+$	3.16×10^{-1}	0.50

The table is *not* intended to be representative of expected and quantitative pH values due to the variability noted above but is intended to reveal how dramatically the anticipated pH can vary from the "measured" value over a range of pH values and with different organic mobile phase constituents. The important conclusions to be drawn from these discussions are twofold. First, the most effective buffer is one that has a pK_a value close to the intended pH. Second, for a buffered mobile phase, one should generate the buffer in the aqueous phase at a proportionately increased concentration such that, upon dilution with the organic constituent, the correct final concentration is obtained.

Even though buffers and additives are present in the mobile phase at very low levels, their purity is still of major concern to the proper functioning of the

TABLE 1.4   "Apparent" pH of Mobile Phase Buffer Solutions

Aqueous pH[a]	"pH" in 50/50 Methanol/Buffer	"pH" in 50/50 Acetonitrile/Buffer
1.70	2.32	2.02
2.21	2.66	2.72
2.90	3.91	3.71
3.46	4.10	4.43
4.38	5.02	5.51
5.34	6.18	6.42
5.62	6.45	6.04
6.91	7.85[b]	7.33[c]

[a] Buffers at pH 1.70, 2.21, 2.91, 5.62, and 6.91 were prepared from phosphate and those at pH 3.46, 4.38, and 5.34 from acetate.
[b] This solution was very cloudy.
[c] This solution was hazy.

chromatographic system. An example is given in Figure 1.5c. Here, triethylamine (TEA) is compared as fresh and aged for more than 1 year. Note that the aged TEA has strong absorbance in the 250–280 nm region, perhaps an indication of the extraction of phenolic constituents from the bottle cap. Once again, proper rotation of materials is needed to ensure good and reproducible chromatography.

As noted above, acetonitrile/water mobile phases are commonly used in RP separations. As a consequence, a standard mobile phase buffer preparation often includes the preparation, filtration, and addition of an aqueous phosphate buffer solution to acetonitrile to produce the working mobile phase. The acetonitrile/aqueous phosphate buffer pair was explicitly chosen to illustrate the following point. When acetonitrile nears the 50/50 v/v level and the phosphate buffer approaches 50 mM (this concentration refers to the singly-charged moiety, $H_2PO_4^-$, and decreases as the charge increases on HPO_4^{2-} and PO_4^{3-}), a precipitate will form. The most critical aspect of this is that the formation of the precipitate may not be immediate. In essence, the chromatographer has created a wonderful recrystallization solvent and slow crystal formation results. As a consequence, the mobile phase is often put into use before the trouble is realized. Later, a fine layer of white precipitate is found coating the bottom of the solvent reservoir. If the precipitate is there, then it is also present in the pump head, abrading the piston and piston seal, in the column frits, constricting flow, and on the support material, precipitating in the pores and modifying the support surface. Therefore, buffer identity, concentration, charge, and solvent composition are all key variables requiring consideration when preparing mobile phases.

Precipitates are very difficult to remove from chromatographic systems. To purge a system in which a precipitate has formed, a mobile phase whose composition is identical to the buffer system is prepared leaving out the buffer. This is pumped through the system until equilibrium is reached. Next a solvent in which the buffer is considerably more soluble, yet compatible with the system, is pumped through the system at a slow flow rate, usually for many hours (or preferably overnight). If a sufficient volume is used (to keep the eluted buffer very dilute) the solvent may be recycled during this process. Unfortunately, the original column performance may never be regained.

Even more subtle is the gradient system in which both solvents A and B have different buffer compositions and precipitation or phase separation occurs as the gradient is formed. If any of the mobile phase component concentrations in the mixing chamber or *on the column* exceed their solubility in the solution, then phase separation or precipitation occurs. Here it is critical to remember that the adsorbed surface layer on the support material is *not* necessarily identical to the mobile phase. This adsorbed layer is usually enriched with the solvent components most similar to the support.

Resolubilization of the column-deposited immiscible liquid phases is time-consuming but can typically be accomplished through the use of a "universal solvent" (e.g., acetone or isopropyl alcohol). As discussed above, redissolving solid precipitates is at best difficult, and typically not possible, especially if the precipitate has formed in the pores of the support material or inlet frit.

To avoid solubility-related problems, it is recommended that mobile phases routinely be prepared at least the day prior to use. In this way equilibrium for the solvent is reached prior to use and observation of any phase separation is easily seen. The absence of precipitates or multiple phases is a strong indication that a usable mixture has been prepared.

1.10   VOLATILITY

The volatility of mobile phase components was not a major chromatographic problem in the past. The only time volatility came into play was in the initial solvent preparation step. To prevent the introduction of gas bubbles into a chromatographic system, with ultimately disastrous results when the bubbles collect in the pump head or the detector, a number of methods are used to degas solvent: sonication (5–10 minutes is common), application of a vacuum (pull vacuum until just before the solvent boils; make sure this is a gentle vacuum so that the glass container is not compromised and imploded), and/or sparging with an inert gas such as helium (see manufacturer's recommendations). Reviews of these techniques are found in the literature [14,25]. Regardless of the technique used, solvent degassing is done for a short time immediately prior to use. By minimizing the time spent in the degassing procedure, the critical composition of the solvent is unaltered.

Recent changes in HPLC pump designs have led manufacturers to recommend a continuous helium sparge of solvents. When pure solvents are used in separate reservoirs and mixed appropriately by the pump system, problems are seldom encountered (but keep in mind solubility issues). Frequently, however, a solvent mixture is prepared in a single reservoir. A continuous sparge leads to gradual and continual changes in composition, with the most volatile components removed to the greatest extent. With high levels of organic modifier in the mobile phase (>25%), significant changes in the chromatography typically do not result even over extended periods of time.

This is not the case when low levels of volatile modifiers are used. The greatest problems arise with the use of extremely volatile low-level mobile phase modifiers (MPMs), such as trifluoroacetic acid and triethylamine in reversed-phase solvent systems and ethyl acetate or dichloromethane components in normal-phase solvent systems. These low-level MPMs are typically used at the 0.1–1.0% v/v level. A continuous sparge over the course of the day will greatly reduce their concentrations via volatilization, and dramatic changes in peak retention and peak shape can result (see Fig. 1.13).

Volatilization effects are minimized if the sparge line is removed from the solvent reservoir after the initial degassing is complete. To prevent resorption of air into the solvent, the sparge line is left in the solvent reservoir but is positioned above the

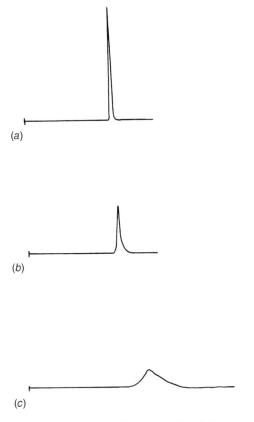

(a)

(b)

(c)

FIGURE 1.13   (a) Original peak using freshly prepared mobile phase. This result was generated from an amine standard and a 50/50/0.1 water/methanol/trifluoroacetic acid mobile phase. This mobile phase is continuously and vigorously sparged. (b) A chromatogram of the same standard injected after three hours of sparging. (c) A chromatogram of the same standard injected after eight hours of sparging.

solvent and the sparge flow is reduced. This creates a blanket of air-free sparge gas above the solvent. Volatilization will still occur since fresh sparge gas is not saturated with solvent vapor. However, the rate of volatilization will be markedly diminished since the surface area of contact is greatly reduced (i.e., the top of the solvent in the reservoir versus the total surface area of hundreds of bubbles rising through the solvent).

If the exact solution composition is critical, then the sparge gas may be presaturated by placing the sparge line in a solvent reservoir filled with mobile phase. The effluent gas from this bottle is then used to sparge the actual in-use solvent. The "sparge solvent" should be replaced frequently.

The problem of preferential volatilization always increases as the volume of the solvent in the solvent reservoir decreases. To virtually eliminate the problem of changing solvent composition, a blanket sparge with frequent replacement of solvent with fresh is recommended.

Temperature is a frequently overlooked chromatographic variable. Retention time is a strong function of temperature. Typically, as temperature increases, retention time decreases and peaks become sharper and more symmetric. Temperature effects are discussed under volatility because solvent volatility increases with increasing temperature. Special care should be taken to avoid working at temperatures at or above the boiling point of major solvent components of the mobile phase. The mixing of lower-boiling-point solvents with higher-boiling-point solvents (e.g., methanol [b.p.= 65°C] with water [b.p.= 100°C]) may extend the working temperature range, but rapid volatilization of methanol can be a significant problem unless specially vented and pressurizable reservoirs are used.

In normal HPLC work, temperatures of 60°C and above are considered high. To skirt the reservoir temperature issues, lengths of column tubing are sometimes placed in heaters and used to raise the solvent temperature. It should be noted that water at elevated temperatures is very aggressive and can cause rapid and irreproducible damage to silica-based columns and other HPLC components. Therefore, regardless of what technique is used, review of the pump, injector, and detector operation manuals is critical to make sure that each and every instrument component is compatible with high-temperature use.

1.11 SOLVENTS: INSTABILITY, REACTIVITY, AND DENATURANTS

Some classes of solvents are inherently unstable. Final breakdown products and/or intermediates may themselves be chemically reactive or alter the chromatographic properties of the solvent enough to be a cause of concern to the analyst. To further illustrate this, two classes of solvents (ethers and chlorinated solvents) and acetone will be discussed. Ethanol, due to its highly regulated status in its neat (or 200 proof) form, is often available in a denatured form. There are many denaturants for ethanol and they will be presented in this section. Finally, problems associated with water will be considered with regard to the pluses and minuses of manufactured versus in-house-produced sourcing.

If questions arise with regard to the stability of any given solvent, refer to any of a number of safety manuals [26,27] or call the solvent manufacturer directly.

1.11.1   Ethers

As mentioned previously, tetrahydrofuran (THF), a cyclic ether, readily forms peroxides. This phenomenon is true in general for ethers whose chemical structures include hydrogen bonded to a primary or secondary α-carbon. Tertiary carbons and methyls groups are much less prone to peroxide formation since a tertiary carbon has no active hydrogen and the methyl radical is not favored energetically. Figure 1.14 shows structures of a number of common ether solvents. From these structures and the information above, ethyl ether, THF, isopropyl ether, and dioxane should be active peroxide-forming compounds (they are) and methyl *t*-butyl ether should resist peroxide formation (it does).

Ethers are typically packaged under dry nitrogen to exclude water and oxygen, two substances that facilitate peroxide formation. Peroxide formation is also initiated by ultraviolet light. As a consequence, ethers are bottled in either metal (tin or aluminum) or amber glass containers. Manufacturers often add stabilizers (a.k.a., preservatives) to the peroxide-forming ethers.

THF is available unstabilized, or stabilized with 25 ppm (typical for GC use) to 250 ppm (typical for HPLC use) butylated hydroxytoluene (BHT; 2,6-di-*t*-butyl-*p*-cresol), which is added to scavenge the peroxide breakdown products of THF. Ethyl ether can be purchased unstabilized or stabilized with ethanol (~2–3%), BHT (1–10 ppm), or a blend of ethanol, water, and BHT (1.5–3.5%, 0.2–0.5%, and 5–10 ppm, respectively). Isopropyl ether is available unstabilized or stabilized with 0.01% hydroquinone or 5–100 ppm BHT. Dioxane is available unpreserved or with 25–1500 ppm BHT as preservative. From the above, it is important that the correct solvent be chosen for use, since most are available unstabilized or stabilized and with a range of stabilizers and stabilizer concentrations.

Peroxides are potent oxidizing agents and are chemically very reactive. Consequently, the use of ethers should be carefully considered in the context of the various chemistries present in the chromatographic system. Obviously, the use of unstable ethers should be avoided for analytes that are readily oxidized by or are chemically reactive with peroxides.

Another part of the chromatographic system susceptible to reaction with peroxides is the support material. Often considered as "inert," the bonded phase or the actual support material itself may be irreversibly modified through the reaction with peroxides. Peroxides under routine operating conditions chemically modify both silica-based aminopropyl and diol phases. Polymeric supports such as poly-(styrene-divinylbenzene) also react with peroxides. In any case, the use of ethers in conjunction with these supports should be avoided and preserved solvents should be used.

In the case of ethyl ether, an interesting historical sidelight has functional implications today. Decades ago it was determined that ethanol helps inhibit the formation of peroxides. Hence, ethanol was added as a preservative at ~2% v/v.

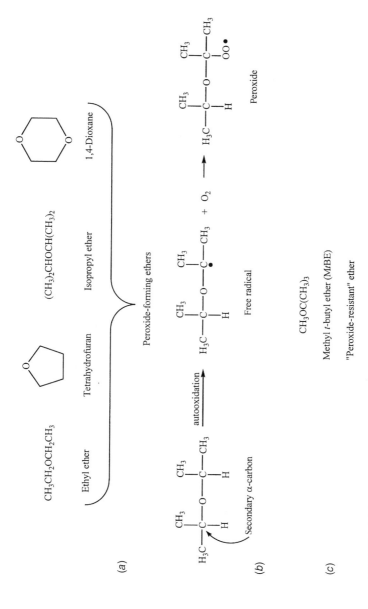

FIGURE 1.14 (*a*) Structures of some commonly used peroxide-forming ethers. (*b*) Scheme for the autooxidative production of a peroxide. (*c*) The structure of a "peroxide-resistant" ether.

33

Numerous extraction and separation methods were developed using this solvent blend. Subsequently, it was determined that ethanol was not really an inhibitor of peroxide formation or an active scavenger of peroxides. Resistance to the removal of the ethanol from ethyl ether arose not because of its lack of efficacy as a preservative but because of the significantly different separation properties that the "preserved" ethyl ether has in comparison with unpreserved ethyl ether. Today, both preserved and unpreserved ethyl ether are in common use and the technique must state specifically which type of ether is to be used.

From a safety point of view, ethers are no different from any other chemical—they should be handled with care at all times and disposed of properly. However, special concerns arise because of the formation of peroxides. In general, peroxides are not visible to the eye. Even though manufacturers usually guarantee peroxide levels below 1 ppm upon packaging, packaging integrity may be compromised by improper storage practices during shipment or after receipt. Storage at elevated temperatures and failure to store the solvent away from light will hasten peroxide formation. Also, if the seal breaks and the bottle cap loosens, then air will leak into the bottle and peroxides will form more rapidly. Since the chromatographer has little or no control over non-laboratory storage, it is good practice to routinely check ethers for peroxide levels prior to use.

With regard to peroxide screening, two classes of colorimetric tests exist for monitoring peroxide levels. The first includes variations of two rapid qualitative tests using potassium iodide [28,29]: Take ~1 mL of the sample to be tested and mix with an equal volume of freshly prepared glacial acetic acid that contains ~0.1 g of KI (potassium iodide) and shake well, *or* add 1 mL of a freshly prepared aqueous 10% KI solution to 10 mL of the sample to be tested and shake well. The oxidation of iodide (colorless) to triiodide (yellow/brown) indicates the presence of peroxides. The intensity of the color indicates the approximate peroxide level. A faint yellow tint correlated with very low levels of peroxide, whereas a brown color indicates potentially dangerous levels of peroxide.

Often a qualitative indication of peroxide level is sufficient to determine whether the solvent is suitable for use. However, if a quantitative peroxide level determination is needed, then a spectrophotometric titanium tetrachloride test is quantitative down to the low-ppm level. The method is quite involved and the reader should refer to the literature method for further details [30].

Peroxide levels above 250 ppm are considered hazardous. (Again, note that considerably lower levels of peroxide may be hazardous if methods requiring ether volume reduction—i.e., evaporation—are used.) If a crystalline deposit is ever observed inside an ether bottle or around the bottle cap or neck, *do not move the bottle.* Crystalline peroxides are particularly prone to explosion on contact. Notify your safety officer for proper disposal.

1.11.2   Chlorinated Alkanes

The most commonly used HPLC solvents in this class are dichloromethane, chloroform, and carbon tetrachloride. All of these solvents, as well as most highly

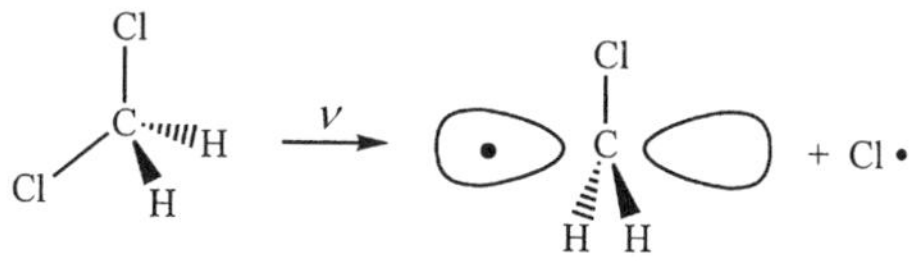

FIGURE 1.15   Free-radical formation. The free-radical product of dichloromethane is extremely reactive.

chlorinated non-aromatic solvents, break down in the presence of water and/or light to form reactive free-radical organic intermediates and hydrochloric acid (see Fig. 1.15). The free radicals ultimately react with other free radicals and reactive compounds to form a wide array of chlorinated by-products. The level and number of by-products, as well as the acidity of the solvent in general, increase as the solvent ages. As a consequence, as with ethers, chlorinated solvents require special attention. Proper storage (cool, dry, dark), judicious rotation of material and appropriate on-hand quantities (3–6 months' working stock maximum) will help ensure solvent integrity.

Because of their rapid breakdown process, chlorinated solvents are almost invariably sold containing a stabilizer. Dichloromethane has ~25 ppm amylene (2-methyl-2-butene), cyclohexene, 400–600 ppm methanol, or a methanol/amylene blend as stabilizer. These alkenes act as chemical "sinks" that react with hydrochloric acid. Cyclohexane has also been used as a preservative but its efficacy is questionable.

Chloroform stabilizers include ethanol (0.5–1%), amylene (50–150 ppm), and an ethanol/amylene blend.

Carbon tetrachloride is an exception and is commonly produced without a preservative. The reason for this is that carbon tetrachloride is frequently used as an aprotic solvent for infrared analyses. Infrared quantitation of oil and grease levels through the intensity of the C—H stretch can be done using carbon tetrachloride. An alkene preservative would render it unsuitable for use.

At ambient temperature the solubility of water in chlorinated solvents ranges from ~80 ppm (carbon tetrachloride) to ~560 ppm (chloroform) to ~2500 ppm (dichloromethane). Manufacturers typically produce these solvents with water levels ≤100 ppm. In these solvents, commonly used in normal-phase HPLC, the support materials (e.g., silica or alumina) are deactivated by water. As the surface-adsorbed water level on silica increases (typically as a result of increases in the water content of the mobile phase), a dramatic loss of resolution occurs, specifically because water in the solvent prevents the active surface sites from playing a role in the retention process.

Reduction of the water content in the mobile phase is achieved through the use of drying agents such as molecular sieves or a silica precolumn. It should be noted that these drying agents are excellent general-use sorbents and will absorb airborne compounds, organic as well as water, during preuse storage. Therefore, these drying agents must be scrupulously cleaned, often by heating, before use in the solvent.

Unfortunately, even after cleaning, significant levels of particulates and high-molecular-weight residues are often present and may desorb into the solvent. Frit clogging and blockage of the column inlet by particulates causes increased system backpressure. The presence of residues can modify the support and change the chromatography.

1.11.3   Ethanol

Perhaps the most difficult aspect of using ethanol in an HPLC system is finding exactly what it is called! A list of synonyms includes alcohol, denatured ethanol, dehydrated alcohol, reagent alcohol, and ethyl alcohol. Now that the solvent can be found, a second confusing issue arises. Ethanol is not readily available in its neat form. This is because the production and use of neat ethanol is subject to close government scrutiny and a seemingly endless paper trail including taxes and tariffs. To avoid this, components called denaturants are added to ethanol so as to remove it from governmental auspices.

Some of the denaturants in laboratory use ethanol are cyclohexane (1%); isopropyl alcohol (~5%); diethyl phthalate (1%); toluene (1–2%); 2-butanone (methyl ethyl ketone) (2%); a mixture of 2-butanone and 4-methyl-2-pentanone (methyl isobutyl ketone) (2% and 0.5%, respectively); methanol (5%), designated SDA Formula 3; ethyl acetate and gasoline (5% and 1%, respectively), designated SDA Formula 1; ethyl acetate, 4-methyl-2-pentanone, and gasoline (each at 1%), designated SDA Formula 1-1; ethyl acetate and methanol (1–2% and 3–5%, respectively); 4-methyl-2-pentanone and kerosene (4% and 1%, respectively), designated Formula CDA 19; and isopropyl alcohol and methanol (5% each). Note that many other denatured ethanols (e.g., bitrex and menthol denatured) are available, typically for use in personal care and cosmetic products.

Again, the proper selection of ethanol is crucial to the reproducibility of an analysis and should be clearly designated in a method. Also note that ethanol composition, if used in as part of a mobile phase, will vary lot-to-lot due to the approximate denaturant level. This means that if peak resolution is small these effects could be critical to the separation.

1.11.4   Water

Nearly all RP separations use water as part of the mobile phase. Because water is a ubiquitous solvent familiar to and used by almost every chromatographer, it is often assumed to present little or no problem in chromatography. This assumption has led to nightmares for many chromatographers.

Water is purified in such a manner that both organic and inorganic contaminants are removed. When water is properly produced and packaged, it has a pH of ~7, low levels of inorganic and organic contaminants, and no bacteria. Most manufacturers monitor these aspects via the pH response, low-UV gradient profile, resistivity measurements, and bacterial tests. The final pH and metal content of water depend not only on the manufacturing process but also on the packaging techniques and contact time with air (i.e., introduction of carbon dioxide). When a sealed bottle of water is opened, airborne bacteria and bacterial nutrients enter. Bacteria start to grow and multiply. Unless the water is mixed with significant levels of organic modifiers (>15% acetonitrile, >20% methanol, etc.), the bacteria continue to multiply. The solvent becomes contaminated with living and dead bacteria, bacterial cell-wall fragments, and peptidic fragments from the interior proteins of the bacteria cell. When this solvent is pumped through an HPLC column, either contaminants collect on the column until it is overloaded and breakthrough occurs, or the solvent becomes strong enough to elute contaminants. In either case, unanticipated and unacceptable non-sample-related peaks result. Bacterial growth can usually be prevented if the water is mixed with an organic modifier (as mentioned above), with a very low-pH buffer (e.g., acetic acid pH < 4), or with a very high-pH buffer (e.g., borax pH > 9). Consequently, when possible, the weak solvent in an RP separation should not be pure water that is then mixed by the pump with the organic component. Rather, the highest possible percentage of organic component in water that is compatible with the chromatographic requirement should be placed in the weaker solvent reservoir.

Above and beyond these problems are the troubles created by the fact that, in bacteria-laden water, bacteria have found their way into every nook and cranny in the HPLC system. The components with the largest surface areas (inlet reservoir filters; in-line and column frits) serve as active multiplication sites for more bacteria. Results of this type of bacterial growth are seen chromatographically as spurious, irreproducible peaks. Correction of the problem is time-consuming. Each stainless steel component needs to be cleaned with a solution of dilute nitric acid followed by water and methanol rinses. (Make sure that each component in the HPLC system is compatible with nitric acid. Components that are not should be cleaned according to the manufacturer's instructions or replaced.)

Some manufacturers add sodium azide to prevent bacterial growth in water. The presence of the azide will affect ion and ion-pair chromatography separations. Since the sodium azide is ionic in aqueous solutions, it may also have subtle but important effects on the RP chromatography of protic analytes, charged analytes, and more complicated analytes such as peptides and proteins. Also, sodium azide poses safety and health risks. Explosive heavy-metal azides may form when incorrect disposal methods are used (e.g. if azides come into contact with metallic plumbing fixtures when discharged down the sink). Beyond this, sodium azide is highly toxic and carcinogenic. Therefore, great care must be exercised when it is used.

Water is the solvent that probably has the greatest number of documented grades of specifications. These include ACS (American Chemical Society), USP (United States Pharmacopoeia), ASTM (American Society for Testing and Materials), and SEMI (Semiconductor Equipment and Materials Institute) specifications as well as more general specifications such as "pyrogen-free." It is critical to ensure that the water used in a method meets the specifications demanded by the method. Since many of these organizations include tests that are non-standard with respect to high-purity HPLC solvent manufacturers, it is important to contact the manufacturer whenever it is unclear whether or not their water is suitable for the intended use.

For those laboratories producing in-house LC-grade water it is important that the proper filters/resins (e.g., carbon, reverse osmosis [RO], ion exchange) are installed to yield the correct level of purity. Many different systems are commercially marketed, and a detailed list of laboratory needs and uses will help the manufacturer of the system assemble the correct configuration (i.e., the order of the filters as well as the type).

The water source feeding the purification system must also be tested. If it contains high levels of ionic and/or organic contaminants, then a special pretreatment station may need to be installed between the source and the "polishing" unit. It should be realized that water sources often originate from city treatment plants and the quality will vary with the season since more bleaching additives are required during the summer months than during the winter months.

A strict preventive maintenance schedule for the exchange of the filters/resins and connecting tubing as well as a routine quality control protocol for finished water purity verification is essential to laboratories producing in-house water.

1.11.5 Acetone

Acetone has superior solubilizing characteristics and is often used as the final rinse to remove water and organic residues from glassware. As wonderful as this characteristic may be, chromatographers must realize that acetone is chemically reactive and undergoes self-condensation to yield diacetone alcohol and higher oligomers. Levels of diacetone alcohol in the low-ppm range are typical in aged acetone. The level is usually of little concern in an HPLC separation. The major concern centers on the use of acetone in conjunction with potentially reactive solvent, solutes or stationary phases. For example, acetone, with time, will cause permanent chemical modification of an aminopropyl support material. Therefore, even though acetone is chromatographically acceptable as a solvent for sugar analyses (which often use an aminopropyl support), it is unacceptable for long-term system compatibility.

A major feature of acetone noted previously is that, along with isopropyl alcohol (IPA), it is considered a "universal" solvent. This means that acetone and IPA are fully miscible with a wide range of compounds within solvent classes such as alkanes, chlorinated alkanes, ketones, and—most importantly—water. Consequently, for those support materials that are used as either NP or RP supports (e.g., cyanopropyl, aminopropyl, diol), acetone (or IPA) may be used as a conversion solvent. For example, to change a cyanopropyl column from a NP hexane/dichloromethane mobile phase to a RP acetonitrile/water system, acetone (or IPA) is equilibrated with the column as an intermediate step. Note that buffers should never be present in any solution during the conversion step. If a buffer is in the mobile phase, then an identical mobile phase without the buffer must be equilibrated with the column prior to the conversion step.

When considering acetone or IPA, the two most important characteristics are as follows. Acetone is a strong absorber in the UV region up to 340 nm; therefore, flushing acetone from the system when using a UV detector may not be ideal.

Conversely, IPA has low absorbance up to 240 nm but is extremely viscous. Slow flow rates are typically used when IPA is the conversion solvent. IPA is also "nonreactive" under normal operating conditions.

In conclusion, solvent instability or unique solvent properties may result in situations where the sample, solvent, column, or system integrity may be jeopardized. An understanding of the solvent and its potential chemical reactivities will save the chromatographer hours of lost time and avoid inaccurate or incorrect results.

1.12 SAMPLE STABILITY IN SOLVENTS

A completely separate yet extremely important aspect of sample/solvent compatibility is that of sample stability in the solvent. There are three distinct scenarios here.

First, consider the overall stability of the sample in the preparation solvent. Note that this solvent may or may not be identical to the mobile phase (although it is generally recommended that the preparation solvent and mobile phase be the same so as to avoid system peak generation, etc.) Samples (this includes standards) spend a large amount of time interacting with the preparation solvent. The time from completion of the preparation to the actual analysis is often hours. In some instances this has been shown to be crucial in cases of protein denaturation in solvents [31] and on columns [32]; decomposition of derivatized analytes [33]; time, temperature, and light effects in the case of ascorbic acid [34]; and pesticide residue decomposition on support media such as graphitized carbon and C_{18} silica [35].

Second, when a sample is not made up in the mobile phase itself, there may be equilibrium issues when the sample is injected into the mobile phase. For example, if the sample is prepared in a solvent that is stronger than the mobile phase, then solubility may be exceeded when the injection plug is diluted by the mobile phase. This type of non-equilibrium sample overload can result in peak fronting (see Fig. 1.16).

The third point, and related to the second point, is when samples are prepared in solvents weaker than the mobile phase. A classic example involves proteins. Here, the protein is often dissolved in an aqueous buffer system and subsequently injected into a mobile phase containing a significant level of organic modifier. When the level of organic is too high (e.g., acetonitrile >20% v/v), the protein may denature and precipitate out of the mobile phase onto the column support material [36].

From the above it is easily concluded that a working knowledge of sample/solvent interaction (above and beyond chemical reactions) is important. One practical method for determining stability is to prepare a standard solution at the beginning of the method validation process. Use this sample as a check standard throughout the validation process. Plotting the recovery versus time will give an accurate estimate of the acceptable "shelf-life" of the analyte in the preparation solvent.

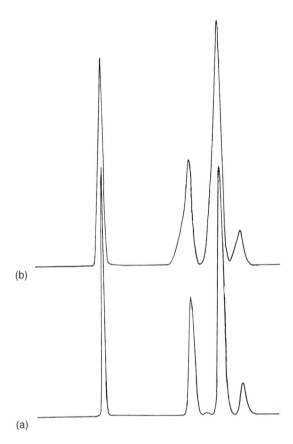

FIGURE 1.16 (*a*) A chromatogram of four UV-blocking agents dissolved in the mobile phase. Note the very good peak symmetry. (*b*) The same agents dissolved in methanol. Predicting when this will and will not be a problem is very difficult.

1.13   MOBILE PHASE INTERACTION WITH STATIONARY PHASE

The mobile phase plays a number of interactive roles in a chromatographic separation. First and foremost, it is in dynamic equilibrium with the stationary phase. For base silica (and other unmodified supports), there exists an adsorbed layer of mobile phase components, not necessarily identical to the mobile phase composition. This is why water deactivates a silica column; it forms strong hydrogen bonds with the residual silanol groups (the active adsorption sites on the surface) and renders them unavailable for adsorptive interactions with the solute.

For bonded phases, such as a C_{18}, the less polar components wet the surface and expand the bonded phase, and thereby provide more partitioning volume. A critical minimum level of organic is needed to accomplish the wetting (typically 2% by volume is considered an absolute minimum). If the bonded phase is not wetted, it collapses onto the surface and causes a dramatic decrease in the available interaction sites. Once the bonded phase is wetted, solutes partition into/onto the phase and displace the adsorbed mobile phase constituents.

Finally, the mobile phase is a potentially destructive element for the stationary phase because the bonded phase or stationary phase is, to greater or lesser degree, soluble in the mobile phase. This destructive aspect takes the form of deactivation (as noted with silica above) or of actual dissolving of the stationary phase and/or removal of the bonded phase. The rate of dissolution increases when the pH is above 7 or below 2. It also increases with increasing mobile phase ionic strength (e.g., higher buffer concentrations) and increasing temperature. In order to maximize the optimal column performance lifetime it is often suggested that a precolumn be used when a particularly aggressive mobile phase is required. The precolumn is inserted between the pump and the injector. An analytical column is also protected through the use of a guard column. The guard column is placed between the injector and the analytical column and also provides protection from excipients found in the sample.

It should be noted that special bonding technologies have led to manufacturers claiming long-term support stability for pH ranging from 1 to 12 and wettability at extremely high water mobile phase content. For pH stability some manufacturers have made use of bulky silanes such as diisopropyloctadecyl- and diisobutylocta-decylsilanes in order to make the silica surface less accessible to aggressive solvents and thereby increase column stability. Other manufacturers have actually created a polymeric "coating" that is bound to the surface.

For wettability at water content >97%, polar functional groups are either interspersed along the bonded phase backbone or used as "end capping" agents. Be sure to read the manufacturer's literature carefully in order to confirm this and determine whether any special considerations need to be taken into account.

1.14 INHERENT CONTAMINANTS

One of the most difficult problems any manufacturer of high-purity solvents faces is the variability of the raw material stock used in solvent production. Solvent raw materials can be generated in many ways, including production via fractionation of large-scale distillations, as by-products of petroleum cracking processes, and as a result of synthetic processes or by-products thereof [37].

Each production method will introduce vastly different levels and types of impurities. As an example, consider the solvent methyl *t*-butyl ether (M*t*BE). When it is the product of a petroleum-based source, numerous ethers, aldehydes, ketones, and alcohols are typically present. Synthesis of M*t*BE from methanol and isobutylene reduces the potential contaminant range to residual methanol and isobutylene and alcoholic contaminants in the methanol and alkene contaminants in the isobutylene. The suitability of either source of M*t*BE for final analytical use depends strictly on the requirement of the end user.

More important, from an end user point of view, is the subtle variability of the raw material over time and the resultant effects on the finished high-purity solvent product. An example of this is methanol. A number of years ago, raw material methanol contained a high and variable level of low-molecular-weight amines. At that time, the manufacturing process did not consistently remove the amines and so the high-purity product often had a "fishy" odor. Beyond the malodor, which was aesthetically displeasing, the very low levels of amines typically presented little problem for the separation. However, in cases where derivatizations were run or in gradient work at low UV settings, these low-level contaminations proved disastrous.

Review of common and uncommon contaminants in high-purity solvents is beyond the scope of this section. The important point here is that the presence of a contaminant in a high-purity solvent is often totally unexpected and unanticipated by both the manufacturer and the end user. One must realize that it is impossible for any manufacturer to develop tests and specifications for unanticipated contaminants in the raw materials. It should be emphasized that feedback from end users to a manufacturer is critical. Feedback makes a manufacturer aware of problems that may then be addressed and eliminated in the future.

1.15   SOLVENT EFFECTS ON FLUORESCENCE

The use of fluorescence as the method of detection is often preferred in HPLC because it usually offers both the lowest detection limits and the highest sensitivities of commonly used detectors. Enhanced specificity is offered because the number of solutes that fluoresce is limited. These properties give the chromatographer the capability of discriminating between low concentrations of compounds of interest that fluoresce from high concentration of non-fluorescent interferents. When interferents fluoresce, further discrimination is possible via the detector by "fine tuning" the excitation wavelength and judiciously choosing the emission wavelength for the analysis.

A wide range of fluorescent derivatization reagents and techniques have been developed to utilize the analytical advantages that fluorescence offers. These derivatization reagents are usually specific to functional groups (e.g., amine, hydroxy, thiol), and their specificity offers yet another opportunity for discriminating against interferents. For excellent reviews, the reader is referred to references [38] and [39].

The choice of a fluorescent tag depends on the type of derivatization (e.g., precolumn or postcolumn) because the stability of analyte–tag chemical bonds varies greatly. Therefore, for precolumn derivatization it is important to determine the time delay between derivatization and analysis as well as the stability of the reaction product in the mobile phase prior to HPLC method development. For postcolumn derivatization, compatibility of the reaction solvent with the chromatographic system as well as the rate of reaction are variables that must be considered.

If a fluorescent derivatizing reagent that is non-fluorescent when unreacted (and therefore detector-transparent) but fluoresces when it becomes the tag (i.e., the part of the derivatizing reagent that is chemically bonded to the analyte) cannot be found, then the elution time of the excess unreacted reagent is important since it is often an extremely large peak compared to the analyte of interest.

Solvent–solute interactions are important in fluorescence analyses. These interactions have large effects on the reproducibility, detection limits, and sensitivity of a method. First, for $n \rightarrow \pi^*$ transition compounds (e.g., ketones) in nonpolar and aprotic solvents (e.g., alkanes and ethers), little or no fluorescence occurs. However, in protic solvents (e.g., methanol) these compounds fluoresce. Conversely, $\pi \rightarrow \pi^*$ transition compounds (e.g., polycyclic aromatic hydrocarbons) fluoresce readily in nonpolar solvents and less in protic solvents.

The presence of heavy atoms (e.g., halogens) in solvents generally tends to decrease fluorescence intensity. The same is true with increasing temperature, where a 1°C increase in temperature decreases fluorescence intensity by ~2% [40].

Special care should be taken when trying to extrapolate results from one solvent system to another, since even minor changes in the polarity or pH of the solvent often have a dramatic effect on the fluorescent behavior of the analyte. These changes occur because of shifts and/or changes in the λ_{ex} and λ_{em} spectra due to stabilization or destabilization of the electronic structure of the analyte (e.g., protonated vs. deprotonated) or changes in the polarity of the surrounding solvent (e.g., hexane vs. methanol).

Impurities in the solvent may also be capable of fluorescing. For example, acrylonitrile, acrolein, and acrylic acid impurities and their adducts with mobile phase components will fluoresce under the proper conditions. This also happens when trifluoroacetic acid is a mobile phase modifier and amine impurities are in the mobile phase. The result is the appearance of spurious chromatographic peaks. As in UV absorbance techniques, these peaks are of variable size when a gradient is used and the size of the peak is directly proportional to the amount of solvent pumped onto the column between gradients. If such solvent combinations are used, then it is recommended that the fluorescence output of the solvent mixture be monitored prior to analytical use.

1.16 PARTICULATES AND SOLVENT FILTERING

Particulate materials in HPLC solvents are the potential source of many problems. These materials block the solvent reservoir inlet filters, score pistons, wear piston seals, and plug column inlet frits. Therefore, particulates must be removed from the solvent prior to use.

Into the late 80's solvent filtering prior to use in any HPC system was a routine requirement. Today, most high-purity solvents are prefiltered before final packaging. Although this does not ensure that the solvent will be particulate-free, particulate levels are orders of magnitude lower than in the past.

At least one manufacturer (Burdick & Jackson; B&J) recommends that end users *never* filter B&J solvents if they are used neat or in mixtures with other B&J solvents. Why? Because the commercially available filter media used to remove particulate material from the solvent often *adds* 10–100 times more particulates than were present in the unfiltered solvent. In the worst-case scenario, the filter medium not only adds particulates but may also add extractable (or adsorbed) organic materials into the solvent. This is very dependent on the solvent and filter, but these extractables invariably present themselves as unexpected and variable chromatographic peaks. Therefore, time, effort, and potential difficulties can be avoided when the appropriate starting solvents are purchased.

Exceptions to the "never filter" rule occur when mobile phases contain MPMs (particularly salts) added to the solvent. When this is the case, the solvent containing the salts should be prepared and filtered independently and then mixed with the other high-purity solvent components. For example, consider a 50/50 v/v methanol/water (0.05 M acetate buffer at pH 4). Here the aqueous acetate buffer is prepared from glacial acetic acid and sodium acetate, filtered, and then added in equal volumes to methanol.

1.17   MANUFACTURERS AND TESTING PROTOCOLS

A major published testing procedure compendium for HPLC solvents is that of the American Chemical Society: *Reagent Chemicals*, 9th edition [41]. Therein are described two procedures, one for testing water for suitability of use in the testing of HPLC solvents. The actual test method is a gradient run on a C_{18} column and applies only to water/acetonitrile and water/methanol gradients. The elution profile is monitored at 254 nm and a maximum peak (not plateau) value is <0.005 AU. Note that the choice of 254 nm means that the test method focuses on phenyl-containing and carbonyl-containing contaminants.

Individual manufacturers have developed their own set of test protocols and you should contact them if you have questions.

2

METHOD OPTIMIZATION

The use of literature methods as a resource is powerful and time-saving. Separations of compounds that are structurally similar to those involved in the proposed separation are often available. Unfortunately, the separation cited is frequently not exactly what is needed and, as a consequence, some method adjustments are required to achieve the desired results.

When the analyst is not so fortunate and no comparable published method is available, a completely new method may need to be developed. In either instance, the process of method optimization is necessary. Note that the analyst determines the definition of the term "optimization." To begin the optimization process, criteria [i.e., minimal (value must be $>X$), maximal (value must be $<Y$), or bounded (between X and Y)] are established for any or all of the following: peak shape, peak height, peak separation, run time, and overall analysis time (and therefore sample throughput). For the last listed criterion, remember that in gradient separations the re-equilibration time must be considered as part of the analysis time since a subsequent injection cannot occur until after the system has reached equilibrium with the initial solvent.

There are many approaches to method optimization. The goal of each is to minimize the time and effort needed to generate an optimal separation. Five commonly used optimization techniques will be discussed here: eluotropic/solvent strength parameters, triangulation and related algorithmic methods, scout gradients, window diagrams, and the "brute-force" method. All of these techniques generate results that either produce an "optimal" or acceptable method or are then used to help narrow the choices for the next set of chromatographic experiments. None of these techniques generates quantitative *predictions* (in fact, at this time there exists no "universal" predictive retention model). Each technique will be presented and described in detail below.

2.1   ELUOTROPIC SERIES AND SOLVENT STRENGTH PARAMETERS

2.1.1 Eluotropic Series

These parameters were developed to provide the chromatographer with a rapid reference to the relative elution strengths of solvents. There are essentially two solvent elution strength scales: the eluotropic series and solvent strength parameters.

The eluotropic series was one of the first sets of data used for estimating the strength of a solvent by its effect on solute retention. The earliest chromatographic supports were underivatized (e.g., base silicas and aluminas) and as a consequence the eluotropic series was initially defined for these supports. To generate distinct values for each solvent, the average retention time was obtained from a large number of solutes on a specific support material. From these retention times each solvent was assigned a distinct eluotropic value, ε° [42]. It should be noted that ε° values are based on adsorption processes so that the use of ε° has its greatest effectiveness in the area of liquid–solid separations.

Recently, these parameters have been expanded to include a limited number of solvents for the reversed-phase octadecyl (C_{18}) support. It should be realized however, that ε° is based on *adsorption* equilibrium and that adsorption (except in very specific instances) is not a predominant retention mechanism in reversed-phase separations. That having been stated, regardless of what system the ε° parameter is used in conjunction with, it is at best an estimate of the solvent strength for an average solute on a generic support.

Table 2.1 shows the ε° values for a wide range of solvents on a number of stationary phases. In general, the relationship between ε° and retention is that as ε° increases in value, the retention of a solute decreases. In chromatographic nomenclature, when comparing the elution characteristics of two solvents, the solvent that produces the longer retention time is referred to as the *weak* solvent whereas the solvent that produces the shorter retention time is referred to as the *strong* solvent. For silica and alumina $0.0 < \varepsilon^\circ < 1.0$. Therefore, on a silica support, hexane, with a low ε° value (0.01), is a very weak solvent while methanol, with a large ε° value (0.95), is a very strong solvent.

More specifically, an increase in ε° of 0.05 (making a stronger solvent) decreases k' by a factor of 2 to 4. [Retention, monitored as capacity factors, $k' (k' \equiv (t_r - t_0)/t_0$, where t_r is the retention time of the solute and t_0 is the system void volume) is a strong logarithmic function of solvent composition.] Conversely, a decrease in ε° of 0.05 (making a weaker solvent) increases k' by a factor of 2–4 [49]. Finally, conversion from a silica-based ε° value to an alumina-based ε° value can be done using the approximate conversion of $1.4\varepsilon^\circ_{(SiO_2)} \approx \varepsilon^\circ_{(Al_2O_3)}$ [43].

The reason for the wide range in effect between alumina and silica is that the ε° value is not only a function of the mobile phase but is dependent upon the solute chemical structure and the support material as well. As a general rule of thumb, the more closely a solvent chemically resembles the support, the stronger an eluent it is (solubility issues withstanding!). For example, hexadecane (C_{16}) is a very strong solvent on an octadecyl (C_{18}) support, whereas ethanol (CH_3CH_2OH) is a very

TABLE 2.1 Eluotropic Strength of Solvents on Various Sorbents[a]

Solvent	$\varepsilon^{\circ}_{(Al_2O_3)}$[b]	$\varepsilon^{\circ}_{(SiO_2)}$[b]	$\varepsilon^{\circ}_{(C_{18})}$[b]	P'
Pentane	$\equiv 0.00$[c]	$\equiv 0.00$	–	0.0
Hexane	0.00–0.01	0.00–0.01	–	0.1
Iso-octane	0.01	0.01	–	0.1
Cyclohexane	0.04	0.03	–	0.2
Carbon tetrachloride	0.17–0.18	0.11	–	1.6
1-Chlorobutane	0.26–0.30	0.20	–	1.0
Xylene	0.26	–	–	2.5
Toluene	0.20–0.30	0.22	–	2.4
Chlorobenzene	0.30–0.31	0.23	–	2.7
Benzene	0.32	0.25	–	–
Ethyl ether	0.38	0.38–0.43	–	2.8
Dichloromethane	0.36–0.42	0.30–0.32	–	3.1
Chloroform	0.36–0.40	0.26	–	4.1
1,2-Dichloroethane	0.44–0.49	–	–	3.5
Methyl ethyl ketone	0.51	–	–	5.7
Acetone	0.56–0.58	0.47–0.53	8.8	5.1
Dioxane	0.56–0.61	0.49–0.51	11.7	4.8
1-Pentanol	0.61	–	–	–
Tetrahydrofuran	0.45–0.62	0.53	3.7	4.0
Methyl t-butyl ether	0.3–0.62	0.48	–	2.5
Ethyl acetate	0.58–0.62	0.38–0.48	–	4.4
Dimethyl sulfoxide	0.62–0.75	–	–	7.2
Diethylamine	0.63	–	–	–
Acetonitrile	0.52–0.65	0.50–0.52	3.1	5.8
1-Butanol	0.70	–	–	3.9
Pyridine	0.71	–	–	5.3
2-Methoxyethanol	0.74	–	–	5.5
n-Propyl alcohol	0.78–0.82	–	10.1	4.0
Isopropyl alcohol	0.78–0.82	0.60	8.3	3.9
Ethanol	0.88	–	3.1	–
Methanol	0.95	0.70–0.73	$\equiv 1.0$	5.1
Ethylene glycol	1.11	–	–	–
Dimethyl formamide	–	–	7.6	6.4
Water	–	–	–	10.2

[a] Compiled from references [42–55].
[b] Values represented as ranges indicate multiple sources.
[c] The "$\equiv$" symbol means that these are defined values.

strong solvent on a silica (SiOH) support. (Note that water is such a strong solvent on silica or alumina that it effectively deactivates the material; i.e., neutralizes all the active retention sites on the surface through strong hydrogen-bond interactions.)

The greatest utility for the ε° values is in the estimation of an approximate mobile phase composition needed to generate acceptable retention times. Unfortunately, the equations used for predicting an appropriate mobile phase composition are complicated and time-consuming to use [50]. To simplify the process, numerous equivalence charts have been generated such as the nomogram shown in Figure 2.1. To use Figure 2.1, connect the identical $\varepsilon°$ values on the top and bottom $\varepsilon°$ scales. Where the resultant line intersects a binary solvent scale, the percentage of each solvent needed is read directly. If such an intersection does not occur, the solvent pair is unsuitable for use in that separation.

As stated above, $\varepsilon°$ is a representation of the average retention for a large set of solutes, generally on one support. Since every solute has a different and unique retention vs. solvent composition response, as consequence selectivity—the ability of a system to resolve components within a mixture—cannot be strictly determined from $\varepsilon°$ values. In NP separations selectivity is often maximized when the stronger component of the mixture (that component having the highest $\varepsilon°$ value) is at levels < 5% or ≫ 50%. Because high $\varepsilon°$ solvents typically have higher water level content, and water deactivates unmodified silica and alumina supports, the lower level of modifier is almost always used.

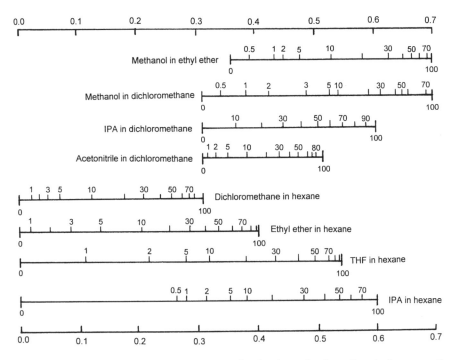

FIGURE 2.1   Eluotropic strength nomogram for the determination of equivalent-strength mobile phases. To use the nomogram: (1) Find the current mobile phase composition; (2) connect the same $\varepsilon°$ values at the top and bottom of the nomogram; and (3) read off the equivalent mobile phase composition from where the line intersects the mobile phase bar. If the line does not intersect the mobile phase composition of interest, then there is no equivalent mobile phase (e.g., 1% THF in hexane has no IPA/dichloromethane equivalent).

For example, three test solutes [chlorobenzene (CB), anisole (A), fluoronitro-benzene (FNB)] were separated on a silica column using mobile phase mixtures with an ε° value of 0.08 (see Fig. 2.1). These mobile phases included 95/5 hexane/dichloromethane, 97.5/2.5 hexane/ethyl ether, 99.25/0.75 hexane/THF, and 99.85/0.15 hexane/isopropyl alcohol. Figure 2.2 shows the separation of the three compounds using these solvents. It is evident that, even though ε° "predicts" equivalency, there are distinct and significant differences not only on the overall retention times but also in the resolution of the first two eluting peaks. For example, the mobile phase with THF generates the longest retention time for FNB but only an intermediate resolution of A from CB. Clearly, the most effective mobile phase (greatest resolution with shortest overall retention time) is 95/5 hexane/dichloromethane. A slight increase in dichloromethane and/or a higher flow rate would increase sample throughput (shorten analysis time).

Depending upon the chemical structure of the solute, the addition of hydrogen bond donating (methanol) or hydrogen bond accepting (ethyl ether) solvents to the mobile phase may improve selectivity (i.e., enhance resolution) due to unique interactions with the solute and/or modification of the silica surface [51,52]. Therefore, ε° cannot predict any enhancement in selectivity that one solvent pair may provide over another since this is often a result of specific mobile phase/solute functional group interactions.

In addition, another limitation to the effectiveness of the ε° values is that commercially available silicas and aluminas vary substantially in physical structure and chemical composition. Surface area, hydrogen activity (often reported as the pH of a 1% slurried suspension), and impurity type/level vary dramatically between producers. These differences have a huge affect on a separation. As a result, the ε° scales offer the analyst as basis from which to decrease the guesswork as to what solvent composition may be an appropriate starting point in the development, or refinement, of a method. The greatest utility results when the comparison is carried out on one column or, barring that, on columns packed from the same lot of packing material.

In conclusion, eluotropic values, although derived from complex mathematical relationships and difficult to generate, offer a rapid means of efficiently investigating different solvents. The use of nomograms is key for multicomponent mobile phases.

2.1.2 Solvent Strength Parameters

A separate approach used in predicting solute retention is the use of polarity parameters, P' [53]. The P' values for a wide range of solvents appear in Table 2.1. Unlike ε°, P' values are derived from partition equilibrium values and, therefore, have little direct application to adsorption-driven separations. Rather, P' values are effectively used in separations involving reversed phases (e.g., octadecyl, octyl, and phenyl). For these systems, the partitioning equilibrium is the major retention process.

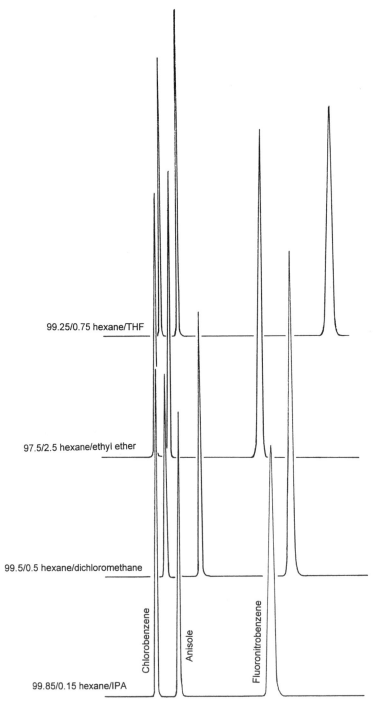

FIGURE 2.2   Normal-phase separation of three analytes on a silica column using mobile phases of equal $\varepsilon°$. Note that both the overall retention times and the resolution between peaks are significantly different.

The relationship between P' and an experimentally determined chromatographic parameter is represented as:

$$\log\left(\frac{k'_2}{k'_1}\right) = \frac{P'_2 - P'_1}{2} \tag{2.1}$$

where, k' is the relative retention of a given solute in two different mobile phases, 1 and 2. In addition, as a rough first approximation, P' for a solvent containing two or more components is

$$P' = V_1 P_1 + V_2 P_2 + \cdots + V_n P_n \tag{2.2}$$

where V_n is the volume fraction of the nth component P_n. Further details may be found in references [44–45].

Use of Eq. 2.2 can give approximate starting mobile phase compositions for method development. For example, to create solvents that have a P' value of 6.9, 25/75 v/v water/acetonitrile ($P'_{H_2O} \times V_{H_2O} + P'_{ACN} \times V_{ACN} = 10.2 \times 0.25 + 5.8 \times 0.75 = 6.9$) or 48/52 v/v water/IPA ($P'_{H_2O} \times V_{H_2O} + P'_{IPA} \times V_{IPA} = 10.2 \times 0.48 + 3.9 \times 0.52 = 6.9$) could be used.

Figure 2.3 shows the utility and limits of this approximation method. Here, four aromatic hydrocarbons are eluted on a C_{18} column. Mobile phases were prepared so that the elution profiles were as closely matched as possible. Calculation of P' values give the following results: 50/50 water/2-methoxyethanol, $P' = 7.85$; 65/35 water/ IPA, $P' = 8.00$; 56/44 water/acetonitrile, $P' = 8.26$; 71/29 water/n-propyl alcohol, $P' = 8.40$; 62/38 water/THF, $P' = 7.84$. (Note the refractive index peak/trough just after the injection of the sample on the water/THF solvent system. This effect is not present on any other chromatogram). From the ranges of P' values obtained, 7.84–8.40, it is evident that the use of this parameter will generate approximate conditions only.

Figure 2.4 vividly shows the real limitations associated with the use of eluotropic solvent parameters. In this separation, water/methanol/acetonitrile and water/2-methoxyethanol/acetonitrile gradients were run as follows: A = 95/2.5/2.5 water (20 mM phosphate buffer pH 2)/organic/acetonitrile, B = 80/10/10 water (buffer)/organic/acetonitrile. The dotted-line profile gives the gradient shape from $t = 0$ to 60 minutes. Note the dramatic difference in the chromatographic profiles, peak shapes and resolution. In fact, the 2-methoxyethanol gradient separated the 11 lignans *and* one degradation compound! These results cannot be anticipated or predicted through the use of solvent strength parameters alone.

However, judicious use of the $\varepsilon°$ and P' parameters will save significant amounts of time and solvent waste. As a final comment, both the $\varepsilon°$ and P' parameters reflect the gross retention mechanisms in adsorption and partition chromatography. Any subtle, but usually critical, selectivity offered by the particular solvent cannot be extracted from these parameters.

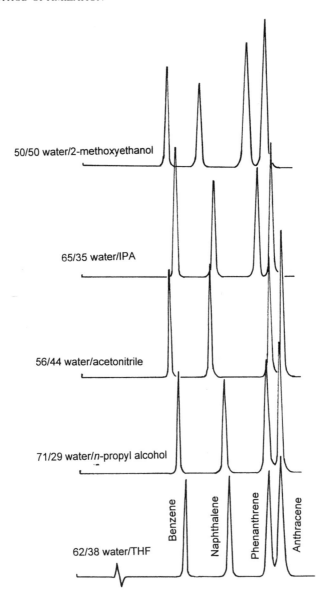

FIGURE 2.3 Chromatograms of benzene, naphthalene, phenanthrene, and anthracene using various water/organic mobile phases. See text for discussion.

2.1.3 Mathematical Representation of Solvent Strength

As discussed earlier, solvent strength is a parameter that predicts the gross chromatographic effects that a solvent has on solute retention. Solvent selectivity, the ability of a solvent to resolve two or more peaks on a given stationary phase, is a vastly more complicated subject.

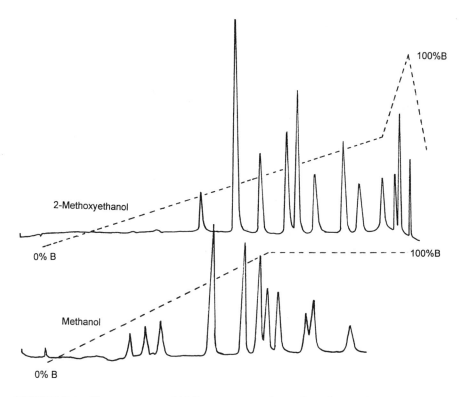

FIGURE 2.4 Chromatograms of 11 lignans separated on a C_{18} column using water/organic gradients.

Here, solvents are classified as to their hydrogen bond acidity (α), hydrogen bond basicity (β) and polarizability/dipolarity (π^*). Although a mathematical representation of the relationship between k' and α, β, and π^* exists as [56–58]

$$\log k' = SP_0 + mV_2 + S(\pi_2^* - d\delta_{kt}) + a\alpha_2 + b\beta_2 \qquad (2.3)$$

the function is complex since the coefficients m, S, d, a and b change with solvent composition. Consequently, the absolute predictive ability of this model, while theoretically powerful, is practically limited.

2.2 TRIANGULATION AND OTHER ALGORITHMIC METHODS

There are two basic types of triangulation methods. One is based on a mathematical representation of solvent selectivity and the other is based on general chromatographic knowledge. Nonetheless, these optimization processes use the same overall strategy: simultaneously alter three variables that affect retention (here the composition of a three-solvent mobile phase), and use the results to guide the choice of the next series of experiments. The initial goal is to achieve acceptable resolution between peaks. The universal goal often expands this to include a minimization of the analysis time.

Consider a more general approach to this optimization scheme for a reversed-phase column (e.g., C_{18}) as shown in Figure 2.5. Each apex of the triangle is designated with a unique mobile phase composition. Here the points are 100% methanol, acetonitrile, or water. The points along the sides connecting any two components represent mixtures of those two components. The points marked 1, 2, and 3, represent 50/50 mixtures of methanol/water, water/acetonitrile, and methanol/acetonitrile, respectively. All points in the interior of the triangle are composites of all three mobile phases. Thus, the point marked 4 represents a 1/1/1 methanol/acetonitrile/water mobile phase.

The optimization process generates the actual separation at each point (i.e., mobile phase composition) and eliminates those that produce the "least acceptable" results. In many cases the initial decision parameter is resolution. As a consequence, the three mobile phases generating the "best" resolution between all peaks are used as the apices of a new triangle. From those three the next series of experiments is constructed. For instance, in the above example, if the results at points 4, 2 and acetonitrile were most acceptable, then the next set of experiments include points 5, 6, 7, and 8 (see Figure 2.5b). Successive iterations of this process will lead to the "best" separation and the separation is optimized. This entire process can be visualized by through a 3D plot as shown in Figure 2.6. The higher the Z value, the "better" the separation [59].

This optimization scheme is used for any mutually miscible trio of solvents. Note that in the above example two of the apices are very strong solvents and retention times will most likely be extremely low. Conversely, the water apex represents such a weak solvent that the analytes may never elute. A basic chromatographic understanding leads to the use of blended mobile phases (e.g., 20/80 methanol/water vs. 100% water) instead of neat solvents at the apices. This approach infuses general common sense into the experimental design and dramatically decreases the number of iterations needed to optimize the separation. The use of computer algorithms [60] can also decrease the optimization process and an analogous tetrahedron (Fig. 2.7) may be used to generate optimal quaternary solvent systems [61]. In fact, some instrument manufacturers have "automated" optimization programs that streamline the process.

In all cases, these approaches are used to generate an isocratic solvent system that provides the optimal resolution between all peaks. Baseline or acceptable resolution is not guaranteed, since this optimization method does not predict column capacity (i.e., the largest number of peaks that may be minimally resolved in a given time). Hence, an acceptable separation, as defined by the chromatographer, may be possible only with a gradient elution. A representative example of the triangulation approach is shown in Figure 2.8. Here the basic optimization iteration is derived from the lower right-hand corner of the triangle shown in Figure 2.5a (i.e. Fig. 2.5b). Each chromatogram represents the separation generated using the mobile phase composi-

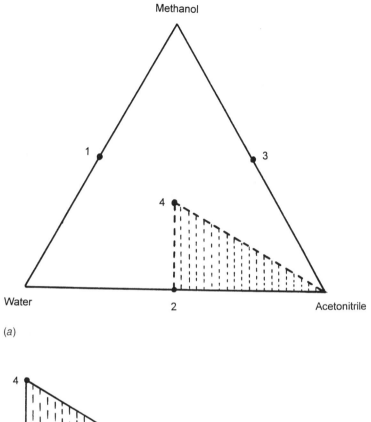

(a)

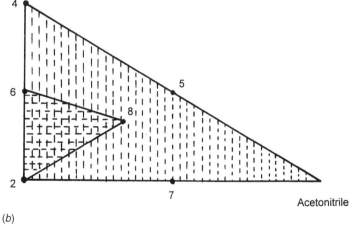

(b)

FIGURE 2.5   (a) An example of a mobile phase triangle with pure solvents as apices. Seven experiments are used in the initial optimization: each apex and points 1 to 4. (b) The second iteration of the optimization process. Four experiments are run: points 5 to 8.

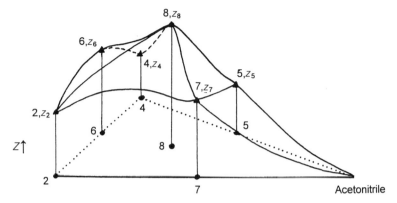

FIGURE 2.6 A plot of the results for the second iteration. Here Z may represent the separation factor, α, between any set of peaks or a parameter that combines the separation with overall retention time. For example, after designating a minimum acceptable α, the parameter $Z = \alpha \times (1/t_r)$, where t_r is the retention time of the latest-eluting peak. Z increases as α increases (resolution gets better) and/or total elution time decreases (sample throughput increases).

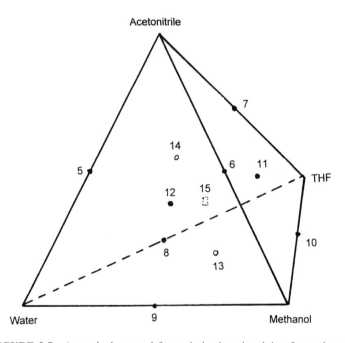

FIGURE 2.7 A tetrahedron used for optimizations involving four solvents.

tion at the indicated point. The point that produced the best separation is point 2. Note that mobile phases associated with points 4, 6, and 8 result in chromatographic elution profiles that are very similar, only shifted in overall retention, whereas 100% acetonitrile generates one peak—all compounds co-elute. From the conditions represented at point 2, small changes in the mobile phase composition can be made to produce full baseline resolution.

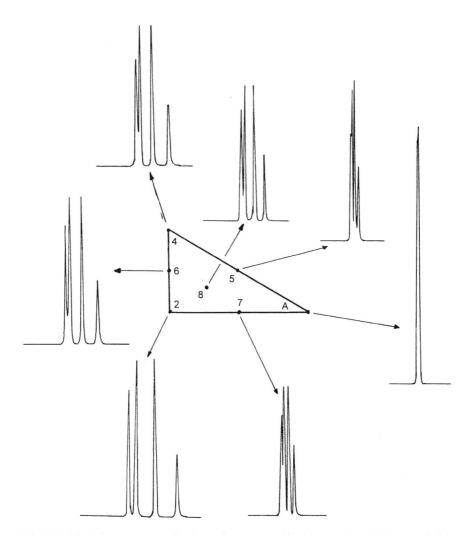

FIGURE 2.8 Chromatograms for the various test mobile phases: A = 100% acetonitrile; 2 = 50/50 acetonitrile/water; 4 = 40/33/27 water/acetonitrile/methanol; 5 = 16/68/16 water/acetonitrile/methanol; 6 = 42/42/16 water/acetonitrile/methanol; 7 = 25/75 water/ acetonitrile; 8 = 35/59/8 water/acetonitrile/methanol.

2.3   SCOUT GRADIENTS

A general drawback to any of the above optimization techniques is that the chromatographer is never *assured* that all the peaks will elute, or if they do that they will not all elute in the void volume. It should be noted that even though this method is gradient by nature, it is used in order to ascertain whether an isocratic mobile phase composition exists that will meet the analysts separation requirements.

Schoenmakers and others [62–65] anticipated the uncertainty in elution by using a rapid gradient elution scheme to aid in the choice of operating conditions. Starting with a weak solvent (e.g., 80/10/10 water/acetonitrile/methanol in a RP system) and rapidly altering the composition to the strongest anticipated mobile phase (e.g., 20/40/40 water/acetonitrile/methanol), all peaks are eluted in a reasonable period and approximate bounds for another gradient are generated by matching the weak solvent composition with the earliest-eluting peak and the strongest solvent composition with the latest-eluting peak (see Fig. 2.9). The weak solvent should initially be weaker than the composition needed to elute the first peak so that elution in the void volume does not occur. Similarly, the strongest mobile phase will be stronger than that needed to elute the last peak to ensure reasonable run times. If there is complete uncertainty as to the retention characteristics of the analytes, then a gradient from 100% weak mobile phase to 100% strongest mobile phase should be run initially. Again, the choices of solvents used are limited only by solubility considerations since each reservoir may be a mixture of solvents. This procedure allows for the rapid assessment of a solvent utility in any given separation.

What process is followed after generating the separation using the scout gradient (Fig. 2.9)? When the gradient profile is superimposed on the chromatogram, an interesting result is obtained. [Note that it is critical that the dwell volume (or time), i.e., the volume in the system between the point of the gradient creation (in the pump) and the head of the columnbre accounted for. The gradient profile must be shifted to compensate for this dwell time.] From the elution time of each peak and the composition of the mobile phase at the time of the elution, an estimate of the solvent strength needed to elute the analyte is obtained. By generating the mobile phase composition for all peaks, it is possible to obtain an approximate mobile phase composition for which the resolution of all peaks may be generated. This point is the weakest mobile phase needed to elute *the first peak*. Unfortunately, this mobile phase composition does not give the shortest overall elution time. As a consequence, experiments with isocratic conditions between the weakest and strongest compositions as dictated by the scout gradient are tried. Also, it is important to remember that a series of scout gradients that incorporate mobile phase components that can offer unique selectivity (e.g., THF) should be included in the optimization protocol.

The scout gradient process is particularly suited to analyses where a large number of analytes need to be separated. The rapid ramping of the mobile phase up to a very strong mobile phase causes the elution of all compounds in a short period of time. A linear gradient should be used since it is the easiest from which to extract the isocratic conditions.

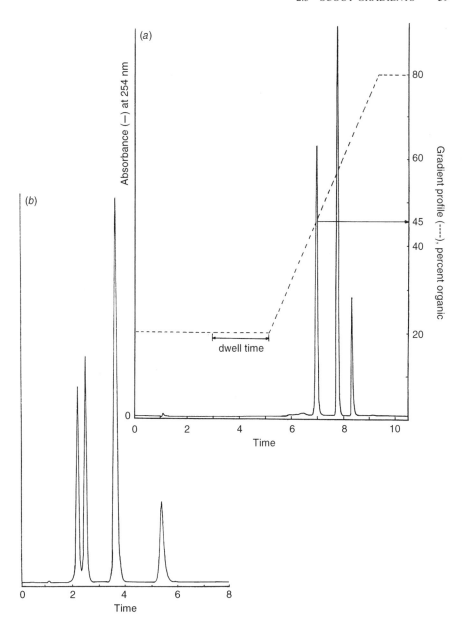

FIGURE 2.9 (*a*) Scout gradient (---) used to estimate the optimal mobile phase composition. Use of the mobile phase composition corresponding to the elution of the first peak generates the longest retention times. The gradient used was 80/10/10 (hold 3 minutes) → 20/40/40 (at 7 minutes, hold 5 minutes) water/acetonitrile/methanol. Note that the entire gradient profile is shifted by the length of the dwell time. (*b*) Elution profile generated at 55/22.5/22.5 water/acetonitrile.

For complex samples, an isocratic separation may not yield an optimal separation. In these cases, the scout gradient serves as the basis for gradient optimization. For areas that have many peaks co-eluting, a plateau in the gradient followed by another gradient may be needed. Nevertheless, these scout gradients generate a large amount of information and provide guidance for further optimization steps.

2.4   WINDOW DIAGRAMS

A window diagram is constructed from the measure of the separation of pairs of compounds from one another, i.e., using a parameter such as the separation factor, α. This approach has been particularly successful for weak organic acids, as shown by Deming [66,67]. In these cases α is plotted against the solvent pH. The "windows" created beneath these curves give the optimum resolution available for all sets of pairs of solutes. Reasonable retention times, etc., must also be taken into account to give the optimal chromatographic working conditions.

In this example four protic solutes are considered: benzoic acid (BA), phenyl-acetic acid (PAA), phenylsuccinic acid (PSA), and 4-hydroxybenzoic acid (HBA). For these optimizations, knowing the pK_a value for each solute is helpful. Here the values are [68]: 4.20 for BA, 4.58 for HBA, 4.31 for PAA, and 3.78/5.55 for PSA (diprotic). The pH of the mobile phase (at least the aqueous component) is made to cover the range of at least ± 1 pH unit of the pK_a extremes (here 1.7 and 6.9). Table 2.2 lists the k' values generated at each pH and then lists all the k'_1/k'_2 ratios calculated therefrom. Note that the ratio is always set up so that the first calculated value is > 1 (i.e., solute 1 is better retained than solute 2). That does not mean that the ratio will remain > 1 throughout the experimental pH range. In fact, looking at the k' ratio values in Table 2.2 shows two instances where the ratio goes to < 1. Where this happens, a reversal of elution order has occurred.

Figure 2.10 is the plotted result of α vs. pH ranging from 1.7 to 5.6. The portion of the graph under the composite curve (i.e., the lowest α value at any pH) indicates that no analyte pair co-elutes under those conditions since $\alpha \neq 1$. Good resolution is often correlated with α values greater than 1.2. In this example, the best resolution occurs at pH 4.1, where α is 1.55.

Two other considerations need to be pointed out here. First, the maximal α does not necessarily correspond to the *optimal* working condition. In essence, α only is a measure of separation not of total elution time. Thus, working at pH 4.1 may take 25 minutes for complete elution, whereas working at pH 2.1 ($\alpha = 1.2$) may require only 15 minutes. When a large number of analyses are pending, this difference could be extremely important.

2.5   BRUTE-FORCE METHOD

The brute-force method does not rely on mathematical functions/relationships, algorithmic iterations, or gradient profiling. It is the original and classic method

TABLE 2.2 Data for Generation of Window Diagram

Solute	pH $\cong$ 1.7	2.2	2.5	3.5	4.4	5.3	5.6
			k' value[a]				
Benzoic acid (BA)	5.13	5.37	5.09	4.53	3.68	0.70	0.63
4-hydroxybenzoic acid (4OH)	1.33	1.39	1.35	1.20	0.83	0.21	0.15
Phenylacetic acid (PAA)	4.05	4.24	4.09	3.53	2.22	0.90	0.86
Phenylsuccinic acid (PSA)	3.41	3.54	3.18	2.56	1.11	0.33	0.29

[a] The mobile phase is 40/60 methanol/water (50 mM buffer) at the indicated pH. Acetic acid was used for pH 3.5, 4.4, and 5.3, whereas phosphate was used for all other buffers.

Solute 1/solute 2[b]			k' ratio[a]				
BA/4OH	3.86	3.86	3.77	3.78	4.43	3.33	4.20
BA/PAA	1.27	1.27	1.24	1.28	1.66	0.78	0.73
BA/PSA	1.50	1.52	1.60	1.77	3.32	2.12	2.17
PAA/4OH	3.05	3.05	3.03	2.94	2.67	4.29	5.73
PAA/PSA	1.19	1.20	1.29	1.38	2.00	2.73	2.97
PSA/4OH	2.56	2.55	2.36	2.13	1.34	1.57	1.93

[a] The k' ratio, k'_1/k'_2, is also commonly referred to as the separation factor, α.
[b] Note that the ratio is always initially generated using the most retained (i.e., highest k') value. This gives an initial ratio greater than 1.

for optimization. This method is very effective for moderately complex separations and relies on the experience of the analyst, and is the way to proceed if an HPLC system does not have gradient capability.

In essence, the brute-force method starts with an intentionally too strong mobile phase ... this is key. This ensures that all analytes will elute in a reasonably short time. In fact, under these conditions, many or all analytes may co-elute. The first mobile phase composition tested may be 90/10 acetonitrile/water. The next step is to sequentially weaken the mobile phase until an acceptable separation is achieved. For example, the next mobile phase may be 80/20 acetonitrile/water. Assessment of the optimal conditions (including separation, elution time, and peak shapes) dictates the subsequent tests. For example, severe tailing indicates that a buffer system (e.g., acetate) or mobile phase modifier (e.g., THF) be added in the next iteration. This flexibility is not an inherent aspect of the triangulation optimization method. The aspect of experience comes into play with the choice and level of buffers and mobile phase modifiers. Sequential modifications to the mobile phase are made until the optimal separation is achieved.

2.6 OTHER OPTIMIZATION TECHNIQUES AND RETENTION PARAMETERS

Other parameters have been used to help predict solute retention with varying mobile phases. Solubility parameters, δ [69,70], octanol/water partition coefficients,

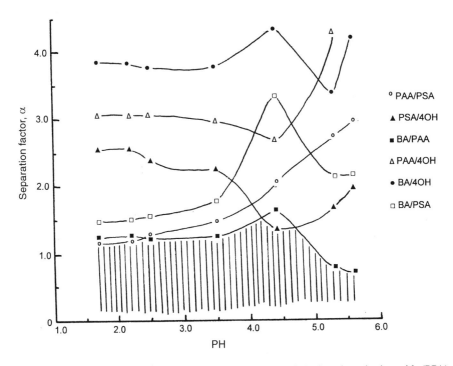

FIGURE 2.10 Window diagram for the separation of 2-phenylpropionic acid (PPA), phenylsuccinic acid (PSA), 4-hydroxybenzoic acid (4OH), and benzoic acid (BA). The highest point under the composite curve offers the best resolution. This occurs at a pH $\sim$4.1 with an α value of approximately 1.5.

P_{oct}, [71–73] and fragmental constants, π, [74,75] are the most common. Their practical use has been somewhat limited due to the complex relationships between the parameters and chromatographic retention. The reader is encouraged to refer to the cited literature for further details. For basic detailed discussions on general method development, the reader should refer to citations [76–79].

3

METHOD VALIDATION AND ONGOING PERFORMANCE EVALUATION

The optimization process as described in Chapter 2 is a critical one for the analyst. Successful method optimization means that targeted, defined criteria (e.g., resolution, elution time, sensitivity) have been met through the use of an experimentally determined set of variables (flow rate, mobile phase composition, system temperature, detector settings, column type, etc.).

However, in many instances, the mere establishment of a separation that meets these criteria is not enough. More critically, and often neglected, omitted, or forgotten, is the fact that the true value of a separation is *not* that the separation was successful once but that the separation will generate comparable results when the method is run exactly as described by *a scientist trained in the field*. This involves a three-step process: instrument qualification, method development/validation, and ongoing system performance checks (known as system suitability tests).

The initial instrument qualification process is detailed and comprehensive and includes both software (computer programs used in data acquisition, analysis, and archiving) and hardware (pumps, detectors, injectors, etc.). Once the instrument is qualified then methods themselves are developed and validated. The development process includes the initial evaluation of the appropriateness of the test, which is often determined by the ability of the method to meet targeted method requirements or mandates (linear range, linearity, accuracy, quantitation limits, resolution, etc.). Once these fundamental requirements are met, the cost effectiveness and overall efficiency are reviewed. Upon acceptance, a developed method then goes through a validation process.

It should be noted that the method that is derived from the development process is captured in a method document. A method document suggests the general operational parameters. Some examples are: sample preparation technique, column type (C_{18}, C_8, silica), typical mobile phase composition, and choice of detector. Thus, a method is the blueprint, or guideline document, that supplies sufficient information that a trained analyst can readily implement and successfully generate acceptable results. On the other hand, a method *does not exactly specify all operating parameters*. This is done deliberately in order to make the method flexible enough to include a wide range of choices in the use of instruments and columns from different manufacturers. Methods that require specific and exacting columns and conditions are not robust and are avoided.

A method is distinctly different from a standard operating procedure (SOP). An SOP is the document that details the exact operating procedures that were derived from the method and are often laboratory specific. An SOP will define sample size/weight and preparation, diluent and dilution volumes, standard curve concentrations, flow rates, detector/data acquisition settings, etc. In essence, the SOP is the master document used in the quality assurance/control laboratory.

Initial validation is done by one laboratory, most often the development laboratory. Next the validation is done by multiple analysts within the development laboratory *on one sample*. The results are reviewed and, if they are acceptable, the method is intralaboratory validated. This means that the development laboratory is capable of generating valid results when using this method. Intralaboratory validation often includes testing the method for the effects generated through small controlled incremental changes (method robustness) in operational parameters such as mobile phase composition, pH, flow rate, column batch, etc. This is often conducted in the development laboratory. If the method is robust, then interlaboratory validation proceeds.

Interlaboratory validation includes the results generated by multiple laboratories, instruments, and analysts. The statistical results are then compared against an existing set of minimum operational specifications (e.g., relative standard deviation is <5%, linearity as described by the correlation coefficient, $r > 0.99$). Once the method has successfully performed in the interlaboratory validation, then the method becomes a validated method ready for general use.

Beyond this validation process, the final requirement for method performance is an ongoing guarantee that the entire system is performing properly. The validation process gives statistical assurance that the method *can* be run successfully by multiple laboratories and analysts. However, a valid method does not *guarantee* that any given run sequence will yield valid results. Therefore, this run-to-run assurance of acceptable method performance is captured in the generation and monitoring of system suitability parameters. Each of these processes will be discussed below.

3.1 QUALIFICATION PROCESS

3.1.1 Instrument Qualification

The instrument qualification process tests the functioning of each component of the instrument individually, compares the results against preset manufacturer's specifications, and produces archivable results. The fully assembled and interconnected instrument is also tested and the final documentation states that the system meets all the needs and specifications initially determined by the laboratory and accepted by the manufacturer.

Instrument qualification is a necessary prerequisite to method development and validation. In essence, if the instrument is not functioning properly, then the results cannot be considered "good." Instrument validation includes four steps: design qualification (DQ), installation qualification (IQ), operational qualification (OQ), and performance qualification (PQ) [80]. These will be presented separately.

Today most instruments are computer driven. The computer can control any or all of the following: instrument operation, data acquisition, data storage/processing, and report generation. As a consequence, software validation is another important aspect of instrument qualification. However, software qualification is currently a much-debated area and is beyond the scope of this presentation.

3.1.2 Design Qualification

Design qualification is performed during the initial assessment of the instrument. It derives directly from the needs of the laboratory. From an assessment of intended method requirements, the minimum specifications of all software and hardware components are generated. These requirements may be either very demanding (e.g., a complex multicomponent multistep gradient separation with very low detection limits) or "routine" (e.g., a purity assay for a single-substance formulation).

Required software capabilities include such things as data acquisition and processing, time/date stamping, access control, and archiving. Hardware requirements cover operational aspects such as the reproducibility and accuracy of the flow rate, injection volume, and gradient formation. Functional ranges from the flow rate, injection volumes, and detector response must also meet the laboratory requirements. A sometimes overlooked aspect of design qualification is the footprint issue. An instrument and its peripherals (computer, printer, etc.) require a predefined bench space: its footprint. Some instruments are designed horizontally and require significant bench space. Many are stackable, i.e., built up vertically. These require the least bench space but also are often the hardest to access for maintenance and troubleshooting.

The laboratory-derived requirements are then compared against system specifications set by manufacturers. If system specifications meet all of the laboratory requirements, then the system meets the design qualification and is considered for use in the analysis.

3.1.3 Installation Qualification

Installation qualification consists of a series of unique steps that are taken by both the laboratory and the manufacturer. The laboratory must prepare the bench space and supply adequate ventilation, electrical connections, and computer interfaces (if any). The manufacturer must assemble the instrument in the prescribed fashion and make sure that any options are correctly integrated/installed. After the instrument is assembled, the manufacturer conducts routine tests at the factory to assure that all components are functioning properly and that the interfaces between components are correct and operational. Once the factory testing is complete and the instrument is approved, it is boxed and shipped, with all appropriate paperwork, to the user.

The next step for the laboratory is to receive the package(s) and visually inspect them for damage. If there is no damage, then the laboratory proceeds to follow the manufacturer's instructions. If this includes removing the components from the shipping package(s), it is critical that each component is scrutinized for identity (against the packing slips) and for damage. If any parts are missing or damaged, the laboratory must notify the manufacturer immediately.

The next step for the manufacturer often is on-site installation. The installation visit is set up once the laboratory has fully prepared the space for the instrument and has verified shipment status (complete, correct, and visually undamaged). Installation qualification includes generating a documented list of components and their serial numbers, establishing physical connections between all components, and generating a record of where and when the installation took place and by whom it was conducted.

3.1.4   Operational Qualification

Once the installation qualification is complete, the instrument is then tested against operational specifications. This often includes testing each component individually. For example, the pump is tested for the accuracy and precision of flow rate; a UV-visible detector has its wavelength settings calibrated, and the linear range, accuracy, and precision are verified (the last three often in conjunction with a test of the injector and pump). These test results are compared against the manufacturer's specifications using qualified standards and test components.

For gradient systems, a check of the gradient formation capability of the system may be performed. This is often done using a water/methanol/acetone gradient from $100/0 \rightarrow 0/100$ starting with a nonabsorbing mobile phase, say 50/50/0 methanol/acetonitrile/acetone, to an absorbing mobile phase, say 50/49/1 methanol/acetonitrile/acetone, monitored at 230 nm. At this time the gradient profile is confirmed and the dwell volume of the system [the volume (flow rate×time) between when the pump starts to change mobile phase composition and when the gradient appears at the detector]. The level of acetone is adjusted so that the plateau of the gradient is approximately 0.5 AU.

3.1.5   Performance Qualification

In reality the OQ and PQ steps often blur. In order to establish that the instrument is functioning up to the requirements established by the laboratory, a series of test samples are typically run in conjunction with the standard/test components. However, a clear distinction occurs when an instrument is purchased for dedicated use in one analysis. In this case, the performance qualification will definitely extend beyond the typical operational qualification and necessarily include running the method itself.

Outside the typical scope of these qualifications is the interface with Laboratory Information Management System (LIMS). In these instances, and where appropriate, archival file systems are established and the appropriate security access parameters are set up. Proper communication with the LIMS is verified. Often, a sample with known characteristics is run and the results are compared against those generated by a validated system.

An instrument qualification will generate documentation that includes at least the following for each test performed: the serial number (or ID number) of each component qualified along with pertinent manufacturer documents, chromatograms or other data/information generated in the qualification process, the person(s) who conducted the qualification and the date of each qualification, and a tabulation of the results generated. These should be kept in hard-copy form in a binder that is specific to each instrument. These results, if possible, should also be electronically archived.

Instrument qualification (here OQ and PQ) is typically conducted once a year or after any major incident requiring repair or replacement (such as replacement of a detector lamp or exchange of a pump). With each qualification, another set of test results is documented and archived.

3.2 METHOD VALIDATION

Once an instrument is qualified and a method meets all established optimization criteria, the next step is method validation. To this point, one analyst (or one group of analysts) has developed a method that seems to "work." Unfortunately, and especially in literature papers, this is the extent of any verification. No information exists that the method may be readily transferable to another laboratory.

To help guarantee that a method is readily utilizable by any trained analyst, a process termed *method validation* has been defined by a number of scientific and regulatory bodies. Some documents were generated in close affiliation with governmental agencies [e.g., the United States Pharmacopoeia (USP) and the United States Food & Drug Administration (FDA)] and some are the result of international cooperation between organizations [e.g., International Conference on Harmonization (ICH) and International Standards Organization (ISO)]. The intention of all the documents generated by these organizations is to give guidance to those analysts involved in the validation of a method. This guidance is meant to produce statistically verifiable and testable results while at the same time allowing for as much scientific "flexibility" as possible.

Common to most documents dealing with method validation are definitions of the following working parameters: accuracy, precision, linearity, linear (or working) range, specificity, detection limit, quantitation limit, ruggedness, robustness, and system suitability. Each of these parameters is generated from the statistical analysis (or the comparison of the parameter to an existing statistical limit) of the results generated during the validation run. Each of the parameters will be discussed below.

It should be noted that many of these parameters have pre-established confidence limit ranges that will capture the needs of the industry. For example, regulated pharmaceutical formulations may have a much more restrictive range than non-regulated nutritional supplements or herbal formulations. In the latter cases, method validation is often done for the support of label claims. Exact requirements should be derived directly from the organization involved.

3.2.1 Accuracy

Accuracy is the measure of how closely the method generates the "true" result. Note that an accurate method need not be a precise method. To establish the method accuracy, the validation is often run against a standard reference material. For example, pharmaceutical compounds are commonly procured from the USP as certified standards. National Institute of Standards and Technology (NIST)-traceable standards are available as well.

For the determination of method accuracy, the initial step is to generate a response vs. concentration curve using the standard material. The concentration range should ideally cover 50–150% of the target value. For purity assays this may not be practical or useful and so a tighter range might be employed. Note that the accuracy of any method is a measure of the instrumental performance as well as of any sample preparation steps.

When the analysis involves the determination of a compound in a mixture or formulation, accuracy is typically determined from the results generated by spiking a matrix blank (i.e., the formulation comprising all components except for the compound(s) of interest) with a predetermined aliquot of the standard material. If a matrix blank is unavailable (as is the case for many natural products), then accuracy may be estimated from a series of standard additions of the compound to the sample itself.

The establishment of accuracy over the critical working range of the method is often done during the method validation process. This might entail three levels of concentration: the target level of the analyte (i.e., 100% of the anticipated level), a low level (e.g., 50–90% of target), and a high level (e.g., 110–150% of target). The recovery of the analyte at each level is compared against the expected level. *Note that the actual acceptable variability of the method from the "true" level is established by the laboratory to match the needs of the analysis.* This accepted variability limit must be established before the analytical results are generated.

3.2.2 Precision

Precision is a measure of the ability of a method to generate the "same" result for multiple analyses of the same sample. Conversely, from the above case with accuracy, a precise method need not be an accurate one. Instrument bias, loss of analyte in the sample preparation, or analyte instability in solvents, may all lead to inaccuracies. However, if the error is constant and reproducible, a very precise result is obtained.

Short-term precision (e.g., replicate analyses during the same day) is often referred to as the *repeatability* of a method. Long-term precision (e.g., over a week by the same laboratory or between laboratories) is called *reproducibility.*

The precision of the method is often established over the anticipated working range of the method and is often determined concomitantly with the accuracy. In such cases, a minimum of triplicate analyses of the low concentration, target concentration, and the high concentration are run. Alternatively, six replicate analyses of the target concentration might be used. Again, the protocol for determination of the precision must be established prior to analysis.

The precision of an analysis is often expressed in terms of the relative standard deviation (in percent, %RSD) or the coefficient of variation (COV). These values are calculated from the standard deviation, s, and mean, x, of the data set:

$$\%\text{RSD} = \text{COV} = 100\frac{s}{x}. \tag{3.1}$$

Note that the smaller the %RSD the less variability there is in the data set.

3.2.3 Linearity

A response vs. concentration plot (see Fig. 3.1) is a visual representation of the relationship between the detector response and the analyte concentration. For a linear relationship the mathematical equation for a straight line is used:

$$y = mx + b \tag{3.2}$$

where y is the concentration, m is the slope of the line (detector response/concentration), and b is the y-intercept. Ideally, the y-intercept is 0 (i.e., at $x = 0$). When the y-intercept is not zero, the method has an offset or bias. An offset (bias) derives from such factors as residual uncompensated mobile phase background absorbance or an improperly "zeroed" detector. Visual inspection of the response vs. concentration plot readily shows this (as would the mathematical equation describing the line via the b value).

The sensitivity is directly derived from a response vs. concentration plot. The sensitivity is the slope of the line: Δdetector response/Δconcentration. A very sensitive method has a large slope, so that a small change in concentration results in a large shift in the response (see Fig. 3.1). A highly sensitive method readily distinguishes between very small differences in sample concentration. However, this limits overall working range. Conversely, a method with low sensitivity shows a small change in response for a similar change in concentration. The working range is often very broad, but the ability to differentiate samples with slight variations in concentration is more difficult. Sensitivity is often "adjusted" through choosing detectors (e.g., fluorescence having very high sensitivity vs. refractive index having a low sensitivity) or detector operational settings (e.g., choice of excitation/emission wavelength settings).

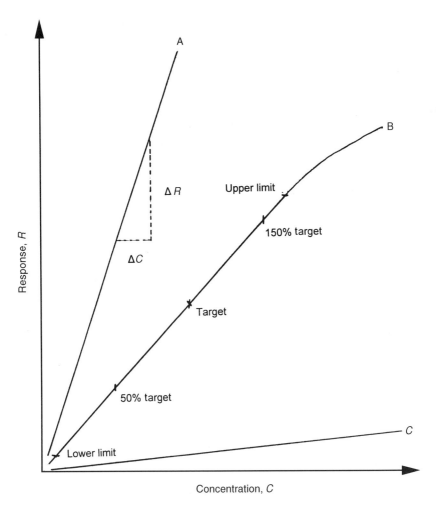

FIGURE 3.1 Three response versus concentration curves with, A having the highest sensitivity (largest response per unit concentration, $\Delta R/\Delta C$) and C the lowest. B has a linear range between the points labeled "lower limit" and "higher limit". Above the high limit concentration the response curve is distinctly nonlinear. The working curve here is defined as running from 50% to 150% of the target (anticipated) analyte level.

The linearity of the relationship is determined from regression analysis of the data and is often reported as either the correlation coefficient, r, or the coefficient of determination, r^2. The mathematical relationship is [81]

$$r = \frac{\sigma_{xy}}{\sigma_x \sigma_y} \tag{3.3}$$

where σ_{xy} is the covariance of x and y, and the standard deviations for x and y are

$$\sigma_{xy} = \frac{1}{N_p}\Sigma(x_i - x)(y_i - y), \qquad \sigma_x = \sqrt{\frac{1}{N-1}\Sigma(x_i - x)^2}, \qquad \sigma_y = \sqrt{\frac{1}{N-1}\Sigma(y_i - y)^2}$$

$$(3.4)$$

where N_p is the number of pairs of data points and N is the number of x or y data points. If a value of $+1.0$ or -1.0 is ever obtained, then there is a perfectly proportional correlation between the response and the concentration. For many analyses, a minimum specification for r or r^2 is typically >0.98. However, some critical analyses may require r or $r^2 > 0.995$. The operational specification limit is determined by the requirements of the laboratory.

Note that in quantitative work the analyst *assumes that there is a correlation* between the response and the concentration, whereas the correlation coefficient actually indicates *only* how confident that analyst may be that there is such a correlation.

3.2.4   Linear and Working Ranges

A well-defined relationship between detector response and analyte concentration is crucial for quantitative analyses. In the past, the working range of an analysis was often defined by the linear portion of the response vs. concentration curve. Although a linear relationship is desirable, it is no longer a requirement since powerful curve-fitting data analysis programs are readily available. The important criterion is that the response–concentration relationship is constant and reproducible.

The linear range is experimentally established and is defined by an upper and lower value. All points within this range meet the required accuracy, precision, and linearity specifications set for the method. The working range implies that the points between the limits are those that are actively and routinely used in the analytical determination. In many instances the linear and working ranges are identical. However, there may be times when the working range is a subset of the linear range. This is acceptable. For example, during method development, the linear range for the method might be established between 10 µg/mL and 500 µg/mL, whereas the required working range might only cover 50–150 µg/mL (i.e., 50–150% of a 100 µg/mL target). Conversely, the working range must never extend beyond the upper and lower limits of the established linear range. Once the linear range is defined, the resulting response vs. concentration plots are often referred to as standard curves, calibration curves, working curves, or linear curves.

The linear range will vary depending on the purpose of the method. There are three distinct scenarios to consider. First, there are those samples of high purity (or neat) for which the expected level of analyte is at or near 100%. Second, formulations may contain analytes at levels that vary anywhere from 0.1% to $>90\%$. Third, the analyte may be an impurity or degradation compound where the analyte levels are trace.

For purity determinations, the range should have the target concentration as the midpoint of the range. Obviously, this is not the case for high-purity samples, i.e., those near 100% purity. In these cases the range may run from an equivalent 80% to 110% of the absolute purity.

For formulation assays, the range of the technique should whenever possible cover from 50% to 150% of the target level. For example, for a sun block lotion containing a 5% level there should be established a linear range from 4.5% to 5.5%.

Finally, for degradation studies, the range from 0 to 100% may be mandatory, whereas for impurities (or degradation products), the linear range may be defined from 0 to X% of the maximum acceptable impurity level, where X is the highest anticipated level of the impurity.

3.2.5 Specificity

Specificity refers to the ability of a method to accurately determine the analyte level in the presence of all other components in the sample matrix. Care should be taken to not confuse this term with selectivity, which refers to the ability of a method to separate (resolve, deconvolve) a pair of compounds from one another (not to generate the unique identification of the analyte from all compounds as implied by specificity). Ion selective electrodes provide an excellent example to illustrate the difference. These electrodes are designed to respond to one ion (e.g., a fluoride selective electrode responds to F^-) but show response to others (e.g., hydroxide and chloride ions in this example) and therefore are selective but not specific.

The specificity of a method often is confirmed through the use of one or more of the following means: independent analysis of a matrix blank, secondary analysis of the analyte peak by a complementary technique (e.g., LC/UV/MS), or matrix spiking with known contaminants or degradation products and monitoring of the effect on the chromatographic results.

3.2.6 Detection and Quantitation Limit

Detection and quantitation limits are also determined experimentally. A detection limit is defined as the lowest concentration level above the background signal that can be determined. Note that detection limits can refer to the instrument alone (instrument detection limits, IDLs) or method detection limits (MDLs), often referred to simply as a detection limit (DL) or limit of detection (LOD). IDLs are generated under ideal conditions where the background signal due to mobile phase or sample matrix is minimized or eliminated and only the operational noise of the instrument is considered. Method detection limits must take into account any noise generated by the mobile phase, sample matrix, and instrument. The difference between the two include all extractable matrix components, different pump and detector noise when used under working conditions (rather than ideal conditions) including gradients, signal-generating mobile phases (albeit low level), and injector-related noise (e.g., due to mis-matched sample/mobile phase composition).

A quantitation detection limit (QDL), often simply called the quantitation limit (QL) or limit of quantitation (LOQ), is defined as the lowest *reportable* analyte level. In all cases IDL ≤ DL ≤ QL. Regardless of the limit under consideration, the key differentiation between them is the signal-to-noise (S/N) ratio value used to define them. To determine the system noise, the instrument is run for a long time (often three times the duration of the analysis) at a very sensitive detector setting. Two things are monitored: the signal noise and the detector drift. Signal noise is due to the detector and pump design and the mobile phase and sample solvent composition. Although there are many statistically valid ways of determining the correct value to use for system noise, a maximum noise level can readily be determined from calculating the noise from the maximum and minimum response levels, R. Once the ΔR is determined, the S/N ratio is calculated. In many instances the S/N for IDL is 1.5–2, for DL is 2–3, and for QL is 3–10. Again, as with other method-related parameters, the analyst ultimately defines the DL and QL.

3.2.7 Robustness

The robustness of a method is typically determined during the method development stage and is a measure of how consistently a method generates the same analytical result when small deliberate changes in operating parameters are made. Many times it is part of the intralaboratory development/validation process. For example, changeable parameters could include organic level in mobile phase, pH of mobile phase, concentration of mobile phase modifiers, and column.

Designed experiments such as the Plackett–Burman design [82] allow the laboratory to rapidly assess the key variables using the least number of experiments. This design uses a maximum–minimum (high–low, A–B, on–off) level approach. A seven-factor design (e.g., pH, temperature, column) requires eight experiments (as compared with a $8 \times 7 = 56$ one-parameter-at-a-time approach). An 11-factor design requires 12 experiments. This and other designs that may also be used are discussed in detail elsewhere [83]. If the design factors include multiple analysts, instruments, and/or laboratories, the resulting "robustness test" is often termed a ruggedness test.

3.2.8 Ruggedness

The concept of ruggedness includes that of robustness but includes the reproducibility of a method when different analysts, laboratories, and instruments are used. Ruggedness determination is almost invariably an interlaboratory result. In many cases, a well-defined collaborative effort is used in which eight or more laboratories analyze the same sample (or a set of samples). It is evident that the potential variability in the method conditions will be significantly greater than in the controlled-design robustness testing. In ruggedness testing not only are the system parameters tested but also the manner of sample handling, sample and mobile phase preparation, and, in areas of the method not explicitly described and defined, analyst interpretation.

A method that has been proven to be both robust and rugged will produce sound statistically supported results by anyone properly trained in the field *when coupled with the final topic of this chapter: system suitability.*

3.3   SYSTEM SUITABILITY

All of the validation parameters above deal with method performance during the critical phases of validation and method review. Unfortunately, no matter how rigorous the validation protocol is for a method, validation cannot preclude or anticipate the effect of instrumentation that is not working properly or analyst error. It is for this reason that an on-going method-specific test (or set of tests) is needed. This is the reasoning behind the use of system suitability parameters.

System suitability refers to a unique set of performance specifications that is directly linked to a method. These specifications are not accuracy, precision, linearity, etc., which are measures of method performance and are used to support product release. These data are generated after the analysis set is complete. Conversely, a system suitability parameter directly ties the immediate performance of the method for a *specific portion of an analytical sequence.* Some parameters that are commonly used as system suitability parameters include theoretical plates and/or peak symmetry for the analyte peak (N), resolution (R_s or α for multiple analyte peaks in a determination), or check sample results (analysis of a previously analyzed sample).

These parameters can be ranges (both upper and lower control limits) as in the case of an analyte level in a check standard; maximum values (upper control limit only) as in the case of peak asymmetry; or minimum values (lower control limit only) as is the case of resolution or plate number. *Parameters that do not directly monitor the actual performance of the method (e.g., mobile phase pH, flow rate determination), although functionally important, are inadequate and inappropriate for system suitability determinations.*

Often, tracking of system suitability parameters is done through the use of control charts. These charts monitor the day-to-day, analysis-to-analysis results and are important for spotting trends, helping identify potential emerging problems, and thereby assisting the analyst in anticipating problems leading to proactive preventive maintenance.

4

ALCOHOLS

Alcohols are a frequently used class of LC solvents. They are used both as major constituents in reversed-phase (RP) mobile phases and as low-volume constituents in normal-phase (NP) separations. They offer the following set of unique properties:

1. They have a hydroxyl functional group, $-OH$, which acts as a hydrogen bond donor *and* acceptor.
2. They are readily available as high-purity liquids over a significant molecular weight range (i.e., methanol, C_1, to octanol, C_8) *and* isomeric forms (e.g., 1-butanol, 2-butanol, isobutyl alcohol [1-methyl-2-propanol] and *t*-butyl alcohol [2-methyl-2-propanol]).
3. They are chemically stable and therefore need no preservative.
4. They pose minimal health and safety hazards when handled properly but are highly flammable.
5. They are typically miscible with a wide range of solvent classes.
6. They are comparatively inexpensive.

Of the alcohols, methanol has been the most widely used, followed by isopropyl alcohol (IPA). Where chemically feasible, almost all reversed-phase separations have at least been attempted using methanol.

Neat ethanol has found limited use, not because it does not offer interesting and useful chromatographic properties, but because of the artificially high cost due to strict government control over its use and dispensation. Denatured ethanol, commonly called reagent alcohol, is readily available in many forms. However, only those with either a hydrocarbon at $\sim$1% levels or ones containing methanol/

IPA mixes at the 1–5% level are compatible with UV work. Note that the potential variability in the level of added denaturant poses potential reproducibility problems for the chromatographer.

In addition to the above, benzene was formerly used as an ethanol-denaturing agent. However, due to the carcinogenic nature of benzene, this type of ethanol is not the typical offering today. Methyl ethyl ketone (MEK) and ethyl acetate are alternatives but themselves are not compatible with most analyses that use a UV detector since MEK and ethyl acetate absorb significantly up to 330 nm. RP gradient work is also impractical since the MEK and ethyl acetate (or the hydrocarbon in the above-mentioned ethanol) is loaded onto the column during the weak solvent portion of the gradient and eluted (as a peak whose size is directly related to the loading volume of the weak solvent between gradient runs) during the strong solvent portion of the gradient.

From a functional chromatographic point of view, the quality of reagent alcohol (or denatured ethanol) suitable for HPLC—that containing IPA and methanol—is more consistent. The amounts of methanol and/or IPA added are typically controlled well enough that the overall strength of the resulting solvent blend does not vary markedly from lot to lot. In addition, the level of denaturant has not been standardized among manufacturers. Neat ethanol is commercially available but must be specifically requested. Unfortunately, the mountain of paperwork and exorbitant tax structure deter its use. Consequently, ethanol has only recently begun to receive the attention that it merits.

IPA is an underutilized solvent. It is known as a "universal solvent" in that it is miscible with water and a wide selection of polar and nonpolar water-immiscible solvents (e.g., hexane, DMSO, dichloromethane). This property allows the chromatographer to create ternary mixtures of water/IPA/organic solvent that offer unique chromatographic selectivities. Because of the ability of IPA to also solubilize a wide range of solutes, it is effectively used to clean up and "regenerate" RP chromatography columns. In general, where column and system compatibility allows, this may be done by back flushing the column with a shallow gradient from 99.8/0/0.2 water/IPA/TFA to 70/30/0.2 IPA/water/TFA at a slow flow rate. The TFA is present to assist in the protonation and solubilization of protic solutes. Make sure that the effluent is shunted directly to waste, not through the detector.

n-Propyl alcohol (NPA) is very similar to IPA in its RP chromatographic properties but the ~40% premium in cost compared to IPA limits its use. NPA's greatest use is in the separation of proteins. NPA is also used as a low-percentage mobile phase modifier in normal-phase separations, where it offers significant chromatographic advantages over other alcohols.

The butanols, 1-pentanol, and 1-octanol are typically used as low volume percent mobile phase additives. They often are used to increase the mobile phase solubility of solutes and confer unique selectivity properties on the mobile phase.

Tables 4.1–4.4 list some important chemical, physical and chromatographic properties as well as general manufacturing specifications and safety parameters for the alcohols [84–92]. Figure 4.1 shows the chemical structures of the solvents listed in Tables 4.1–4.4.

TABLE 4.1 Physical Properties of Alcohol Solvents[a]

	MeOH	EtOH	NPA	IPA	NBA	SBA	IBA	TBA	PA	EG
Molecular weight	32.04	46.07	60.09	60.09	74.12	74.12	74.12	74.12	88.15	62.07
Density (g/mL)	0.7913	0.7893	0.8037	0.7854	0.8097	0.808	0.803	0.786	0.811	1.1088
Viscosity (cP)	0.55	1.194	2.3	2.4	2.98	4.21	6.68	4.31[b]	3.68	21
Solubility in water (%)	100	100	100	100	7.8	20	8.5	100	1.7	100
Water sol. in solvent (%)	100	100	100	100	20.07	37	16.4	100	9.2	100
Boiling point (°C)	64.70	78.0	97.2	86.26	117.5	99.5	107.8	83	137.8	198
Melting point (°C)	−97.68	−114.1	−126.2	−88.0	−89.2	−114.7	−108	24.8	−78	−13
Refractive index (n_D)	1.3284	1.3600	1.3856	1.3772	1.3993	1.3970	1.3959	1.3870	1.4090	1.4310
Dielectric constant	32.35	25.3	20.81	18.62	17.84	17.26	17.93	12.6	14.5	38.66
Dipole moment	2.87	1.69	3.09	1.68	1.66	1.8	1.79	1.7	1.8	2.20
Surface tension (dyne/cm)	22.5	22.8	23.7	21.7	24.6	23.5	23.8	20.7	25.6	48.4

[a] All values at 20°C (except boiling and melting points) unless otherwise noted.
[b] At 25°C.

Abbreviations: MeOH, methanol, methyl alcohol; EtOH, ethanol, ethyl alcohol; NPA, 1-propanol, *n*-propyl alcohol; IPA, isopropyl alcohol, 2-propanol; NBA, 1-butanol, *n*-butyl alcohol; SBA, 2-butanol, *sec*-butyl alcohol; IBA, 2-methyl-1-propanol, isobutyl alcohol; TBA, 2-methyl-2-propanol, *t*-butyl alcohol; PA, 1-pentanol, *n*-pentyl alcohol; EG, ethylene glycol.

TABLE 4.2 Chromatographic Parameters of Alcohol Solvents

	MeOH	EtOH	NPA	IPA	NBA	SBA	IBA	TBA	PA	EG
Eluotropic strength $\varepsilon°$ on Al_2O_3	0.95		0.82	0.82	0.7					
Eluotropic strength $\varepsilon°$ on SiOH	0.70			0.60						
Eluotropic strength $\varepsilon°$ on C_{18}	1.00		10.1	8.3						
Solvent strength parameter, P'	5.1		4.0	3.9	3.9	4.0	4.0			
Hildebrandt solubility parameter, δ	14.5	13.0	12.0	11.5	11.3	10.8		11.4	11.1	10.7
Hydrogen bond acidity, α	0.93	0.83	0.78	0.76	0.79			0.68		0.90
Hydrogen bond basicity, β	0.62	0.77		0.95	0.88			1.01		0.52
Dipolarity/polarizability, $\pi*$	0.60	0.54	0.52	0.48	0.47			0.41		0.92

Abbreviations: MeOH, methanol, methyl alcohol; EtOH, ethanol, ethyl alcohol; NPA, 1-propanol, *n*-propyl alcohol; IPA, isopropyl alcohol, 2-propanol; NBA, 1-butanol, *n*-butyl alcohol; SBA, 2-butanol, *sec*-butyl alcohol; IBA, 2-methyl-1-propanol, isobutyl alcohol; TBA, 2-methyl-2-propanol, *t*-butyl alcohol; PA, 1-pentanol, *n*-pentyl alcohol; EG, ethylene glycol.

TABLE 4.3 Common Manufacturing Quality Specifications of Alcohol Solvents[a]

	MeOH	EtOH[b]	NPA	IPA	NBA	SBA	IBA	TBA	PA	EG
UV cutoff (nm)	205	205	210	205	215	260	220		220	
Percent water (maximum)	0.05	0.1	0.05	0.06	0.03	0.05	0.05	0.1	0.3	0.2
Available as ACS tested[c]	ABE[d]FJM	JM		ABEFJM	AFJM	n.a.[e]	AFJM	AJM		n.a.
Available as HPLC-grade[c,h]	ABEFJM	AEF	ABEM	ABEFJM	ABF		ABJ	A		AF[g]JM
Available through[f]			FJ		E	AFJM	E	F	AFJM	

[a] Abbreviations: MeOH, methanol, methyl alcohol; EtOH, ethanol, ethyl alcohol; NPA, 1-propanol, n-propyl alcohol; IPA, isopropyl alcohol, 2-propanol; NBA, 1-butanol, n-butyl alcohol; SBA, 2-butanol, sec-butyl alcohol; IBA, 2-methyl-1-propanol, isobutyl alcohol; TBA, 2-methyl-2-propanol, t-butyl alcohol; PA, 1-pentanol, n-pentyl alcohol; EG, ethylene glycol. n.a. = not available.

[b] Denatured ethanols are not for use in HPLC. They often contain methyl ethyl ketone, ethyl acetate, and a hydrocarbon, each at the ~1% level. Reagent alcohol or alcohol is the typical designation for HPLC-quality ethanol. It contains methanol and IPA each at the 3–6% range and the ethanol is present at the 89–95% range. Refer to manufacturer's specifications for details.

[c] Manufacturer's code: A = Aldrich; B = Burdick & Jackson; E = EM Science; F = Fisher; J = JT Baker; M = Mallinckrodt.

[d] Not tested for formaldehyde or acetaldehyde.

[e] Not available since an ACS test does not exist.

[f] Available as a high-purity solvent but not specifically designated as ACS tested or HPLC grade. This does not mean a lesser quality solvent, just that it is not specifically tested for these applications. If these manufacturers produce either ACS or HPLC solvent, they are not listed under this heading.

[g] Designated ACS even though a set of ACS tests does not exist for ethylene glycol.

[h] Burdick & Jackson sells ACS tested solvents in 5 gal. cans and 55 gal. drums. B&J HPLC grade solvents meet all ACS specifications.

TABLE 4.4 Safety Parameters of Alcohol Solvents[a]

	MeOH	EtOH	NPA	IPA	NBA	SBA	IBA	TBA	PA	EG
Flash point[b] (TCC)(°C)	11	13	23	12	37	24	28	11	33	111
Vapor pressure (mm)(°C)	25	19	20	20	5	12.1	8.8	24.5	53	0.06
Threshold limit value (ppm)	200	1000	200	400	50	100	50	100		
CAS number	67-65-1	64-17-5	71-23-8	67-63-0	71-36-3	78-92-2	78-83-1	76-65-0	71-44-0	107-21-1
Fire[c]	3	3	3	3	3	3	3	3	3	1
Reactivity[c]	0	0	0	0	0	0	0	0	0	0
Health[c]	1	1	1	1	2	1	1	1	1	1

[a] *Abbreviations*: MeOH, methanol, methyl alcohol; EtOH, ethanol, ethyl alcohol; NPA, 1-propanol, *n*-propyl alcohol; IPA, isopropyl alcohol, 2-propanol; NBA, 1-butanol, *n*-butyl alcohol; SBA, 2-butanol, *sec*-butyl alcohol; IBA, 2-methyl-1-propanol, isobutyl alcohol; TBA, 2-methyl-2-propanol, *t*-butyl alcohol; PA, 1-pentanol, *n*-pentyl alcohol; EG, ethylene glycol.
[b] TCC = TAG closed cup.
[c] According to National Fire Protection Association ratings [92]:

Fire: 4 = Materials that vaporize at room temperature and pressure and burn readily.
3 = Liquids or solids that can ignite under room conditions.
2 = Materials that ignite with elevated temperature or with moderate heat.
1 = Materials that must be preheated before they ignite. 0 = Materials that will not burn.

React: 4 = Materials that, by themselves, can detonate or explode under room conditions.
3 = Materials that can detonate or explode but require an initiator (e.g., heat).
2 = Materials that undergo violent chemical reactions at elevated temperatures or pressures or react with water.
1 = Materials that are, by themselves, stable but that may become unstable at elevated temperatures and pressures.
0 = Materials that are stable even under fire conditions and do not react with water.

Health: 4 = Short exposure times to these materials are lethal or cause major residual injury.
3 = Short exposure times to these materials cause temporary and/or residual injuries.
2 = Lengthy (but not chronic) exposure to these materials may cause temporary incapacitation and/or minor residual injury.
1 = Materials that, upon exposure, cause irritation but only minor residual injury.
0 = Materials that, upon exposure under fire conditions, offer no more hazard than ordinary combustible materials.

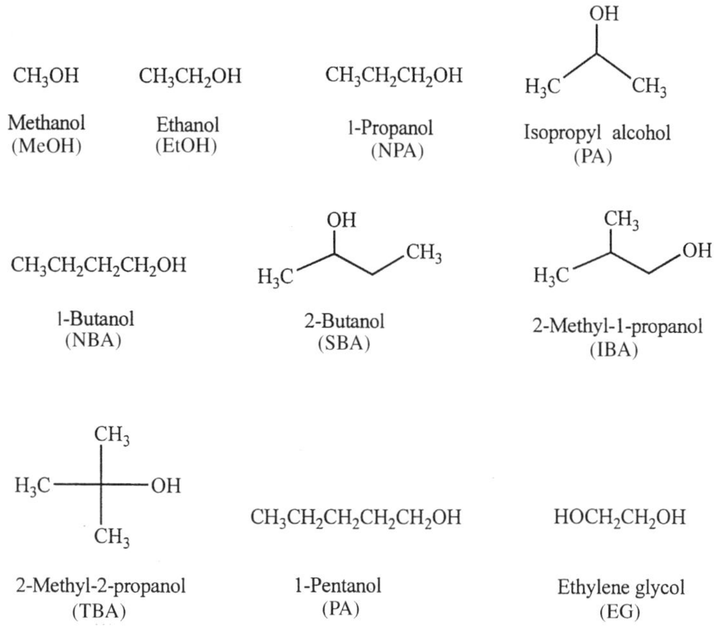

			OH H_3C CH_3
CH_3OH	CH_3CH_2OH	$CH_3CH_2CH_2OH$	
Methanol (MeOH)	Ethanol (EtOH)	1-Propanol (NPA)	Isopropyl alcohol (PA)

$CH_3CH_2CH_2CH_2OH$

1-Butanol
(NBA)

2-Butanol
(SBA)

2-Methyl-1-propanol
(IBA)

2-Methyl-2-propanol
(TBA)

$CH_3CH_2CH_2CH_2CH_2OH$

1-Pentanol
(PA)

$HOCH_2CH_2OH$

Ethylene glycol
(EG)

FIGURE 4.1

4.1 IMPURITIES

Alcohols tend to contain other alcohols and related ethers, aldehydes, ketones, and carboxylic acids as impurities. This is especially true for longer chain-length alcohols. The type and level of impurity depends upon the methods used for manufacture and cleanup in the production of the feedstock material, the raw material for the high purity solvent, and the manufacturing process of the high-purity product itself.

Methanol frequently contains ethanol, acetone, formaldehyde and acetaldehyde at ppb (parts per billion) levels [93]. These low levels usually have a minimal effect on chromatographic performance. Methyl ethyl ketone is often present at higher levels, which is seen as a higher-wavelength UV cutoff value. Although now uncommon, methylamine was in the past found in high-purity methanol. Its presence gave methanol a uncharacteristic "fishy" odor and could result in marked alteration of the chromatography of basic amine compounds. The presence of any primary amine is a potential problem when the derivatization of amines is a part of the analysis protocol. The result is a least one unanticipated peak (more if the methylamine is a distribution of substituted compounds, i.e., monomethyl-, dimethyl-substituted or longer chains such as ethyl or propyl).

Ethanol is a peculiar case. High-purity ethanol (200 proof) is difficult to obtain for laboratory use because of rigid federal regulations, high tax levels, and strict record keeping surrounding its use. Consequently, the trace levels of methanol and IPA left after processing are dwarfed by the 1–5% levels of each added to produce unregulated HPLC ethanol, termed "denatured" or "reagent" alcohol. Additionally, acetone, formaldehyde, acetaldehyde, and traces of ethyl acetate may be present in the finished product.

IPA and NPA often contain various butanol isomers, ethers, and aldehydes/ketones as impurities. Since IPA and NPA are both relatively strong solvents both for RP and NP separations, such low-level impurities often have no practical effect on separations. The UV cutoff and absorbance–wavelength curves may, however, be significantly affected.

The butanols often contain a wide variety of other alcohols, alkenes, ketones, aldehydes, ethers and acids as impurities. Some isomeric forms of butanol have similar boiling point ranges, making their production as high-purity solvents relatively difficult and the end product expensive. Although not typically considered for most LC uses, 2-butanol (*sec*-butyl alcohol) is soluble in water up to approximately 20% and water is soluble in 2-butanol to almost 40%. Because of their miscibility with numerous organic solvents and high solubility levels with water, butanols may be considered for use as co-solvents in solubilizing water/organic mixtures that are immiscible by themselves. Unique selectivities are obtained through their use as a result of the different steric configurations of the branched butanols.

More information on impurities and general properties may be found in references [94] and [95].

4.2   GENERAL ANALYTES

4.2.1   Simple Substituted Benzene Analytes

Alcohols are used with great success in the separation of compounds capable of undergoing strong hydrogen bond interactions: phenols, carboxylic acids, ketones, and substituted amines. The hydrogen-bond donor/acceptor duality of the alcohols is often considered the key to successful separations of these analytes. In the case of weakly hydrogen bonding compounds (e.g., phenols and ketones) the alcohols preferentially interact with residual surface silanol groups and thereby generate more symmetric peaks. This is not necessarily true for strong hydrogen bond acceptors such as amines. In these cases, a mobile phase modifier specifically chosen to prevent the surface interactions is often needed as well.

Urinary phenol, cresol isomers (eluting as one peak), p-aminophenol, and p-nitrophenol were baseline separated in <10 min using a 30/70/0.1 methanol/water/H_3PO_4 mobile phase and a C_{18} column ($\lambda = 215$ nm). Detection limits of 1 mg/L (S/N = 2) were reported for all analytes [96]. In a separate study [97], the 1,2- and 1,3-dihydroxy- and 1,3,5-trihydroxybenzene isomers were analyzed on a porous graphite column (electrochemical detector at +1.1 V potential vs. Ag/AgCl) using a 40-min 25/75 → 100/0 methanol/water (50 mM $HClO_4$/$LiClO_4$ at pH 4) gradient. A 0.1 µg/L detection limit was reported.

Busto et al. [98], conducted a series of studies on phenol and substituted phenols in drinking water samples. In this study an optimum methanol/water (acetic acid to pH 2.7) gradient was generated for the baseline separation of phenol and 11 methyl- and chloro-substituted phenols in <40 min on a C_{18} column ($\lambda = 280$ nm). A detailed multiparameter optimization protocol for the generation of the gradient profile and system temperature was also given. From this optimization work a 70°C system temperature and an initial 40/60 methanol/buffer mobile phase were chosen. A 1.85% increase per minute in methanol ramp was used to elute the phenols. All peak shapes were excellent and the latest-eluting peak, pentachlorophenol, eluted in 25 min. Analyte concentration levels ranged from 30 to 200 ppm. To obtain an isocratic separation the authors [99] re-examined the above separation using a computer optimization version of the solvent triangle approach described in Chapter 2. The apices were acetonitrile–water, THF–water, and methanol–water mixtures. An interesting and informative resolution map was generated within the triangle using 10 mobile phase combinations. The recommended mobile phase composition was 28.7/4/3.8/63.5 methanol/acetonitrile/THF/water on a 50°C C_{18} column ($\lambda = 280$ nm). Detection limits using this method were reported to be up to 100 times lower (200 ppb) than in the comparable gradient separation.

Benzoic acid and benzoyl peroxide were isolated from flour during the bleaching process [100]. Separation and quantitation were done on a C_{18} column ($\lambda = 227$ nm) using an 80/20 methanol/water mobile phase for benzoyl peroxide (elution at 6 min) and a 95/5 water (30 mM phosphate buffer at pH 6.5)/methanol mobile phase for benzoic acid (elution at 6 min). It is interesting that good retention of the acid is achieved at a pH above the pK_a of the acid since the analyte is negatively charged at that pH. In many cases, negatively charged analytes are poorly retained or unretained. Combination of the analyses could be accomplished through the use of a gradient, but the detection wavelength might need to be increased to the 250–280 nm range to avoid severe baseline shifts during the gradient. Regardless, standard curves from 5 to 150 ppm were reported.

o-, *m*-, and *p*-Nitrobenzoic acid isomers were separated and analyzed on a C_{18} column ($\lambda = 254$ nm) using an 80/20/0.4 water/IPA/acetic acid (pH 2.99) mobile phase [101]. Linear ranges of 30–1100 μg/mL and detection limits of 5 μg/mL were reported. Baseline resolution was achieved in a run time of <10 min. As noted above, the key here is to make sure the acid level in the mobile phase is always high enough to keep these low-pK_a analytes fully protonated.

Benzoyl peroxide

Chloroaniline was photochemically decomposed and the degradation products (cyclopenta-1,3-diene-1-carbonitrile, 2-aminophenoxazine-3-one, and three *o*-phenyl-disubstituted anilines such as 2-chloro-3-aminophenol) were extracted from water and separated on a C_{18} column ($\lambda = 280$ nm). A 60/40 methanol/water (with 1.8 g/L ammonium acetate) mobile phase resolved all compounds in <15 min [102].

The premise of this study was to evaluate a subcritical water extraction process for phenol, chlorophenols, and chloroanilines [103]. In the process, excellent separation of each of these classes of compounds was obtained on a C_{18} column (no wavelength cited but a typical value would be 250–280 nm). Phenol and chlorophenols (4-chloro; 2,3-, 2,4-dichloro; 2,3,6-, 2,4,5-trichloro) were baseline resolved in 30 min using a 60/40 methanol/water mobile phase. Aniline (extremely tailed use acid in the mobile phase to correct) and chloroanilines (4-chloro; 2,3-, 2,5-dichloro; 2,3,6-, 2,4,5-trichloro) were resolved using a 25-min 40/60 → 90/10 methanol/watergradient.Workingconcentrationsrangedfrom2to40µg/g.

Metol (*N*-methyl-*p*-aminophenol) oxidation products were extracted from water and separated on a C_{18} column ($\lambda = 271$ nm) using a 15-min 95/5 → 100/0 water (0.68 M formic acid/ammonia buffer at pH 2.5)/methanol gradient. Trihydroxy-benzene, hydroquinone, and numerous hydroxylated dimers were identified [104].

Screening methods for 44 nitro-, chloro-, and alkyl mono- and disubstituted phenolic compounds in water were developed on three columns: cyanopropyl, C_{18}, and diphenyl ($\lambda = 220$ nm). A 60-min 50/50 → 90/10 methanol/water (10 mM acetate buffer at pH 4.8) gradient was used. Not all 44 compounds were baseline resolved, but a subset of 28 were resolved on the C_{18} column using the above gradient. Detection limits were reported individually for all components and ranged from 0.5 to 1.7 ppm [105].

2-Aminophenoxazine-3-one

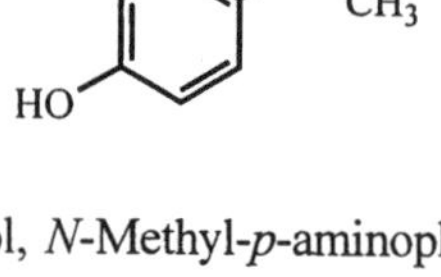

Metol, *N*-Methyl-*p*-aminophenol 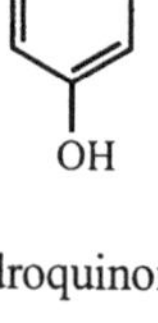   Hydroquinone

The retention characteristics of acetanilide and 23 monosubstituted acetanilides (alkyl, C_1–C_4 and isomers; halogenated; hydroxy) were studied with methanol/water (66.6 mM phosphate buffer at pH 7.4) mobile phases ranging from 30% to 65% methanol [106]. Two C_{18} columns ($\lambda = 235$ nm) were used in the study in order to generate a predictive method for the separation of substituted acetanilides. Although k' values are not explicitly given, they may be calculated from the equations presented. Interesting selectivity differences were found, especially with the hydrogen bond accepting substituents (e.g., chloro, bromo, and nitro) where methanol was much more effective than THF or acetonitrile in generating the separation.

The capacity factors (k') for 32 benzene, fused benzene, and phenol derivatives (including alkyl-, chloro-, bromo-, and nitro-substituted) were obtained under various isocratic elution conditions on a C_{18} column ($\lambda = 254$ nm or 282 nm) over the range of 70–100% methanol (in 10% increments) in water [107]. The k' values for a smaller set of these analytes were tabulated for mobile phase compositions ranging from 60/40 to 0/100 methanol/water. A complete set of tabulated results was presented and provides an excellent basis from which to begin method development.

Knox et al. [108] studied the retention characteristics of 54 aromatic hydrocarbons (benzene—hexamethylbenzene and selected isomers, toluene—decylbenzene and selected isomers, indane, indene, tetrahydronaphthalene, and naphthalene—fluorene and terphenyls) on three C_{18} columns using 70/30, 80/20, and 90/10 (w/w) methanol/water mobile phases. (Note that the v/v ratios are approximately 74/26, 83/17 and 92/8, respectively.) Capacity factors were tabulated for each column and each mobile phase composition when the analytes eluted with a $k' = 40$. The k' values ranged from $\sim$1 (benzene) to $\sim$39 (1,2,3-triisopropylbenzene) for 70/30 and $\sim$0.2 (benzene) to $\sim$20 (n-tridecylbenzene) for 90/10 methanol/water.

A 100% methanol mobile phase was used in the study of the retention characteristics for over 50 alkyl-substituted benzenes on a porous graphite column [109]. Retention times varied from 1.8 min ($k' = 0.2$) for benzene (just slightly more than the void volume) to 58 min ($k' = 37$) for pentamethylbenzene. A distinct advantage of this system over a normal C_{18} system was the enhanced selectivity of the graphite column towards both isomeric forms as well as methyl and methylene homologs. The k' values and retention times for all analytes are tabulated. The retention differences between polymethylbenzenes and alkyl-substituted benzenes of the same carbon number (e.g., trimethylbenzenes vs. n-propylbenzene) were studied in detail on C_{18} and phenyl columns using an 80/20 methanol/water mobile phase.

Acetanilide

Indene

Fluorene

Isomeric differentiation, as defined by chromatographic resolution, was insignificant on reversed-phase supports when compared with that on graphite support.

Armstrong et al. [110] studied 19 sets, with two to five compounds in each set, of structural isomers (e.g., cresols and xylenes to methylindoles and prostaglandins) on a β-cyclodextrin column ($\lambda = 254$ nm or 280 nm). Good peak shapes and resolution were obtained for all sets using mobile phases ranging from 30/70 methanol/water (for cresols) to 90/10 methanol/water (for vitamin D_2). It was noted that retention for disubstituted benzenes (e.g., the hydroxy/methyl, dimethyl, nitro/hydroxy, nitro/amine and amine/carboxy substituent pairs) usually decreased in the order *para* > *ortho* > *meta* on the β-cyclodextrin column.

Six aromatic aldehydes (benzaldehyde, isovanillin, vanillin, 4-methoxy-, 4-methyl- and 3,4-dihydroxybenzaldehyde) were derivatized with 2-amino-4,5-ethylenedioxyphenol and eluted on a C_{18} column ($\lambda = 330$ nm, ex; 390 nm, em) using a 70/30 methanol/water mobile phase [111]. Separation was complete in <30 min and detection limits of 5 pmol (S/N = 3) were reported.

Fifteen ketones (methyl ethyl ketone to diisobutyl ketone and phenylethyl ketone to phenylisobutyl ketone) were eluted from a C_{18} column ($\lambda = 254$ nm) using isocratic conditions and mobile phases consisting of 0.4–1.0 mole fraction methanol in water [112]. Plots of $\log k'$ versus mole fraction methanol are presented along with subsequently derived retention equations.

Phenolic acids (ferulic, syringic, *p*-coumaric, vanillic, caffeic, protocatechuic, and *p*-hydroxybenzoic) were extracted from soil and analyzed on a C_{18} column ($\lambda = 245$ nm) using a 38-min 70/30 → 30/70 (2/97.25/0.5/0.25 methanol/water/acetic acid/ethyl acetate)/(80/17/2/1 methanol/water/acetic acid/ethyl acetate) gradient. Detection limits were reported as <1 µg/g for all compounds [113]. A working concentration curve of 1–10 µg/mL was also generated.

Fifteen biliary metabolites of 1,3-dichlorobenzene (e.g., *S*-2,4-dichlorophenylcysteine, mercapturic acid, glutathione, *trans*-2,4-dichloro-6-[glutathione-*S*-2,4-dien-1-ol]) were extracted from bile, fractionated, and subsequently separated on a

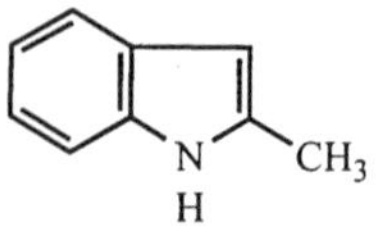

2-Methylindole

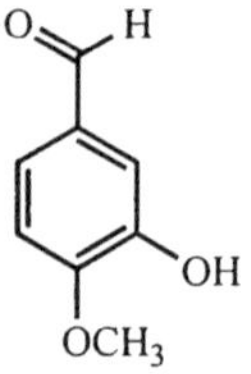

Isovanillin

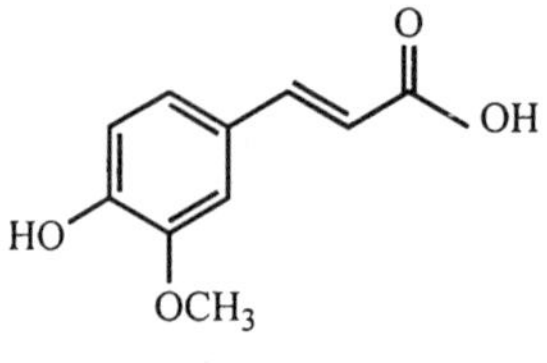

Ferulic acid

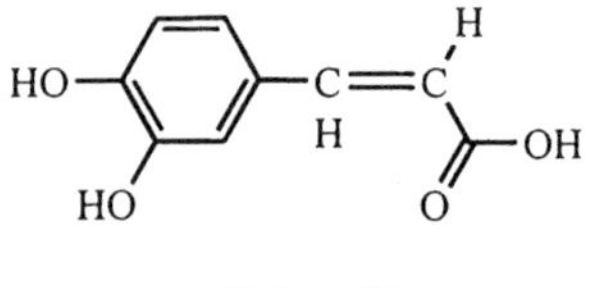

caffeic acid

C_{18} column ($\lambda = 275$ nm). The ratios of methanol/water (80 mM H_3PO_4 with 1.9–2.3% sodium dodecyl sulfate [SDS]) used in the separation ranged from 66/34 to 45/55 depending on which fraction was processed. A particularly impressive chromatogram documented the excellent resolution of 12 metabolites using a 2/1 methanol/water (80 mM H_3PO_4 with 2.3% SDS) mobile phase. Elution was complete in <40 min [114].

Two biological urinary metabolites of styrene (mandelic acid and phenylglyoxylic acid) were quantitated down to the 1 mg/L range using a 10/90 methanol/water (0.5% acetic acid to pH 4.5 with diethylamine) mobile phase and a C_{18} column ($\lambda = 254$ nm). A linear working range over two orders of magnitude was obtained and peak shapes were excellent [115]. Baseline resolution of the two analytes as well as hippuric acid and o-methylhippuric acid (an internal standard), was achieved in <10 min.

A series of 113 alkylbenzenes, chlorobenzenes, chlorotoluenes, chloronaphthalenes, and chlorobiphenyls were studied with respect to retention on a C_{18} column ($\lambda = 254$ nm). Four isocratic ethanol/water mixtures (80/20, 85/15, 90/10, and 95/5) were used as mobile phases. The study of the elution behavior of individual chlorobiphenyl isomers gave extremely useful comparative retention data. Estimates of the effects of methylene- and chloro-substitution on retention are also described in detail [116].

The retention characteristics of over 40 small polar molecules (phenols, phenyl alcohols, imines, phenylcarboxylic acids, phenyl esters, phenyl ethers, phenyl ketones, nitro- and cyanophenols) were studied on C_4 and C_{18} stationary phases [117]. Experimental results explain the advantages of methanol over other organic mobile phase components (e.g., acetonitrile) in the separation of aliphatic alcohols, phenols, and carboxylic acids. Methanol, through its reciprocal hydrogen bond donor/acceptor character forms stable complexes with these solutes *in the stationary phase*, giving enhanced selectivity for these solute types. The k' values for these 40 solutes are tabulated for 20/80, 25/75, and 50/50 methanol/water mobile phases.

Propellant residues from discharged single base (nitrocellulose) and double base (nitrocellulose/nitroglycerin) smokeless powders were examined by LC [118]. Five analytes were identified and separated (2,4-dinitrotoluene, N-nitrosodiphenylamine

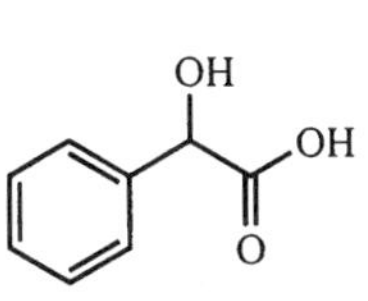

Mandelic acid 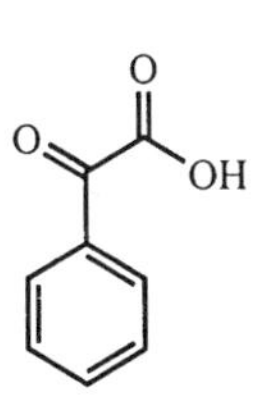   Phenylglyoxylic acid 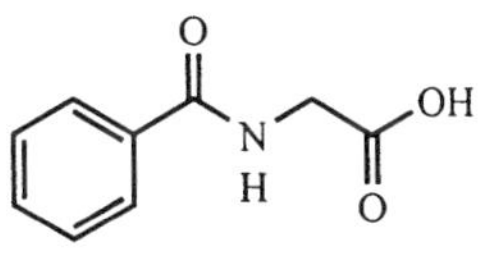   Hippuric acid

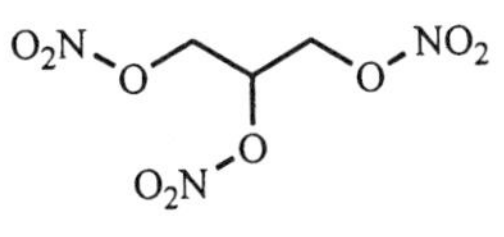

Nitroglycerine 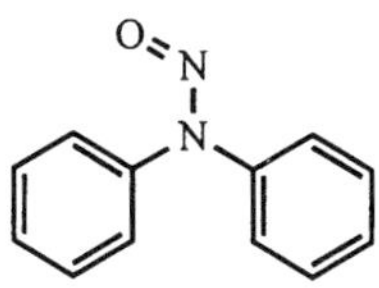 N-Nitrosodiphenylamine

[NNDA], 4-nitrodiphenylamine [4DNPA], 2,2′- and 2,4-dinitrodiphenylamine, and 2-nitrodiphenylamine [2NDPA]) on a C_{18} column ($\lambda = 385$ nm) using a 70/30 methanol/water mobile phase. Other unidentified analytes were seen in the chromatographic profiles, which were used for the characterization of different powders. Approximately 50–100 µg of propellant residue was injected. Similarly, diphenylamine was separated from NNDA, 4DNPA, and 2NDPA on a C_8 column ($\lambda = 292$ nm) using an isocratic 19.4/9.6/71.0 IPA/acetonitrile/water (50 mM chloroacetic acid to pH 2.7) mobile phase [119]. Peak shapes were excellent and good resolution was obtained. Levels below 100 pmol were detected and elution was complete in 15 min.

It has been found that the selectivity generated by methanol/water solvents can be dramatically enhanced for some analytes through the addition of THF to the mobile phase. Such analytes include benzene, benzylalcohol, 3-phenylpropanol, 2,4-dimethylphenol, and diethyl phthalate [120]. Optimal resolution was achieved with a C_{18} column ($\lambda = 254$ nm) using a 10/25/65 methanol/THF/water mobile phase. Elution was complete in 22 min.

Markus and Kwon [121] separated the metabolites of benzonitrile (i.e., cyanophenols) on a C_{18} column ($\lambda = 254$ nm) using a 60/20/920 methanol/acetic acid/water mobile phase. In addition, benzylnitrile, o-cyanobenzyl alcohol, and phthalide were separated in 20 min using the mobile phase above as the initial conditions and a gradient to 100% methanol.

Stronger hydrogen bond accepting species, such as primary amines, have been separated using methanol as the organic modifier [122]. Putrescine, spermidine, spermine, cadaverine, and a series of their acetyl amine analogs were separated and quantitated as their benzoylated derivatives using a C_{18} column ($\lambda = 254$ nm). Interestingly, even through the derivatized solutes are secondary amines, an

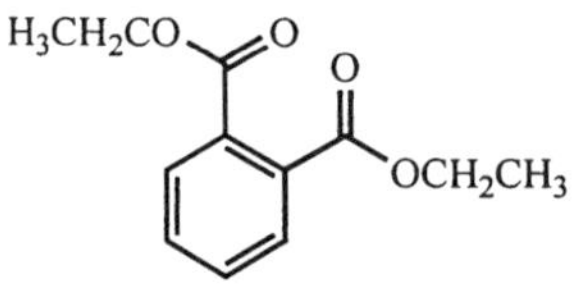

Diethylphthalate

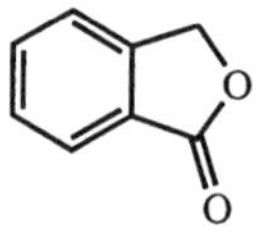

Phthalide

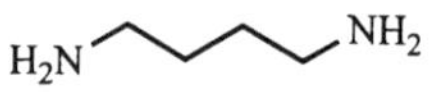

Putrescine; 1,-4-diaminobutane

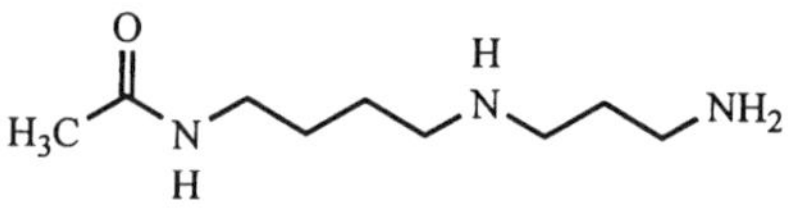

N^1-Acetylspermidine

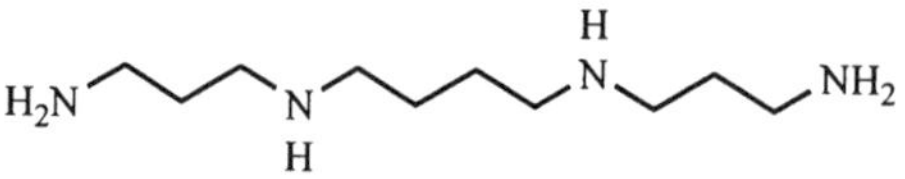

Spermine

Cadaverine; 1,5-diaminopentane

unbuffered 62/38 methanol/water solvent generated baseline resolution, excellent peak shape, and elution in 10 min. A linear calibration curve from 1.25 to 25 μM was reported. In an extension of this work, Schenkel et al. [123] expanded the above test set to include 1,3-diaminopropane and 1,6-diaminohexane. These analytes were extracted from cancer cells and converted to their benzoylated amine derivatives. Excellent peak shape was obtained on a C_{18} column ($\lambda = 234$ nm) using a 55/45 methanol/water mobile phase. Detection limits of 2–12.5 nmol injected were reported (analyte dependent). Elution was complete in <40 min and peaks were well resolved except for the N^1-acetylspermidine/1,6-diaminohexane pair. The authors noted that unless special treatment of the sample occurs after derivatization but prior to analysis, benzoyl chloride by-products such as benzoic acids and methyl benzoate, will co-elute with analytes of interest.

The retention of pyridine and 15 mono- and disubstituted alkylpyridines was studied on a 20°C C_{18} column using a series of methanol/water (50 mM phosphate buffer at pH 7.0) mobile phases [124]. Standard concentrations were prepared at the 100 μg/mL level. Whereas the use of a lower-viscosity mobile phase is usually preferable (e.g., acetonitrile/water), in this instance acetonitrile/water gave rise to peaks exhibiting a greater degree of tailing. The k' values (when <15) for all analytes were tabulated for isocratic mobile phases ranging from 65/35 to 25/75 methanol/buffer.

The separation of 5-aminosalicylic acid (as a bulk chemical) from 13 of its common production impurities was achieved using a C_{18} column ($\lambda = 215$ nm) and an 85/11/4 water (80 mM phosphate buffer at pH 2 with 5 mM heptanesulfonic acid and 70 mM NaCl)/methanol/THF mobile phase. Impurities at levels of <0.1% were readily detected [125]. Excellent peak shape was obtained for all analytes and separation was complete in <40 min.

Smith and Burr [126] studied the retention of 73 disubstituted benzenes (functional groups included alkyl, nitro, bromo, chloro, carboxyl, nitrile, amide, amino, hydroxyl, methoxy, and phenyl) on a C_{18} column ($\lambda = 254$ nm) using a series of methanol/water (1.37 g NaH_2PO_4 with 1.58 g Na_2HPO_4 to pH 7) mobile phases. The k' values for these compounds were tabulated for mobile phases ranging in composition from 40/60 to 80/20 methanol/buffer.

4.2.2   Other Organic Compounds

Nine perhalogenated compounds (CCl_4, CBr_4, CCl_2Br_2, CCl_3Br, C_2Br_4, C_2Br_6, C_2Cl_4, C_2Cl_6, $C_2Br_2Cl_4$) were quantitated on a C_{18} column ($\lambda = 220$ nm) using an 80/20 methanol/water mobile phase [127]. Elution was complete in 17 min but

p-Aminosalicylic acid

baseline resolution between all peaks was not achieved. Linear ranges of 0.01–20 mg/mL and detection limits of 1–5 mg/L (analyte dependent) were reported.

Nakajima et al. [128] used 5-(4-pyridyl)-2-thiophenemethanol to derivatize four sets of carboxylic acids: acetic (C_2) to valeric (C_5) and heptanoic (C_7); dodecanoic (C_{12}), tetradecanoic (C_{14}), hexadecanoic (C_{16}) and octadecanoic (C_{18}); and even bromobenzoic acid isomers and anisic acid isomers. A C_{18} column ($\lambda = 300$ nm, ex; 360 nm, em) was used for all separations with the following methanol/water mobile phases: $30/70 \rightarrow 100/0$ for the short-chain aliphatics; 100% methanol for the long-chain aliphatics; and 95/5 for the remaining two sets. Detection limits ranged from 5 fmol for C_{18} to 5800 fmol for C_2, with linear ranges of up to three orders of magnitude reported. Peak shapes and resolution were excellent.

Although Barceló et al. [129] focused on the performance of solid phase extraction materials, the analyte base of 22 aromatic sulfonates (e.g., benzenesulfonate, 2-amino-1,5-naphthalenedisulfonate, 1,3-benzenedisulfonate, 1-hydroxy-6-amino-3-naphthalenesulfonate, 2-napthalenesulfonate, diphenylamine-4-sulfonate) is extensive. Separation is achieved on a C_{18} column using a complex 30-min $100/0 \rightarrow 25/75$ water (5 mM triethylamine with 5 mM acetic acid to pH 6.5)/methanol gradient. Elution of all peaks was complete in <20 min and so, for selected analytes, a shorter analysis time may be possible. Standards of 50 µg/L were used and excellent peak shape and baseline resolution for all but three pairs of analytes was achieved. In those cases partial resolution was obtained.

Mucoaldehyde, a suspected breakdown product of benzene, was resolved from its corresponding disubstituted acid/aldehyde, diacid, alcohol/aldehyde, and alcohol/acid forms on a C_{18} column ($\lambda = 265$ nm) using a 25-min $0/100 \rightarrow 10/90$ methanol/water (1% acetic acid) gradient. Elution was complete in 30 min but baseline resolution between all compounds was not obtained [130].

Fifteen carboxylic acids and phenolics were extracted from white wines and analyzed on a C_{18} column ($\lambda = 210$ nm for tartaric, malic, lactic, and acetic acids; 278 nm for gallic, protocatechuic, caftaric, p-hydroxybenzoic, vanillic, caffeic,

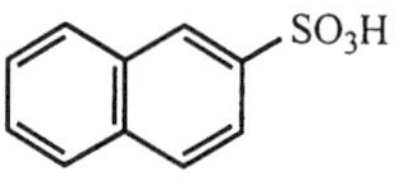

2-Naphthalenesulfonic acid

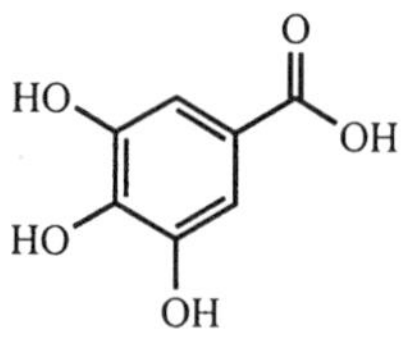

Gallic acid

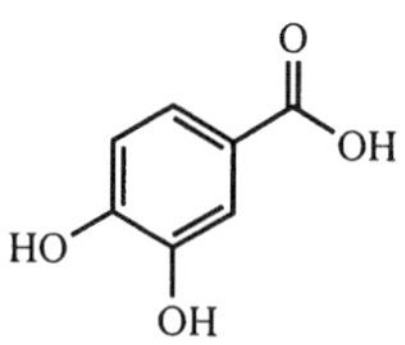

Protocatechuic acid

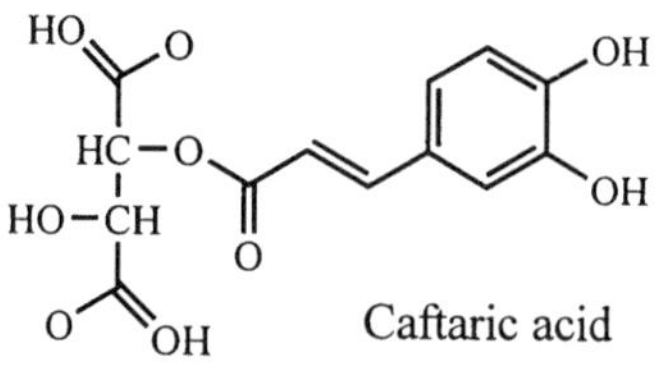

Caftaric acid

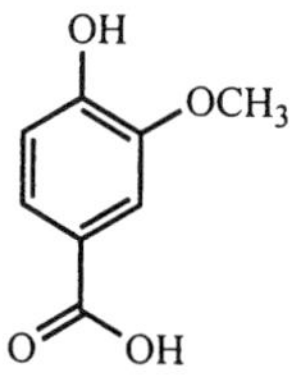

Vanillic acid

synergic, coumeric, ferulic, sorbic acids, and catechin). A complex 40-min $0/100 \rightarrow 50/50$ methanol/water (add H_2SO_4 to pH 2.5) gradient fully resolved all but vanillic acid from caffeic acid. Acetonitrile was also successfully used as the organic component of the mobile phase [131].

Peroxycarboxylic acids (C_8, C_9, C_{10}, and C_{12}) were baseline resolved in 12 min on a C_{18} column (amperometric detector; -0.27 V vs. Ag/AgCl) using a 70/30 methanol/water (15 mM phosphate at pH 6) mobile phase [132]. Detection limits of 50 µM were claimed. Interestingly, UV detection at 206 nm was also reported but, not surprisingly, excessive background (remember that the UV cutoff for methanol is typically 203–205 nm) precluded its effective use.

Dinitrophenylhydrazine (DNPH) derivatives of formaldehyde, acetaldehyde, acrolein, acetone, methyl ethyl ketone, propionaldehyde, butyraldehyde, and benzaldehyde were separated on a C_{18} column ($\lambda = 365$ nm) using a 75/25 methanol/water mobile phase [133]. Detection limits were reported as 1.5 pmol for all analytes. Separation was very good and elution was complete in <15 min. The authors noted that acetonitrile was not effective as a mobile phase constituent. Similarly, DNPH derivatives of formaldehyde, acetaldehyde, propionaldehyde, butyraldehyde, acrolein, and benzaldehyde were analyzed from automobile and stack exhaust gases using a C_{18} column ($\lambda = 360$ nm) and a 73/27 methanol/water mobile phase [134]. Peak shapes and separations were good. Sample detection limits of 10–60 ppb were reported, whereas on-column injection amounts of 2–50 ng were readily detected.

Fatty alcohols (C_8 to C_{20}) are often used in cosmetic formulations. Katayama et al. [135] used 2-(4-carboxyphenyl)-5,6-dimethylbenzimidazole as a fluorescent tag. This tag was found to be compatible with normal "wet" solvents (reducing the need for scrupulously anhydrous ones) and reacted to form the final product at room temperature. The separation of saturated C_{12}, C_{14}, C_{16}, C_{18}, and C_{20} fatty alcohols was optimized on a C_{18} column ($\lambda = 338$ nm, ex; 428 nm, em) using an 85/15 methanol/IPA mobile phase. Separation was good enough to allow baseline separation of even and odd carbon chain lengths. All compounds eluted in <20 min. Detection limits of 0.2–0.4 pg/20 µL injected (S/N = 3) were reported.

Various vegetable oils were characterized as to their saturated fatty acid profiles for C_8–C_{24} and selected unsaturated fatty acids: 16 : 1, 18 : 1, 18 : 2, 18 : 3 and 20 : 1. The fatty acids were converted to their methyl esters and subsequently converted to their hydroxamic acid derivatives [136]. The products were then separated on a C_{18} column ($\lambda = 213$ nm) using a 70/30 $\rightarrow$ 95/5 methanol/water (20 mM phosphate buffer at pH 3) gradient. The background absorbance shift due to methanol was

2,4-Dinitrophenylhydrazine

Acrolein

ingeniously counteracted by the addition of 125 μL of 1% w/v NaNO$_3$ to each liter of mobile phase. It should be kept in mind that the decrease in baseline drift is offset by an increase in background absorbance, resulting in a decrease in sensitivity and higher detection limits.

A series of nine alkyl and alkylphenol ether carboxylates were separated on a C$_{18}$ column ($\lambda = 276$ nm or RI detector) using a 75/15/10 methanol/water/acetonitrile mobile phase containing 4 mM tetrabutylammonium hydrogensulfate and 1 mM tetrabutylammonium hydroxide [137]. A linear range of 0.01–10 mg/mL for UV detection and 0.1–10 mg/mL for RI detection was reported. Although baseline resolution was not achieved for all analytes, various mixtures used to create surfactant formulations could be readily identified and characterized using this method. Elution for the longest-retained analyte, oleth-6-carboxylic acid, occurred in <25 min.

The retention of 13 organophosphorus acids (e.g., dimethylthiophosphoric, diisopropylphosphoric, ethylmethylthiophosphinic, cyclohexylmethylphosphinic, and pinacolylmethylphosphinic acid) was studied on a C$_{18}$ column (thermospray MS) using a 40/60 methanol/water (0.1 M ammonium acetate buffer at pH 5.0 with 1 mM or 5 mM tetrabutylammonium hydroxide [TBAH]) mobile phase [138]. The retention time for each compound was tabulated for both the 1 mM and 5 mM TBAH concentrations. Elution times ranged from 3 to 33 min for these compounds. With selected ion monitoring, 100 pg detection limits were reported. The authors noted that methylphosphonic acid eluted as an extremely broad peak at any pH between 3.0 and 6.0. They speculated that this was due to their working below the pK_a of the acid (6.5) and stated that working at a higher pH would have damaged the silica-based column. The use of a C$_{18}$ precolumn, which acts as a saturated column to protect the analytical column, would have allowed for the use of pH >7 mobile phases.

In a very interesting application, α-, β-, γ-cyclodextrin (CD), 6-O-glucosyl-β-CD, and 6-O-β-maltosyl-CD were resolved on a C$_{18}$ column ($\lambda = 355$ nm) using a 94/6 water (0.025% I$_2$ with 0.05% KI) mobile phase [139]. Elution was complete in 20 min. Plots of the effects of percent methanol and iodine concentration on retention were presented. Linear ranges of 0.05–0.25 mM were reported along with detection limits of 5–70 μM (analyte dependent).

Four positional isomers of 6^1,6^n-di-O-(t-butyldimethylsilyl)cyclomaltooctaose were well resolved and eluted in <30 min on a C$_{18}$ column (RI detector) using a 70/30 methanol/water mobile phase [140]. Similarly, four di-O-tritylcyclomalto-octaoses were baseline resolved in 20 min on a C$_{18}$ column ($\lambda = 240$ nm) using a 75/25 methanol/water mobile phase. Peak shapes were all excellent.

Cyclohexylmethylphosphinic acid

Homologous series of linear polysulfides ($R_2N-S_n-NR_2$, where $n = 1-11$ sulfide linkages) were analyzed as substituted dipropyl-, dibutyl-, and dipentylamines using a C_{18} column ($\lambda = 254$ nm) and 100% methanol mobile phase [141]. This study was conducted to generate predictive retention models and so separations were not optimized. However, adequate separations (each homolog was clearly distinguishable) resulted. Elution times ranged from 50 min (propyl with $n = 12$) to 75 min (pentyl with $n = 9$). The $\ln k'$ results for ethyl ($n = 2-12$) through hexyl ($n = 2-9$) are tabulated.

Ethanol and IPA have gained widespread use as polar mobile phase components in chiral separations (see also Pharmaceutical Analytes, Section 4.7). Here the choice of solvent is critical since the steric structure of the solvent mediates the solute interaction with the stationary phase chiral binding site through either (1) direct interaction with the chiral site, or/and (2) interaction with an achiral site near the active chiral site, thereby altering the steric configuration of the site [142].

Enantiomeric separation of nonpharmaceutical compounds include N-alkyl-N-methylaniline N-oxides (ethyl to butyl plus isomers) on a Chiralcel OD column ($\lambda = 210$ nm) using 1% to 3% ethanol in hexane [143]. Carrea et al. [144] separated the enantiomers of various substituted chromium and magnesium tricarbonyl metallocenes (η^6-benzene and η^5-cyclopentadiene) on a Chiralcel OD column ($\lambda = 315$ nm) with an isocratic mobile phase that varied from 1% to 10% ethanol in hexane depending on the enantiomeric pair involved. Chromium tricarbonyl compounds complexed with a variety of η^6-arenes were separated on a Whelk-O column ($\lambda = 315$ nm) using a 20/80 IPA/hexane as the mobile phase [145].

4.2.3 Organometallic Compounds and Metal–Ligand Complexes

Di- and trimethyllead, di- and triethyllead, and methyl- and ethylmercury were resolved from each other and rainwater contaminants on a C_{18} column ($\lambda = 235$ nm) using a 40/60 methanol/water (0.1 M citric acid at pH 5.9 with 2.2 mM methyl thioglycate) mobile phase. Elution was complete in 50 min and detection limits of 280–800 ng/L (analyte dependent) were reported [146].

Nine organomercury compounds (e.g., methyl-, methoxyethyl-, phenyl-, tolyl- and benzoyl-mercury; nitromersol and mersalylic acid) were baseline resolved on a C_{18} column ($\lambda = 230$ nm) using a 25 min 30/70 $\rightarrow$ 50/50 methanol/water (50 mM ammonium acetate buffer at pH 5.0 with 0.1 mM 2-mercaptoethanol) gradient [147]. Note that the mercaptoethanol has a strong stench and so the system should be located in a well-ventilated area. Detection limits from 7 to 95 µg/L were reported (analyte dependent).

Nitromersol

Four butyltin and phenyltin compounds (diphenyltin, dibutyltin, triphenyltin, tributyltin) were separated on a C_8 column ($\lambda = 425$ nm, ex; 496 nm, em) using an 84/15/1/0.03 methanol/water acetic acid/triethylamine mobile phase [148]. Separation was complete in 14 min. Monophenyltin and monobutyltin required acetic acid levels of up to 10% for elution. In these cases where particularly aggressive mobile phases are needed, a precolumn should be used to protect the analytical column. Plots of k' vs. % water and % triethylamine were presented. Linear ranges of 0.5–800 μg/L Sn and limits of detection and quantitation of 0.1–5 μg/L Sn (S/N = 3) and 0.5–11 μg/L Sn (S/N = 10, analyte dependent), respectively, were reported.

Metal–ligand complexes have been separated using methanol-based solvents [149]. 8-Quinolinethiol (QT) metal complexes were used to monitor specific metal ion levels (Co(II), Zn(II), Hg(II), Ni(II), Cu(II), and Fe(III)) during the stainless steel manufacturing process. These complexes were baseline resolved on a C_{18} column (photodiode array detector, $\lambda = 350$–800 nm) in 25 min using an 82/18 methanol/water (5 × 10^{-5} M EDTA) mobile phase. The excess QT and its oxidation products were also resolved from the species of interest. Ethanol at ~65% also gave acceptable results. Plots of k' versus percent alcohol for each species were given for methanol over the 80–90% range and for ethanol over the 60–70% range. Linear working curves from 2 × 10^{-5} to 1 × 10^{-3} M were reported. On-column injections of 0.1 μg were easily detected. The authors noted that the use of acetonitrile caused significant peak broadening and was unsuitable in this separation.

Separation of the ions Cd(II), Pb(II), Ni(II), Co(II), Cr(III), Cu(II), and Hg(II) was achieved as their diethyldithiocarbamate complexes. A C_{18} column ($\lambda = 270$ nm) used in conjunction with a 70/30 methanol/water mobile phase (containing 0.1 mM carbamate). The analysis was 28 min long and a linear range of 0.01–2 mg/mL was reported. Peaks were acceptable and good resolution was obtained [150]. Various other carbamates were tested as chelating agents but with less success than the one above.

A series of ions (Pd(II), Be(II), Ga(III), Fe(III), Al(III) and Cu(II)) were extracted from water as their acetylacetone complexes and baseline resolved on a C_{18} column ($\lambda = 330$ nm) using a 58/35/6/1 methanol/water/dichloromethane/acetic acid mobile phase [151]. Elution was complete in 40 min and detection limits of 1–5 ppb were reported. Peak shapes were uniformly good. The authors used this technique to determine Cu(II) and Fe(III) levels in tap water. It should be noted that other β-diketone complexing agents, used like ion pair reagents, should also be effective here.

8-Quinolinethiol

Saitoh et al. [152] separated seven rare-earth ions (Nd(III), Gd(III), Tb(III), Dy(III), Ho(III), Er(III), Lu(III)) as their tetraphenylporphine complexes using a C_{18} column ($\lambda = 555$ nm) and a 90/10 methanol/water (0.5% acetylacetone with 0.68% triethylamine) mobile phase. Injections of 10 μL of 0.1 mM metal-complex solutions were made. The Nd(III) complex was stable for less than one hour in any of the solvents methanol, acetone, acetonitrile, or dichloromethane. Elution was complete in <15 min. Similarly, the tetraphenylporphine complexes of VO(IV), Cu(II), Ni(II), Zn(II), and Pd(II) were resolved on a C_{18} column ($\lambda = 420$ nm) using a methanol/octane mobile phase where octane was present at less than 0.1 mole fraction [153]. (It should be noted that 0.1 mole fraction octane is equivalent to ~21% in methanol.) A 1 μL injection of a standard containing 4×10^{-6} M metal complex was readily detected.

2-(6-Methyl-2-benzothiazolylazo)-5-diethylaminophenol was used as the complexation reagent in the separation of Os(IV), Ir(IV), Pt(II), Ru(III), Co(II), and Ni(II) on a 35°C C_8 column ($\lambda = 575$ nm). Baseline separation was achieved in <30 min and detection limits of 65 pg/mL to 50 ng/mL were obtained (ion dependent). Linear working ranges up to 3000 ng/mL were reported [154].

Pd(II), Rh(III), Ru(III), and Pt(II) were baseline resolved as their 4-(5-nitro-2-pyridylazo)resorcinol complexes using a C_{18} column ($\lambda = 536$ nm) and a 50/10/40 methanol/ethyl acetate/water (10 mM acetate buffer at pH 4.0 with 10 mM tetrabutylammonium bromide and 0.01 mM sodium EDTA) mobile phase [155]. Separation was complete in 20 min and detection limits of ~1 ng/mL were reported. A plot of k' versus percent methanol ranging from 40% to 90% with no ethyl acetate present was given. The k' for some complexes exceeded 9 when the mobile phase contained <50% methanol, so this is probably not an effective system to use. Ethyl acetate was used to enhance the selectivity of the separation. Levels of 4% to 20% ethyl acetate were used to generate baseline separation.

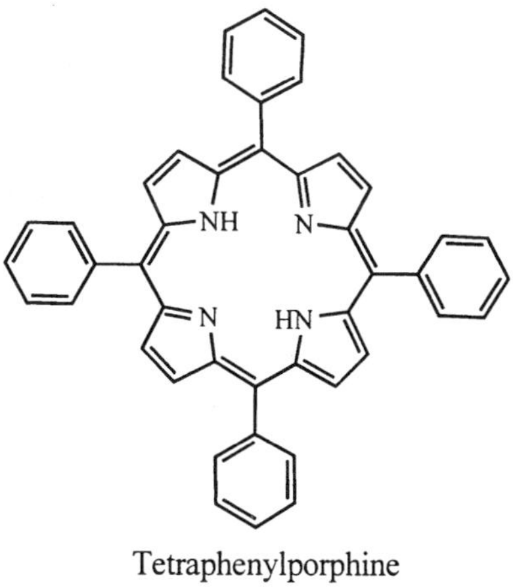

Tetraphenylporphine

4.2.4   Summary

It is evident that alcohols in the mobile phase provide an extremely powerful means for separating analytes containing hydrogen bond accepting (e.g., nitro or carboxy) or donating (e.g., hydroxyl) functional groups. These separations were successfully carried out in both the reversed-phase and normal-phase modes. Therefore, the manner in which the sample is prepared (i.e., the final sample solvent) should be taken into consideration when developing the quantitative analysis. Another major use for alcohols is as the mobile phase modifier in conjunction with chiral bonded phase separation of enantiomers.

4.3   ENVIRONMENTALLY IMPORTANT ANALYTES

4.3.1   PAHs, Substituted PAHs, and Related Analytes

A seemingly ubiquitous and commonly analyzed group of LC analytes is the polycyclic aromatic hydrocarbons (PAHs). The standard promulgated EPA methods (e.g., SW-846 Method 8310 and Method 610) utilize a 25-min approximately $40/60 \rightarrow 100/0$ acetonitrile/water gradient to separate 16 PAHs (from naphthalene to benzo[g,h,i]perylene). Since these are long-standing promulgated methods, they are well-established and reliable. Both fluorescence ($\lambda = 280$ nm, ex; 389 nm, em) and UV ($\lambda = 254$ nm) detection are permissible. Programmed fluorescence detectors have also been used to yield lower detection limits.

   The addition of low-level mobile phase modifiers was studied with regard to their effect on retention and selectivity for 16 PAHs on a C_{18} column ($\lambda = 254$ nm) using aqueous sodium dodecyl sulfate (SDS) or cetyltrimethylammonium bromide and 5–15% IPA or methanol or 2–7% 1-butanol as mobile phases [156]. A chromatogram of baseline resolved components was shown using a mobile phase of 0.1 M SDS in 85/15 water/IPA at 60°C. Standard solutions of ~200 ng/mL and 20 µL injections were used. Excellent information concerning the effects of 1-butanol on selectivity was also presented.

   Thirty-one extremely large PAHs (from pyrene with four fused rings to tetra-benzo[a,cd,f,lm]perylene with nine fused rings) were studied using a polymeric C_{18} column (photodiode array detector, $\lambda = 250$–600 nm). Methanol/dichloromethane mobile phases were used to solubilize and elute these compounds [157]. Plots for k'

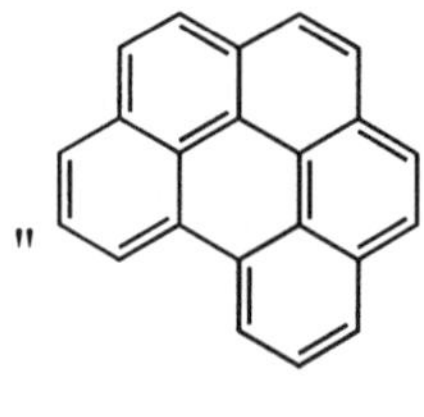

Benzo[g,h,i]perylene

versus percent dichloromethane in methanol were shown for all solutes. The effect of dichloromethane on solute spatial conformation and resulting interaction with the bonded phase is discussed in detail. These changes manifest themselves as reversals in retention orders as the dichloromethane level changes.

Anigbogu et al. [158] studied the effects of methanol, t-butyl alcohol (TBA), and cyclopentanol (CP) on anthracene and pyrene retention on a C_{18} column ($\lambda = 255$ nm) using 50% to 70% methanol in water containing 3 mM β-cyclodextrin and 1% TBA or CP. On the basis of retention effects, the authors speculate that TBA and CP assist in the formation of a cyclodextrin/pyrene complex and conclude that TBA and CP may be effectively used as mobile phase modifiers in these fused-ring systems. Schuette and Warner [159] conducted a similar study on the effects of 1-pentanol on β- or γ-cyclodextrin/PAH complexes. A solution of 0.1 M 1-pentanol with 5 mM γ- or β-cyclodextrin increased the fluorescence emission intensity markedly in the 430–490 nm range. Chromatographic selectivity and efficiency were enhanced and the detection limits were lowered by nearly an order of magnitude.

Sample stability in solvents is a key consideration for any analysis but is particularly important in PAH analyses [160]. The PAHs acenaphthene (458), fluoranthene (94), benz[a]anthracene (4.3), benzo[k]fluoranthene (76), benzo[a]pyrene (9), and dibenz[a,h]anthracene (36), all EPA priority pollutants, photodegrade in methanol (time in minutes for 50% decomposition is given in parentheses). The presence of various solvent/solute photodegradation adducts was confirmed by GC/MS. Therefore, the proper choice of solvent in tandem with careful sample manipulation is essential for reproducible and accurate results.

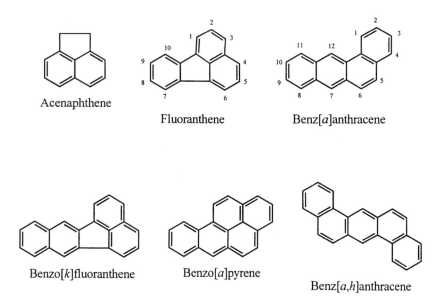

Acenaphthene

Fluoranthene

Benz[a]anthracene

Benzo[k]fluoranthene

Benzo[a]pyrene

Benz[a,h]anthracene

Pyrene, and mono-, di-, and trichloropyrenes were separated as distinct peaks on a C_{18} column ($\lambda = 320$ nm, ex; >385 nm, em) using a 98/2 methanol/water mobile phase [161]. Separation was complete in 30 min. The authors also interfaced the LC effluent to a Fourier transform infrared (FTIR) spectrometer in order to positively identify the positional isomers of these compounds. The lowest amount injected from which positive confirmation was claimed by FTIR was 10 ng.

Eight perchlorinated polycyclic aromatic hydrocarbons (benzene, acenaphthylene, naphthalene, biphenyl, anthracene, fluoranthene, phenanthrene, pyrene) were separated on a C_{18} column ($\lambda = 250$ nm) using an 80/20 methanol/hexane mobile phase [162]. The authors noted that other mobile phase combinations such as ethanol/cyclohexane, acetonitrile/THF, and ethanol/hexane were not suitable for this separation. Elution was complete in 42 min. The pyrene and acenaphthylene peaks coeluted. A standard of $\sim$1 mg/mL of each component generated detectable peaks.

The retention of fluoranthene and 13 of its liver metabolites was studied using a C_{18} column ($\lambda = 260$ nm) and a 40-min 10/90 → 100/0 methanol/water gradient [163]. The metabolites separated and identified included the 2,3-diol epoxides; tetradiols and nonvicinal tetradiols; and the 1-, 3-, and 8-hydroxyls. Fluoranthene was retained longest and eluted at $\sim$43 min. Peak shapes were uniformly excellent and resolution was good throughout.

Cyclopenta[c,d]pyrene (CPP) and a series of five hydroxy to tetrahydroxy metabolites were baseline resolved on a C_{18} column ($\lambda = 275$ nm) using a complex 40-min 10/90 → 100/0 methanol/water gradient [164]. Excellent peaks shapes resulted and resolution was good between all pairs of compounds. Elution was complete in 34 min with the unmetabolized nonpolar CPP eluting last.

5,6-Dimethylchrysene, its 1,2-dihydroxy and 1-hydroxy metabolites, and 1-hydroxy-5-(hydroxymethyl)-6-methylchrysene were resolved on a C_{18} column ($\lambda = 254$ nm) using a complex 30/70 → 100/0 methanol/water gradient [165]. Similarly, dibenz[a,j]anthracene (DBA) and three metabolites (5,6-dihydroxy-, 3,4-dihydroxy- and 1,2,3,4-tetrahydroxy-DBA) were resolved on a C_{18} column ($\lambda = 254$ nm) using a complex 60-min 50/50 → 100/0 methanol/water gradient [166].

An excellent separation of $trans$-3,4-dihydroxy-3,4-dihydrodibenz[a,h]anthracene from 13 liver metabolites was achieved by Platt and Schollmeier [167] on a C_{18} column ($\lambda = 280$ nm) using a 60-min 60/40 → 100/0 methanol/water (10 mM ammonium carbonate at pH 7.8) gradient. Elution was complete in 50 min. Proposed structures for the metabolites were depicted.

Perchlorobenzene   Perchlorobiphenyl   Chrysene

Six urinary metabolites of benz[*j*]aceanthrylene (such as, the 1,2-dihydrodiol-8-hydroxy-, 1,2-dihydrodiol-10-hydroxy-, dihydrodiol-phenol-, and *trans*-1,2-dihydrol-substituted analogs) were isolated and resolved on a C_{18} column ($\lambda = 290$ nm) using a 50-min 64/36 $\rightarrow$ 0/100 water (10 mM ammonium acetate buffer at pH 5.3)/methanol (10 mM ammonium acetate) gradient. Components were well resolved from one another and co-extracted materials [168].

In a separate study [169], metabolites of benzo[*a*]pyrene (three hydroxyls, three diols and three quinones) were separated on a PAH column ($\lambda = 254$ nm or 295 nm, ex; 430 nm, em) using a complex 60-min 60/40 $\rightarrow$ 100/0 methanol/water gradient. Complete resolution of all pairs of analytes was not achieved.

Benz[*a*]anthracene and six of its metabolites (7,12-dihydroxymethyl-, 7-formyl-, 7-formyl-12-methyl-, 7-hydroxymethyl-, 7-hydroxy-12-methyl-, and 7-methyl-benz[*a*]anthracene) were extracted from rat tissue and separated on a C_{18} column ($\lambda = 254$ nm) using a 100% methanol mobile phase [170]. The formyl-substituted metabolites showed strong fronting behavior, indicating that they were potentially chemically unstable, at too high a level in the sample (unlikely), or affected by a sample solvent/mobile phase mismatch (most likely). All other metabolites exhibited excellent peak shapes and good overall resolution was obtained. Elution was complete in 1.5 h. Peak identification was made by mass spectrometry.

Phenalene, methyl- and dimethylphenalene, and benzanthracene levels were studied in various diesel fuel distillates [171]. Baseline separation was achieved in <20 min when a C_{18} column (electrochemical detector, +0.6 V) and an isocratic 85/15 methanol/water (0.33 M sodium nitrate) mobile phase were used. Detection limits of 10 mg/L and a working concentration range of up to 40 mg/L were reported. The pH of the mobile phase had a strong effect on the sensitivity of the method. As the pH increased from 1.2 to 9.0 so did the detector response. Unfortunately, the response from interferents (indoles and carbazoles) increased faster than did that of the PAHs. Therefore, the use of a protic buffers was avoided and the sodium nitrate electrolyte was used instead.

Tjioe and Hurtubise [172] and Jansen et al. [173] generated baseline separations of four tetrol isomers of benzo[*a*]pyrene. Both groups used C_{18} columns ($\lambda = 254$ nm and $\lambda = 246$ nm, ex; 396 nm, em; respectively) and isocratic conditions: 55/45 methanol/water and 60/40 methanol/water, respectively. Tjioe reported detection limits of ~150 pg and complete elution in 23 min. Jansen claimed detection limits of ~3 pg and complete elution in 13 min. Tjioe also quantitated the tetrols using solid-matrix room temperature fluorescence and achieved 1–10 fmol detection limits (isomer dependent). Rozbeh and Hurtubise [174] increased the number of hydroxysubstituted benzo[*a*]pyrenes to 14 and studied their retention on a

Benz[*j*]aceanthrylene

Phenalene

Carbazole

C_{18} column ($\lambda = 254$ nm) using either an 80/20 or a 75/25 methanol/water mobile phase with 0–2.5 mM γ- or β-cyclodextrin added. The γ-cyclodextrin additive generated better resolution and shorter elution times.

Fu and co-workers [175] studied 30 amino- and acetylamino-PAHs (naphthalene to dibenz[a,h]anthracene) on three C_{18} columns ($\lambda = 254$ nm). The retention times for all analytes were tabulated using a 90/10 methanol/water mobile phase. Similarly, they studied 46 nitro-PAH compounds [176]. The backbone PAHs in this study again ranged from naphthalene to dibenz[a,h]anthracene. Here, some of the compounds were partially saturated. Structures and nomenclature were presented for all the analytes. Retention times were tabulated for all compounds on a C_{18} column ($\lambda = 254$ nm) using a 90/10 methanol/water mobile phase and ranged from 2 to 30 min. A detailed discussion of the suspected retention mechanism was given.

4.3.2 Nitrated and Chlorinated Nonpesticide/Herbicide Pollutant Analytes

The retention behavior of 34 mononitrated and polynitrated PAH compounds (e.g., nitroquinolines, nitronaphthalenes, nitrofluorenones, nitrobiphenyls, nitroanthracenes) was studied on a C_{18} column using methanol/water mobile phases of 50/50, 60/40, and 70/30 at temperatures ranging from 35 to 65°C [177]. Each compound was injected at the 1 µg level. Retention and selectivity effects were discussed in detail and k' values for each compound at each mobile phase composition/temperature setting were tabulated.

MacCrehan et al. [178] developed a separation for 12 nitro-substituted PAHs (e.g., 2-nitrofluorene, 9-nitroanthracene, 3-nitrofluoranthrene, 6-nitrochrysene, and 1,3-, 1,6- and 1,8-dinitropyrene) extracted from diesel exhaust and air particulates. A C_{18} column (electrochemical detector, -0.50 V vs. Ag/AgCl at a gold amalgam electrode; and fluorescence, $\lambda = 360$ nm, ex; 430 nm, em) and a 17-min $91/9 \rightarrow 60/40$ (at 5 min hold 12 min) (64/5/31 methanol/IPA/water)/acetonitrile gradient was used. Detection limits of 1–4 ng/mL for the electrochemical detector and 0.5–2 ng/mL for the fluorescence detector were reported. It should be noted that a very large oxygen (O_2) peak was observed near the void volume in the electrochemical detector-generated chromatogram. Otherwise peak shapes and resolution were good for both detection methods.

A series of 16 polar ammunition plant by-products such as nitrobenzoic acids (2-, 3-, 4-nitro and 2,4-dinitro), nitrophenols (3- and 4-nitro, 2,6- and 3,5-dinitro, 2,4,6-trinitro, 4-methyl-2-nitro, 3-methyl-4-nitro, 4-methyl-2,6-dinitro, 2-methyl-4,6-dini-

2-Nitroquinoline   3-Nitrofluorenone   2-Nitrofluorene

tro), RDX (hexahydro-1,3,5-trinitro-1,3,5-triazine), HMX (octahydro-1,3,5,7-tetra-nitro-1,3,5,7-tetrazocine), and dipicrylamine (2,2',4,4',6,6'-hexanitrodiphenylamine) were separated on a 27°C C_{18} column ($\lambda = 254$ nm) using a 40-min 53/47 (hold 20 min) → 15/85 (at 40 min) water (5 mM H_2SO_4 at pH 2)/methanol gradient. Excellent resolution was obtained except for the 3-nitrobenzoic acid/3-nitrophenol pair [179]. Detection down to 100 ng/L was reported.

Seven priority pollutant nitroaromatic (nitrobenzene and nitro-, nitroamino-, and dinitrotoluene) and nitramine residues (i.e., HMX, RDX, TNBA, TNB, DNB, Tetryl, TNT) were extracted from soil and separated on a C_{18} column ($\lambda = 254$ nm) using a 50/50 methanol/water mobile phase [180]. Retention times ranged from 2 to 15 min. Detection limits varied from 1 to 30 ng/g (analyte dependent). The linear working range was reported as 10–10,000 µg/g. This study is extremely important in that it also describes in detail the effects of numerous operating parameters (e.g., sample and standard stability, matrix effects).

Ethyleneglycol dinitrate, pentaerythritol tetranitrate, nitroglycerin, Tetryl, RDX, HMX, DNT and TNT were separated on a C_{18} column ($\lambda = 254$ nm) using a 50/50 methanol/water mobile phase. Elution was complete in <10 min [181]. To enhance sensitivity, a postcolumn Griess reaction was run (sulfanilimide and N-[naphthyl-(1)]ethylenediamine photolyzed with sample). Peak shapes and resolution were excellent. Detection limits of 100 pg were reported.

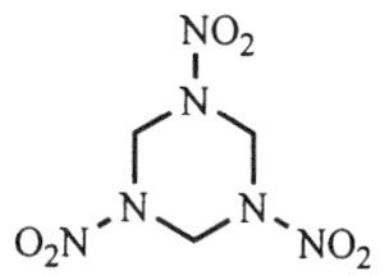

RDX

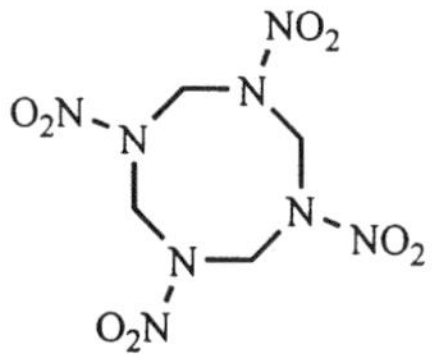

HMX

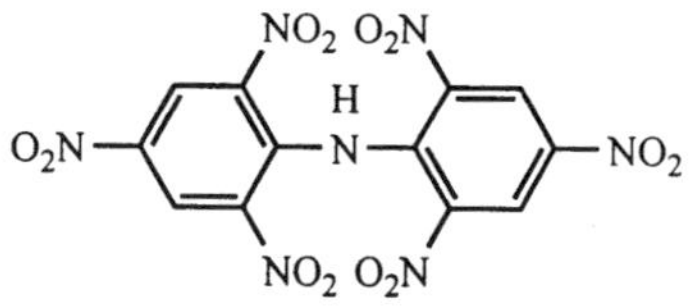

Dipicrylamine

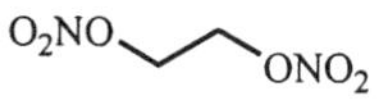

Ethyleneglycol dinitrate

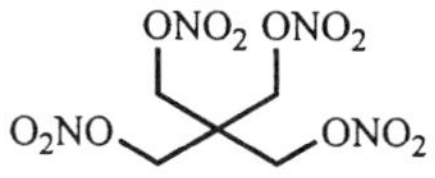

Pentaerythritol tetranitrate

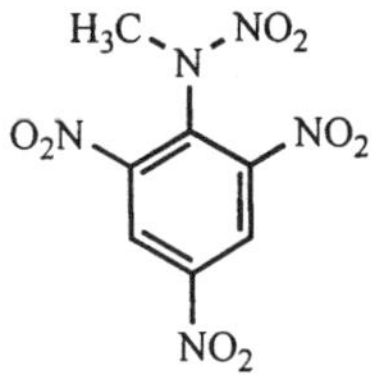

Tetryl

2,4,6-Trinitrotoluene (TNT) and its biotransformation products (2,6-diamino-4-nitrotoluene, 2,4-diamino-6-nitrotoluene, 4-amino-2,6-dinitrotoluene, and 2-amino-4,6-dinitrotoluene) were baseline resolved on tandem C_{18}/cyanopropyl columns ($\lambda = 230$ nm) using a 60.5/25/14.5 water/methanol/THF mobile phase. Elution was complete in <20 min [182].

Twenty-one nitro-substituted compounds (including HMX, RDX, Tetryl, nitroguanidine, styphnic and picric acids, nitrobenzene, and glyceryl and glycol dinitrates) were separated on a 40°C C_{18} column using a 100/86 methanol/water (phosphate buffer at pH 3) mobile phase [183]. Elution order was dependent upon the methanol level in the mobile phase, and marked retention time variation was noted for different batches of packing materials from the same manufacturer as well as between manufacturers. However, a general correlation between retention on all materials was described that allowed for data comparison and method transfer between packing materials.

Three dinitropyrenes (1,3-, 1,6-, and 1,8-) were isolated from soil samples and separated on a C_{18} column ($\lambda = 375$ nm, ex; 450 nm, em) using an 80/20 methanol/water mobile phase [184]. Elution was complete in 25 min and peaks were baseline resolved. Standards were run from 0.04 to 3.7 ng/g (analyte dependent). Detection limits of about 2 pg injected (S/N = 3) were reported.

Kimata et al. [185] studied the retention behavior of over 50 polychlorinated dibenzo-p-dioxins using C_{18}, pyrenyl, nitrophenyl and nitrophenoxy bonded phase columns ($\lambda = 235$ nm). Selectivity in terms of bonded phase/mobile phase choice was described in detail along with the presentation of over 30 chromatograms and complete tabulated results. Mobile phases were either 90/10 or 80/20 methanol/water. The authors noted that the dioxin standards were made up in nonane and had to be solvent-exchanged prior to injection to prevent the generation of poorly shaped chromatographic peaks due to the immiscibility of nonane with the mobile phase. Components of a subset of the analytes listed in the above work, the tetrachlorodibenzo-p-dioxins [186], were chromatographed on a C_{18} column ($\lambda = 235$ nm). The k' values for 22 of the isomers were tabulated using a 100% methanol mobile phase. Detection down to 2 ng injected was reported.

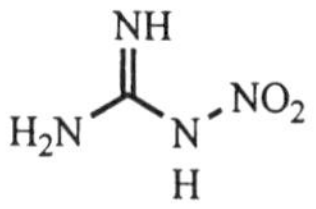

Nitroguanidine

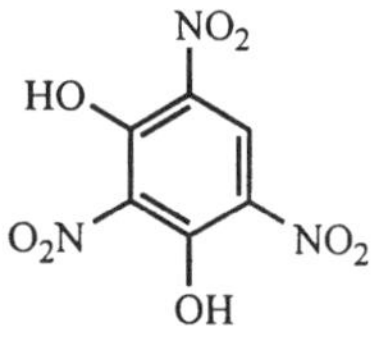

Styphnic acid

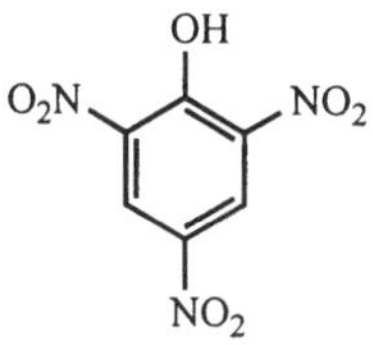

Picric acid

4.3.3 Pesticides, Herbicides, and Fungicides

4.3.3.1 Pesticides

Thiophanate-methyl and its metabolites (2-aminobenzimidazole, carbendazim) were isolated from water and separated on a C_{18} column ($\lambda = 270$ nm). A 50/50 methanol/water with 0.6% ammonia mobile phase generated elution in 10 min [187]. A linear working range of 0.25–10 µg/L with a detection limit of 0.25 µg/L (S/N = 3) was reported.

Cheung et al. [188] studied 2-{4-[(7-chloro-2-quinoxalinyl)oxy]phenoxy}-propanoic acid and its decomposition products (e.g., the ethyl ester of 2-[4-hydroxyphenoxy]propanoic acid, 4-[7-chloroquinoxalinyloxy]phenol, 2-hydroxy-7-chloroquinoxaline) were resolved on a C_{18} column ($\lambda = 245$ nm) using an 80/20/0.1 methanol/water/TFA mobile phase. Elution took 15 min. Linear ranges were reported as 80–290 µg/mL with a detection limit of 1.5 ng per injection (S/N = 3).

Coumaphos, fluvalinate, and bromopropylate (and its degradation compound 4,4'-dibromobenzophenone) were isolated from honey and separated in 8 min on a capillary C_{18} column ($\lambda = 313$ nm, 254 nm, 233 nm, 313 nm, respectively) using a

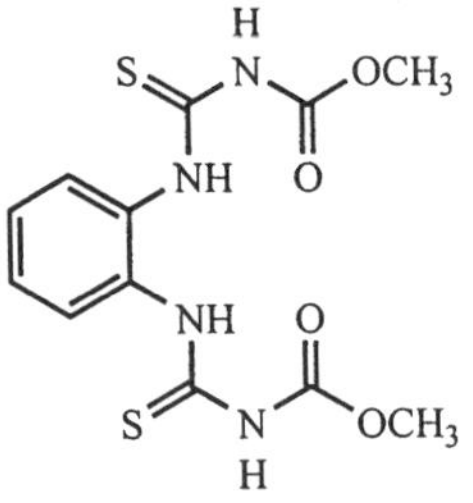

Thiophanate-methyl

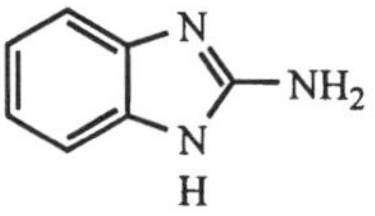

2-Aminobenzimidazole

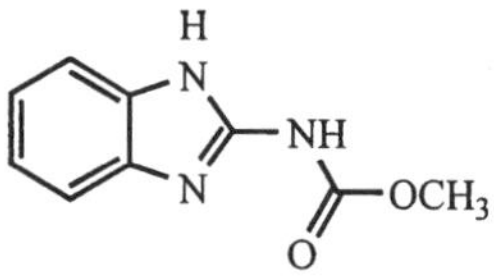

Carbendazim

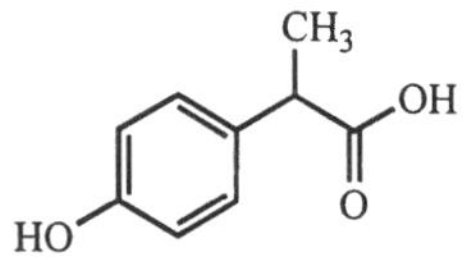

2-(4-Hydroxyphenoxy)propionic acid

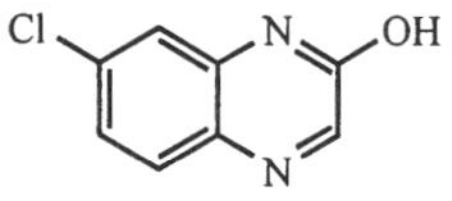

2-Hydroxy-7-chloroquinoxaline

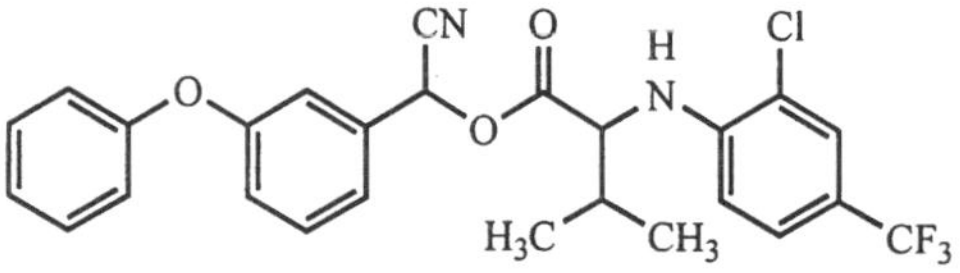

Fluvalinate

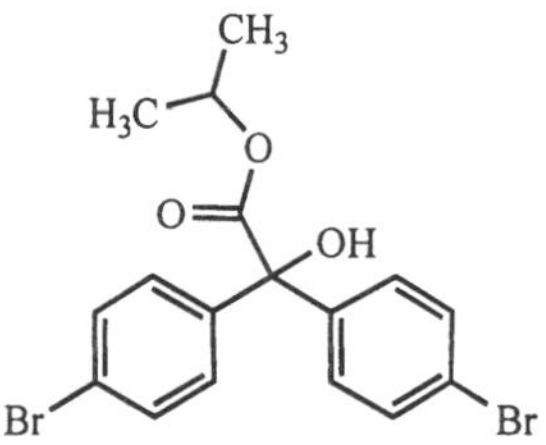

Bromopropylate

90/10 methanol/water mobile phase [189]. A plot of the effect of changing the percent water on $\log k'$ is presented. Spikes of 40 to 70 µg/kg were used.

Cyromazine and its decomposition product residue, melamine, were isolated from soil and separated on a strong ion exchange column ($\lambda = 214$ nm) using a 25/75 methanol/water (2.04 g/L KH_2PO_4 to pH 3 with H_3PO_4) mobile phase [190]. Elution was complete in 16 min. A linear range of 0.05–3 µg/mL was established and detection limits of 50 pg injected (10 ppb soil) were reported.

Paraoxon, guthion, fenitrothion, and methyl- and ethyl-parathion were extracted from fruit and separated in 8 min using a 75/25 methanol/water (25 mM acetic acid) mobile phase and a C_{18} column ($\lambda = 260$ nm). Detection limits of 0.3 ng injected were reported and calibration standards from 3 to 200 ng were used [191]. These values corresponded to <10 ng/L in water samples and 50 µg/kg in soil extracts.

Fifteen chlorinated alicyclic pesticides (e.g., chlordane, isodrin, heptachlor, endrin, dieldrin, and various epoxides) were separated using 70/30 methanol/water water (0.01% H_3PO_4) as the mobile phase on a C_{18} column ($\lambda = 210$ nm). Due to the low UV wavelength required for monitoring analyte elution and the use of methanol it should not be too surprising to find that an elevated detection limit of 1.5 µg injected was reported [192]. Strong consideration should be given to replacing some of the methanol content with acetonitrile in order to reduce the background absorbance generated by the mobile phase.

The organophosphorus pesticides oxydemeton-methyl, trichlorfon, dimethoate, dichlorvos, demeton-s-methyl, fenitrooxon, and fenamiphos (and two metabolites)

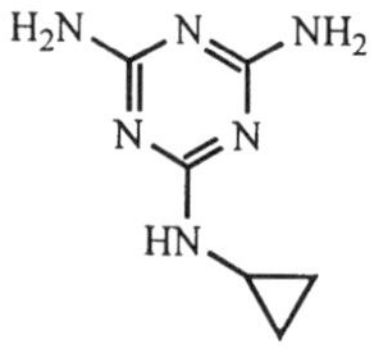

Cyromazine

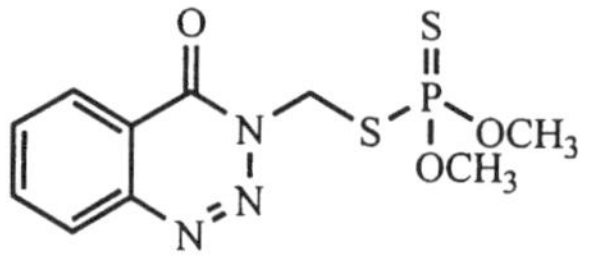

Guthion

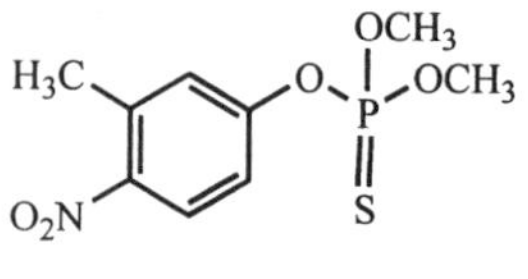

Fenitrothion

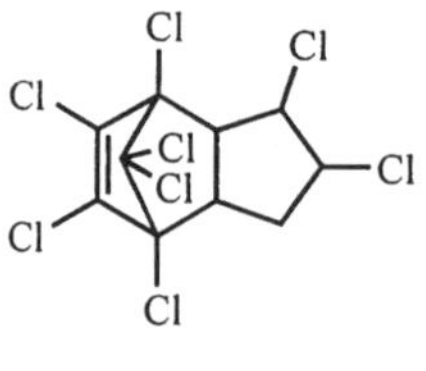

Chlordane

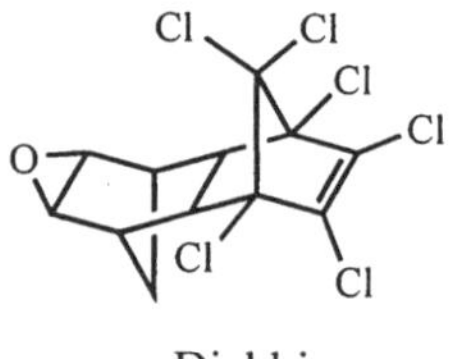

Dieldrin

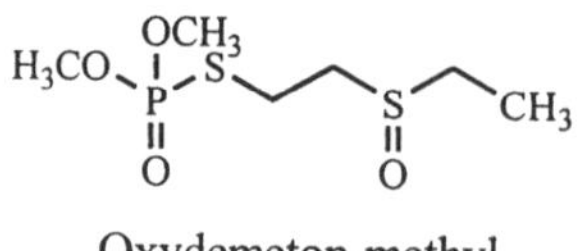

Oxydemeton-methyl

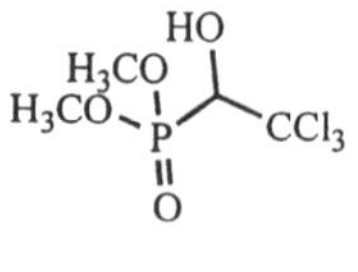

Trichlorfon

were extracted from water samples and separated on a cyanopropyl column (electrospray MS) using a 45-min $28/72 \rightarrow 60/40$ methanol/water gradient [193]. Excellent separation and complete elution were achieved in <35 min. Good sensitivity resulted (0.01 µg/L).

Phosmethylan and six production contaminants (e.g., 2-chlorobutyroanilide, di-[(2-chlorobutyroanilido)methyl]sulfide, O,O,S-trimethylphosphorodithioate) were resolved using a C_{18} column ($\lambda = 230$ nm) and an 80/20 methanol/water mobile phase. Elution was complete in 15 min, but the peak shape of the dimer-form (sulfide) contaminant was poor. Contaminant levels of 1–10%, down to 0.4 µg injected, were quantitated [194].

Wilkins [195] extracted diflubenzuron, dihydrocapsaicin, and capsaicin from fruits and vegetables and separated them on a C_{18} column ($\lambda = 254$ nm) using a 75/25 methanol/water (50 mM ammonium acetate) mobile phase. Elution was complete in <3 min. A detection limit of 0.25 mg residue per kg of foodstuff was reported.

In a much broader study, a total of 25 organophosphorus pesticides, anilides, carbamates, and phenylureas were separated and quantitated by thermospray MS using a C_{18} column and 45-min $10/90 \rightarrow 90/10$ methanol/water (100 mM ammonium acetate) gradient [196]. Very good separation was achieved and for each analyte a mass of 180 ng injected was easily detected. Again using a thermospray MS detector, 11 high-polarity pesticides (e.g., fenuron, metoxuron, isoproturon, diuron, linuron, chloroxuron) were baseline resolved on a C_8 column (also $\lambda = 245$ nm) in 45 min using a $20/80 \rightarrow 95/5$ methanol/water gradient. Ammonium acetate was added postcolumn and had a final concentration of 70 mM. Detection limits of 1–100 ng/L were obtained. Peak shapes were excellent [197].

Carbamate pesticides can be monitored in the UV at ~220 nm, by postcolumn derivatization with o-phthalaldehyde and fluorescence detection, or by thermospray MS [198]. Following US EPA Methods 531.1 and 531.5 for drinking water (C_{18} column with $\lambda = 330$ nm, ex; 465 nm, em and a 30-min $0/100 \rightarrow 0/100$ methanol/water gradient), baseline separation of 12 N-methyl and N-methyloxime carbamates and selected oxidation products was achieved with detection limits of <0.6 ppb [199].

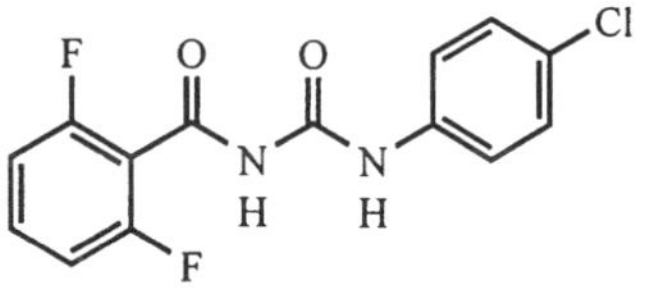

Diflubenzuron

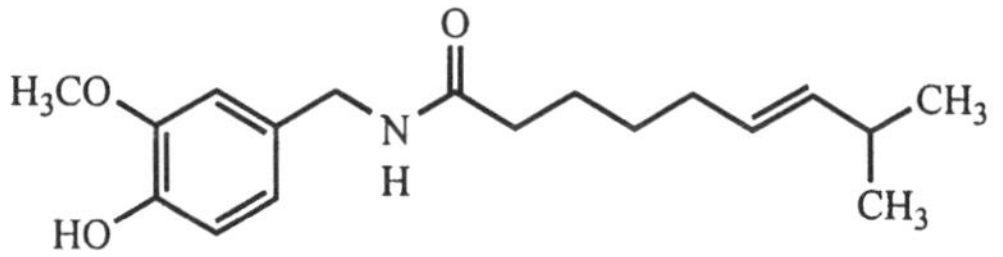

Capsaicin

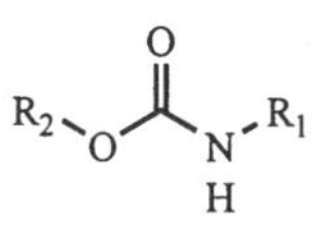

Carbamate backbone

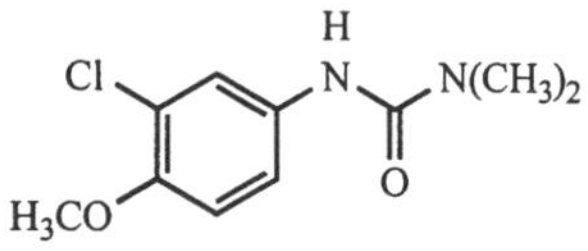

Metoxuron

Individual pesticide formulations, such as technical grade aldicarb, were successfully fractionated (aldicarb sulfoxide and sulfone and propionaldoxime) using a 30/70 methanol/water mobile phase on a C_{18} column ($\lambda = 240$ nm) when 200 µL of a 2.3 mg/mL sample was injected [200].

Two chlorophenoxyacetic acids, 2,4-D and dicamba, were extracted from soil and water and analyzed on a C_{18} column ($\lambda = 236$ nm) using a 50/50 methanol/water (1% acetic acid) mobile phase [201,202]. Detection limits of 0.1 µg/g in soil or 1.0 µg/mL in water were reported. In a similar fashion, eight chlorophenoxyacid residues (e.g., mecoprop, 2,4-D, dichlorprop, fenoprop) were extracted from water samples and analyzed on a C_{18} column ($\lambda = 228$ nm) using a 60/40 methanol/water (30 mM phosphate buffer at pH 3.0) mobile phase [203]. The authors noted that 0.05% trifluoroacetic acid (TFA) used in place of the phosphate buffer as the mobile phase modifier was just as effective and produced significantly different retention times for the analytes. Molar absorptivities were tabulated for the analytes at 205 nm, 228 nm, and 280 nm. Detection limits of 0.5–2 µg/L were reported.

Crescenzi and co-workers [204] analyzed water samples for 20 acidic pesticide compounds (e.g., dicamba, bentazone, bromoxynil, warfarin, dinoterb, pentachlorophenol, coumafuryl, and 4,6-dinitro-*o*-phenol). A C_{18} column (electrospray MS or $\lambda = 220$ nm) was used in conjunction with a 40-min 30/70 → 75/25 (85/15 methanol/acetonitrile)/water (both phases 0.1 M in K_2HPO_4 with 0.2 M tetrabutyl-ammonium fluoride) gradient. Excellent peak shapes resulted, but five pairs of

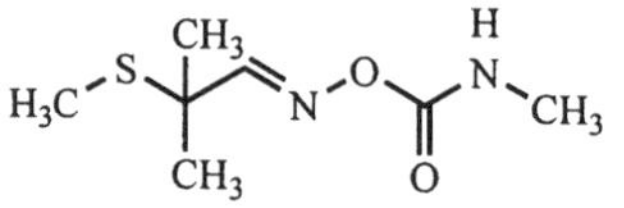

Aldicarb

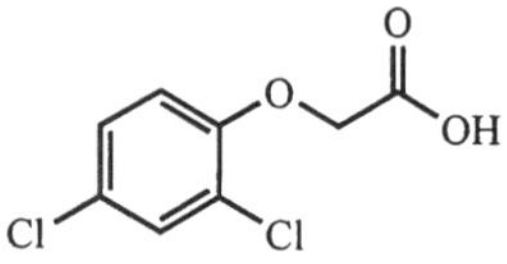

2,4-D 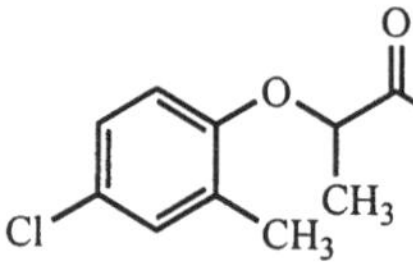   Mecoprop 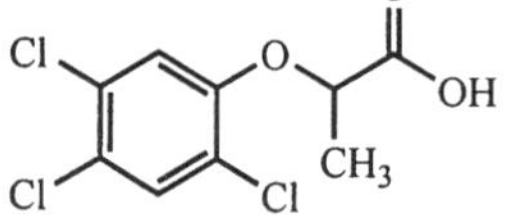   Fenoprop

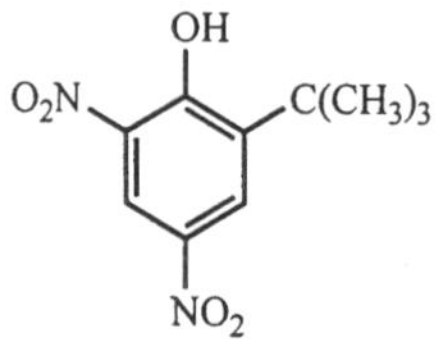

Dinoterb 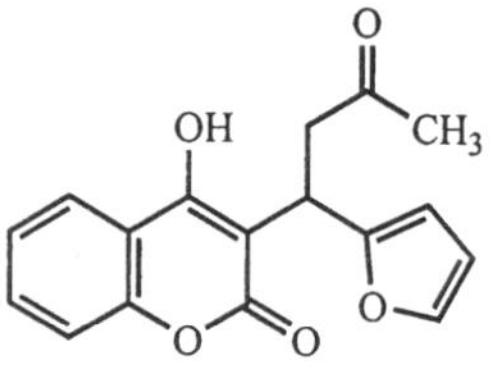   Coumafuryl

analytes were poorly resolved. Detection limits of ~2 ng injected (S/N = 3) were reported.

Sixty-eight pesticides (e.g., phenylamides, phenylcarbamates, aliphatic carbamates, methylthiotriazines, phosphothiolates) were chromatographed using a C_{18} column and a $30/70 \rightarrow 77/23$ methanol/water (10 mM NaCl) gradient [205]. Of these, 19 pesticides (e.g., barban, norflurazon, disulfoton sulfone, prometryn, disulfoton, fenamiphos, chlorobenzilate, alachlor, propoxur) were separated as described above and quantitated using photolysis and electrochemical detection (electrochemical detector with glassy carbon electrode, +1.0 V vs. Ag/AgCl). Detection limits of <10 ng were reported for all analytes and baseline separation was achieved in <35 min. The author noted that acetonitrile could not be used in this separation because it generated electroactive compounds during the photolysis process, which polarized the electrode.

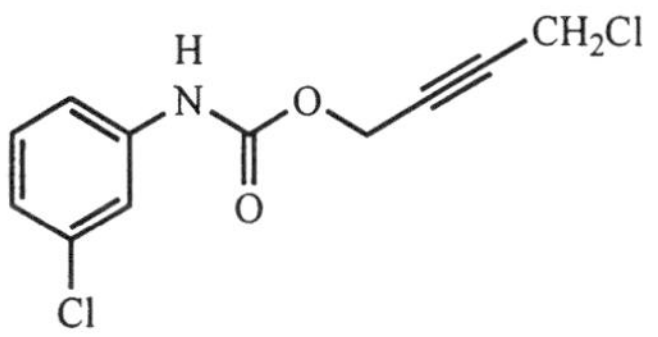

Barban

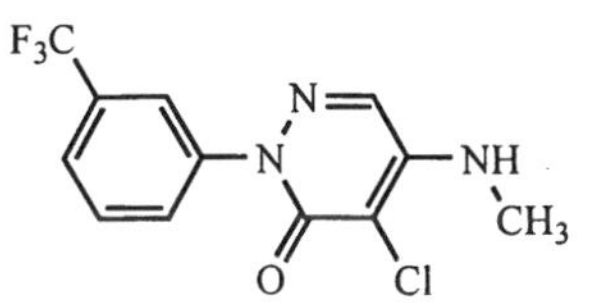

Norflurazon

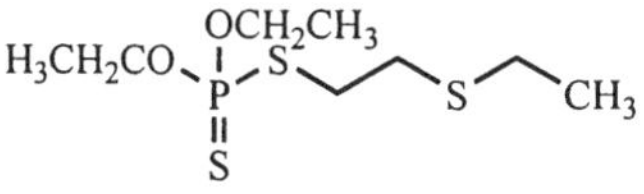

Disulfoton

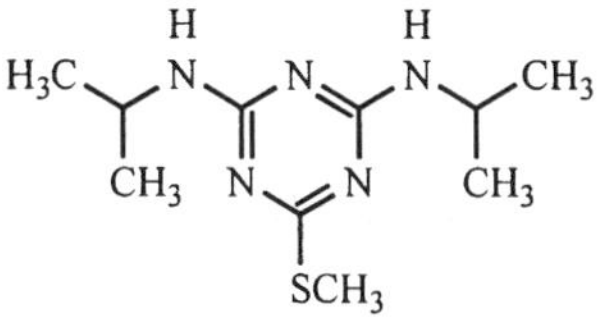

Prometryn

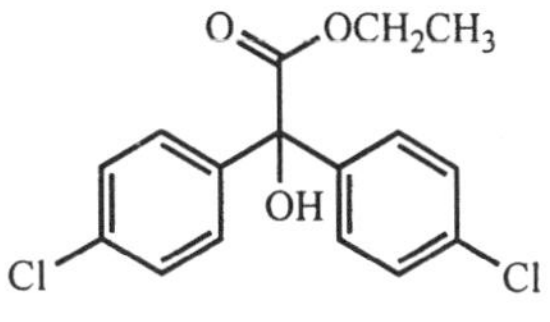

Chlorobenzilate

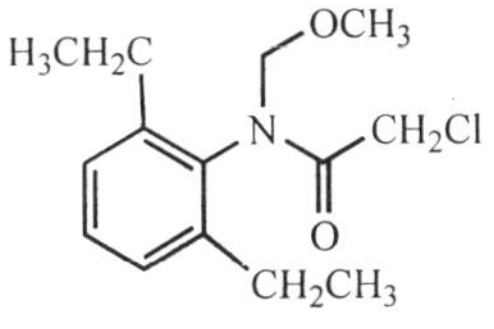

Alachlor

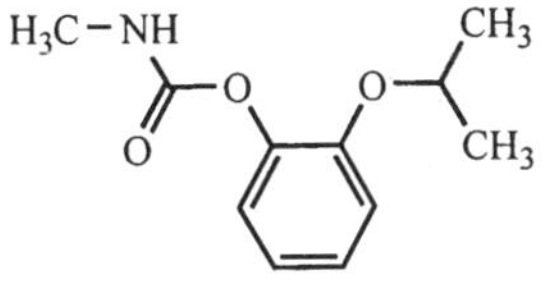

Propoxur

4.3.3.2 Herbicides

Five sulfonylurea herbicides (cinosulfuron, chlorsulfuron, thifensulfuron-methyl, metsulfuron-methyl, sulfometuron-methyl) were microwave-extracted from soil and analyzed on a 30°C C_{18} column ($\lambda = 226$ nm) using a 45/55 methanol/water (0.1% phosphoric acid) mobile phase [206]. The authors used spiked soil samples ranging from 20 to 1000 µg/kg and reported a detection limit of 2 ng injected (S/N = 3). Although the analytes were well resolved, the system exhibited a strong background absorbance decrease throughout the 10-min analysis.

Six aryloxyphenoxypropionic acids (clodinafop, fluazifop, haloxyfop, diclofop-methyl, fenoxaprop, quizalofop) were extracted from drinking water and spring water and separated on a C_{18} column ($\lambda = 240$ nm) with a complex 25-min $42/58 \rightarrow 12/88$ water (0.1% TFA)/(70/30/0.025 methanol/acetonitrile/TFA)

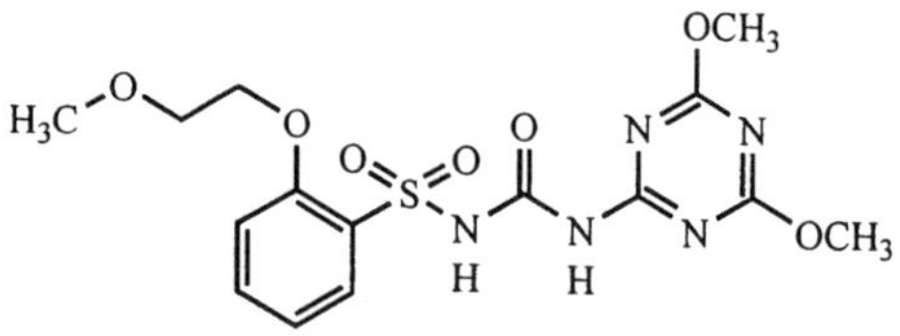

Cinosulfuron

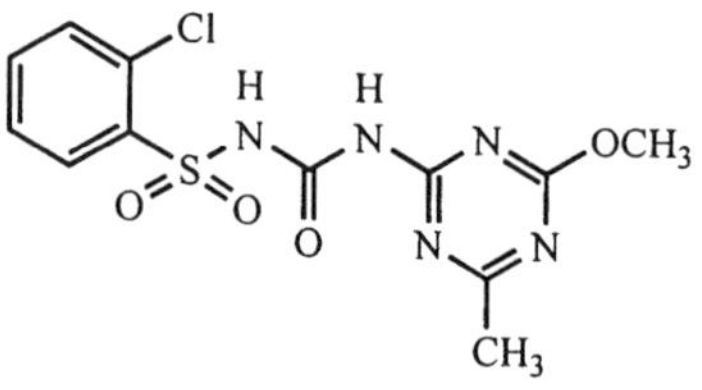

Chlorsulfuron

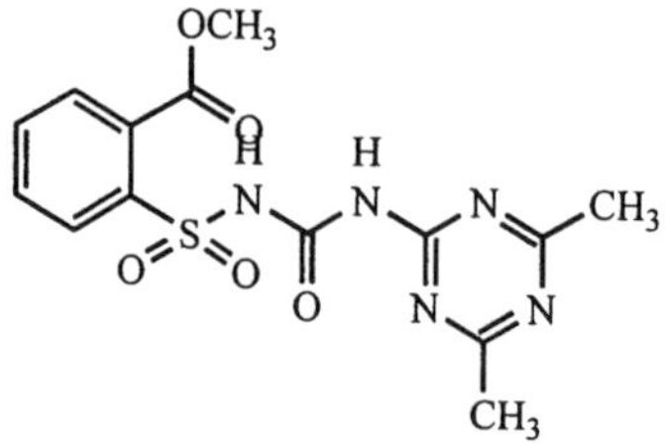

Sulfometuron-methyl

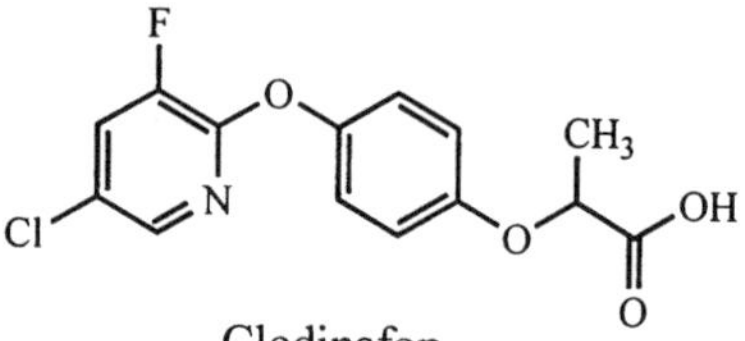

Clodinafop

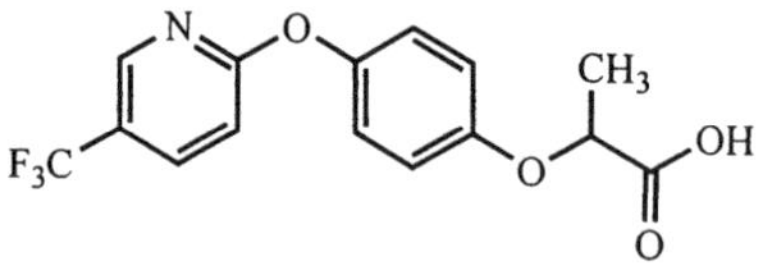

Fluazifop

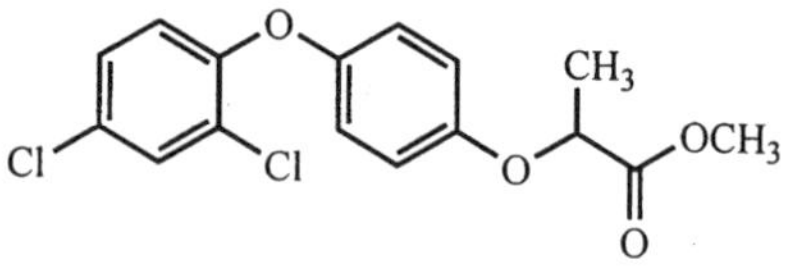

Diclofop-methyl

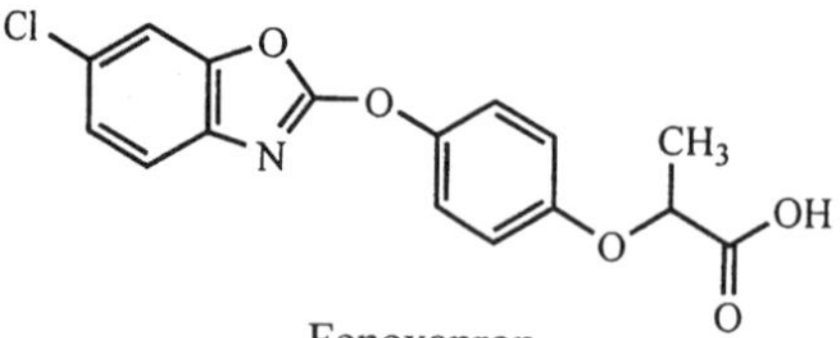

Fenoxaprop

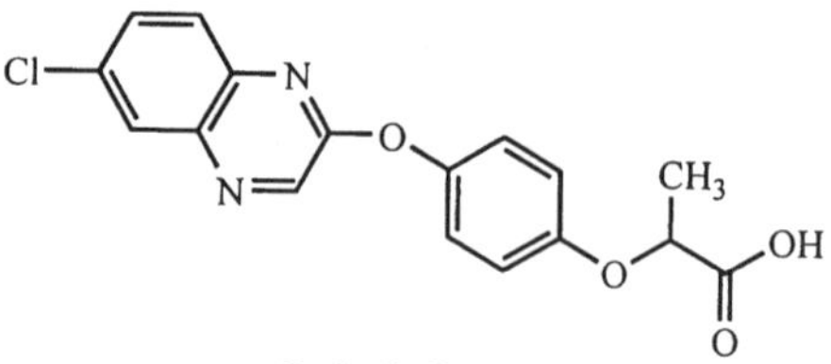

Quizalofop

gradient [207]. The analytes were well resolved but eluted on a strongly concave baseline generated by an early-eluting large co-extraction peak and a prominent upward baseline shift halfway through the elution. This may have been due to a very small signal of analyte (since no scale is given but accompanying PDA-derived spectra have signals of 0.001–0.0015 AU). Working solutions of 1–2 µg/mL were used. Detection limits ranged from 7 to 20 ng/L.

Barceló and co-workers [208] used negative ion electrospray MS as the detector in the analysis of eight acidic herbicides (benazolin, bentazone, 4-chloro-2-methyl-phenoxyacetic acid, 4-[4-chloro-2-methylphenoxy]butyric acid, 6- and 8-hydroxy-bentazone) in water. A C_{18} column was used with a 30-min $20/80 \rightarrow 80/20$ methanol/water (formic acid at pH 2.9) gradient. The detection limits were ~ 0.03 µg/L. Co-elution of 6- and 8-hydroxybentazone occurred but these analytes were quantitated separately through the use of single ion monitoring (SIM). Excellent peak shapes were obtained and all other compounds were well resolved.

In an interesting separation developed by Sanchis-Mallols et al. [209] emphasis is placed on the importance of solvent specificity in a separation: seven herbicides [dicamba, bentazone, 2,4-dichloro-, 4-chloro-2-methyl-, and 2,4,5-trichlorophen-oxyacetic acid, 4-chloro-2-methylphenoxybutyric acid, 2-(2,4-dichlorophenoxy)- and 2-(2-methyl-4-chlorophenoxy)propionic acid] were fully resolved in <10 min using a C_{18} column ($\lambda = 230$ nm) and a 50/42/8 methanol/water (10 mM NaH_2PO_4 to pH 2.5 with H_3PO_4)/pentanol mobile phase. Neither methanol nor acetonitrile alone could cleanly separate all analytes. Chromatograms resulting from the use of methanol/buffer/pentanol using various mobile phase compositions were presented. Linear ranges of 0.1–0.5 mg/mL and detection limits of 5–100 µg/L were reported.

The retention characteristics of nine phenylureas (e.g., phenuron, hydroxy-methoxuron, bis-N,N'-(3-chloro-4-methylphenyl)urea, neburon, and N-phenylurea) were studied on a cyanopropyl column ($\lambda = 254$ nm) under normal-phase conditions of hexane/IPA (90/10 to 10/90), and reversed-phase conditions of IPA/water (10/90 to 90/10). Complete resolution of all nine herbicides was not achieved, but six were well resolved and eluted within 15 min using an isocratic 90/10 hexane/IPA mobile phase [210]. Extensive tables of k' values for a wide range of isocratic mobile phases for both the NP and RP separations were presented.

Bentazone

Benazolin

N-Phenylurea

Thirteen phenylureas (e.g., linuron, metoxuron, monuron, chlorturon, chloroxuron) extracted from various crops were baseline resolved in 35 min using a C_{18} column ($\lambda = 242$ nm) and a $34/6/60 \rightarrow 68/12/20$ methanol/acetonitrile/water gradient [211]. Excellent resolution and peak shapes were obtained. With the documented cleanup procedure, routine detection limits down to 10 ppb for various fruits, vegetables, and cereals were reported. Brinkman and co-workers [212] studied a similar set of 15 phenylureas in drinking water and surface water using a thermospray MS detector. An 18-min $40/60 \rightarrow 80/20$ methanol/water (0.1 M ammonium acetate) gradient on a C_{18} column resolved all but one pair of analytes (difenoxuron from isoproturon). Peak shapes were excellent and the other analyte pairs were adequately resolved. Detection limits were reported to range between 5 and 20 ng/L for drinking water samples (analyte dependent).

Linuron and four of its metabolites (e.g., 3,4-dichloroaniline, *N*-[3,4-dichlorophenyl]-*N'*-methylurea) were extracted from water and soil samples and quantitated using an amperometric detector (+1.3 V vs. Ag/AgCl). Good resolution was obtained with complete elution in 15 min using a C_{18} column and a $38/10/52$ methanol/ethyl acetate/water (2 g KNO_3/L of water) mobile phase. Detection limits of 0.5 ng injected and working curve ranges up to 1 µg injected were reported [213]. The amperometric detector was five times more sensitive than a UV detector at $\lambda = 247$ nm.

Water samples containing 3 ppm levels of analytes yielded detectable peaks. Simazine, atrazine, and propazine were separated and quantitated using a C_{18} column (thermospray MS) and a 50/50 methanol/water (1% acetic acid) mobile phase. Here, peaks of 100 ng injected were detected [214]. In a separate study, particle beam MS detection was also used with a C_{18} column and a 70/30 methanol/water (1.7 mg/L phenoxyacetic acid and 1% acetic acid) mobile phase. A 50 ng injected detection limit was reported. An excellent 10-min separation was achieved with this method.

In an extremely diverse survey, Bagheri et al. [215] generated a separation for 39 carbamates, triazines, phenylureas, and organophosphoros pesticides using a C_{18} column (thermospray MS) with a $10/90 \rightarrow 90/10$ methanol/water (0.1 M ammonium acetate) gradient. Excellent overall resolution between all compounds and

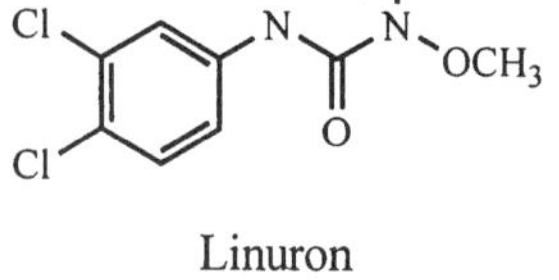

Linuron 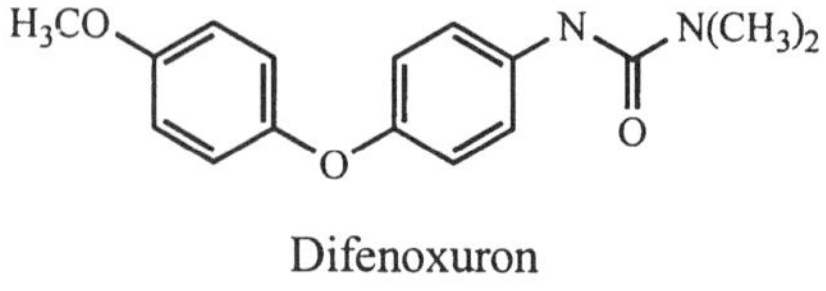   Difenoxuron

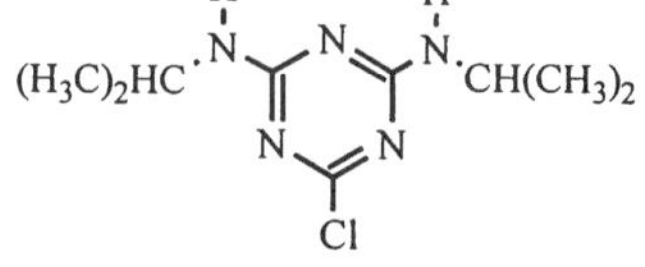

Propazine

complete elution was obtained in <50 min. Calibration curves from 0.1 to 10 μg/L were generated for each analyte. .

Seven chlorotriazines (simazine, cyanazine, hydroxy- and deethyl-atrazine, atrazine, deisopropylatrazine, chlorodiamino-s-triazine) were baseline resolved in <14 min using a C_8 column (thermospray MS) and a 20/80 → 55/45 methanol/water (50 mM ammonium acetate with 1% formic acid) gradient [216]. Excellent peak shapes and good resolution were obtained. Detection limits in the 20 ng injected range were reported. The authors noted that these detection limits could not be achieved on a C_{18} column.

4.3.3.3   *Fungicides*

Five thiurams (N,N,N',N'-tetramethyl-, N,N,N',N'-tetraethyl-, N,N,N',N'-di-methyldiphenyl-, N,N,N',N'-tetrabutyl-, and N,N,N',N'-tetrabenzylthiuram disulfide, N,N,N',N'-tetramethylthiourea, and N,N,N',N'-tetramethyl- and N,N,N',N'-tetraethylthiuram monosulfide) were baseline resolved on a C_{18} column ($\lambda = 254$ nm) using a 50-min two-ramp 40/60 → 88/12 (at 32 min) → 100/0 (at 50 min) methanol/water gradient [217]. Standard concentrations at 100 mg/L were readily detected. Ten degradation products were also resolved but not identified.

Four dithiocarbamate residues (ziram, mancozeb, propineb, metam-sodium dihydrate) were isolated from fruits and vegetables and separated on a 30°C C_{18}/polyvinylalcohol column ($\lambda = 286$ nm and electrochemical (EC) detector, +0.6 V vs. Pt). Samples were incompletely resolved using a 15-min 90/10 → 40/60 water (10 mM sodium EDTA with 10 mM Na_2HPO_4 to pH 11 with NaOH)/methanol gradient [218]. Note that the very high pH of the mobile phase is compatible with this column. A plot of the retention vs. percent methanol is presented. Standards ranged from 25 to 500 μg/L. Detection limits of ~5 μg/L and quantitation limits of ~10 μg/L were obtained from both UV and electrochemical (EC) detectors. However, the authors noted that the EC detector required a 2-day equilibration prior to use.

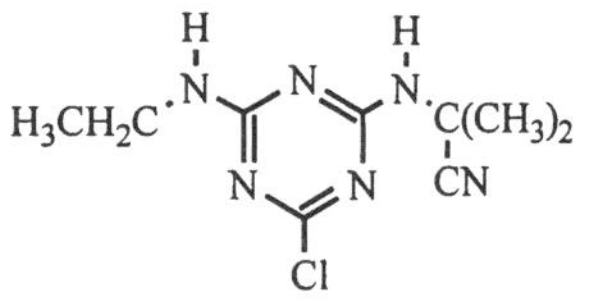

Cyanazine

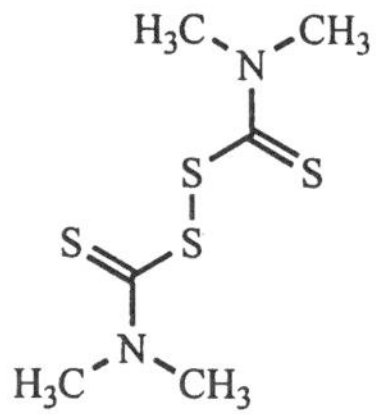

Thiuram (tetramethyl)

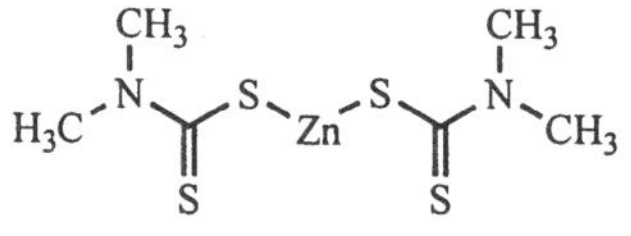

Ziram

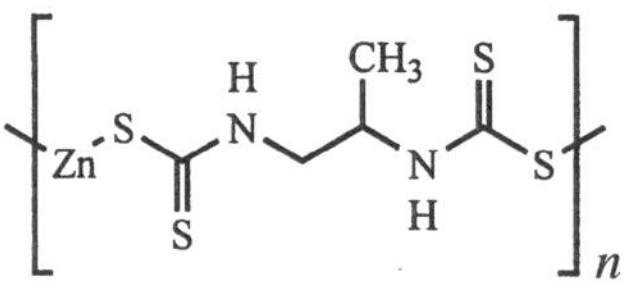

Propineb ($n > 1$)

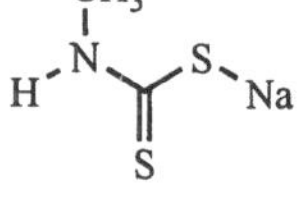

Metam-sodium

Two benzimidazoles (thiabendazole, carbendazim) were recovered from oranges and grapes and separated in <4 min on a 55°C C_{18} column ($\lambda = 285$ nm) using a 50/50 methanol/water mobile phase [219]. The authors noted that, along with increasing retention time, peaks broadened significantly but remained symmetric with a decreasing percent methanol in the mobile phase. Linearity was established over the 0.5–1000 ng injected range. Detection and quantitation limits of 0.5 ng injected (S/N = 3) and 2 ng injected (S/N = 9) were reported, respectively.

The transformation products of vinclozolin were isolated from a fungus incubate and analyzed on a C_{18} column ($\lambda = 254$ nm) using a 50-min 30/70 → 95/5 methanol/ water gradient [220]. The carbamic acid product peak shape is extremely poor as compared with the others. The authors do not note or explain the reason for this, but this is the only acid present, it elutes prior to all other peaks, and no buffer is present in the mobile phase to keep the solute protonated.

The fungicides based on complexes of dimethyldithiocarbamate (DMDC; $Zn[DMDC]_2 \equiv$ Ziram; $Fe[DMDC]_3 \equiv$ Ferbam), ethylenebis(dithiocarbamate) (EBDC; $Na_2[EBDC] \equiv$ Nabam; $Zn[EBDC] \equiv$ Zibeb; $Mn[EBDC] \equiv$ Maneb; $(Mn/Zn)[EBDC] \equiv$ Mancozeb), and the flotation agent ethylxanthogenate were isolated from water and analyzed as their copper(II) complexes [221]. A C_{18} column ($\lambda = 260$ nm) and a 70/30 water (2 mM sodium acetate at pH 4.5 with 2.5 mM hexanesulfonate and 100 µm Cu(II))/methanol mobile phase generated the elution of $Cu(DMDC)_2$ in 10 min. The authors noted that higher Cu(II) levels increased background noise considerably and at 750 µM a blue precipitate formed. Analyte response was unaffected by the concentration of hexansulfonate and pH. Significant system peaks were present but a linear range of 40–500 nM and a detection limit of 20 nM (S/N = 3) were reported. The other analytes listed above were baseline resolved on the same column ($\lambda = 287$ nm) but a 45/55 methanol/ water (3.25 mM hexansulfonate with 2.5 mM acetate at pH 4 and 130 µM Cu(II)) mobile phase was used. Detection limits here were 30 nM.

Ethanolic sesame seed extracts were purified to obtain a taurine transport inhibitory substance (TTIS) using a two-step method. Fractions were collected

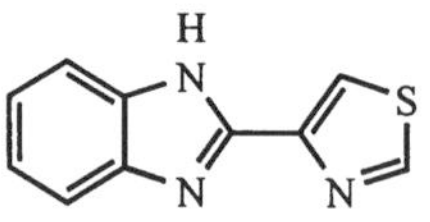

Thiabendazole

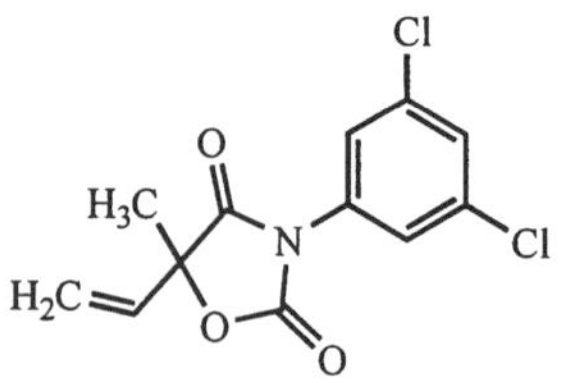

Vinclozolin

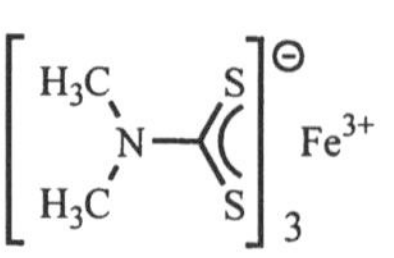

Ferbam

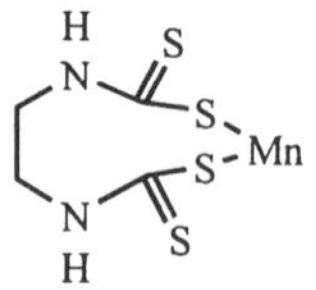

Maneb

from a C_{18} column ($\lambda = 210$ nm) using a 40-min 84/16 (hold 8 min) → 100/0 (at 20 min hold 20 min) methanol/water gradient and then analyzed on a C_8 column using a 30-min 85/15 (hold 5 min) → 100/0 (at 20 min hold 10 min) gradient [222]. In the analysis step, the TTIS eluted at ~9 min with no other peak elution after 11 min. Therefore, this analysis could be dramatically shortened.

Feruloylagmatine was extracted from winter wheat and analyzed on a C_{18} column ($\lambda = 325$ nm). A 60-min 72/25 → 65/35 water (0.05% TFA)/methanol gradient generated excellent peak shape and elution in ~27 min. A related compound, *p*-coumaroylagmantine was detected (at 21 min) as well [223].

4.3.4   Summary

Alcohols show unique chromatographic properties for highly oxygenated compounds: phenols, carboxylic acids, nitro-substituted, ketones and aldehydes. Extremely complex mixtures could be resolved due to the unique selectivity that alcohols provide for these strongly hydrogen bond interactive solutes. Even weakly hydrogen bonding analytes such as the chlorinated pesticides were effectively analyzed with methanolic mobile phases. The addition of a small volume percent of ethyl acetate, THF, and/or acetonitrile (all hydrogen bond acceptors) often enhances peak sharpness. It is important to note that methanol can be effectively used as the mobile phase organic constituent in APCI, thermospray, particle beam and electrospray MS detection systems.

4.4   INDUSTRIAL AND POLYMER ANALYTES

4.4.1   Surfactant and Additive Analytes

Surfactants (surface active agents) are very important constituents of many industrial formulations. In these formulations, it is often not just one compound that is of interest. Rather, the overall identity, as determined by the presence and distribution of the individual components, is critical. Kondoh et al. [224] developed a method for determining non-ionic surfactants containing ester groups, such as sorbitan and sucrose fatty acid esters and polyoxyethylene fatty acid esters, as their *o*-nitro-phenylhydrazine derivatives. To remove the residual free fatty acid fraction, 6 mM triethylamine (TEA) was added to the 85/15 methanol/water starting mobile phase. The free fatty acids then eluted in the void volume and the separation of the analytes of interest was conducted on a C_{18} column ($\lambda = 550$ nm). Elution and identification up to the penta-ester resulted when a 50-min 0/85/15 → 75/25/0 ethanol/methanol/water (6 mM TEA) gradient was used.

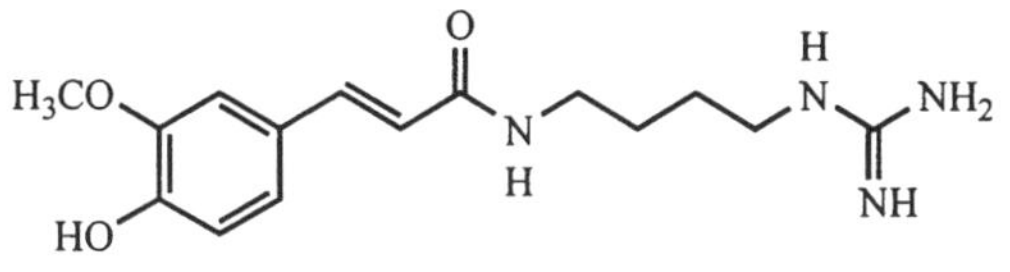

Feruloylagmatine

Levels of ethoxylated alcohol surfactants, $C_nH_{2n+1}(OC_2H_4)_xOH$ (where $n = 12$ and $x = 1$ to 30), in water were monitored through extraction and derivatization with phenyl isocyanate. Successful characterization was achieved using a 30-min $80/20 \rightarrow 100/0$ methanol/water gradient and a C_{18} column ($\lambda = 240$ nm). Detection limits of 0.1 ppm were reported [225]. Crescenzi and co-workers [226] used electrospray MS as the detector for analyzing aliphatic ethoxylate alcohols in water samples. In this study a C_8 column and a 20-min $80/20 \rightarrow 100/0$ methanol/ water (0.1 mM TFA) gradient was used. Excellent resolution of C_{12} to C_{18} ethoxylates was obtained and a detection limit of 20 pg injected was reported. Detection limits of 2 µg/L and 0.2 µg/L for river water and drinking water, respectively, were reported.

Five octylglycoside non-ionic surfactants (e.g., 1-*O*-octyl-β-D-galactopyranoside, 1-*S*-octyl-β-D-thioglucopyranoside, 1-*C*-octyl-1-deoxy-β-D-glucopyranoside) were separated in 8 min on a porous graphitized carbon (PCG) column (ELSD, 30°C tube temp., 2.2 bar air pressure) using a 95/5 methanol/water mobile phase [227]. Peak shapes were excellent and the peaks were baseline resolved. The authors noted that gradient use on the PCG column required a re-equilibration time of approximately 30 min! Standard concentrations ran from 100 to 500 mg/L.

Tsuda et al. [228] studied three alkylphenol polyethoxylates (4-nonylphenol, and nonylphenolmono- and nonyldiethoxylate) and three alkylphenols (4-*t*-butylphenol, 4-octylphenol, bisphenol A). These were isolated from fish and shellfish and separated on a phenyl column ($\lambda = 275$ nm, ex; 300 nm, em). A 25-min $40/60 \rightarrow 20/80$ water/methanol gradient was used to generate baseline resolution of all peaks. A linear range of 0.05–2.0 µg/mL and a detection limit of 2 ng/mL (S/N = 3) were reported.

Linear alkyl sulfonates (C_{10}–C_{14}) and sulfophenoxycarboxylic acids (C_2–C_{11}) were isolated from marine samples and simultaneously separated on a C_8 column

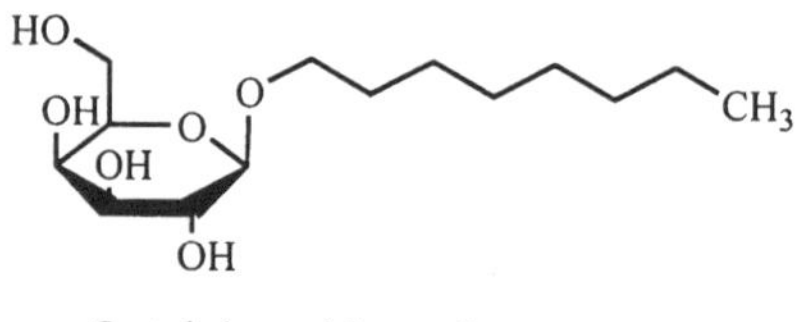

Octylglycoside surfactant
1-*O*-octyl-β-galactopyranoside

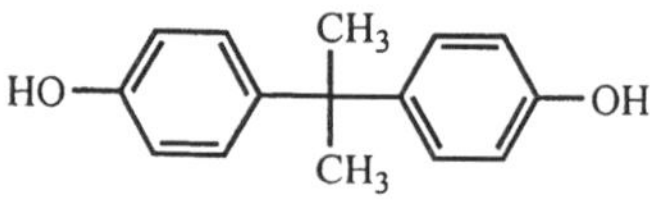

Bisphenol A

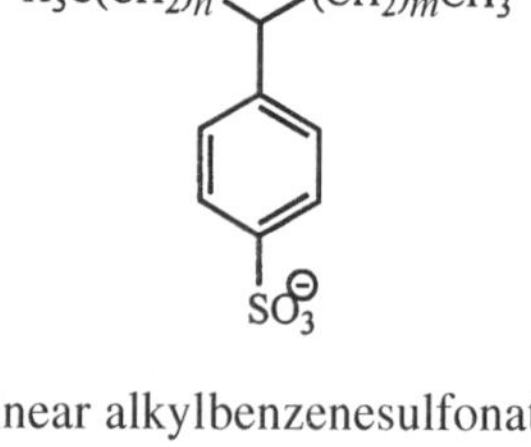

Linear alkylbenzenesulfonate

($\lambda = 225$ nm, ex; 295 nm, em). A complex 55-min $90/10 \rightarrow 0/100$ water/(8/2 methanol/water with 1.25 mM tetramethylammonium hydrogensulfate) gradient generated excellent peak shapes and separation for all compounds [229]. A linear working range of 10–450 μg/L and a detection limit of 0.2 μg/L was reported.

Detergents, linear alkylbenzenesulfonates (C_9–C_{13}), were separated on a C_8 column ($\lambda = 225$ nm, ex; 290 nm, em) using an 80/20 methanol/water (0.1 M $NaClO_4$) mobile phase [230]. Peak shapes were excellent and elution was complete in <12 min. River water extracts were studied, with a 4 ng detection limit claimed.

Optical brighteners are detergent additives. Twenty-three brighteners in laundry detergents were analyzed on a C_8 column ($\lambda = 330$ nm) using a 50-min 40/40/20 methanol/acetonitrile/water (50 mM tetrabutylammonium bromide) to 90/10 IPA/dichloromethane (10 mM tetrabutylammonium bromide) gradient [231]. Retention times for all peaks were tabulated (typically between 6 and 25 min) and actual quantitative results for 10 marketed detergents were presented. Peak shapes were excellent.

Dimethyl to didecyl phthalates were separated using a C_{18} column and a 60/40 IPA/water mobile phase [232]. When a gradient was employed, phthalates up to didodecyl substituted were resolved in <50 min. Plots of $\log k'$ versus chain length for various IPA/water mobile phases provide a predictive basis for further method development.

4.4.2 Polymers and Polymer Additives

The use of reversed-phase HPLC for the characterization of oligomers, polymers and polymer additives has received much attention recently. Twelve polymer additives (Irganox 245, 259, 565, 1010, 1035; Tinuvin P, 234, 320, 326, 327, 328; Anox 3114) were well resolved on a 40°C C_{18} column ($\lambda = 254$ nm) using a 92/8 methanol/water mobile phase [233]. Elution was complete in 90 min. Confirmation of peak identity was obtained by the use of a photodiode array detector. The k' values for eight selected solutes were tabulated for methanol/water compositions ranging from 85/15 to 95/5.

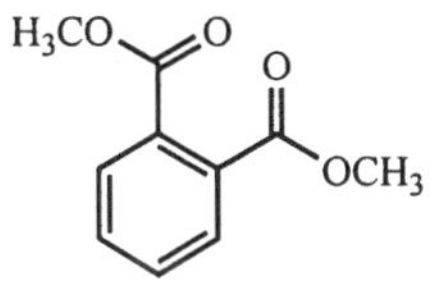

Dimethyl phthalate

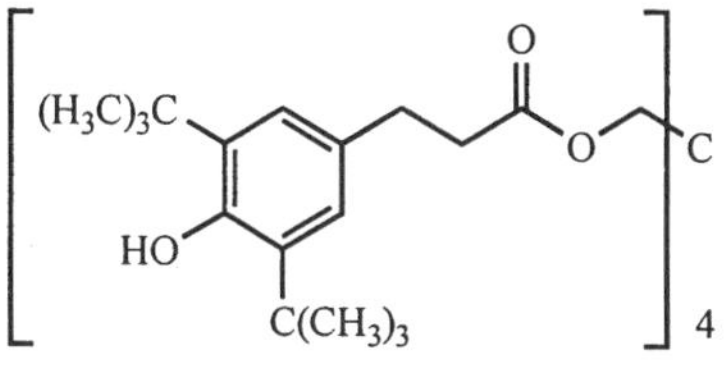

Irganox 1010

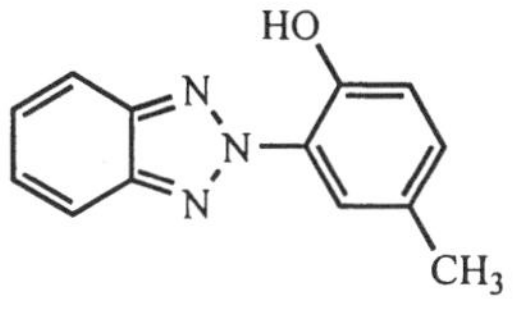

Tinuvin P

A polyethylene resin stabilizer, Cyasorb UV 1084 ([2,2'-thiobis(4-*t*-octylphenol-ato)]-*n*-butylamine Ni(II)), and its degradation products were monitored using a C_{18} column ($\lambda = 285$ nm) and a 94.5/5.0/0.5 IPA/water/acetic acid mobile phase [234]. Ten other polymer additives (e.g., BHT, BHEB, UV 531, Isonox 129, Ethyl 330, Irgafos 168) were monitored as well. Elution times were reported and fell in the range 2–12 min. Concentration ranges of 10–500 mg/L were used.

Isodecyl end-capped propylenediol adipate polyester plasticizers were character-ized using a pentafluorophenyl column ($\lambda = 210$ nm and IR = 1790–1680 cm^{-1}) and a 92/8 methanol/water mobile phase. Sample elution was complete in 25 min [235]. Sample concentrations were 1.2 g/10 mL (10 µL injection).

Novolac resins (phenol/formaldehyde condensates) were characterized on a 50°C C_{18} column ($\lambda = 280$ nm) using a 40-min 40/60 → 100/0 methanol/water gradient [236]. Unreacted monomers and oligomers up to five units long were separated. For the analysis, 10 µL aliquots of samples containing 0.5% resin were injected. Peak shapes were excellent.

Polyethylene glycols, PEG (MW = 200–1000), were analyzed on a C_8 column (RI detector) using a 30/70 methanol/water mobile phase for MW 200 samples and a 54/46 methanol/water mobile phase for MW 1000 samples [237]. Excellent resolution of the oligomer peaks was obtained and the analyses were complete in <30 min. Detection limits of 5 µg/mL (S/N = 3) were reported. Similarly, a 40-min 10/90 → 80/20 methanol/water gradient on a C_{18} column (evaporative light scattering detector) provided excellent separation of polyethylene glycol (MW = 200–1000) oligomers. Samples were 2% w/v polymer in methanol and 10 µL aliquots were injected [238].

Two polyethylene glycol (PEG) polymer distributions were characterized with respect to their molecular weight. A PEG 300 was analyzed in 10 min on a C_{18} column with a 30/70 methanol/water mobile phase [239]. Similarly, a PEG 600 sample was eluted in 25 min. Good oligomer resolution was achieved for both polymers. Sample concentrations ranged from 5 to 15 g/L. An evaporative light scattering detector (ELSD) was used.

DNA, RNA, hemoglobin, and protein complexes of polyethylene glycol were chromatographed on a C_{18} column ($\lambda = 260$ nm). Elution used reagent alcohol as the mobile phase. (Note: Care should be taken to order the correct reagent alcohol.) Excellent resolution resulted for each class of complexes [240]. For example, three DNA/PEG peaks were resolved: one for the number of base pairs <30, one for base pairs between 30 and 70, and one for base pairs between 70 and 130. In this case

Irgafos 168

elution was complete in 30 min. Protein (lysozyme, protease) and hemoglobin complexes were resolved in 45 min.

Polybutylene glycols (MW = 650–3000) were analyzed on a C_4 column (evaporative light scattering detector) using a 1.5-h methanol/water (0.4 mL/L acetic acid) gradient [241]. Peak shapes were somewhat broad and the authors note that potential precipitation of the sample was possible when methanol was used on C_{18} columns. To prevent this, a good choice for an organic solvent is probably IPA or n-propyl alcohol due to the increased solubility of the analyte in these solvents. A 10 µL injection of a 2% solution was made for each sample. A 25/75 → 75/25 acetonitrile/water gradient gave generally better peak shapes and precipitation was not documented in this solvent system.

Polystyrene (MW = 650,000) was characterized on a C_{18} column (photodiode array detector) using a 30-min 80/20 → 10/90 methanol/dichloromethane gradient [242]. Methanol precipitated the sample on the column and dichloromethane eluted the polymer according to its molecular weight. Effects of sample load (12.5–400 µg injected) on the chromatographic elution profile were discussed as well as alternative normal-phase analyses.

Block co-polymers of styrene/t-butyl methacrylate (MW = 240,000) were analyzed on C_{18} and phenyl columns (photodiode array detector) using a 15-min 90/10 → 0/100 methanol/THF gradient [243]. Useful MW distribution characterization resulted when known polystyrene standards were utilized as the molecular weight calibrators. Injections containing 4 µg of polymer were used. Methanol has been used as the polar mobile phase constituent for co-poly(styrene/acrylonitrile) [244] and styrene/ethyl methacrylate co-polymer analyses [245,246]. Ethanol provided increased solubility and was used for polymethyl methacrylate-graft-polydimethyl siloxane co-polymers [247].

The kinetics of an epoxy-amine polymerization reaction were followed via LC on a C_{18} column ($\lambda = 254$ nm) using a 20-minute methanol/water gradient [248]. A 20 µL injection of a 1 mg/mL solution was typical. Peaks were fronted. Peak shapes may be improved through the addition of a silanol-blocking amine to the mobile phase. It should be noted, however, that fronting could also be due to continued chemical reaction occurring during the chromatographic process so that only quenching the reaction will help. Nevertheless, peaks were well enough resolved to give the qualitative results the authors sought.

Chang [249] analyzed (methoxymethyl)melamine resins on a C_{18} column ($\lambda = 235$ nm) using a 50-min 25/75 → 100/0 methanol/water (50 mM ammonium acetate) gradient. A 0.2% solution of each polymer was prepared of which 10 µL injections were made. Peak shapes were excellent. Thirty-three resin fragments were separated (20 monomeric and 13 dimeric) ranging in MW from 200 to 750. These components were positively identified by thermospray MS.

Methoxymelamine resins, namely CYREZ 963 and CYREZ 350 commercial resins, were analyzed using a C_{18} column ($\lambda = 225$ nm) and a 50-min 25/75 → 100/0 (at 35 min hold 15 min) methanol/water gradient. Peak shapes were excellent [250]. Distinct and characteristic profiles for these polymers were obtained. A series of 10 µL injections of 0.05% solutions was made in these analyses.

The C_4, C_6, C_8, C_{11}, and C_{12} alkyl esters of biphenyl-4,4′-dicarboxylic acid were separated in <12 min using a C_{18} column ($\lambda = 254$ nm) and a 92/8 methanol/water (add acetic acid to pH 2.8) mobile phase [251]. With 10 μL injections of samples at concentrations of 40 μg/mL, the C_{12} alkyl ester was noticeably tailed. A shallow gradient and the addition of a stronger surface modifier (e.g., TFA) would help decrease elution times and produce sharper peaks.

The preparation of bisphenol A from the condensation of acetone with phenol often results in the production of many by-products [252]. Fourteen byproducts [e.g., 2,2,4-trimethyl-2*H*-chromen, 4-*t*-butylphenol, 10,10-dimethylxantan, dimethylhydroxybisphenyl; 1,1,3-trimethyl-3-(4-hydroxyphenyl)-5-indanol; 1,1,-spiro-bis(3,3-dimethylindanol-5)], along with phenol and bisphenol A, were separated on a C_{18} column (photodiode array detector, $\lambda = 210$ nm to 600 nm with 254 nm, 280 nm, and 305 nm most commonly used) using a 17-min 60/40 → 100/0 methanol/water gradient. The sample was prepared as 0.5 g/20 mL of ethylene glycol with impurity levels running from 0.1 to 2.5% (w/w) of bisphenol A.

The kinetics of the formation of phenolic resol resin was followed using a 35°C C_{18} column ($\lambda = 280$ nm). An 180-min 80/20 → 20/80 water/methanol gradient separated such reaction components as 2-hydroxymethylphenol, 2,6-dihydroxymethylphenol, 2,6′dihydroxymethyl-*p*, *p*′-dihydroxydiphenylmethane, etc., which were followed over the course of the reaction [253].

4.4.3 Sunscreen Agents

Scalia [254] extracted five sunscreen agents (2-ethylhexyl-*p*-dimethylaminobenzoate, 2-hydroxy-4-methoxybenzophenone, 2-ethylhexyl-*p*-methoxycinnamate [octyl

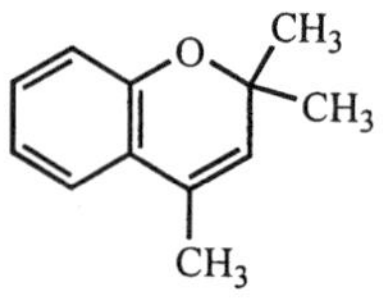

2,2,4-Trimethyl-2*H*-chromen

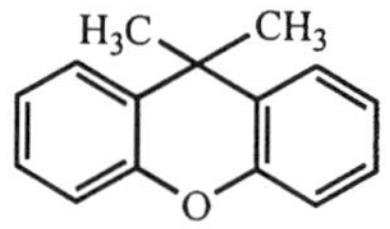

10,10-Dimethylxantan

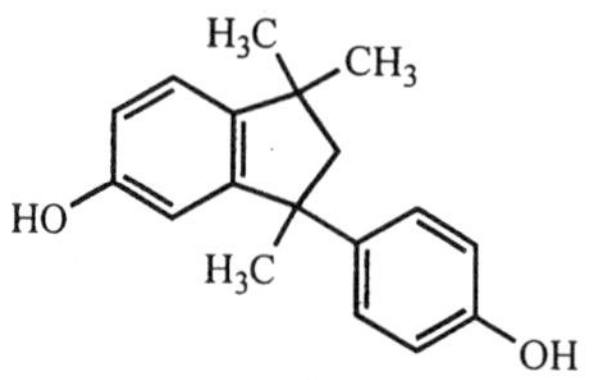

1,1,3-Trimethyl-3-(4-hydroxyphenyl)-5-indanol

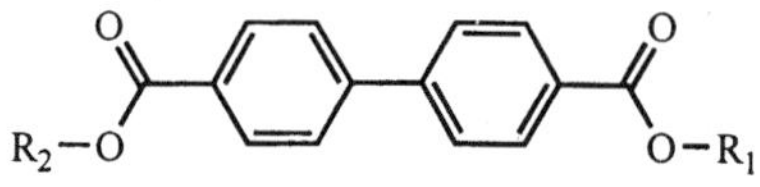

Biphenyl-4,4′-dicarboxylic alkyl ester

methoxycinnamate], 4-methylbenzylidine camphor, 4-*t*-butyl-4′-methoxydibenzoyl-methane) from cosmetic formulations and resolved them on a 30°C C_{18} column ($\lambda = 230$ nm). Baseline resolution and complete elution were achieved in <12 min with a 45/10/10/35/0.5 methanol/acetonitrile/THF/water/acetic acid mobile phase. Linear ranges from 45 to 80 μg/mL or 450 to 800 μg/mL were used.

Seven benzophenone sunscreen agents (benzophenones 1, 2, 3, 4, 5, 8, 10; often hydroxylated and/or methoxylated benzophenones) were extracted from suncreams, lipsticks, and other personal care products and separated on a C_{18} column ($\lambda = 286$ nm). A 30/30/40 methanol/acetonitrile/water mobile phase generated good peak shapes and resolution in 40 min [255]. All detection limits were similar and were reported as about 1 mg/L.

Five sunscreens (benzophenone-3 [2-hydroxy-4-methoxybenzophenone, Uvinul M40], PEG-25 *p*-aminobenzoic acid [Uvinul P25], 3-[4-methylbenzylidine]camphor [Eusolex 6300], 4-*t*-butyl-4′-methoxydibenzoylmethane [Eusolex 9020], octyldi-methyl-*p*-aminobenzoic acid [Eusolex 6007]) stabilities to photodegradation were studied through chromatographic assay [256]. Baseline resolution of the five compounds was achieved when a C_8 column (PDA wavelength range of 200–400 nm) and a 20-min 80/20/1 → 100/0/1 (at 10 min hold 10 min) methanol/water/acetic acid gradient was used. Linear ranges of 0.5–50 μg/mL and detection limits of ~0.01 μg/mL (analyte dependent, S/N = 3) were reported.

A series of nine sunscreen additives (Uvinuls) were separated and quantitated on a C_{18} column ($\lambda = 292$ nm) using a two-ramp 2/3/95 → 35/35/30 (at 50 min) → 100/0/0 (at 60 min) methanol/acetonitrile/water (1% w/w acetic acid) gradient [257]. The compounds resolved included 2-hydroxy-4-methoxy-, 2,2′-dihydroxy-4,4′-dimethoxy-, 2,4-dihydroxy-, and 2,2′,4,4′-tetrahydroxybenzophe-

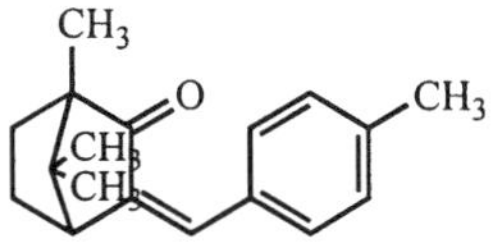

4-Methylbenzylidine camphor

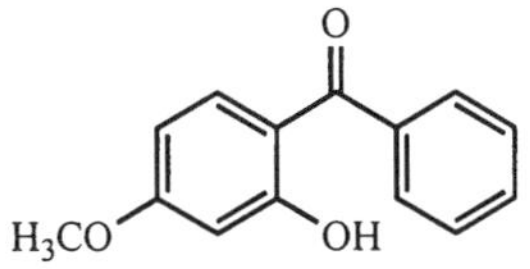

Eusolex 4360, Uvinul M40, benzophenone 3, 2-hydroxy-4-methoxybenzophenone

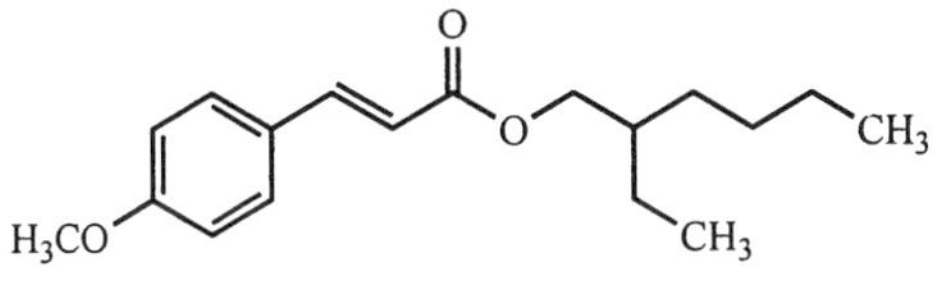

Octyl methoxycinnamate

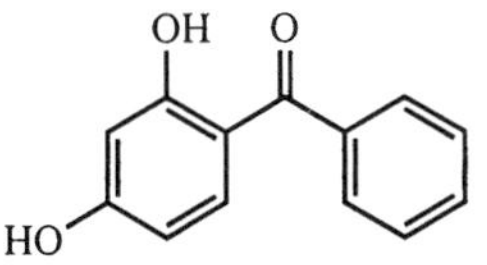

Benzophenone-1; 2,4-Dihydroxybenzophenone

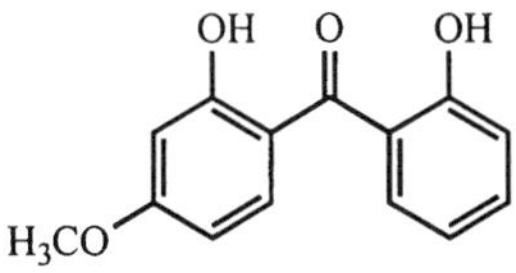

Benzophenone-8; 2,2′-Dihydroxy-4-methoxybenzophenone

none and 2-cyano-3,3-diphenylacrylic acid ethyl ester. Detection limits of 10 ng injected and linear concentration ranges up to 5 µg injected were reported.

4.4.4 Dyes

In an excellent and study by Pel et al. [258], a total of 15 direct dyes (2-nitro-1,4-phenylenediamine and 4-nitro-1,2-phenylenediamine), primary intermediates (3-aminophenol, resorcinol), primary intermediates (1,4-phenylenediamine, 2-amino-phenol, 4-aminophenol), and other intermediates (hydroquinone, 2,6-diaminopyr-idine, metol, 3,4-diaminotoluene, 2-amino-5-nitrophenol, 4-chlororesorcinol, 4-chloroaniline, hexylresorcinol) used in hair dye formulations were isolated and separated on a 48°C base deactivated C_8 column ($\lambda = 235$ nm). A multiramp 45-min $100/0 \rightarrow 20/80$ water (50 mM acetic acid to pH 5.9 with ammonia)/methanol gradient generated good resolution between all peaks. A table of an additional 35 components was listed with their retention times for the above gradient profile. Since many of these compounds are light- and heat-sensitive, plots of stability over time of irradiation and heat with and without the antioxidant ascorbic acid were presented. Working concentration ranges of 5–200 µg/mL and detection limits in the neighbor-hood of 1 µg/mL (S/N = 3) were reported.

Four dimers of catechin (8-8, 6-6, 8-6a, and 8-6b) were separated on a C_{18} column ($\lambda = 280$ nm) using a 30-min $90/10 \rightarrow 0/100$ water (1% formic acid)/ (80/20/0.2 methanol/water/formic acid) gradient. These compounds were fraction-ated and collected and then dehydrated and oxidized to yield yellow pigments. These pigments were separated in the same manner as above and monitored at $\lambda = 440$ nm. A number of by-product reaction pathways are show in the article [259].

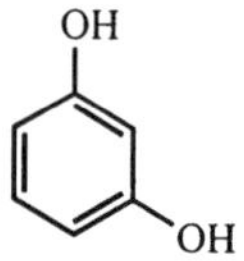

Resorcinol

p-Phenylenediamine

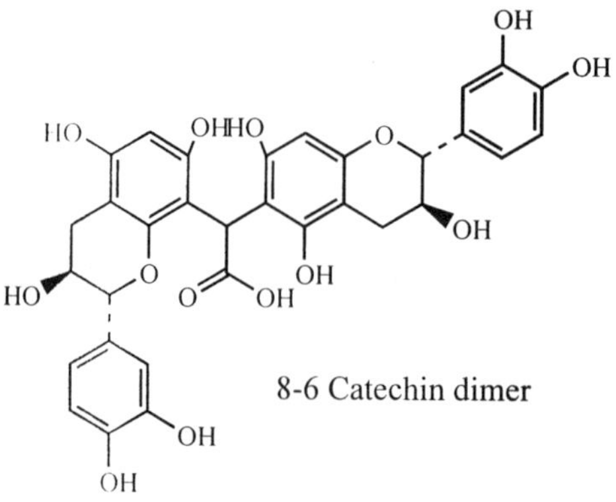

8-6 Catechin dimer

Four sandocryl textile dyes (Yellow [natural yellow 28], Blue [vat blue 41], Green [basic green 4], Red [basic violet 16]) were separated on 29°C C_8 column ($\lambda = 205$ nm). A 20-min $40/60 \rightarrow 0/100$ water/methanol gradient generated baseline res-olution [260]. The authors noted that an analogous acetonitrile gradient was not effective.

Novotná et al. [261] studied a set of eight historical anthroquinone and naphthoquinone dyes (carminic acid, laccaic acid A and B, lawsone, juglone, alizarin, lapachol, emodin). They were well resolved in 12 min on a C_{18} column ($\lambda = 270$ nm) using a 15-min exponential gradient from $40/60 \rightarrow 95/5$ methanol/ water (0.1 M citrate pH 2.5). Separately, alizarin and purpurin were resolved on a C_{18} column ($\lambda = 245$ nm) using a 72/25 methanol water (0.2 M acetate pH 4.3) mobile phase. Detection limits of 0.6–12 ng injected (analyte dependent) were reported. Linear curves up to the solubility limit of ~1 mM were also used.

Six food dyes (E-110 [Sunset Yellow FCF, FD & C Yellow 6]; E-123 [Amaranth, FD & C Red 2]; E-124 [Ponceau 42]; E-122 [Carmosine]; E-127 [Erythrosine, FD & C Red 3]; E-120 [Carminic acid]) were fully resolved on a C_{18} column ($\lambda = 520$ nm) using an 8-min $60/40 \rightarrow 50/50$ (at 1 min) $\rightarrow 0/100$ (at 3 min hold 5 min) methanol/

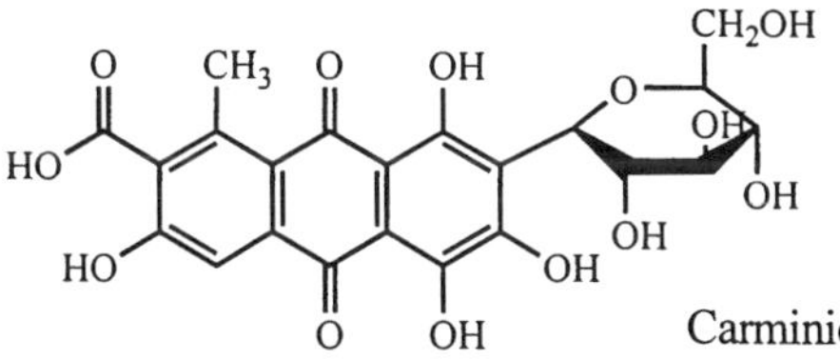

Carminic acid

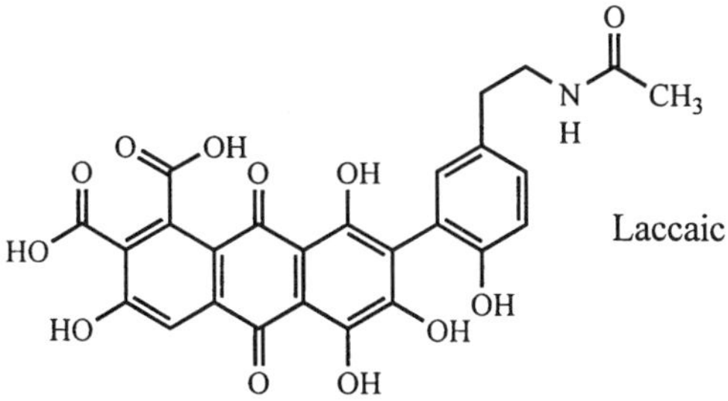

Laccaic acid A

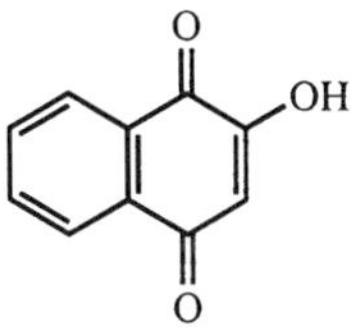

Lawsone 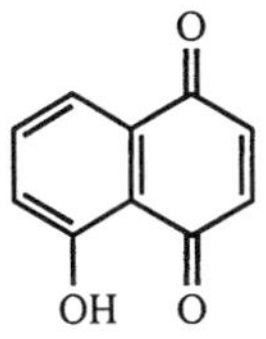 Juglone 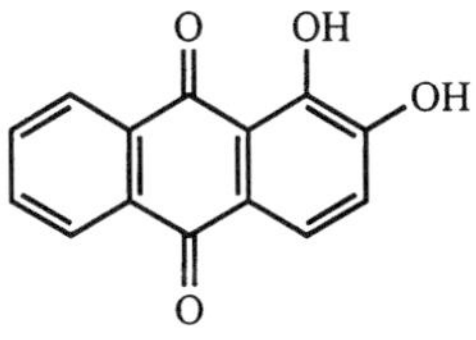 Alizarin

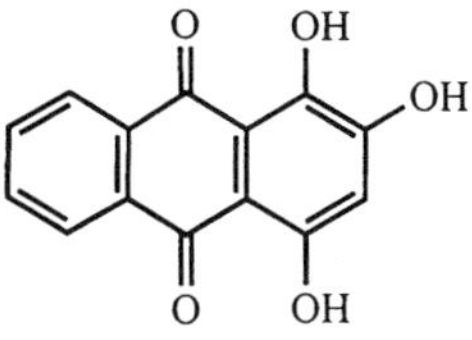

Purpurin

water with 5 mM tetrabutylammonium hydroxide (TBAH) in 0.1 M NaH$_2$PO$_4$ pH 7 mobile phase [262]. Plots detailing the effects of changing percent methanol, pH, and TBAH concentration were presented. Linear ranges of 2–40 mg/mL and detection limits of 50–400 ng/mL (analyte dependent) were reported.

Eight sulfonated azo dyes (Acid Red 1, 8, 29, and 106, Acid Violet 5, Reactive Red 2 and 4, Reactive Orange 16) were individually characterized with respect to major components using a base deactivated C$_{18}$ column ($\lambda = 254$ nm). Isocratic mobile phase compositions of 35/65 to 50/50 methanol/water (50 mM ammonium acetate) were used (dye specific). Typical run times were 40–60 min

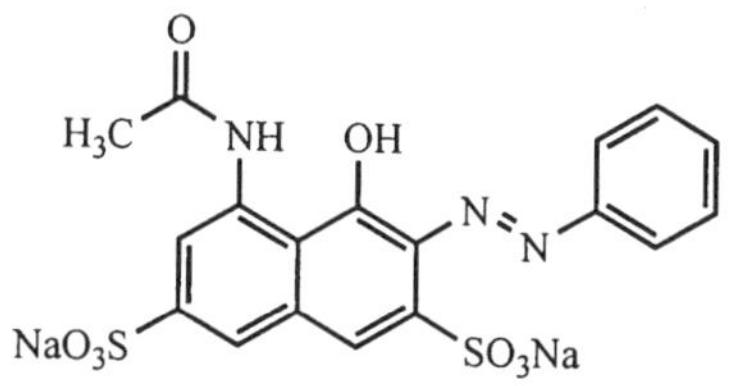

Acid Red 1

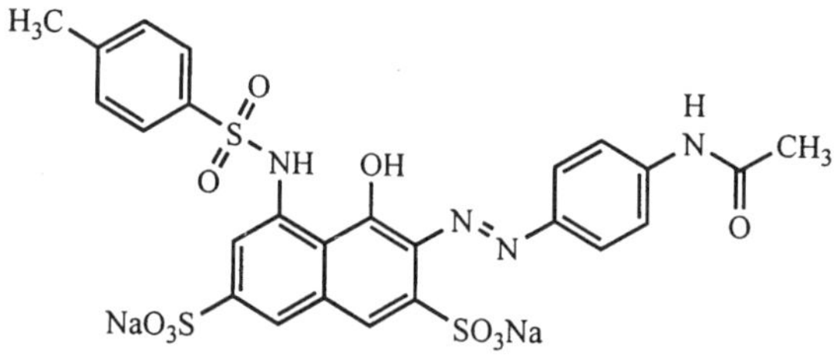

Acid Violet 5

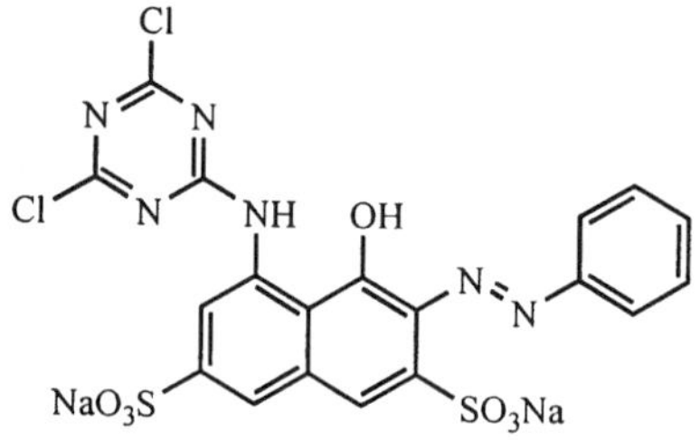

Reactive Red 2

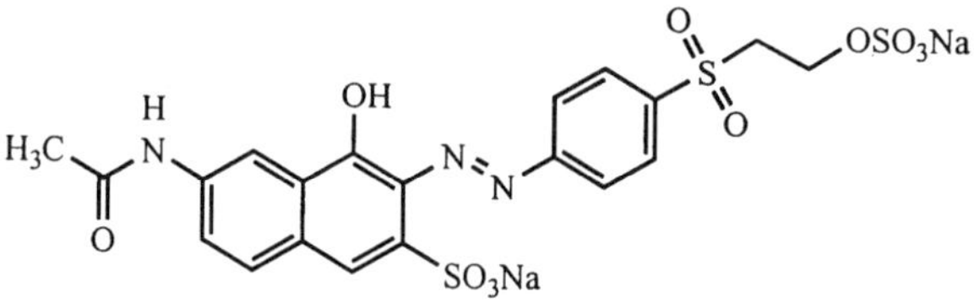

Reactive Orange 16

[263]. Acetonitrile did not lead to improved chromatography. A plot of the affect of ammonium acetate concentration on retention, asymmetry, and theoretical plates for Acid Red 1 was presented. Identification of eluting peaks was done through preparative LC fractionation, collection, and subsequent analysis by LC/MS.

Dyes were extracted from polyester and diacetate textile fibers and analyzed [264]. Fifteen commercial dyes (including three Disperse Yellows, five Disperse Oranges, three Disperse Reds, two Disperse Blues and one each Disperse Brown, Basic Yellow and Vat Blue) were separated on a C_{18} column (thermospray MS) using a 25-min 50/50 → 100/0 (at 5 min hold 20 min) methanol/water (0.1 M ammonium acetate) gradient. These results were developed for use in forensic analyses. Structures for all the analytes in this study are presented in this work.

The oxidative degradation products of Uniblue A were separated from the dye on a C_{18} column ($\lambda = 254$ nm) using a 70/30 methanol/water (10 mM phosphate buffer at pH 6.7) mobile phase [265]. Elution was complete in <10 min and overall resolution was good but not baseline. Dye standard solution concentration was 4 mM.

4.4.5 Other Industrial Analytes

Alcohols have also been used as modifiers in the study of fullerenes in at least four recent and separate studies [266–269]. The reader is referred to Chapter 5 (Section 5.4) for more details.

A protocol for the identification of polybrominated flame retardant compounds was developed using a 23°C C_{18} column. A 97/3 methanol/water (0.15 g KH_2PO_4 and 0.25 g Na_2PO_4/100 mL) mobile phase was used. Elution was complete in 18 min [270]. The retention times for compounds such as tetrabromobisphenol A, 2,4,6-tribromophenol, tetrabromophthalic anhydride, and decabromodiphenyloxide as well as their UV maxima (ranging from 203–280 nm) were tabulated. For raw or technical grade materials and their brominated mixtures, such as for octabomodiphenyloxide and hexabromobiphenyl, a table of the elution times for all constituents

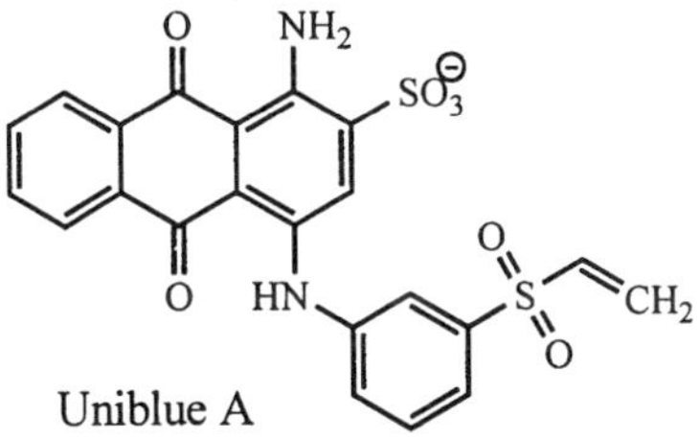

Uniblue A

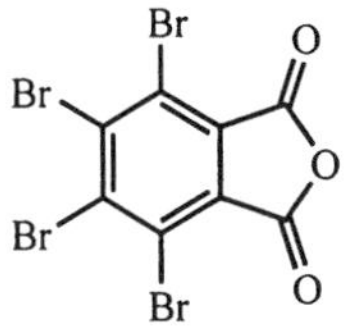

Tetrabromophthalic anhydride

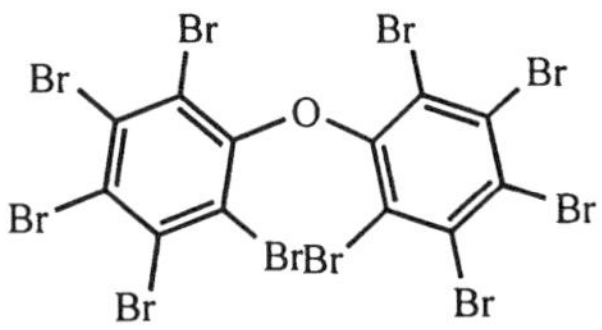

Decabromodiphenyl oxide

is given along with the corresponding individual λ_{max} values (210–225 nm). The authors noted that cyanopropyl and phenyl columns were ineffective for all materials. The buffer system used here is probably not the most effective since the pH is well above the pK_a for some compounds (e.g., 2,4,6-tribromophenol). This is evidenced by the chromatography, where these acidic compounds elute at or near (or are even excluded from the pore volume and elute *prior* to) the system void volume (e.g., a retention time of 1.55 min on a 250 × 4.6 mm column at a 1 mL/min flow rate). A lower-pH buffer would offer retention since the acids would then be protonated (i.e., uncharged).

Sulfur (S_8), tetramethylthiuram mono- and disulfide, 2-mercaptobenzothiazole, tetramethylthiuram polysulfides, and 2-benzothiazole-2,2'-disulfide were resolved on a C_{18} column ($\lambda = 280$ nm) using an 85/15 methanol/water mobile phase [271]. Sulfur was last eluted at 25 min.

Löfgren et al. [272] studied a variety of printing inks using a C_{18} column ($\lambda = 510$ nm brown and red inks, 570 nm blue inks or fluorescence $\lambda = 350$ nm, ex; 550 nm, em) and a 30-min 50/50 → 100/0 methanol/water (10 mM phosphate buffer at pH 3) gradient. Various document-extracted ink chromatograms are shown along with those of standard pigments and dyes.

4.4.6 Summary

The use of alcohols to aid in the solubilization of lower molecular weight polymers is very effective when the polymer is very polar (polyethylene glycols, methoxy-melamines, etc.). In the case of higher molecular weight polymers or less polar polymers, such as polystyrene, a stronger solvent such as dichloromethane or THF is needed for solubilization and chromatogram generation. Methanol was particularly effective in commercial dye separations and was again found to be compatible with thermospray MS detectors.

4.5 BIOLOGICAL ANALYTES

4.5.1 Carboxylic Acid Analytes

One major class of analytes that has been studied repeatedly over the years is that of the fatty acids. They are conveniently separated into three classes: short-chain ($<C_{12}$ chain lengths), long-chain (C_{12}–C_{24}), and very long-chain ($>C_{24}$).

Mono-, di-, and trichloroacetic acid and mono- and dibromoacetic acid levels were monitored in water [273]. Separation was generated on a C_{18} column

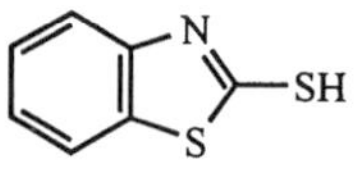

2-Mercaptobenzothiazole

($\lambda = 210$ nm) using a 50/50 methanol/water (50 mM tetrabutylammonium chloride pH 5.0) mobile phase. Monobromoacetic acid eluted on the shoulder of the nitrate peak under these conditions. Elution was complete in 15 min. Plots and tables of the effect of percent methanol, pH, and ion pair reagent were presented. Detection limits ranged from 0.15 to 3 mg/L.

Because of the low molar absorptivities for fatty acids above 205 nm and, conversely, the limited number of solvents with low enough absorbtivities below 205 nm, short-chain carboxylic acids and their long-chain counterparts, the fatty acids, and other saturated compounds are commonly derivatized either prior to or during the chromatographic process. Alternatively, a "universal" detector (e.g., refractive index [RI] or evaporative light scattering detector [ELSD]) is used. The advantage of derivatization lies in the fact that chromophoric derivatives can give orders of magnitude lower detection limits than RI or ELSD methods.

Carboxylic acids in wine (e.g., lactic, acetic, tartaric, malic, succinic, citric) were analyzed as their phenacyl bromide derivatives on a 30°C C_{18} column ($\lambda = 254$ nm). These compounds were baseline resolved in <18 min using a 30/70 $\rightarrow$ 90/10 methanol/water gradient [274]. Excellent peak shapes were obtained. The study went on to further optimize the separation using the triangulation method detailed in Chapter 2. Here mixtures of methanol/water, acetonitrile/water and THF/water were used as the apices. A quaternary solvent generated optimum results. The same authors [275] optimized the derivatization and separation of the above acids plus acetone and methylmalonic acid. A C_{18} column ($\lambda = 254$ nm) and a 20-min 40/60 $\rightarrow$ 75/25 methanol/water gradient gave the best results. Other derivatization compounds studied were 4'-phenylphenacyl bromide and 4'-bromophenacyl bromide.

Twenty-five naphthacyl derivatized of fatty acids ($C_{7:0}$ to $C_{22:6}$) were separated on a 30°C C_{18} column ($\lambda = 246$ nm) using a 45-min 80/10/10 $\rightarrow$ 86/10/4 (at 30 min) $\rightarrow$ 90/10/0 (at 40 min hold 5 min) methanol/acetonitrile/water gradient [276]. Peak shapes were excellent and the resolution was exceptional throughout the analysis. For example, $C_{14:0}/C_{16:1}$, $C_{16:0}/C_{18:1}$, and $C_{18:3}/C_{14:0}/C_{22:6}/C_{16:1}$ groups were well resolved. Linear ranges from 2 pmol to 10 nmol injected and detection limits of 0.1 ng injected were claimed.

Tartaric acid

Malic acid   Succinic acid

Four annatto compounds (natural color compounds), α- and β-bixin and α- and β-norbixin, were extracted from high-fat dairy products and resolved on a C_{18} column ($\lambda = 500$ nm). Peaks were unresolved and extremely fronted when a 100% methanol mobile phase was used [277]. However, the use of a 90/10 methanol/water (2% acetic acid) mobile phase produced good peak shape and excellent resolution. The addition of acetic acid obviously kept the analytes fully protonated, thereby generating symmetric peak shape. The analysis was complete in <10 min. Detection limits of 5 ng/g and calibration ranges up to 450 μg/g were reported for the all analytes.

Mycolic acids, clinical diagnostic markers for the mycobacterium tuberculosis group, were studied as the p-bromophenacyl bromide (BPB) and 4-methyl-6,7-dimethoxycoumarin (MDC) derivatives. Two fluorescent tags were also studied. A 20-min 60/40 → 6/94 methanol/IPA gradient and a C_{18} column ($\lambda = 254$ nm) produced elution profiles that were reproducible and indicative of individual bacteria [278]. A fine review details the use of a wide range of other potential derivatizing agents for carboxylic acid residues, namely the fatty acids [279].

Eleven chlorogenic acids (e.g., p-coumaroylquinic acids, caffeoylquinic acids, feruloylquinic acids) were extracted from coffee and analyzed on a C_{18} column

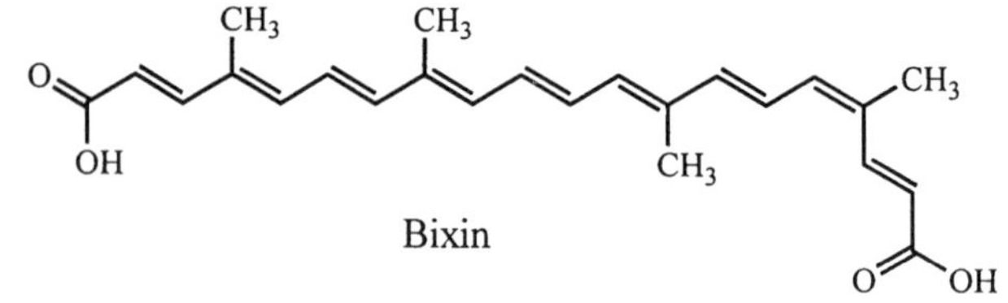

Bixin

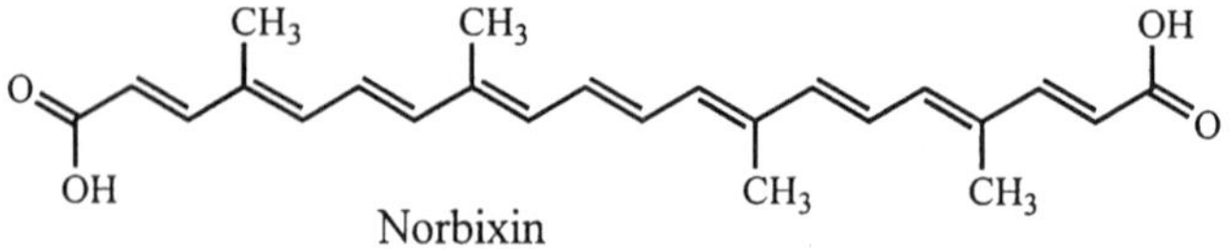

Norbixin

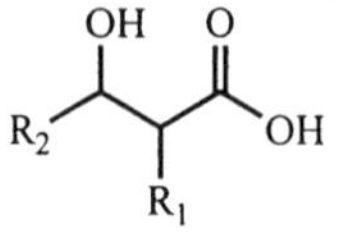

Mycolic acids
($R_1 = C_{20}$–C_{24} linear alkane; $R_2 = C_{30}$–C_{60} complex structure)

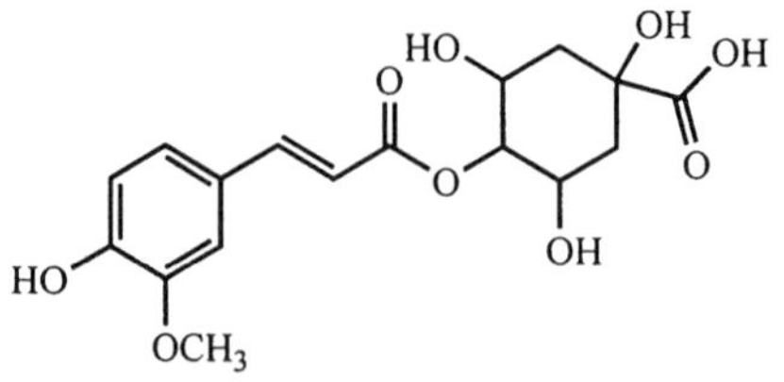

Feruloylquinic acid

($\lambda = 325$ nm) using a 15/85 methanol/water (8.47 g citric acid with 4.5 g KOH and 2.2 g HCl, all per liter at pH 3) mobile phase [280]. Excellent peak shapes resulted and at least 20 acid compounds, some unidentified, were separated under these conditions. Elution was complete in <30 min.

Ribotta et al. [281] studied a series of six catecholamine metabolites (vanilmandelic acid, homovanillic acid, 3,4-dihydroxymandelic acid, 3,4-dihydroxy- and 4-hydroxy-3-methoxyphenylacetic acid, vanillic acid). They were well resolved on a C_{18} column (electrochemical detector, 0.75 V). A 15/85 methanol/water (2 mM tetrabutylammonium bromide at pH 4.5) mobile phase eluted all compounds in <10 min. A linear range of 3×10^{-7} to 5×10^{-5} M and detection limits of 2×10^{-8} (S/N = 3) were reported. A considerable amount of supportive data such as retention time vs. pH, % methanol, and tetrabutylammonium bromide concentration is also given. Also, detector response versus potential from 0.3 to 0.8 V is shown.

Six 1,3-thiazolidinecarboxylic acids were baseline resolved as their dansyl derivatives on a C_{18} column ($\lambda = 265$ nm or 365 nm, ex; 550 nm, em) using a 30-min 43/57 $\rightarrow$ 75/25 methanol/water (9 mM NaH_2PO_4 with 30 mM Na_2HPO_4 phosphate at pH 7.4) gradient [282]. Excellent peak shapes were obtained. Linear concentration curves from 200 ng to 20 µg injected were generated.

Eleven phenolic acids present in olive oils (e.g., tyrosol, hydroxytyrosol, vanillic acid, syringic acid, caffeic acid, o- and p-coumaric acid, p-hydroxybenzoic acid) were studied using a C_{18} column and a 70-min 6/94 $\rightarrow$ 100/0 methanol/water (3% acetic acid) gradient [283]. Acetonitrile adversely affected chromatographic performance and required lengthy re-equilibration times between injections. Molar absorptivities are tabulated for all compounds and detection limits ranged from 1 to 6 ng injected (analyte dependent).

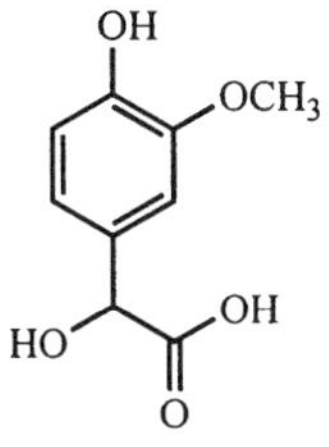

Vanilmandelic acid 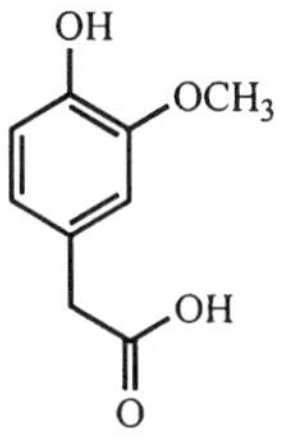   Homovanillic acid 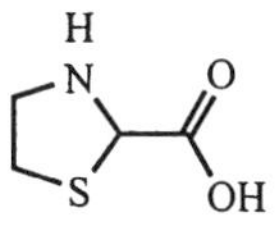   1,3-Thiazolidine-2-carboxylic acid

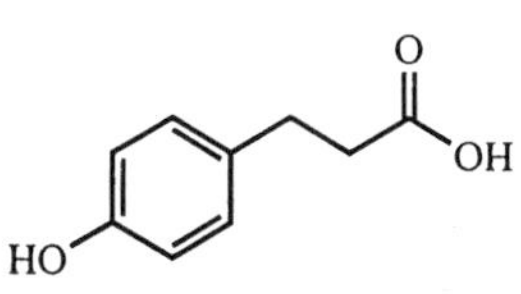

Tyrosol 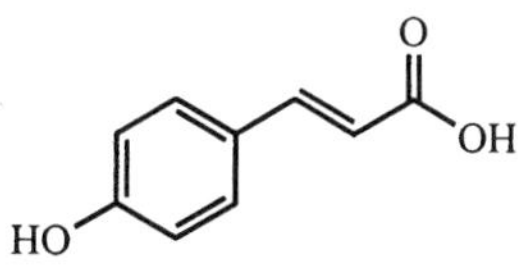   p-Coumaric acid

Six bile acids (cholic, chenodeoxycholic, deoxycholic, ursodeoxycholic, urso-cholic, lithocholic) levels were determined in ursodeoxycholic acid pharmaceutical preparations (for treatment of cholesterol gallstones). Excellent resolution and a 40-min elution time were achieved on a C_{18} column (ELSD, drift tube 70°C, N_2 nebulizing gas 2.1 bar) using a 66/22/18 methanol/acetonitrile/water (acetic acid to pH 4) mobile phase [284]. Linear ranges from 0.1 to 5.6 µg injected and detection limits of 30–60 ng injected (analyte dependent) were reported. Similarly, a set of seven bile acids were separated on a C_{18} column (RI detector) using a methanol/water (0.1 M acetic acid) mobile phase [285]. Taurocholic, glycocholic, chenodeoxy-cholic, deoxycholic, and lithocholic acids were well resolved. Cholic and glyco-chenodeoxycholic acids were poorly resolved. Elution was complete in 45 min. Detection limits of 0.5 nmol and a working range of 0.5–75 nmol were reported. The authors also noted that storage of samples in glass containers could lead to as much as a 25% loss of analyte per hour. They followed chromatographically the complete conversion of glycocholic acid to cholic acid over 59 h. Capacity factors were also tabulated for the same mobile phase conditions but using a C_6 column. No additional benefits as to separation were evident.

Various combinations of lipoxins were separated on a C_{18} column. LXB_4, LXA_4, 14S-LXB_4, and 6S-LXA_4 (electrochemical detector, +1.2 V vs. Ag/AgCl) were resolved in 10 min using 65/35 methanol/water (1 mM TFA) mobile phase [286]. Detection limits of 15–30 fmol (S/N = 3) were reported (analyte dependent). Similarly, a 10-min analysis of 10 ng injected LTC_4, PGB_2, LTD_4, LTE_4, and LTB_4 ($\lambda = 278$ nm) using a 45/30/25 water (0.1% EDTA with 10 mM sodium acetate at pH 5.5)/methanol/THF mobile phase was reported [287]. Detection limits of 500 pg injected were reported.

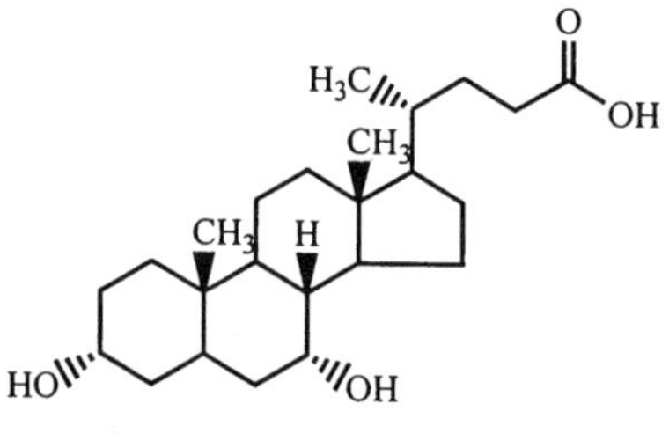

Chenodeoxycholic acid

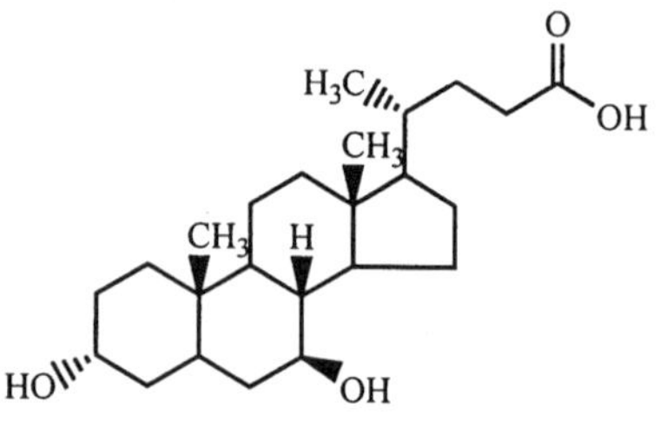

Ursodeoxycholic acid

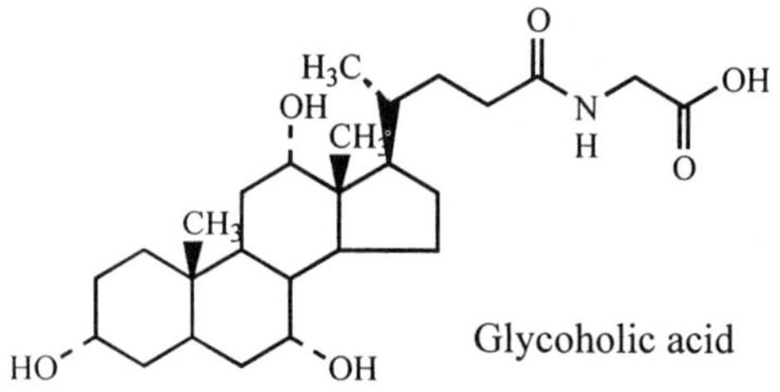

Glycocholic acid

The urinary and hepatocyte metabolites of tobacco-specific nitrosamines (e.g., 4-[methylnitrosamino]-1-[3-pyridyl]-1-butanone and butanol, 4-hydroxy- and 4-oxo-4-[3-pyridyl]butyric acid, 4-[methylnitrosamino]-1-(N-oxy-3-pyridyl)-1-butanone and butanol) were well resolved on a C_{18} column ($\lambda = 254$ nm) using a complex 75-min $0/100 \rightarrow 30/70$ methanol/water (20 mM sodium phosphate buffer at pH 7) gradient [296]. Excellent peak shapes were obtained and elution was complete in 70 min. Electrospray MS detection was also used with a nearly identical setup: the sodium phosphate buffer was replaced with a 10 mM ammonium acetate buffer at pH 6.7.

Norharman (β-carboline), harman (1-methyl-β-carboline), and 1-ethyl-β-carboline (as internal standard) were separated in 10 min on a C_{18} column (thermospray MS) using a 23/77 methanol/water (0.1 M ammonium formate with 0.1 M formic acid buffer at pH 3.4) mobile phase [297]. Alternatively, fluorescence detection ($\lambda = 300$ nm, ex; 433 nm, em) and 32/68 methanol/water (0.1 M potassium phosphate buffer at pH 3.0) were used. Detection limits of 5–10 ng/L were reported (analyte dependent).

Kotzabasis et al. [298] used a C_{18} column ($\lambda = 254$ nm) and a 23-min $55/45 \rightarrow 84/16$ methanol/water gradient to separate the benzoyl derivatives of putrescine, cadaverine, spermidine, spermine, and agmatine extracted from leaves. Levels as low as 0.2 nmol were easily detected. Peak shapes were dramatically dependent upon the sample solvent chosen. Samples dissolved in 100% methanol generated extremely tailed peaks, whereas samples dissolved in 80/20 methanol/water generated peak doublets. Odd peak shapes are not uncommon when a sample solvent–mobile phase mismatch occurs. A sample solvent similar to the mobile phase composition almost invariably gives the best peak shape and the most reproducible results. Whenever possible, a sample solvent identical in composition to the mobile phase is generally recommended.

Seventeen biogenic amines were separated as their o-phthalaldehyde derivatives in <25 min [299]. Methylamine to hexylamine, along with their isomers, putrescine, phenethylamine, and cadaverine were baseline resolved on a 60°C C_{18} column ($\lambda = 330$ nm, ex; 445 nm, em) using a $60/0.4/39.6 \rightarrow 80/0.2/19.8$ methanol/THF/water (0.03% triethanolamine) gradient. Detection limits of 25 µg/L and linear ranges of 100 µg/L to 10 mg/L were reported. Excellent peak shapes were generated.

Svendsen [300] used a C_{18} column (electrochemical detector with a 14-electrode array, one electrode poised at +60 mV then each successive electrode poised sequentially +40 mV higher to the highest potential of +840 mV) and a 20-min $1/99 \rightarrow 40/60$ methanol/water (0.1 M phosphate buffer at pH 3.4) gradient to

Norharman   Harman   Agmatine

separate and identify 30 electroactive compounds. These included L-dopa, epineph-rine, guanine, hydroxyphenols and hydroxybenzoic acids, melatonin, octopamine, salsolinol, xanthine, uric acid, and similar compounds. Approximate working electrode oxidation potentials are tabulated for each compound along with its retention time. Excellent selectivities and positive sample identification were achieved using this technique.

Argpyrimidine (AP) is a Maillard reaction product formed from maltose in beer. It was isolated on a 25°C wide-pore C_{18} column ($\lambda = 320$ nm, ex; 398 nm, em) using a complex 75-min 90/10 → 0/100 water/(70/30 methanol) with 1.2 mL/L hepta-fluorobutyric acid gradient. The fraction containing AP was analyzed on a C_{18} column using a 30-min 98/2 (hold 15 min) → 0/100 (at 20 min hold 10 min) water (1 g NaH_2PO_4 to pH 6.5)/(6/4 propanol/water [1 g NaH_2PO_4 to pH 7.0]) gradient. AP eluted in 14 min but co-extracted compounds eluted over the entire time the strong solvent was pumped through the column. At 30 min, the baseline had not yet returned to zero. Samples spiked with 450 fmol of AP generated a readily detectable peak [301].

Benzidine and six metabolites (e.g., N-acetylbenzidine, N-[deoxyguanosin-8-yl]-N-acetylbenzidine) were extracted from liver tissue and resolved in <20 min using a C_{18} column ($\lambda = 280$ nm) and a 23-min 35/65 → 80/20 methanol/water (20 mM ammonium acetate buffer at pH 6.7) gradient [302]. Peaks were somewhat broad, but resolution was good. Increased buffer concentration and an increase in temperature would help peak shape considerably.

3,3'-Dichlorobenzidine and monoacetyl-3,3'-dichlorobenzidine were extracted from hemoglobin and analyzed using a C_{18} column (electrochemical detector at 0.8 V) and a 50/50 methanol/water (20 mM sodium phosphate buffer at pH 5.0) mobile phase [303]. Monoacetyl-3,3'-5,5'-tetramethylbenzidine was used as a

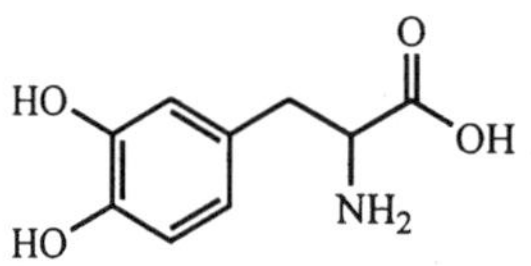

L-Dopa; levodopa

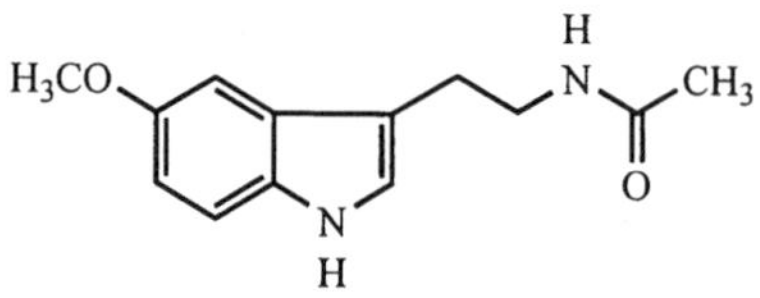

Melatonin

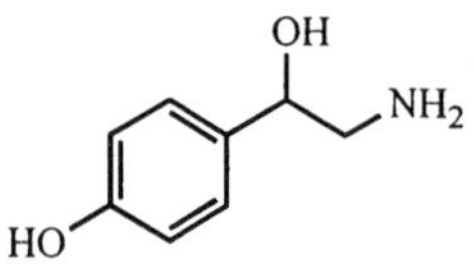

Octopamine

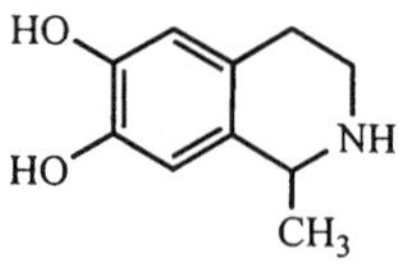

Salsolinol

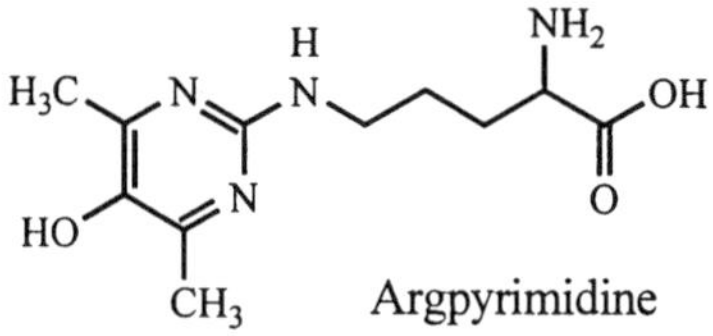

Argpyrimidine

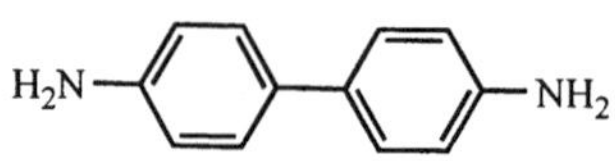

Benzidine

surrogate to determine recoveries, which ranged from 92% to 98%. Elution was complete in ~32 min and detection limits of 6 ng/g hemoglobin were reported.

Indole-3-acetic acid and indole-3-acetaldehyde are the products of enzymatic action on tryptophan by some rhizobia nodules. These three analytes were separated on a 40°C C_{18} column ($\lambda = 280$ nm) using a 67/33 methanol/water (5% acetic acid) mobile phase [304]. Elution was complete in 10 min and good peak shapes and resolution were achieved.

Six indole alkaloids (tryptamine, bufotenin, serotonin, tryptophan, 5-hydroxy- and 5-hydroxy-N-methyltryptophan) were separated in 15 min using a 10/90 ethanol/ water (0.1 M citrate/phosphate buffer at pH 2.8) mobile phase on a C_{18} column ($\lambda = 267$ nm). Excellent chromatographic results were obtained and detection limits down to "tens of nanograms" were obtained [305].

Studies by Ikarashi [306] and Leung [307] have documented the separation of large sets of neurochemicals (29 and 26, respectively). The first study used a C_{18} column (fast atom bombardment MS) and a 15-min 0/100 → 40/60 methanol/water (1% glycerol and 0.2% TFA) gradient. (Note that glycerol was described as a necessary component for creating a successful LC interface with the FAB MS detector.) Although baseline resolution was not obtained for every pair, all pairs were at least substantially resolved. The second study utilized electrochemical detection at a carbon paste electrode (+0.75 V vs. Ag/AgCl) and separation on a C_{18} column using a 10/90 methanol/water (40 mM sodium acetate with 10 mM citric acid, 91 mg/L sodium octyl sulfate, 50 mg/L EDTA, and 13 mM NaCl) mobile phase. Detection limits of 5 pg were reported. Separation was complete in <50 min. With this high level of salt in the mobile phase it is critical that the mobile phase never be allowed to remain in the system for extended periods of time, especially with no flow.

4.5.3 Aflatoxins, Mycotoxins, and Other Toxic Analytes

Aflatoxin B_1 and two hydroxylated derivatives (aflatoxin P_1 and Q_1) were extracted from urine and analyzed on a C_{18} column ($\lambda = 365$ nm, ex; 466 nm, em for Q_1 and

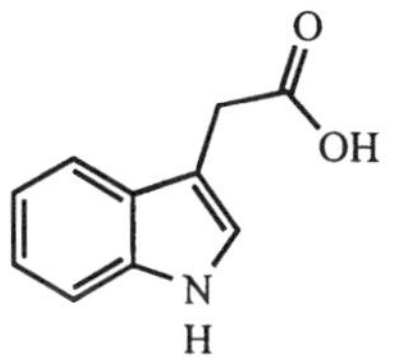

Indole-3-acetic acid

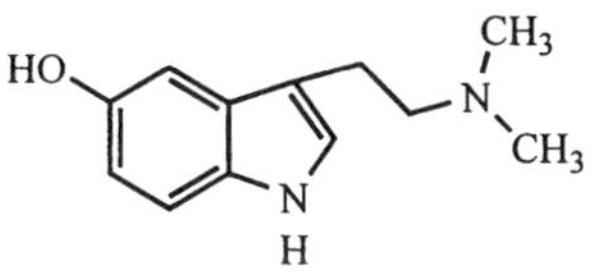

Bufotenin

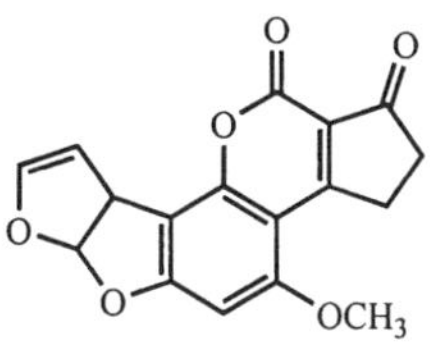

Aflatoxin B_1

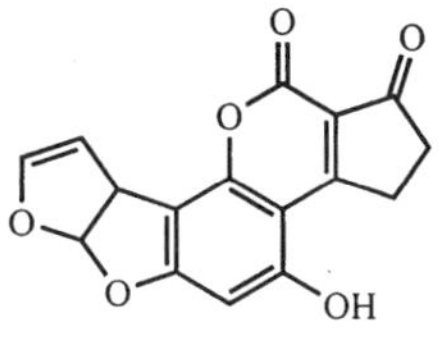

Aflatoxin P_1

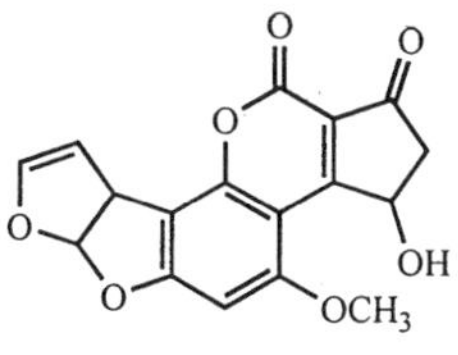

Aflatoxin Q_1

504 nm, em for P_1; 360 nm, ex; 435 nm, em for B_1) using a 60/40 water/acetonitrile mobile phase [308]. Postcolumn addition of heptakis-di-O-methyl-β-cyclodextrin (10 mM aqueous solution) generated detection limits of 0.5 ng/mL (S/N = 3) and linear ranges of 2–60 ng/mL. Elution was complete in 30 min and peak shapes were acceptable for a postcolumn analysis.

Aflatoxins B_1, G_1, M_1, Q_1, and aflatoxin B_1-8,9-epoxide glutathione conjugate were extracted from liver tissue and baseline resolved on a C_{18} column ($\lambda = 362$ nm) using a complex 20-min 0/100 → 90/10 (at 20 min hold additional 3 min) (95/5 methanol/THF)/water (0.1% ammonium phosphate at pH 3.5) gradient [309]. Elution was complete in 23 min and peak shapes were excellent.

In some studies, methanol and methanol-containing solvents have been found to be incompatible with aflatoxins [310] and aflatoxin hemiacetals [311]. Methanol-generated aflatoxin decomposition products, seen chromatographically as either severely shouldered peaks and/or distinct peak doublets, form very gradually as the methanol ages. Interestingly, replacement of the aged methanol with "fresh methanol" (presumably meaning an unopened methanol solvent bottle) rectifies the problem [312]. In fact, the small amount of methanol (0.5–2%) used in chloroform as a stabilizer introduced enough reactive material into chloroform-based NP solvents that aflatoxin degradation was seen in these cases as well. Any solvents preserved with added methanol should therefore be noted when used for aflatoxin separations.

Ginkgotoxin (4'-O-methylpyridoxine) and its 5'-glucoside were extracted from *Ginkgo biloba* and separated on a C_{18} column ($\lambda = 280$ nm, ex; 370 nm, em) using a 10/90 methanol/water (50 mM KH_2PO_4 to pH 3 with H_3PO_4) mobile phase [313]. Good peak shapes, near baseline resolution, and a <8 min elution time was obtained. A linear range of 10–100 ng injected was reported.

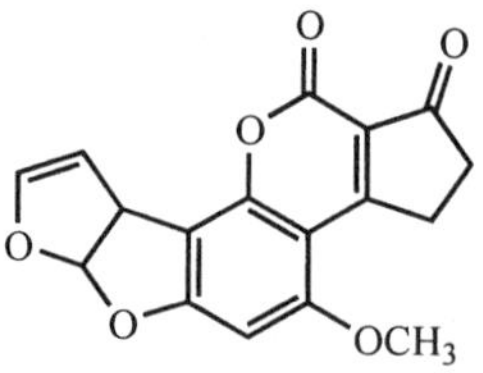

Aflatoxin M_1 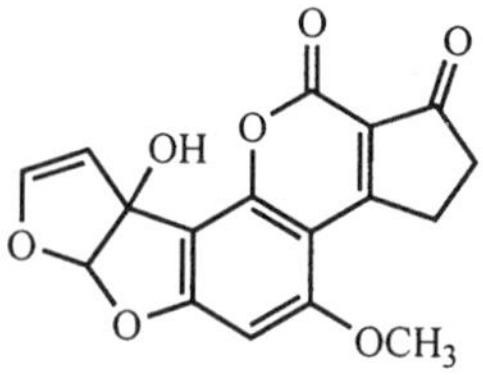   Aflatoxin M_1

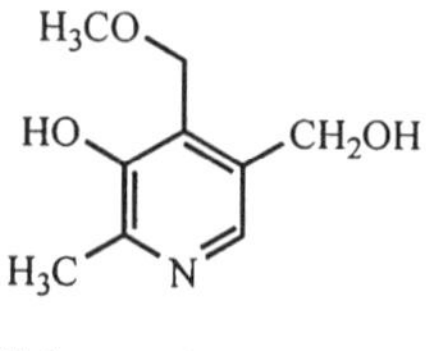

Ginkgotoxin

   Five shellfish toxins (okadaic acid, 7-*O*-acetylokadaic acid, and dinophysistoxin-1, -2 and -3) were baseline resolved and eluted with excellent peak shapes using either a C_4 or a C_{18} column (APCI-MS) with a 70/30 methanol/water (0.1% TFA) mobile phase [314]. The use of methanol as organic modifier lowered detection limits 10-fold (0.4 µg/mL) as compared with acetonitrile. The authors cite increased solubility of the toxins as the explanation for the decrease in detection limits.

   Zearalenone, α- and β-zearalenol (zeralanone as internal standard [IS]) were isolated from beer and quantitated by LC/MS/MS [315]. A C_8 column was used with a 65/35 methanol/water with 15 mM ammonium acetate mobile phase. Elution was complete in 14 min. The authors note that the negative ion mode was more sensitive that the positive ion mode. MS parameters were given as $T = 400°C$; N_2 collision gas; needle current 4 µA; voltage 30 eV. A linear range of 0.15–500 µg/L, a 0.07–0.15 µg/L detection limit, and a 0.03–0.06 limit of quantitation (analyte dependent) were reported.

   Stander et al. [316] studied a series of 11 ochratoxins (e.g., ochratoxin α and β, ochratoxin A and B, (4*R*)- and (4*S*)-4-hydroxyochratoxin A, bromoochratoxin B, citrinin). These were isolated from *Aspergillus ochraceus* and resolved on a C_{18} column ($\lambda = 250$ nm, ex; 454 nm, em) using a 50/60/2 water/methanol/acetic acid mobile phase. Elution was complete in 25 min. No detection limit was given but 0.05 mg/kg was considered a trace amount.

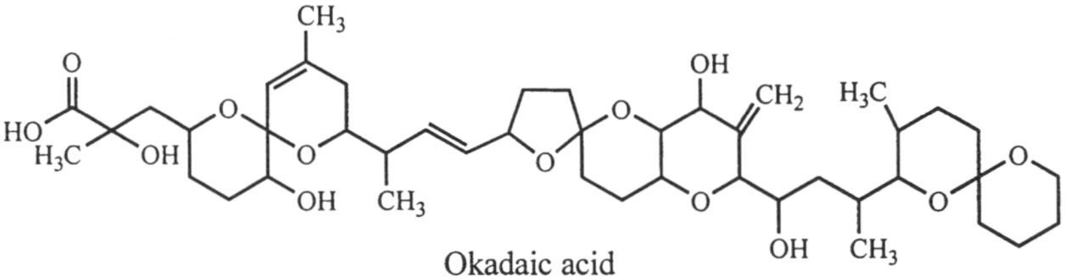

Okadaic acid

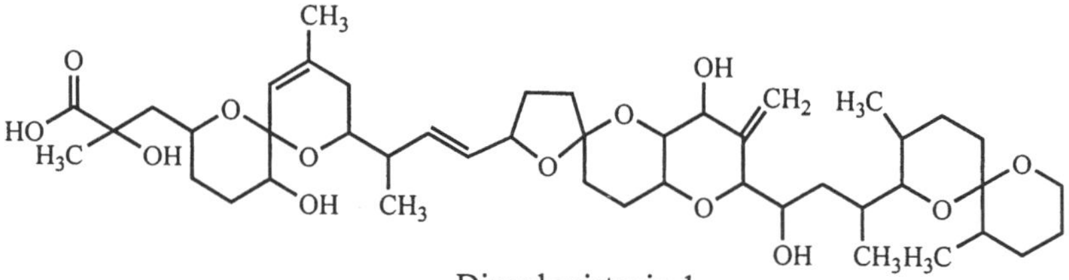

Dinophysistoxin-1

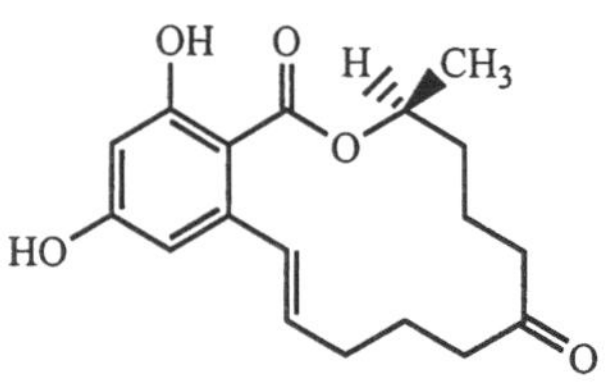

Zearalenone

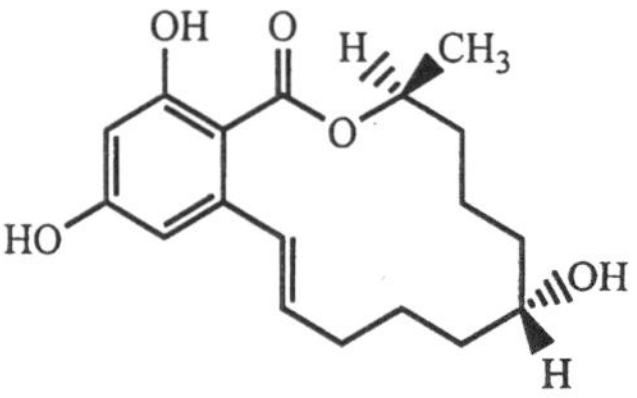

α-Zearalenol

Lantadene A, a hepatotoxin, was isolated from bacteria and analyzed on a C_{18} column ($\lambda = 210$ nm) using a 71/20/9/0.01 methanol/acetonitrile/water/acetic acid mobile phase. Elution was complete in 17 min [317].

Six *Alternaria* mycotoxins (altenuene, alternariol and its methyl ether, altertoxin-I and -II) were extracted from foodstuffs and analyzed [318]. Excellent peak shape and good resolution resulted when a C_{18} column (photodiode array detector, $\lambda = 210–450$ nm) and a 20-min $50/50 \rightarrow 85/15$ methanol/water (H_3PO_4 to pH 3) gradient were used. Detection limits varied from 0.2–10 ppm (analyte dependent).

Brevetoxins (PbTx-1, PbTx-2 and PbTx-9) were separated on a C_{18} column (electrospray MS or $\lambda = 215$ nm) using an 85/15 methanol/water mobile phase. Elution was complete in slightly more than 30 min with detection limits of 1 pmol or less (S/N = 3) reported [319].

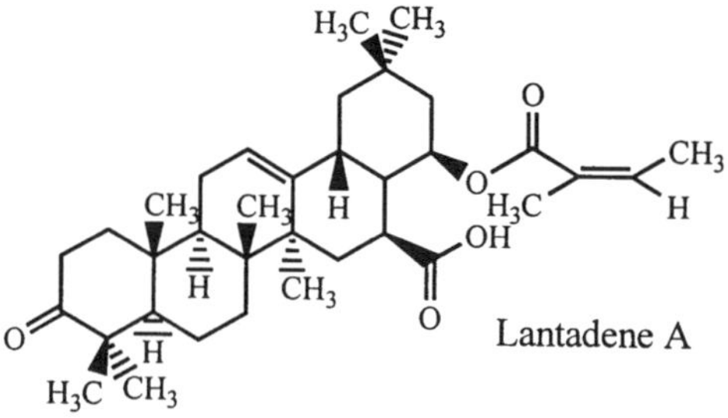

Lantadene A

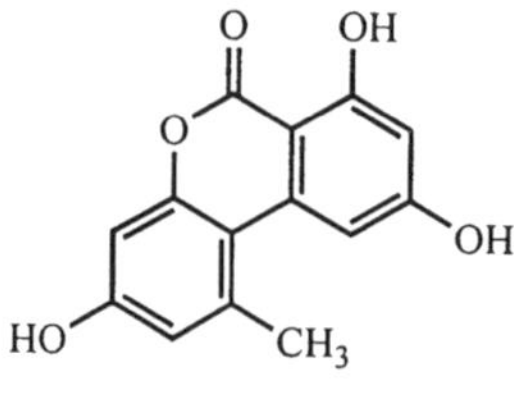

Altenariol

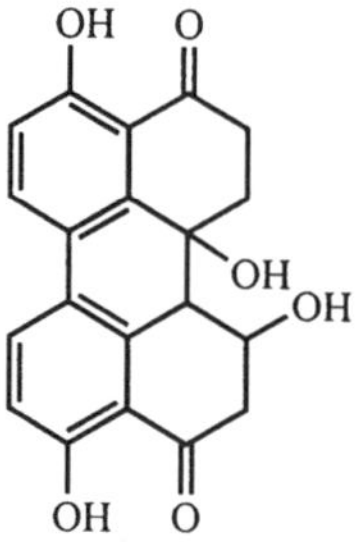

Altertoxin-I

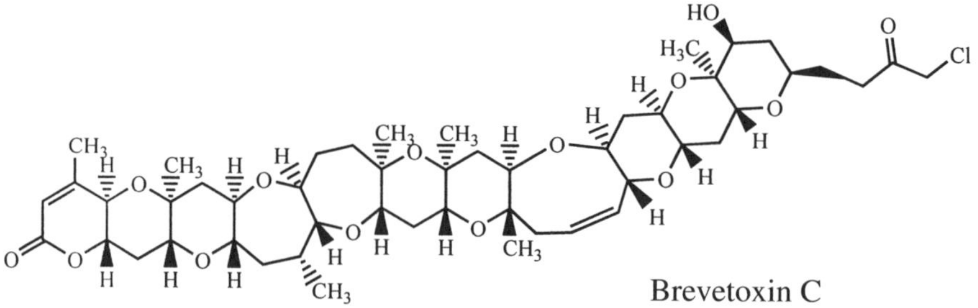

Brevetoxin C

Ten coumarins (scopoletin, umbelliferone, xanthotoxin, isopimpinellin, angelicin, bergapten, oxypeucedanin, imperatorin, ostruthol, isoimperatorin) were extensively studied using a 50°C C_{18} column ($\lambda = 320$ nm) and isocratic 54% methanol, 36% ethanol or 24% n-propyl alcohol/water mixtures [320]. Three elution order changes occurred when the organic modifier was changed from methanol to ethanol to n-propyl alcohol (e.g., the elution order in methanol was bergapten [B] < (+)-oxypeucamin, [+O] < imperatorin [I] < ostruthol [O]; in ethanol it was +O < B < O < I; in n-propyl alcohol it was +O < B < I < O). For all mobile phases the analytes eluted in <25 min. It is interesting to note that solute selectivities can vary markedly when solvents that have such similar chemical structures and properties are used. These selectivities can be used as powerful tools, especially in difficult separations.

Coumarin and 10 of its hepatic metabolites (3-, 4-, 5-, 6-, 7- and 8-hydroxy-coumarin, o-coumaric acid, 3,7- and 6,7-dihydroxycoumarin, and o-hydroxyphenyl-acetic acid) were resolved on a 40°C C_{18} column ($\lambda = 280$ nm) using a 25-min $10/90 \rightarrow 65/35$ methanol/(95/5 water/formic acid) gradient [321]. Standards containing ~ 20 ng of each component were readily detected. Peak shapes were uniformly excellent and elution was complete in 24 min.

Eight anticoagulant rodenticides from serum (warfarin, coumafuryl, coumatetra-lyl, brodifacoum, bromadiolone, chlorophacinone, diphacinone, difenacoum) were baseline resolved on a C_{18} column ($\lambda = 285$ nm) using a 20-min $52/48 \rightarrow 87/13$

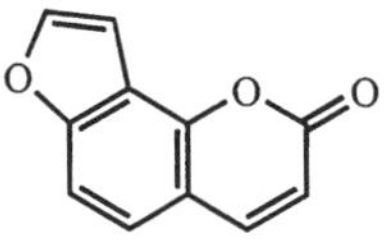

Angelicin

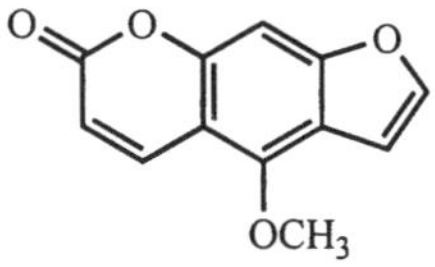

Bergapten

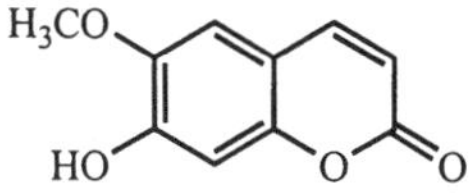

Scopoletin

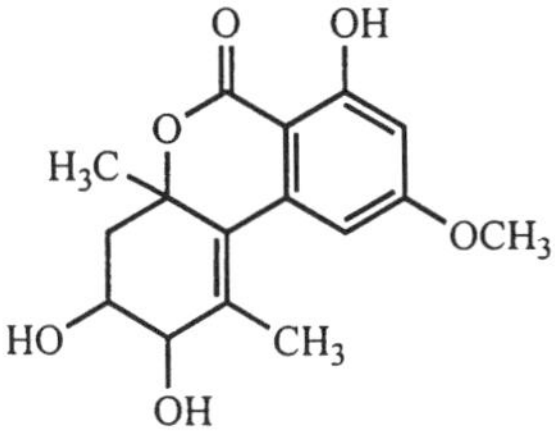

Altenuene

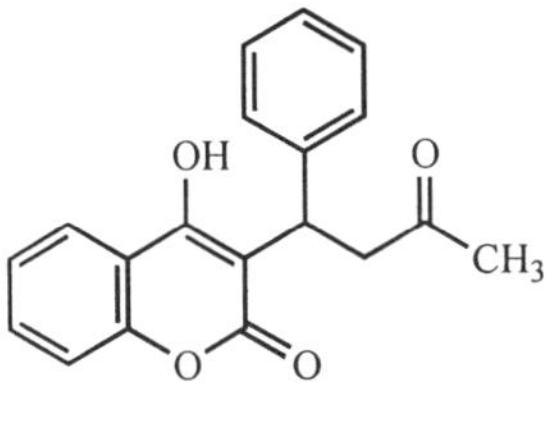

Warfarin

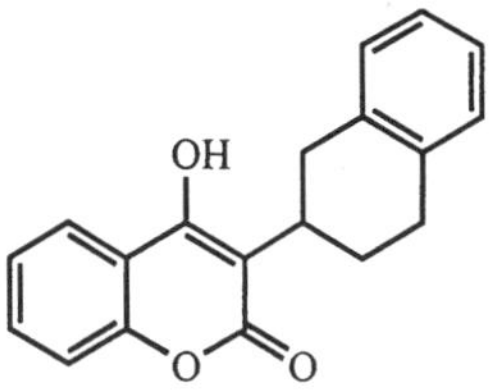

Coumatetalryl

methanol/water (30 mM tetrabutylammonium hydroxide pH 6) gradient. Working ranges were found to be 5–500 ng/mL [322].

The o-phthalaldehyde derivatives of fumonisins A and B were well resolved on a C_{18} column ($\lambda = 335$ nm, ex; 440 nm, em) using an 80/20 methanol/water (0.1 M sodium phosphate at pH 3.35) mobile phase. Good peak shape was reported [323]. For corn extracts, detection limits of 50 ng/g were reported.

Methanol extracts of neem oil were concentrated and injected onto a preparative C_{18} column ($\lambda = 215$ nm). Ten separate fractions (e.g., azadirachtin I, H, A, D, B, nimbin, salannin) were isolated using a two-ramp 60/40 (hold 60 min) $\rightarrow$ 70/30 (at 120 min) $\rightarrow$ 80/20 (at 130 min) methanol/water gradient. A 5 g aliquot of the concentrate was injected and processed [324].

Five thapsigargins (nortriobolide, trilobolide, thapsigargin, thapsitranstigan, thapsigargicin) were extracted from the *Thaspia garganica* plant and were well resolved on a C_{18} column ($\lambda = 230$ nm). An 83/17 methanol/water mobile phase generated the separation in <10 min [325].

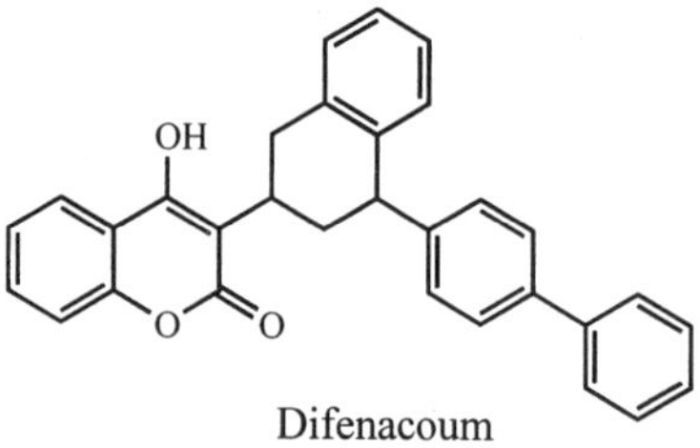

Difenacoum

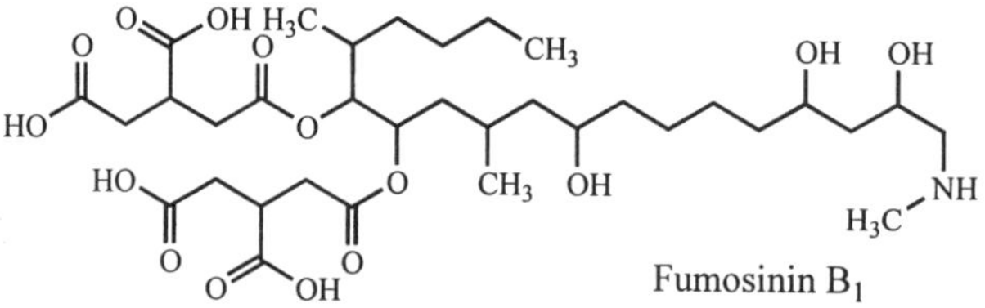

Fumosinin B_1

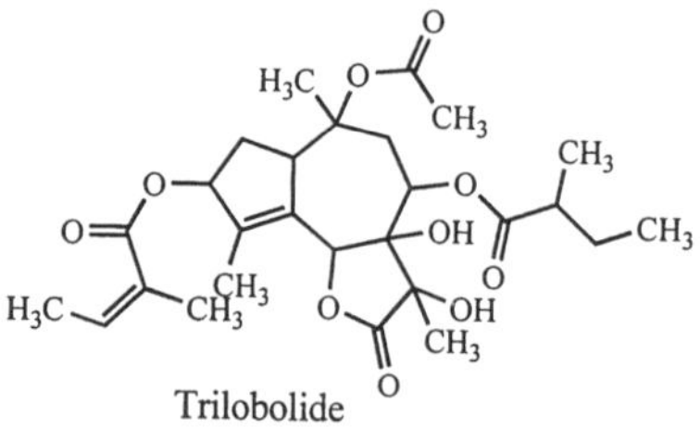

Trilobolide

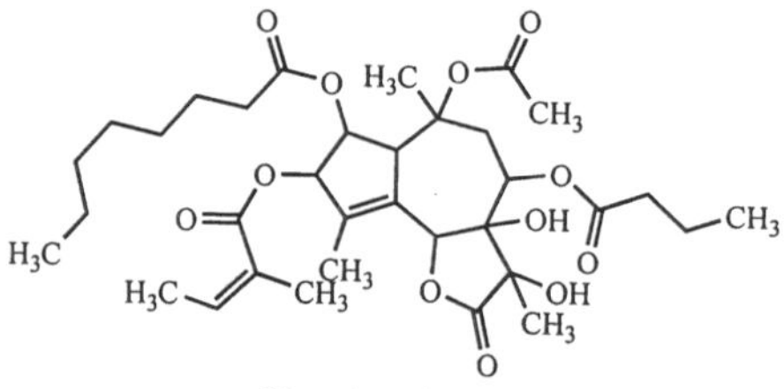

Thapsigargin

4.5.4 Vitamins and Related Analytes

4.5.4.1 Water-soluble Vitamins and Related Compounds

Eight water-soluble vitamins (nicotinamide, thiamine, riboflavin, pyridoxine, pyridoxal, pyridoxamine, cyanocobalamine, folic acid) were extracted from powdered infant milk and separated using ion-pair chromatography [326]. A mobile phase consisting of 15/85 methanol/water (5 mM octanesulfonic acid and 0.5% triethylamine to pH 3.6 with triethylamine or acetic acid) and a C_{18} column (photodiode array detector ranging from 250–370 nm) generated baseline resolution in 50 min. Note that thiamine, the last-eluting peak, appeared 30 min after the closest-eluting peak. A gradient could be used effectively here. In general, the working range was 0.05–5 mg/L and detection and quantitation limits fell in the 0.02–1 μg/mL and 0.03–0.25 μg/mL ranges, respectively.

Pyridoxine, pyridoxal, pyridoxamine, pyridoxal phosphate, pyridoxamine phosphate, and 4-pyridoxic acid were baseline resolved and detection limits of $\ll 1$ pmol were obtained on a C_{18} column ($\lambda = 328$ nm, ex; 393 nm, em) using a $0/100 \rightarrow 100/0$ (at 10 min hold 10 min) IPA/water (33 mM phosphate buffer pH 2.2 with 10 mM octanesulfonic acid) gradient [327]. Excellent quantitative results from spinal fluid samples were also obtained.

Homocysteine, cysteine- and homocysteine-*S*-bimane, and cysteine-*S*-bimane-glycine were extracted from plasma and serum samples were baseline resolved in 10 min using a C_8 column ($\lambda = 390$ nm, ex; >418 nm, em) and a $5/95 \rightarrow 100/0$ methanol/water (2.5 mL acetic acid to pH 3.40) gradient. Detection limits of ~2 pmol (S/N = 3) and linear response ranges up to 200 pmol were obtained [328].

Eleven folylmonoglutamates were separated on a phenyl column and quantitated using electrochemical detection (+900 mV vs. Ag/AgCl), fluorescence detection ($\lambda = 295$ nm, ex; 365 nm, em), or photodiode array detection ($\lambda = 200$–400 nm). A 15/85 methanol/water (50 mM phosphate buffer pH 3.5) mobile phase generated

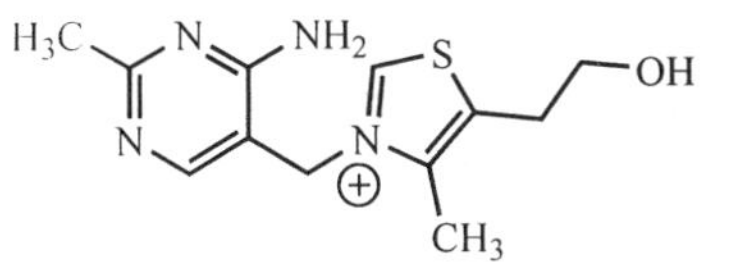

Thiamine

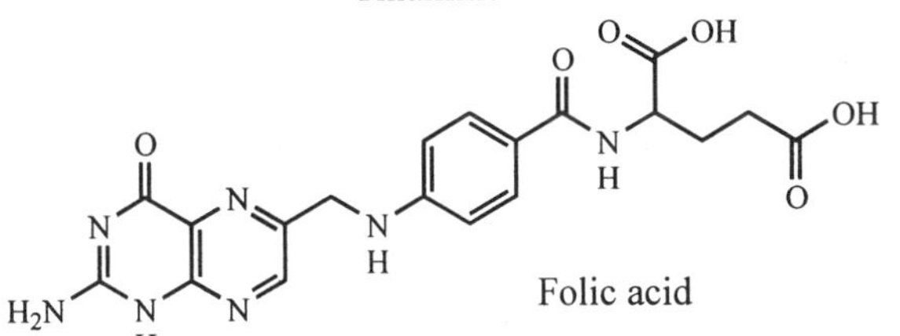

Folic acid

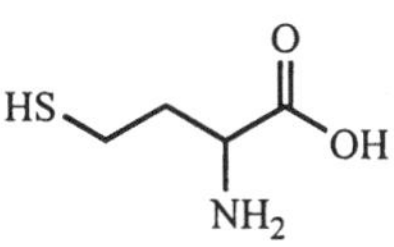

Homocysteine

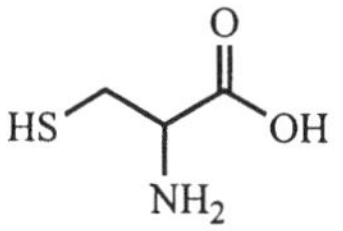

Cysteine

baseline resolution in <25 min [329]. Levels of 5 ng-injected were detected by UV. UV spectra from 200 to 400 nm were plotted for all the compounds analyzed in the study. For compounds showing response, electrochemical detection and fluorescence detection gave detection limits an order of magnitude lower than UV.

Four cobalamins (hydroxo- cyano-, adenosyl-, methylcobalamin) and two cobinamides (cyano-α- and cyano-β-cobinamide) were baseline resolved on a C_{18} column ($\lambda = 278$ nm) using a 30-min 90/10 → 60/40 water (25 mM ammonium acetate to pH 4 with acetic acid)/methanol gradient [330]. Peak shapes were excellent. A linear range of 10–100 µg/mL and a detection limit of 2 µg/mL were reported.

Pyroxidine, thiamin hydrochloride, cyanocobalamin, and folic acid from tablets were resolved on a 35°C β-cyclodextrin column ($\lambda = 254$ nm) using a 20/80 methanol/water (50 mM phosphate buffer pH 7) mobile phase [331]. Elution was complete in <7 min. Peaks shapes were somewhat broad. A lower pH and a more concentrated buffer would most likely help correct this problem.

Biotin and dethiobiotin were derivatized with panacyl bromide and separated on a C_{18} column ($\lambda = 380$ nm, ex; 470 nm, em) using a 15-min 60/40 → 30/70 water/methanol gradient [332]. Baseline resolution was achieved and elution was complete in <8 min. Linear response was obtained over the concentration range of 10–1000 pmol injected.

Fish plasma samples were extracted for isolation of melatonin. The extracts were then analyzed on a C_{18} column ($\lambda = 268$ nm, ex; 352 nm, em) using a 60/40 methanol/water mobile phase [333]. Melatonin was eluted in <10 min. The working curve covered the range 3–600 pg/mL. A detection limit of 3 pg/mL was reported.

4.5.4.2 Fat-soluble Vitamins and Related Compounds

Vitamin E_1 (α-tocopherol) was extracted from feed and analyzed at the levels of 8–23 µg/mL on a C_{18} column ($\lambda = 295$ nm, ex; 330 nm, em) with a 95/5 methanol/water mobile phase [334]. Arnaud et al. [335], chromatographically studied serum extracts for retinol, α-tocopherol, and β-carotene on a C_{18} column ($\lambda = 450$ nm, 325 nm or 292 nm) and published the retention times of these compounds in pure methanol and 17 other binary and ternary mobile phases including methanol/acetonitrile/hexane, methanol/acetonitrile/cyclohexane, methanol/hexane, and methanol/acetonitrile/dichloromethane. Detection limits of 0.1–1.5 µM

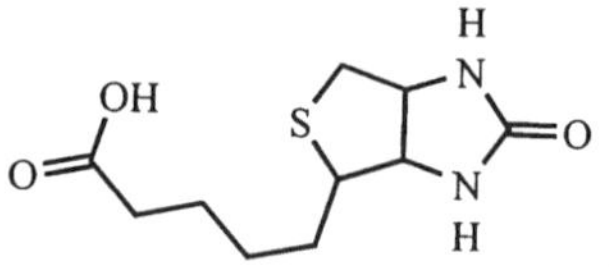

Biotin

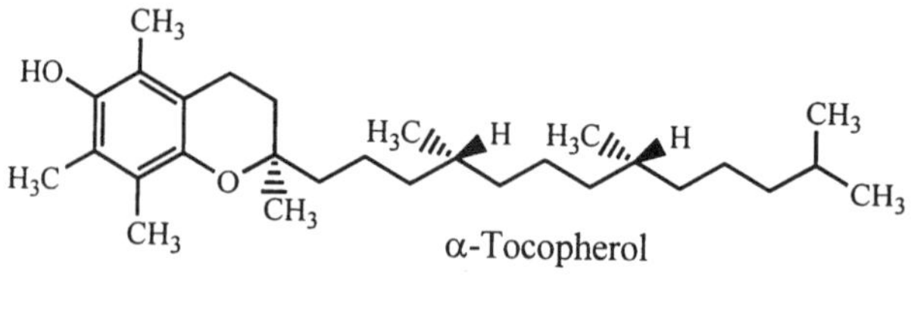

α-Tocopherol

(S/N = 3) were reported. Analyses were complete in 13 min. These results are valuable reference points for the development of separations for more complicated analyte mixtures.

Vitamin E, coenzyme Q, and related compounds (e.g., tocotrienols, ubiquinols) were extracted from tissues and analyzed on a C_{18} column (in-line UV at $\lambda = 275$ nm and amperometric detector set at 500 mV) using a 28-min 39/61 (hold 16 min) $\rightarrow$ 0/100 (at 18 min hold 10 min) (8/2 methanol/water)/ethanol with 0.2% $LiClO_4$ gradient. Note that the ethanol is designated as 95/5 ethanol/IPA IPA and samples are "preserved" from oxidation through the addition of BHT. As expected, the amperometric detector generates signals for all but ubiquinones 9 and 10, whereas UV detection of the tocotrienols is poor. A very large system peak occurs in the UV chromatogram, causing an off-scale negative peak from 21 to 24 min. Standards ranged from 50 to 200 µM [336].

Antioxidant levels (ubiquinone-10, ubiquinol-10, α- and γ-tocopherol) in neonate and infant plasma were determined using a C_{18} column (electrochemical detector, electrode 1 = -150 mV, electrode 2 = $+600$ mV) and an 800/180/65 methanol/ethanol/IPA with 31.7 mM ammonium formate mobile phase [337]. Elution was complete in 30 min. A standard solution containing 1–35 pmol of each material gave readily detected peaks.

The retention behavior of four tocopherols (α, β, γ, and δ) and 5,7-dimethyltocol was studied in detail using 0.5–2% levels of ethanol, IPA, n-propyl alcohol, 1-butanol and t-butyl alcohol in hexane ($\lambda = 298$ nm, ex; 345 nm, em) on an aminopropyl column [338]. Significant selectivity differences were noted, with the butyl alcohols giving the best results.

α-Tocomonoenol, a vitamin E analog, was extracted from salmon eggs and analyzed on a C_{18} column (electrochemical detector, $+600$ mV vs. Ag/AgCl) using a methanol (50 mM $NaClO_4$) mobile phase [339]. Elution was complete in 14 min and peak shape was excellent.

In a novel approach to the separation of carotenoids, Tai and Chen [340], used a C_{30} column ($\lambda = 450$ nm) and an 89/1/10 methanol/dichloromethane/IPA mobile phase. Daylily (*Hemerocallis disticha*) extracts were characterized. Sixteen compounds (e.g., neoxanthin, lutein-5,6-epoxide, β-cryptoxanthin, 9-*cis*-β-carotene)

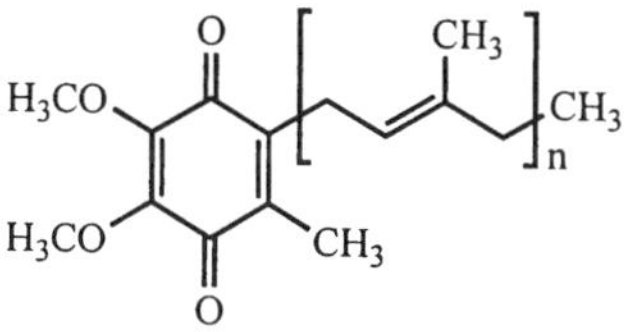

Ubiquinones; coenzyme Qs

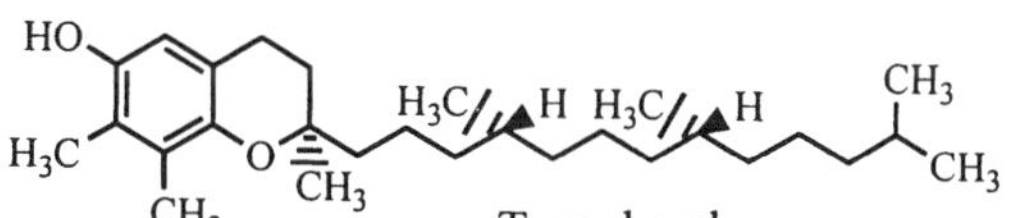

γ-Tocopherol

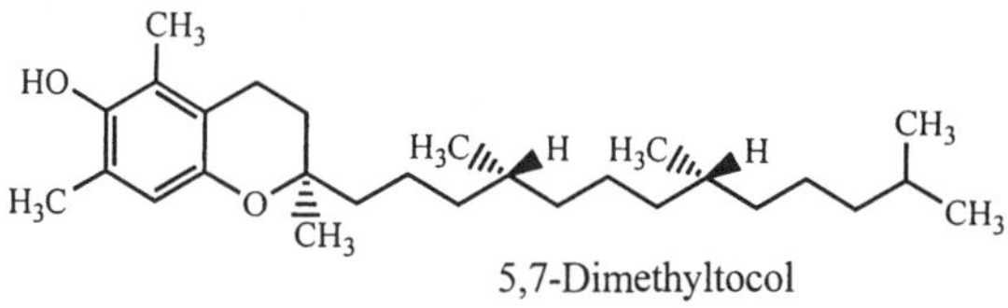

5,7-Dimethyltocol

were identified. The authors noted that the "classical" C_{18} separation did not give as good resolution as this method. Peak shapes were excellent and overall resolution was very good. Elution was complete in 55 min.

A series of six carotenoids (e.g., capsanthone, capsanthin, cucurbitaxanthin) were isolated from plasma after ingestion of paprika juice. Analysis was performed on a C_{18} column ($\lambda = 452$ nm) using a 10/7/2/3.3 methanol/acetonitrile/dichloromethane/water mobile phase [341]. Elution was complete in 30 min and lutien and zeaxanthin were not fully resolved.

Indyk and Woollard [342] provided an extremely detailed and complete 38-laboratory collaborative study in which vitamin K_1 (phylloquinone) was extracted from infant formula and quantitated using a C_{18} column (for all isomers) or a C_{30} column (for *cis/trans* isomer separation) and a hand-packed 20×4 mm zinc powder postcolumn reactor column used to reduce all components and generate a more highly fluorescent product. Detection was by fluorescence, 243 nm, ex; 430 nm, em. Elution was complete in <15 min using a 90/10 methanol/dichloromethane (containing 5 mL/L of methanolic solution of 0.274 g $ZnCl_2$/mL with 82 mg sodium acetate/mL and 60 mg acetic acid/mL). Such a high salt content will require frequent system flushes. A working curve from 6.25 to 31.25 ng/mL was used.

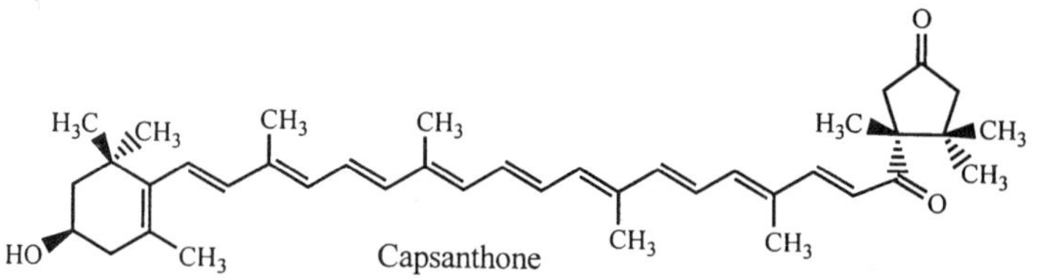

Capsanthone

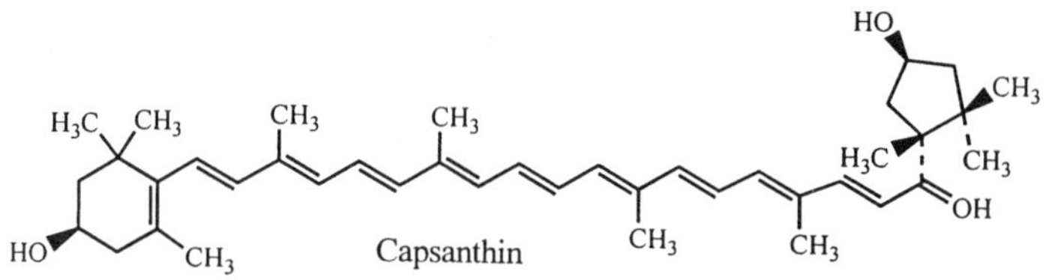

Capsanthin

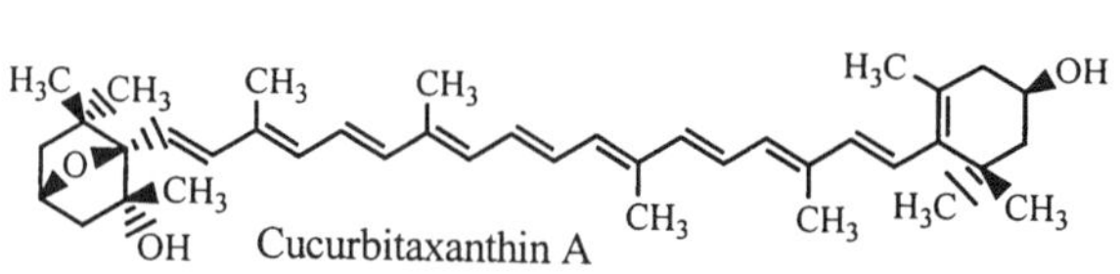

Cucurbitaxanthin A

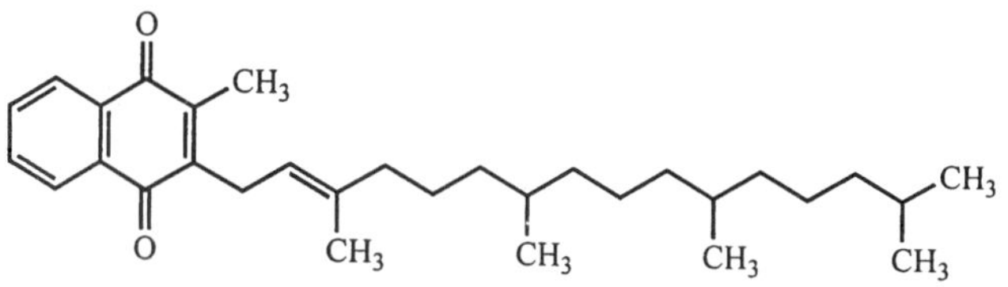

Phylloquinone, vitamin K_1

Vitamin K_1 was also isolated from emulsified nutritional supplements [343]. Quantitation was accomplished using post-column reduction on a platinum oxide catalyst. A 50/50 ethanol/methanol mobile phase generated a 10-min retention time on a 40°C C_{18} column ($\lambda = 320$ nm, ex; 430 nm, em). The reported detection limit was 0.1 pg injected (S/N = 3) with a linear range of 0–2 pg injected. It is important to remember that many types of HPLC-grade ethanol are available and, although this is probably not important in this separation, care should be taken to ensure that this information is provided.

Vitamin K (phylloquinone), 2′, 3′-dihydrophylloquinone, and menaquinones (where n = 4, 5, 6) were extracted from milk and infant formula and analyzed on a C_{18} column ($\lambda = 243$ nm, ex; 430 nm, em). All compounds were well resolved and eluted in 15 min using a 90/10/0.5 methanol/dichloromethane/methanol (with 10 mM zinc chloride, 5 mM sodium acetate, and 5 mM acetic acid) mobile phase [344]. Note that this level of salts in the mobile phase could precipitate in various parts of the LC system. Flush the system regularly. The linear working curve extended from 2 to 50 ng/mL with a detection limit of 1.5 ng/mL. The authors noted that this compared favorably to a UV detection limit (at $\lambda = 269$ nm) of 50 ng/mL.

A series of vitamin K_2 menaquinones (MK4–MK10) were isolated from bowel contents and analyzed on a C_{18} column ($\lambda = 338$ nm, ex; 425 nm, em) using post-column reduction with ethanolic 0.1% $NaBH_4$ [345]. A 35-min isocratic 95/5 ethanol/water elution gave good resolution.

Andreoli et al. [346] studied the separation of six fat-soluble vitamins (vitamin A, vitamin A acetate and palmitate, vitamin D_3, and vitamin E and vitamin E acetate)

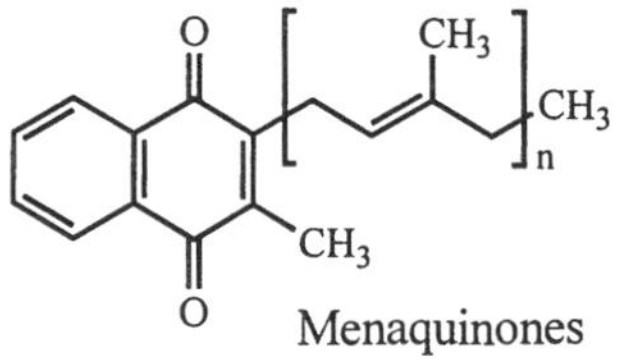

Menaquinones

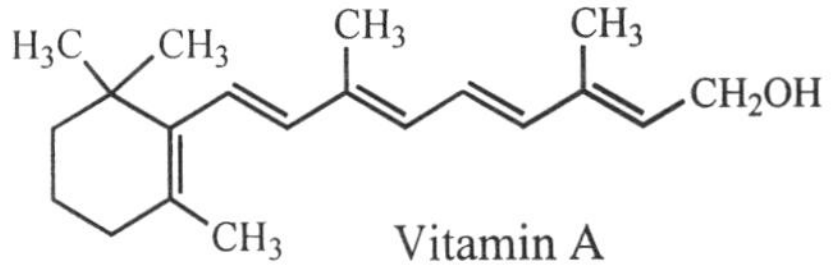

Vitamin A

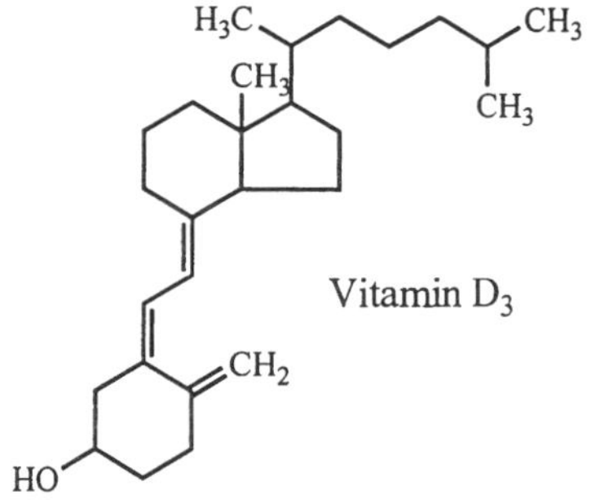

Vitamin D$_3$

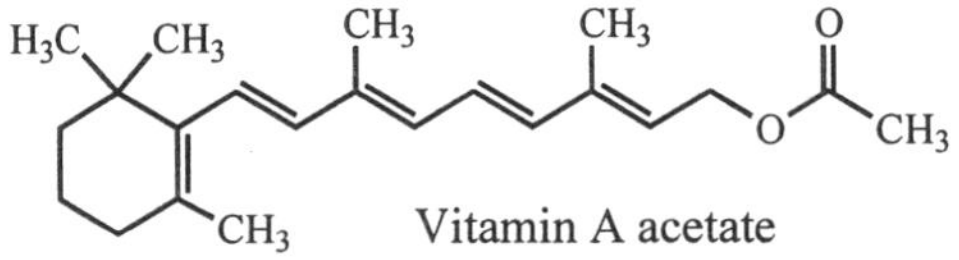

Vitamin A acetate

on a C_8 column ($\lambda = 325$ nm for all vitamin As; $\lambda = 265$ nm for D_3, and $\lambda = 295$ nm for vitamin E and vitamin E acetate) using a 98/2 methanol/water mobile phase. Baseline resolution was achieved in 13 min. The authors presented an interesting comparison of detection limits for a narrow-bore (2 mm ID) and an analytical (4 mm ID) column: 0.001–0.7 ng injected for narrow-bore and 0.09–2.5 ng injected for analytical (analyte dependent). Linear ranges for the narrow-bore were 0.05–50 through 6–1200 ng injected (analyte dependent).

Vitamins A, D, E, and pro-vitamin D_2 were extracted from animal feed and analyzed on a C_{18} column ($\lambda = 290$ nm) using methanol as the mobile phase. Vitamin A was poorly retained and eluted on the shoulder of a large early-eluting interferent peak; the other peaks were well resolved. Elution was complete in <10 min. The linear range extended from 10 ng to 50 µg injected [347]. The reported detection limit was 10 ng/g feed.

An effective separation of 10 fat-soluble vitamins and provitamins (retinal, retinyl palmitate and acetate, ergosterol, 7-dehydrocholesterol, tocopherol acetate, and Vitamin D_2, D_3, E, and K_1) was generated in <35 min. Milk or butter samples were extracted and analyzed on a C_{18} column ($\lambda = 325$ nm for retinyls, 264 nm for vitamin Ds, and 280 nm for all others) using a complex 33-min $99/1/0 \rightarrow 70/0/30$ methanol/water/THF gradient [348]. Peak shapes and resolution were very good. Linear ranges of 15–1000 ng injected and detection limits of 1 ng injected (analyte dependent) were reported.

Eleven retinols (e.g., 4-oxoretinoic acid, 4-oxoretinol, retinoyl-β- and retinyl-β-glucuronide, retinoic acid, 5,6-epoxyretinol, retinol) and retinyl acetate, linoleate,

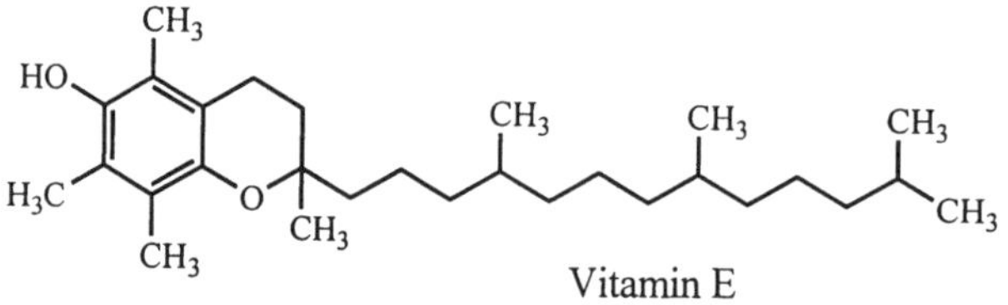

Vitamin E

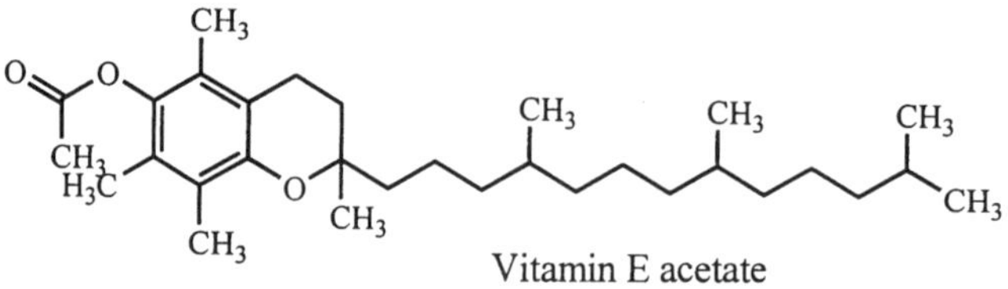

Vitamin E acetate

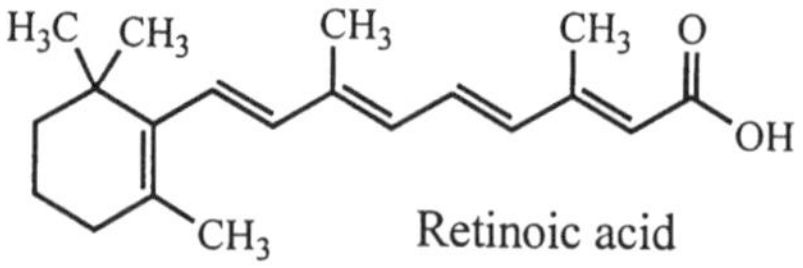

Retinoic acid

palmitate, and stearate were extracted from serum samples and resolved on a C_{18} column ($\lambda = 340$ nm) with a 20-min 68/32/0 → 75/0/25 methanol/water (10 mM ammonium acetate)/dichloromethane gradient [349]. Samples contained 0.1% w/v BHT to help limit oxidation due to exposure. A sample load range from 0.5 to 20 ng injected was reported. Excellent peak shapes and baseline resolution were obtained. Retinyl stearate eluted last at 29 min.

A series of 14 retinoids, carotenoids, and tocopherols from very polar (e.g., all-*trans*-4-oxo-retinoyl β-glucuronide, all-*trans*-4-oxo-retinoic acid, all-*trans*-5,6-epoxyretinoic acid) to slightly polar (e.g., all-*trans*-retinol, all-*trans*-retinoic acid) to nonpolar (e.g., all-*trans*-lycopene, all-*trans*-β-carotene) were resolved on a Microsorb-MV column ($\lambda = 330$ nm) using a 100/0 → 0/100 (at 15 min hold 19 min) (3/1 methanol/water with 10 mM ammonium acetate)/(4/1 methanol/chloroform) gradient [350].

Seventeen retinoic acids, retinoyl-β-glucuronides, retinol, retinal, and retinyl fatty acids were resolved with a C_{18} column ($\lambda = 340$ nm) using a 50-min 70/30/0 → 88/12/0 → 83/0/17 methanol/water (10 mM ammonium acetate)/chloroform gradient [351]. Samples and standards were all stored in solvents containing BHT (10 μg/mL). Excellent separation and good peak shapes were obtained and working curves from 5 to 200 ng were generated and the results were tabulated.

In an excellent presentation, Deli et al. [352], studied the carotenoid composition of extracts of paprika. A total of 34 carotenoids were identified and quantitated. The chromatographic run was completed in 50-min using a C_{18} column ($\lambda = 450$ nm) and an 88/0/12 → 94/0/6 → 50/50/0 methanol/acetone/water gradient. Baseline resolution between major components was achieved. Peak identities were confirmed by GC/MS.

Eight carotenoids (capsorubin, capxanthin, zeaxanthin, apo-8′-carotenal, β-cryptoxanthin, α-carotene, β-carotene) were extracted from various *Capsicum* species and analyzed on a C_{18} column ($\lambda = 470$ nm). The gradient used was a 25-min 25/75 → 75/35 (at 15 min) → 0/100 (75/20/5/0.05 acetonitrile/methanol [50 mM ammonium acetate]/dichloromethane/triethylamine)/methanol with 0.1% BHT acting as a preservative [353]. Good resolution was achieved for all but one pair of compounds. Athough no standard concentrations were given, samples containing anything from 3 to 1300 μg/100 g sample were detected.

all-*trans*-Retinal

11-*cis*-Retinal

Lutein, astaxanthin, β-cryptoxanthin, and α- and β-carotene were separated on a C_{30} column (electrospray MS) using a unique 60-min 85/15 $\rightarrow$ 10/90 methanol/ MtBE (1 mM ammonium acetate) gradient. Postcolumn derivatization with 2,2,3,3,4,4,4-heptafluoro-1-butanol dramatically increased the sensitivity of the method [354]. Detection limits 100-fold lower than that obtained with UV detectors (i.e., 1–2 pmol injected) were reported for α- and β-carotene. The isomerization of *trans*-astaxanthin to 9-*cis*- and 13-*cis*-astaxanthin was monitored through separation on a 25°C C_{18} column ($\lambda = 480$ nm) using an 85/5/5.5/4.5 methanol/ dichloromethane/acetonitrile/water mobile phase [355]. Separation was good and elution was complete in <10 min.

In an excellent article by Bell et al. [356], the retention of 12 carotenoids (zeaxanthin, lutein, echinenone, β-cryptoxanthin, and α-, 9-*cis*-α-, 15-*cis*-α-, 9'-*cis*-α-, 13-*cis*-α-, β-, 13-*cis*-β-, and 15-*cis*-β-carotene) was studied with respect to temperature (277 K to 323 K) on C_{18}, C_{30}, and C_{34} columns ($\lambda = 450$ nm). Methanol was used as a mobile phase on the C_{18} column and 95/5 methanol/methyl t-butyl ether on the C_{30} and C_{34} columns. The authors noted that acetonitrile was a potential mobile phase modifier but that high acetonitrile levels often led to decreases in recoveries. The use of dichloromethane was discouraged since residual HCl due to natural solvent degradation was implicated in poor recoveries as well. The latter is supported by the fact that low molecular weight alkenes are often used as preservatives in dichloromethane. A series of van't Hoff plots (essentially $\ln k'$ vs. $1/T$) were presented where C_{18} phase showed near linearity and C_{30} and C_{34} phases exhibited nonlinear relationships for most carotenoids.

The carotenoid profiles (17 compounds: α-, β-, ξ-carotene, α-, β-cryptoxanthin, auroxanthin A and B, mutatoxanthin A and B, trollichrome, antheraxanthin, (see structures on page 147) *cis*-antheraxanthin, neoxanthin, *cis*-violaxanthin, lutein, zeaxanthin/isolutein [co-elute], phytofluene) of orange juices were generated to determine the point of origin for the juices. A C_{30} column ($\lambda = 350$ nm, 430 nm, and 486 nm) and a complex 60-min 90/5/5 $\rightarrow$ 50/50/0 methanol/methyl t-butyl ether/water gradient were used and resulted in good separation. Standards used at 20 mg/L generated good response [357].

Twenty-three carotenoids (e.g., capsorubin, capsanthin and its 5,6-epoxide, auroxanthin, mutatoxanthin, cryptoxanthin, cryptocapsin, β-carotene) were extracted from *Asparagus officinalis* and identified through separation on a C_{18} column ($\lambda = 450$ nm). A complex 44-min $12/88/0 \rightarrow 9.6/80.4/10 \rightarrow 0/50/50$ water/ methanol/acetone gradient was used. Many sets of peaks were not fully resolved, but were well enough separated for assignment of chemical identity [358].

In an intriguing paper, Lee [359] generated an objective measure of the red grapefruit juice color through the separation and quantitation of phytofluene,

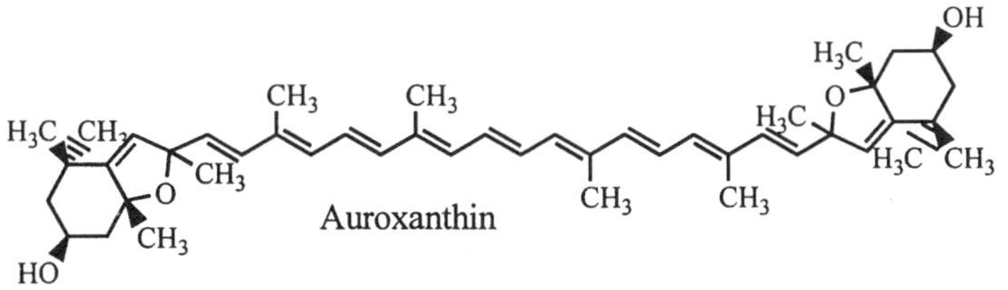

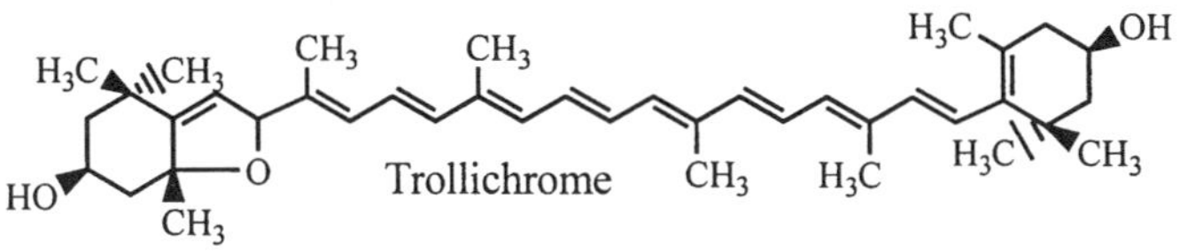

Auroxanthin

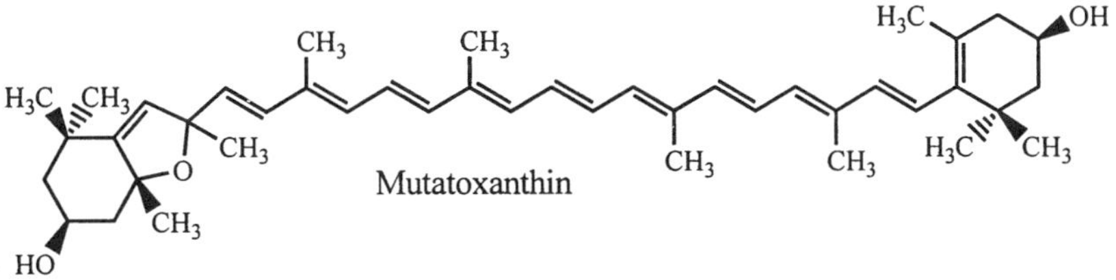

Mutatoxanthin

Trollichrome

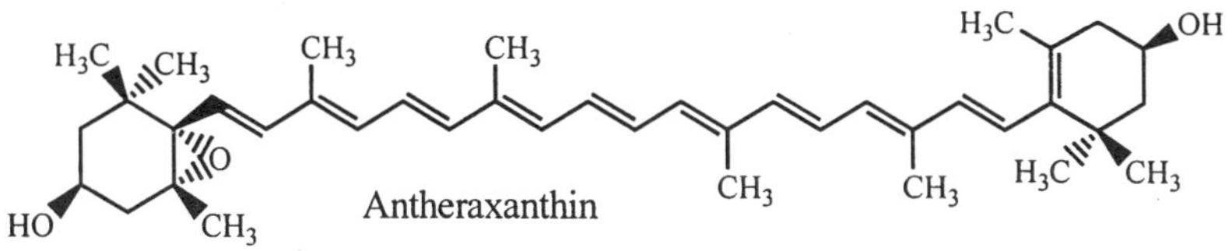

Antheraxanthin

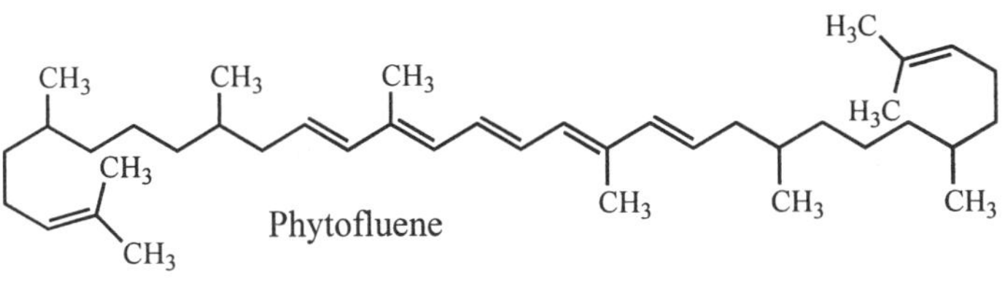

Phytofluene

ζ-carotene, β-carotene isomer, β-carotene, and lycopene. The analysis was run on a C$_{30}$ column (λ = 450 nm) using a 75/25 → 65/35 (at 10 min) → 45/55 (at 20 min hold 5 min) methanol/methyl *t*-butyl ether with 0.05% triethylamine and 0.01% BHT gradient.

4.5.5   Terpenoids, Flavonoids, and Other Naturally occurring Analytes

4.5.5.1   *Terpenoids*

Several sesquiterpene lactones (parthenin, coronopilin, tetraneurin A, hysterin, 1α-hydroxy-4β-*O*-acetylpseudoguaian-6β,12-olide, 4α-*O*-acetylpseudoguaian-6β-olide)

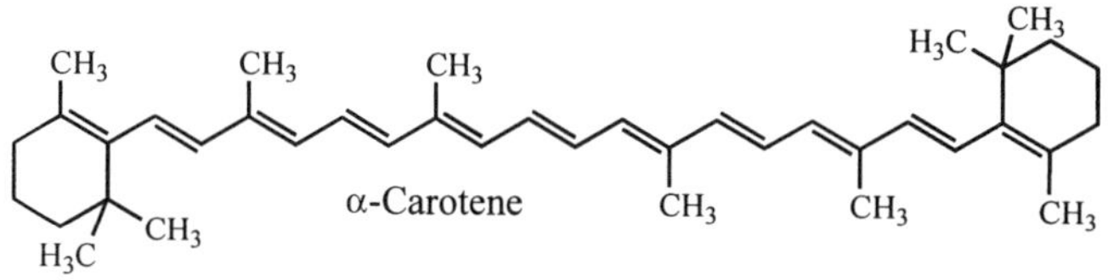

α-Carotene

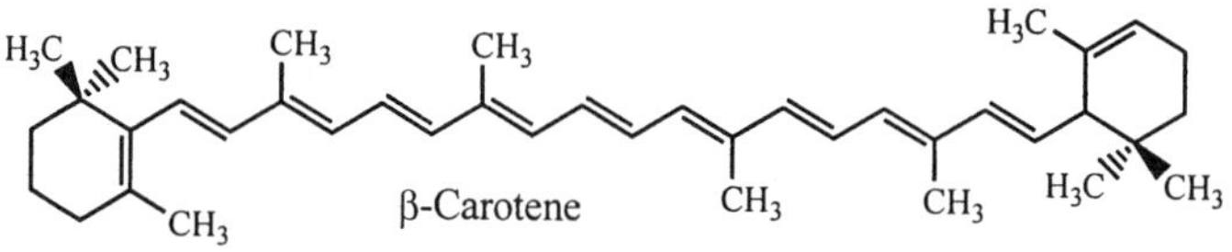

β-Carotene

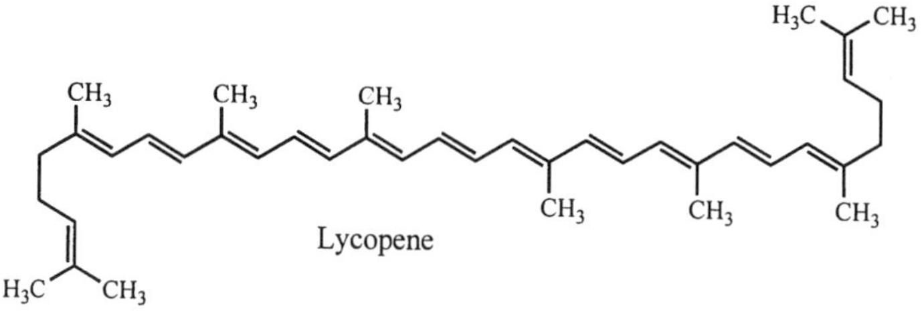

Lycopene

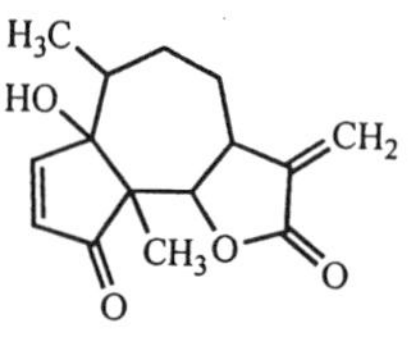

Parthenin

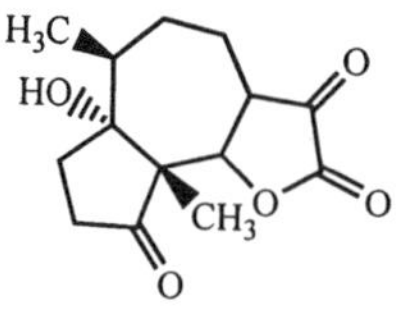

Coronopilin

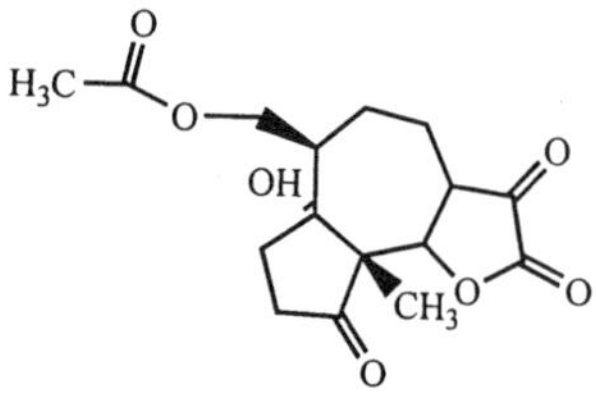

Tetraneurin A

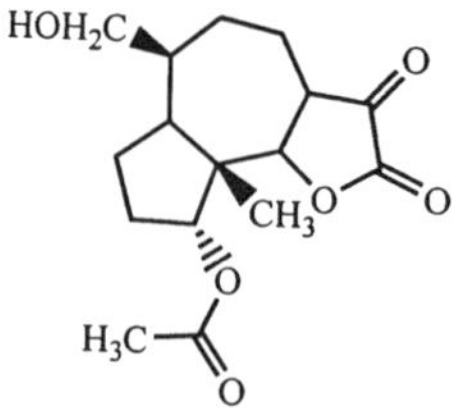

Hysterin

were extracted from *Parthenium hysterophorus* and analyzed on a C_{18} column ($\lambda = 210$ nm). Five of these compounds (excluding the olides) were resolved in 45 min using a 94/4 water/*n*-propyl alcohol mobile phase. Coronopilin, hysterin, and the two olides were separated using a two-step 10/90 (hold 20 min) → 22/78 (at 38 min) → 37/63 (at 53 min) *n*-propyl alcohol/water gradient. The effects of solvent background absorbance are striking in this example. With each step, the baseline deflects nearly 30% of full scale! Also, an extremely large and reproducible peak occurs at the second step, most likely a solvent contaminant. Finally, it should be noted that the backpressure of this system will go through a 2.7-fold increase over the course of the analysis [360].

Triterpenes were extracted from *Cimicifuga racemosa* and separated on a Discovery C_{18} column ($\lambda = 200$ nm and ELSD, nozzle $T = 40°C$, N_2 pressure = 2 bar). The authors noted that other C_{18} and phenyl columns could not resolve the peaks: cimiracemoside A, 26-deoxyactein and actein, and five minor unidentified triterpenes. The sensitivity at 200 nm was severely compromised due to the use of a 35-min 58/21/21 → 52/14/34 water/acetonitrile/"alcohol" (where alcohol = 90.6/4.5/4.6 ethanol/methanol/IPA) gradient. Remember that alcohols have UV cutoff values from about 203 to 206 nm. For the ELSD, the concentration range of 8–500 μg/mL was represented in a log response vs. log concentration plot. A detection limit of 20 μg/mL was reported [361].

Two diterpenes, cafestol and kahweol, were isolated from brewed coffees and separated on a C_{18} column ($\lambda = 230$ nm and 290 nm) using a 20-min 30/70 → 5/95 water/methanol gradient [362]. Peak shapes were excellent and baseline resolution was obtained. Detection limits of 50 μg/L were reported and standards of 40–300 μg/g were easily detected.

Six terpenes (asiaticoside, asiatic acid, madecassic acid, oleanolique acid, glycyrrhetinic acid, hederagenine) were extracted from *Centella asiatica* and base-line resolved with excellent peak shape on a C_{18} column ($\lambda = 206$ nm). A 65/5/35 methanol/acetonitrile/water (50 mM acetate at pH 3) mobile phase eluted all peaks in 50 min [363]. Plots of the effect of changes in percent methanol and pH on retention were plotted. The influence of phosphate and formate buffers was also examined. The authors note that a small percentage of acetonitrile added to the mobile phase had a strong positive sharpening effect on the peak shape. A linear range from 0.5–7.5 mg/mL with detection limits of 5 μg/L was reported.

Actein

Hederargenine

Ten diterpenes (serradiol, linearol, conchitriol, foliol, isofoliol, tobarrol, andalusol, sidol, lagascatriol, siderol) were extracted from various *Sideritis* species and analyzed on a C_{18} column ($\lambda = 220$ nm) using a 30/70 water/methanol mobile phase [364]. A chromatogram of the separation involving three analytes (linearol, conchitiol, isofoliol) was presented but the k' and α values for all 10 diterpenes were also tabulated. Dry mass of these compounds was reported as 0.004–0.19%.

Chen et al. [365] analyzed *Ginkgo biloba* extracts for diterpenes (ginkgolides A, B, and C) and the sesquiterpene (bilobalide) content using a C_{18} column (RI detector) and a 23/77 methanol/water mobile phase. Peaks were somewhat broad (perhaps due to the RI), but good resolution was achieved and elution was complete in 24 min. A plot of k' vs. percent methanol is shown. A linear range of 4–60 µg

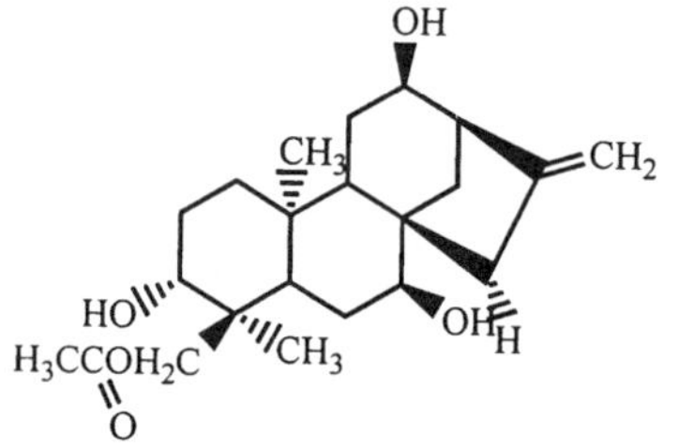

Linearol

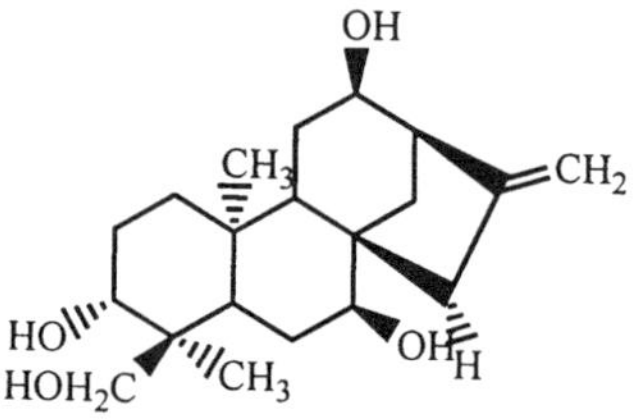

Foliol

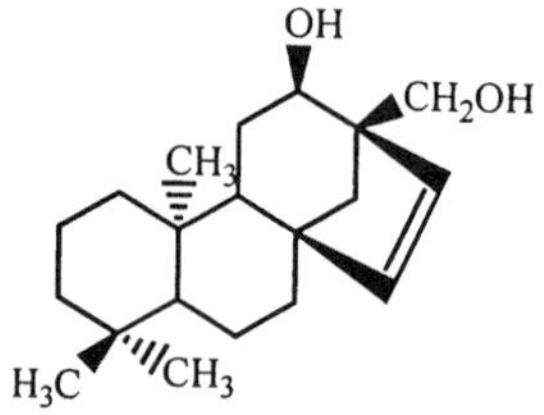

Tobarrol

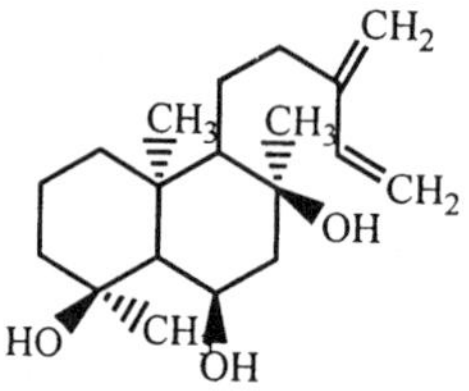

Andalusol

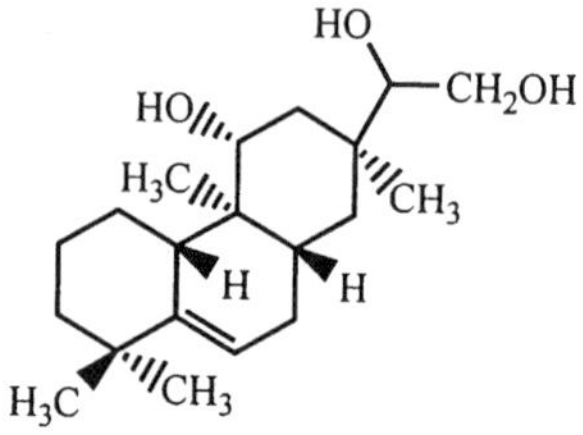

Lagascatriol

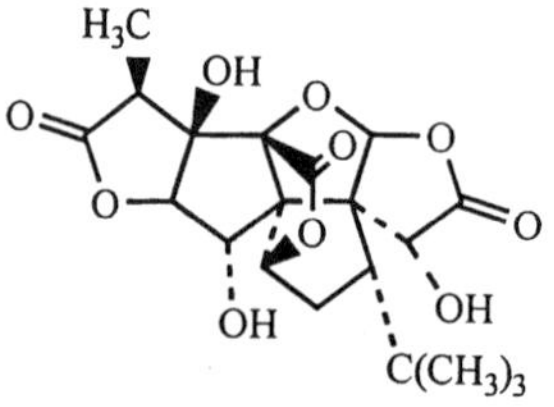

Ginkgolide B

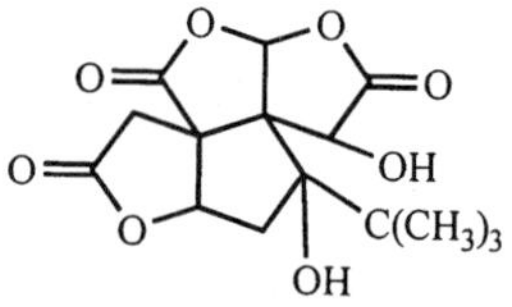

Bilobalide

injected and a detection limit of 0.5 μg injected were reported. For best results the system should be at constant temperature to yield steady RI output.

Sharma et al. [366] studied a series of six lantadenes (pentacyclic terpenoids with C_{22} side chains; lantadene A, B, C, D and reduced A and B. They were extracted from *Lantana camara* leaves and well resolved on a C_{18} column ($\lambda = 210$ nm) using an 85/15/0.1 methanol/water/acetic acid mobile phase. Elution was complete in <50 min. Linear ranges of 0.05–20 μg injected (analyte dependent) were reported.

4.5.5.2 Flavanoids

Two isoflavanoids and one triterpenoid (formononetin, isoliquiritigenin, 18-β-glycyrrhetinic acid) were used as compliance marker compounds for a patient on an anticancer drug extracted from licorice roots [367]. A C_{18} column (particle beam MS) and a rapid 1-min gradient from 40/55/5 to 95/0/5 methanol/water/formic acid (hold 10 min) generated the separation.

Eucalyptus honey extracts were characterized for flavonoid content. Six flavonoids (myricetin, tricetin, luteolin, quercetin and its 3-methyl ether, kaempferol),

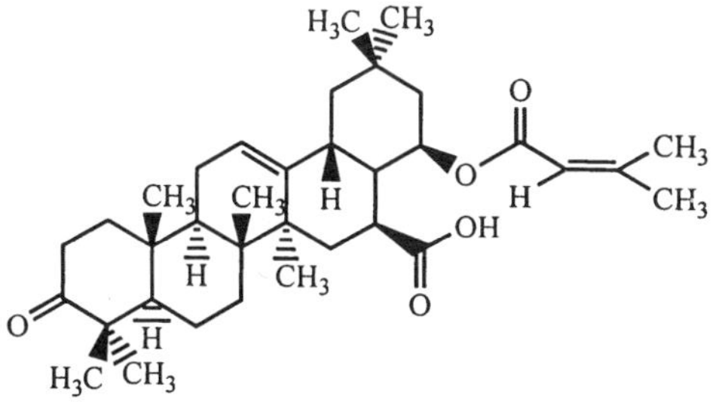

Lantadene B

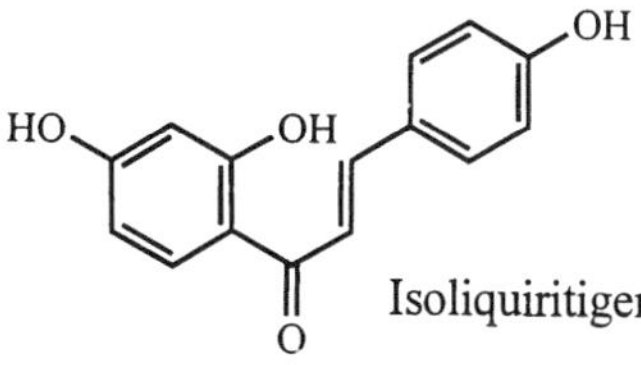

Isoliquiritigenin

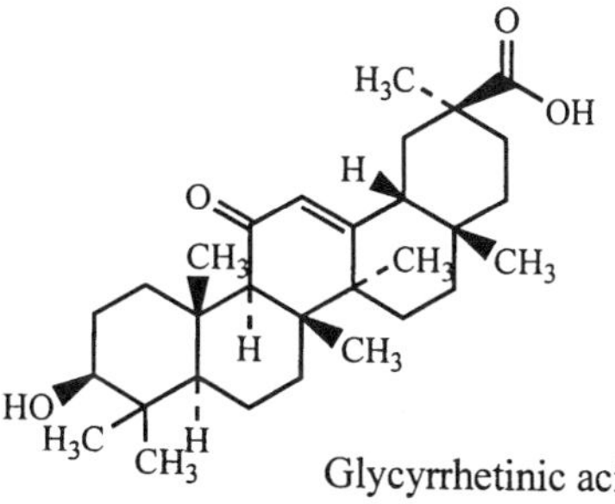

Glycyrrhetinic acid

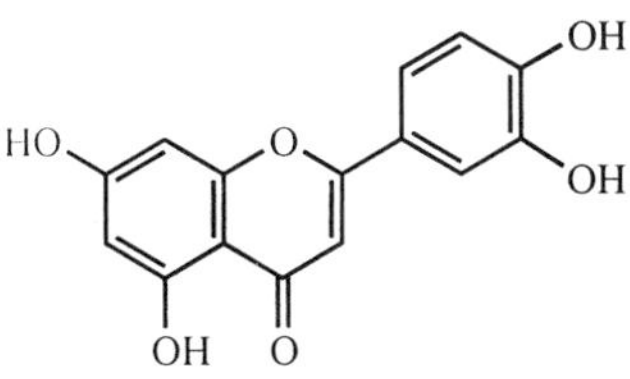

Luteolin

pinobanskin, and chrysin were resolved on a C_{18} column ($\lambda = 340$ nm) using a complex 60-min 70/30 → 20/80 (95/5 water/formic acid)/methanol gradient [368]. Excellent peak shapes and good resolution were achieved. Reported levels of these compounds ranged from 1 to 500 µg/10 g of sample.

7,8-Benzoflavone (BF) and five of its liver metabolites (6- and 7-hydroxy-BF, 5,6- and 7,8-dihydroxy-BF and BF-5,6-oxide) were extracted from tissue and separated on a C_{18} column ($\lambda = 280$ nm) using a 70/30 methanol/water mobile phase [369]. Excellent peaks shapes and resolution were obtained. Elution was complete in 20 min.

The retention behavior of a set of 12 isoflavanoids and two coumaronochromes (e.g., 2'-hydroxygenistein, wighteone, luteone, licoisoflavone A, lupalbigenin) was studied on a C_{18} column ($\lambda = 254$ nm) using an 80-min 45/55 → 100/0 methanol/water (0.5% acetic acid) gradient. Good overall resolution was obtained and a linear concentration range from 10 ng to 5 µg was reported [370].

Diastereomers of four flavinoids (naringin, prunin, neohesperidin, narirutin) were resolved on a β-cyclodextrin column ($\lambda = 280$ nm) using a 20-min 95/5 → 50/50 (90/10/0.5 methanol/water/acetic acid)/(95/5 methanol/water) gradient [371]. Nine flavanoids [372] extracted from *Ginkgo biloba* leaves (quercetin, apigenin,

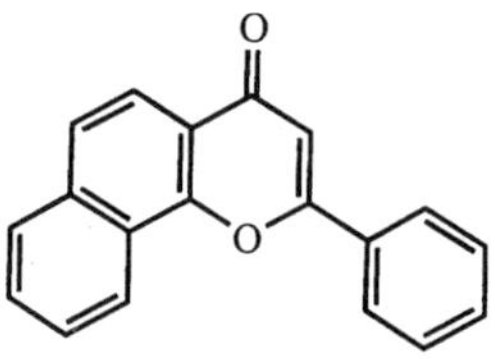

7,8-Benzoflavone

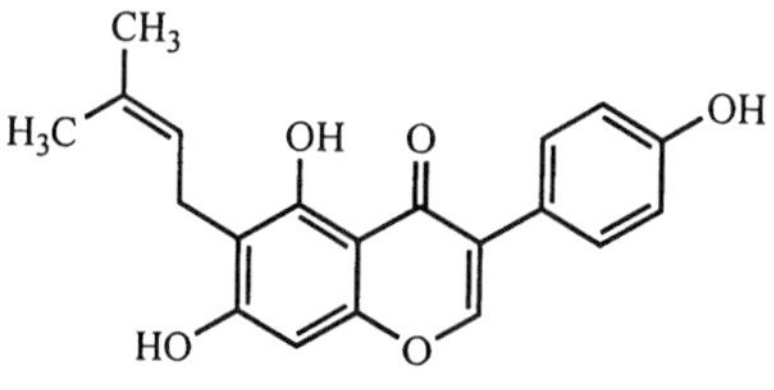

Wighteone

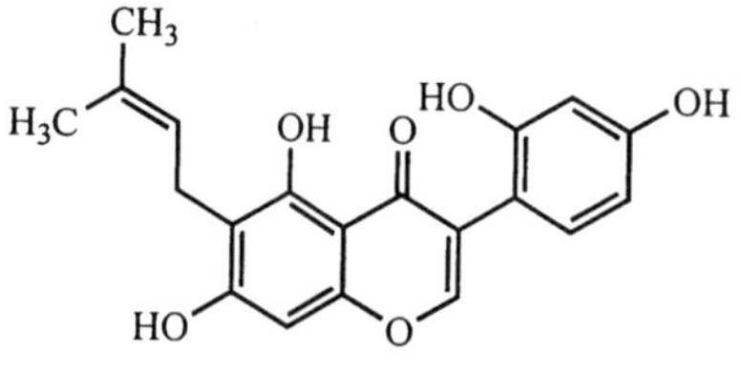

Luteone

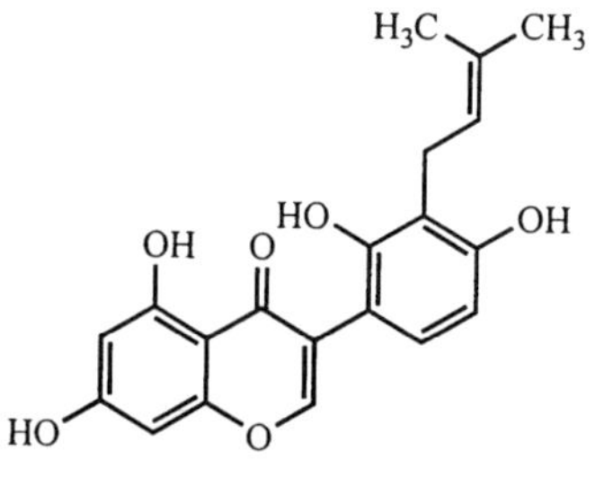

Licoisoflavone A

isorhamnetin, kaempferol, amentoflavone, bilobetin, ginkgetin, isoginkgetin, and sciadopitysin) were baseline resolved on a C_{18} column ($\lambda = 370$ nm) using a 20-min $22.5/22.5/55.0 \rightarrow 30/30/40$ methanol/THF/water (0.5% H_3PO_4) gradient.

The urinary metabolites of the flavonoid glycoside eriocitrin (eriodictyol, homoeriodictyol) were separated from the parent compound on a 40°C C_{18} column ($\lambda = 333$ nm and negative ion electrospray MS) using a 30-min $20/80 \rightarrow 70/30$ methanol/water (5% acetic acid) gradient [373]. Baseline resolution was achieved. Although no standard concentrations were given, recovered sample concentrations were listed from 30 to 400 ng/mL.

4.5.5.3 *Caffeine and related compounds*

Seven chlorogenic and hydroxycinnamic acids (caffeic, *p*- and *o*-coumeric, ferulic, 4-methoxy-, 3,4-dimethoxy-, 3,4,5-trimethoxycinnamic) were extracted from green coffee and baseline resolved on a C_{18} column ($\lambda = 320$ nm) using a complex 47-min $15/85 \rightarrow 80/20$ methanol/water (1% acetic acid) gradient [374]. Peak shapes were excellent. Linear ranges of 4–400 μg/mL with detection limits of 0.1 μg/mL were reported.

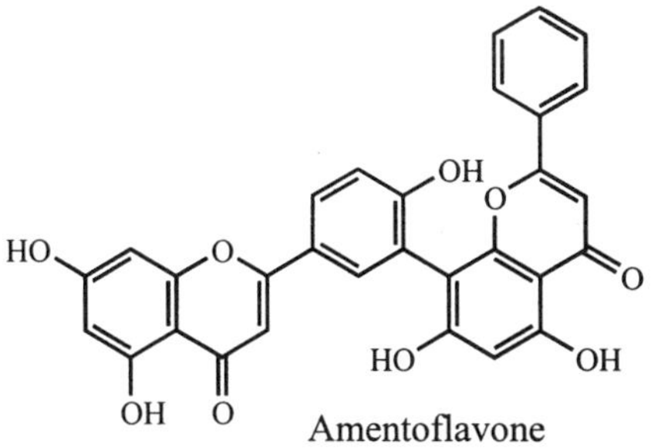

Amentoflavone

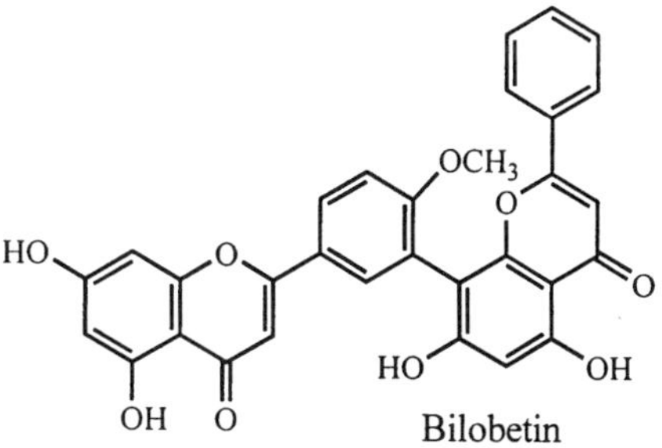

Bilobetin

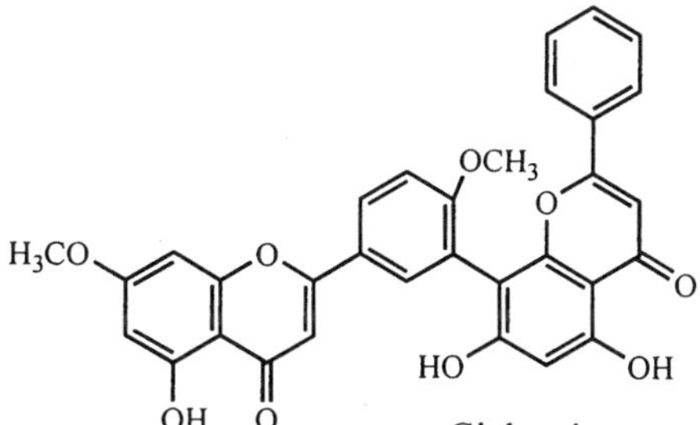

Ginkgetin

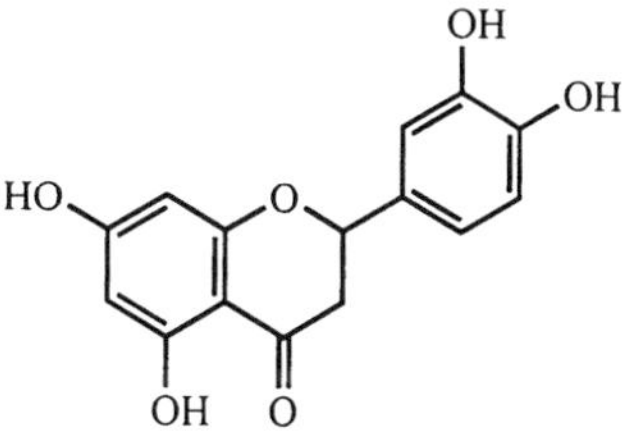

Eriodictoyl

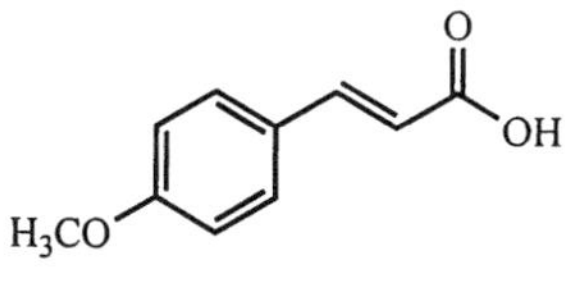

4-Methoxycinnamic acid

Methyluric acids are common decomposition products of methylxanthines such as caffeine. Five such compounds (1- and 7-methyluric acid, 1,7- and 1,3-dimethyl-uric acid, and 1,3,7-trimethyluric acid) were extracted from urine and separated on a C_8 column ($\lambda = 280$ nm). A two-ramp $95/5 \rightarrow 80/20$ (at 8 min) $\rightarrow 70/30$ (at 15 min) water (acetate pH 3.5)/methanol gradient generated excellent separation and peak shapes [375]. The linear range was 25 µg/mL to 3 mg/mL with reported detection limits of 10 ng/mL.

Five caffeine metabolites (5-acetylamino-6-formylamino-3-methyluracil, 1-methyl- and 1,7-dimethylxanthine, 1-methyl- and 1,7-dimethyluric acid) were extracted from urine and separated on a C_{18} column ($\lambda = 280$ nm). In addition, nine other related compounds (e.g., 3- and 7-methyluric acid, 3- and 7-methyl-xanthine, and 1,3,7-trimethylxanthine) were also studied with these compounds using a 12/1/87 methanol/acetonitrile/water (0.05% acetic acid) mobile phase [376]. Concentration ranges of 4–100 µM were used. Peak shapes were very good and most peaks were well resolved. Elution was complete in <35 min.

Caffeine and 14 metabolites (e.g., paraxanthine, 5-acetylamino-6-amino-3-methyluracil, dimethyluric acids, methylxanthines, and methyluric acids) were extracted from urine and separated on a C_{18} column ($\lambda = 280$ nm) using a 45-min $92.5/7.5 \rightarrow 60/40$ water (0.05% acetic acid)/methanol gradient [377]. Linear ranges were reported as 0.5–20 µg/mL with detection limits of ~2 ng/mL.

Nine xanthines (xanthine, 1-, 7-, and 3-methylxanthine, isocaffeine [IS], theobromine, paraxanthine, theophylline, caffeine) were extracted from serum and urine separated on a 32°C C_8 column ($\lambda = 270$ nm) using a 20-min $90/10 \rightarrow 70/30$ water (50 mM ammonium acetate)/methanol gradient [378]. Theobromine and para-xanthine were incompletely resolved. A linear range from 0.2 to 20 µg/mL and detection limit of 2 µg/mL were reported.

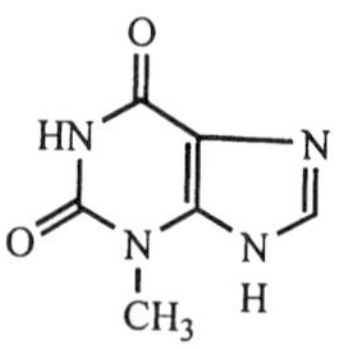

3-Methylxanthine

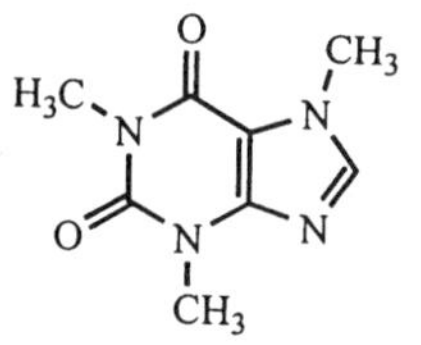

Caffeine

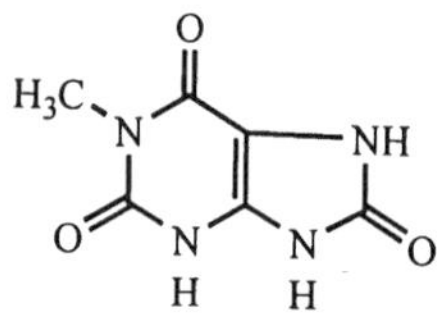

1-Methyluric acid

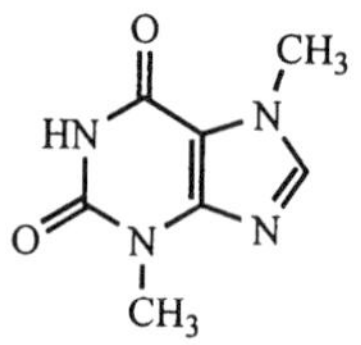

Theobromine

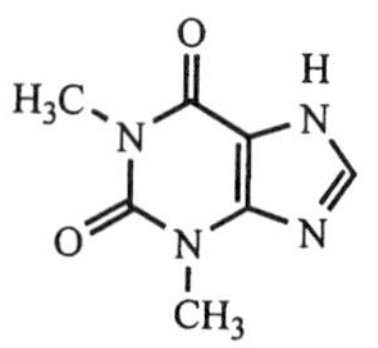

Theophylline

The purity of commercial saffron spice was assessed through the analysis of extracts on a 30°C C_{18} column ($\lambda = 250$ nm, 310 nm, and 440 nm) using a 50-min $80/20 \rightarrow 30/70$ water/methanol gradient. Along with the desired picricrocin, other metabolites were found (2,6,6-trimethyl-4-hydroxy-1-carboxaldehyde-1-cyclohexene, 3-gentiobiosil-kaempferol, α-crocin, safranal, and crocins 2, 3, 4, 5, and 6). Excellent peak shapes and resolution were obtained. In addition, though not shown, common adulterant k' values (tabulated for tartracin, Ponceau S, 4-nitroaniline, and methyl orange) were presented. Linear ranges of 0.25–1.5 mg/mL were reported [379].

4.5.5.4   *Other Compounds*

Egallic acid (gallic acid dimer) was extracted from fruit samples and isolated from other extracted components on 40°C C_{18} column ($\lambda = 360$ nm) using a 62.4/37.45/0.15 water/methanol/H_3PO_4 mobile phase. Elution was complete in 30 min due to the elution of other materials (egallic acid eluted at 10 min). A linear range of 0.1–100 µg/mL, detection limits of 0.015 µg/mL, and quantitation limits of 0.05 µg/mL were reported [380]. An alternative mobile phase of 82/12 water (5 mM KH_2PO_4 to pH 2.5)/acetonitrile was also used.

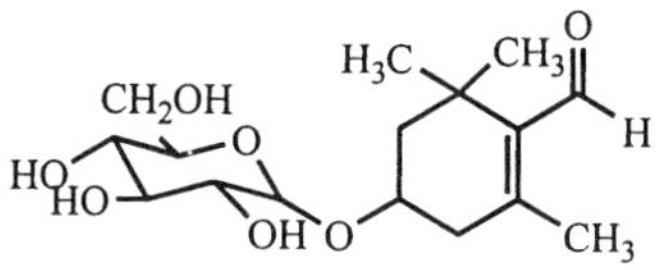

Picricrocin

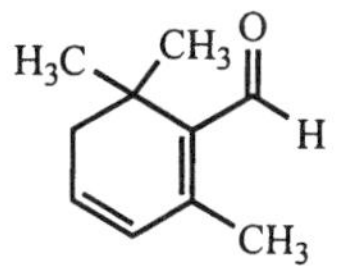

Safranal

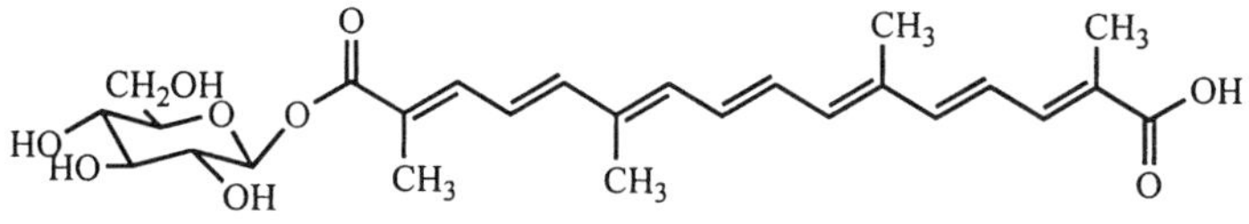

Crocin 4

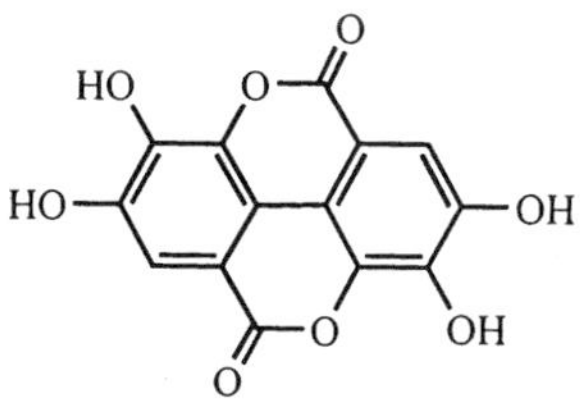

Egallic acid

Gil et al. [381] characterized pomegranate juice for its antioxidant activity by determining the levels of galloylglucose, punicalagin, hydrolysable tannin, delphinidin and cyanidin 3-glucosides, and egallic acid. A C_{18} column ($\lambda = 280$ nm) and a three-ramp 100/0 (hold 5 min) $\rightarrow$ 88/12 (at 15 min hold 3 min) $\rightarrow$ 54/46 (at 38 min) $\rightarrow$ 0/100 (at 43 min hold 2 min) water/methanol gradient were used. Reported standard levels ranged from 6 to 550 mg/L.

Five common phenolic antioxidants (propyl, octyl, and dodecyl gallate, butylated hydroxyanisole and butylated hydroxytoluene) used in bakery products were extracted and analyzed on a 21°C C_{18} column ($\lambda = 280$ nm). An 18-min 50/50 $\rightarrow$ 15/85 (at 3 min hold 15 min) (95/5 water/acetic acid)/(95/5 acetonitrile/acetic acid) gradient was used to generate baseline resolution and good peak shapes. A standard curve of 2–100 µg/mL was used [382]. It should be noted that 5% acetic acid generates a fairly aggressive mobile phase. Consideration should be given to the use of a precolumn or a lower level or an alternate (e.g., TFA) modifier.

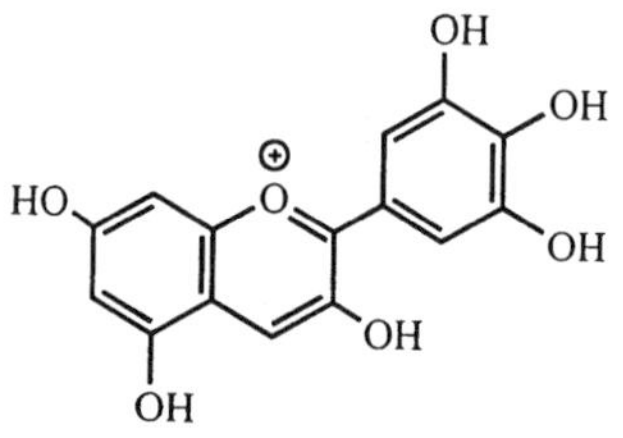

Delphinidin

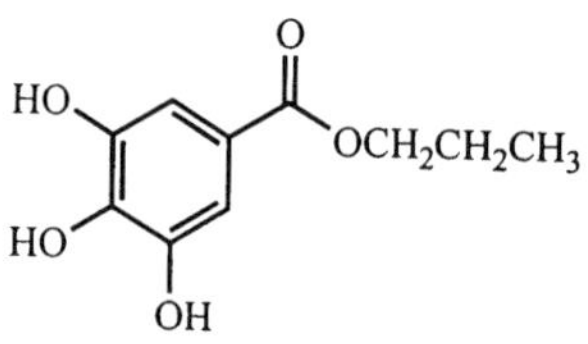

Propyl gallate

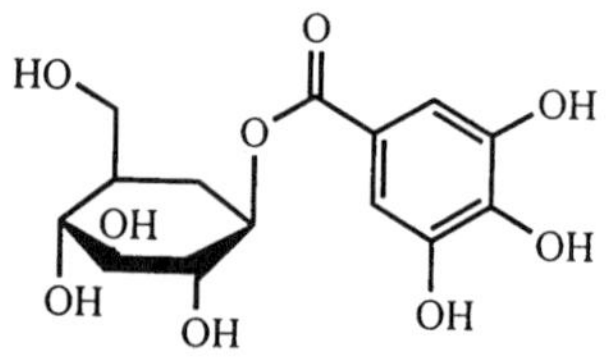

Galloylglucose

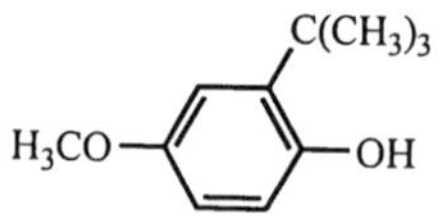

BHA; butylated hydroxyanisole

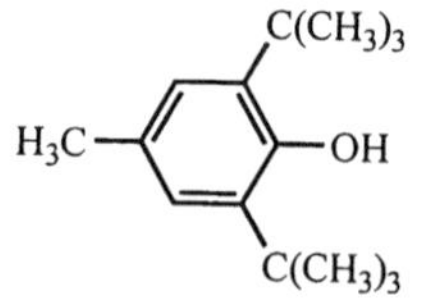

BHT; butylated hydroxytoluene

Thirteen phytoestrogens and their metabolites (daidzein and genistein and their 7-*O*-glucosides, secoisolariciresinol, dihydrodaidzein and dihydrogenistein, enterodiol, matairesinol, equol, enterolactone, *O*-desmethylangolensin, anhydrosecoisolariciresinol) were isolated from plasma and separated on a 37°C C_{18} column (two four-potential coulometric array detectors [ESA 580] with settings of 200 mV, 420 mV, 490 mV, 500 mV, 575 mV, 600 mV, 670 mV and 720 mV). A complex 62-min $80/20 \rightarrow 0/100$ (80/20 water [50 mM sodium acetate buffer at pH 5]/methanol)/(40/40/20 water [50 mM sodium acetate buffer at pH 5]/methanol/acetonitrile) gradient was used [383]. Peaks that eluted close to one another were deconvolved through the detector potential settings. Detection limits of ~1 ng/mL and linearities of up to 61 ng injected were reported. The authors also reported that samples made up in methanol did not negatively affect the separation if the injection volume was 10 μL or less.

Four crocins (1, 2, 4, and 4), crocetin, and *cis*-crocin were extracted from saffron and analyzed on a C_{18} column ($\lambda = 420$ nm) using a complex 30-min $60/40 \rightarrow 10/90$ water (1% acetic acid)/methanol gradient. Peak shapes were excellent and all peaks were baseline resolved [384]. The last analyte eluted at 23 min (the internal standard eluted at 28 min). Linear ranges were reported as 5–50 μg/mL to 30–300 μg/mL, whereas detection limits were in the 0.2–3 μg/mL range (analyte dependent).

Aristolochic acids I and II were extracted from *Coptidis rhizoma* and analyzed on a 40°C C_{18} column ($\lambda = 310$ nm) using a 15-min $60/40 \rightarrow 0/100$ water (1% acetic acid)/methanol gradient [385]. Peak shapes were excellent but the acids were only partially resolved from other extracted compounds. A linear range of 0–100 mg/L and quantitation limits of ~0.25 mg/L were reported.

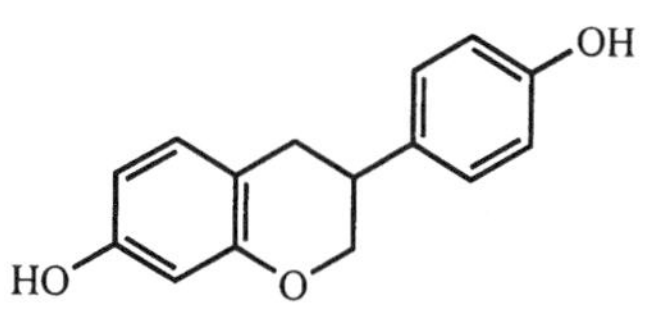

Equol

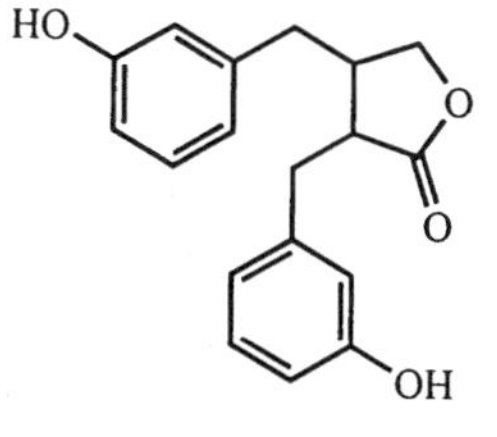

Enterolactone

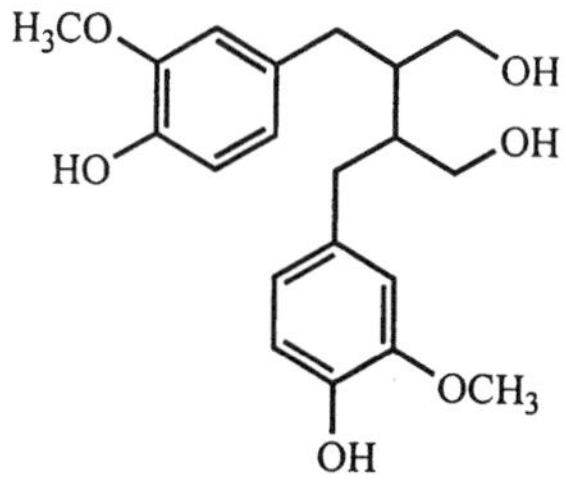

Secoisolariciresinol

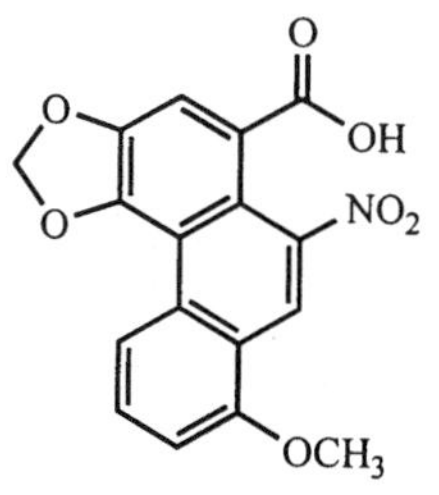

Aristolochic acid I

Racemic jasmonic acid (from *Petunia*) was reacted with a variety of amines (e.g., methylpropylamine, 2-phenylamine, (*S*)-3-methylbutylamine, tryptamine, (*S*)-2-methylbutylamine, tyramine) to form a series of natural-occurring *N*-jasmonoyl-amine conjugates. These products were separated on a C_{18} column ($\lambda = 210\,nm$) using a 50/50 methanol/water (0.2% acetic acid) mobile phase. 3-Methylbutylamine and 2-phenylethylamine conjugates co-eluted, but good resolution was obtained for the other conjugates and elution was complete in 60 min. The standard concentration was 10 mg/mL [386].

Zhang et al. [387] isolated five steroidal saponins from *Anemarrhena asphode-loides*: anemarrhenasaponins I and Ia. They were separated in 40 min on a C_{18} column with an 80/20 methanol/water mobile phase. Timosaponins B-I, B-II, and B-III were resolved on the same column but in 55 min with a 65/35 methanol/water mobile phase. The detector wavelength was not given, but compounds of similar structure, due to the lack of strongly absorbing chromophores, are usually run at $\lambda = 220\,nm$. Note, however, that the UV cutoff for methanol is typically around 205 nm and so work below 210 nm would not be recommended.

Three pigments (aloesaponarin II, 5-hydroxyaloesaponarin II, and their precursor 3,8-dihydroxy-1-methyl-9,10-anthraquinone-2-carboxylic acid) were isolated from *Streptomyces coelicolor* on a deactivated base C_{18} column ($\lambda = 430\,nm$, ex; 530 nm, em) using a 40-min 100/0 (hold 2 min) $\rightarrow$ 30/70 (at 10 min) $\rightarrow$ 0/100 (at 40 min) (60/40 methanol/water [0.5% acetic acid])/(90/10 methanol/acetonitrile) gradient [388]. Baseline resolution and excellent peaks shapes were obtained. It should be noted that the colors of these compounds were acid/base sensitive so that keeping the mobile phase solution pH stable might be important if quantitative work is attempted. Under these conditions, the "pH" will change dramatically throughout the analysis.

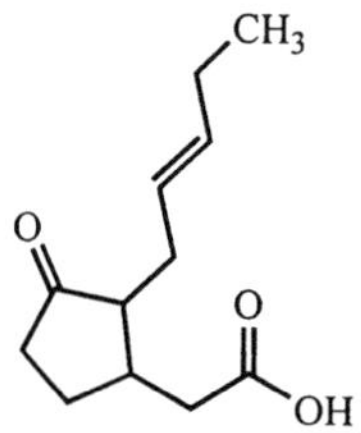

Jasmonic acid

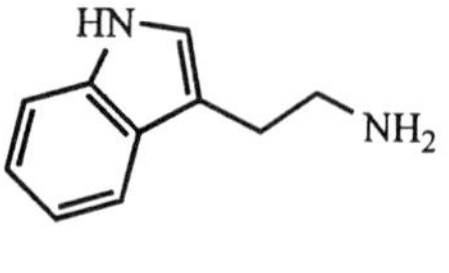

Tryptamine

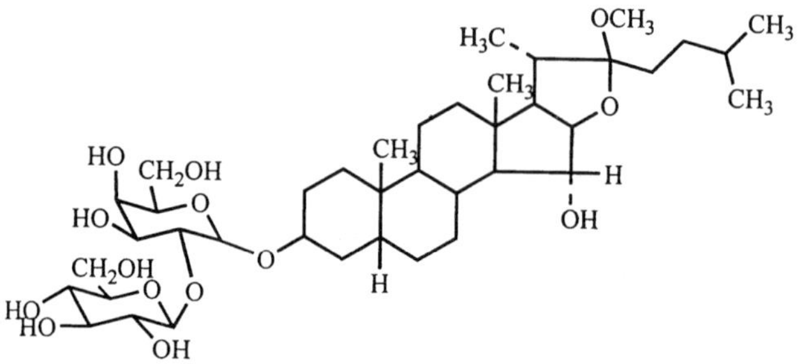

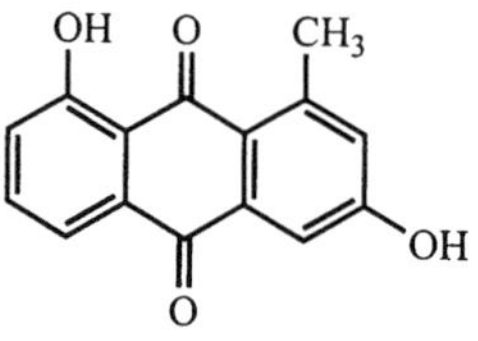

Aloesaponarin II

Anemarrhenasaponin Ia

Four procyanidins (dimers B2, B5, B4, and trimer C1) were isolated from the leaves and flowers of *Crataegus* spp. and resolved on a C_{18} column (photodiode array detector, $\lambda = 220\,nm$, $260\,nm$, $280\,nm$, $290\,nm$) with a complex 60-min $82/18 \rightarrow 0/100$ water (0.5% phosphoric acid)/methanol gradient [389]. The last peak of interest eluted at 40 min. The rest of the gradient was run to remove other extracted components. During the gradient, procyanidin C1 was incompletely resolved from epicatechin (an extracted flavonoid). Standard concentrations ranged from ~5 to 150 µg/mL (analyte dependent).

This method [390] describes the characterization of vanilla extracts, synthetic precursors (e.g., ethyl vanillin) and adulterants (e.g., coumarin) on a C_{18} column ($\lambda = 275\,nm$). In addition to vanillin and the two compounds listed above, 12 other characterization compounds included anisic, syringic, vanillic and *p*-hydroxybenzoic acids, piperonal, guaiacol, eugenol, *p*-hydroxybenzyl alcohol, *p*-anisaldehyde, *p*-hydroxybenzaldehyde, caffeine, and syringaldehyde. Excellent peak shapes and baseline resolution for all peaks were achieved using a two-ramp $100/0/0 \rightarrow 70/30/0.2$ (at 40 min hold 10 min) $\rightarrow 0/100/0.2$ (at 60 min) water (0.05% acetic acid)/methanol/THF gradient. A linear range of 0.3–150 ppm and a detection limit of 0.2 mg/L were reported.

Liu et al. [391] extracted nine major active components from *Hypericum perforatum* (rutin, hyperoside, isoquercitrin, quercetin, quercitrin, psuedohypericin,

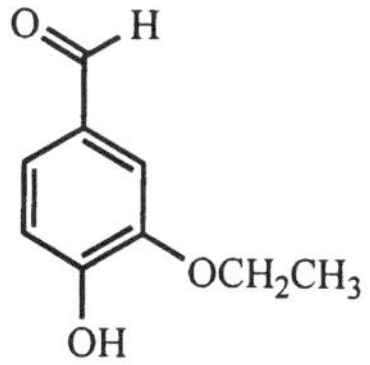

Ethylvanillin

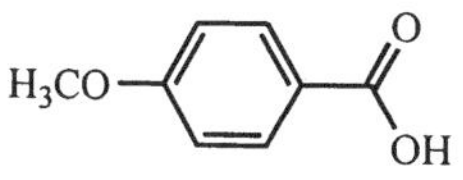

p-Anisic acid

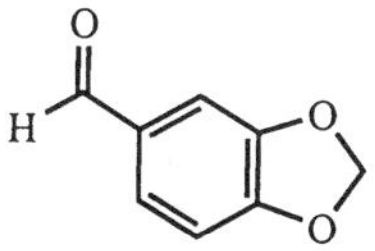

Piperonal

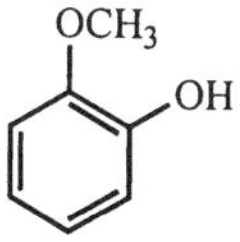

Guaiacol

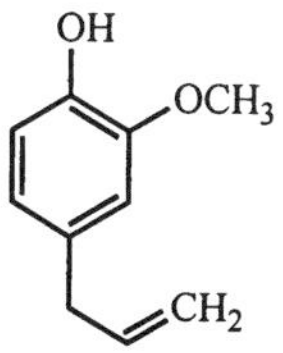

Eugenol

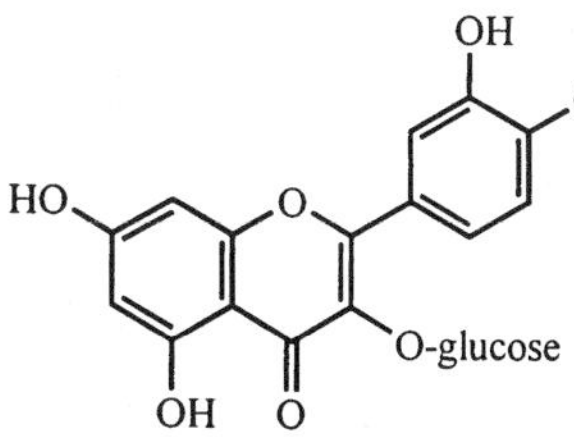

Isoquercitrin

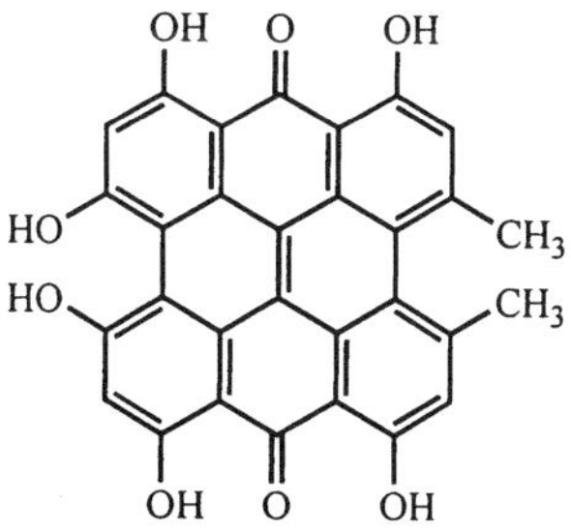

Hypericin

hypericin, hyperforin, adhyperforin) and separated them in 80 min using a C_{18} column ($\lambda = 257$ nm, 584 nm, 276 nm) and a complex $90/10/0.5 \rightarrow 0/100/0.5$ water/(13/7 methanol/acetonitrile)/TFA gradient. Resolution between all compounds was not achieved. Working standard concentrations varied in the range from 0.1 to 4 mg/mL (analyte dependent).

Protopseudohypericin, pseudohypericin, protohypericin, and hypericin were extracted from *Hypericum perforatum* blossoms and baseline resolved on a C_{18} column ($\lambda = 590$ nm) using a 12-min $70/30 \rightarrow 90/10$ (at 8 min hold 4 min) (5/4 methanol/acetonitrile)/(water [0.1 M triethylammonium acetate]) gradient [392]. Excellent peaks shapes were obtained. The total injected mass of the four compounds was approximately 12 µg.

Notopterol and isoimperatorin, two coumarins extracted from *Notopterygium forbessi* (Chinese Qianghuo) and were used for species identification. The

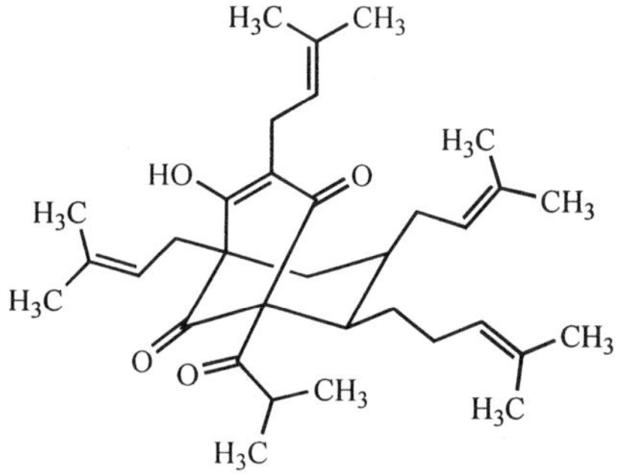

Hyperforin

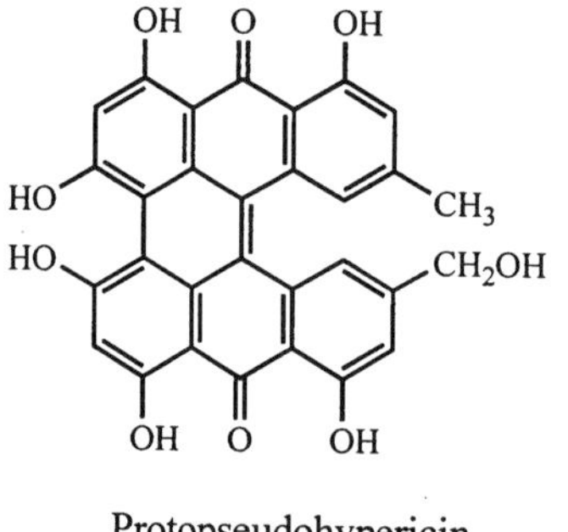

Protopseudohypericin

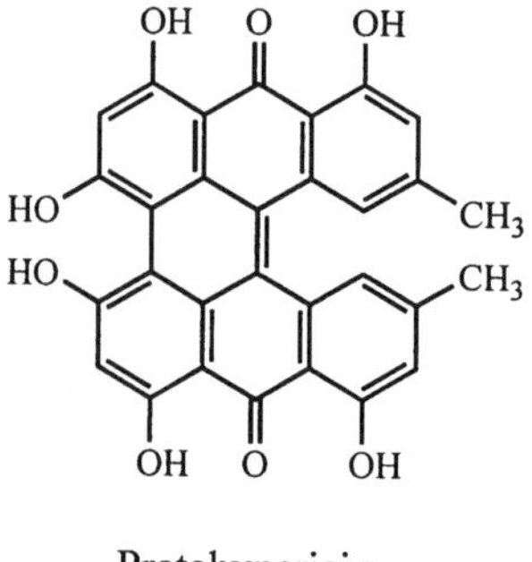

Protohypericin

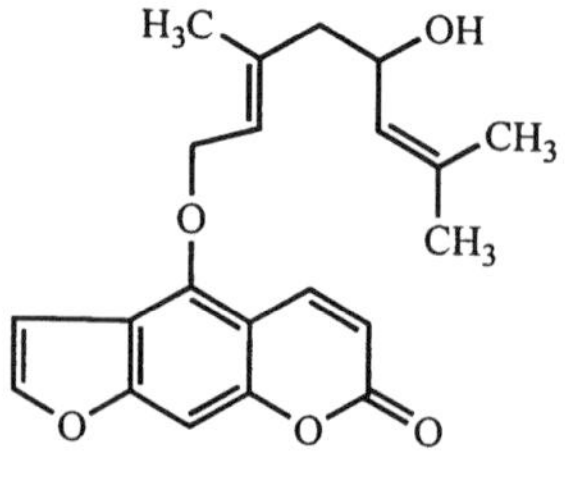

Notopterol

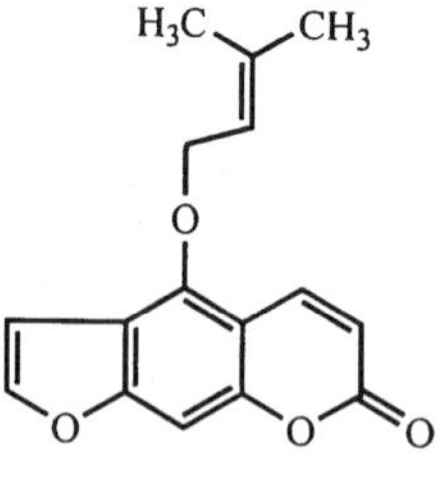

Isoimperatorin

compounds were baseline resolved on a 40°C C_{18} column ($\lambda = 254$ nm) using a 40/30/30 water/methanol/acetonitrile mobile phase [393]. Elution was complete in 20 min.

Boswellia resin was extracted and analyzed for four boswellic acids (acetyl-11-keto-β-, acetyl-11-hydroxy-β-, acetyl-11-methoxy-β-, 3-O-acetyl-9,11-dihydro-β-boswellic acid) on a 33°C C_{18} column ($\lambda = 210$ nm). Elution was complete in 18 min using a 90/10 methanol/water with 0.01% TFA. The keto and hydroxy forms were not completely resolved under these conditions [394].

Two constituents of karanja oil (karanjin, pongamol) were isolated and separated on a C_{18} column ($\lambda = 300$ nm and 350 nm) using an 85/13.5/1.5 methanol/water/acetic acid mobile phase [395]. Elution was complete in 20 min and peaks were baseline resolved. Calibration curves from 2 to 10 µg/mL and detection limits of 0.1 µg/mL were reported. The authors noted that an 85/15 methanol/water mobile phase generated a poor peak shape for pongamol, a β-diketone. This is common for these types of compounds since they are very interactive as complexing agents and can produce strong hydrogen bond interactions with residual surface silanol groups. The addition of acid generally prevents these interactions and improves peak shape.

Homogentisic acid was isolated from strawberry-tree honey and analyzed in <15 min on a C_{18} column ($\lambda = 292$ nm). A 10/90 methanol/water (10 mM H_2SO_4) mobile phase generated good peak shape. A linear range of 50–550 mg/kg was reported [396].

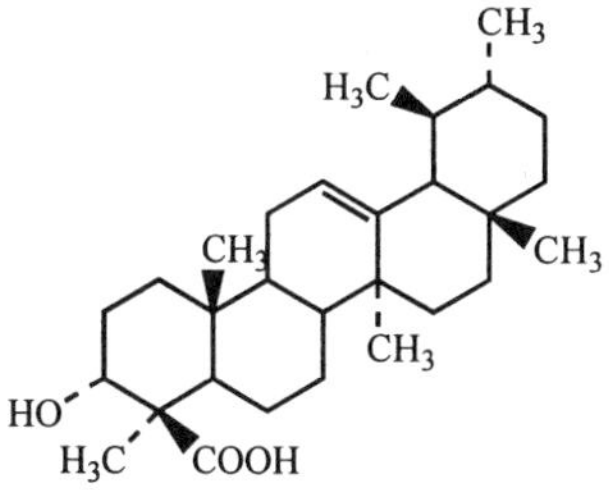

β-Boswellic acid

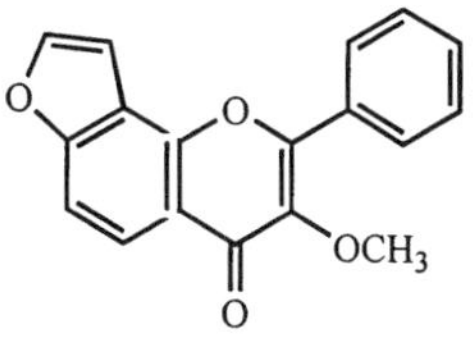

Karanjin

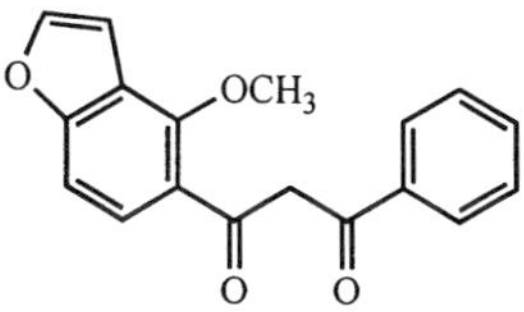

Pongamol

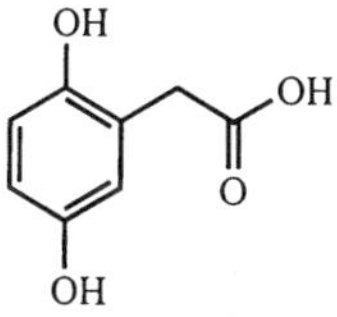

Homogentisic acid

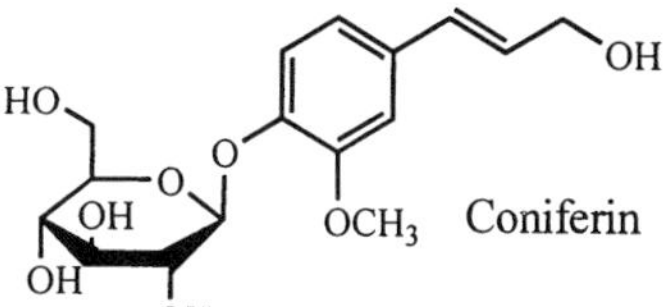

Coniferin

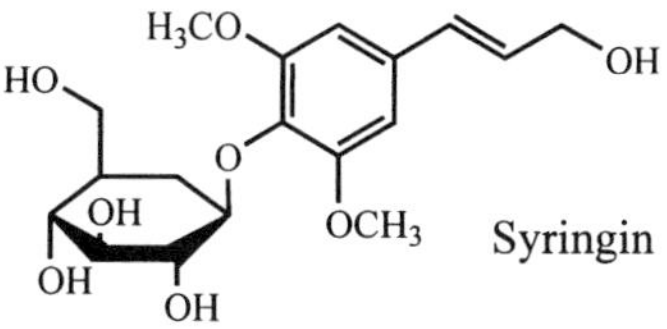

Syringin

Three phenylpropanoid glucosides (coniferin, kalopanaxin, syringin) were used to identify subspecies of *Viscum album*. These were extracted from powdered leaves and stems and separated on a C_{18} column ($\lambda = 264$ nm). Baseline resolution and complete elution were achieved in 25 min using a 73.5/20/6.5 water/methanol/ water (0.1 M sodium acetate) mobile phase [397]. Calibration curves from 300 to 750 ng/mL were used.

Powdered samples of *Radix astragali* were extracted and analyzed for astragaloside IV concentration [398]. The analyte was derivatized with benzoyl chloride and quantitated on a C_{18} ($\lambda = 230$ nm) column using a 90/4/6/0.2 methanol/THF/ water/triethylamine mobile phase. Elution occurred at 25 min and a linear range of 4–80 µg/mL was reported along with a detection limit of 2 µg/mL.

Abietic acid and dehydroabietic acid were extracted from ointments and separated in <7 min on a C_{18} column ($\lambda = 225$ nm, ex; 285 nm, em) using an 87/13/0.02 methanol/water/H_3PO_4 mobile phase [399]. The authors noted that detection at 200 nm or 239 nm was ineffective due to a large number of co-eluting peaks. Fluorescence offered excellent specificity. Peak shapes were excellent and baseline resolution was obtained. The linear range was found to be 0.05–5 µg/mL and detection limits were 0.05 µg/mL (S/N = 3).

Trichothecenes (nivalenol, deoxy- and 15-acetyldeoxynivalenol) were extracted from bananas and analyzed on a C_{18} column ($\lambda = 225$ nm). A 35/65 methanol/ water mobile phase generated a 10 min baseline-resolved separation [400]. Standards were run from 2 to 20 µg/g and detection limits of 5 ng/g were reported. Similarly, in the same study, α- and β-zearalenol and zearalenone were well resolved using a 65/35 methanol/water mobile phase. Standard levels were 2–20 µg/g and detection limits of 9 ng/g were reported.

He et al. [401] extracted six lignans (schisandrin, schisanhenol, deoxy- and (γ-schisandrin, schisantherin A and B) from *Schisandra chinensis* and analyzed them on a C_{18} column ($\lambda = 225$ nm) using a 15-min 60/40 → 100/0 methanol/water gradient. Baseline resolution was achieved, except for schisantherin A and B. Nine additional peaks were tentatively identified by positive ion electrospray MS. Reference standards were run at 150 µg/mL and were readily detected.

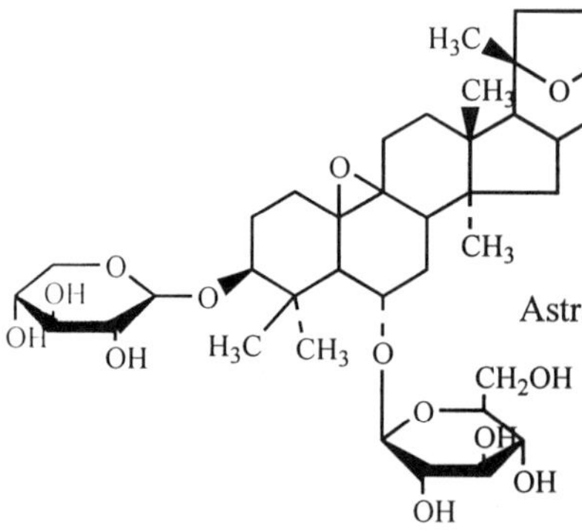

Astragaloside IV

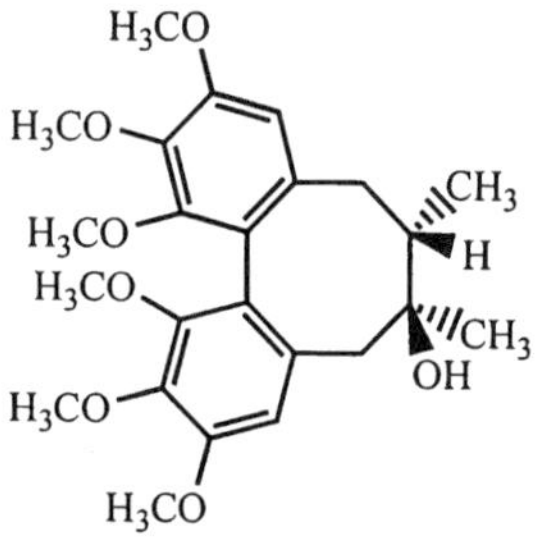

Schisandrin

Six kava lactones (methysticin, dihydromethysticin, kavain, dihydrokavain, yangonin, demethoxyyangonin) were isolated from *Piper methysticum* roots and analyzed on a 40°C base deactivated C_{18} column ($\lambda = 220$ nm) using a 20/20/60/0.1 methanol/acetonitrile/water/acetic acid mobile phase [402]. Separation with baseline resolution was achieved in 34 min. The authors noted that C_8, phenyl, or generic C_{18} columns did not fully resolve all peaks. For the described separation, acetic acid was reported as leading to significantly improved resolution and sensitivity. Calibration standards were run from 5 to 500 µg/mL. Detection limits of 1 µg/mL (S/N = 3) were reported.

Ginkgo biloba leaves were extracted and characterized by their phenolic acid profile (protocatechuic, *p*-hydroxybenzoic, vanillic, caffeic, isovanillic, *cis*- and *trans-p*-coumaric, *cis*- and *trans*-ferulic, sinapic acid). A C_{18} column ($\lambda = 254$ nm) and a 73/35 water/methanol mobile phase were used to elute all compounds in 30 min. In general, peak shapes were good, as was overall resolution [403].

A set of nine polyphenols found in wines (e.g., catechin, *cis*- and *trans*-polydatin, rutin, *cis*- and *trans*-resveratrol) were baseline resolved on a C_{18} column ($\lambda = 280$ nm or 305 nm) using a 40-min $80/15/5 \rightarrow 75/20/5$ (at 5 min) $\rightarrow 50/45/5$ (at 30 min hold 10 min) water/methanol/water (33% acetic acid) gradient [404]. Peak shapes were excellent and linear ranges varying from 10–200 mg/L to 0.5–10 mg/mL, depending on the analyte, were reported.

Daidzein and genistein and their metabolites, equol and *O*-desmethylangolensin, were extracted from milk and analyzed on a C_{18} column ($\lambda = 260$ nm) using a complex 25-min $95/5 \rightarrow 5/95$ (90/10 water/acetic acid)/(10/5/1 methanol/acetonitrile/dichloromethane) gradient [405]. Excellent peak shapes and good resolution were achieved. Calibration curves from 0 to 150 µM were generated with detection limits of 25–165 nM (analyte dependent) reported.

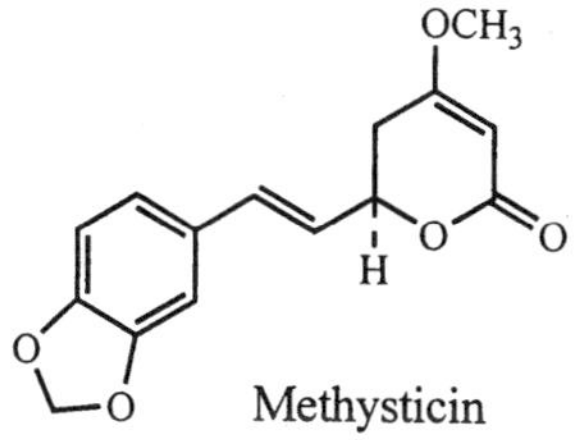

Methysticin

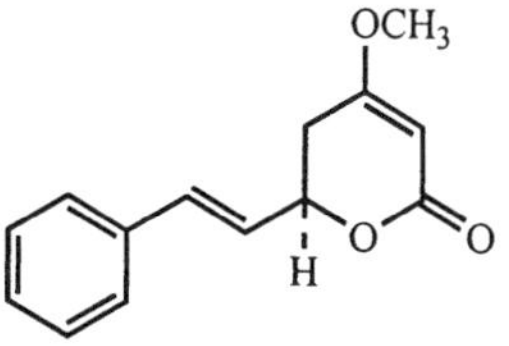

Kavain

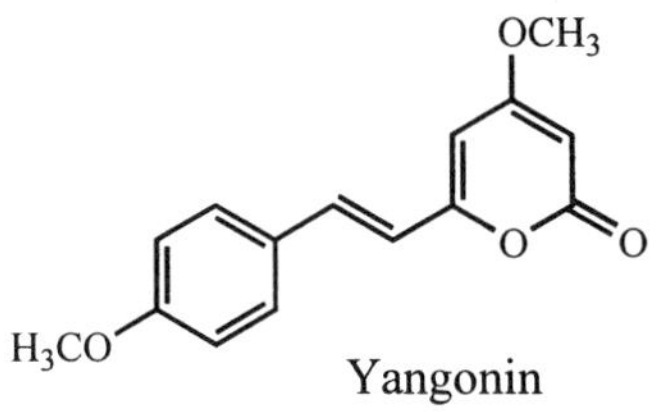

Yangonin

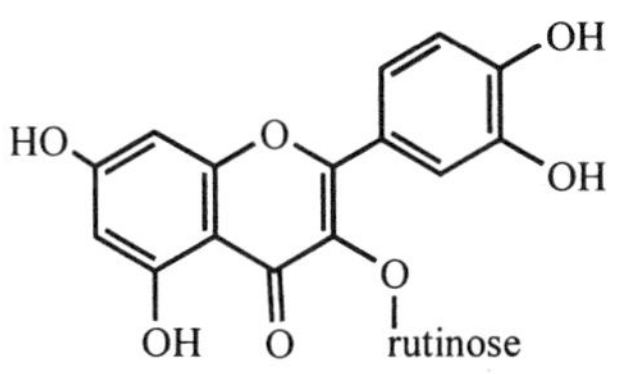

Rutin

Peroxisomicine A1 was extracted from *Karwinskia* and analyzed on a C_{18} column ($\lambda = 269$ nm and 410 nm) using a 74/26 methanol/water (made as 795 mL 0.1 M citric acid with 205 mL Na_2HPO_4 to pH 3) mobile phase [406]. The analyte eluted in 14 min, well resolved from other matrix extracts. A concentration range of 0.126–12.6 μg/mL was used and quantitation limits of 30 ng/mL (at 410 nm with S/N = 10) were reported.

Ten iridoid glucosides (monotropein, scandoside, asperulosidic acid and four analogs, geniposidic acid, asperuloside, daphylloside) were extracted from *Gallium* species and studied using a C_{18} column ($\lambda = 233$ nm). The choice of mobile phase, 85/15 water/methanol, led to co-elution of five peaks, all retained at about 5 min [407]. The latest-eluting peak had a retention time of 45 min. This method could be improved through the use of an acidic mobile phase modifier (to protonate the acids that elute near 5 min) and a shallow gradient starting with a weaker initial mobile phase (e.g., 95/5/0.2 water/methanol/TFA). This would increase retention of the polar species and decrease retention of the late-eluting peak.

Five hydroxyanthroquinones (rhein, emodin, aloe-emodin, chrysophanol, physcion) were isolated from *Rhuem officinale* and separated on a 40°C C_{18} column ($\lambda = 410$ nm). Good peak shape, baseline resolution, and complete elution in 17 min were achieved with an 80/20 methanol/water (0.1% H_3PO_4) mobile phase [408].

Four naphthaquinones were extracted from *Arnebia densiflora* roots and analyzed on a C_{18} column ($\lambda = 520$ nm). A 95/5/0.1 methanol/water/formic acid mobile phase generated a 10-min separation [409]. Teracrylalkannin and β,β-dimethylacrylalkannin were fully resolved but the α-methyl-n-butylalkannin co-eluted with isovalerylalkannin. A gradient starting with a considerably weaker initial mobile phase should effectively separate all peaks within a reasonable timeframe. Linearity was reported over the range of 25–200 μg/mL.

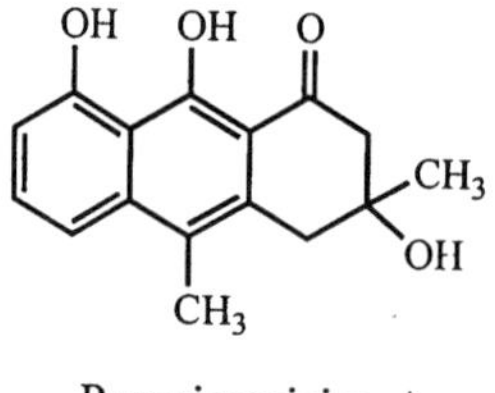

Peroxisomicine A$_1$

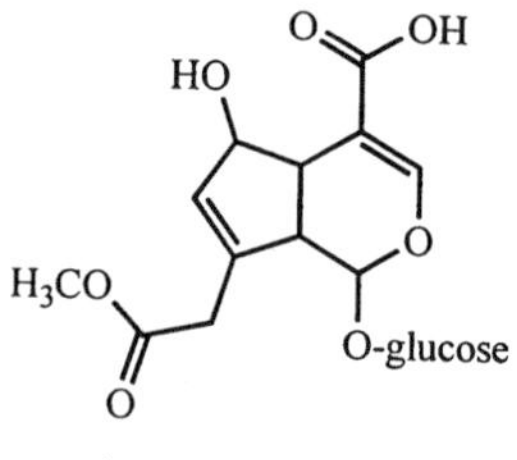

Asperulosidic acid

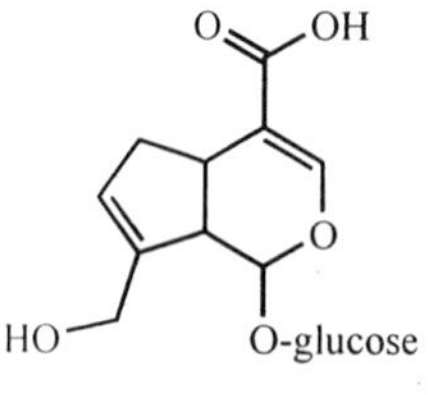

Geniposidic acid

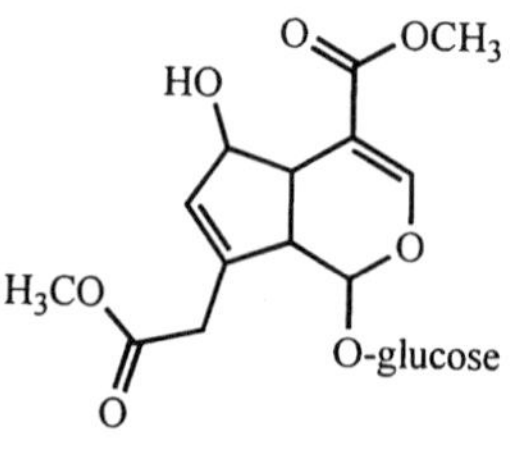

Daphylloside

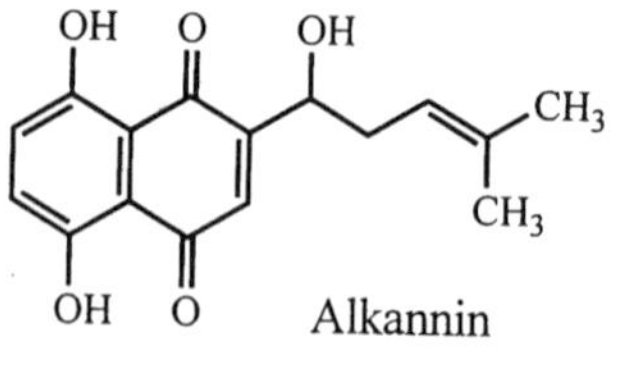

Alkannin

In a very nice study, a set of 20 compounds were extracted from the rhizome of *Angelica sinensis* (umbelliferone). Twelve were identified as phthalides (e.g., senkyunolide, butylphthalide, (Z)-ligustilide, (Z,Z'-6.8', 7.3'-diligustilide, angelicide, levistolide). All were separated on a 45°C C_{18} column ($\lambda = 270$ nm) using a 40-min 63/35 → 0/100 water (0.25% acetic acid)/methanol gradient [410]. Positive ion elestrospray MS was used for identification (N_2 drying gas $T = 360°$ at 40 mL/min, nebulizing pressure 80 psi).

Agruello et al. [411] extracted six glucosinolates and thioglucosides (sinigrin, progoitrin, glucoiberin, glucoraphanin, glucotropaeolin, glucobrassin) from cabbage. A 100/0 → 40/60 (at 17 min) → 20/80 (at 20 min hold 3 min) water (50 mM NaH_2PO_4 buffer at pH 3.2)/methanol (50 mM NaH_2PO_4 at pH 3.2 [must be apparent pH here!]) gradient on a C_{18} column ($\lambda = 228$ nm) generated enough separation between analytes and co-extracted compounds to allow for identification. The authors noted that a low pH was necessary to protonate the surface silanol groups and thereby increase retention. Increased ionic strength of the mobile phase (i.e., increase in buffer concentration) led to increased retention.

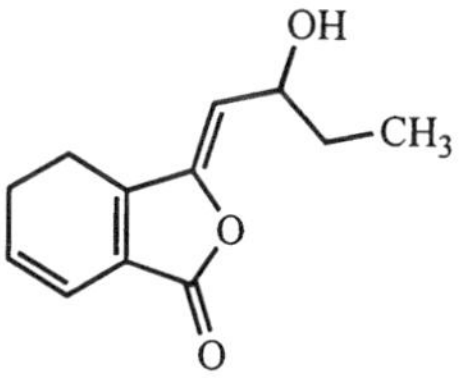

Senkyunolide F

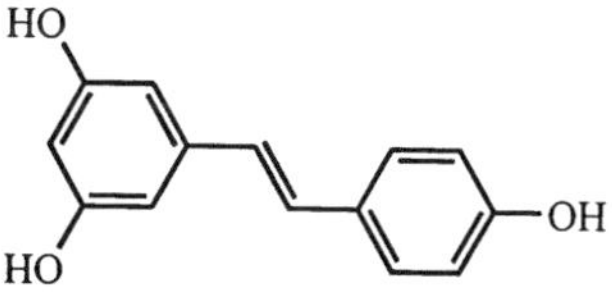

trans-Resveratrol

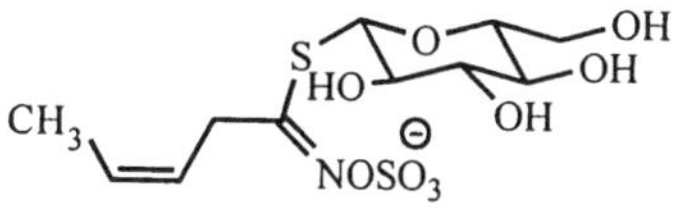

Sinigrin

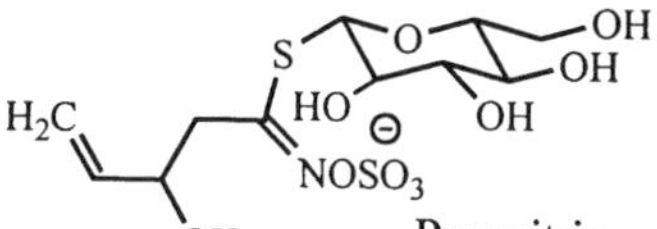

Progoitrin

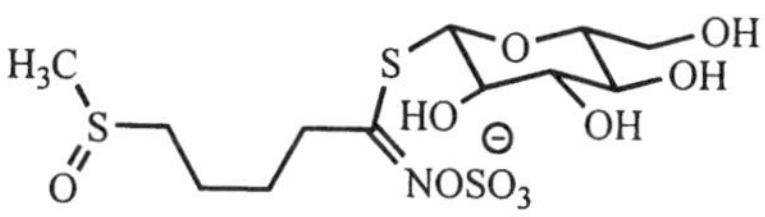

Glucoraphanin

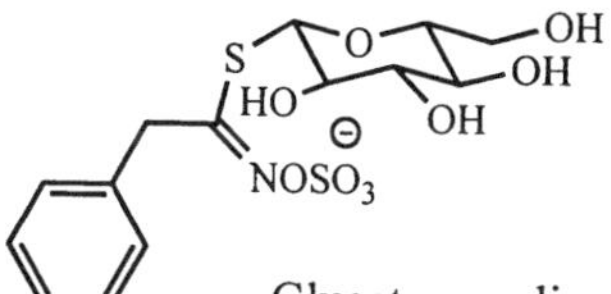

Glucotropaeolin

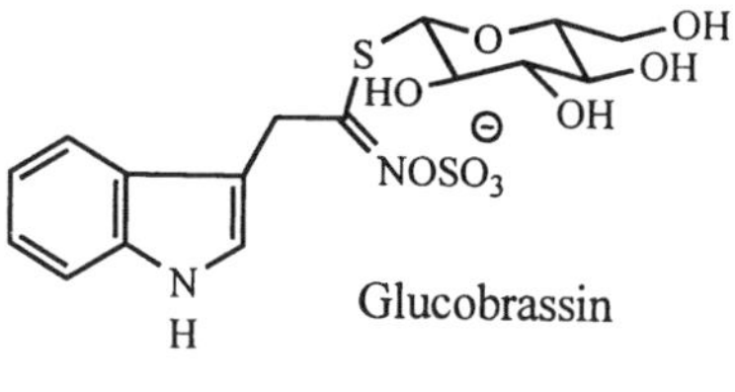

Glucobrassin

The determination of 18α- and 18β-glycyrrhizin isolated from plasma was achieved on a 50°C C_{18} column ($\lambda = 254$ nm). A 45/55/0.5 water/methanol/60% 60% $HClO_4$ mobile phase generated good separation in 12 min [412]. A linear range of 1–100 μg/mL was reported.

Aerial material from *Centella asiatica* was extracted and characterized for asiaticoside content [413]. A 70/30 methanol/water (1% TFA) mobile phase and a C_{18} column ($\lambda = 220$ nm) provided separation of the analyte from numerous other compounds. Elution was complete in <10 min. A linear range for asiaticoside was reported as 1–30 μg injected.

Thirteen phenolic compounds (e.g., anthrones such as aloins A and B, 8-O-methyl-7-hydroxyaloins A and B, 10-hydroxyaloin A; phenyl pyrones such as aloenin; chromones such as aloesin, isoaloeresin D) were extracted from *Aloe vera* and separated on a C_{18} column ($\lambda = 293$ nm). A complex 60-min $30/70 \rightarrow 70/30$ methanol/water gradient generated excellent peaks shapes and resolution. Calibration curves for all 13 compounds were generated from 0.002–2 mg/mL [414].

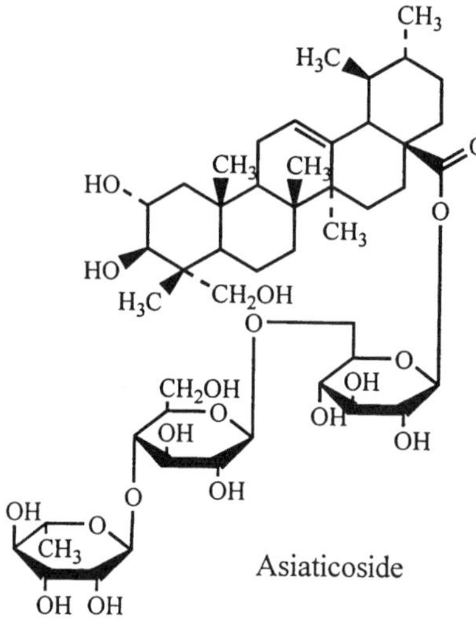

Asiaticoside

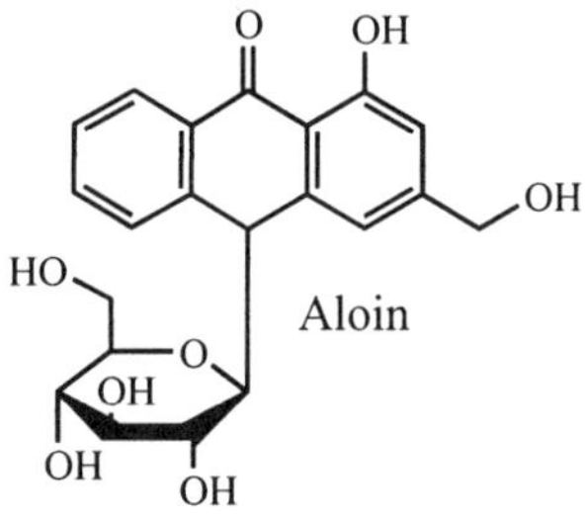

Aloin

Chromones

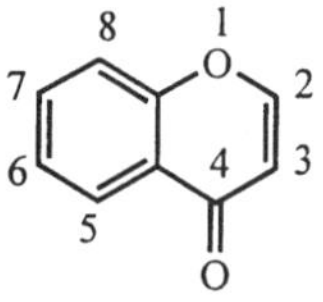

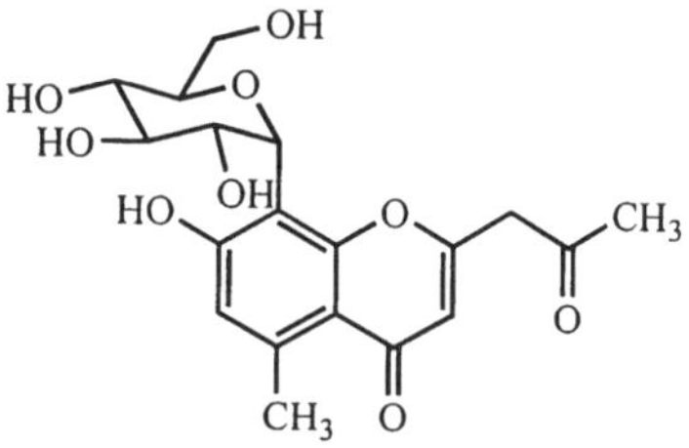

Aloesin

Various fungi from the genus *Phoma* were extracted with ethyl acetate and characterized by their chromatographic profiles using a C_{18} column (photodiode array detector from $\lambda = 200–400$ nm, monitor at $\lambda = 230$ nm) with a 60-min $100/0 \rightarrow 0/100$ water/methanol mobile phase gradient. Marine versus terrestrial strains were differentiated. Two compounds positively identified by thermospray MS were mellein and hydroxymellein [415].

Various classes of steroidal compounds have been effectively analyzed [416]. Kraan et al. [417] separated the 6α- and 6β-hydroxylated derivatives of corticosterone, 11-dehydrocorticosterone, and 11-deoxycortisol isocratically on a C_{18} column ($\lambda = 254$ nm) using a 35/65 methanol/water (phosphoric acid to pH 3.4) mobile phase. Varin et al. [418] extracted spironolactone and four isomers from plasma and urine and separated them in 8 min on a C_{18} column ($\lambda = 254$ nm) using a 65/35 methanol/water (phosphate buffer at pH 3.4) mobile phase. Calibration curves of 6–400 µg/mL in plasma and 30–2000 ng/mL in urine were used.

The metabolites of estradiol (estriol, and 2-hydroxy-, 4-hydroxy-, 6α-hydroxy-, and 15α-hydroxyestradiol) were well resolved on a 40°C C_{18} column (electrochemical detector, 0.8 V vs. Hg/Hg_2Cl_2) using a 60/40 methanol/water (50 mM ammonium phosphate buffer at pH 2.5) mobile phase. Elution was complete in 35 min [419]. Calibration curves of 0–40 pmol injected were used. Detection limits ranged from 10–50 fmol injected (S/N = 3, analyte dependent).

An outstanding separation of eight cholesteryl esters was achieved on a C_{18} column ($\lambda = 210$ nm) using a 50/50 IPA/acetonitrile mobile phase [420]. Peak shapes were excellent and elution was complete in 70 min. Detection limits of 15–60 nmol were reported.

A set of five benzoate esters of cholesterol oxides (7-keto-, epoxy-, 7-hydroxy-, 25-hydroxycholesterol and cholestanetriol) were baseline resolved on a C_{18} column ($\lambda = 230$ nm) using an 85/15 IPA/water mobile phase. Elution was complete in 20 min [421]. Detection limits of 500 ng injected were reported and a linear working range of 0.8–10 µg injected was generated.

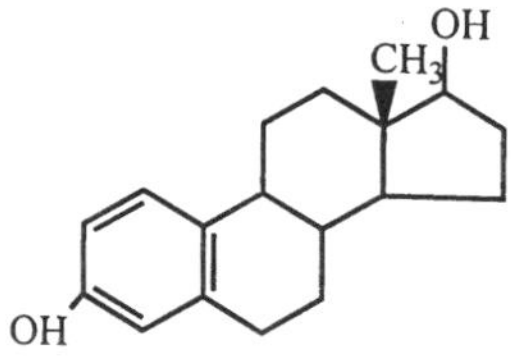

Estradiol

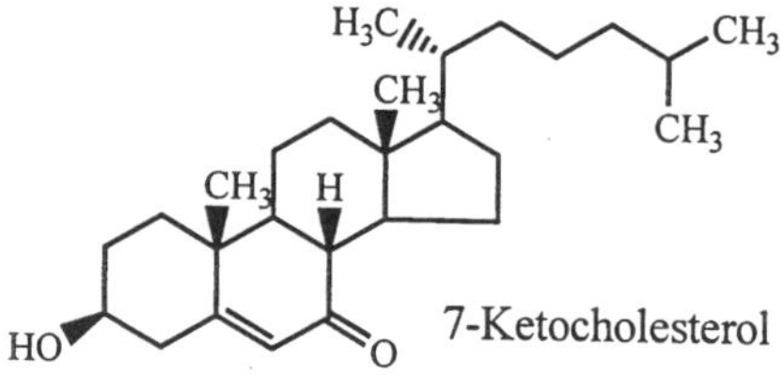

7-Ketocholesterol

Five 20-oxosteroids (pregnenolone, 3β-hydroxy-5α-, 3β-hydrpxy-5β-, 3α-hydropy-5α- and 3-hydroxy-5β-pregnan-20-one) and their sulfates were separated as their 4-(N,N'-dimethylaminosulfonyl)-7-hydrazino-2,1,3-benzoxadiazole derivatives on a C_{18} column ($\lambda = 450$ nm, ex; 550 nm, em). The mobile phase was a 3/1 methanol/water (4 mM γ-cyclodextrin). Good resolution was obtained and elution was complete in 25 min [422].

Twenty-five ecdysteroids, derived from 20-hydroxyecdysone, were studied in both normal-phase (silica column) and reversed-phase (C_{18} column) modes. IPA/dichloromethane/water (30/125/2), IPA/cyclohexane/water (40/400/3), or IPA/iso-octane/water (30/100/2) was selected as NP mobile phase. Here is an instance where the water concentration in a immiscible matrix (e.g., iso-octane) is increased due to the presence of a mutually miscible solvent, IPA. Methanol/water (50/50), ethanol/water (30/70), or IPA/water (18/72), all containing 0.1% TFA, were studied as reversed-phase mobile phases [423]. Methanol proved particularly effective in the RP separation of analytes that varied by the degree of unsaturation (i.e., number of double bonds). IPA, as a RP solvent, was superior in resolving 5α–5β pairs. IPA was extremely effective in the separation of 20-hydroxyecdysone and polypodine B mixtures. These results are not surprising since ecdysones are polyhydroxylated.

Eight proanthocyanidins (catechin, gallocatechin, epicatechin, epigallocatechin and their phloroglucinol adducts) were separated on a C_{18} column ($\lambda = 280$ nm and photodiode array detector) using 1 mg/mL standard solutions [424]. Analysis was complete in 50 min using a $0/100 \rightarrow 60/40$ methanol/water (1% acetic acid) gradient. Standards of ~ 10 µg injected were easily detected. Peak shapes were

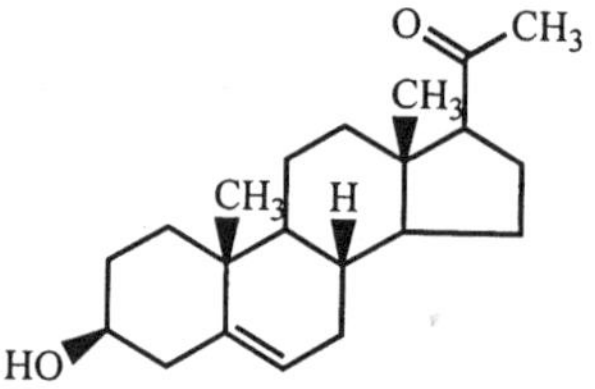

Pregnenolone

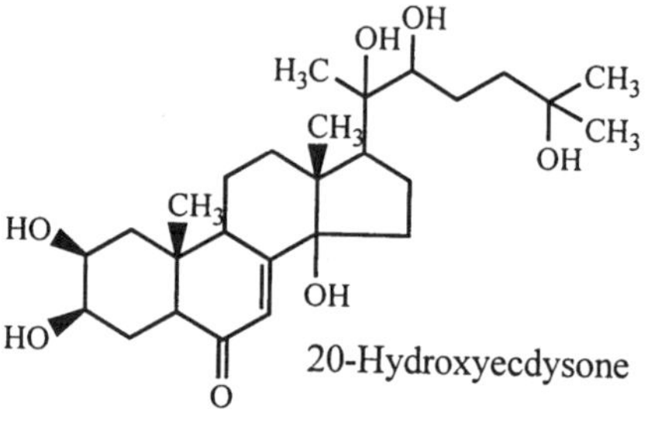

20-Hydroxyecdysone

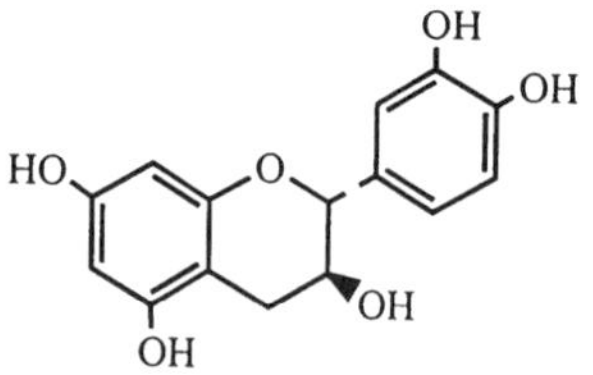

Catechin

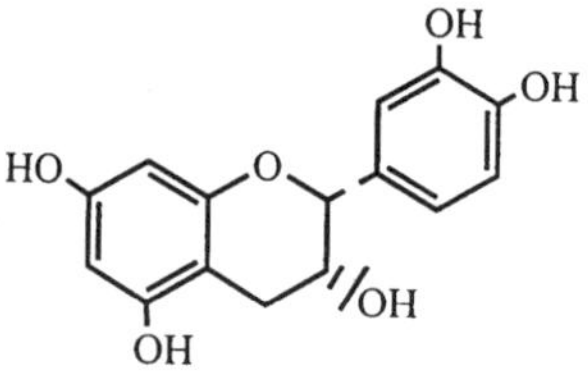

Epicatechin

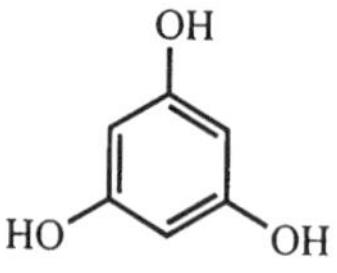

Phloroglucinol

excellent but separation of epigallocatechin from catechin-4-phloroglucinol was not complete.

4.5.6 Analytes Derived from Oils and Fats

Triglycerides have been separated in various fats and oils using both reversed-phase and normal-phase systems [425]. Alcohols have been used effectively as the low-volume mobile phase modifier for NP work (the reader is referred to Chapters 5 and 6 for details).

The hydroperoxide by-products of the reaction of unsaturated triacylglycerol (TAG) (e.g., stearoyl : oleoyl : stearoyl, palmitoyl : palmitoyl : oleoyl) with t-butyl hydroperoxide were studied on a C_{18} column (ELSD, drift tube 85°C) using a 30-min $80/20 \rightarrow 20/80$ methanol/IPA gradient [426]. Up to 16 peaks were generated in the product chromatogram. Electrospray MS was used to identify some of the peaks. Common products included TAG hydroperoxide, epoxide, and combined peroxide/epoxide products.

The products of the action of sphingosine kinase on sphingosine 1-phosphate isolated from serum and plasma were derivatized with o-phthalaldehyde and separated on a C_{18} column ($\lambda = 340$ nm, ex; 455 nm, em). Five products (sphinga-nine, its 4-D-hydroxy and 4-D-hydroxy-1-phosphate analogs; sphingosine and its 1-phosphate analog) were well resolved in 40 min using a $83/16/1$ methanol/water (10 mM KH_2PO_4 buffer to pH 5.5)/water (1 M tetrabutylammonium dihydrogen-phosphate) mobile phase [427]. Linearity for sphingosine 1-phosphate over the range from 5 to 500 pmol injected was reported.

Twelve phosphatidylcholine molecular species (e.g., 18 : 2/20 : 4, 16 : 0/20 : 4, 18 : 0/20 : 4, 18 : 0/18 : 2) were obtained from lever microsomes and separated on two 250×4.6 mm C_{18} columns in series (ELSD, N_2 nebulizer at 1.8 L/min, drift tube $T = 100$°C). Good resolution was achieved with 45-min $40/48/12/0 \rightarrow 40/42/8/8$ (at 15 min hold 30 min) hexane/IPA dichloromethane/water gradient [428].

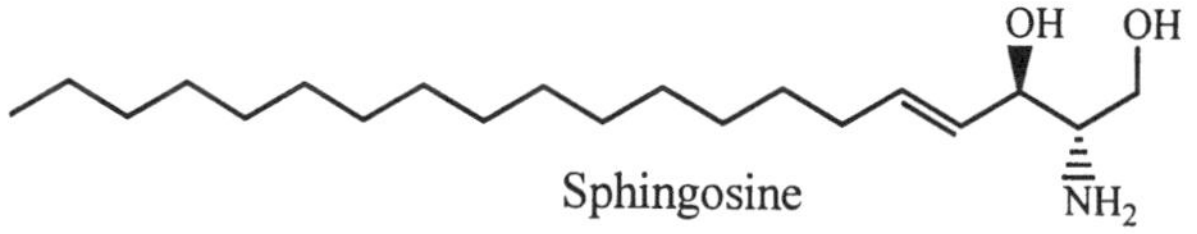

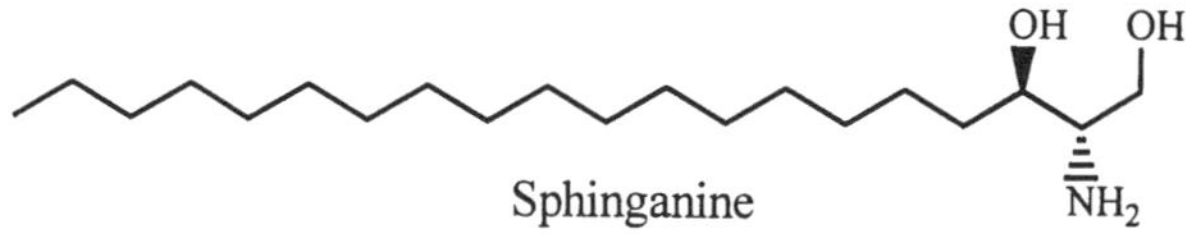

Mounts et al. [429] separated the phosphatidylethanolamine (PE), phosphatidy-linositol (PI), phosphatidic acid (PA) and phosphatidylcholine (PC) fractions in soybean oils using a complex 25-min $100/0 \rightarrow 0/100$ ($50/50$ $CHCl_3/THF$)/ ($1/92/7$ $CHCl_3$/methanol/ammonium hydroxide) gradient on a silica column (ELSD). It should be noted that this is an extremely aggressive mobile phase (containing NH_4OH) for the silica column. A precolumn is recommended for protection of the analytical column. Using this method, decomposition of phospho-lipids was followed over the course of 10 days. This group of analytes was also separated isocratically on a cyclodextrin column (ELSD) using a $35/32.7/26.8/5.5$ hexane/IPA/ethanol/water (5 mM tetramethylammonium phosphate buffer at pH 6.3) mobile phase [430]. The sensitivities for each compound class under these conditions were tabulated but no detection limits were given.

PE, PI, lyso-PE, and phosphatidylserine (PS) phospholipids were extensively studied on a silica column (ELSD) using a 35-min $446/415/104/30/5 \rightarrow 546/216/154/80/4$ IPA/iso-octane/chloroform/water/THF gradient. A response versus concentration plot was presented for each class of compounds, with PE having the highest sensitivity approximately twice that for PS. Detection limits of 10 µg injected and a linear range to 250 µg were reported [431]. Electrospray MS was used as the detector for PC, PE, and PS using a C_{18} column and a 17-min $88/0/12 \rightarrow 88/12/0$ methanol/hexane/water (0.5% ammonium hydroxide) gradient. The detection limits were 5 fmol (including a 1/100 split ratio) for PC and PE and >100 fmol for PS [432].

Markello et al. [433] developed a separation on a silica column that gave class separation of nonpolar lipids, glycolipids, phospholipids, and sphingolipids in 60 min. This work was based on the seminal work of Christie [434, 435] and used a silica column and a complex 60-min hexane/THF $(99/1) \rightarrow$ chloroform/IPA $(1/4) \rightarrow$ IPA/water $(1/1)$ gradient (ELSD). Results were used to assist in the diagnosis of metabolic disease in humans. Retention times for all 13 classes of compounds (fatty acids, PE, PI, PS, PC, ceramides, etc.) were tabulated. The authors noted that retention of CE, TAG and phytol was extremely sensitive to the water level in the mobile phase (i.e., creation of a hydrated and deactivated silica surface). The authors claim that the key here was the use of a saturation column (a column placed in the solvent flow path prior to the injector) and an extended column re-equilibration period after each analysis.

	X
Phosphatidic acid	-H
Phosphatidylethanolamine	
Phosphatidylserine	$-CH_2CH_2N$
Phosphatidylcholine	$-OCH_2CH(NH_2)COOH$
	$-OCH_2CH_2N^+(CH_3)_3$
Phosphatidylinositol	

R_1 and R_2 are commonly C16-20 alkyl or alkenyl chains.

Phosphatidic acid and analogs

The activity of ceramidase was monitored through its action on ceramide. Ceramide was derivatized with 4-chloro-7-nitrobenzo-2-oxa-1,3-diazole, NBD) and the resulting NBD-derivatized ceramide, sphingomyelin, and free fatty acids were separated on a C_{18} column ($\lambda = 455$ nm, ex; 530 nm, em) using an 85/15/0.15 methanol/water/H_3PO_4 mobile phase [436]. Baseline resolution was achieved and complete elution in <15 min.

4.5.7 Nucleotides, Nucleosides, and Related Analytes

The levels of 13 urinary nucleosides (e.g., dihydrouridine, pseudouridine, 1-methyladenosine, inosine, guanosine, 5'-deoxy-5'-methylthioadenosine) were monitored at 260 nm and 280 nm using a 30°C C_{18} column and a 40-min 100/0 → 40/60 water (25 mM KH_2PO_4 buffer at pH 4.7)/(3/2 methanol/water) gradient [437]. Peak shapes were excellent as was resolution. Concentrations were reported as nmol/μmol creatinine (in the range 2–300 nmol/mL).

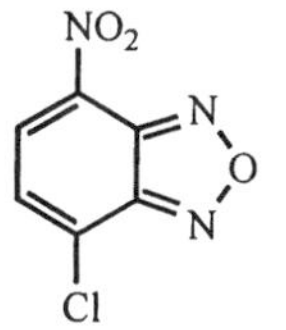

NBD-chloride;
4-chloro-7-nitro-2-oxa-1,3-diazole

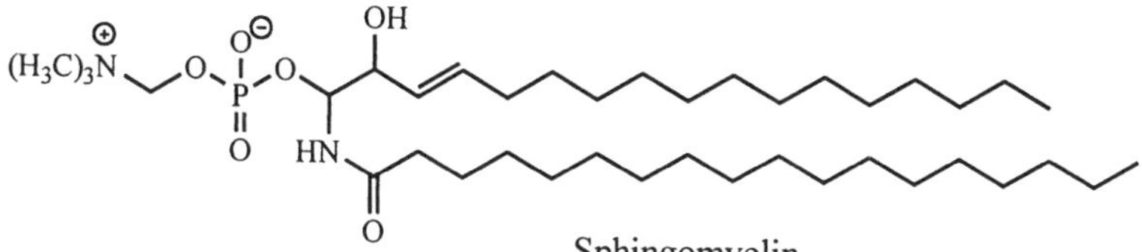

Sphingomyelin

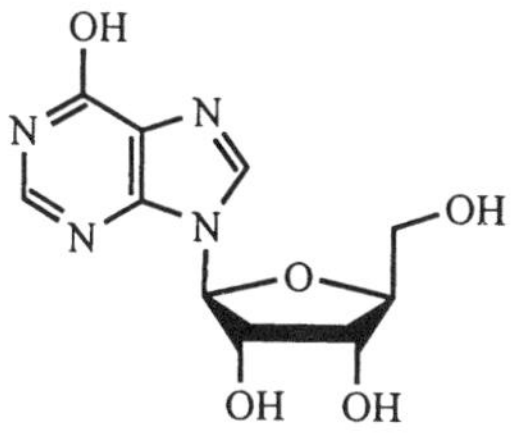

Inosine

The $\alpha \to \beta$ anomerization forms of NAD, NADP, NADH, and NADPH were extracted from erythrocytes and analyzed on a Knauer IGY column ($\lambda = 340$ nm). A 100/0 (hold 3 min) $\to$ 96/4 (at 7 min hold 2 min) $\to$ 90/10 (at 9 min hold 8 min) (80/20 water [0.2 M KH_2PO_4 buffer at pH 6.4 for oxidized analytes and pH 7.6 for reduced analytes])/methanol gradient generated good separations [438]. Reported analyte levels ranged from 0.4 to 55 nmol/mL red blood cells.

Micheli and Sestini [439] report on the study of NAD/NAPD precursors, metabolites, and catabolites (e.g., nicotinate and its mononucleotide, nicotinamide and its mononucleotide, AMP, inosine monophosphate) in erythrocytes. Samples were extracted from red blood cells and analyzed on a C_{18} column ($\lambda = 260$ nm and 280 nm). A 93/7 (hold 5 min) $\to$ 70/30 (at 6 min hold 4 min) water (0.1 M KH_2PO_4 buffer at pH 5.5 with 8 mM tetrabutylammonium sulfate)/(70/30 water [0.1 M KH_2PO_4 buffer at pH 5.5 with 8 mM tetrabutylammonium sulfate]/methanol) gradient generated excellent peak shapes and resolution. Reported erythrocyte levels ranged from 10 to 50 nmol/mL.

Normal and modified urinary nucleosides were monitored on a 30°C C_{18} column ($\lambda = 260$ nm and 280 nm) using a complex 30-min 100/0 $\to$ 40/60 water (25 mM KH_2PO_4 at pH 4.7)/(60/40 methanol/water) gradient [440]. Out of the 16 compounds monitored (e.g., uridine, 6-methyladenosine, 5-methyluridine, 2-methylguanosine, xanthosine) four compounds (pseudouridine, 1-methyladenosine, 3-methyluridine, 1-methylinosine) were clearly elevated-concentration markers for cancer. Standard concentrations ran from 1 to 4 µM. Overall resolution was exceptional as were peak shapes.

The k' values for 41 ribonucleotides, deoxynucleotides, cyclic nucleotides, and deoxycyclic nucleotides were generated on a C_{18} column ($\lambda = 254$ nm) using a 4/96

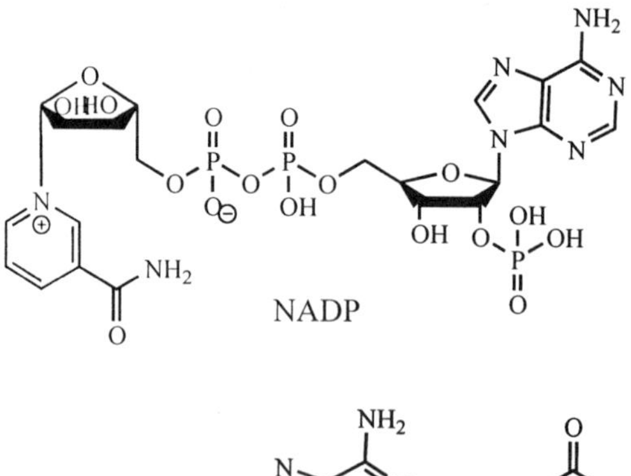

NADP

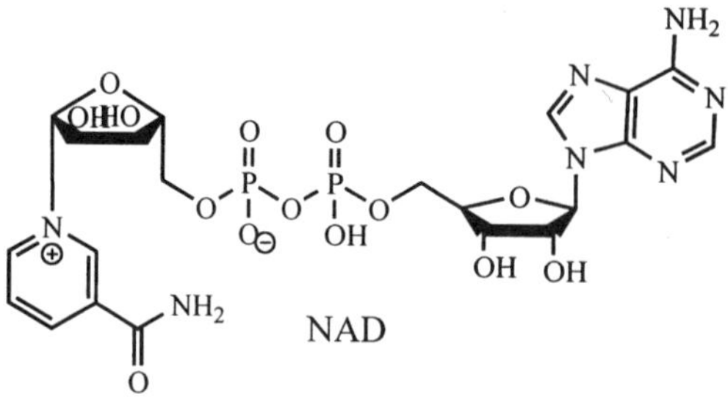

NAD

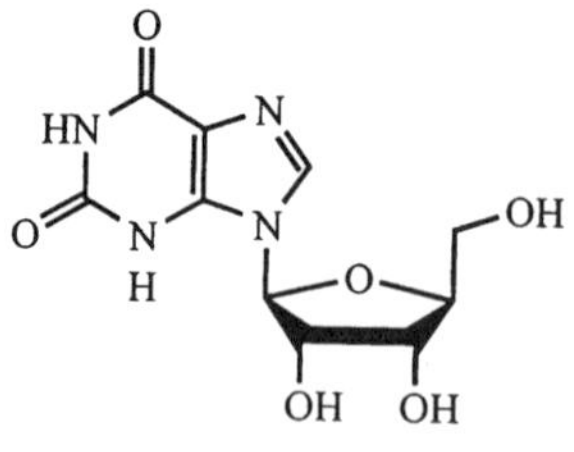

Xanthosine

methanol/water (83.3 mM triethylammonium phosphate pH 6.0) mobile phase [441]. A complete tabulation of these k' values shows cytidine 5'-monophosphate least retained ($k' = 0.4$) and adenosine 3':5'-cyclic monophosphate the most retained ($k' = 38$). The effects of increasing methanol concentration from 5% to 7% on k' are tabulated for 12 select analytes. Chromatograms show the elution of the ribonucleotides and cyclic nucleotides (elution complete in 2 h) separately from the deoxynucleotides/deoxycyclic nucleotides (elution complete in 80 min). Overall resolution and peak shapes were good.

An excellent separation of five DNA nucleotides (2'-deoxyguanosine, 2'-deoxy-5-methylcytidine, 2'-deoxycytidine, 2'-deoxyadenosine) and their five sulfur mustard adducts (products from alleged mustard gas exposure) was generated on a C_{18} column using a complex 40-min 0/100 → 48/52 methanol/water (25 mM ammonium carbonate buffer at pH 8) gradient [442]. Excellent peak shapes and resolution were obtained.

Mucochloric acid reacts with nucleosides to form ethenocarbaldehyde derivatives [443]. The derivatives of adenosine and cytidine were resolved on a C_{18} column ($\lambda = 325$ nm) using a 20-min 10/90 → 30/70 methanol/water (10 mM KH_2PO_4 buffer at pH 4.6) gradient. Adenosine, ethenoadenosine, and the ethenocarbaldehyde derivative were resolved and eluted in 20 min, whereas cytidine and ethenocytidine were resolved and eluted in 16 min. Mucochloric acid eluted at ~8 min and was badly fronted (i.e., more of the peak eluted prior to the peak maximum). This could be due to the fact that the pH of the mobile phase did not result in a fully protonated form of the acid. Conversely, the acid is also capable of reversible cyclization and this peak fronting may be an indication of an equilibrium process that occurs on the same timescale as the separation. A change in the pH of the solvent of the system temperature would alter the equilibrium and hence the peak shape and provide useful information as to how to further modify the system to produce good chromatography.

Thirteen purines (e.g., hypoxanthine, xanthine, methylxanthines, dimethylxanthines, methylguanines) were extracted from urine and separated on a C_{18} column ($\lambda = 254$ nm) using a 45-min 100/0 → 80/20 water (10 mM KH_2PO_4 buffer at pH 5.5)/methanol gradient. Excellent resolution was obtained even though the peaks were noticeably tailed. Interestingly, neither the pH (4.0–6.5) nor the ionic strength (10–100 mM) had any effect on k'. A linear range of 0.1–10 nmol injected with a detection limit of 20 pmol was reported [444].

2'-Deoxycytidine

Mucochloric acid

The retention characteristics of 21 deoxyuridine derivatives were studied on an alumina column ($\lambda = 260$ nm) using a 90/10/0.5 methanol/dichloromethane/TFA mobile phase [445]. The k' values were tabulated for all analytes and ranged from 0.5 to 11. The authors noted that for IPA/water solvents, the k' decreased as the IPA content increased from 0 to 40% but then increased as the %IPA rose to 90%. Similar results were obtained with methanol/dichloromethane solvents as the methanol content rose above 60%.

Two sets of *syn-* and *anti*-enantiomers of four 5,6-dimethylchrysene-1,2-dihydrodiol-3,4-epoxide adducts of both deoxyadenosine and deoxyguanosine were baseline resolved in 1.5 h using a C_{18} column ($\lambda = 262$ nm) and a $40/60 \rightarrow 56/44$ methanol/water gradient [446]. Excellent peak shapes were obtained.

Three neuraminic acids (*N*-glycoyl-, *N*-acetyl-, and *N*-propionylneuraminic acids) were derivatized with 1,2-diamino-4,5-(methylenedioxy)benzene and separated on a 40°C C_{18} column ($\lambda = 373$ nm, ex; 448 nm, em). A 100/15/10 water (10 mM acetate buffer at pH 5)/methanol/acetonitrile mobile phase generated a 9-min baseline resolved chromatogram [447]. Solutions of 0.1–1 nmol analyte were used in the procedure.

Nicotinic acid, nicotinamide, inosine monophosphate, adenosine 5′-monophosphate (AMP), adenosine 5′-diphosphate (ADP) and adenosine 5′-triphosphate (ATP) were extracted from red blood cells and baseline resolved on a C_{18} column ($\lambda = 280$ nm or 254 nm) using a complex 22-min $4/96 \rightarrow 30/70$ methanol/water (100 mM KH_2PO_4 with 8 mM tributylamine at pH 5.5) gradient. Peak shapes were excellent [448].

4.5.8   Other Analytes

Chlorophylls (see structure on page 175) and their corresponding pigments (chlorophyll *a* and *b*, chlorophyllides *a* and *b*, pheophytin *a* and *b*, pheophoride *a* and *b*) were extracted from cherimoya fruit. Separation on a C_{18} column ($\lambda = 660$ nm or 440 nm, ex; 660 nm, em) was accomplished using an $80/20/0 \rightarrow 80/0/20$ (at 15 min hold 10 min) methanol/water (1 M ammonium acetate)/acetone gradient [449]. The authors note that the use of 0.5–1 M ammonium acetate was necessary to generate good overall selectivity, efficiency, and resolution.

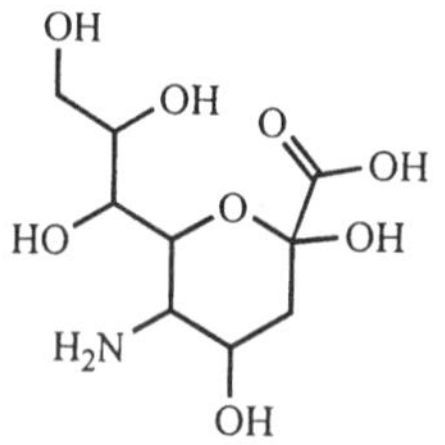

Neuraminic acid 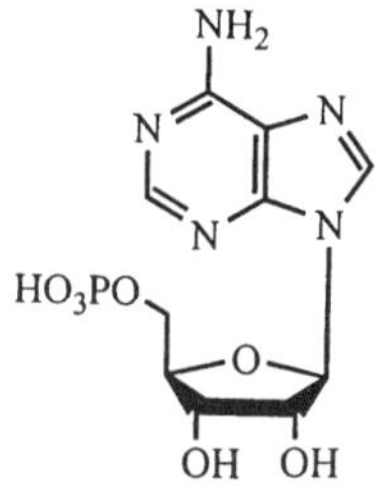   Adenosine monophosphate

Seven *n*-alkylhydroperoxides (ranging from C_6 to C_{18}, even chain) were baseline resolved and eluted in 20 min using a C_{18} column ($\lambda = 295$ nm, ex; 415 nm, em) and an $80/20 \rightarrow 95/5$ (at 6.5 min hold 14 min) methanol/water (H_3PO_4 to pH 3.5) gradient. Detection was generated through a postcolumn reaction of horseradish peroxidase/*p*-hydroxyphenylacetic acid (20 mg/4 mg to 250 mL of aqueous 10 mM KH_2PO_4 at pH 7 solvent) and 30 mM NaOH with the column effluent. The authors noted that acetonitrile had a negative impact on sensitivity. Detection limits of 0.4–1 µM (analyte dependent) were reported [450].

Octaethylporphyrin and etioporphyrin as well as their Ni and VO complexes were separated on a C_{18} column ($\lambda = 386$–401 nm) using a $94/6 \rightarrow 100/0$ (at 22 min hold 2 min) (1/1 methanol/acetonitrile)/water gradient [451]. Excellent separation and peak shapes were achieved. Linear ranges of 30–8000 ng/mL were reported.

Coproporhyrin isomers I and III were isolated from urine and feces and separated on a 29°C C_8 column ($\lambda = 406$ nm or $\lambda = 405$ nm, ex; 620 nm, em). A 46-min $100/0 \rightarrow 35/65$ (at 30 min) $\rightarrow 10/90$ (at 46 min) (90/10 water [1 M ammonium acetate buffer at pH 5.5]/acetonitrile)/methanol gradient baseline resolved the analytes. A ramp to 100% methanol was run after the gradient to elute others materials extracted from the sample [452]. Levels of 40–2000 µmol/24 h (urine) and 70–2200 nmol/g (feces) were reported.

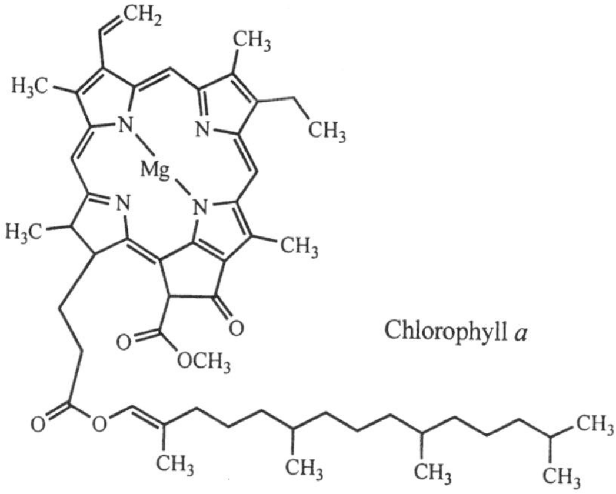

Chlorophyll *a*

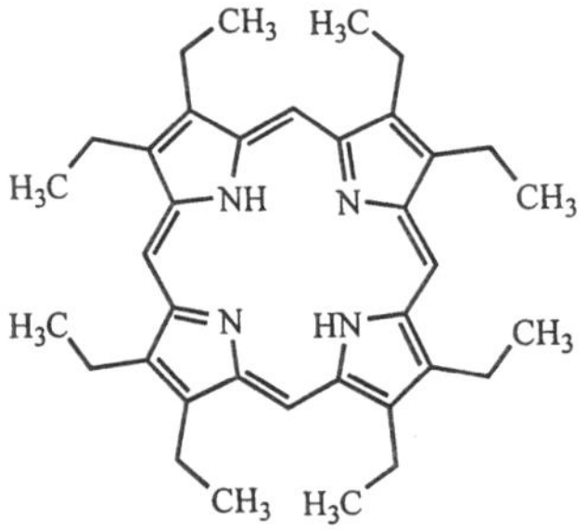

Octaethylporphyrin

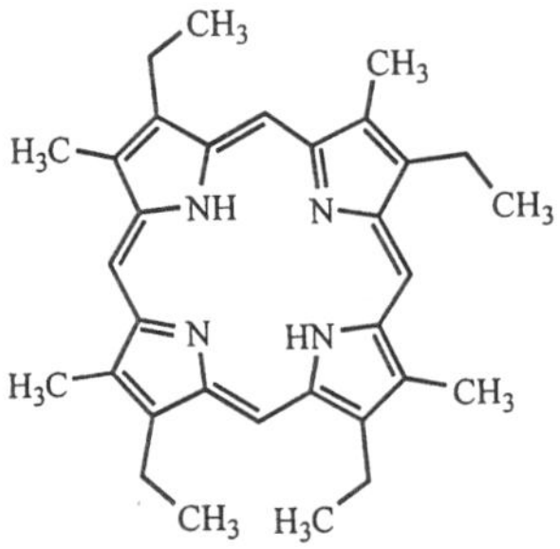

Etioporphyrin

Porphyria associated with a rat model of Wilson's disease included six porphyrins (uro-, hepta-, hexa-, penta-, copro-, and mesoporphyrin). Baseline resolution and excellent peak shapes were generated on a 35°C C_{18} column ($\lambda = 395$ nm, ex; 620 nm, em) using an 85/15 (hold 5 min) $\rightarrow$ 0/100 (at 15 min hold 5 min) water (0.1 M sodium phosphate at pH 3.5)/methanol gradient [453]. For urine extracts, standards of 10–1000 nM with detection limits of 2 nM were reported.

Triiodothyronine (T_3) and thyroxine (T_4) were separated from the analogous mono-, di- and tribromothryronine analogs on a C_{18} column ($\lambda = 295$ nm or amperometric detector, 1.2 V vs. Ag/AgCl) using a 47/10/43 methanol/acetonitrile/water (0.06% TFA) mobile phase [454]. Detection limits were found to be 40 ng/L with a nonlinear working curve up to 1200 ng/L. Peak shape was difficult to discern since complete resolution of the peaks, under the stated optimal conditions, was not obtained.

Five iodoamino acids (3-iodotyrosine [MIT], 3,5-diiodotyrosine [DIT], T_3, T_4 and 3,3',5'-triiodothyronine [rT_3]) were separated on a C_{18} column (ICP/MS) using a

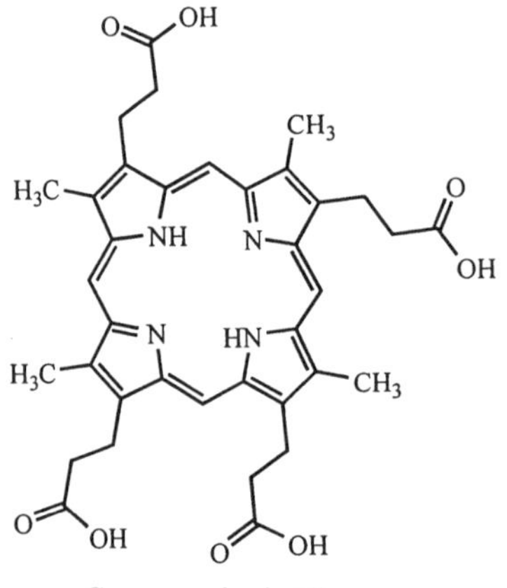

Coproporphyrin III

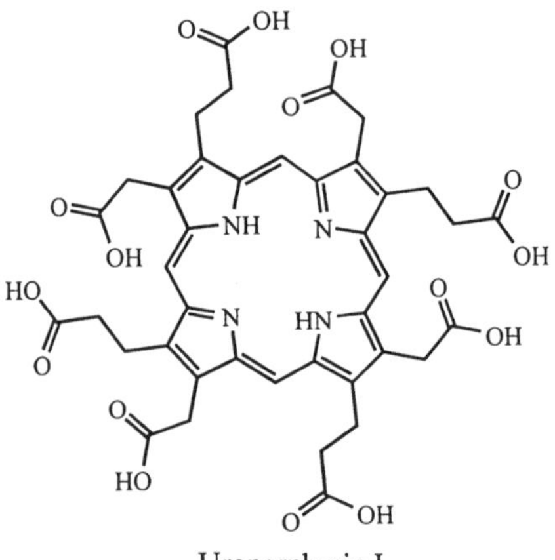

Uroporphyrin I

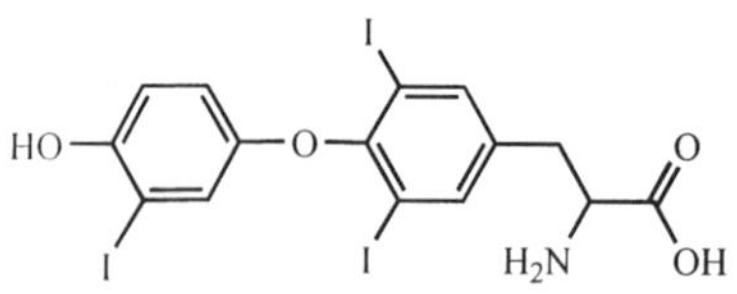

Triiodothyronine

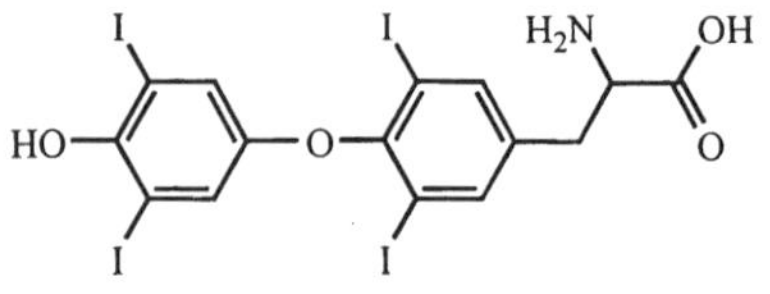

Thyroxine

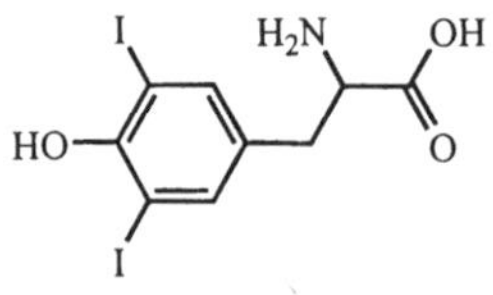

3,5-Diiodotyrosine

10/90 methanol/water mobile phase to quantitate MIT and DIT and a 50/50 methanol/water mobile phase to quantitate MIT/DIT, T_3, T_4 and rT_3 [455]. Detection limits, based on iodine equivalents, were reported as $\sim$100 pg. Linear response was recorded over four orders of magnitude of concentration. Excellent resolution was obtained and total elution was completed in <15 min.

Hill et al. [456] separated, in <20 min, hydroquinone from four of its S-substituted glutathione derivatives on a C_{18} column (electrochemical 15-channel detector from −250 to +450 mV using glassy carbon) using a 6/94 methanol/water (4 mM citric acid in 8 mM ammonium acetate buffer at pH 4) mobile phase. Successful analyte detection down to the 1 μmol level was reported. Positive identification of the derivatives was done using a C_{18} column (fast atom bombardment MS) and a 5/95 (93/7 methanol/glycerol)/([93/7 water/glycerol] with 100 mM ammonium acetate buffer at pH 5) mobile phase.

Thirty-nine chlorophyll transformation products were separated on a C_{18} column ($\lambda = 400$ nm) using a 40-min gradient from 90/10 methanol/water to 90/5/5 acetone/methanol/water [457]. Positive identification of methylated compounds within fractions was made by thermospray MS. Pheophorbide *a* methyl ester, zeaxanthin, pyropheophorbide *a* methyl ester, isorenieratene, pheophytin *a* and *b* and pyropheophytin *a* and *b* were confirmed.

Malachite green and crystal violet (leuco and chromatic, reduced and oxidized forms, respectively) were separated on a C_{18} column ($\lambda = 600$ nm) using an 81/19

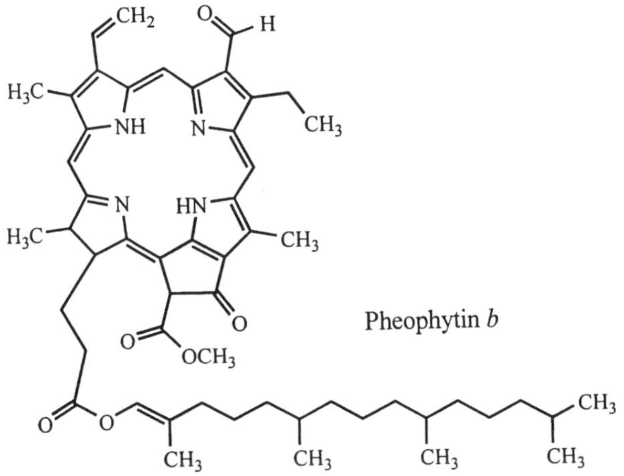

Pheophytin *b*

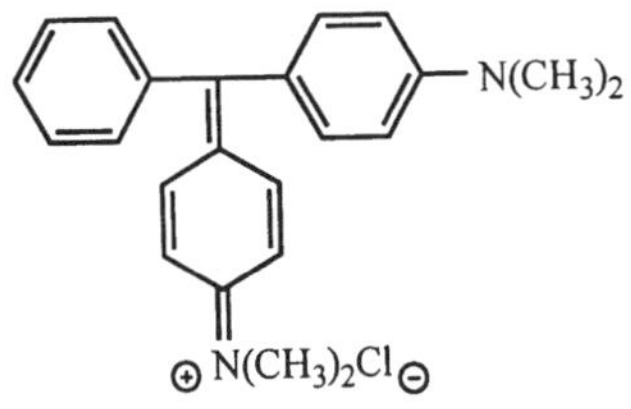

Malachite green

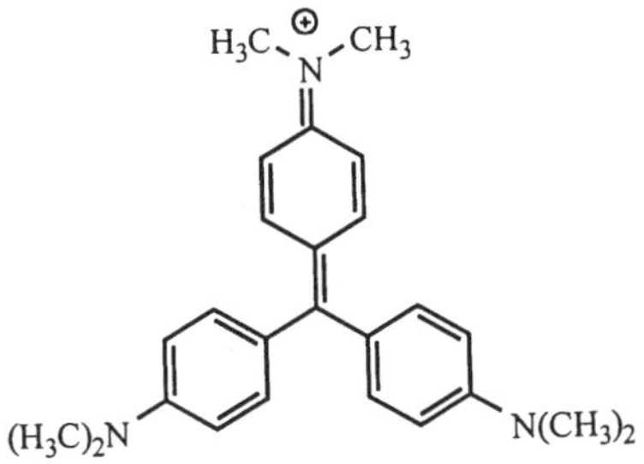

Crystal violet

methanol/water (0.1 M sodium acetate buffer at pH 6) mobile phase [458]. Detection limits were reported as ~0.2 ng with linear working curve ranges of 1–100 ng. Separation was complete in <10 min with peaks being slightly tailed.

4.5.9   Summary

Alcohols can be extremely effective when used in separations involving compounds with hydroxyl, carboxy, and/or carboxylic acid functional groups. The presence of the alcohol in the mobile phase almost invariably leads to sharper peak shapes. Part of this enhancement is due to the competitive hydrogen bond interaction with the analyte at the expense of analyte–surface interactions (as determined by hydrogen bonding with residual silanol groups). However, an even greater role for the alcohol is the enhanced solubility in the mobile phase that should assist mass transfer of the analyte from the surface to the mobile phase. Alcohols, even at low percentages (10–20%), should be considered for separations of analytes containing the functional groups detailed above.

Conversely, alcohols should not be used in the separation of sugars. First, the resultant peak shapes generated on the typical column of choice, aminopropyl, are considerably worse than those generated using acetonitrile. Second, the reaction of the reducing sugars with the aminopropyl bonded phase, via a Schiff base formation, is accelerated by alcohols [459].

4.6   AMINO ACID, PEPTIDE, AND PROTEIN ANALYTES

4.6.1   Amino Acid Analytes

Amino acid analysis is important not only from a clinical aspect—amino acid metabolism disorders can be fatal if not diagnosed early—but also as the basis for protein and peptide sequencing. Due to the combination of the small amounts of material used in an actual analysis and the lack of chromophores on most amino acid residues, derivatization is common practice. Derivatization may be done prior to either separation or detection.

Hearn and co-workers [460,461] have published a series of results relating amino acid hydrophobicity to reversed-phase retention of amino acids and peptides. The authors used wide-pore C_4 or C_{18} columns and a gradient of IPA/acetonitrile/water (0.1% TFA) in generating results. A total of 1738 peptides covering 12 previously generated amino acid hydrophobicity scales were part of the study. Predicted and actual retention times of overlapping heptamers in myohemerythrin were presented. This study and previous studies cited therein offer excellent theoretical and experimental bases for the predictive chromatography of amino acids and peptides.

The isoaspartic acid released during the deamidation of asparginine from peptides and proteins was monitored on a C_{18} column ($\lambda = 254$ nm) using an $85/15 \rightarrow 50/50$ (at 10 min hold 5 min) $\rightarrow 0/100$ (at 18 min hold 3 min) (90/10 water [25 mM KH_2PO_4 with octanesulfonic acid to pH 3.2]/methanol)/methanol gradient [462]. Isoaspartic acid, S-adenosyl-L-methionine, and S-adenosylhomocys-teine were resolved in <16 min. Standard concentrations ranged from 0.02 to 2 mM.

The simultaneous analysis of glutathione (tripeptide), γ-glutamylcysteine, and 16 amino acids was achieved using two separate HPLC derivatization methods. o-Phthalaldehyde (OPA) was used for all compounds and monobromobimane (MBB) for the thiols compounds only. Both methods used a C_{18} column (fluorescence detection using OPA filters). The OPA derivatives were separated very well using a 100-minute 100/0 (hold 5 min) → 70/30 (at 45 min) → 0/100 (at 90 min hold 10 min) (90/10 water [20 mM NaH_2PO_4 at pH 6.8]/methanol)/(40/60 water [20 mM NaH_2PO_4 at pH 6.8]/methanol) gradient [463]. Plots of individual standard curves from 0.01 to 1 nmol injected were presented along with a table of retention times and relative responses for all compounds. The glutathione, γ-glutamylcysteine, and cysteine were separated in 12 min using a 100/0 → 88/12 water (90/10 water [0.25% acetic acid to pH 4.3 with NaOH]/methanol)/(10/90 water [0.25% acetic acid to pH 4.3 with NaOH]/methanol) gradient.

Plasma amino acids were separated as their o-phthalaldehyde derivatives in <20 min on a C_{18} column ($\lambda = 230$ nm, ex; 389 nm, em) using a 98/2 → 0/100 (50/50 water/water [9 mM KH_2PO_4 with 0.5 M TEA to pH 6.9])/(35/15/50 methanol/acetonitrile [9 mM KH_2PO_4 with 0.5 M TEA to pH 6.9]) gradient [464]. Linear plots of fluorescence emission versus concentration from 5 to 800 μmol/L were obtained. It should be noted that separation was complete with baseline resolution in <12 min but significant column reequilibration time between injections was necessary.

Eleven dansylated amino acid isomers were separated on a β-cyclodextrin column ($\lambda = 254$ nm) using methanol/water (0.2 M ionic strength phosphate buffer at pH 6.5) mobile phases [465]. The level of methanol needed to generate a separation varied from 15% to 25%. At 15% methanol, four sets of enantiomers (glutamine, serine, norvaline, norleucine) are partially resolved in <30 min. The authors claim that water is an essential component of the mobile phase, but the next study clearly argues against this statement.

Armstrong and co-workers [466] studied the retention of the DL isomers of dansyl-substituted amino acids on both (R)- and (S)-(naphthylethyl)carbamoyl-modified β-cyclodextrin column. The k' and α values for nine amino acids with a wide range of substituents added to the terminal amine group were tabulated. The mobile phase compositions used were analyte-dependent but ranged from 99/1 ethanol/acetic acid to 25/75 ethanol/acetonitrile (+1% acetic acid) with a 50/50 ethanol/acetonitrile (+1% acetic acid) composition being most frequently used.

Glutathione   o-Phthalaldehyde   Glutamine

9-Fluorenylmethylchloroformate (FMOC) was used in the precolumn derivatization technique for amino acids. Separation of 30 residues was completed in 23 min using a C_{18} column ($\lambda = 263$ nm, ex; 313 nm, em) and an $18/82 \rightarrow 99/1$ methanol/water (25 mM ammonium phosphate at pH 6.5) gradient. Amino acid levels down to 100 pmol were readily detected [467].

Naphthalenedialdehyde was used as a precolumn derivatization reagent in the determination of desmosine, isodesmosine, and 17 other amino acid residues [468]. More stable complexes (as compared with o-phthalaldehyde derivatives) were cited as the analytical advantage. Detection limits of 100 fmol (S/N = 2) were reported and the analysis was completed in <35 min. A C_{18} column ($\lambda = 420$ nm, ex; 490 nm, em) and a $10/5/85 \rightarrow 63/1/36$ methanol/THF/water (5 mM sodium citrate) gradient were used. Additionally, the amino acids were monitored electrochemically (+750 mV vs. Ag/AgCl). Peak shape was good, but complete separation of all residues was not achieved.

4.6.2   Peptide Analytes

A set of imidazole dipeptides (carnosine, anserine, balenine, histidine, 1-methylhistidine, 3-methylhistidine, homocarnosine) were isolated from equine muscle, derivatized with o-phthaladehyde immediately prior to analysis, and separated on a C_{18} column ($\lambda = 340$ nm, ex; 450 nm, em). A complex 45-min $100/0 \rightarrow 0/100(995/5$ water [12.5 mM sodium acetate at pH 7.2]/THF)/(55/35/15 water [12.5 mM sodium acetate at pH 7.2]/methanol/acetonitrile) gradient was used. Note that the buffer preparation uses the addition of isomolar amounts of sodium acetate (1.026 g/L) to acetic acid (0.751 g/L). The resulting pH of the aqueous solution should be the pK_a of acetic acid (4.76). Peak shapes were excellent and only 1-methyl- and 3-methylhistidine peaks lacked baseline resolution. Working standards were

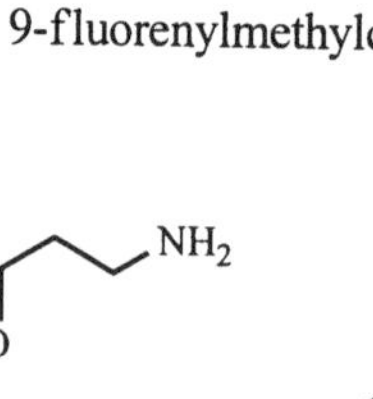

FMOC; 9-fluorenylmethylchloroformate

Carnosine   Anserine

prepared over the range 0.01–1 mM [469]. Detection limits were 0.01 mmol/kg (S/N = 3).

Twenty-five sets of benzyloxycarbonyl-glycine-X-Y-OMe epimers were chosen to study the racemization process that occurs during peptide synthesis [470]. These tripeptides were resolved on a 40°C C_{18} column ($\lambda = 245$ nm) using a number of isocratic methanol/water mobile phases. The inserted amino acid residues, X and Y, included combinations of alanine, valine, leucine, isoleucine, and phenylalanine. Mobile phase compositions ranged from 45% to 65% methanol (analyte dependent), capacity factors from 3 to 14, and α values from 1.11 to 1.39.

In a similar fashion, the retention behavior of 23 tripeptide epimers (with the basic structure Z-Ala-X-Val-OMe, where X is either the D or L form of the inserted amino acid) was studied on a C_{18} column ($\lambda = 220$ nm) using a series of isocratic methanol/water mobile phases [471]. The k' values for all the LLL and LDL forms and the resulting α values are tabulated for each epimer. Reasonable elution times were achieved, with mobile phases ranging from 60% to 80% methanol. The LDL-epimers were consistently more retained than the LLL-epimers. The authors attribute this to a larger surface area presented to the packing material in the LDL conformation.

Four cyclic heptapeptide toxins (extracted from *Microceptis auruginosa*) consisting of D-alanine, *erythro-β*-methylaspartic acid, glutamic acid, L-leucine, L-alanine, N-methyldehydroadenosine, and 3-amino-9-methoxy-2,6,8-trimethyl-10-phenyl-deca-4,6-dioenoic acid were baseline resolved in 20 min using a C_{18} column ($\lambda = 238$ nm) and a 58/42 methanol/water (50 mM phosphate buffer at pH 3) mobile phase [472]. Excellent peak shapes were obtained.

4.6.3 Protein Analytes

Alcohols were studied as denaturants for various proteins by Herskovits et al. [473]. The amount of organic solvent in water needed to denature cytochrome *c*, myoglobin, or chymotrypsinogen decreased in the order ethylene glycol ≫ methanol > ethanol > IPA > *n*-propyl alcohol ≫ 2-butanol > 1-butanol. Interestingly, ethylene glycol at the 60% level did not denature any of these proteins, whereas ~50% methanol denatured them all. These results were supported by Bull [474], who used methanol, ethanol, *n*-propyl alcohol, and 1-butanol, and seven proteins. In this study 30% methanol denatured all proteins, whereas less than 10% *n*-propyl alcohol caused the same degree of denaturation.

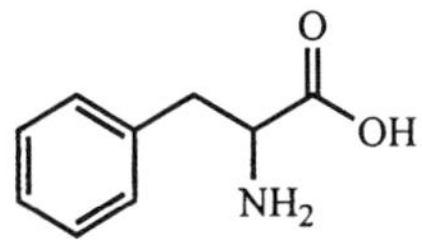

Phenylalanine

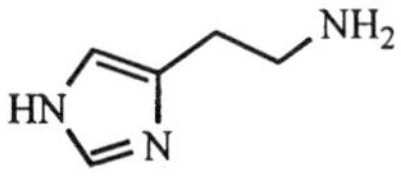

Histamine

The extremely complex and important relationship between peptide/protein retention and the percentage of organic in the mobile phase has received a great deal of attention. In a classic study, Hearn and Grego [475] studied the retention of analytes composed of 1 to 5 amino acid residues on a C_{18} column using isocratic mobile phases containing various levels of methanol (i.e., alcohol/water [20 mM phosphate buffer at pH 2.25 or 15 mM triethylammonium phosphate at pH 2.95]). As expected, retention decreased as the alcohol level increased from $0 \rightarrow 50\%$. However, after 50%, the retention of these analytes *increased* as the concentration of methanol increased. The rate of increase was dependent upon both the buffer used and the composition of the analyte. This behavior was *not* observed for the isocratic chromatography of myoglobin or apomyoglobin on a C_8 column using IPA or ethanol [476]. Eleven to fifteen residue polypeptides showed this behavior to an even greater extent with methanol and to a lesser degree with IPA.

Simpson et al. [477] studied the retention behavior of six proteins (ribonuclease, cytochrome c, α-lactalbumin, bovine serum albumin, trypsin inhibitor, ovalbumin) on four C_{18} columns using isocratic mobile phases containing 20–95% IPA (with 0.1% TFA). Plots of k' vs. percent IPA show sharp drops in k' starting at between 20% and 50% IPA (protein dependent and column dependent). The k' remained low (<1) until it rose sharply for all proteins between 60% and 80% IPA. The reasoning behind the U-shaped k' versus organic modifier plots focuses on the denaturation of the protein. Retention decreases as expected when the organic level increases, but low levels cause little or no change in protein conformation. As the protein denatures, its conformation becomes more random and formerly internal hydrophobic amino acid residues come into contact with the mobile phase. As this process begins to occur, the increased hydrophobic nature of the protein and the increased strength of the mobile phase offset one another and k' is constant over a range (analyte and organic identity dependent) of mobile phase compositions. At higher organic levels, protein denaturation occurs to such a great extent that k' actually increases. It should be noted that protein denaturation leads to irreversible adsorption or very slow desorption of the protein from the surface of the support material. This is a prime factor leading to low protein recoveries in LC work.

A number of human serum proteins were analyzed using a wide range of bonded phases (C_{18}, phenyl, C_8, diphenyl) and either a $10/90 \rightarrow 70/30$ or a $0/100 \rightarrow 100/0$ gradient of n-propyl alcohol/water (triethylamine acetate buffer at pH 4.0). Serum characterization was complete in <60 min [478]. C_{18} bonded phases gave the lowest absolute recovery of proteins with C_8, phenyl, and diphenyl yielding recoveries $>80\%$ at the 10 µg total mass injected level. Recovery increased to $>95\%$ at the 100 µg level.

Protamine (nuclear proteins of reproductive cells) separations were optimized on a C_{18} column ($\lambda = 220$ nm) using a 0 to 7% ethanol gradient in water with 0.2% TFA. The authors found that both the mass of protein injected and the flow rate (here 0.5–1.0 mL/min) had a significant effect on the separation, whereas the effects of changes in the injection volume were minimal [479]. These findings are totally consistent with a stationary phase-induced denaturation process.

The *Bacillus* and *Aspergillus* protease hydrolysates of α_5-casein were fractionated by gel electrophoresis. Each fraction was further characterized by separation on a C_{18} column ($\lambda = 220$ nm). A 60-min $100/0 \rightarrow 30/70$ (at 55 min) $\rightarrow 9/91$ water (0.06% TFA)/methanol gradient was used. This method readily differentiated the protease hydrolysates as well as the fractions within each hydrolysate.

Two studies by Karger and co-workers [480, 481] studied in detail the denaturing effect of *n*-propyl alcohol on recombinant human growth hormone (rhGH) and its *N*-methionyl variant. The effects of changes in temperature, pH, and percent *n*-propyl alcohol on the chromatography using a 45°C C_4 column ($\lambda = 295$ nm, ex; 360 nm, em) were presented. A detection limit of $10\,\mu g/mL$ (S/N = 2) was reported. Optimal separation of rhGH from its variant was achieved when a 78/22 *n*-propyl alcohol/water (0.1 M phosphate buffer at pH 6.5) mobile phase was used. Elution was complete in 35 min.

An interesting study found that the addition of an alcohol "booster" to IPA led to a lower total organic solvent needed for elution, a decrease in the extent of denaturation observed in the eluted protein, and an increase in overall recovery [482]. Both 1-butanol and 2-methyl-1-butanol were highly effective "booster" alcohols. The column contained a diphenyl packing and the mobile phase had either 0.1% TFA or 0.1 M H_3PO_4 present (e.g., 33/4/63 IPA/2-methyl-1-butanol/water [0.1 M H_3PO_4 to pH 2.3]). Twelve proteins having molecular weights of 11,000 (insulin) to 160,000 (γ-globulin) were studied. When *n*-propyl alcohol or *n*-propyl alcohol/1-butanol mobile phases were used, recoveries were all >80%, most in the 95–100% range.

Fernández and Sinanoglu [483] found that not only does the organic solvent cause protein denaturation but the solvent will also offer a point of maximal protein stabilization in its native conformation. For example, it was found that lysozyme was most stable in an 8/92 methanol/water solvent. This kind of information is of potentially great importance to the chromatographer but is very rarely found or employed.

Armstrong and co-workers [484] introduced a novel variation in the protein elution scheme. A reverse gradient was run from high organic to water—specifically, 90/10 acetonitrile/water (1% TFA) to 60/40 acetonitrile/water (1% TFA). Since a general plot for proteins of $\log k'$ versus percent acetonitrile in the mobile phase is U-shaped with a minimum k' value between 30% and 45% acetonitrile, elution will occur irrespective of whether the gradient goes from a weaker to stronger solvent (traditional reversed-phase gradient) or from a stronger to weaker solvent. The protein is initially in its denatured form, so a major consideration is whether the protein needs to be in its native conformation for further analysis after separation or will be analyzed in its denatured form. Chromatographic results should be more consistent because the protein is already denatured. Note that simple molecules cannot be analyzed in this fashion since the plot of $\log k'$ versus percent acetonitrile for them is a continuous decrease in retention as the acetonitrile level increases.

Methanol and IPA and their effects on peptide and protein conformation were studied from both kinetic and chromatographic viewpoints [485]. Kinetics plays an extremely important role in the conformation and retention mechanisms of proteins.

Multiple peaks may occur when the retention time is on the same order as the kinetics of the conformational change. Conversely, for small cyclic dipeptides, retention may be predicted on the basis of preferred conformation and resultant surface area. Although the details are complex, these general concepts must be kept in mind when working with proteins so that multiple peaks due to conformational equilibria are not mistaken for separate compounds.

Oxidized human growth hormone (hGH) variants were preparatively separated on a PLRP-S column using a 100-min 34/66 → 39/61 n-propyl alcohol/water (25 mM ammonium acetate buffer at pH 7.5) gradient. Excellent resolution was obtained for the oxidized and native hGH using a 15 mg load [486]. Further identification of the fractions was done by LC/MS.

4.6.4   Summary

In the initial stages of protein and peptide separations, methanol, ethanol, IPA [487], and n-propyl alcohol were widely and effectively used. As the requirements for lower UV cutoffs became more demanding, the 205–210 nm cutoff range for the alcohols became unacceptable. Even though the alcohols were less prone to denature proteins than other solvents, only one alternative reversed-phase organic solvent remained: acetonitrile. Alternative detectors (e.g., ELSD and MS) should lead to a renewed interest in the alcohols for protein separations.

4.7   PHARMACEUTICAL ANALYTES

4.7.1   Drug Surveys and Screening Procedures

The monitoring of impurities in pharmaceutical formulation raw materials and final product formulation compositions as well as possible metabolites of these pharmaceutical compounds is an extremely important, highly regulated area. An analytical method that is submitted in a new drug application is literally mandated and run in perpetuity thereafter. This is because of the exorbitantly high cost of validating a new method (not to mention the inevitable questions as to what was done incorrectly in the first method). Unfortunately, method development timetables are usually so compressed that truly optimal processes (especially with regard to long-term ruggedness) are difficult to obtain. Consequently, a sound understanding of all options available to the analyst is critical.

Extremely valuable in this regard are those papers dealing with a large number of pharmaceutical compounds and their retention behavior in methanolic mobile phases. Over 300 compounds were used in a systematic HPLC toxicological survey [488]. A 30°C C_{18} column (photodiode array detector, $\lambda = 190–800$ nm with the peak maxima tabulated for each compound) and an isocratic 65/5/30 methanol/THF/water (10 mM phosphate buffer at pH 2.6) mobile phase were used. Blood or plasma samples were extracted and the resulting residue was analyzed. Peaks were identified by retention time and confirmed by their UV spectrum. This review is also important in that it collates, incorporates, and cites the results from 24 other less extensive studies.

Smith et al. [489] studied the retention of 28 basic drug compounds (e.g., nitrazepam, papaverine, cocaine, propranolol, morphine, amphetamine, prolintane, pipazethate, ephedrine, strychnine) on a silica column ($\lambda = 254$ nm) using a 90/10 methanol/water (80 mM each of 3-[cyclohexylamino]-2-hydroxy-1-propane sulfo-nate and 3-[cyclohexylamino]-1-propanesulfonic acid to pH 10.0) mobile phase. Injections of 1 µL of standards ranging from 40 to 1000 µg/mL were made. The k' for each analyte was tabulated and, under these conditions, ranged from 0.3 to 10. The authors noted a gradual but continuous decrease in retention during system operation. A significant void was found at the head of the column. This reempha-sizes the fact that basic solvents (pH > 7) and acidic solvents (pH < 2.5) rapidly dissolve a silica support. A precolumn with a packing material matching the functionality of the analytical column should be used to presaturate the mobile phase with silica and thereby protect the analytical column.

De Leenheer and co-workers [490] used a polybutadiene-modified alumina support (photodiode array detector, $\lambda = 225$–350 nm) in a separate study. Alumina is stable in basic solutions, so that hydrogen bonding effects on basic compound retention are effectively neutralized by the presence of sodium hydroxide. Retention results were tabulated for 134 drugs that are commonly included in toxicological screens. A 10/90 $\rightarrow$ 90/10 methanol/water (0.0125 M NaOH) gradient was run and complete elution was obtained in <30 min. Peak shapes were very good. Obviously, separation of all peaks was not critical since this study was used for screening purposes only. However, this work is an excellent resource for generating initial mobile phase conditions in method development.

The evaluation of the retention of 18 imidazol(in)e derivative drugs was done on an alumina-based C_{18} column, and two silica based C_{18} columns ($\lambda = 254$ nm)

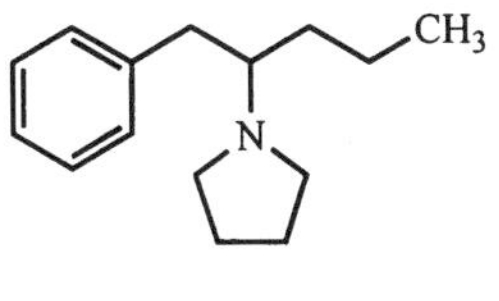

Prolintane

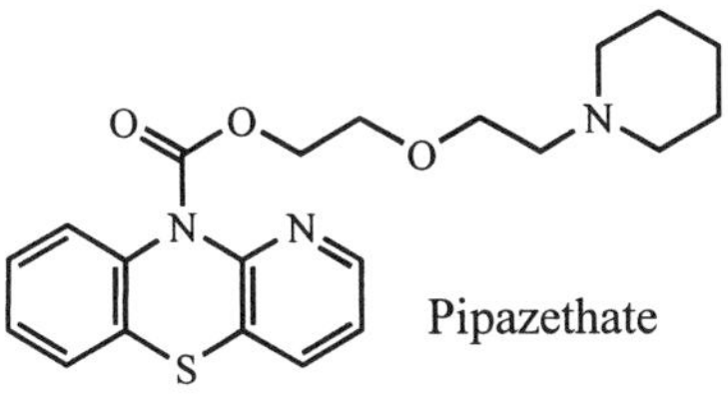

Pipazethate

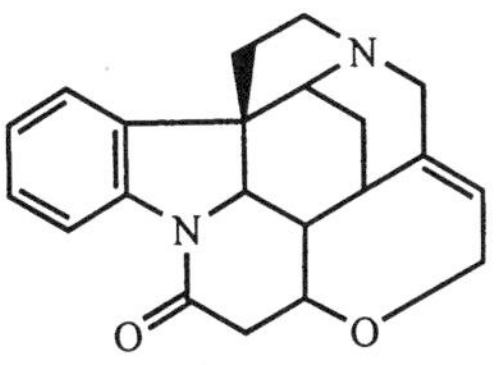

Strychnine

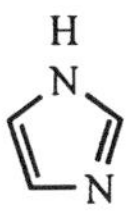

Imidazole

using isocratic mobile phases consisting of 20–80% methanol in water [491]. A 40 mM acetate/phosphate/borate buffer at pH 2.9 (for silica) or pH 10.9 (for alumina) with 0.25% 1-octanol was part of the mobile phase. It is interesting to note that a significant number of analytes exhibited a U-shaped k' vs. percent methanol plot on the silica-based C_{18}, but few exhibited this curvature on the alumina-based C_{18} column. As mentioned in the previous section, this type of behavior is not typical for smaller molecules. However, due to the potential multiple analyte/surface hydrogen bond interactions for many drug compounds, these effects must be kept in mind when developing a method.

Koves [492] developed a rapid screening method for barbiturates, benzodiaza-pines, phenytoin, salicylic acid, acetaminophen, and theophylline. An isocratic (50/50 or 40/60) methanol/water mobile phase in conjunction with a C_{18} column ($\lambda = 229$ nm and photodiode array detector, $\lambda = 210$–367 nm) proved particularly effective for this purpose.

4.7.2   Retention Mechanisms for Drug Compounds

Gustafsson et al. [493] showed that multiple peak formation is possible for drug compounds (similar to dipeptides with an L-proline residue) that undergo a *cis–trans* isomerization whose transition rate is comparable to the chromatographic timescale. Ramipril and ramiprilate were chosen for study on a C_{18} column ($\lambda = 220$ nm) using a 55/45 methanol/water (0.1 M ionic strength phosphate buffer at pH 2.0) mobile phase. Under these conditions, changing the temperature altered the *cis/trans* isomerization ratio and, at 50°C, one sharp peak was eluted. Changes in mobile phase pH, organic constituent level, buffer concentration, and bonded phase also have strong effects on the chromatography. Awareness of such possible isomerization equilibria can save many hours of wasted and/or counterproductive work.

Cox and Stout [494] studied thiamin, caffeine and morphine retention on a silica column. A mobile phase was used consisting of $x/100 - x$ methanol/water (50 mM phosphate buffer at pH 4.6) where $x = 15$–75. As expected, k' decreased as x increased from 15% to 40%. Unexpectedly, k' for thiamin and caffeine *increased* as x increased from 40% to 75%. The explanation presented by the authors is that the effective mobile phase pH increased from 4.6 to 6.4 as the methanol level increased from 15% to 75%. This leads to increased silanol ionization (the "pK_a" for surface silanol groups usually falls between 4 and 6). An alternative, or perhaps comple-mentary, explanation is that at higher solution pH phosphate is more highly charged and begins to precipitate on the silica surface due to its limited solubility in highly methanolic solvents. This process increases the overall surface charge and results in increased retention times. Limited-solubility buffers present in a mobile phase must be used with care.

4.7.3 NSAIDs and Analgesic Drugs

Caffeine and five arylpropionic nonsteroidal anti-inflammatory drugs (NSAIDs) (indoprofen, ketoprofen, naproxen, fenbufen, ibuprofen) were extracted from blood and analyzed on a C_{18} column ($\lambda = 254$ nm). A two-ramp $90/10 \rightarrow 45/55$ (at 10 min) $\rightarrow 0/100$ (at 30 min) water (0.1 M formate pH 3)/methanol gradient generated excellent peak shape and baseline resolution [495]. The last peak eluted at 18 min. This raises the question of the utility of the last 12 min of the gradient. If this is a flush step, then a step gradient immediately after the last-eluting peak will dramatically shorten the overall analysis time. The reported linear range is 5–500 mg/L with a detection limit of 2 mg/L (S/N = 3) and a quantitation limit of 8 mg/L (S/N = 10).

Diclofenac and six metabolites (e.g., hydroxydiclofenac, dihydroxydiclofenac sulfate, 4-amino-3,5-dichlorobenzene sulfonic acid) were isolated from urine and separated on a C_{18} column ($\lambda = 282$ nm). A $95/5 \rightarrow 40/60$ (at 20 min) $\rightarrow 0/100$ (at 22 min hold 4 min) water (10 mM ammonium formate at pH 7.0)/methanol gradient gave excellent separation. Time-of-flight MS was used to confirm peak identity [496].

Clonixin was extracted from plasma and urine and quantitated on a C_{18} column ($\lambda = 290$ nm) using a $36/28/36$ methanol/acetonitrile/water (0.3% acetic acid) mobile phase [497]. It is interesting to note that the elution time for a urine extract was 7.1 min, whereas that for the plasma sample was 9.8 min. No change in mobile

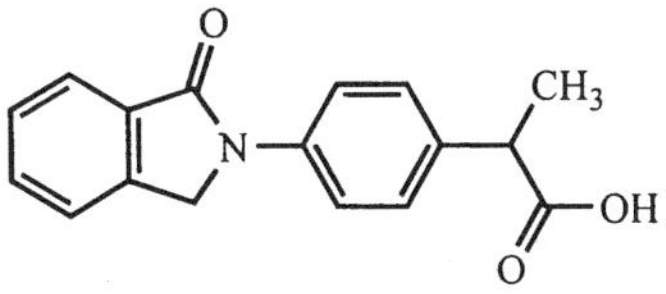

Indoprofen

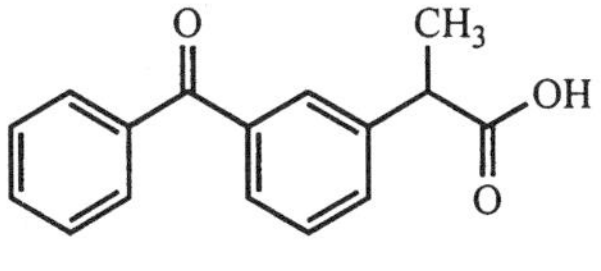

Ketoprofen

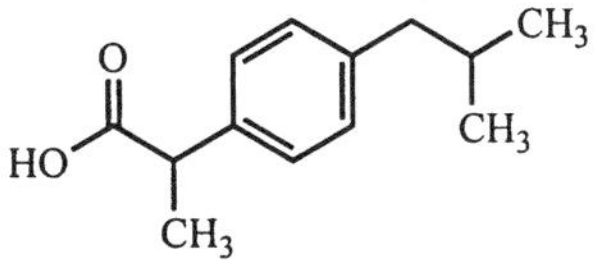

Ibuprofen

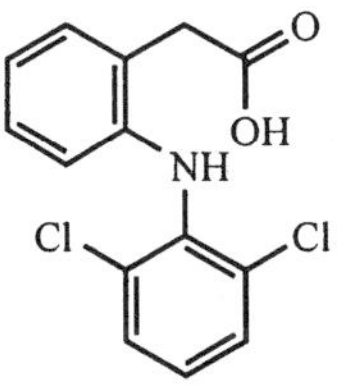

Diclofenac

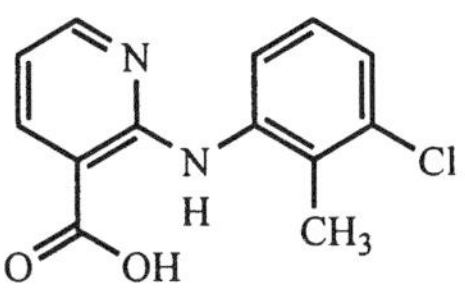

Clonixin

phase composition was mentioned. A linear concentration range of 0.01–2.0 µg/mL and a detection limit of 10 ng/mL (plasma) and 20 ng/mL urine were reported.

In a unique method Portier et al. [498] extracted fentanyl and midazolam from plasma and separated them on a silica column ($\lambda = 200$ nm) using 0.02% of 70% $HClO_4$ in methanol mobile phase. Separation was complete in 5 min but the analytes eluted on the shoulder of a large and tailed peak that eluted starting at the system void volume. This separation is amazing in two regards. First, it uses a silica column that, under these conditions, is fully deactivated. Second, the wavelength chosen for detection is well below the UV cutoff for methanol. Regardless of the conditions used, a fentanyl standard curve from 200 to 2000 pg/mL and one for midazolam from 50 to 400 ng/mL were reported; quantitation limits were 200 pg/mL and 10 ng/mL, respectively.

Ibuprofen and its hydroxy and carboxy metabolites were extracted from plasma and urine and separated on a C_{18} column ($\lambda = 220$ nm) using a mobile phase consisting of 65/35 methanol/water (1 mL H_3PO_4/L). Separation was complete in <15 min [499]. Recoveries from spiked samples were found to be greater than 95% for plasma (1 to 500 µg/mL) and greater than 85% for urine (50 to 500 µg/mL).

Weak solvents are sometimes used for samples in order to generate a chromatographically advantageous "compression" of the peak when it enters the column. However, the loss of flufenamic acid, indomethacin, and naproxen on supposedly inert PEEK (polyetheretherketone) injector tubing raises the question of the prudence of this practice. Hambleton et al. [500] used a narrow-bore 25°C C_{18} column ($\lambda = 254$ nm) and a 58/42 methanol/water (20 mM phosphate buffer at pH 7.0) mobile phase. A 20 µL PEEK injection loop was used and overfilled. When the analyte was in the aqueous buffer only (a weak solvent), peak carryover generated an increase in peak area until the sixth replicate injection. The peak area at this time was over 80% higher than the initial injection peak area. Increasing the methanol concentration in sample solvent dramatically decreased but did not eliminate this effect. Increasing the volume overloaded, increasing the solute solvent strength, and switching injector loops to stainless steel or fused silica all reduced the problem.

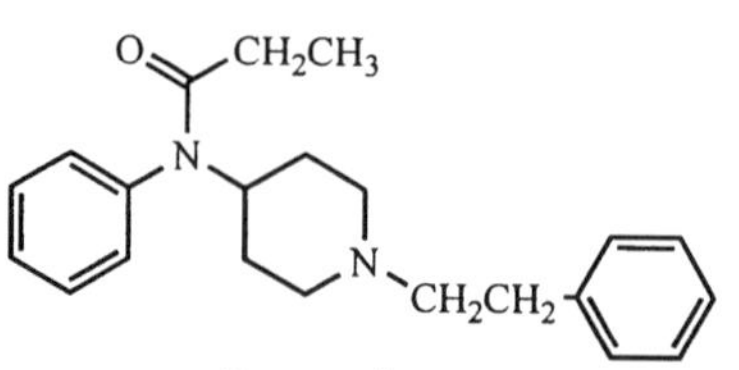

Fentanyl

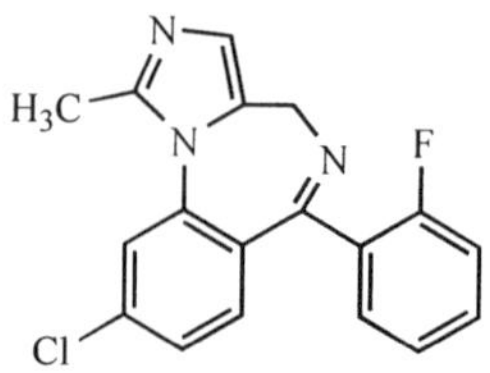

Midazolam

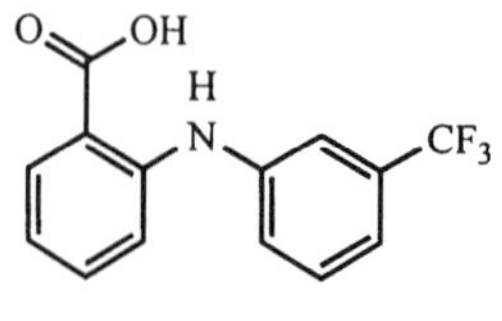

Flufenamic acid

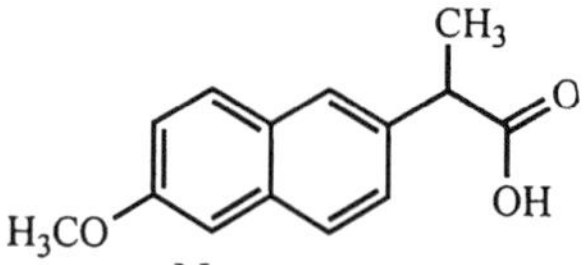

Naproxen

Acetaminophen, acetylsalicylic acid (aspirin), caffeine, and propoxyphene hydrochloride were well resolved on a C_{18} column ($\lambda = 254$ nm) using a 15/85 methanol/water (10 mM sodium acetate buffer at pH 4.1) mobile phase. Elution was complete in <12 min and 500 ng injections were easily detected [501].

4.7.4 Antibiotic Drugs

Yun et al. [502] separated 11 cephalosporin antibiotics (cefsulodin, cefadroxil, cefuroxime, cefoxitin, cefotaxime, cefazolin, cefaclor, cephalexin, cephradine, cephaloglycin, cephalothin) on a C_{18} column (pulsed amperometric detector). A complex 50-min 87/11/2 → 87/2/11 water (10 mM acetate buffer at pH 4.7)/methanol/acetonitrile gradient was used. Detection limits were reported as 30 ppb.

The bone marrow metabolites of chloramphenicol (CP) succinate were studied using a C_{18} column ($\lambda = 280$ nm) and a 70/30 water (10 mM NaH_2PO_4 at pH 6)/methanol mobile phase [503]. Along with the title compound, the retention times

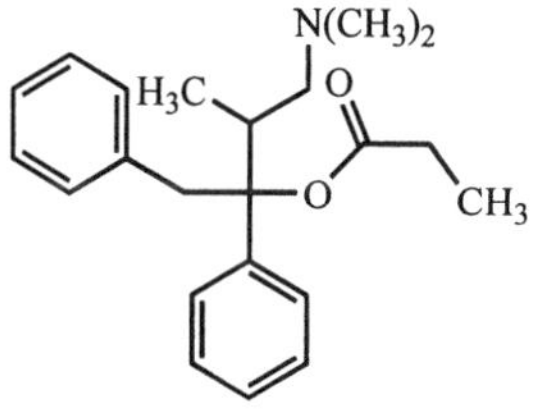

Propoxyphene

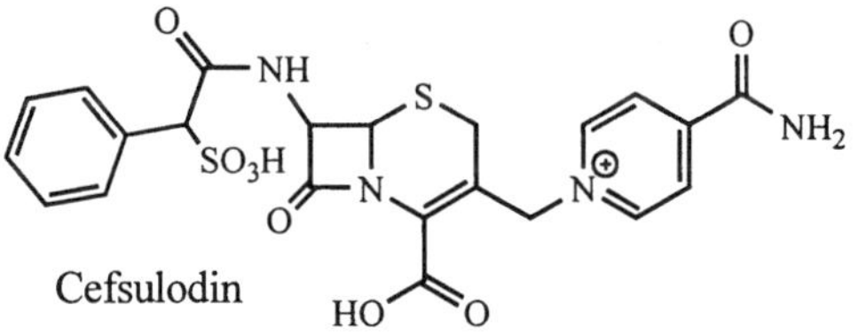

Cefsulodin

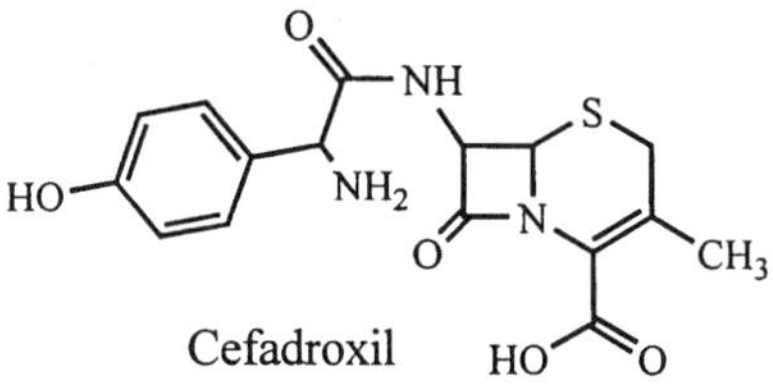

Cefadroxil

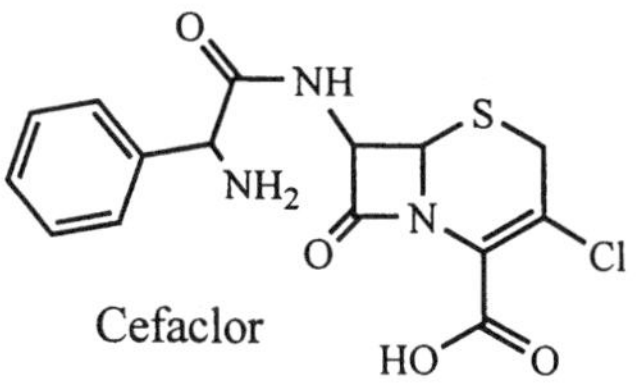

Cefaclor

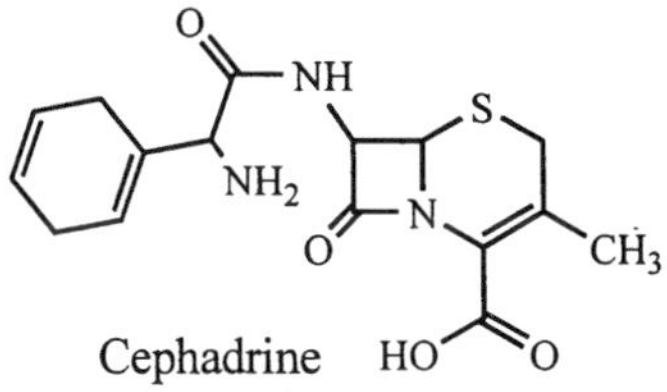

Cephadrine

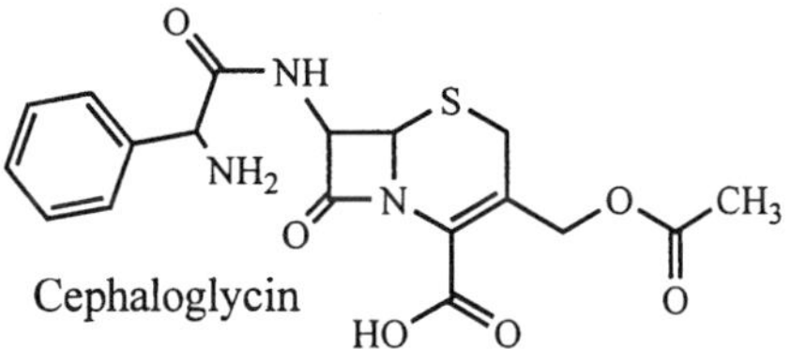

Cephaloglycin

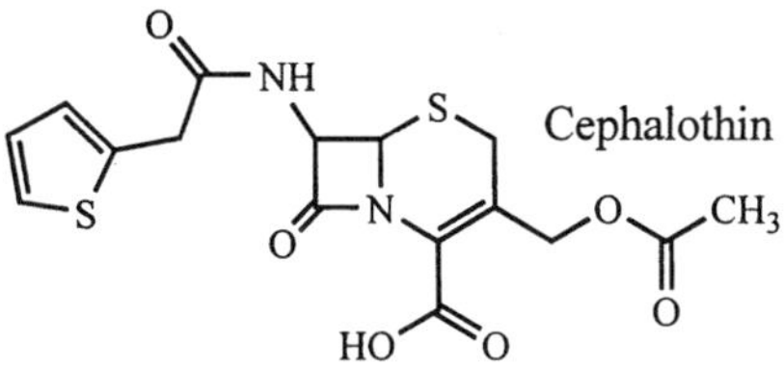

Cephalothin

for the CP oxamic acid, amine, aldehyde, glycerol and glucuonide compounds, the CP succinates I and II, and nitroso-substituted chloramphenicols were tabulated. For the five analogs actually shown in chromatograms, the overall peak shapes were somewhat broad but excellent resolution was achieved in 15 min. Reported standard concentrations of 5 µg/mL were used. In another study CP and 10 metabolites (e.g., CP-glucuronide, DP-diacetate, CP-3-acetate) were extracted from excreta and separated on a C_{18} column ($\lambda = 280$ nm) using a complex 60-min methanol/water gradient (20/80 [0.1 M citrate buffer at pH 4.4]) → (50/50 [0.1 M phosphate buffer at pH 5.3]) → (80/20 [0.1 M ammonium acetate buffer at pH 4.4]). Good resolution was obtained [504].

Four components of gentamicin (termed C_1, C_2, C_{1a}, C_{2a}) were extracted from bovine milk, derivatized with o-phthalaldehyde, and separated on a C_{18} column ($\lambda = 340$ nm, ex; 430 nm, em). An 82/18/0.1 water (11 mM pentanesulfonic acid with 5.6 mM sodium sulfate)/methanol/acetic acid mobile phase generated baseline resolution and elution in 20 min [505]. Standards ran from 30 to 240 ng/mL and detection limits were reported as 0.4 ng/mL.

Four polyether antibiotics (lasalocid, monensin, saliomycin, narasin) were extracted from feed, derivatized with 2,4-dinitrophenylhydrazine (DNPH), and quantitated on a C_{18} column ($\lambda = 305$ nm for lasalocid and 393 nm for all other compounds). A 90/10 methanol/water (1.5% acetic acid) mobile phase generated baseline resolution, good peak shape, and elution in <15 min. It should be noted that

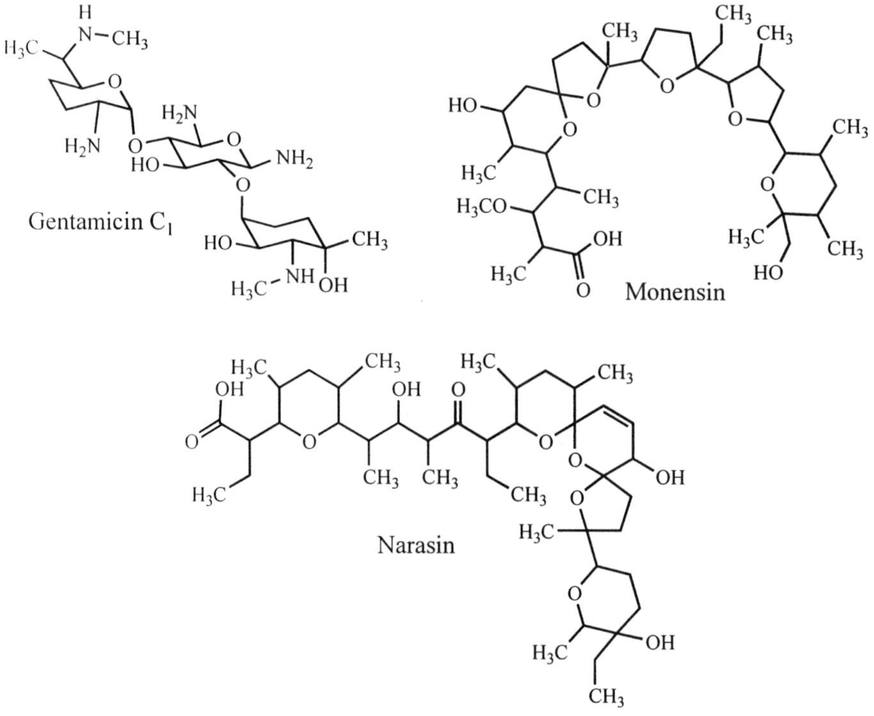

Gentamicin C₁

Monensin

Narasin

samples were extracted with methanol and directly derivatized. Samples spiked with 50–150 mg/kg were used and peaks were easily detected at the lowest concentration [506].

Doxycycline and its epimer, 4-epidoxycycline, were extracted from turkey tissue and separated on a PRLP-S column ($\lambda = 406$ nm, ex; 515 nm, em through post-column addition of 5% m/v zirconyl chloride). An 80/15/5 (hold 5 min) → 40/20/40 (at 20 min) water (10 mM oxalic acid)/acetonitrile/methanol gradient did not generate good resolution between the analytes, most likely due to the extra band broadening caused by the postcolumn system. A linear range of 100–500 ng/g and detection limits of 1 ng/g (S/N = 3) were reported [507].

Sulfaguanidine, sulfadiazine, and succinyl sulfathiazole from a pharmaceutical powder were baseline resolved on a C_{18} column ($\lambda = 270$ nm) using a 20/80 methanol/water (50 mM ammonium acetate) mobile phase [508]. Elution was complete in 5 min and peak shapes were excellent. Twelve sulfonamides were well resolved on a C_{18} column ($\lambda = 254$ nm) using an isocratic 6/94 IPA/water (20 mM sodium dodecylsulfate [SDS] with phosphate buffer at pH 3.0) mobile phase. Elution was complete in 15 min. The effects of changing % IPA and SDS concentration on k' and α values were presented [509]. This study provides excellent background information for method development. Detection limits of ~1 µg/mL were reported.

Penicillin G, ampicillin, amoxicillin, cloxacillin, and cephapirin were separated on a C_{18} column ($\lambda = 230$ nm or positive ion electrospray MS) using three mobile

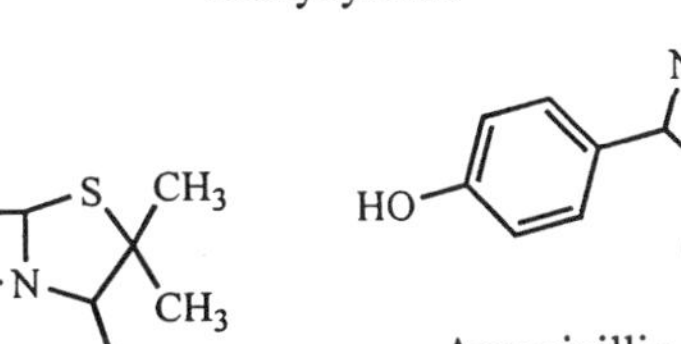

Doxycycline

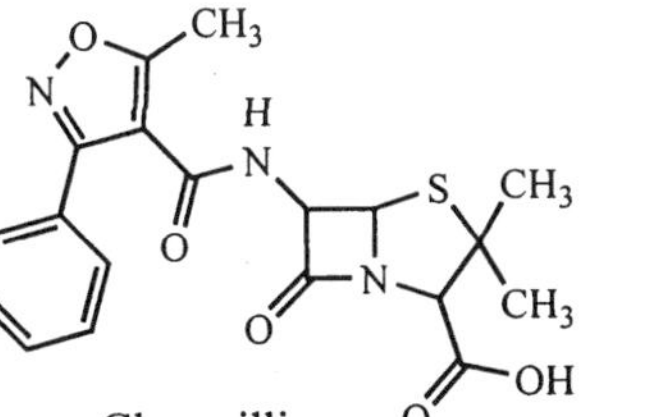

Ampicillin

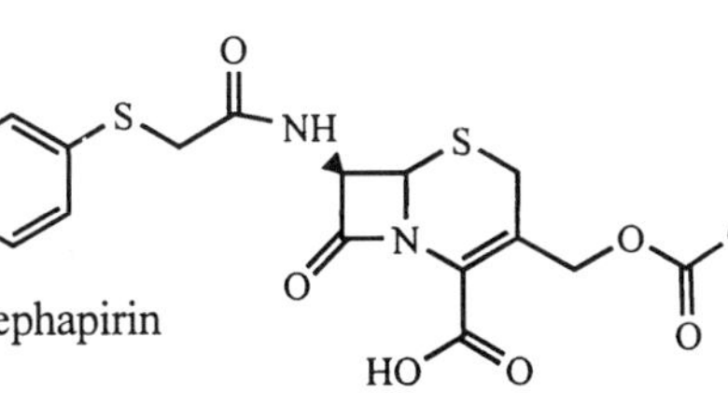

Amoxicillin

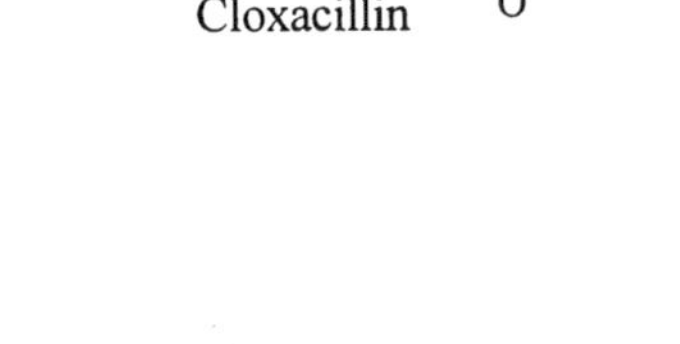

Cloxacillin

Cephapirin

phases: 80/20, 50/50, 20/80 methanol/water [510]. Ion response was determined with a 50/50 methanol/water mobile phase with various individual mobile phase modifiers: 0.1% TFA (pH 2.2), 10 mM heptafluorobutyric acid (pH 2.2), 0.1% acetic acid (pH 3.4), 10 mM tetrabutylammonium hydroxide (pH 8.4), and 50 mM ammonium hydroxide (pH 10.5). Previous studies by these authors showed cloxacillin to be unstable in all storage solvents studied [511]. However, a 25/25/50 ethanol/acetonitrile/water storage solvent showed only 5% degradation over 8 weeks of storage, whereas 100% methanol, 50/50 methanol/water, and 50/50 acetonitrile/water (0.2% phosphoric acid) showed >80% degradation of cloxacillin in 2 weeks. The degradation products (cloxacillinpenicilloic acid, cloxacillinpenilloic acid) were separated using a C_{18} column (thermospray MS) and a 12/2.5/85.5 IPA/acetic acid/water mobile phase.

Degradation of samples is always an important consideration for method development. A series of mobile phases may need to be tested to find one compatible with the sample. Time studies for storage of samples and standards should also be run. Derivatized products typically have limited stability and analytical results could be extremely dependent on the time between the derivatization event and analysis. Also, the use of pure water as the sample solvent should be avoided since bacterial growth presents a major problem for the chromatographer.

Erythromycin A was resolved from 10 related impurities (e.g., erythromycin B, C, E and F, pseudoerythromycin A enol ether, anhydroerythromycin A) in 40 min using a 70°C PLRP-S column ($\lambda = 215$ nm). A 16.5/3/5/75.5 t-butyl alcohol/acetonitrile/water (0.2 M phosphate buffer at pH 9.0)/water mobile phase was used [512]. The linear range for erythromycin A was 100–400 µg injected, with impurity levels down to 0.03% (300 ppm) still detectable. Note that the elevated operating temperature, while not necessarily precluding the use of methanol, is above the boiling point of pure methanol (65°C) and could lead to rapid compositional changes and thereby irreproducible results. At such elevated temperatures, the injection of the room-temperature samples could also result in erratic results—a fact that is often overlooked. Solute stability at these elevated temperatures is also critical, since samples are held at these temperatures for many hours in a sample queue.

4.7.5 Anticancer Drugs

Forgács and Cserháti [513] studied 21 anticancer drugs with the intention of determining the relative hydrophobicities of these drugs. Retention of these compounds (e.g., vinblastine, paraplatin, doxorubicin, mitomycin C, methotrexate,

Mitomycin C

Methotrexate

mitolactol, paclitaxel) was monitored on a C_{18} column ($\lambda = 215$ nm) using isocratic solvents whose composition ranged from 0 to 90% methanol in water (25 mM KH_2PO_4). The correlation of k' with change of methanol concentration over the range is reported and is very valuable for method development.

Eibl et al. [514] characterized a series of alkylphosphocholines (APCs)—potential antineoplastic agents—based on alkyl chain length. Linear APCs (C_{14}–C_{22}, even only) were baseline resolved in 12 min on a trimethylsilyl column (RI at 35°C) using an 85/15 methanol/water mobile phase. Similarly, $C_{21:1}$, $C_{23:1}$, $C_{24:1}$, $C_{25:1}$ and erucyl-PC were resolved in 12 min (UV detection at $\lambda = 206$ nm). This operating wavelength is problematic, being so close to the UV cutoff for methanol. Indeed, significant baseline drift is observed over the course of a single chromatogram. Finally, oleyl- and elaidyl-PC were resolved in 25 min on a C_8 column (refractive index at 35°C) using an 80/20 methanol/water mobile phase. Linear ranges for these analytes were reported as 5–160 nmol injected (for RI) and 5–80 nmol (UV).

Eight liver metabolites of porfiromycin (e.g, 2-methylamino-7-aminomitosene, 1,2-*cis*- and 1,2-*trans*-1-hydroxy-2-methylamino-7-aminomitosene, and phosphorylated and decarbamoylated analogs) were isolated and separated on a 50°C C_{18} column ($\lambda = 250$ and 550 nm). A two-ramp $100/0 \rightarrow 90/10$ (at 15 min) $\rightarrow 60/40$ (at 40 min) gradient of water (50 mM K_2HPO_4 to pH 5.0)/methanol was used. Elution was complete in <30 min and no further washout peaks were observed, so that a shorter gradient may be appropriate [515].

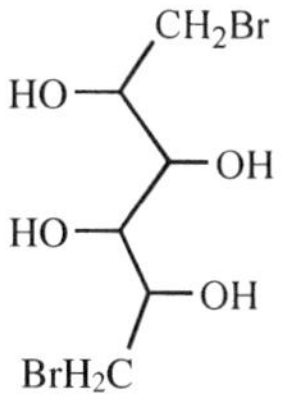

Mitolactol

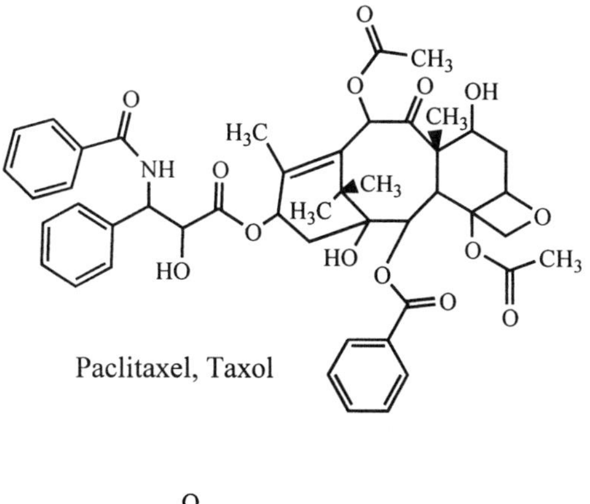

Paclitaxel, Taxol

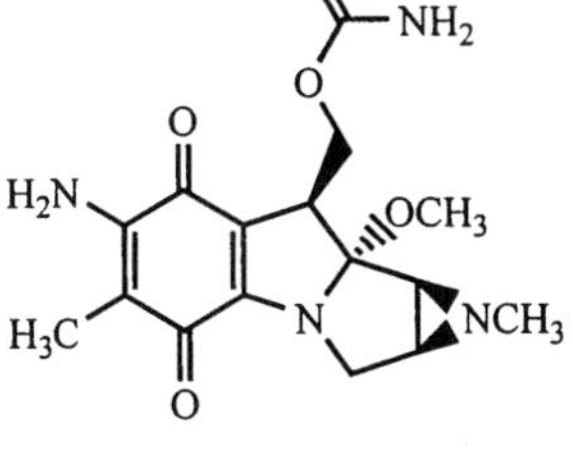

Porfiromycin

A water-soluble peptidic-mitomycin C antitumor antibiotic, 7-*N*-[(2-γ-L-glutamyl-amino)ethyldithioethyl] mitomycin C, and two metabolites, the methyl sulfide and the symmetrical disulfide, were extracted from plasma and analyzed on a C_{18} column ($\lambda = 375$ nm). A 15-min $58/42 \rightarrow 67/33$ methanol/water (50 mM KH_2PO_4 to pH 6.5 with NaOH) gradient generated baseline resolution. The working concentration range was reported as $10–1000$ ng/mL with detection limits of 10 ng/mL [516].

The leukemia treatment drug 6-mercaptopurine and seven metabolites (e.g., 6-methylmercaptopurine, 6-mercaptopurine riboside, 6-thioguanosine, 6-thioxanthine) were extracted from plasma and separated on a C_{18} column ($\lambda = 295$ nm and 330 nm). A 50-min $90/10$ (hold 10 min) $\rightarrow 0/100$ (at 50 min) 98/2 water [30 mM ammonium phosphate pH 3]/methanol)/(60/40 water [30 mM ammonium phosphate pH 3]/methanol) gradient was used [517]. Detection limits were reported as 20–50 nM (analyte dependent).

6-Mercaptopurine and two metabolites (6-thioguainine, 6-methylmercaptopurine) were extracted from red blood cells and analyzed on a 40°C C_{18} column ($\lambda = 289$ nm and 332 nm) with a $100/0$ (hold 4 min) $\rightarrow 86/14$ (at 14 min hold 6 min) water (0.1% heptanesulfonic acid with 20 mM phosphate at pH 6.4)/methanol gradient [518]. The analytes were well resolved from a sizable number of other extracted compounds. Linear ranges of 20–2000 pmol/25 mg hemoglobin were reported. The detection limits were cited as 20 pmol/25 mg hemoglobin.

8-Chloroadenosine and two serum metabolites (8-chloroinosine and 8-chloro-adenine) were isolated and analyzed on a C_{18} column ($\lambda = 263$ nm) using an $86/11/3/0.1$ water (1% acetic acid)/methanol/acetonitrile/tetrabutylammonium bromide mobile phase [519]. The 8-chloroadenosine peak was severely tailed and eluted last at 21 min. A more effective mobile phase modifier should be investigated (e.g., a TFA/triethylamine mixture). A linear range of 0.1–10 µg/mL and detection and quantitation limits of 0.5 µg/mL (S/N = 2) and 4 ng/mL (S/N = 10), respectively, were reported.

5-Bromo-2′-deoxyuridine and five hydrolysis products (bromouracil, uracil, 5-hydroxy-2′-deoxyuridine, 5-methyl-3(2*H*)-furanone, 2′-deoxyuridine) were separated in 14 min on a C_{18} column ($\lambda = 205$ nm) using a $90/10$ water (10 mM KH_2PO_4 buffer at pH 4.8)/methanol mobile phase [520]. Note that the system is operated at a wavelength very close to the cutoff for methanol. Detection limits and sensitivity will undoubtedly be adversely affected. Standards with concentrations of 10 mg/mL gave strong responses. Peak shapes were very good, as was overall resolution.

6-Mercaptopurine   6-Thioxanthine   5-Bromo-2′-deoxyuridine

Tamoxifen and two major metabolites (4-hydroxy- and desmethyltamoxifen) were extracted from breast tumor tissues and separated in 10 min on a C_{18} column ($\lambda = 265$ nm) using an 89/11 methanol/water (1% triethylamine) mobile phase [521]. The chosen internal standard was incompletely resolved from the tamoxifen. A linear curve was generated from 2 to 2000 ng injected with a detection limit of 40 pg injected claimed.

Auraptene, an anticancer phytochemical candidate, was extracted from a wide range of citrus fruits and analyzed on a 40°C C_{18} column ($\lambda = 325$ nm). A 75/25 methanol/water mobile phase generated elution in 6 min. A detection limit of 0.1 µg/g was claimed [522].

Taxol and seven other major taxoids (10-deacetylbaccatin III, baccatin IV, 1-hydroxybaccatin I, 2-acetoxybrevifoliol, brevifoliol, 2′-deacetoxydecinnamoyltaxinine J, 2′-deacetoxytaxinine J) were extracted from *Taxus wallichiana* and characterized on a C_{18} column ($\lambda = 228$ nm). An 80/20 → 60/40 (at 20 min) → 47/53 (at 26 min) → 20/80 (at 31 min) → 0/100 (at 37 min hold 8 min) (20/5/75 methanol/acetonitrile/water)/(30/35/35 methanol/acetonitrile/water) gradient generated excellent peak shapes and baseline resolution [523].

Paclitaxel and six related taxanes [524] were separated on a pentafluorophenyl column ($\lambda = 227$ nm and 273 nm) using a 50-min 76/24 → 50/50 (6/94 reagent alcohol/water)/acetonitrile gradient (where alcohol refers to reagent alcohol and the manufacturer's lot specification was 90.6% ethanol, 4.5% methanol, 4.9% IPA). Excellent peak shape and resolution were achieved.

4.7.6 Antiepileptic Drugs

Vidal et al. [525] extracted the anticonvulsant lamotrigine from serum and analyzed it in <2 min on a silica column ($\lambda = 280$ nm) using a 94/5.96/0.04 methanol/water/water (1 M ammonium phosphate) to pH 5 with H_3PO_4 mobile phase. The analyte was cleanly resolved from all other peaks. Note that the silica column, with the high water level in the mobile phase, is fully deactivated. The linear range was 0.5–20 µg/mL, with a reported quantitation limit of 0.35 µg/mL.

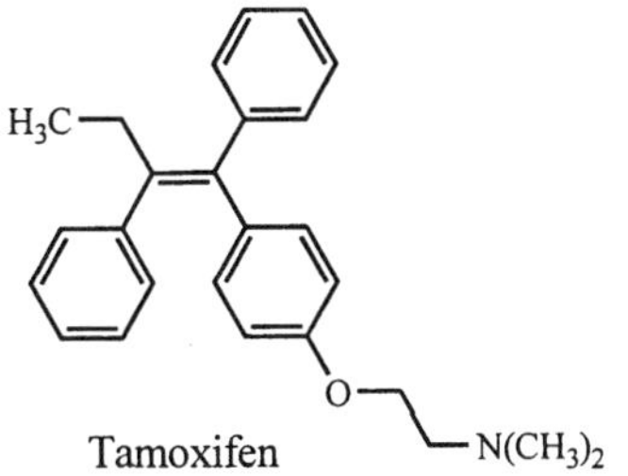

Tamoxifen

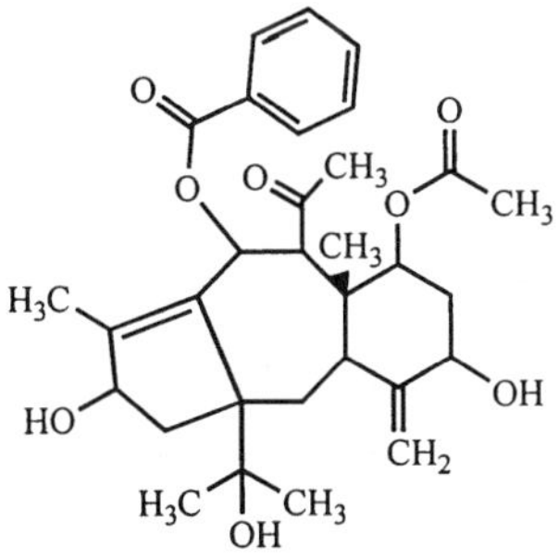

Brevifoliol

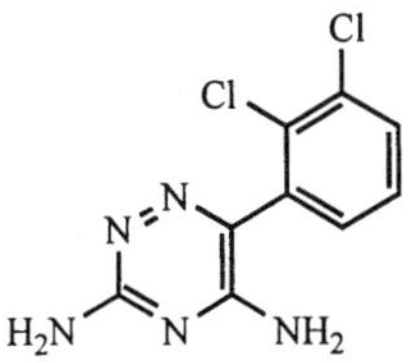

Lamotrigine

Urine levels of carbamazepine and five metabolites (carbamazepine epoxide and acridan, carbamazepine-2- and -3-hydroxide, and carbamazepine-10,11-*trans*-diol) were determined on a C_{18} column ($\lambda = 240$ nm to 280 nm at 18 min) using a 15/15/70 (hold 12 min) → 15/35/50 (at 24 min hold 6 min) acetonitrile/methanol/water (9.5 mM KH_2PO_4 buffer at pH 7) gradient [526]. Peak shapes were excellent and well resolved. A calibration curve was generated from 0.1 to 1000 µg/mL. Quantitation limits were reported as 0.2 µg/mL.

Valproic acid from serum (with an undecylenic acid internal standard) was analyzed as its 4-bromomethyl-7-methoxycoumarin derivative on a 40°C C_{18} column ($\lambda = 322$ nm, ex; 695 nm, em) using an 80/20 methanol/water mobile phase [527]. The authors noted that a 1.6-fold increase in sensitivity was obtained by changing λ_{ex} to 228 nm, but more interferences were then detected (i.e., selectivity decreased). The use of methanol rather than acetonitrile resulted in a 2-fold increase in sensitivity. The linear working range was reported as 6–200 µg/mL.

Barbital, pheno-, buto-, amo- and heptabarbital, and secobutobarbitone were resolved on a C_{18} column ($\lambda = 235$ nm) using a 26/5/29/40 ethanol/ IPA/ methanol/water (0.01 M acetate buffer at pH 5.5) mobile phase. Detection limits of 0.09–0.24 ng were reported. Linear ranges up to ~150 ng were achieved [528]. The analysis was completed in 10 min.

4.7.7   Steroidal Drugs

Testosterone and six cell microsome metabolites (6β-, 16α-, 16β, 7α-, 2α-hydroxytestosterone, and androstenedione) were well resolved and eluted in <35 min on a 40°C C_{18} column ($\lambda = 242$ nm) using a 50/38.5/11.5 water (0.1 mM Na_2HPO_4 at pH 6)/methanol/acetonitrile mobile phase [529]. The authors make special note that the use of THF is to be avoided due to chances of its reacting with the analytes. Linear ranges of 280–4600 ng/mL for metabolites and 20–40 µg/mL for testosterone were reported.

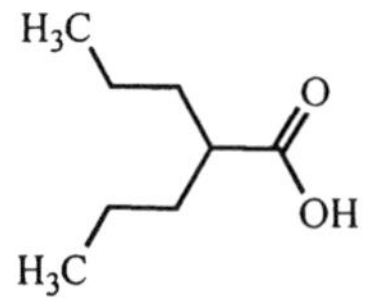

Valproic acid 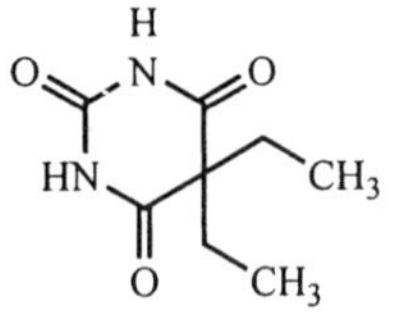   Barbital

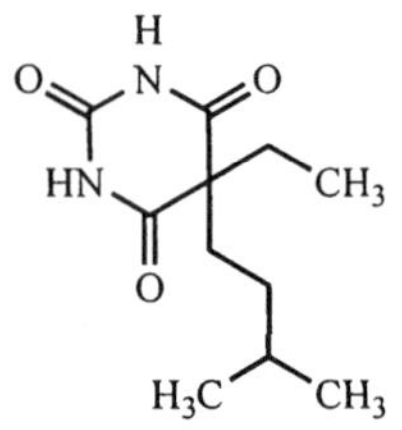

Amobarbital

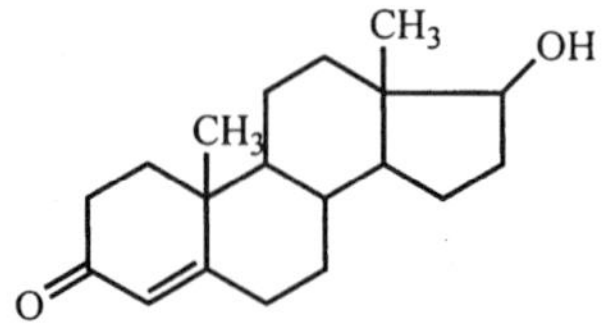

Testosterone

Three corticosteroids (dexamethasone, flumethasone, triamcinolone acetonide) were recovered from feedstock and separated on a C_{18} column ($\lambda = 242$ nm) using a 50/50 methanol/water mobile phase [530]. Baseline separation was achieved and elution was complete in 14 min. Detection limits of 100 µg/mL were reported.

Ethinylestradiol, norgestrel, and norethisterone acetate from tablets were resolved on a β-cyclodextrin column ($\lambda = 280$ nm) using a 40/60 methanol/water (50 mM phosphate buffer at pH 7.0) mobile phase [531]. Peak shapes were excellent and 25 ng injected was easily detected.

Prednisolone, prednisone, and four 20-reduced metabolites were analyzed on a C_{18} column ($\lambda = 242$ nm) using a 57.5/43.5 methanol/water mobile phase. Detection limits of 10 ng/mL and a linear working range to 1 µg/mL were reported [532]. Excellent peak shapes were generated and elution was complete in 30 min. Ten potential steroid and 10 common drug interferents were also analyzed; cortisol was the only co-eluting peak (with prednisolone).

Hirata et al. [533] developed a method to separate prednisolone, prednisone, cortisol, cortisone, and deflazacort metabolites II & III. A C_{18} column ($\lambda = 254$ nm) and a 10/90 $\rightarrow$ 30/70 IPA/water (0.05 M acetate buffer at pH 4.5) gradient gave good peak shapes and baseline resolution. Detection limits of 10 ng/mL were reported.

A 7/93 1-pentanol/water (0.1 M SDS) mobile phase was used to separate seven anabolic steroids (e.g., methyltestosterone, progesterone, testosterone propionate and enanthate) in under 15 min on a C_{18} column ($\lambda = 246$ nm). The effects of 1-pentanol and SDS levels were studied and described [534]. The rapid elution and low UV background absorbance (due to the low levels of pentanol present) offer great advantages over the use of other alcohols (since they would need higher levels and

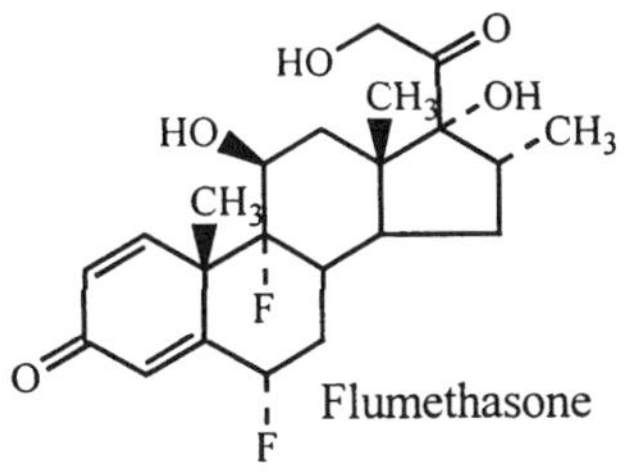

Flumethasone

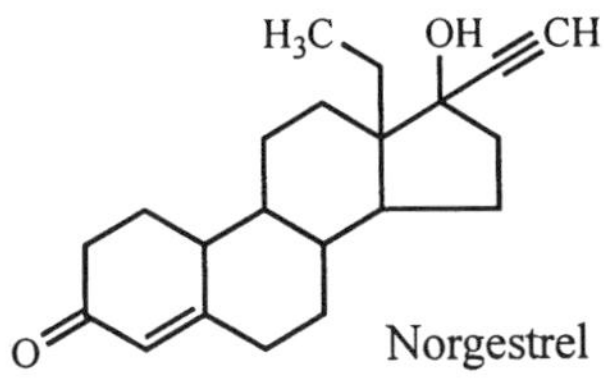

Norgestrel

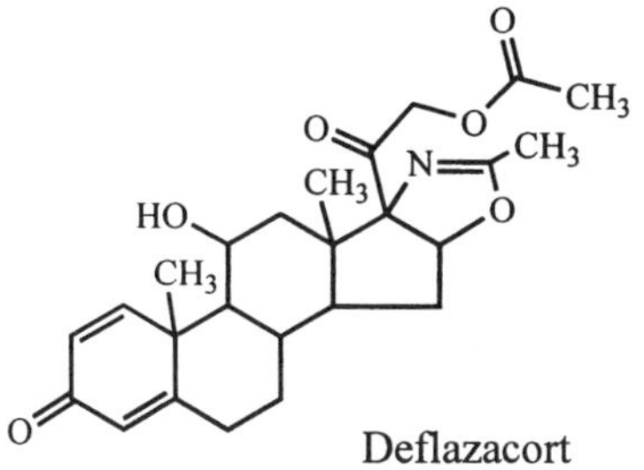

Deflazacort

would generate a higher background signal and different selectivities). Detection limits of 5 µg/mL (20 µL injection) were reported.

4.7.8   Anthelmintics

Mebendazole and its degradation product, 2-amino-5-benzoylbenzimidazole, were determined in tablets and suspensions C_{18} column ($\lambda = 290$ nm) using a 5/3/2 water (50 mM KH_2PO_4)/methanol/acetonitrile mobile phase [535]. Elution was complete in <8 min. The authors note that the removal of phosphate from the mobile phase leads to severe peak tailing. The linear range of the parent compound was 40–160 µg/mL and of the degradation compound 0.4–20 µg/mL. Detection limits were reported as 3 µg/mL and 15 ng/mL and quantitation limits as 9 µg/mL and 120 ng/mL, respectively.

Febantel is an anthelmintic predrug that is rapidly converted to fenbendazole and oxfendazole. These two compounds and seven related liver metabolites (e.g., oxfendazole amine, 4-hydroxyfendazole, oxfendazole sulfone) were isolated and analyzed on a C_8 column ($\lambda = 290$ nm). A two-ramp 65/35 → 56/44 (at 18 min) → 20/80 (at 23 min hold 7 min) water (0.1 M ammonium carbonate)/methanol gradient was used. Complete resolution of all peaks was not obtained, but none co-eluted [536].

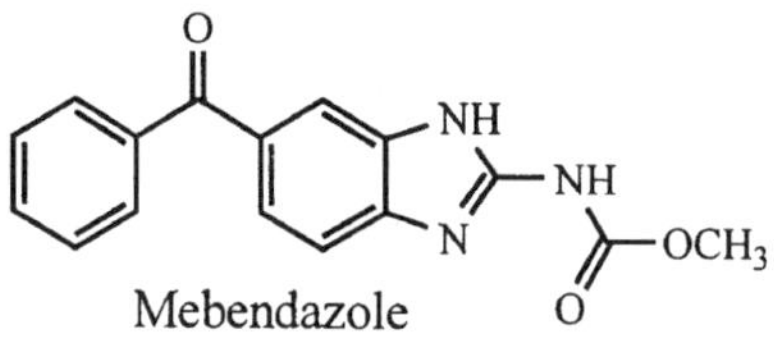

Mebendazole

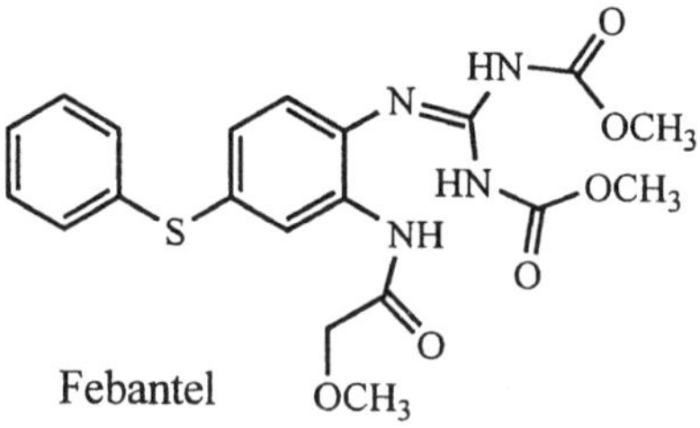

Febantel

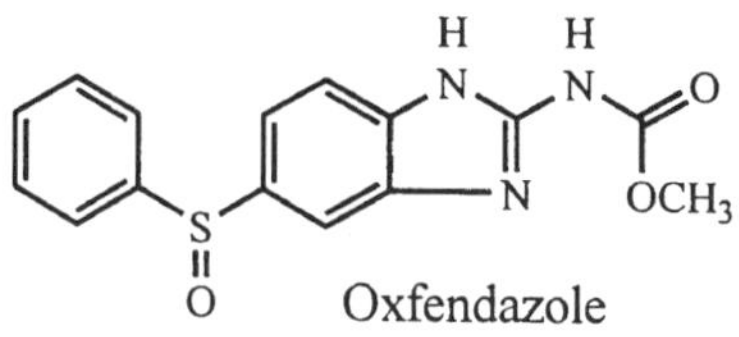

Oxfendazole

Sher Ali et al. [537] extracted abamectin, eprinomectin, moxidectin, ivermectin, and doramectin from liver and derivatized them with 1-methylimidazole/trifluoroacetic anhydride prior to separation and analysis on a C_{18} column ($\lambda = 365$ nm, ex; 465 nm, em). Baseline resolution and excellent peak shapes were obtained with a 97/3 methanol/water mobile phase. Elution was complete in <10 min. A standard curve from 25 to 100 ppb was used.

4.7.9 Illicit Drugs

Morphine and three metabolites (morphine-3- and -6-glucuronide, normorphine) were resolved on a porous graphitized carbon (PGC) column ($\lambda = 220$ nm) in 18 min using a 60/40 methanol/water (50 mM ammonium acetate buffer at pH 10.6) mobile phase. It should be noted that a PGC column is stable under these high pH conditions whereas many silica supports are not. Linear ranges of $10–100\,\mu g/mL$ were reported [538]. The effects of pH, percent methanol, column temperature, and ion pairing agent on retention were presented graphically.

Morphine, dilaudid, naloxone, and naltrexone were extracted from urine and separated on a C_{18} column (electrochemical detector at +0.85 V vs. Ag/AgCl) using a 10/90 IPA/water (0.1 M ammonium phosphate at pH 4.5 with 0.02% sodium octylsulfate) mobile phase [539]. Baseline resolution was obtained. More positive potentials (e.g., +0.95 V) lead to increased sensitivity but faster poisoning of the electrode. A linear response range from 50 to 5000 ng was obtained.

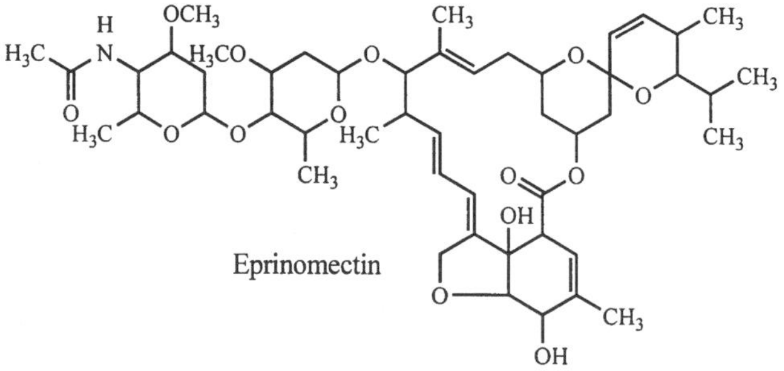

Eprinomectin

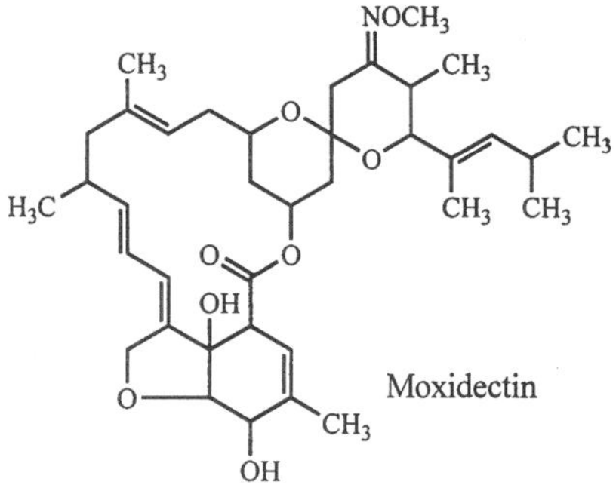

Moxidectin

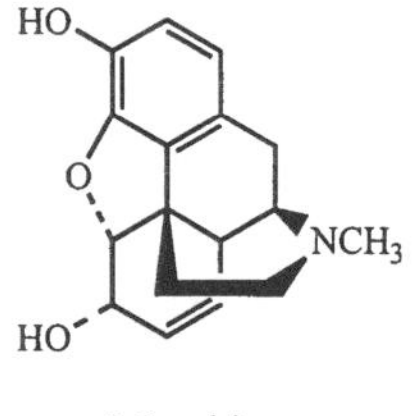

Morphine

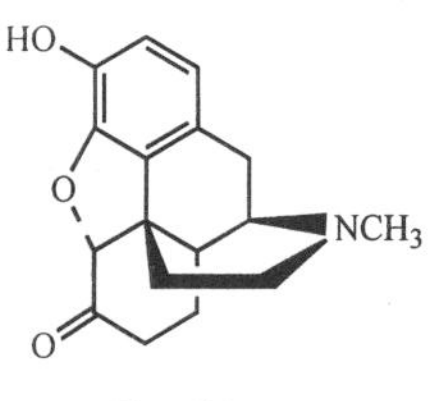

Dilaudid

In an extremely powerful technique, Dams et al. [540] separated 18 opiates and their derivatives (morphine, codeine, naloxone, hydrocodone, papaverine, dextromethorphan, noscapine, bupremorphine, methadone, heroin, thebacone, ethylmorphine, 6-monoacetylmorphine, acetyldihydrocodeine, acetylcodeine, normethadone, normorphine, norcodeine) with baseline resolution in <12 min! This was accomplished on a "fast" LC phenyl column 53 × 7 mm ($\lambda = 280$ nm) using a simple $100/0 \rightarrow 50/50$ (at 10 min) $\rightarrow 0/100$ (at 12 min hold 1 min) (90/5/5 water/methanol/acetonitrile with 50 mM ammonium acetate)/(50/50 methanol/acetonitrile with 50 mM ammonium acetate) gradient. The mobile phase components were adjusted to pH_{app} 4.5 with formic acid prior to use. The phenyl column was chosen over the conventional C_8 or C_{18} columns because of the increased selective interaction with analytes containing the planar phenyl functional group. Detection limits from 50 to 450 ng/mL (analyte dependent) were reported.

γ-Hydroxybutyrate and γ-butyrolactone levels were determined in illegal preparations using a C_{18} column ($\lambda = 214$ nm) and a 70/30 water (10 mM KH_2PO_4 to pH 3.0 with H_3PO_4)/methanol mobile phase [541]. Excellent peak shape and baseline resolution were obtained in <5 min. Calibration curves from 0.5 to 30 μg injected were used. The authors noted that the high on-column concentrations of γ-hydroxybutyrate resulted in peak splitting, perhaps a result of sample solvent mismatch with the mobile phase. Also, significant baseline drift was evident at very low analyte concentration, a consequence of the methanol in the mobile phase. Detection limits of ~150 ng injected were reported.

Four *Cannabis sativa* constituents (co-eluting acidic cannabinoids, cannabidiol, cannabinol, Δ^9-tetrahydrocannabinol) were extracted from hashish powder and analyzed on a C_{18} column ($\lambda = 220$ nm and electron impact MS, desolvation $T = 45°C$, ionization energy 70 eV, ion source $T = 250°C$, filament current 200 μA). An 80/20 methanol/water mobile phase generated baseline resolution in 20 min [542]. The reported calibration range was 0.25–50 μg/mL with detection limits of 0.2 μg/mL (S/N = 3). These authors later modified this method in order to be able to determine hashish sources. The major difference is that a 30-min gradient from $70/30 \rightarrow 90/10$ methanol/water was used. This enables the analysis of the above analytes plus cannabichromene [543]. Here calibration standards ranged from 0.04 to 0.6 μg/mL.

Methadone

Thebacone

γ-Butyrolactone

Arachidonylethanolamine (anandamine, a cannabinnoid) was extracted from various culture media, dansylated and analyzed on a C_{18} column ($\lambda = 255$ nm). Heptadecanoylethanolamine was used as an internal standard. A complex 35-min $40/60 \rightarrow 99.5/0.5$ methanol/water (0.1% acetic acid) gradient resolved the analyte of interest from co-extracted materials and the IS in <29 min [544]. The linear range for the analysis was reported as $0.5-100$ µg/mL with a detection limit of 4 nmol ($S/N = 3$).

4.7.10 Antihistimines

The antihistamines synopen and ephedrine were separated and eluted in <10 min on a diol column ($\lambda = 254$ nm) using a 30/70 methanol/water (10 mM triethylamine to pH 4 with H_3PO_4) mobile phase. For the most reproducible results the authors correctly noted that the pH of the solution should be monitored in the aqueous phase and then added to the organic. An extensive table of the effects of changing triethylamine concentration, pH, and percent methanol on the retention of the analytes is also presented [545].

Ranitidine, ranitidine N-oxide, ranitidine S-oxide, and desmethylranitidine were baseline resolved in 8 min on a 40°C C_{18} column using a 46/5/49 methanol/acetonitrile/water (35 mM phosphate buffer at pH 7.0) mobile phase [546]. Often, as in this case, small amounts of acetonitrile added to the mobile phase will yield sharper peaks and, therefore, overall better efficiency. Here, excellent peak shapes were obtained and responses were linear between 10 and 1000 ng.

Methapyrilene, methapyrilene N-oxide, and mono-N-desmethyl methapyrilene were extracted from urine and baseline resolved on a C_{18} column ($\lambda = 254$ nm) using a 40-min $10/90 \rightarrow 100/0$ methanol/water (10 mM phosphate buffer with 200 µL/L triethylamine to pH 7.0) gradient [547]. Extrapolation from the absorb-

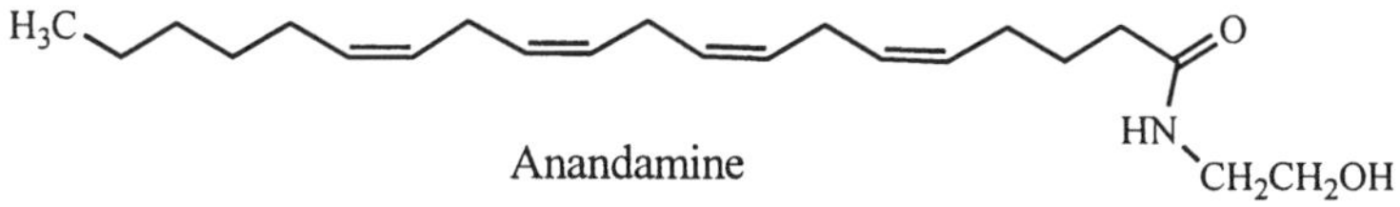

Anandamine

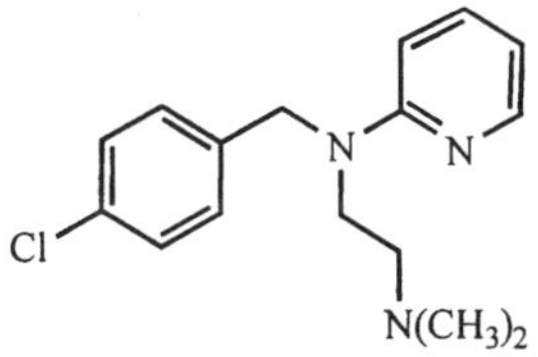

Synopen

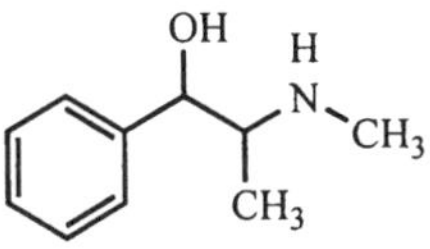

Ephedrine

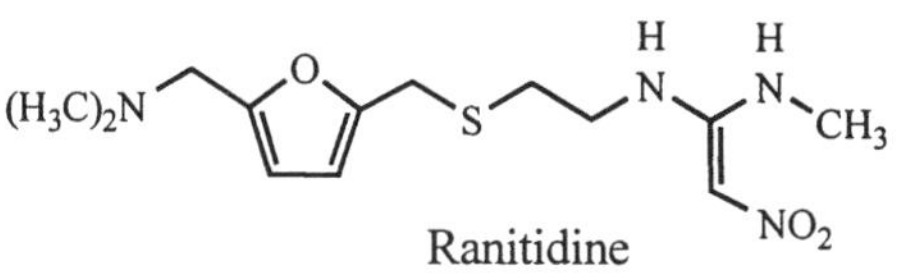

Ranitidine

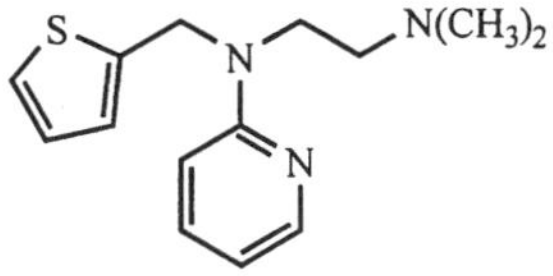

Methapyrilene

ance obtained from a 5 µg standard injection gave an approximate detection limit of 50 ng injected. Good resolution and peak shapes were obtained.

4.7.11   Anti-HIV Drugs

The separation of hydrocinnamoyl chloride (also known as 3-phenylpropionyl chloride, a precursor to indinavir), from related production impurities (α-chlorocinnamic acid, hydrocinnamic acid) and aniline derivatives (3-phenylpropionamide, N-hydrocinnamoyl anilide, N-benzyl anilide, N-cinnamoyl anilide, N-α-chlorocinnamoyl anilide, N-3-cyclohexylpropionyl anilide) was accomplished in 25 min on a C_8 column ($\lambda = 215$ nm). The anilides were derivatized with benzyl amine prior to analysis. A complex 30-min methanol/water (0.1% H_3PO_4 at pH 2.2) gradient from 55/45 → 75/25 was used [548]. Considerable baseline drift was observed due to the absorbance of methanol at 215 nm. A linear range of 0.2–2.0 mg/mL and detection limits of 200 µg/mL (S/N = 5) were reported.

Iododoxorubicin and three urinary metabolites (doxorubicin, (13R)- and (13S)-iododoxorubicinol) were separated on a C_{18} column ($\lambda = 470$ nm, ex; 522 nm, em) using a 60/40 methanol/water (60 mM phosphate buffer at pH 3) mobile phase [549]. Elution was complete in 80 min and detection limits of 6 ng/mL were reported.

Thymidine, 5-methoxypsoralen, and their photoadducts were separated on a C_{18} column ($\lambda = 254$ nm) using a 35/65 methanol/water mobile phase [550]. The cis-syn and cis-anti monocyclo-adducts were well resolved from the linear adducts and the psoralen dimers. Elution was complete in 45 min.

4.7.12   Antianxiety and Antipsychotic Drugs

Khedr and Sakr [551] separated buspirone hydrochloride and four degradation products (buspirone acid hydrochloride, a bispyrimidinylpiperazinebutane, a chlorobutylazaspirodecanedione, and an azaspirodecanedione) on a C_{18} column ($\lambda = 244$ nm and 210 nm) using a 65/35 (hold 5 min) → 46/54 (at 10.5 min hold

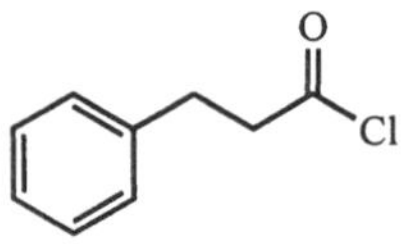

Hydrocinnamoyl chloride

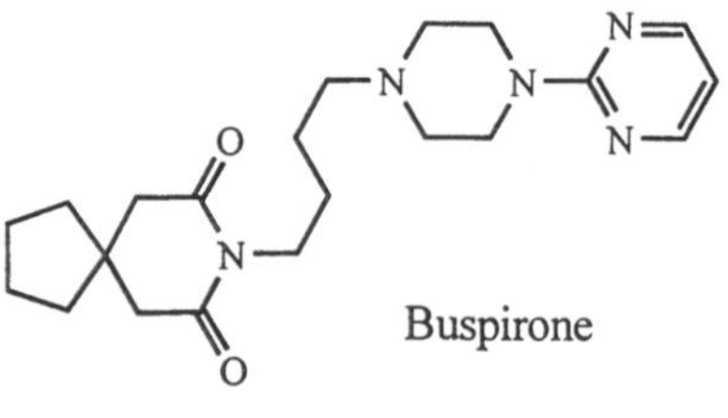

Buspirone

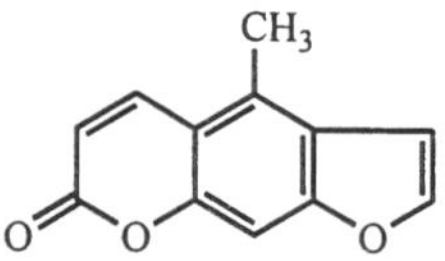

5-Methylpsoralen

20 min) water (1.36 g KH$_2$PO$_4$/L to pH 6.9 with NaOH)/(17/13 methanol/acetonitrile) gradient. Standard concentrations ran from 5 to 50 μg/mL, detection limits were reported as 0.5 μg/mL (S/N = 3) and quantitation limits as 3 μg/mL (S/N = 10). The gradient at 210 nm shows a marked baseline shift due to the increased level of methanol throughout the gradient. Elution was complete at 30 min. Peak shapes were good and baseline resolution was achieved.

Diazepam, three metabolites (N-desmethyldiazepam, temazepam, oxazepam) and clonazepam [IS] were isolated from plasma or urine and separated on a base deactivated C$_{18}$ column (λ = 232 nm). A 50/10/40 methanol/acetonitrile/water (50 mM KH$_2$PO$_4$ pH 3.5) mobile phase generated baseline resolution in 12 min [552]. Calibration standards ran from 10 to 2000 ng/mL with detection limits of 10 ng/mL reported. Similarly, diazepam and three metabolites (nordiazepam, oxazepam, temazepam) were baseline resolved in 12 min on a C$_{18}$ column (λ = 257 nm) using a 60/35/5 methanol/water/acetonitrile mobile phase [553].

Nine benzodiazapines (medazepam, chlordiazepoxide, tofisopam, nitrazepam, estazolam, triazolam, oxazepam, chlotiazepam, diazepam) were baseline resolved in <25 min using a C$_{18}$ column (λ = 241 nm) and a 60/35/5 methanol/water/acetic acid mobile phase [554]. Excellent peak shapes were obtained and detection limits of 1 ng injected (S/N = 3) were reported. A linear working curve from 10 to 1000 μg/L was reported. Additionally, ten benzodiazapines were resolved on a C$_{18}$ column (λ = 254 nm) using various isocratic methanol/water mobile phases (50/50 to 80/20), flow rates (from 0.6 to 1.0 mL/min) and temperatures (from 26°C to 50°C) to optimize the separation. A series of 20 μL injections of 10–80 mg/mL standards were used. The optimal results were obtained with a 50/50 methanol/water mobile phase, a flow rate of 0.82 mL/min and a temperature of 50°C. Elution was complete in 18 min [555].

The degradation of lorazepam in acidic solution was followed on a base deactivated C$_8$ column (λ = 230 nm). A 35/20/45 methanol/acetonitrile/water (0.1 M ammonium acetate with 5 mM KH$_2$PO$_4$ to pH 6 with acetic acid) mobile phase resolved lorazepam from 2-amino-2,5-dichlorobenzophenone and 6-chloro-4-(2-chlorophenyl)-2-quinazolinecarboxaldehyde in <17 min. A linear range of 0.5–1.8 μg/mL and detection limits of 0.2 μg/mL were reported [556].

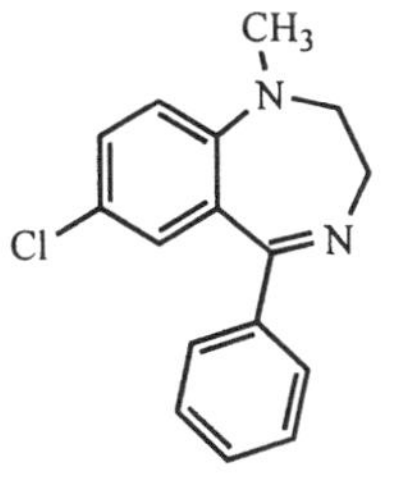

Medazepam

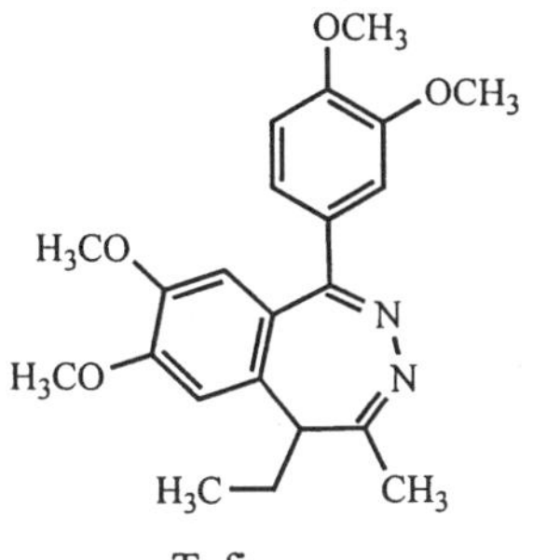

Tofisopam

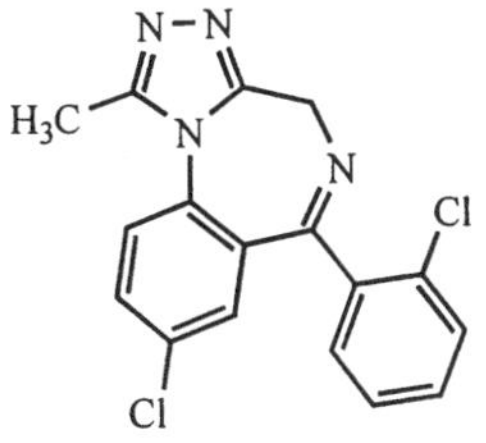

Triazolam

4.7.13   Other Drug Analytes

The antifilarial methyl-[5-(α-amino-4-fluorobenzyl)benzimidazol-2-yl]carbamate (MAFBC) and four metabolites (e.g., flubendazole and decarbamoylated-MAFBC) were extracted from plasma and separated on a C_{18} column ($\lambda = 291$ nm) using a $70/30 \rightarrow 45/55$ (at 10.25 min hold 2.75 min) $\rightarrow 30/70$ (at 17.56 min hold 3 min) water (50 mM phosphate to pH 4.0 with H_3PO_4)/(75/25 methanol/acetonitrile) gradient [557]. Good peak shapes and resolution were obtained. The calibration range used was 10–500 ng/mL and quantitation limits of 10 ng/mL (parent) and 20–1000 ng/mL for the metabolites (analyte dependent) were reported.

Five antidiabetic drugs (tolbutamide, chlorpropamide, glipizide, gliclazide, glibenclamide) were isolated from blood or serum and separated on a C_{18} column ($\lambda = 225$ nm). A 25-min 58/42 (hold 11 min) $\rightarrow$ 71/29 (at 15 min hold 10 min)

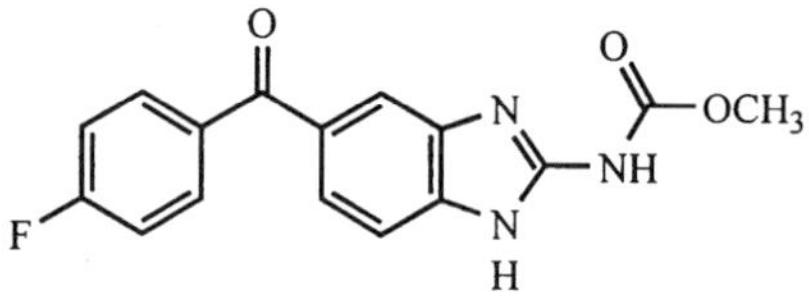

Flubendazole

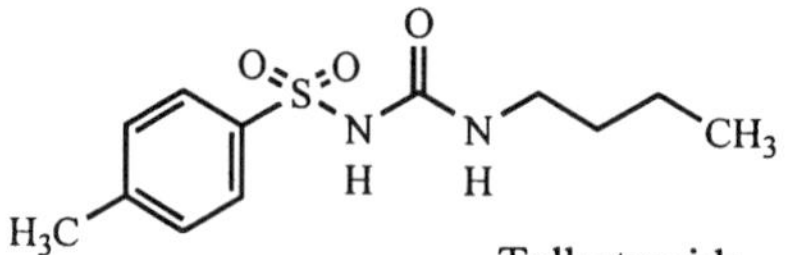

Tolbutamide

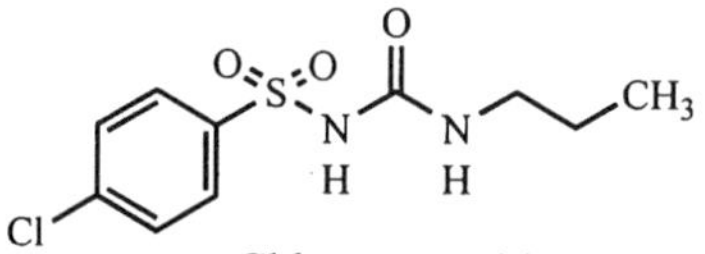

Chlorpropamide

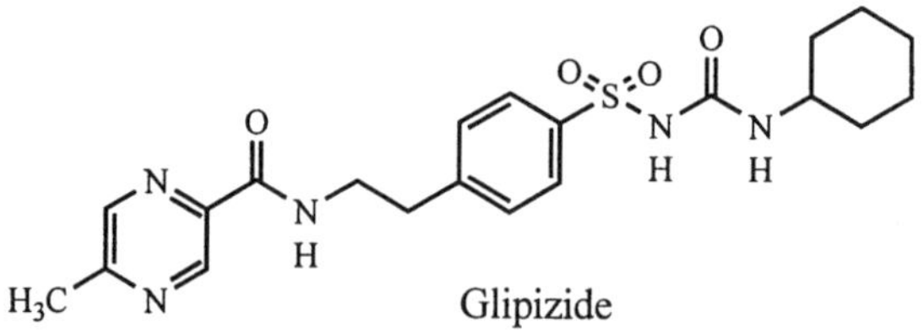

Glipizide

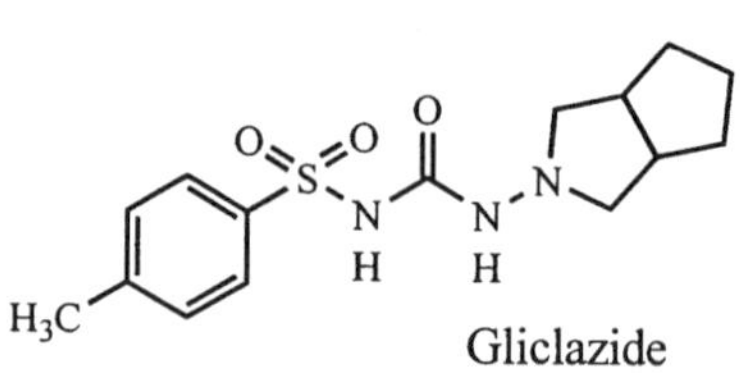

Gliclazide

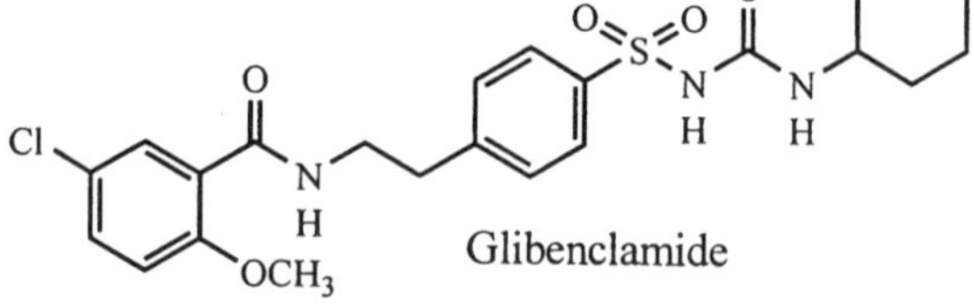

Glibenclamide

methanol/water (10 mM KH_2PO_4 buffer at pH 3.5) gradient generated baseline resolution. A significant baseline shift occurred during the gradient due to the absorbance of methanol at 225 nm. Acetonitrile might be an effective option. A 2 µg/mL sample was readily detected [558].

The antimicrobial thimerosal is used as a preservative in hepatitis vaccines. Two degradation compounds (dithiosalicylic acid, thiosalicylic acid) were separated from their parent on a C_{18} column ($\lambda = 226$ nm) using a 65/35/0.9 methanol/water/ H_3PO_4 mobile phase [559]. Elution was complete in 12 min. A calibration curve from 20 to 80 g/mL was used and a 5 g/mL detection limit (S/N = 3) was reported.

A new fasciolicidic agent, 6-chloro-5-(1-naphthyloxy)-2-methylthiobenzimida-zole and two metabolites (the sulfone and sulfoxide) were extracted from sheep urine and analyzed on a C_{18} column ($\lambda = 304$ nm) using a 40/30/30 methanol/ acetonitrile/water mobile phase [560]. Good peak shapes and resolution were achieved along with complete elution in less than 12 min. Linearity was obtained over the range 0.2–12.5 µg/mL.

Pimobendan and its liver microsomal metabolite O-desmethylpimobendan were isolated from an incubate and separated on a 40°C C_{18} ($\lambda = 338$ nm, ex; 405 nm, em) using a 3/1/4 methanol/acetonitrile/water (0.6% ammonium acetate) mobile phase [561]. Good peak shapes and resolution were achieved and elution was complete in 6 min. A linear range of 0.1–200 ng injected and a detection limit of 0.1 ng injected were reported.

Chloroquine (an antibacterial) and its mono- and bisdesethylchloroquine meta-bolites were extracted from human microsomes and separated on a C_1 column ($\lambda = 250$ nm, ex; 380 nm, em) using an isocratic 70/30/1 methanol/water/ triethylamine mobile phase [562]. Elution was complete in 35 min, all peaks were broad, and chloroquine was tailed. Since these are basic amine compounds, the addition of TFA could help peak shape. Also, it should be noted that C_1 columns are not resistant to basic solutions, the bonded phase being readily cleaved under such conditions. Therefore, a precolumn or the use of a C_8 or C_{18} column should be considered. Two separate linear curve regions were cited: 78–1250 nM and 1250–20000 nM with a quantitation limit of 78 nM (S/N = 3).

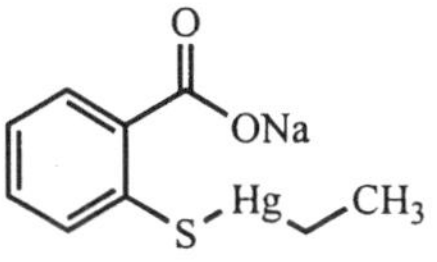

Thimerosal

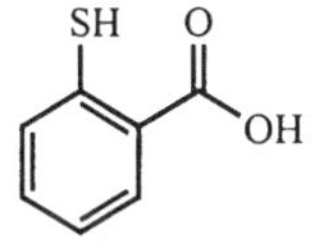

Thiosalicylic acid

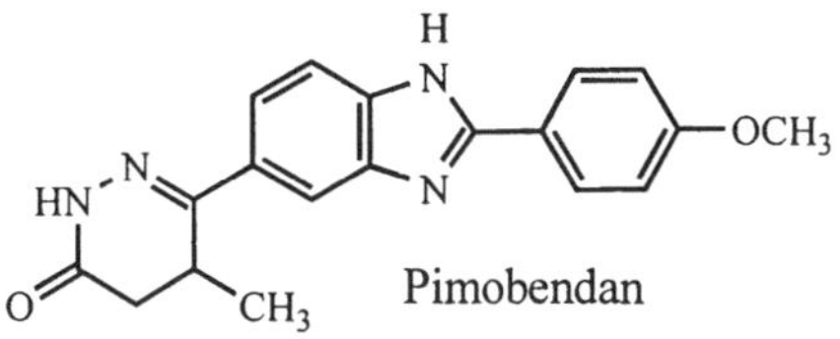

Pimobendan

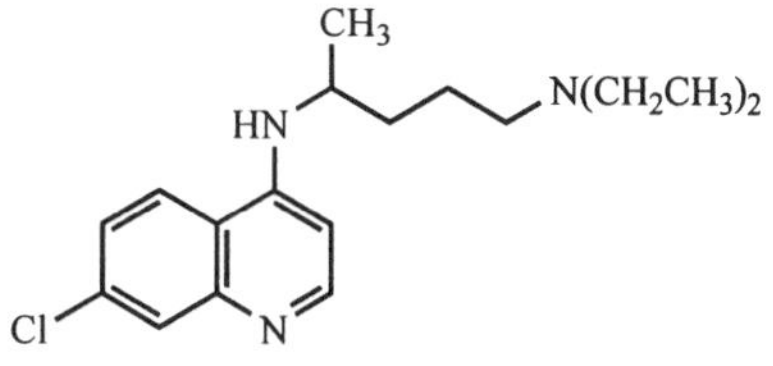

Chloroquine

Severe chronic pain is sometimes treated with co-administered ketamine/bupivacaine. These compounds were isolated from plasma and analyzed on a cyanopropyl column ($\lambda = 215$ nm). A 718/220/80/2 water (10 mM NaH_2PO_4 buffer at pH 2.34)/methanol/acetonitrile/H_3PO_4 mobile phase generated baseline separation in <8 min. The peaks were quite tailed. Introduction of TFA and/or a TFA/triethylamine buffer system in place of the phosphate system may remedy this. Calibration curves from 10 to 400 ng/mL (ketamine) and 125 to 4000 ng/mL (bupivacaine) were stated, with the lowest concentration serving as the quantitation limit. Detection limits (S/N = 4) of 1 ng injected and 0.8 ng injected, respectively, were reported [563].

Salicylic acid, three impurities (4-hydroxybenzoic acid, phenol, 4-hydroxyisophthalic acid), and two metabolites (gentisic acid, salicylglycine) were baseline resolved in 12 min on a phenyl column ($\lambda = 235$ nm) using a mobile phase of 40/60/1 methanol/water/H_3PO_4 with 2.5 g β-cyclodextrin/L to "pH 2" [564]. The author notes that a C_{18} column was ineffective. Linear ranges of 0.1–100 µg/mL and detection limits of 0.1 µg/mL (S/N = 3) were reported.

Detajmium, an antiarrhythmic drug, and its parent compound ajmaline were extracted from serum and analyzed on a 40°C C_{18} column ($\lambda = 247$ nm, ex; 353 nm, em) using a 75/22/3 water (0.1 M phosphate at pH 3.5)/methanol/acetonitrile mobile phase [565]. Baseline resolution and elution in 9 min was achieved. A linear range of 1–200 ng/mL and a quantitation limit of 1 ng/mL were reported.

The antidepressant venlafaxine and its metabolite oxydesmethylvenlafaxine were extracted from plasma and separated on a C_{18}/cyanopropyl mixed mode column (electrochemical detector at 0.65 V and 0.98 V). These compounds have UV chromophores as well (phenyl groups). A 10-min analysis was achieved with a 30/70 water (50 mM KH_2PO_4 at pH 4.8)/methanol mobile phase. Note that the buffer has a very low capacity here (nearly equidistant from the first two phosphate pK_a values). A standard range from "0" to 200 ng/mL was developed and detection limits of 0.5 ng/mL (S/N = 3) were reported [566].

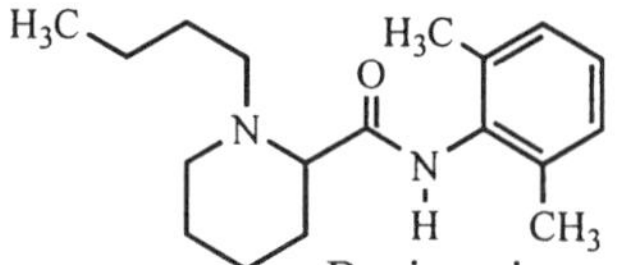

Bupivacaine

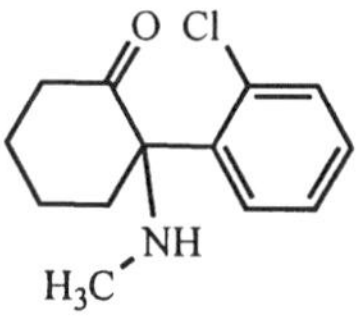

Ketamine

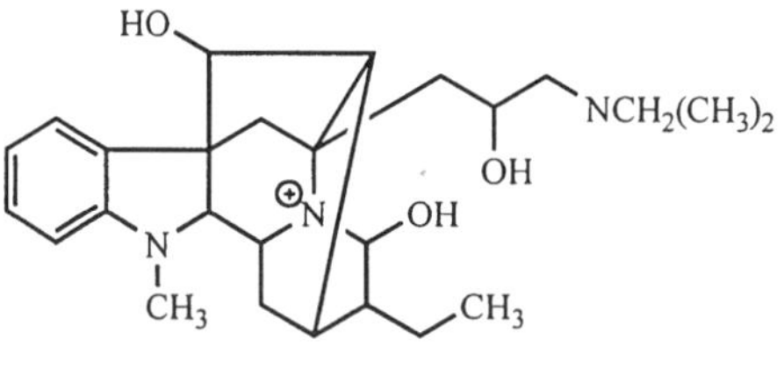

Detajmium

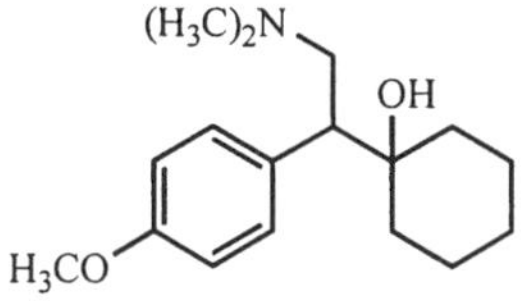

Venlafaxine

Fluoxetine (Prozac) and its metabolite (norfluoxetine) are racemic. Both compounds were isolated from plasma and urine and separated in 55 min on a 30°C acetylated β-cyclodextrin column ($\lambda = 214$ nm). Good resolution of all four compounds was achieved using a 25/75 methanol/water (0.3% triethylamine) mobile phase [567]. A standard curve from 25 to 200 μg/mL was established for each enantiomer.

Rifampicin/isoniazid formulations are used in resistant tuberculosis cases. These and the 3-formylrifampicin and isonicotinyl hydrozone decomposition products were separated on a C_{18} column ($\lambda = 229$ nm) using a 45/55 methanol/water mobile phase [568]. Degradation profiles for ~2 mg/mL samples were analyzed, giving all four peaks at levels that were easily detected.

Benzocaine and its N-acetylbenzocaine metabolite were extracted from fish fillet and separated on a 40°C C_{18} column ($\lambda = 289$ nm for benzocaine, 271 nm for the metabolite). A 10-min elution time was generated using a 55/45 methanol/water mobile phase [569]. Calibration curves from 20 to 5000 ng/g were used and detection limits of 20 ng/g were reported.

The separation of the R and S enantiomers of the bactericide ofloxacin was achieved on a C_{18} column ($\lambda = 330$ nm, ex; 505 nm, em) using an 86/14 water (6 mM L-phenylalanine with 3 mM copper sulfate)/methanol mobile phase [570]. Separation was complete in 18 min with good resolution. Linearity was reported over the range 11–45 μM with detection and quantitation limits of 0.6 μM and 6 μM.

Two photochemical decomposition compounds of lomefloxacin were separated and analyzed on a C_{18} column ($\lambda = 280$ nm, ex; 430 nm, em) using a 70/30/0.1 water/methanol/H_3PO_4 mobile phase [571]. Baseline resolution and excellent peak shapes were obtained. Elution was complete in 10 min. A standard 10 μg/mL parent

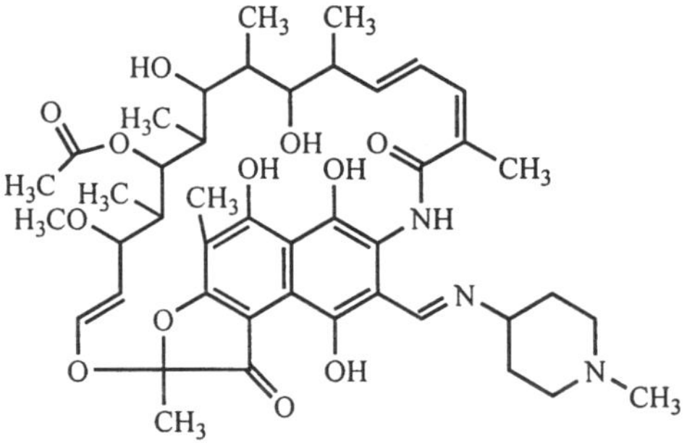

Rifampicin

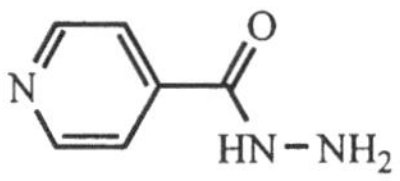

Isoniazid

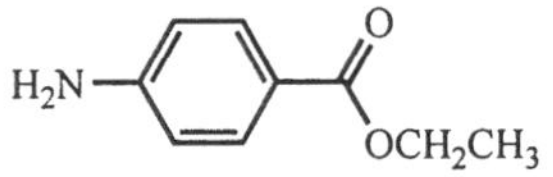

Benzocaine

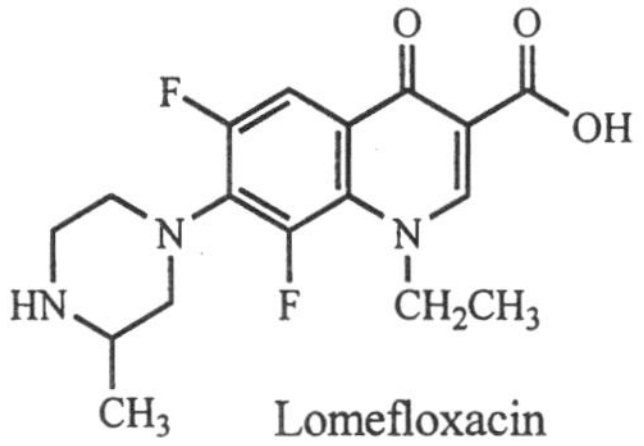

Lomefloxacin

solution was degraded and peaks at 80% decomposition were readily detected for all compounds.

Cisapride and its metabolite, norcisapride, were extracted from urine and separated on a C_8 column ($\lambda = 295$ nm, ex; 300 nm, em) using a 45/55 methanol/water (20 mM NaH_2PO_4 to pH 7) with 1 g/L triethylamine mobile phase [572]. Baseline resolution was obtained and elution was complete in 15 min. A linear range of 50–2000 ng/mL and a quantitation limit of 50 ng/mL were reported.

Undecylenic acid was extracted from ointments and powders, derivatized with 4'-nitrophenacyl bromide, and analyzed on a C_8 column ($\lambda = 265$ nm). A 50/30/20 methanol/acetonitrile/water mobile phase generated a 7 min elution [573]. A table of the effect on retention times of changing phenacyl derivatives (e.g., 4'-methyl, 4'-phenyl, 4'-bromo) and the column to a C_{18} is presented. The linear range was reported as 12.5–300 µg injected.

Gestrinone (contraceptive) was extracted from serum and analyzed on a C_{18} column (positive ion electrospray MS, source $T = 200°C$, ion spray voltage +5.0 kV, nebulizer N_2 gas 1 L/min, curtain gas 1.25 L/min, oriface voltage 45 V, collision pressure 3×10^{-5} Torr) using a 0.2% formic acid in methanol mobile phase [574]. Elution time was approximately 3 min. A linear range of 4–180 ng/mL with a detection limit of 0.8 ng/mL was reported. Formic acid greatly enhanced method sensitivity, whereas acetonitrile caused a decrease. The authors noted that detection could be achieved at $\lambda = 340$nm. Here the sample was resolved from all interferences with a 70/30 methanol/water mobile phase but the analysis time was 30 min.

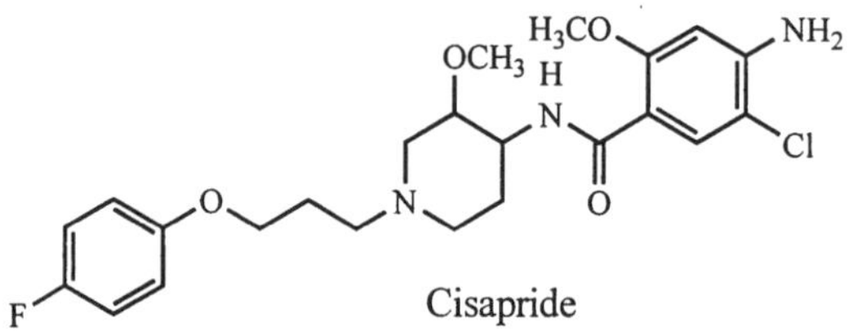

Cisapride

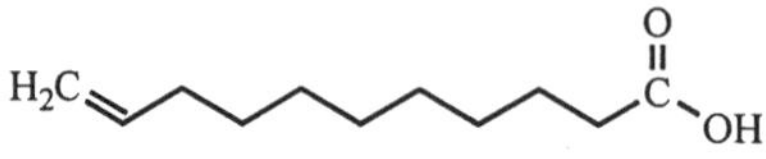

Undecylenic acid

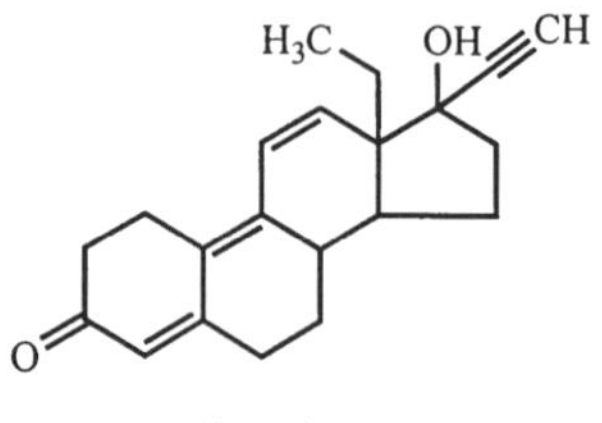

Gestrinone

Six 1,4-dihydropyridines (nifedipine, nimodipine, nisoldipine, nicardipine, laci-dipine, felodipine) were extracted from plasma and separated on a C_{18} column (electrochemical detector 1000 mV vs. Ag/AgCl). Baseline separation was achieved in 16 min using a 70/30 methanol/water (2 mM sodium acetate pH 5) mobile phase [575]. Baseline resolution and good peak shapes were obtained. The authors noted that at pH > 6 the ionizable nicardipine became unacceptably tailed. Also, the background noise increased to an unacceptable level when the buffer concentration exceeded 5 mM. Methanol/water was used due to the lower background noise generated as compared with analogous acetonitrile/water mobile phases. A plot of retention time versus percent methanol was shown. Standards ranged from 50 to 1000 ng/mL and detection limits of ~30 ng/mL (S/N = 3) were reported.

Shervington [576] analyzed cough syrups containing pseudoephedrine, guaife-nesin, and dextromethorphan on a C_{18} column ($\lambda = 257$ nm and 280 nm) using a 54/45/1 water/methanol/formate buffer (pH 4.3) prepared as 35 mL ammonia to which 30 mL water and 30 mL formic acid are carefully added, then diluted to 100 mL with water). Elution was complete in 19 min, but the last-eluting peak (dextromethorphan) had poor peak shape and eluted 10 min after guaifenesin (the next nearest eluting peak). A gradient or a stronger mobile phase should be considered. Also, the guaifenesin calibration curve is distinctly nonlinear and inspection of the chromatogram gives a potential explanation: the peak absorbance is >1.6 absorbance units. Better selection of working wavelengths could alleviate this problem.

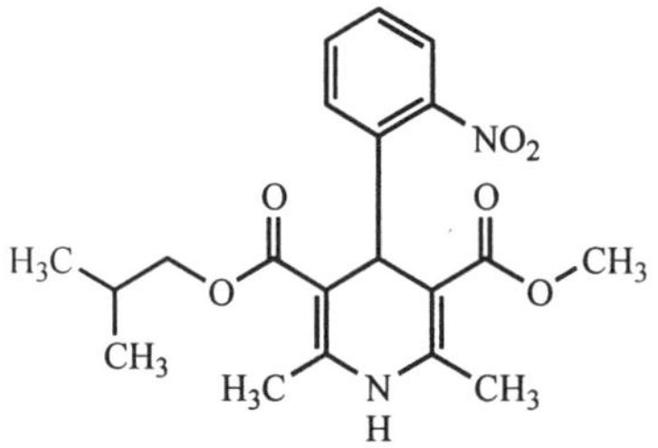

Nisoldipine

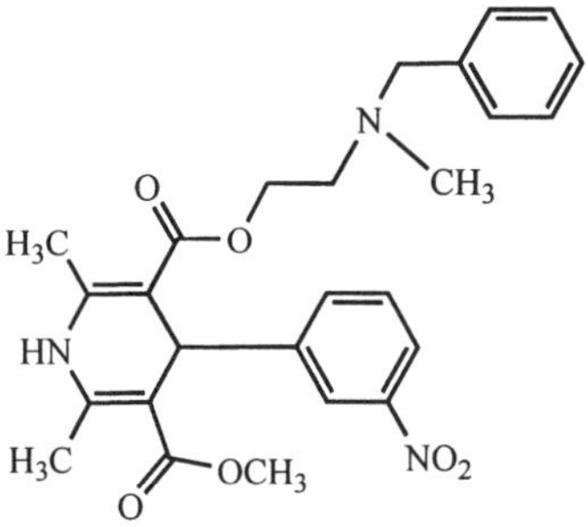

Nicardipine

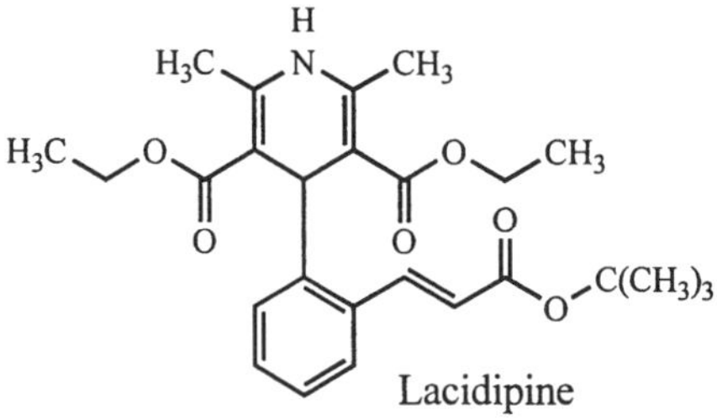

Lacidipine

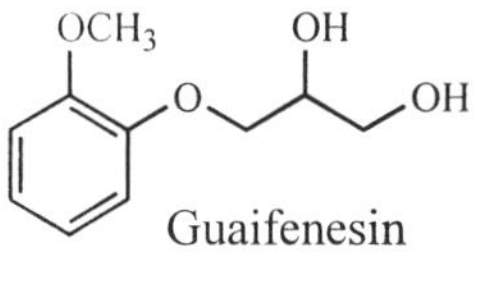

Guaifenesin

Four antibacterial compounds (dimetridazole, ronidazole, metronidazole, hydroxymetridazole) were isolated from poultry meat and separated on a C_{18} column (positive ion electrospray MS, capillary $T = 220°C$; quadrapole $T = 70°C$; no auxiliary gas; offset voltage -0.2 V; spray voltage 5kV, N_2 sheath gas 80 psi; collision voltage 10V). An 81/13/6 water (formic acid)/methanol mobile phase generated complete elution in <8 min. A linear range of 0.025–0.04 mg/L and detection limits of 2–5 µg/kg were reported [577].

Isoniazid (illegal for use in treatment of tuberculosis) was extracted from milk samples and analyzed on a C_{18} column ($\lambda = 330$ nm). A 47/30/23 water (1% ammonium acetate pH 5.6)/methanol/acetonitrile mobile phase gave good resolution between the analyte and other matrix interferences [578]. Standards from 0.1 to 100 mg/L and detection limits of 0.2 ng/mL were reported.

The separation of four calcium channel blocker drugs (diltiazem, verapamil, nifedipine, nitrendipine) and five metabolites (e.g., demethyldiltiazem, norverapamil) was accomplished on a C_{18} column ($\lambda = 230$ nm) using a 2/2/1/0.04 methanol/water (40 mM ammonium acetate pH 7.9)/acetonitrile/triethylamine mobile phase. Elution was complete in 11 min. A plot of the effect of pH from 6.9 to 7.9 on retention is shown. Linear ranges were generated from 10 to 5000 ng/mL and detection limits of 8 ng/mL were reported [579].

Tramadol, an opioid-type analgesic, metabolites ($1R,2R$ and $1S,2S$ diastereomers of O-demethyl-, -N,O-didemethyl- and N,N,O-tridemethyltramadol glucuronide), were isolated from urine and separated on a C_{18} column ($\lambda = 270$ nm, ex; 300 nm, em) using a 16.7/83.3 methanol/water (20 mM nonanesulfonic acid to pH 4.5) mobile phase [580]. Elution was complete in 28 min and peaks were well resolved. Standard concentrations ranged from 0.35 to 4.5 µg/mL and detection limits were reported as 10 ng/mL (S/N = 3).

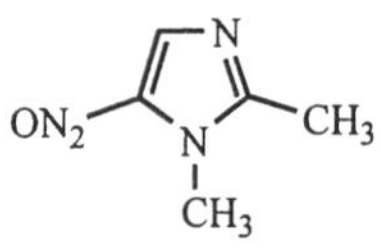

Dimetridazole 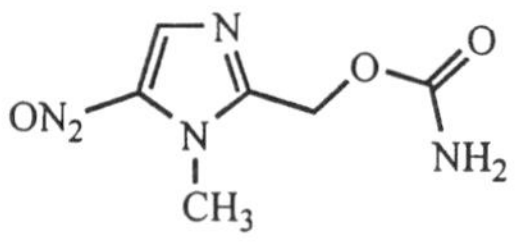   Ronidazole

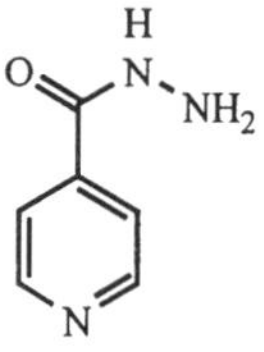

Isoniazid

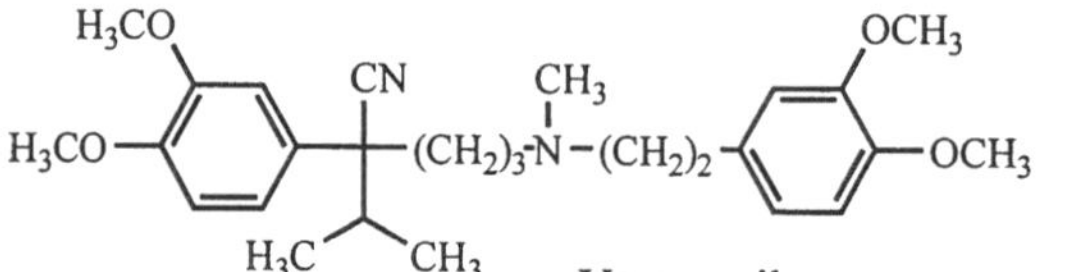

Verapamil

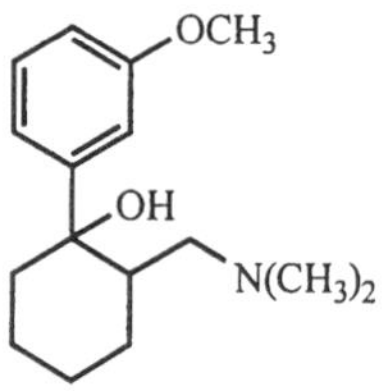

Tramadol

Nalmefene (opioid antagonist), and its metabolites nornalmefene and nalmefene glucuronide were isolated from liver tissue and analyzed on a C_{18} column ($\lambda = 230$ nm) using a 15-min $80/20 \to 50/50$ water (50 mM NaH_2PO_4 + 0.2% triethylamine [TEA] to pH 4.3 with H_3PO_4)/methanol gradient [581]. Good peak shapes and excellent resolution resulted. A consistent and negative baseline drift was observed for all chromatograms presented. This is most likely due to the decreasing TEA concentration during the gradient since TEA has a strong absorbance up to 260 nm (see Chapter 1, Fig. 1.5). A potential remedy is to add 0.2% TEA to the methanol, thereby equalizing the concentration in both mobile phase components.

Ivermectin and abamectin were extracted from cattle feces and chromatographed as their 1-methylimidazole derivatives [582]. Separation was complete in 15 min on a C_{18} column ($\lambda = 365$ nm, ex; 470 nm, em) using a 95/5 methanol/water mobile phase. The detection limits were reported as 2 ng/g (S/N = 50) and the response vs. concentration curve was found to be linear up to 2 µg/g.

Disulfiram is reduced to diethyldithiocarbamate (DDTC) in the blood and subsequently esterified to the DDTC-methyl ester [583]. From this compound, the DDTC-methyl sulfoxide, DDTC-methyl sulfine and S-methyl-N,N-diethylthiocarbamate metabolites are formed. These four converted compounds were extracted from liver tissue and baseline resolved on a C_{18} column using a 50/50 methanol/water mobile phase. Peak shapes were good and elution was complete in 15 min.

The enantiomers of azepinoindole and its N-desmethyl metabolites were well resolved on a β-cyclodextrin-modified silica column ($\lambda = 231$ nm). A 1/99 t-butyl alcohol/water (10 mM sodium phosphate buffer at pH 7 with 15 mM β-cyclodextrin and 2 mM triethylamine) mobile phase was used to generate the chiral separation. The assay calibration range covered 100–500 ng/mL. Peaks were somewhat tailed. Elution was complete in 25 min [584].

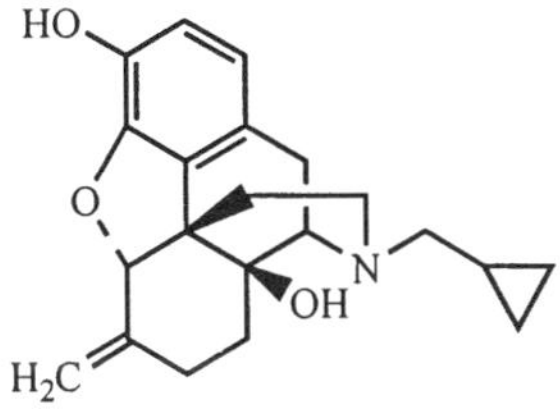

Nalmefene

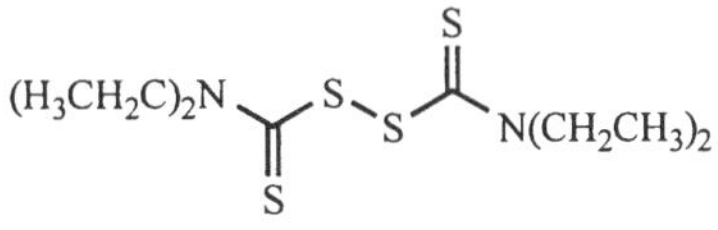

Disulfiram

Separation of amitriptyline from seven of its metabolites was accomplished in a unique fashion [585]. A silica column and a 90/10 methanol/water (65 mL 25% ammonia with 11 mL acetic acid buffer at pH 9.1) mobile phase was used. Consequently, the silica was fully deactivated and modified by the buffer. Although baseline resolution was not achieved, good peak shape was obtained and the elution was complete in <10 min. Using the same conditions, 14 basic drugs (e.g., tetracaine, amiloride, betahistine, diphenhydramine, chlorpheniramine, quinacrine, strychnine, pyrantel) were separated in under 20 min. Finally, eight beta-blockers (e.g., epanolol, practolol, atenolol) were separated under identical conditions in <10 min. Whereas baseline resolution was not achieved for all pairs of compounds in these three separations, the short run times under isocratic conditions used here suggest that "mixed mode" combinations (i.e., normal-phase support and a reversed-phase solvent) provide many options for consideration in the separation of very basic compounds.

The retention characteristics of 31 β-adrenergic blocking drugs were extensively studied by Massart and co-workers [586] on a C_{18} column ($\lambda = 220$ nm) using isocratic methanol/water (phosphate buffer at pH 4.0 at an ionic strength of 0.1) mobile phases varying from 30% to 60% in methanol. Peak shape was dramatically improved through the addition of 2 mM dioctylamine to the mobile phase (with a concomitant decrease in k'). These results indicate that the dioctylamine eliminates or modifies a source of strong retention for these analytes; most probably residual surface silanol groups. It obviously does not present itself as an ion pair reagent or the retention time for the analyte would have increased. Examples like these give important insights into the critical factors governing retention.

Amitriptyline

Betahistine

Quinacrine

Pyrantel

Epanolol

Practolol

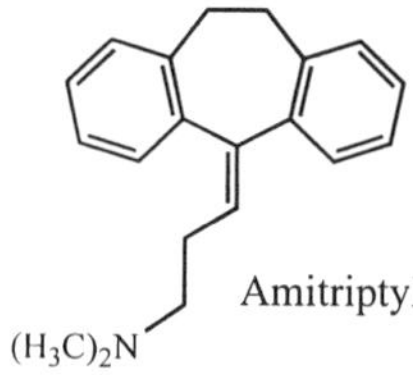

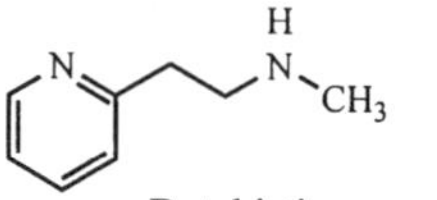

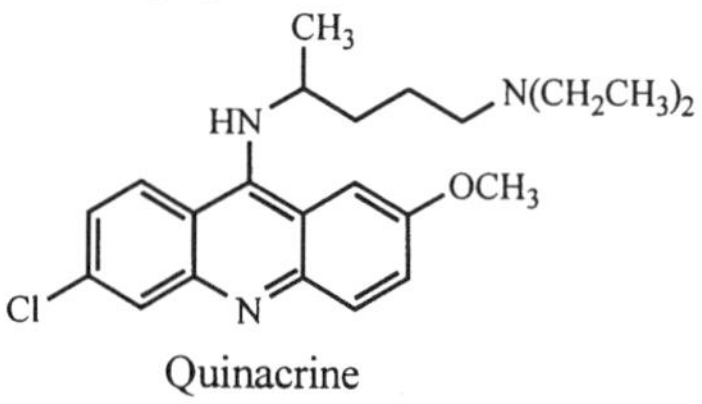

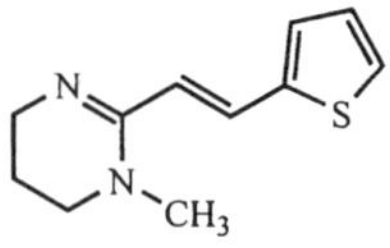

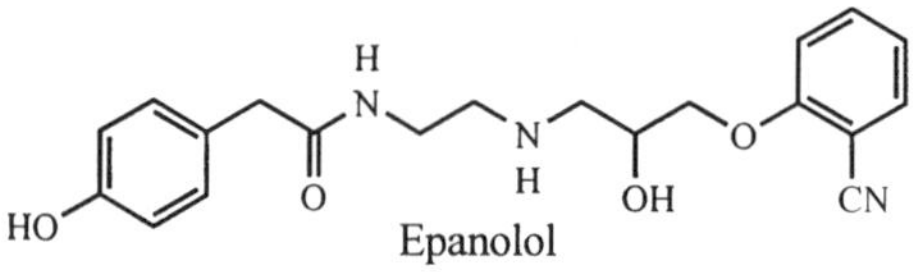

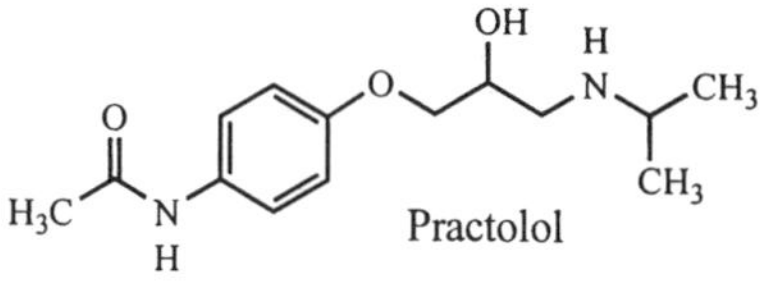

The enantiomers of metoprolol and its acidic urine metabolites were baseline resolved on a C_{18} column ($\lambda = 223$ nm, ex; 340 nm, em) using a complex 25-min $75/25 \rightarrow 90/10$ methanol/water (0.1 M phosphate buffer as 13.8 g NaH_2PO_4 and 1.59 g propylamine hydrochloride/L to pH 3.2 with H_3PO_4) gradient [587]. Concentration curves were generated over the range from 7.5 to 225 μg injected.

Theophylline and four of its liver metabolites (1-methyluric acid, 1,3-dimethyluric acid, 3-methylxanthine, uric acid) were extracted from tissue and baseline resolved on a C_{18} column ($\lambda = 273$ nm) using a 13/87 methanol/water (5 mM tetrabutylammonium hydrogensulfate with 10 mM sodium acetate at pH 4.75) mobile phase [588]. Peak shapes were excellent and elution was complete in 20 min.

Spironolactone, 7α-thiomethylspirolactone, 7α-thiospirolactone, and canrenone were resolved on a C_{18} column ($\lambda = 254$ nm) using a 30-min $65/35 \rightarrow 100/0$ methanol/water gradient [589]. Peaks were tailed but baseline resolution was achieved nonetheless. The addition of a mobile phase modifier such as triethylamine and/or trifluoroacetic acid would most likely greatly reduce this problem.

Table 4.5 lists some USP methods for a variety of analytes that use alcohol mobile phases [590].

4.7.14   Summary

Many drug compounds and metabolites that have multiple hydroxyl, carboxyl, and amine functionalities may be analyzed using mobile phases containing alcohols. Often, in the case of basic amines, a mobile phase modifier, such as dioctylmethylamine, is present to sharpen peaks and therefore increase overall column peak capacity. Compatibility of alcohols with a number of MS interfaces is an important factor to keep in mind when developing methods that have the potential of ultimately being transferred to newer instrumentation. Care must be taken when phosphate buffers are used due to their limited solubility in methanolic mobile phases. Stronger consideration of longer-chain alcohols in drug analysis should occur since unique selectivity and favorable retention properties often result.

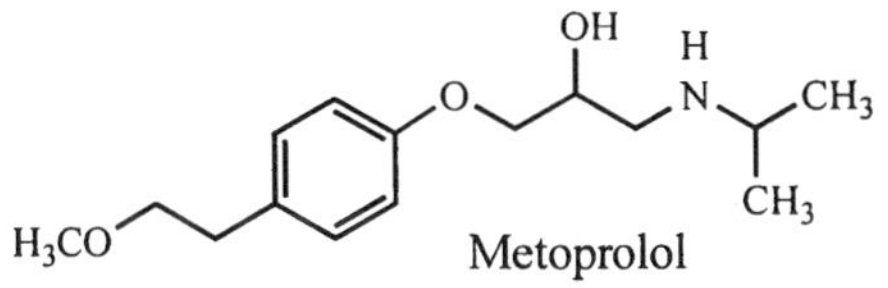

Metoprolol

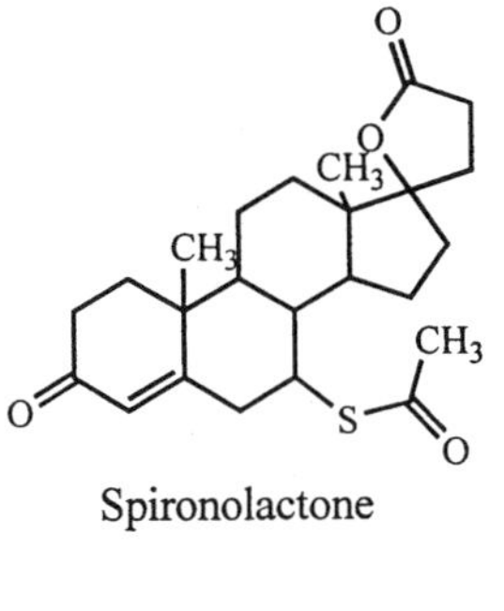

Spironolactone

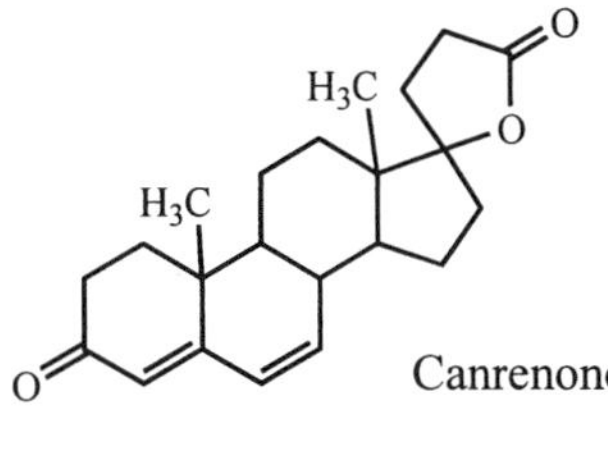

Canrenone

TABLE 4.5 USP Methods[a]

Analyte(s)	Co-analyte (Internal Standard)	Column	Mobile Phase	Detection (nm)	USP Page
Amprolium	2-Picoline	C_1	500/450/50 water (6 g sodium heptanesulfonate with 12 mL acetic acid and 2 mL triethylamine/methanol/acetonitrile	254	141
Baclofen (tablets)		C_{18}	550/440/20 water (0.3 N acetic acid)/ methanol/water (0.36 N sodium pentanesulfonate)	265	195
Bumetanide (injection)	4-Ethylbenzaldehyde	C_{18}	50/45/5/2 methanol/water/THF/acetic acid	254	254
Captopril (tablets)	Captropril disulfide	C_{18}	55/45 methanol/water (with 0.5 mL H_3PO_4)	220	296
Chlorambucil (tablets)	Propylparaben	C_{18}	50/50 alcohol/water (with 1 mL acetic acid)	254	372
Cimetidine (assay)		C_{18}	80/20 water/methanol (with 0.3 mL H_3PO_4)	220	412
Dibucaine		C_{18}	70/30 methanol/water (1.2 g sodium lauryl sulfate, 0.2 g sodium acetate, 2 mL triethylamine to pH 5.6 with acetic acid)	254	542
Droperidol	4-Fluoroacetophenone	C_{18}	70/28/2 methanol/water/water (31 g boric acid/L water to pH 7 with NaOH)	280	614
Flutamide (capsules)	Testosterone	C_{18}	7/4 methanol/water (50 mM KH_2PO_4)	254	751
Fluxinin meglumine (paste)	Sodium benzoate	C_{18}	70/30/1 methanol/water/acetic acid	254	724

	Internal standard	Column	Mobile phase	RI	
Guanadrel sulfate (assay)	Ethylparaben	C$_{18}$	47/53 methanol/water with 6.25 g camphor sulfonic acid and 0.8 g NH$_4$NO$_3$ to pH 5 with acetic acid		797
Iodohippurate sodium (injection)		Phenyl	75/25/1 water/methanol/acetic acid	265	892
Isradipine		C$_{18}$	50/40/10 water/methanol/THF	326	941
Miconazole (injection)		C$_8$	50/30/20 methanol/acetonitrile/water (5 g ammonium acetate)	230	1112
Mometasone furoate	Beclomethasone dipropionate	C$_8$	65/35 methanol/water	254	1124
Padimate O (lotion)	Ethylparaben	C$_{18}$	85/15/0.5 methanol/water/acetic acid	308	1252
Quazepam (tablet)		C$_8$ (40°C)	70/30 methanol/water	254	1452
Sulconazole nitrate		C$_{18}$	70/30 methanol/water (1.9 g sodium pentane sulfonate to pH 3.8 with 2 NH$_2$SO$_4$	230	1559

^{a} Reference [590].

5

ALKANES AND ALKYL AROMATICS

Alkanes are nonpolar solvents. This property drastically limits the ability of the alkanes to solubilize even moderately polar analytes. Alkyl aromatics have similar properties, but their solubilization capability is increased significantly due to their polarizability; i.e., the π electrons in the ring system enhance interaction with more polar compounds.

A distinct advantage of these nonpolar compounds is their excellent chemical stability and overall low reactivity. The alkanes are available in high-purity form from pentane to iso-octane. Selected higher chain homologs, for example hexa-decane, can also be obtained in high-purity form. As with all homologous series of solvents, the cost of the solvent increases dramatically as chain length increases, and the overall purity decreases (due to the presence of many isomeric forms) as chain length increases. For example, n-hexane is available at a purity of >99% (as n-hexane) and is 2–3 times more expensive than hexanes (>85% n-hexane and >98% hexanes).

Another advantage is the immiscible nature of alkanes with water. This property allows the alkanes to be used as extraction solvents for aqueous solutions. If the sample can be extracted, concentrated, and injected, at least two sample preparation steps and potential major sources of sample loss—namely, either concentration to dryness and reconstitution of residue or concentration and solvent exchange—are eliminated. These considerations during method development are rarely considered and the opportunity is lost.

A disadvantage to the use of alkanes is their very low viscosity, which leads to potential instrumentation problems. The low backpressure generated in the system often fails to seat outlet check valves, leading to erratic flow rate characteristics and, consequently, variable retention times. Therefore, a flow restrictor may be needed in these cases to produce an artificial backpressure on the check valve. The restrictor is positioned between the check valve and the injector and prevents backflushing of the mobile phase back into the reservoir.

A concern associated with the commonly used alkanes, hexane and toluene, is that they have very high vapor pressures at room temperature. Because of this volatility, efforts must be made to make sure that samples dissolved in these solvents are stored in tightly sealed vials so that volumes stay constant. For LC systems that require continuous solvent sparging, frequent replacement of the solvent or a sparging reservoir is necessary. Finally, appropriate ventilation is essential due to the extremely flammable nature of these solvents and their vapors.

Pentane is infrequently used in LC methods due to its high volatility (pentane is ~4 times more volatile than hexane). A major advantage pentane offers is that its UV vs. absorbance curve does not go above 0.25 AU even at 190 nm. Therefore, quantitation at extremely low UV wavelengths is readily accomplished using pentane.

It should be noted that the group of solvents known as "petrols" belongs to the class of alkyl hydrocarbon solvents. A distillation fraction from 30 to 70°C produces petroleum ether. The final product is predominantly a mixture of pentanes and hexanes, but a wide range of other compounds is present at the 0.5% range. Due to the greater chance of significant compositional variability, petroleum ether (and the higher boiling range fractions termed "technical petrols") is rarely used in quantitative LC work but is more frequently used in preparative LC and as an extraction solvent. High-purity petroleum ether is considerably less expensive than either pentane or hexane.

Hexane is the most commonly used alkane in LC work. It is available in both UV and nonspectrophotometric (NS) forms. The UV material is intended for most LC work, whereas the NS form is typically used for extractions followed by GC work or LC work using a refractive index (RI) detector or an evaporative light scattering detector (ELSD). Unlike pentane, hexane may be purchased for LC use as either >85% n-hexane and >99.9% total saturated hexane isomers, >96% n-hexane and >99.8% saturated total hexane isomers or >99% n-hexane, and >99.9% total saturated hexane isomers. A premium is paid for every increase in purity as n-hexane (~20% for 95% n-hexane and ~100% for 99% n-hexane). Little, if any, practical chromatographic difference has been reported as long as the remaining impurities are predominantly hexane isomers.

Heptane is available both as >96% n-heptane and >99.9% as saturated n-heptane isomers. The 30% premium paid for heptane over the cost of hexane is seldom warranted, except in particularly demanding separations. A major advantage heptane offers is a vapor pressure that is approximately 1/4 that of hexane (and 1/12 that of pentane).

Iso-octane (2,2,4-trimethylpentane) is a more commonly used nonpolar so vent than heptane. Due to its branched-chain nature, iso-octane usually has a higher solubility level in polar solvents than does n-heptane. The cost of LC-quality iso-octane is typically less than that of n-heptane and more than that of hexane. For these reasons, iso-octane is frequently the next alkane solvent of choice after hexane.

Cyclohexane and cyclopentane are sometimes used in preference to hexane and pentane due to their unique molecular shapes and the possible selectivity enhancements they can offer. The significantly higher cost (30%) of the cyclic alkanes as compared with their linear analogs usually effectively precludes their use since it is a rare instance in which a cyclic alkane *must* be utilized in order to achieve a separation.

The use of benzene, especially in the US marketplace, has been drastically curtailed due to its carcinogenic nature. Often used for liquid/liquid extractions, it has typically been replaced by toluene, cyclohexane or hexane.

Toluene is used in LC work when analytes are large, fairly nonpolar, and have a chromophore with a $\lambda_{max} > 280\,nm$ or have a solution concentration high enough with a refractive index different enough from the solvent to be analyzed with an RI detector. An ELSD is particularly suited for use with solvents such as these since absorance is not an issue (volatility is the key). Due to its polarizable nature (π-electrons), toluene is much more effective at solubilizing polarizable and moderately polar compounds. This is a great advantage in many separations.

Tables 5.1–5.4 list some important chemical, physical, and chromatographic properties as well as general manufacturing specifications and safety parameters for the alkanes and alkyl aromatic solvents [84–92]. Figure 5.1 shows the chemical structures for the solvents listed in Tables 5.1–5.4.

5.1 IMPURITIES

In pentane, 2-methylbutane is typically the major impurity and is present at the high-ppm level. Much lower levels of higher molecular weight alkanes and alkenes are also present. Petroleum ether, by definition, has no impurities since it is defined by a distillation range rather than by a compositional specification. This leads to considerably more lot-to-lot variability.

Hexane, as mentioned above, is often manufactured as the mixture of saturated isomeric *n*-hexanes. Grossly elevated levels of alkenes and their reaction products shift the UV cutoff to higher wavelengths and often cause significant background absorbances up into the 210 nm range. These include 2-methylpentane, 3-methyl-pentane, and methylcyclopentane. Cyclohexane and small quantities of alkenes are also typically present. Hunchak and Suffet [591] stated that diethylphthalate, octyl adipate, glycerol acetate, ethyleneglycol diacetate, and decanoic acid were found in 300-fold concentrated hexane. Previous comments on stability not withstanding, alkanes will, over very extended periods and exposure to oxygen, heat, and light, degrade as in the set of reactions shown below [592].

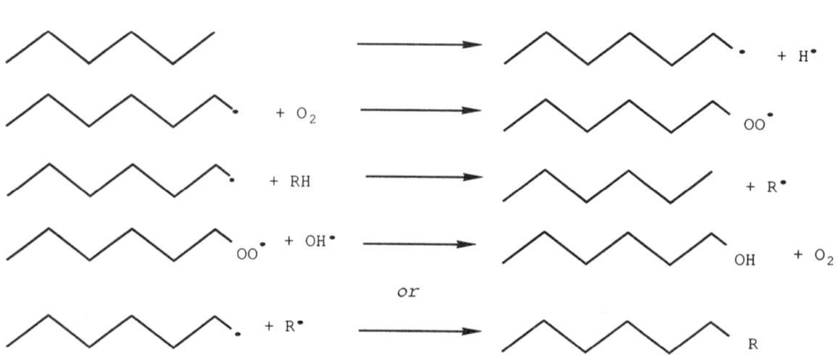

TABLE 5.1 Physical Properties of Alkane and Aromatic Solvents[a]

	Pent	cPent	Hex	cHex	Hept	TMP	Dec	Benz	Tol
Molecular weight	72.15	70.14	86.18	84.16	100.21	114.23	138.26	78.11	92.14
Density (g/mL)	0.6262	0.7454	0.6594	0.7785	0.6837	0.6919	0.8865	0.874	0.8669
Viscosity (cP)	0.23	0.44	0.31	1.0	0.41	0.50	2.415	0.652	0.59
Solubility in water (%)	0.04	0.01	0.014	0.006	0.0003[b]	0.0002[b]	< 0.02	0.07	0.074
Water sol. in solvent (%)	0.009	0.01	0.01	0.01	0.01[b]	0.006	0.0063	0.07	0.05
Boiling point (°C)	36.07	49.26	68.7	80.72	98.43	99.24	191.7	80.1	110.62
Melting point (°C)	−129.7	−93.87	−95.3	6.54	−90.6	−107.4	−124	5.5	−95
Refractive index (n_D)	1.3575	1.4064	1.3749	1.4262	1.3876	1.3914	1.4758	1.5010	1.4969
Dielectric constant	1.84	1.97	1.89	2.02	1.92	1.94[b]	2.15	2.28	2.33
Dipole moment (D)	0.0	0.0	0.0	0.0	0.0	0.0	0.0	0.0	0.31
Surface tension (dyne/cm)	16.0	22.4	18.4	25.0	20.3	18.8	29.4	28.9	28.5

Abbreviations: Pent, pentane; cPent, cyclopentane; Hex, hexane; cHex, cyclohexane; Hept, heptane; TMP, iso-octane, 2,2,4-trimethylpentane; Dec, Decalin®, decahydronaphthalene; Benz, benzene; Tol, toluene.

[a] All values at 20°C (except boiling and melting points) unless otherwise noted.

[b] At 25°C.

TABLE 5.2 Chromatographic Parameters of Alkane and Aromatic Solvents[a]

	Pent	cPent	Hex	cHex	Hept	TMP	Dec	Benz	Tol
Eluotropic strength $\varepsilon°$ on Al_2O_3	0.00	0.05	0.01	0.04	0.01	0.01		0.32	0.29
Eluotropic strength $\varepsilon°$ on SiOH	0.00		0.00		0.01				0.22
Eluotropic strength $\varepsilon°$ on C_{18}									
Solvent strength parameter, P'	0.0	0.1	0.1	0.2	0.1	0.1		2.7	2.4
Hildebrandt solubility parameter, δ	7.0	8.7	7.3	8.2	7.4	6.9	8.8	9.2	8.9
Hydrogen bond acidity, α	0.00		0.00	0.00	0.00			0.00	0.00
Hydrogen bond basicity, β	0.00		0.00	0.00	0.00			0.10	0.11
Dipolarity/polarizability, $\pi*$	−0.087		−0.08	0.000	−0.08			0.59	0.54

Abbreviations: Pent, pentane; cPent, cyclopentane; Hex, hexane; cHex, cyclohexane; Hept, heptane; TMP, iso-octane, 2,2,4-trimethylpentane; Dec, Decalin[R], decahydronaphthalene; Benz, benzene; Tol, toluene.

Cyclohexane contains significant levels of the dimethylpentanes (all isomers) and methylcyclohexane. Linear hexane isomers and benzene are usually present at low-ppm levels. Benzene, at these levels, will produce a nonzero background absorbance in the 250–280 nm range on the absorbance vs. wavelength spectrum.

Heptane commonly contains high-ppm levels of methylcyclohexane and various heptane isomers, low-ppm levels of other alkanes and traces of various alkenes. Odd-numbered linear alkanes are notoriously difficult to purify. In addition, as the number of carbons in the compounds increases, so does the number of possible isomers. Many of these isomers are not separated from the n-alkane during large-scale manufacture. Hence, overall purity, as strictly defined by the n-alkane content, usually decreases while the number and level of impurities increase as the alkane size increases.

Iso-octane usually contains saturated isomeric octane forms as well as other linear, branched, and cyclic alkanes and trace levels of alkenes. Since iso-octane is a branched-chain molecule, the presence of low levels of positional isomers is usually of little or no chromatographic consequence.

Benzene, a commonly used solvent at one time, is rarely used today because of its carcinogenic nature. When available, benzene usually contains ppm levels of toluene and xylenes and lower levels of C_5–C_7 alkanes. Polar sulfur-containing compounds such as thiophene and carbon disulfide are also present.

Toluene is frequently used as a mobile phase in gel permeation chromatography (GPC). In GPC, trace levels of benzene, xylenes, linear and cyclic alkenes, thiophene, and carbon disulfide are of little consequence unless they can react with analytes of interest. However, in normal-phase (NP) HPLC, the polar impurities can be of critical importance since they modify the polar surface and will significantly alter the resulting chromatography. Therefore, a high-purity toluene should be acquired for HPLC use.

TABLE 5.3 Common Manufacturing Quality Specifications of Alkane and Aromatic Solvents[a]

	Pent	cPent	Hex[b]	cHex	Hept	TMP	Dec	Benz	Tol
UV cutoff (nm)	190	200	195	200	200	215	200	278	284
Percent water (maximum)	0.01	0.01	0.01	0.01	0.01	0.01	0.01	0.03	0.03
Available as ACS tested[c,h]	n.a.[d]	n.a.	ABFJM	AFJM	n.a.	AFJM	n.a.	AFJ	ABFJM
Available as HPLC-grade[c]	ABEFJ	ABE	ABEFJM	ABEFJ	ABEFJM	ABEFJM	B	AEJ	ABEFM
Available through[e]	M		$A^f E^f J^{f,g} M^{f,g}$				AFJ		

[a] *Abbreviations*: Pent, pentane; cPent, cyclopentane; Hex, hexane; cHex, cyclohexane; Hept, heptane; TMP, iso-octane, 2,2,4-trimethylpentane; Dec, Decalin®, decahydronaphthalene; Benz, benzene; Tol, toluene.

[b] Unless otherwise noted, this entry column refers to 85% *n*-hexane, >99% hexanes.

[c] Manufacturer's code: A = Aldrich; B = Burdick & Jackson; E = EM Science; F = Fisher; J = JT Baker; M = Mallinckrodt.

[d] Not available since an ACS test does not exist.

[e] Available as a high-purity solvent but not specifically designated as ACS or HPLC grade. This does *not* mean a lesser quality solvent, just that it is not specifically tested for these applications. If these manufacturers produce either ACS or HPLC solvent, they are not listed under this heading.

[f] Available as >95% *n*-hexane.

[g] Available as >99% *n*-hexane.

[h] Burdick & Jackson sells ACS tested solvents in 5-gal. cans and 55-gal. drums. B&J HPLC grade solvents meet all ACS specifications.

TABLE 5.4 Safety Parameters of Alkane and Aromatic Solvents[a]

	Pent	cPent	Hex	cHex	Hept	TMP	Dec	Benz	Tol
Flash point[b] (TCC) °C	-40	-37	-22	-20	-4	4	58	-11	4
Vapor pressure (Torr @ 20°C)	420	400	124	77.5	35.5	41	<1.0	74.6	28.5
Threshold Limit Value (ppm)	600	600	50	300	400	75	25	10	100
CAS number	109-66-0	287-92-3	110-54-3	110-82-7	142-82-5	540-84-1	91-17-8	71-43-2	108-88-3
Flammability[c]	4	3	3	3	3	3	2	3	3
Reactivity[c]	0	0	0	0	0	0	0	0	0
Health[c]	1	1	1	1	0	0	2	2	2

[a] *Abbreviations*: Pent, pentane; cPent, cyclopentane; Hex, hexane; cHex, cyclohexane; Hept, heptane; TMP, iso-octane, 2,2,4-trimethylpentane; Dec, Decalin®, decahydronaphthalene; Benz, benzene; Tol, toluene.

[b] TCC = TAG closed cup.

[c] According to National Fire Protection Association ratings [92]:

Fire: 4 = Material that vaporize at room temperature and pressure and burn readily.
3 = Liquids or solids that can ignite under room conditions.
2 = Materials that ignite with elevated temperature or with moderate heat.
1 = Materials that must be preheated before they ignite.
0 = Materials that will not burn.

React: 4 = Materials that by themselves, can detonate or explode under room conditions.
3 = Materials that can detonate or explode but require an initiator (e.g., heat).
2 = Materials that undergo violent chemical reactions at elevated temperatures or pressures or react with water.
1 = Materials that are, by themselves, stable but that may become unstable at elevated temperatures and pressures.
0 = Materials that are stable even under fire conditions and do not react with water.

Health: 4 = Short exposure times to these materials are lethal or cause major residual injury.
3 = Short exposure times to these materials cause temporary and/or residual injuries.
2 = Lengthy (but not chronic) exposure to these materials may cause temporary incapacitation and/or minor residual injury.
1 = Materials that, upon exposure, cause irritation but only minor residual injury.
0 = Materials that, upon exposure under fire conditions, offer no more hazard than ordinary combustible materials.

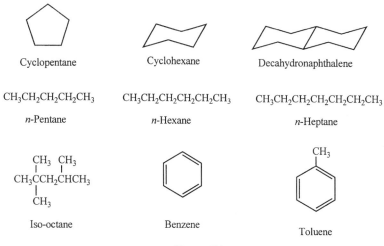

Figure 5.1

5.2 GENERAL ANALYTES

5.2.1 Simple Substituted Aromatic Analytes

One of the traditional uses of low-polarity or nonpolar solvents is their use as solubilizing and supporting solvents in the normal-phase separation of both homologous series [593] and positional isomers [594]. Recently, nonpolar solvents have found major use in the separation of enantiomeric compounds [595].

For small solutes, much study has centered on the generation of data for classes of compounds [596]. The retention behavior of 38 alkyl-substituted benzenes was studied on silica and alumina columns ($\lambda = 200–300\,$nm) using hexane as the eluent. A strong *ortho*-substituent effect was noted. This effect is common in adsorption chromatography and will be discussed again below. Retention, given as $\log k'$ values, was tabulated for all compounds used in the study.

The retention of 46 alkylbenzenes was studied on an alumina column ($\lambda = 254\,$nm) at 25°C using a pentane mobile phase [597]. Retention times and k' values were tabulated for mobile phases with "0.5% and 1.0% water." At "0.5% water" the k' values ranged from 0.2 (*sec*-butylbezene, time $= 2.4\,$min) to 3.2 (hexamethylbenzene, time $= 8.6\,$min). The authors used a moisture control system set at 0.5% and 1.0% water. What value this really translates to is unclear since water is soluble in pentane only to the 0.009% level. Perhaps the water level refers to that on the alumina itself. Regardless of this, the authors state that the reproducibility of the retention decreases with decreasing water content due to increased heterogeneity of the alumina.

Chlorobenzenes, chloronaphthalenes, and chlorobiphenyls were studied on a silica column using n-octane, n-decane, iso-octane, and cyclohexane [598]. Capacity factors were tabulated for all compounds. For these compounds, as solvent chain length increased (n-hexane $\rightarrow$ n-octane $\rightarrow$ n-decane), retention time increased. Branched chain (n-octane $\rightarrow$ iso-octane) or cyclic (n-hexane $\rightarrow$ cyclohexane) solvents yielded longer retention times since they were less effective than their straight-chain analogs at solubilizing the analyte.

Sixteen alkylaromatic compounds (e.g., o-xylene, ethylbenzene, n-amylbenzene, 1-ethylnaphthalene and 2-n-hexylnaphthalene) were chromatographed on a silica column ($\lambda = 254$ nm) using hexane and 0.5–2.0% dichloromethane, chloroform, carbon tetrachloride, or ethyl bromide as the mobile phase modifier [599]. For the stronger solvent modifiers (chloroform and dichloromethane) linear decreases in the reciprocal of the adjusted retention time (i.e., $1/[(\text{retention time}) - (\text{void volume})]$) versus mole fraction of modifier in the mobile phases resulted. However, for the weaker modifiers (carbon tetrachloride and ethyl bromide) the plot was distinctly nonlinear at low mole fractions of modifier (0.01–0.02) and linear at higher mole fractions of the modifier (>0.05).

The purpose of this experiment was to show the applicability of normal-phase gradients to LC separations. In what was described as a "random" selection of 10 analytes (2-phenylethylbromide, 1,4-diphenylbutane, phenetole, nitrobenzene, chlorostilbene oxide, Sudan red 7B, 4-chlorobenzophenone, veratrole, acetophenone, phthalic acid bis-2-ethylhexylester) all peaks were well resolved with good peak shape on a silica column ($\lambda = 254$ nm). A two-ramp 100/0 (hold 2 min) $\rightarrow$ 50/50 (at 8.5 min hold 5 min) $\rightarrow$ 0/100 (at 20 min) hexane/dichloromethane gradient was used [600]. Methyl t-butyl ether was used in place of dichloromethane but resolution was not as good.

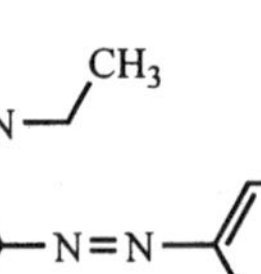

Phenetole 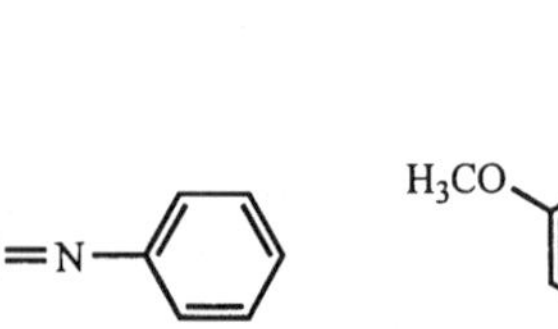   Chlorostilbene oxide

Sudan red 7B   Veratrole

Smith and Cooper [601] studied the retention of three nonpolar solutes (phenanthrene, chrysene, perylene) and four polar solutes (nitrobenzene, 1,2-dinitrobenzene, phenol, aniline) in hexane and hexane/x mobile phases (where $x =$ chloroform, methyl t-butyl ether [MtBE], and dichloromethane at the 5%, 10%, 15%, and 20% levels) on cyanopropyl, aminopropyl, and diol columns. From this work, the solvent strength of each mixture was determined for use in predicting chromatographic retention. More importantly, complex solvent/solute/adsorbed solvent/stationary phase interactions were described highlighting important and unique selectivities offered by these combinations. For example, altering the mobile phase composition from 3% MtBE in hexane to 12% MtBE in hexane (on a cyanopropyl support) leads to a decrease in the retention of phenol and aniline. What is unexpected is the concomitant reversal of the elution order (phenol/aniline to aniline/phenol). This type of reversal of elution order is rare in reversed-phase separations (ion-pair systems notably excluded) but may be a considerable advantage in normal-phase separations.

Langer and co-workers [602] studied the effects on retention of 43 substituted benzenes, phenols, cresols, naphthalenes, and anilines on a silica or an Fe(III)-modified silica ($\lambda = 254$ nm) using a 92/8 hexane/MtBE mobile phase. The retention times for the hydroxyl-, amino-, and nitro-substituted solutes increased (Fe(III)-modified silica vs. silica), whereas those for esters, aldehydes, and ketones decreased. A chromatogram of four selected polar hydroxylated solutes (diethyl phthalate, 3-methoxyphenol, p-cresol, and *trans*-cinnamic acid) showed good peak shapes on the base silica and broadened peaks on the Fe(III)-modified silica. It should be noted that hydroxyquinone was converted to benzoquinone on the Fe(III)-modified silica. The completeness of the conversion was monitored chromatographically, but the kinetics of the reaction was not followed (i.e., the degree of decomposition vs. contact time on the sorbent with adjustment of the flow rate). This study emphasizes the many and varied interactions that may occur between a solute and the chromatographic support.

The *ortho*-, *meta*-, and *para*-isomers of chloro-, bromo-, iodo-, nitro-, methyl-, hydroxyl-, and aminophenols were studied on a silica column ($\lambda = 270$ nm) with various mixtures of 1-butanol (1–2.5%) in hexane or heptane [603]. The elution order in terms of substituent and increasing retention times was: Cl < CH$_3$ < NO$_2$ < H < OH < NH$_2$. The authors noted that reversals in solute elution times occurred as the 1-butanol level changed. Absolute retention times and substituent-related α

Phenanthrene

Perylene

values were given over the range of 1-butanol/hexane mobile phase mixtures. For example, retention times increased from *ortho* < *meta* < *para* for chloro-, methyl- and nitrophenols, whereas retention increased from *para* < *meta* < *ortho* for hydroxyphenols. The authors attribute these elution differences to the hydroxy group being a strong electron donor whereas the chloro and nitro groups are electron acceptors.

Cooper and Hurtubise [604] worked with a set of 34 compounds that included substituted phenols, dihydroxyaromatic compounds, and larger polyaromatic compounds (e.g., naphthalene, anthracene and pyrene). For the monohydroxy compounds, retention (as $\log k'$) on an aminopropyl column ($\lambda = 254$ nm or 280 nm) in isocratic mobile phases consisting of 90/10 to 50/50 heptane/IPA was tabulated. The range of $\log k'$ values was found to be -0.2 for 1,2,3,4-tetrahydro-1-naphthol and 1.2 for 1-hydroxypyrene at 90/10; -0.9 for 1-indanol and 0.3 for 1-hydroxypyrene at 50/50. Dihydroxyl compounds were similarly studied but with mobile phase compositions of 60/40 to 20/80 heptane/IPA. Here $\log k'$ values were found to be 0.5 for 3,5-dihydroxytoluene and 1.3 for 2,3-dihydroxynaphthalene at 60/40; -0.1 for 2,6-dihydroxytoluene and 0.9 for 2,3-dihydroxynaphthalene at 20/80. A series of 1–6 µL injections of 1–12 mg/mL standards were used to generate the results. These results were used to generate predictive models for solute retention, but are also valuable for use in method development.

The retention of eight hydroxy-substituted naphthalenes on silica ($\lambda = 280$ nm) was studied using isocratic heptane/IPA (99.8/0.2 to 95/5) and heptane/ethyl acetate (98/2 to 80/20) mobile phases [605]. A detailed explanation of the solvent modifier effect on retention is given and basically results from the hydrogen-bonding capabilities of IPA (hydrogen bond acceptor and donor) vs. ethyl acetate (hydrogen bond acceptor only). Both surface and solution interactions are considerably different for these two modifiers and therefore selectivity and retentivity were significantly influenced.

Snyder and co-workers [606] studied the retention of benzyl alcohol, *m*-nitroacetophenone, 10 substituted naphthalenes, chrysene, and perylene on a 30°C diol column using a series of isocratic dichloromethane/hexane (0/100 to 35/65) mobile phases. Retention results for all compounds at various isocratic mobile phase compositions are tabulated. Five steroids (prednisone, corticosterone, adrenosterone, 4-androstene-17α-ol-3-one, and 4-androstene-17β-ol-3-one) were similarly studied but at ranges of dichloromethane from 13% to 80%. Also presented in this work is an equation from which the eluotropic strength of an A + B solvent mixture, i.e.,

1,2,3,4-Tetrahydro-1-naphthol   1-Indanol   Adrenosterone

ε°_{AB}, may be calculated. This is a very valuable reference in terms of both data and general chromatographic interest.

Twenty-one high-polarity di- and trisubstituted hydroxyl-, chloro-, nitro-, and amino-benzenes, naphthalenes, and pyridines were chromatographed on silica and Florisil$^{®}$ columns ($\lambda = 254$ nm). Three mobile phases were used: 87.5/12.5 heptane/IPA, 60/40 heptane/dioxane, and 60/40 heptane/THF [607]. Although the overall *average* retention was similar for all three solvents, marked variation on absolute retention and selectivity occurred as the modifier was changed. Plots of $\log k'$ vs. mobile phase composition are shown for both silica and Florisil$^{®}$ columns. The number of comparative results such as these for large sets of high-polarity solutes is limited, making this an especially valuable reference.

The retention times of six methylanilines, four methylphenols, five methylquinolines, two aminonaphthalenes, thymol, 2-naphthol and 1-nitronaphthalene were generated for a range of isocratic hexane/THF mobile phases (98/2 to 80/20) on a Florisil$^{®}$ column ($\lambda = 254$ nm). Mobile phases of 20% dioxane, 5% IPA or 15% THF in heptane all gave similar overall average retention times for the compounds above, but when they were compared in terms of their functional group selectivities they were markedly different [608]. Tabulated results are presented for reference.

Aniline and seven methyl-, chloro-, methoxy-, and nitroanilines were separated on a γ-cyclodextrin column ($\lambda = 254$ nm) using a 92.5/7.5 heptane/IPA mobile phase [609]. The k' values were tabulated for these compounds for solvents containing 7.5 to 100% IPA and then to 90/10 water/IPA.

Hadley et al. [143] separated six N-alkyl-N-methylaniline N-oxides (ethyl to butyl and isomers) and their enantiomers on a Chiralcel OD column ($\lambda = 210$ nm) using 1.5–3.0% ethanol in hexane as the mobile phase. Results are tabulated for each solute and its enantiomer. Nearly complete resolution of all 12 compounds was achieved in one chromatographic run, but only after extremely long chromatographic run times (>240 min). Separation of individual enantiomeric pairs was more effective with most analyses completed in <40 min.

Numerous autoxidation products (e.g., 1,3- and 1,4-dihydroxybenzene, 1,3- and 1,4-di-[1-methylethyl]benzene dihydroperoxide, 1-methyl-1-phenylethanol) were separated from their 1,3- and 1,4-diisopropylbenzene parent compounds on a silica column ($\lambda = 257$ nm) with a 98/2 hexane/IPA mobile phase [610]. Each set of compounds (i.e., the 1,3 vs. the 1,4) was studied separately. Lower limits of

6-Methylquinoline

Thymol

detection were in the range 0.5–1 µg injected. Excellent peaks shapes were obtained as well as good overall resolution.

The retention characteristics of 29 aza-arenes (e.g., pyridine, acridine, quinoline, benz[a]acridine, and numerous substituted analogs) were studied on a diol column using a 97.5/2.5 iso-octane/ethanol mobile phase [611]. Dimethylbenz[a]acridine was least retained ($k' < 3$) and indole had the greatest retention ($k' > 9$). The study also worked with 29 phenols (e.g., numerous alkyl-substituted phenols, nitrophenols, and halogenated phenols). They were studied in detail on the diol column using a 50/50 iso-octane/dichloromethane mobile phase. Appropriate selection of solvent composition provided baseline resolution of isomeric groups (e.g., dimethylphenols using hexane/ethyl acetate).

A comprehensive study of over 300 compounds, covering the most highly volatile aroma components of foods and plants, has been conducted by Lübke et al. [612]. Mobile phases ranging from 100% pentane up to 50/50 pentane/ethyl ether at 15°C were used in conjunction with a diol phase. The various classes of compounds studied included the pyrroles, pyridines, aliphatic esters, aliphatic ketones, aromatic aldehydes, phenols, alkyl alcohols, unsaturated alcohols and hydrocarbons. The addition of a hydroxyl substituent lead to the greatest increase in solute retention. In general substituent effect on retention decreases in the order hydroxyl > nitrile > ketone > ester > aldehyde > ether > hydrocarbon. Thiols, furans, thiophenes, and thioethers were poorly retained using this system. Carboxylic acids and amines could be eluted, but slow kinetic desorption of these analytes from the silanophilic sites on the column caused erratic retention behavior unless the column was completely re-equilibrated between injections. This study is very important because the enormous amount of retention information generated may be effectively used as the basis for initial separation conditions in new method development.

Mono- and disubstituted 4,4'-chalcones were the focus of two extensive studies [613]. A total of 48 *E-S-cis-* and *Z-S-cis*-chalcones were chromatographed on aminopropyl, diol, cyanopropyl, C$_{18}$, and C$_8$ columns ($\lambda = 270$ nm) using a 97/3 heptane/THF mobile phase. Capacity factors for all compounds on all columns are

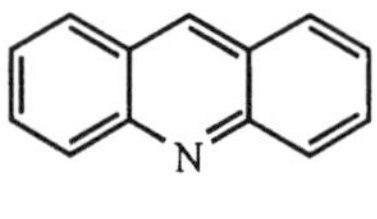

Acridine

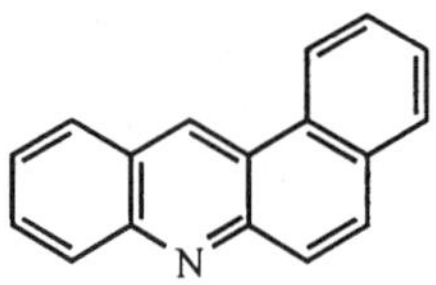

Benz[a]acridine

Thiophene

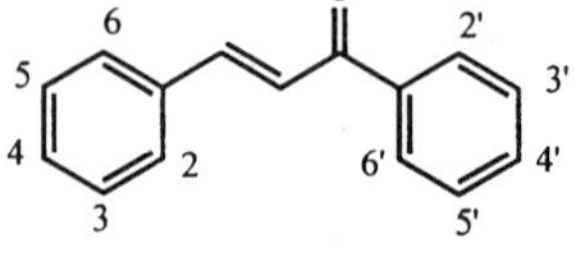

Chalcones

tabulated. These results were used to generate a predictive retention model, but their value in method development should not be underestimated. The authors expanded the range of mobile phase modifiers to include dioxane, ethanol, propanol, octanol, and dimethylformamide [490]. The effects of all modifiers at the 0.5% level in heptane were studied individually. An aminopropyl column was used. Capacity factors are given for each compound and every mobile phase modifier. These studies examine both isomeric and functional group effects on retention in NP work and provide excellent experimental data.

5.2.2 Organometallics and Metal–Ligand Complexes

Hexamethylpropyleneamine oxime (HMPAO) stereoisomers and their technetium-99m complexes were resolved on a 40°C Chiralcel OD column [614]. The d- and l-uncomplexed isomers were separated in 20 min using a 97/3 hexane/IPA (0.01% diethylamine) mobile phase. The ^{99m}Tc complexes were also resolved but with an 85/15 hexane/IPA mobile phase. Temperature increases from 20°C to 40°C greatly improved peak shape and resolution. A table of resolution for the ^{99m}Tc complexes of $meso$-, d- and l-HMPAO was generated for 90/10 to 0/100 hexane/IPA mobile phases. Retention times for 65/35 to 85/15 hexane/IPA were also tabulated. For all these mobile phases the retention times were under 15 min.

Tan et al. [615] successfully resolved the group VI (Cr(VI), W(VI), and Mo(VI)) metal–decacarbonyl (diphenylphosphinyl)alkane (ethyl to hexyl) bridge complexes on a mixed mode cyanopropyl/aminopropyl column ($\lambda = 254$ nm). A 97/3 hexane/chloroform mobile phase was used. Plots of k' versus alkyl bridge length were shown with the k' values varying from 4 to 12. Individual k' values were tabulated for all compound combinations.

Enantiomeric forms of carboxy-, acetyl-, hydroxymethyl- and aldehyde-substituted cyclopentadienyl and benzyl complexes of Cr, Mn, and Fe are separated on Chiralcel OC, OD, OT, and OB columns ($\lambda = 220$ nm) using hexane/IPA and hexane/ethanol mobile phases (at the 1–3% level). Better peak resolution was obtained when ethanol was used in the modifier [616]. For both modifiers, resolution increases rapidly as percent organic modifier decreases. Capacity factors of 5–20 were obtained depending upon the solvent/solute combination chosen.

The Ni(II) and VO(II) complexes of etioporphyrins and deoxophylloerythroetioporphyrins from fuel and crude oils were studied on an aminopropyl column ($\lambda = 400$ nm, ex; 620 nm, em) using a 25-min $45/55/0 \rightarrow 20/30/50$ toluene/hexane/dichloromethane gradient [617]. A chromatogram of four neat metal–porphyrin standards and four crude oil extracts are shown. Due to the very large number of possible homologs of the porphyrins, a manifold of peaks was obtained. These distributions were distinctly different for each crude oil sample. Tentative identification, as related to the retention of the neat standards, is given.

5.2.3 Summary

Alkane solvents as the major solvent component are extremely useful when either positional isomer or enantiomeric separations are needed. The selection of support material should not be limited to silica alone. Cyanopropyl, aminopropyl, and diol columns have exhibited useful and unique retention properties when used in conjunction with the alkane-based solvent system. The addition of small percentages (typically <10%) of a more polar modifier often results in a decrease in retention, overall better peak shapes and, when chosen properly, an enhancement of selectivity.

5.3 ENVIRONMENTALLY IMPORTANT ANALYTES

5.3.1 PAHs, Substituted PAHs, and Related Analytes

Polycyclic aromatic hydrocarbons (PAHs) are nonpolar analytes regulated by the EPA. One standard method, US EPA SW-846 Method 8310, dictates that an acetonitrile/water gradient on a C_{18} column be used. However, PAH solubility decreases rapidly as the ring number increases from naphthalene (two fused rings) to benzo[g,h,i]perylene (six fused rings). Such solubility limitations are not a problem when alkane solvents are used.

Fernandez and Bayona [618] developed a method for the fractionation of polyaromatic hydrocarbons from dichloromethane extracts of air particulates and marine sediment. The resulting fractions were quantitated by GC/MS. A silica column ($\lambda = 254$ nm or 254 nm, ex; 390 nm, em) generated the following fractions: PCB, PAH, NO_2-PAH/2°-amine-PAH, keto-PAH, keto-PAH/quinones, quinones, 3°-amine-PAH, and 3°-amine-PAH/hydroxy-PAH. The separation required a 70-min $100/0 \rightarrow 0/100$ hexane/dichloromethane gradient. Levels of 47 compounds were monitored, typically well below the µg/g range.

Alkylbenzenes and alkylnaphthalenes were studied on a silica column using hexane, hexane/1-chlorobutane, hexane/1-bromobutane, or hexane/IPA as the mobile phase [619]. Modifier levels ranged from 0.005% to 10%. Capacity factors versus carbon number were plotted for each solvent mixture. Selectivity decreased for all solvent modifiers *except* 1-chlorobutane, for which selectivity *increased* as the level increased from 2% to 8%. The authors attribute this to the formation of π-complexes between the 1-chlorobutane and the PAH solutes. Selectivity decreased, as expected, for the alkylnaphthalenes when 1-bromobutane was used. Selectivity was lost rapidly as the level of IPA increased from 0.01% $\rightarrow$ 0.05% indicating that at low IPA concentrations IPA (or the water contained in the IPA) readily modifies or deactivates the silica support.

Zawadiak et al. [620] studied the retention of three oxidation products of 2-isopropylnaphthalene (IPN), 1-(2-naphthyl)ethanone, 2-(2-naphthyl)-2-propanol, 1-methyl-1-(2-naphthyl)ethyl hydroperoxide, and six of 2,6-diisopropylnaphthalene (DIPN) (e.g., 2-isopropyl-6-[1-hydroxy-1-methylethyl]naphthalene, 1-[6-isopropyl-2-naphthyl]ethanone, 1-methyl-1-[2-naphthyl]ethyl hydroperoxide) on a 25°C silica column ($\lambda = 220$ nm for IPN analytes and 227 nm for DIPN analytes). A 99/1 hexane/IPA mobile phase resolved all IPN products in 4 min. Linear working ranges were reported as 0.05–5 µg injected. A 99/1 (hold 3 min) $\rightarrow$ 97/3 (at 11 min) hexane/IPA gradient baseline resolved and eluted all DIPN compounds in 10 min.

Corresponding working ranges were reported as 0.05–1 µg injected. The effect of changing the IPA level on retention was plotted for all compounds.

Metabolites of 7,9- and 7,10-dimethylbenz[c]acridine (e.g., 7-[hydroxymethyl]-9-methylbenz[c]acridine, 7,9-dimethylbenz[c]acridine-5,6-oxide) were separated in 80 min using a silica column ($\lambda = 270$ nm) and 95/5 petroleum ether/ethanol or 90/10 petroleum ether/ethyl acetate, respectively. At least 23 metabolites were positively identified [621]. The retention behavior of benz[c]acridines and 11 methyl, dimethyl, and trimethyl analogs were studied on silica and aminopropyl columns [622]. Individual solute retention times were tabulated. For the silica column a 95/5 hexane/ethanol mobile phase was used and retention times ranged from 2.5 to 33 min. Incomplete resolution of three isomers resulted. Hexane and the aminopropyl column produced retention times of 8–26 min with at least partial resolution obtained for all isomers.

A series of triol, triolhydroxyethylthioether, and methoxytriol-benzo[a]pyrene (BP) enantiomer pairs were resolved on an (S)-N-(3,5-dinitrobenzoyl)leucine (DNBL) column ($\lambda = 247$ nm) using a series of hexane/(2/1 ethanol/acetonitrile) mobile phases [623]. The ratio of hexane/modifier ranged from 90/10 to 70/30 depending upon the analytes. Retention times were documented for all compounds along with their resolutions. The authors note that there is a striking difference in the chromatography between the DNBL column and a related (R)-N-(3,5-dinitroben-zoyl)phenylglycine (DNBPG) column. For example, the chromatograms for the enantiomeric separation of 7,8,9-triol-10-hydroxyethylthioether-BP using an 80/20 hexane/(2/1 ethanol/acetonitrile) mobile phase show very tailed peaks resulting on the DNBPG column with elution times over 35 min. On the DNBL column, however, peak shapes are excellent and elution is complete in <20 min.

Some dichlorodihydro-PAHs exhibited surface-catalyzed degradation on silica support materials. Nilsson and Colmsjö [624] found that >90% of the injected amount (level not reported) of 4,5-dichloro-4,5-dihydropyrene degraded to 4-chloropyrene and pyrene while in contact with the silica for <8 min. Dry hexane was used as the mobile phase. Deactivation of the silica through the saturation of hexane with water decreased this degradation to <20%. The use of an aminopropyl column in place of a silica column led to less than 5% degradation, whereas the use of a cyanopropyl column resulted in less than 0.5% degradation. The authors concluded that the surface silanol groups were the key factors causing the degradation since deactivation of the surface with water or derivatization of the silanol groups invariably decreased the amount of degradation. With pentane as the mobile

7,9-Dimethylbenz[c]acridine

phase, the cyanopropyl column exhibited a higher selectivity for the dichlorodihy-dro-PAHs than the aminopropyl column. Twenty-seven chloro-, dichlorodihydro-, and parent PAH compounds were studied and k' values were tabulated for the cyanopropyl column/pentane solvent system.

Seventeen selected dihydrodiol isomer pairs of three- and four-ring PAHs were resolved on a silica column ($\lambda = 254$ nm) as their O-methyl ethers [625]. The parent compounds included 9,10-dihydrodiol-1-methylphenanthrene, 5,6-dihydrodiol-benz[a]anthracene, and 5,6-dihydrodiol-7-methylbenz[a]anthracene. Successful eluents ranged from the highest solvent polarity of 94.6/5/0.2/0.2 hexane/ethyl acetate/methanol/THF (for 4,5-dihydrodiolbenzo[a]pyrene) to the lowest solvent polarity of 99.5/0.66/0.33 hexane/ethanol/acetonitrile (for 5,6-dihydrodiolben-zo[c]phenanthrene). Elution times ranged from 4 to 23 min.

Diesel exhaust emissions were monitored through fractionation on a silica column using a 100/0/0 (hold 10 min) → 0/100/0 (at 40 min hold 10 min) → 0/0/100 (at 70 min hold 20 min) hexane/dichloromethane/acetonitrile gradient [626]. The collected PAH fraction (eluting from 8 to 28 min) was further fractionated on an aminopropyl column using a 20-min gradient from 100% hexane to 100% dichloromethane. The PAHs with 4- and 5-membered rings eluted between 15 and 20 min. Individual peak identification was done by RP chromatography.

Heavy petroleum was characterized with respect to its aromatic (e.g., benzene, 1-methylnaphthylene), and nonaromatic (e.g., tetradecane) content. These results were comparable to the lengthier ASTM D2549-91 method. Two 300 × 4.1 mm 35°C aminopropyl columns (RI detector) and a 95/5 hexane/IPA mobile phase generated complete elution in <15 min [627].

5.3.2   Nitrated and Chlorinated Nonpesticide/Herbicide Analytes

Excellent peak shape and baseline resolution of o-, m-, and p-dinitrobenzene were obtained on a silica column ($\lambda = 254$ nm) using a 99.5/25/0.5 pentane/dichloromethane/IPA mobile phase [628]. Detection limits of <10 μg/mL and a linear concentration range up to 160 μg/mL were reported. The analysis was complete in <10 min.

Two dinitrotoluenes (2,4- and 2,6-DNT) and two trinitrotoluenes (2,4,5- and 2,3,4-TNTs) were baseline resolved from "finished product" TNT (2,4,6-trinitro-toluene) in <20 min on a silica column ($\lambda = 235$ nm). An 80/20 heptane/dichloromethane mobile phase was used. Elevated detection limits were found because dichloromethane has a UV cutoff of ~233 nm [629]. A 98.5/1.5 heptane/THF mobile phase gave better detection limits and a lower background absorbance, but some resolution was lost. When sequential injections were made using the heptane/dichloromethane solvent system, resolution was gradually lost due to contaminant buildup on the column. A routine rinse of the column with 100% dichloromethane restored column performance to near original performance. Although not stated in the paper, such rinses are most effective if the column is back-flushed. Remember that any flush solvent should be shunted directly to waste to prevent contamination of the detector. Nonetheless, the required detection of

<0.5% 2,4-dinitrotoluene in TNT was easily achieved. Similarly, Zou et al. [630], resolved TNT and its biodegradation compounds, 2,4-DNT, 4-amino-2,6-DNT, 2-amino-4,6-DNT, 2,6-diamino-4-NT, and 2,4-diamino-6-NT on a silica column ($\lambda = 254$ nm) using 79/21 hexane/IPA as a mobile phase. Peak shape was excellent and complete resolution in <10 min was achieved.

THF/hexane (3/97) was the mobile phase used to study the retention of 46 nitro-PAHs [176]. Retention times ranged from 4 to 14 min on a silica column ($\lambda = 254$ nm). A "Pirkle-type" chiral phenylglycine column was also used to separate some compounds using a 27/2/1 hexane/ethanol/acetonitrile mobile phase. Elution times ranged from 3.2 to 15.6 min. Nineteen parent unsubstituted PAH and partially saturated PAH compounds were also chromatographed under identical conditions, yielding retention times in the 4–7 min range for silica and 4.1–9.3 min for the chiral phase. Interesting effects due to positional isomers were documented and correlated to the position of the nitro group with respect to the aromatic ring plane (i.e., parallel or perpendicular). These tabulated results form a good basis from which to develop specific methods.

Dinitro- and trinitrophenanthrenes were fractionated using NP techniques. Crude dinitrophenanthrene products were collected in nine fractions using a silica column ($\lambda = 280$ nm) and 50/50 hexane/dichloromethane as a mobile phase [631]. A rapid change to 70/20/10 hexane/dichloromethane/ethyl acetate eluted a tenth fraction. Elution was complete in <20 min. Similar techniques were used for the trinitrophenanthrenes. Further fractionation was carried out on late eluting collected fractions using different mobile phases such as 60/35/5 hexane/dichloromethane/ethyl acetate. Structural confirmation was done by GC/MS.

5.3.3   Pesticide, Herbicide, and Fungicide Analytes

The levels of fungicides fentin acetate and fentin hydroxide were determined in formulations using a silica column ($\lambda = 286$ nm) and a 95/5 hexane/acetic acid mobile phase [632]. Elution was complete in 5 min and peaks were resolved. Calibration solutions from 0.1 to 10 µg/mL were used.

Technical grade toxaphene was chromatographically fractionated and 20 chloro-boranes (hexa to nona) and five chlorocamphenes (hexa to octa) were identified [633]. Two normal-phase columns, silica and aminopropyl, were used with a hexane mobile phase ($\lambda = 220$ nm). Elution fractions were collected over a 20-min period

Fentin hydroxide

Toxaphene (example), mix of polychlorinated camphene where average number of chlorines is 8 and further characterized. For further separation, a reversed-phase separation on a C_{18} column was run with an 86/14 acetonitrile/water mobile phase. The authors note that these conditions are not optimal due to limited sample solubility.

Nine carbamates were baseline resolved and exhibited excellent peak shapes on both a cyanopropyl column (7–10% IPA in hexane gradient) and on a silica column ($\lambda = 220$ nm) using a 10-min $7/93 \rightarrow 9/91$ IPA/hexane gradient [634]. Striking retention differences occurred. For example, pirimicarb eluted ~3 min on the cyanopropyl column and at ~10 min on the silica column. Peak shapes and overall resolution was also significantly better on the cyanopropyl column as compared with the silica column.

Seven pyrethroid pesticides used in common formulations were studied using a silica column ($\lambda = 275$ nm) with a 97.3/0.7 iso-octane/ethyl acetate mobile phase [635]. Determination of the total number of impurities was verified by using a mobile phase with a different selectivity: 99.4/0.6 iso-octane/dioxane. No change in the number of peaks occurred with this change. Detection limits of 150 µg/mL and a linear concentration working curve up to 3000 µg/mL were obtained.

Kutter and Class [636] completely resolved isomers and partially separated enantiomers of cypermethrin in 16 min on a silica column ($\lambda = 220$ nm) using a 99.5/0.5 hexane/THF mobile phase. Partial resolution of the four pair of enantiomers was achieved in 42 min through the use of a Pirkle chiral column and a 99.9/0.05/0.05 hexane/TFA/IPA mobile phase.

The enantiomers of pyriproxyfen were studied in a similar fashion (hexane with added alcohol mobile phase modifiers) on a methylbenzoate modified cellulose support ($\lambda = 254$ nm). In this study, the capacity factor (k') and separation (α) were monitored at 95/5 hexane/alcohol (methanol or ethanol) or 90/10 hexane/alcohol (n-propyl alcohol, IPA, 1- and 2-butanol, iso- and t-butyl alcohol, 1-, 2-, and 3-pentanol, and 1- and 2-hexanol). Similar increases in retention were recorded as the modifier was changed from methanol to 1-pentanol [637]. In this study, however, IPA gave lower retention times than n-propyl alcohol. It is also interesting to note that the elution order for the S versus the R enantiomer was S then R for methanol,

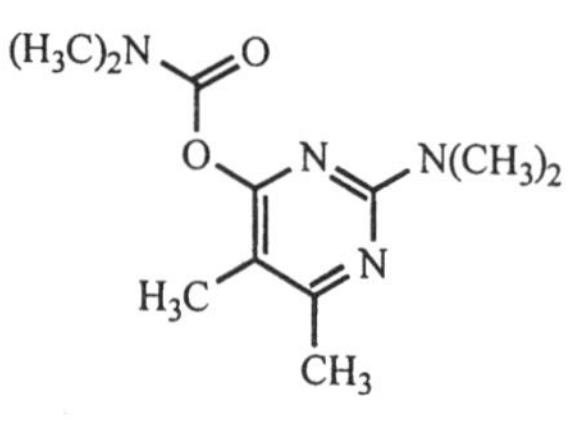

Pirimicarb

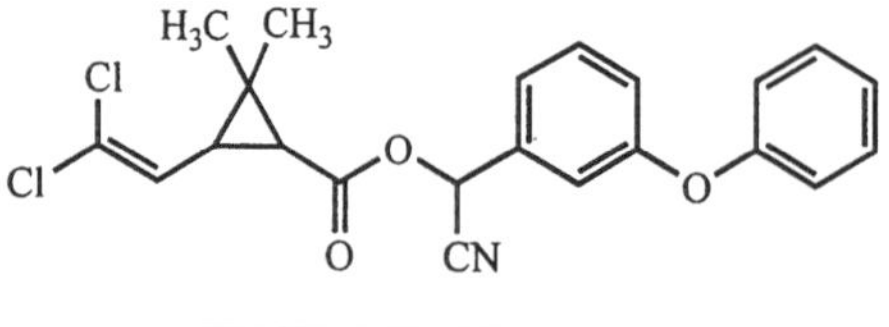

Cypermethrin

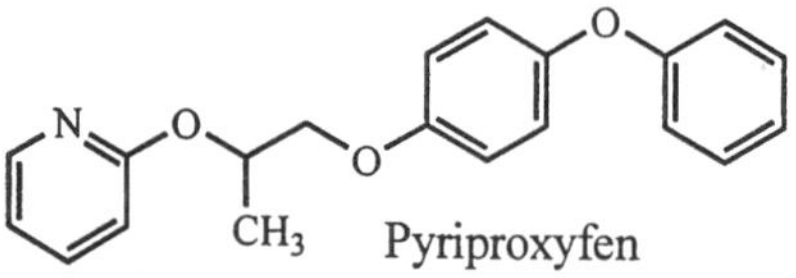

Pyriproxyfen

ethanol, IPA and *t*-butyl alcohol, but was reversed for all other alcohol modifiers (with no separation achieved for 2-butanol). Steric effects were cited as the cause for the retention reversals.

Amitrole, diuron, and simazine were separated, eluted in 8 min and quantitated on a cyanopropyl column ($\lambda = 205$ nm or amperometric detection at $+165$ mV vs. Ag/AgCl) using a 62/36/2 hexane/IPA/water (with 2.5 g/L LiClO$_4$ for UV or 2 g/L trichloroacetic acid for amperometry) mobile phase [638]. Good peak shapes were obtained and detection limits of 2.5 ng (UV) and 200 pg (amperometry) were claimed. A major advantage of the amperometric detector compared with the UV detector was the selectivity for amitrole. Working range concentrations of up to 500 ng injected were reported.

Fischer and Jandera [210] studied the retention of nine phenylurea herbicides (fenuron, desfenuron, deschloromethoxuron, neburon, isoproturon, *N*-phenylurea, hydroxymethoxuron, bis-*N*,*N*′-(3-chloro-4-methylphenyl)urea, and *N*-butyl-*N*′-phenylurea) on a cyanopropyl column ($\lambda = 254$ nm) using various ratios of hexane/IPA as mobile phases. Good resolution was obtained when a 90/10 hexane/IPA solvent system was used. Elution was complete in 20 min. However, *N*-phenylurea tailed badly. (The addition of a low level of acetic acid or triethylamine would very likely help sharpen this peak.) The k' values for all these pesticides were tabulated for mobile phases ranging from 90/10 to 0/100 hexane/IPA. A similar study was done using an aminopropyl column. Only seven analytes were resolved in the study (at 80/20 hexane/IPA). Capacity factors for all pesticides were tabulated for this column from 95/5 to 0/100 hexane/IPA. The information in this study is excellent for those needing assistance in developing new methods for the separation of urea-based analytes.

Four metabolites of *N*-methylcarbazole (carbazole, 3-hydroxycarbazole, 3-hydroxy-*N*-hydroxymethylcarbazole, *N*-hydroxymethylcarbazole) were separated on a silica column using a 25/1 hexane/IPA mobile phase [639]. Peak shapes were good and elution was complete in 42 min. Analysis times could be greatly decreased through the use of a gradient, since the last two peaks eluted at 18 and 42 min.

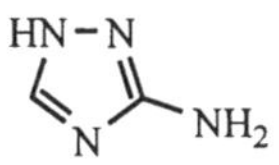

Amitrole

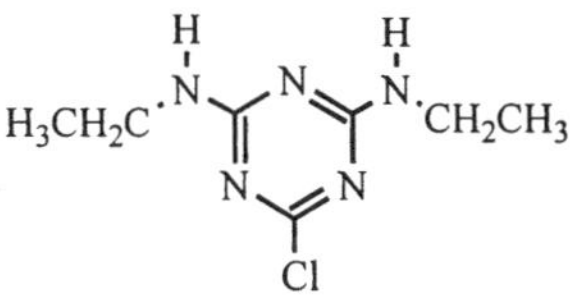

Simazine

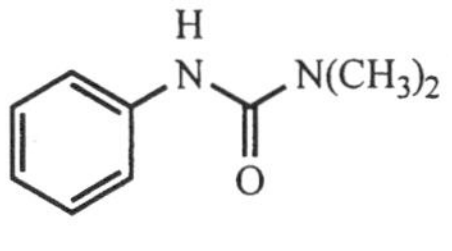

Fenuron

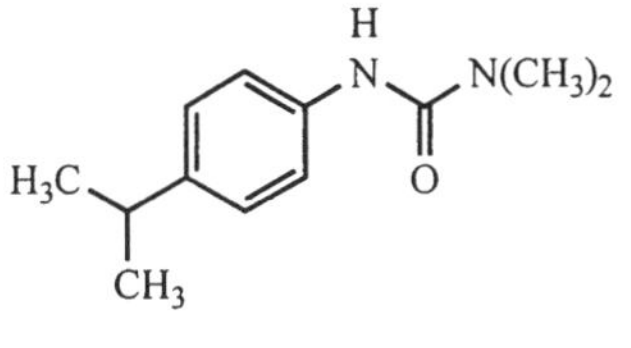

Isoproturon

5.3.4   Summary

Once again, isomeric and enantiomeric forms of polar compounds can be effectively separated using NP supports and alkane-based mobile phases. When anhydrous alkanes are used as the mobile phase, active silanol groups can strongly interact with solutes. These interactions can cause peak broadening or can catalyze chemical reactions. Therefore, the choice of mobile phase modifier is critical—the modifier governs the overall surface interactions and dictates the selectivity of the separation. The only concern in using alkane-based solvent systems is the limited solubility some analytes will have in the mobile phase used.

5.4   INDUSTRIAL AND POLYMER ANALYTES

5.4.1   Surfactant and Additive Analytes

Polyethoxylated surfactants were separated based on the number of ethylene oxide oligomers (EON). Two methods were utilized. A silica column ($\lambda = 270$ nm) and a 30-min $100/0 \rightarrow 70/30$ (90/5/5 n-heptane/chloroform/methanol)/(50/50 chloroform/methanol) gradient was able to produce a well-structured profile for EON = 1–25. An aminopropyl column was used for an EON 1–30 polymer. A 45-min $100/0 \rightarrow 80/20$ (90/5/5 n-heptane/chloroform/methanol)/(50/50 chloroform/metha-nol) gradient gave good profile definition. Samples were made up to a concentration of 0.05 mol/L [640].

Miszkiewicz et al. [641] used a 55-min gradient, $100/0 \rightarrow 40/60$ hexane/(95/5 IPA/water) and a 50°C aminopropyl column ($\lambda = 277$ nm or ELSD, drift tube 50°C, 1.5 L/min N$_2$) for the characterization of alcohol (AE) and alkylphenolethoxylates (APE). The authors noted that for those APEs with a low number of ethylene oxide residues, the ELSD skews the molecular weight distribution to higher molecular weights. This is due in part to the volatility of those species (and therefore a lower than expected signal). Conversely, the UV absorptivity must be corrected for the decrease in molar absorptivity as the chain length increases. The latter fact is critical if one pure standard is used for quantitation of all components. Overall, peak shapes were excellent and the authors claimed excellent reproducibility.

Ethoxylated alkylphenols are used widely as detergents, emulsifiers, and surface-active agents. Oligomers of up to 10 ethylene oxide units were separated on a silica column ($\lambda = 270$ nm) using a 15-min $90/5/5 \rightarrow 70/15/15$ heptane/chloroform/methanol gradient [642]. A gradient to 10% (50/50 chloroform/methanol)/90% (70/10//0 heptane/chloroform/methanol) reduced the analysis time with minimal loss of resolution. This study also presented the effects of mobile phase composition (80/10/10 to 90/5/5 heptane/chloroform/methanol) on an aminopropyl column for

Ethylene oxide

oligomers in the 10–30 unit range. The elution times varied from 15 to 32 min as the oligomer length increased. Tributylphenolethoxylates were also characterized on an aminopropyl column using a 20-min $90/5/5 \rightarrow 0/50/50$ heptane/chloroform/methanol gradient. Sample concentrations of around 50 mM were used and 5 µL aliquots were injected.

The same set of compounds were extracted from water and analyzed using a cyanopropyl column ($\lambda = 229$ nm, ex; 310 nm, em) and a complex 20-min $99/1 \rightarrow 58/42$ (20/80 hexane/THF)/[10/90 water/IPA) gradient [643]. Excellent peak shape and resolution were obtained for oligomers with up to 18 monomer units. Detection limits were found to be about 0.1 µg/L for each oligomer. Levels of 1–3 µg/L were detected and resolved into complete oligomer distributions (1–18 ethoxy units). The authors note that in the extract concentration step the sample (3 µg) was lost because of adsorption to the glass. Losses were minimized when the sample extract was spiked with 5 µL of an alcohol methoxylate (e.g., C_{12-14} alcohol-3EO trade-named SURFONIC$^{(R)}$ L24-3) prior to concentration. The sample also needed protection from air and dissolved oxygen.

Jandera et al. [644] separated ethoxylated alkylphenols (the alkyl was methyl to pentadecyl) using either a $25/75 \rightarrow 85/15$ ethanol/heptane gradient on a silica column ($\lambda = 254$ nm or 230 nm) or a $60/40 \rightarrow 90/10$ n-propyl alcohol/heptane gradient on a cyanopropyl column. These methods were used to characterize a number of commercial formulations. Most analyses were complete in less than 40 min. In general, excellent peak shapes and elution profiles were generated with these systems. Analyses were complete in 30–40 min.

Ethoxylated alcohol distributions from 2 to 40 ethylene oxide units (e.g., Brij 76) were characterized on a diol phase (ELSD) using a 25-min $90/10 \rightarrow 20/80$ hexane/(98/2 chloroform/IPA) gradient [645]. Good separation throughout the homologous series resulted. Peak shapes were acceptable as well. Tables of retention times for hexadecylalcohol-ethoxylated compounds (1–6 ethoxy groups) and for C_{10}–C_{18} hexaethoxylated alcohols were presented.

In an excellent and comprehensive study, Martin [646] generated exceptionally well-resolved chromatograms of fatty acid and alcohol adducts of ethylene oxide (EO) and propylene oxides. An aminopropyl column (ELSD) and a 30-min $76/19/5 \rightarrow 56/14/30$ hexane/chloroform/methanol gradient were used. Separate chromatograms were also generated for EO condensates of fatty alcohols with $n < 10$ and $n > 10$ ($n =$ number of carbons in the linear alcohol). EO condensates of fatty acids were characterized using the same chromatographic conditions as for the alcohols but using a $75/20/5 \rightarrow 50/12.5/37.5$ gradient.

Ethoxylated alcohols with the alkyl chain length of 12 and the number of ethoxylate chains varying from 2 to 20 were extracted from influent and effluent wastewaters and were well resolved as their phenylisocyanate derivatives on an aminopropyl column ($\lambda = 240$ nm). A 30-min $100/0 \rightarrow 65/35$ (35/15 hexane/ethylene dichloride)/(185/65 acetonitrile/IPA) with 800 µL acetone/L added gradient was used. Detection limits down to 100 ppb were reported [225].

4-Nonylphenolpolyglycol ether surfactants with 2, 4, 6, 8, 10, 13, and 15 ethoxy units were resolved on a diol column ($\lambda = 275$ nm). A 12-min $70/30 \rightarrow 40/60$ (95/5 hexane/dichloromethane)/(50/40/10 hexane/dichloromethane/IPA) gradient generated excellent separation [647]. This procedure was also used to identify aqueous surfactant solutions such as Arkopal N-20, N-40, N-60, and N-100. Detection limits (for each nonylphenolpolygycol ether constituent) ranged from 50 to 130 ng.

5.4.2 Other Polymeric Analytes

Styrene/ethyl or methyl methacrylate co-polymer elution profiles were studied on a phenyl column ($\lambda = 259$ nm) using a 98/0/2 (for 0.5 min) → 3/35/62 (at 0.5 min) iso-octane/THF/methanol step gradient. Samples were prepared in THF containing 250 ppm BHT. The resulting chromatography for mixtures of well-defined co-polymer ratios was detailed and discussed [648]. The author notes that gradient work at 230 nm was impossible with THF due to the large baseline shift generated by the THF. The interesting three-component solvent system is therefore proposed and described in detail.

Styrene/ethyl methacrylate (S/EMA) co-polymer elution profiles were character-ized on a cyanopropyl column (evaporative light scattering detector) using various heptane/methanol/dichloromethane gradients [649]. For example, a mixture of five S/EMA co-polymers, present at the 10 μg injected level, were analyzed in 10 min using a step gradient from 100% heptane to 80/20 heptane/dichloromethane at 0.1 min followed by a linear gradient reaching 30/50/20 heptane/methanol/ dichloromethane at 10 min.

Methyl and butyl methacrylate copolymers were analyzed using a 30-min 98/2 → 0/100 toluene/methyl ethyl ketone gradient on a silica column (ELSD). Molecular weights of polymers from 9.8 K to 40 K were characterized. 100 μL aliquots of 1–10 μg/mL poly(methyl methacrylate) standards were used [650]. Poly(methyl methacrylate)-graft polydimethylsiloxane co-polymers were character-ized using a silica column (ELSD) and a 60/4/36 → 0/10/90 cyclohexane/ ethanol/toluene gradient [247]. The authors noted that the sample was not soluble in the initial mobile phase and so a stronger solvent was used. The retention process therefore is one of initial precipitation of the sample onto the head of the column followed by redissolving and subsequent elution. The results from the HPLC method were graphically compared with size exclusion/solvent evaporation FTIR results and excellent correlation was obtained.

5.4.3 Fullerenes

Solubility is an issue for fullerenes especially in semipreparative and preparative work. However, the very limited solubilities in hexane (13 μg/mL C_{70}) and iso-

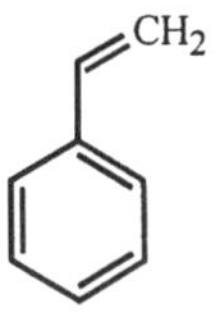

Styrene

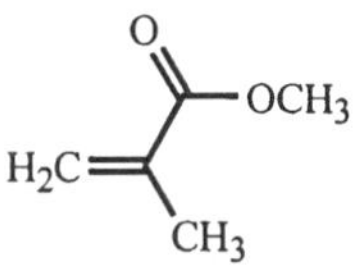

Methyl methacrylate

octane (26 μg/mL C_{60}) also cause chromatographic problems as noted above. Dichlorobenzene (36 mg/mL C_{70}), CS_2 (10 mg/mL C_{70}), and dichloromethane (254 μg/mL C_{60}) are solvents in which the fullerenes have the highest solubilities but have major drawbacks chromatographically (too strong as eluents, too difficult to remove to reclaim sample). Contrast this with <10 μg/mL C_{60} solubility in methanol and <2 μg/mL in acetone [651].

Fullerene separations have been studied extensively [269]. Due to their high molecular weights and nonpolar character, the alkanes and alkylbenzenes have been widely used as the solvents of choice. C_{60} and C_{70} fullerenes were resolved on a dinitroanilinepropyl column ($\lambda = 380$ nm) using either a 30/70 dichloromethane/ hexane or a 50/50 hexane/benzene mobile phase [652]. Solutions containing 1 mg/mL of sample were prepared and 5 μL injections were made. The authors note that ethyl ether/hexane solvents generated broad asymmetric peaks most likely due to the precipitation of the sample.

Ohta et al. [653] examined the retention of a series of fullerenes: C_{60} to C_{84}. The effects of bonded phase (C_8, C_{18}, C_{30}), mobile phase (toluene/aceonitrile 100/0 to 60/40), and temperature (0°C to 60°C) were investigated. One successful 80-min baseline separation was for C_{60}, C_{70}, C_{76}, $C_{78-2v'}$, C_{78-2v}, and C_{78-D3} fullerenes on a 30°C C_{30} column ($\lambda = 312$ nm) using a 60/40 toluene/acetonitrile mobile phase. Similarly, C_{60}, C_{70}, C_{76}, C_{78}, C_{82}, and C_{84} fullerenes were separated on a 60°C C_{30} column ($\lambda = 312$ nm) in 40 min using a 60/40 toluene/acetonitrile mobile phase.

Hexane and a C_{18} column [654] were conditions also used for this separation. A triphenyl column ($\lambda = 320$ nm) and a hexane mobile phase were used to study the temperature effects on the retention of fullerenes. For the range of 30–70°C, little or no change in the retention profile occurred [655]. Unlike the study presented above, mixtures of 5–15% ethyl ether in pentane on a phenyl column ($\lambda = 330$ nm) successfully separated C_{60} and C_{70} fullerenes but could not separate the higher molecular weight species [656]. Plots of $\ln k'$ versus percent ethyl ether in pentane were shown for C_{60} and C_{70} fullerenes.

Temperature affects the separation of fullerenes quite dramatically when monomeric or polymeric C_{18} phases are used. With hexane as mobile phase ($\lambda = 312$ nm), the chromatographic changes due to temperatures ranging from 20°C down to -70°C were studied [654]. Increased resolution of C_{76} to C_{84} fullerenes was obtained as the temperature was decreased from 20°C down to -20°C. Below this temperature, all resolution was completely lost regardless of the monomeric or polymeric nature of the C_{18} phase. The authors speculate that the bonded phase becomes rigid or "frozen" at temperatures below -20°C, which precluded effective interaction with the solutes. The compatibility of alkanes with very low-temperature work is rarely utilized and could be advantageous for many other separations.

Diederich and Whetten [657] separated C_{60}, C_{70}, C_{76}, and C_{84} fullerenes in 20 min on a C_{18} column ($\lambda = 310$ nm) using a 50/50 toluene/acetonitrile mobile phase. Jinno et al. [658], baseline resolved C_{60}, C_{70}, C_{76}, $C_{78}(C_{2v'})$, $C_{78}(C_{2v})$, C_{80}, and C_{90} fullerenes in 160 min on monomeric and polymeric C_{18} columns ($\lambda = 325$ nm) using a 45/55 toluene/acetonitrile mobile phase. On the C_{18} columns, a very definite decrease in retention occurred as the temperature increased from

10°C to 70°C: polymeric C_{18}, from 240 to <10 min; monomeric C_{18}, 480 to 40 min. The best resolution occurred between 20°C and 40°C.

The four 3-methylcyclohexanone [2 + 2] photoadducts of C_{60} were baseline resolved on a chiral bonded phase ($\lambda = 300$ nm) using a 2/1 toluene/hexane mobile phase [659]. In this study, a 100 μL aliquot of a 5 mg/mL standard was injected. Three monoadducts of the reaction of C_{70} fullerene with 1,2-bis(bromomethyl)-4,5-dimethoxybenzene were resolved on a Buckyclutcher I column ($\lambda = 310$ nm) using either a 2/3 toluene/hexane or 2/3 hexane/dichloromethane mobile phase [660]. Peak shapes were good but resolution was incomplete. Elution was complete in 70 min.

5.4.4   Summary

Alkanes and alkylbenzenes are extremely important components of mobile phases in the analyses of polymers and other common industrial analytes due to the high solubility levels they provide. Modification by low levels (typically <10%) of a more polar component provides reasonable retention times and important selectivity enhancements as well as increased solubility of many solutes.

5.5   BIOLOGICAL ANALYTES

Many biologically important analytes are nonpolar, have large nonpolar functional groups, or are moderately polar in nature. These compounds often have limited solubility in water but large solubilities in organic solvents. Accordingly, a number of normal-phase separations have been developed to take advantage of these properties.

5.5.1   Carboxylic Acid Analytes

The separation of the positional isomers of acetic and butyric acid-substituted diacylglycerols (palmitoyl and stearoyl) from butteroil was accomplished as their dinitrophenyl isocyanate derivatives on a YMC-Pack 3,5-(R)-(+)-1-(1-naphthyl)-ethylamine column ($\lambda = 226$ nm) at 25°C using a 40/10/1 hexane/1,2-dichloro-ethane/ethanol mobile phase [661]. Injections were 10–20 μg and elution was complete in <30 min.

The hydroperoxidation products of linoleic acid were studied by Wu and co-workers [662]. Ten products (e.g., 13-hydroxy-(9Z-11E)-octadeca-9,11-dienoic acid, 9-hydroxy-(10E-12E)-octadeca-10,12-dienoic acid, ketooctadecadienoic acid, hydroxyoctadecadienoic acid) were resolved in <30 min on a silica column ($\lambda = 234$ nm) using a 98/2/0.05 hexane/IPA/acetic acid mobile phase. Peak shapes were excellent and good resolution was obtained. The same conditions were used in semipreparative work. Structures were confirmed by GC/MS.

Similarly, eight hydroperoxidase products of octadecadienoic and octadeca-trienoic acids (e.g., hydroperoxyoctadecatrienoic acid, 13(S)-hydroperoxy-(9Z,11E,15Z)-octadecatrienoic acid) were resolved on a silica column using a

70/30/0.5 hexane/ethyl ether/acetic acid mobile phase. Peak shapes and resolution were very good. Elution was complete in 30 min [663].

Gérard et al. [664] studied the retention of 16 hydroxy and epoxy fatty acids. Cyanopropyl, diol, and silica columns were tested in conjunction with various solvent gradients. Results using cyclohexane, iso-octane, or hexane modified with an alcohol/acid combination (chosen from IPA, butanol or isoamyl alcohols and sulfuric, formic or acetic acids) were presented. The optimal separation was obtained with a silica column (ELSD) using a 40 min $99/1 \rightarrow 0/100$ (99.3/0.5/0.2 hexane/ IPA/acetic acid)/(79.8/20/0.2 hexane/IPA/acetic acid) gradient. Excellent resolution between hydroxyoctadecanoic acid positional isomers (e.g., 2-hydroxy, 4-hydroxy, 10-hydroxy, and 12-hydroxy) was obtained and peak shapes were very good. Linear concentration vs. detector response plots were generated from 5 to 200 µg injected and 1 µg detection limits were reported.

5.5.2 Vitamins and Related Analytes

In three separation schemes, retinyl, retinal, and retinol palmitate isomers were analyzed [665]. A silica column was used for all separations. Seven retinal isomers (7-, 9-, 11-, 13-*cis*, 7,9-, 11,13-di-*cis*, and all-*trans*) were eluted and baseline resolved in 18 min with a 97/3 *n*-heptane/methyl *t*-butyl ether ($\lambda = 371$ nm) mobile phase. The seven retinol isomers (9-, 11-, 13-*cis*, 7,9-, 7,13-, 9,13-, 11,13-di-*cis*, 9,11,13-tri-*cis*, and all-*trans*) were incompletely resolved and eluted in 22 min using a 94/6 *n*-heptane/methyl *t*-butyl ether ($\lambda = 325$ nm) mobile phase. Finally, retinyl isomers (9-, 11-, 13-*cis*, 7,9-9,13-di-*cis*) were incompletely resolved in 5 min using a 99/1 *n*-heptane/methyl *t*-butyl ether ($\lambda = 325$ nm) mobile phase. For this study 0.1 ng injections were made.

Although this paper is ostensibly to differentiate packing materials, it also provides excellent information regarding the separation of tocopherols (α, β, γ, and δ) and tocotrienols (α, β, γ, and δ). Excellent separation of all compounds was achieved in 20 min on a silica column ($\lambda = 294$ nm, ex; 326 nm, em) using a 96/4 hexane/1,2-dioxane mobile phase [666]. A diol column was also effective when a 94/4 hexane/methyl *t*-butyl ether mobile phase was used. Aminopropyl columns were less effective overall.

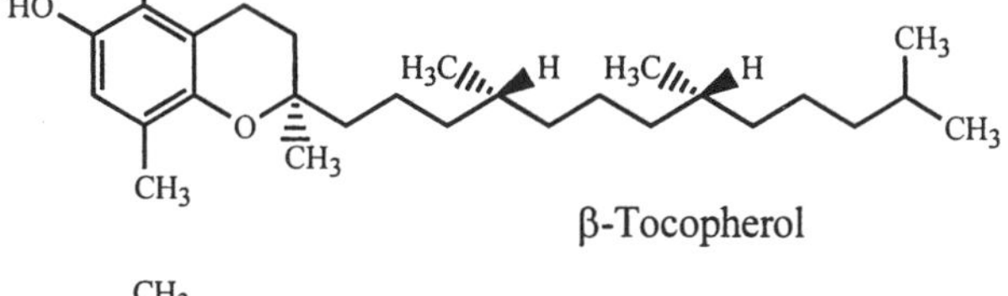

β-Tocopherol

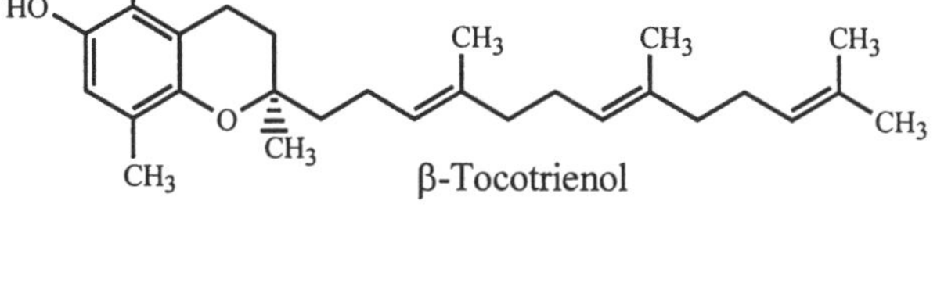

β-Tocotrienol

A series of tocotrienols (α-, β-, γ- and δ-tocotrienol) and two related compounds (desmethyl- and didesmethyltocotrienol) were extracted from rice bran and analyzed on a silica column ($\lambda = 295$ nm, ex; 330 nm, em) using a 99.8/0.2 hexane/IPA mobile phase [667]. Good resolution was obtained for all but the α/γ pair. Elution was complete in 26 min. Standards ranged from 0.25 to 1000 μM (analyte dependent).

Abidi and Mounts [338] studied α-, β-, γ, and δ-tocopherol and 5,7-dimethyltocol retention on β- and γ-cyclodextrin columns ($\lambda = 298$ nm, ex; 345 nm, em) using both cyclohexane and hexane mobile phases modified with alcohols (ethanol, IPA, n-propyl alcohol, 1-butanol, and 2-methyl-2-propanol), ethers (dioxane, THF, diisopropyl ether, MtBE, or tetrahydropyran) or ethyl acetate. A k' versus percent hexane and percent cyclohexane plot was shown for each modifier. Selected chromatograms and extensive tables of k' and α values are presented. Most elutions were complete in less than 45 min. In general, peak shapes were excellent with the notable exception of when MtBE was used as the modifier. In this instance, very broad peaks were generated. Why this occurred for MtBE and not for diisopropyl ether is not explained, nor is it readily explainable. Also observed was a significant decrease in fluorescence intensity when ethyl acetate was the mobile phase modifier. The authors ascribed this result to the decreased solubility of the analytes in the solvent, since the effect was not observed with any other solvent system. Another possibility is that ethyl acetate may effectively quench the fluorescence (it is the only carbonyl-containing solvent used in the study).

The α-, β-, γ-, and δ-tocopherols and tocotrienols were baseline resolved on a silica column ($\lambda = 290$ nm, ex; 330 nm, em) using a complex 24-min (99.5/0.5 iso-octane/ethyl acetate) $\rightarrow$ (97.3/1.8/0.9 iso-octane/ethyl acetate/acetic acid) $\rightarrow$ (98.15/0.9/0.85/0.1 iso-octane/ethyl acetate/acetic acid/2,2-dimethoxypropane) gradient [668]. The 2,2-dimethoxypropane reacts with the silica-bound water to form acetone and methanol that are readily flushed from the system. The use of this compound as an additive to the last mobile phase component in the sequence produces a continually "regenerated" active silica surface. Evidence of the effectiveness of this approach was tabulated as a nearly 50% loss in the amount of tocotrienols eluted when no 2,2-dimethoxypropane was used as compared with a $\ll$1% loss when it was present. In general, the use of 2,2-dimethoxypropane is an effective method of reactivating a water-deactivated silica. α-, β-, and γ-Tocotrienols and β-tocopherol in stillingia oil were baseline resolved on a silica column ($\lambda = 295$ nm, ex; 330 nm, em) using a 95/5 hexane/dioxane mobile phase. Elution was complete in 20 min and excellent resolution and peaks shapes were generated [669].

β-Carotene, 6,13-cis-retinol, all-$trans$-retinol, and α-, β-, γ-, and δ-tocopherol extracted from cheeses were baseline resolved using a silica column ($\lambda = 450$ nm, carotene; 280 nm, ex and 325 nm, em, tocopherols; 325 nm, ex and 475 nm, em,

$$H_3C \diagdown \diagup CH_3$$
$$H_3CO \diagup \diagdown OCH_3$$

2,2,-Dimethoxypropane

retinols) using an 18-min 50/50 → 10/90 hexane/(99/1 hexane/IPA) gradient [670]. Samples were preserved using pyrogallol as the antioxidant. Excellent peak shapes and resolution were obtained. Detection limits of 0.16 ng to 0.9 ng were reported.

Retention of nine all-*trans*-retinal and seven retinol isomers were studied on a silica column ($\lambda = 325$ nm) using various hexane/M*t*BE (3–7%), hexane/dioxane (6–7%), and heptane/M*t*BE (6–7%) mobile phases [671]. The retinals were adequately resolved in <15 min using a 97/3 hexane/M*t*BE mobile phase ($\lambda = 371$ nm). The best resolution of the retinols occurred with a 93/7 hexane/M*t*BE solvent ($\lambda = 325$ nm). Analysis time was slightly longer than 20 min. This same separation, with slightly poorer resolution, was accomplished in <12 min using a 93/7 heptane/M*t*BE mobile phase. A detection limit for all-*trans*-retinol of 0.1 ng was reported.

Both *cis*- and *trans*-retinoic acids were extracted from plasma and separated from retinol on a silica column ($\lambda = 350$ nm) using a 200/0.7/0.135 hexane/IPA/acetic acid mobile phase [672]. Elution was complete in 25 min. Detection limits of 0.5 μg/L for acids and 10 μg/L for retinols were obtained. Linear responses were generated up to 2 mg/L. Petroleum ether/1,2-dichloroethane/IPA (80/19.3/0.7) and petroleum ether/acetonitrile/acetic acid (99.5/0.2/0.3) mobile phases were also used for the separation of retinols and retinoic acids, respectively [673].

Fourteen keto- and dihydroxycarotenoids (e.g., all-(*E*)-3′-epilutein, (9*Z*)- and (13*Z*)-lutein, (9*Z*)-, (13*Z*)-, and (15*Z*)-zeaxanthin, ε,ε-carotene-3-dione, 3-hydroxy-β, ε-caroten-3′-one) were extracted from plasma and separated on a nitrile column ($\lambda = 325$ nm and 495 nm) using a 74.62/20.00/0.25/0.10 hexane/dichloromethane/methanol/diisopropylamine mobile phase [674]. A table of absorbance maxima is given for the analytes in the mobile phase. (The nonpolar carotenoids all eluted in the first 10 min and were successfully separated in a reversed-phase method described Chapter 9.) Elution of the polar carotenoids was complete in <30 min. Good peak shapes were generated.

Five carotenoids found in marine shellfish (lycopene, cantaxanthin, β-cryptoxanthin, lutein, and zeaxanthin) were separated on a silica column ($\lambda = 458$ nm) in <20 min using an 82.5/17.5 hexane/acetone mobile phase. These analytes were collected in highly purified form by using a 500 × 9.4 mm silica column and a 79/21 hexane/acetone mobile phase. Peak shapes were excellent and resolution was good in both analytical and preparative modes [675]. Isaksen and Francis [676] preparatively separated up to seven carotenoids (up to 50 mg total load) from broccoli and paprika using a silica column ($\lambda = 435$ nm). Depending on the extract, elution took anywhere from 80 min to 3 h with a 65/35 petroleum ether/acetone mobile phase.

Zeaxanthin

Hara et al. [677] studied the retention behavior of 10 fat-soluble vitamins and vitamin homologs (retinol, retinal, ergocalciferol, cholecalciferol, α-, β-, γ-, and δ-tocopherol, menadione, phylloquinone) on a silica column ($\lambda = 254$ nm or 292 nm) using hexane modified with ethyl acetate, THF, or IPA. Ethyl acetate and THF gave similar plots of capacity factor (0.1–10) vs. percent organic modifier (5–20%) [678]. Resolution and asymmetry were also similar. IPA proved to be a much stronger solvent, most likely due to its competitive hydrogen bond donating properties. Capacity factors from 0.2 to 8 resulted as %IPA was changed from 5 to 1. Asymmetry was found to be considerably better in the IPA-modified solvent and it was therefore recommended for use.

5.5.3 Terpenoids, Flavonoids, Steroids and Related Compounds

Herrera et al. [679] used an interesting approach to the analysis of four 5-hydoxy-7-methoxyflavones (5-hydroxy-7-methoxyflavone, 5-hydroxy-7,4′-, 5,3′-dihydroxy-7,4′-dimethoxyflavone, and 5-hydroxy-7,3′,4′-trimethoxyflavone). A normal-phase gradient system was used for generating the separation. The column chosen was aminopropyl ($\lambda = 346$ nm) and a complex 53-min 85/0/15 → 0/100/0 → 0/40/60 hexane/chloroform/acetonitrile gradient generated baseline resolution.

Six polymethoxylate flavones (tangeretin, nobiletin, sinensetin, tetra-O-methyscutellarein, 3,3′,4′,5,6,7,8-hepta- and 3,3′,4′,5,6,7-hexamethoxyflavone) were isolated from mandarin oils and analyzed on a silica column ($\lambda = 315$ nm) using a 95/5 hexane/ethanol mobile phase [680]. Quantitation was made against a coumarin IS. Concentration levels ranged from 0.01–0.40 g analyte/100 g oil. Peaks were well resolved and elution was complete in 24 min.

Both normal-phase and reversed-phase separations were performed on a series of six mono-, di-, tri-, and pentamethoxyflavones extracted from *Flos primula veris* flowers [681]. The normal-phase separation generated baseline resolution of all compounds in <15 min using a 92/8 hexane/IPA mobile phase and silica column ($\lambda = 330$ nm, ex; 440 nm, em). Although resolution was good, the pentamethoxyflavone peak was so broad that it was nearly indiscernible from the baseline. The authors also noted that periodic flushing of the column with IPA was needed to avoid irreversible adsorption and buildup of highly polar compounds on the column. For

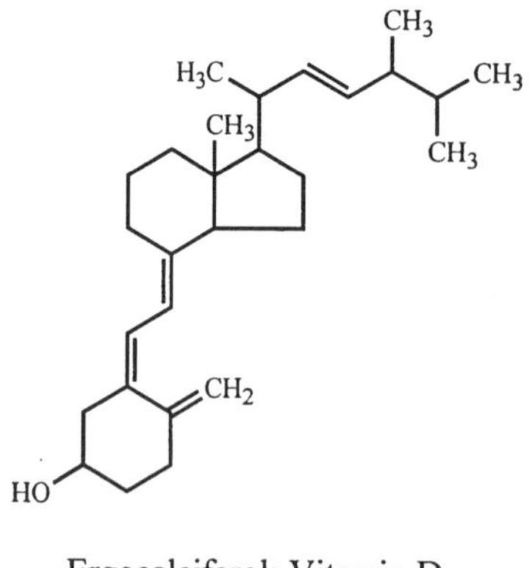

Ergocalciferol; Vitamin D$_2$

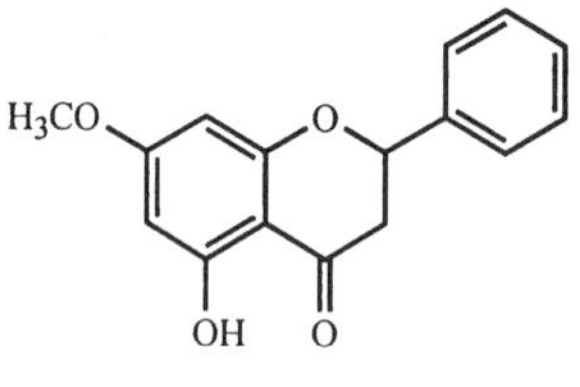

5-Hydroxy-7-methoxyflavone

the reversed-phase separation a C_{18} column was used along with a 60/40 water/ acetonitrile mobile phase. Excellent peak shapes were obtained, but two mono-methoxyflavones were poorly resolved and the di- and pentamethoxyflavones co-eluted. Analysis time was 25 min.

Fucosterol, bifurcadiol and eleganediol extracts from algae were baseline resolved with excellent peak shapes in <12 min using a silica column (RI detector) and a 60/40 iso-octane/ethyl acetate mobile phase [682]. Between 12 and 36 μg of sample were detected (analyte dependent).

Three γ-lactones (parthenolide, marrubiin, artemisinin) from plant extracts were quantitated on a silica column ($\lambda = 210$ nm or 225 nm) using either an 85/15 or a 90/10 hexane/dioxane mobile phase [683]. Good resolution of the analytes of interest from other extracted components was obtained and linear absorbance vs. concentration plots were obtained for the 0.2–5 mg/mL range. Analysis time was 15 min.

The optical isomers of seven flavanones (flavanone, 5-methoxy-, 4'-methoxy-, 5-methoxy-, 2'- and 6-hydroxy-, 6-methoxyflavanone, pinostrobin) were resolved on a Chiralcel OD column ($\lambda = 254$ nm) using a 90/10 hexane/IPA mobile phase [684]. Capacity factors ranged from 1.3 to 2.9 and α-values from 1.16 to 1.60. A 100/4 hexane/dioxane mobile phase used in conjunction with a ChiraSpher column ($\lambda = 254$ nm) resolved three sets of enantiomers (flavanone, 6-hydroxy-, and 4'-methoxyflavanone) in <25 min.

Two 3-flavanols, ten 4-flavanones, four flavones, and four 3-flavonols were resolved as individual sets as their peracetylated derivatives on a nitrile column ($\lambda = 278$ nm) using a 60/40 hexane/ethyl acetate mobile phase [685]. The 4-flavanone separation took 30 min, whereas the other sets were resolved in 17 min or less. A complex 30-min 60/40 → 55/45 hexane/ethyl acetate gradient resolved 17 of the entire set of 20 compounds.

Tanshinone I and IIA and crypto- and dehydrotanshinone were quantitated over the 0.1–100 nmol concentration range using a silica column ($\lambda = 285$ nm) and a

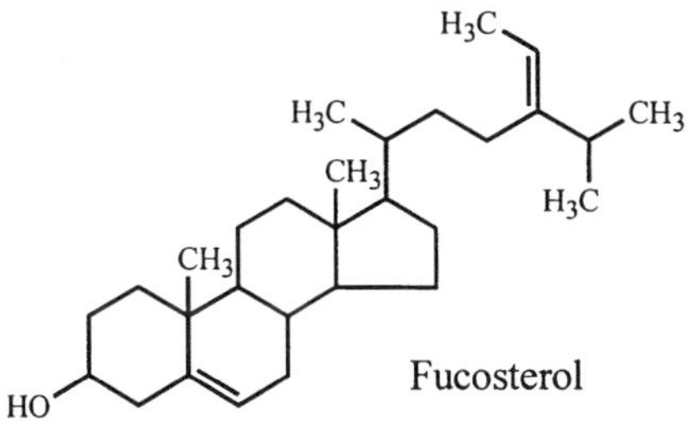

Fucosterol

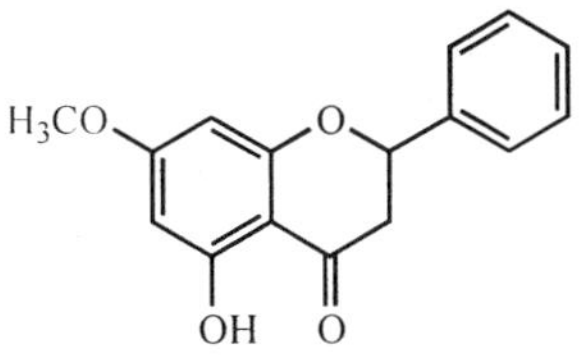

Pinostrobin

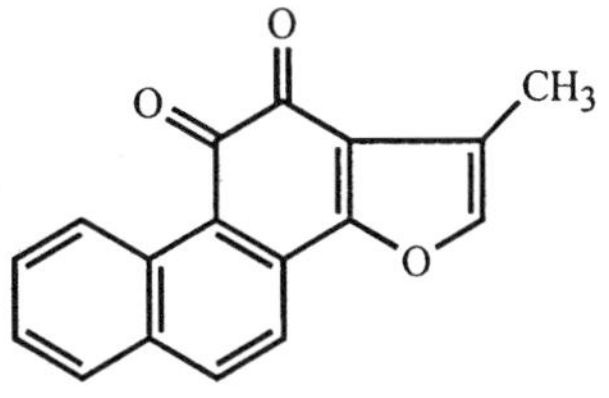

Tanshinone I

92/8 hexane/dioxane mobile phase. Elution was complete in 20 min. Good peak shapes were obtained [686].

Five bile acids (glyco- and tauro-3α,6α-dihydroxy-5β-cholanoic acid, chenodeoxycholic acid, hyocholic acid, 6-ketolitocholic acid) were extracted from bile and analyzed as their methyl esters. Baseline resolution was obtained on a cyanopropyl column (RI detector) in 25 min when an 89/6/5 hexane/IPA/dichloromethane (+1% 1-pentanol and 1% water) mobile phase was used [687].

5.5.4   Analytes Derived from Oils and Fats

In a nice presentation, a series of nonvolatile triglyceride oxidation products were studied using both UV and MS detection [688]. Chromatograms and spectra of 1,3-dipalmitoyl-2-linoleoylglycerol, as well as its hydroxyl, hydroperoxy, epoxy, and oxo oxidation products were generated with the UV first in series with the MS. Separation was conducted on a silica column and elution was done using a complex 57-min 97/3 → 50/50 hexane/(9/1 hexane/IPA) gradient. UV monitoring was done at $\lambda = 235$ nm, but the epoxy and oxo products generated no response. Prior to the MS, a 0.15 mM NaI solution in 1/1 ethanol/methanol was added to the UV effluent. Positive ion electrospray was used with a capillary voltage of 3.5 kV and a fragmentor voltage of 80 V. Nebulizer pressure was 20 psi; gas $T = 350°$C; drying gas flow rate 3 L/min. Calibration curves on the MS were distinctly nonlinear over the 1–500 ng injected range.

Four autoxidation products of methyllinoleate (i.e., methyl-13-hydroperoxy-X-octadecadienoate, where X = *cis*-9-*trans*-11, *trans*-9-*trans*-11, *cis*-10-*trans*-12, and *trans*-10-*trans*-12) were separated, but not baseline resolved, on a silica column ($\lambda = 234$ nm) using an 88/12 heptane/diethyl ether mobile phase. Elution was complete in <10 min [689].

Twelve conjugated linoleic acids (three groups of four each: *trans/trans*, *cis/trans*, or *trans/cis*, *cis/cis*) were converted to their methyl esters and analyzed on a silver-impregnated silica column ($\lambda = 233$ nm) using a 99.9/0.1 hexane/acetonitrile mobile phase [690]. The three groups were baseline resolved from one another but the isomers within each group were only partially resolved from one another. Elution was complete in 30 min. The separation was extremely sensitive to changes in the acetonitrile level.

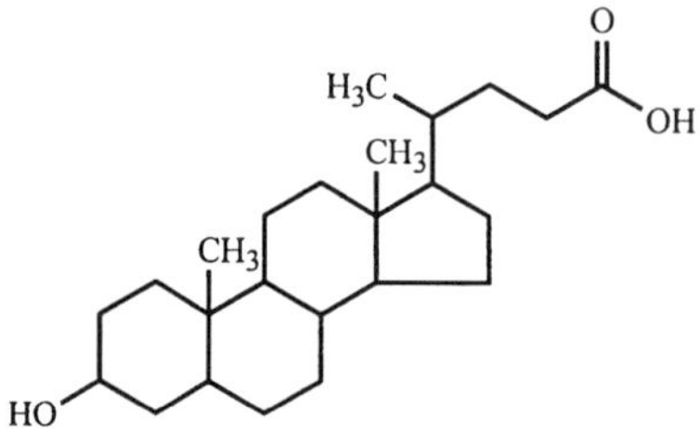

Lithocholic acid

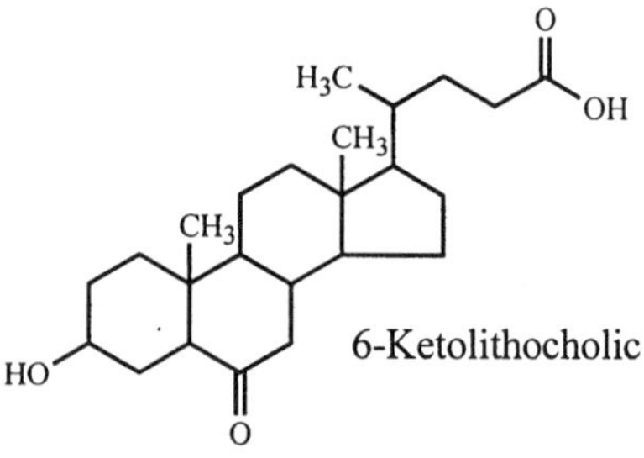

6-Ketolithocholic acid

$$HOOC(CH_2)_7CH=CHCH_2CH=CHCH_2CHCHCH_2CH_3$$

Linoleic acid

The 9- and 13-hydroperoxides of methyl linoleate were separated on a silica column ($\lambda = 205$ nm) using a 99.5/0.5 hexane/IPA mobile phase [691]. For the systems that contained vitamin E, only the two peaks were obtained (all the same isomer). However, when the oxidation product was generated without vitamin E present, the appearance of *cis-trans* and *trans-trans* isomers occurred. Complete resolution of all isomers was not achieved. In all cases elution was complete in 10 min. Similarly, the peroxidation products of linoleic acid, 9- and 13-hydroperoxy-linoleic acids, were separated on a silica column ($\lambda = 234$ nm) using a 98/1.9/0.1 hexane/ethanol/acetic acid mobile phase [692]. Elution was complete in <11 min. The peaks were incompletely resolved. The stock solutions were 4 mM.

Hydroperoxides of triacylglycerols (trilinolein, triolein) and cholesterol esters (cholesteryl linolate, oleate, arachidonate) were separated as their diphenyl-1-pyrenylphosphine derivatives on a silica column ($\lambda = 352$ nm, ex; 380 nm, em)] using a 24-min $100/0 \rightarrow 99.1/0.9$ hexane/n-butanol gradient [693]. Samples were dissolved in solvent containing 0.5 g/L BHT to prevent their oxidation. Separation was complete in <30 min. Working concentration curve ranges from 5 to 1000 pmol were reported and detection limits of 2 pmole (S/N = 3) were claimed.

Glycerol diacetate monopropionates were purified by preparative HPLC using a silica column (RI detector) and an isocratic 80/20 hexane/MtBE mobile phase [694]. Elution was complete in <20 min. Similarly, glycerol monohydroxystearate was purified on a cyanopropyl column ($\lambda = 230$ nm) using a 95/5 hexane/IPA mobile phase [695]. Separation was complete in <30 min.

The stereoisomeric pairs of ricinoleate, methyl isoricinoleate, and methyl esters of three additional bis-homoallylic hydroxy fatty acids were separated as their α-naphthylethyl isocyanate derivatives [696]. A silica column and a 95/5/1 hexane/ethyl acetate/THF mobile phase generated the separation. Elution was generally complete in <30 min and resolution was excellent in all cases.

Mono- and diacylglycerides were studied using a silica column (ELSD). An 8-min $98/2 \rightarrow 2/98$ hexane/(80/10/10/1 hexane/IPA/ethyl acetate/10% aqueous formic acid) gradient separated eight glyceride analytes (cholesteryl myristate, tripalmitin, palmitic acid, 1,3-dipalmitin, cholesterol, 1,2-dipalmitin, 1-mono and 2-monopalmitin). A 20 µL aliquot of samples containing 0.1–1 mg/mL of peanut oil was injected [697].

The resolution of several asymmetric triacylglycerols was accomplished using a 25°C Chiracel OD column ($\lambda = 210$ nm). For example, a 200/1 n-hexane/IPA mobile phase almost fully resolved 1-eicosapentaenoyl-2,3-dicapryloyl-sn-glycerol (ECC) from its CCE analog. Elution time was 50 min. Similarly, 1-docosahexaenoyl-

$$HOOC(CH_2)_7CH=CHCH=CHCH(OOH)CH_2CH_2CHCH_2CH_3$$

13-Hydroperoxylinoleic acid

CH$_2$OH		CH$_2$OR$_1$
CHOH	+ 3 fatty acids =	CHOR$_2$
CH$_2$OH		CH$_2$OR$_3$

Glycerol   Triacylglycerols; TAGs

2,3-dicapryloyl-*sn*-glycerol (DCC) was resolved from CCD by 22 min using the same conditions [698].

Separation of interesterified fats from various oils into their free fatty acid and mono-, di-, and triglyceride fractions was rapidly achieved on a silica column (ELSD) using a 90/10 iso-octane/IPA mobile phase [699]. Baseline resolution was obtained along with complete elution in under 15 min. Tallow, sun oil, and sunflower oil elution profiles are shown.

Diacyl-*sn*-glycerols were studied as their (*S*)-(+)-(1-naphthyl)ethylurethane derivatives on a silica column ($\lambda = 280$ nm). Baseline resolution was obtained in approximately 1 h for each of the 1,3-dioleoyl-2-palmitoyl-, 1,2-dioleoyl-2-palmitoyl-, and 1,3-dipalmitoyl-2-oleoylglycerols using a 99.6/0.4 iso-octane/*n*-propyl alcohol (2% in water) mobile phase [700]. (Alumina, C_{18}, diol, aminopropyl, and cyanopropyl columns did not successfully resolve the analytes.) The authors noted that constant retention times were difficult to maintain over the course of a day. Retention times continually lengthened unless 2% water in *n*-propyl alcohol was initially added to the mobile phase. The final recommendation of the authors was to make up enough mobile phase material at one time to be able to analyze all samples while frequently exchanging the solvent reservoir with fresh solvent.

Good resolution of the 3,5-dinitrophenylurethane derivatives of *sn*-1,2- and *sn*-2,3-diacylglycerols with the total acyl carbon number ranging from 32 to 44 was obtained using a 150/20/1 hexane/1,2-dichloroethane/ethanol mobile phase on a Sumichiral OA-4100 column ($\lambda = 254$ nm). The column temperature was maintained at $-20°$C! Elution was complete in just under 2 h (flow rate 0.5 mL/min). Separation of the SS, SU, and UU enantiomers of distearic- and dioleicglycerol derivatives was also demonstrated using the same column, but a 170/10/1 hexane/1,2-dichloroethane/ethanol mobile phase ($T = -10.5°$C). Separation took 5 h (flow rate 0.25 mL/min). No explanation is given why such slow flow rates were used [701].

A series of 1,2-diacylbenzylglycerides (C_8, C_{16}, $C_{18:1}$) were synthesized from racemic glycerol acetonide [702]. Seventeen intermediates and final products were identified via chromatography on a silica column (RI detator) using a hexane/IPA mobile phase (2–4% IPA depending upon the compound). Retention times for all compounds were tabulated. A 20 µl aliquot of 1% w/v samples was injected. Enantiomeric separation of the benzylglyceride and the 1- or 2-acyl-3-benzylglycerols was accomplished on a Chiralcel OB column coated with cellulose tribenzoate. A mobile phase of hexane/IPA (1–4% IPA depending upon analyte) was used. Capacity factors ranged from 1.4 to 28.3 and were tabulated along with α values for each enantiomeric pair.

Conjugated linoleic acid isomer determination in beef fat was conducted on a series of three 250×4.6 mm Ag^+-silica columns ($\lambda = 233$ nm) using a 99.9/0.1 hexane/acetonitrile mobile phase [703]. Sixteen isomers were clearly delineated in 55 min, e.g., *trans*-12,*trans*-14, *cis*-12,*trans*-14, *cis*-11,*trans*-13, *trans*-10,*trans*-12, *cis*-9,*trans*-11, *trans*-8,*cis*-10, *trans*-7,*cis*-9, and *trans*-6,*trans*-8. Individual isomer levels ranged from 0.02 to 1.95 mg/g fat.

In a nice study by Ye et al. [704] the comparative performance of narrow-bore and analytical silica columns on the separation of α-, β-, γ-, and δ-tocopherols was determined. Temperature was held at 29°C and peaks were detected at $\lambda = 298$ nm. A 99.2/0.8 hexane/IPA mobile phase generated baseline resolution and elution in 14 min. Linear ranges of 0.33–15.5 ng injected was established and quantitation limits (analyte dependent) were 14–51 pg injected and 72–307 pg injected for narrow-bore and analytical, respectively.

Indyk [705] studied the tocopherols, stigmasterol, and cholesterol extracted from milk powder on a C_{18} column ($\lambda = 295$ nm, ex; 330 nm, em for tocopherols, or 212 nm UV detection for all other analytes). Elution was complete in 30 min when a 99.9/0.1 hexane/IPA mobile phase was used. Detection limits were given as 0.04 µg/injection and 4 mg/kg sample (cholesterol). The author noted that the use of isopropyl ether as the extraction solvent caused problems with the chromatography. Baseline "disturbances" were reported to occur early in the chromatogram, in the region where the tocopherols eluted, most likely due to the elution of diisopropyl ether and diisopropyl ether degradation components.

An excellent separation of lipid fractions (cholesterol ester [CE], triacylglycerol [TAG], cholesterol [C], phosphatidylethanolamine [PE], phosphatidylinositol [PI], phosphatidylserine [PS], phosphatidylcholine [PC], sphingomyelin [SM], lysophosphatidylcholine [LPC], cardiolipin [CP]) was developed by Redden and Huang [706]. A two-ramp 37-min $99/1/0/0/0 \rightarrow 0/0/80/20/0 \rightarrow 0/0/50/0/50$ isooctane/THF/IPA/chloroform/water gradient was used. Superb resolution was obtained on a 35°C silica column (mass detector). Sample concentrations ranging from 1.1 µg/mL to 6.2 µg/mL were used to create a standard working curve (mass injected was adjusted by increasing the injection volume from 2 to 40 µL).

The PE, PI, phosphatidic acid (PA), PS, and PC lipid classes from soybean oils were analyzed by NP chromatography [707]. A silica column ($\lambda = 206$ nm) and a 20-min $56/42/2 \rightarrow 38/51/11$ hexane/IPA/water gradient gave good separation. Detection limits of 100 ppm for these phosphorus-containing compounds were reported.

A similar method separated the following lipids from human fibroblasts into the following groups: CE, TAG, phytols, C, diglycerides, fatty acids, ceramides, PE, PS, SM, and LPC on a silica column (ELSD). Working curves were shown for CE, TAG, C, ceramide, PC, and SM fractions. A two-ramp 60-min $99/1/0/0/0 \rightarrow 0/0/20/80/0 \rightarrow 0/0/0/50/50$ hexane/THF/chloroform/IPA/water gradient was used. Samples containing 30 nmol total lipid phosphorus were used [433].

A slightly different set of lipid fractions (mono- and digalactosyldiacylglycerol, PC, PE, TAG, LPC, PI, and PA) was studied on a diol column (ELSD) using a hexane/IPA/water gradient containing one or more acids or bases (acetic acid, ammonium acetate, or triethylamine) and one or more organic modifiers (THF, ethyl acetate, iso-octane, or 1-butanol). Various gradients were run to optimize the separation. A 30-min $81.5/17/1.5/0.08/0 \rightarrow 0/84.5/1.5/0.08/14$ hexane/IPA/

acetic acid/triethylamine/water gradient gave the best results. An injection of 4–8 µg was made for each group. Elution was complete in 30 min [708].

Amniotic fluid phospholipids were separated on a silica column (ELSD) using a 45-min 50/50 → 0/100 (at 30 min hold 15 min) (60/40 hexane/IPA)/(300/200/29 hexane/IPA/water) gradient. Fractions studied included diphosphatidylglycerol, PE, PG, PI, PS, LPE, PC, lysolecithin, phosphatidylmonomethyl- and phosphatidyldimethylethanolamine, and SM [709]. Samples down to 1 µg injected were used. Nonlinear response vs. concentration plots were generated in the 10–40 µg injected range.

PE, PI, PA, and PC in lecithins were separated on a silica column ($\lambda = 206$ nm) in 10 min using an 80/80/10 hexane/IPA/water (0.2 M acetate buffer at pH 4.2) mobile phase [710]. Precision and accuracy data were presented for the results with standards covering the range 1–4 mg/mL (10 µL injections). Note that the use of acetate as the buffer with a detection wavelength of 206 nm will generate an extremely high background absorbance.

It is evident that the hexane to IPA to water theme is common for the fractionation of complex lipid mixtures in the procedures listed above. This is reasonable. Molecules with such vastly different polarities would need a mobile phase shift from nonpolar (hexane) to very polar (water) to generate elution. Keep in mind that for a silica column, significant reequilibration times (usually 10–15 min) will be necessary to take the column back to initial conditions, especially when water is involved. Also, note that IPA is mandatory for the gradient to work since hexane and water, by themselves, are immiscible.

5.5.5 Other Analytes

Seven chlorophyll b allomers (e.g., 13^2-methoxy-, 13^2-hydroxy-, and 13^2-hydroxy-10-methoxychlorophyll b, and $3^1,3^2$-didehydro-7^1-oxorhodochlorin-15-glycolic acid $13^1,15^2$-dimethyl-17^3-phytyl ester Mg^{2+} salt) were baseline resolved on a silica column ($\lambda = 440$ nm) using an 88/12 hexane/THF mobile phase [711]. Elution was complete in 35 min.

Chlorophyll a and b, pheophytin a, pyropheophytin a, and the zinc complexes of pyropheophytin a and pheophytin a were isolated from fresh and frozen processed pea presamples [712]. A normal-phase separation using a silica column ($\lambda = 658$ nm) and a 22-min complex 98.3/1.7 → 97/3 hexane/IPA gradient generated excellent resolution and peak shapes.

Chlorophyll a and b, pheophytin a and b, and β-carotene from olive oil extracts were separated and quantitated on a silica column ($\lambda = 409$ nm, 430 nm, or 452 nm) using a 98.5/1.5 hexane/IPA mobile phase. Detection limits of 0.5 ng total injected were reported and amounts from 0.1 to 45 µg/g were quantitated by this method [713].

Drexler and Ballschmiter [714] separated pheophytin a and b, chlorophyll a, a', and b, and xanthophyll in <10 min using a diol column ($\lambda = 425$ nm) and a 100/2 hexane/ethanol mobile phase. In the same work, a description of the separation of

Mg(II) and Zn(II) chelates of mesoporphyrin IX-dipropyl ester and didodecyl ester was presented. The diol column ($\lambda = 403$ nm) along with a 100/0.5 hexane/ethanol mobile phase eluted the six compounds in 20 min. Additionally, the Cu(II), Ni(II), Zn(II), Mg(II), and 2H$^+$ complexes of mesoporphyrin-IX dihexyl esters were resolved in 25 min with a 100/0.375 hexane/ethanol mobile phase on the same diol column ($\lambda = 405$ nm).

The bacteriochlorophyll 663 (Δ2,6-phytadienol chlorophyll *a*) was isolated from green sulfur bacteria and separated from bacteriochlorophyll *a* and *a'* on a silica column ($\lambda = 400$ nm) using a 100/1.5/0.2 hexane/IPA/methanol mobile phase [715]. Peak shapes were excellent and elution was complete in 25 min. However, co-extracted bacteriochlorophyll *c* eluted at 72 min.

The reduced and oxidized forms of coenzymes Q$_9$ and Q$_{10}$ were separated from one another and ubichromenol in <8 min on a C$_{18}$ column ($\lambda = 275$ nm). A 10/90 hexane/methanol mobile phase was used. Detection limits of 2 ng injected were reported. Working linear concentration ranges of 0.2–100 µg/mL were used. Hexane undoubtedly plays a pivotal role in eluting these compounds that have long alkene substituents [716].

Ceramide types III and IV were isolated from stratum corneum lipid extracts and separated from cholesterol, free fatty acids, and triglycerides on a silica column (ELSD) in <15 min using a 19/1 hexane/ethanol mobile phase [717]. Related compound classes (hydroxy fatty acid ceramide, C2 ceramide, sphingolipid, phyto-sphingosine) were separated on a silica column using a 90/10/1/1 chloroform/ethanol/triethylamine/formic acid mobile phase. In this case elution was complete in 10 min.

Sulfatide (ceramide galactosyl-3'-sulfate) is present in many tissues. Accumulation of sulfatide and related compounds resulting from the storage disorder metachromatic leukodystrophy was followed by a normal-phase separation on a silica column ($\lambda = 229$ nm) using a 16-min 99/1 → 80/20 hexane/dioxane gradient [718]. Good resolution between sulfated and six related compounds was obtained and the last analyte of interest eluted in <10 min.

The benzoic anhydride derivatives of ceramide-related compounds (fatty acids, non-hydroxy fatty acid ceramide, dihydroceramide, sphingosine, sphinganine) were separated on a silica column ($\lambda = 230$ nm) using a 99.44/0.45 cyclohexane/IPA mobile phase [719]. Elution was complete in 8 min. Good resolution was achieved.

Sphingomeylin diastereomers (naturally occurring D-*erythro*-(2S,3R)-from L-*threo*-(2S,3S)-sphingomyelin) were separated on a 45°C diol column (ELSD, drift tube, 110°C, gas flow 2 L/min) using a complex 25-min 82/17/0/1/0.8 → 0/85/14/1/0.8 hexane/IPA/water/acetic acid/triethylamine gradient [720].

Sulfatide

Baseline resolution was achieved and plots of the effect of retention time versus temperature (25–45°C) were presented. Peaks of 2 nmol injected were detectable.

Five urinary corticosteroids (tetrahydrodeoxycorticosterone, tetrahydro-11-deoxycortisol, tetrahydrocortisol, allo-tetrahydrocortisol, tetrahydrocortisone) were isolated, derivatized with 9-anthroyl cyanide, and separated on a silica column ($\lambda = 370$ nm, ex; 470 nm, em). A $98/2 \rightarrow 93/7$ (at 20 min hold 20 min) hexane/IPA gradient generated good peak shape. Standards of 10 ng injected were easily detected [721].

Caude and co-workers [722] used Pirkle-type tyrosine linked dinitrobenzene stationary phases ($\lambda = 254$ nm) to study the resolution of the enantiomers of alkyl-N-arylsulfinamoyl esters. To optimize the separation, various ratios of 98/8 hexane/ethanol and 50/50 hexane/chloroform were mixed to form a ternary eluent. An informative plot of capacity factor and separation factor is presented for one compound and a table of retention and selectivity is given for various hexane/polar solvent mixtures (polar solvent = ethanol, IPA, chloroform, or dichloromethane).

5.5.6   Summary

The alkanes play a very important role in the chromatography of mid-sized nonpolar or slightly polar molecules. With the alkanes, analyte solubility is increased and, due to the low UV cutoff, the lowest detection limits at the lowest operating wavelengths are possible. The addition of small percentages of polar mobile phase modifiers facilitates elution and offers unique selectivity properties.

5.6   AMINO ACID AND PEPTIDE ANALYTES

Since the 1980s there has been an exponential growth in the number of separations based on the resolution of enantiomers. Due to the nearly universal finding that one enantiomeric form of a compound has a thousandfold or greater potency or, in some cases, toxicity, production and monitoring of the individual components of an enantiomeric mixture has become important.

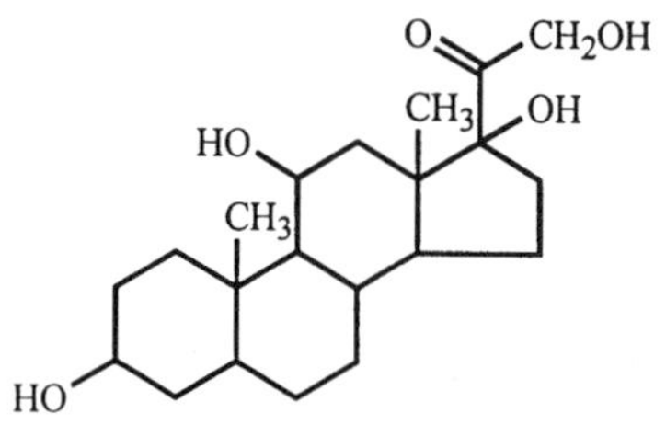

Tetrahydrocortisol

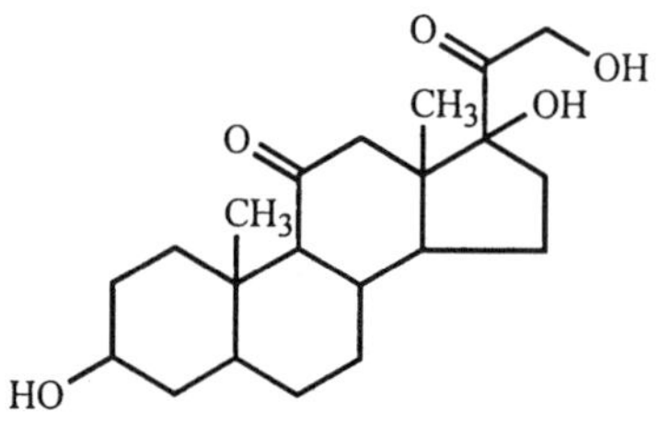

Tetrahydrocortisone

As early as 1980, studies involving the resolution of *d*- and *l*-amino acid enantiomers on a Pirkle-type chiral stationary phase (CSP) were being described. Hara and co-workers [723] used an *N*-acylation/*O*-alkylation procedure to study 10 leucine derivatives and 16 amino acid enantiomers on an *N*-formyl-L-valinylaminopropyl chiral phase. For the leucine enantiomers a hexane/IPA mobile phase was used ($\lambda = 230$ nm). The IPA level varied from 0.5% to 6% depending on the compound. Resolution (α) was at least 1.09 in all cases. The *N*-acetyl amino acid *t*-butyl esters were most successfully resolved when either a 20/80 hexane/ethyl ether or a 70/30 hexane/dichloromethane mobile phase was used. IPA was not effective in these separations. The authors suggest that the residual silanol group stationary phase/analyte hydrogen bond interaction is enhanced when an aprotic solvent system is used, whereas protic solvents, because of their overwhelming concentration in the mobile phase, preferentially interact with these sites.

Pirkle et al. [724] tabulated the k', α, and R_s values for 10 α- and β-amino acids as their *N*-(2-naphthyl) derivatives on a *N*-(2-naphthyl)alanine or a *N*-(2-naphthyl)valine column ($\lambda = 254$ nm or 280 nm). Isocratic mobile phase compositions ranged from 95/5 to 80/20 hexane/IPA. With these chiral stationary phases, resolution depends upon the differential interaction of the solute with the π–π^* electrons in the naphthyl residue and/or the hydrogen bond interaction with the spacer group by which the naphthyl residue is attached to the silica support. Therefore, aggressive hydrogen bonding mobile phase components (e.g., amines) are usually avoided. IPA is frequently used since it is an excellent solubilizing agent and is miscible with the common chiral separation solvents hexane, heptane and iso-octane.

Ôi et al. [725] studied the retention of six amino acids (alanine, valine, leucine, methionine, phenylglycine, and phenylalanine) as their *N*-acetyl methyl esters and *N*-(3,5-dinitrobenzoyl) derivatives on an (*R*)- or (*S*)-1-(α-naphthyl)ethylamine stationary phase ($\lambda = 230$ nm or 254 nm). Capacity factors and α values are tabulated for all pairs of amino acid derivatives. A 40/10/1, a 100/20/1, or a 200/20/1 hexane/1,2-dichloroethane/ethanol mobile phase was used to generate the separation.

The absolute configurations of α-chiral amines as their *N*-succinimidyl-α-methoxyphenylacetates were determined on a silica column ($\lambda = 254$ nm) using hexane/ethyl acetate mobile phases [726]. The composition of the mobile phase varied from 75/25 to 50/50 hexane/ethyl acetate depending upon the identity of the analyte. The authors claim that their method can be used without prior separation, purification, and identification of standards.

5.6.1 Summary

Due to the very limited solubility of underivatized peptides and proteins in nonpolar solvents, it is unlikely that NP separations for peptides and proteins will ever become an important technique. However, derivatized amino acids and short-chain peptides (seven amino acid residues or fewer) have been successfully chromatographed. Diastereomeric separations are done easily on the many chiral stationary phases currently available. In these cases, IPA is frequently used as the low-level mobile phase modifier in hexane, heptane, or iso-octane due to its great solubilizing qualities.

5.7 PHARMACEUTICAL ANALYTES

As was seen in earlier sections, alkane-based solvents are well suited for the separation of large nonpolar molecules (e.g., retinoids and carotenes). Small molecules having positional isomers are effectively separated when a strict alkane-based solvent/normal phase sorbent combination is used. These areas of use have been greatly expanded since the mid-1980s through the use of alkane-based solvents in the separation of chiral compounds—mostly involving small polar molecules. The importance of the analysis of enantiomers has grown as it has become increasingly apparent that for pharmaceuticals often only one enantiomer is therapeutically active, whereas the other enantiomer or other diasteriomers may have severely reduced activity, be totally ineffective or, worse yet, be toxic or lethal. Consequently, complete resolution and identification of each species in a mixture is critical.

5.7.1 Cardiac Glycosides

Digoxin and five metabolites (digoxigenin, digoxigenin mono- and bisdigitoxoside, and (20S)- and (20R)-dihydrodigoxin) were extracted from urine and feces and were separated as their 1-naphthyl derivatives on a silica column ($\lambda = 217$ nm, ex; >340 nm, em) using a 6/1/1 hexane/dichloromethane/acetonitrile mobile phase [727]. Peak shapes were excellent and all compounds were baseline resolved. Detection limits of 5–125 ng/mL in urine and 10–250 ng/200 g feces were reported. Elution was complete in 15 min.

β-Methyldigoxin was extracted from plasma and separated from digoxin and digoxigenin (and its mono- and bisdigitoxosides) on a silica column ($\lambda = 220$ nm) using a 69/20/10/1 heptane/IPA/dichloromethane/water mobile phase [728]. The analysis was complete in <25 min, baseline resolution was achieved, and excellent peak shapes were recorded. A linear working concentration range of 0.02–50 ng/mL was reported. Recoveries at 0.5 and 3.0 ng/mL were tabulated.

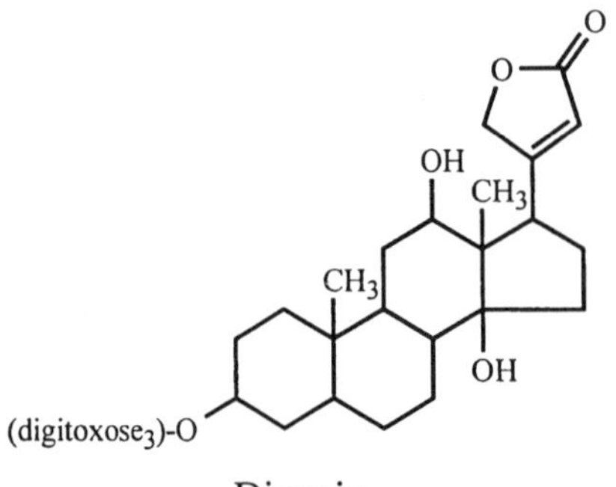

Digoxin 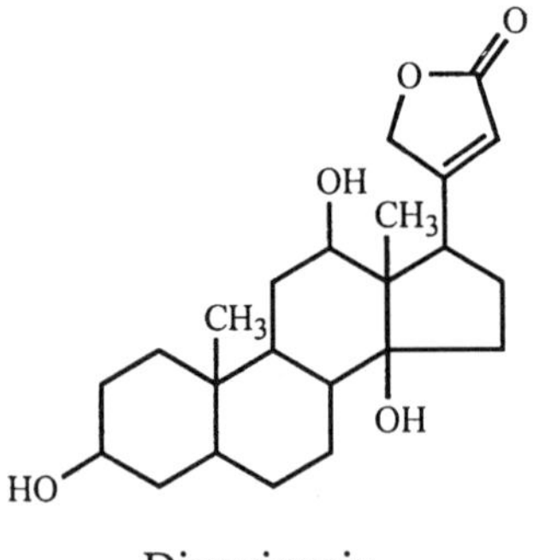   Digoxigenin

5.7.2 NSAIDs and Analgesic Drugs

Enantiomers of six NSAIDs (ibuprofen, pirprofen, ketoprofen, benoxaprofen, carprofen, protizinic acid) were resolved as their benzylamine derivatives on a Chiralcel OJ column ($\lambda = 230$ nm) using an 80/20/0.5 hexane/IPA/acetic acid mobile phase [729]. A 20 μL injection of 500 μg/mL standard was used in the study. Elution was complete in <30 min while obtaining good resolution and good peak shapes.

The enantiomers of tiaprofenic acid and its 3-isomer impurity were resolved and separated from two additional production impurities (5-benzoyl-2-acetylthiophene and 5-benzoyl-2-ethylthiophene) [730]. Impurity levels of 0.2–4.0 μg/mL vs. 50–200 μg/mL levels for tiaprofenic acid were studied. A Chiralcel OD column ($\lambda = 296$ nm) was used with a 94/6/0.1 hexane/IPA/TFA mobile phase. Complete elution required 80 min.

5.7.3 Benzodiazepines

Diazepam, prazepam, and three of their metabolites—nordiazepam, (3R)- and (3S)-hydroxyprazepam—were baseline resolved on an (R)-(3,5-dintirophenylbenzoyl)-glycine chiral phase ($\lambda = 232$ nm) using a 77/20/3 hexane/dichloromethane/IPA mobile phase [731]. Elution was complete in <20 min and excellent peak shapes were obtained.

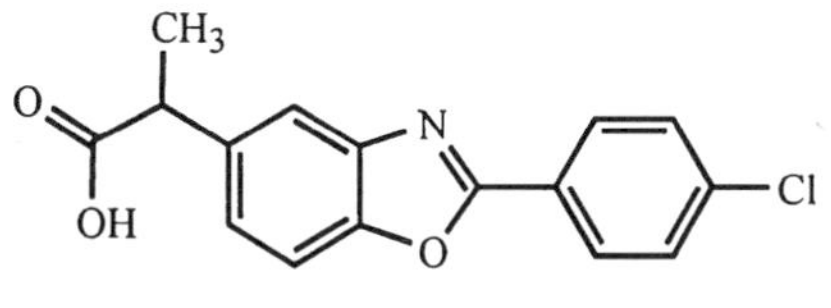

Benoxaprofen

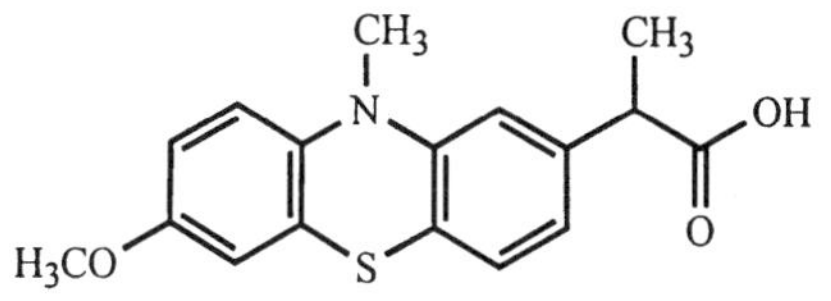

Protizinic acid

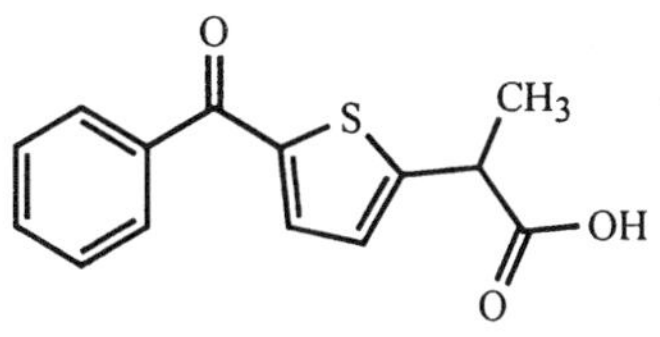

Tiaprofenic acid

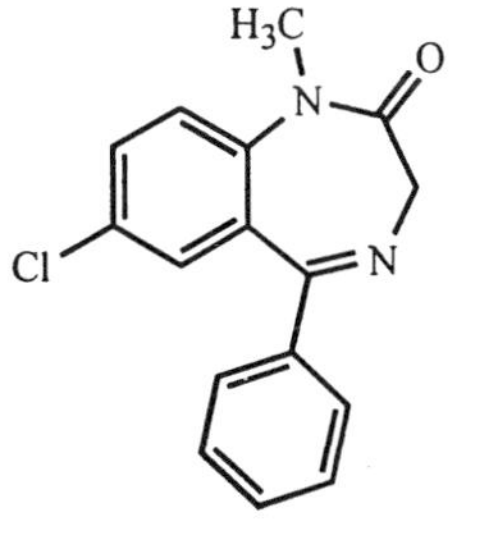

Diazepam

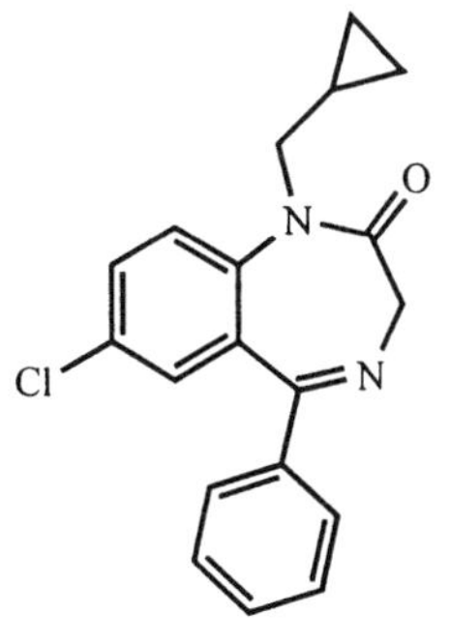

Prazepam

Lorazepam and its 3-*O*-acyl, 1-*N*-acyl-3-*O*-acyl-, and 3-*O*-methyl derivatives were baseline resolved on a silica column ($\lambda = 230\,\text{nm}$) using a 90/10 hexane/(2/1 ethanol/acetonitrile) mobile phase [732]. Elution was complete in <30 min. The lorazepam peak was quite tailed. The retention of the enantiomers of each of these compounds was determined on six Pirkle-type chiral columns. Eluent composition ranged from 90/10 hexane/IPA to 95/5 → 91.5/8.5 hexane/(2/1 ethanol/acetonitrile) to 77/20/3 → 70/20/10 hexane/IPA/1,2-dichloroethane. Retention times ranged from 13.5 to 56 min. Each pair was adequately resolved under one or more sets of conditions. All results are tabulated. One topic not often addressed in such studies is compound racemization half-lives. Here, neat solvents and experimental mobile phases were used. Water and neat alcohols resulted in 50% racemization in <30 min. The hexane/1,2-dichloroethane/IPA mobile phases yielded racemization times of >50 min. This time increased as the level of IPA decreased. Essentially no racemization occurred ($t_{1/2} > 5000\,\text{min}$) in neat 1,2-dichloroethane or acetonitrile.

5.7.4   Other Analytes

Odapipam (dopamine receptor antagonist) and five metabolites (e.g., *N*-desmethyl, hydroxylated, dehydrogenated analogs) were extracted from microsomal incubates and separated on a silica column ($\lambda = 280\,\text{nm}$ and 295 nm) using a 900/100/2/1 *n*-heptane/IPA/water/water (25% ammonia) mobile phase [732a]. All peaks were well resolved and elution was complete in <12 min. Although chromatographic concentrations were not given, the incubate was 100 μM in odapipam and that peak, as well as the metabolites, is readily detectable.

As mentioned above, the alkanes are heavily used for enantiomeric separations, especially when Pirkle-type or cyclodextrin packings are used [142]. Cleveland [733] studied 70 racemic pharmaceuticals on up to seven different Pirkle-type packings (photodiode array) using various isocratic mobile phase mixtures ranging from 40/35/25 to 92/5/3 hexane/1,2-dichloroethane/ethanol (0.1% TFA). Adequate resolution was achieved only 10–45% of the time an any selected phase. Therefore, an appropriate choice of bonded phase is critical in maximizing its utility.

Oxcarbazepine was extracted from plasma samples and fully resolved from the enantiomeric forms of its metabolites, 10,11-dihydro-10-hydroxycarbamazepine and

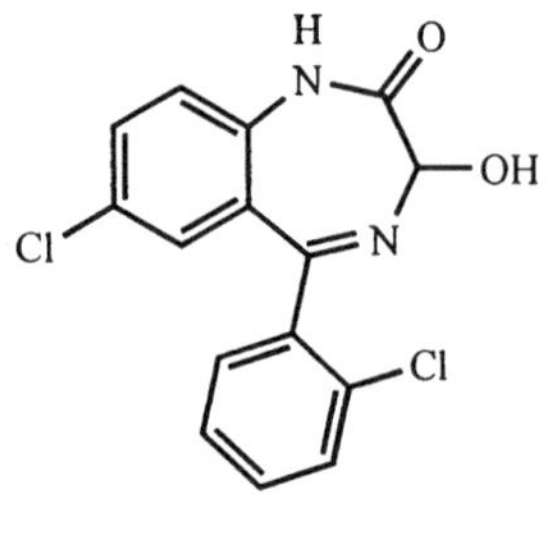

Lorazepam

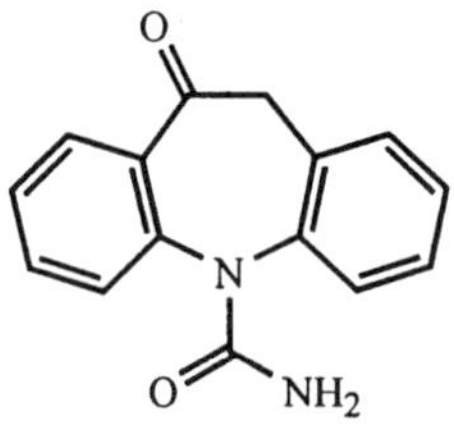

Oxcarbazepine

trans-10,11-dihydroxycarbamazepine, on Chiralcel OD/ODH columns in series ($\lambda = 220$ nm). Detection limits (S/N = 3) of 5 ng/mL were reported and baseline resolution was obtained in <30 min using a 70/30 hexane/ethanol mobile phase. Standard concentration ranges of 10–500 ng/mL were used [734].

Hooper et al. [735] separated the enantiomers of 1-hydroxytacrine on a Chiralcel OF column ($\lambda = 325$ nm) using a 70/30/0.1 hexane/IPA/diethylamine mobile phase. Analysis was complete in 40 min. The urinary metabolites 1-hydroxytacrine and 2-hydroxytacrine were baseline resolved from their parent compound in under 30 min on a cyanopropyl column ($\lambda = 325$ nm) using the same mobile phase as above. Working curves of 0.1–10 ng/mL were used. Detection limits of 0.01 µg injected were reported.

Propranolol and five metabolites were separated on a Chiracel OD column ($\lambda = 240$ nm, ex; 320 nm, em) using a 91/9/0.1 hexane/ethanol/diethylamine mobile phase. Good peak shapes and resolution resulted [736]. Elution was complete in 30 min. Standards of ~20 µmol/L were injected (50 µL).

Betaxolol enantiomers were baseline resolved in <10 min on a Chiralcel OD column ($\lambda = 273$ nm) using an 87/13/0.05 hexane/IPA/diethylamine mobile phase. Detection of 2.5 µg/mL was easily achieved [737]. Omission of the diethylamine component of the mobile phase led to extremely tailed peaks. Very similar results were obtained for cicloprolol [738], where separation was achieved in 10 min using an 80/15/0.05 hexane/IPA/diethylamine mobile phase.

Ponder et al. [739] studied the retention of promethazine, ethopropazine, trimeprazine, and trimipramine enantiomers on nine different chiral phases

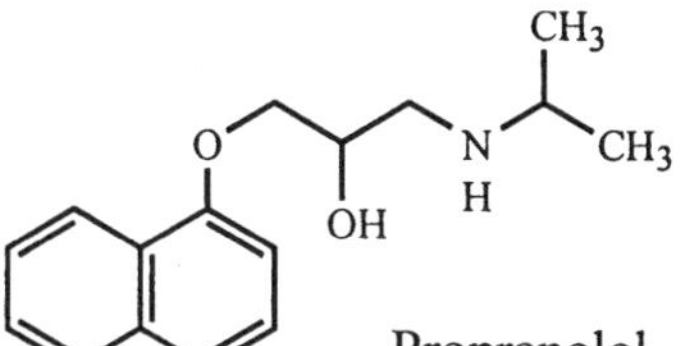

Propranolol

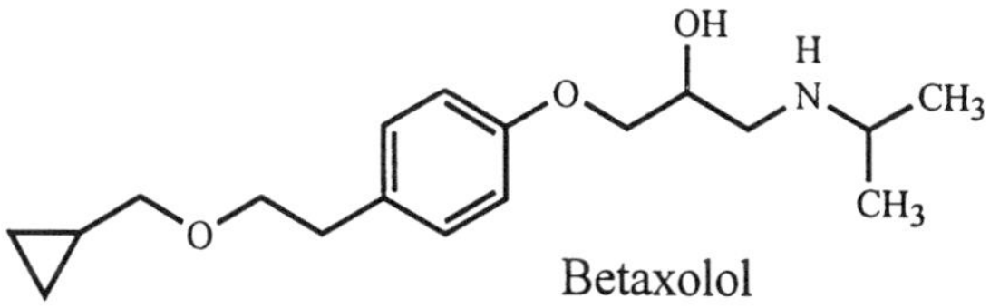

Betaxolol

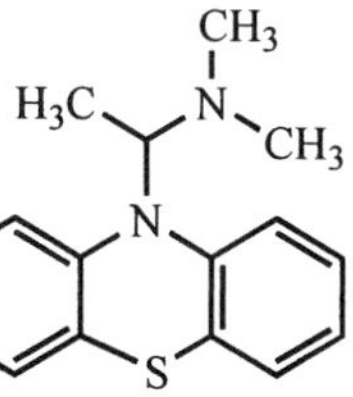

Promethazine 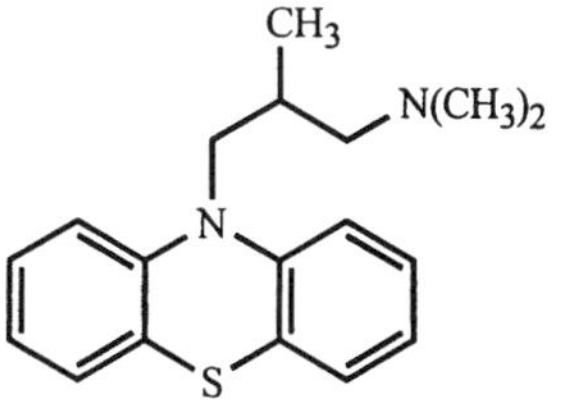   Trimeprazine 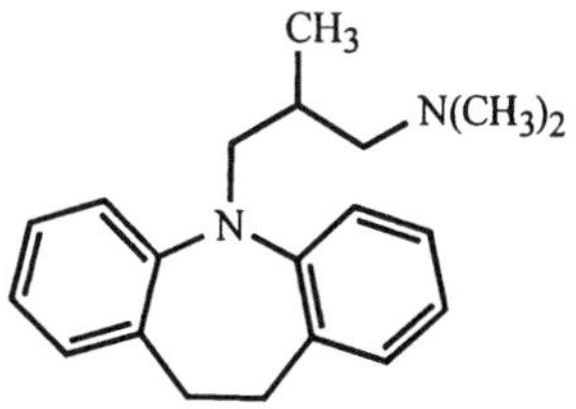   Trimipramine

TABLE 5.5 USP Methods[a]

Analyte(s)	Co-analyte (Internal Standard)	Column	Mobile Phase	Detector (nm)	Page
Anthralin	o-Nitroaniline	Silica	82/12/6 hexane/dichloromethane/acetic acid	354	148
Calcifediol	Testosterone	Silica	6/6/5/3 heptane/heptane (water-saturated)/ethyl acetate/dichloromethane	254	274
Dapsone		Silica	7/1/1/1 pentane/IPA/acetonitrile/ethyl acetate	254	496
Dihydrotachysterol	Preergocalciferol	Silica	100/1 iso-octane/IPA	254	569
Levmetamfetamine (limit of methamphetamine in) as the naphtylchloroformate derivative		3,5-dinitrobenzoyl-L-phenylglycine	98/1.5/0.5 hexane/IPA/acetonitrile	274	959
Phytonadione (E and Z-isomers)	Cholesteryl benzoate	Silica	2000/1.5 hexane/n-amyl alcohol	254	1330
Probucol (related compounds)	USP system suitability std	Silica	4000/1 hexane/alcohol	254, 420	1395
Tolbutamide	Tolazamide	Silica	475/475/20/15/9 hexane/hexane (water-saturated)/THF/alcohol/acetic acid	254	1676
Tretinoin (assay)	Isotretinoin	Silica	99.65/0.25/0.1 iso-octane/IPA/acetic acid	352	1683

[a] Reference [590].

($\lambda = 254$ nm). Hexane/ethanol/TFA gave the best results on cyclodextrin columns, whereas hexane/1,2-dichloroethane/ethanol/TFA and/or hexane/dichloro-methane/ethanol/TFA mobile phases gave acceptable R_s values and k' values of <15 for most enantiomers. (It should be noted that acetylated and γ-cyclodextrins gave no resolution of enantiomers under any conditions because the cyclodextrin cavity size is too small to accept the analyte.) Generalizations about appropriate maximal solvent strengths needed to effect separation were presented for most columns. The effects of retention and separation due to chlorinated mobile phase components (dichloromethane, chloroform, and 1,2-dichloroethane) are also tabulated.

Thioridazine enantiomers were extracted from serum and separated on a phenyl-methylurea column ($\lambda = 263$ nm) using a 45/45/10/0.0075 hexane/dichloro-methane/methanol/1 M methanolic ammonium acetate mobile phase [740]. Resolution was good and peaks eluted in <10 min. Analyte peaks were well removed from serum components and detection limits of $\sim$50 ng/mL were reported. Additionally, thioridazine and eight of its metabolites (e.g., thioridazine N-oxide, northioridazine, and thioridazine 5-sulfate and disulfone) were baseline resolved in <12 min using a silica column ($\lambda = 254$ nm) and an 8/1/1 iso-octane/methanol/dichloromethane (0.036% methylamine) mobile phase [741].

The United States Pharmacopoeia has many well-established protocols for the analysis of drug compounds. See Table 5.5 for details [590].

5.7.5   Summary

Many hydroxylated pharmaceutical compounds are effectively separated and quantitated on NP support materials using alkane-based solvents. Polar modifiers are added to the alkane to cause more rapid elution and enhance selectivity. Hexane and heptane are the mainstay solvents for use in the enantiomeric separations that utilize Pirkle-type support materials. Often alcohols are used as the modifier in such separations. The major disadvantage to the use of alkane mobile phases is the often-limited solubility of the analyte therein. It should be noted, however, that the separation of enantiomeric compounds still utilizes these solvents, especially if per-analysis derivatization is done.

6

CHLORINATED ALKANES AND CHLORINATED BENZENES

Chlorinated solvents have been extensively used as extraction solvents. As is shown in Table 6.1, these solvents have very limited solubilities in water (1.6% for dichloromethane to 0.0025% for 1,2,4-trichlorobenzene). Conversely, these solvents all offer low water solubilities, even when water saturated (0.31% for 1,2-dichlorobenzene to 0.008% for carbon tetrachloride). The chlorinated solvents also offer a wide range of polarities (in order of decreasing polarity): n-butyl chloride, dichloromethane, chloroform, and carbon tetrachloride. Most chlorinated solvents are still readily available in high-purity form, but environmentally driven legal restrictions are beginning to take effect and will, ultimately, severely limit or totally prohibit their production and use.

The chlorinated solvents offer the following set of unique properties:

1. They have one or more chloro group substituents, −Cl, which act as weak hydrogen bond acceptors.
2. They are immiscible with water.
3. Some chloroalkanes readily degrade to HCl and a variety of other chlorinated by-products and, therefore, contain a variety of additives designed to either prevent degradation and interact or react with the degradation products.
4. They are miscible with a wide range of organic solvents.
5. They readily dissolve a wide variety of polar and nonpolar compounds.

Of the chlorinated solvents, dichloromethane (DCM) is the most commonly used HPLC solvent followed, to a much lesser extent, by chloroform. The high volatility of DCM and chloroform is an added bonus for the extraction chemist since sample

260

concentration (by extraction solvent volume reduction) or solvent exchange (total removal of the extraction solvent followed by the subsequent dissolution of the sample in another solvent) can be efficiently done.

Even though DCM and other chlorinated solvents are highly effective extraction solvents, they have somewhat limited chromatographic utility. One major reason is that the property that makes them excellent extraction solvents (water immiscibility) precludes or severely limits their use in traditional reversed-phase separations. The increased use of ternary solvents with the use of an "intermediary" mutually miscible solvent, such as IPA, has enabled the use of chlorinated solvents as RP components. Keep in mind, however, that the UV cutoffs for the alkyl chlorinated solvents are quite high, ranging from 220 nm to 260 nm.

In NP separations, the chlorinated alkanes have been utilized effectively to enhance selectivity in separations and/or significantly increase the solubility of the analyte in the mobile phase. Solubility is a critical limiting factor in many polymer and fullerene analyses as well as purification methods. Chlorinated solvents typically provide the highest solubilities, compared with other common LC solvents, for these analytes. Therefore, chlorinated solvents are frequent components of the mobile phase in these analyses.

Chlorinated benzenes, because of their much higher UV cutoffs (>285 min versus 233 nm for DCM) and high cost (more than twice the cost of DCM), are rarely used in HPLC standard methods. However, when systems in which temperatures in excess of 70°C are needed (in order to solubilize analytes) and a refractive index detector or methods that employ UV detection where a high wavelength is used, the temperature stability of the chlorinated benzenes offers a practical alternative.

Also, standard HPLC bonded phases are cleaved from the surface faster when the system operating temperatures increase. This is exacerbated by any HCl degradation by-product present in the chlorinated solvent. Therefore, when the required operating temperature reaches 70°C, a solvent-compatible polymeric resin or an underivatized support (zirconia, alumina or silica) should be considered.

High-purity chlorinated solvents are produced in a manner that initially removes the vast majority of contaminants (both decomposition and raw material processing). Unfortunately, as dicussed above, some chlorinated alkane solvents are inherently unstable and decompose into chlorinated free radicals and their by-products [742]:

$$RCl + hv \rightarrow R\cdot + Cl\cdot$$
$$Cl\cdot + HR\cdot \rightarrow HCl + R: \qquad (6.1)$$
$$R\cdot + R\cdot \rightarrow RR\cdot$$

Light, heat, and O_2 initiate the formation of these free redicals. Prevention of their reappearance depends largely upon maintaining proper storage. For example, exposure to heat and light accelerates decomposition, so that storage in a cool dark place is a key factor in maintaining solvent integrity.

Manufacturers of high-purity chlorinated alkanes often use a preservative (or stabilizer) that acts as either an inhibitor to free radical formation or as a scavenger for the decomposition. DCM preservatives include ppm levels of either alkenes (cyclohexene, amylene [2-methyl-2-butene]), alkanes (cyclohexane), or alcohols (methanol). Chloroform is preserved with either ppm levels of amylene or 1% ethanol and sometimes a combination of both. The alkenes react with hydrochloric acid to form chloroalkanes, whereas the alcohols apparently form adducts with the HCl through strong hydrogen bond interactions. The utility of cyclohexane, compared with the other preservatives, is questionable. Regardless of the preservative used, most manufacturers print an expiration date on each bottle label. The printed date is usually 9 months to 1 year after the date of bottling. This gives the user information that serves as an initial warning to check for potential problems of an aged solvent.

The formation of chlorinated free radical compounds makes it extremely important to guarantee that the analytes of interest are neither affected by nor reactive toward these decomposition compounds. The use of recently produced high-purity solvents is preferable since the post-production removal of breakdown components by the analytical laboratory is time-consuming and expensive. After purchasing the solvent, the best method for preventing potential solvent-related problems is to make sure that the solvent is stored properly (cool, dry, out of direct light, in a vented storage area) and used prior to the manufacturer's expiration date. Conversely, the laboratory solvent supply should be rotated and limited to a maximum of one month's stock to ensure solvent integrity through its timely use.

A concern with the use of chlorinated solvents is that chloroform, carbon tetrachloride, and 1,2-dichloroethane are confirmed carcinogens. Dichloromethane is a suspected carcinogen. Therefore, these solvents should be handled with care and in a well-vented area as much as possible. Resevoirs and waste containers should be covered to prevent solvent vapor from getting into the laboratory area. Finally, the waste solvent generated by chlorinated solvent-containing mobile phases should be collected in separate waste receptacles since chlorinated solvents require different disposal methods from nonchlorinated HPLC solvents.

Tables 6.1–6.4 list some important chemical, physical and chromatographic properties as well as general manufacturing specifications and safety parameters for the chlorinated alkanes and chlorinated benzenes [84–92]. Figure 6.1 shows the chemical structures for the solvents listed in Tables 6.1–6.4.

6.1 IMPURITIES

Carbon tetrachloride is relatively stable and at room temperature is, under normal ambient laboratory conditions, unaffected by light or oxygen. Most contaminants found in CCl_4 are by-products of the chlorination process used for its manufacture. These contaminants include ppm levels of phosgene, ($Cl_2C=O$), CS_2, HCl, and low-molecular weight aldehydes.

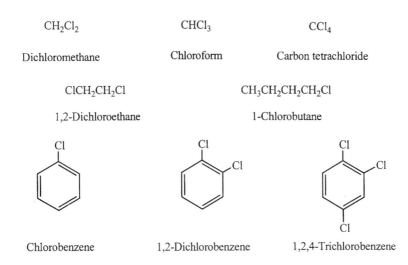

Figure 6.1

Chloroform is the least stable of the chlorinated solvents used in HPLC. Within an hour or two of its exposure to light, moisture, and oxygen, detectable levels of phosgene, HCl, and Cl_2 are formed. The presence of 1% ethanol greatly stabilizes the chloroform, presumably due to the alcohol's inhibition of the initiators for chlorinated free radical reactions. Alkenes, e.g. amylene, are added to $CHCl_3$ to remove the reactive decomposition compounds that do form. Impurities found in $CHCl_3$ due to the raw material manufacture process are similar to those found in CCl_4: phosgene, CCl_4, aldehydes, Cl_2, and HCl.

Dichloromethane (DCM) is not affected by moisture at room temperatures but does decompose, albeit much more slowly than $CHCl_3$, when stored in the light and/or at elevated temperatures. Decomposition products include chloroform, 1,1,2,2-tetrachloroethane, HCl, and phosgene. The manufacturing process creates $CHCl_3$, CH_3Cl, HCl, and Cl_2 as the major impurities in DCM. Various alkenes and methanol are currently used in the high-purity solvent as preservatives for DCM. Prior to choosing the proper DCM, tests should be run on standards with each type of stabilizer (e.g., amylene versus methanol) to ensure that no reactive products or intermediates are present that react with or chromatographically interfere with the detection of the analyte(s) of interest.

It should be remembered that although the preservatives are added to the solvent to prevent degradation, these additives themselves are also impurities. Alkenes, such as amylene or cyclohexene, and cyclohexane are typically present at the 25–250 ppm range. The level of preservative is typically increased in proportion with the instability of the solvent. All degradation products (e.g., chlorocyclohexane when cyclohexene is used) become impurities that may "appear" in an analysis as the solvent ages. These impurity levels increase with time and can reach the 100 ppm

TABLE 6.1 Physical Properties of Chlorinated Solvents[a]

	DCM	CHCl₃	CCl₄	EDC	BuCl	CB	DCB	TCB
Molecular weight	84.93	119.38	153.82	98.96	92.57	112.56	147.00	181.46
Density (g/mL)	1.326	1.4892	1.594	1.253	0.8862	1.1058	1.3058	1.454
Viscosity (cP)	0.44	0.57	0.97	0.79	0.45	0.80	1.32[b]	
Solubility in water (%)	1.60	0.815	0.08	0.81	0.11	0.05	0.013	0.0025
Water solubility in solvent (%)	0.24	0.056	0.008	0.15	0.08	0.04	0.31[c]	0.020
Boiling point (°C)	39.75	61.15	76.65	83.48	78.44	131.69	180.48	213.5
Melting point (°C)	−95.14	−63.55	−22.95	−35.66	−123.1	−45.58	−17.01	16.9
Refractive index (n_D)	1.4241	1.4458	1.4601	1.4448	1.4021	1.5248	1.5514	1.5717
Dielectric constant	9.08	4.81	2.24	10.42	7.39	5.59	10.2	2.24
Dipole moment (D)	1.60	1.04	0.00	1.80	2.05	1.69	2.50	
Surface tension (dyne/cm)	28.16	27.16	26.75	32.23	23.75	33.28	26.84	

Abbreviations: DCM, dichloromethane, methylene chloride; CHCl₃, chloroform, trichloromethane; CCl₄, carbon tetrachloride, tetrachloromethane; EDC, 1,2-dichloroethane, ethylene dichloride; BuCl, *n*-butyl chloride, 1-chlorobutane; CB, chlorobenzene; DCB, 1,2-dichlorobenzene, *o*-dichlorobenzene; TCB, 1,2,4-trichlorobenzene.

[a] All values except boiling and freezing points at 20°C unless otherwise noted.
[b] At 25°C.
[c] At 24°C.

TABLE 6.2 Chromatographic Parameters of Chlorinated Solvents[a]

	DCM	CHCl$_3$	CCl$_4$	EDC	BuCl	CB	DCB	TCB
Eluotropic strength $\varepsilon°$ on Al$_2$O$_3$	0.42	0.40	0.18	0.49	0.26	0.30		
Eluotropic strength $\varepsilon°$ on SiOH	0.30	0.26	0.11					
Eluotropic strength $\varepsilon°$ on C$_{18}$								
Solvent strength parameter, P'	3.1	4.1	1.6	3.5	1.0	2.7	2.7	
Hildebrandt solubility parameter, δ	9.7	9.3	8.6	9.9	8.4	9.7	10.0	
Hydrogen bond aciditivity, α	0.30	0.44	0.00	0.00	0.00	0.00	0.00	
Hydrogen bond aciditivity, β	0.00	0.00	0.00	0.00	0.00	0.07		
Dipolarity/polarizability, $\pi*$	0.82	0.58	0.28	0.81	0.39	0.71	0.67	

[a] *Abbreviations*: DCM, dichloromethane, methylene chloride; CHCl$_3$, chloroform, trichloromethane; CCl$_4$, carbon tetrachloride, tetrachloromethane; EDC, 1,2-dichloroethane, ehtylene dichloride; BuCl, n-butyl chloride, 1-chlorobutane; CB, chlorobenzene; DCB, 1,2-dichlorobenzene, o-dichlorobenzene; TCB, 1,2,4-trichlorobenzene.

TABLE 6.3 Common Manufacturing Quality Specifications of Chlorinated Solvents[a]

	DCM	CHCl₃	CCl₄	EDC	BuCl	CB	DCB	TCB
UV cutoff (nm)	233	245	263	228	220	287	295	308
Percent water (maximum)	0.01	0.02	0.01	0.01	0.02	0.03	0.02	0.01
Available as ACS tested[b]	ABEFJM	ABEFJM	AJ	AFJM	n.a.[c]	AJM	n.a.	n.a.
Available as HPLC-grade[b]	ABEFJM	ABEFJM	A	ABE	ABEF	BM	ABJ	BEFJ
Available through[d]	ABEFJM	ABEFJM	A	ABE	JM	F	FM	A

[a] *Abbreviations*: DCM, dichloromethane, methylene chloride; $CHCl_3$,chloroform, trichloromethane; CCl_4, carbon tetrachloride, tetrachloromethane; EDC, 1,2-dichloroethane, ehtylene dichloride; BuCl, *n*-butyl chloride, 1-chlorobutane; CB, chlorobenzene; DCB, 1,2-dichlorobenzene, *o*-dichlorobenzene; TCB, 1,2,4-trichlorobenzene.

[b] Manufacturer is code: A = Aldrich; B = Burdick & Jackson; E = EM Science; F = Fisher; J = JT Baker; M = Mallinckrodt

[c] No ACS test exists for this solvent.

[d] Available as a high-purity solvent but not specifically designated as ACS or HPLC grade. This does *not* mean a laser quality solvent, just that it is not specifically tested for these applications. If the manufacurers produce either ACS or HPLC solvent, they are not listed under this heading.

TABLE 6.4 **Safety Parameters of Chlorinated Solvents**[a]

	DCM	CHCl$_3$	CCl$_4$	EDC	BuCl	CB	DCB	TCB
Flash point[b] (TCC) (°C)	NBT	NBT	NBT	13	−10	29	69	105
Vapor pressure (Torr @ 20°C)	350	158.4	89.55	83.35	80.1	8.8	1.2	1
Threshold limit value (ppm)	500	2	2	1		75	50	5
CAS number	75-09-2	66-67-3	56-23-5	107-07-2	109-96-3	108-90-7	95-50-1	120-82-1
Flammability[c]	1	0	0	3	2	3	2	1
Reactivity[c]	0	0	0	2	0	0	0	0
Health[c]	2	2	3	2	3	2	2	2

[a] *Abbreviations*: DCM, dichloromethane, methylene chloride; CHCl$_3$, chloroform, trichloromethane; CCl$_4$, carbon tetrachloride, tetrachloromethane; EDC, 1,2-dichloroethane, ethylene dichloride; BuCl, *n*-butyl chloride, 1-chlorobutane; CB, chlorobenzene; DCB, 1,2-dichlorobenzene, *o*-dichlorobenzene; TCB, 1,2,4-trichlorobenzene.

[b] TCC=TAG closed cup; NBT indicates no flash point generated by TAG closed cup

[c] According to National Fire Association ratings [92]:

Fire: 4 = Materials that vaporize at room temperature and pressue and burn readily.
  3 = Liquids or solids that can ignite under room conditions.
  2 = Materials that ignite with elevated temperature or with moderate heat.
  1 = Materials that must be preheated before they ignite
  0 = Materials that will not burn.

React: 4 = Materials that by themselves, can detonate or explode under room conditions.
  3 = Materials that can detonate or explode but require an initiator (e.g., heat).
  2 = Materials that undergo violent chemical reactions at elevated temperatures or pressures or react with water.
  1 = Materials that are, by themselves, stale but that may become unstable at elevated temperatures and pressures.
  0 = Materials that are stable even under fire conditions and do not react with water.

Health: 4 = Short exposure times to these materials are lethal or cause major residual injury.
  3 = Short exposure times to these materials cause temporary and/or residual injuries.
  2 = Lengthy (but not chronic) exposure to these materials may cause temporary incapacitation and/or minor residual injury.
  1 = Materials that upon exposure, cause irritation but only minor residual injury.
  0 = Materials that, upon exposure under fire conditions, offer no more hazard than ordinary combustible materials.

level if solvent storage (or the initial production) is improper. Chloroform is also available with ~1% ethanol. This may be considered a low level, but it produces large changes in NP retention times and extraction equilibria and therefore needs to be monitored.

Butyl chloride is quite stable compared with chloroform or DCM. Additives are not typically needed to prevent degradation. Higher cost, without dramatic enhancement of the chromatographic results, and higher levels of impurities have limited its use.

The chlorinated benzenes are stable in light and water even at elevated temperatures. Impurities that come from production steps include benzene, chlorinated benzenes, toluene, chlorotoluenes, HCl and Cl_2.

Even though many of the high-purity solvents offered today are pretested to guarantee low HCl levels (and, hence, low degradation compound concentrations), it is always wise to keep in mind that the manufacturer and distributor have no control over storage after the solvent leaves their facilities. Most truck trailers are not temperature controlled, so that solvents may be exposed to high temperatures during delivery. This greatly facilitates solvent degradation. If questions arise as to the integrity of a chlorinated alkane solvent, two quick and simple tests can save hours of frustration, lost samples and/or damaged columns: (1) Extract chloroform or DCM with an equal volume of water and test the aqueous layer with dilute $AgNO_3$ (a precipitate, AgCl, indicates the presence of Cl^-); (2) Extract or DCM with an equal volume of an aqueous KI solution; Cl_2 will oxidize I^- to I_3^- and turn the organic (bottom) layer purple [743]. To remove these impurities, the chlorinated solvent may be extracted with acidic, basic and/or neutral water solutions. Although this procedure removes the acid and dissolved gas contaminants, it also creates a water-saturated solvent.

6.2   GENERAL ANALYTES

In most cases, RP separations have been the dominant means by which chromatographers achieve separation of small solutes. In NP separation and some other specialized cases, however, the chlorinated alkanes have given the chromatographer a unique approach to a separation.

6.2.1   Simple Substituted Benzene Analytes

Snyder and co-workers [606] tabulated the results for the retention of 13 substituted aromatic compounds on a diol column. The effect of isocratic hexane/ dichloromethane (DCM) mobile phases in which the DCM systematically increased from 5% to 35% (in 5% increments) was studied: multiple substituent effects (e.g., 1-nitronaphthalene vs. 1,5-dinitronaphthalene), positional isomer effects (e.g., 1- vs. 2-naphthyl acetate and 1- vs. 2- acetonaphthalene), and functional group effects (e.g., 1-nitro- vs. 1-cyano- vs. 1-acetonaphthalene). Five steroids (prednisone, corticosterone, adrenosterone, 4-androsten-17β-ol-3-one, 4-androsten- 17α-ol-3-one) were also studied with the hexane/DCM mobile phase adjusted so that log k' values of $\sim$1, $\sim$0.75, $\sim$0.5, and $\sim$0.1 were obtained for the solutes. The percent DCM needed to achieve a log k' value of 1.0 varied from 13 for adrenosterone to 59 for prednisone. Similarly, 36% DCM resulted in a log k' of 0.41 for 4-androsten-17β-ol-3-one, but prednisolone had a log k' of 0.29 even at 100% DCM. This large difference in retention behavior may be explained by the prednisolone having the opportunity to undergo more hydrogen bonding with the silica than the androstenolone, prednisolone having one additional $-$C=O and two extra $-$OH groups.

The retention of eight carboxylic acids (hippuric, homovanillic, 4- and 3-hydroxybenzoic, 4-hydroxycinnamic, 4-aminobenzoic, 2-hydroxyphenylacetic, and 5-hydroxyindole-3-acetic acid) on a silica column was thoroughly studied using a series of DCM/methanol/water mobile phases [744]. Acetate buffer was added to keep the acids in their fully protonated and uncharged states. As the percent water rose from <0.1% to 1% in an 80/20 DCM/methanol (10 mM acetic acid with 10 mM potassium acetate) mobile phase, an approximate doubling in k' resulted for all compounds. A subsequent increase in water level from 1% to 2.5% water yielded a much larger increase in the k' values. This sudden increase in retention was attributed to water in the solvent completely filling the pores and covering the surface of the packing material as the water level rose past 1%. This water layer, rather than deactivating the silica as is common for other analytes, actually provides a preferential partition layer for these very hydrophilic compounds. As expected, the percentage of methanol in the mobile phase had a strong effect on retention. As percent methanol increased, retention decreased. Resulting k' values for mobile

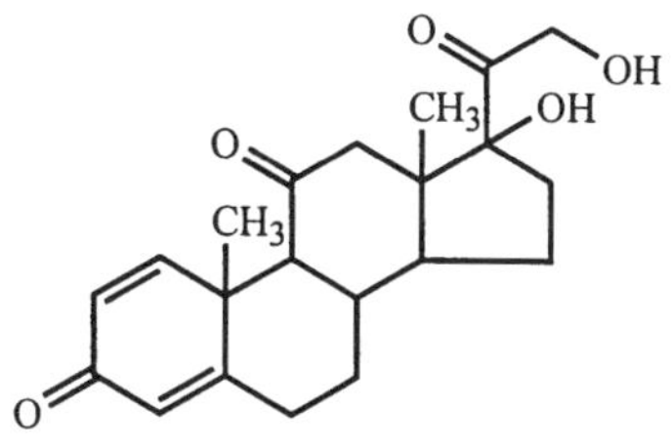

Prednisone

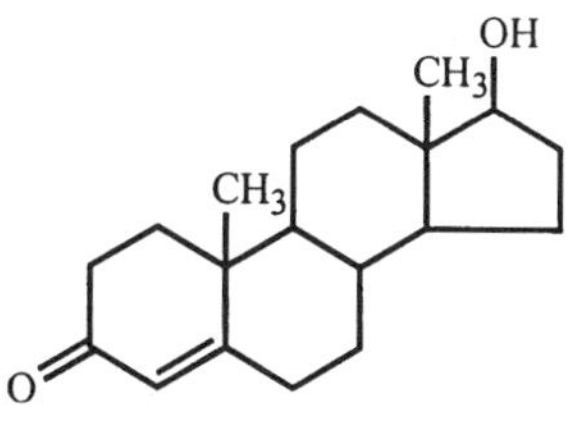

4-Androsten-17-ol-3-one

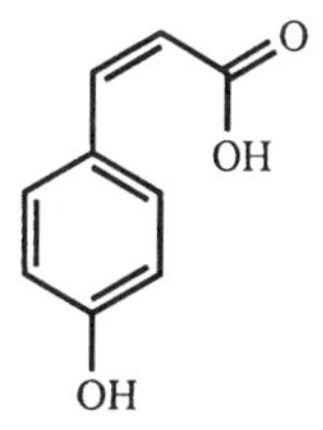

4-Hydroxycinnamic acid

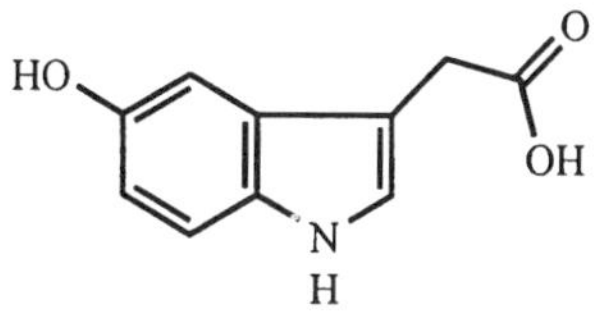

5-Hydroxyindole-3-acetic acid

phases containing 10–40% methanol are given. Other significant variables that were studied included the concentration of buffer (k' increases rapidly from no buffer to 5 mM then gradually and slightly decreases from 5 mM to 960 mM buffer); the pH (k' sigmoidally increases from pH 3 to pH 7 with the k' vs. pH plot results resembling individual acid titration curves); and the buffer counterion (k' decreases as counterion becomes "harder", $K^+ > Na^+ > NH_4^+ > Li^+$). Complete separation of the above eight analytes as well as 3,4-dimethoxybenzoic acid and benzoic acid was achieved in 12 min. Peak shapes were uniformly excellent. Four nucleotide bases (thymine, uracil, 5-methylcytosine, cytosine) are baseline resolved and eluted in 7 min using an 80/20/2 DCM/methanol/water (100 mM acetic acid with 10 mM potassium acetate buffer at pH 3.75) mobile phase. Six aromatic sulfonic acids (anthraquinone-2-, benzosuberone-4-, indane-4-, benzene-, 1-naphthol-, and 6-hydroxynaphthalene-2-sulfonic acid) were baseline resolved and eluted in <10 min using the same conditions. Finally, sorbic acid and benzoic acid were analyzed in DCM extracts of breads and fruit. Separation was complete with baseline resolution in 4 min using 80/20/2.3 DCM/methanol/water (10 mM acetic acid with 10 mM potassium acetate pH 4.75).

A series of twenty-nine substituted phenols (e.g., alkyl, phenyl, fluoro, chloro, bromo, nitro, and multisubstituted) was studied using a diol column and a 50/50 chloroform/iso-octane mobile phase [611]. In general, retention increased as the polar nature of the solute increased (e.g., in order of increasing retention: 2,4,6-tributylphenol < 4-amylphenol < 4-methylphenol < phenol < 4-chlorophenol < 4-nitrophenol). As might be expected, this system showed good isomer selectivity (k' for 3-nitrophenol = 4.20, for 4-nitrophenol = 5.50). However this selectivity was not as pronounced as that found using an underivatized silica support.

The effects of water levels in chloroform and 1,2-dichloroethane on solute retention is an important factor when using silica or alumina supports [745]. Ethyl benzoate, dibutyl, diethyl, and dimethyl phthalate, and o- and p-nitroaniline were used as test solutes to monitor the magnitude of the effect of water level on retention. For example, k' values for these solutes on the alumina support were initially between 1 (for dimethyl phthalate) and 8 (for p-nitroaniline) when water levels were <60 ppm in the chloroform mobile phase. When water levels reached approximately 150 ppm, all k' values had fallen below 2. For water levels >300 ppm all k' values = 1. This reemphasizes the critical need for water levels in nonaqueous NP separations to be strictly controlled.

A second important issue raised in this study is that of viscosity changes in very low viscosity solvents. The authors claim a 30% decrease is seen in overall system efficiency (as measured by number of theoretical plates, N) when the less viscous solvent, chloroform (0.57 cP) is replaced by 1,2-dichloroethane (0.79 cP). Unfortunately, viscosity is not the only variable affected when the solvents are changed (note water level in the solvent, different eluotropic strengths and selectivities, etc.) and so the actual loss due to the change in viscosity is hard to estimate. Nevertheless, factors such as this must be kept in mind when developing a successful NP separation.

6.2.2 Organometallics and Metal–Ligand Complexes

Prior to use of $CHCl_3$ as a low-level mobile phase modifier (75/20/5 methanol/ water/chloroform) in the separation of Hg(II), Co(II), Fe(II), Cu(II), Zn(II), Pb(II), and Ni(II) complexes of 8-quinolinethiols (8QTs), the chloroform was pretreated with three successive washes: concentrated H_2SO_4, then 4 M NaOH, then distilled water [149]. The final treated $CHCl_3$ was then placed in a sealed bottle and stored in the refrigerator, where it was of usable quality for one week. Even with this pretreatment procedure, chloroform extracts of aqueous metal-8QTs showed anomalous behavior as seen by decreased sensitivity for Co, Ni, and Zn complexes and peak splitting for the Co complex. Although the authors present no explanation for this observation, it should be noted that metal-chelate complexes are often extremely pH-sensitive. Small pH changes will alter the metal-chelate bonding interactions and thereby change the complexation equilibrium and the absorbance spectrum—in this instance to a lower molar absorptivity. This observation explains both the decrease in sensitivity and the anomalous peak formation. The presence of a weakly complexing buffer system may help give more consistent results.

6.2.3 Summary

Chlorinated alkane solvents offer significantly different specificities from other solvent classes. This, coupled with their solvating capabilities, makes them effective mobile phase modifiers, not only in NP separations but in RP separations when a solvent, such as IPA, can create a miscible ternary mixture (e.g., water/IPA/ chlorinated alkane).

6.3 ENVIRONMENTALLY IMPORTANT ANALYTES

The analysis of many priority pollutants in soil, sludge, and water matrices begins with an extraction of the sample by DCM. Consequently, any method that is compatible with a sample already preconcentrated in DCM should be of particular interest to the practicing environmental analyst.

6.3.1 PAHs, Substituted PAHs, and Related Compounds

The retention behavior of 31 large polyaromatic hydrocarbons (PAHs) consisting of four fused rings (pyrene) to nine fused rings (tetrabenzo[*de,hi,op,st*]pentacene) was

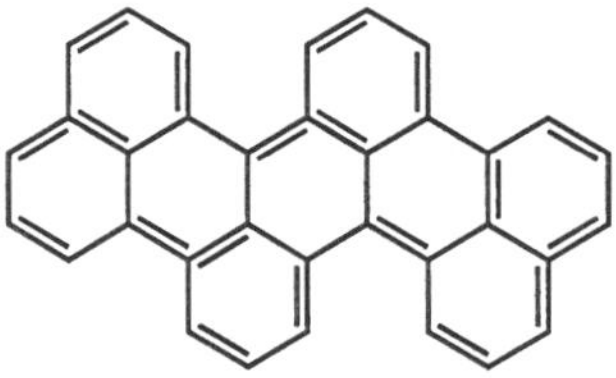

Tetrabenzo[*de,hi,op,st*]pentacene

studied on a C_{18} column (photodiode array detector) using isocratic mobile phase mixtures of methanol and DCM ranging from 100% DCM to 100% methanol [157]. The resulting plots of $\log k'$ vs. %DCM were made for each compound. Interesting and unexpected nonlinearities occurred in these plots that were not seen in typical NP separations on silica. The authors concluded that this effect was due to the considerably higher DCM used in the RP mode compared with that used for a silica column. The higher DCM levels supposedly caused a solvent-induced nonplanarity of the solutes and hence unexpected retention behavior. This explanation was supported by the fact that elution orders changed dramatically for some sets of compounds when the %DCM was changed. For example, dibenzo[cd,lm]perylene (DBP) eluted significantly before benzo[lm]phenanthro[4,5,6,abcd]perylene (BPP) at DCM levels <50%. However, BPP eluted significantly before DBP when the DCM level was >80%. This type of retention behavior adds an additional and powerful selectivity factor for separation that goes beyond normal solute/solvent/support interactions.

An interesting 100% hexane → 100% DCM → 100% acetonitrile → 100% DCM gradient on a silica column ($\lambda = 254$ nm, ex; 390 nm, em) fractionated sample extracts into eight classes of PAHs/PCBs and quinones in 70 min [613]. This was particularly effective since the samples (urban particulates and marine sediments) were extracted with DCM prior to cleanup and the subsequent fractionation procedure detailed above. Detection limits in the range 0.1–25 μg/g of solid (analyte dependent) were reported. Individual components in the collected fractions were then quantitated by GC-FID and GC-MS since the HPLC solvents used are compatible with these techniques.

In an attempt to fractionate eight test nitrogen heterocyclic compounds (e.g., acridine, phenazine, carbazole, 1,2-bis(2-pyridyl)ethylene) from seven test PAHs (acenaphthene, perylene, decacyclene, coronene), Ruckmick and Hurtubise [746] used a silica column ($\lambda = 280$ nm) and a 99.975/0.025 CCl_4/DMSO or a 95/5 CCl_4/$CHCl_3$ mobile phase. The CCl_4/DMSO solvent almost succeeded in resolving all. All nitrogen heterocycles were well resolved. On the other hand, the PAHs were poorly resolved. The CCl_4/$CHCl_3$ mobile phase rapidly eluted and resolved all the PAHs, but could not elute the nitrogen heterocyclic compounds. A potential method would involve a gradient from 95/5 CCl_4/$CHCl_3$ to 99.975/0.025 CCl_4/DMSO, thereby obtaining the best elution and resolution characteristics of both systems.

Dibenzo[cd,lm]perylene

Decacyclene

6.3.2   Nitrated and Chlorinated Nonpesticide/Herbicide Analytes

A number of heptane/polar modifier mobile phase combinations on a silica column ($\lambda = 235$ nm or photodiode array detector) were evaluated [629] during the development of a reproducible and sensitive method for the determination of the manufacturing source of 2,4,6-trinitrotoluene (TNT) found at explosion sites. This is done by identification and quantification of trace manufacturing impurities resulting in distinct "fingerprints" for individual TNT manufacturers. Of these mobile phase modifiers, DCM and THF gave the most useful separations. Whereas a 98.5/1.5 heptane/THF mobile phase had a lower UV cutoff than the comparative 80/20 heptane/DCM mobile phase, the overall chromatographic resolution was not as good when the THF solvent was used. The heptane/DCM mobile phase gave excellent resolution of 2,4-dinitro-, 2,3,4-trinitro- and 2,4,5-trinitrotoluene from TNT in <20 min.

6.3.3   Pesticide and Herbicide Analytes

Parathion and two of its hepatic metabolites, 4-nitrophenol and paraoxon, were baseline resolved on a silica column ($\lambda = 254$ nm) using a 93/7/0.02 DCM/acetonitrile/acetic acid mobile phase [747]. Peak shapes were excellent and the analysis was complete in <10 min.

Ethylenethiourea (ETU) was extracted from fruit and vegetable peels and analyzed on a cyanopropyl column ($\lambda = 240$ nm) using a 48/48/4 chloroform/hexane/methanol mobile phase [748]. ETU eluted in 12 min as a peak that was well resolved from a series of earlier-eluting and later-eluting extracted components. Total analysis time was 20 min. The authors make special note of the instability of ETU in methanol, reporting a >50% loss in 24 h when samples were extracted with and stored in methanol. Extraction followed by immediate analysis was recommended. Concentrations down to 1 μg/g were detectable.

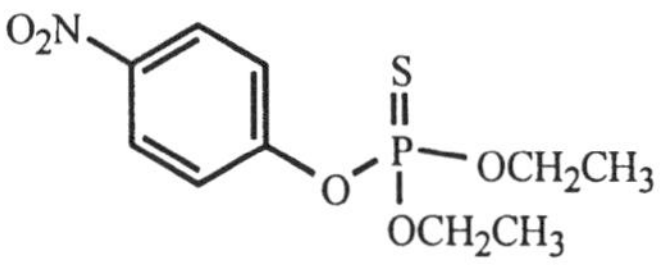

Parathion

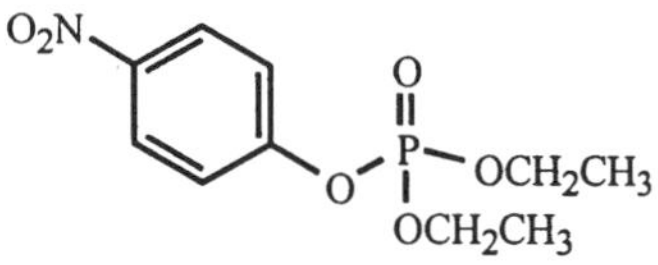

Paraoxon

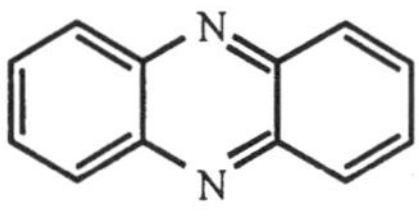

Phenazine

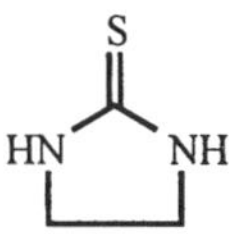

Coronene

Ethylenethiourea

6.3.4   Summary

Chlorinated solvents are extremely useful when large or high molecular weight compounds are to be analyzed. The ability of DCM and chloroform to solubilize materials having a wide range of polarities makes them particularly attractive extraction solvents. This property is generally overlooked in LC method development either because of immiscibility with water or high UV cutoff issues. Today, the next best thing to a moratorium on carbon tetrachloride and chloroform is underway. Procurement and disposal costs (waste stream separation of chlorinated solvents from others) is heightening the awareness of using alternative solvent systems for methods. Although the use of chlorinated solvents in the analytical laboratory may never be fully excluded, the increased of use of these solvents is unlikely.

6.4   INDUSTRIAL AND POLYMER ANALYTES

Chlorinated benzenes have been infrequently used in RP and NP separations. Their major use has been as solvents for the analysis of ultrahigh molecular weight (UHMW) polymers when solubility in normal GPC solvents at room or slightly elevated temperatures is limited. For complete and consistent solubilization, the UHMW polymers often require temperatures in excess of 130°C. This makes handling samples difficult, regardless of the solvent. 1,2-Dichlorobenzene and 1,2,4-trichlorobenzene are stable at this temperature and can successfully solubilize these analytes. This, therefore, is their major area of use.

6.4.1   Surfactant and Additive Analytes

Chloroform has been used as a mobile phase modifier in methods involving non-ionic ethoxylated alkylphenols and alcohols. Commonly used mobile phases are hexane or heptane with methanol and chloroform typically in ratios of about 70/10/20 [642, 645]. These methods have been described in detail in Chapter 4.

The characterization of several polymer additives was conducted on an aminopropyl column (ELSD). An isocratic 10/90 methanol/DCM mobile phase was used for Tinuvin 327 and Tinuvin 770 [749]. A $99.4/0.6 \rightarrow 97/3$ chloroform/methanol gradient was used to characterize Atmer 129, Radiamuls 142 and the components of technical-grade glyceryl monostearate. Approximately 0.1 µg injections were made. Irgafos P-EPQ and Irgafos 168 were analyzed on a C_{18} column ($\lambda = 220$ nm) using a 65/35 acetonitrile/chloroform mobile phase. This is a particularly interesting choice of mobile phase since the working wavelength is at least 25 nm below the UV cutoff for chloroform! The authors cite no difficulties arising from this choice of solvent at this wavelength.

6.4.2 Polymeric Analytes

The chlorinated alkanes have proven useful for solubilizing lower molecular weight polymers and oligomers. As detailed in the alkane and alcohol chapters, dichloromethane (DCM) has been used in conjunction with methanol and heptane gradients for the characterization of polystyrenes [272] and styrene/ethyl methacrylate co-polymers [649] and with heptane for co-poly (styrene/acrylonitrile) materials [244, 527].

Methanol/DCM mobile phases (ranging from 25 to 75% DCM) were used to characterize polystyrene polymers ranging in molecular weight from 3600 to 2.7×10^6 on C_{18} columns with base silicas of varying pore diameter. Injections of 0.5 μg sample were used. This study was conducted to deconvolve adsorption from solubilization effects from size exclusion contributions to retention [750]. Peak retention shifted from elution with 65 to 72% DCM as the sample load increased from 0.5 to 80 μg injected. These effects were attributed to sample solubility effects.

A number of polydisperse polystyrenes (MW ranges from 56,000 to 358,000) were characterized using a C_{18} column ($\lambda = 262$ nm) and a 35-min 46/54 → 66/34 DCM/acetonitrile concave gradient. This method was shown to be less sensitive to concentration (0.5 μg → 25 μg in 5 μL) and to offer better resolution than the classical size exclusion methodology [751].

Sato et al. [752] did a similar study on poly(styrene/methyl methacrylate) co-polymer (PMMA). A 100/0 → 60/40 acetonitrile/DCM gradient was used with a polystyrene column. As pore size decreased, significant band broadening occurred as the retention mechanism became less dominated by adsorption while size exclusion began to play an equally important role in the separation process.

Zimina et al. [753] studied PMMA co-polymers and poly(styrene/t-butyl methacrylate) (PtBMA) co-polymers on a 300 Å silica column under "critical conditions"—conditions in which one component of an A-B block co-polymer is eluted whereas the alternate components have no effect on its elution. PMMA polymers with molecular weights of 27,700, 60,000, and 107,000 were chromatographed under critical (57.0/43.0 DCM/acetonitrile) and near-critical (57.2/42.8, 56.9/43.1 and 56.7/43.3 DCM/acetonitrile) conditions. In those separations on either side of the critical condition where near-critical conditions were used, peak shape and elution order changed dramatically. A 1 μL injection of 0.1% w/v standards was used. Similar results were obtained for an 80,000 MW PtBMA co-polymer where critical conditions occurred at a mobile phase consisting of 90.7/9.3 DCM/acetonitrile.

Ethyl/butyl methacrylate co-polymers were analyzed on a 60°C silica column ($\lambda = 233$ nm) using a 20-min 99/1 → 90/10 1,2-dichloroethane/ethanol gradient [754]. Co-polymers with ratios ranging from 100% polybutyl methacrylate to 100% polyethyl methacrylate and 75/25, 50/50 and 25/75 mixtures were used. All five peaks were resolved. The authors discuss the trade-off between the sensitivity lost by working at 233 nm (rather than at a lower wavelength where the methacrylates absorb more strongly) and the higher background introduced by the 1,2-dichloroethane (UV cutoff = 228 nm) and their choice of 233 nm. Elution was complete in 10 min.

A 30-min 99/1 → 93/7 chloroform/ethanol gradient was used with a silica column ($\lambda = 254$ nm) to characterize styrene/methyl and ethyl methacrylate co-polymers [755]. That the ethanol content was critical was shown through a series of chromatograms for a 50/50 styrene/methyl methacrylate co-polymer and a 35/65 styrene/ethyl methacrylate co-polymer. For 25 μL injections of 0.1% w/v samples, the 50/50 co-polymer completely eluted with a 97/3 chloroform/ethanol mobile phase but was completely adsorbed to the silica at 99/1. Similarly, the 35/65 co-polymer eluted at 95/5 chloroform/ethanol and did not elute at 98/2. Temperature effects (40–70°C) on the level of ethanol needed for elution were tabulated for these co-polymers as well.

Mori [756] characterized styrene-methyl, styrene-ethyl, and ethyl-*n*-butyl methacrylate co-polymers on a silica column ($\lambda = 233$ nm). The polymethacrylates have strong UV absorption at 230 nm but little or no absorbance above 250 nm. Consequently, the authors did not consider chloroform as a potential solvent due to its UV cutoff of 245 nm. Interestingly, THF was also eliminated from use because the authors cited a high UV cutoff. This is true only for preserved THF that contains BHT. THF UV has a cutoff of <220 nm and would therefore be acceptable for use as long as it is chemically compatible with the solutes. From this reasoning, a 1,2-dichloroethane/ethanol gradient was used in this study. A gradient was run (0 → 100% ethanol) over 15 or 20 min depending upon the polymer studied. Sample concentrations were 0.01%.

Styrene/acrylonitrile co-polymers were characterized on a C_{18} ($\lambda = 260$ nm) column using a 28-min 100/0 → 0/100 heptane/DCM gradient or on a silica column using a 40-min 20/80/0 → 0/2.75/97.25 heptane/DCM/methanol gradient [757]. Comparisons of the results for the two methods with respect to percent acrylonitrile in the co-polymer were tabulated. A linear plot of retention time vs. percent acrylonitrile was obtained using the silica column method. The authors noted that co-polymers containing in excess of 20% acrylonitrile were strongly absorbed to both NP and RP supports. Elution was achieved only with DCM mobile phases containing more than 20% methanol.

Mori [758] used a 20-min 98/2 → 85/15 1,2- dichloroethane/ethanol gradient to characterize poly(styrene/vinyl acetate) block co-polymers on a 30°C silica column ($\lambda = 254$ nm). A 100 μL injection of 0.1% w/v samples was used in the analysis. Samples were completely eluted in 10 min.

6.4.3 Fullerenes

DCM and chloroform are solvents in which fullerenes have an acceptably high solubility. Accordingly theses solvents have been used frequently in chromatographic systems that separate, analyze, and/or purify the fullerenes. For the lower molecular weight fullerenes, (e.g., C_{60} and C_{70}) DCM has been used in hexane at the 20–50% range to obtain separation [652, 759]. However, a significantly higher level of DCM was necessary for the separation of isomers or higher molecular weight fullerenes [269].

For example, Ettl et al. [760] isolated the chiral (D_2) allotrope of C_{76} from a prefractionated "C_{70}" cut fullerene mixture. A 70/30 DCM/acetonitrile mobile phase on a C_{18} column successfully isolated this allotrope. Similarly, a series of six C_{60} and C_{70} fullerene isomers were baseline resolved on a C_{18} column (photodiode array detector, $\lambda = 200$–600 nm) using a 45-min 0/100 → 100/0 DCM/acetonitrile gradient [761].

6.4.4   Summary

Chlorinated solvents offer unique solubility and selectivity properties that no other solvent class has. Unfortunately, their incompatibility with the most commonly used detectors—UV and fluorescence—has dramatically limited their use. Although these solvents may be used with an RI detector, the high detection limits for the RI detector still limit the use of chlorinated solvents. The increased availability and use of more sensitive detectors that are compatible with chlorinated solvents—ELSD and MS—may lead to an increased interest in the chlorinated solvents, especially since detection limits for the ELSD are at least 10-fold lower than for RI detectors and for MS are even lower.

6.5   BIOLOGICAL ANALYTES

6.5.1   Carboxylic Acid Analytes

An analysis of the positional distribution of fatty acids (as their methyl esters) was done for triacylglycerols in butterfat. Approximately 200 μg of butterfat sample was injected [762]. Excellent separation and peak shapes for a series of individual standards (e.g., $C_{4:0}$–$C_{20:0}$, even; $C_{18:1}$–$C_{22:1}$, even; $C_{18:2}$ and $C_{18:3}$) were obtained using a C_{18} column (ELSD) and a 45-min 20/80 → 50/50 chloroform/acetonitrile gradient. The total weight of fractionated triacylglycerols was 20–40 μg with the lowest reported detectable amount of ∼2 μg.

The *cis* and *trans* isomers of linolenic acid were analyzed as their phenacyl derivatives on a silver-impregnated silica column ($\lambda = 242$ nm). The method for the preparation of the column is discussed. A 49.75/49.75/0.5 1,2-dichloroethane/ DCM/acetonitrile mobile phase was used and elution was complete in under 40 min [763]. Peaks were well resolved and peak shapes were good. A linear concentration range of 0–200 μg was reported.

6.5.2   Vitamins and Related Analytes

Biotin and dethiobiotin were derivatized with panacyl bromide and separated on a silica column ($\lambda = 380$ nm, ex; 470 nm, em) using a 95/5 dichloromethane/

methanol mobile phase [764]. Baseline resolution was achieved and elution was complete in <7 min. Linear response was obtained over the concentration range of 10–1000 pmol injected. It is interesting to note that the sensitivity for these analytes is much higher than when using an analogous separation reversed-phase system.

The separation of *cis* and *trans* isomers of various carotenoid isomers has been well studied and is covered in a review [765]. Low levels of chlorinated solvent (5–25%) in acetonitrile or methanol mixtures (70/10 to 47/47) have given adequate separations on C_{18} columns [766, 767]. Better separations were obtained on silica-based polymeric C_{18} supports than analogous monomeric C_{18} supports.

In an excellent review of variables influencing the analysis of carotenoids by HPLC, Scott [768] noted that some parameters had adverse effects on the system. First, stainless steel frits seemed to lead to lowered responses for β-cryptoxanthin and α- and β-carotene. Replacement of the frits with "metal-free" frits increased responses, but the replacement of stainless steel column tubing and injector tubing did not increase response. This is most likely a result of the large disparity in contact surface areas between the frits and the tubing. Storage of lycopene samples in ~0.1% BHT (antioxidant)-preserved chloroform vs. chloroform without BHT showed considerably less degradation with time. The same results were found for BHT-preserved THF vs. unpreserved THF. Optimization of extraction procedures is also addressed in this paper.

Schoefs et al. [769] extended the above methods to include the analyses of chlorophylls and protochlorophylls extracted from kidney bean, wheat, and cucumber seeds along with the *cis–trans* carotenoids. A C_{18} column ($\lambda = 437$ nm) and a complex 23-min $100/0 \rightarrow 80/20$ (70/30 acetonitrile/methanol)/DCM gradient gave adequate separation of 23 compounds, although not all pairs were baseline resolved. To address this problem, a photodiode array detector was used to deconvolve and identify overlapping peaks.

The α-, β-, and γ-tocopherol levels in a wide variety of foods (e.g., cheeses, fish, nuts, tomato products, vegetables, and teas) were determined using a C_{18} column ($\lambda = 290$ nm, ex; 330 nm, em) and a 30/70/5 DCM/acetonitrile/methanol (0.001% TEA) mobile phase [770]. Results are tabulated for over 40 products. Detection limits of 100–500 µg/kg were reported.

An interesting isocratic solvent system, 22/71/4/2/1 DCM/acetonitrile/methanol/water/propionic acid, was used to elute carotenoids, retinol, retinal, and their dehydro analogs (21 compounds total) on a C_{18} column ($\lambda = 352$ nm or 450 nm) from fish serum and eggs [771]. A table was generated that contained retention times and k' values for all compounds. Elution was complete in just under 60 min, with retinyl palmitate and α- and β-carotene the only analytes with retention times of >20 min. Standards of 2 µg/mL were used and easily detected. Baseline resolution was not achieved between all analyte pairs.

Vitamin D_3 and 11 metabolites were chromatographed on a silica column using a 28-min $99.98/0.02 \rightarrow 94/6$ DCM/methanol gradient [772]. Baseline resolution was not obtained between all pairs of analytes, but excellent peak shapes and complete elution in 30 min were obtained.

6.5.3   Analytes from Fats and Oils

In an excellent and detailed study by Gaudin et al. [773], the effects of mobile phase composition on elution and mobile phase modifier on ELSD response in the analysis of four ceramides (N-stearoyl-D-sphingosine, N-palmitoyl-D-sphingosine, N-palmitoyl-DL-dihydrosphingosine, and ceramide III) were presented. Separation was generated on a porous graphitized carbon column held at 50°C (ELSD drift tube $T = 35$°C, N_2 pressure 1 bar). For a 55/45/0.1/0.1 chloroform/methanol/formic acid/triethylamine mobile phase, a 6-min separation generated four excellent peak shapes and baseline resolution. Removal of the modifiers caused peaks to decrease in intensity and split. For an 81/190.1/0.1 THF/methanol/formic acid/triethylamine mobile phase, elution time was still 6 min, but resolution between ceramide III and the stearoyl compound was incomplete. It should be noted that mobile phase combinations utilizing acetone, ethyl acetate, and n-propyl alcohol did not elute the analytes. Ceramide standard concentrations varied from 0.3 to 1.0 mg/mL.

A large range of glycolipids obtained from edible plant sources was characterized using a silica column (ELSD, temperature 60°C, 2.0 bar N_2 nebulizer gas) and a $99/1 \rightarrow 57/25$ (at 15 min) $\rightarrow 10/90$ (at 20 min hold 5 min) chloroform/(95/5 methanol/water) gradient [774]. A mixture of acylated steryl glucoside, monogalactosyldiacylglycerol, steryl glucoside, ceramide monohexoside, and digalactosyldiacylglycerol was baseline resolved in 12 min. Detection limits of 0.2–0.5 µg injected (S/N = 3) were reported. Pumpkin extract was a plant example. A Floch extract yielded 10 compounds, the five standards listed above plus trigalactosyldiaclyglycerol free fatty acids (all co-eluted), and phosphatidylinositol, phosphatidylcholine, and phosphatidylethanolamine.

The triacylglycerol (TAG) autoxidation products of canola oil were studied using two 250×4.6 mm C_{18} columns (ELSD, drift tube $T = 140$°C, N_2 nebulizer gas flow at 2 L/min and a complex 85-min $70/30 \rightarrow 30/70$ acetonitrile/dichloromethane gradient [775]. Along with the triglycerols (e.g., ranging from linoliec : oleic : oleic to stearic : oleic : stearic), hydropexoxides (e.g., linoleic : oleic : oleic hydroperoxide) and epoxides (e.g., palmitic : oleic : oleic epoxide) were identified.

Winter butterfat was characterized with regard to triacyglycerol (TAG) fractions on a silver-impregnated silica column (ELSD) using a complex 35-min $100/0 \rightarrow 0/100$ (4/1 DCM/1,2-dichloroethane)/acetone gradient [661, 776]. A standard composed of 1 mg/10 µL of purified TAGs was injected. From these fractions, identification of 33 separate fatty acid residues was made on a C_{18} column using a complex 80-min $30/70 \rightarrow 65/35$ (4/1 DCM/1,2-dichloroethane)/acetonitrile gradient. From these chromatograms, a compositional table was generated.

Palmer and Palmer [777] separated and identified 33 TAGs in various edible oils (e.g., soybean and olive oil). These TAG compounds were comprised of various combinations of linoleic, linolenic, palmitic, palmitoleic, stearic, oleic, and arachidic acids. A C_{18} column (ELSD) and a 60-min $30/70 \rightarrow 70/30$ DCM/acetone gradient were used. Injections of samples were 20 µL of a 1.0% w/v solution, whereas purified standards were run at 0.05%. Neff et al. [778] used a C_{18} column and a 120-min $30/70 \rightarrow 60/40$ DCM/acetonitrile gradient for TAG analysis but used an FID detector. Here 1–2 mg of each individual TAG was injected. Héron et al. [779] studied borage oil to monitor γ-linolenic acid (18 : 3 Z6, Z9, Z12) on a C_{18} column (ELSD) using an isocratic 32/68 DCM/acetonitrile mobile phase. Peak identification could not be obtained for all peaks by LC/MS (even though excellent separation was obtained) because of the high degree of similarities between the MS peaks for many of these compounds. Since standards are not available for γ-linolenic acid TAGs, the authors then describe a method of sequential sample spiking with known TAGs and use of the resulting retention times as predictive markers.

Vaghela and Kilara [780] separated cerebrosides, PI, PE, PS, PC, sphingomeylin, and lysophosphatidylcholine in whey protein concentrates using a narrow-bore silica column (ELSD) and a 25-min $60/34/6/0.25 \rightarrow 80/20/0/0$ chloroform/methanol/water/ammonium hydroxide (20%) gradient. Detection limits were 2–20 ng (S/N = 2) and a linear concentration range from 0.25 to 3 µg injected was used. Ammonium hydroxide was used since the effluent must be volatile and be compatible with an ELSD. However, the authors note that NH_4OH is very aggressive toward and readily dissolves silica. Very low NH_4OH concentrations should be used because this dissolved silica could be redeposited in the ELSD. The authors also note the advantages of narrow-bore columns were achieved in this study as well: reduction of up to 80% of solvent consumption and 10-fold improvement in detection limits.

Four sphingosines were analyzed as their N-naphthimide derivatives on a silica column ($\lambda = 260$ nm, ex; 370 nm, em) using a 35/25/20 hexane/chloroform/ethyl acetate mobile phase [781]. The *threo-*, 2H-*threo-*, 2H-*erythro-*, and *erythro-*sphingosines were baseline resolved in 23 min. Picomole detection limits were claimed.

Five lipids (ceramide IV, sphingosine, sphingosine phosphate, and dihydro- and phytosphingosine) were isolated from cell cultures and analyzed on a silica column (ELSD, drift tube $T = 70°$, N_2 nebulizer gas at 1.6 bar) using a 90/01/1/1 chloroform/ethanol (200 proof)/triethylamine/formic acid mobile phase [782]. Good resolution and peak shapes were achieved and elution was complete in 16 min. A linear response from 1.7 to 17 µg injected was generated and a detection limit of 100 ng injected were reported.

6.5.4   Other Analytes

An uncommon study of sugar retention was conducted by Herbreteau [459]. Ribose, erythritol, rhamnose, xylose, arabinose, xylitol, sorbose, fructose, mannose, mannitol, glucose, and galactose were eluted on a silica column with an 80/19.8/0.2 DCM/methanol/water mobile phase and a diol column using an 84/16 DCM/methanol mobile phase. All k' values were <4 on the diol column and <16 on the silica column. These k' values were compared with the k' values obtained on an aminopropyl column using an isocratic 80/20 acetonitrile/water mobile phase. An important advantage of these alternative systems is that the possible Schiff base reaction between a reducing sugar and the aminopropyl functional group is eliminated.

Procyanidins (e.g., monomeric catechin and epicatechin) can form oligomers up to dodecamers. These compounds were extracted from apples and separated on a 37°C silica column ($\lambda = 280$ nm or 276 nm, ex; 326 nm, em). A complex 70-min $82/14/4 \rightarrow 10/86/4$ dichloromethane/methanol/(1/1 acetic acid/water) gradient generated good resolution and peak shapes [783]. Cinnamon and peanut skin extract chromatograms showed the presence of dimers through pentamers. Electrospray MS was used to confirm peak identity.

Procyanidins from cocoa were treated under conditions mimicking gastric juices and analyzed for their rate of decay. Dimers through pentamers were monitored using a silica column ($\lambda = 230$ nm, ex; 310 nm, em) and a complex 60-min $82/14/4 \rightarrow 10/86/4$ dichloromethane/methanol/water (2% acetic acid) gradient [784]. However, elution of the pentamer occurred in <40 min. All analytes were well resolved from one another.

Plant growth regulator (G_1, G_2, and G_3) substances were extracted from eucalyptus leaves and baseline resolved on a silica column ($\lambda = 240$ nm) using a 65/35 DCM/iso-octane mobile phase [785]. The peaks eluted from a 20 μL injection of a 2.4 mg/10 mL purified extract were excellent and readily detected. Elution was complete in <8 min.

Sotolon, a flavor component in French sherry, was analyzed using a diol column ($\lambda = 232$ nm) and a 60/40 DCM/hexane mobile phase. Soloton eluted at ~17 min and was nearly baseline resolved from a large number of co-extracted analytes [786]. Results from various samples were tabluated with levels down to 3 μg/L reported.

Four paspalitrem-type tremorgenic mycotoxins (paxilline, paspaline, paspalinine and paspalicine) extracted from molds were baseline resolved on a silica column ($\lambda = 230$ nm) using a 90/10 DCM/ethyl ether mobile phase. Peak shapes were excellent and elution was complete in 5 min. Linear working curves from 2 to 300 ng/μL and detection limits of 0.5 μg/mL were reported [787].

Thirty-six mycobacteria species were characterized by their mycolic acid profiles (as the *p*-bromophenacyl esters) using a 10-min $98/2 \rightarrow 70/30$ methanol/DCM gradient on a C_{18} column ($\lambda = 260$ nm). A set of 276 samples were analyzed, characterized, and identified with >91% accuracy [788]. Even though the chromatograms were very complex, clearly defined sets of related compounds were eluted using this method, allowing a multivariate analysis of the chromatograms to be developed that increased the accuracy of the identification procedure to a >95% level.

Sotolon

Paxilline

Vanadium(V) extracted from clam tissue was determined as its N-phenylbenzo-hydroxamic acid (PBHA) complex on a silica column ($\lambda = 430$ nm) using a 97/3 chloroform/methanol (0.9 mM PBHA) mobile phase [789]. Absorption spectra are presented for the PBHA–V(V) complexes in chloroform/methanol mobile phases ranging from 5% to 30% methanol. Choice of mobile phase is important not only because of its effects on the spectral characteristics of the complex, but also because of its effects on complex stability. Linear concentration curves were obtained up to 200 µg/L, with detection limits reported as 8 µg/L.

6.5.5   Summary

Chlorinated solvents are very beneficial when analyses of solutes covering a wide range of solubilities need to be chromatographically determined in one run. Gradients from water-rich to strictly organic solvents (or vice versa) are particularly effective. Chloroform and DCM have been used routinely as mobile phase constituents for carotenoid, retinoid, and tocopherol analyses.

6.6   AMINO ACID AND PEPTIDE ANALYTES

Most proteins have only a limited solubility in chlorinated solvents and so chlorinated solvents are not generally used for the analyses of these compounds.

Diastereomeric mixtures of cyclic thiodipeptides (cyclothioalaninethioalanine, cyclothioalaninethioproline) were resolved as their 1-fluoro-2,4-dinitrophenyl-5-L-alanine amides on a silica column ($\lambda = 250$ nm, 254 nm, 270 nm, 279 nm, or 281 nm) using a 98/2 DCM/ethyl acetate mobile phase [790]. The separation, when scaled up to semipreparative levels, required a 95/5 DCM/ethyl acetate mobile phase. Capacity factors of <3.0 and separation factors greater than 1.2 were achieved.

The t-butyl ester derivatives of the d- and l-N-acetyl amino acids, S- benzylcys-teine, N-t-butyltryptophan, phenylglycine, and phenylalanine were well resolved ($\alpha > 1.3$) with k' values <2.5 using 70/30 hexane/DCM or 70/30 hexane/chloro-form mobile phases on a (N-formyl-L-valinylamino)propyl bonded phase ($\lambda = 230$ nm). The use of aprotic mobile phase modifiers, instead of the more commonly used alcohols, resulted in greater separation between enantiomers (as determined by

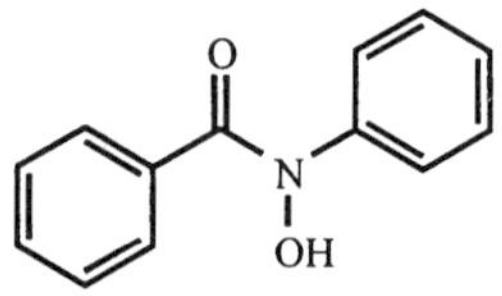

N-Phenylbenzohydroxamic acid

increased α values). The authors attribute this to the strongest part of these enantiomeric separations being the result of solute/stationary phase hydrogen bonding [716].

The p- and n-conformers of the Cys^3- and Cys^7-N-methylated analog of triostin A were studied using a silica column ($\lambda = 239$ nm) and a 12/7/1 chloroform/ acetonitrile/IPA mobile phase [791]. The equilibrium between the *cis* and *trans* positions of two methylcysteine groups of the octadepsipeptide result in the formation of two distinct peaks with the n-conformer eluting first. As the temperature is raised from 15°C (two resolved peaks eluting at 3 and 4 min) to 50°C, the peaks merge into one symmetrical peak at just <4 min retention. From these results, computer simulation leads to conversion rates of $k_{n-p} = 0.04/s$ and $k_{p-n} = 0.02/s$ at 25°C. Whenever unexpected peak splitting is obtained chromatographically, it must be determined whether the splitting is a result of inappropriate chromatographic conditions, column or sample load problems or possible solute comformer equilibria. It should be noted that similar peak splitting was also seen in the RP mode and will be discussed in the acetonitrile section.

Ten peptide fractions from the cyanogen bromide cleavage of bacteriorhodopsion were resolved on an aminopropyl column ($\lambda = 272$ nm) using a 45-min $7/1/1 \rightarrow 1/7/1$ chloroform/methanol/isopropylamine gradient [792]. Peak shapes were good and separation was adequate.

To summarize, chlorinated solvents have received limited use in amino acid and peptide work, mostly because of the very limited solubilities of the analytes in these solvents. However, many of these compounds may be derivatized prior to injection to lower detection limits and increase method sensitivity. Often these reagents are nonpolar fluorescent molecules whose fluorescent quantum yield is dramatically decreased by the presence of heavy atoms (i.e., chlorine). Hence chlorinated solvents will probably always have limited use for this class of compounds.

6.7 PHARMACEUTICAL ANALYTES

The most consistent and common use of 1-chlorobutane has been as a major component in the United States Pharmacopoeia analyses of steroidal compounds. Table 6.5 shows the analytes and conditions, the vast majority being steroidal in nature [590]. It should be noted that the number of analyses utilizing these solvents has decreased as they are replaced using routine RP solvent systems. The mobile phases are not simply water-saturated. Rather, the water content of the mobile phase (and therefore the amount of water on the silica surface in equilibrium with the mobile phase) is strictly controlled by mixing portions of water-saturated mobile phase constituents (i.e., 1-chlorobutane or DCM) with "dry" solvent. It should be emphasized that for 1-chlorobutane, a 50% water-saturated solution is $\sim$400 ppm in water. Methanol and THF have normal water levels in the 50–1000 ppm range. Therefore, for truly consistent results not only should the water level in 1-chlorobutane be well established but the water level in the other mobile phase constituents must be monitored as well.

Maytansine (an antileukemia agent) and nine maytansinoid homologs (e.g., maytanbutine, maytanprine, maysenine, maysine, maytanacine, normaysine) were separated on a silica column ($\lambda = 254$ nm) using a 95/2.5/2.5 dichloromethane/dioxane/IPA mobile phase [793]. Incomplete separation of compounds resulted, partly due to the very broad peaks. A small amount of strong acid (e.g., TFA) may sharpen peaks considerably. Elution was complete in 18 min.

Quinine was quantitated from serum, plasma, and red blood cell samples [794]. A silica column ($\lambda = 254$ nm) and a 100/9/0.4 DCM/methanol/water (1 M perchloric acid) mobile phase were used. Elution was complete in <15 min and detection limits of <250 ng/mL were reported. The linear concentration range was 0.25–20 µg/mL. The k' values for potential interferents were tabulated (e.g., quinidine, dihydroquinine, chloroquine, primaquine, dapsone, pyrimethamine). Further discrimination between other antimalarial drugs and quinine was achieved through the use of fluorescence detection ($\lambda = 350$ nm, ex; 418 nm, em). The authors noted that an increase in either the DCM or the perchloric acid concentration led to an increase in the k' for quinine.

Five kanamycin analogs were separated from kanamycin on a silica column ($\lambda = 254$ nm). A 35-min 95/5 → 65/35 chloroform/ethyl acetate gradient gave good peak shapes and resolution [795].

Whereas chloroform [722, 723] and 1,2-dichloroethane [733, 739] have been used extensively as mobile phase modifiers in the separation of enantiomers, they can also be used as the major mobile phase constituent. For example, chloroform- and DCM-based mobile phases were used to separate the enantiomers of the N-3,5-dinitrobenzoyl-α-amino phosphonates of metoprolol, oxprenolol, propanolol, pronethalol, pindolol, and bufuralol [796]. An (R)-N-(3,5-dinitrobenzoyl)phenylgly-

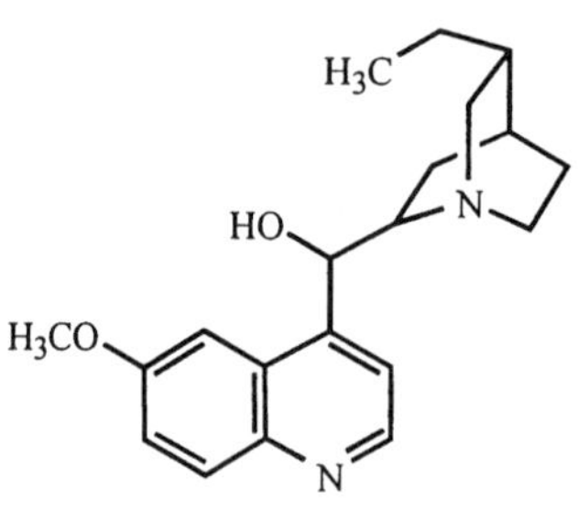

Dihydroquinine

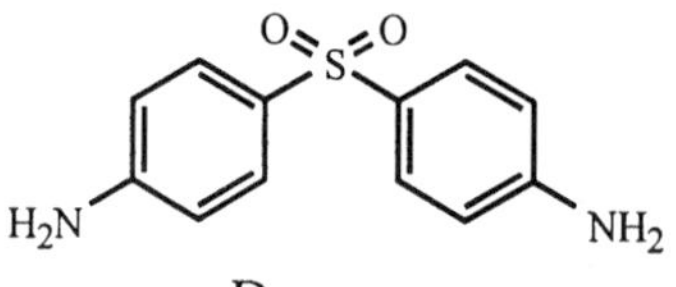

Dapsone

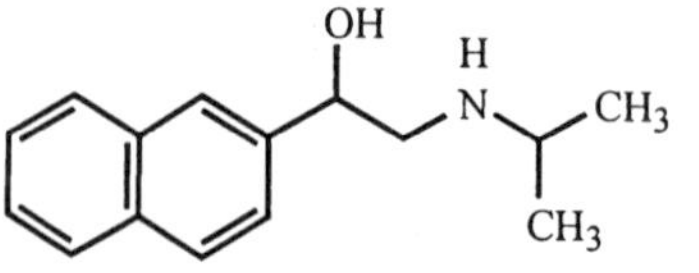

Pronethalol

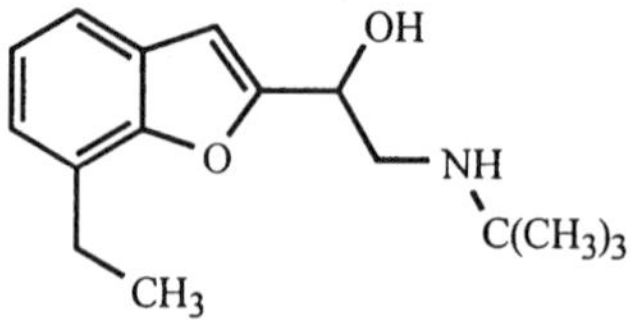

Bufuralol

TABLE 6.5 USP Methods[a]

Analyte(s)	Co-analyte (Internal Standard)	Column	Mobile Phase	Detector (nm)	USP Page
Diflorasone diacetate	Isoflupredone	Silica	350/125/15/10 1-chlorobutane/dichloromethane (water saturated)/acetic acid/THF	254	557
Ergocalciferol (oral solution)	Preergocalciferol	Silica	Chloroform (alcohol-preserved)	254	653
Hydrocortisone	Prednisone	Silica	890/56/28/24/0.4 1-chlorobutane/THF/methanol/acetic acid/water	254	823
Hydrocortisone acetate	Prednisone	Silica	475/475/70/35/30 1-chlorobutane/1-chlorobutane (water-saturated)/THF/methanol/acetic acid	254	828
Medroxyprogesterone acetate (injectable suspension)	Progesterone	Silica	700/300/80 1-chlorobutane (water-saturated)/hexane (water-saturated)/acetonitrile	254	1029
Methylprednisolone	Prednisone	Silica	475/475/70/35/30 1-chlorobutane/1-chlorobutane (water-saturated)/THF/methanol/acetic acid	254	1090
Prednisolone	Betamethasone	Silica	95/95/14/7/6 1-chlorobutane/1-chlorobutane (water-saturated)/THF/methanol/acetic acid	254	1380
Prednisone		Silica	98/2 chloroform/methanol	254	1387
Sulindac (tablets)		Silica	38/5/1 chloroform/ethyl acetate/acetic acid	332	1582

[a] Reference [590].

cine column ($\lambda = 254$ nm) was used. The mobile phase was either 19/1 chloro-form/ethanol (0.5 g/L ammonium acetate) or 19/1 DCM/ethanol (0.5 g/L ammonium acetate). System temperature was set at 21°C, 0°C, and −24°C. For all pairs of enantiomers, the separation (α) increased significantly as the temperature decreased. The k' and α values are tabulated for the DCM/ethanol mobile phase.

Phenolphthalein and aloin levels in tablet formulations were quantitated using a silica column ($\lambda = 249$ nm) and an 88/12/0.5 chloroform/methanol/acetic acid mobile phase [797]. Good peak shapes resulted and elution was complete in 3 min! A potential problem with this analysis is that the phenolphthalein elued very near the void volume. Any baseline disturbance due to the injection could therefore create difficulties for reproducibly quantitating phenolphthalein. A plot of k' versus percent chloroform (from 80% to 96%) gives the reader excellent information as to the choice of a mobile phase that would be more likely to generate a robust method.

In summary, chlorinated solvents have been effectively used in the separation of enantiomers. Their excellent solubilizing character makes it possible to use them with solutes having a wide range of polarities. These solvents also have been effectively used in the analysis of steroids and steroidlike compounds. Although the use of chlorinated solvents as main mobile phase constituents is decreasing, there are many areas in which they are effectively used as mobile phase modifiers to enhance selectivity or increase solubility of analytes.

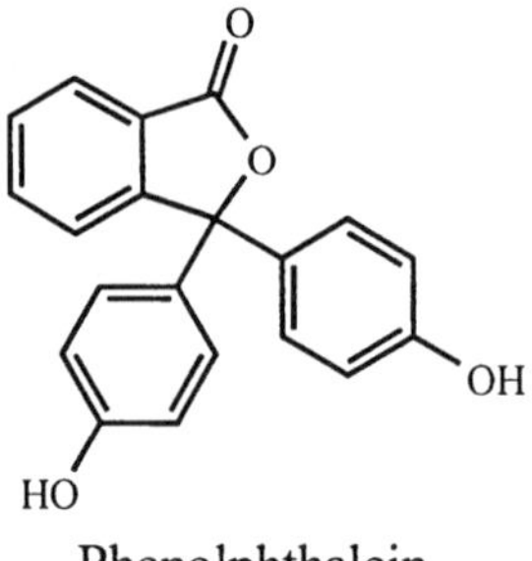

Phenolphthalein

7

ETHERS

Ethers are solvents of low to moderate polarity. They are proton acceptors and in general have dipole moments much lower than those of alcohols. Ethers readily solubilize nonpolar to moderately polar solutes. Ethers also run the gamut of solubility with water: THF and dioxane are miscible, whereas ethyl ether and methyl t-butyl ether (MtBE) are immiscible. THF and dioxane are therefore commonly used as the organic component in a binary solvent for RP separations. Ethyl ether and MtBE are used as the polar constituents in NP separations or in ternary aqueous solvents (with a mutually miscible third component such as IPA) for RP separations.

Ethers tend to be very reactive, and the less stable ones (ethyl ether and THF) decompose readily to form peroxides and more highly oxidized compounds. The decomposition process is accelerated by the solvent being exposed to light, heat, and/or air.

Ethyl ether can be purchased unpreserved, with a 1–2% ethanol stabilizer, with ~1 ppm of the antioxidant BHT, or with a combination of both ethanol and BHT. Note that the presence of even small quantities of ethanol greatly alters the eluotropic strength of the NP mobile phase, whereas the addition of BHT swamps a UV detector set at $\lambda < 280$ nm. These additives drastically modify the silica surface and thereby change retentivity of any compound and the resolution between compounds.

Anhydrous ethyl ether may also be used as an LC solvent. In NP work, the difference between "anhydrous" (<50 ppm water max.) and "HPLC grade" (<300 ppm water max.) is enough to markedly alter the chromatographic behavior of the column. Therefore, when a NP method is developed using ethyl ether, the exact specifications for the ethyl ether should be documented.

Ethyl ether is not a commonly used RP solvent. When it is used, as stated above, it is almost invariably present with water and a solubilizing solvent, often IPA. Nevertheless, even in RP separations the type and level of stabilizer in the ethyl ether is important and to guarantee rugged and reproducible chromatographic results the specifications for the ethyl ether need to be documented.

If unpreserved ethyl ether is used as a mobile phase constituent, it must be routinely tested for the presence of peroxides. A visual ferrous ammonium sulfate test is conducted as follows [798]: 5 mL 1% ferrous ammonium sulfate (freshly prepared), 0.5 mL 1 N sulfuric acid, and 0.5 mL 0.1 N ammonium thiocyanate are mixed (and decolorized with a trace of zinc dust if necessary) and shaken with an equal quantity of the ether to be tested. If peroxides are present, a red color develops. This method is significantly more sensitive than the KI test described in Chapter 1. However, improper storage can drastically increase the rate of peroxide formation and present a safety hazard to the user. This is especially true when pre- or postanalysis sample concentration steps involve ether (e.g., extractions or preparative LC).

The decomposition products of ethyl ether (peroxides) are very reactive. Because of this reactivity, it should be remembered that the concentration of any percent level of mobile phase constituent is usually orders of magnitude larger than the concentration of the injected analyte. Therefore, even if only a small fraction of the solvent decomposes, the resulting concentration of decomposition product can rapidly reach and exceed the concentration of the analyte. When solvent/analyte reactions do occur, the observed chromatographic result ranges from a shifted retention time and altered detector response factor, to tailed and/or split peaks, to total loss of analyte.

M*t*BE is quite stable and therefore forms peroxides very slowly. This fact has led chromatographers to start using M*t*BE more frequently as an organic mobile phase constituent. The major drawback to M*t*BE is that even the highest purity available contains a plethora of low-level low molecular weight contaminants. These contaminants include aldehydes, ketones, and carboxylic acids. These compounds often have a strong effect in NP work, cause elevated absorbance levels in low-wavelength UV detection work, and influence the background in fluorescence analyses.

THF presents similar stability problems to those described for ethyl ether. However, in the case of THF, BHT is almost exclusively used as a preservative. The normal ~250 ppm level of BHT present in the THF do not usually affect the chromatography as much, only because the UV detector instantly shows off-scale absorbance levels once the preserved THF enters the detector cell. Obviously, RI detectors and ELSDs are less affected, but if preserved THF is used in NP work the same modification of the silica surface will occur as described above for ethyl ether.

Unpreserved THF poses the same set of stability problems as does unpreserved ethyl ether. Peroxides will form even with the presence of a preservative. The rate of formation is dependent upon storage conditions and the time between manufacture and use. One issue not noted above is that the ethers are typically packaged under nitrogen. The nitrogen is used to remove oxygen from the manufacturing and packaging system, thereby preventing oxygen from reacting with the ethers to assist in the formation of peroxides. Once the bottle is opened, oxygen is reintroduced into the solvent and rapid formation of peroxides begins again.

As noted above, ethers are typically used as one of two or more components in a mobile phase, so once the mixture has been made, the potential for significant levels (in terms of safety hazards) of peroxides to build up in solution is small. Regardless of this, it is extremely important that the samples and injection solvents are compatible with ethers, primarily for safety reasons, but beyond that to ensure rugged and reproducible chromatographic results as noted above.

Ethers tend to have high vapor pressures, especially ethyl ether and M*t*BE. Work with these solvents should always be done in a well-ventilated area or under a hood. Ethers are extremely flammable and have low flash points. Care should be taken to keep them away from open flame or spark conditions. This also means that when a large volume of these solvents is transferred, both the container and receiver must be grounded.

Ethyl ether and M*t*BE have very low viscosities that could lead to instrumentation problems if these solvents are used in conjunction with other low-viscosity solvents (such as hexane). The problem of poorly seating check values and cavitation is more likely under this condition and therefore a flow restrictor may be needed to correct the problem.

The viscosity problems are the opposite for THF and dioxane. When mixed with water for RP separations, these mixtures generate viscosities over twice that of pure water (i.e., >2 cP). Therefore, gradients using significant volume percentages of THF ($>30\%$) need to be monitored so that backpressure limits are not exceeded.

THF and dioxane have been used in the separation of low-polarity to moderately polar analytes (retinols, steroidal compounds, etc.) These solvents have been shown to generate superior selectivity for those low polarity solutes that have accessible $-OH$ groups. It should be noted that dioxane is a known carcinogen and should be handled appropriately.

Tables 7.1–7.4 list some important chemical, physical, and chromatographic properties as well as general manufacturing and safety parameters for the ethers [84–92]. Figure 7.1 shows the structures for those solvents listed in Tables 7.1–7.4.

7.1 IMPURITIES

The most important fact that needs to be conveyed about ether compounds has been raised numerous times already: they form peroxides, which, at a high enough concentration, can detonate. A typical reaction scheme for peroxide formation is shown below for ethyl ether [799]:

$$H_3CH_2C \diagup \!\!^{O} \!\!\diagdown CH_2CH_3 + O_2 \longrightarrow H_3CHC \diagup \!\!^{O} \!\!\diagdown_{OOH} CH_2CH_3 \tag{7.1}$$

From the hydroperoxide, peroxides, acids, aldehydes, and numerous adducts may form. Alcohols, such as ethanol in the case of ethyl ether, are also present as a result of the initial manufacturing and purification process.

TABLE 7.1 Physical Properties of Ether Solvents[a]

	EE	IPE	M*t*BE	THF	Diox	2ME
Molecular weight	74.12	102.18	88.14	72.11	88.11	76.10
Density (g/mL)	0.7133	0.6433	0.741	0.888	1.0336	0.9646
Viscosity (cP)	0.24	0.33	0.27	0.55	1.37	1.72
Solubility in water (%)	6.89	0.94	4.8	100	100	100
Water solubility in solvent (%)	1.26	0.55	1.5	100	100	100
Boiling point (°C)	34.55	68.5	55.2	66.0	101.32	124.6
Melting point (°C)	−117.4	−85.5	−108.6	−108.5	11.80	−85.1
Refractive index (n_D)	1.3524	1.3682	1.3689	1.4072	1.4224	1.4021
Dielectric constant	4.33	3.95		7.58	2.25	16.93
Dipole moment (D)	1.15	1.13	1.32[b]	1.75[b]	0.00	2.36
Surface tension (dyne/cm)	17.1	17.8	19.4[c]	26.4[b]	33.7	31.3

Abbreviations: EE, ethyl ether, diethyl ether; IPE, isopropyl ether, diisopropyl ether; M*t*BE, methyl *t*-butyl ether, *t*-butyl methyl ether; THF, tetrahydrofuran; Diox, 1,4-dioxane; 2ME, 2-methoxyethanol, methyl cellosolve, ethylene glycol monomethyl ether.
[a] All values except boiling and freezing points at 20°C unless otherwise noted.
[b] At 25°C.
[c] At 24°C.

7.2 GENERAL ANALYTES

7.2.1 Simple Substituted Benzenes and Related Analytes

Ethers have been very effectively used as low-percentage mobile phase constituents in both NP and RP separations. Unique selectivity was generated in RP by the ether

TABLE 7.2 Chromatographic Parameters of Ether Solvents[a]

	EE	IPE	M*t*BE	THF	Diox	2ME
Eluotropic strength $\varepsilon°$ on Al_2O_3	0.38	0.28	0.3–0.4	0.45	0.56	0.74
Eluotropic strength $\varepsilon°$ on SiOH	0.43		0.48	0.53	0.51	
Eluotropic strength $\varepsilon°$ on C_{18}				3.7	11.7	
Solvent strength parameter, P'	2.8		2.5	4.0	4.8	5.5
Hildebrandt solubility parameter, δ	7.4	7.0		9.0	10.0	11.4
Hydrogen bond acidity, α	0.00	0.00		0.00	0.00	
Hydrogen bond basicity, β	0.47	0.49		0.55	0.47	
Dipolarity/polarizability, $\pi*$	0.27	0.27		0.58	0.27	0.71

Abbreviations: EE, ethyl ether, diethyl ether; IPE, isopropyl ether, diisopropyl ether; M*t*BE, methyl *t*-butyl ether, *t*-butyl methyl ether; THF, tetrahydrofuran; Diox, 1,4-dioxane; 2ME, 2-methoxyethanol, methyl cellosolve, ethylene glycol monomethyl ether.
[a] All values except boiling and freezing points at 20°C unless otherwise noted.

TABLE 7.3   Common Manufacturing Quality Specifications of Ether Solvents[a]

	EE	IPE[b]	M/BE	THF	Diox	2ME
UV Cutoff (nm)	215		210	212	215	210
Percent water (maximum)	0.03	0.2	0.02	0.03	0.05	0.08
Available as ACS tested[c]	ABFJM	JM	n.a.[d]	ABFJM	AJFM	A
Available as HPLC-grade[c]	ABJFM		ABEFJM	ABEFJM	ABE	AB
Available through[e]		AF				FJM

[a] *Abbreviations*: EE, ethyl ether, diethyl ether; IPE, isopropyl ether, diisopropyl ether; M/BE, methyl *t*-butyl ether, *t*-butyl methyl ether; THF, tetrahydrofuran; Diox, 1,4-dioxane; 2ME, 2-methoxyethanol, methyl cellosolve, ethylene glycol monomethyl ether.
[b] Stabilized with hydroquinone.
[c] Manufacturer code: A = Aldrich; B = Burdick & Jackson; E = EM Science; F = Fisher; J = JT Baker; M = Mallinckrodt.
[d] No ACS test exists for this solvent.
[e] Available as a high-purity solvent but not specifically designated as ACS or HPLC grade. This does *not* mean a lesser quality solvent, just that it is not specifically tested for these applications. If the manufacturers produce either ACS or HPLC solvent, they are not listed under this heading.

acting as a hydrogen bond acceptor to the solute. In NP separations this type of interaction of the ether with the surface, in particular the silanol group on the silica surface, was critical. Beyond this, how the sample solvent affects chromatographic results can be just as important as is noted below.

Hydroquinone and the 4-methoxyphenol-, 3-ethoxyphenol-, and 4-benzyloxy-phenol ethers were extracted from cosmetic formulations and analyzed on a phenyl column ($\lambda = 295$ nm) using a 45/55 THF/water mobile phase. Elution was complete in 9 min. Baseline resolution and excellent peak shapes were obtained. This is an interesting result since THF is known to produce peak splitting at elevated levels for some analytes. Standard curves ranging from 50 to 400 mg/L were used. Detection limits of $\sim$2 mg/L were reported [800].

Dzido and Engelhardt [117] noted that a combined THF interaction with the solutes (chalcones for Walczak and phenols, alcohols, ketones, and nitro compounds for Dzido) in tandem with a THF interaction with the stationary phase yielded an apparent solvent strength weaker than expected on RP supports. Whether this is due to the THF creating a more "structured" and less accessible bonded phase, to the presence of enriched (compared with the mobile phase THF concentrations) pockets of THF on the surface covering residual silanol group, or to the partitioning of a THF/solute associated pair, or to a combination of all three is not known. What is important is that this effect, while not unique to THF, is an important and common result when THF is used.

Ethyl ether can also produce such effects. Phenol and *o*-, *m*-, and *p*-nitrophenols were baseline resolved with excellent peak shapes on a silica column ($\lambda = 270$ nm) using a 93/6/1 hexane/ethyl ether/IPA mobile phase [603]. Neither separation nor

TABLE 7.4 Safety Parameters of Ether Solvents[a]

	EE	IPE	M*t*BE	THF	Diox	2ME
Flash point[b] (TCC)(°C)	−45	−28	−27	−14	12	39
Vapor pressure (Torr @ 20°C)	442	148	240[c]	142	29	6.2
Threshold limit value (ppm)	400			200	25	25
CAS number	60-29-7	108-20-3	1634-04-4	109-99-9	123-91-1	109-86-4
Flammability[d]	4	3	3	3	3	2
Reactivity[d]	1	1	0	1	1	0
Health[d]	1	1	2	2	2	2

Abbreviations: EE, ethyl ether, diethyl ether; IPE, isopropyl ether, diisopropyl ether; M*t*BE, methyl *t*-butyl ether, *t*-butyl methyl ether; THF, tetrahydrofuran; Diox, 1,4-dioxane; 2ME, 2-methoxyethanol, methyl cellosolve, ethylene glycol monomethyl ether.
[a] All values except boiling and freezing points at 20°C unless otherwise noted.
[b] TCC = TAG closed cup.
[c] 25°C.
[d] According to National Fire Protection Association ratings [92]:
Fire: 4 = Materials that vaporize at room temperature and pressure and burn readily.
   3 = Liquids or solids that can ignite under room conditions.
   2 = Materials that ignite with elevated temperature or with moderate heat.
   1 = Materials that must be preheated before they ignite.
   0 = Materials that will not burn.
React: 4 = Materials that, by themselves, can detonate or explode under room conditions.
   3 = Materials that can detonate or explode but require an initiator (e.g., heat).
   2 = Materials that undergo violent chemical reactions at elevated temperatures or pressures or react with water.
   1 = Materials that are, by themselves, stable but that may become unstable at elevated temperatures and pressures.
   0 = Materials that are stable even under fire conditions and do not react with water.
Health: 4 = Short exposure times to these materials are lethal or cause major residual injury.
   3 = Short exposure times to these materials cause temporary and/or residual injuries.
   2 = Lengthy (but not chronic) exposure to these materials may cause temporary incapacitation and/or minor residual injury.
   1 = Materials that, upon exposure, cause irritation but only minor residual injury.
   0 = Materials that, upon exposure under fire conditions, offer no more hazard than ordinary combustible materials.

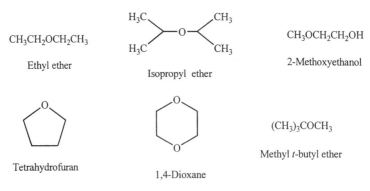

CH₃CH₂OCH₂CH₃

Ethyl ether

Isopropyl ether

CH₃OCH₂CH₂OH

2-Methoxyethanol

Tetrahydrofuran

1,4-Dioxane

(CH₃)₃COCH₃

Methyl *t*-butyl ether

Figure 7.1

good peak shapes were obtained when a hexane/alcohol mobile phase (without the ethyl ether) was used.

Dioxane and THF interactions with a silica or Florisil$^{(R)}$ surface and the resulting effects on retention were found to be markedly different [607, 608]. Whereas THF deactivated and shielded the surface through hydrogen bond formation, dioxane interacted concomitantly with the surface and the solute through its pair of diametrically opposed ether oxygens. This resulted in either retention by co-adsorption of hydrogen bond donating solutes or displacement of hydrogen bond accepting solutes. Di- and trifunctional solutes were also studied. A graphical comparison of k' values for 17 test solutes (including dimethylanilines, cresols, and methyl quinones) using three mobile phases (THF/hexane 15/85, dioxane/hexane 20/80, and IPA/hexane 5/95) clearly showed special selectivities obtained through the use of dioxane vs. THF or IPA as a mobile phase constituent. Excellent and thorough discussions of solute/solvent interactions and their effects on the retention process were presented in this work.

The retention characteristics of 23 substituted benzenes (e.g., anisole, nitrobenzene, chlorobenzene, benzonitrile, benzaldehyde) were studied on C_{18} columns [801]. The k' values were tabulated for the following mobile phase compositions: 60/40 THF/water, 35/35/30 THF/methanol/water, and 37.5/12.5/50 THF/methanol/water. From these data, a predictive retention model based on principal solute component factors was developed. Defining three factors led to a model that reproduced the analyte retention to within 2.2%.

In summary, the eluotropic strengths of the ethers, in both NP and RP modes, often preclude their use as the major constituent of a mobile phase. However, the ability of the ethers to interact with the silica surface or with solutes as hydrogen bond acceptors leads to the generation of unique and valuable selectivity and retention properties. Frequently, the addition of ethers to the mobile phase increases resolution and results in a concomitant decrease in the overall retention time. The ability of ethers to interact with the silica surface or solutes as hydrogen bond acceptors leads to unique or valuable selectivity and retention properties. Frequently, resolution is improved and retention is decreased.

7.3 ENVIRONMENTALLY IMPORTANT ANALYTES

7.3.1 PAHs, Substituted PAHs, and Related Analytes

Thirteen PAH compounds were baseline resolved on a PAH column ($\lambda = 350$ nm, ex; 430 nm, cm) using a 20-min 100/0 $\rightarrow$ 70/30 methanol/THF gradient [802]. Elution was complete after a 10-min final hold time. Detection limits ranged from 0.5 to 5 ng injected. Peak shapes were excellent.

Sander and Wise [803] reported the unique separation of the 16 EPA priority PAH pollutants on a C_{18} column in <30 min using a 20-min 80/20/0 $\rightarrow$ 80/0/20 acetonitrile/water/MtBE gradient (hold 10 min) at $-8°$C! Peak shapes were excellent and baseline resolution was excellent. The authors also used this same system to baseline resolve all six methylchrysene isomers [804].

The retention behavior of 26 azaarenes (e.g., quinoline, acridine, benz[a]acridine, quinazoline, 2,2′-biquinoline, quinoxaline, and various methyl substituted analogs) was studied on a C$_8$ column ($\lambda = 254$ nm) using a series of isocratic THF/water and dioxane/water mobile phases [805]. Although methanol/water gave the best results for the correlation of k' with molecular weight, dioxane and THF gave enhanced selectivities (as compared with methanol/water mobile phases) between analytes containing hindered and nonhindered nitrogen atoms. Interestingly, dioxane proved to be a much weaker solvent for the elution of this class of compounds than predicted from solvent strength parameters.

7.3.2 Pesticide and Herbicide Residue Analytes

A 100% THF mobile phase was used with an α-cyclodextrin column ($\lambda = 238$ nm) in the separation of the *syn* and *anti* isomers of 1,3,5-trinitroso-1,3,5-hexahydro-triazine [806]. When the system temperature was 20°C, these isomers co-eluted at 5 min in such a manner as to appear as a single chromatographic peak. At -20°C, however, a nearly complete baseline resolution of the isomers is obtained, with retention times being 6 and 7.5 min. The trade-off is the severe peak tailing that occurs at the lower temperature. Levels of the analytes were ~0.5 μg/μL.

Cserháti and Forgács [807] studied the retention of phenoxyacetic acid and a series of 10 mono-, di, and trichlorinated analogs on a porous graphitized carbon (PGC) column ($\lambda = 230$ nm). The capacity factors for these compounds were determined using isocratic mobile phases of dioxane/water ranging from 30% to 85% dioxane (in 5% increments). A linear relationship between log k' and percent dioxane in the mobile phase was generated and is useful for potential method development. Peak shapes were improved through the addition of 50 mM acetic acid to the mobile phase (which protonated the acids). The same authors [808] studied the

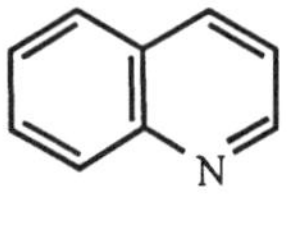

Quinoline 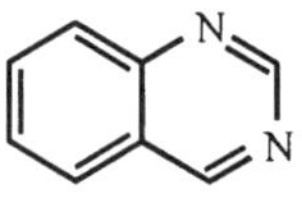   Quinazoline

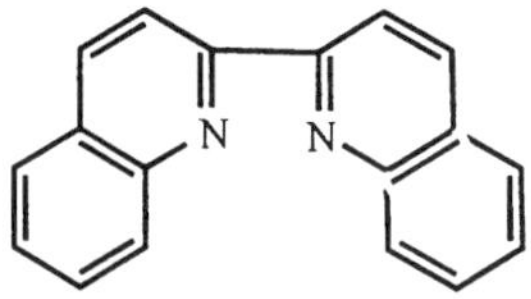

2,2′-Biquinoline

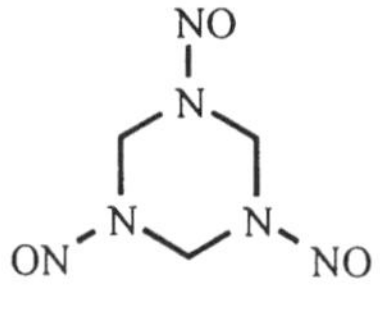

1,3,5-Trinitroso-1,3,5-tetrahydrotriazine

retention characteristics of 30 pesticides (e.g., linuron, atrazine, iodofenphos, bromoxynil, lenacil, ethofumesate, diphenamid, *p,p'*-DDT) on the PGC column ($\lambda = 240$ nm). Dioxane/water mobile phases were used. The dioxane level was varied from 65% to 90% in 5% increments. A 20 μL injection of a 0.5 mg/mL standard was used. The k' values for each solvent could be readily calculated from linear regression curves of log k' vs. percent dioxane plots.

Five organotin compounds used as pesticides or PVC stabilizers (fenbutatin oxide, triphenyltin chloride, tricyclohexyltin hydroxide, di-*n*-butyltin dichloride and diphenyltin dichloride) were baseline resolved on a cyanopropyl column ($\lambda = 430$ nm, ex; 495 nm, em) using a 96/2/2 hexane/THF/acetic acid mobile phase [809]. Detection limits of 1–2 ng injected for standards, 20–30 ng/L for surface water, and 200–300 ng/g for soils were obtained when a postcolumn irradiation/morin (3,5,7,2′,4′-pentahydroxyflavone) reaction was used.

Since ethers do not enhance the separation of non-hydrogen bond donating compounds, they do not find much use in separations of aromatic and chlorinated compounds. However, interesting applications have been developed for carboxylic acids or organotin compounds and for low-temperature work on PAHs.

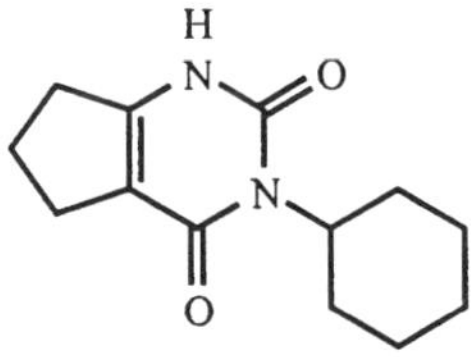

Lenacil

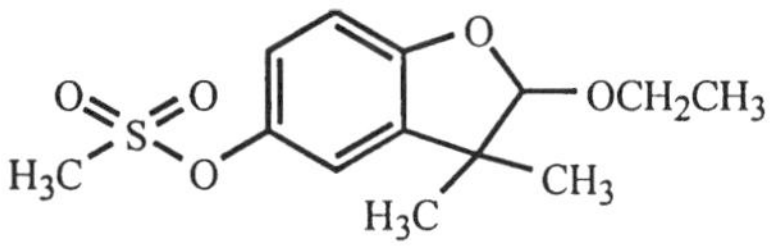

Ethofumesate

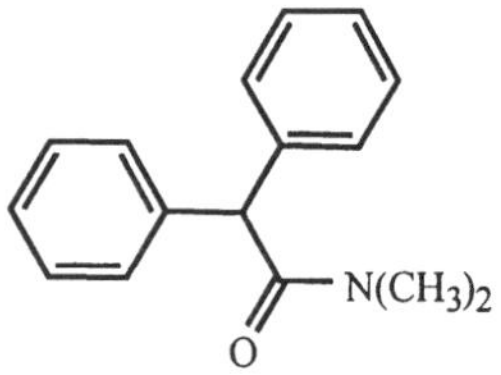

Diphenamid

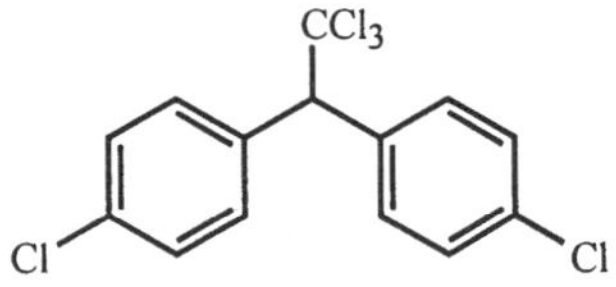

p,p'-DDT

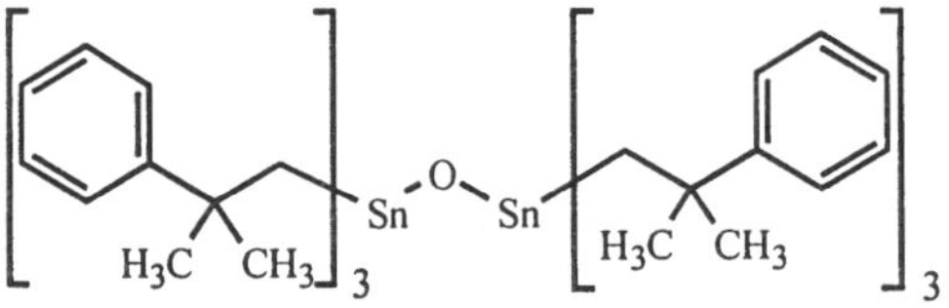

Funbutatin oxide

7.4 INDUSTRIAL AND POLYMER ANALYTES

7.4.1 Surfactant and Additive Analytes

Five polypropylene food packaging additives (1,3,2,4-di-p-methylbenzilidene sorbitol, 2,4-di-t-butylphenol, Tinuvin 326, Ultranox 626, Irganox 101, Irgafos 168) were extracted from samples and separated on a C_8 column ($\lambda = 280$ nm) and using a 35-min $40/10/10/40 \rightarrow 40/30/30/0$ THF/acetonitrile/methanol/water gradient. The authors reported detection limits for Irganox, Ultranox, and Irgafos of 5 mg/mL and linear ranges are estimated to be 10–20 times this level [810].

Although this is more of a theoretical paper presenting information of the GPC characterization of polyesters, Philipsen et al. [811] present an excellent separation of polyester PE7 components generated on a 35°C C_{18} column ($\lambda = 277$ nm) using a 50-min $35/65 \rightarrow 85/15$ THF/water with 0.2 mL/L acetic acid gradient. The sample concentration of 100 mg/mL gave absorbance readings of 0.2 to 1.0 AU readings for most of the oligomers.

Alkylether sulfate surfactants can be separated from the parent alcohol through the use of a 4-min $90/10 \rightarrow 40/60 \rightarrow 0/100$ water/THF gradient on a C_{18} column (ELSD). Detection limits of about 20 nmol for the sulfate and 5 nmol for the parent alcohol were reported [812].

A series of linear alkylbenzenesulfonates (LAS) were extracted from influent, effluent and river water samples. Baseline resolution was achieved on a C_1 column ($\lambda = 225$ nm, ex; 290 nm, em) using a 45/55 THF/water (0.1 M sodium perchlorate) mobile phase [813]. Peaks were somewhat broad but complete resolution of C_9–C_{15} analogs was achieved. Detection limits (S/N=3) were reported as 1.5 ng per component injected and 7–10 ppb total LAS (S/N=10). Elution was complete in 15 min.

7.4.2 Polymeric Analytes

Traditionally, gel permeation chromatography (GPC) or size exclusion chromatography (SEC) have been used for the characterization of polymers. THF has been a mainstay solvent in GPC because it readily solubilizes many polymers, both nonpolar and polar. The THF used in GPC is usually preserved with BHT to prevent the irreversible damage that peroxides can inflict on the poly(styrene–divinylbenzene) gels. Often BHT creates no problems with detection since many polymers have no UV chromophores and, therefore, RI detection is used. Recently,

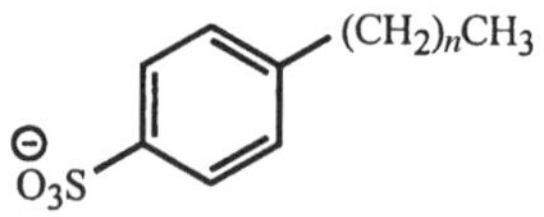

Alkylbenzenesulfonate (linear)

RP and NP separations have been used in the characterizations of polymers. Yet, the same need to solubilize the polymer remains and consequently THF has received much attention as a solvent for use in separations of this type.

Partially hydrolyzed poly(vinyl alcohol) polymer samples were separated using size exclusion chromatography and then characterized with respect to their degree of hydrolysis on a PLRPS column (ELSD, drift tube $T = 100°$C, N_2 nebulizer gas 12.5 L/min) using an 8-min 98/2 → 30/70 water/THF gradient [814]. Elution times and profiles were clearly different for each collected fraction and plots of percent hydrolysis versus retention time were presented.

Glöckner and Wolf [243] characterized a styrene/t-butyl methacrylate co-polymer ($StBM$; MW 240,000) on both C_{18} and phenyl columns ($\lambda = 259$ nm). A 12-min 90/10 → 30/70 methanol/THF gradient followed by an 8-min re-equilibration step was used for the analysis. Each injection contained 0.4 μg of $StBM$ and a polystyrene standard. The authors noted that the $StBM$ co-polymer in THF stock solution sometimes became turbid, even when preserved THF was used. Not surprisingly, elution curves changed when the turbidity was noted. No comment was made as to the possible origin of the turbidity. When gradient elution is used for polymer analyses and the initial solvent is a weak solvent, precipitation is very likely to occur. A balance between adsorption, precipitation, and redissolution needs to be struck in order to generate meaningful chromatographic elution profiles.

Glöckner et al. [815] examined this balance in the characterization of styrene/ethyl and styrene/methyl methacrylates (SEM and SMM, respectively) on a C_{18} column ($\lambda = 260$ nm). In these separations a step gradient was used from 100% acetonitrile to 20–50% THF at 0.01 min followed by a 10-min gradient to 50% heptane (with the percent acetonitrile decreasing inversely to the heptane). A 5–20 μg of sample was injected for each run. Excellent resolution of 32%, 55%, and 68% ethyl methacrylate co-polymers was achieved in 8 min.

The chemical compositional distribution for poly(methyl methacrylate)-*graft*-polystyrene was determined on a C_{18} column ($\lambda = 254$ nm) using 100 μL injections of 300 μg/mL samples [816]. Elution profiles were generated using a complex 21-min 90/10 → 0/100 acetonitrile/THF gradient with a 10-min hold and 15-min re-equilibration step. Molecular weights of 86,000 to 292,000 were characterized. The authors note that the chemical composition distribution was determined by RP-HPLC, but the results had significant uncertainty when compared with theoretical computation. The same authors [817] noted that when strong adsorption or precipitation of a polymer occurs, significant residual levels of sample elute on successive gradients. Although graft co-polymers of methyl methacrylate and styrene were the focus of the study, the authors warn that this is a general phenomenon that has practical implications involving the generation of accurate polymer compositional data.

Novalac resins made from p-cresol condensation with formaldehyde were characterized on a 50°C C_{18} column ($\lambda = 280$ nm) using a complex 55-min 0/50/50 → 50/50/0 THF/methanol/water gradient. Samples were made up as 0.5% w/v solutions of the polymer (10 μL injected). The p-cresol is separated from the oligomers and excellent resolution was obtained up to the octomer [236].

Polyester resins having average molecular weights of 11,400–16,000 were characterized with respect to their elution profiles on a 25°C C_{18} column ($\lambda = 277$ nm or ELSD, 80°C, N_2 at 3 bar) using a 23-min 70/30/0.02 → 0/100/0.02 water/THF/acetic acid gradient [818]. Typical sample concentrations were 20 mg/mL.

In a review by Podzimek [819] the following polymers were all characterized on a 40–50°C C_{18} column ($\lambda = 254$ nm or 280 nm) using THF/water or THF/methanol/water gradients: epoxy resin based on bisphenol A (oligomers up to 20), epoxy resin based on bisphenol A/tetrabromobisphenol A, unsaturated polyester resin, and low and high molecular weight cresols. Analysis of extracts from epoxy resin prepregs for dicyandiamide, bisurea, N,N,N',N'-tetraglycidyl-4, 4'-diaminodiphenylmethane, and the diglycidyl ether of bisphenol A. Generally, gradients from 30–40% THF in water to 80% THF in water over 60 or 80 min were used. Injections were 10 μL of 0.3–1.5% THF standard solutions.

Sheih and Benton [820] present an excellent review of the analyses and characterizations of a series of epoxy resins. The authors note that a C_{18} column ($\lambda = 230$ nm, 254 nm, or 280 nm) and an acetonitrile/water gradient, a THF/water gradient, or a ternary acetonitrile/THF/water gradient can successfully generate the required separations. Further, acetonitrile/water mobile phases (and an approximate 40/60 → 100/0 gradient) are only compatible for use with lower molecular weight resins because of the limited solubility of epoxy resins in this solvent system. For higher molecular weight resins, THF/water (an approximate 30/70 → 100/0 gradient) is recommended. Additionally, a ternary system of acetonitrile/THF/water offers the advantages of both systems. Chromatographic results are shown for advanced solid epoxy resins (up to MW 12000, 20 oligomers), advanced phenolic resins, novolac epoxy resins, and precursors and p-(t-butyl)phenol-modified solid epoxy resins.

Many fundamental retention topics are covered in another review [821]. Examples cited included isocratic 83/17 dioxane/water and C_{18} for polystyrene oligomers; isocratic 60/40 THF/water and C_{18} for polystyrene oligomers; and 0/100 → 100/0 THF/methanol gradient and C_{18} for styrene/ethyl methacrylate co-polymers. A critical parameter often overlooked is potential injection overload that creates a temporary precipitation event on the column. This condition is similar to the instances above where the sample precipitated on the column because of its limited solubility in the mobile phase. This review presents many more specific and important details that need to be considered when the analyst is developing polymer characterization methods.

7.4.3   Fullerenes and Other Industrial Analytes

Fullerenes were separated on a C_{18} column using a series of solvents containing ethyl ether and either pentane or methanol [656]. Ethyl ether (100%) separated C_{60}, C_{70}, C_{76}, C_{82}, C_{84}, and C_{90} in <10 min. A plot of k' vs. percent ethyl ether in pentane is presented. Interestingly, a large change in mobile phase composition (from 0 to 50% ethyl ether in pentane) resulted in little or no change in the k' of the fullerenes. With methanol in ethyl ether, however, k' values for C_{60} ranged from approximately -1 (for 100% ethyl ether) to >1.5 (for 80/20 methanol/ethyl ether). These differences in retention behavior are described as being due to solubility considerations; the fullerenes have a much higher solubility in pentane or ethyl ether as compared with methanol.

Harwood and Mamantov [822] separated C_{60} and C_{70} fullerenes in 15 min on a C_{18} column ($\lambda = 254$ nm) using a 60/40 THF/acetonitrile mobile phase. The k' values for C_{60} and C_{70} of 0.5 and 0.6 using an 80/20 THF/acetonitrile mobile phase and of 2.2 and 3.8 using a 60/40 THF/acetonitrile mobile phase are given. Peak fronting occurred as the injection volume increased from 10 to 35 μL. The authors do not report what the sample concentration is, so that their conclusion that the peak shape distortion results from a solubility overload cannot be assessed. The samples are dissolved in DCM, an excellent solvent for fullerenes. Therefore, a complex "desolvation" process, perhaps coupled with the solubility effects, may be responsible for the fronting behavior. This is supported by the fact that when a 70/30 THF/water mobile phase is used (definitely a poor solvating system), peak splitting (not fronting) is observed and so complex chromatographic behavior is occurring.

Stearic and 12-hydroxystearic acids were extracted from lubricating greases and then separated on a C_{18} column (refractive index detector) using a 70/30 THF/water (0.1% TFA) mobile phase [823]. Elution was complete in <8 min. Four grease samples were analyzed and the chromatograms were shown. The resulting sample contents (3–13% by weight) were tabulated.

Ethers are excellent solubilizing solvents for a wide range of polymers and high molecular weight compounds. Ethers are also relatively strong solvents in both RP and NP separations. These properties coupled with their hydrogen bond accepting property often lead to unique selectivities and shorter retention times as compared with comparable alcohol replacements.

7.5 BIOLOGICAL ANALYTES

7.5.1 Carboxylic Acid Analytes

Evans and McGuffin [824] studied the effect of injecting a 1 μL aliquot of a 5×10^{-4} M 4-(bromomethyl)-7-methoxycoumarin-derivatized C_{10} fatty acid dissolved in 100% THF onto a C_{18} column eluted with a 90/10 methanol/water mobile phase. A split peak resulted. The authors could not definitively explain the effect but noted that the best solution to the potential problem is to dissolve the sample in the mobile phase itself or in a solvent that is only slightly weaker than the mobile phase. This split peak effect is not uncommon when high percent THF solute solvents are used.

Four hydroxycinnamic acids (sinapinic, caffeic, ferulic, and o-coumaric acids) from orange juice were baseline resolved on a C_{18} column ($\lambda = 324$ nm or $\lambda = 335$ nm, ex; 435 nm, em) using a 12/5/83 → 35/0/65 (at 6 min hold 22 min) THF/acetonitrile/water (2% acetic acid) gradient [825]. Fluorescent detection gave detection limits of 1 ppm. During method development, the authors found that resolution was greatest and retention shortest when the organic modifier was THF, followed in terms of decreasing efficiency by acetonitrile and then methanol. These compounds were also baseline resolved from narirutin and hesperidin, two potential HPLC interferent compounds commonly present in orange juice.

The 9-anthracenemethyl ester derivatives of aliphatic carboxylic acids (C_6–C_{20}) were determined in spent engine oils using a postcolumn bis(2,4-dinitrophenyl)oxalate/H_2O_2 chemiluminescence system [826]. A C_8 column and a complex 23-min 40/30/30 → 40/60/0 acetonitrile/THF/water gradient resulted in excellent baseline separation and peak shapes for nine such compounds. Detection limits were ~30 pmol injected with a linear concentration range of 0–4 nmol/mL reported. The authors note that methanol is incompatible with this technique because of its effective quenching of the chemiluminescence process.

7.5.2 Vitamins and Related Analytes

The retention behavior of retinol, retinal, ergocalciferol, cholecalciferol, α-, β-, γ-, and δ-tocopherol, menadione, and phylloquinone was studied on a silica column ($\lambda = 254$ nm or 292 nm) using a THF-modified hexane mobile phase [677]. A plot of k' versus percent THF in hexane (from 5% to 20%) is shown and is a good resource for method development work. THF provided superior selectivity for these solutes as compared with IPA. Conversely, IPA yielded sharper more symmetric peaks, especially for those solutes that have accessible hydroxyl functional groups.

A RP method for the determination of α-, β-, γ-, and δ-tocopherol in vegetable oils was developed by Mounts and Warner [827]. A C_{18} column (ELSD) and a 60/25/15 acetonitrile/THF/water mobile phase baseline resolved the tocopherols in

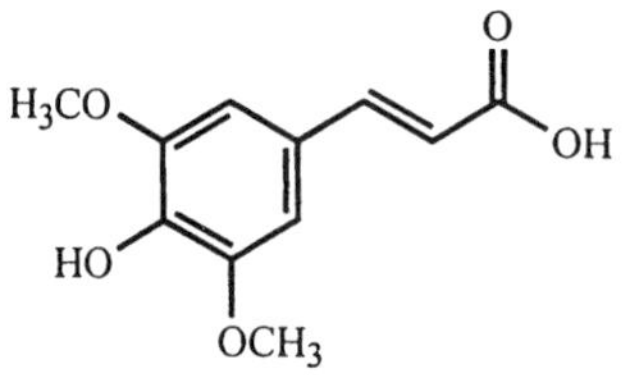

Sinapinic acid

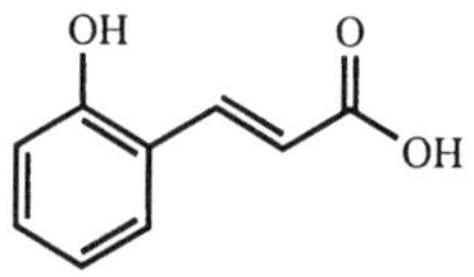

o-Coumaric acid

30 min. The authors claimed that the RP separation of β- and γ-tocopherol was "not common" and attributed their success to the use of hexane as the injection solvent. Since hexane is immiscible with the mobile phase, the authors contended that the tocopherols were deposited at the head of the column, which resulted in enhanced resolution. Care must be exercised, however, since the hexane will dynamically modify the support surface. If the system is not fully re-equilibrated between injections, the residual hexane left on the column will adversely affect the reproducibility of the technique. It should be noted that this technique yielded broad but symmetric peaks (e.g., δ-tocopherol eluted at 13 min but the peak was 3 min in width).

Besler et al. [671] studied retinal and retinol isomer retention on a silica column ($\lambda = 325$ nm or 371 nm) with hexane/MtBE (97/3 and 93/7), hexane/dioxane (93/7 and 94/6), and heptane/MtBE (94/6 and 93/7) mobile phases. The purpose of this work was to find an acceptable replacement for dioxane, namely, MtBE. The best overall chromatography resulted when 94/6 hexane/dioxane resolved 11,13-di-cis-, 13-cis-, 9,11-13-tri-cis-, 9,13-di-cis-, 11-cis-, 7,11-di-cis, 9-cis-, 7,9-di-cis, and all-$trans$ retinol in 15 min. Broader peaks but better resolution were obtained for the same solute set when 93/7 hexane/MtBE was used (elution was complete in 21 min). MtBE in hexane also gave excellent resolution of 13-cis-retinoic acid and all-$trans$ retinoic acid methyl esters on a silica column ($\lambda = 340$ nm) in <5 min, whereas dichloromethane/hexane (35/65) required 17 min and toluene/hexane (45/55) required 10 min. It was found that for the toluene/hexane mobile phase, very small changes in the water content of the system dramatically and adversely affected the chromatography. The MtBE/hexane mobile phase provided more overall stability with respect to these changes.

Carotenoids (astaxanthin, α- and β-carotene, lutein, and β-cryptoxanthin) were separated on a C_{30} column (photodiode array detector, $\lambda = 250$–600 nm in line with positive ion electrospray MS) using a 60-min $85/15 \rightarrow 10/90$ methanol/MtBE all with 1.0 mM ammonium acetate gradient [354]. To increase the MS response, a postcolumn oxidation step using a 50/50 methanol/MtBE solution containing 2% 2,2,3,3,4,4,4-heptafluoro-1-butanol was added to the effluent stream at 50 μL/min. Detection limits of 1 pmol were reported.

Sander et al. [828] separated seven xanthophylls (e.g., lutein, zeaxanthin, capsanthin, echinenone) and seven carotenoids (e.g., α-, β-, and γ-carotene, lycopene, 15-cis-β-carotene) on a C_{30} column ($\lambda = 450$nm) using a 90-min $81/15/4 \rightarrow 6/90/4$ methanol/MtBE/water gradient. Excellent resolution between all 14 components and peak shapes were obtained. The authors note that a nonaqueous methanol/MtBE gradient resolves all compounds except for the β-cryptoxanthin/echinenone pair. As noted above, the addition of 4% water resolves this pair and retains separation of all other compounds as well.

7.5.3   Terpenoids, Flavanoids, Steroids, and Related Analytes

Su et al. [829] analyzed huangqin roots extracts and analyzed for flavonoid antioxidant content. Baicalin, baicalein, ganhuangenin, ganhuangemin, wogonin, and oroxylin A were all identified. A C_{18} column ($\lambda = 275$ nm) and a 66/14.5/12.5/5/2/0.01 water/THF/dioxane/methanol/acetic acid mobile phase generated good resolution and peak shape and an elution time of 50 min. Although no standard curve information was given, spiking concentrations of baicalein ranged from 50 to 400 ppm.

Curcumin levels were determined using a C_{18} column ($\lambda = 420$ nm) with a 40/60 THF/water (1% citric acid to pH 3.0 with KOH) mobile phase [830]. A linear curve from 0.2 to 20 μM and detection limits of 5 ng/mL were reported. In the same study the concentration of tetrahydrocurcumin (THC) was determined using a C_{18} column

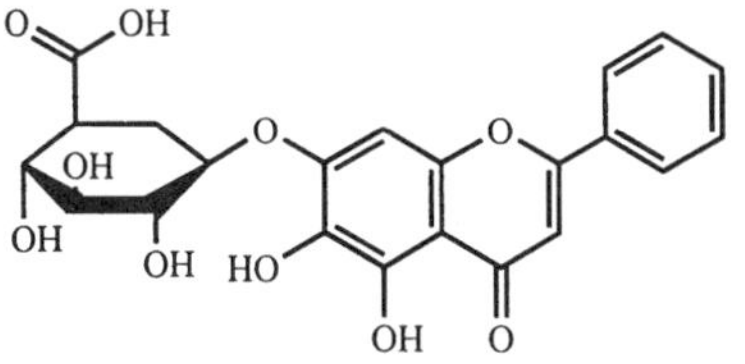

Baicalin

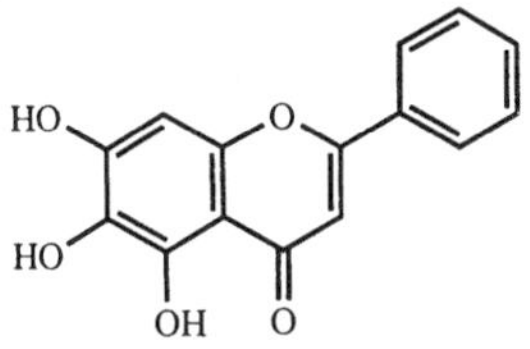

Baicalein

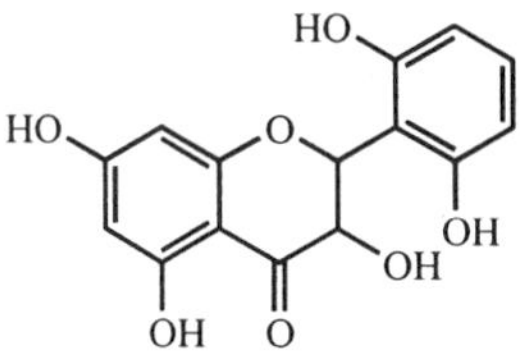

Ganhuangemin

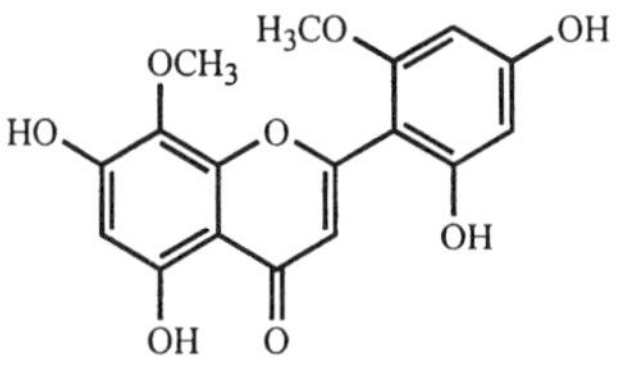

Ganhuangenin

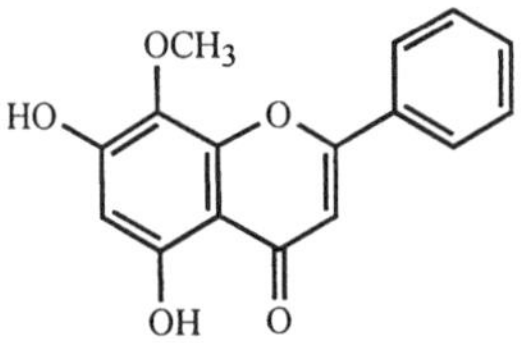

Wogonin

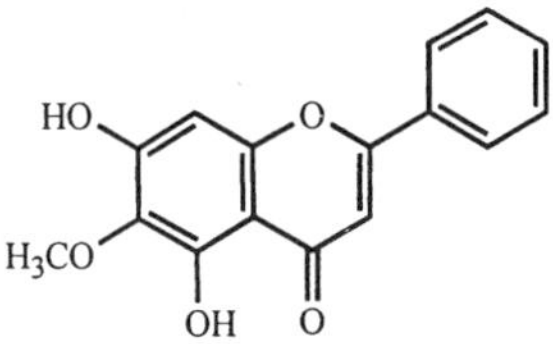

Oroxylin A

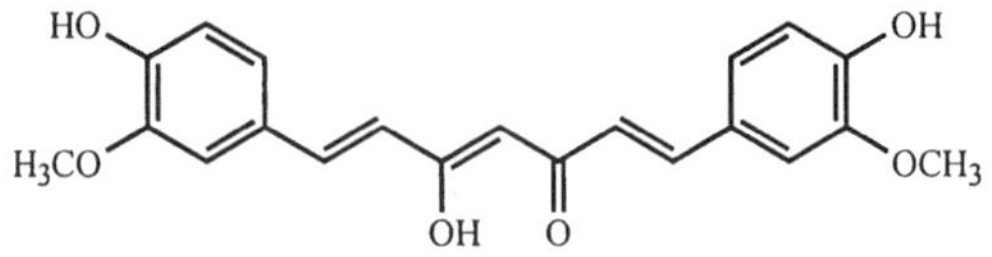

Curcumin

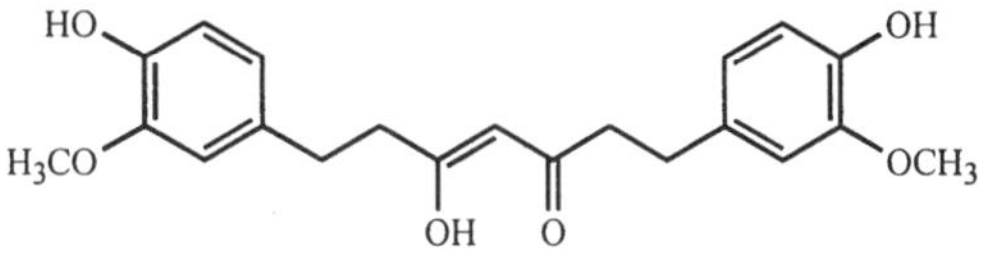

Tetrahydrocurcumin

($\lambda = 280$ nm) and a 10/30/70 THF/acetonitrile/water (1% acetic acid pH 3–7) mobile phase. The peak was broad regardless of the pH used, but the best results were at pH 3. This is most likely due to ineffective shielding of the surface from the strong hydrogen bonding β-enol–enol tautomer. Note that curcumin peak shape was excellent but that the strong hydrogen bond acceptor THF is present at high levels in the mobile phase. Finally, plasma samples were treated and the resulting dehydro-curcumin and THC as well as curcumin and THC glucuronides were resolved in 25 min with a C_{18} column ($\lambda = 280$nm) using a 10/30/70 THF/acetonitrile/water (1% acetic acid pH 3) mobile phase.

Four glycoalkaloids (α-chaconine, α-tomatine, demissine, α-solanine) were extracted from potatoes. A 15-min separation was generated on an aminopropyl column ($\lambda = 208$ nm) using a 50/30/20 THF/acetonitrile/water (25 mM KH_2PO_4) mobile phase [831, 832]. It is interesting to note that even though the detection wavelength is well below the UV cutoff value for THF, the baseline was quite stable. However, the overall sensitivity and working range are undoubtedly compromised. No working standard concentrations were cited.

Four *Aconitum* alkaloids (aconitine, mesaconitine, hypaconitine, jesaconitine) were extracted from blood and separated on a 40°C C_{18} column ($\lambda = 260$ nm for jesaconitine and 235 nm for all others). A 14/86 THF/water (0.2% TFA) mobile phase generated baseline resolution and elution in 15 min. Calibration curves from 100 to 10000 ng/mL and detection limits of 50 ng/mL were reported [833].

Seven tashinones IIA (abietane-type compounds) extracted from roots were separated on a silica column ($\lambda = 285$ nm) using a 92/8 hexane/dioxane mobile

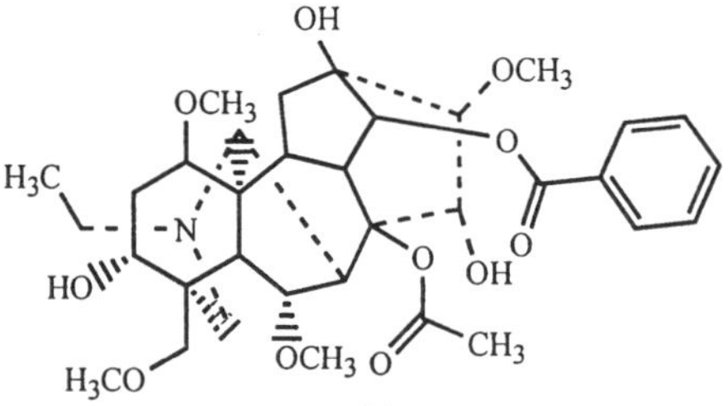

Aconitine

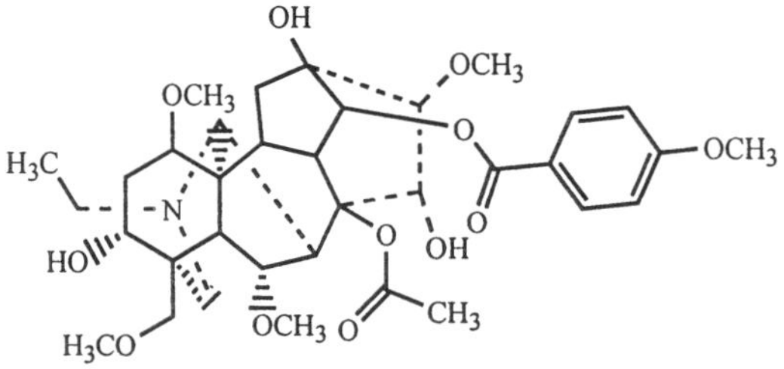

Jesaconitine

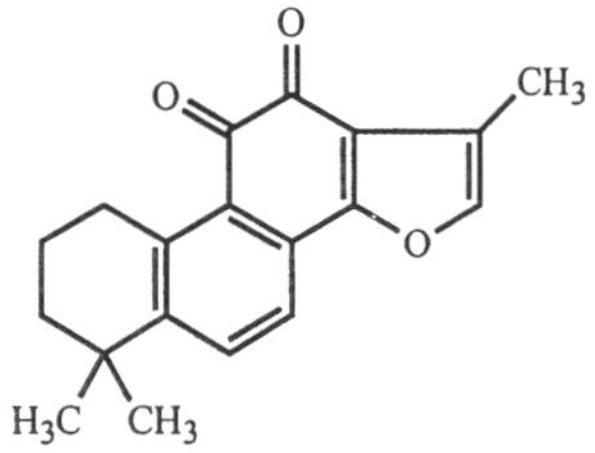

Tanshinone IIA

phase [686]. Linear absorbance versus concentration plots were obtained over the 30 pmol to 10 nmol range. Elution was complete within 20 min. Ferruginol co-eluted with an impurity and was eluted as an independent peak using the silica column and a 97.6/2.4 hexane/dioxane mobile phase. Elution was complete in 15 min and a working concentration range of 0.1–100 nmol was reported.

Thirteen kaempferol and quercetin rhamnodiglucosides (flavinol glucosides) in tea were separated as their benzoyl derivatives using a silica column ($\lambda = 260$ nm and 354 nm) and a 150/90/30 iso-octane/ethyl ether/acetonitrile mobile phase [834]. Similar underivatized compounds were also separated on a C_{18} column ($\lambda = 354$ nm) using a 10/13/77 dioxane/methanol/water (2% acetic acid) mobile phase. Again, 13 were separated with detection limits of $\sim$0.04 g/kg tea reported [835].

7.5.4 Analytes Derived from Fats and Oils

In an excellent study of neutral lipids by Foglia and Jones [836], a 100/0/0.4 (hold 5 min) → 20/80/0.4 (at 15 min hold 2 min) hexane/MtBE/acetic acid gradient was used with a cyanopropyl column (ELSD, 40°C drift tube, and $N_2 = 1.5$ L/min). Two separations are shown, one involving neutral lipid standards (cholesterol, oleate, methyl oleate, oleic acid, triolein, 1,3- and 1,2-diolein, 1- and 2-monoolein) and the other IPA-transesterified tallow separated into fractions (isopropylesters, free fatty acids, triglycerides, 1,3- and 1,2-diglycerides, and 1-monoglyceride). Area vs. concentration plots for these analytes were nonlinear over the 0–50 μg/mL range. Overall separations were excellent, as were peak shapes.

Triacylglycerols from various plant fats and oils were separated and identified using a C_{18} column ($\lambda = 215$ nm) and a 30 min 75/25 → 25/75 propionitrile/MtBE gradient [837]. Although baseline resolution was not achieved for all solute pairs, enough resolution resulted that 24 triacylglycerols from soybean oil (e.g., palmitic : oleic : palmitic to stearic : stearic : oleic) could be identified. Each injection contained 5 μg of sample. Other fats (cocoa butter) and oils (olive, palm, rape, peanut, sunflower, linseed) were also studied.

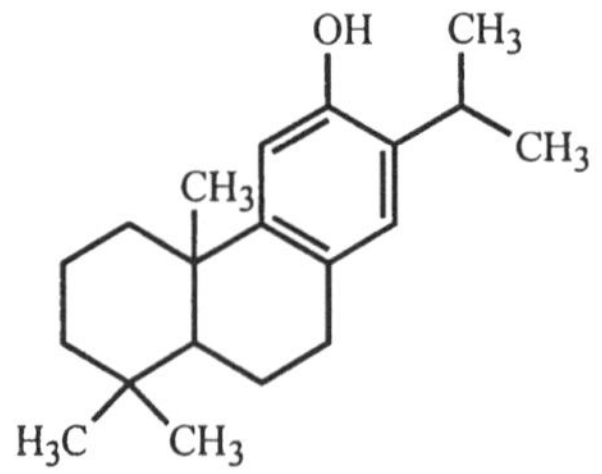

Ferruginol

M*t*BE was used as a mobile phase constituent in a system designed to separate the phospholipase-treated molecular species (phosphatidylcholine [PC] and phosphatidylethanolamine [PE]) found in the liver [838]. A C_{18} column ($\lambda = 205$ nm) and a 72/18/8/2 acetonitrile/IPA/M*t*BE/water mobile phase were used. Within the PC fraction, nine separate fatty acid fractions were separated and identified in 70 min. Similarly for the PE fraction, the fatty acids fractions were resolved in 80 min.

7.5.5 Other Analytes

Seven browning pigments from apricots were analyzed on a C_{18} column ($\lambda = 436$ nm). A complicated 108-min 99/0/1 → 55/5/40 → 0/90/10 water/acetonitrile/THF gradient was used for the separation [839]. A misidentification of the last-eluting peak (at 108 min) may have been made. This peak was considered an analyte of interest, but was considerably larger than the other analyte peaks and was quite variable in size. It should be noted that an important chromatographic event occurred in this particular analysis around the time of the last-eluting peak: the gradient reversed. This is important because reverse-gradient peaks are often generated, as the authors note, when the sample contains compounds of widely different polarities and solubilities. Also, it needs to be realized that the mobile phase loading process (i.e., the buildup of low-level contaminants on the surface during the re-equilibration time between gradients) is very complex and not well understood. However, as a diagnostic tool with all other variables held constant, the size of the peaks associated with this process are usually directly proportional to the volume of weak mobile phase that is pumped through the column between gradients.

THF was used as a low-level mobile phase constituent in the analysis of 5-dimethylaminonaphthalene-1-sulfonyl chloride derivatives of iodothyronines extracted from brain and liver tissue and serum [840]. A 40-min 28/0.5/71.5 → 90/10/0 acetonitrile/THF/water (3 mM H_3PO_4 with 0.2 mM TEA) gradient and a C_{18} column ($\lambda = 7$–51x excitation filter; 3–72M emission filter, both from Gilson) and were used for the separation. Seven monitored compounds were baseline resolved: 3-monoiodo- and 3,5-diiodo-L-tyrosine; L-thyronine; 3,5-diiodo-, 3,5,3′-triiodo-, reverse 3,3′,5′-triiodo-, and 3,3′,5,5′-tetra-iodo-L-thyronine (T_4). Peak shapes were excellent, even at the 40-min mark (T_4 elution). Standard injections of 0.5 pmol were readily detected.

Amine compounds in wine (histamine, tryptamine, tyramine, phenylethylamine, putrescine, cadaverine, 1,6-diaminohexane, and tryptophan) were quantitated as their *o*-phthalaldehyde derivatives on a C_{18} column (electrode array, with electrodes set at +80 mV, +160 mV, +240 mV and +320 mV vs. Ag/AgCl) using a complex 38-min 85/15 → 0/100 (12.5/5/82.5 acetonitrile/THF/water [0.1 M sodium acetate at pH 6.5])/([25/30/45 acetonitrile/THF/water [0.1 M sodium acetate at pH 6.5]) gradient [841]. The range in poised potentials allowed for discrimination between overlapping peaks (e.g., tryptophan and tryptamine). Detection limits were approximately 15 μg/L for all analytes.

A series of 10 coumarins (e.g., umbelliferone, xanthotoxin, imperatorin, ostruthol) were studied with respect to retention behavior using a C_{18} column ($\lambda = 320$ nm) and either 42/58 dioxane/water or 28/72 THF/water mobile phases [320]. The k' values are tabulated for the 10 analytes under these conditions. To compare solvent effects, the percentages of THF and dioxane were chosen to generate similar retention times for the last-eluting peak. It is interesting to note that an elution order reversal occurred (imperatorin and ostruthol) when changing from THF to dioxane as the organic modifier. The ostruthol has a hydroxyl functional group that the imperatorin does not. The increased probability for hydrogen bonding to the dioxane vs. THF may explain the longer retention time observed in the dioxane system.

The retention of three extracted γ-lactones (parthenolide, marrubiin, artemisinin) was studied on a silica column ($\lambda = 210$ nm and 225 nm) using hexane/dioxane (90/10 or 85/15) mobile phases to isolate these compounds from other extractants [683]. Elution times were 15 min or less. The linear concentration range was reported as 0.2–5 mg/mL.

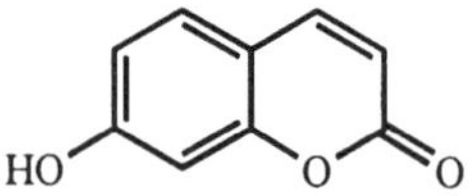

Umbelliferone

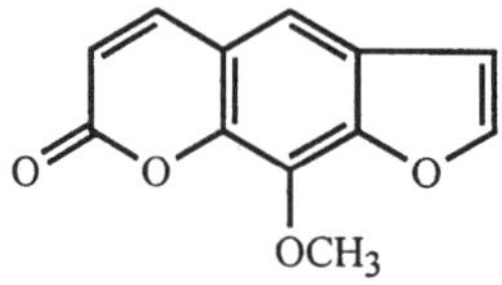

Xanthotoxin

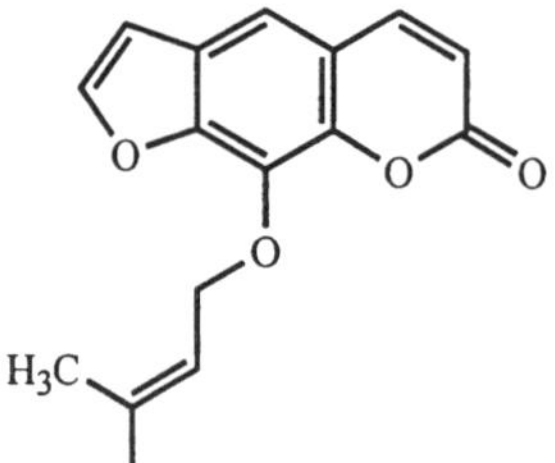

Imperatorin

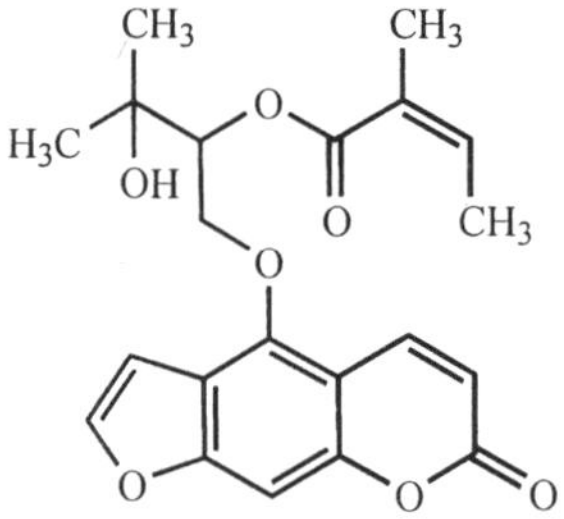

Ostruthol

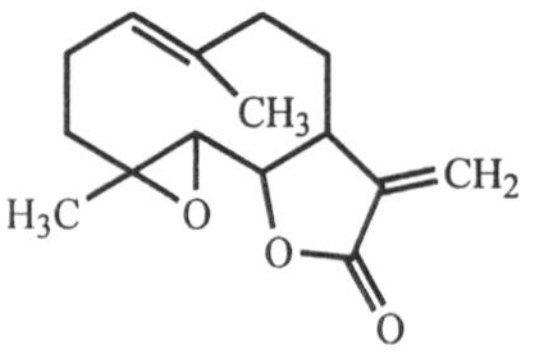

Parthenolide

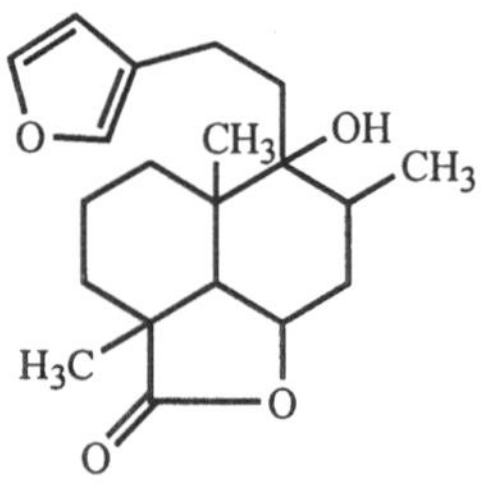

Marrubiin

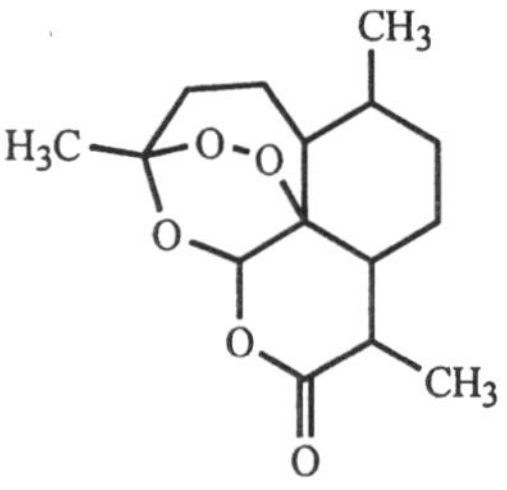

Artemisinin

Shepherd [312] compared the selectivity effects of various mobile phase organic components on the separation of aflatoxins B_1, B_2, G_1, and G_2. The column was C_{18} ($\lambda = 360$ nm) and the mobile phases were 60/40 methanol/water, 40/60 acetonitrile/water, and 20/80 THF/water. All mobile phases eluted the aflatoxins in <15 min, but the acetonitrile- and methanol-containing mobile phases gave incomplete resolution of the analytes. The THF-modified mobile phase not only baseline resolved all the aflatoxins but also gave better peak shape.

Ethers have been used to generate unique selectivities for compounds that are hydrogen bond donors (e.g., alcohols and other hydroxylated compounds). For steroidal and large low-polarity compounds (e.g., retinols, tocopherols) the hydrogen bond interaction often leads to enhanced selectivity. Coupled with the solvent strength effects of ethers, this yields resolution equivalent to or better than other solvent systems but with shorter overall analysis time.

7.6 AMINO ACID AND PEPTIDE ANALYTES

Individual sets of enantiomers of N-acetyl amino acid t-butyl esters of leucine, valine, norleucine, norvaline, alanine, isoleucine, serine, asparagine and glutamine, and O-acyltyrosine were baseline resolved on an (N-acyl-L-valyl-amino)propyl column ($\lambda = 230$ nm) using an 80/20 ethyl ether/hexane mobile phase [723]. The separation factor, α, for these separations was always between 1.2 and 1.4. The k' values ranged from 2 to 9. The authors note that a change to a protic solvent (e.g., IPA) led to a significant decrease in α, whereas similar α values were obtained with other aprotic solvents (dichloromethane and chloroform), indicating that hydrogen bond interaction between the solute and the support is a key factor in the resolution of these analytes.

When 2-methoxyethanol was used as a mobile phase constituent in the separation of insulin, cytochrome c, bovine serum albumin, myoglobin, and ovalbumin on a diphenyl phase [482], good peak shapes were obtained and reasonable separation, but other solvents (alcohols) were far superior. Thus, ether solvents are rarely used in protein work.

In summary, ethers have received limited use in the separation of amino acids and proteins because they are strongly denaturing in the case of proteins. For both amino acids and peptides, superior separations have generally been achieved with the use of alcohols and acetonitrile.

7.7 PHARMACEUTICAL ANALYTES

THF is the ether most commonly used as a mobile phase component in the analysis of pharmaceutical compounds. It is rarely used as the only organic component of an aqueous binary mobile phase but is commonly present at low levels (1–15%) in addition to another organic component (e.g., acetonitrile or methanol). THF at these levels tends to dramatically improve peak shapes and concomitantly decrease retention time.

Four quinolone residues (flumequine, nalidixic acid, oxolinic acid, and piromidic acid) were extracted from catfish muscle and separated on a 45°C styrene divinyl benzene polymer column ($\lambda = 280$ nm for piromidic acid; 325 nm, ex, 365 nm, em for all others) using a 720/120/60 water/THF/acetonitrile with 0.213 mL H_3PO_4 mobile phase [842]. Elution was complete in 17 min. A linear range of 10–200 ng/g sample and a quantitation limit of around 5 ng/g were reported.

Spironolactone and three aqueous degradation products (7α-thio-, 7α-thiomethyl-spirolactone, and 6β-hydroxy-7α-thiospirolactone) were analyzed on a phenyl column ($\lambda = 238$ nm for spironolactone and 280 nm for all others). Octyl and C_{18} columns were also used, but the phenyl gave acceptable results at considerably shorter analysis times. The final 45-min separation was generated with a 21/79 THF/water mobile phase [843]. Acetonitrile, methanol, and mixtures thereof did not generate similar selectivity. Concentration ranges of 0.5–4 μg/mL (degradation products) and 5–30 μg/mL for spironolactone were reported.

Six steroid hormones (cortisol, 11-deoxycortisol, prednisolone, methylpredniso-lone, prednisone, dexamethasone) were isolated from serum and urine and analyzed on a C_{18} column ($\lambda = 254$ nm) using a 72/25/3 water/THF/methanol mobile phase [844]. Excellent peak shapes and baseline resolution were obtained over the course of the 30-min separation. Linear ranges of 25–3000 nmol/L and detection limits of 5 nmol/L (S/N=1) were reported. Note here that S/N=1 means that the signal is

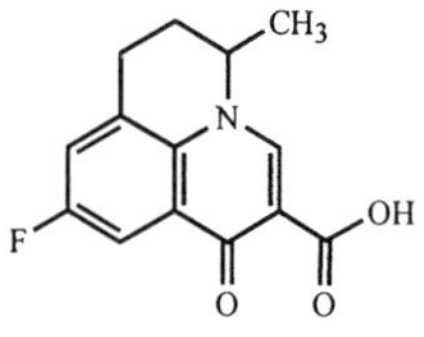

Flumequine

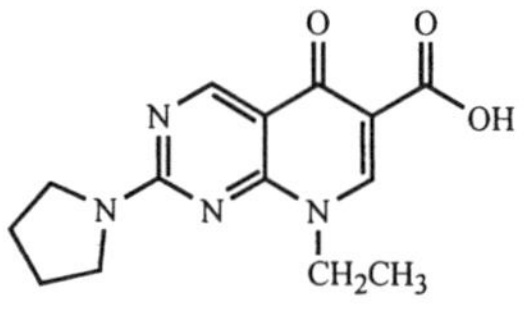

Piromidic acid

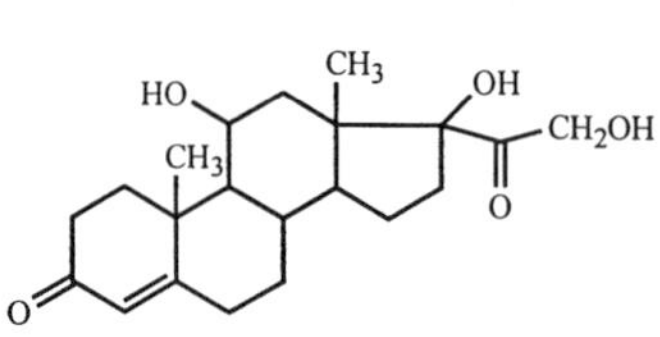

Cortisol

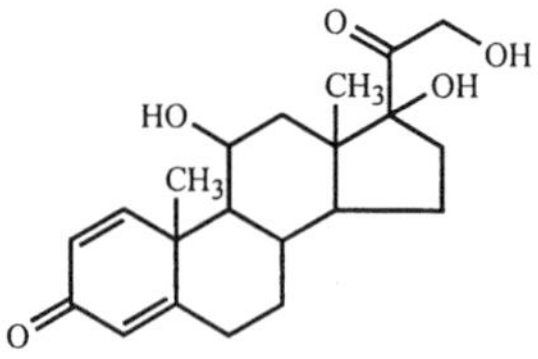

Prednisolone

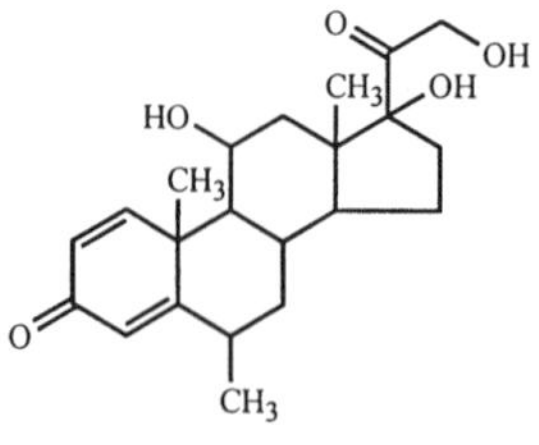

Methylprednisolone

essentially the same size as the noise and the quantitation is realistically based on retention time alone.

Nalbuphine (narcotic agonist-antagonist analgesic) and six degradation compounds (e.g., 2,2′-bisnalbuphine, *p*-hydroxybenzoic acid, 10-ketonalbuphine) were baseline resolved on a C_8 column ($\lambda = 280$ nm) using a 36-min two-ramp 96/4 $\rightarrow$ 72/28 (at 21 min hold 4 min) $\rightarrow$ 0/100 (at 26 min hold 10 min) (95/5/0.05 water/acetonitrile/TFA)/(70/25/5/0.05 water/THF/acetonitrile/TFA] gradient [845]. Peak shapes were excellent. A linear range of 0.5–20 µg/mL along with detection and quantitation limits of 0.01 µg/mL (S/N=3) and 0.05 µg/mL (S/N=10), respectively, were reported.

Caffeine, theophylline, theobromine, and paraxanthine were extracted from brain tissue and serum and baseline resolved on a C_{18} column ($\lambda = 273$ nm) using a 3/97 THF/water (10 mM phosphate buffer at pH 6.5). The elution was complete in 8 min and no matrix interferents co-eluted near the analytes of interest [846]. Standard curves were run from 0.1 to 10 µg/L and detection limits (S/N=2) were reported as 1.6 ng injected.

Aspartame and saccharin were quantitated in dietary formulations by separation on a C_{18} column ($\lambda = 210$ nm for aspartame and 270 nm for saccharin). Separation from other matrix interferents (e.g., acesulfame, 4-sulfamoylbenzoic acid, resorcinol (internal standard), *p*- and *o*-toluenesulfonamide) was done using a 5/5/90 THF/methanol/water (80 mM triethylamine phosphate buffer at pH 3.0) in <20 min [847]. Excellent resolution and peaks shaped were observed. Standards were in the 5–100 µg/mL range. It should be remembered that both methanol and THF have significant absorbance at 210 nm but are used effectively in this case because they are present at very low levels.

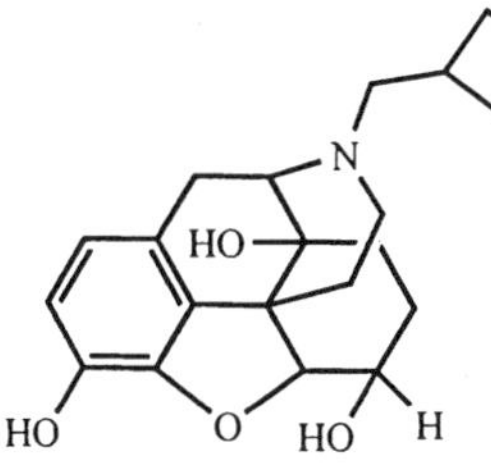

Nalbuphine

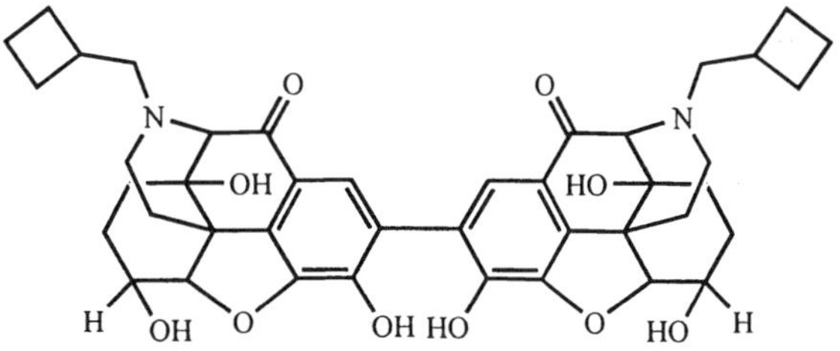

2,2'-Bisnalbuphine

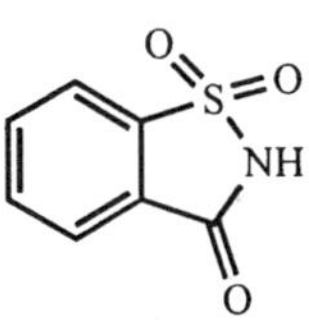

Saccharin

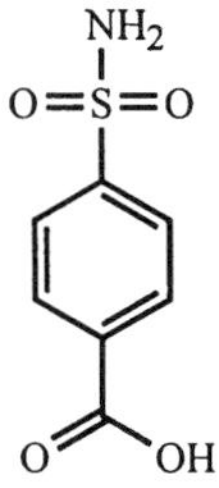

4-Sulfamoylbenzoic acid

Drug compounds often contain one or more basic amine functional groups and THF has the distinct advantage of being able to act as a hydrogen bond acceptor to both the residual surface silanol groups and the basic amine group as well. This prevents analyte/silanol group interactions that commonly lead to severe peak tailing or kinetically slow desorption of the analyte from the surface. Consequently, THF has been used as a mobile phase constituent with levels of 10% or less in separations described in earlier sections: 5-aminosalicylic acid and impurities [125], ramipril and ramiprilate [493], and taxol and related compounds [524]. Representative USP methods that have been developed with THF as a component are listed in Table 7.5 [590].

Other separations also use low-level THF. Trazodone was separated from 10 related impurity compounds on a C$_8$ column ($\lambda = 248$ nm) using a 15/5/15/67.5 THF/methanol/acetonitrile/water (0.5% TFA) mobile phase [848]. The trazodone level was 10 μg injected and the other compounds were ~0.2% of this amount (20 ng). Impurity levels down to 0.01% were quantitated. Good separation was obtained between peaks and elution was complete in 38 min.

Benzalkonium chloride (C$_{12}$, C$_{14}$, and C$_{16}$ homologs) in eye care products was quantitated using a cyanopropyl column ($\lambda = 215$ nm) and a 2500/1500/20 water/THF/triethylamine mobile phase [849]. Baseline resolution and complete elution were achieved in 12 min. Recoveries ranged from 91% to 102% at the 50 μg/mL concentration level.

Cortisol, cortisone, corticosterone, prednisone, and prednisolone were extracted from serum and separated on a C$_{18}$ column ($\lambda = 215$ nm) at 50°C using an 80/20 water/THF mobile phase [850]. Peak shapes were good and baseline resolution was achieved. Elution was complete in 20 min. Detection limits of 0.4 μg/dL with a linear concentration range up to 60 μg/dL were reported. The authors also presented a very useful table of relative retention times under identical conditions for 15 additional potentially interfering steroids.

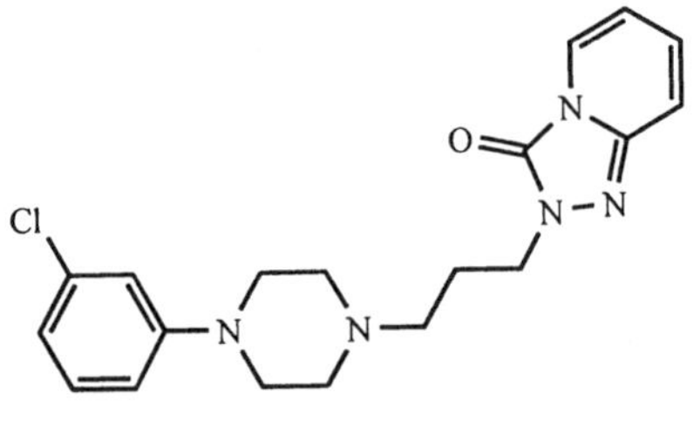

Trazadone

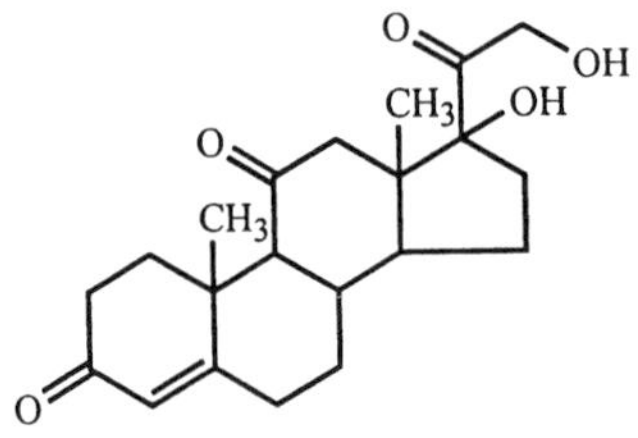

Cortisone

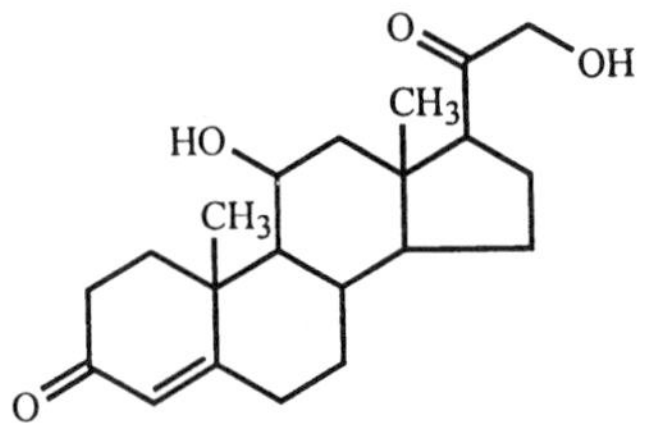

Corticosterone

TABLE 7.5 USP Methods[a]

Analyte(s)	Column	Mobile Phase	Detector (nm)	US Page
Acetaminophen and codeine phosphate capsules	C_{18}	600/360/40/1 methanol/water/THF/H_3PO_4 (with 4.44 g docusate sodium/L)	280	34
Fluocinolone acetonide	C_{18}	77/13/10 water/acetonitrile/THF	254	726
Griseofulvin	CN	60/35/5 water/acetonitrile/THF	254	788
Nitrofurantoin (limit of nitrofurazone)	C_{18}	10/90 THF/water (6.8 g KH_2PO_4/L to pH 7.0)	375	1185
Terbutaline sulfate	C_8 (40C)	750/140/110 water (1.1 g sodium octanesulfonate)/methanol/THF	280	1605

[a] Reference [590].

311

Ibuprofen, flurbiprofen, and methoxyphenylacetic acid enantiomers were resolved as their naphthalenemethylamine derivatives on an (R)-N-(3,5-dinitroben-zoyl)phenylglycine column ($\lambda = 214$ nm) using a 92.5/7.5 hexane/dioxane mobile phase [851]. This mobile phase gave optimal performance when compared with IPA/hexane or methanol/hexane mobile phases. It should be noted that although dioxane has a UV cutoff of 215 nm its use at such low levels in the mobile phase gave an acceptably low background absorbance level.

Diastereomeric α-ketoamide calpain inhibitors were resolved on an (S)-leucine-(R)-1-(α-naphthyl)ethylamine column ($\lambda = 210$ nm) using a 78/17/5 or an 86/11/3 hexane/dioxane/acetonitrile mobile phase [852]. Good resolution was obtained and elution was complete in 25–30 min. Results at 60°C and 0°C were shown. Good resolution was obtained at the elevated temperature and poor resolution at the lower temperature. When the lower temperature was used, the retention time increased from 22 to 25 min, or more than 10%.

Cocaine, and its metabolites (benzoylecgonine, benzoylnorecgonine, norcocaine, bupivarcaine [internal standard]) were baseline resolved and separated from acepro-mazine, ketamine, and atropine [853]. A C_8 column ($\lambda = 215$ nm or 235 nm) and a 96.75/3.25/0.0025 water (2.5 mM phosphate buffer at pH 2.75)/THF/TEA mobile phase gave complete elution in 25 min. Samples extracted from serum gave detection

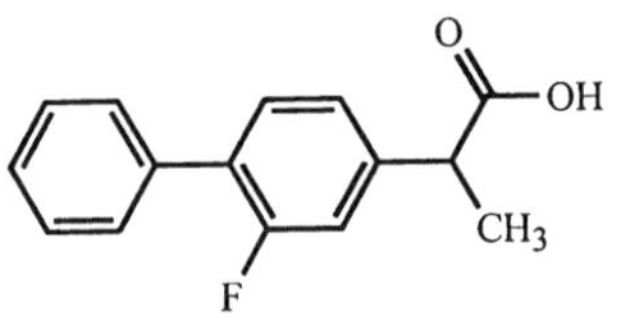

Flurbiprofen

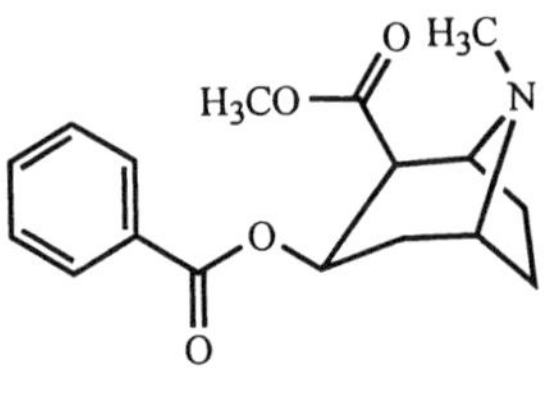

Cocaine

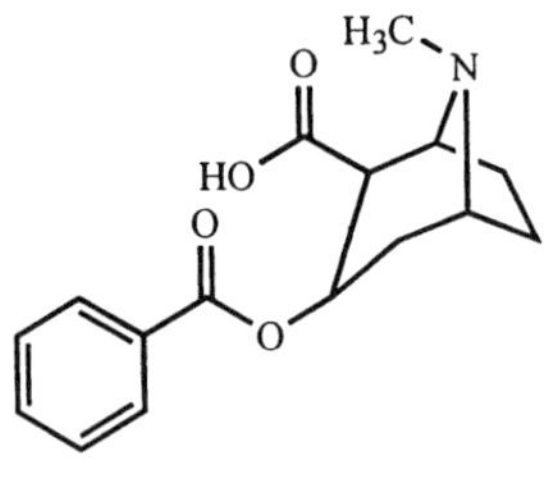

Benzoylecgonine

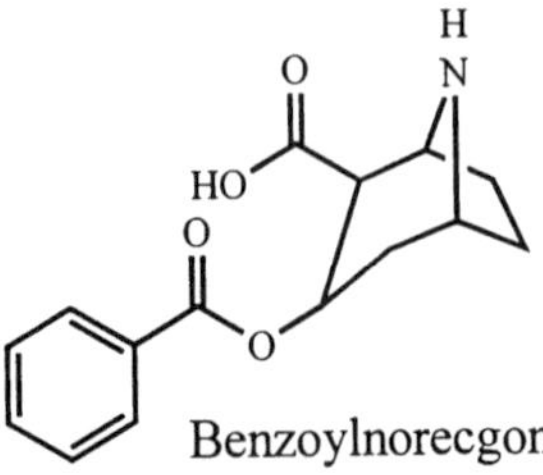

Benzoylnorecgonine

limits of 1–5 ng/mL (S/N = 6, analyte dependent). A calibration curve was generated over the 50 ng/mL to 1 μg/mL range.

7.7.1 Summary

THF and dioxane offer significant improvement in overall chromatographic efficiency for many hydroxylated and amine-containing compounds. Low levels of these solvents appear to enhance the separation through decreasing the hydrogen bond interaction of the solute with the support surface. At such low levels (often <10%), detection at wavelengths <210 nm can still be effectively conducted.

8

KETONES AND ESTERS

Ketones are solvents of moderate polarity. Like ethers, they are hydrogen bond acceptors and have dipole moments intermediate between those of ethers and alcohols. Ketones have a wide range of solubilities with water, from acetone being miscible, to methyl ethyl ketone (2-butanone) being soluble at 27%, to methyl propyl ketone (2-pentanone) being soluble to the 6% level. The higher molecular weight ketones are soluble to the 2% level or less. Ethyl acetate is soluble at the 9% level.

Acetone is often viewed (at least from the chromatographer's reference point) as a "universal" solvent. It is miscible with nearly all commonly used LC solvents. Therefore, it may be effectively used to change columns (e.g., cyanopropyl) from RP to NP working modes.

As an example, consider a mobile phase system on a cyanopropyl column of 50/50 methanol/water (50 mM phosphate buffer pH 3.0). If the analyst now wishes to conduct a NP separation using a 98/2 hexane/ethyl acetate mobile phase, a solution of the NP solvent should not be directly pumped through the system. Since the methanol/water mixture is immiscible with the hexane/ethyl acetate mixture, the methanol/water in the pores will be trapped and the phosphate buffer will precipitate in and around the silica pores, on the packing and frit surfaces, and in the detector cell. It is extremely difficult to redissolve and resolubilize this complex system and often the original performance of the column is never recovered. The correct way of switching solvents is to flush the column thoroughly with the 50/50 methanol/water solvent *without* the buffer. This make take 20–30 column volumes. After this process is complete, 100% acetone is pumped through the column. Approximately 10–25 column volumes are usually adequate. This is followed by the hexane/ethyl acetate mobile phase. If a UV detector is used, the absorbance at, say, 254 nm may be monitored to follow the flush-out of the acetone. Once a stable baseline is obtained,

314

the system can then be used for analysis. The exact opposite sequence is followed in order to switch from the NP to the RP system.

This process can also be conducted by substituting IPA for acetone. The major disadvantage associated with the use of IPA instead of acetone is that IPA produces a very high-viscosity solvent in aqueous solutions and therefore generates a high system backpressure.

Ketones are chemically reactive and undergo acid- and base-catalyzed self-condensation reactions. For acetone the base-catalyzed self-condensation reaction yields a product of diacetone alcohol [854]:

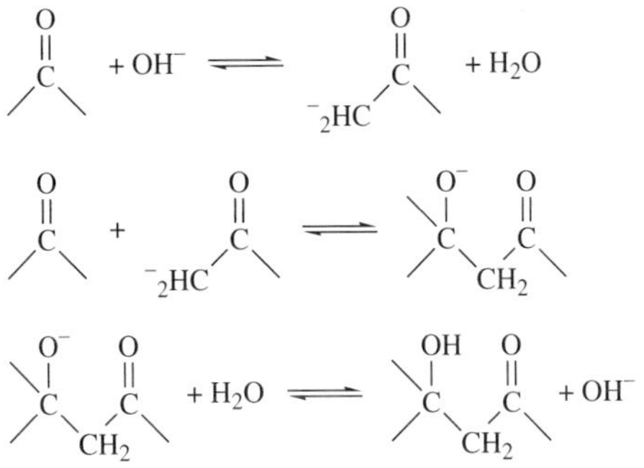

$$(8.1)$$

This reaction can proceed one step further to produce the acetone trimer. A condensation with elimination reaction for acetone yields mesityl oxide and, via a subsequent step, pherone:

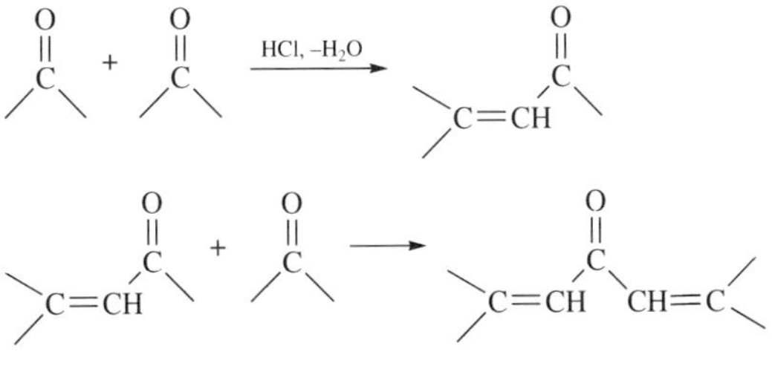

$$(8.2)$$

Other ketones also undergo these reactions, but the reactions products are much more varied because, in addition to the condensation itself, rearrangement and elimination steps can give unsaturated products as well as the simple dimer. Ethyl acetate does not react in such a fashion.

Ketones tend to have solvent strengths similar to those of the alcohols in RP phase separations but are considerably weaker solvents than the alcohols in NP systems. Regardless, extensive use of these ketone solvents has not occurred because the UV cutoff is extremely high (>320 nm). (Ethyl acetate has a UV cutoff of ~256 nm and therefore has been used more in conjunction with UV detection.) The high UV cutoff limits the use of ketones to analyses that (1) do not depend on absorbance for detection (e.g., RI detectors and ELSDs), (2) monitor for analytes that have a chromophore with a significant ε above 360 nm, or (3) use low volumes of the ketones in the mobile phase.

Acetone and ethyl acetate are widely available in high-purity form that is suitable for HPLC work. Many of the higher molecular weight ketones are also available in high-purity form, but because of their limited solubility in water their use in LC has been very limited.

The ketones have high vapor pressures at room temperature and should be used in a well-ventilated area. Ketones should be kept away from sparks and open flames since they are also flammable. When transferring these materials it is prudent to ground both containers to each other and an earth ground during the transfer process. Other than these normal precautions for the handling of flammable materials, ketones present no major safety hazards as far as they or their normal breakdown products are concerned.

Acetone has a very low viscosity. This usually does not present an instrumentation problem (i.e., improper check valve seating as with hexane) because it is rarely used above the 50% level in a mobile phase mixture due to its being a strong solvent. A problem often overlooked in the use of acetone is its high water content. High-purity acetone has a typical water level of >500 ppm that, as was discussed in Chapters 5 and 6, will chemically modify a silica or alumina surface and dramatically change the chromatography.

Tables 8.1–8.4 list some important chemical, physical and chromatographic properties as well as general manufacturing and safety parameters for the ketones and ethyl acetate [84–92]. Figure 8.1 shows the structures of the solvents listed in Tables 8.1–8.4.

8.1   IMPURITIES

The classic methods for producing acetone are from the propylene fraction of the petroleum cracking process, as a fermentation by-product of butyl alcohol, and as a cumene by-product of phenol [855]. Considering the different and varied sources, it readily follows that acetone and the other ketones usually contain significant and variable levels of higher ketone homologs, alcohols, various aldehydes (positional isomers of the ketone homologs), esters, unsaturated alkenes (elimination products) and carboxylic acids. The identity and level of impurities is dependent upon the manufacturing process.

Most of these impurities are removed by various process techniques during the manufacture of high-purity solvents. In general, these resulting high-purity solvents are "free" of these impurities since many manufacturers certify them to meet ACS (or other) specifications that include tests for acid and base character and residual aldehyde and alcohol content.

TABLE 8.1 Physical Properties of Ketone and Ester Solvents[a]

	ACT	MEK	MPK	MIBK	MBK	MIAK	MAK	EA
Molecular weight	58.08	72.11	86.13	100.16	100.6	114.19	114.19	88.1
Density (g/mL)	0.7900	0.8049	0.8082	0.8008	0.8113	0.888	0.8154	0.9006
Viscosity (cP)	0.36	0.43	0.51	0.59	0.638	0.8	0.8	0.45
Solubility in water (%)	100	24.0	5.9	1.9	1.75	0.54	0.44	8.7
Water solubility in solvent (%)	100	10.0	3.3	1.9[b]	2.12	1.3	1.3	3.3
Boiling point (°C)	56.29	79.64	102.4	116.5	127.2	144.9	150.4	77.11
Melting point (°C)	−94.7	−86.69	−77.8	−84	−56.9	−73.9	−26.9	−84.0
Refractive index (n_D)	1.3587	1.3788	1.3901	1.3957	1.4007	1.4072	1.4087	1.3724
Dielectric constant	21.45	18.51	15.45	13.11[b]	12.2		11.95	64
Dipole moment (D)	2.69	2.78			2.7			1.78
Surface tension (dyne/cm)	23.32	24.5	24.82	23.64	25.43		26.6	23.75

Abbreviations: ACT, acetone; MEK, methyl ethyl ketone, 2-butanone; MPK, methyl propyl ketone, 2-pentanone; MIBK, methyl isobutyl ketone, 2-methyl-4-pentanone; MBK, methyl butyl ketone, 2-hexanone; MIAK, methyl isoamyl ketone, 2-methyl-5-hexanone; MAK, methyl amyl ketone, 2-heptanone; EA, ethyl acetate.
[a] All values except boiling and freezing points at 20°C unless otherwise noted.
[b] At 25°C.

TABLE 8.2   Chromatographic Parameters of Ketone and Ester Solvents[a]

	ACT	MEK	MPK	MIBK	MBK	MIAK	MAK	EA
Eluotropic strength $\varepsilon°$ on Al_2O_3	0.56	0.51		0.43				0.58
Eluotropic strength $\varepsilon°$ on SiOH	0.53							0.48
Eluotropic strength $\varepsilon°$ on C_{18}	8.8							
Solvent strength parameter, P'	5.1	4.7	4.5	4.2		4.0		4.4
Hildebrandt solubility parameter, δ	9.6	9.3			8.3	8.4	9.0	9.1
Hydrogen bond acidity, α	0.08	0.06						
Hydrogen bond basicity, β	0.48	0.48						
Dipolarity/polarizability, $\pi*$	0.71	0.67						

[a] *Abbreviations:* ACT, acetone; MEK, methyl ethyl ketone, 2-butanone; MPK, methyl propyl ketone, 2-pentanone; MIBK, methyl isobutyl ketone, 2-methyl-4-pentanone; MBK, methyl butyl ketone, 2-hexanone; MIAK, methyl isoamyl ketone, 2-methyl-5-hexanone; MAK, methyl amyl ketone, 2-heptanone; EA, ethyl acetate.

TABLE 8.3   Common Manufacturing Quality Specifications of Ketones and Esters[a]

	ACT	MEK	MPK	MIBK	MBK	MIAK	MAK	EA
UV cutoff (nm)	330	329	331	334	330	330		256
Percent water (maximum)	0.50	0.05	0.05	0.05	0.05	0.05		
Available as ACS tested[b]	ABEFJM	AM	n.a.[c]	AJFM	n.a.	n.a.	n.a.	ABEFJM
Available as HPLC-grade[b]	ABEFJM	ABEJM	BE	ABEM		BE		ABEFJM
Available through[d]		JA			AJ	AE	AJ	

[a] *Abbreviations:* ACT, acetone; MEK, methyl ethyl ketone, 2-butanone; MPK, methyl propyl ketone, 2-pentanone; MIBK, methyl isobutyl ketone, 2-methyl-4-pentanone; MBK, methyl butyl ketone, 2-hexanone; MIAK, methyl isoamyl ketone, 2-methyl-5-hexanone; MAK, methyl amyl ketone, 2-heptanone; EA, ethyl acetate.
[b] Manufacturer's code: A = Aldrich; B = Burdick & Jackson; E = EM Science; F = Fisher; J = JT Baker; M = Mallinckrodt.
[c] No ACS test exists for this solvent.
[d] Available as a high-purity solvent but not specifically designated as ACS or HPLC grade. This does *not* mean a lesser quality solvent, just that it is not specifically tested for these applications. If the manufacturers produce either ACS or HPLC solvent, they are not listed under this heading.

8.2   GENERAL ANALYTES

Olsen and Hurtubise [856] studied the retention of 1- and 2-naphthol, 1,2,3,4- and 5,6,7,8-tetrahydronaphthol, 1- and 2-naphthalenemethanol, and 1- and 2-naphthalene-ethanol on a silica column ($\lambda = 280$ nm) using a series of hexane/ethyl acetate mobile phases. Ethyl acetate content was varied from 2% to 20%, solutes were eluted under isocratic conditions, and a table of the $\log k'$ values for these analytes with ethyl acetate levels at 2% increments was generated. Well-correlated linear $\ln k'$ versus percent ethyl acetate relationships for individual solutes indicate that for

TABLE 8.4 Safety Parameters of Ketones and Esters[a]

	ACT	MEK	MPK	MIBK	MBK	MIAK	MAK	EA
Flash point[b] (TCC)(°C)	−20	−9	7	18	25	36	39	−4
Vapor pressure (Torr @ 20°C)	184.5	74	11.5	16		19.9	35.2	73
Threshold Limit Value (ppm)	750	200	200	50		50		400
CAS number	67-64-1	78-93-3	107-87-9	108-10-1	591-78-6	110-12-3	110-43-0	141-78-6
Flammability[c]	3	3	3	3	3	2	2	3
Reactivity[c]	0	0	0	1	0	0	0	0
Health[c]	1	1	2	2	2	1	1	1

[a] *Abbreviations*: ACT, acetone; MEK, methyl ethyl ketone, 2-butanone; MPK, methyl propyl ketone, 2-pentanone; MIBK, methyl isobutyl ketone, 2-methyl-4-pentanone; MBK, methyl butyl ketone, 2-hexanone; MIAK, methyl isoamyl ketone, 2-methyl-5-hexanone; MAK, methyl amyl ketone, 2-heptanone; EA, ethyl acetate.

[b] TCC = TAG closed cup.

[c] According to National Fire Protection Association ratings [92]:

Fire: 4 = Materials that vaporize at room temperature and pressure and burn readily.
3 = Liquids or solids that can ignite under room conditions.
2 = Materials that ignite with elevated temperature or with moderate heat.
1 = Materials that must be preheated before they ignite.
0 = Materials that will not burn.

React: 4 = Materials that, by themselves, can detonate or explode under room conditions.
3 = Materials that can detonate or explode but require an initiator (e.g., heat).
2 = Materials that undergo violent chemical reactions at elevated temperatures or pressures or react with water.
1 = Materials that are, by themselves, stable but that may become unstable at elevated temperatures and pressures.
0 = Materials that are stable even under fire conditions and do not react with water.

Health: 4 = Short exposure times to these materials are lethal or cause major residual injury.
3 = Short exposure times to these materials cause temporary and/or residual injuries.
2 = Lengthy (but not chronic) exposure to these materials may cause temporary incapacitation and/or minor residual injury.
1 = Materials that, upon exposure, cause irritation but only minor residual injury.
0 = Materials that, upon exposure under fire conditions, offer no more hazard than ordinary combustible materials.

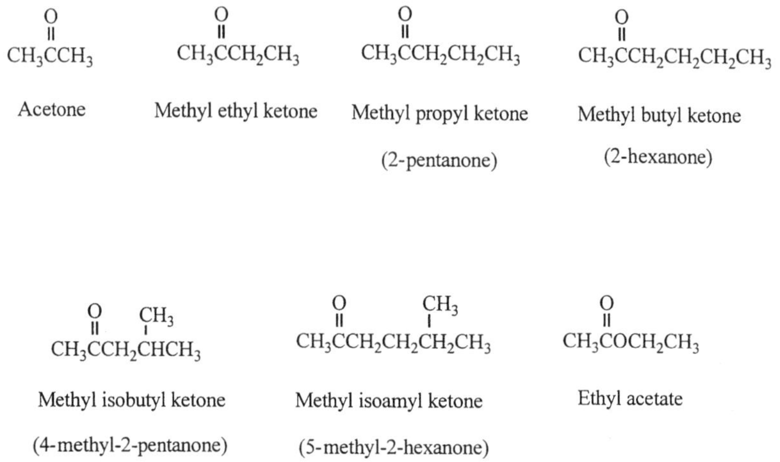

$$
\underset{\text{Acetone}}{\overset{\overset{\displaystyle O}{\|}}{CH_3CCH_3}}
\qquad
\underset{\text{Methyl ethyl ketone}}{\overset{\overset{\displaystyle O}{\|}}{CH_3CCH_2CH_3}}
\qquad
\underset{\substack{\text{Methyl propyl ketone} \\ \text{(2-pentanone)}}}{\overset{\overset{\displaystyle O}{\|}}{CH_3CCH_2CH_2CH_3}}
\qquad
\underset{\substack{\text{Methyl butyl ketone} \\ \text{(2-hexanone)}}}{\overset{\overset{\displaystyle O}{\|}}{CH_3CCH_2CH_2CH_2CH_3}}
$$

$$
\underset{\substack{\text{Methyl isobutyl ketone} \\ \text{(4-methyl-2-pentanone)}}}{\overset{\overset{\displaystyle O \quad CH_3}{\| \quad |}}{CH_3CCH_2CHCH_3}}
\qquad
\underset{\substack{\text{Methyl isoamyl ketone} \\ \text{(5-methyl-2-hexanone)}}}{\overset{\overset{\displaystyle O \quad\;\; CH_3}{\| \qquad |}}{CH_3CCH_2CH_2CH_2CH_3}}
\qquad
\underset{\text{Ethyl acetate}}{\overset{\overset{\displaystyle O}{\|}}{CH_3COCH_2CH_3}}
$$

FIGURE 8.1

hydroxylated aromatic compounds retention can reasonably be predicted. Three solutes, 2,6-, 2,5-, and 3,5-dimethylphenol, were baseline resolved on a diol column using a 90/10 hexane/ethyl acetate mobile phase [611].

The separation of Rh(III), Ru(II), Pd(II), and Pt(II) as their 4-(5-nitro-2-pyridy-lazo)resorcinol complexes was achieved on a C_{18} column ($\lambda = 536$ nm) using a 50/10/40 methanol/ethyl acetate/water (10 mM acetic acid buffer pH 4.0 with 10 mM tetrabutylammonium bromide and 10 mM sodium EDTA) mobile phase [155]. Elution was complete in <20 min. Peaks were baseline resolved. The detection limits were reported to be between 0.5 and 2.6 ng/mL (ion dependent) and linear working concentrations were generated from ~2 to 100 ng/mL for all analytes. The authors note that ethyl acetate levels between 4% and 20% were effective for this separation but do not state that ethyl acetate is responsible for the selectivity obtained.

8.3 ENVIRONMENTALLY IMPORTANT ANALYTES

Ethyl acetate was used as a low-level mobile phase constituent in the separation of seven phenolic acids (protocatechuic, p-hydroxybenzoic, vanillic, syringic, caffeic, p-coumaric, ferulic) from soil [113]. A C_{18} column ($\lambda = 254$ nm) and a complex 42-min 70/30 → 30/70 (2/97.25/0.5/0.25 methanol/water/acetic acid/ethyl acetate)/ (80/17/2/1 methanol/water/acetic acid/ethyl acetate) gradient generated good resolution of the peaks of interest. Although ethyl acetate is a low-level constituent in the mobile phase, the authors note that the reproducibility of the method suffered greatly when a helium sparge into the solvent was used and the ethyl acetate was preferentially volatilized. As a result, the retention times changed enough to cause the co-elution of three analytes (vanillic, syringic, and caffeic acids). The authors also note that the solvent system used, while not giving the most optimal separation, is one that presents minimal health and safety hazards.

Linuron and three metabolites were resolved on a C_{18} column ($\lambda = 247$ nm or amperometric detection at +1.3 V vs. Ag/AgCl) using a 38/52/10 methanol/water/ethyl acetate mobile phase all with 2 g/L KNO_3 as the supporting electrolyte [213]. Elution was complete in 15 min and 5 ng injected of each metabolite was detectable by amperometry. This method was found to be five times more sensitive than the UV detection method for metabolites. Linuron, however, had a lower detection limit using UV detection at 247 nm.

The levels of seven pyrethroids were determined in various pesticide formulations using a silica column ($\lambda = 275$ nm) and a 99.3/0.7 iso-octane/ethyl acetate mobile phase [635]. Peak shapes were excellent and the elution time for the longest-retained was ~13 min. The authors warned that the sample precipitated from solution if made up in iso-octane or the mobile phase but that an 80/20 iso-octane/dichloromethane solvent was acceptable. An interesting chromatogram was presented of a sample containing residual acetone (from sample preparation). Acetone, in this setup, appeared as a very broad peak that eluted over the 18–22 min range. Thus, if acetone is used in the sample preparation procedure and is present as a residual component, the analysis time will be increased by at least 50%!

To summarize, as chromatographers begin to search in earnest for "safer" (i.e., posing fewer health hazards) mobile phases, ethyl acetate may become more heavily utilized, especially as a low-level mobile phase constituent. As seen above, ethyl acetate offers some unique chromatographic and chemical properties, especially when solutes that are reactive with protic solvents are separated.

8.4 INDUSTRIAL AND POLYMER ANALYTES

Because of their limited water solubility, a series of polymer additives were characterized using various nonaqueous mobile phases [749]. Hostanox SE-10 (dioctadecyl disulfide) and DSTDP were analyzed on a C_{18} ($\lambda = 220$ nm or ELSD) column using a 100% acetone mobile phase. Note that acetone does have a low absorbance "window" around the 220 nm region in its UV spectrum that allows it to be used effectively in this analysis. Sumilizer BHT, Irganox 1010, and Irganox 1076 were analyzed on a C_{18} column with a 35/65 acetone/acetonitrile mobile phase. These analyses were typically complete in <30 min and generated excellent peak shapes and resolution.

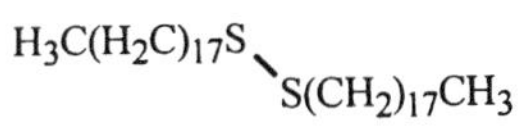

$$H_3C(H_2C)_{17}S{\diagdown}S(CH_2)_{17}CH_3$$

Hostanox SE-10; dioctadecyl disulfide

Methyldibromoglutaronitrile (1,2-dibromo-2,4-dicyanobutane, BCB) is a preservative added to various cosmetic products [857]. Over 30 shampoos and creams were analyzed for their BCB content. BCB was extracted from samples and quantitated on a C_8 column (electrochemical detector at -0.6 V vs. Ag/AgCl or UV at $\lambda = 220$ nm) with a 40/60 acetone/water (20 mM sodium sulfate and 2 mM sodium chloride) mobile phase. The BCB level was determined and baseline resolved from Bronopol and Bronidox. Separation was complete in <20 min. Peak shapes were excellent. Detection limits (S/N= 3) for BCB were 0.5 ppm by electrochemical detection and the linear working range was reported as 0.5–40 ppm with 50 ppm detection limits and a 50–300 ppm, linear range, for UV detection.

Poly(methyl methacrylate) and polytetrahydrofuran polymers were studied at the "critical point of adsorption." This critical point of adsorption occurs where the retention of a given polymer is governed strictly by the number and types of functional groups on the polymer [858]. The authors show plots of log MW vs. retention time for various mobile phase compositions on a given column. The critical point is reached when the retention time becomes independent of the molecular weight of the polymer. For poly(methyl methacrylate) that point was reached on a silica column (RI detector) with a 73/27 methyl ethyl ketone/cyclohexane mobile phase. For polytetrahydrofuran, the silica column and a 95/5 acetone/hexane mobile phase created the critical conditions. This approach has enabled the individual blocks within the co-polymer to be studied (i.e., the portion of the polymer that can make contact with the support surface).

Methyl/butyl methacrylate block co-polymers were characterized on a silica column (ELSD) using a 30-min $98/2 \to 0/100$ toluene/methyl ethyl ketone gradient [659]. Co-polymers of differing molecular weights (9800 to 40,000) and co-polymer percentages were readily differentiated using this method. Samples were made up at the 1–10 mg/mL level and 25 µL was injected.

8.5 BIOLOGICAL ANALYTES

The *cis* and *trans* diarylpropenamine isomers were isolated from feces, derivatized with 3-(4'-bromo-[1,1'-biphenyl]-4-yl)-3-(phenyl-*N*,*N*-dimethyl-2-propen-1-amine), and analyzed on a silica column ($\lambda = 272$ nm). Baseline resolution and excellent peak shapes were achieved along with complete elution in <10 min with a 59/20/1 ethyl acetate/hexane/triethylamine mobile phase [859]. A linear range of 10–150 µg/mL and with a detection limit of 2 µg/mL were reported.

The levels of retinol, α-tocopherol, and retinyl palmitate in infant plasma samples were determined using a C_{18} column ($\lambda = 305$ nm) and a $100/0 \to 40/60$ (at 9 min hold 6 min) (90/10 acetonitrile/water)/(10/90 IPA/ethyl acetate) gradient [860]. Excellent separation and peaks shapes were obtained. All compounds eluted within 12 min. Detection limits were reported as 0.1–1.0 µg/mL (S/N = 3). The resulting working curve was linear up to 8 µg/mL.

Five retinol isomers, three *syn*-retinal oxime isomers, and a total of nine *anti*-retinal oxime and retinol isomers were baseline resolved using an 85.4/11.2/2/1.4

hexane/ethyl acetate/dioxane/1-octanol mobile phase and a silica column ($\lambda = 325$ nm). Elution was complete in <25 min and peak shapes and resolution were excellent [861].

Hara et al. [677] studied the effect of changing the mobile phase composition on the retention of 10 fat-soluble vitamins (e.g., *trans*-retinol, retinal, ergocalciferol, cholecalciferol, menadione, phylloquinone). A silica column ($\lambda = 254$ nm or 292 nm) was used. The ethyl acetate level in hexane was systematically changed from 5% to 20% and the results were plotted in a ln k' vs. percent ethyl acetate format. As expected, when the percent ethyl acetate increased, the overall retention of the analytes decreased. Interesting, however, was the fact that the changes in retention of both menadione and retinal were so different from the other analytes (e.g., α-, β-, γ-, and δ-tocopherol) that reversals in retention order occurred. These data present an excellent basis from which to develop a new method for fat-soluble vitamins.

The retention of six paprika and broccoli carotenoids (including β-carotene, lutein, antheraxanthin, violaxanthin, and *cis*-neoxanthin) was studied on a silica column ($\lambda = 435$ nm) using a 35/65 acetone/petroleum ether mobile phase for analysis [676]. Preparative-scale separation of β-carotene, cryptoxanthin and its epoxide, capsanthin, and capsorubin was achieved on a silica column using the same mobile phase. Chromatograms showing the influence of 1 mg vs. 40 mg loads were presented. The separation went from baseline resolution for the five compounds with the 1 mg injection to the presence of separate but merged peaks throughout the entire 80-min chromatogram. The authors claim that up to 50 mg of sample can be treated per injection with the 40 mg injection.

A polymeric C_{18} column ($\lambda = 450$ nm) and an 80/20 methanol/ethyl acetate mobile phase were used to baseline resolve α- and β-carotene and lycopene in 15 min [828]. A monomeric C_{18} gave reversal of retention for the components, an end-capped C_{18} gave co-elution of all components, and finally the polymer C_{18} gave complete separation of all analytes.

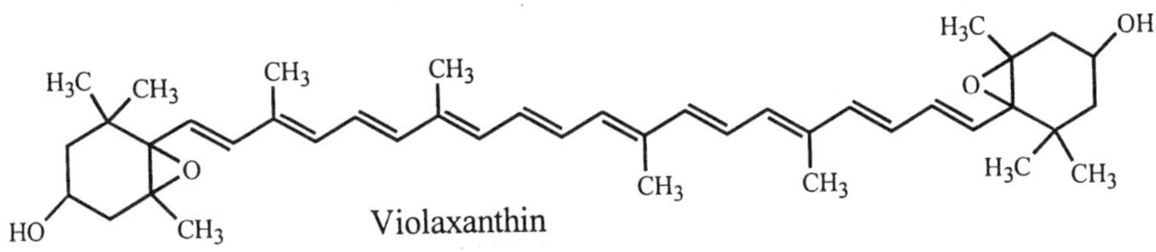

Violaxanthin

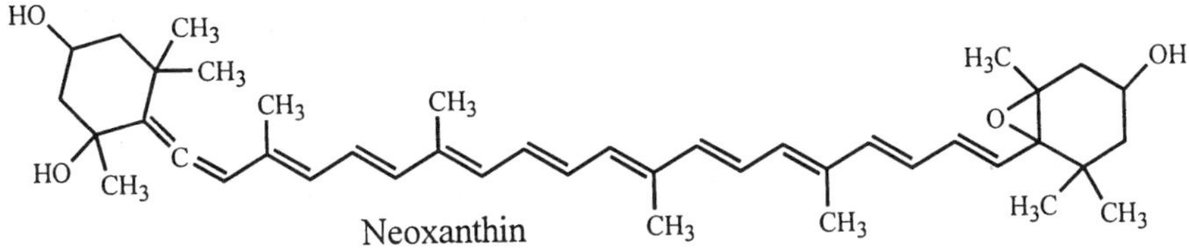

Neoxanthin

Fifty-eight distinct carotenoid components of black paprika were detected on a C_{18} column (photodiode array detector, $\lambda = 340$–480 nm) using a complex 44-min $88/0/12 \rightarrow 100/0/0 \rightarrow 50/50/0$ methanol/acetone/water gradient [353]. About 50% of the peaks were identified and their levels were tabulated. The overall composition of the extract was also calculated from the LC run.

Various mixtures of acetone/acetonitrile have been used in the separation and characterization of triglycerides from a number of plant and food extracts. Hierro et al. [862] studied the composition of avocado oil using a C_{18} column (ELSD) and a complex 120-min $35/65 \rightarrow 70/30$ acetone/acetonitrile gradient. Trilinolein, tricaproin, tricaprylin, tricaprin, trilinolenin, trimyristin, tripalmitin, triolein, tristearin, and numerous mixed triglycerides were separated and identified by their equivalent carbon number. Twenty-four distinct peaks were generated and a linear regression analysis based on standards generated a regression equation from which peak identifications were made. Injections of 10–70 μg of pure triglyceride standards were used to establish a response curve.

In a series of two articles, the composition of palm olein and the transesterification of palm olein by nonspecific lipases were determined chromatographically using a C_{18} column (RI detector) and a $63.5/36.5$ acetone/acetonitrile mobile phase [863, 864]. The generation of a profile of end products was complete in 40 min. It is interesting to note that the retention time for the palm olein varied at least 1.5 min from injection to injection (29–30.5 min). No comment was made as to the source of this variability. Nevertheless, the profiles were consistent and 11 transesterification products were separated from the original components and quantitated.

The triacylglycerol (TAG) composition of the seeds of five *Amaranthus* accessions were determined from the fat residue from a Soxhlet extraction. Two 250×4.6 mm $30°C$ C_{18} columns (RI detector) in series and a $60/40$ acetone/acetonitrile mobile phase were used in the analysis. A 10 μL aliquot of a 5% fat extract was injected. The acyl substituents of the TAG ranged from linolenic : linolenic : linoleic to palmitic : oleic : steric. Squalene was also separated and identified [865]. Elution was complete in 75 min. Pili nut (*Canarium ovatum*) oil was similarly analyzed. In this case the TAGs ranged up to stearic : stearic : oleic and elution took 95 min [866].

Numerous other papers have addressed the separation of triglycerides in a similar fashion: cottonseed oil, $60/40$ acetone/acetonitrile, C_{18}, (RI detector) [867]; olive oil, $93/7$ acetone/acetonitrile, C_{18}, RI detector [868]; determination of the commercial grade of olive oils, $63.6/34.4$ acetone/acetonitrile, C_{18}, RI detector [869]; peanut and cottonseed oil, $55/45$ or $60/40$ acetone/acetonitrile, C_{18}, RI detector [870]. All analyses required between 35 and 45 min to complete. All gave adequate separation and peak shapes.

In a collaborative study, a general guideline for separation and quantitation of triglycerides was developed by Firestone [871]. Six different vegetable oils (soybean, almond, sunflower, olive, rapeseed and palm) were characterized using a high carbon load C_{18} column (RI detector) and a $50/50$ acetone/acetonitrile mobile phase mixture. A presentation of a separation factor versus the number of double bonds in the triglyceride is made and is of great benefit to chromatographers needing to generate new methods of separation for triglycerides.

Triglycerols were determined in human milk in a similar fashion. A C_{18} (RI detector) column and 64/36 acetone/acetonitrile mobile phase generated the separation [872]. Components with equivalent carbon numbers ranging from 36 to 54 were eluted in 45 min. Greater resolution was obtained by placing two C_{18} columns in series and increasing the analysis time to 90 min. Thirty-nine peaks were clearly delineated and identified.

Five novel triacylglycerols (TAGs) were enzymatically manufactured to contain arachidonoyl, stearoyl, and palmitoyl substituents (e.g., triarachidonin, 1,3-diarachi-donoyl-2-stearoyl). Baseline separation was generated on two 40°C 250 × 4.6 mm C_{18} columns in series (RI detector) using a 67/33 acetone/acetonitrile mobile phase. Elution was complete in <40 min [873].

A method for monitoring the adulteration of olive oil in terms of its equivalent carbon number (ECN) distribution was developed [874]. The ECN = 42 peak in olive oil is roughly 1% of the total triglyceride, whereas it is ~25% for corn oil, soybean oil, and sunflower seed oil. The separation of ECNs from 42 to 52 (even) was achieved on a C_8 column (RI detector) using a 70/30 acetone/acetonitrile mobile phase. Excellent resolution of the peaks and good peak shapes were obtained. Elution was complete in less than 15 min.

Fatty alcohol ethoxylates were extracted from plant materials. The C_{12} portion was preparatively purified into individual $n = 8$ to $n = 13$ oligomeric fractions. A silica column (RI detector) and a 98/2 butanone/water mobile phase were used [875]. A 0.5 mL injection of a 40% sample solution (using the mobile phase as the solvent) produced excellent peak resolution and good peak shapes. Elution was complete in 22 min.

Nine chlorophyll derivatives (chlorophyll *a* and *b*, chlorophyllides *a* and *b*,

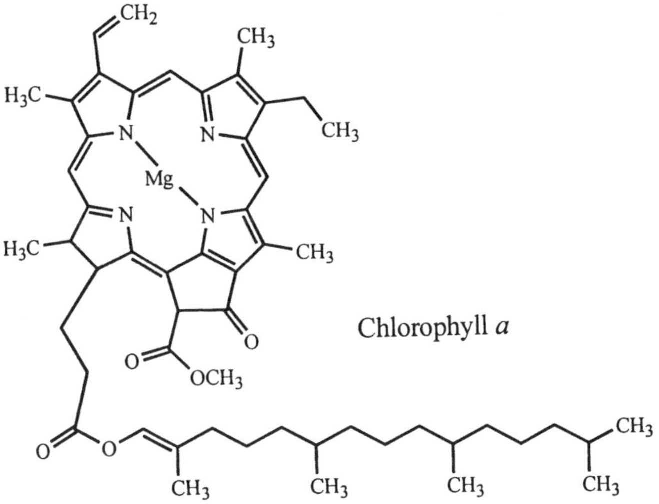

Chlorophyll *a*

pheophytins *a* and *b*, pheophorbide *a*, and pyropheophytins *a* and *b*) from spinach were separated and analyzed on a C_{18} column (positive ion FAB-MS) using a 20-min $15/65/20/0.5 \rightarrow 60/30/10/0.5$ ethyl acetate/methanol/water/glycerol gradient. Sample loadings were 15 µg injected [876].

Li and Inoue [877] generated both preparative and analytical separations of manganese(III) chlorophyll *a* and *b*. These analytes were preparatively separated on a C_{18} column ($\lambda = 420$ nm) using a 50/50 acetone/acetonitrile mobile phase. Analytical work using a 40/60 acetone/acetonitrile mobile phase generated peaks that were terribly tailed. Switching the mobile phase to 90/10 acetone/acetonitrile (5 mM sodium acetate) gave excellent peak shape, excellent resolution, and elution of both compounds in <6 min. These analytes had a reported detection limit of ~0.1 µg/mL and a linear working curve from 0 to 20 µg/mL.

Taylor and McDowell [878] separated 28 distinct pigment components from a tea leaf extract using a C_{18} column (photodiode array detector, $\lambda = 200$–750 nm) and a 20-min $100/0 \rightarrow 50/50$ (at 10 min hold 10 min) (90/10 acetonitrile/water)/ethyl acetate gradient. Twenty-three peaks were identified by retention time and UV-vis absorption spectra.

A smaller set of eight xanthophyll pigments (e.g., chlorophylls, β-carotene, neoxanthin, lutein, zeaxanthin) were baseline resolved in <15 min using a C_{18} column ($\lambda = 440$ nm) and a step gradient from 94.75/1.75/1.75/1.75 acetonitrile/water/dichloromethane/methanol to 50/50 ethyl acetate/acetonitrile at the 5 min mark after injection [879]. Peak shapes were excellent. Concentrations ranging from 0.1 to 10 nmol/mL were used as standards.

Twenty pigments extracted from algal samples were well separated in 30 min using a C_{18} column ($\lambda = 440$ nm) and a complex 15-min $45/45/10/0 \rightarrow 50/0/0/50$ acetonitrile/methanol/acetone/water (3.75 g tetrabutylammonium

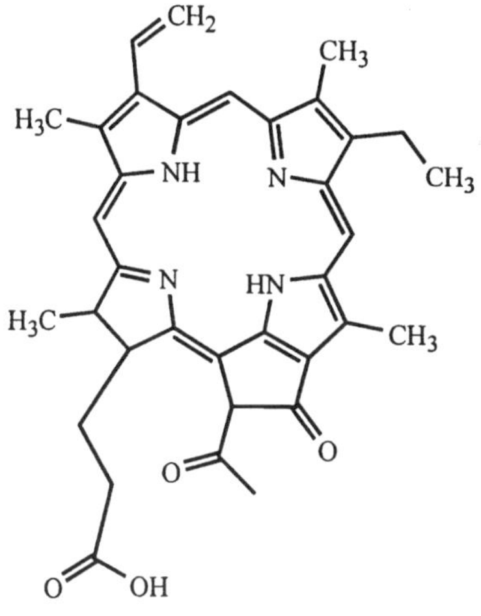

Pheophorbide *a*

acetate with 19.25 g ammonium acetate/250 mL water) gradient [880]. An interesting aspect of this separation is that after the gradient reaches its final composition at 15 min, this composition is held for 5 min and then a 1-min reverse gradient to the original conditions is followed by a 15-min hold. Peaks of interest elute in up to 27 min using this gradient scheme! Detection limits ranged from 0.4 to 2 μg/L. Less common pigments monitored included oscillaxanthin, fucoxanthin, myxoxanthophyll, diadinoxanthin, alloxanthin, and diatoxanthin.

8.7 PHARMACEUTICAL ANALYTES

Testosterone acetate, methyltestosterone, testosterone, bolasterone, and progesterone were baseline resolved on a silica column with postcolumn treatment using 30 mM $Tb(NO_3)_3$ (fluorescence detection of complex, $\lambda = 245$ nm, ex; 547 nm, em) using a 60/40 ethyl acetate/cyclohexane mobile phase [881]. Elution was complete in <20 min. Detection limits of 0.5 ng/mL and working concentration ranges of 1–1000 ng/mL were reported. Peak shapes were very good. Results for the NP separation were better (in both detection limits and working curve concentrations) than the RP separation presented in the same paper.

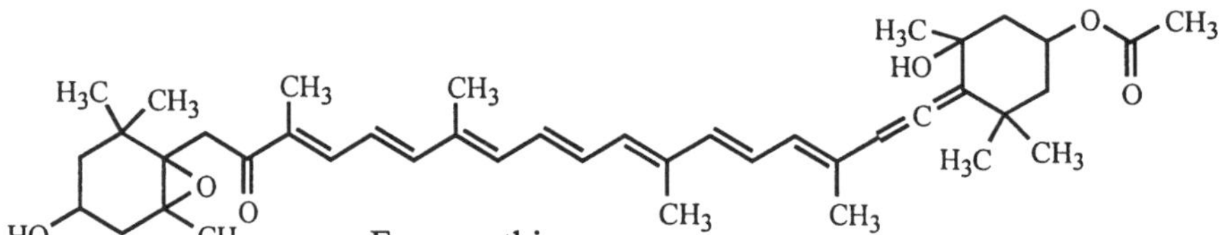

Fucoxanthin

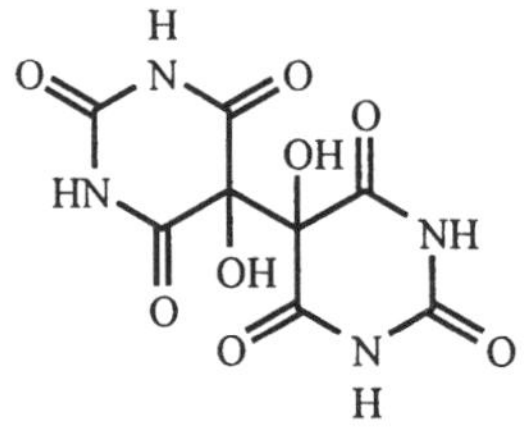

Alloxanthin

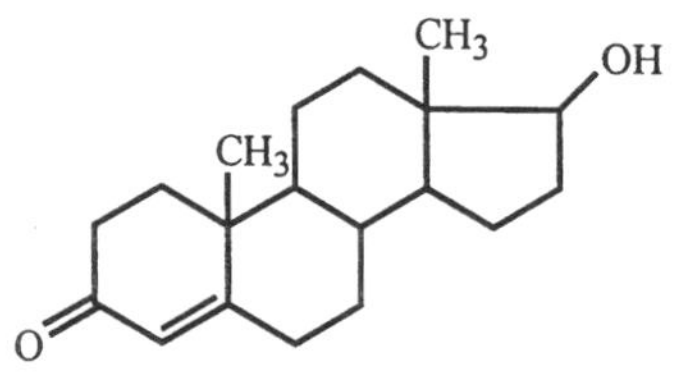

Testosterone

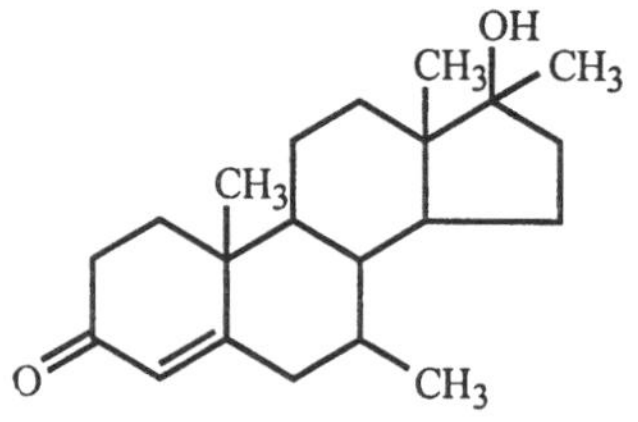

Bolasterone

The isomers of 4-amino-3-(4-chlorophenyl)butyric acid were separated on a preparative scale using a silica column ($\lambda = 260$ nm) and 80/20 ethyl acetate/hexane mobile phase [882]. Up to 375 mg of sample was injected and baseline resolution was achieved in 6 min.

Mesitylate esters are common intermediates in chemical synthesis and are very unstable. In order to monitor the extent of a reaction, it is often necessary to monitor all reaction species: the precursor, mesylate intermediates, and end products. The mesylation reaction of 3-(2-isopropylphenyl)-1-hydroxy-1-R-propane (where R is a bulky chlorinated hydrocarbon) was followed on a diol column ($\lambda = 320$ nm) using a 15-min 15/85 → 5/95 (75/25 hexane/toluene)/ethyl acetate gradient [883]. Intramolecular ring closure of the analyte occurred on the silica support at room temperature. Chromatography yielded a large, tailed cyclic ether peak followed by the mesylated material followed by the unreacted diol compound. Elution was complete in 14 min. The silica-facilitated reaction was greatly reduced when the system temperature was taken down to 10°C and was prevented when the system temperature was −30°C.

In summary, acetone and ethyl acetate are very effective solvents when used with large slightly polar solutes. The ability of these solvents to keep a wide range of solutes in solutions, their moderate solvent strengths, and their low viscosities make them unique. Only IPA has similar properties. The trade-offs are that IPA is a high viscosity solvent, and acetone and ethyl acetate have higher UV cutoff values.

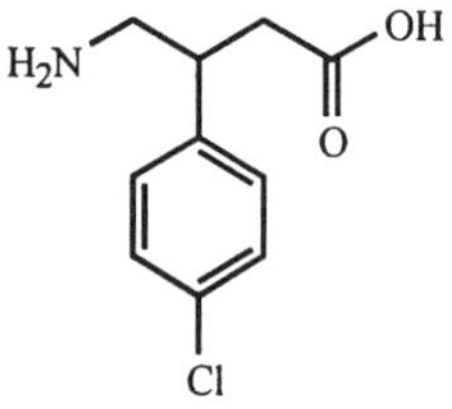

4-Amino-3-(4-chlorophenyl)butyric acid

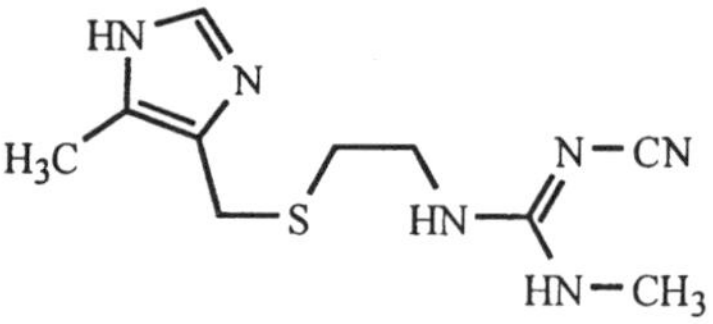

Cimetidine

9

NITRILES AND NITROGENOUS SOLVENTS

By far the most commonly used organic mobile phase component is acetonitrile. It offers a unique set of properties that set it apart from the other LC solvents:

1. It is moderately polar.
2. It has a midrange solvent strength.
3. It is, overall, an excellent solubilizing solvent and typically generates sharp well-defined chromatographic peaks.
4. It is miscible with a wide range of organic solvents as well as water, and its mixtures with water are of low viscosity when compared with analogous alcohol or ether mixtures.
5. It is a weak hydrogen bond acceptor.
6. It has a very low UV cutoff.

The areas of widest application for acetonitrile include pharmaceutical, amino acid, peptide and protein, and various biological compound analyses. Acetonitrile has found limited use in the analysis of polymers.

It should be noted, especially in light of all the positive properties that acetonitrile offers for use in LC, that acetonitrile presents at least four drawbacks: (1) Phosphate buffers, especially under multiply charged buffer pH conditions, have very limited solubility in acetonitrile; (2) acetonitrile is fairly unstable and reactive with strong acids (e.g., sulfuric acid); (3) it is one of the most expensive solvents used in large quantities in HPLC; and (4) when compared with the alcohols it presents an increased health hazard. However, with proper care and handling, acetonitrile is basically a safe and effective chromatrographic solvent.

Propionitrile and butyronitrile have been used in very limited and specific cases—namely, in the analysis of lipids when acetonitrile does not effectively solubilize the sample. The major drawbacks to the use of these solvents are lack of availablility in a high-purity grade (comparable to acetonitrile) and the increased health hazards associated with them as compared with acetonitrile.

Most of the nitrogenous compounds used as solvents in HPLC are polar strong hydrogen bonding species. The most common are triethylamine (TEA) and diethylamine. Pyridine may be included in this group; however, its high UV cutoff (330 nm) severely limits its utility. Some of the major advantages to their use are:

1. They have reasonably low UV cutoffs when fresh (not pyridine).
2. They are very volatile, so that a sample can be recovered easily.
3. They are strong hydrogen bonding compounds, making them ideal for use as silanol group blocking agents.
4. They are soluble in most solvent systems at a useful level (<2%).

TEA and diethylamine are very strong bases. Therefore, they will aggressively dissolve silica-based support materials. A precolumn should be used to protect the analytical column from a rapid and irreversible loss of bonded phase. A precolumn, which is positioned after the pump head and before the injector, contains packing material that matches the analytical column in functionality (e.g., octyl, octadecyl, underivatized base silica) but is made from large-particle (40–60 mesh) material. These columns are effectively and efficiently dry-packed in the laboratory but need frequent regeneration (i.e., all packing material is completely removed from the precolumn and replaced).

Anyone who has worked with these compounds will immediately be able to relate a major drawback to their use: They are noxiously odiferous (i.e., they stink!). Their characteristic "fishlike" odor, once experienced, is quite unforgettable. They also degrade (oxidize) readily, so they must be opened and used over a short period of time (weeks to months).

Dimethylacetamide and dimethylformamide are extremely polar hydrogen bond accepting compounds. They are such strong solvents that they have found little use in HPLC. They do offer unique selectivity properties and therefore should at least be kept in mind when developing complex separations of highly polar compounds.

9.1 IMPURITIES

Acetonitrile is typically a by-product of the large-scale production of acrylonitrile (from ammonia and propylene) and contains a wide range of very low-level impurities [884]; these include acrylonitrile, allyl alcohol, acrylic acid, and acetic acid. Because of the widespread use of acetonitrile in many synthetic and chromatographic applications, much attention has been given to the development of purification methods that essentially eliminate the presence of these impurities. Since the vast majority of acetonitrile use in LC is in reversed-phase separations and almost invariably one of the components is water, one "impurity" that is of little or no concern is water. The reason for mentioning this at all is that most manufacturers now offer a low-water-content premium-priced acetonitrile (<10 ppm water as compared to 30–100 ppm typical) that is intended for use in water-sensitive biosynthetic work. Purchase of this premium-cost material is rarely warranted for RP work.

Propionitrile and butyronitrile contain all the alcohol and acid impurities of acetonitrile plus their isomeric and unsaturated analogs. They are not readily available at purities of >97%. These solvents are expensive when compared with acetonitrile.

The impurities in diethylamine and triethylamine involve propyl and methyl substituents in place of an ethyl group. Depending on the method of production and purification, oxidized forms may also be present. These oxidation products are responsible for the increase in UV cutoff as the solvent ages (see Chapter 1). The presence of inorganic acids can produce salt formation and precipitates.

Pyridine is typically synthetically produced using any of a number of different starting compounds (e.g., crotonaldehyde, ammonia, water and formaldehyde). One advantage of pyridine is that it is difficult to oxidize [885]. Two corresponding disadvantages are the high UV cutoff (>300 nm) and the unpleasant odor. Pyridine often contains the methylated analogs (i.e., lutidines and picolines) and other saturated cyclic nitrogen-containing compounds. As with acetonitrile, pyridine is available as a premium-priced low-water solvent that is not necessary for ordinary chromatographic work.

Pure dimethylformamide (DMF) and dimethylacetamide are colorless and practically odorless. DMF slowly hydrolyzes to formic acid and dimethylamine when in contact with water [886]. The high UV cutoffs for these solvents (>260 nm) limit their utility.

Tables 9.1–9.4 list some important chemical, physical, and chromatographic properties as well as general manufacturing and safety parameters for the nitriles and nitrogenous solvents [84–92]. Figure 9.1 shows the structure of the solvents listed in Tables 9.1–9.4.

9.2 GENERAL ANALYTES

9.2.1 General Sample Solvent Considerations

Mobile phase/sample solvent mismatch is a common problem. As seen in the methanol example in Figure 1.16, peak shape can be grossly affected. Papers have been published dealing with the effect that the injection solvent has on peak shape in acetonitrile mobile phases as well. Hoffman et al. [887] used benzyl alcohol, tryptophan, cimetidine, and phenylalanine as test solutes. A C_{18} column was used

TABLE 9.1 Physical Properties of Nitriles and Nitrogeneous Solvents[a]

	ACN	PRN	DMF	DMA	DEA	TEA	Pyr
Molecular weight	41.05	55.08	73.10	87.12	73.14	101.19	79.10
Density (g/mL)	0.7822	0.7774	0.9487	0.9366[b]	0.707	0.726	0.9832
Viscosity (cP)	0.38[c]	0.29[b]	0.92	0.838[b]	0.388	0.394	0.95
Solubility in water (%)	100	10.3	100	100	100	100	100
Water solubility in solvents (%)	100		100	100	100	100	100
Boiling point (°C)	81.60	97.4	153.0	166.1	55	88.8	115.25
Melting point (°C)	−43.8	−92.8	−60.4	−20	−50	−115	−41.55
Refractive index (n_D)	1.3441	1.3655	1.4305	1.4384	1.3861	1.4000	1.5102
Dielectric constant	37.5	29.7	36.71[b]	37.78[b]	3.6	2.42	12.4
Dipole moment (D)	3.44	4.05	3.86[b]	3.72[b]	0.92	0.77	2.37[b]
Surface tension (dyne/cm)	19.10	26.75[b]	36.76	32.42[d]	20.4	20.7	36.88

[a] Abbreviations: ACN, acetonitrile, methyl cyanide; PRN, propionitrile, ethyl cyanide; DMF, N,N-dimethylformamide; DMA, N,N-dimethylacetamide; DEA; diethylamine; TEA, triethylamine; Pyr, pyridine.
[b] At 25°C.
[c] At 15°C.
[d] At 30°C.

332

TABLE 9.2 Chromatographic Parameters of Nitriles and Nitrogeneous Solvents[a]

	ACN	PRN	DMF	DMA	DEA	TEA	Pyr
Eluotropic streingth $\varepsilon°$ on Al_2O_3	0.65						0.71
Eluotropic strength $\varepsilon°$ on SiOH	0.52						
Eluotropic strength $\varepsilon°$ on C_{18}	3.1		7.6				
Solvent strength parameter, P'	5.8		6.4	6.5		1.9	5.3
Hildebrandt solubility parameter, δ	11.9	10.7	11.8	10.8		0.00	10.6
Hydrogen bond acidity, α	0.19		0.00	0.00		0.00	0.00
Hydrogen bond basicity, β	0.31	0.37	0.69	0.76		0.71	0.64
Dipolarity/polarizability, $\pi*$	0.75	0.71	0.88	0.88		0.14	0.87

[a] *Abbreviations:* ACN, acetonitrile, methyl cyanide; PRN, propionitrile, ethyl cyanide; DMF, *N,N*-dimethylformamide; DMA, *N, N*,-dimethylacetamide; DEA; diethylamine; TEA, triethylamine; Pyr, pyridine.

in the studies. Benzyl alcohol was chromatographed using an 20/80 methanol/water mobile phase and 0/100, 50/50, 67/33, and 100/0 acetonitrile/water mixtures as the solute solvent. Peak shape deteriorated as the acetonitrile level increased. At 100% acetonitrile, a distinct double peak was formed. Phenylalanine and tryptophan exhibited similar behavior when the same solute solvents were used but with an 8/92 acetonitrile/water (5 mM phosphate buffer at pH 3.5) mobile phase. Finally, cimetidine exhibited no peak doubling behavior under these conditions. Sample injection size was also studied. Vukmanic and Chiba [888] generated similar results for acetonitrile/water solute solvents using a C_{18} column and methyl-2-benzimidazole carbamate and 3-butyl-2,4-dioxo[1,2-*a*]-*s*-triazinobenzimidazole (STB) as test solutes. Their findings emphasize the effect of injection volume. For example, a 10 µL injection of STB yielded one well-shaped peak for any injection solvent composition of acetonitrile/water 50/50. However, the peak was considerably broader at 50 µL injected and a double peak was generated at 100 µL.

Another effect of the mobile phase/sample solvent mismatch is a change in the sensitiviy ($\Delta R/\Delta C$) of the method for an analyte. As an example, Perlman and Kirschbaum [889] prepared a series of solutes (e.g., captopril, nadolol, *o*-nitroaniline, triamcinolone acetate, methylparaben) in neat solvents: acetonitrile, methanol, DMSO, and dichloromethane. These solutions were then injected onto a C_{18} or phenyl column ($\lambda = 214$ nm or 270 nm) and eluted with 50/50 methanol/water or 38.8/1.1/960 ethanol/water/dichloromethane mobile phases. Significant differences in the peak areas resulted for some but not all analytes. Deterioration of peak shapes was also common. Prediction of these changes was nearly impossible. For example, *o*-nitoaniline (in methanol) exhibited an increased peak area in methanol/water, whereas *p*-nitroaniline was unaffected. An awareness of the unexpected and unpredictable effects the sample solvent has on both the quantitative results and the overall separation is critical when developing a method.

TABLE 9.3   Common Manufacturing Quality Specifications of Nitriles and Nitrogeneous Solvents[a]

	ACN	PRN	DMF	DMA	DEA	TEA	Pyr
UV cutoff (nm)	190		268	268			330
Percent water (maximum)	0.01		0.03	0.03			0.01
Available as ACS tested[a]	ABEFJM	n.a.[c]	AFJM		n.a.	n.a.	ABJM
Available as HPLC-grade[a]	ABEFJM		ABEJM	ABEM		BF	ABEJM
Available through[d]		AE			AEFJM	AJEM	

[a] *Abbreviations:* ACN, acetonitrile, methyl cyanide; PRN, propionitrile, ethyl cyanide; DMF, N,N-dimethylformamide; DMA, N, N,-dimethylacetamide; DEA; diethylamine; TEA, triethylamine; Pyr, pyridine.
[b] Manufacturer's code: A = Aldrich; B = Burdick & Jackson; E = EM Science; F = Fisher; J = JT Baker; M = Mallinckrodt.
[c] No ACS test exist for this solvent.
[d] Available as a high-purity solvent but not specifically designated as ACS or HPLC grade. This does *not* mean a lesser-quality solvent, just that it is not specifically tested for these applications. If the manufacturers produce either ACS or HPLC solvent, they are not listed under this heading.

9.2.2   Simple Substituted Hydrocarbon and Benzene Analytes Retention Studies

Virgin olive oil is characterized in terms of its complex phenolic compound content (elenolic acid, oleuropeine aglycone and dialdehyde, p-hydroxyphenylethanol, ligstroside dialdehyde, 3,4–dihydroxyphenylethanol) through extraction and analysis on a C_{18} column ($\lambda = 225$ nm). A complex 40-min $85/15 \rightarrow 66/34$ water (1 mM H_2SO_4)/acetonitrile gradient was used [890]. Baseline resolution was not obtained for multiple components, but discrete peaks are generated for each.

Maillard reaction products (5-hydroxy-2-methylfurfural, furosine, methylfurfural, furfural [2-furaldehyde], 2-furanal methyl ketone) were isolated from milk and separated on a C_{18} column ($\lambda = 280$ nm). An 80/20/0.2 water (5 mM sodium heptanesulfonate)/acetonitrile/formic acid mobile phase generated baseline resolution in 28 min [891]. The calibration curve for furosine covered the range of 0.1–0.8 µg/mL. The detection limit was reported as 0.63 µg/mL. Other furfurals were quantitated in the range 0.05–0.5 µg/mL.

The degradation process for a series of organic anhydrides (dimethylmaleic anhydride and acid, maleic anhydride and acid, crotonic acid, methylacrylic anhydride and acid, citroconic anhydride and acid) commonly used in industry were followed using a C_{18} column ($\lambda = 220$ nm) and a 70/30 water (10 mM ammonium phosphate at pH 2.5)/acetonitrile mobile phase [892]. Each pair (anhydride and acid) was chromatographed separately and all were well resolved under these conditions. Elution took from 5 to 22 min depending on the analytes. Injections of 25 nmol generated peaks about 0.002 AU.

S-Phenyl- and S-benzylmercapturic acids (metabolites of benzene and toluene) were extracted from urine, derivatized with monobromobimane, and analyzed on a

TABLE 9.4 Safety Parameters of Nitriles and Nitrogeneous Solvents[a]

	ACN	PRN	DMF	DMA	DEA	TEA	Pyr
Flash point[b] (TCC)	5.6	36	58	63	−28	6	20
Vapor pressure (Torr @ 20°C)	88.8[c]	47.22[c]	2.7	1.3[c]		54	20.73[c]
Threshold limit value (ppm)			10	10			
CAS number	7.5-05-8	107-12-0	68-12-2	127-19-5	109-89-7	121-44-8	110-86-1
Flammability[d]	3	3	2	2	3	3	3
Reactivity[d]	0	1	0	0	0	0	0
Health[d]	2	4	1	2	3	3	2

[a] *Abbreviations:* ACN, acetonitrile, methyl cyanide; PRN, propionitrile, ethyl cyanide; DMF, N,N-dimethylformamide; DMA, N, N,-dimethylacetamide; DEA; diethylamine; TEA, triethylamine; Pyr, pyridine.
[b] TCC = TAG closed cup.
[c] At 25°C.
[d] According to National Fire Protection Association ratings [92]:
Fire: 4 = Materials that vaporize at room temperature and pressure and burn readily.
           3 = Liquids or solids that can ignite under room conditions.
           2 = Materials that ignite with elevated temperature or with moderate heat.
           1 = Materials that must be preheated before they ignite.
           0 = Materials that will not burn.
React: 4 = Materials that, by themselves, can deteriorate or explode under room conditions.
           3 = Materials that can detonate or explode but require an initiator (e.g., heat).
           2 = Materials that undergo violent chemical reactions at elevated temperatures or pressures or react with water.
           1 = Materials that are, by themselves, stable but that may become unstable at elevated temperatures and pressures.
           0 = Materials that are stable even under fire conditions and do not react with water.
Health: 4 = Short exposure times to these materials are lethal or cause major residual injury.
           3 = Short exposure times to these materials cause temporary and/or residual injuries.
           2 = Lengthy (but not chronic) exposure to these materials may cause temporary incapacitation and/or minor residual injury.
           1 = Materials that, upon exposure, cause irritation but only minor residual injury.
           0 = Materials that, upon exposure under fire conditions, offer no more hazard than ordinary combustible materials.

35°C Supelcosil DP column ($\lambda = 375$ nm, ex; 480 nm, em). The elution was actually achieved in 15 min using an 85/15/0.1 water/THF/TFA mobile phase, but a 60/40/0.1 acetonitrile/water/TFA was necessary to rapidly elute all other extracted compounds [893]. Peaks were not baseline resolved from one another but were well resolved from all other extracted peaks. The linear range was reported as 10–250 µg/L with detection limits of 1 µg/L.

Thirteen fluorinated benzoic acids (e.g., 2-fluoro-, 2,6-difluoro-, 2,4,5-trifluoro-, 4-[trifluoromethyl]benzoic acid) were well resolved on a 35°C C_{18} column ($\lambda = 270$ nm) using an 84/16 (hold 7 min) → 20/80 (at 8 min hold 3 min) water (5 mM phosphate at pH 4.3)/acetonitrile gradient [894]. Linear curves were

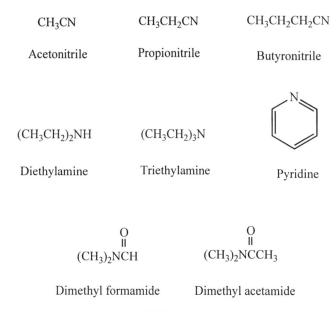

FIGURE 9.1

generated from 10 to 600 µg/L (analyte dependent) and detection limits of 10 µg/L were reported.

Hanai et al. [895] generated k values for 43 alkanes, alkylbenzenes, chloro and polychlorinated benzenes, and 41 alkylphenols, chlorophenols, and bromophenols on a C_{18} column using 60/40 to 90/10 acetonitrile/water mobile phases. Once again such a comprehensive tabulation of k values is extremely valuable for method development work.

In an excellent and thorough study, 15 chlorinated phenols were well resolved on a C_{18} column ($\lambda = 260$ nm) using a 31/79 acetonitrile/water (50 mL citrate buffer at pH 4) mobile phase [896]. Elution was complete in 90 min. The late-eluting peaks were sufficiently well resolved that a gradient would have been very effective at reducing analysis times. Plots of log k' vs. percent acetonitrile (30–46%) at pH 4 and 6 were generated. A further plot of the minimum resolution between all analytes versus pH (3–7) and percent acetonitrile (30–64) was presented.

Fourteen chloro-, chloromethyl-, and dichloroanilines were extracted from water and analyzed on a C_{18} column ($\lambda = 240$ nm) using a 36/64 acetonitrile/water (10 mM phosphate buffer at pH 7) mobile phase [897]. The samples were preserved with hydrazine sulfate. Without a preservative, up to 70% of the anilines were degraded within 24 h of collection. All components were eluted in 22 min and overall resolution was quite good, especially considering that a number of unknown peaks eluted as well. Detection limits "well below 0.1 μg/L" were reported.

In an extensive study of the amide metabolites of N-benzylanilines, Ulgen et al. [898], tabulated the retention times for 43 compounds on a C_{18} column ($\lambda = 254$ nm). A series of isocratic acetonitrile/water mobile phases were used. Typically, sets of four or five compounds were separated. For example, benzaldehyde, 2,4,6-trimethylaniline, N-benzoyl-2,4,6-trimethylaniline, N-(2,4,6-trimethylphenyl)-α-phenylnitrone, and N-benzyl-2,4,6-trimethylaniline were baseline resolved in 30 min using a 40/60 acetonitrile/water (phosphate buffer at pH 7) mobile phase. The results for six other sets of analytes were tabulated as well. Typical acetonitrile/water (phosphate buffer at pH 7) mobile phase compositions ran from 40/60 to 50/50. Elution times varied between 2.5 and 36 min. If the other separations are anything like the one seen in the only chromatogram shown, this is a very valuable and thorough set of results.

In an interesting study, McCalley [124] studied the effect of the organic mobile phase constituent (acetonitrile, methanol, or THF) on the peak shapes of 16 basic amine compounds such as pyridine and monoalkyl- and dialkyl-substituted pyridines on a C_{18} column. The mobile phase compositions were adjusted so that the k' values were at least comparable. It was uniformly shown that a 40/60 acetonitrile/water (37.5 mM phosphate buffer at pH 7.0) mobile phase resulted in slightly lower k' values but consistently higher asymmetry values (i.e., peaks exhibited greater tailing) than either a 55/45 methanol/water (50 mM phosphate buffer at pH 7) or a 25/75 THF/water (30 mM phosphate buffer at pH 7) mobile phase. The author did not point to the solvent directly as the cause for such differences; rather he postulated that the solvent effect on the protonation of the solute was the driving factor. Hence, if the solute is more likely protonated in the acetonitrile mobile phase, then the solute will undergo strong hydrogen bond interactions with the surface silanol groups. The solvent system giving the best results (in terms of peak symmetry) was that containing THF. This should not be surprising since THF is well known to sharpen peaks when added at low volumes (2–10%) to mobile phases, most likely due to its being a strong hydrogen bond acceptor (and therefore interacting strongly with residual silanol groups).

An extensive table of $\log k'$ values for 32 phenols and substituted phenols (nitro, chloro, bromo, alkyl), benzene and substituted benzenes (alkyl, chloro, bromo), and naphthalene, anthracene, pyrene, and chrysene was generated by Bosch et al. [107]. Mobile phase compositions covered the entire 100% water to 100% acetonitrile range. No values were generated for $\log k'$ values >2. Although these data were used to help generate a predictive retention scheme, the results are an excellent resource for use in method development.

Aminophenols (*ortho*, *meta*, *para*), dihydroxybenzenes (pyrocatechol, resorcinol, hydroquinone), and *p*-phenylenediamine are typical hair dye components [899]. These were isolated from product and separated on a polystyrene column ($\lambda = 280$ nm) using a 25/75 acetonitrile/water (0.3 M ammonium phosphate at pH 5.15) mobile phase. Peaks are somewhat tailed, but good resolution is achieved nonetheless. Plots showing the effects of changing buffer and acetonitrile concentration are also presented. Standard ranges from 0.1 to 50 µg/mL and detection limits of 0.05–0.16 µg/mL (analyte dependent) were reported.

Sodium benzoate and potassium sorbate preservative levels in foods (juices, soda, sauces) were determined using a C_{18} column ($\lambda = 225$ nm for benzoate, 255 nm for sorbate) and a 90/10 water (0.3 g ammonium acetate/L to pH 4.2 with acetic acid)/acetonitrile mobile phase [900]. Peaks were somewhat broad but well resolved and eluted in <15 min. Linear curves were reported as 2.5–200 mg/L with detection limits of 1 mg/L.

Five hydroxyaromatic compounds (1- and 2-hydroxy-, and 2,7-dihydroxynaphthalene, 5-hydroxy-1,4-naphthoquinone, 4,4'-dihydroxybiphenyl) were extracted from water and baseline resolved on a phenyl column ($\lambda = 254$ nm). Elution was complete in 15 min using a 22/78 acetonitrile/water (0.1 M acetate at pH 4.7) mobile phase [901]. The stated linear range was 0.002–10 mg/L with detection limits of 0.7–3.2 µg/mL (analyte dependent).

Four furans (2-furaldehyde, 2-acetyl furan, 2-methyl furan, 5-(hydroxymethyl)-2-furfural) and nine phenolics (e.g., gallic acid, 3,4,5-trimethoxyphenylacetic acid, vanillin, coniferaldehyde) were extracted from distilled spirits (whiskey,

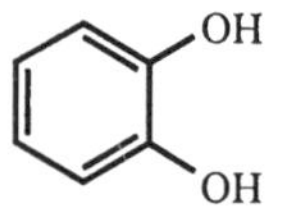

Pyrocatechol

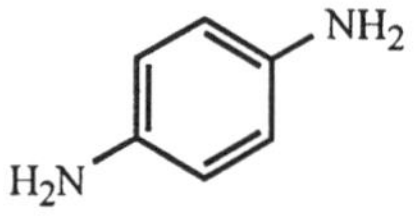

p-Phenylenediamine

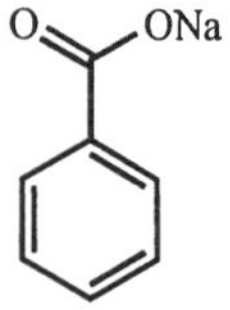

Sodium benzoate

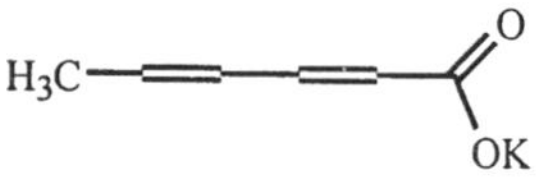

Potassium sorbate

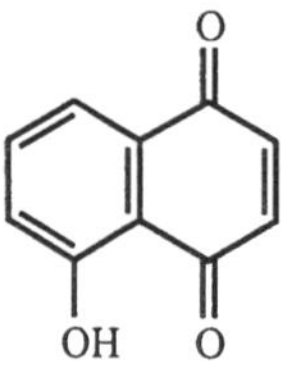

5-Hydroxy-1,4-naphthoquinone

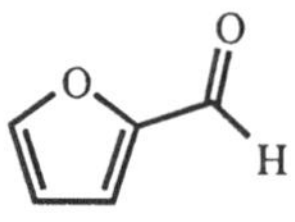

2-Furaldehyde

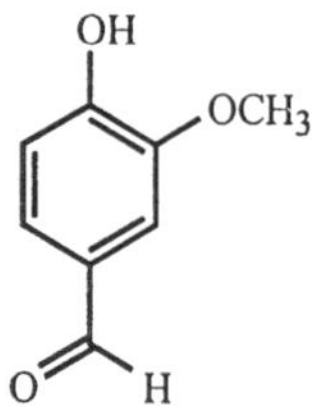

Vanillin

brandy, rum) and analyzed on a 25°C C_{18} column (photodiode array detector varied to match individual analyte peak absorbance which ranged from 255 to 341 nm). A complex nonlinear 155-min 90/10 → 50/50 water (0.2 mL H_3PO_4 with 2 mL acetic acid/L to pH 2.1)/acetonitrile (0.2 mL H_3PO_4 with 2 mL acetic acid/L to pH 2.4) gradient gave good separation for all peaks of interest. (Note that the pH for acetonitrile is an estimate due to the nonaqueous nature of the solvent.) The reported linear ranges for the analytes were given as 0.2–500 mg/L (except egallic acid at 0.03–100 mg/L) and the detection limits as ≥ 0.01 mg/L [902].

9.2.3   Other Compounds

Methyl-, ethyl-, propyl-, and butylparabens were extracted from various pharmaceutical formulations and quantitated using a C_{18} column ($\lambda = 230$ nm) and a 40/60 acetonitrile/water (6.5 mM NaH_2PO_4 buffer at pH 5.5) mobile phase [903]. Elution was complete in <7 min and complete resolution was obtained. Injections of 10 μL aliquots of standards ranging from 10 to 30 μg/mL were made. Detection limits of ~1 μg/mL (S/N = 3) were reported.

The flavor additive *trans*-4-phenyl-3-buten-2-one, the -2-ol analog, and 4-phenyl-2-butanone were extracted from liver microsomes and cytosol and baseline resolved on a C_{18} column ($\lambda = 260$ nm) using a 40/60 acetonitrile/water mobile phase [904]. Elution was complete in <25 min.

The reaction of soybean peroxidase on methyl *N*-methylanthranilate produces methylanthranilate [905]. This conversion was followed by separation on a C_8 column ($\lambda = 332$ nm) using a 50-min 100/0/0.1 → 50/50/0.1 water/acetonitrile/TFA gradient. Elution was complete in <30 min, peaks were well resolved, and peak shape was excellent. No interferents were observed, and so a shorter, steeper gradient seems advisable.

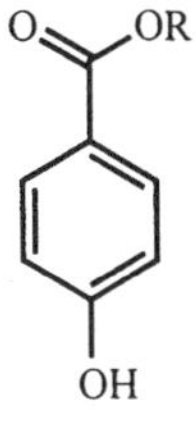

Paraben;
R = methyl, ethyl, etc

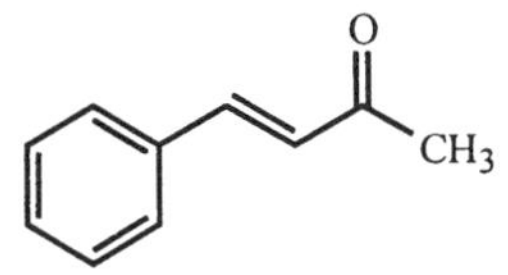

trans-4-Phenyl-3-buten-2-one

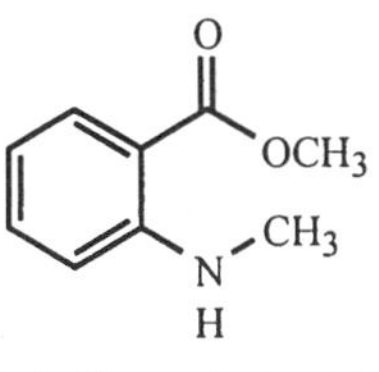

Methyl *N*-methylanthranilate

Nitric oxide is a by-product of the action of the enzyme NO-synthetase on L-arginine in tissues. The NO product is reacted with 3-amino-4-(N-methylamino)-2′,7′-difluorofluorescein and analyzed on a 30°C C_{18} column ($\lambda = 500$ nm, ex; 515 nm, em) using a 94/6 water (10 mM Na_2HPO_4 buffer at pH 7.2)/acetonitrile mobile phase [906]. Elution was complete in 14 min. A linear range of 2–200 nM was claimed.

N-Acyl- and N-glycolylneuraminic acids (sialic acids) were isolated from serum and tissue samples, per-O-benzoylated, and analyzed on a C_{18} column ($\lambda = 231$ nm) using a 67/33 water/acetonitrile mobile phase [907]. Good resolution was achieved and elution was complete in <20 min. Calibration standards of 0.05–50 μg injected were used and detection limits of 10 ng injected were reported.

Benzothiazole, 2,2′-(dithiobis)benzothiazole, and 2-mercapto-, 2-(methylthio)- and 2-(thiocyanomethylthio)benzothiazole were baseline resolved on a C_{18} column ($\lambda = 250$ nm, 380 nm, or 325 nm) using a complex 20-min 63/37 → 90/10 acet-onitrile/water (4 mM NaH_2PO_4 buffer at pH 4.5) gradient. Peak shapes were excellent. Extraction from industrial wastewater samples gave detection limits in the low (0.1–10) ng injected range [908].

Sodium tetraphenylborate (STB) degradation products (tri-, di-, and mono-phenylborate and phenol) were recovered from river water and analyzed on a C_{18} column ($\lambda = 219$ nm). The polyphenyl borates were separated in 15 min using a 36/28/36 acetonitrile/methanol/water (0.1% diammonium hydrogenphosphate) mobile phase [909]. The STB peak shape was extremely fronted. This could be due to co-eluting impurities or could possibly be due to charge effects, and a different buffer might produce better chromatography. Phenol and phenylborate ($\lambda = 217$ nm) were resolved and eluted in 10 min with a 30/70 acetonitrile/water

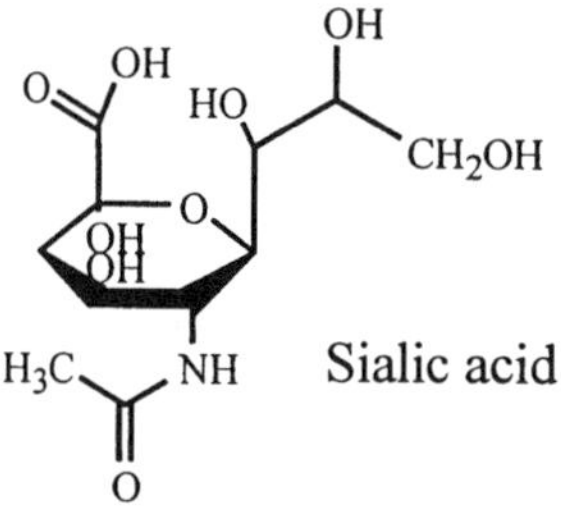

Sialic acid 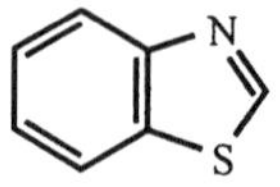   Benzothiazole

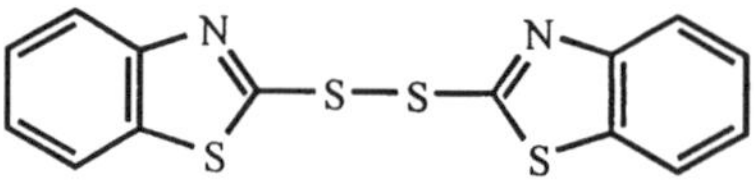

2,2′-(Dithiobis)benzothiazole 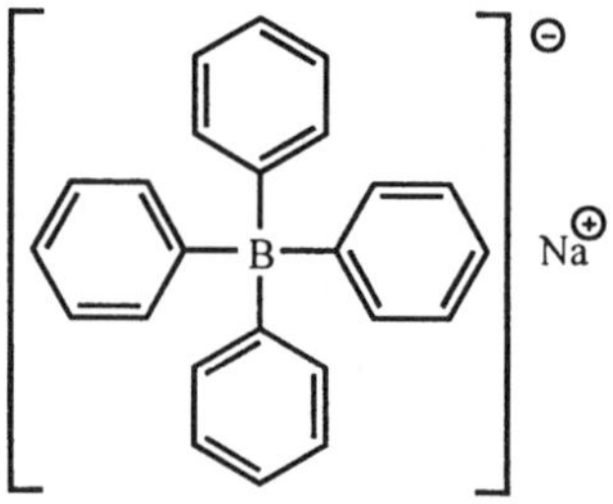   Sodium tetraphenylborate

mobile phase. Peak shapes were good. Spiked sample concentrations ranged from 50 to 500 mg/L.

In a novel approach to the separation of thiosulfate, thiocyanate, and polythionates (tri-, tetra-, penta-, and hexathionate) Miura and Kawaoi [910] achieved baseline separation on a C_{18} column ($\lambda = 230$ nm) using an 80/20 water/acetonitrile mobile phase that contained 6 mM tetrapropylammonium hydroxide (TPAH) adjusted to pH 5.0 with acetic acid. Elution was complete in 22 min. The authors tabulated UV and conductivity detector sensitivity and found that the UV detection gave better results. Also, a lower UV wavelength did not increase sensitivity because of the increase of background absorbance due to the acetic acid and TPAH. Detection limits were reported in the 10–60 nM range (S/N = 3).

Six aromatic oligo-guanylhydrazones (N,N'-bis[3,5-diacetyl]decanetetrakis-, 2,6-diacetylaniline-, 1,4-diphenyl-1,4-butanedione-bis-, N,N'-bis[3,5-diacetyl-phenylhexanediamide]-tetrakis-, 3,3'-[trimethylenedioxy]-di-p-anisaldehyde-bis-, N-[4-acetylphenyl]-N'-[3,5-diacetylphenyl]urea-trisamidinohydrazone]) were separated on a C_8 column ($\lambda = 265$ nm). A 30-min $100/0 \rightarrow 75/25$ water (10 mM heptane sulfonate with 10 mM tetramethylammonium chloride and 4.2 mM H_3PO_4)/ acetonitrile gradient generated good resolution [911]. Linear ranges of 0.1-50 μg/mL and quantitation limits of ~0.8 μg/mL were reported.

A series of seven p-t-butylcalixarenes (p-t-butylcalix[n]arene where $n = 4$–10) and bishomooxacalix[4]arene (condensation products of p-t-butylphenol and formaldehyde) were resolved on a C_{18} column ($\lambda = 288$ nm) using a 15-min $90/10/0/0.1 \rightarrow 40/0/60/0.1$ acetonitrile/methanol/ethyl acetate/TFA gradient [912]. Peaks for 5 ppm standards were easily detected.

Asakawa et al. [913] studied the separation of three synthetic interlocked macrocycles ([2]catenane, [3]catenane, [2]rotaxane) on a C_{18} column ($\lambda = 260$ nm) using a $100/0/0.1$ (hold 8 min) $\rightarrow 5/95/0.1$ (at 28 min hold 14 min) water/acetonitrile/TFA gradient. Each of these products were produced through the cyclization of a polyether through a cyclobis(paraquat-p-phenylene) or a cyclobis(paraquat-4,4'-biphenylene) macrocycle. As a consequence, the separation included the precursor molecules as well as the desired product. In each case, baseline resolution was achieved in <30 min. Stock solution concentrations of 1 mM were used and readily detected.

Two metabolites of $trans$-4-phenyl-3-buten-2-one (4-phenyl-2-butanone, $trans$-4-phenyl-3-buten-2-ol) were extracted from blood and separated on a C_{18} column

A simple guanylhydrazone;
methylglyoxal bis-guanylhydrazone

Calix[6]arene

($\lambda = 2660\,\text{nm}$) using a 40/60 acetonitrile/water mobile phase [914]. Excellent peak shapes and resolution along with a 25-min elution time were achieved.

This study dealt with the separation and analysis of eleven 2,4-dinitrophenyl-hydrazine (DNPH) derivatized aldehydes and ketones (formaldehyde, acetaldehyde, acetone, hexanal, propanal, crotonaldehdye, butanal, pentanal, acrolein, benzalde-hyde, *p*-tolualdehyde). A small nonporous C_{30} column, $33\times4.6\,\text{mm}$, ($\lambda = 360\,\text{nm}$) and a $75/25 \rightarrow 38/62$ water/acetonitrile gradient generated a complete elution in 3 min. Acrolein and propanal co-eluted. Similar results were obtained on a C_{18} column ($150\times4.6\,\text{mm}$) using a 15-min $51/49 \rightarrow 20/80$ water/acetonitrile gradient [915]. Hexanal, heptanal, octanal, and nonanal were isolated from air and also derivatized with DNPH and separated in 5 min on a C_{18} column ($\lambda = 360\,\text{nm}$) using an 80/20 acetonitrile/water mobile phase [916]. Linear curves were generated from 50 to 1000 ng/mL with detection limits reported at 20–80 ng injected (analyte dependent).

A series of alkanals and alkenals, degradation products of polyunsaturated fatty acids, were separated as their dabsylhydrazine derivatives on a C_{18} column ($\lambda = 436\,\text{nm}$) using a 60-min $40/60 \rightarrow 80/20$ acetonitrile/water gradient [917]. Acrolein, propionaldehyde, crotonaldehyde, butyraldehyde, 4-hydroxy-2-nonenal, malonaldehyde, *trans*-2-pentenal, hexanal, heptanal, *trans*-2-octenal, octanal, *trans*-2-nonenal, and nonanal were baseline resolved. When malonaldehyde and trans-2-hexenal were included in the set, they co-eluted with pentanal. Peak shapes were excellent and detection limits of 5 ng injected (S/N = 4) were reported.

Acrolein, crotonaldehyde, and methacrolein were separated as their anthrone derivatives on a C_{18} column ($\lambda = 405\,\text{nm}$, ex; 480 nm, em) using a 60/40 acetonitrile/water mobile phase [918]. Elution was complete in 10 min and peak shapes were very good. Detection limits of 5 ppb and linear concentration ranges from 20 ppb to 14 ppm were reported.

The retention characteristics of 4,4'-disubstituted benzanilides and benzamides (chloro, methyl, methoxy, nitro) were conducted on a C_{18} column ($\lambda = 254\,\text{nm}$) using a series of acetonitrile/water mobile phases [919]. The contribution of the substituent to retention was tabulated and a predictive model for the calculation of $\log k'$ values was presented. Mobile phase compositions from 35/65 to 70/30

2,4-Dinitrophenylhydrazine

Dabsylhydrazine

Benzanilide   Benzamide

acetonitrile/water were used and a plot of $\log k'$ values (calculated) versus $\log k'$ (actual) was also presented.

9.2.4 Organometallics and Metal–Ligand Complexes

Selenols (RSeH), diselenides (RSe-SeR) and selenyl sulfides ($R'S$-SeR) were separated on a C_{18} column (electrochemical detector at -1.10 V vs. Ag/AgCl). A 5/95 acetonitrile/water (5 mM phosphate buffer at pH 2.9 with 40 mg/L sodium octylsulfate) mobile phase was used [920]. A plot of k' versus percent acetonitrile for five such compounds was presented for the range of 1–8%. The k' values varied from 4–50 at 1% acetonitrile to <0.2–4 at 8% acetonitrile. Selectivity of detection was achieved by setting the potential to -0.55 V (for diselenides and selenyl sulfides) or $+0.15$ V (for selenols only).

Vanadium(V), nickel(II), zinc(II), copper(II), cobalt(III), aluminum(III), iron(III), and manganese(III) were separated in 26 min as their 8-hydroxyquinoline complexes using a C_{18} column ($\lambda = 390$ nm) and a 40/60 acetonitrile/water (20 mM sodium acetate pH 7.5) with 5 mM 8-hydroxyquinoline mobile phase [921]. Good overall resolution was obtained. Peaks were somewhat broad. The standard concentration was 50 ng/mL. The effect of changes in percent acetonitrile and 50/50 organic/buffer (where organic was 1,4-dioxane, acetone, or THF) were plotted. However, no chromatograms generated under these conditions were presented.

Trace levels of vanadium(V) were extracted from water and coal fly ash samples and analyzed as the 2-(8-quinolylazo)-5-(dimethylamino)phenol complex [922]. A C_{18} column ($\lambda = 550$ nm) and a 50/50 acetonitrile/water (10 mmol/kg tetrabutyl-ammonium [TBA] bromide) mobile phase generated the separation. A number of potentially co-eluting metal ions were studied as well: Ni, Co, and Fe. Detection limits of 3 pg were reported. Vanadium(IV) and (V), Pb(II), Cu(II)and Cr(VII) were resolved on a C_8 column ($\lambda = 245$ nm) with a 12/78 acetonitrile/water (50 mM TBA with 2 mM EDTA to pH 6) mobile phase [923]. Elution was complete in <20 min and peak shapes were good. Detection limits of 1 ng and linear concentration ranges of 1–30 μg/mL were reported.

The carbonyl complexes of group VI (Cr, W, Mo) as $M(CO)_6$ and Group VII (Mn$_2$, Re$_2$, MnRe) as $M_2(CO)_{10}$ were resolved by Li et al. [924] on a C_{18} column. A 10-min group VI separation was achieved using a 70/30 acetonitrile/water mobile phase, whereas a 10-min Group VII separation required an 80/20 acetonitrile/water mobile phase. In the same paper, a series of alkylphosphines ($\phi 2P\text{-}(CH_2)_n\text{-}P\phi_2$, where $n = 1$–6) were also separated on a C_{18} column ($\lambda = 254$ nm). Elution was generated in 25 min using a 90/10 acetonitrile/water mobile phase.

8-Hydroxyquinoline

Two texaphyrin (aromatic pentadentate ligands) complexes, motexafin gadolinium and lutetium, were extracted from plasma and separated on a 55°C C_{18} column ($\lambda = 470$ nm) using a 59/21/20 water (0.1 M ammonium acetate to pH 4.3 with acetic acid)/acetonitrile/methanol mobile phase. Peaks were well resolved and elution was complete in 20 min. Linear ranges from 0.01–30 μM and quantitation limits of 0.01 μM gadolinium and 0.1 μM lutetium were reported [925].

A series of eight substituted 3-hydroxypyridin-4-one (e.g., *N*-methyl-2-methyl-, *N*-ethyl-2-ethyl-, *N*-methyl-2-amide-6-methyl-) iron chelates were separated on a PRLP-S column ($\lambda = 280$ nm) using a 20-min $98/2 \rightarrow 65/35$ water (5 mM sodium heptanesulfonate to pH 2.0 with HCl)/acetonitrile gradient [926]. Note that HCl is very aggressive toward stainless steel components and another acid should be tried to replace it here. Baseline resolution and excellent peak shapes were obtained. The retention times for 22 other analogs are also tabulated.

Gold(I) was quantitated as its cyano complex by HPLC using a C_{18} column ($\lambda = 214$ nm) and a 32/68 acetonitrile/water (5 mM TBA). Note that cyanide ion is extremely toxic and its use poses a great health hazard. Make sure that the solution pH is always extremely basic and that the effluent is disposed of as a segregated hazardous waste. The complex eluted in 8 min. Pd(II) and Pt(II) were resolved from Au(I) when the acetonitrile level was decreased to 23%. In this case elution was complete in 35 min. The detection limit for Au(I) was reported as 0.4 ppb [927].

Four organomercury compounds (methoxyethyl, ethyl, phenyl, and methyl) were extracted from water and separated from inorganic mercury (Hg^{2+}) as their pyrrolidinethiocarbamate (PDTC) complexes [928]. A C_{18} column (λ not given) and a 60/40 acetonitrile/water (5 mM sodium PDTC to pH 5.5 with ammonium acetate) mobile phase resolved these compounds in <10 min. Detection limits of 5 ng were reported.

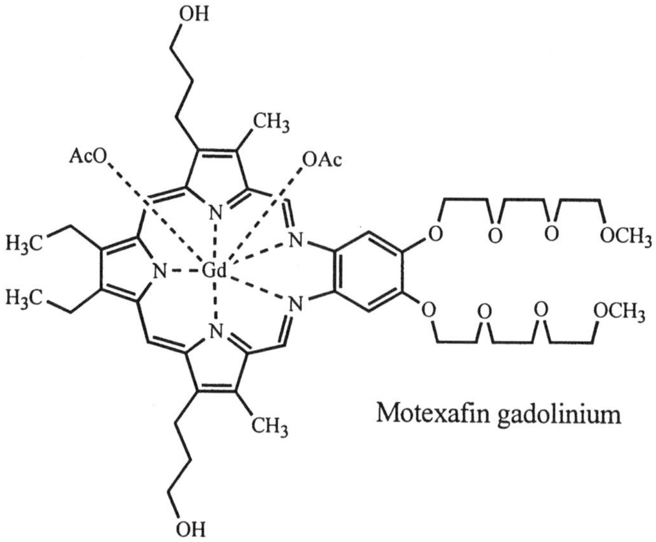

Motexafin gadolinium

9.2.5 Summary

Acetonitrile is successfully used for the separation of a wide range of analytes, both polar and nonpolar. Typically its inclusion in the mobile phase results in sharper peaks and shorter retention times as compared with alcohols. Acetonitrile is the most commonly used RP solvent because of these facts and the unique set of properties it brings to HPLC: low UV cutoff and low viscosity mixtures when used with water.

9.3 ENVIRONMENTALLY IMPORTANT ANALYTES

9.3.1 Substituted Benzenes and Related Analytes

3-Methyl-4-trifluoromethylaniline and four metabolites (e.g., 2-trifluoromethyl-5-acetamidobenzoic acid and 3-methyl-4-trifluoromethylaniline-N-glucuronide) were isolated from urine and separated on a base deactivated C_{18} column ($\lambda = 254$ nm). Note that this experiment utilized an in-series NMR detector and so deuterated solvents were used. If UV is the detector, this expense is not necessary. A 30-min 95/5 → 50/50 water (50 mM KH_2PO_4 to pH 2.5 with H_3PO_4)/acetonitrile gradient generated good separation [929].

Alarcón and co-workers [930] successfully separated nine priority pollutant substituted phenols (e.g., 2,4-dimethyl, 2-chloro, 2- and 4-nitro, 2,4,6-trichloro) from phenol using a C_{18} column ($\lambda = 280$ nm) and a 34/10/56 acetonitrile/ methanol/water (30 mM ammonium acetate buffer at pH 5.0 with 0.15 mM cetyltrimethylammonium bromide [CTAB]) mobile phase. Good resolution was obtained but it was noted that this result was very dependent on the C_{18} column/CTAB concentration combination used. Elution was complete in <15 min.

The separation of nine phenolic degradation products (e.g., gallic acid, acetovanillone, anisic acid, acetosyringone) found in foundry sands was optimized on a C_{18} column ($\lambda = 275$ nm) using a 96/3/1 (hold 8 min) → 72/3/25 (at 48 min hold 10 min) water/acetic acid/acetonitrile gradient [931]. Baseline resolution was not achieved between all peaks, but all peaks were readily identifiable. The use of a concave gradient did not give any better resolution. This is an interesting paper in that the thought process for optimizing the separation is presented in detail.

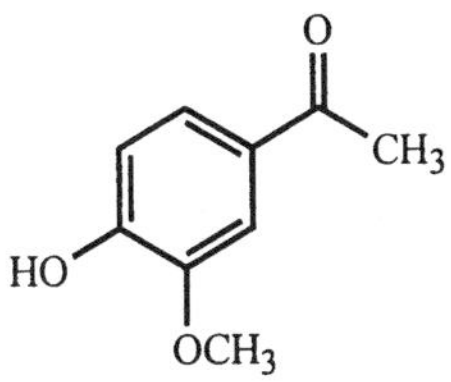

Acetovanillone 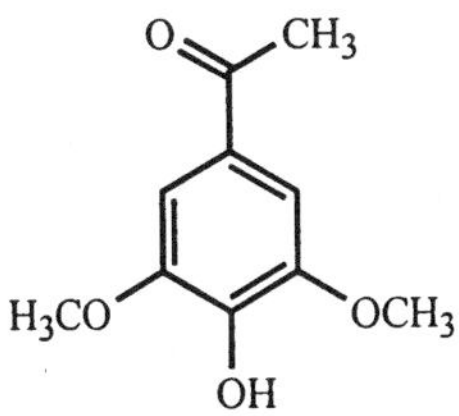   Acetosyringone

Guaiacol, bromoxynil, ioxynil, trichlorosyringol, *o*-chlorovanillin, dinoseb, and 11 chloro-, nitro- and methyl-substituted phenols were baseline resolved on a C_{18} column using a 24-min 33/64 → 75/25 (90/10 acetonitrile/methanol with 0.015% TFA)/water (0.05% TFA) gradient. The monitored wavelength was changed from 280 to 230 nm at 7.8 min. Peak shapes were uniformly excellent. Detection limits were <20 ng injected for each analyte. With a preextraction process described in the paper, overall detection limits were reported as 0.1 μg/L [932].

Thirteen chloroanilines (two chloroaniline isomers, five chloromethylaniline isomers, five dichloroaniline isomers, and 5-chloro-2-hydroxyaniline) and aniline were extracted from river and drinking water samples [933]. They were baseline resolved on a C_{18} column ($\lambda = 245$ nm or electrochemical detection at +0.95 V vs. Ag/AgCl) using a complex 60-min 26/74 → 40/60 acetonitrile/water (50 mM acetate buffer at pH 4.9) gradient. Excellent resolution and peak shapes were obtained. Detection limits were reported as 15–25 ng/L (S/N = 5) with UV detection and 3–5 ng/L for electrochemical detection. A table of analyte λ_{max} and pK_a values was presented.

9.3.2 PAHs, Substituted PAHs, and Related Analytes

A standard PAH separation is EPA method 610 (or SW-846 8310). This calls for the use of a C_{18} column ($\lambda = 254$ nm or 280 nm, ex; 389 nm cutoff, em) and a 25-min 40/60 → 100/0 acetonitrile/water gradient to separate 16 priority pollutants (naphthalene to indeno[1,2,3-*cd*]pyrene). Fluorescence detection limits of less than 0.1 μg/L are cited. Wise et al. [934] reviewed PAH analyses. Both monomeric and polymeric C_{18} columns were studied with respect to retention and separation characteristics. A typical method was a 30-min 40/60 → 100/0 acetonitrile/water gradient with UV detection at $\lambda = 254$ nm. A large number of references that include slight variations of this and the EPA 610 scheme are cited within this review.

A series of methylated chrysenes, picenes, and perylenes were separated on a 28°C C_{18} column ($\lambda = 265$ nm, ex; 365 nm, em) using an 80/20 acetonitrile/water (for chrysenes) or a 95/5 acetonitrile/water (for picenes and perylenes) mobile phase [935]. The authors noted that, against expectations, some of the methylated solutes eluted *prior* to their unsubstituted analogs. A chromatogram supporting this showed that 5-methyl- and 6-methylchrysene eluted before chrysene on a Bakerbond C_{18} column. This behavior was attributed to a combination of the length : width ratio of the solute, the position of the methyl substituent (both on the rings and with

Bromoxynil

3,4,5-Trichlorosyringol

Dinoseb

respect to the ring plane), and the characteristics of the support material (pore size, carbon loading).

Animal bedding was analyzed for 9,10-dimethyl-1,2-benzanthracene levels [936]. Extracts were characterized using a C_{18} column ($\lambda = 254$ nm) and a 90/10 acetonitrile/water mobile phase. A co-extracted peak was not fully resolved from the analyte, therefore a weaker mobile phase should be considered. Elution was complete in 3 min. A calibration curve from 1 to 20 ppm was used.

The retention of 34 nitrated PAH compounds (e.g., quinolines, naphthalenes, biphenyls, anthracenes, fluorenes) was studied on a C_{18} column using mobile phases consisting of 50/50, 60/40, and 70/30 mixtures of acetonitrile/water at 35°C, 45°C, 55°C, and 65°C [179]. Aliquots (1 μL) of samples containing 1 mg/mL of each compound were injected. The log k' value was tabulated for each compound under each mobile phase/temperature combination. This is an excellent reference for those seeking detailed information on the retention behavior of nitro-PAHs.

Four nitropyrenes (1-nitropyrene, 1,3-, 1,6-, and 1,8-dinitropyrene) were isolated from airborne particulates and separated in 17 min on a 40°C C_{18} column (electrochemical detector, 0.60 V vs. Ag/AgCl) using a 70/30 acetonitrile/water (10 mM sodium monochloroacetate at pH 4.7) mobile phase [937]. A plot of response versus pH showed a distinct maximum at 4.7. Working curves from 0.006 to 1.6 μg/mL (analyte dependent) and detection limits of ~20 pg injected were reported.

In an impressive set of separations Kettrup et al. [938] analyzed 15 azaarenes (e.g., 4-azafluorene, acridine, 4-azapyrene, benzo[h]quinoline, dibenz[c,h]acridine) and seven amino-polyaromatic hydrocarbons (amino-PAH) (e.g., 1-aminonaphthalene, 2-aminofluorene, 6-aminochrysene) on a PAH column. The azaarenes were baseline resolved in 40 min using a 30.7°C system with a two-ramp 40/60 → 70/30 (at 16 min) → 75/25 (at 43 min) acetonitrile/water (20 mM tris(hydroxymethyl)-aminomethane to pH 6.9 with HCl) gradient. Fourteen separate fluorescent excitation/emission combinations were utilized to optimize response. All excitation wavelengths were 246–290 nm and all emission wavelengths were 330–463 nm. A standard of 1 ng injected produced readily detectable peaks. For the amino-PAHs, the same column and system temperature was used but the gradient was changed to

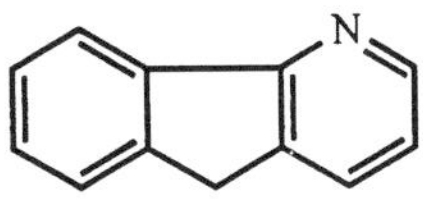

4-Azafluorene

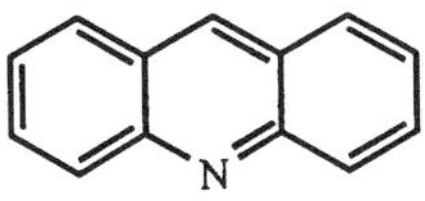

Acridine

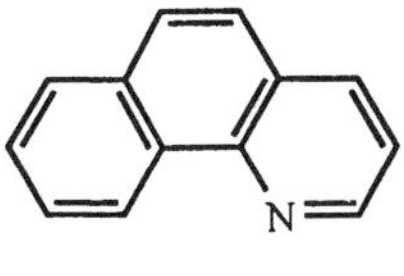

Benzo[h]quinoline

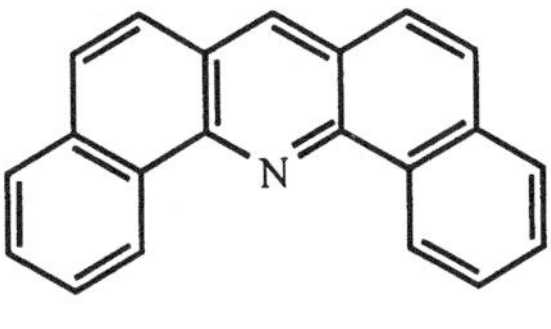

Dibenz[c,h]acridine

30/70 → 80/20 acetonitrile/water (tris buffer as above). Here, standards of 1.25 ng injected were used. The authors reported detection limits ranging from 7 to 160 pg injected (analyte dependent).

A set of five carcinogenic heterocyclic amines (2-amino-3-methylimidazo[4,5*f*]quinoline, 2-amino-3,8-dimethyl-, 2-amino-3,4,8-trimethyl-, and 2-amino-3,4,7,8-tetramethylimidazo[4,5*f*]quinoxaline,   2-amino-1-methyl-6-phenylimidazo [4,5*b*]pyridine) were analyzed on a C_8 column [939]. The quinoline and di- and trimethylquinoxalines were baseline resolved in 15 min using a 15/85 acetonitrile/water (1.4 mL triethylamine/L with H_3PO_4 to pH 3.2) mobile phase. The quinoline and dimethylquinoxaline co-eluted but all other analytes were resolved using a 30/70 acetonitrile/water (1.4 mL triethylamine/L with H_3PO_4 to pH 3.2) mobile phase. The reported linear range was 5 ng/mL to 10 μg/mL with detection limit of ~20 pmol injected (S/N = 2).

Eight heterocyclic aromatic amine carcinogens were extracted from cooked meats. A C_{18} column and a complex 30-min 95/5 → 45/55 water (10 mM triethylamine phosphate at pH 3.6)/acetonitrile resolved 2-amino-1-methyl-, 2-amino-3,8-dimethyl-, and 2-amino-3,4,8-trimethylimidazo[4,5-*f*]quinoxaline with monitoring by UV ($\lambda = 262$ nm) and fluorescence ($\lambda = 307$ nm, ex; 370 nm, em). A 40-min 95/5 → 20/80 water (25 mM triethylamine phosphate at pH 3.6)/acetonitrile gradient on the same column resolved 2-amino-1,6-dimethylimi-

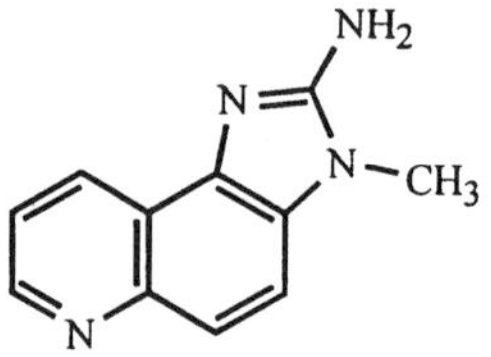

2-Amino-3-methylimidazo[4,5*f*]quinoline

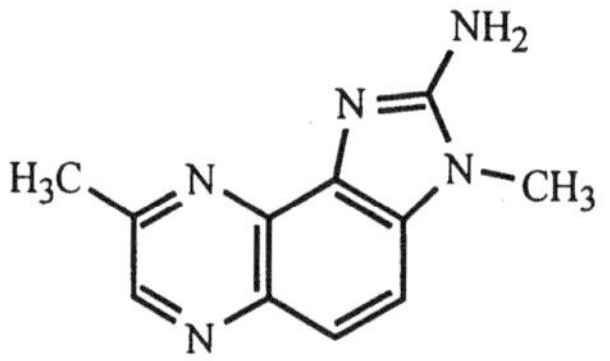

2-Amino-3,8-dimethylimidazo[4,5*f*]quinoxaline

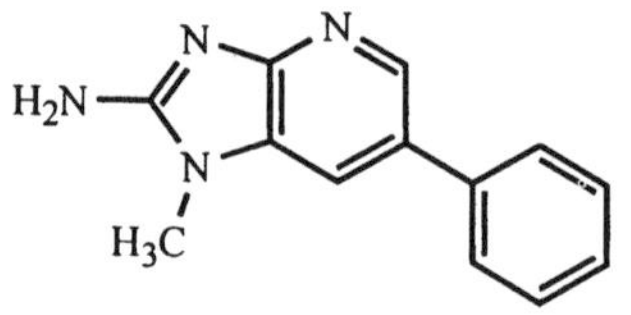

2-Amino-1-methyl-6-phenylimidazo[4,5*b*]pyridine

dazo[4,5*b*]pyridine, 2-amino-3-methylimidazo[4,5*f*]quinoxaline, 1,5,6-trimethyl- and 3,5,6-trimethylimidazopyridine. Conentrations ranged from 1 to 300 ng/g [940].

Five mutagenic heterocyclic aromatic amines (2-amino-3-methyl-, 2-amino-3,8-dimethyl-, 2-amino-3,4,8-trimethyl-, 2-amino-1,7,9-trimethylimidazo[4,5*f*]-quinoxaline, 2-amino-1-methyl-6-phenylimidazo[4,5*b*]pyridine) were isolated from cooked meat and separated on a C_{18} column ($\lambda = 265$ nm, atmospheric pressure chemical ionization MS; capillary $T = 250°$C, vaporizer $T = 500°$C, N_2 sheath gas pressure 20 psi , collision voltage 13 eV). A complex 16-min $90/10 \rightarrow 0/100$ (95/5 water [25 mM ammonium acetate pH 8.5)/acetonitrile)/acetonitrile gradient generated good resolution. A linear range of 10–30,000 pg injected and detection and quantitation limits of 15 and 45 ng/kg, respectively, were reported [941].

Proflavin (3,6-diaminoacridine) and three metabolites (3-*N*-acetyl-, 3-*N*-8-D-glucuronosyl-, 3-*N*-8-D-glucuronosyl-6-*N*-acetylproflavin) were extracted from trout tissue and separated on a C_{18} column ($\lambda = 262$ nm) using a 100/0 (hold 5 min) $\rightarrow$ 75/25 (at 13 min hold 7 min) (93.23/5.0/1.4/0.37 water/acetonitrile/acetic acid/triethylamine)/(4.91/95.0/0.07/0.02 water/acetonitrile/acetic acid/triethylamine) gradient [942]. Good peak shapes and resolution were achieved.

Six hepatic metabolites of 7,9- and 7,10-dimethylbenz[*c*]acridine (e.g., 7-[hydroxymethyl]-10-methylbenz[*c*]acridine, 7,9-dimethylbenz[*c*]acridine-5,6-oxide) were separated from the parent compounds on a C_{18} column ($\lambda = 270$ nm) using a complex 80-min $24/76 \rightarrow 100/0$ acetonitrile/water gradient [621]. Analytes eluted from 35 min to 75 min. Peak shapes were excellent, even at the long retention times. Peaks of interest were well separated from other extracted compounds.

The deoxyadenosine and deoxyguanosine adducts of *syn*-benzo[*g*]chrysene-11,12-dihydrodiol-13,14-epoxide were baseline resolved on a C_{18} column ($\lambda = 264$ nm) with a 50-min $25/75/0 \rightarrow 12/60/28$ acetonitrile/water/methanol gradient [943]. The *R* and *S* conformational adducts were separated individually using a 24/76 acetonitrile/water mobile phase. Elution required nearly 120 min in all cases and peaks were quite tailed.

The isomers of tetrachlorodibenzo-*p*-dioxin were separated on both monomeric and polymeric C_{18} columns ($\lambda = 235$ nm) using an 88/12 acetonitrile/water mobile phase [186]. Eleven discernible peaks were generated on the polymeric phase (which gave superior separation). Each sample mixture contained ~3 ng of each isomer. Peak shapes were good and elution was complete in 35 min.

Proflavin

7,9-Dimethylbenz[*c*]acridine

9.3.3   Nitro-, Nitroso-, and Chlorinated Nonpesticide/herbicide Pollutant Analytes

The anaerobic metabolites of 2,4-dinitrotoluene and 2,6-dinitrotoluene (2,6-diaminotoluene, 2-hydroxylamino-6-nitrotoluene, 2-nitroso-6-nitrotoluene, 2-amino-6-nitrotoluene, 3,3'-diamino-2,2'-dimethylazoxybenzene, 2,2'-dimethyl-3,3'-dinitroazoxybenzene) were baseline resolved on a C_{18} column ($\lambda = 250\,nm$) using a complex 65-min 15/85 → 60/40 acetonitrile/water gradient [944]. Detection limits of 5 ng injected were reported. A linear concentration range of 5–50 ng injected was generated.

The biological degradation of 2,4,6-trinitrotoluene (TNT) was followed by Ahmad and Roberts [945] using a C_{18} column (photodiode array detector, $\lambda = 200$–$600\,nm$) and a complex 18-min 10/90 → 100/0 acetonitrile/water ($50\,\mu L/L\ H_3PO_4$ to pH 3.2) gradient. Phloroglucinol (1,3,5-benzenetriol), pyrogallol (1,2,3-benzenetriol), methyl phloroglucinol, p-cresol, 2,2',6,6'-tetranitro-4,4'-azoxytoluene, 4,4',6,6'-tetranitro-2,2'-azoxytoluene, and the typical reduced-form nitrotoluenes (e.g., 2-amino-4,6-dinitrotoluene) were separated from TNT and RDX (hexahydro-1,3,5-trinitro-1,3,5-triazine) in under 20 min. Detection limits of 10 ng injected (for the benzenetriols) and <1 ng injected for the nitro-substituted compounds were reported. Although the overall chromatography was reported to be independent of pH (at least around pH 3.2), the stability of 2,4,6-triaminotoluene (TAT) was such that it eluted as five peaks at pH 6.0, four peaks at pH 4.0, and two peaks (eluting near the void volume of the system) at pH 3.2. The stability of the TAT was then examined in the standard solutions and it was found that the TAT could be eluted as one peak when the standard solution was made up in 50/50 acetonitrile/methanol.

Four bacterial metabolites of 2,6-dinitrotoluene (2-amino-6-nitrotoluene, 2-nitroso-6-nitrotoluene, 2-hydroxylamino-6-nitrotoluene, 2,2'-dimethyl-3,3'-dinitroazoxybenzene) were baseline resolved on a C_{18} column ($\lambda = 250\,nm$) using a complex 70-min 15/85 → 65/35 acetonitrile/water gradient [946]. Linear response ranges from 10–150 ng injected were obtained. Peak shapes were very good and elution was complete in 60 min.

Some of the most volatile priority pollutants that have been quantitated by HPLC are the N-nitrosoamines. The dimethyl- to dibutyl-, piperidyl-, and pyrrolidyl-substituted N-nitroso compounds were well resolved as their 4-(2-phthalimidyl)ben-

3,3'-Diamino-2,2'-dimethoxyazoxybenzene   Pyrogallol

zoyl chloride derivatives [947]. A C_{18} column ($\lambda = 299$ nm, ex; 426 nm, em) and a 48/52 acetonitrile/water mobile phase were used for the separation. Elution was complete in <20 min. Detection limits of ~1 pmol injected (S/N = 3) for each analyte and working ranges of 0.16–100 nmol/L were reported. A plot of k' versus percent acetonitrile (40–70%) was shown for these analytes. The k' values for the dipropyl and dibutyl analytes were well over 5 when the acetonitrile level dropped below 50%.

The methyl, hydroxy, methoxy, chloro, cyano, and nitro derivatives of N-methyl-N-nitrosoaniline were separated on a C_{18} column ($\lambda = 290$ nm) using a 30/70 acetonitrile/water (10 mM phosphate buffer at pH 5.6) mobile phase [948]. The methyl and chloro derivative peaks were noticably fronted. Elution was complete in 30 min. The same separation was achieved in <15 min with excellent peak shapes, when the mobile phase was changed to acetonitrile/methanol/water/80% H_3PO_4 (at pH 3.2). A mobile phase pH low enough to keep the amine analytes fully protonated seems to be necessary to prevent unfavorable analyte interaction with the silica support. Plots of log k' versus percent acetonitrile (from 20% to 50%) were nonlinear. The detection limits were reported as 3×10^{-6} M. Linear ranges were found to cover the range 5×10^{-6} to 2×10^{-4} M.

The syn and $anti$ conformers of N-nitrosopipecolinic acid, N-nitrosonipecotic acid, N-nitrosothiazolidine-4-carboxylic acid, and N-nitrososarcosine were individually baseline resolved on an α-cyclodextrin column ($\lambda = 238$ nm) using a 90/10 acetonitrile/water (10 mM triethylammonium acetate) mobile phase [806]. A 10 μL aliquot of 0.5 μg/μL standards was injected. No resolution of conformers was achieved at room temperature, but when the system temperature was decreased to $-8°$C, good resolution was obtained. Peak shapes were good and elution times ranged from 6 to 20 min.

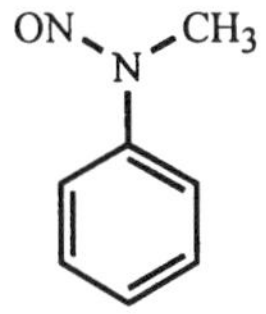

N-Methyl-N-nitrosoaniline 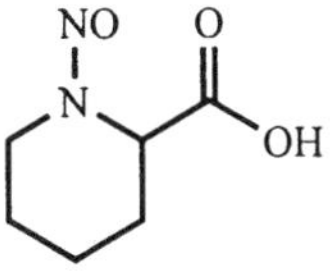 N-Nitrosopipecolinic acid

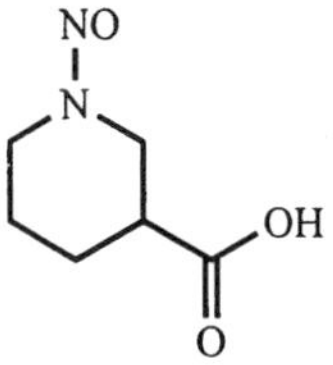

N-Nitrosonipecotic acid 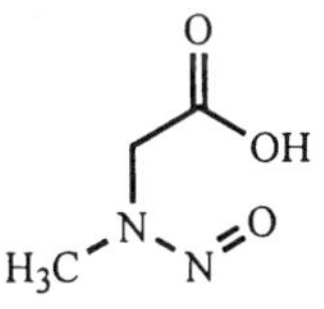 N-Nitrosarcosine

9.3.4 Pesticide, Herbicide, and Fungicide Analytes

9.3.4.1 *Pesticides*

The pesticides (folpet, procymidone, triazophos) represented three poorly resolved components out of 22 analyzed (e.g., metomyl, dichloran, iprodione, α-endosulfan, vinclozolin, carbophenothion). The separation was conducted on a C_{18} column (photodiode array detector, $\lambda = 200$ nm to 280 nm). A 56/27/17 (hold 2 min) → 5/90/5 (at 20 min) water/acetonitrile/methanol gradient was used. Calibration curve ranges were 0.2–12 µg/mL. Quantitative values for the poorly resolved peaks were attempted through the use of multivariate calibration techniques [949].

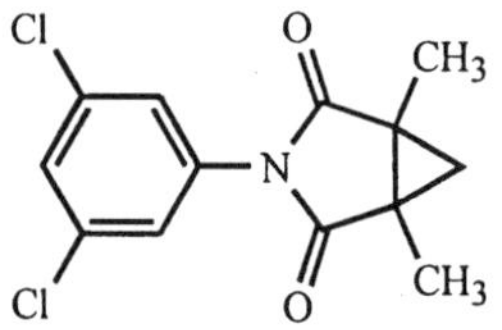

Procymidone

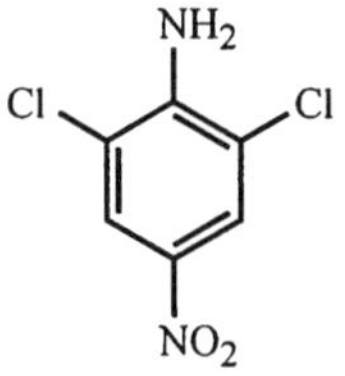

Dichloran

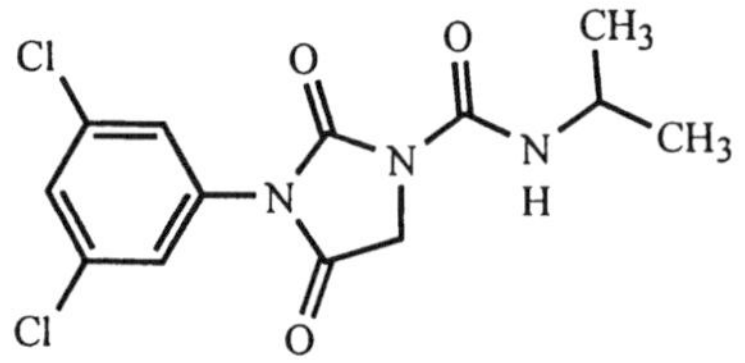

Iprodione

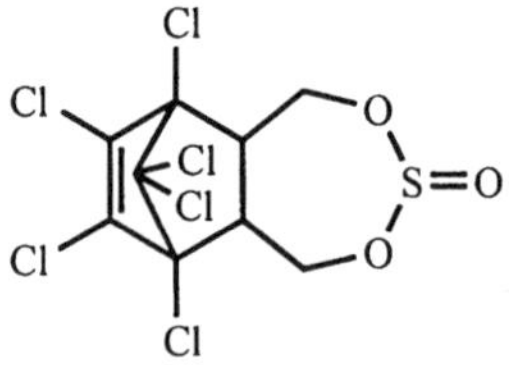

Endosulfan

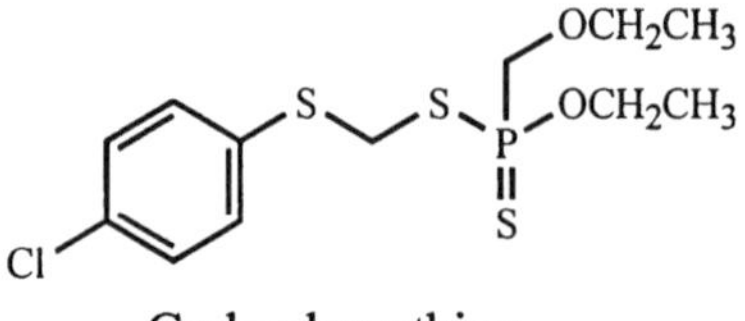

Carbophenothion

In a very nice study Parrilla and Martinez Vidal [950] analyzed twenty-one pesticides (metomyl, dimethoate, aldicarb, dichlorvos, atrazine, carbofuran, diuron, methiocarb, folpet, triazophos, iprodione, vinclozolin, chlorfenvinphos, chlorpyrifos, tetradifon, chlropyrofos-m, α- and β-endosulfan, endosulfan-s, carbophenothion). They were recovered from ground water and well resolved on a C_{18} column ($\lambda = 212$ nm) using a 22-min $56/27/17 \rightarrow 5/90/5$ water/acetonitrile/methanol gradient. A general linear range was reported as 0.50–4 µg/L with detection limits of from 0.1 to 0.1 to 4 µg/L.

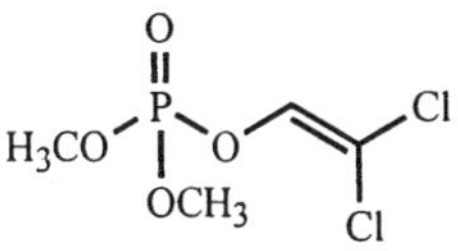

Dichlorvos

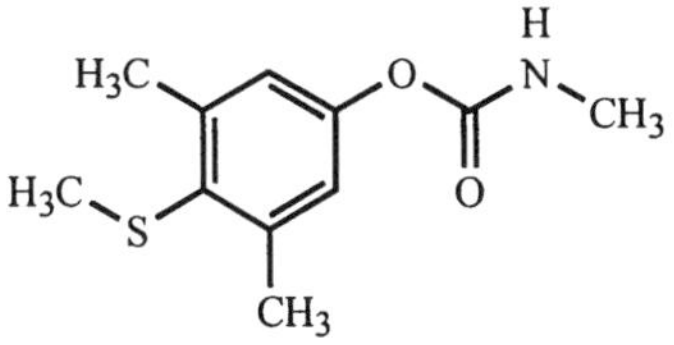

Methiocarb

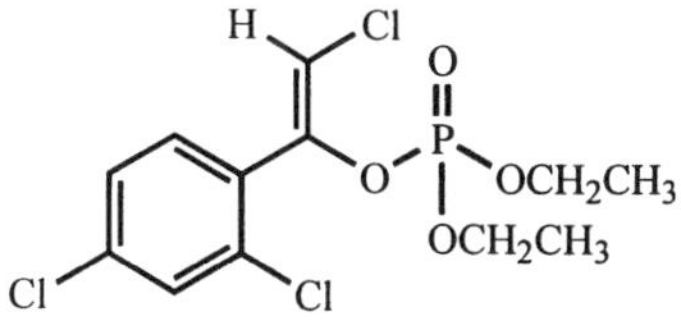

Chlorfenvinphos

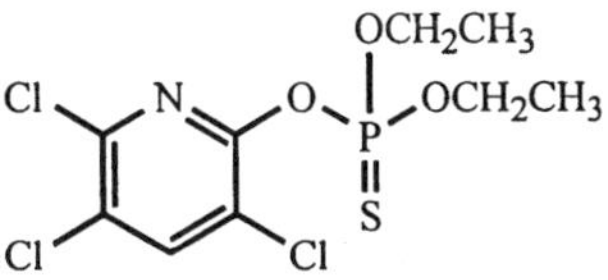

Chlorpyrifos

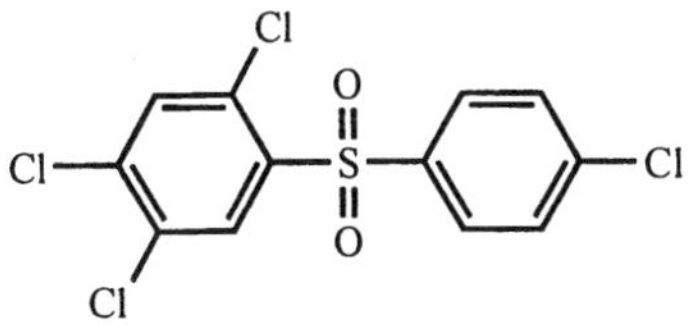

Tetradifon

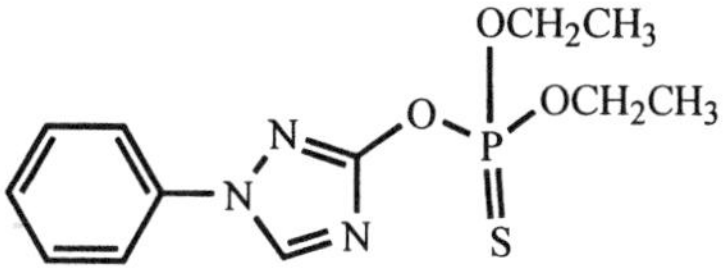

Triazophos

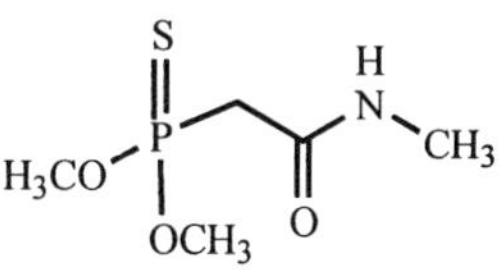

Dimethoate

Six antifouling pesticides (chlorothalonil, Irgarol 1051, Sea-nine 211, diuron, dichlofluanid, TCMTB) were extracted from seawater and separated on a C_{18} column ($\lambda = 254$ nm or positive ion electrospray MS, drying gas at 10 L/min and 350°C, nebulizer pressure 40 psi, capillary voltage 2.4 kV). A 100/0 (hold 2 min) → 0/100 (at 32 min) water (50 mM ammonium formate)/acetonitrile gradient generated excellent peak shapes and resolution [951]. For MS the linear range extended from 5 to 5000 ng/L with detection limits of 1–25 ng/L reported (analyte dependent).

The diasteriomers of metolachlor were separated on a 10°C graphitized carbon column ($\lambda = 210$ nm). The authors noted that the low temperature was necessary for improved resolution. A 1/1 acetonitrile/water mobile phase generated baseline resolution with complete elution in 17 min [952]. Standard solutions of 1–10 mg/mL were used.

Although Parilla et al. [953, 954] reported the quantitation of fenamiphos and folpet on a C_{18} column (photodiode array detector, $\lambda = 210$ nm, 224 nm, or 252 nm) the method developed actually baseline resolves 22 pesticides (e.g., metomyl, aldicarb, dichlorvos, diuron, methiophos, iprodione, chlorfenvinphos, α- and β-endosulfan, carbophenothion) in 25 min. A 27/17/56+ → 90/5/5 (at 20 min hold 5 min) acetonitrile/methanol/water gradient was used. Peak shapes were excellent and representative chromatograms plus tabulated results of retention times for each pesticide were presented. Linear working ranges and detection limits are tabulated for all the pesticides used in the study.

An excellent separation of 28 pesticides was generated by Huen et al. [955]. All compounds were baseline resolved on a C_{18} column ($\lambda = 234$ nm) using a 40-min 10/90 → 90/10 acetonitrile/water (1 mL H_3PO_4/L) gradient. The test compounds

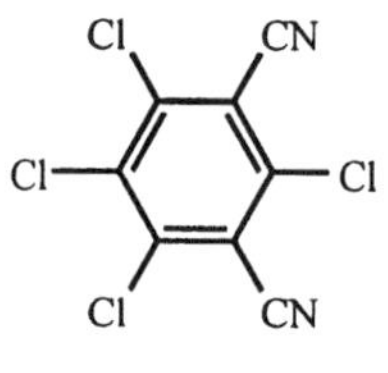

Chlorothalonil

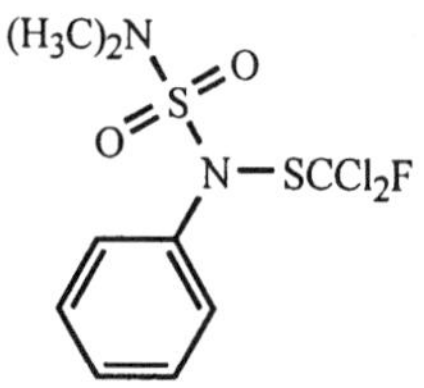

Dichlofluanid

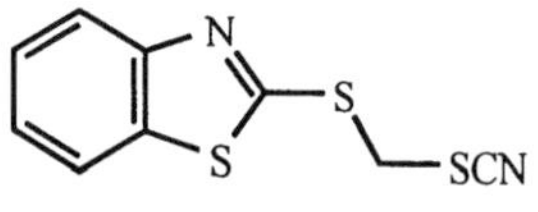

TCTMB

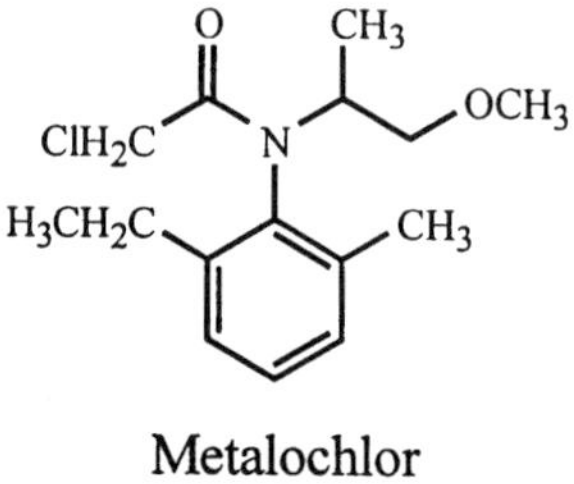

Metalochlor

included: metamitron, diuron, atrazine, metazachlor, neburon, pyrazaphos, chlorpyriphos, trifluralin, and fenpropathrin. Detection limits for drinking water samples were reported to be 0.01 μg/L.

Thirty-three pesticides (including triazines, phenylureas, dinitroanilines, acetanilides, thiocarbamates, and phosphothioates) were extracted from water and resolved on a C_{18} column ($\lambda = 220$ nm). A nonlinear 85-min 20/80 → 85/15 acetonitrile/water (1 mM ammonium acetate) gradient was used [956]. Concentrations of the analytes varied from 250 mg/mL to 1.1 μg/mL (20 μL injected). Detection limits ranged from 6 ng/mL for napropamide to 163 ng/mL for phenylurea (S/N = 6). In the same study, a set of nine acidic phenolic pesticides (thifensulfuron-methyl, metsulfuron-methyl, dicamba, MCPA (4-chloro-2-methylphenoxyacetic acid), MCPB (4-chloro-2-methylphenoxybutyric acid), bromoxynil, dichlorprop, ioxynil, bifenox, haloxyfop) was also studied. Here, a C_{18} column ($\lambda = 229$ nm) was also used and a 16-min 48/52 → 60/40 acetonitrile/water (17 mM H_3PO_4) gradient generated a baseline resolution of the analytes. Detection limits of 25–80 ng/L were reported (analyte dependent).

Seventeen organophosphate pesticides (e.g., famphur and famphur oxon, paraoxon, fenthoxon, ronnel, 3-chloro-4-methyl-7-hydroxycoumarin, stirofos, parathion, coumaphos) were extracted from beef tissues and baseline resolved on a C_{18} column (photodiode array detector, $\lambda = 190$–350 nm). A complex 34-min 55/45 → 70/30 acetonitrile/water gradient was used [957]. Excellent peak shapes were obtained. A

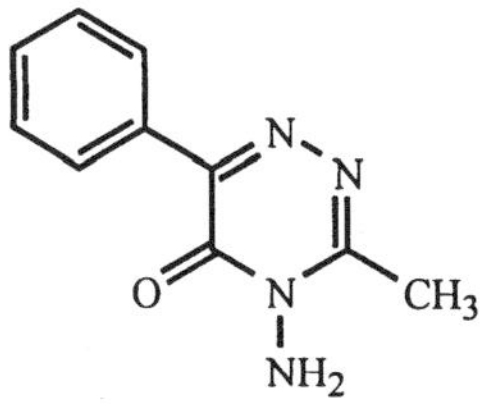

Metamitron

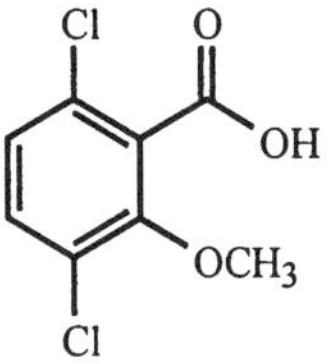

Dicamba

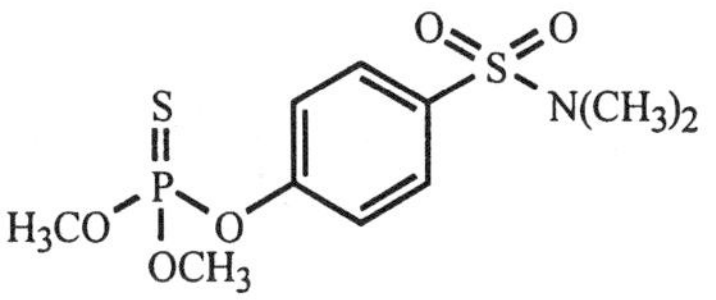

Famphur

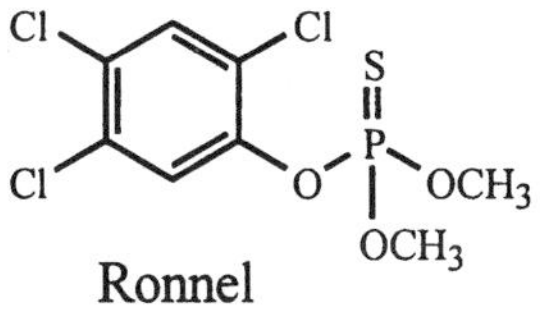

Ronnel

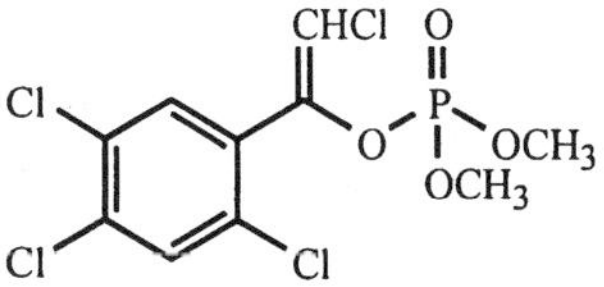

Stirofos

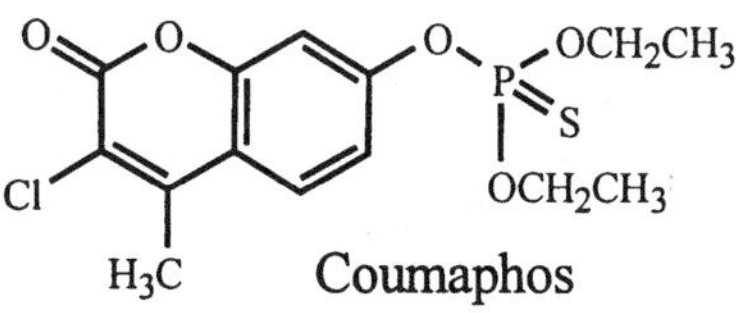

Coumaphos

table of ε_{max}, recoveries and detection limits (0.25–1.0 ppm, analyte dependent) was presented for all analytes.

The levels of *cis*- and *trans*-monocrotophos along with three decomposition products (4-hydroxy-*N*-methyl crotonamide, monomethylacetoacetamide and its chlorinated analog) were determined in technical-grade product. An 87/13 water/ acetonitrile mobile phase and a C_{18} column ($\lambda = 218$ nm) generated baseline resolution and complete elution in <10 min. Standard ranges of 1–50 µg/mL for decomposition products and 1–5 mg/mL from the active compounds were reported. Detection limits of 5–10 ng injected (S/N = 4) were tabulated [958].

Amitraz and five of its hydrolysis products (e.g., 2,4-dimethylphenylformate, 2,4-dimethylaniline, methylaminedialdehyde) were baseline resolved on a C_{18} column ($\lambda = 210$ nm) using a 16.5-min 100/0 → 0/100 (30/70 acetonitrile/water [10 mM triethylamine with 0.75 M H_3PO_4 at pH 6.1])/(90/10 acetonitrile/water [10 mM triethylamine with 0.75 M H_3PO_4 at pH 6.1]) gradient [959]. A 20 µg/L amitraz solution was used to produce the degradation products, but no quantitative level was presented.

Pirimicarb and its desmethyl and desmethylformamide metabolite residues were isolated from fruit and vegetables and separated on a carbamate column (positive ion electrospray MS, lens voltage 0.3 kV; capillary voltage 3 kV; cone potential 25 V; collision gas argon at 0.001 mbar; collision energy 30 eV). An 80/20 → 10/90 (at 9 min hold 4 min) water (0.1% ammonium acetate with 0.1% acetic acid)/ acetonitrile (0.1% acetic acid) gradient was used. A calibration range of 2.5–500 ng/mL was reported [960]. This was nonlinear and fitted with a quadratic equation (pirimicarb shown).

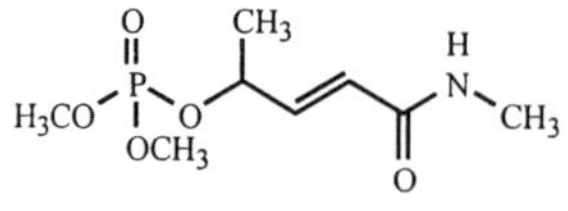

Monocrotophos

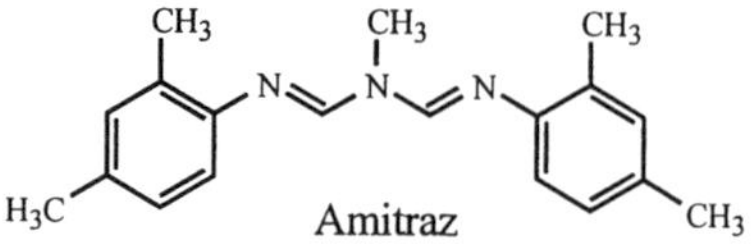

Amitraz

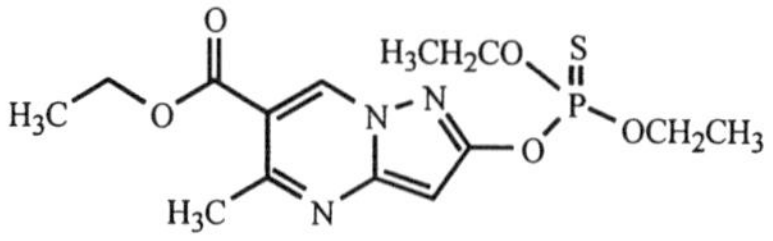

Pyrazophos

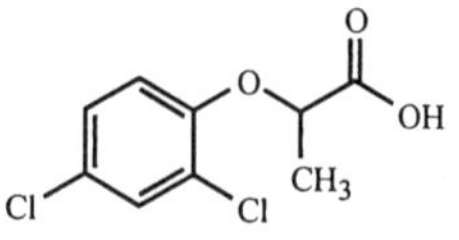

Dichlorprop

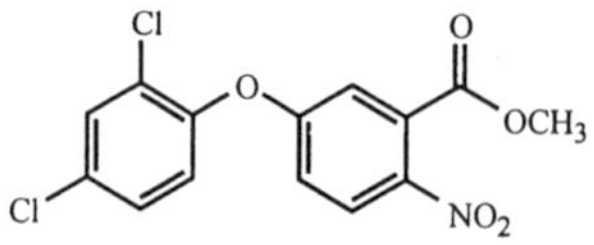

Bifenox

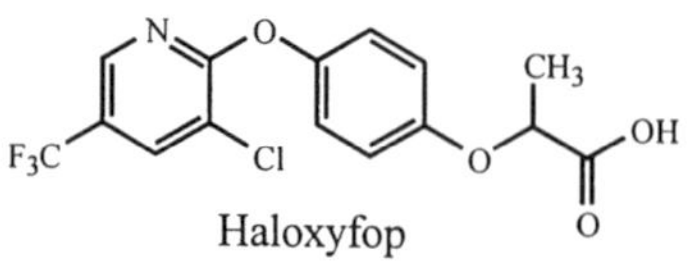

Haloxyfop

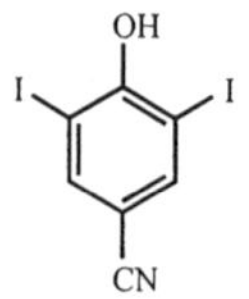

Ioxynil

Five pesticide residues (carbendazim, thiabendazole, thiophanate-methyl, benomyl, imidacloprid) were isolated from fruits and vegetables and separated on a C_8 column (positive ion electrospray MS, drying gas flow 10 L/min and $T = 325°C$; nebulizer pressure, 50 psi; capillary voltage 2.5 kV; corona current, 4 µA fragmentor voltage 50 V; vaporizer temperature 400°C). Good resolution and complete elution were obtained in 12 min using a 95/5 water (ammonium formate to pH 4 with formic acid)/acetonitrile mobile phase [961]. Linear calibration was obtained over the range of 0.01–5 µg/mL and limits of detection were reported in the 0.01 µg/mL range for full-scan and 0.002 µg/mL for single-ion monitoring.

Abamectin, doramectin, ivermectin, moxidectin, and two homologs were extracted from milk and derivatized with trifluoroacetic anhydride. Separation was

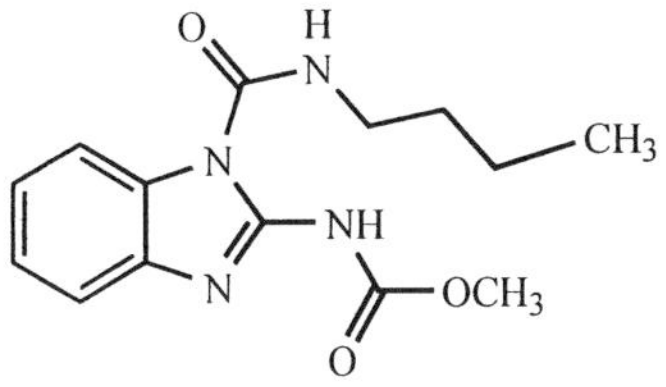

Benomyl

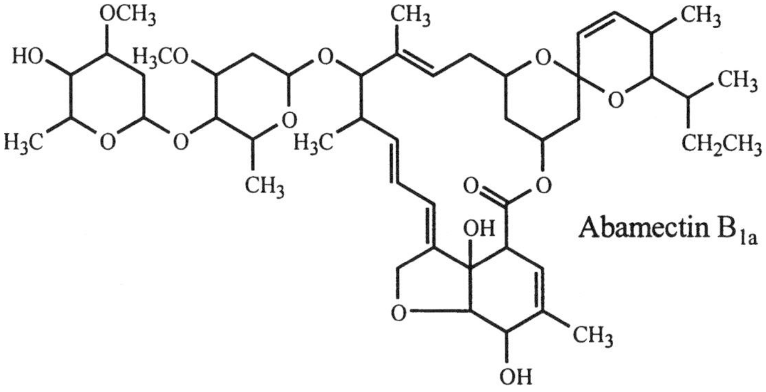

Abamectin B$_{1a}$

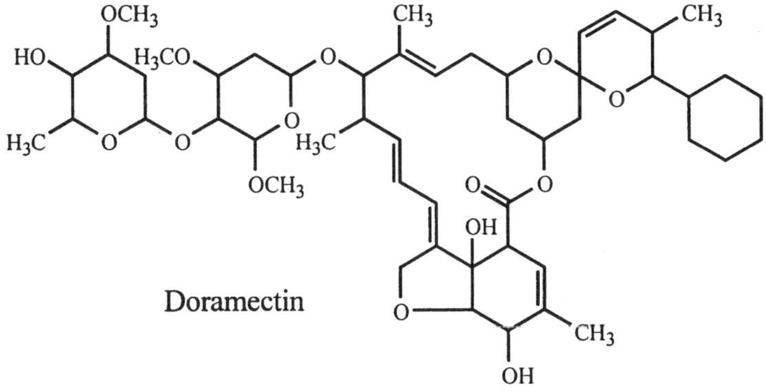

Doramectin

performed at 30°C on a C_{18} column ($\lambda = 364$ nm, ex; 475 nm, em) in <20 min with a 90/6/4 acetonitrile/THF/water mobile phase [962]. Fortified samples ranged from 1 to 30 ng/mL. Detection limits were reported as 0.3 ng/mL (S/N = 3). Authors noted rapid degradation of the derivatized analyze when exposed to light.

The simultaneous determination of permethrin, N,N-diethyl-m-toluamide, and pyridostigmine and their degradation compounds (m-toluamide, m-toluic acid, N-methyl-3-hydroxypyridinium bromide, m-phenoxybenzyl alcohol, m-phenoxybenzoic acid) were extracted from urine and plasma and analyzed on a C_{18} column ($\lambda = 208$–230 nm). A two-ramp 99/1 → 25/75 (at 6 min) → 1/99 (at 11 min hold 4 min) water (to pH 3.2 with acetic acid)/acetonitrile gradient was used to generate baseline resolution and excellent peak shapes [963]. Linear curves from 100 to 5000 ng/mL and quantitation limits of 150 ng/mL were reported.

Chlorophacinone and diphacinone were isolated from commercial rodenticides and analyzed on an aminopropyl column ($\lambda = 324$ nm) using a 70/30/0 → 70/0/30 (at 5 min hold 7 min) acetonitrile/water (1% acetic acid)/water (10 mM ammonium acetate) gradient [964]. The authors noted that the addition of ammonium acetate was crucial to the separation. Standards were run from 1 to 300 µg/mL. A detection limit of 2 ng/mL (S/N = 2.5) was reported.

Rodriguez et al. [965] isolated diflubenzuron and three metabolites (2,6-difluorobenzamide, 4-chlorophenylurea, 4-chloroaniline) from pine needles and separated them in 8 min on a C_{18} column (photodiode array detector, 190–400 nm) using a 50/2/48 acetonitrile/methanol/water mobile phase. Excellent resolution and peak shapes were obtained, even though these were acids. Calibration standards ran from 5 to 10,000 ng/mL and detection limits of 0.3–30 ng/mL were reported (analyte dependent).

Flupyrazofos and two microsomal metabolites (flupyrazofos oxon, 1-phenyl-3-trifluoromethyl-5-hydroxypyrazole) were resolved on a C_{18} column ($\lambda = 230$ nm) using a 60/40 acetonitrile/water (10 mM KH_2PO_4 at pH 5.0) mobile phase. Elution was complete in 25 min with an 18-min separation between the first set of two peaks

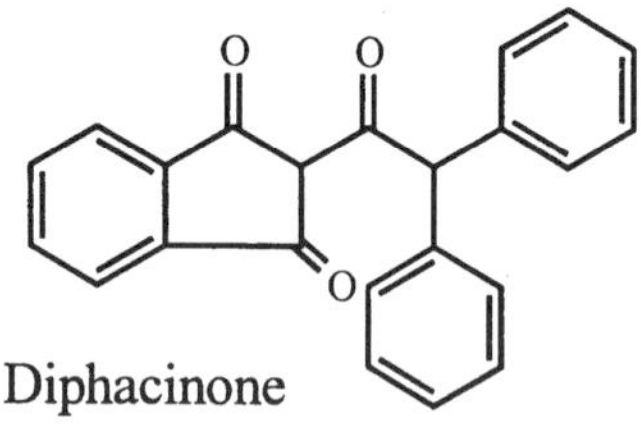

Diphacinone

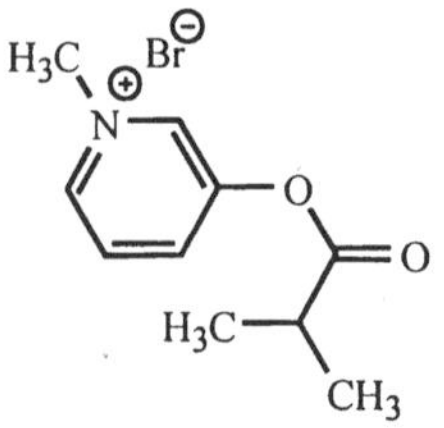

Pyridostigmine bromide

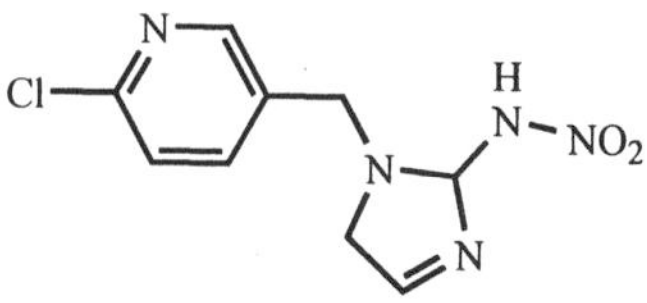

Imidacloprid

and the last-eluting peak. To markedly decrease analysis time, strong consideration (barring extractable matrix interferences) should be given to a sharp gradient beginning as the second peak elutes [966].

Oxytetracycline was extracted from shrimp tissue and analyzed on a C_{18} column ($\lambda = 365$ nm) using a 70/27.5/2.5 water (20 mM oxalic acid)/acetonitrile/methanol mobile phase [967]. Elution was complete in 6 min. The peak showed significant tailing and so additional or different mobile phase modifier should be used. A linear curve from 0.05 to 3 µg/mL and a detection limit of 0.5 µg/g were reported.

Niclosamide residue was extracted from homogenized fish and analyzed on a C_{18} column ($\lambda = 335$ nm or 360 nm) using a complicated 30-min 70/30 → 0/100 water (58 mM sodium acetate at pH 3.8)/acetonitrile gradient [968]. Elution of the niclosamide occurred at 26 min with no evident peaks eluting from 4 to 24 min. Although it was stated that other metabolites were possible, they were not observed here and so a much faster gradient should be considered to reduce analysis time. A detection limit of 10 ng/g sample was reported.

The residues of four pesticides (diazinon, carbaryl, malathion, fenitrothion) were recovered from sesame seeds and baseline resolved on a C_{18} column ($\lambda = 225$ nm) using a 50/50 acetonitrile/water (0.1% acetic acid) mobile phase [969]. Excellent separation and elution were achieved in 21 min. Linearity was obtained over the range 2–4500 ng/mL with detection limits of 5–50 ng/mL (analyte dependent).

Pentachlorophenol and three of its biodegradation products (pentachloroanisole, tetrachloro-*p*-hydroquinone, tetrachloro-*p*-benzoquinone), were baseline resolved in 18 min on a C_{18} column ($\lambda = 254$ nm) using an isocratic 60/40 acetonitrile/water (0.1% acetic acid) mobile phase [970]. This was considerably more efficient than a methanol/water gradient, not because of better resolution or shorter elution times but because there was no re-equilibration time needed between injections for the acetonitrile/water system.

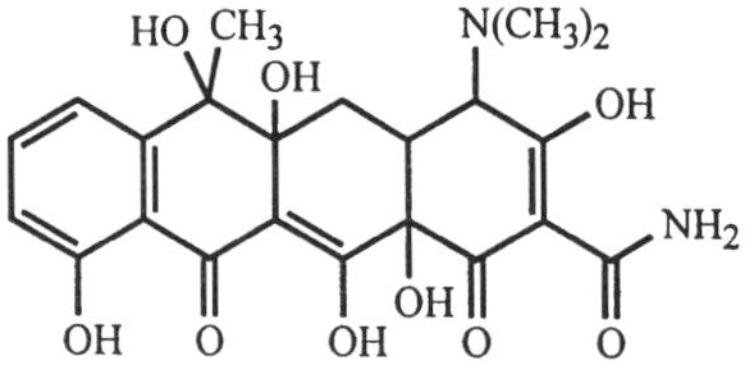

Oxytetracyeline

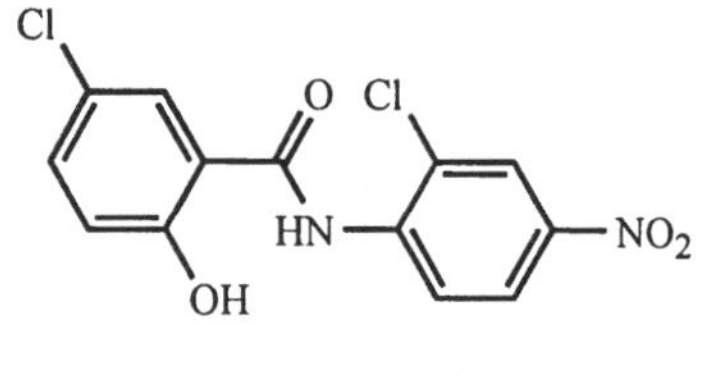

Niclosamide

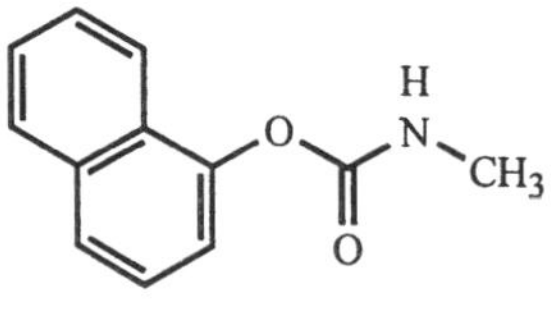

Carbaryl

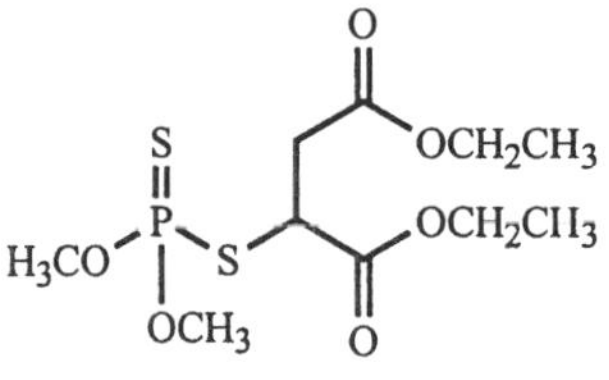

Malathion

The phenyl sulfone, sulfoxide, and sulfonic acid metabolites of fenamiphos were extracted from soil and analyzed using a C_{18} column ($\lambda = 240$ nm) and a $0/100 \rightarrow 50/50$ (at 30 min) $\rightarrow 100/0$ (at 50 min) acetonitrile/water (1 g/L acetic acid) gradient [971]. Detection limits of 10 ppb and a linear concentration curve up to 500 ppb were reported.

Twenty-six methyl carbamate residues and metabolites (e.g., aldicarb-sulfoxide and -sulfone, aldicarb, carbaryl, methiocarb and its sulfoxide and sulfone, etrofolan, baycarb, butocarboxim, thiofanox) were extracted from plant and soil matrices and analyzed as their o-phthalaldehyde derivatives using a C_{18} column ($\lambda = 340$ nm, ex; 455 nm, em). A 15-min $20/80 \rightarrow 80/20$ acetonitrile/water gradient was used [972]. Peaks corresponding to 10 ng injected were readily detected. Under these conditions, carbofuran, propoxur, and bendiocarb co-eluted but were resolved using a separate shallow 15-min $35/65 \rightarrow 40/60$ acetonitrile/water gradient.

Ethyl N-phenylcarbamate, 4,4'-methylenebis(ethyl phenylcarbamate), 4,4'-methylenebis(phenyl isocyanate) and 10 decomposition products were separated as their urea derivatives on a C_{18} column ($\lambda = 254$ nm) using a 75/25 acetonitrile/water (1% TEA to pH 3.0 with H_3PO_4) mobile phase [973]. Very good resolution and peak shapes were obtained. Elution was complete in 30 min and detection limits were reported to be 1–7 ng injected (analyte dependent).

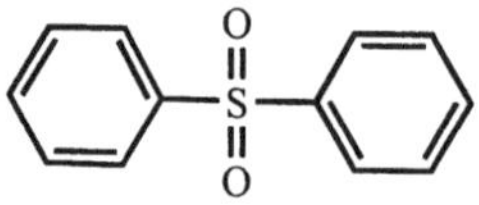

Phenyl sulfone

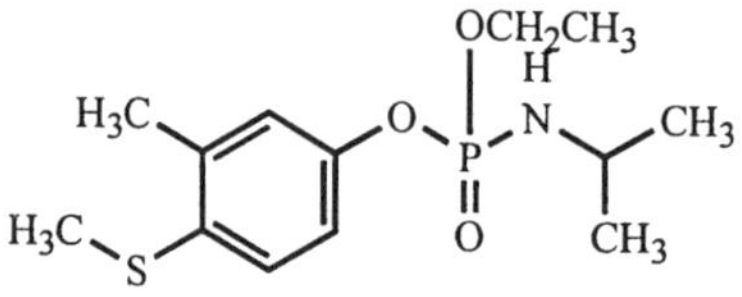

Fenamiphos

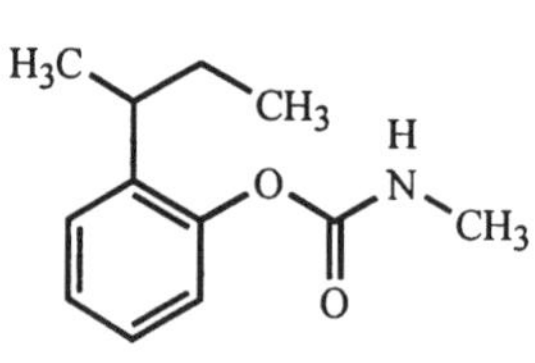

Baycarb, BPMC, Fenobucarb

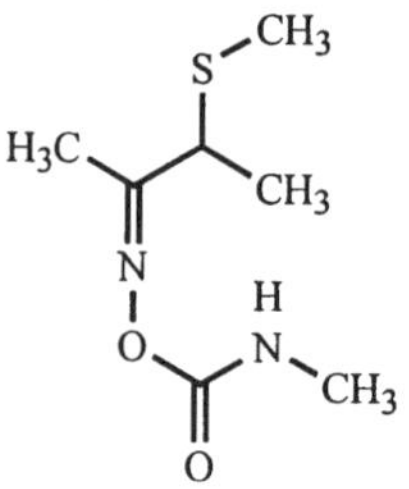

Butocarboxim

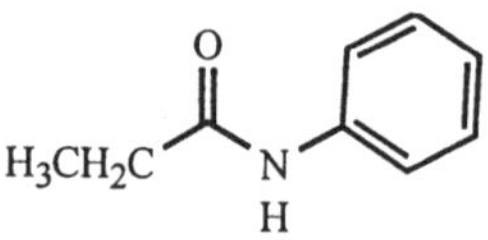

Ethyl N-phenylcarbamate

Newsome et al. [974] studied the chromatography of 11 carbamate pesticides (e.g., aldicarb, aldicarb sulfoxide and sulfone, oxamyl, carbofuran, 3-hydroxy- and 3-ketocarbofuran) on a C_8 column. A postcolumn o-phthalaldehyde derivatization technique ($\lambda = 336$ nm, ex; 440 nm, em) was compared with an atmospheric pressure ionization MS detector. A 20-min $12/78 \rightarrow 70/30$ acetonitrile/water gradient generated good resolution and peak shapes. Detection limits for all compounds were tabulated and were ~0.1 ng for fluorescence and 1 ng for APCI/MS.

1-Naphthylamine, naptalam (N-naphthylphthalamic acid), and antu [1-naphthylthiourea) were extracted from river water samples and quantitated on a C_{18} column ($\lambda = 220$ nm) using a 45/55 acetonitrile/water [0.1% acetic acid) mobile phase [975]. Elution was complete in <5 min and good peak shapes were obtained. Retention times and resolution factors for isocratic mobile phases ranging from 20% to 50% acetonitrile were also tabulated. The authors noted that significant levels of noise were obtained when the detection wavelength was set at 215 nm, most likely due to the acetic acid. A detection limit of ~30 ppb was reported.

Seven pyrethroid insecticides (allethrin, cypermethrin, tetramethrin, bifenthrin, permethrin, fenpropathrin, fenvalerate) extracted from surface water samples were

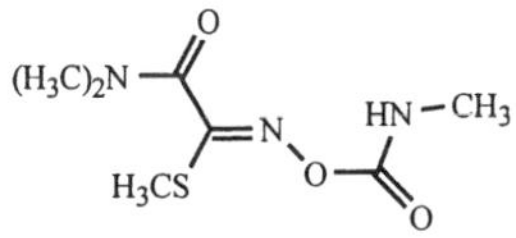

Oxamyl

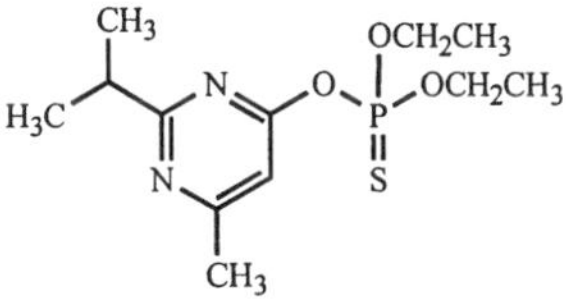

Diazinon

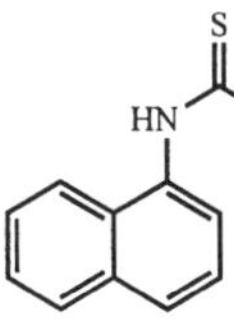

Antu

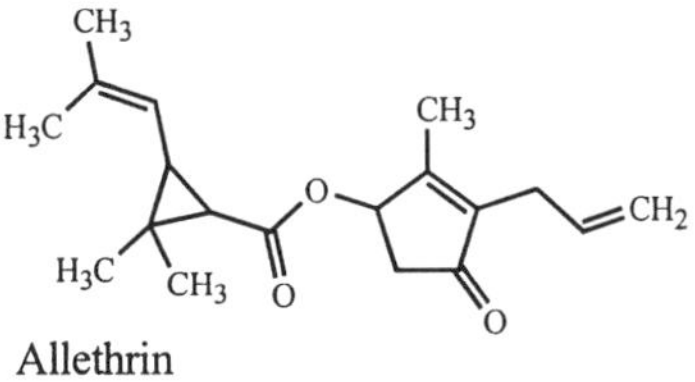

Allethrin

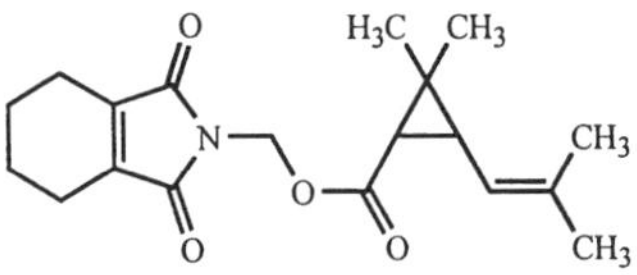

Tetramethrin

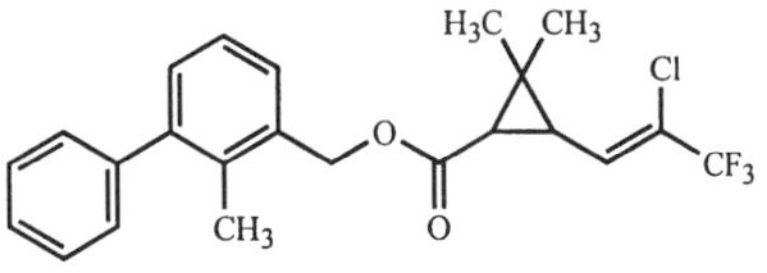

Bifenthrin

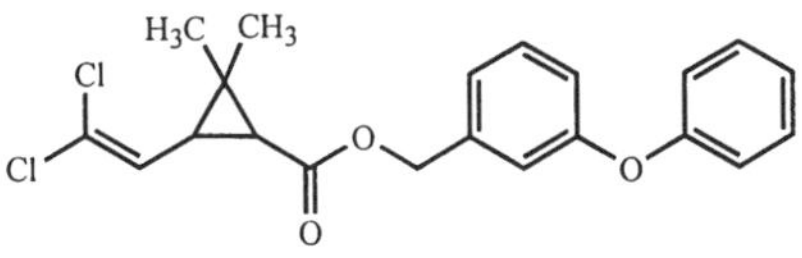

Permethrin

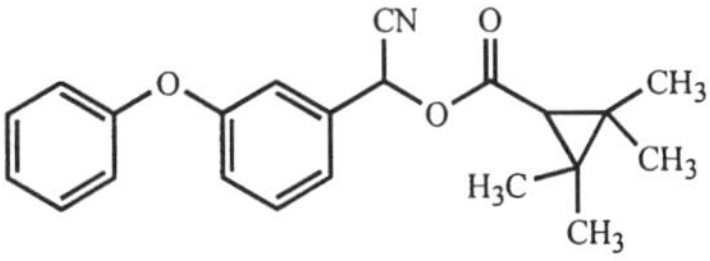

Fenpropathrin

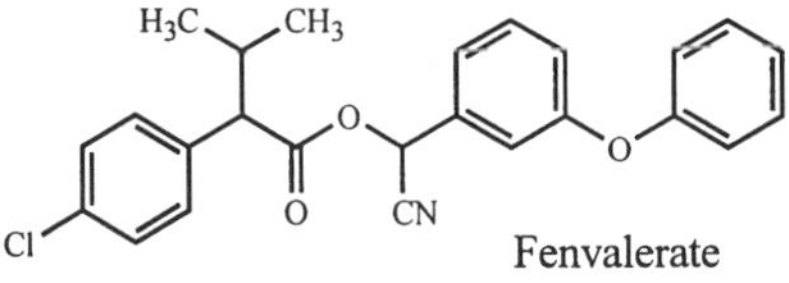

Fenvalerate

baseline resolved on a 40°C PAH column ($\lambda = 235$ nm) using a 35/65 (hold 12.5 min) $\rightarrow$ 80/20 (at 20 min) acetonitrile/water gradient [976]. The authors note that the analytes can adsorb to storage vessel walls, and this loss was avoided through the addition of 1×10^{-3} M Brij-35. Too high a level of Brij-35 caused breakthrough problems during the chromatography. Detection limits between 0.05 and 0.4 µg/L (S/N = 3] were reported (analyte dependent).

Methoxychlor and monohydroxy-, bishydroxy-, ring hydroxylated- and tris-hydroxymethoxychlor were extracted from liver tissue and resolved on and eluted from a C_{18} column ($\lambda = 228$ nm) in 16 min using a 50/50 acetonitrile/water mobile phase [977]. Good resolution and peak shapes were obtained.

Five superwarfarin rodenticides (chlorophacinone, bromadiolone, difethialone and *cis*- and *trans*-brodifacoum) were extracted from serum and separated on a C_8 column ($\lambda = 285$ nm or 265 nm, ex; 400 nm, em). Very good peak shapes were generated with a 45/25 acetonitrile/water (20 mM acetate buffer at pH 4.7 with 1 mM tetrabutylammonium hydroxide) mobile phase [978]. Detection limits for UV were 20–75 ng/mL, whereas for fluorescence they were 3–12 ng/mL (however, chlorophacinone and difethialone did not fluoresce). Concentration curves were linear from 100 to 1000 ng/mL.

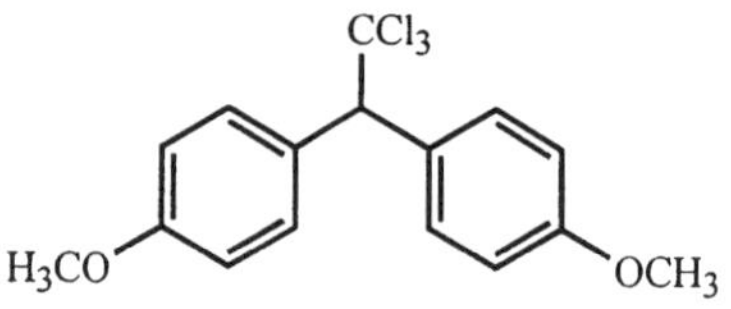

Methoxychlor

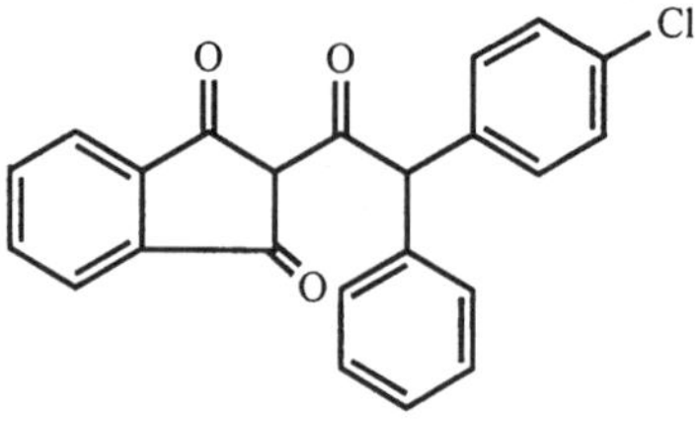

Chlorophacinone

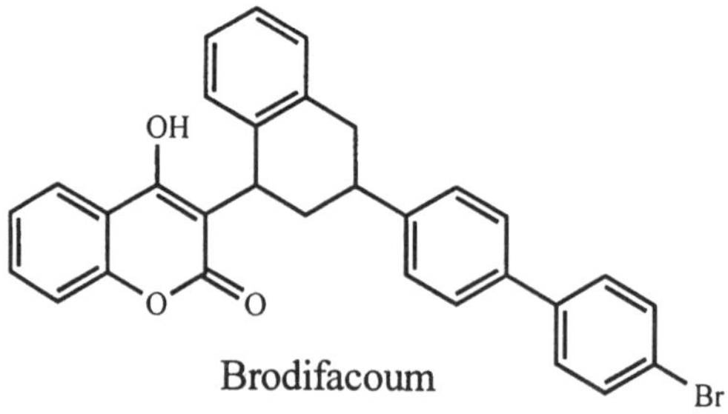

Brodifacoum

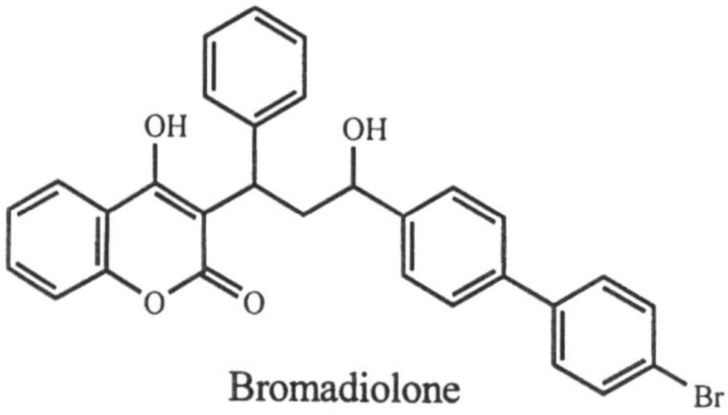

Bromadiolone

Gill disease in fish is treated with chloramine-T. The residue marker compound, *p*-toluenesulfonamide, is extracted from fish meat and analyzed on a 50°C C_{18} column ($\lambda = 226$ nm) using a 77/23 water/acetonitrile mobile phase [979]. An impressive list of retention times for 20 potentially interfering compounds was tabulated. Elution was complete in 11 min and excellent peak shape was achieved. A concentration range of 20–1000 ng/g and detection and quantitation limits of 4 ng/g and 13 ng/g, respectively, were reported.

Destruxins A and B were isolated from *Metarhizium anisopliae* fermentation broth and separated on a C_{18} column ($\lambda = 215$ nm) using a dual ramp $100/0 \rightarrow 73/27$ (at 20 min) $\rightarrow 10/90$ (at 25 min hold 5 min) water/acetonitrile gradient [980]. Although the A and B forms were identified, at least six other unidentified peaks were present. The authors speculate that one or more of these may be other known destruxins (e.g., C, D, A_2, etc.). Standard curves covering the range 0.05–0.2 mg/mL for A and 0.1–1.75 mg/mL for B were presented.

Three mosquitocidal compounds (isomethyl eugenol, costunolide, parthenolide) were extracted from *Magnolia salicifolia* and separated on a C_{18} column ($\lambda = 217$ nm and 222 nm) using a 70/30 acetonitrile/water mobile phase [981]. Baseline resolution was achieved and elution was complete in <10 min. A linear range of 4–125 µg/mL was reported.

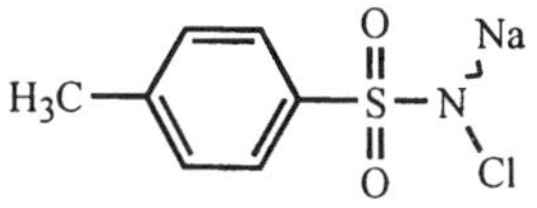

Chloramine-T

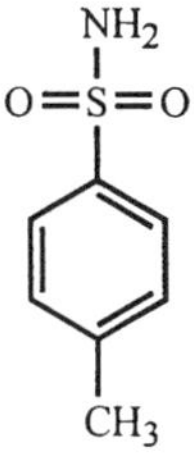

p-Toluenesulfonamide

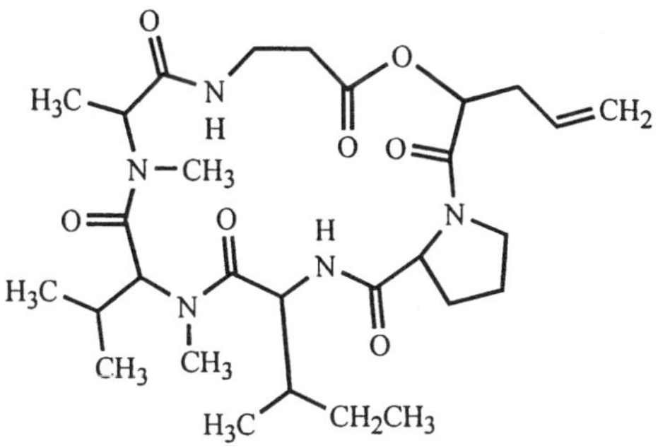

Destruxin A

Six pyrethrins (jasmolin I and II, cinerin I and II, pyrethrin I and II) were extracted from the flower heads of chrysanthemums and separated on a C_{18} column ($\lambda = 230$ nm). A complex 32-min $58/42 \rightarrow 100/0$ acetonitrile/water gradient generated excellent resolution and peak shapes for all components [982]. Standards were prepared at $100 \mu g/mL$ and were easily detected.

The xanthine dyes, phloxine B and uranine, are used as pesticides. These dyes were extracted from guava fruit and separated on a C_{18} column ($\lambda = 493$ nm for uranine and 546 nm for phloxine B) using a 15-min $80/20 \rightarrow 20/80$ water (0.5 M ammonium acetate)/acetonitrile gradient [983]. Separation and peaks shapes were excellent. Samples were spiked with 0.05–$5 \mu g/g$ with limits of detection reported as $0.02 \mu g/g$.

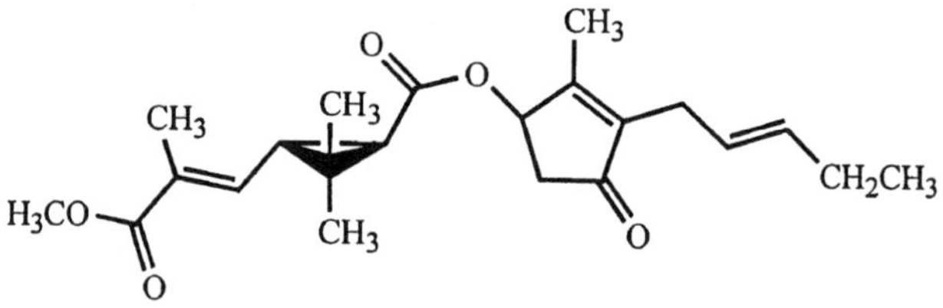

Jasmolin I

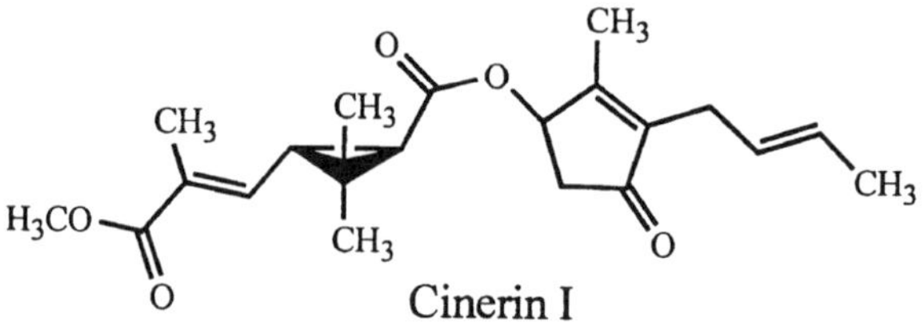

Cinerin I

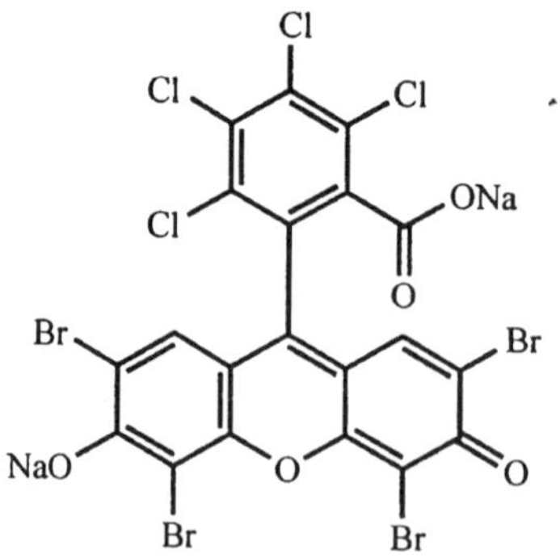

Phloxin B

Spinosads are naturally occurring insect control agents found in *Saccharopoly-spora spinosa* bacteria. After extraction, five compounds (spinosyn A, D, K, and B, and *N*-demethyl spinosad D) were baseline resolved on a 30°C C_{18} column ($\lambda = 250$ nm) using a 44/44/12 acetonitrile/methanol/(67/33 water [2% ammonium acetate]/acetonitrile) mobile phase [984]. Elution was complete in 25 min. Standard concentrations of 0.03–10 μg/mL were used and detection and quantitation limits were reported as 0.02 μg/g and 0.006 μg/g, respectively.

9.3.4.2 Herbicides

The separation of 12 phenylurea hebicides (metabenzthiazuron, fenuron, metoxuron, monuron, chlortoluron, fluometuron, isoproturon, diuron, chloroxuron, chlorbromuron, buturon, neburon) extracted from drinking water was attempted on a base deactivated C_8 column ($\lambda = 245$ nm) using a 35/65 acetonitrile/water [10 mM

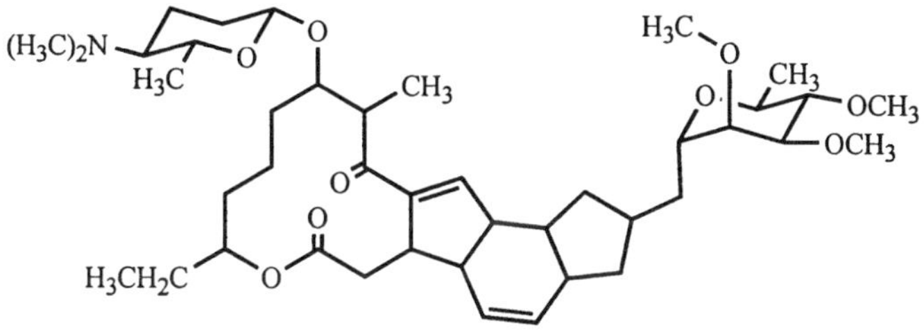

Spinosyn A

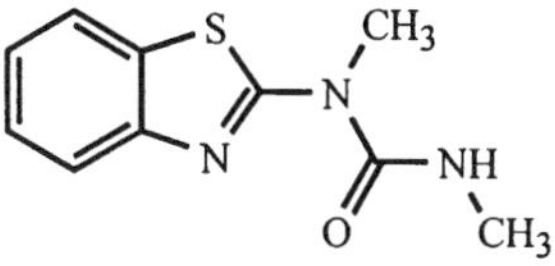

Metabenzthiazuron

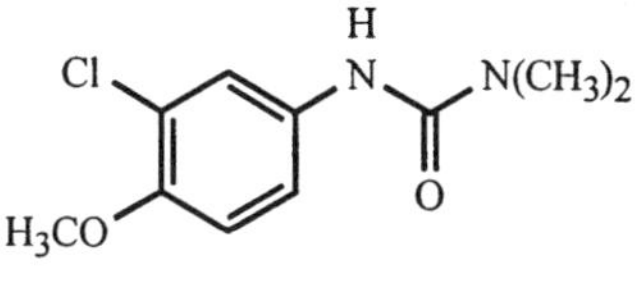

Metoxuron

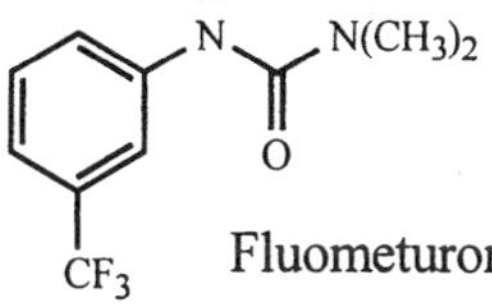

Fluometuron

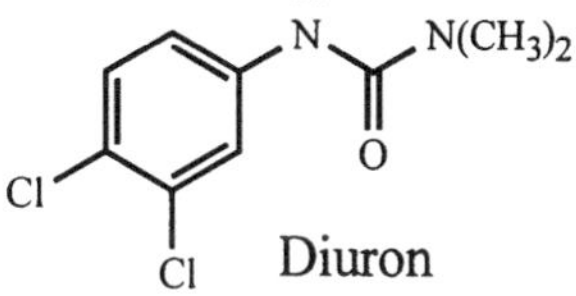

Diuron

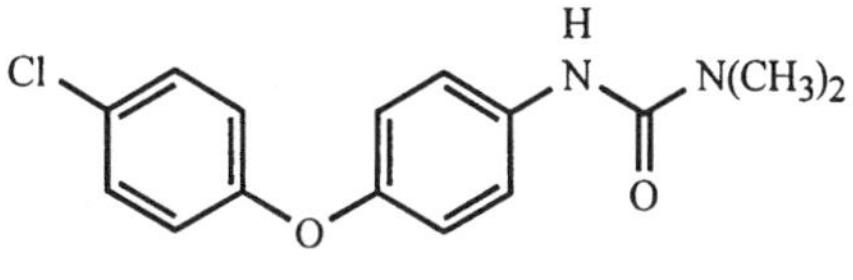

Chloroxuron

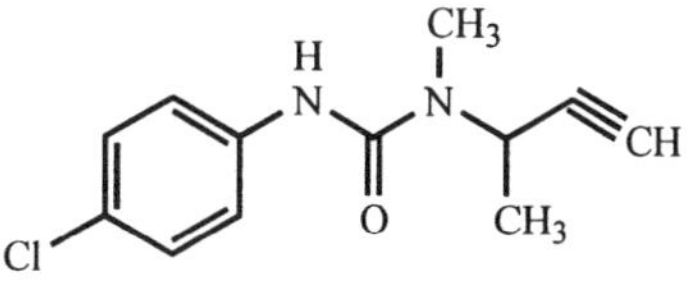

Buturon

Na$_2$HPO$_4$ to pH 7 with H$_3$PO$_4$) mobile phase. Baseline resolution was achieved for all but the fluormeturon/isoproturon pair. Elution was complete in 55 min and the late-eluting peaks were well separated and quite broad. This method would be an ideal candidate for a gradient elution. This would offer better resolution for the early-eluting peaks and narrower peaks for the late-eluting ones. As is, a linear range of 100–400 µg/L and detection limits of ~10 µg/L were reported [985].

Flumetsulam, nicosulfuron, metosulam, aclonifen, and bifenox were extracted from maize and separated on a C$_{18}$ column (positive ion ion spray MS) using a 90/10 → 80/20 (at 10 min) → 95/5 (at 20 min) water/acaetonitrile all with 2 mM ammonium acetate gradient [986]. Samples were spiked at the 100 ng/g level and method detection limits of 0.4–4 ng/g were reported.

Marek et al. [987] studied alloxydim, clethodim, sethoxydim, and two oxidation products (clethodim and sethoxydim sulfoxides). These were resolved on a 30°C C$_8$ column ($\lambda = 254$ nm) using a 68.6/30/1.4 acetonitrile/water/acetic acid mobile

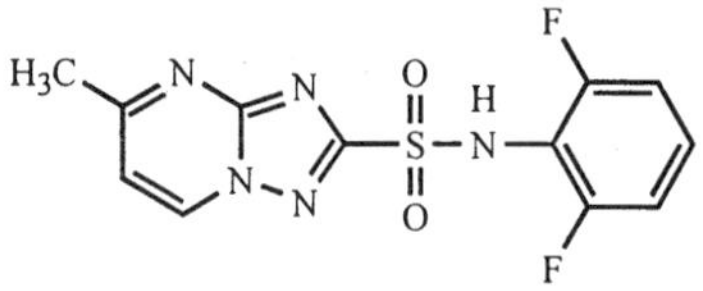

Flumetsulam

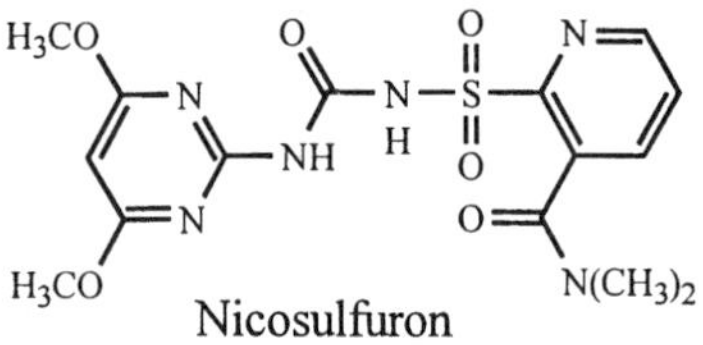

Nicosulfuron

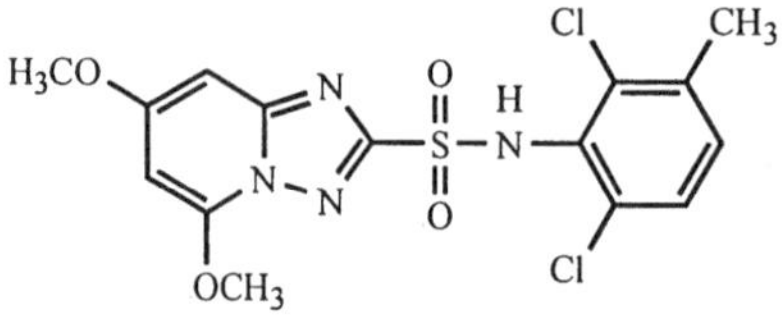

Metosulam

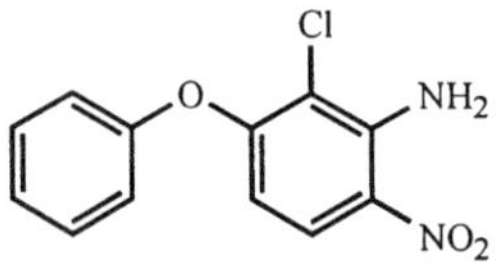

Aclonifen

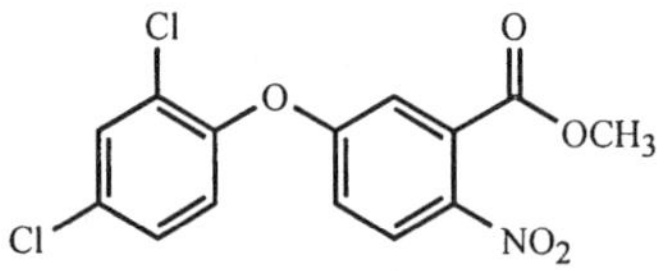

Bifenox

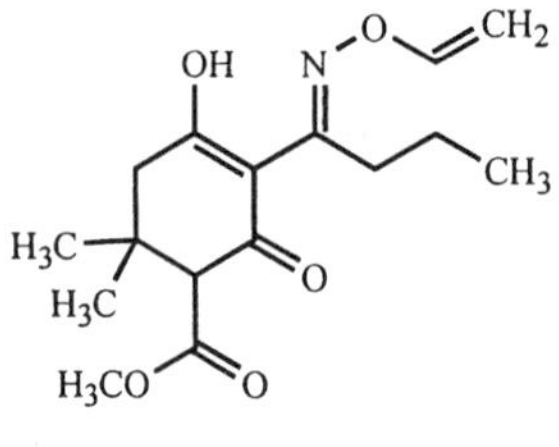

Alloxydim

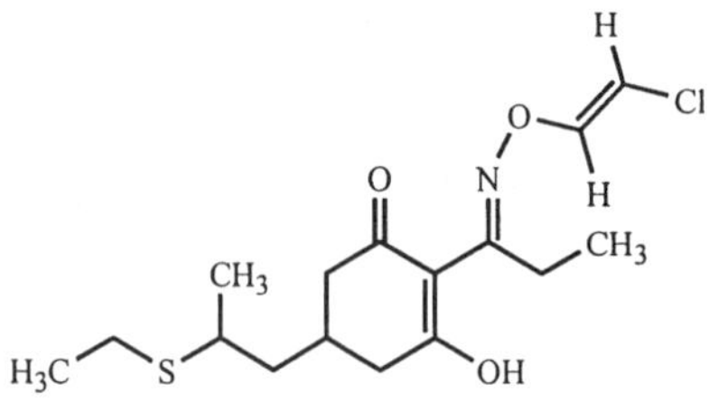

Clethodim

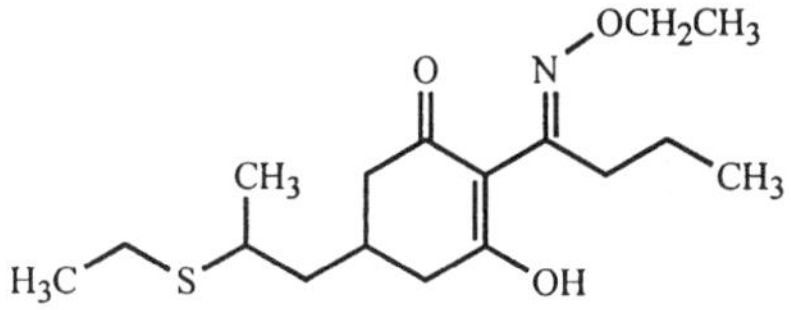

Sethoxydim

phase. Much better sensitivity (quantitation limits were reported as 1 mg/L for UV and 0.08 µg/L for ES/MS) was provided by positive ion electrospray MS (multiplier voltage, 2.5 V; quad $T = 120°C$). For MS work the C_8 column was used but a 12.5-min $30/70 \rightarrow 90/10$ acetonitrile/water (0.1% formic acid) gradient was used.

Imazosulfuron, a sulfonylureic herbicide, was isolated from drinking water and soil and analyzed on a C_{18} column ($\lambda = 238$ nm) using a 45/55 acetonitrile/water (0.1% acetic acid) mobile phase [988]. The elution time was about 13 min and the analyte was well resolved from co-extracted materials. The analyte peak appeared somewhat broad. Additional acid in the mobile phase or TFA at 0.1% might be more effective at generating a sharper peak. A calibration curve was run from 0.05 to 1 µg/mL with a quantitation limit reported as 5 mg/kg (soil) and 0.05 µg/mL (water).

In a very simple and elegant method by Perkins et al. [989], hexazinone was analyzed in ground water through direct injection on a C_8 column ($\lambda = 247$ nm). A 60/25/15 water/acetonitrile/methanol mobile phase generated elution in 8 min. The linear range was 0.3–33 µg/L and the quantitation limit was 10 µg/L.

Four sulfonylurea herbicides (thifensulfuron methyl, metsulfuron methyl, chlorsulfuron, tribenuron methyl) were extracted from water and soil samples and

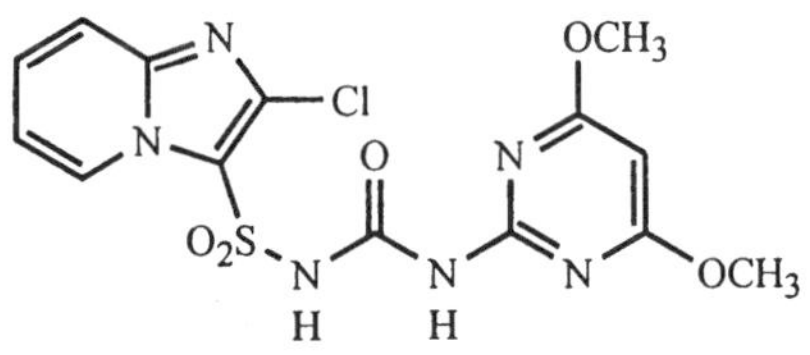

Imazosulfuron

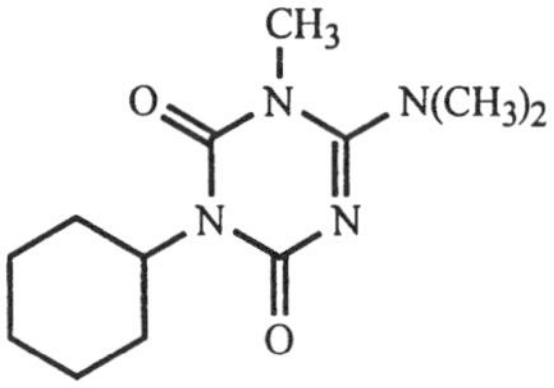

Hexazinone

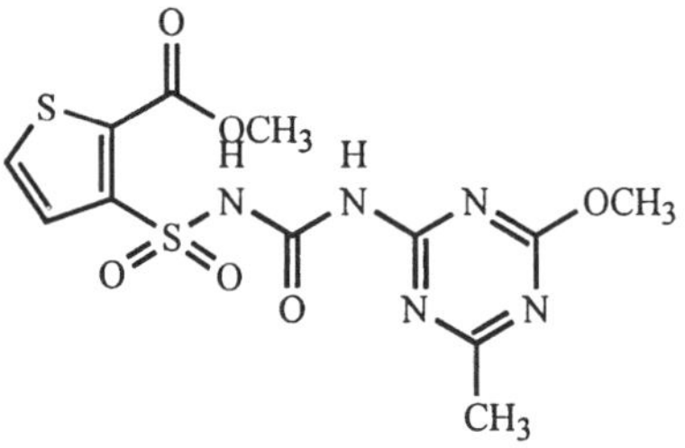

Thifensulfuron methyl

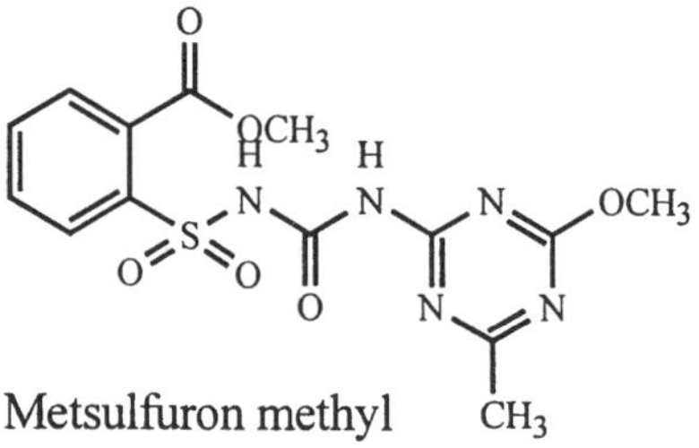

Metsulfuron methyl

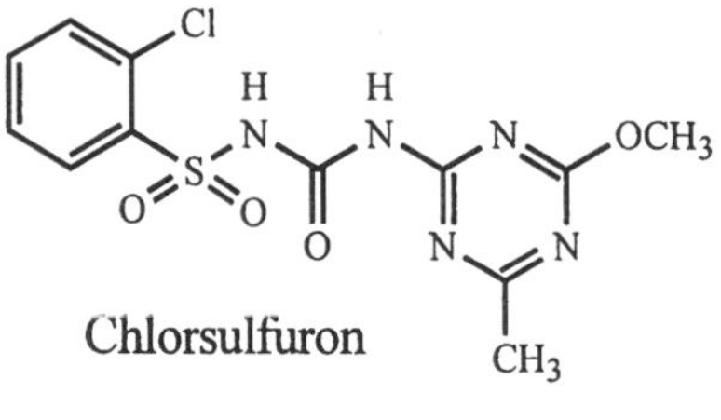

Chlorsulfuron

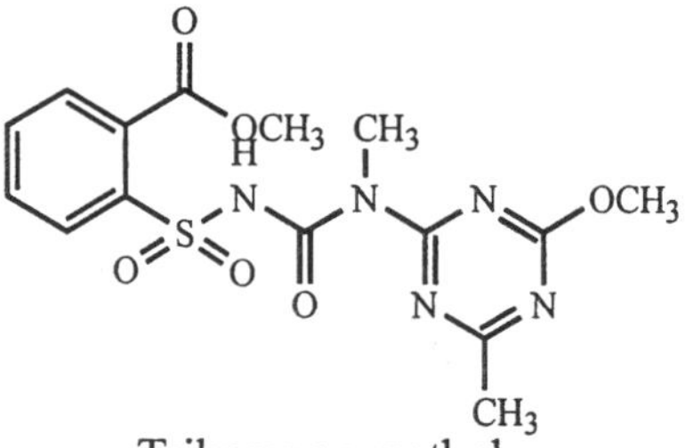

Tribenuron methyl

separated on a C_{18} column ($\lambda = 225$ nm). A 70/30 (hold 4 min) → 20/80 (at 29 min) water (20 mM NaH_2PO_4 to pH 3.4 with H_3PO_4)/acetonitrile gradient gave baseline resolution [990]. A 10 µg/mL working standard was used and quantitation limits of 0.02 µg/L were reported.

Berger [991] extracted 10 phenylurea herbicides (fenuron, monuron, monolinuron, chlorotoluron, diuron, isoproturon, linuron, metobromuron, chlorbromuron) and six degradation products (e.g., 3-[3,4-dichlorophenyl]-1- and 3-[4-chlorophenyl]-1-methoxyureas and methylureas) from soil. These were analyzed using two coupled (i.e., 2 × 25 cm) C_{18} columns ($\lambda = 240$ nm). Excellent resolution and peak shapes were obtained from all 16 compounds using a 35/65 (hold 45 min) → 50/50 (at 75 min hold 10 min) acetonitrile/water gradient. Standards of 0.1–10 µg/mL were used and a quantitation limit of 0.1 µg/mL (0.02 mg/kg soil) was reported.

Dinitroaniline herbicides (dinitramine, ethalfluralin, trifluralin, pendimethalin, isopropalin) were extracted from soil and water matrices and baseline resolved on

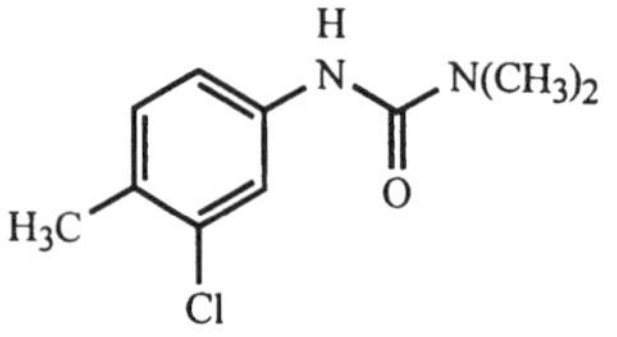

Chlortoluron

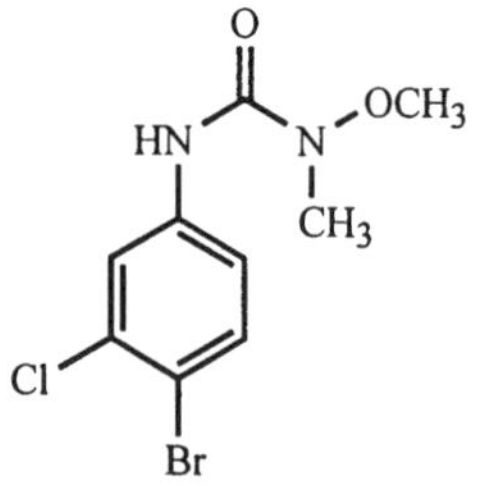

Chlorbromuron

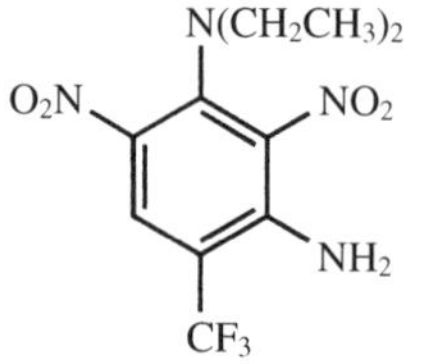

Dinitramine

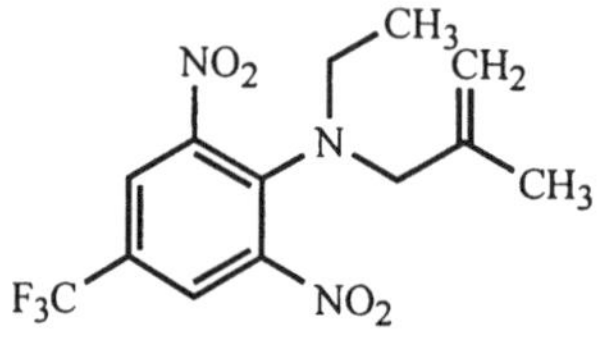

Ethalfluralin

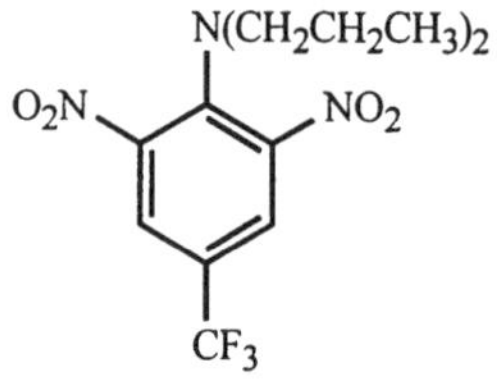

Trifluralin

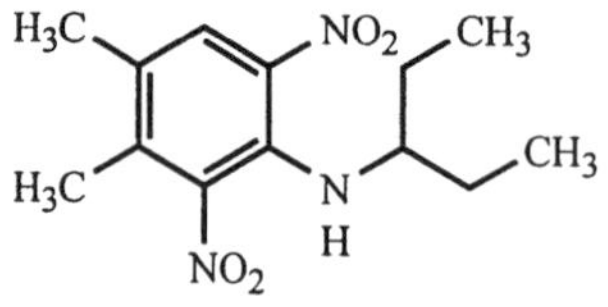

Pendimethalin

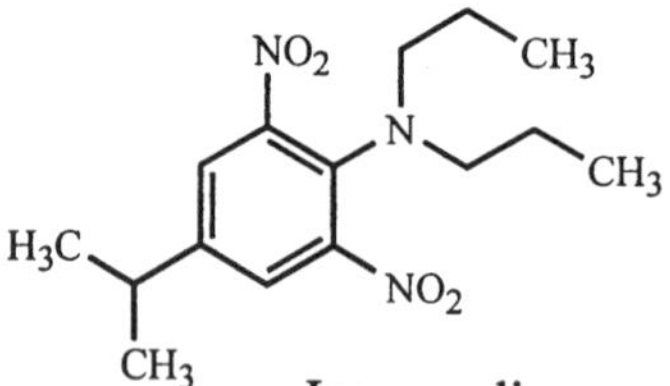

Isopropalin

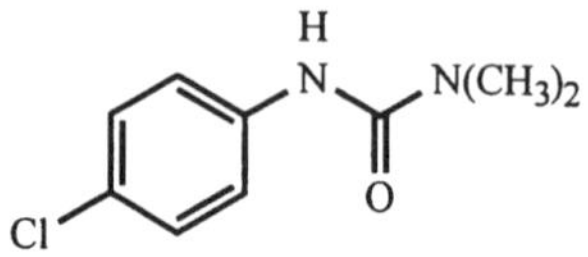

Monuron

several RP columns including C_{18}, C_8, C_6 and C_1 ($\lambda = 220$ nm). Good resolution was obtained using a C_{18} column and a 55/45 acetonitrile/water mobile phase [992]. Similar success was achieved with a 25/75 acetonitrile/water mobile phase and a C_1 column. Both methods gave complete elution in <20 min. Detection limits were reported as 20 µg/kg in soil samples and 0.5 µg/L in water samples.

Simazine, atrazine, propazine and terbuthylazine were extracted from water, baseline resolved and separated from 3-chloro-4-methoxyaniline, 3-chloro-4-methyl-aniline, 4-isopropylaniline, and 3,4-dichloroaniline on a C_{18} column ($\lambda = 230$ nm) using a complex 60-min 18/85 → 46/54 acetonitrile/water (50 mM $HClO_4$/$LiClO_4$ mixture at pH (4) gradient. Peak shapes were excellent and detection limits of less than 10 ng/L were reported [993]. Propazine, atrazine, simazine and six metabolites were baseline resolved on a C_{30} column (photodiode array detector, $\lambda = 200$– 356 nm) using a 32-min 15/85 → 70/30 acetonitrile/water (1 mM phosphate buffer at pH 7.0) gradient. Concentrates of water extracts gave <10 ng/L detection limits [994].

Steinheimer [995] studied atrazine and related compounds, deisopropylatrazine, deethylatrazine, and terbuthylazine, in soil and water samples. These compounds were resolved using a C_{18} column ($\lambda = 220$ nm) and a complex 23-min 10/0/90 → 100/0/0 → 0/100/0 acetonitrile/methanol/water gradient. The last peak of interest eluted at 17-min. Acetonitrile and methanol flushes were done to clean the system between injections. Calibration curves are shown for four of the analytes and were linear over the 0.2–50 ppm range. Detection limits of extracted samples were reported as 0.4 µg/L (water) and 40 µg/kg (soil). Solubility in water was given for the components, the lowest being terbuthylazine at 8 ppm.

Atrazine and four metabolites generated from topical applications [2-hydroxy-4-ethylamino-6-isopropylamino-s-triazine, 2-chloro-4-ethylamino-6-amino-s-triazine, 2-chloro-4-amino-6-isopropylamino-s-triazine, 2-hydroxy-4,6-diamino-s-triazine) were resolved on a C_{18} column ($\lambda = 223$ nm). A 14-min 20/80 → 50/50 aceto-nitrile/water gradient was used [996]. The assay was linear from 2 to 120 µg/mL (20 µL injections). Peak shapes and resolution were good.

Hexazinone and six soil and vegetation metabolites (desisopropylatrazine, hydroxyatrazine, desethylatrazine) were baseline resolved in 30 min using a 40°C C_{18} column (photodiode array detector, $\lambda = 200$–254 nm, and thermospray MS). A complex 50-min 8.5/91.5 → 70/30 acetonitrile/water (50 mM ammonium acetate) gradient produced excellent peak shapes [997]. Detection limits of 50–400 ppb were reported (analyte dependent).

Atrazine   Terbuthylazine

Paraquat and diquat extracted from high-moisture crops were analyzed on a silica column ($\lambda = 257$ nm for paraquat and 310 nm for diquat). A 40/60 acetonitrile/water (5.0 g NaCl to pH 2.2 with HCl) mobile phase was used [998]. The use of methanol in place of acetonitrile produced very broad and tailed peaks. A comparison of this separation was made with a separation system that consisted of an aminopropyl column and a ternary acetonitrile/methanol/water (NaCl/HCl system) mobile phase. The latter system took >2 hours to equilibrate, whereas the silica system was ready for use in 15 min. A plot of k' versus percent acetonitrile (from 10% to 70%) was U-shaped and excessive peak tailing occurred at levels of <30% and >50% acetonitrile. The authors make a special note that they minimized the NaCl/HCl concentrations used because of the corrosive effects chloride salts and acids have on LC hardware. Detection limits of 10 ppb and linear working ranges from 2 to 500 ng injected were reported.

9.3.4.3 *Fungicides*

2-Hydroxy-3-aminophenazine and 2,3-diaminophenazine levels in carbendazim formulations were determined using a separation generated on a C_{18} column ($\lambda = 254$ nm) and a 75/25 water (25 mM K_2HPO_4 pH 7.5 with H_3PO_4)/acetonitrile acetonitrile mobile phase. The aminophenazine was badly tailed and the peaks were not fully resolved. Elution time was 14 min. A better choice of mobile phase might have included TFA or THF as mobile phase modifiers. Standard solutions from 0.1 to 10 ppm were used and a limit of quantitation of 0.71 ppm and limit of detection of 0.6 ppm were reported [999].

Bitertanol (fungicide) residues were recovered from strawberries and quantitated on a 40°C C_{18} column ($\lambda = 265$ nm, ex; 325 nm, em) using a 70/30 acetonitrile/water mobile phase [1000]. Peak shapes were excellent and elution was complete in 5 min. Standards ranged from 0.05 to 2 µg/mL. A detection limit of 0.01 µg/mL (S/N = 3) was reported.

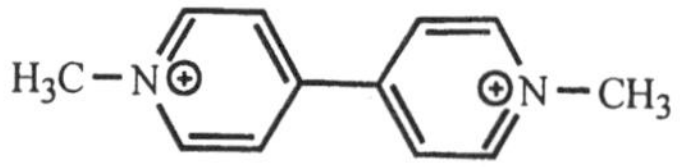

Paraquat

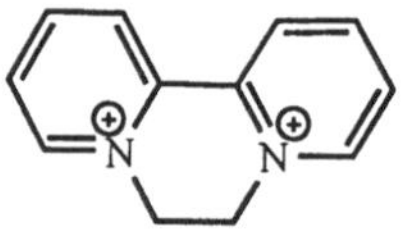

Diquat

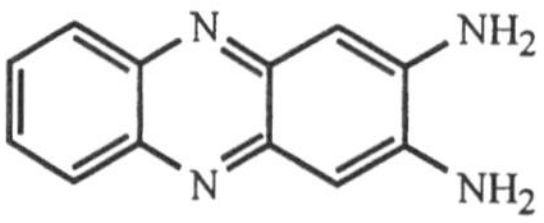

2,3-Diaminophenazine

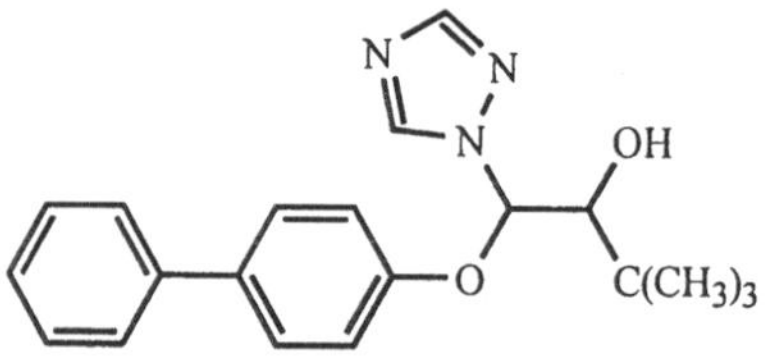

Bitertanol

Phenazine-1-carboxylic acid, 2-hydroxyphenazine, and 2-hydroxyphenazine-1-carboxylic acid were extracted from *Pseudamonas aureofaciens* cultures and analyzed on a C_{18} column ($\lambda = 257$ nm) using a 25-min 65/35/0.1 → 0/100/0.1 water/acetonitrile/TFA gradient [1001]. Elution of the components is complete in 15 min and the two carboxylic acids are not resolved using this gradient system. Therefore, it seems prudent to change to a shallower gradient or have an initial isocratic hold period prior to a gradient in order to force a separation.

Benomyl, diphenyl, *o*-phenylphenol, and thiabendazole were extracted from whole oranges. Separation on a 40°C C_{18} column ($\lambda = 285$ nm, ex; 315 nm, em) was achieved with a 40/40/20 water (2.88 g *n*-dodecylsulfate/400 mL to pH 2.5 with H_3PO_4)/acetonitrile/methanol mobile phase in 15 min. Benomyl and thiabendazole were incompletely resolved. Samples were spiked at the 0.1–90 µg/g range (analyte dependent) and detection limits of 0.05 µg/g were reported [1002].

Malachite green, gentian violet, leucomalachite green, and leucogentian violet (antifungal for fish) residues were isolated from trout and catfish and separated on a C_{18} column ($\lambda = 588$ nm). Note that the leuco forms need to be oxidized to exhibit a UV response. This was done through a postcolumn electrochemical cell poised at +0.9 V. Elution was complete in <9 min using a 55/45 acetonitrile/water (0.4 g ammonium acetate with 1 mL triethylamine to 450 mL then to pH 3.6 with acetic acid) mobile phase [1003]. The malachite green and leucogentian violet were quite tailed but fully resolved. Standards of 80 ng/mL were easily detected.

9.3.5 Summary

Acetonitrile has been used as the organic mobile phase constituent for nearly every separation of environmentally important analytes. Often, acetonitrile not only gives the best peak shape but the shortest retention times and best resolution as well.

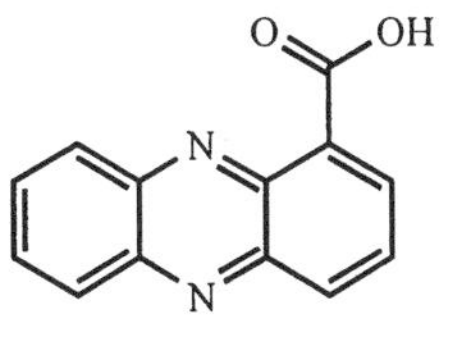

Phenazine-1-carboxylic acid

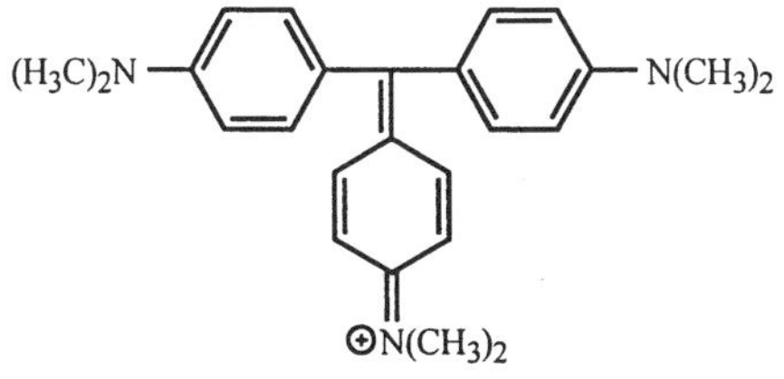

Gentian violet

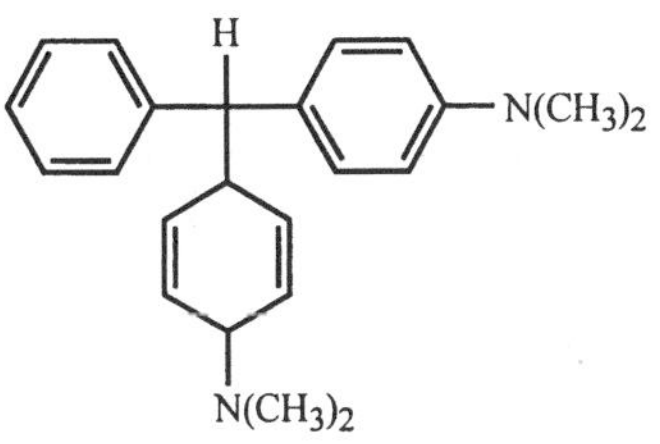

Leucomalachite green

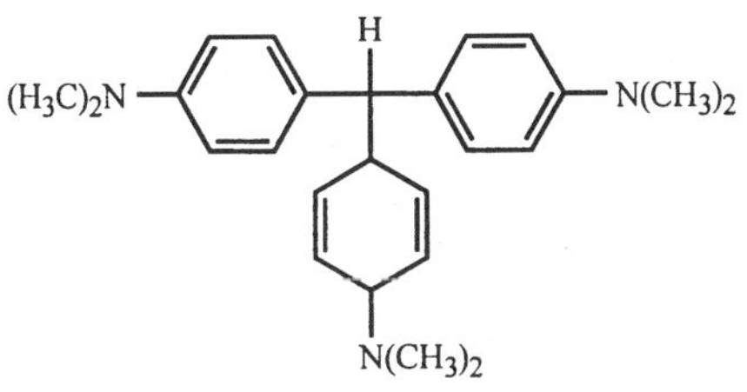

Leucogentian violet

The one case where this is not true was in the separation of hydroxylated compounds (e.g., phenol). With these compounds, the addition of methanol almost always improves the overall efficiency of the separation.

9.4   INDUSTRIAL AND POLYMER ANALYTES

9.4.1   Surfactant and Additive Analytes

Polymer additives Irgafos P-EPQ and Irgafos 168 were analyzed on a C_{18} column ($\lambda = 220\,nm$) using a 65/35 acetonitrile/chloroform mobile phase. Elution was complete in 10 min. Irganox 1010, Irganox 1076, and Sumilizer BHT were analyzed similarly but with a 65/35 acetonitrile/acetone mobile phase. A set of four antioxidants (Santowhite, Irganox 3114, Irganox 1010, Irganox 1330) were baseline

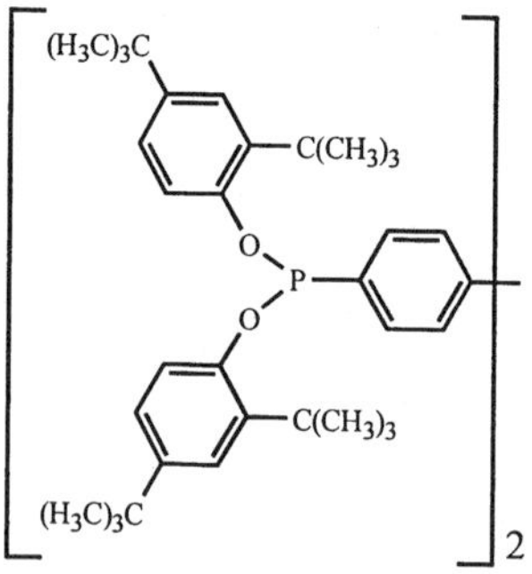

Irgafos P-EPQ

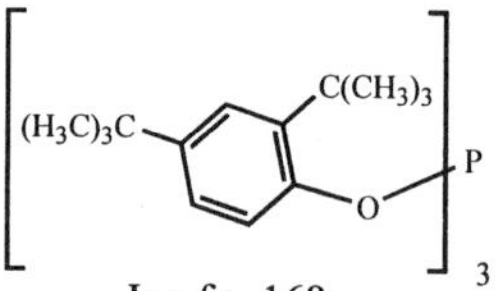

Irgafos 168

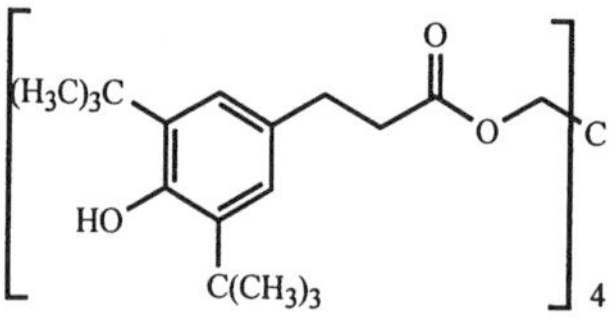

Irganox 1010

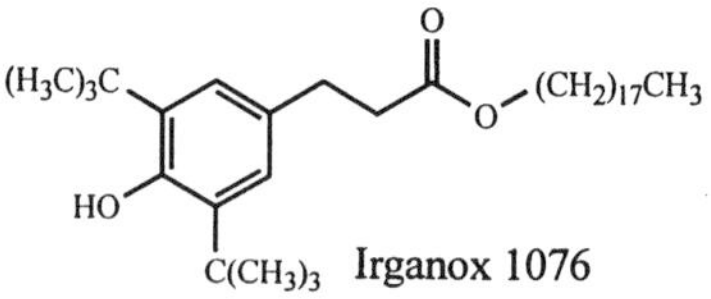

Irganox 1076

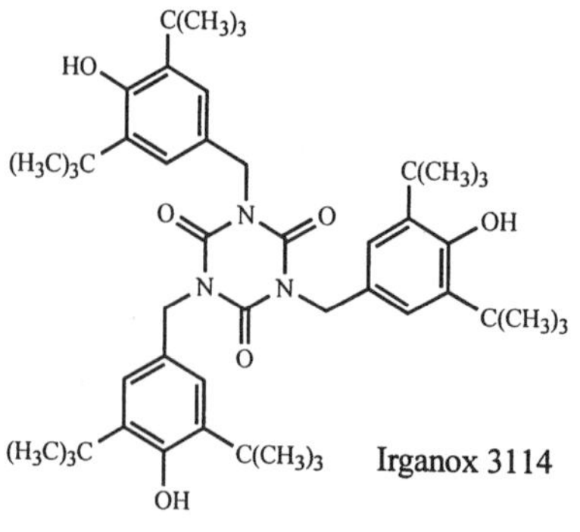

Irganox 3114

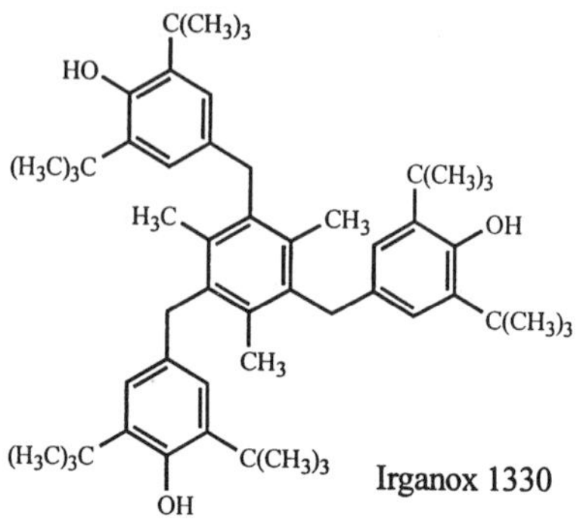

Irganox 1330

resolved on a C_{18} column ($\lambda = 280$ nm) using a 100% acetonitrile mobile phase [1004]. A 20 μL injection of 5 μg/mL standards gave readily detected peaks. Elution was complete in 12 min. Peaks shapes were excellent.

Three fluorescent whitening agents (4,4′-bis[2-sulfostyryl]biphenyl [BSB], 4,4′-bis[4-chloro-3-sulfostyryl]biphenyl, 4,4′-bis[{4-anilino-6-morpholino-1,3,5-triazin-2-yl}amino]stilbene-2,2′-disulfonate [BMS]) were isolated from sediment and surface water and baseline resolved on a C_{18} column ($\lambda = 350$ nm, ex; 430 nm, em). A 22-min $30/70 \rightarrow 60/40$ (60/40 acetonitrile/methanol)/water (0.1 M ammonium acetate pH 6.5) gradient eluted all peaks in 17 min (the internal standard selected eluted at 21 min). Isomers of the BSB and BMS were also generated. Concentration ranges were reported from 2 to 4000 ng/L (analyte dependent) with limits of quantitation of ∼0.4 ng/L [1005].

Stilbene-, biphenyl stilbene-, pyrazoline-, oxazole-, and coumarin-based brighteners used in detergent formulations were separated on a C_8 column and monitored by UV ($\lambda = 340$ nm) and fluorescence ($\lambda = 340$ nm, ex; 390 nm, ex), and API MS [1006]. A table of relative ion intensities for the MS detector showed that a 95/5 acetonitrile/water mobile phase gave the highest ion intensity, whereas a 50/50 acetonitrile/water mobile phase caused a 16–95% reduction in ion intensity and a 50/50 acetonitrile/water (5 mM ammonium acetate) mobile phase resulted in a range from a 10-fold increase to an 85% reduction in ion intensity. A set of samples ranging from 3 to 15 ppm (20 μL injected) was used for working curve generation.

An alkylamide arginine surfactant (where alkyl = C_{10}, C_{12}, C_{14}, C_{16}) was studied on a cyanoporpyl column ($\lambda = 210$ nm) using a 42/58 water (0.075% TFA)/(80/20 acetonitrile/water with 0.1% TFA) mobile phase [1007]. All four components were baseline resolved in <13 min. Excellent peak shape resulted for a 1 μg injection. The column was severely overloaded at 25 μg injected. Linear ranges of 0.05–6 mM and detection limits of 0.007 mM (S/N = 3) were reported.

Nonylphenolethoxylate surfactants were extracted from wastewater samples and analyzed on both a cyanopropyl and a C_{18} column ($\lambda = 229$ nm). For oligomers of $n = 1$ to 3 (ethoxy units) a 98/2 acetonitrile/water mobile phase and the C_{18} column were used [1008]. Peaks were broad, especially the trimer, most likely due to slow mass transfer effects. Elution was complete in 15 min. For oligomers $n = 4$ to 14, a cyanopropyl column and a 95/5 acetonitrile/water mobile phase produced very good separations and peak shapes. Detection limits of 1 μg/L for individual surfactants were reported.

Cretier et al. [1009] analyzed aliphatic alcohol ethoxylates (non-ionic surfactants). These were separated according to alkyl chain length C_{10}–C_{16}, and, when collected and combined, were anlayzed as to their degree of ethoxylation. The

Alkylamide arginine surfactant

prefractionation of the various alkyl lengths fraction was done using a 50-min $75/25 \rightarrow 100/0$ methanol/water gradient on a C_{18} column (ELSD; drift tube 50°C, nebulizer pressure 1.5 bar). From this, even and odd chain length fractions were combined. A 100-min linear gradient from $50/50 \rightarrow 100/0$ acetonitrile/water was then run to resolve the ethoxylated components. Ethoxylation ran from 1 to 30 units. Samples concentrations were 5 g/L.

Thirty-one separate peaks were generated during the analysis of non-ionic octylphenol polyether alcohol surfactants on a C_{18} column (photodiode array detector). The use of an isocratic 25/75 acetonitrile/water mobile phase produced excellent peak shapes [1010]. The chromatogram exhibited two definite groups of peaks. The first 15-peak group eluted between 2 and 14 min and the second 16-peak group between 18 and 80 min. No individual peak identification was given.

Surfactant-containing products such as disodium hexadecyldiphenyloxide sulfo-nate (Dowfax MADS) and the dihexadecyl analog (Dowfax DADS), and sodium dihexylsulfosuccinate (Aerosol 80-I) are used in the remediation of trichroloethene-contaminated aquifers [1011]. The materials, along with residual trichloroethene, 1,1-dichloroethene, and cis- and trans-dichoroethene, were recovered from water and analyzed on a 30°C C_{18} column ($\lambda = 210$ nm for the chlorinated materials and 238 nm for the surfactants). A two-step 10/90 (hold 1 min) $\rightarrow$ 40/60 (at 4 min) $\rightarrow$ 95/5 (at 6 min) acetonitrile/water gradient eluted all analytes in <7 min. The chlorinated compounds were baseline resolved and linearity was established from 0.05 to 15 mg/L, whereas the surfactants were linear over a range of 0.1–20 mg/L (MADS) and 0.01–2.5 mg/L (DADS). Quantitation limits were reported as 0.05 mg/L for chlorinated and 0.1 mg/L for surfactants (S/N = 10). Aerosol 80-I was eluted using a complex $20/80 \rightarrow 70/30$ acetonitrile/water (3 mM NaCl) gradient. The calibration curve was reported as 20 to 120,000 mg/L (!) with a quantitation limit of 20 mg/L. Interestingly, isopropyl alcohol was a quantitatable peak in the chromatogram of Aerosol 80-I.

In a nice presentation by Sarrazin et al. [1012], eleven biodegradation products of 1-(p-sulfophenyl)dodecane were extracted from sea water. Some of these compounds include the following acids: 3-phenylpropionic, phenylacetic, 3-(p-hydroxyphenyl)-propionic, p-hydroxylphenylacetic, p-hydroxybenzoic, p-sulfobenzoic, p-sulfo-phenylacetic. A 35-min $95/5 \rightarrow 60/40$ water (8 mM KH_2PO_4 to pH 2.2 with H_3PO_4)/acetonitrile gradient on a C_{18} column ($\lambda = 215$ nm) generated good separation for most compounds. The authors correctly noted that the pH of the mobile phase needed to remain significantly below the pK_a of the acids in order to

Dowfax MADS

p-Sulfobenzoic acid

keep the acids fully protonated and thereby uncharged (and retained). A linear range of 0.03–0.35 μg injected was reported.

Dequalinium chloride (a bisquaternary ammonium surfactant), its dimer, and two positional isomer impurities were analyzed on a C_{18} column ($\lambda = 322$ nm) using a 30-min 90/10 → 20/80 water [0.1 $NaClO_4$ to pH 3.0 with $HClO_4$)/acetonitrile gradient [1013]. Linear curves from 0.2 to 5 μg/mL (parent compound) and 10 to 100 μg/mL for the impurities were generated with detection limits of 2 ng injected cited for all compounds.

9.4.2 Polymeric Analytes

Low-density polyethylene samples were analyzed for their additive content. Five additives (butylated hydroxytoluene, butylhydroxyethylbenzene, Isonox 129, Irganox 1076, Irganox 1010) were separated on a C_{18} column ($\lambda = 200$ nm) using a 15-min 75/25 → 100/0 acetonitrile/water gradient. Peak were resolved and elution was complete in 11 min [1014]. Levels of 70–1000 ppm were found.

Poly(ethylene glycols) (PEGs) of molecular weight 200–1560 were analyzed on a C_{18} column using various acetonitrile/water mobile phases [1015]. An interesting and ingenious departure from the norm was presented in this method: the addition of a visualizing reagent that deliberately generated a well-controlled nonzero background absorbance (i.e., indirect detection at $\lambda = 210$ nm). This precluded the need for the analyte of interest to have a chromophore or be derivatized prior to analysis since the visualizing reagent was displaced in proportion to the amount of analyte present and was monitored. However, both positive and negative peaks appear in the chromatogram and so a data acquisition system capable on handling this is necessary. For PEG standards of molecular weight 200, 400, and 1000, the chromatographic conditions were as follows: 10/90 acetonitrile/water (0.8 mM caffeine), $\lambda = 210$ nm, 15 min elution; 15/85 acetonitrile/water (0.8 mM p-hydroxybenzoate), $\lambda = 255$ nm, 30 min; 20/80 acetonitrile/water [0.8 mM p-hydroxybenzoate), $\lambda = 255$ nm, 30 min, respectively. This method separated up to the 35-mer in a PEG 1540 sample. Peak shapes were very good.

The anaerobic degradation products of PEGs were analyzed on a C_{18} column ($\lambda = 190$ nm) using a 41-min 4/96 → 72/28 (50/50 acetonitrile/water [1 mM

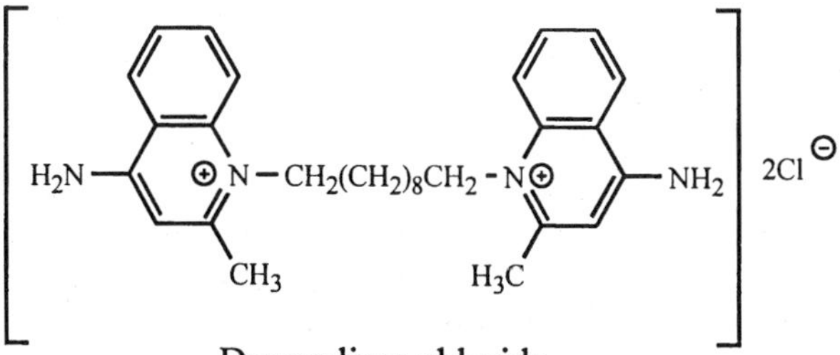

Desqualium chloride

H_3PO_4 with 3 ppm HNO_3])/water (1 mM H_3PO_4 with 3 ppm HNO_3) gradient [1016]. This analysis was possible at 190 nm because the UV cutoff for acetonitrile is usually $\lambda = 188$ nm or below. The addition of nitric acid was a nifty trick used to artificially produce a background absorbance in the water so that the change in background absorbance during the gradient would be acceptable. The resulting chromatogram of a 50 mM mixture of PEGs with molecular weights 200, 300, 400, 600, 1000, and 1500 generated excellent peak shapes and profiles up to $n = 50$.

Four polyesters (Alftalat 3258 and 3352, Crylcoat 430 and 801] were character-ized through the elution profile generated on a 1.5 μm nonporous C_{18} column ($\lambda = 280$ nm) using a 30/70/0/0.2 → 70/30/0/0.2 (at 20 min) → 88/2/10/0.2 (at 50 min hold 25 min) acetonitrile/water/THF/TFA gradient [1017]. Good definition of low → midrange oligomers was obtained. The fingerprint detection limits were 5000 ppm for the Alftalat polymers and 10,000 ppm for the Crylcoat polymers. Samples were prepared and analyzed at the 1%, 2.5%, 5%, and 10% w/v levels.

Mengerink et al. [1018] quantitated the first six linear and cyclic oligomers of polyamide-6 (i.e., caprolactam and its linear analogs) on a C_{18} column. Linear oligomers were analyzed by UV detection ($\lambda = 200$ nm) whereas the cyclic oligomers were analyzed using o-phthalaldehyde/3-mercaptoproprionic acid post-column derivatization with fluorescence detection ($\lambda = 330$ nm, ex; 420 nm, em). A 22-min 99/1 → 49.5/50.5 water (10 mM H_3PO_4)/acetonitrile gradient was used. Peak shapes were excellent and the 1–6 oligomers were well resolved. As would be expected, the baseline for the UV method exhibited noticeable baseline shifting over the course of the gradient since acetonitrile has a significant absorbance there (~0.05 AU). The fluorescence-derived chromatogram exhibits no drifting. Detection limits of 100 mg/kg for linear oligomers and 20 mg/kg for cyclic oligomers were reported.

Residual monomers of epoxy resins (i.e., m-xylylenediamine and bisphenol A diglycidyl ether) were extracted from cured epoxy resins and analyzed on a C_{18} column ($\lambda = 275$ nm, ex; 300 nm, em for m-xylylenediamine) using a complex 20-min 30/70 → 75/25 acetonitrile/water gradient [1019]. Excellent resolution of bisphenol F, bisphenol A, three bisphenol F diglycidyl ether isomers, and bisphenol A diglycidyl ether resulted. Peak shapes were also excellent. Linear concentration curves from 20 to 1000 μg/L were obtained. The m-xylylenediamine was derivatized with fluorescamine and analyzed with the same column and gradient as for bisphenol (but with different excitation/emission wavelengths) and linear curves from 10–800 μg/L were obtained.

A series of three preservatives that release formaldehyde upon degradation (Germall-115, Glydant, Dowicil-75] were resolved and separated from formaldehyde as their 3,5-diacetyl-1,4-dihydrolutidine derivatives [1020]. A PRP-1 column ($\lambda = 410$ nm) and a 30/70 acetonitrile/water mobile phase were used to generate the 15-min separation. Detection limits of 400 ppb and a linear range of two orders of magnitude were reported. The authors note that standards made from pure preservatives show substantial formaldehyde peaks upon analysis, i.e., degradation occurs in the short time between standard preparation and analysis. No comment on the possible reason for the degradation (or source of contamination) was made.

The photochromic spirooxazine compound deposited in polyurethane films (i.e., 1,3-dihydro-1,3,3-trimethylspiro[2H-indole-2,3']-[3H]naphth[2,1-b][1,4]-oxazine) was photodegraded and the products, substituted indole and naphthalene fragments, were monitored by HPLC [1021]. A C_{18} column (photodiode array detector, $\lambda = 200$–450 nm) and a 40-min 40/60 → 100/0 acetonitrile/water gradient generated a baseline separation between the original spirooxazine and four identified degradation compounds.

Bisphenol-A-diglycidyl ether and the *ortho/ortho-*, *para/para-*, and *ortho/para-*bisphenol-F-diglycidyl ethers were extracted from fish and separated in <10 min on a 40°C C_{18} column ($\lambda = 225$ nm, ex; 310 nm, em). Use of a 20-min 60/40 → 100/0 acetonitrile/water gradient resulted in the generation of good resolution and peak shapes [1022]. Linear ranges of 0.5–4 mg/L and detection limits of 0.056 mg/L (S/N = 3) were reported.

Methyl methacrylate, *N*,*N*-dimethyl-*p*-toluidine, and benzoyl peroxide were extracted from various methyl methacrylate polymers and analyzed on a C_{18} column ($\lambda = 235$ nm) using a 52/48 acetonitrile/water mobile phase [1023]. Elution was complete in 30 min. Bisphenol A and 4,4′-dichlorophenylsulfone were similarly extracted and resolved on a C_{18} column ($\lambda = 235$ nm) using a 50/50 acetonitrile/water mobile phase. Elution was complete in 22 min.

Residual isocyanate monomeric extracts from 19 commercial polyurethanes were analyzed as their 9-(methylaminomethyl)anthracene derivatives on a 45°C C_{18} column ($\lambda = 254$ nm, ex; 412 nm, em). An 80/20 acetonitrile/water (3% TEA to pH 3 with H_3PO_4) mobile phase was used [1024]. Analytes included 2,4- and 2,6-toluenediisocyanate, 2,4-toluene diisocyanate dimer, diphenylmethane-4,4′-diisocyanate, hexamethylene diisocyanate, cyclohexyl-, phenyl-, and octadecylisocyanate and, isoperhone diisocyanate. The effect of varying the level of acetonitrile (70–85%) on retention (as k') was presented as a plot for six different C_{18} materials. Detection limits were reported as 0.03 mg/kg.

9.4.3 Dyes and Related Analytes

The ancient dyes berberine, palmatine, and jatrorrhizine were recovered from paper and analyzed on a C_{18} column ($\lambda = 345$ nm). The 40/10/49.5/0.5 ethanenitrile/methanol/water [50 mM TFA]/sodium lauryl sulfate (no concentration given so assumed by weight) mobile phase gave baseline resolution and excellent peak shapes [1025]. Separation was complete in <15 min and complete resolution of all peaks was obtained.

Berberine

Palmatine

Angelino et al. [1026] studied three food dyes, amaranth, new coccine, and chromotrope. These were baseline resolved in <10 min using a C_{18} column (photodiode array detector, $\lambda = 190$–600 nm) using a 63/27 (27 not a typo!) water (5.0 mM octylammonium phosphate at pH 6.0) mobile phase. Plots showing the effect of ion-pair reagent concentration on retention show final elution times of >40 min at the 7.5 mM level! Plots of temperature and pH effect on retention are also included. Two additional dyes, sunset yellow and tartrazine were added to the above list and a 28-min baseline resolved separation generated using a 66/24 water (5.0 mM octylammonium phosphate at pH 6.0)/acetonitrile mobile phase.

Three red dyes (carmoisine [acid red 14, food red 3], amaranth [acid red 27, food red 9], Ponceau 4R [acid red 18, food red 7]) were isolated from confectionaries and separated in 25 min on a 25°C C_{18} column ($\lambda = 520$ nm) using a 70/30 water/acetonitrile with 5 mM octylamine to pH 6.4 with H_3PO_4. Note that the nonaqueous pH is irreproducible [1027]. Calibration standards ranged from 0.05 to 0.2 mg/mL with detection limits reported at 12 µg/mL (S/N = 3).

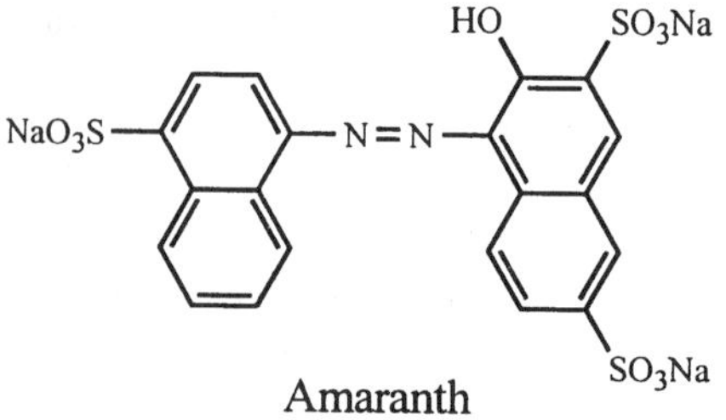

Amaranth

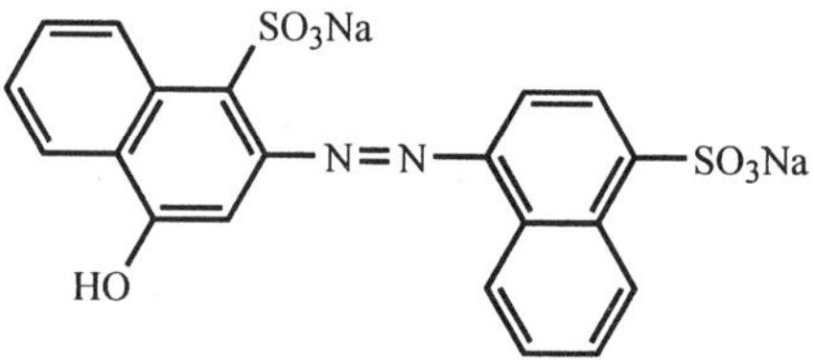

Chromotrope

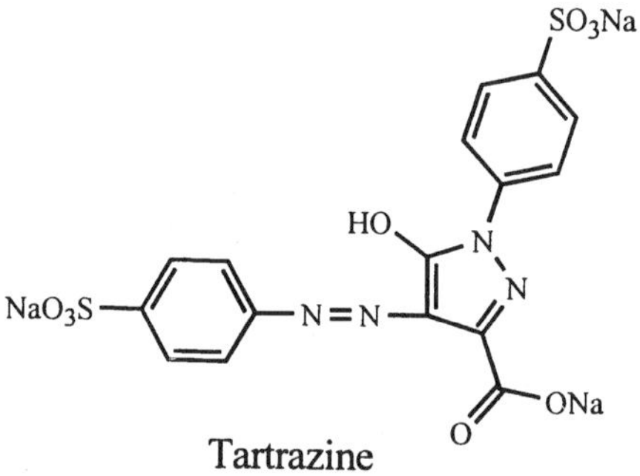

Tartrazine

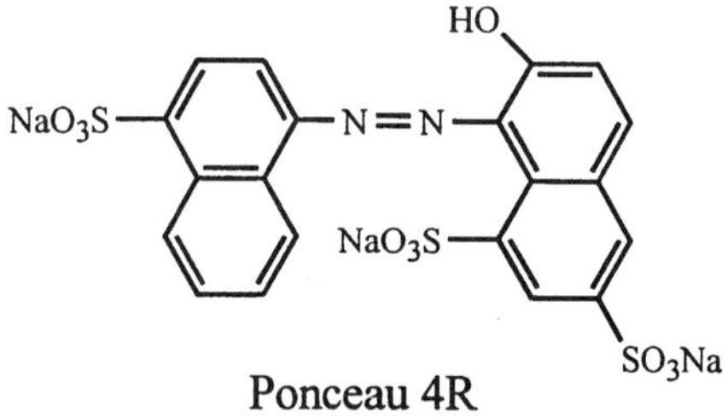

Ponceau 4R

Eight sulfonated dyes (Acid Red 2, 27, 88, Acid Orange 7, 12, 52, Acid Yellow (1) were well resolved on a 25°C PAH column ($\lambda = 460$ nm) using an ion-pair reagent system: 73/27 water (phosphate at pH 6.7)/acetonitrile with 2.4 mM butylamine. Acceptable peak shapes resulted and elution was complete in 16 min [1028]. Plots of the effects of pH, percent acetonitrile, and butylamine concentration on ln k' were presented. Linear ranges ran from 15 to 500 μg/L, detection and quantitation limits were 15 μg/L and 25 μg/L, respectively.

A set of nine temporary and direct dyes used in hair coloring products (e.g., Disperse Blue 1, 3, 26, and 99, Disperse Violet 1, and N^1, N^4, N^4-tris-(β-hydroxyethyl)-1,4-diamono-2-nitrobenzene [HDN]) were isolated from product by Scarpi et al. [1029]. These were separated on a base deactivated C_8 column ($\lambda = 450$ nm and 630 nm) using a 70/30 (hold 10 min) → 20/80 (at 20 min) water (5 mM heptanesulfonic acid)/acetonitrile gradient. The example shown has Disperse Blue 1 and 3, Disperse Violet 1, and HDN eluted in 16 min. Although this is not unacceptable, a modified gradient and the addition of an acidic mobile phase modifier would help the resolution and peak shapes. A linear range of 10–100 mg/mL was reported.

Sunset Yellow FCF (Food Yellow No 5), six subsidiary colors, and four raw materials/intermediates were separated on a 40°C C_{18} column ($\lambda = 232$ nm) using a 20-min 100/0 → 65/35 water (0.04% ammonium carbonate)/acetonitrile gradient

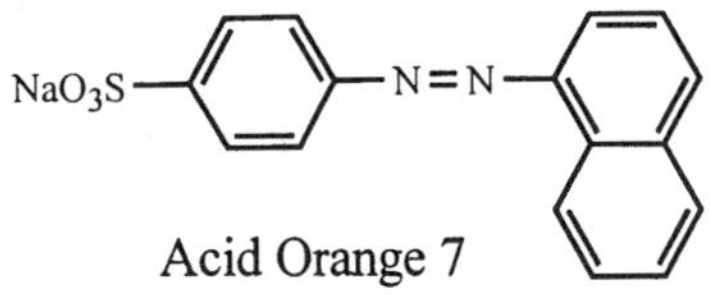

Acid Orange 7

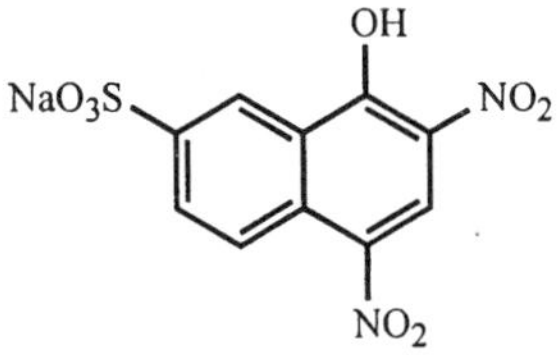

Acid Yellow 1

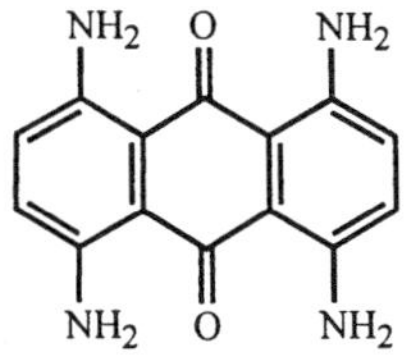

Disperse Blue 1

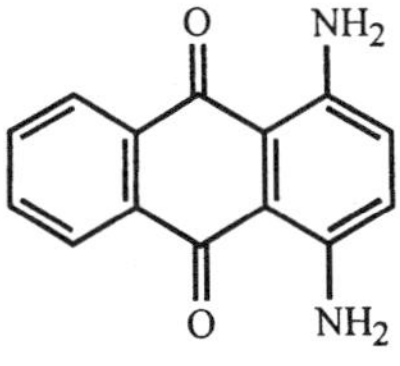

Disperse Violet 1

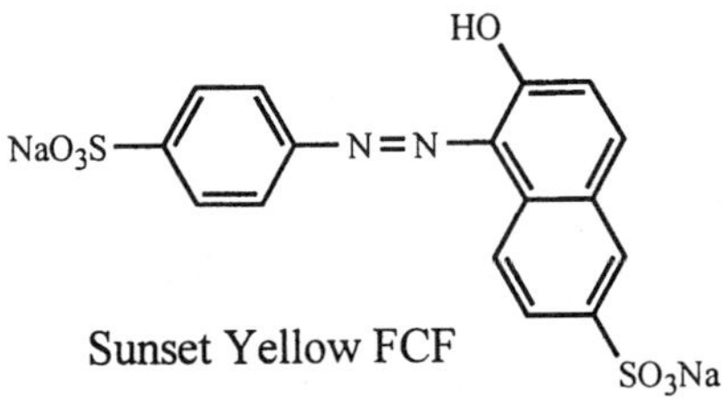

Sunset Yellow FCF

[1030]. Subsidiary compounds include structures such as 6-hydroxy-5-phenylazo-2-naphthalenesulfate and 4-(2-hydroxy-1-naphthylazo)benzenesulfate. Raw materials and intermediates included sulfanilic acid and 6,6′-oxybis(2-naphthalenesulfate). Detection limits of 0.01–0.1% were reported.

Sixty-seven lots of Sunset Yellow FCF (FD&C Yellow No.6) were tested for combined benzidine on a C_{18} column ($\lambda = 500$ nm) using a 50-min $0/100 \rightarrow 20/80$ acetonitrile/water (with 1.5% ammonium acetate and 0.5% acetic acid) gradient [1031]. Detection limits for combined benzidine were reported as 10 ng/g sample.

Seventeen blue non-ball inks were characterized using a C_{18} column (photodiode array detector, $\lambda = 200$–800 nm) and an 80/20 acetonitrile/water [5 mM heptanesulfonic acid with acetic acid to pH 4.7) mobile phase [1032]. Analysis was complete in 20 min. A set of retention versus wavelength versus intensity plots were given for different ink samples and a number of ways of analyzing the data were presented. The authors noted that the acetonitrile/water mobile phase gave optimal results as compared with methanol/water mobile phases. Red printing inks were analyzed in a similar fashion using a C_{18} column ($\lambda = 254$ nm, 510 nm, and 350 nm, ex; 550 nm, em) and a $30/70 \rightarrow 100/0$ acetonitrile/water (10 mM $KClO_4$ to pH 3 with $HClO_4$) gradient [272]. Chromatographic profiles were presented and detection limits of $\sim$50 ng were reported.

In an excellent study, White and Catterick [1033] characterized and analyzed 40 acidic dyes according to their composition, shape, and retention on a PLRP-S column (photodiode array detector, $\lambda = 190$–600 nm). A 50/50 acetonitrile/water (0.7 g/L citric acid with 3.4 g/L tetrabutylammonium hydrogen sulfate adjusted to pH 9.0 with ammonia) mobile phase was used. Five classes of dyes were included in the study: phenyl-N=N-monosulfonated naphthols, naphthyl-N=N-naphthols, phenyl-N=N-p-hydroxylated naphthyls, monosulfated phenyl-N=N-naphthols, and phenyl-N=N-disulfonated naphthols. Tables were generated in which the relative retention times, relative absorbances (vs. 500 nm) for 250–450 nm (50 nm intervals), and UV maximum absorbances were listed. From positional/spatial arguments and data presented in the study, information about the chemical structure of an analyte could be determined with as little as 100 ng of sample. Structures were presented for all compounds.

In a very nice separation by Zhou et al. [1034], 10 arylamines (aniline, benzidine, o-anisidine, quinoline, o-toluidine, 4,4-methylenedianiline, 3,3′-dimethoxy- and 3,3′-dimethylbenzidine, 4-aminobiphenyl, N,N-dimethylaniline) commonly found in a dye process plant were extracted from water and baseline resolved and exhibited excellent peak shapes on a base deactivated C_{18} column ($\lambda = 254$ nm). A 30/70 (hold 8 min) $\rightarrow$ 90/10 (at 16 min hold 4 min) acetonitrile/water gradient eluted all compounds within 15 min. A linear range of 0.1–100 mg/L was reported.

9.4.4   Other Industrial Analytes

Because fullerenes are sparingly soluble in acetonitrile, it was not an effective solvent on its own. Consequently, it was used in conjunction with other solvents such as toluene [269, 657] and dichloromethane [658]. When too high a level of acetonitrile was present in the mobile phase, the fullerenes precipitated onto the column packing. The resolubilization process was controlled by the rate at which the mobile phase passed over the precipitate. Therefore, these samples often elute as severely tailed or double peaks due to the dual "retention" mechanism process: precipitation/resolubilization followed by partition/adsorption.

Diesel oils were characterized by fractionating a sample into basic and neutral fractions and subsequently analyzing each fraction individually [1035]. The basic fraction was then analyzed for $C_2–C_4$ alkylbenzoquinone content and the neutral fraction for $C_1–C_5$ alkylcarbazole content. Both fractions were analyzed on a C_{18} column (particle beam MS) using a 40-min 40/60 → 100/0 acetonitrile/water gradient.

Gasoline was tested for adulteration by kerosene by monitoring the naphthalene content (associated with kerosene) of the gasoline sample [1036]. The sample was prepared to be 1% v/v gasoline in methanol. An 80/20 acetonitrile/water mobile phase and a C_{18} column ($\lambda = 285$ nm) separated the analytes of interest: naphthalene, 2-methylnaphthalene and 2,6-dimethylnaphthalene. Elution was complete in <15 min.

9.4.5 Personal Care and Cosmetic Analytes

Nine different commercially available ceramides (Ceramide III, IIIB, and VI; N-palmitoyl-, N-stearoyl-, N-oleoyl- and N-nervonoyl-D-sphingosine, and N-palmitoyl- and N-lignoceroyl-DL-dihydrosphingosine) were analyzed on a 30°C C_{18} column (ELSD), drift tube 35°C, N_2 at 1 bar) using a 15-min 95/5/0 → 35/5/60 acetonitrile/THF/n-propyl alcohol (all phases made 10 mM in both triethylamine and formic acid) gradient. The authors studied the effects of mobile phase components (methanol and dichloromethane offered no advantages). Peak area was shown to be a strong positive function of mobile phase modifier concentration from 1 to 10 mM, whereas the overall retention was unaffected [1037].

Three sunscreen agents (Eusolex 2292, octylmethoxycinnamate; Eusolex 4360, 2-hydroxy-4-methoxybenzophenone; Eusolex 6300, 3-(4-methylbenzylidene)camphor) were extracted from cosmetics and separated on a C_{18} column ($\lambda = 300$ nm) using a 90/10 acetonitrile/water mobile phase [1038]. Elution was complete in <4 min. Linearity spanned from 0.2 to 8.0 µg/mL and detection limits of 3 ng/mL (S/N = 3) were reported. A plot of retention time versus percent acetonitrile was presented.

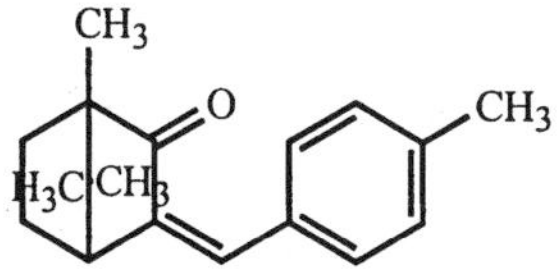

Eusolex 6300,
3-(4-Methylbenzylidine)camphor

Rhodamine B is a colored additive used in cosmetics that has been banned for use in Europe. This method describes a way to isolate and assay for rhodamine B using a C_8 column ($\lambda = 525$ nm) and a 15-min $50/50 \rightarrow 70/30$ acetonitrile/water (0.1 M $NaClO_4$ to pH 3.0 with $HClO_4$) gradient [1039]. A calibration curve from 2 to 400 µg/mL was reported and a 0.4 µg/mL standard has an absorbance of over 0.03 units.

The phenolphthalein level in cosmetic products was determined after extraction using a 25°C C_8 column ($\lambda = 230$ nm and 280 nm) and a 30-min $90/10 \rightarrow 10/90$ water (1% acetic acid)/acetonitrile gradient. This method also separated methyl-, ethyl-, propyl- and butylparaben and triclosan preservatives, allowing for their quantitation as well. Standards at the 66 µg/mL phenolphthalein and 25 µg/mL preservative were readily detectable. Detection limits of 5 ng injected ($S/N = 3$) were reported [1040].

Allantoin, urea, and lysine pyroglutamate were extracted from cosmetic formulations and analyzed by hydrophilic interaction on a polyhydroxyethyl A column ($\lambda = 200$ nm). An 80/20 acetonitrile/water (75 mM triethylamine phosphate at pH 2.8) mobile phase generated a 7-min separation and good resolution [1041]. A plot of the effect of changing percent acetonitrile and triethylamine phosphate concentration on retention time was presented. Linear ranges from 6 ng/mL to 420 µg/mL, and limits of detection of 0.7–13 µg/mL (analyte dependent) were reported.

9.4.6 Summary

Acetonitrile has been widely used in the separation and quantitation of small industrial-related compounds. However, many polymers and large hydrocarbonaceous molecules have low solubility in acetonitrile, limiting its use.

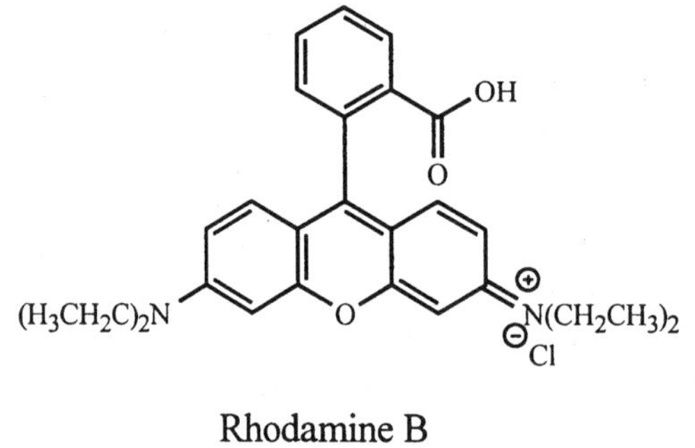

Rhodamine B

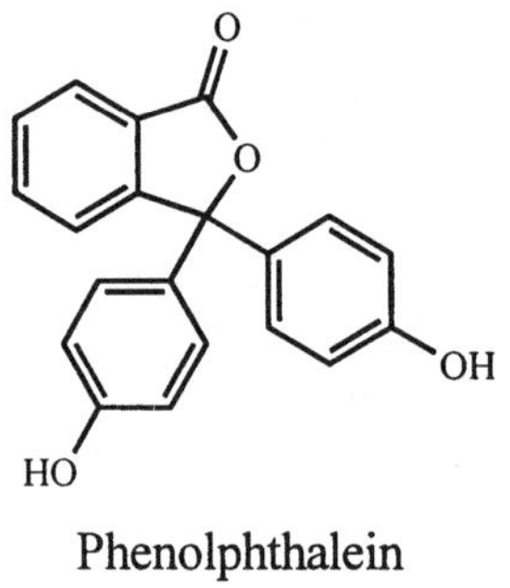

Phenolphthalein

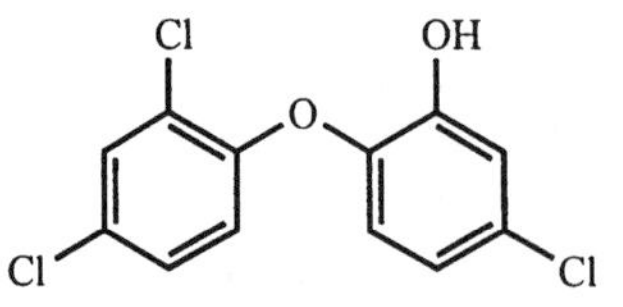

Triclosan

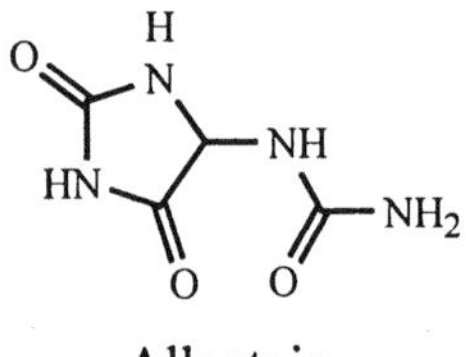

Allantoin

9.5 BIOLOGICAL ANALYTES

9.5.1 Carboxylic Acid Analytes

The fatty acid composition of *Agathis robusta* seeds was determined from the analysis of extracts that were converted to their picolinyl esters and separated on a $20°C$ C_8 column (ELSD) using an acetonitrile mobile phase [1042]. Elution was complete in 55 min. The following fatty acids were found: 18:4 and 20:5 co-eluted; 18:3 and 20:4 co-eluted; 18:2 and 20:3 co-eluted; 18:1; 16:0; and 18:0. A 1 mg sample size generated reasonable peak intensities.

Urinary 8-isoprostaglandin $F_{2\alpha}$ and two additional F-isoprostaglandins [9β, 11α- and 9α, 11β-isoprostaglandin F_2) were isolated and well resolved on a C_8 column (negative ion electrospray MS, desolvating plate $T = 80°C$, ring lens voltage 90 V, orifice voltage 50 V, orifice 1 $T = 200°C$) using a 70/30/0.1 water/acetonitrile/ acetic acid mobile phase [1043]. Elution time was 37 min. A linear curve from 20 to 5000 pg/mL was established along with a detection and quantitation limits of 2 pg/mL and 100 pg/mL, respectively.

A series of perfluorinated carboxylic acids (from perfluorohexanoic to perfluoro-decanoic) were extracted from liver, derivatized with 3-bromoacetyl-7-methoxycou-marin, and analyzed on a C_{18} column ($\lambda = 366$ nm, ex; 419 nm, em). Baseline resolution, excellent peak shape, and complete elution were achieved in 15 min using a 3/1 acetonitrile/water mobile phase [1044]. The linear range was 0.1– 10 nmol injected and detection limits were reported as 50 pmol/50 mg sample ($S/N = 3$).

In a very informative paper by Tindall et al. [1045], a series of aliphatic acids (acetic, propionic, butyric) and their corresponding anhydrides were analyzed on 12 different sorbents ranging from silica to C_8 to C_{18}. The authors reported accomplishing the separation of all six on a C_{18} column ($\lambda = 210$ nm) using a 90/10 (hold 1 min) $\rightarrow$ 55/45 (at 8 min hold 7 min) water (0.02% H_3PO_4)/acetonitrile gradient. Excellent peak shapes and resolution were achieved. The anhydride degradation rate was also monitored with respect to column type (minimal influence) and presence/ absence of H_3PO_4 (large influence on some). Detection limits of 1 ppm were claimed.

Twenty-two carboxylic acids and carbonyl compounds (e.g., oxalacetic acid, glyoxylic acid, glyceraldehyde, glyoxal, acetone, isobutyraldehyde, benzaldehyde, cyclohexanone) were separated as their 2,4-dinitrophenylhydrazine (DNPH) derivatives on a $55°C$ C_{18} column ($\lambda = 360$ nm). A complex $100/0 \rightarrow 0/100$ (1/1.4/6 acetonitrile/methanol/water [5 mM KH_2PO_4 buffer at pH 6.7])/(3/2 acetonitrile/ water [5 mM KH_2PO_4 buffer at pH 6.7]) gradient was used [1046]. In this study both mobile phase composition and flow rates were changed throughout the course of the separation. Complete resolution was not achieved between all 22 compounds but the authors note that the presence of a significant level of methanol ($\sim$25%) was required to generate adequate resolution between most peaks. Detection limits from 15 to 500 ng/L were reported.

A set of five acids (dehydroabietic, linolenic, abietic, linoleic, oleic) and seven less polar compounds (e.g., ergosterol, campesterol, β-sitosterol, trilinolein, triolein) were extracted from Scots pine and resolved on a C_{18} column (ELSD, tube temperature 40°C, nebulizer pressure at 2 bar). Excellent peak shapes and baseline resolution was achieved for all compounds using a 90/10/0.1/0/0 (hold 8 min) → 60/0/0/8/40 (at 23 min hold 27 min) acetonitrile/water/acetic acid/methanol/dichloromethane gradient [1047]. Typical standards ranges were reported as 4–1000 µg/mL. The authors noted that the methanol was critical to producing good peak shapes for the sterols and that UV detection would not have allowed for the analysis of all compounds (both from a chromophore and solvent absorbance point of view).

Benzoic acid extracted from orange juice was the subject of a collaborative study that used a PRP-1 column ($\lambda = 230$ nm) and a 40/60 acetonitrile/water (50 mM phosphate buffer at pH 2.3) mobile phase [1048]. Elution was complete in <8 min and a detection limit of 0.5 ppm (S/N = 10] was reported. The working range was 0.5–10 ppm.

Oxolinic acid was extracted from fish serum and analyzed on a 24°C C_{18} column ($\lambda = 327$ nm, ex; 360 nm, em) using a 68/22/10 water (1 mM H_3PO_4 at pH

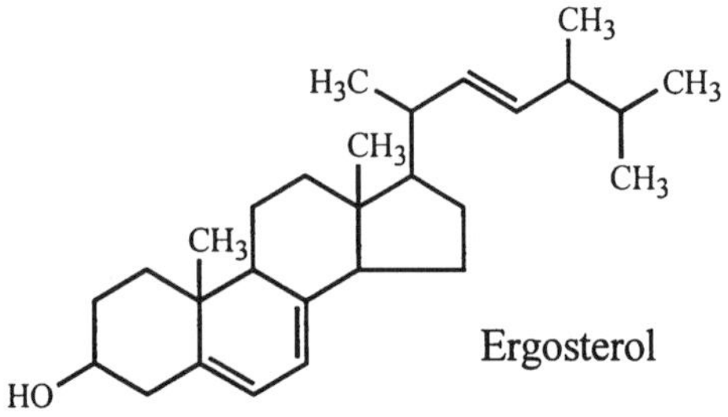

Ergosterol

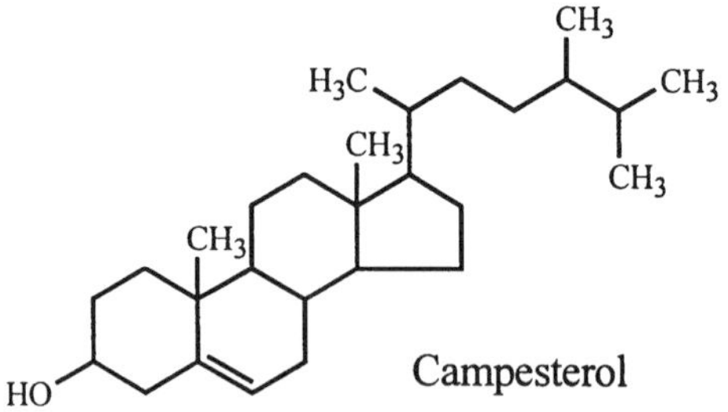

Campesterol

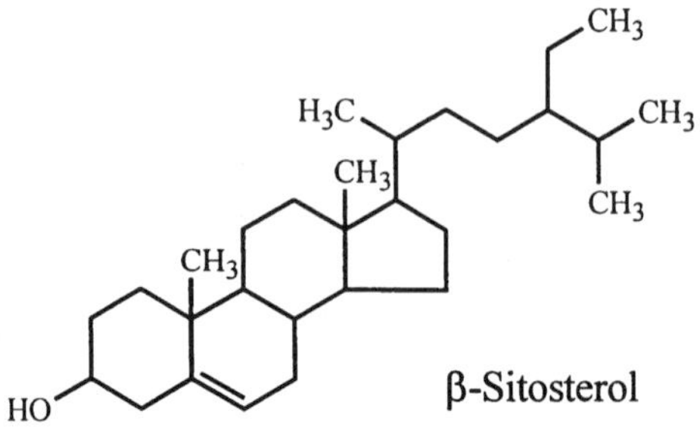

β-Sitosterol

3.4)/acetonitrile/THF mobile phase [1049]. Elution was complete in 6 min. Changing the H_3PO_4 concentration from 1 to 20 mM had only minor effects on retention and peak shape. A linear range of 0.02–2.5 µg/mL and detection and quantitation limits of 5 ng/mL and 15 ng/mL, respectively, were reported.

Peroxyacetic acid (the reaction product of acetic acid and hydrogen peroxide) and hydrogen peroxide levels were simultaneously determined in brewery cleaning-in-place effluents as their methyl-p-tolyl sulfide and triphenylphosphine derivatives, respectively. Baseline resolution of the derivatives and the excess derivatizing reagents was achieved in <5 min on a C_8 column ($\lambda = 225$ nm). A step gradient from $40/60 \rightarrow 100/0$ acetonitrile/water was performed at 3 min. A baseline response to the gradient was evident at approximately 4 min, meaning that the total "dwell" time of the system cannot be more than about 1 mL (a very short column was used here). Linear curves from 2.5 to 1000 µmol for the acid and 7.5 to 300 µmol for the peroxide were reported [1050]. Detection limits were in the low µmol range (S/N = 3).

N-Acyl- and N-glycosylneuraminic acids (sialic acids) were isolated from serum and tissue samples, per-O-benzoylated, and analyzed on a C_{18} column ($\lambda = 231$ nm) using a 67/33 water/acetonitrile mobile phase [1051]. Good resolution was achieved and elution was complete in <20 min. Calibration standards of 0.05–50 µg injected were used and detection limits of 10 ng injected were reported.

Caffeic acid and caffeic acid o-quinone, were resolved from 12 polyphenol oxidase products on a C_{18} column ($\lambda = 280$ nm) using a 40-min $4/96 \rightarrow 36/74$ acetonitrile/water (2.5% acetic acid) gradient [1052]. A working curve from 0.01 to 1.0 mM caffeic acid was used. The analytes of interest were baseline resolved from one another and the other products.

Ginkgolic acid is a mixture of several 2-hydroxy-6-alkylbenzoic acids, two of which (alkyl = 1-octene or 3-decene) were baseline resolved on a 45°C C_{18} column ($\lambda = 308$ nm) with a 92/7/1 acetonitrile/water/acetic acid mobile phase [1053]. Elution was complete in 6 min and peaks of interest were well resolved from all extracted material. In a separate study, Pietta et al. [1054] isolated six metabolites (hippuric acid, 4-hydroxy- and 3-methoxy-4-hydroxyhippuric acid, 4-hydroxy-, 3,4-dihydroxy-, and 3-methoxy-4-hydroxybenzoic acid) of *Ginkgo biloba* leaf extracts from urine and blood samples and analyzed on a C_{18} column (photodiode array detector). An 88/5/5/2 water/acetonitrile/methanol/acetic acid mobile phase generated good resolution and complete elution in 15 min. The standard used was at the 1 mg/mL level.

Ginkgolic acid I

Vanillylamine, L-phenylalanine, caffeic acid, coumaric acid, ferulic acid, cinnamic acid, capsaicin, and dihydrocapsaicin were extracted from green peppers and resolved on a C_{18} column ($\lambda = 236$ nm) using a 35-min $0/100 \rightarrow 100/0$ aceto-nitrile/water (pH $= 3.0$, buffer not given) gradient [1055]. The phenylalanine peak was severely tailed, but still did not co-elute with other compounds of interest. The choice of a better buffer or a higher buffer concentration might decrease the tailing. No detection limits were given but a table of results listed most analyte levels down to 50 µg/g.

The temperature-dependent equilibrium between the two anomers, 11-α and 11-β forms, of 11-hydroxythromboxane B_2 was studied on a C_{18} column ($\lambda = 192$ nm) using a 35/65 acetonitrile/water (pH 3.5 with TFA/NaOH solution) mobile phase [1056]. Note here that the TFA will absorb strongly at 195 nm since the chosen wavelength is only 4 nm higher than the cutoff for high-quality acetonitrile! Fully resolved peaks eluting at 3 and 6 min were obtained at a system temperature of 0°C. This separation was lost to a highly-tailed split peak at 10°C, a single highly-tailed peak at 30°C, and a broad single peak at 40°C. The influence of mobile pH on the equilibrium was shown in a plot of the ratio of the two peaks (at 0°C).

A series of 24 prostaglandins (PG) (e.g., thromboxane, TX; leukotriene, LT; hydroxyeicosatetraenoic acid, HETE) were analyzed on a C_{18} column ($\lambda = 192$ nm, 237 nm, or 280 nm) using a 30-min $59/41/0/0.07 \rightarrow 0/0/100/0.03$ acetonitrile/ methanol/water/formic acid gradient [1057]. The elution order for selected compounds was: 20-COOH-LTB$_4$ < TXB$_2$ < estradiol (3.5 min) < PGB$_2$ < 6-trans-LTB$_4$ < LTB$_4$ (7.5 min) $\ll$ progesterone < 15-HETE < 11-HETE < 12-HETE [20 min] < LTC$_4$ < LTE$_4$ < LTD$_4$ (30 min). Peak shapes were excellent and good resolution was obtained.

Seventeen hydroperoxy polyunsaturated fatty acids (e.g., 5-(S)-hydroperoxyeico-satetraenoic acid, 5-(S)-hydroperoxyeicosatetraenoic acid) were extracted from brain tissue and derivatized with acetic anhydride [1058]. The resulting solution was injected onto a C_{18} column ($\lambda = 235$ nm) and eluted with a 15/8/2 acetonitrile/ water (0.1 M ammonium formate)/0.1 M formic acid mobile phase. The internal standard (12-hydroxyoctadecadienoic acid-d_8) was made up at the 600 pmol level and stored in a solution containing BHT to prevent its oxidation. Thermospray MS was also used as a detector and was most sensitive when a 15/6/4 acetonitrile/0.1 M acetic acid/0.1 M ammonium acetate mobile phase was used. In this example, detection limits of 0.1–0.5 pmol were obtained by single ion monitoring (SIM).

Vanillylamine

Thromboxane A$_2$

In an exceptional study by Miwa [1059], three sets of fatty acids were separated, all as their 2-nitrophenylhydrazide derivatives. Ten short-chain acids (lactic, acetic, propionic, isobutyric, n-butyric, isovaleric, n-valeric, 2-ethylbutyric [IS], isocaproic, n-caproic) were separated on a 30°C C_8 column ($\lambda = 400$ nm) using a 30/20/50 acetonitrile/methanol/water (to pH 4–5 with HCl) mobile phase. Elution was complete in 24 min and peaks of 150 pmol were readily detected. In a separate experiment, 29 fatty acid derivatives (e.g., caprylic, capric, myristic, eicosapentenoic, α-linolenic, di-homo-γ-linolenic, elaidic, docosatrienoic, linoelaidic, erucic) were separated on a C_{18} column using a 86/14 acetonitrile/water (to pH 4–5 with HCl) mobile phase. Elution was complete in 22 min and, for the high peak capacity, resolution was outstanding. Peaks of 150 pmol were readily detected. Finally, a mixture of hydroxylated carboxylic acids was analyzed (citric, tartaric, malic, succinic, fumaric, glycolic, L-pyroglutamic, lactic, acetic) were separated on a C_8 column using an 80/10/10 water/acetonitrile/methanol with 5 mM triethylamine mobile phase. Elution was complete in 28 min and overall resolution was very good. Overall limits of detection were reported as 500 fmol to 4 pmol injected (S/N = 2).

A set of three chromatograms detailed the separation of short-chain (C_4–C_8, methyl-C_3, methyl-C_4), medium-chain (C_5–C_8, C_{11}–C_{14}), and long-chain (C_{12}–C_{18}, even) fatty acids on a C_{18} column ($\lambda = 205$ nm) using three different parabolic acetonitrile/water gradients: $10/90 \rightarrow 85/15$, $20/80 \rightarrow 100/0$, and $70/30 \rightarrow 100/0$, respectively [1060]. Linear concentration curves from 60 to 800 nmol were obtained and detection limits of ∼50 nmol were reported.

Seven fatty acids ($C_{14:0}$, $C_{16:0}$, $C_{16:1}$, $C_{18:0}$, $C_{18:1}$, $C_{18:2}$, and $C_{20:4}$) were extracted from cardiac tissue and analyzed as their 9-anthryldiazomethane derivatives [1061]. Excellent peak shapes and baseline resolution were obtained on a C_{18} column ($\lambda = 365$ nm) using a 93/7 acetonitrile/water mobile phase. The total analysis time was 60 min. The authors noted that small changes in the mobile phase composition or minor deviations from a flow rate of 0.6 mL/min destroyed the resolution between the $C_{14:0}$, $C_{16:1}$, $C_{18:2}$, and $C_{20:4}$ compounds. The detection limits were reported as 15 fmol injected (S/N = 4). Linear working curves are shown for up to 400 pmol injected.

Lipoic and dihydrolipoic acids were isolated from urine, derivatized with 2-(4-aminophenyl)-6-methylbenzothiazole, and separated on a C_8 column ($\lambda = 343$ nm, ex; 423 nm em). Elution was complete in 10 min using an 80/20 acetonitrile/water mobile phase. Excellent peak shape and resolution were obtained. A linear working range of 1–250 ng/mL and detection limits of 0.5 ng/mL (S/N = 3) were reported [1062].

Dicarboxylic acids (C_6–C_{16}, even) were analyzed as their mono-coenzyme A, mono-carnitine, and 4-nitrobenzyl esters [1063]. The mono-coenzyme A esters were baseline resolved on a C_{18} column (photodiode array detector, $\lambda = 200$–300 nm) using a complex 45-min $5/95 \rightarrow 50/50$ acetonitrile/water (50 mM KH_2PO_4 buffer at pH 5.3) gradient. Peak shapes were excellent. The mono-carnitine esters were analyzed as their 4-bromophenacyl derivatives on a C_8 column ($\lambda = 260$ nm) using a complex 40-min $60/38/2 \rightarrow 95/0/5$ acetonitrile/water/water (0.15 M triethylamine phosphate buffer at pH 5.6) gradient. The 4-nitrobenzyl derivatives were resolved on a C_{18} column ($\lambda = 265$ nm) using a complex 40-min $40/60 \rightarrow 100/0$ acetontrile/water gradient. Peak shapes were excellent.

Arachidonic, palmitoleic, linoleic, eicosatrienoic, oleic, palmitic, and stearic acids were extracted from venous blood and baseline resolved as their 4-bromomethyl-7-methoxycouramrin derivatives in 40 min on a C_{18} column ($\lambda = 325$ nm, ex; 398 nm, em) using an 85/15 acetonitrile/water mobile phase [1064]. Working curves were generated for concentrations of 50–500 µmol/L. Detection limits of ~10 µmol/L (S/N = 3) were reported. Samples were diluted in solutions containing BHT to prevent oxidation of the acids.

Ten steryl ferulate and *p*-coumarate esters were extracted from corn and rice and were identified using a C_{18} column ($\lambda = 325$ nm) and an 82/3/2/13 acetonitrile/1-butanol/acetic acid/water mobile phase. Good resolution was obtained and the elution was complete in 35 min. Chromatograms showing the effects of varying water level (1–21%) on peak shape and resolution were shown. Levels from 70 µg to 26 mg/g sample were tabulated [1065].

Thirteen free and conjugated bile acids (e.g., cholic, taurocholic, deoxycholic, lithocholic, taurolithocholic, glycodeoxycholic acids) were baseline resolved on a 35°C PLRP-S column (pulsed amperometric detector at +0.03V vs. Ag/AgCl). A 55-min $20/60/30 \rightarrow 29/51/20$ acetonitrile/water/water (0.5 M NaOH) gradient was used [1066]. The presence of 0.1 M hydroxide is essential for the amperometric detection of the hydroxyl functional group on the bile acids. However, the authors

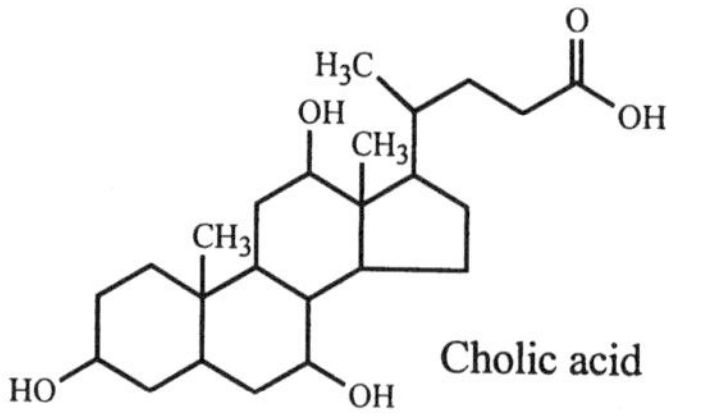

Cholic acid

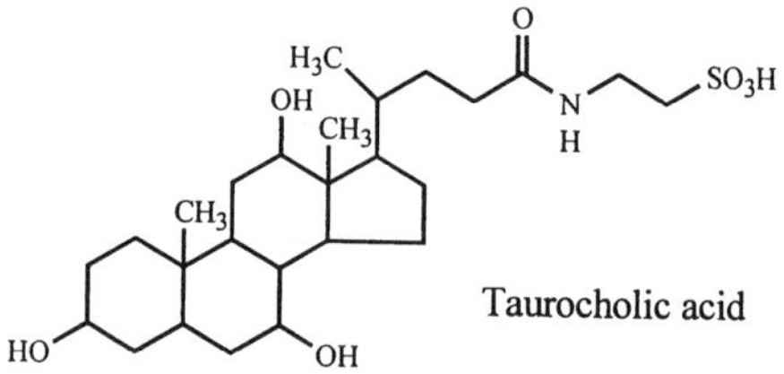

Taurocholic acid

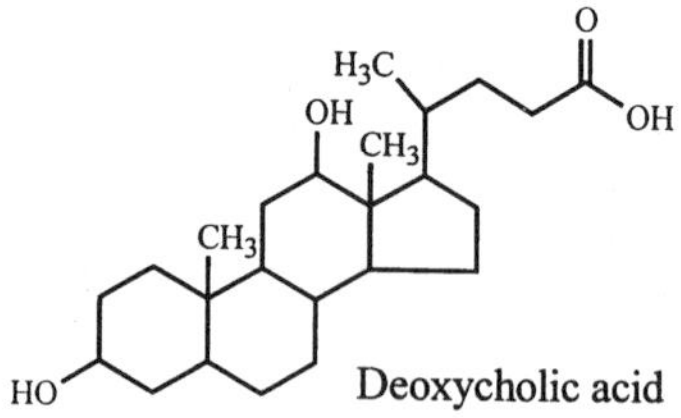

Deoxycholic acid

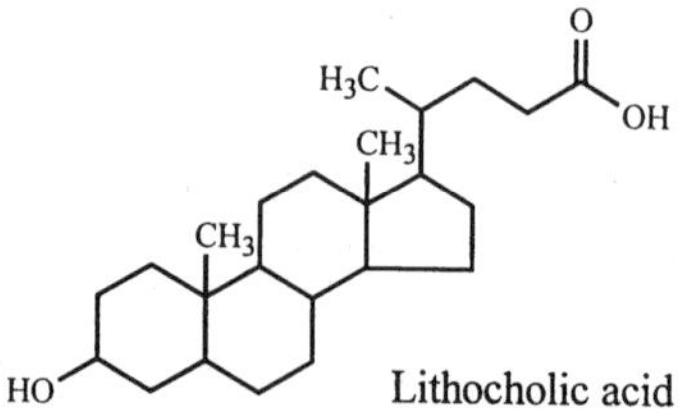

Lithocholic acid

note that acetonitrile slowly decomposes to ammonium and acetate ion under these basic conditions. Accordingly, postcolumn addition of the hydroxide is recommended. Peak shapes were excellent and detection limits of 60 pmol injected (S/N = 3) were reported. A linear concentration range of 0.6–14 μmol was obtained.

Seven of the major bile acids (e.g., taurocholic, α- and ω-tauromuricholic, glycohydrodeoxycholic) were extracted from bile and serum and analyzed on a C_{18} column ($\lambda = 200$ nm and fast atom bombardment MS) using a complex 45-min $20/80 \rightarrow 90/10$ (97/3 acetonitrile/glycerol)/(95/5 water [19 mM ammonium carbonate buffer at pH 4.0]/glycerol) gradient [1067]. Detection limits for the UV work were reported as 0.1 nmol. Peak shapes were good, but one pair co-eluted (taurocholate and taurohyodeoxycholate) and were successfully monitored by MS.

Iida et al. [1068] studied the separation of 15 free, glycine-aminated, and taurine-aminated bile acids (lithocholic, chenodeoxycholic, ursodeoxycholic, deoxycholic, cholic acids) on a C_{18} column ($\lambda = 205$ nm). A number of acetonitrile/water (phosphate buffer at pH 3.5–7.5) mobile phase combinations were examined. The optimal resolution occurred with a 23/77 acetonitrile/water (0.3% phosphate buffer at pH 7.0) mobile phase. Eleven peaks eluted in <40 min but four sets of peaks co-eluted. The remaining peaks were eluted in 20 min using a 28/72 acetonitrile/water (0.3% phosphate buffer at pH 7.0) mobile phase. Based on these results, a gradient starting at a lower percent acetonitrile and ramping to a higher percent acetonitrile finish would be particularly effective at increasing resolution while keeping analysis times reasonable.

9.5.2 Basic Amine Analytes

Furopyrroleopyridine, the highly fluorescent blue product of a crosslink reaction with 3-deoxyglucosone and N-α-acetyllysine, was isolated from the reaction mixture using a C_{18} column ($\lambda = 270$ nm or 370 nm) and a complex 35-min $100/0 \rightarrow 0/100$ water (0.1% TFA)/acetonitrile (0.1% TFA) gradient. Four distinct peaks were separated, but only one major peak was isolated and was identified as the title compound [1069].

Five polyamines (1,3-diaminopropane, putrescine, cadaverine, spermidine, spermine) were extracted from freeze-dried plant and animal tissue and baseline resolved as their dansyl chloride derivatives. A C_{18} column ($\lambda = 252$ nm, ex; 500 nm, em) and an acetonitrile/water rapid gradient (70/30 for 4 min to 100/0 in 1 min) generated the separation [1070]. Detection limits of 0.5 pmol and a linear concentration range up to 35 pmol were reported. Elution was complete in <10 min. The dansyl derivatives of the above compounds and N^1-acetylspermidine were resolved on a C_{18} column ($\lambda = 340$ nm, ex; 510 nm, em) using a two-tier 5.5-min $40/60 \rightarrow 100/0$ acetonitrile/water (10 mM heptanesulfonate at pH 3.4) gradient [1071]. Excellent resolution and peaks shapes were achieved and elution was complete in <8 min. Detection limits of 1 pmol injected and linear ranges of up to 200 pmol injected were reported.

Yang and Thyrion [1072] studied peak shape and retention on a wide range of basic amine-containing compounds: procaine, adiphenine, drofenine, nafronyl, tetracaine, meclofenoxate, 4-aminobenzoic acid, and caffeine. The basic mobile phase was 65/35 acetonitrile/water (20 mM acetate buffer at pH 4.5) and the column a C_{18} ($\lambda = 260$ and 280 nm). The effect of the identity (i.e., di-*n*-butylamine and triethylamine) and concentration of mobile phase modifier from 0 to 0.3% was studied. Retention times from 2 to 20 min resulted and asymmetry factors in general were <2.5.

Melatonin levels in commercial tablet and capsule products were determined using an electrochemical detector (+850 mV vs. Ag/AgCl), a C_{18} column, and an 80/20 water (15 mM $NaClO_4$ with 40.6 mM sodium citrate, 2.15 mM sodium octanesulfonate, and 10 mM diethylamine hydrochloride, and 27 μM Na_2EDTA)/ acetonitrile mobile phase [1073]. Elution was complete in 3.5 min. A linear range of 0.1–1 mg injected and a detection limit of 3 pg injected ($S/N = 3$) was reported.

The urinary metabolites of 5-hydroxyindole (5-hydroxytryptophan, 5-hydroxy-tryptamine, 5-hydroxyindol-3-ylacetic acid, 5-hydroxytryptophol, *N*-acetyl-5-hy-

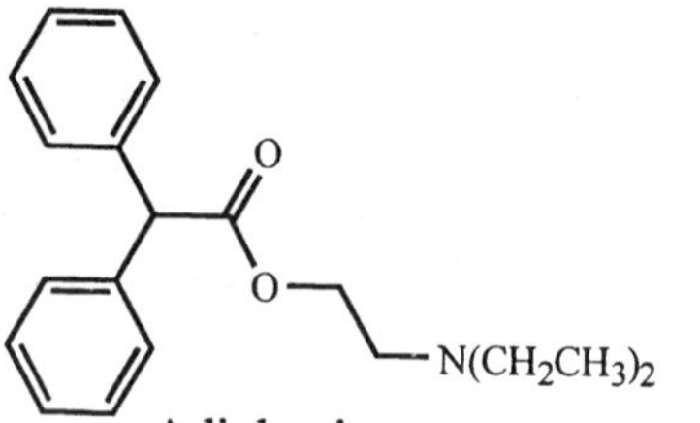

Adiphenine

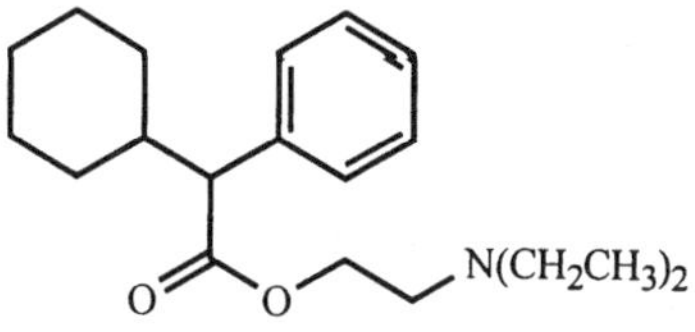

Drofenine

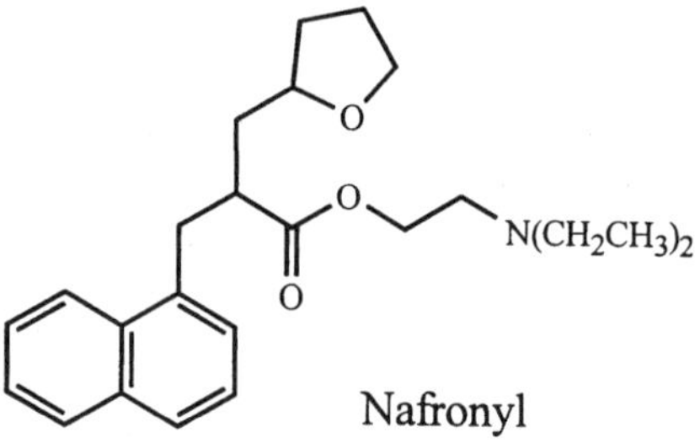

Nafronyl

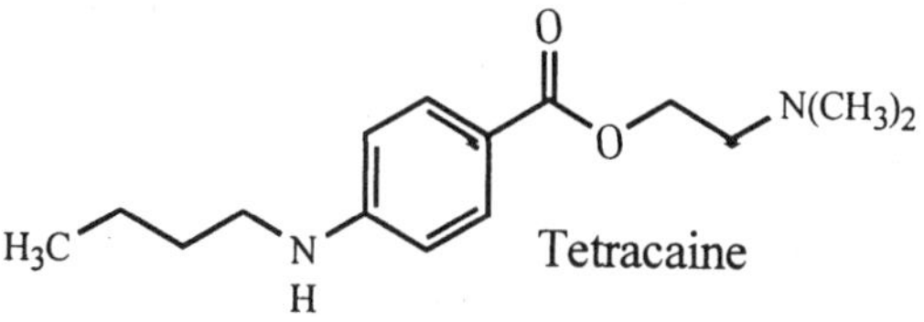

Tetracaine

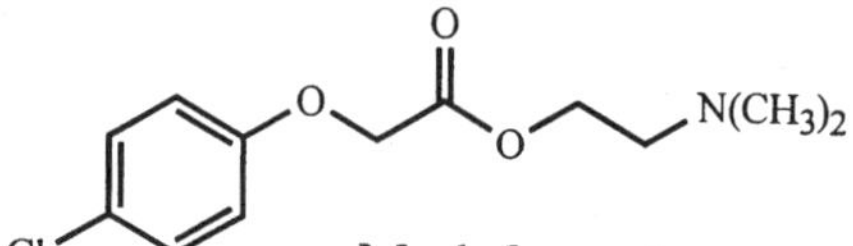

Meclofenoxate

droxytryptamine) were baseline resolved on a C_{18} column using a 5/95 acetonitrile/water (10 mM acetate buffer at pH 4.7) mobile phase [1074]. The compounds were postcolumn derivatized with a benzylamine/potassium hexacyanoferrate(III) mixture ($\lambda = 345$ nm, ex; 481 nm, em). Elution was complete in 40 min. Detection limits of 100–500 fmol injected (S/N = 3) were reported (analyte dependent). Peak shapes were excellent.

An optimization procedure for the separation of epinephrine bitartrate, L-DOPA, 3,4-dihydroxyphenylacetic acid, norepinephrine·HCl and dopamine·HCl (with 3,4-dihydroxybenzylamine·HCl as internal standard), was described by He et al. [1075]. A C_{18} column was used in conjunction with an electrochemical detector (+0.6 V vs. Ag/AgCl). A window diagram of relative retention times for adjacent eluting solute pairs (i.e., t_{R2}/t_{R1}) resulted in three acceptable solvent composition windows. The optimal solvent conditions were found to be 2.5/97.5 acetonitrile/water (0.23% sodium acetate with 0.02% EDTA and 0.066% sodium heptanesulfonate adjusted to pH 3.9 with monochloroacetic acid). Elution was complete in <30 min and all peaks were well resolved. Detection limits for dog or human plasma samples were reported as ~10 pg/mL for epinephrine and norepinephrine.

Serotonin (and 5-hydroxyindoleacetic acid) and a total of eight catecholamines and metabolites including norepinephrine and dopamine (and their major metabolites e.g., 4-hydroxy-3-methoxyphenylethylene glycol, homovanillic acid) were separated from epinephrine, tyrosine and tryptophan on a 30°C C_{18} column ($\lambda = 230$ nm ex; 320 nm em or amperometric detection at +0.7 V vs. Ag/AgCl on glassy carbon) using a 19/181 acetonitrile/water (12.16 mM citric acid with 11.60 mM diammonium phosphate, 2.54 mM sodium octyl sulfate, 3.32 mM dibutylamine phosphate, and 1.11 mM sodium EDTA) mobile phase [1076]. Standards and samples were stored at −70°C and only used for one day. Brain tissue extracts were studied and detection limits of between 75 pg injected (17 ng/g) for tyrosine and 520 pg injected (120 ng/g) for tryptophan were reported for fluorescence detection. 5-Homovanillic acid and 3,4-dihydroxyphenylacetic acid were not detected by this method. With amperometric detection, all analytes gave a response and the detection limits ranged from 25 pg injected (6 ng/g) for dopamine and 50 pg injected (12 ng/g) for 3-methoxytyramine. Elution of all components was complete in <20 min and resulting peaks shapes were excellent.

Thymine and six photoirradiation products generated in the presence of p-aminobenzoic acid (5,6-dihydrothymine, 5-hydroxymethyluracil, cis-anti-, cis-syn-, trans-anti- and the trans-syn-cyclobutane dimers of thymine) were separated on a C_{18} column (photodiode array detector, $\lambda = 200$–400 nm) using a 30-min 0/100 → 50/50 acetonitrile/water [75 mM phosphate buffer at pH 4.4] gradient [1077]. Elution was complete in 25 min and good peak shapes were obtained.

Caffeine and four metabolites—theobromine, paraxanthine, theophylline, and 1,3,7-trimethyluric acid—were extracted from liver tissue and resolved on a C_{18} column ($\lambda = 250$ nm) using an 85/15 → 5/95 (at 30 min hold 25 min) (895/95/7/3 water/acetonitrile/THF/acetic acid)/(695/295/7/3 water/acetonitrile/THF/acetic acid) gradient [1078]. Samples were reconstituted in the initial mobile phase and phenacetin and 4-hydroxyanilide were also added. Resolution and peak shapes were excellent and elution was complete in 55 min.

Nicotinic acid and nicotinomide were extracted from cooked sausage and separated on a 35°C C_{18} column ($\lambda = 261$ nm) using a 75/25 water (5 mM heptanesulfonic acid to pH 3.3 with H_3PO_4)/acetonitrile mobile phase [1079]. Peaks were tailed, so the addition of TFA or a TFA/triethylamine mobile phase modifier might lead to better results. Elution was complete in <4 min and baseline resolution was achieved. Calibration curves from 20 to 300 μg/mL and detection limits of 0.30 mg/100 g were reported.

The tobacco alkaloid myosmine 3-(1-pyrrolin-2-yl)pyridine, N-nitrosonornico-tine, and 4-hydroxy-1-(3-pyridyl)-1-butanone were baseline resolved on a C_{18} column ($\lambda = 233$ nm and 254 nm) using a 39-min 100/0 → 50/50 water (20 mM phosphate buffer at pH 6.5)/acetonitrile gradient [1080]. No concentration data were given.

Nicotine and seven related compounds (cotinine, nornicotine, myosmine, β-nicotyrine, nicotine *trans*-N-oxide, anatabine, anabasine) were separated on a C_{18} column ($\lambda = 254$ nm) using a complex 10-min 100/0 → 50/50 (water [0.024 M K_2HPO_4 to pH 8.5]/methanol)/(95/5 acetonitrile/water) gradient [1081]. Excellent peak shapes and good resolution was obtained. Standards of 50 μg/mL nicotine and 5 μg/mL all other components gave easily detected peaks.

Ten pair of racemic nicotine analogs (e.g., N'-[2,2-difluoroethyl]nornicotine, N-benzyl-N-α-dimethyl-3-pyridinemethanamine, 1-benzyl-2-phenylpyrrolidine, N'-[2-methylbenzyl]nornicotine, 5,6-cyclohexenonicotine) were individually resolved on a β-cyclodextrin column ($\lambda = 254$ nm) using various isocratic acetonitrile/water [1% TEA to pH 7.1 with acetic acid] mobile phases [1082]. The mobile phase composition (ranging from 5/95 to 30/70), resulting k' values and separation factors (α) are tabulated for each pair of compounds. A 20-min gradient of acetonitrile/water (buffer as above) from 10/90 → 70/30 led to the baseline resolution of three racemic analyte pairs: N'-benzylnornicotine, N'-(2,2,2-trifluoroethyl)nornicotine and N'-(2-naphthylmethyl)nornicotine. A series of 5–10 μL injections of 0.1–0.5% samples were used. An interesting study of the retention of racemic N'-benzylnor-nicotine in both methanol and acetonitrile/buffer mobile phases (both organic levels at 20–100%) was presented. For methanol, k' decreases, as expected, with increasing percent methanol. For acetonitrile, however, the k' decreases to 40% acetonitrile, stays constant to 80% acetonitrile, and then increases from 80% to 100% acetoni-

Nicotinic acid   Myosmine   N-Nitrosonornicotine

Cotinine   β-Nicotyrine

trile. The authors speculate that the retention increase is caused either by increased hydrogen bond interaction with the surface or another surface complexation process. The former explanation is consistent with the differences between acetonitrile (weak hydrogen bond acceptor) and methanol (strong hydrogen bond acceptor and donor).

Seven pungent principles (e.g., piperine, piperettyline, piperoleines A and B, piperettine, piperylene) were extracted from *Piper nigrum* seeds and analyzed on a C_{18} column ($\lambda = 254$ nm, 280 nm, 343 nm, and 364 nm) using a 48/52 acetonitrile/ water (1% acetic acid) mobile phase [1083]. Elution was complete in 20 min. Peak shapes were acceptable but incomplete resolution was obtained for three analytes. Piperine standard concentration ranged from 0.025 to 0.1 g/L.

Kuronen et al. [1084] separated eight steroid glycoalkaloids (solanidine, α-solanine, α-chaconine, tomatidine, α-tomatine, solasodine, dehydrotomatidine, dehy-drotomatine) on a 50°C C_{18} column ($\lambda = 205$ nm) using a complex 25-min $80/20 \rightarrow 0/70$ water (25 mM tetraethylammonium phosphate buffer at pH 3.0)/acetonitrile gradient. Surprisingly, the baseline shift over the course of the gradient was minimal (note the low wavelength and the use of the ammonium modifier). The effects of changing column temperature on retention were plotted. Stock standards of 1 mg/mL were used and 50 μL injections generated easily detected peaks.

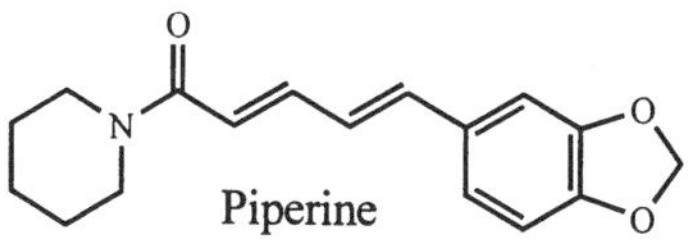

Piperine

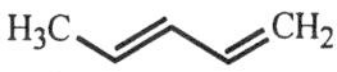

Piperylene

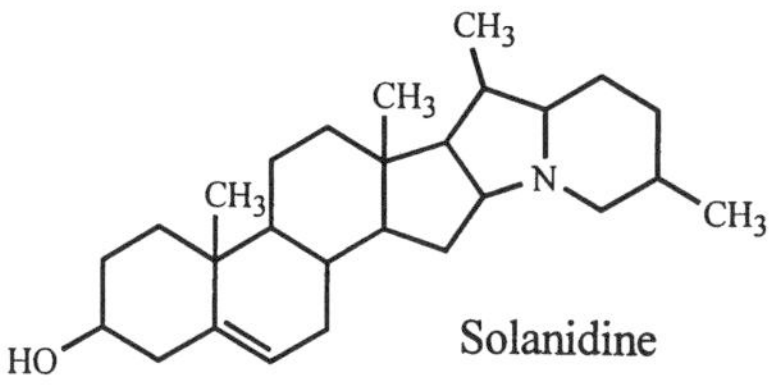

Solanidine

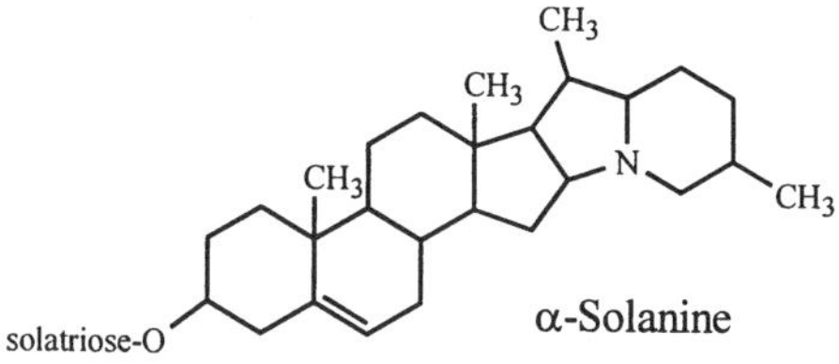

α-Solanine

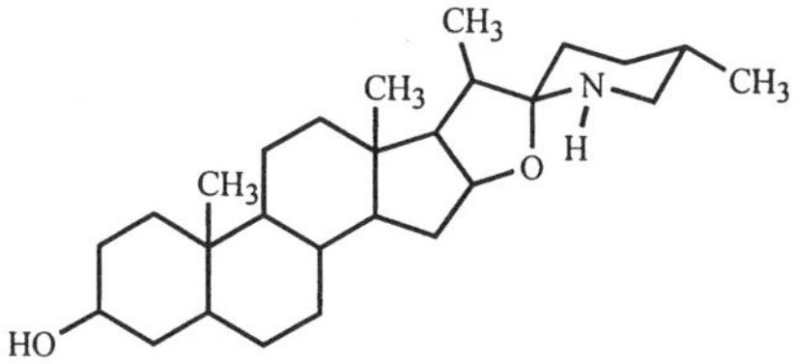

Tomatidine

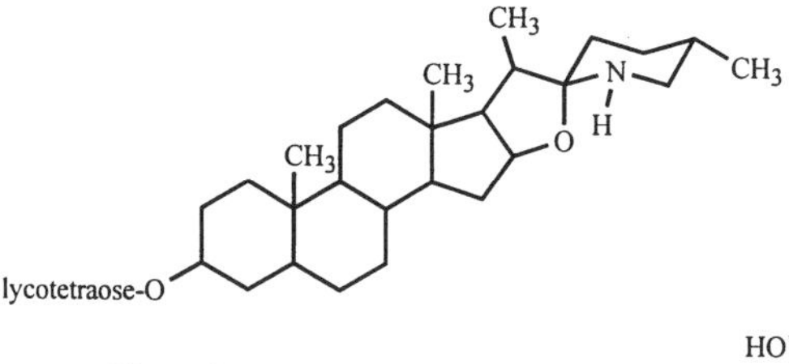

α-Tomatine

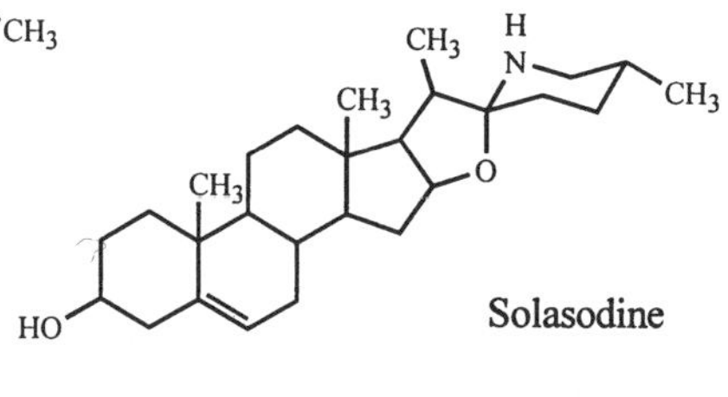

Solasodine

Various varieties of potatoes were characterized by their α-solanine and α-chaconine levels. Extracts were analyzed on a C_{18} column ($\lambda = 200$ nm) using a 30/70 acetonitrile/water (50 mM ammonium phosphate buffer at pH 6.5) mobile phase. Elution was complete in 15 min. The analyte peaks were somewhat broad but were baseline resolved [1085]. A better choice of pH (i.e., pH < analyte pK_a) and mobile phase modifier (e.g., TFA) would likely produce sharper peaks. A linear range from 0.01 to 0.3 mg/mL was reported as well as detection limits of 0.01 mg/mL.

Tomatine and dehydrotomatine were isolated from tomato plants and separated on a C_{18} column ($\lambda = 210$ nm). A 14-min 90/10/0.1 → 50/50/0.1 water/acetonitrile/TFA gradient generated good separation. The content of leaf and green fruit extracts were compared. Spike standards of 1–12.5 µg injected were cited [1086].

Allantoin was extracted from milk and analyzed on an aminopropyl column ($\lambda = 214$ nm) using a 90/10 acetonitrile/water mobile phase [1087]. The authors note an increase in retention time with increased acetonitrile in the mobile phase. This is most likely due to decreasing allantoin solubility as the acetonitrile level increases. Elution was complete in <4 min and the analyte peak was well resolved from other extracted materials. A linear range of 10–50 µM and a detection limit of 1 µM were reported.

Allantoin, uric acid, xanthine, and hypoxanthine were extracted from ruminant blood and urine. Samples were treated with and without DNPH [2,4-dintriphenylhydrazine) prior to analysis to derivatize the allantoin. Two 25 cm C_{18} columns were connected in series ($\lambda = 205$ nm and 254 nm for nonderivatized sample; 254 nm and 360 nm for derivatized sample). A complex 45-min 100/0 → 80/20 water (2.5 mM $(NH_4)H_2PO_4$ to pH 3.5 with H_3PO_4)/acetonitrile gradient was used [1088]. It should be noted that for the underivatized allantoin case, elution is at or very close to the void volume. This is potentially a major problem since many times co-extractant will elute near the void volume. Conversely, the derivatized case has allantoin eluting at 37 min, nearly 20 min after the last analyte and 15 min after the last co-extracted peak. A gradient to a stronger solvent could markedly reduce the analysis time. For the derivatized samples, a linear range of 3–500 µM and detection limits of 5 nmol (200 nmol for hypoxanthine) were reported.

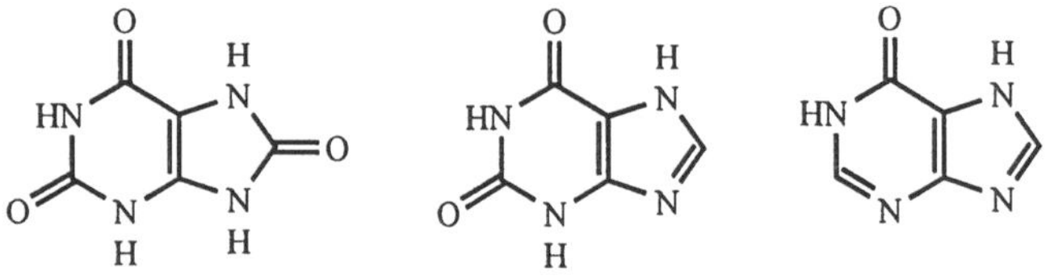

Uric acid   Xanthine   Hypoxanthine

Sanguinarine is a marker analyzed for in the adulteration of mustard oil [1089]. The oil is diluted and separation is generated on a C_{18} column ($\lambda = 280$ nm) using a 55/21/20/4 acetonitrile/methanol/water/THF mobile phase. Elution was complete in 10 min with good resolution from argemone and other mustard-related compounds. All peaks were somewhat broad. This is not unexpected for sanguinarine, a compound that has a quaternary (permanently charged) amine group. The use of a mobile phase modifier such as a TFA/triethylamine combination may be effective here. A linear range of 0.01–1 mg/g and detection limit of 5 μg/g were reported.

Two tricyclic indole derivatives, 1,2,3,4-tetrahydro- and 1-methyl-1,2,3,4-tetrahydro-β-carboline-3-carboxylic acid, were extracted from cooked and raw fish and analyzed on a 40°C C_{18} column ($\lambda = 270$ nm, ex; 343 nm, em). A two-ramp $100/0 \rightarrow 68/32$ (at 8 min) $\rightarrow 10/90$ (at 18 min) water (50 mM ammonium phosphate buffer at pH 3)/(20/80 water [50 mM ammonium phosphate buffer at pH 3]/acetonitrile) gradient generated excellent peak shapes and resolution [1090]. Detection limits in the low ng/g range were reported.

Kim *et al.* [1091] extracted a series of nine alkamides (e.g., undeca-(2Z,4E)-diene-8,10-diynoic acid isobutylamide, dodeca-(2E,4E,8Z)-, (10Z)-tetraenoic acid isobutylamide, dodeca-(2E,4Z)-diene-8,10-diynoic acid 2-methylbutylamide) from

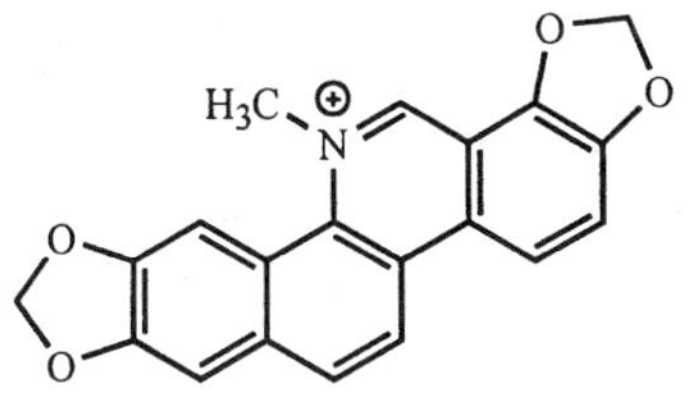

Sanguinarine

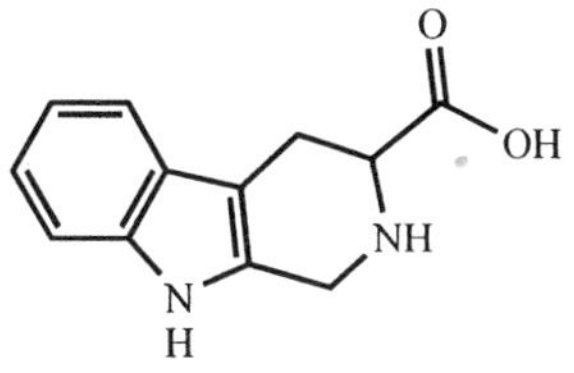

1,2,3,4-Tetrahydro-β-carboline-3-carboxylic acid

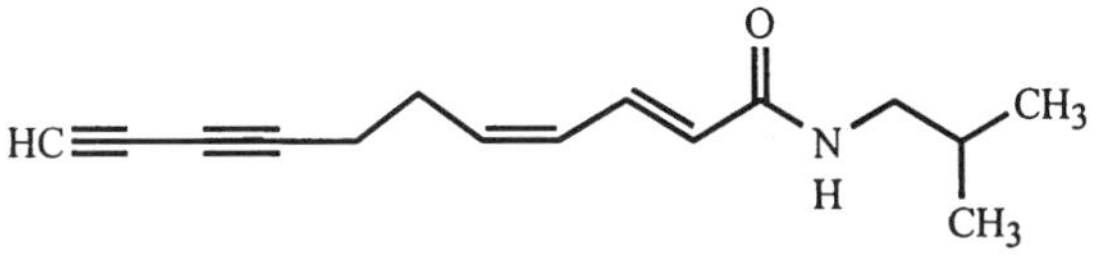

Undeca(2E,4Z)diene -8,10-diynoic acid isobutylamide

Echinacea purpurea and studied them on a C_{18} column ($\lambda = 254$ nm). A 30-min $60/40 \rightarrow 30/70$ water/acetonitrile gradient was used. Complete resolution of only two components was achieved and complete co-elution of two sets of two compounds resulted. All compounds eluted prior to 20 min. A C_8 column, the replacement of some acetonitrile with methanol, and a shallower gradient may be more successful in this separation. Standard mixtures from 0.25 to 1 µg injected were used.

Alkamides from the roots of *Echinacea purpurea* were isolated and separated on a 45°C C_{18} column ($\lambda = 263$ nm). A complex 35-min $55/45 \rightarrow 0/100$ water/ acetonitrile gradient was used. Two major components were identified: undeca-(2E,4Z)-diene-8,10-diynoic acid isobutylamide and dodeca-(2E,4E,8Z,10Z)-tetraenoic acid isobutylamide [1092]. At least nine other related compounds were isolated and identified as alkamides. Detection limits for the major components were reported as 15 ng injected (S/N = 5).

An excellent separation of 14 heterocyclic aromatic amines (e.g., aminoimidazoazaarenes such as 2-amino-3,8-dimethylimidazo[4,5f]quinoxaline; glutamic acid pyrolysates such as 2-amino-6-methyldipyrido[1,2-a:3'2'-d]imidazole; carbolines such as 3-amino-1,4-dimethyl-5H-pyrido[4,3f]indole) extracted from fried or broiled meats was developed on a C_{18} column ($\lambda = 263$ nm). A complex 55-min $5/95/0 \rightarrow 15/85/0 \rightarrow 15/0/85 \rightarrow 55/0/45$ acetonitrile/water (10 mM TEA adjusted to pH 3.2 with H_3PO_4)/water (10 mM TEA adjusted to pH 3.6 with H_3PO_4) gradient was used [1093]. The analytes of interest eluted in <30 min, peak shapes were uniformly excellent, and detection limits of "low ng/g" were reported.

A series of ten opines (mannopine, mannopinic acid, cucumopine, octopine, allo-octopine, octopinic acid, nopaline) and opine analogs (normannopine, galactopine,

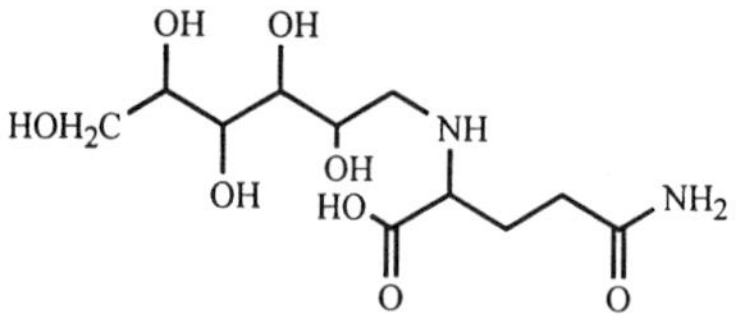

Mannopine

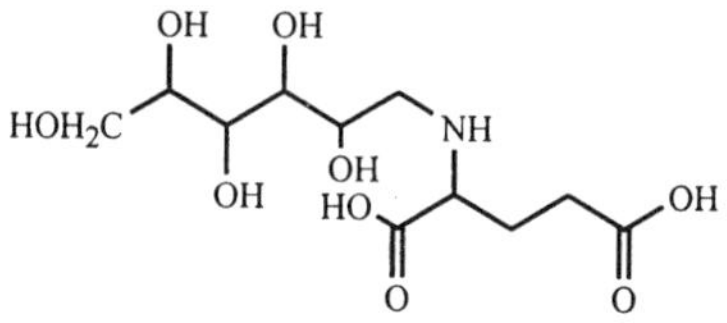

Mannopinic acid

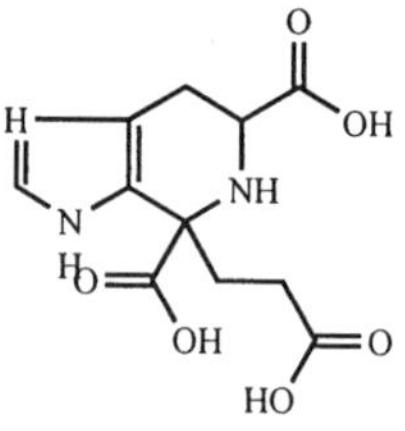

Cucumopine

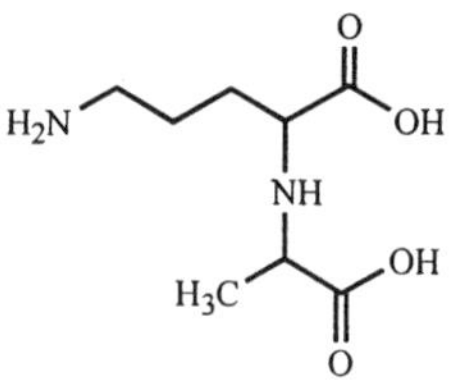

Octopinic acid

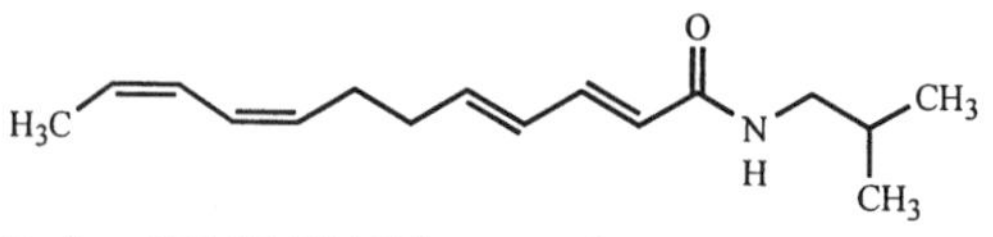

Dodeca(2E,4E,8Z,10Z)tetraenoic acid isobutylamide

glucopine) were derivatized with NBD-F [4-fluoro-7-nitrobenzoxadiazole) and separated on a C_{18} column ($\lambda = 470$ nm, ex; 530 nm, em). An 84/16/0.12 water/ acetonitrile/TFA mobile phase generated an excellent separation in 25 min. Standard solutions of 0.75–37.5 nmol were used and detection limits of 1 pmol were reported [1094].

Five biotransformation products of N-methylbenzylamidine (benzamide, benzamidine, N-methylbenzamide, N-hydroxy-N-methylbenzamidine and N-methylbenzamidoxime) were baseline resolved in <30 min on a C_{18} column ($\lambda = 229$ nm) using 12/88 acetonitrile/water (0.08% 85% H_3PO_4 and 0.001% sodium octanesulfonate) mobile phase. Peak shapes were very good and a working curve of 5–300 pmol injected was reported [1095].

Putrescine, cadaverine, spermidine, and spermine were extracted from plant materials and derivatized with 9-fluorenylmethylchloroformate (FMOC). Separation was achieved on a C_8 column ($\lambda = 265$ nm, ex; 310 nm, em) using a complex 35-min 50/50 → 100/0 (20/0.5/79.5 acetonitrile/THF/water [0.1 M sodium acetate buffer at pH 4.2])/(79.5/0.5/20 acetonitrile/THF/water [0.1 M sodium acetate buffer at pH 4.2]) gradient [1096]. These compounds were well resolved from other compounds that reacted with the FMOC reagent such as methanol, aspartic acid (used to react with the excess FMOC reagent), and ornithine. Peak shapes were good and detection limits were reported as 7–28 pg (analyte dependent).

9.5.3 Aflatoxins, Mycotoxins, and Other Toxic Analytes

Two reviews [312, 1097] focus on the separation of aflatoxins B_1, B_2, G_1 and G_2. Often isocratic acetonitrile/water or acetonitrile/methanol/water mobile phases are used in conjunction with C_{18} columns ($\lambda = 360$ nm). Normally acetonitrile levels of 25–40% are used. If methanol is used it is approximately at the 5–15% level and directly replaces an equivalent amount of acetonitrile. Most importantly, TFA is almost invariably added at the 0.1–0.5% level. TFA forms a hemiacetal product with the aflatoxins that increases sensitivity by at least a factor of 3. Complete baseline resolution is usually achieved in <20 min. Fluorescence detection has been used, with 360 nm excitation and 440 nm emission wavelengths chosen. Detection limits of 1–10 pg were commonly reported. Postcolumn derivatization with iodine has also been used in conjunction with fluorescence detection to increase sensitivity.

Ten microcystins (algal hepatotoxins) were extracted from water samples and separated on a C_{18} column (photodiode array detector, $\lambda = 200–300$ nm) using a complex 50-min 30/70 → 0/100 acetonitrile/water (0.05% TFA) gradient.

Benzamidine

Spermidine

Ornithine

Excellent peak shapes and baseline resolution for all analytes were obtained. The microcystins included in the study were: microcystin-RR, -YR, -LR, -FR, -LA, -LY, -LW, -LF, and nodularin. Working curve concentration ranges of 34 ng to 9 μg/L were used [1098].

Murata et al. [1099], separated three microcystins (RR, YR, and LR) as their dansylated derivatives. Peroxyoxalate chemiluminescence detection was chosen using a 0.5 mM bis[4-nitro-2-(3.6.9-trioxadesyloxycarbonyl)phenyl] oxalate with 50 mM hydrogen peroxide postcolumn reaction. This method gave a detection limit of 15 fmol (S/N = 10). Linear working ranges were reported as 15–1670 fmol. Separation was achieved on a C_{18} column (Jasco 885-CL chemiluminescence detector was optimal) with a 40/60 acetonitrile/water (0.05% TFA) mobile phase. Peak shapes and resolution were excellent and elution was complete in <10 min.

Two trichothecene mycotoxins (nivalenol and deoxynivalenol) and two related esters (3-acetyl- and 15-O-acetyl-4-deoxynivalenol) were extracted from wheat flour and separated on a C_{18} column ($\lambda = 220$ nm). A complex 36-min 92/8 → 0/100 (9/10 water/acetonitrile)/acetonitrile gradient gave good resolution and peak shape [1100]. Standards containing 3 ppm of each compound were easily detected.

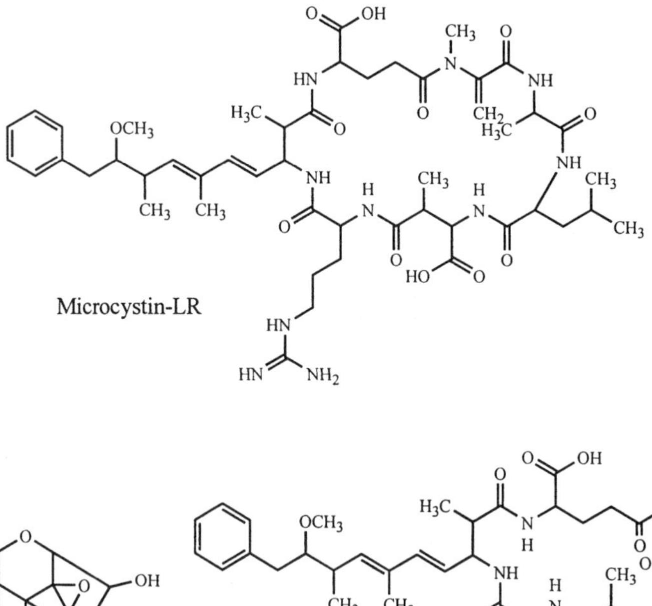

Microcystin-LR

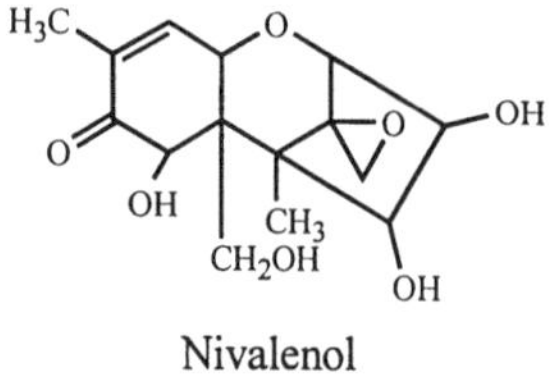

Nivalenol

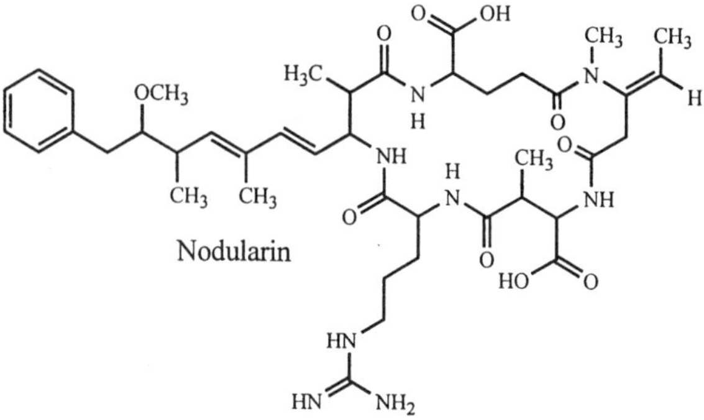

Nodularin

In an excellent study seven mycotoxins (deoxynivalenol, patulin, diacetoxyscir-penol, HT-2 toxin, T-2 toxin, zearalenone, ochratoxin A) were baseline resolved on a 35°C C_{18} column (photodiode array detector, $\lambda = 210$–340 nm) using a 7-min $40/60 \rightarrow 70/30$ acetonitrile/water (phosphoric acid to pH (3) gradient [1101]. Peak shapes were excellent and detection limits of 5–20 µg/20 g wheat were reported. Linear concentration curves from 40 to 800 ng injected were obtained. The UV spectra for all analytes in a 30/40 acetonitrile/water were presented. Luf et al. [1102] extracted nivalenol and deoxynivalenol from wheat and separated them on a 50°C C_{18} column (negative ion APCI; N_2 carrier gas 250 L/h, sheath gas 125 L/h, source $T = 100$°C, vaporizing $T = 400$°C, cone voltage 20 V) using an 82/9/9 water/acetonitrile/methanol mobile phase. Plots of peak area vs. cone voltage and vaporization temperature are given. The authors note that the addition of ionic mobile phase modifiers (e.g., ammonium or acetate) suppressed the analyte signal. Elution is complete in 7 min. Linear curves from 12.5 to 250 ng injected and quantitation limits of 300 µg/kg were reported.

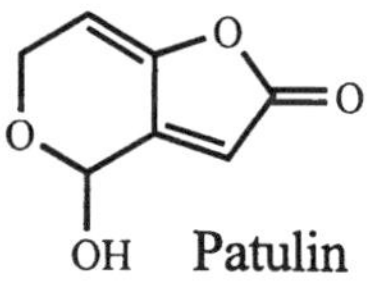

Patulin

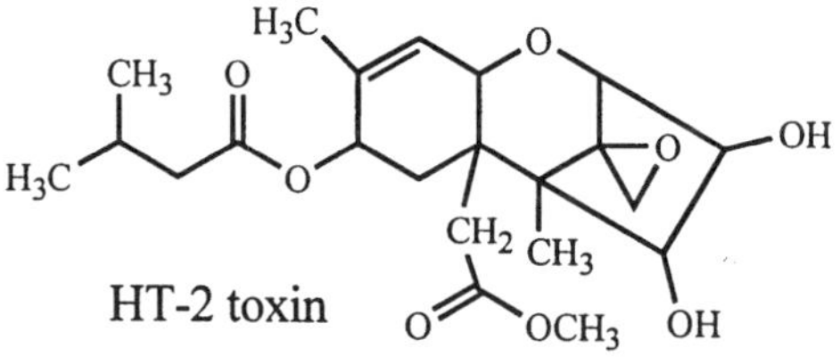

HT-2 toxin

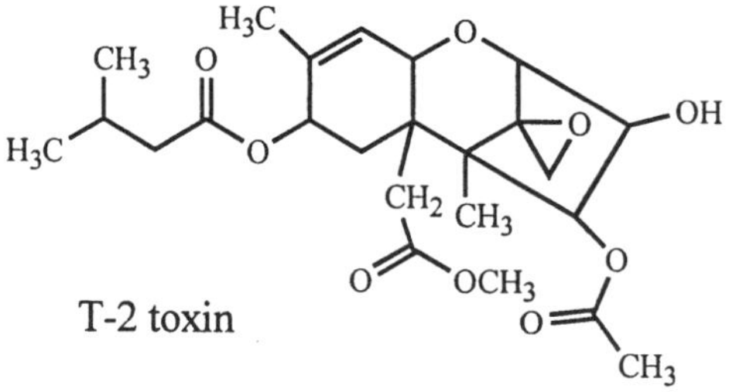

T-2 toxin

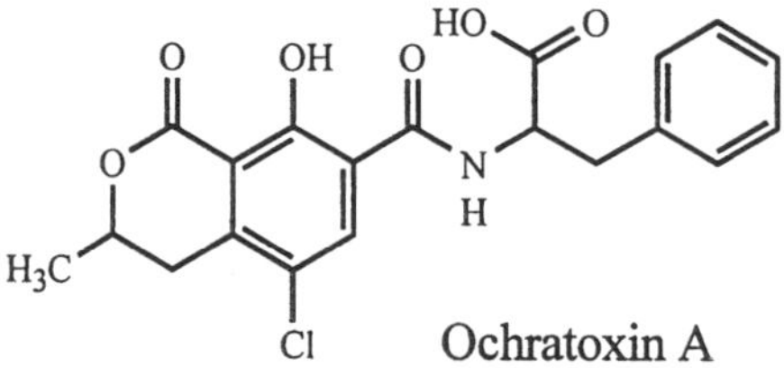

Ochratoxin A

Fusaproliferin (FUS) and beauvericin (BEA), were extracted from *Fusarium subglutinans* and analyzed on a C_{18} column (positive ion electrospray MS; source voltage 5.0 kV, capillary $T = 250°C$, sheath gas 80 units, auxiliary gas 10 units, capillary voltage 41 V) using a 30/70 (hold 5 min) → 0/100 (at 8 min hold 7 min) (90/10/0.1 water/acetonitrile/formic acid)/(10/90/0.1 water/acetonitrile/formic acid) gradient [1103]. Standards ranging from 0.1 to 5 μg FUS/mL and 0.001 to 1 μg BEA/mL were used. Detection limits of 2 ng FUS injected and 20 pg BEA injected (S/N = 2) were reported.

The mycotoxin citrinin was extracted from corn and analyzed on a 40°C C_{18} column ($λ = 450$ nm, ex; 620 nm, em). A 40/60 acetonitrile/water (2% formic acid) mobile phase eluted the citrinin in 10 min. However, the peak exhibited significant tailing [1104]. The addition of an amine (e.g., triethylamine) should be considered to help correct this. A citrinin standard of ~40 μg/mL was used and detection limits of 200 ng/g were reported.

Anatoxin-*a*, homoanatoxin-*a* and four degradation products (dihydroanatoxin-*a*, dihydrohomoanatoxin-*a*, epoxyhomoanatoxin-*a*, epoxyanatoxin-*a*) were isolated from blue-green algae, derivatized with NBD-F (4-fluoro-7-nitro-2,1,3-benzoxadiazole) , and separated on a 35°C C_{18} column ($λ = 470$ nm, ex; 530 nm, em). The use of a 55/45 water/acetonitrile mobile phase generated baseline resolution, excellent

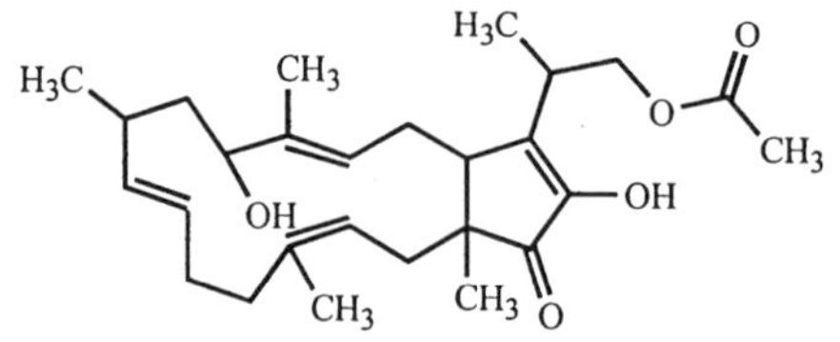

Fusaproliferin

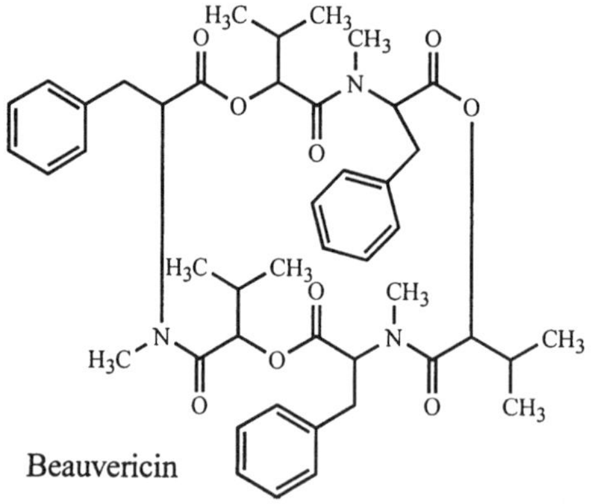

Beauvericin

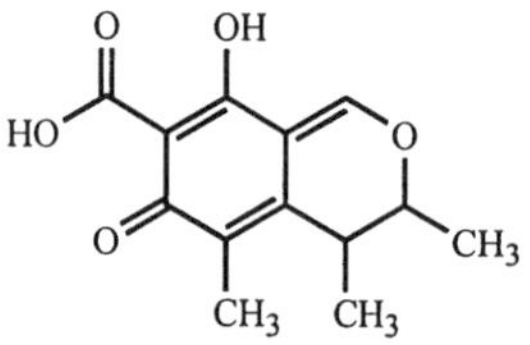

Citrinin

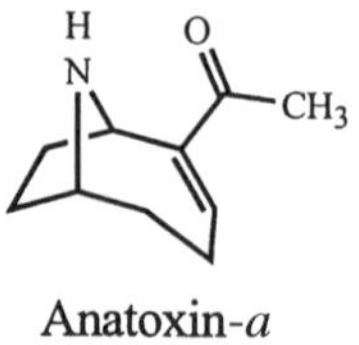

Anatoxin-*a*

peak shape, and complete elution in <30 min. Loads of 10 μg/mL were easily detected and the authors reported a 10 ng/L detection limit [1105].

The polypeptidic toxin α-amanitin was isolated from urine and analyzed on a C_{18} column (electrochemical detector +500 mV) using a 90/10 water (5 mM phosphate buffer at pH 7.2)/acetonitrile mobile phase [1106]. Elution was complete in 16 min. A linear curve from 10 to 200 ng/mL was generated and a detection limit of 2 ng/mL (S/N = 5) and a quantitation limit of 10 ng/mL (S/N = 25) were reported. Peaks eluted free from interferences in 15 min.

Altersolanols A, B, C, D, E and F were extracted from *Alternaria solani* and baseline resolved on a 25°C C_{18} column ($\lambda = 270$ nm) using a complex 23-min 18/82 → 90/10 acetonitrile/water gradient [1107]. Excellent peak shapes were generated. Detection limits of 0.2–0.5 pmol/5 μL injection (S/N = 3) were reported. A limited linear working range of 0.4–1.0 μM was also reported.

Okadaic acid and dinophysistoxin-1, -2, and -2B were extracted from phytoplankton, derivatized with 9-anthryldiazomethane, and separated on a 30°C C_{18} column ($\lambda = 365$ nm, ex; 412 nm, em). An 80/2/15 acetonitrile/methanol/water mobile phase generated baseline resolution, excellent peak shape, and complete elution in 38 min. Okadaic acid standards from 0.3 to 1.3 μg/mL were run and a detection limit of 0.1 ng injected was reported [1108]. Shellfish toxins okadaic acid, dinophysistoxin-1, and pectenotoxin-6 were extracted from scallops and mussels and analyzed on a 35°C C_{18} column (electrospray MS; capillary voltage 4.5 kV, skimmer voltage 74 V, capillary $T = 200$°C, N_2 sheath gas 60 psi , N_2 auxiliary gas 5 units). A 70/30 acetonitrile/water (0.1% acetic acid) mobile phase was used to elute all compounds in 10 min [1109]. Linear ranges of 10 pg to 30 ng injected with detection limits of 5 pg injected were reported.

Okadaic acid and pectenotoxin-2 (P2) were extracted from phytoplankton, derivatized with 9-anthryldiazomethane, and separated in 18 min on a C_{18} column

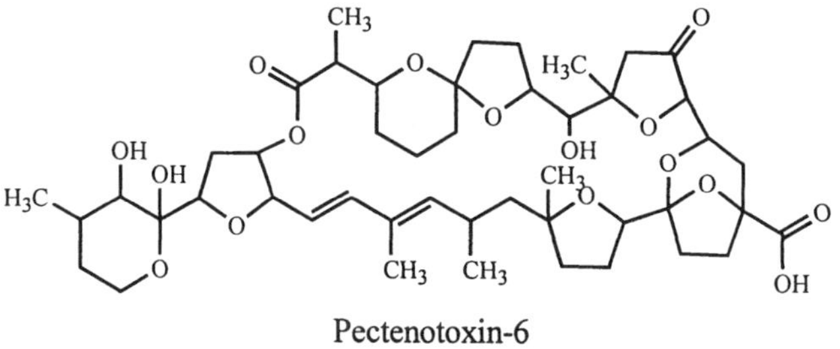

Pectenotoxin-6

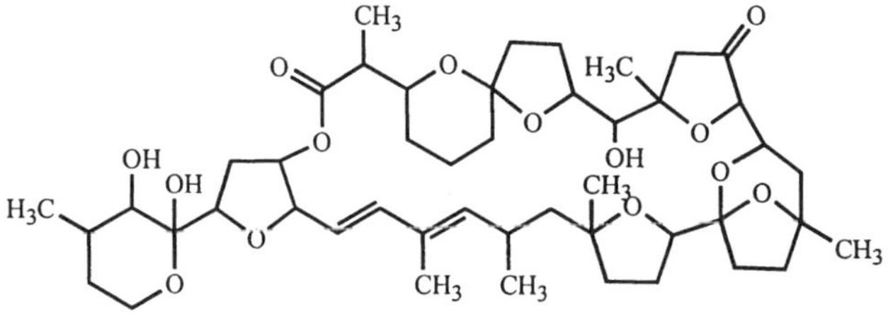

Pectenotoxin-2

($\lambda = 365$ nm, ex; 412 nm, em) using an 80/5/15 acetonitrile/methanol/water mobile phase [1110]. Underivatized P2 and pectenotoxin *seco* acid and its 7-*epi* analog were separated on the same column but with a 60/40/0.1 acetonitrile/ water/TFA mobile phase. In this case, identification was made with an electrospray MS (r.f. mode only, collision energy 30 eV). Excellent resolution and peak shapes along with a 16-min elution time were obtained.

Gymnodimine and gymnodimine B were isolated from *Gymnodinium* spp. and, although this is a semipreparative HPLC technique [1111], it should be readily convertible to an analytical one. A C_{18} column ($\lambda = 215$ nm) and a 55/45 acetonitrile/water (0.8% methanol and 0.005% triethylamine) generated baseline resolution, excellent peak shapes, and elution in <7 min.

Hederasaponins B, C, and D and α-, β- and γ-hederin were extracted from leaves and baseline resolved on a C_{18} column (ELSD) using a 25-min 35/65 $\rightarrow$ 63/37 acetonitrile/water gradient. The last analyte of interest eluted in 20 min. Peak shapes were excellent and detection limits of ~0.1 µg/20 µL were reported [1112].

α-Chaconine and its hydrolysis products β_1-, β_2- and γ-chaconine) were well resolved on a C_{18} column ($\lambda = 200$ nm) using a 35/65 acetonitrile/water (0.1 M

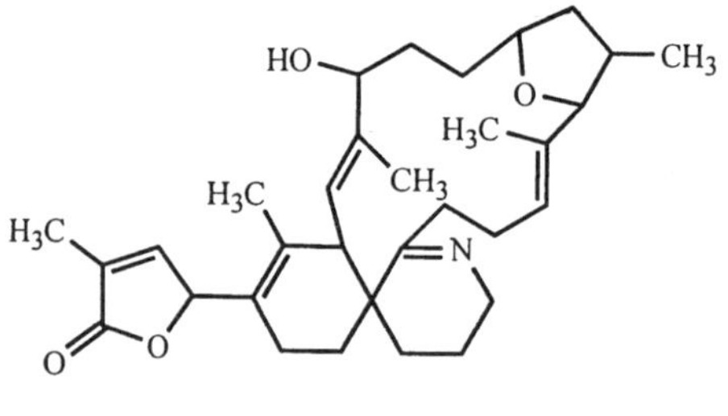

Gymnodimine

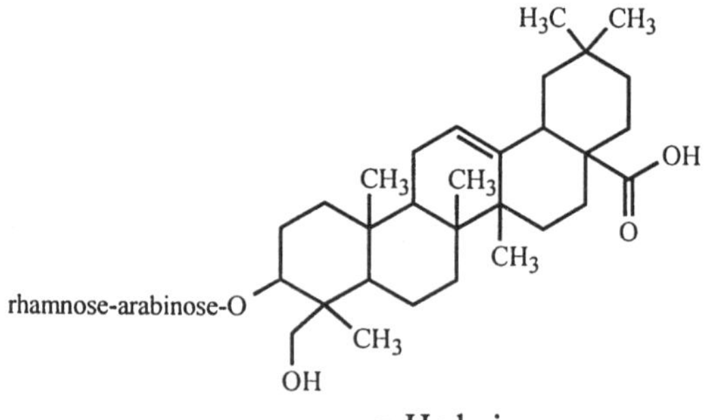

α-Hederin

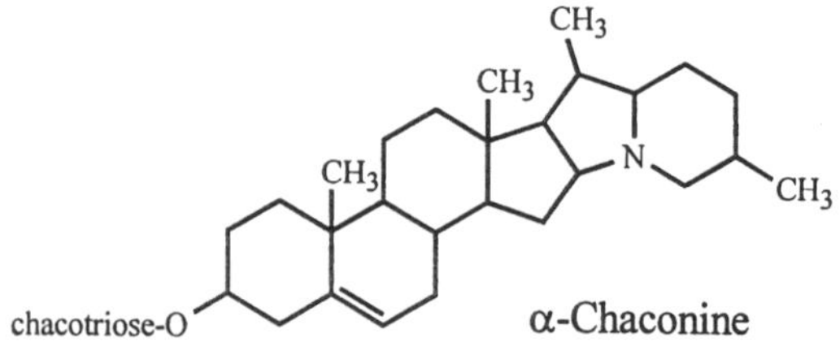

α-Chaconine

ammonium phosphate at pH 3.5) mobile phase [1113]. Peak shapes were good and elution was complete in <15 min. The fully deglycolated product, solanidine, was eluted in <10 min using the same column but a 60/40 acetonitrile/water (10 mM ammonium phosphate at pH 3.0) mobile phase. In both cases the baseline was somewhat noisy. Note that for this work acetonitrile and water are the only truly acceptable mobile phase constituents for UV work at 200 nm.

9.5.4 Vitamins and Related Analytes

9.5.4.1 Water-soluble Vitamins and Related Compounds

In a wonderfully short and effective separation, seven water-soluble vitamins (ascorbic acid, niacin, niacinamide, pyridoxine, folic acid, thiamine, riboflavin) were baseline resolved on a C_8 column using an isocratic 7/93 acetonitrile/water (1% acetic acid and 5 mM sodium heptanesulfonate) mobile phase [1114]. Elution was complete in 6 min.

Riboflavin, flavin mononucleotide, and flavin–adenine dinucleotide, precursors to "skuny" odors in some beverages, were determined from a direct sample injection. These compounds were separated on a C_{18} column ($\lambda = 265$ nm, ex; 525 nm, em) in 9 min using a complex 8-min 95/5 → 5/95 water (50 mM NaH_2PO_4 to pH 3.0 with H_3PO_4)/acetonitrile gradient [1115]. The effect of buffer concentration on the fluorescence intensity of the peaks is shown. The authors note that the use of perchloric acid methanol mobile phase gradients led to unacceptable peak broadening or peak splitting. Reported linear ranges varied from 217 µg/L to 11–225 µg/L with detection limits of 0.5–2 µg/L (S/N = 3) and quantitation limits of 2–7 µg/L (analyte dependent).

Riboflavin and two metabolites (flavin adenine dinucleotide [FAD], flavin mononucleotide [FMN]) were extracted from plasma and baseline resolved on a C_{18} column ($\lambda = 445$ nm, ex; 530 nm, em). An 8-min separation was achieved with a 15/85 acetonitrile/water (15 mM magnesium acetate with 10 mM KH_2PO_4 to pH 3.4 with H_3PO_4) mobile phase [1116]. Standards ranged from 2 to 250 nM with detection limits given as 4 nM.

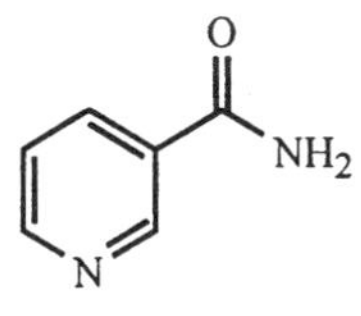

Niacinamide 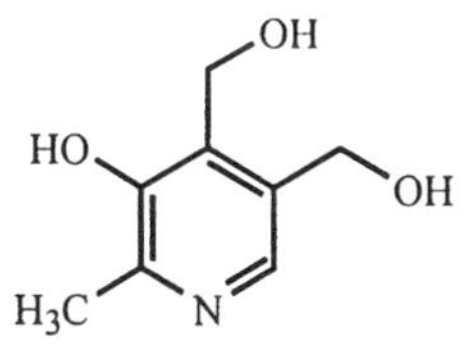   Pyridoxine 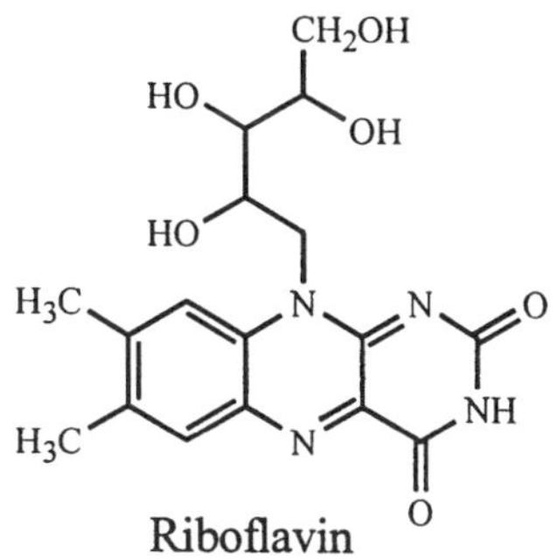   Riboflavin

A separation of pteroylglutamate$_{1-7}$ and numerous reduced forms (e.g., dihydrofolate, tetrahydrofolate) and derivatives (e.g., methyl and formyl) was achieved on a C_{18} column ($\lambda = 280$ nm) using a 90/10 (hold 5 min) $\rightarrow$ 64/36 (at 15 min) $\rightarrow$ 50/50 (at 35 min) $\rightarrow$ 35/65 (at 52 min) water (25 mM NaCl with 0.5 mM dithioerythritol, and 5 mM tetrabutylammonium phosphate [TBAP])/(60/40 acetonitrile with 25 mM NaCl, 0.5 mM dithioerythritol, 5 mM TBAP, and 1 mM KH_2PO_4) gradient [1117]. Injections of 0.5–5 nmol were readily detected.

Folate and five related compounds (tetrahydro-, 5-methyltetrahydro-, 5-formyltetrahydro-, 10-formyl-, and 10-formyldihydrofolate) were extracted from foods and analyzed on a C_{18} column ($\lambda = 280$ nm). A complex 20-min 95/5 $\rightarrow$ 90/10 water (33 mM phosphate buffer at pH 2.1)/acetonitrile gradient was used [1118]. Elution was complete in 18 min. Tetrahydropfolate was incompletely resolved from coextracted materials. Calibration standards ran from 8 to 100 ng/mL.

Dehydroepiandrosterone (DHEA) levels are determined in dietary supplements using a C_{18} column ($\lambda = 292$ nm) and a 60/40 acetonitrile/water (25 mM KH_2PO_4 to pH 3.5 with H_3PO_4) mobile phase [1119]. DHEA was resolved from testosterone, androsterone, and pregnenolone and eluted in <10 min. Calibration standards were run from 2 to 50 μg/mL. Detection and quantitation limits were reported as 0.9 μg/mL and 3.3 μg/mL, respectively.

9.5.4.2 Fat-soluble Vitamins and Related Compounds

Vitamins D_2 and D_3 were extracted from milk and were determined using a C_{18} column ($\lambda = 265$ nm) and a 97/3 acetonitrile/methanol mobile phase [1120]. They were fully resolved and eluted in <25 min. The authors noted that a 10-min re-equilibration time with 97/3 acetonitrile/methanol was needed to generate reproducible results. Whether this was due to late-eluting peaks or some other system

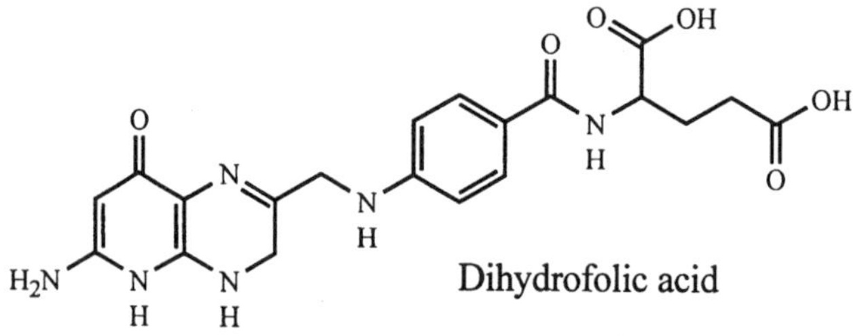

Dihydrofolic acid

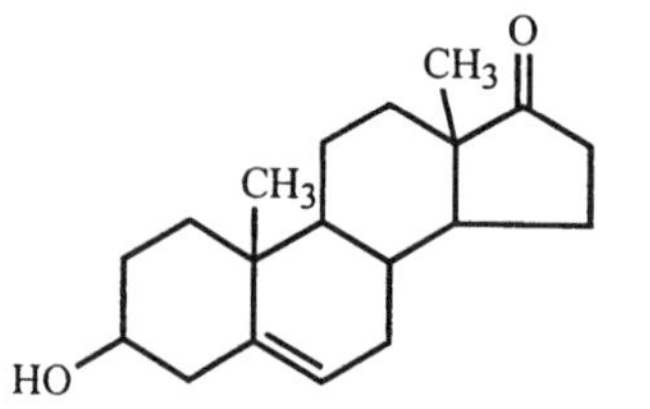

Dehydroepiandrosterone

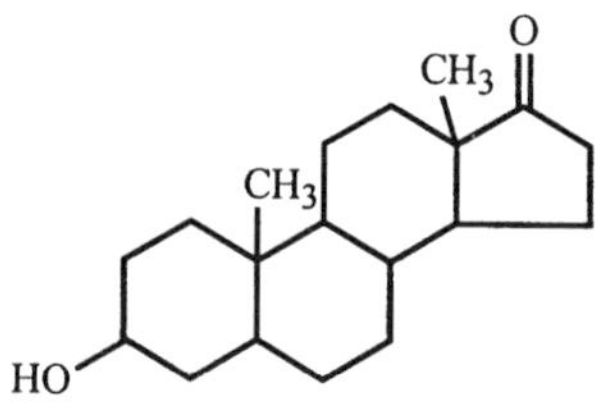

Androsterone

variable was not indicated. Detection limits of 2.5 ng injected and working curve concentrations up to 60 ng injected were reported.

Falecalcitriol is a fluorinated analog of vitamin D_3. It and three hydroxylated metabolites were resolved on a 50°C C_{18} column ($\lambda = 265$ nm) using a $50/50 \rightarrow 0/100$ (at 12 min hold 3 min) water/acetonitrile gradient [1121]. Elution was complete in <15 min and peaks were baseline resolved. Peak identification was generated by EI/MS.

The retention behavior of glucuronides and sulfates of provitamin D, vitamins D_2, D_3, and 25(OH)D_3 as their 4-[4-(6-methoxy-2-benzoxazolyl)phenyl]-1,2,4-trazo-line-3,5-dione derivatives was studied using a C_8 column ($\lambda = 265$ nm or 320 nm, ex; 380 nm, em). The two sulfate conjugates of provitamin D were resolved in 10 min on a C_8 column using a 66/34 acetonitrile/water (2% $NaClO_4$ and 5 mM methyl-β-cyclodextrin), while the two glucuronide conjugates were resolved and eluted in 14 min under the same conditions. Detection limits of 5 fmol (S/N = 5) were reported. The identical conditions produced baseline resolution of the sulfate and glucuronide conjugates of vitamin D as well. Addition of cyclodextrin was critical to the separation [1122]. Without it, the sulfate conjugate peaks co-eluted. Three positional isomers of vitamin 25(OH)D_3 glucuronides were baseline resolved in 15 min using a 50/50 acetonitrile/water (2% $NaClO_4$ to pH 3.0 with $HClO_4$) mobile phase.

Andrikopoulos et al. [1123] developed an excellent separation scheme for an interesting mixture of compounds: phenolic antioxidants, tocopherols, and trigycerides. This was accomplished on a C_{18} column ($\lambda = 280$ nm or photodiode array detector, $\lambda = 210$–330 nm) with a complex 67-min $70/30/0 \rightarrow 0/100/0 \rightarrow 0/40/60$ (water with H_3PO_4 to pH 3.0)/(7/5 acetonitrile/methanol)/IPA gradient. The 10 antioxidants (e.g., BHT, BHA, propyldodecylgallate, t-butylhydroquinone,

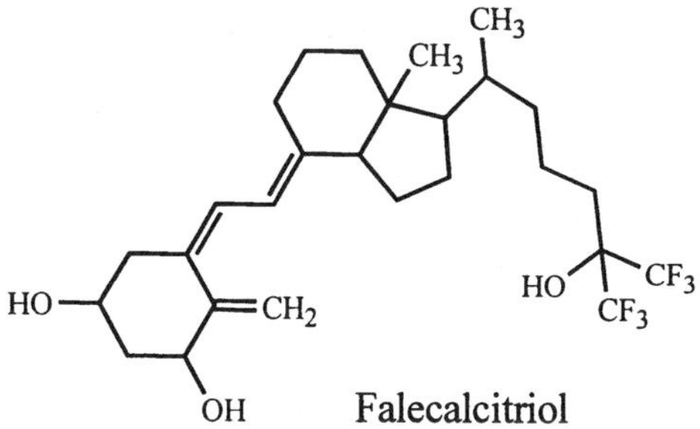

Falecalcitriol

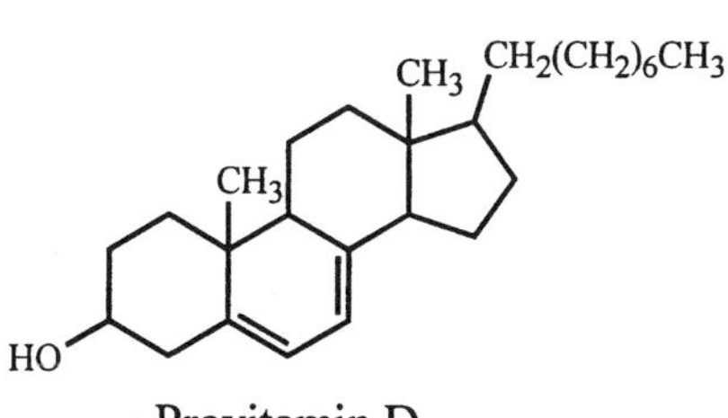

Provitamin D_3

nordihydroguaiaretic acid) were baseline resolved and completely eluted in the first 20 min. Tocopherols (α-, β-, γ-, and δ-) were baseline resolved and eluted between 28 and 34 min. Triglycerides with equivalent carbon numbers from 42 to 50 were eluted between 40 and 50 min. Differentiation of triglycerides was achieved through varying the detector wavelength: 215 nm for saturated and unsaturated triglycerides, 280 nm for conjugated triglycerides, and 230 nm for oxidized unsaturated triglycerides. Individual 20 µL aliquots of working solutions containing 300 mg/mL triglycerides, 50 µg/mL antioxidants, and 300 µg/mL tocopherols were injected. Excellent peak shapes were obtained for all compounds.

Lutein, zeaxanthin, β-cryptoxanthin, lycopene, α-carotene, and β-carotene were extracted from fruits and vegetables and baseline resolved on a C_{18} column ($\lambda = 450$ nm) using a 75/20/5 acetonitrile/methanol/dichloromethane (0.1% BHT and 0.05% TEA) mobile phase [1124]. The authors note that carotenoids degrade on silica-based columns. The addition of TEA was found to stop the degradation process. For example, the average recovery of test analytes increased from 68% to 92% upon the addition of 0.1% TEA to the mobile phase. Linear concentration curves from 0.05 to 4 µg/L and detection limits of $\sim$0.02 µg/mL were reported.

A total of 22 carotenoids were extraced from human plasma, 13 of which were identified (e.g., astaxanthin, lutein, echinenone, cis-lycopene, carotenes). A 45-min separation was achieved on a C_{18} column (photodiode array detector, $\lambda = 300$–600 nm) using a 70/15/5/10 acetonitrile/methanol (50 mM ammonium acetate)/water/dichloromethane mobile phase [1125]. Good (but not necessarily baseline) resolution was obtained for all peaks. A linear range of 10–200 ng injected and detection limits of 5 ng injected were reported.

Khachik et al. [1126] studied the effect of injection solvents on the peak shapes of carotenoids. The study was conducted on a C_{18} column ($\lambda = 450$ nm) and a 30-min 10/85/2.5/2.5 → 10/50/20/20 acetonitrile/methanol/dichloromethane/hexane gradient was used. When the sample was dissolved in THF, dichloromethane, or toluene, double or multiple peaks were formed. Not unexpectedly, increased injection volumes exacerbated the problem. Most likely, this finding is the result of the extremely limited solubility of the sample solvent in the initial mobile phase. The retention process is therefore not partitioning but a kinetically controlled desolvation process. Acetonitrile, acetone, and methanol provided excellent peak shapes (they are all miscible with the initial mobile phase).

Nordihydroguaiaretic acid

9.5.5   Terpenoids, Flavonoids, and Other Naturally Occurring Analytes

9.5.5.1   Terpenoids

The sesquiterpene fungal toxins illudin S and M were extracted from *Omphalotus nidiformis* and analyzed on a 30°C C_{18} column ($\lambda = 320$ nm) using a 10-min $90/10/1 \rightarrow 10/90/1$ water/acetonitrile/acetic acid gradient. Identity was confirmed by APCI/MS/MS. The authors noted that a splitting of the illudin S peak (eluting at $\sim$3.5 min) was due to the sample being dissolved in acetonitrile and correctly recommend the use of a solvent closer to the mobile phase composition to eliminate the problem [1127].

In a excellent study, Es-Safi et al. [1128] studied that reaction products of catechin and furfural or 5-(hydroxymethyl)furfural and their effects on food color change. Separation of the species was achieved using a 30°C C_{18} column (photodiode array detector, $\lambda = 250$–600 nm) and a $95/5 \rightarrow 70/30$ (at 40 min) $\rightarrow 60/40$ (at 50 min) $\rightarrow 0/100$ (at 55 min) (98/2 water/formic acid)/(80/18/2 acetonitrile/water/formic acid) gradient. The 8-8, 6-8, and 6-6 catechin dimer furfuryl adducts and well as xanthene and xanthylium salts were positively identified by electrospray MS (negative ion, orifice voltage -70 V, ion spray voltage -4 kV) using a narrow-bore C_{18} column with the effluent monitored at $\lambda = 280$ nm and a similar gradient profile.

Four triterpenes (asiatic acid, madecassic acid, asiaticoside, madecassoside) were extracted from *Centella asiatica* and baseline resolved on a C_{18} column

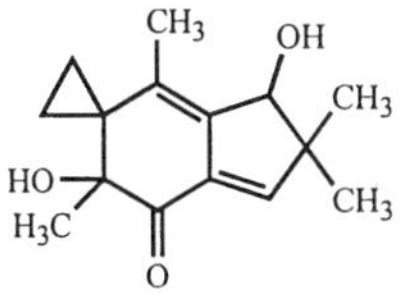

Illudin M

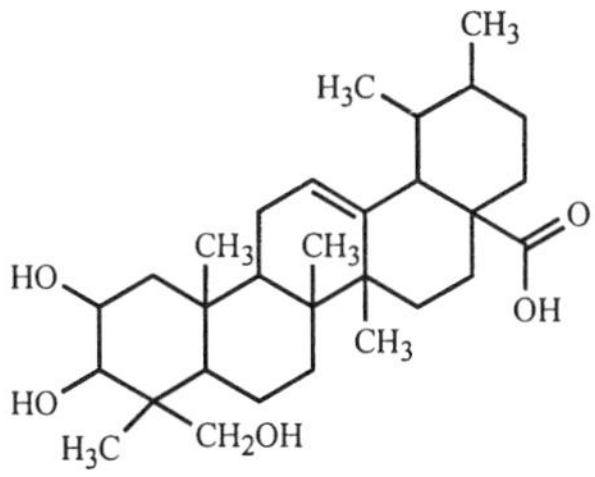

Asiatic acid

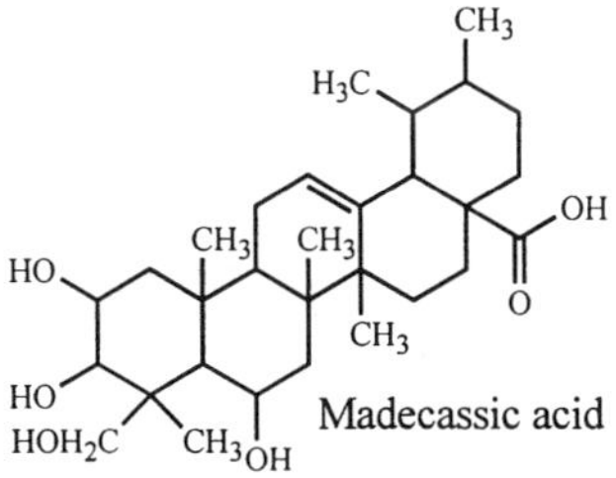

Madecassic acid

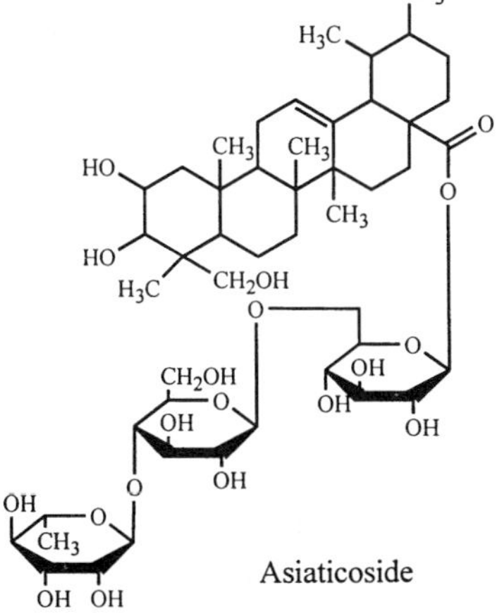

Asiaticoside

($\lambda = 205$ nm) using a three-ramp $90/10/0.05 \rightarrow 72/28/0.05$ (at 15 min) $\rightarrow$ $60/40/0.05$ (at 20 min) $\rightarrow 10/90/0.05$ (at 25 min hold 10 min) water/acetonitrile/H_3PO_4 gradient. Peak shapes were excellent. A linear range of 1–50 µg injected and a detection limit of 0.3 µg injected were reported [1129].

Loganin and the metabolism products of *Lonicera japonica* cell cultures, secologanin and 7-deoxyloganin, were extracted from culture media and resolved on a 40°C C_{18} column ($\lambda = 240$ nm) using a $90/10 \rightarrow 75/25$ (at 30 min hold 5 min) water/acetonitrile gradient [1130]. Elution was complete in 33 min and all peaks were baseline resolved.

Five sesquiterpenes (α-bisabolol, α-bisabolol oxide A and B, chamazulene, β-farnesene) and two polyacetylenes ((E)- and (Z)-ene-yne-dicycloether) were extracted from *Chamomilla recutita* and separated on a C_{18} column ($\lambda = 200$ nm) using a 15-min $19/80/1 \rightarrow 59/40.5/0.5$ acetonitrile/water/phosphoric acid gradient [1131]. Peak shapes were excellent and baseline resolution was achieved. Standard concentrations from 50 to 200 µg/mL were used.

A series of highly oxygenated triterpenes, curcurbitacins, were extracted from various plants and characterized on a C_{18} column ($\lambda = 210$ nm and negative ion

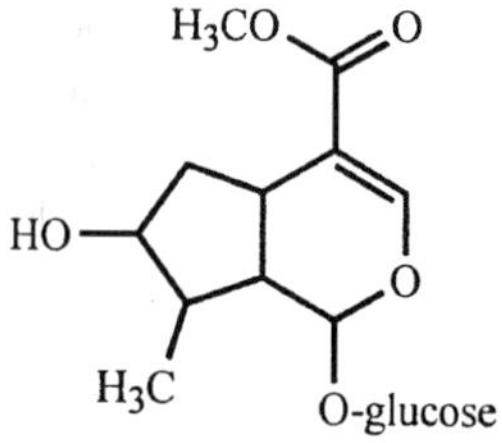

Loganin 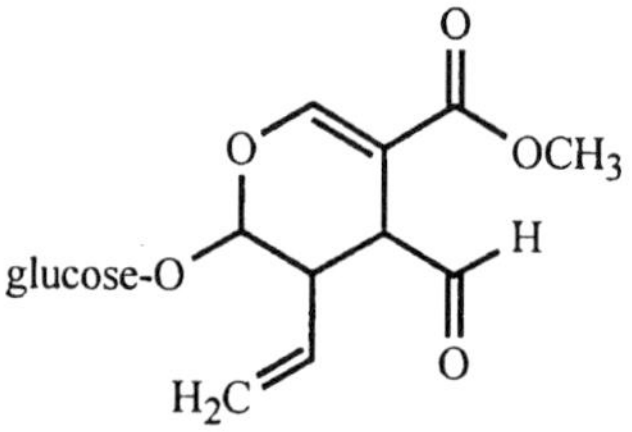   Secologanin 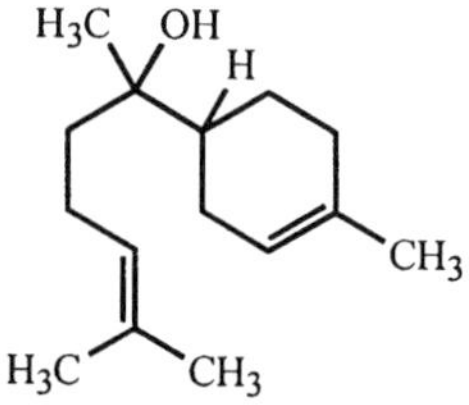   α-Bisabolol

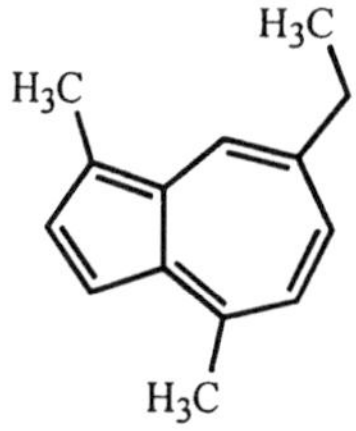

Chamazulene 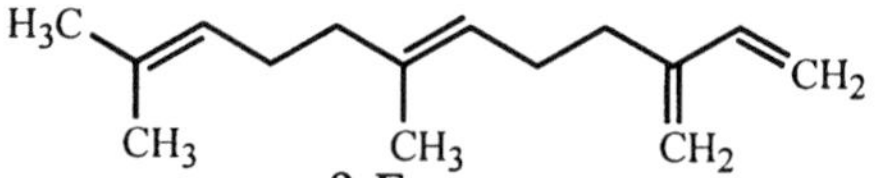   β-Farnesene

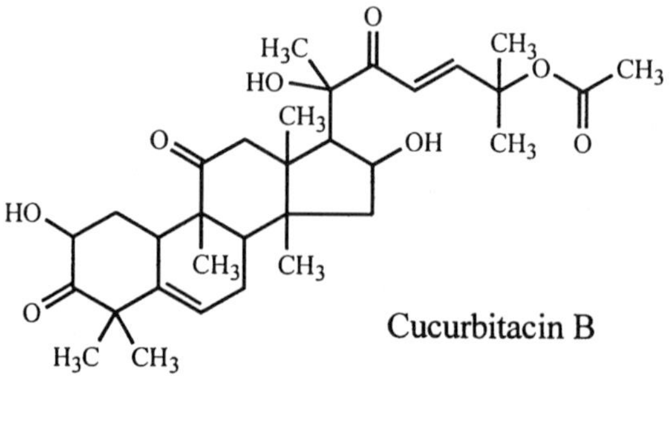

Cucurbitacin B

electrospray MS, spray voltage 4.5 kV; capillary $T = 200°C$; CID offset 5 V; sheath gas 40 psi). The gradient was complex, running from $80/20 \rightarrow 10/90$ water (0.01% TFA)/acetonitrile over 60 min [1132]. Over 35 compounds were identified ranging from curcurbitacin I- and E-glucoside, gratioside, curcurbitacin I, B, D, and L, arvenine I and III, and a large series of methylated, acetylated, glucosylated 19-norlanostadiones and ene-ones.

Neem seeds were extracted and analyzed for triterpenoids (salannin, nimbin, 6-deacetylnimbin) and azadirachtin content [1133]. Separation was achieved on a 30°C C_{18} column (positive ion APCI, probe $T = 400°C$, source block $T = 150°C$, N_2 sheath and drying gas at 300 L/h, argon collision gas at 0.002 mbar, source pin voltage 3.5 kV, cone voltage 25 V) using an 80/20 (hold 5 min) $\rightarrow$ 35/65 (at 20 min hold 2 min) water/acetonitrile (0.1% TFA) gradient. A working curve for azadirachtin was generated from 0.5 to 200 µg/mL with a detection limit of 870 pg/mL.

Extracts of *Leontodon autumnalis* were characterized from their sesquiterpenoids (crepidiaside A and B, 15-glucopyranosyloxy-2-oxoguaia-3,11(13)-dien-1α,5α, 6β,7α,10αH-12,6-olide and four related compounds) using a complex 35-min $88/12 \rightarrow 40/60$ water/acetonitrile gradient on a C_{18} column (phototdiode array detector, $\lambda = 220–260$ nm). Peak shapes were excellent and resolution was good overall [1134].

Wolfender et al. [1135] screened plant extracts for various phenolic and terpene glycosides (e.g., phlominol, sesamoside, lisianthioside, bellidifolin-8-O-glucoside) using an acetonitrile/water (0.05% TFA) gradient with a C_8 or C_{18} columns (thermospray MS). The gradients were typically run from ~5 to 75% acetonitrile

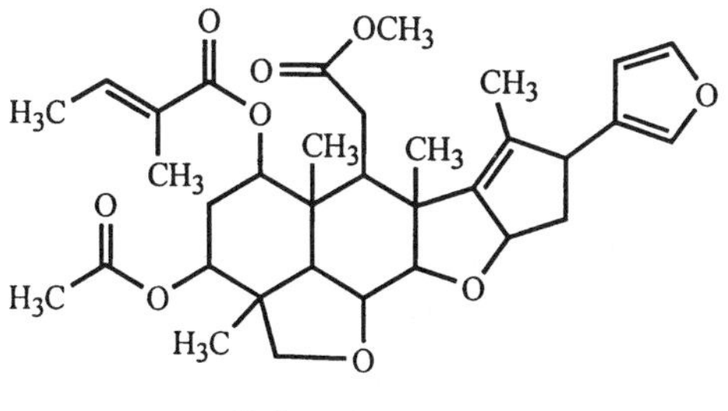

Salannin

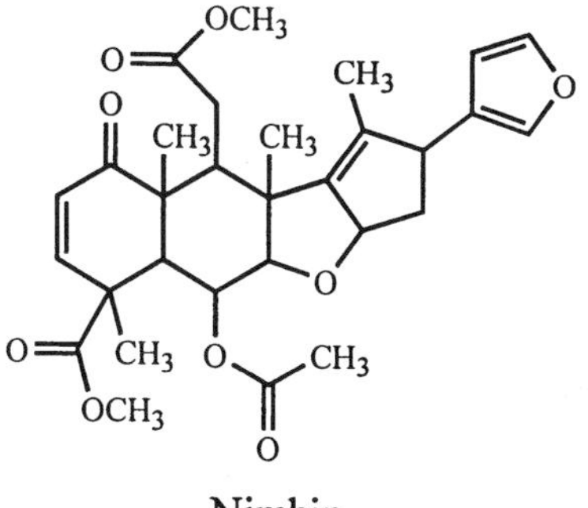

Nimbin

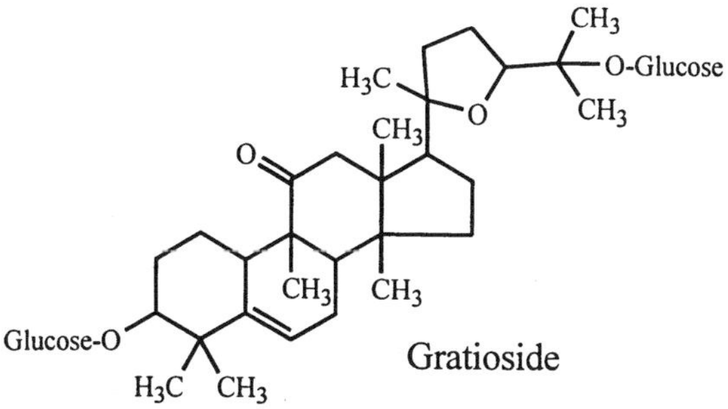

Gratioside

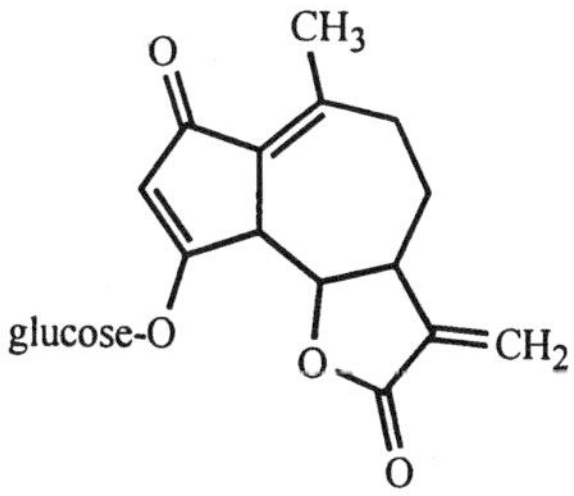

Crepidiaside

and lasted 30–50 min. The same group also studied saponins (e.g., glucose, rhamnose, arabinose, glucuronic acid, glucosamine) from plants using the thermospray MS as the LC detector [1136]. In this case, a 30-min $30/70 \rightarrow 80/20$ acetonitrile/water gradient on a C_{18} column ($\lambda = 206$ nm) was used for generating the separation. The system effluent was made MS-compatible through the post-column addition of 0.5 M ammonium ion. Injections of 2 µg of pure sample were used.

9.5.5.2   Flavanoids and Related Compounds

In an excellent method, Menghinello et al. [1137] resolved 19 flavonoid aglycones (catechin, rutin, naringin, hesperidin, myricetin, morin, quercetin, naringenin, apigenin, kaempferol, isorhamnetin, 3- and 6-hydroxyflavone, and 3,6 and 3,7-dihydroxyflavone, chrysin, acacetin, galangin, flavanone) using a 35-min $90/10/0.1 \rightarrow 50/50/0.1$ water/acetonitrile/TFA gradient on a C_{18} column ($\lambda = 254$ nm and 340 nm). Peaks were sharp. A calibration curve from 200 to 500 µM and detection limits ranging from 135 to 580 µg/mL were reported. Similarly, naringin, neoeriocitrin, hesperidin, and neohesperidin were isolated from orange juices in order to determine whether adulteration with grapefruit juice had occurred [1138]. These compounds were baseline resolved on a C_{18} column ($\lambda = 280$ nm) using an 88/12 water (0.1% acetic acid)/acetonitrile mobile phase. Analysis was complete in 12 min. Linear ranges were reported as 5–100 µg/g and detection limits as 2 ppm.

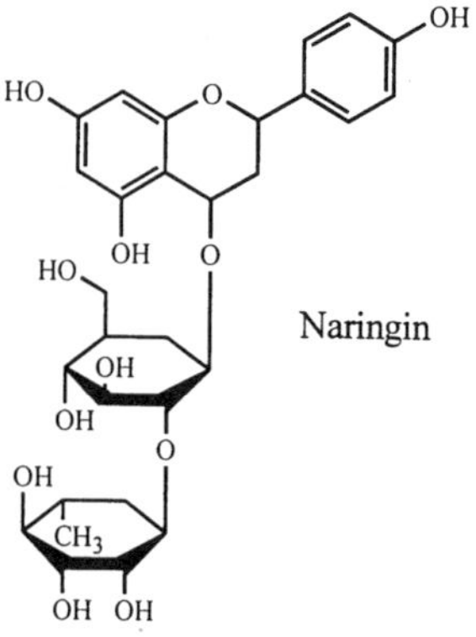

Naringin

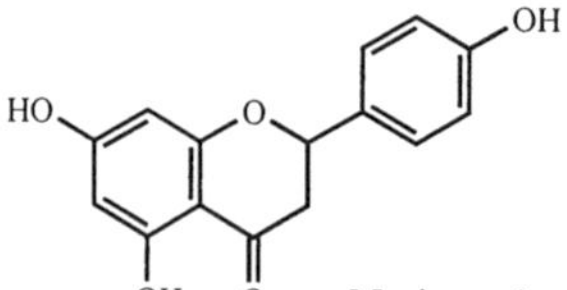

Naringenin

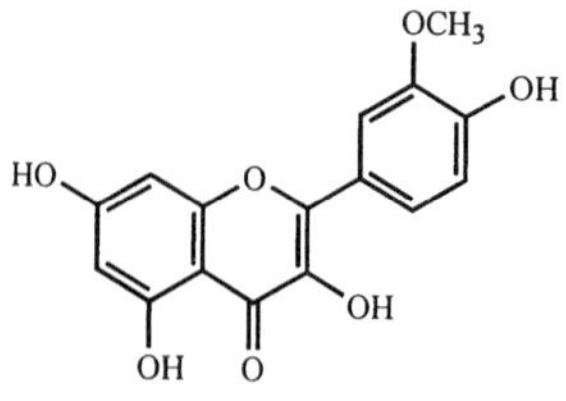

Isorhamnetin

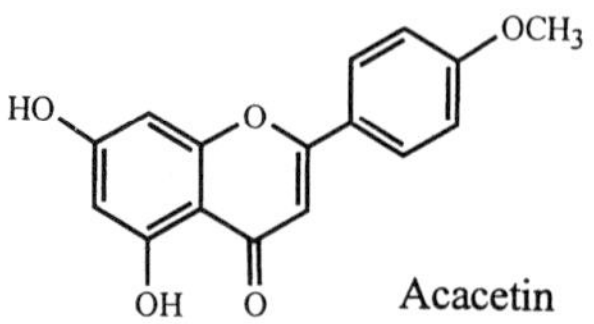

Acacetin

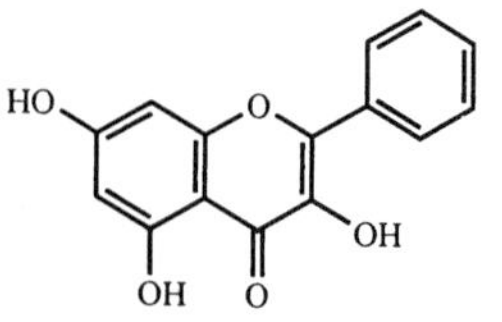

Galangin

Onion skin and wine extracts were analyzed for bioflavonoid content (quercetin, kaempferol, fisetin, rutin, myricetin, morin). A 50°C C$_8$ column ($\lambda = 370$ nm) and a 71/28/1 water/acetonitrile/TFA mobile phase generated a 15-min elution. Peaks were resolved and shapes were excellent. The wine extract had a large early-eluting interfering set of peaks (three analytes eluted on the shoulders of these peaks). A linear range of 0.1–30 µg/mL was reported [1139].

A set of six phenolic materials [3,4-dihydroxyphenylethanol, elenolic acid glucoside, demethyloleuropein, quercetin 3-rutinoside, luteolin 7-glucoside, oleuropein) were extracted from olives and well separated on a C$_{18}$ column ($\lambda = 280$ nm for the phenylethanol and 340 nm for all others). A 20-min 15/85 → 40/60 acetonitrile/water (H$_3$PO$_4$ to pH 2.5) gradient was used to generate the separation [1140]. Analyte levels in samples were reported in the 0.05–3.5 mg/g range.

Seven flavonoids were identified in grapefruit (narirutin, naringin, hesperidin, neohesperidin, didymin, quercetin, poncirin). A complex 60-min 90/10 → 20/80 water/acetonitrile gradient used with a C$_{18}$ column ($\lambda = 280$ nm) resolved all the peaks of interest. However, no sample preparation was done and many matrix peaks were also eluted, generating chromatographic interferences. Preanalysis sample preparation should be considered. Standards from 25 to 100 µg/mL were used [1141].

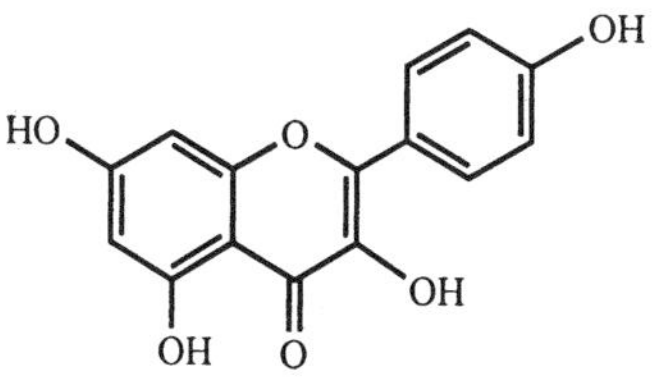

Kaempferol

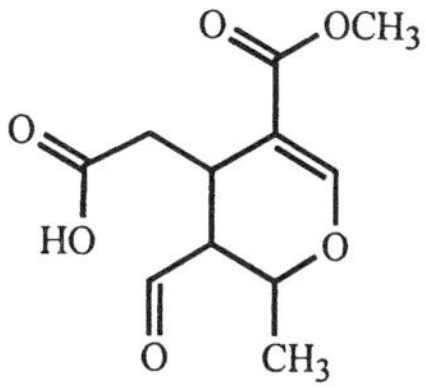

Elenolic acid

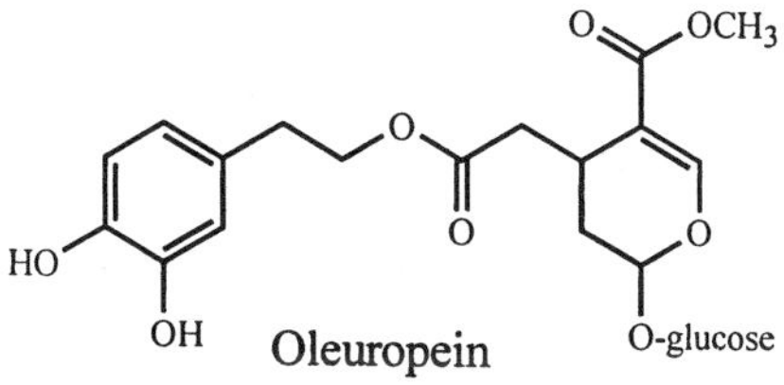

Oleuropein

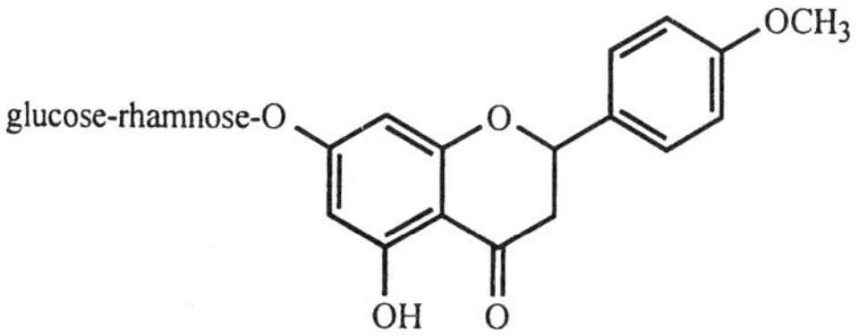

Didymin

Xanthohumol and five other related prenylflavonoids (desmethyl- and isoxantho-humol, 6- and 8-prenylnaringenin, 6-geranylnaringenin) were separated on a C_{18} column ($\lambda = 254$ nm) using a $60/40 \rightarrow 0/100$ (at 15 min hold 5 min) acetonitrile/water (1% formic acid) gradient [1142]. Identification was by positive ion APCI MS, nebulizer interface 500°C, corona discharge voltage 8 kV, discharge current 3 µA, orifice voltage +55 V, collision voltage 30 V. Elution was complete in 18 min. For MS a standard curve of 1–800 µg/L was used (analyte dependent).

Eight flavonoids (e.g., calycosin, formononetin, ononin, and various glucosylated analogs) were extracted from *Astragalus* roots and separated on a C_{18} column ($\lambda = 260$ nm) using a 40-min $82/18/0.025 \rightarrow 58/420.025$ water/acetonitrile/acetic acid gradient [1143]. Peak shapes were excellent and peaks were generally well resolved. The last peak eluted at 33 min.

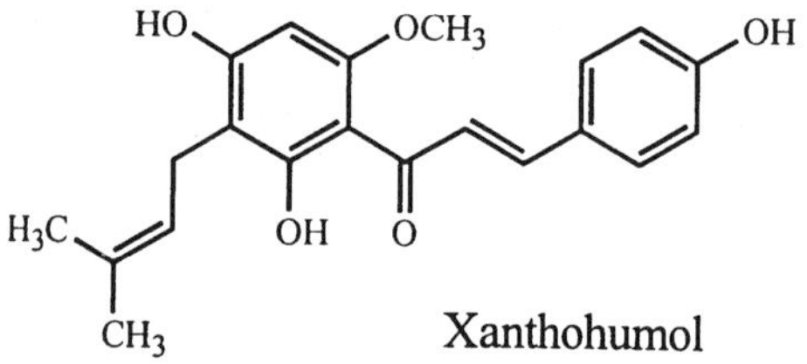

Xanthohumol

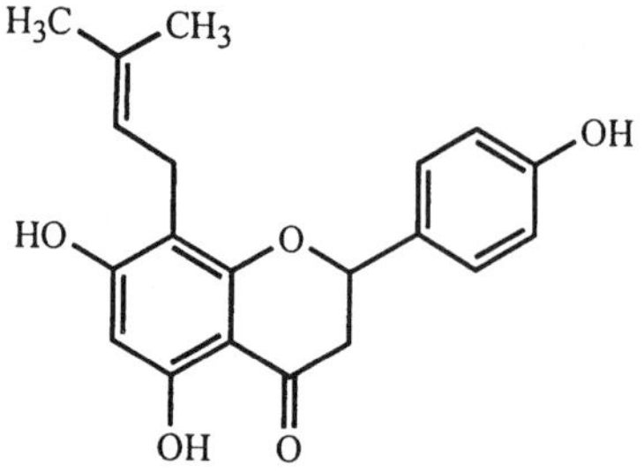

8-Prenylnaringenin

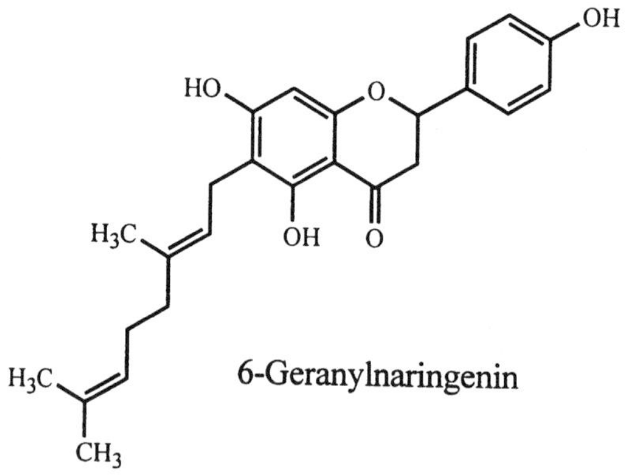

6-Geranylnaringenin

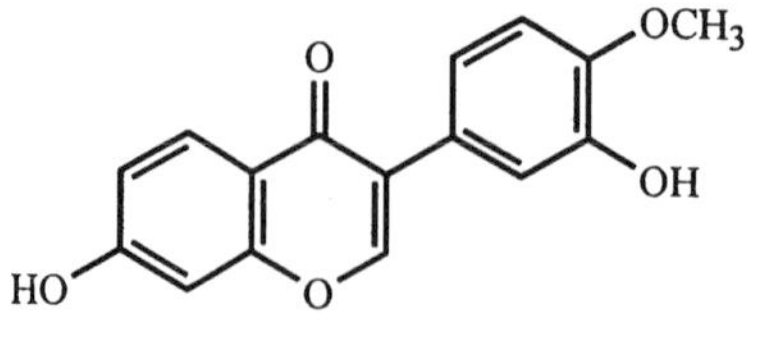

Calycosin

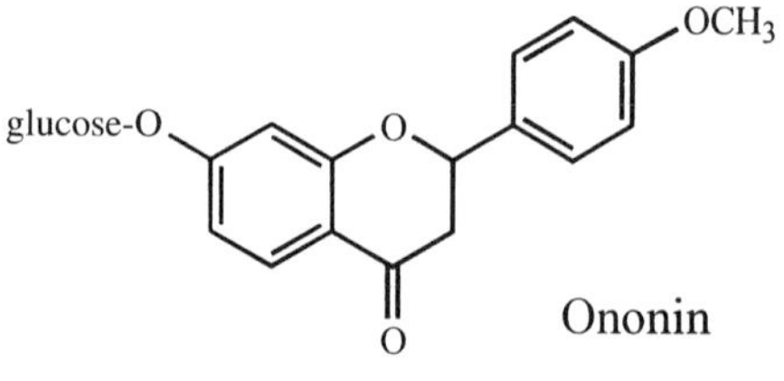

Ononin

The Chinese medicinal preparation ping-wei-san was analyzed for eight marker compounds (glycyrrhizin, hesperidin, nobiletin, tangeretin, honokiol, magnolol, 3,3',4',5,6,7,8-hepta- and 5-hydroxy-3',4',6,7,8-pentamethoxyflavone) on a 30°C C_{18} column ($\lambda = 275$ nm 0–18 min, 250 nm thereafter). A two-ramp 100/0 (hold 15 min) → 65/35 (at 25 min) → 55/45 (at 55 min hold 20 min) (20/80 acetonitrile/water [H_3PO_4 to pH 2.5])/(70/30 acetonitrile/water [H_3PO_4 to pH 2.5]) gradient generated excellent resolution and peak shapes [1144].

Seven flavanones and xanthones (euchrestaflavanone C and B, osajaxanthone, toxyloxanthone C, macluraxanthone, alvaxanthone, 8-prenylxanthone) were isolated from the root bark of osage orange trees and resolved on a C_{18} column ($\lambda = 280$ nm) using a 42-min 60/40 → 40/60 (981.5/18.5/0.005 water/acetonitrile/TFA)/(0/100/0.005 water/acetonitrile/TFA) gradient. Peaks were well resolved from one another and from other extracted compounds. Positive-ion APCI was used for peak identification [1145].

Mouly et al. [1146] isolated a set of six polymethoxylated flavones (sinensetin, hexa- and heptamethoxyflavone, nobiletin, tetramethyl-O-scutellarein, tangeretin) from orange juice and baseline resolved in 23 min on a C_{18} column ($\lambda = 330$ nm). The mobile phase was 60/40 water/acetonitrile. Typical standard concentrations were in the 10–25 mg/L range.

Seventeen flavonoids (three flavones: apigenin, luteolin, tangeretin; six flavanols: galangin, kaempferol, quercetin, myrcetin, rhamnetin, isorhamnetin; three flavonones: naringenin, eriodictyol, hesperetin; five catechins/catechin esters: catechin, epicatechin, epigallocatechin, epicatechin gallate, epigallocatechin gallate) were identified and quantitated in a variety of vegetables, fruits, and berries [1147]. Catechins were determined by UV or electrochemical detection. The latter used an

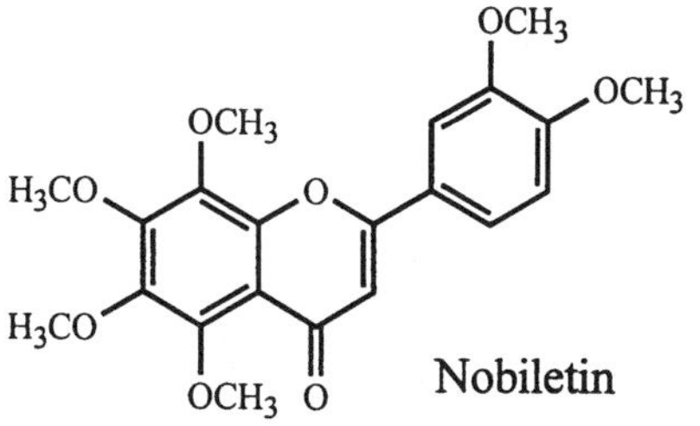

Nobiletin

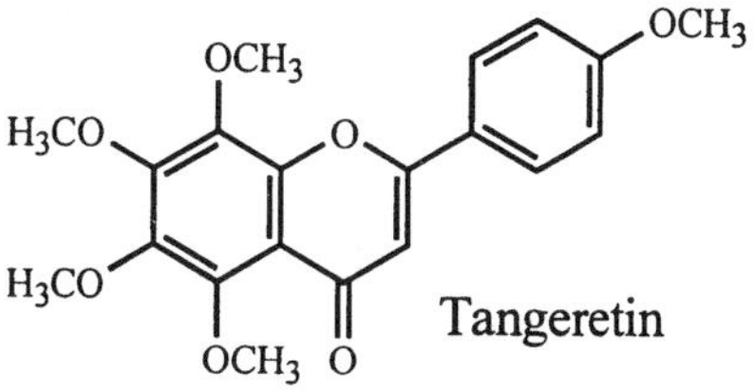

Tangeretin

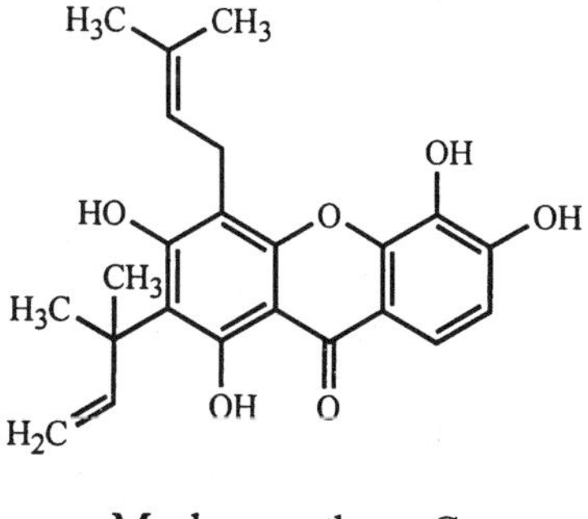

Macluraxanthone C

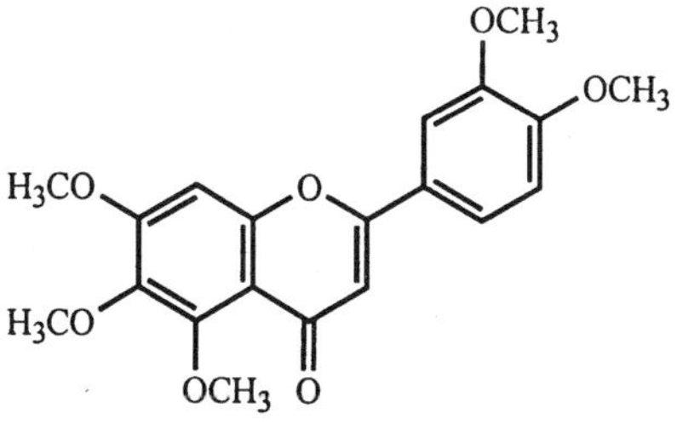

Sinensetin

array detector with potentials ranging from 100 to 800 mV. A 30-min separation was achieved on a C_{18} column using an 86/14 water (50 mM H_3PO_4)/acetonitrile mobile phase. Twenty minutes of baseline separate the latest-eluting peaks. Detection limits (7–100 times lower than for UV detection) ranged from 4 to 40 ng/mL. Diode array detection ($\lambda = 270$–375 nm), a C_{18} column, and a 95/5 (hold 5 min) $\rightarrow$ 50/50 (at 55 min hold 10 min) gradient generated the separation. Detection limits of 80–250 ng/mL were reported.

This study involved the separation and quantitation of flavanones (eriocitrin, neoeriocitrin, narirutin, naringin, hesperidin, neohesperidin) and flavones/flavonols (quercetin, apigenin, kaempferol, chrysin, galangin) isolated from fruit juices. The flavanones were separated on a C_{18} column ($\lambda = 280$ nm) using an 80/20 water (formic acid to pH 2.4)/acetonitrile mobile phase. Elution was complete in 12 min and good separation was obtained. The established linear range was 0.2–50 µg/mL with a detection limit of 0.2 µg/mL (S/N = 3) reported. The flavones/flavonols were separated on the same column ($\lambda = 265$ nm) with a 50/50 $\rightarrow$ 35/65 (at 2 min hold 10 min) water (formic acid to pH 2.4)/acetonitrile gradient. Once again good resolution was obtained. The linear range was 1–100 µg/mL and detection limits of 0.1 µg/mL were reported [1148].

The urinary levels of nine phenolic metabolites of herbal medicines *Daisaiko-to* and *Shosaiko-to* (naringenin, aloe-emodin, rhein, liquiritigenin, davidigenin, medi-

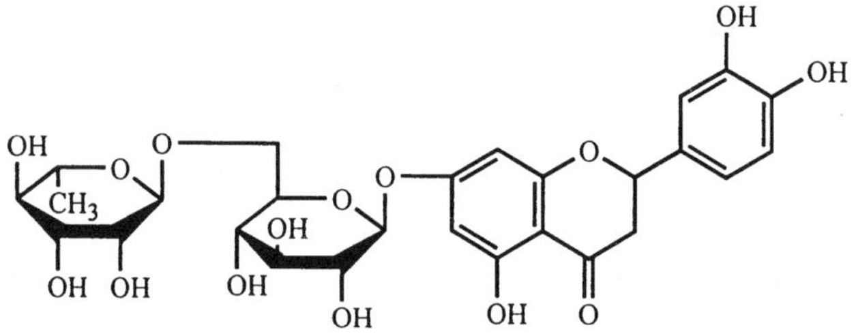

Eriocitrin

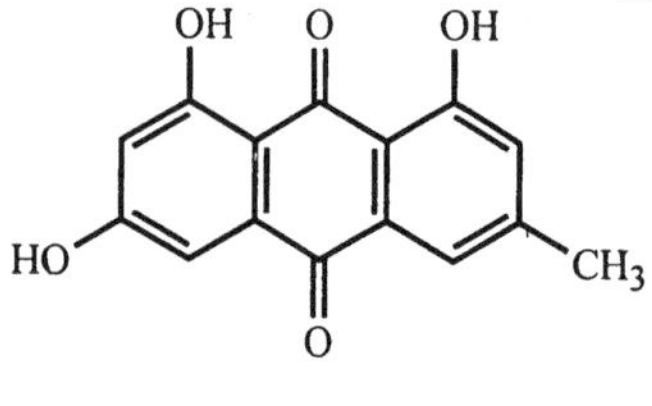

Emodin

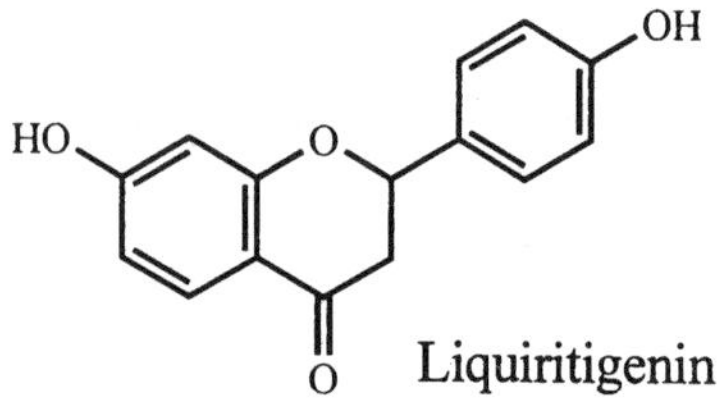

Liquiritigenin

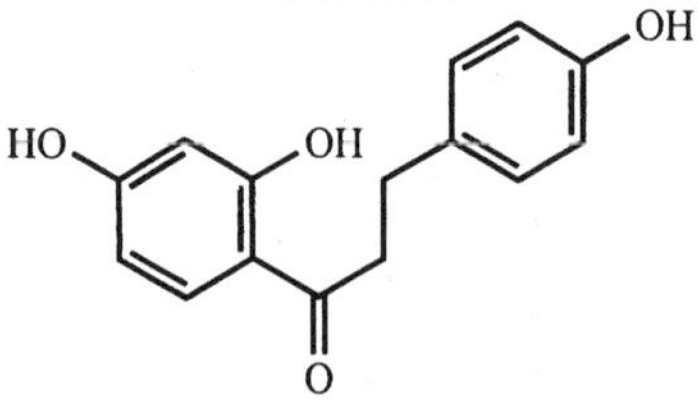

Davidigenin

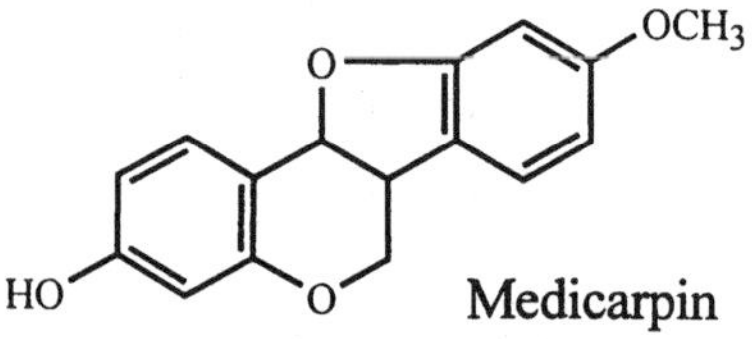

Medicarpin

carpin, baicalein, wogonin, oroxylin-A) were determined using a 67/33 acetonitrile/water (0.5 mM H_3PO_4) mobile phase and 30°C C_{18} column (photodiode array detector, $\lambda = 195$–400 nm). Elution was complete in 30 min and peaks were adequately resolved [1149]. A useful table and plot of the relationship of k' to the percent acetonitrile is also presented.

Four isoflavones (daidzein, genistein, formononetin, biochanin A) and coumestrol were extracted from soybeans and separated on a phenyl column (photodiode array detector, $\lambda = 240$ to 350 nm). Good peak shape and resolution were obtained in 22 min using a 33/67 acetonitrile/water (1% acetic acid) mobile phase [1150]. The authors note that C_8, C_{18}, and cyanopropyl columns were ineffective in the separation. The effect of changing percent acetonitrile on log k' is shown. Linear ranges were reported as 0.02–11 mg/L with detection limits of ~100 nM.

Nine isoflavones (daidzin, genistein, glycitin, genistin [and the corresponding glycitin and genistin malonyl analogs], daidzein, genistin, acetylgenistein) were extracted from soybean cultivars and resolved on a C_{18} column ($\lambda = 254$ nm). A 45-min 85/15/1 → 69/31/1 water/acetonitrile/acetic acid gradient generated excellent resolution and peak shapes [1151].

Quercetin, pinobanksin, kaempferol, pinocembrin, chrysin, galangin, and tectochrysin were extracted from sunflower honey and resolved on a C_{18} column

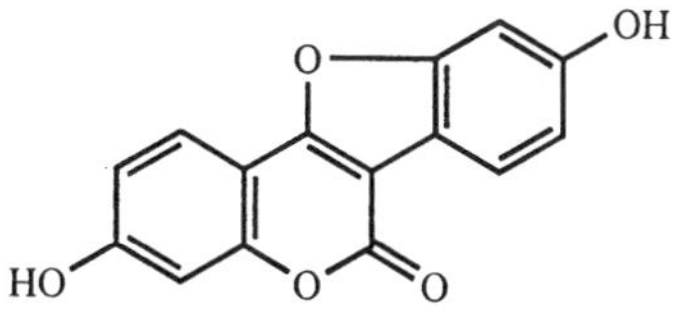

Coumestrol

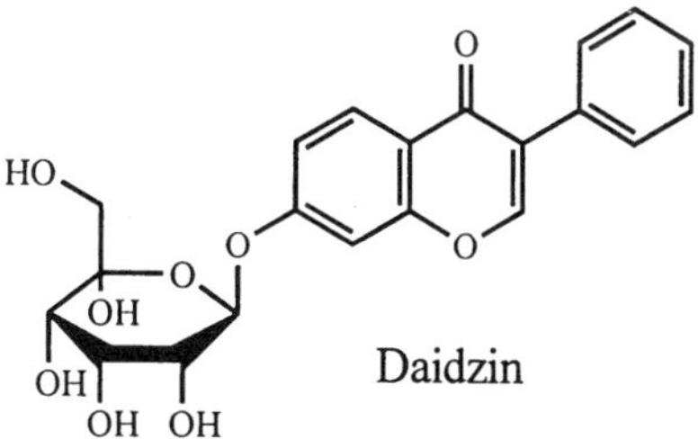

Daidzin

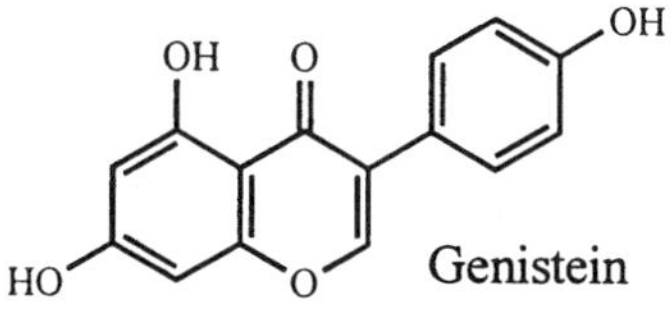

Genistein

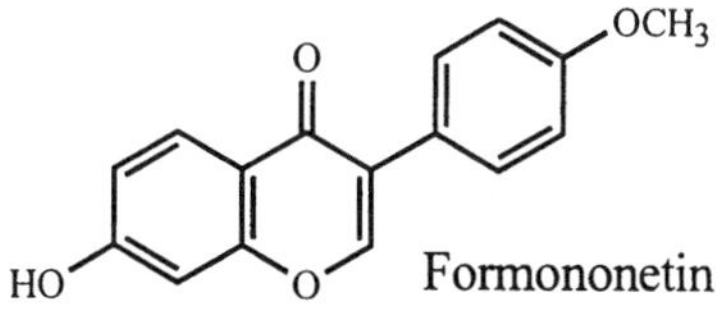

Formononetin

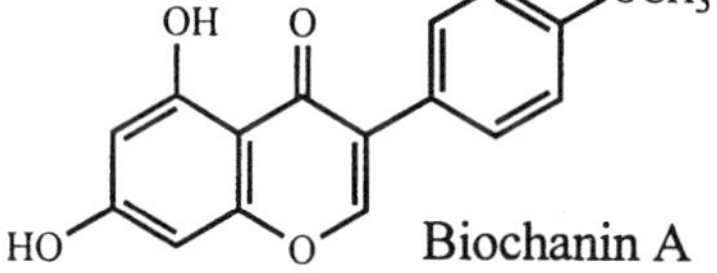

Biochanin A

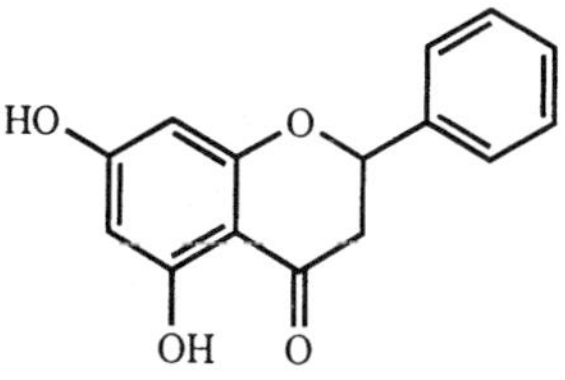

Pinocembrin

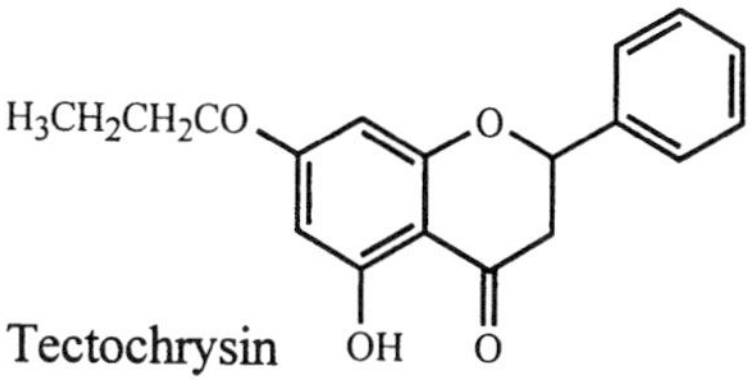

Tectochrysin

($\lambda = 366$ nm) using a complex 60-min $0/100 \rightarrow 70/30$ acetonitrile/water (phosphoric acid at pH 2.6) gradient [1152]. All compounds eluted between 40 and 55 min. Peak shapes were uniformly good and interferent peaks were not a major problem.

Extracts from *Ginkgo bilboa* were fingerprinted using a C_{18} column ($\lambda = 350$ nm) and a complex 30-min $15/1.5/83.5 \rightarrow 0/78/22$ (65/35 THF/IPA)/acetonitrile/water (0.5% H_3PO_4) gradient [372]. Excellent separation of 33 compounds resulted with all peaks identified (e.g., quercetin, kaempferol, myricetin, isorhamnetin, luteolin, and their glycosylated and rhamnosylated analogs, apigenin, gingketin, sciadopitysin). Peak shapes were uniformly excellent. In a similar fashion [834], 13 flavanol glycosides (e.g, myricetin-, kaempferol-, quercetin-3-*O*-glucoside, quercetin-3-*O*-galactoside, quercetin-3-*O*-rhamnoside) were well resolved on a C_{18} column ($\lambda = 354$ nm) using an 85/15 acetonitrile/water (20 mL acetic acid/L) mobile phase. Elution was complete in 40 min.

Seven constituents extracted from a Wuu-Ji-San preparation (liquiritin, hesperidin, cinnamic acid, cinnamaldehyde, glycyrrhizin, honokiol, magnolol) were baseline resolved on a C_{18} column ($\lambda = 254$ nm) using a $5/95 \rightarrow 70/30$ (at 30 min hold 20 min) acetonitrile/water (0.03% H_3PO_4) gradient [1153]. Working concentration ranges from 5 ng–125 µg/mL were used. Co-elution of interferences was negligible and peaks shapes were excellent.

Six flavanone glycosides (eriocitrin, narirutin, hesperidin, neoeriocitrin, naringin, neohesperidin) were extracted from citrus juices and resolved on a C_{18} column

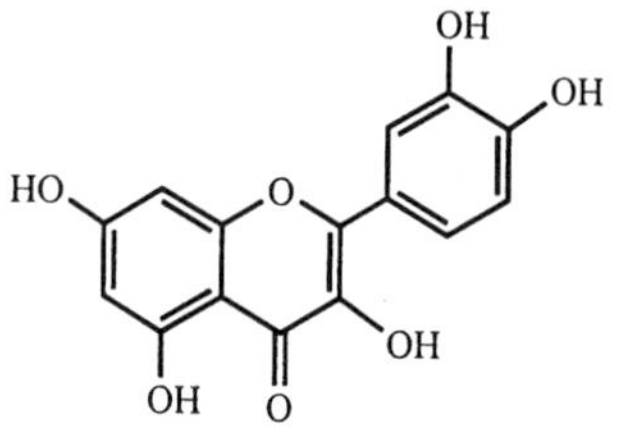

Quercetin

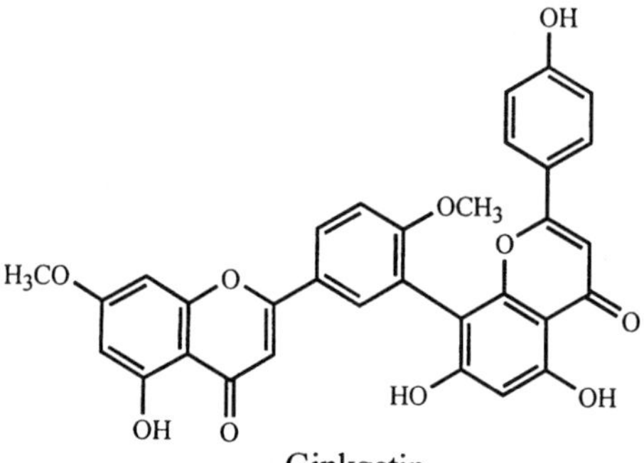

Ginkgetin

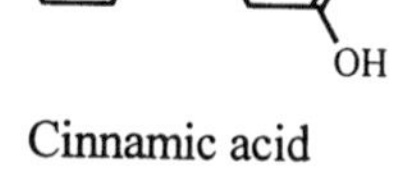

Cinnamic acid

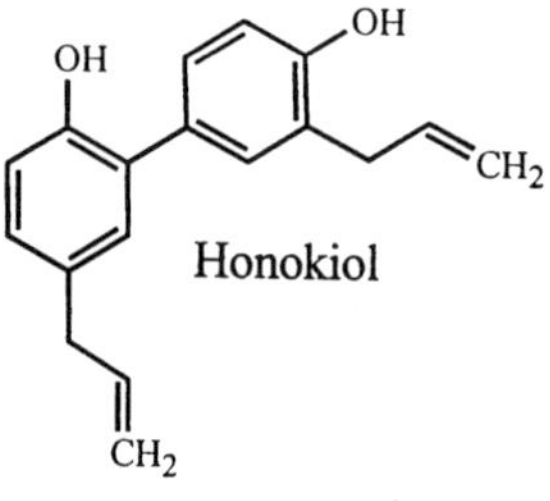

Honokiol

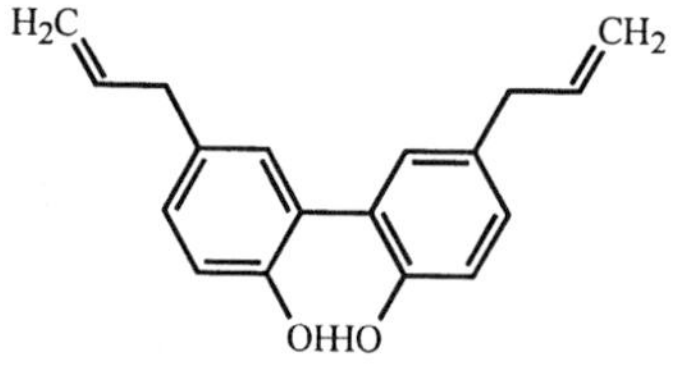

Magnolol

($\lambda = 280$ nm). Elution was complete in 30 min when a 16/3/80/1 acetonitrile/ THF/water/acetic acid mobile phase was used [1154].

Theogallin, gallic acid, epicatechin gallate, catechin, caffeine, epicatechin, epigallocatechin gallate, and epigallocatechin were extracted from tea and separated on a C_{18} column ($\lambda = 278$ nm) using a 157/40/2/1 water/dimethyl formamide/ methanol/acetic acid mobile phase [840]. Baseline resolution was achieved and elution was complete in 25 min.

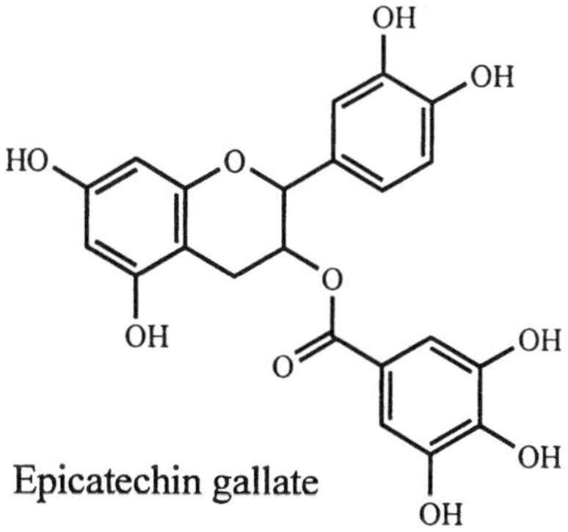

Epicatechin gallate

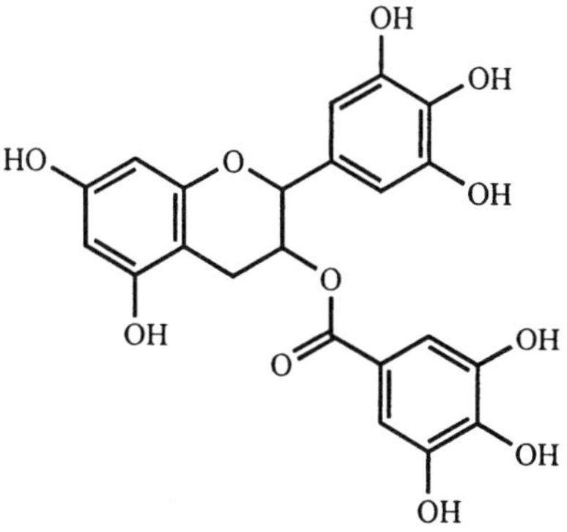

Epigallocatechin gallate

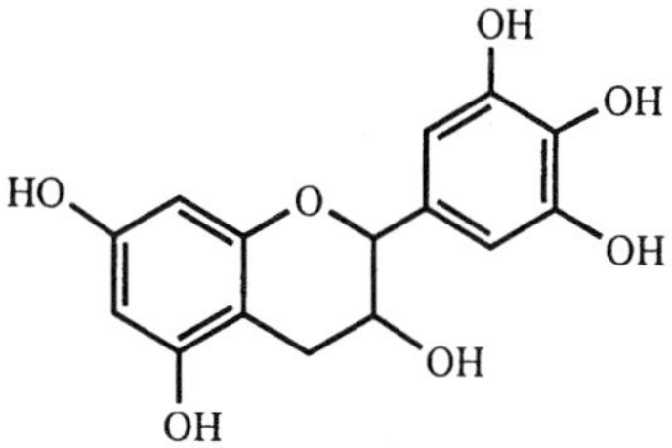

Epigallocatechin

In a superb investigative analysis of polyphenols in barley and beer, Whittle et al. [1155] identified 57 compounds (e.g., gallocatechin dimer, procyanidin pentamers, prodelphinidin trimers, procyanidin tetramers) using a C_{18} column ($\lambda = 280$ nm and positive ion electrospray MS, capillary voltage 2.45 kV, lens 0.12 V, source $T = 120°C$, cone voltage 20 V or 90 V) and a 180-min $98/2/0.1 \rightarrow 0/100/0.1$ water/acetonitrile/acetic acid gradient. In a separate experiment, an electrochemical detector (dual channel $+350$ mV and -650 mV) and a 140-min $97.5/2.50.1 \rightarrow 0/100/0.1$ water/acetonitrile/acetic acid gradient resolved common phenolics (e.g., catechin, gallic acid) and less common compounds such as fureneol, prodelphinidin B3, sinapic acid, tryptophol, and 4-vinylguiacol.

Five catechins (catechin, epicatechin gallate, epicatechin, epigallocatechin, epigallocatechin gallate) and caffeine were extracted from 35 different green, black, and instant teas [1156]. Baseline resolution and overall excellent peak shapes were achieved on a C_{18} column ($\lambda = 275$ nm) using a two-ramp $90/10 \rightarrow 70/30$ (at 10 min) $\rightarrow 20/80$ (at 15 min hold 5 min) ($94.7/4.3/1$ water/ acetonitrile/formic acid)/($49.5/49.5/1$ water/acetonitrile/formic acid) gradient. Standards ranged from 5 to 100 µg/mL with detection and quantitation limits reported as 1 µg/mL (S/N = 3) and 5 µg/mL (S/N = 10), respectively.

Five major tea catechins (e.g., epicatechin, epicatechin gallate, epigallocatechin 3-gallate) were isolated from plasma and baseline resolved on a C_{18} column (coulometric detector at 70 mV) using a complex 44-min $96/4 \rightarrow 67/33$ water (1.75% acetonitrile, 0.12% THF, and 100 mM sodium phosphate buffer at pH 3.35)/(58.5/29/12.5 acetonitrile/water/THF with 15 mM sodium phosphate buffer at "pH 3.45") gradient [1157]. The authors noted that bacterial growth in the weak solvent was found at phosphate levels below 100 mM. They also noted that acetic acid was not compatible with the electrochemical detection system setup used here. Standards from 1 to 10,000 ng/mL were used. Detection limits of 5 ng/mL (S/N = 3) were reported.

Sheu and Chen [1158] baseline resolved eight extracted constituents of tea (e.g., 10,10′:threosennoside B, emodin, wogonin, oroxylin A 7-O-glucuronide, glycyr-rhizin, baicalein) using a C_{18} column ($\lambda = 254$ nm) and a complex 50-min

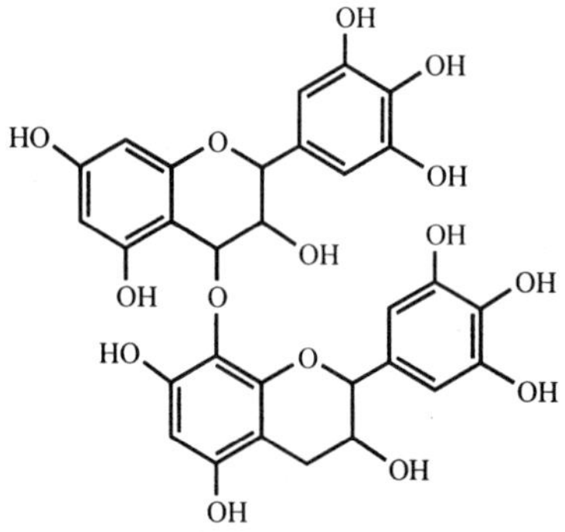

Gallocatechin dimer A

95/5 → 0/100 (8/1/1 water/acetonitrile/methanol [3.0 mM tetrabutylammonium phosphate with 7.3 mM KH_2PO_4 adjusted to pH 4.2 with H_3PO_4]/(1/2/2 water/ acetonitrile/methanol [3.0 mM tetrabutylammonium phosphate with 7.3 mM KH_2PO_4 adjusted to pH 4.2 with H_3PO_4] gradient. Plots of k' versus both KH_2PO_4 and tetrabutylammonium concentration were shown to be complicated and neither linear nor logarithmic in nature. A linear working range of 0.5– 225 µg/mL (analyte dependent) was reported for these compounds. Detection limits of 0.08–0.5 ng injected (S/N = 3) were obtained.

Two major elderberry anthocyanins (cyanidin-3-glucoside, cyanidin-3-sambubio-side) and two minor ones (cyanidin-3,5-diglucoside, cyanidin-3-sambubioside-5-glucoside) were extracted from human serum and separated on a C_{18} column ($\lambda = 512$ nm). A 10-min elution time was achieved using a 350/50/50 water/ formic acid/acetonitrile mobile phase [1159]. This level of formic acid is very aggressive and is not recommended for routine use unless a precolumn is used. Linear ranges of 0.4 ng/mL to 1 µg/mL, detection limits of 0.15 ng/mL (S/N = 3), and quantitation limits of 0.4 ng/mL (S/N = 10) were reported. Note that the level of formic acid is very high.

Berente et al. [1160] developed a method to classify German red wines through the determination of nine anthocyanins (delphinidin-, cyanidin-, petunidin-, paeo-nidin-, and malvidin-3-glucosides, paeonidin-, and malvidin-3-acetylglucosides, and paeonidin- and malvidin-3-coumarylglucosides). A 30-min 90/10 → 55/45 (5/95 acetonitrile/buffer)/(50/50 acetonitrile/buffer) gradient followed by a step gradient to 100% 50/50 acetonitrile/buffer was used. The buffer is 10 mM KH_2PO_4/H_3PO_4 at pH 1.6. The column was C_{18} ($\lambda = 518$ nm), held at 50°C. Excellent peak shapes and resolution were obtained. The linear range was estimated from malvidin-3-glucoside as 5–125 µg/mL.

In an impressive separation of anthocyanic pigments from various juices and syrups, 14 compounds from a blueberry extract (e.g., delphinidin-3-glucose, -galactose, and -arabinose and similarly substituted cyanidin, malvidin, and peonidin aglycones) were well resolved and identified using a C_{18} column ($\lambda = 525$ nm) and an 84/6/10 water/acetonitrile/formic acid mobile phase [1161]. Other juices (namely strawberry, sour cherry, black and red currant, grape, elderberry, and raspberry) were analyzed in a similar fashion. Slight adjustments in the acetonitrile content (e.g., 81/9/10 water/acetonitrile/formic acid) were used to separate other species (e.g., pelargonidin and petunidin) substituted with other sugars (e.g., rutinose, sambubiose). All separations were complete in <60 min. One standard, cyanidin-3-glucoside chloride), was prepared at the 50 mg/L level.

Nine anthocyanin pigments (cyanidin and delphinidin acylated di- and tri-glycolsylates) were extracted from various flower petals and separated on a C_{18} column ($\lambda = 508$ nm) using an 80/20 → 33/67 (at 30 min) → 30/70 (at 33 min) → 0/100 (at 40 min) water (1.5% H_3PO_4)/(20/24/1.5/54.5 acetic acid/ acetonitrile/H_3PO_4/water) gradient. If the 20% acetic acid figure is correct, this is an extremely aggressive solvent system and a precolumn is a necessity. Peak shapes were good as was overall resolution. The last peak of interest eluted in slightly less than 30 min [1162].

Anthocyanin pigment UV spectra were used to characterize and identify various plant extracts [1163]. A C_{18} column ($\lambda = 520$ nm for peak monitoring and photo-diode array detector, $\lambda = 250\text{--}610$ nm for peak characterization) was used with an 85/15 acetonitrile/(85/15 water/acetic acid) acid mobile phase to generate the elution profiles for pepper, eggplant, blueberry, and huckleberry extracts. Elution was complete in 60 min. A second elution protocol was used to separate and identify 12 anthocyanins in huckleberry juice. Here a complex 40-min $12/78 \rightarrow 25/75$ acetonitrile/water (4% H_3PO_4) gradient was used in conjunction with the C_{18} column. The set of compounds identified included delphinidin-3-rutinoside-5-gluco-side, petunidin-3-*p*-coumaroyl-rutinoside-5-glucoside, and malvidin-3-feroylrutino-side-5-glucoside.

9.5.5.3 *Alkaloids and Related Compounds*

The alkaloid composition of *Eschscholzia californica* (poppy) populations was investigated through the extracts. Chelerythrine, sanguinarine, californidine, eschscholzine, α-allocryptopine, protopine and *O*-methylcaryachine were resolved on a C_8 column ($\lambda = 200\text{--}600$ nm). A complex 35-min $76/24 \rightarrow 60/40$ water (10 mM octanesulfonic acid with 150 mM triethylamine to pH 3 with H_3PO_4)/acetonitrile gradient generated good separation and resolution and complete elution in 30 min [1164].

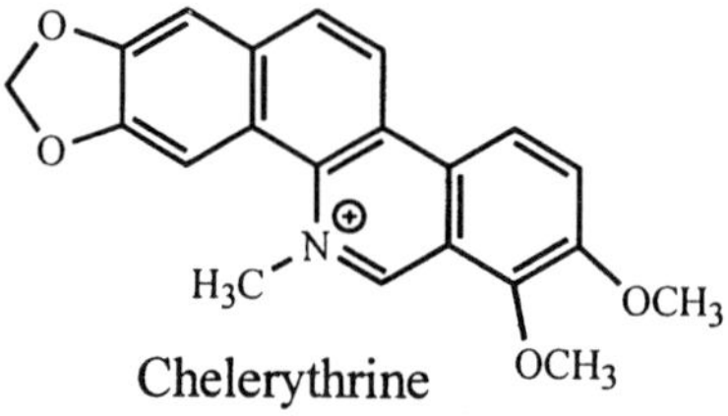

Chelerythrine

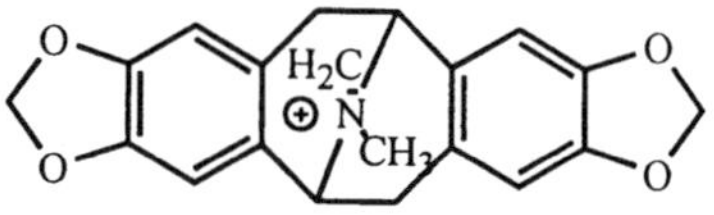

Californidine

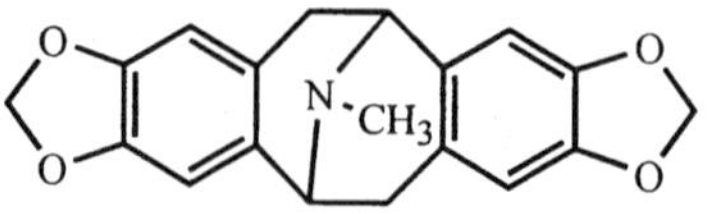

Eschscholzine

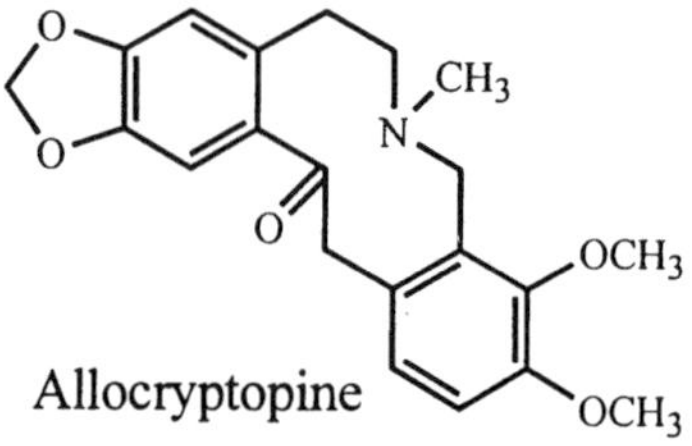

Allocryptopine

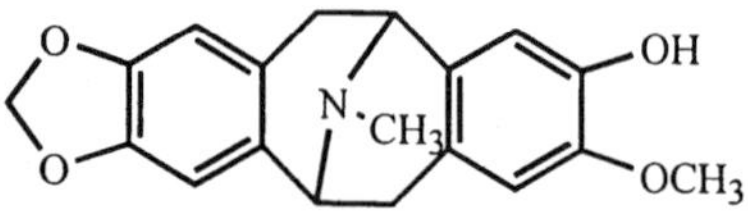

Caryachine

Fourteen isoquinoline alkaloid compounds (e.g., pavine, californidine, caryachine, aporphine, glaucine, protopine, allocryptopine, benzophenanthridine, chelirubine, sanguinarine) were extracted from *Eschscholtzia californica* and analyzed using a C_8 column ($\lambda = 280$ nm and positive ion MS/MS) using a complex 55-min $60/40 \rightarrow 0/100$ water (1 mM sodium dodecylsulfate with 10 mM triethyl amine to pH 2.5 with H_3PO_4)/acetonitrile gradient [1165]. All peaks of interest eluted between 30 and 55 min. Incomplete resolution of all compounds was obtained but closely eluting peaks were differentiated by MS/MS.

Galanthamine and three other alkaloids (*N*-formylnorgalanthamine, tazettine, haemanthamine) were extracted from the roots, leaves, and flowers of *Narcissus confusus* and analyzed on a C_{18} column ($\lambda = 280$ nm) using a 40/60 water (10 mM octanesulfonic acid with H_3PO_4 to pH 3)/acetonitrile (12 mM octanesulfonic acid) mobile phase. Elution was complete in <7 min but peaks were not fully resolved [1166]. Linear ranges from 2 to 32 µg/mL were reported.

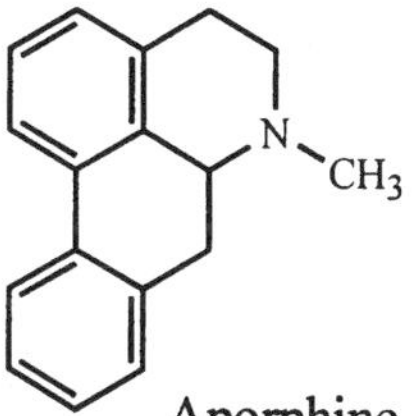

Aporphine

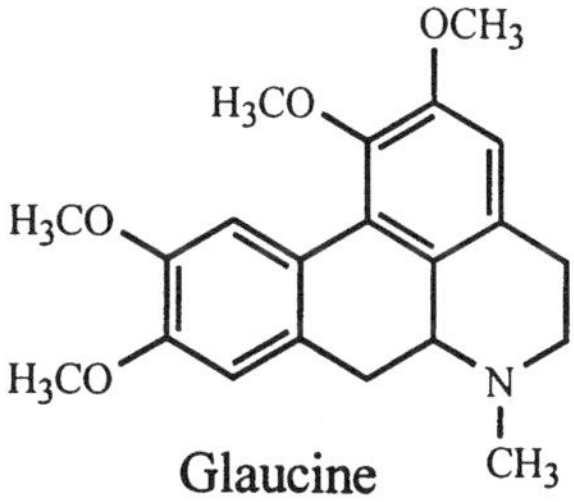

Glaucine

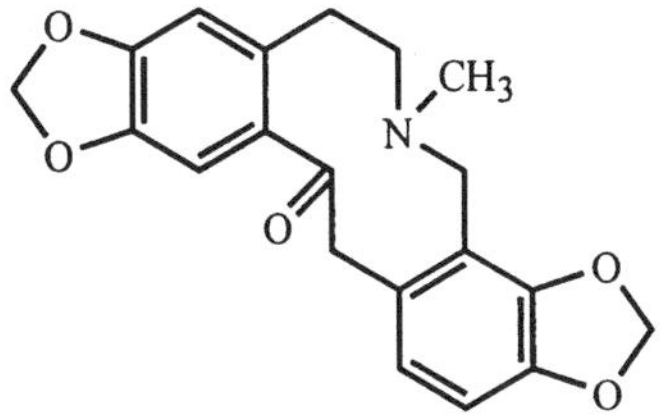

Protopine

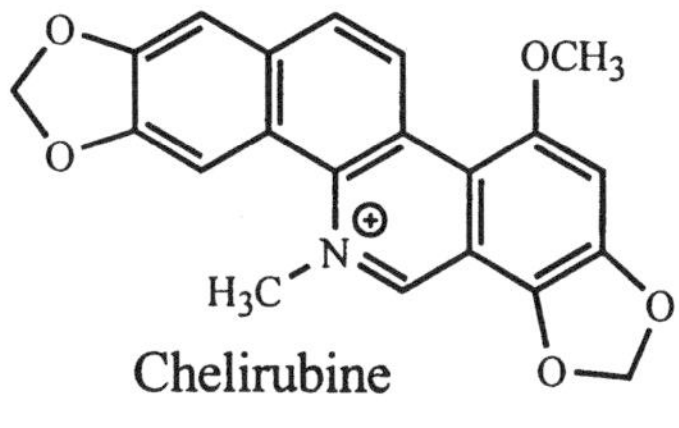

Chelirubine

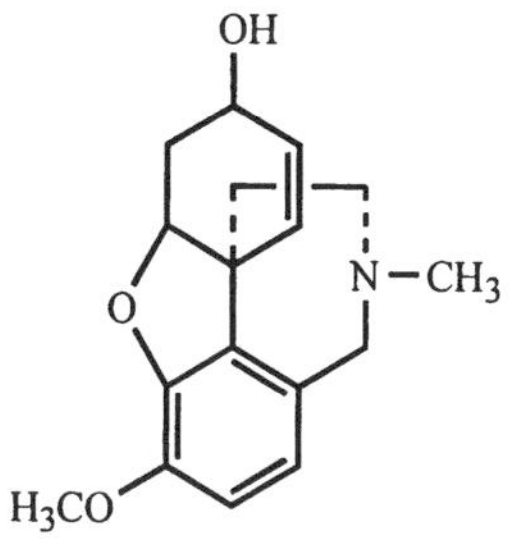

Galanthamine

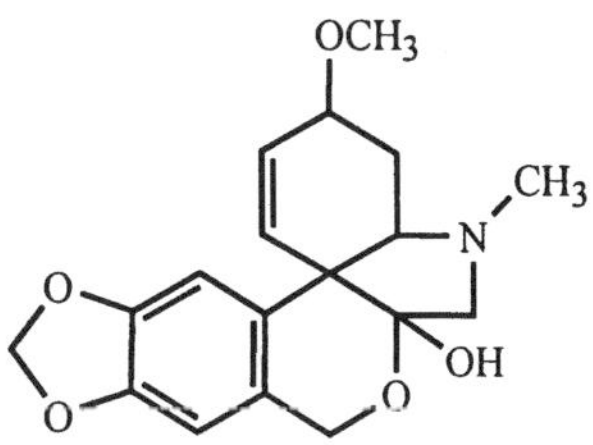

Tazettine

Six ergot alkaloids (ergotamine and an isomer and five hydroxylated metabolites) were extracted from the fungus *Claviceps* and baseline resolved on a C_{18} column ($\lambda = 250$ nm, ex; 370 nm long-pass filter). A 90/10 → 60/40 (at 15 min hold 7 min) (90/10 water [2.6 mM ammonium carbonate]/methanol)/acetonitrile gradient generated excellent peak shapes [1167]. Similarly, ergonovine, ergotamine, ergocornine, α-ergocryptine, and ergocristine were isolated from *Claviceps purpurea* and analyzed on a C_8 column ($\lambda = 333$ nm, ex; 418 nm, em). Excellent resolution and peak shapes were obtinaed using a 50/50 acetonitrile/water (25 mM H_3PO_4 at pH 3.1) mobile phase [1168]. Elution was complete in <15 min. Standards were run from 2 to 400 ng/mL with detection and quantitation limits of 0.5–3 ng/g and 3–18 ng/g (analyte dependent), respectively, reported.

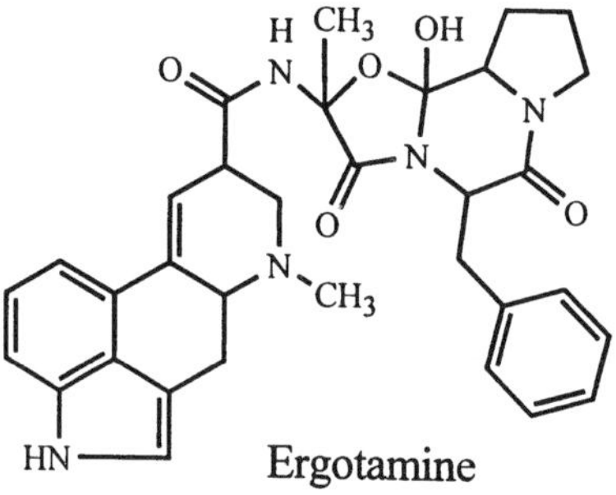

Ergotamine

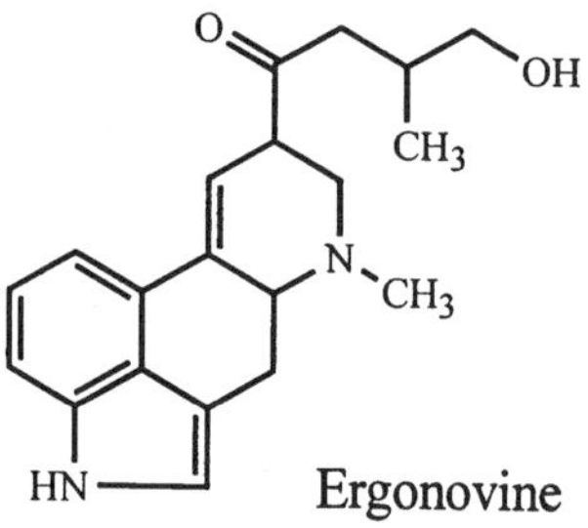

Ergonovine

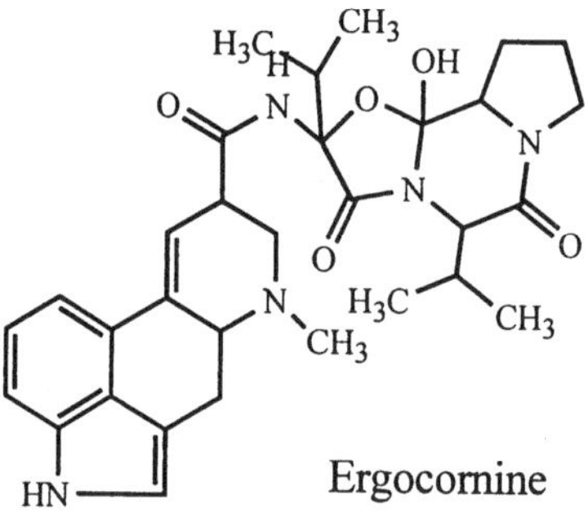

Ergocornine

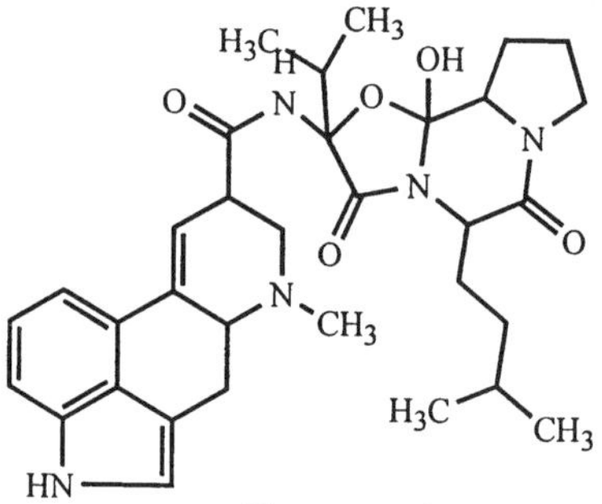

α-Ergocryptine

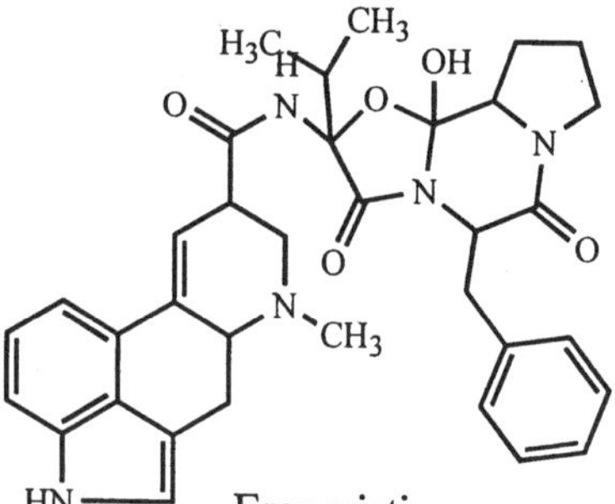

Ergocristine

Six pyrrolizidine alkaloids (jacobine, jacozine, jacoline, jaconine, senecionine, seneciphylline) were extracted from honey and separated on a 65°C PRLP-S column (APCI positive ion MS) using a 30-min 90/10 → 10/90 water (0.1 M ammonium hydroxide)/acetonitrile gradient [1169]. Standards ranged from 0.005 to 10 µg/mL and detection limits of 0.002 mg/kg were reported.

Sun et al. [1170] extracted the bisbenzylisoquinoline alkaloids aromoline, homoaromoline, berbamine, obamegine, colorflammine, thalrugosine, norbaberine,

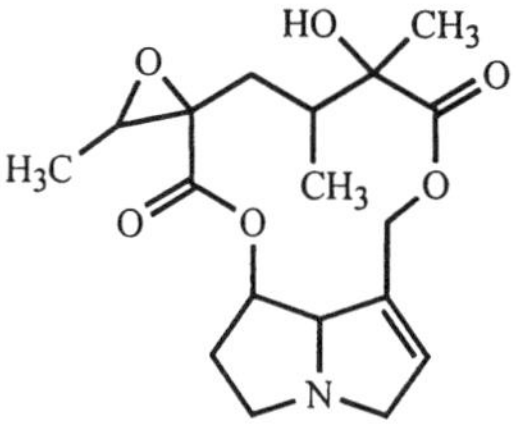

Jacobine

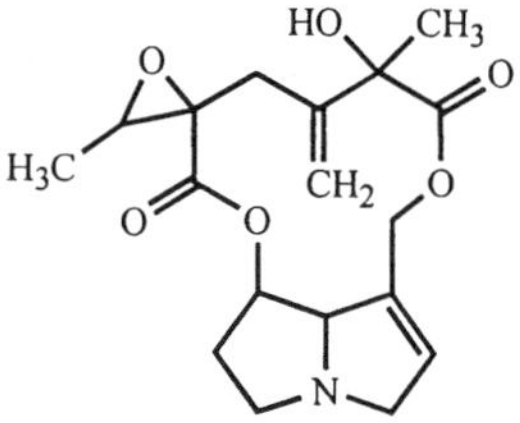

Jacozine

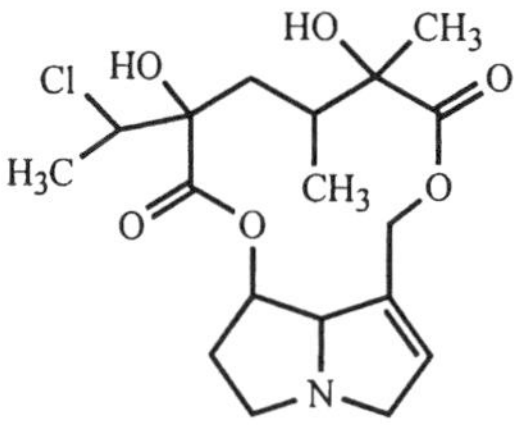

Jaconine

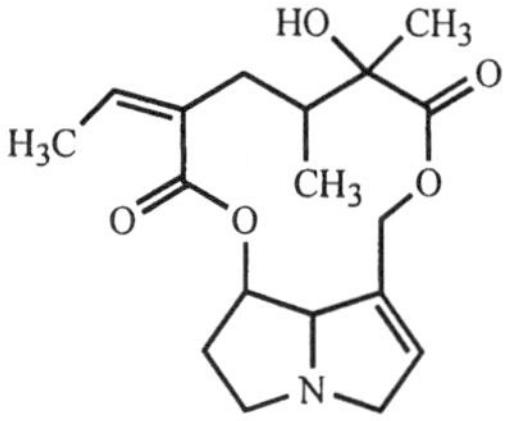

Senecionine

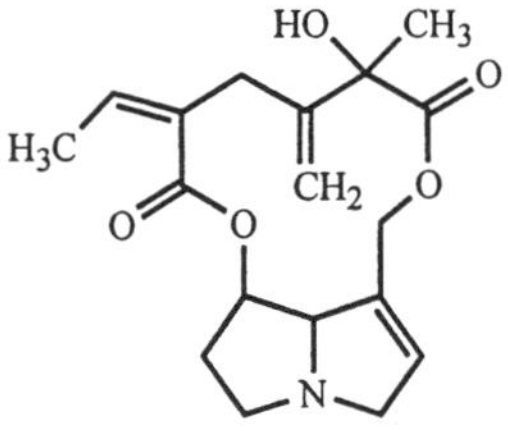

Seneciphylline

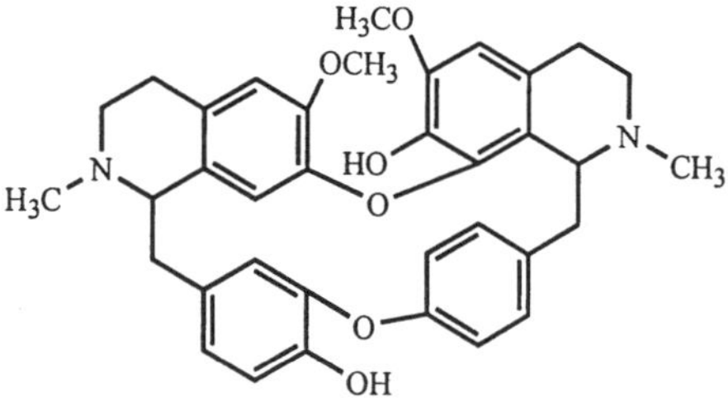

Aromoline

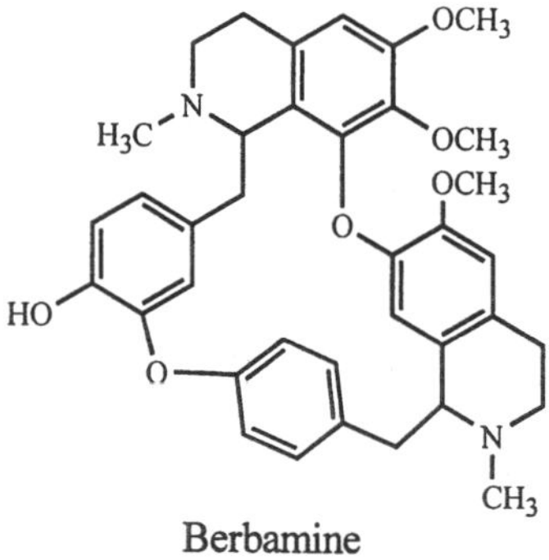

Berbamine

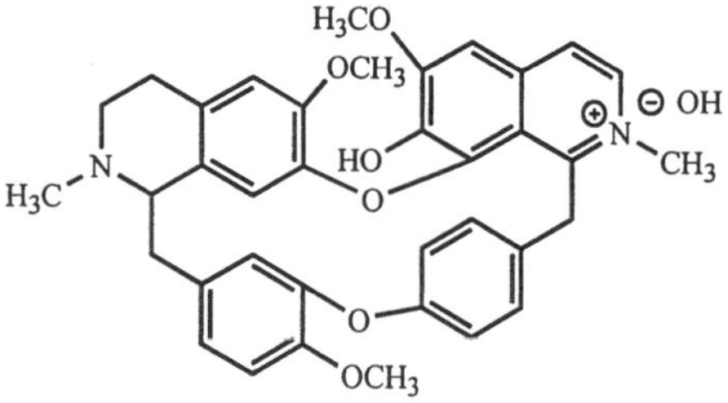

Colorflammine

tetrandrine, and isotetrandrine from the stem wood of *Dehaasia triandra* and separated them on a C_{18} column ($\lambda = 207$ nm) using a 73/27 water (50 mM NaH_2PO_4 at pH 3.0 with 0.1% diethylamine and 2 mM sodium heptanesulfonate)/acetonitrile mobile phase. Peaks were well resolved and elution was complete in 16 min. A plot of retention time versus heptansulfonate concentration was presented. A linear range of 10–200 µg/mL was used and detection limits of 200–1600 pg injected (S/N = 3, analyte dependent) were reported.

Five naphthylisoquinoline alkaloids (dimers michellamine A, B, and C, and momomers korupensamine A and B) were separated on a C_{18} column (positive ion electrospray MS, collision energy 25 eV; collision pressure 2.5 mTorr) using a complex 19-min 85/15 → 60/40 water (0.1% TFA)/acetonitrile gradient [1171]. Test solution concentrations were 20 ng/mL.

Eight isoquinoline alkaloids (e.g., chelidonine, berberine, coptisine, dihydrosanguinarine) were extracted from *Chelidonium majus* and resolved on a C_{18} column ($\lambda = 290$ nm) using a 46/10/44 acetonitrile/methanol/water (0.05 M tartaric acid with 0.5% sodium dodecyl sulfate) mobile phase [1172]. Peaks were slightly tailed. Elution was complete in 40 min. Standards of 0.01–0.28 mg/mL were used and 1 µL aliquots were injected. Capacity factors for all compounds were presented and plotted versus percent acetonitrile (43–47%), over which range the average capacity factor deceased by ~40%.

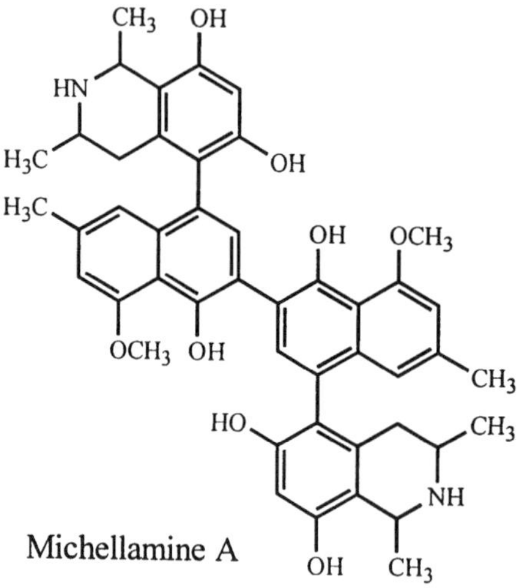

Michellamine A

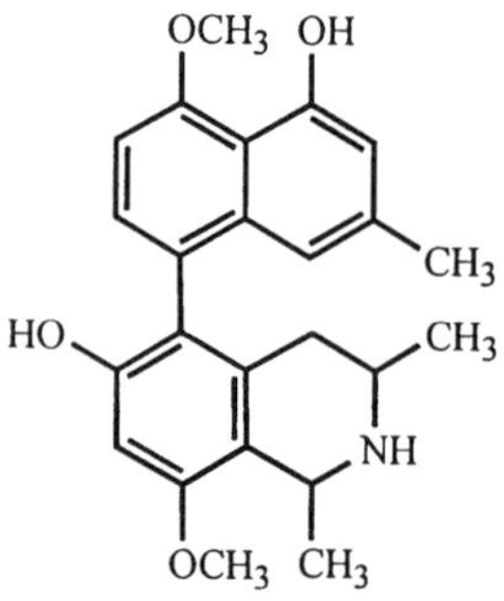

Korupensamine E

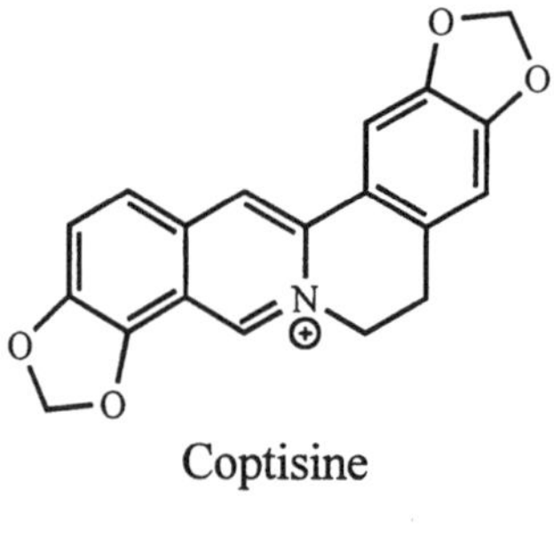

Coptisine

Verotta et al. [1173] studied a series of nine alkaloids (e.g., calycanthine, *meso*-chimonanthine, hodgkinsine, quadrigemine B and C, psychotridine) by extraction from different parts of the *Psychotria colorata* plant and separated on a C_{18} column ($\lambda = 254$ nm). An interesting two-ramp $72/28/0.1 \rightarrow 85/15/0.1$ (at 25 min) $\rightarrow 72/28/0.1$ (at 40 min) methanol/water/diethylamine gradient was used. In general good resolution was achieved.

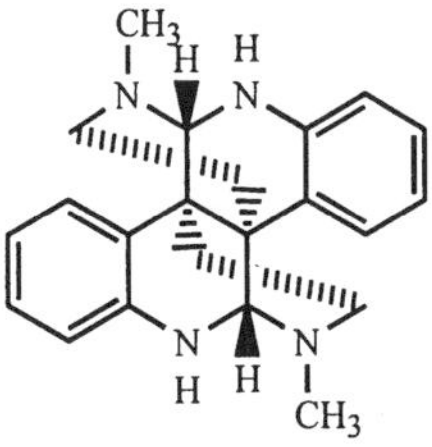

Calycanthine

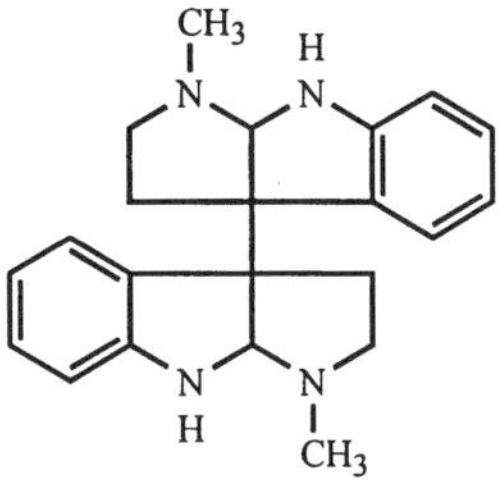

meso-Chimonanthine

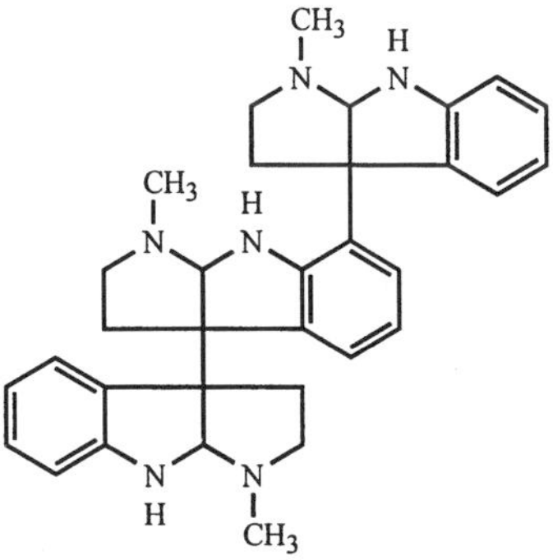

Hodgkinsine

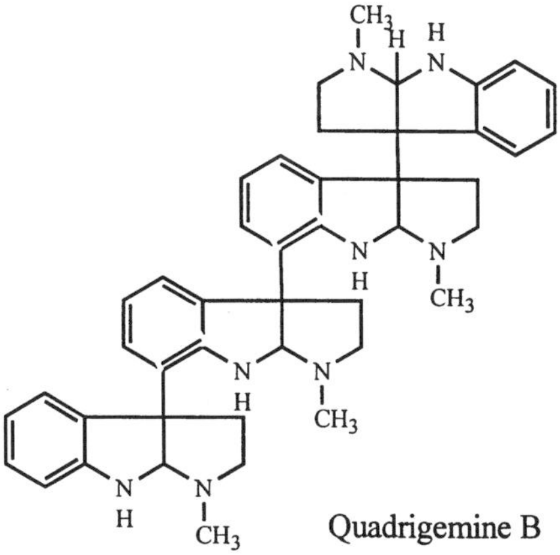

Quadrigemine B

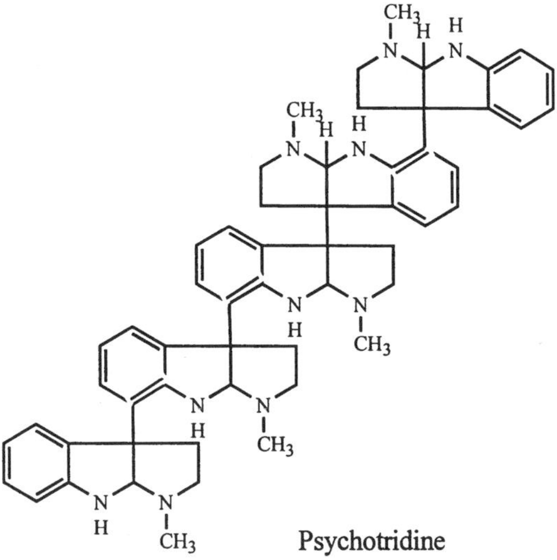

Psychotridine

A series of genus *Colchicum* alkaloids (colchicoside, thiocolchicoside, *N*-deacetyl- and *O*-demethylthiocolchicoside, colchicine, β-lumicolchicine, thiocolchicine, *N*-deacetylthiocolchicine) were isolated and analyzed on a C_{18} column ($\lambda = 380$ nm). A complex 25-min 93.5/0.5/6 $\rightarrow$ 12/82/6 water (5 g KH_2PO_4/L to pH 4.5 with H_3PO_4)/acetonitrile/THF gradient was used. Note that the buffer capacity is poor at this pH. A linear range of 90–900 µg/mL was reported [1174].

Ten β-carboline alkaloids (norharman, harman, harmol, harmine, 3-hydroxy-methylnorharman, harmalol, harmaline, tetrahydronorharman, tetrahydroharman,

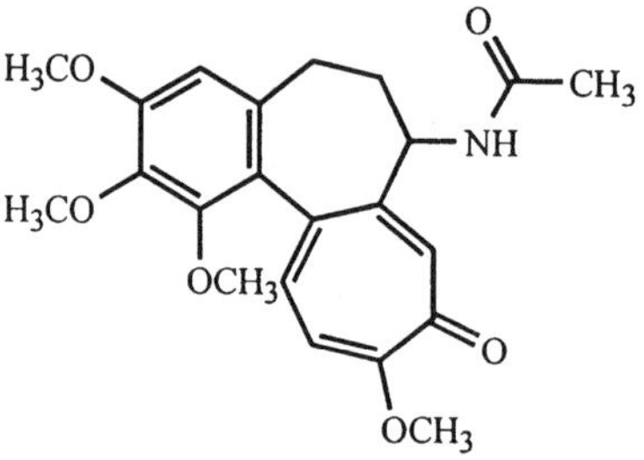

Colchicine

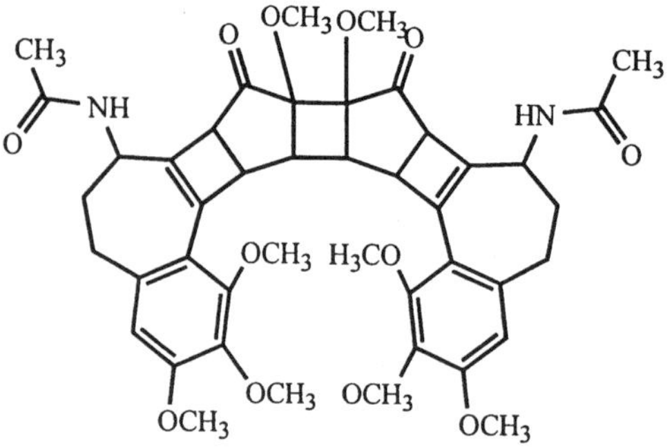

α-Lumicolchicine

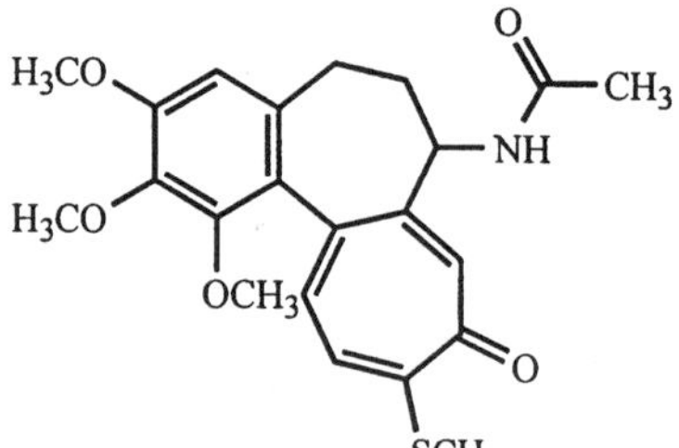

Thiocolchicine

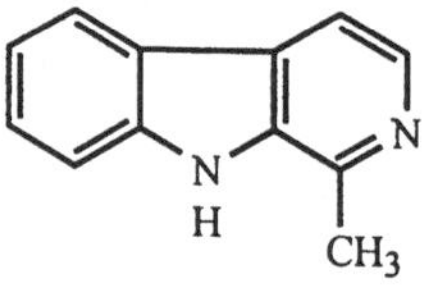

Harman

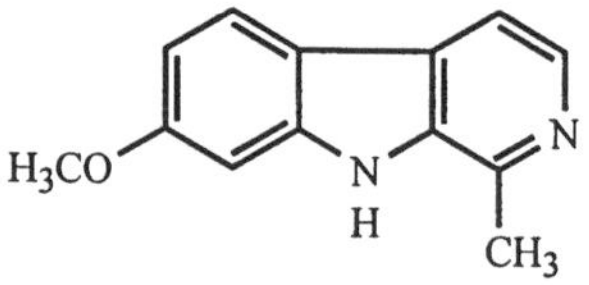

Harmine

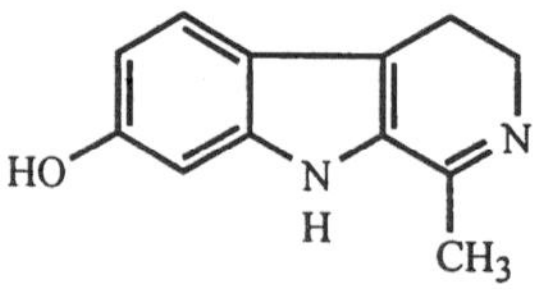

Harmalol

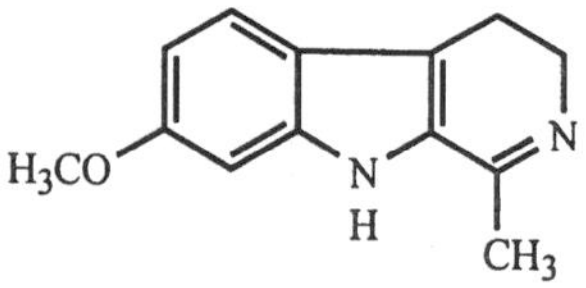

Harmaline

2-ethyltetrahydronorharman) were extracted from a variety of plants. An 81.8/18/0.2 water/acetonitrile/TFA mobile phase used with a 50°C C$_8$ column produced well-resolved peaks and complete elution in 19 min [1175]. Dual fluorescence detectors, one set at $\lambda = 275$ nm, ex; 350 nm, em; the other programmable at 325/417 nm (0–7.5 min), 300/447 nm (7.5–10.5 min), 300/430 nm (10.5–14.5 min), and 330/417 nm (14.5 min to end) were used. Standards from 0.01 to 50 ng/mL and detection limits of 2–20 pg injected (analyte dependent) were reported.

Eight species of *Fumaria* were characterized by Soušek et al. [176] as to their alkaloid composition. Sixteen compounds were extracted (e.g., fumaritine, fumariline, parfumine, adlumidiceine, fumarophycine, sinactine, protopine, coptisine, palmitine), separated, and identified using a C$_{18}$ column ($\lambda = 280$ nm) and a complex 30-min 100/0 → 60/40 water (10.1 mL triethylamine/L to pH 2.5 with H$_3$PO$_4$)/acetonitrile gradient [1176]. Peaks were slightly tailed but overall good resolution was achieved. Standards were prepared at the 10–30 µg/mL range.

Clivorine and four metabolites (e.g., clivoric acid, dehydroretronecine) were isolated from incubates and separated using a PRP-1 guard column to collect the compounds and then elution through a PRP-1 analytical column ($\lambda = 230$ nm). The load period was 5 min using an aqueous 0.2% formic acid to pH 3.4 with ammonia mobile phase. Once the switch occurs, a two-ramp 100/0 → 75/25 (at 30 min) → 70/30 (at 35 min) water (0.2% formic acid to pH 3.4 with ammonia)/ acetonitrile gradient was run. Peaks were baseline resolved. Linear ranges from 5 to 120 µg/mL and detection limits of 0.2–10 µg/mL (analyte dependent) were reported [1177].

9.5.5.4 Fats, Oils, and Related Analytes

Dietary fats and oils were analyzed as their bromophenacyl derivatives on a C$_8$ column (UV, not listed but probably λ between 254 nm and 280 nm). The C$_8$ was preferentially chosen over a C$_{18}$ due to the complete resolution and significantly shorter analysis time [1178]. Various oils gave elution profiles that included the following fatty acids: lauric, linolenic, linoleic, palmitic, oleic, stearic, erucic, and arachidic. A 50-min gradient from 70/30 → 95/5 acetonitrile/water was used. A linear range of 0.2–18 µg injected and a limit of detection of about 0.05 µg injected (S/N = 3) were reported.

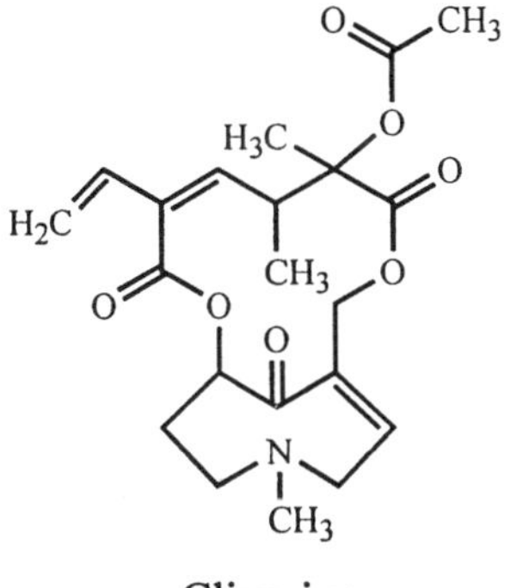

Clivorine

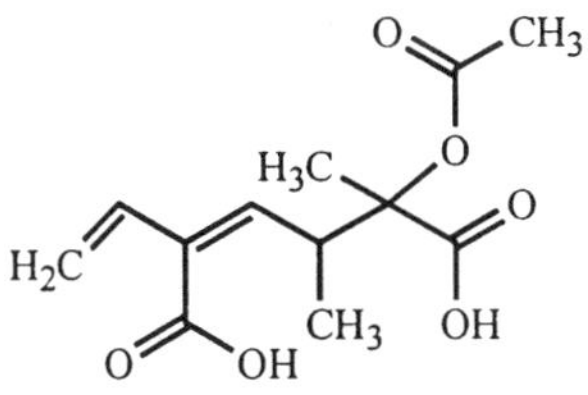

Clivoric acid

Luf et al. [1179] extracted a set of seven cholesterol oxides from processed food (7α-, 7β-, and 25-hydroxycholesterol, $5,6\alpha$- and $5,6\beta$-epoxycholesterol, 7-ketocholesterol, cholestan-$3\beta,5\alpha,6\beta$-triol) and analyzed them by positive ion APCI-MS. UV detection at $\lambda = 206$ nm was not feasible due to the very low concentrations present in the sample. Separation was achieved in 18 min on a C_{18} column using a 60/40 acetonitrile/methanol mobile phase. Detailed studies for the optimization of MS detection included vaporizer temperature (450°C), cone voltage (20 V), and source temperature (200°C). Carrier and nebulizing gas was N_2. Carrier gas flow was set at 250 L/h and the sheath gas at 50 L/h. Linear ranges of 1–500 ng injected and limits of detection at 0.5 ng injected were reported.

Cholesterol and nine 10-cholesteryl esters (e.g., linolenate, laurate, myristate, stearate and arachidate) extracted from blood were separated on a C_{18} column ($\lambda = 206$ nm) using a 60/40 acetonitrile/IPA mobile phase [1180]. Peaks were somewhat broad but overall resolution was good. The addition of more acetonitrile sharpened the peaks, but resolution suffered when the acetonitrile level went above 60%. The authors note that UV detection became impossible at $\lambda < 205$ nm (due to IPA background absorbance) and that special testing of the IPA was necessary prior to use to ensure that the background absorbance would be acceptable. A linear working range from 0.3 to 50 μg injected (in 20 μL) was reported.

The kinetics of lipoxygenase were followed by the relative change in amount of linoleic acid, 1-palmitoyl-2-linoleoyl-*sn*-glycerol-3-phosphoethanolamine, 1-linoleoyl-2-stearoyl-*sn*-glycero-3-phosphocholine, and trilinolein in the sample [1181]. These compounds were baseline resolved on a C_{18} column ($\lambda = 205$ nm) using a $0/86/14 \rightarrow 51/48/1$ (at 13 min hold 8 min) acetonitrile/methanol/water gradient. Elution of the most retained peak, trilinolein, was at 21 min. As an alternative

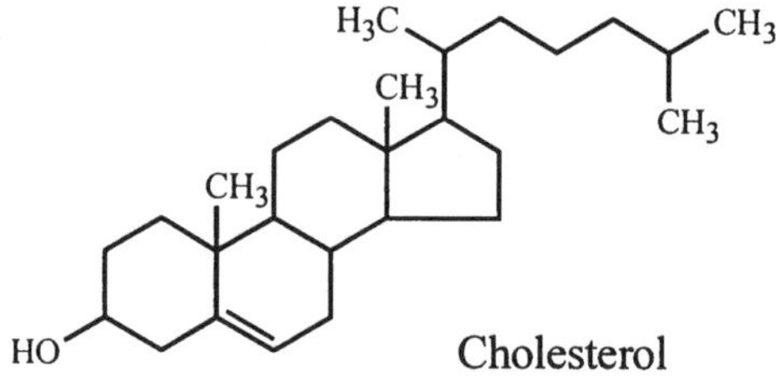

Cholesterol

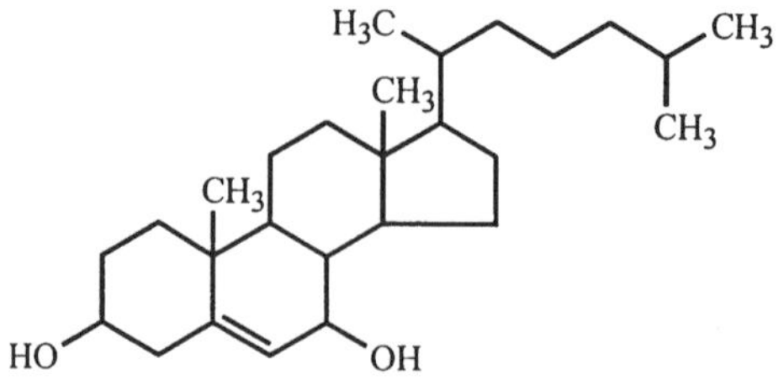

7-Hydroxycholesterol

detection scheme, the authors note that ammonium acetate is compatible with an ELSD whereas choline chloride produced significant baseline noise.

The hydroperoxy oxidation products of trilinoleoyl- dilinoleoyl-oleoyl-, and linoleoyl-dioleoyl glycerols were isolated from sunflower oil and analyzed on a C_{18} column ($\lambda = 246$ nm or 210 nm or ELSD, $T = 130°C$, N_2 pressure 70 mm). Other hydroperoxy product separations were also studied (e.g., palmitoyl, and stearoyl). A complex 60-min $95/5 \rightarrow 60/40$ acetonitrile/methyl t-butyl ether gradient was used [1182]. Peroxide values were reported over the range 0.5–50 meq/kg.

The triglyceride profile of cocoa butter was determined on a series (two of 250×4.6 mm) 30°C C_{18} columns (ELSD, drift tube 100°C, N_2 set at 40 mm flow). A 60-min $80/20 \rightarrow 46/53$ acetonitrile/dichloromethane gradient was used. Seventeen triglycerides were identified (e.g., ranging from 1-palmitoyl-2-lauroyl-3-oleoyl-glycerol to 1,3-distearoyl-2-lauoyl-glycerol to tristearoylglycerol). Samples were 5 mg/mL (10 µL injected). Overall good peak shape and resolution was achieved. The authors noted that for the concentration range of 10–200 µg injected, a strongly nonlinear blot of response versus concentration was obtained [1183].

The retention behavior of phosphatidic acid methyl esters (C_{16}–C_{20}) was studied on a C_{18} column ($\lambda = 208$ nm) using a $70/22/8$ acetonitrile/methanol/water (5 mM tetraalkyltriethylammonium phosphate) mobile phase [1184]. A number of quaternary alkyltriethylammonium phosphates (pentyl, hexyl, heptyl, octyl, and dodecyl) were tested as the mobile phase modifier. The pentyl gave the best separation. Capacity factors were tabulated for all analytes in all mobile phases. Peaks were not well resolved, indicating that a gradient might have given better results.

Phospholipids were extracted from erythrocyte membranes and were analyzed along with their lipoxygenase peroxide products using a 40°C aminopropyl column. Phospholipids (phosphatidylcholine [PC], phosphatidylethanolamine [PE], phosphatidylinositol, phosphatidylserine [PS], and sphingomyelin) were separated using a $67/22/11$ acetonitrile/methanol/water (0.2% TFA) mobile phase ($\lambda = 210$ nm). Elution was complete in 30 min and excellent resolution was achieved. The peroxide forms of PC, PE, and PS were separated at the same time but monitored at $\lambda = 234$ nm. All samples were stabilized with 20 mg/L BHT solutions [1185].

Ramesha et al. [1186], described the subclass separation of diacyl, alkylacyl, and alkenylacylethanolamine phospholipids from brain tissues using the 1-anthroyl derivatives. A C_{18} column ($\lambda = 254$ nm) and a $70/30$ acetonitrile/IPA mobile phase were used to generate the separation. Twenty-one peaks were generated and characterized as to their molecular species identity. The authors reported a detection limit of 0.5 pmol injected (S/N = 5) for individual components and 2 pmol injected for mixtures. Linear concentration ranges up to 200 pmol injected were obtained.

A number of triacylglyceride separations have been presented earlier using C_{18} columns and acetone/acetonitrile [862, 863, 868], propionitrile/MtBE [837] and acetonitrile/chloroform [1187] mobile phases. Other methods using acetonitrile have been developed for TAGs and substituted TAGs.

Hydroxylated triacylglycerols from castor oil and beef tallow were separated according to their equivalent carbon number (ECN) using a 35°C C_{18} column (RI detector) and a 74/26 acetonitrile/THF mobile phase [1188]. A 20 μL aliquot of a 10–20% triacylglycerol solution was injected. The ECN range went from 34 (retention time 3 min for ricinoleoyl/ricinoleoyl/ricinoleoyl (R/R/R)] to 48 (retention time 45 min for stearoyl/stearoyl/stearoyl [SSS]) with palmitoyl, linoleoyl, hydroxystearoyl, and juniperoyl substituents represented as well.

Kuksis et al. [1189], studied the separation of short-chain triacylglycerols (TAGs) in butter oil and long-chain TAGs in menhaden oil using a C_{18} column (negative chemical ionization direct inlet MS) and a 40-min $10/90 \rightarrow 90/10$ acetonitrile/propionitrile (1% dichloromethane) gradient. The dichloromethane is added to produce a chloride attachment source that allowed for the unambiguous identification of the TAGs. The same group identified, quantitated (as a percent of the total sample), and tabulated the results for over 150 component peaks in the butter oil sample [1190]. An RP HPLC method using the C_{18} column (ELSD) and a 40- or 90-min $10/90 \rightarrow 90/10$ acetonitrile/IPA gradient was also presented [1191]. These studies give valuable reference information for triacylglyceride separation and identification.

9.5.6   Nucleotides, Nucleosides, and Related Analytes

Adenosine, inosine, hypoxanthine, xanthine, adenine, guanosine, and β-NAD$^+$ were extracted from placental tissue and baseline resolved on a C_{18} column ($\lambda = 254$ nm) using a complex 12-min $0/100 \rightarrow 20/80$ acetonitrile/water (7.5 mM ammonium phosphate buffer at pH 6.0) gradient. ATP, ADP, AMP, and NADP were also resolved using this method [1192]. The separation of guanosine from hypoxanthine was incomplete in this mobile phase. Standards containing 50–200 pmol/20 μL of each analyte generated easily detected peaks and good peak shapes. Detection limits of low nmoles analyte per gram tissue were reported.

The retention behavior of the 2′-deoxyribonucleosides (adenosine, cytidine, uridine, thymidine, guanosine) were studied on a C_{18} column ($\lambda = 254$ nm) with a series of mobile phases ranging from 100% water to 12.5/87.5 acetonitrile/water [1193]. Capacity factors for each compound were plotted and ranged from 5 to 90 in 100% water to 0.05 to 0.7 in 12.5/87.5 acetonitrile/water.

A series of peptide-oligonucleotide conjugates (POCs) were prepared and analyzed on a PRP-1 column ($\lambda = 258$ nm). A 60-min $100/0 \rightarrow 40/60$ water (0.1 M triethylammonium acetate at pH 9.7)/acetonitrile gradient was used. Peaks such as unreacted DNA, FMOC-(δOrn)$_{12}$-Cys(NH$_2$)-DNA, and (δOrn)$_3$-Cys(NH$_2$)$_2$-DNA were separated. Eight products were produced and identified [1194].

Ally and Park [1195] analyzed creatine, phosphocreatine, four purine bases (e.g., xanthine, adenosine), and seven nucleotides (e.g., IMP, GTP, ATP) from heart tissue extracts using a C_{18} column ($\lambda = 210$ nm for 5 min to monitor creatine and phosphocreatine then switch to 254 nm) and a 22-min $0/100 \rightarrow 50/50$ acetonitrile/water (35 mM KH$_2$PO$_4$ with 6 mM tetrabutylammonium sulfate and 125 mM EDTA buffer at pH 6.0) gradient. The authors noted that due to an unacceptably large baseline shift during the gradient, the 210 nm wavelength was not acceptable for use throughout the gradient, hence the switch to 254 nm. This could very well be related to the use of a sulfate salt in the mobile phase. A tetrabutylammonium phosphate salt might provide a lower background. The authors also note that NAD and NADP can be separated using this method as well (monitor at a $\lambda = 340$ nm). Reported tissue levels ranged from 0.1 to 9 $\mu mol/g$ tissue.

A series of diadenosine polyphosphates (diadenosine diphosphate to diadenosine hexaphosphate) were extracted from platelets and baseline resolved on a 22°C boronate affinity column ($\lambda = 254$ nm). A complex 15-min $100/0 \rightarrow 50/50$ water (10 mM K_2HPO_4 with 2 mM tetrabutylammonium hydrogenphosphate buffer at pH 6.8)/(20/80 water/acetonitrile) gradient [1196] was used. Standards containing 100 ng of each component generated easily detected peaks.

2,3′-Dideoxyadenosine (ddA) and 2,3′-dideoxyinosine (ddI) in plasma samples were resolved from uric acid and hypoxanthine on a C_{18} column ($\lambda = 254$ nm) using a $0/100 \rightarrow 10/90$ (at 10 min hold 10 min) N,N-dimethylformamide/water (20 mM [NH_4]H_2PO_4 with 5 mM tetrabutylammonium phosphate at pH 6.8) gradient [1197]. Elution of samples extracted from plasma was complete in 20 min (the internal standard, N-methyl-2′-deoxyadenosine, eluted last). A 500 ng injection was easily detected. Detection limits of 10 ng/injection or 0.2 $\mu g/mL$ in plasma were reported. Separation of only ddA from ddI was achieved on the same column using an isocratic 7.5% N,N-dimethylformamide in buffer (10 mM [NH_4]H_2PO_4 buffer at pH 6.8) mobile phase.

The metallointercalator–DNA conjugates ([Ru^{3+}][9.10-phenanthrone quinone diimine]$_2$[4-butyric acid-4′-methyl bipyridyl]-DNA, where DNA is a 20-mer oligonucleotide) attached to the 5′ or 3′ terminus, were synthesized and separated on a C_{18} column ($\lambda = 260$ nm or 390 nm). A 50/50 water (50 mM ammonium acetate buffer at pH 6)/acetonitrile mobile phase generated baseline resolution and elution in 35 min [1198].

Kanduc et al. [1199] fractionated liver tRNA into over 60 peaks using a 37°C C_{18} column ($\lambda = 254$ nm) and a 200-min acetonitrile/water (50 mM ammonium acetate at pH 6.6) gradient. The composition was changed at the rate of 1% every 10 min. The areas for each individual peak ranged from 0.001 to 5% of the total peak area.

The reaction products of α-acetoxy-N-nitrosopyrrolidone and α-acetoxy-N-nitrosopiperidine with DNA were characterized on a C_{18} column ($\lambda = 254$ nm) using a 30-min $0/100 \rightarrow 30/70$ acetonitrile/water (10 mM phosphate buffer at pH 7.0)

Dideoxyadenosine

gradient [1200]. The products, N^2-(tetrahydofuran-2-yl)deoxyguanosine and N^2-(3,4,5,6-tetrahydro-2H-pyran-2-yl)deoxyguanosine, were well resolved from DNA fragments.

9.5.7 Sugars and Related Analytes

Underivatized sugars and other polyhydric alcohols have routinely been separated on specialized columns with a mobile phase of modified water (often modified with specific cationic additives such as Ca^{2+}, Pb^{2+} or Ag^+) at elevated temperatures [70–90°C) or on an aminopropyl column using acetonitrile/water mobile phases [1201]. Although solvent systems have been used that replace the acetonitrile, for example, acetone/water and acetone/ethyl acetate/water [1202], it should be noted that the acetone will chemically react and permanently modify the surface of an aminopropyl column. As a result, column lifetime is shortened and retention characteristics will tend to vary slightly but continuously over time.

Sugars are solubility-limited in acetonitrile and so samples are often made up in pure water. This has two important chromatographic ramifications: (1) there will be a peak associated with water in the refractive index detector-based chromatogram, and (2) the retention of sugars increases as acetonitrile levels in the mobile phase increase. This is due to the limited solubility of sugars in acetonitrile. Therefore, even though acetonitrile is considered a strong solvent, solubility dominates the retention process at elevated acetonitrile levels.

Because sugars have no chromophores with significant absorbances above 200 nm, and since the refractive index detector is a universal detector, it is often the detector of choice for underivatized sugars. Typical sample working concentrations for sugars are 0.05% or higher. In order to simplify the chromatography, some sort of sample pretreatment or cleanup is also usually necessary to remove as many contaminants as possible.

Eight sugars (e.g., xylose, glucose, mannose, galactose, sorbitol, mannitol, all D form) found in candy, chewing gum, and juice were derivatized with p-nitrobenzoyl chloride and analyzed on a C_{18} column ($\lambda = 260$ nm) using a 65/53 acetonitrile/water mobile phase [1203]. Peak shapes and resolution were excellent. Elution was complete in 28 min. The effect of percent acetonitrile in the mobile phase on retention was plotted. A linear range of 0.01–100 μg/mL (approximately 100 times more sensitive than RI detection) was reported.

The retention behavior of fructose, glucose, sucrose, maltose, and lactose on various aminopropyl columns (RI detector) was studied using an 80/20 acetonitrile/water mobile phase. Elution for all compounds was complete in 15–20 min with all peaks well resolved. This method, or one very similar to it, is used for the sugar analyses of a wide variety of food products. For example, a food sample that contains elevated salt levels poses a problem for sugar analyses since the chloride ion can co-elute with fructose under normal conditions [1204]. In this work, the optimal mobile phase composition of 70/30 acetonitrile/water (on an aminopropyl column) had to be modified to 75/25 in order to shift the retention of the first-eluting sugar (fructose) past the elution time of chloride (here 4 min or $k' \sim 1$). Standards of ~0.25% w/w were made and 25 μL was injected.

Nojiri et al. [1205] analyzed a series of six sugar alcohols (*meso*-erythritol, D-xylitol, D-glucitol, D-mannitol, maltitol, parachinit) and sucralose, sucrose, and D-mannose (glucose co-eluted) that are found in typical confectioneries as the nitrobenzylated derivatives. A phenyl column ($\lambda = 260$ nm) and a 67/33 acetonitrile/water mobile phase eluted all peaks in 50 min. The authors noted that C_{18}, C_8, aminopropyl, and cyanopropyl columns did not give acceptable results. The effect of changing percent acetonitrile on retention time is plotted. A linear range of 10–250 µg/mL was reported.

A number of saccharides (lactose, maltose, 2′-fucosyllactose, glucose, mannose) and polysaccharides (maltose oligomers 3-mer to 6-mer; glucose tetramer [Gc_4]) were isolated from urine, derivatized with benzoic anhydride/butyl-4-aminobenzoate and separated on a C_{18} column ($\lambda = 304$ nm). Gc_4 is a putative marker for Pompe disease and was eluted in 26 min using a 30-min 92/8 → 86/14 (20/80 acetonitrile/water [10 mM tetrabutylammonium chloride])/acetonitrile gradient. A linear curve from 0.5 to 15 µg/mL was reported. Detection limits of 4.5 pmol per injection were claimed [1206].

A series of six unsaturated disaccharides derived from hyaluronic acid and chondroitin sulfate were derivatized with dansylhydrazine and analyzed on an aminopropyl column ($\lambda = 350$ nm, ex; 530 nm, em). A 90/20 acetonitrile/water (0.1 M acetate buffer at pH 5.6) generated good peak shape and elution in 50 min [1207]. Linear curves from 50 to 500 pmol injected were reported. Samples of chondroitinase ABC digests of trachea and shark cartilage were presented. Detection limits of <50 pmol (20 ng) were cited.

At least 23 different oligosaccharide alditols were identified in egg-jelly mucin from *Xenopus laevis* using an aminopropyl column ($\lambda = 206$ nm) and a 60-min 70/0/25 → 50/50/0 acetonitrile/water (30 mM KH_2PO_4)/water gradient [1208]. An excellent choice of solvents and buffer used in the mobile phase gradient resulted in minimal baseline drift. A table of the structures that correspond to each peak is presented. Overall resolution and peak shapes were very good.

Seven types of oligosaccharides were investigated with respect to the effect of bonded phases on retention [1209]. The degree of polymerization varied from the monomer (e.g., glucose, mannose, maltose), to an intermediate level (such as a dimer or heptamer like cellotetraose, isomaltotriose, xylopentose, maltoheptose, mannohexose) to a maximum octomer (e.g, arabanooctose). The column of choice was a β-cyclodextrin with detection at 195 nm (!) or by refractive index. Separation of the malto-oligosaccharides (70/30 acetonitrile/water) and manno-oligosaccharaides (65/35 acetonitrile/water) is shown and resolution is very good. Other elution patterns can be determined from the provided table of $\log k'$ vs. water plots. Oligosaccharides of maltose (≤ 25 oligomers) and isomaltose (≤ 20 oligomers) were separated on a β-cyclodextrin column (RI detector) using either a 65/35 or 70/30 acetonitrile/water mobile phase, respectively. In both cases elution was complete in about 40 min [1210]. An increase in the acetonitrile level to 70% yielded better resolution between the low molecular weight oligomers, but the higher oligomers (>14) were undetectable (too highly retained). Inulin-derived oligosaccharides (up to 13 oligomers) were well resolved in 20 min using a 70/30 acetonitrile/water mobile phase. Xylose oligosaccharides (≤ 8 oligomers) were poorly resolved even when an 80/20 acetonitrile/water mobile phase was used. The authors worked with a series of other oligosaccharides based on glucose, fructose, and mannose. They compared the retention of these compounds with the normal monosaccharide retention order (pentoses < hexoses < aldohexoses) and found it similar for the oligomers (xylose oligomer < fructose oligomer < glucose or mannose oligomer).

Chitin (partially deacylated β-(1 → 4]-linked N-acetyl-D-glucosamine) was degraded by sonication and the resulting oligosaccharide fragments were analyzed on an aminopropyl column (RI detector). A 65/35 acetonitrile/water mobile phase separated the monomer to heptamer oligosaccharide fragments in 20 min [1211].

Chitosan digosaccharides (polymeric form of linear β-1,4-linked glucosamine residues) was degraded with chitosanase and the resulting degree of polymerization of the chitosan fragments was analyzed on an aminopropyl column (ELSD). A 40/52/8 → 40/42/18 (at 25 min hold 10 min) acetonitrile/water/methanol gradient generated excellent resolution of oligomer fragments ranging from 2 to 8 residues. Peak shapes were excellent [1212].

The inulin level in meat products is determined through extraction, enzymatic hydrolysis, and analysis of the resulting frustose. Fructose, with rhamnose as IS, was well resolved from glucose and sucrose in 15 min on an animopropyl column (RI detector) using an 80/20 acetonitrile/water mobile phase [1213]. The linear range was reported as 0.10–3.25 mg/mL.

Mono-, di-, and trisulfated Δ-disaccharides generated from the action of heparin and heparin sulfate lyases on heparin, heparin sulfate, and Fragmin, were analyzed on a C_{18} column ($\lambda = 232$ nm). The lyase fractions are composed of either D-glucuronic or L-iduronic acids and glucosamine (as the N-acetylated, N-sulfated, or unsubstitued form). In all, 12 disaccharides were resolved using a 20/80 (hold 8 min) → 46.7/53.3 (at 23 min) → 0/100 (at 43 min) water/acetonitrile with 10 mM tetrabutylammonium hydrogensulfate to pH 6.7 gradient. Most peaks were fully resolved and peak shapes were very good [1214]. Calibration curves were generated over the 0.005–50 µg/mL range. Detection limits of 1 ng/mL (S/N = 2) were reported.

Monosaccharides in glycoproteins were separated as their p-aminobenzoic acid ethyl ester derivatives on a Pico-Tag column ($\lambda = 254$ nm) using a 75/25 (water [50 mM acetate buffer at pH 4.5])/(40/40/20 water [50 mM acetate buffer at pH 4.5]/acetonitrile/methanol) mobile phase [1215]. The effects of changing both the temperature (35°C or 45°C) and organic mobile phase content on retention were presented as plots. Excellent resolution between glucosamine, galactosamine, lactose, maltose, galactose, mannose, ribose, xylose, N-acetylglucosamine, N-acetylgalactosamine, fucose, and 2-deoxyglucose derivatives was obtained. Peaks from the injection of 2.5 nmol of each sugar were easily detected.

An interesting separation of four oligosaccharides (cellobiose, di-N-acetylchito-biose) was developed on an aminopropyl column (electrospray MS). A 60-min 100/0 → 0/100 acetonitrile/water gradient was used. This gradient was very effective since the oligosaccharides have very limited solubility in acetonitrile and are therefore concentrated at the top of the column. Peaks are fairly sharp even at a 35–45 min elution time [1216]. This gradient method should be kept in mind for oligosaccharide work when gradient-compatible detectors are used.

Polytosylated cyclodextrins (octakis[6-*O*-tosyl]-2-*O*-tosyl-γ-cyclodextrin and octakis- and heptakis[6-*O*-tosyl]-γ-cyclodextrin) were separated on an aminopropyl column ($\lambda = 220$ nm) using a 50-min $100/0 \rightarrow 70/30$ acetonitrile/water gradient [1217]. Just as with simple sugars, the gradient starts with a high acetonitrile content in which the solutes have limited solubility. Water is added to cause elution.

9.5.8 Other Analytes

Montelukast (leukotriene D_4 receptor antagonist) and five metabolites (three hydroxy- and two sulfoxide-substituted forms) were extracted from liver microsome incubates and baseline resolved on a C_{18} column ($\lambda = 350$ nm, ex; 400 nm, em) using a 45/55 (hold 30 min) $\rightarrow$ 35/65 (at 35 min hold 25 min) water (5 mM heptanesulfonic acid with 50 mM KH_2PO_4 to pH 6.2 with NaOH)/acetonitrile gradient [1218]. Baseline resolution and excellent peak shapes were obtained.

Secoisolariciresinol diglucoside was isolated from flaxseed and analyzed on a 25°C C_{18} column ($\lambda = 280$ nm) using a 30-min $100/0 \rightarrow 70/30$ water (10 mM phosphate at pH 2.8)/acetonitrile gradient [1219]. Peak shape was good as was the overall resolution. Standards of 20–160 µg/mL were used.

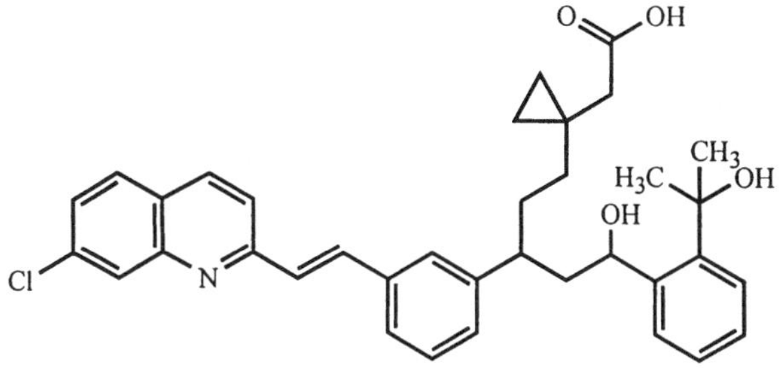

Montelukast

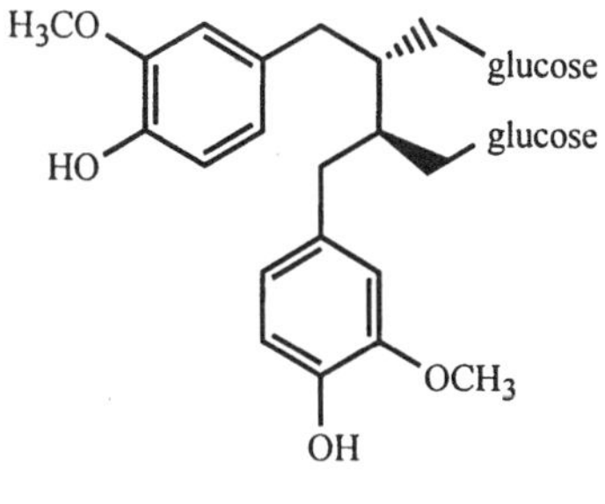

Secoisolariciresinol diglucoside

A set of 19 desulfoglucosinates were extracted from 14 types of seeds (family Brassicaceae) and separated on a C_{18} column ($\lambda = 229$ nm). A 95/5 (hold 2 min) → 59/41 (at 16 min hold 2 min) → 5/95 (at 22 min hold 5 min) water-/acetonitrile gradient was used [1220]. Peak identification was obtained by positive ion electrospray MS using a postcolumn formic acid mixing tee and the following MS settings: capillary voltage 2.7 kV, cone voltage 17 V, extractor 1 V, source block $T = 110°C$, desolvation $T = 200°C$, nebulizer (N_2) flow 76 L/h, desolvation gas flow 430 L/h. Analytes included the desulfo forms of progoitrin, epiprogoitrin, sinigrin, 4-hydroxyglucobrassicin, glucoalyssin, glucobrassin, and glucotropaeolin. A table of results shows the determination of individual compound concentrations down to 0.1 µmol/g sample.

Seven monacolins (e.g., monacolin K, dihydro-, dehyro-monacolin K, monacolin L) and three pigments (monascidin A, monascorubrine, monascorubramine) were isolated from red yeast rice and separated on a C_{18} column ($\lambda = 218$ nm and 237 nm) using a complex 32-min 80/20 → 10/90 water (0.04% H_3PO_4)/acetonitrile gradient [1221]. Excellent peak shapes and resolution were achieved. No concentrations were reported, but the chromatogram shown was of a 0.5 g red yeast rice powder in 10 mL alcohol/water solvent and 0.01–0.2% w/w was commonly found for the individual components.

Two xanthophylls, capsorubin and capsanthin, were extracted from red peppers and analyzed on a C_{18} column ($\lambda = 450$ nm) using an 80/10/10 acetonitrile/

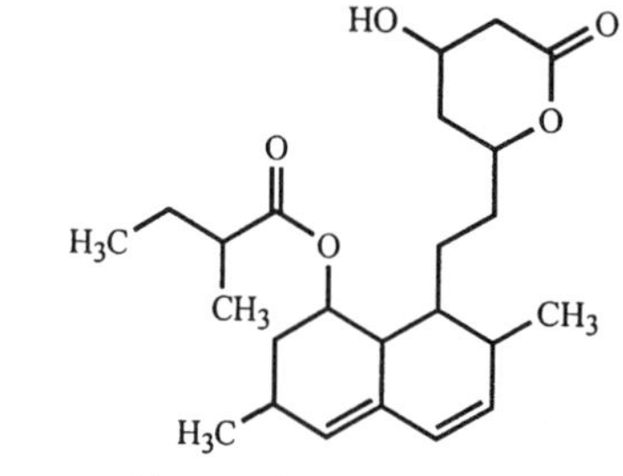

Monacolin K

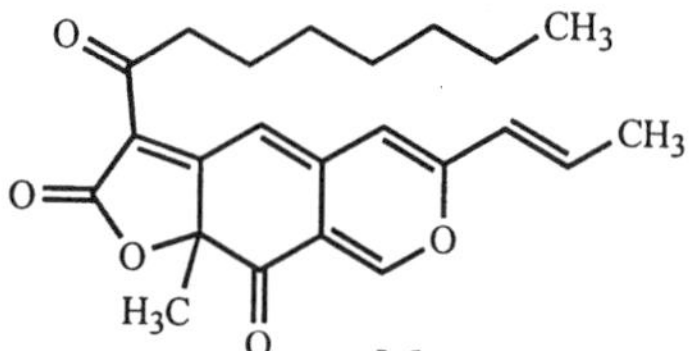

Monascorubrine

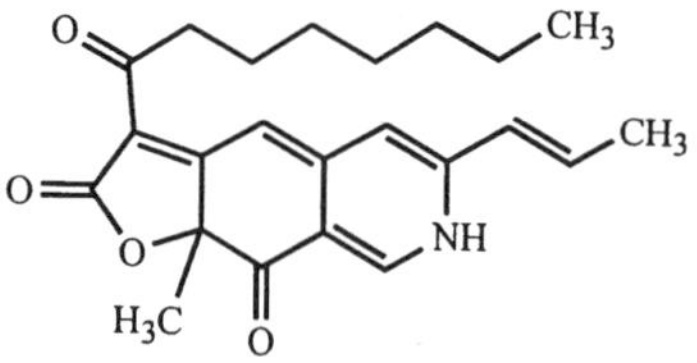

Monascorubamine

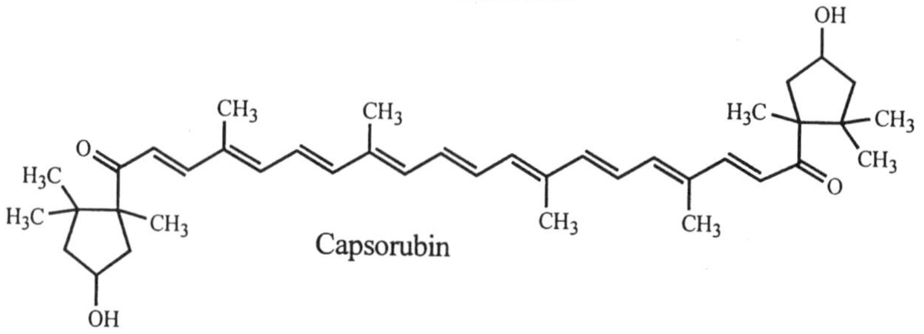

Capsorubin

IPA/ethyl acetate mobile phase [1222]. Elution was complete in <4 min. The authors note that the extremely limited solubility of these compounds in water led to the use of a nonaqueous mobile phase. They further noted that higher concentrations of acetonitrile led to peak splitting perhaps due to on-column solubility issues. Linear curves were generated from 10 to 100 ng injected and detection limits of 30 ng injected were reported.

The hexane extracts of the roots of various *Angelica*, *Ligusticum*, and *Levisticum* species were characterized using a C_{18} column ($\lambda = 210$ nm) using a two-ramp $60/40 \rightarrow 45/55$ (at 15 min) $\rightarrow 5/95$ (at 33 min hold 2 min) water/acetonitrile gradient [1223]. Fourteen compounds were identified (e.g., coniferyl ferulate, butylphthalide, falcarnidiol, imperatorin, osthol, columbianedin, angelol A). Good separation and excellent peak shapes were obtained. Linear ranges of 0.01–1.5 mg/mL were cited for falcarnidiol.

Plumbagin and 7-methyljuglone are naphthoquinones found in the herb known as sundew. They were separated and analyzed on a deactivated base C_{18} column ($\lambda = 425$ nm) using a 38/62 (95/5 acetonitrile/THF)/water (0.2 M acetic acid to pH 3 with triethylamine) mobile phase. Methanol was ineffective in this separation. Good resolution and peak shapes and elution were achieved in 16 min. Calibration curves from 0.08 to 1.0 mg/mL along with detection limits of 0.02 μg/mL were reported [1224].

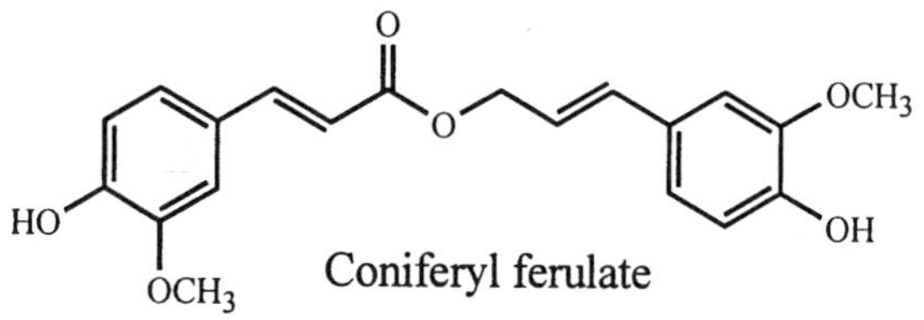

Coniferyl ferulate

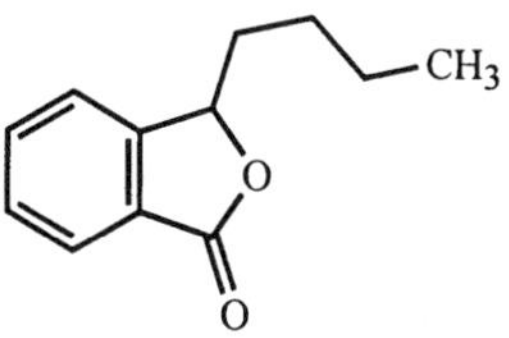

Butylphthalide

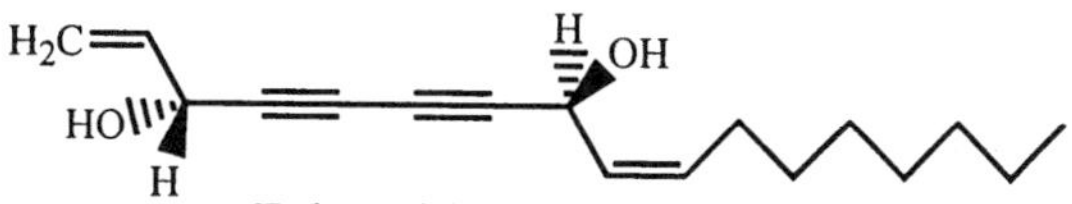

Falcarnidiol

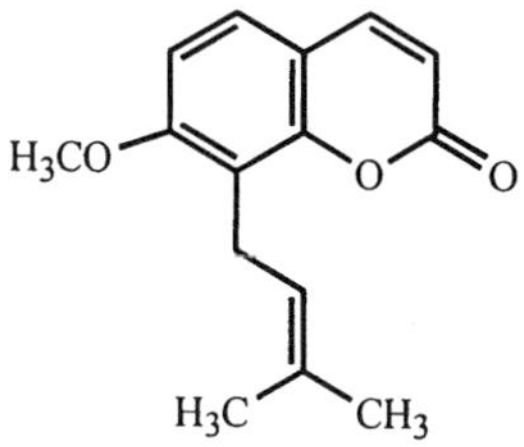

Osthol

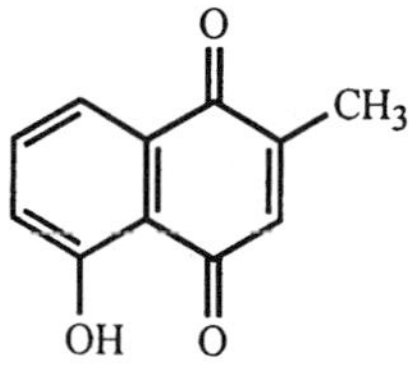

Plumbagin

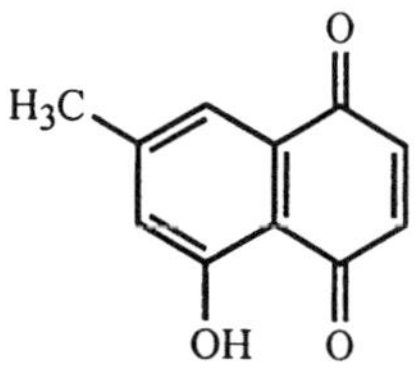

7-Methyljuglone

Seven lignans (secoisolariciresinol, pinoresinol, matairesinol, hinokinin, laricir-esinol, arctigenin, nordihydroguaiaretic acid) were isolated from flaxseed meal and separated on a C_{18} column ($\lambda = 280$ nm) using a 55-min 70/30 → 50/50 water [0.2% acetic acid]/acetonitrile gradient [1225]. Three peaks were not baseline resolved and no concentrations were cited.

Five 1,3-benzodioxanes (fargesin, sesamin, asarinin, 1,3-benzodioxole-5-(2,4,8-triene) isobutyl and methyl nonaoate) were isolated from *Piper mullesua* and separated on a 26°C C_{18} column ($\lambda = 220$ nm) using a 65/35 acetonitrile/water mobile phase [1226]. Elution was complete in 12 min and peak shapes and resolution were good. A calibration curve of 2–20 µg injected was reported.

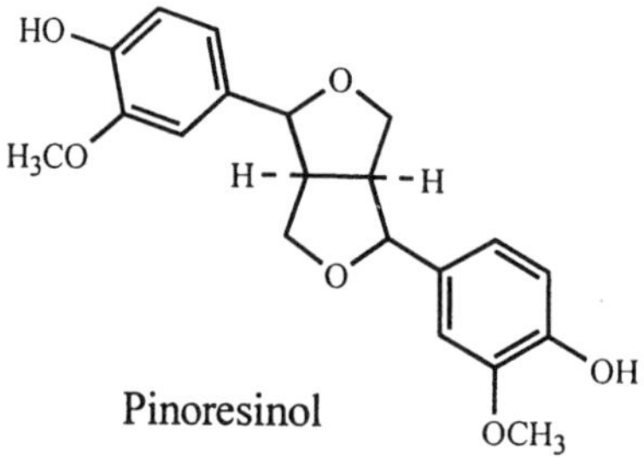

Pinoresinol

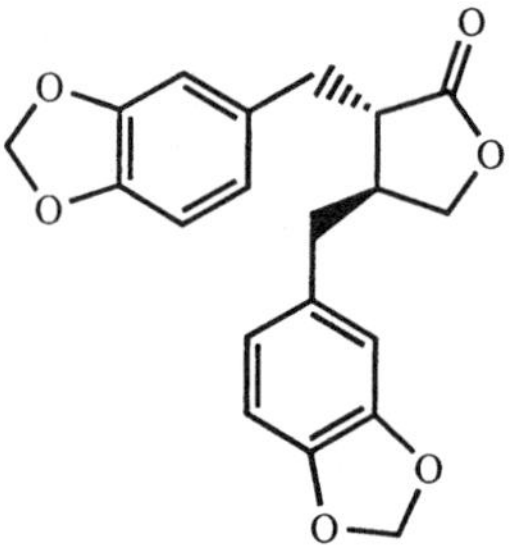

Hinokinin

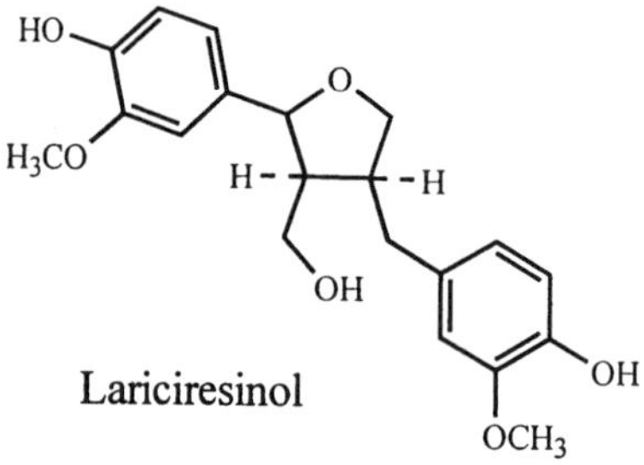

Lariciresinol

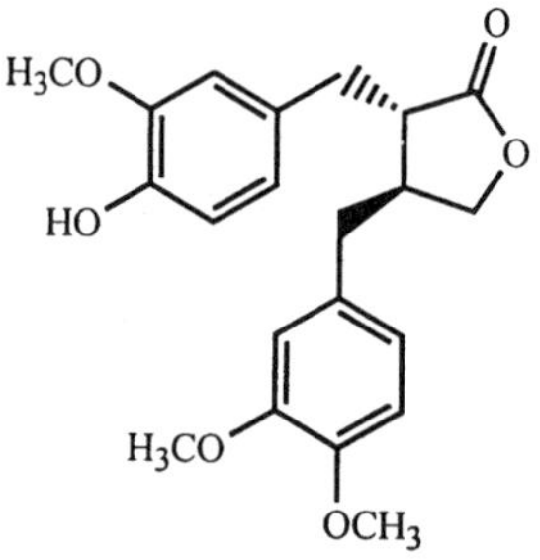

Arctigenin

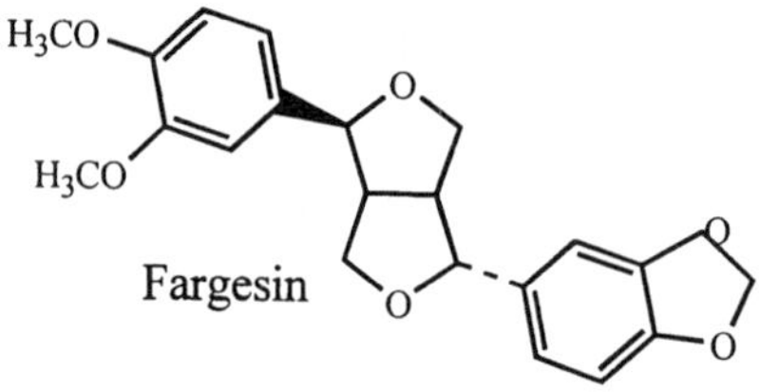

Fargesin

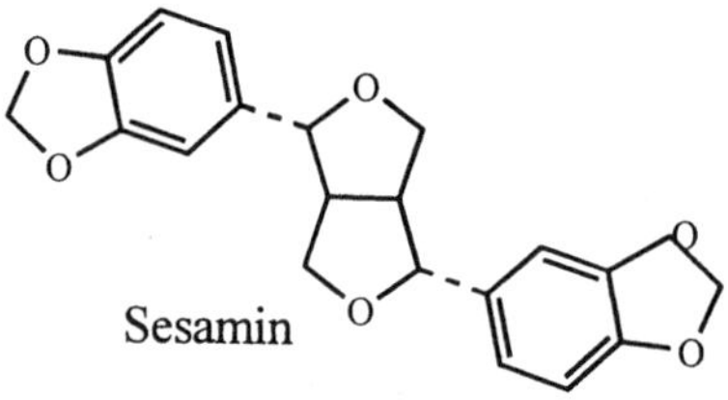

Sesamin

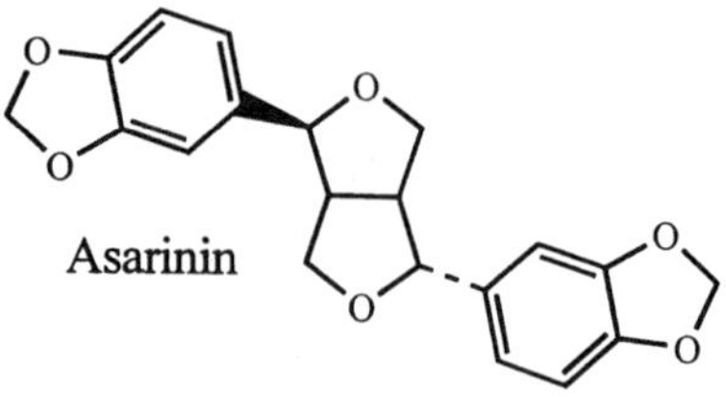

Asarinin

Stevioside and five metabolites (isosteviol, steviabioside, 15α-hydroxysteviol, steviol-16,17α-epoxide, steviol) were isolated from urine and blood and separated on a C$_{18}$ column ($\lambda = 210$ nm) using a 20-min 70/30 → 35/65 water/acetonitrile gradient. Calibration curves covered the 10–60 μg/mL range and detection limits were reported as ~0.5 μg per injection. Overall good peak shapes and resolution were obtained [1227].

Five hydroxyanthroquinone derivatives (aloe-emodin, rhein, emodin, chrysophanol, physcion) were extracted from rhubarb and baseline resolved on a C$_{18}$ column ($\lambda = 254$ nm). A 36/64 (hold 5.5 min) → 80/20 (at 20.5 min hold 5 min) acetonitrile/water (36 mM triethylamine to pH 2.5 with H$_3$PO$_4$) gradient was used. Peak shapes were excellent. The calibration range ran from 0.5 to 40 μg/mL [1228].

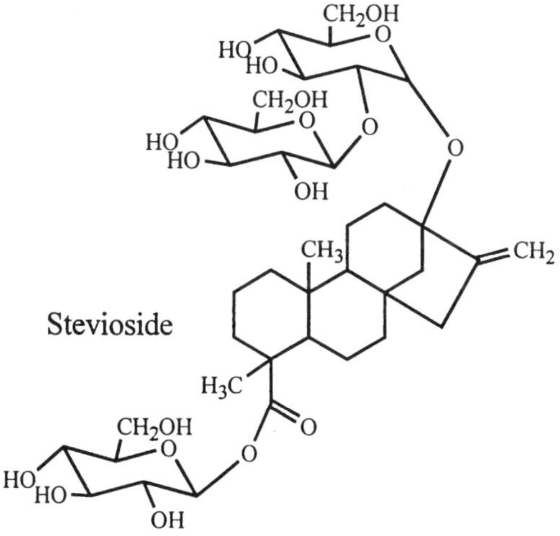

Stevioside

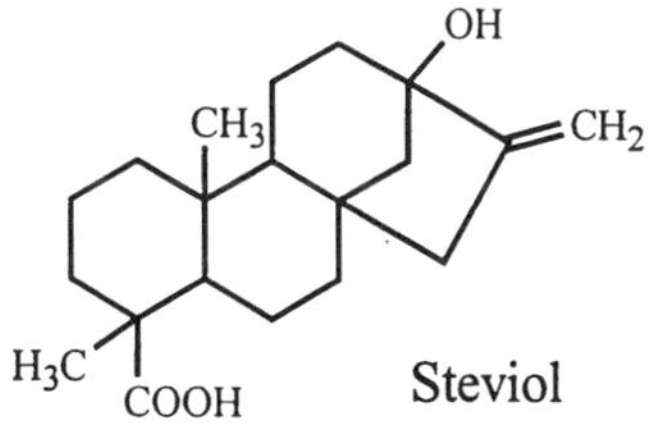

Steviol

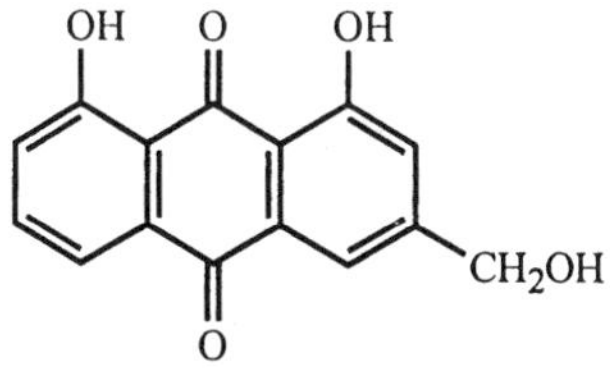

Aloe-emodin

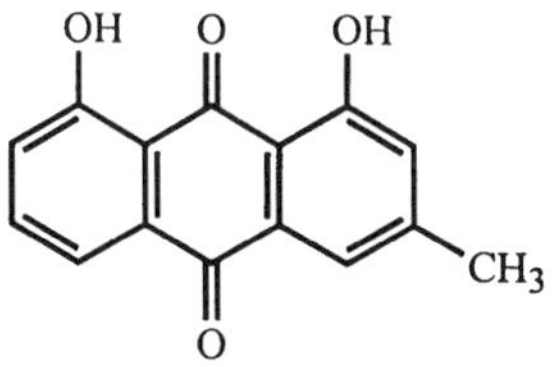

Chrysophanol

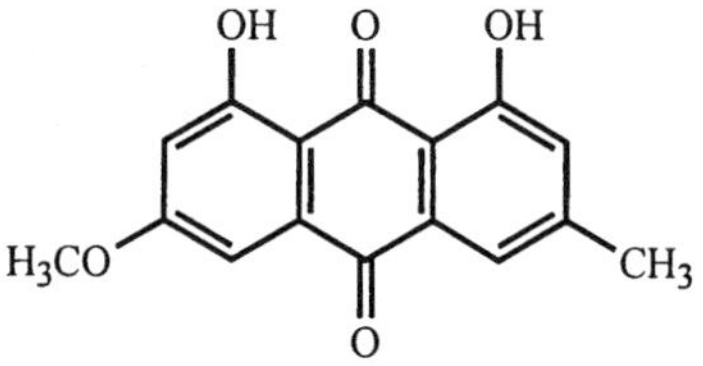

Physcion

A series of six α-pinene hydroxyl radical reaction products were derivatized with 2,4-dintirophenylhydrazine (DNPH) (formaldehyde, acetaldehyde, acetone, pinonaldehyde [both mono- and di-DNPH substituted], campholenealdehyde) and separated on 35°C C_{18} column ($\lambda = 360$ nm) using a 95/5 → 16/84 (at 50 min hold 10 min) gradient [1229]. Peak shapes and resolution were excellent. Peak identities were confirmed by negative ion APCI.

Six anka pigments (rubropunctatin, monascorubrin, monascin, ankaflavin, monascorubramine, rubropunctatamine) were extracted from a *Monascus purpureus* fermentation broth and separated on a C_{18} column ($\lambda = 233$ nm). Elution was complete in 18 min using an 80/20 acetonitrile/water mobile phase. Peak shapes were broad yet well resolved. Positive identification was achieved through positive ion APCI/MS/MS [1230].

Hypericum perforatum (St. John's Wort) was extracted and analyzed for pseuodohypericin, hypericin, hyperforin, and adhyperforin content. Excellent peak shapes

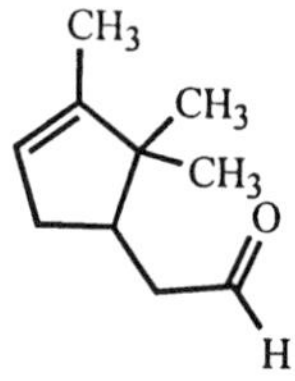

Campholenealdehyde

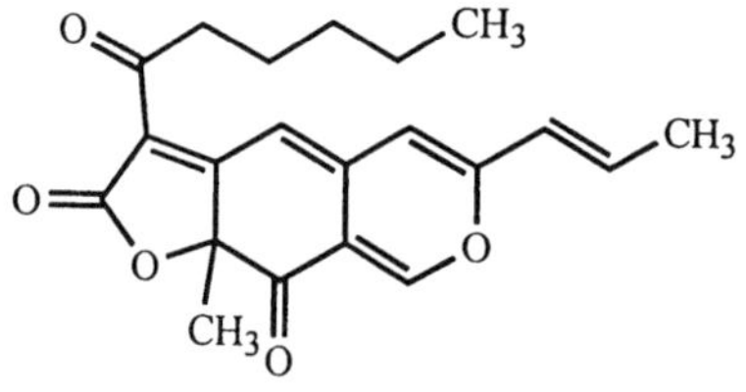

Rubropunctatin

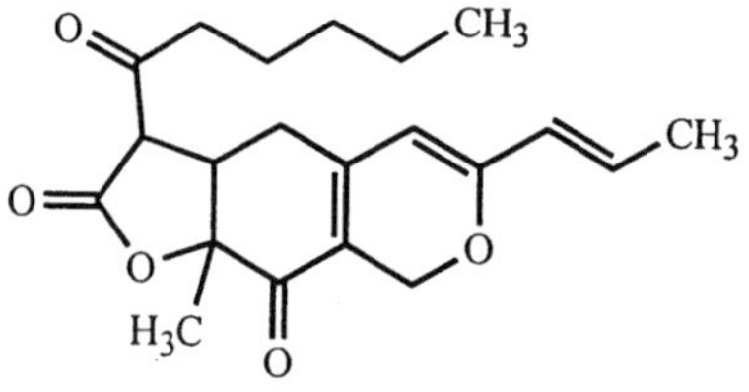

Monascin

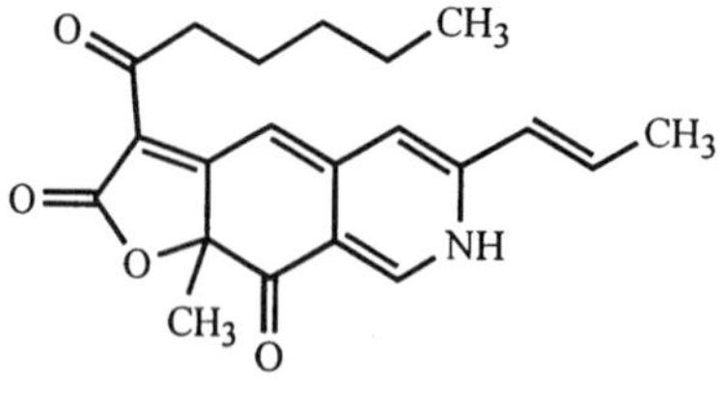

Rubropunctatamine

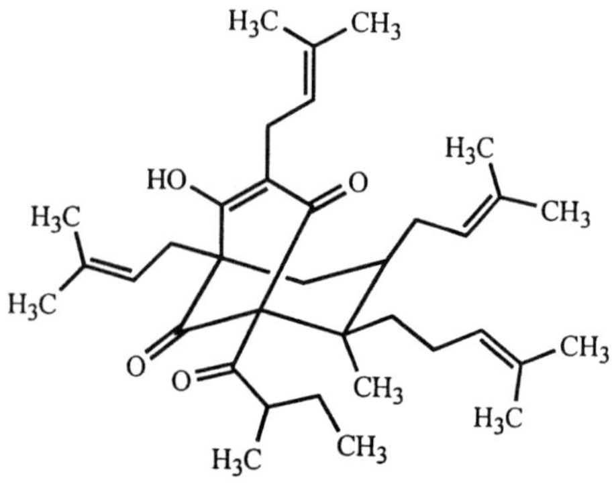

Adhyperforin

and baseline resolution were achieved in 8 min using a deactivated base C_{18} column ($\lambda = 470$ nm ex; 590 nm, em for hypericins and 280 nm for hyperforins) and an 80/20 acetonitrile/water (0.1 M triethylammonium acetate) mobile phase [1231]. Standards of hyperforin: hypericin 20 ppm:2 ppm to 2520 ppm:25 ppm were used. The authors noted the sensitivity of extracted hypericins to light and oxygen exposure and plotted the concentration versus exposure time. Over 90% of the hypericins degrade in 120 min of exposure.

Methanolic *Hypericum perforatum* extracts were characterized on a 30°C C_{18} column ($\lambda = 270$ nm) using a complex 55-min $100/0/0 \rightarrow 5/80/15$ water (0.3% H_3PO_4)/acetonitrile/methanol gradient [1232]. Good peak shapes and resolution were achieved. The following compounds' identities were confirmed positive ion thermospray MS: I3,II8-biapigenin, chlorogenic acid, rutin, hyperoside, pseudohypericin, hyperforin, hypericin, and adhyperforin. In the study numerous additional mobile phase constituents (e.g, THF, TFA, ammonium acetate), columns (C_8 and cyanopropyl), and column temperatures (25–35°C) were evaluated during the optimization process.

Hyperforin was isolated from plasma and analyzed on a mixed-mode C_{18}/cyanopropyl column ($\lambda = 272$ nm) using an 81/19 acetonitrile/water (to pH 4.5 with H_3PO_4) mobile phase [1233]. Elution was complete in 9 min. A linear range of 0.15–3 µg/mL and a detection limit of 4.5 ng injected were reported. Interestingly, no mention was made of analyte instability.

Panax ginseng roots were extracted and 18 ginsenosides were resolved and identified. These included (20S)-protopanaxadiol and (20S)-protopanaxatriol aglycones, glycosolates, and malonylates. Excellent peak shapes and baseline resolution were achieved in 60 min using a 25°C base deactivated C_{18} column (ELSD, nebulizer 70°C; N_2 nebulizing gas 2.5 bar) using an $80/20 \rightarrow 72/28$ (at 15 min) $\rightarrow 68/32$ (at 35 min) $\rightarrow 60/40$ (at 55 min hold 5 min) water (8 mM ammonium acetate to pH 7 with NH_4OH)/acetonitrile gradient [1234]. Linear ranges of 2–2000 µg/mL (analyte dependent) were reported.

Eleven ginsenosides (six major: Rb_1, Rb_2, Rc, Rd, Re, Rg_1; five minor: F_4, Rg_2, Rg_3, Rg_5, Rf) were steam extracted from the roots of *Panax gineng* and separated on

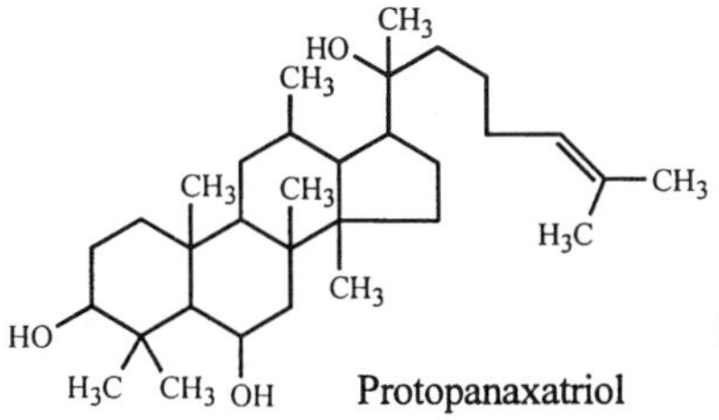

Protopanaxatriol

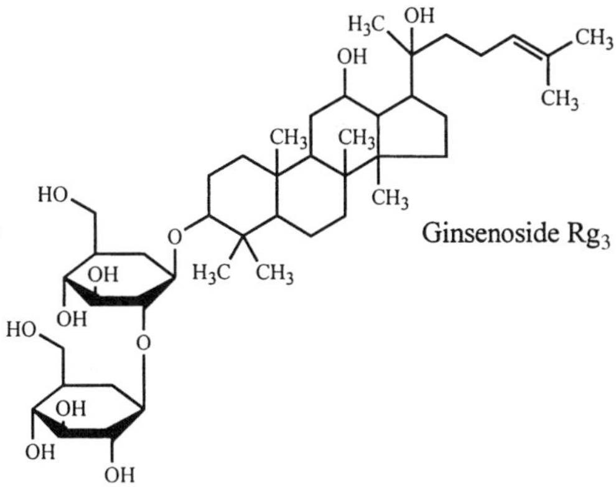

Ginsenoside Rg$_3$

an aminopropyl column (ELSD, drift tube 115°, N_2 nebulizer gas at 2.3 L/min). An 80/20 (hold 7 min) → 10/90 (at 25 min hold 20 min) (80/15/5 acetonitrile/IPA/water)/(80/15/20 acetonitrile/IPA/water) gradient produced excellent peak shapes and overall good resolution [1235]. This is an interesting separation involving glycosides in which the retention is based on increasing water solubility of the analytes. Therefore, as is the case in sugar separations on an aminopropyl column, retention decreases with increasing water levels in the mobile phase. Ginsenoside levels varied from 0.13 to 0.65% w/w.

Ginsenoside Rg_3 was isolated from plasma and analyzed on a C_{18} column ($\lambda = 203$ nm) using a 43/565/1 acetonitrile/water/IPA mobile phase [1236]. The analyte was well resolved from co-extracted compounds and eluted in 17 min. A linear range of 2.5–200 ng/mL and a detection limit of 1.5 ng/mL were reported.

He et al. [1237] separated 13 pungent constituents (gingerols, gingerdiols, shogaols, methylgingerols, gingerdione) that were extracted from ginger on a C_{18} column ($\lambda = 230$ nm and 280 nm) using a complex 38-min 45/55 → 0/100 water/acetonitrile gradient. Each peak was positively identified by positive ion electrospray MS; quadrapole $T = 150°C$, drying N_2 $T = 350°C$ at 40 mL/min, nebulizing N_2 80 psi. All peaks were not cleanly separated from co-extracted compounds, but were clearly discernible by UV detection.

Two gingerdiols ((3S,5S)- and (3R,5S)-6-gingerdiol) and two glucosides were extracted from ginger and separated on a C_{18} column (photodiode array detector, $\lambda = 190$–340 nm). A 100-min 30/70 → 80/20 acetonitrile/water gradient was used, but elution of the last peak occurred well before 30 min. Baseline resolution and excellent peak shapes were obtained [1238].

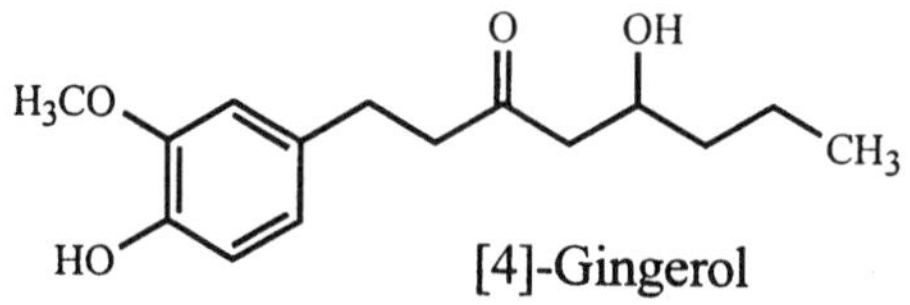

[4]-Gingerol

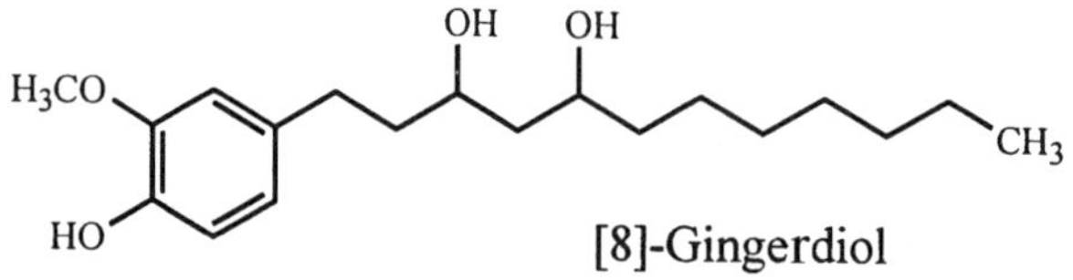

[8]-Gingerdiol

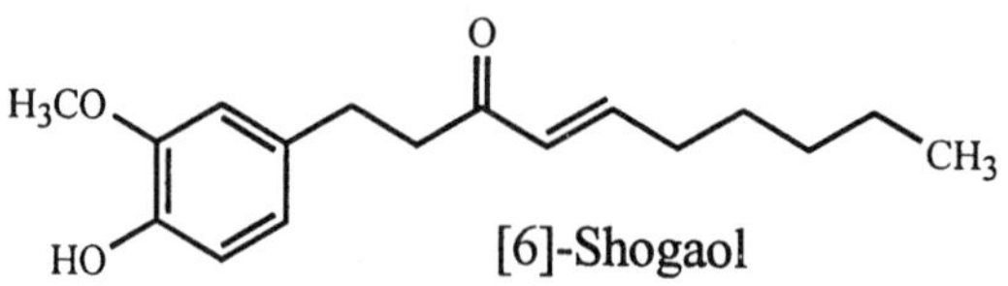

[6]-Shogaol

Eight phytoalexins (*cis*- and *trans*-resveratrol and their glucosides and dehydro-dimers, *trans*-pterostilbene, *trans*-ε-viniferin) were extracted from grapevine leaves [1239]. A 30-min $90/10 \rightarrow 10/90$ water/acetonitrile gradient was run on a C_{18} column ($\lambda = 330$ nm, ex; 374 nm, em). All analytes eluted in <25 min with excellent peak shapes. Resolution was not baseline for all compounds. A calibration curve from 2 to 200 ng injected was reported.

Lucidin primveroside and ruberythric acid were extracted from madder root and analyzed on a C_{18} column ($\lambda = 254$ nm) using a 22/78 acetonitrile/water (20 mM ammonium acetate at pH 4) mobile phase [1240]. The analytes were eluted as sharp peaks in <14 min but were incompletely separated.

Parthenolide was extracted from feverfew powder and resolved from co-extracted components on a C_{18} column ($\lambda = 210$ nm) using a 20-min $50/50 \rightarrow 15/85$ water (50 mM NaH_2PO_4)/(90/10 acetonitrile/methanol) gradient [1241]. Parthenolide eluted at 10 min. Note that significant baseline drift occurred throughout the gradient due to the low wavelength and the presence of methanol in the strong phase. Calibration standards of 25–400 µg/mL were used.

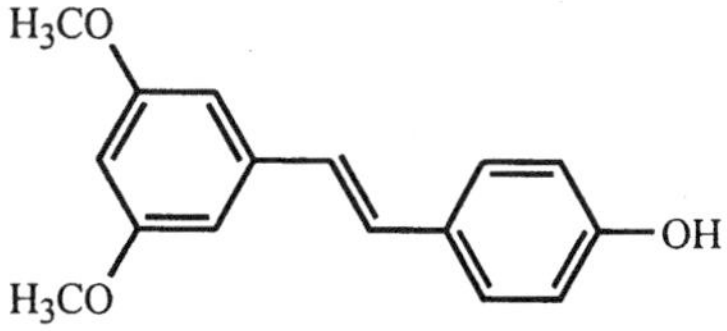

trans-Pterostilbene

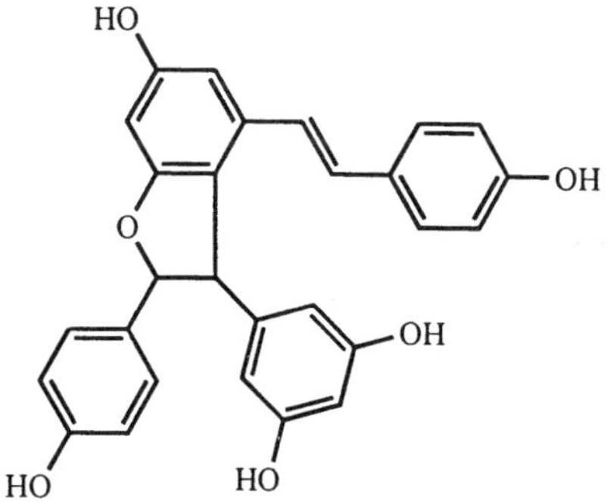

trans-ε-Viniferin

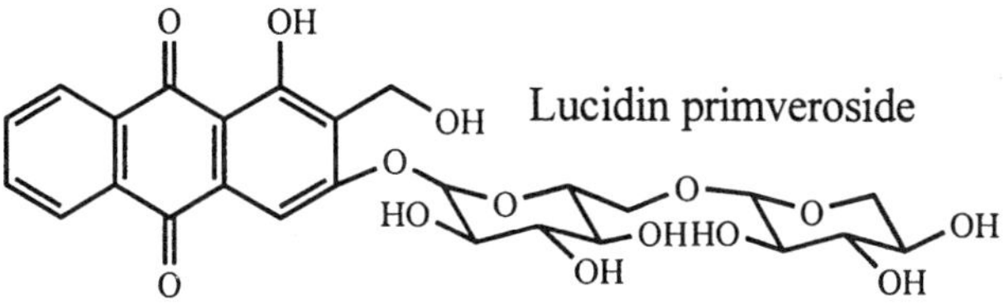

Lucidin primveroside

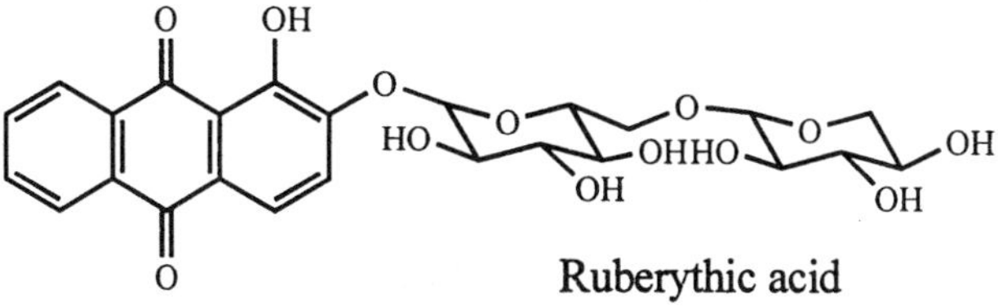

Ruberythic acid

Fennel tea constituents (chlorogenic acid, quercetin-O-glucuronide, p-anisalde-hyde, *trans*-anethole) were analyzed on a C_{18} column ($\lambda = 254$ nm) using a complex 40-min 88/12 → 0/100 water/acetonitrile each formic acid to pH 3.2 gradient. Note that pH in an organic solvent is poorly defined. Peak shapes were excellent and elution was complete in 37 min. The closest-eluting pair of peaks is resolved by over 6 min, so a more aggressive gradient would reduce analysis time considerably [1242]. A linear range of 0.1–2 μg injected was reported.

Herraiz [1243] isolated a series of four tetrahydro-β-carbolines (e.g., 6-hydroxy-1-methyl-1,2,3,4-tetrahydro-β-carboline, 1-methyl-1,2,3,4-tetrahydro-β-carboline-3-carboxylic acid) from cocoa and chocolate and separated them on a 40°C C_{18} column ($\lambda = 270$ nm, ex; 343 nm, em). Serotonin, tryptamine, and tryptophan were also separated and detected using this method. A three-ramp 100/0 → 68/32 (at 8 min) → 10/90 (at 18 min) → 0/100 (at 20 min) water (ammonium phosphate to pH 3 with H_3PO_4)/acetonitrile gradient generated good separation. Detectable levels ranged from 0.03 to 1.8 μg/g sample.

(E)- and (Z)-guggulsterones (hypolipidemic agents) were extracted from *Commiphora mukul* resin and analyzed on a C_{18} column ($\lambda = 245$ nm). A complex 87-min 65/35 → 0/100 water/acetonitrile gradient gave well-resolved peaks that were free other extracted components. A linear range of 15–125 μg/mL was reported [1244].

Glyoxal, methylglyoxal, and diacetyl (2,3-butadione) were extracted from beer and urine and analyzed after derivatization with o-phenylenediamine [1245]. The separation took 11 min on a C_{18} column ($\lambda = 315$ nm) using an 80/20 acetonitrile/water (40 mM acetate at pH 4.5) mobile phase. Not surprisingly, some samples generated interfering peaks. Determination of analyte concentrations in the 3–25 μM range was reported.

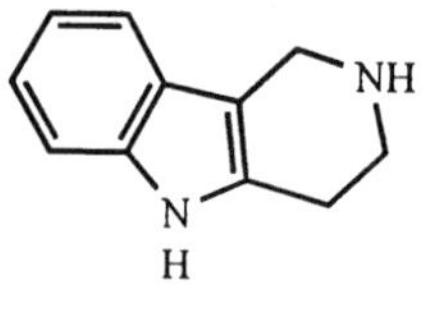

1,2,3,4-Tetrahydro-β-carboline

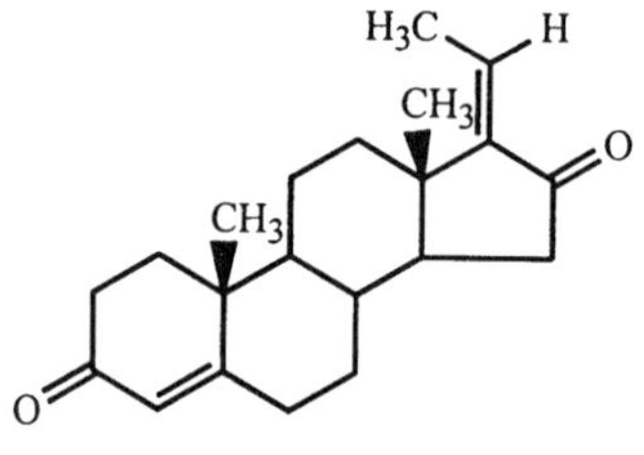

(E)-Guggulsterone

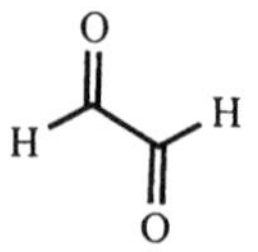

Glyoxal

Dimov et al. [1246] isolated ten iridoid glucosides (e.g., daphylloside, asper-ulosidic acid, geniposidic acid, monotropein, scandoside) from *Galium* plants and analyzed them on a diol column ($\lambda = 233$ nm) using a 98/2 acetonitrile/THF mobile phase. Peaks were somewhat broad, resolution was incomplete for all compounds, and elution was complete in 40 min. The authors note that the use of ion-pair reagents or addition of buffers did not improve the separation. Stock solutions ranged from 0.4 to 0.90 mg/mL.

Parishin and parishin B and C were extracted from Chinese medicinal prepara-tions and analyzed on a C_{18} column ($\lambda = 222$ nm) using a complex 35-min $90/10 \rightarrow 83/17$ water (1% H_3PO_4)/acetonitrile gradient [1247]. Peaks of interest were well resolved from potential interferences. A linear range of 20–400 μg/mL was reported.

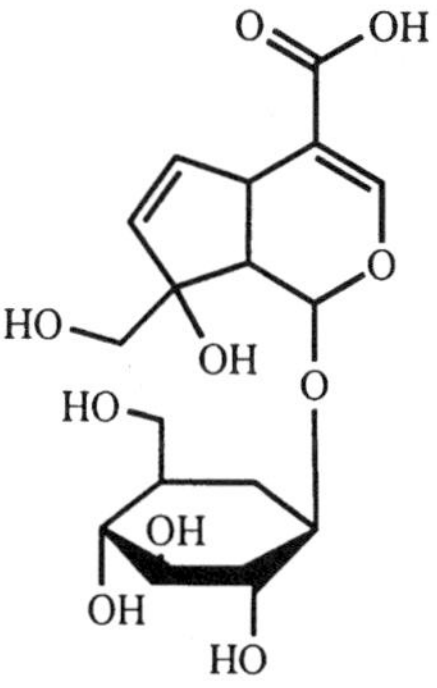

Monotropein

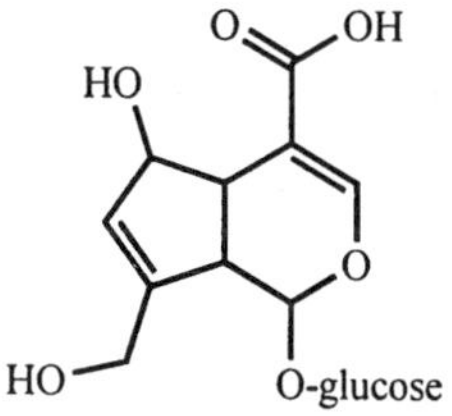

Scandoside

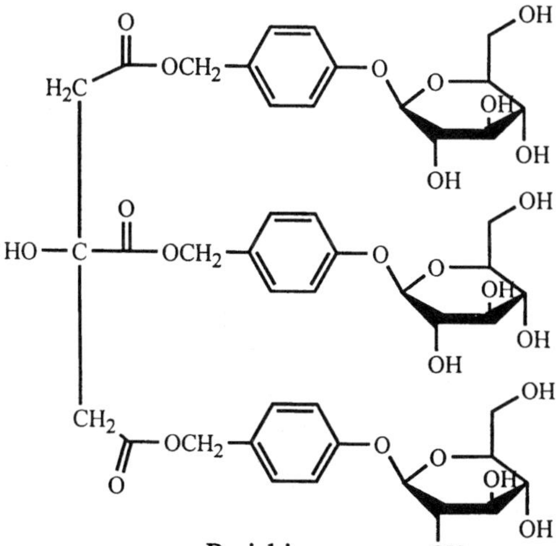

Parishin

Lehtonen [1248] studied phenolic aldehydes (salicylaldehyde, 4-hydroxybenz-aldehyde, protocatechualdehyde, vanillin, syringaldehyde, coniferylaldehyde, and sinapaldehyde) in distilled alcoholic beverages. These compounds were baseline resolved on a C_{18} column (photodiode array detector, $\lambda = 230$–380 nm) using a 30-min $10/90 \rightarrow 30/70$ acetonitrile/water (1 mM H_3PO_4) gradient. The UV spectrum for each analyte as obtained during the chromatographic run was shown. Good peak shapes were generated and levels from 1 to $10\,\mu g/L$ distillate were reported. Detection limits of ~ 0.3 ng injected (S/N = 5) and linear working curves up to 400 ng were reported.

Tyrosine, 3-monoiodotyrosine, 3,5-diiodotyrosine, 3,5,3′-triiodothyronine (T_3), and 3,5′,3′-triiodothyronine (reverse T_3) were baseline resolved on a C_{18} column ($\lambda = 254$ nm) with a 50/49/1 acetonitrile/water/acetic acid mobile phase [1249]. Linear concentration curves from 40 to 800 ng injected were generated. In the same study, a set of 14 thyroacetic acids and iodoamino acids were baseline resolved and eluted in 25 min. Here a C_{18} column and a 60-min $10/90 \rightarrow 75/25$ acetonitrile/water (0.1% H_3PO_4) gradient were used. Some of the additional analytes included 3,5-diiodothyroacetic acid, 3,3′,5-triiodothyroacetic acid, and 3′,5′-diiodothyronine.

Hiserodt et al. [1250] studied the allergen atranorin found in oak moss. It and eight related compounds (chloroatranonin, two atranol isomers, chloroatranol, methyl and ethyl hematommate, methyl and ethyl chlorohemmatomate) were extracted from oakmoss and separated on a C_{18} column (negative ion APCI; vaporizer $T = 300°C$, capillary $T = 250°C$, sheath gas 60 psi, electron multiplier 1.03 kV, corona discharge, $5\,\mu A$ (at 3.5 kV). A $90/10/2 \rightarrow 5/95/2$ (at 30 min hold 10 min) water/acetonitrile/acetic acid gradient gave baseline resolution and excellent peak shapes. The last peak eluted before 32 min. A $250\,\mu g/mL$ standard was easily detectable.

Two limonoid glucosides, obacunone and nomilin glucoside, were extracted from *Citrus tangeria* and analyzed on a C_{18} column ($\lambda = 215$ nm) using a 15/85

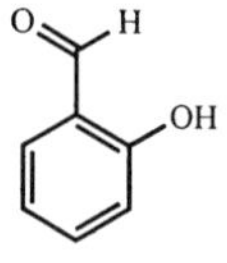

Salicylaldehyde 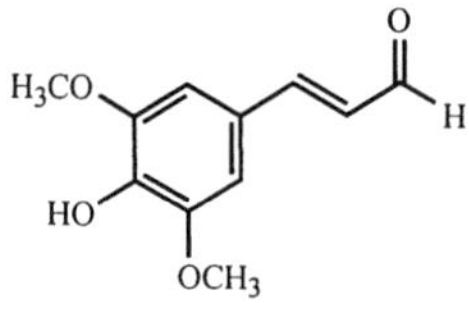   Sinapinaldehyde 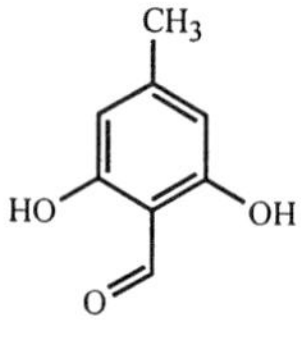   Atranol

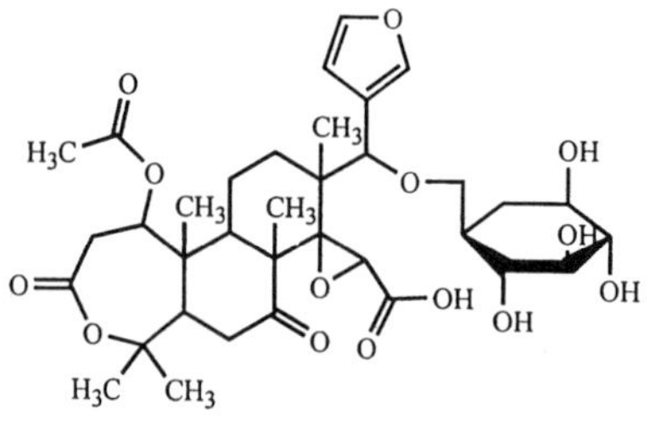

Nomilin glucoside

acetonitrile/water (0.05% TFA) mobile phase [1251]. Separation was complete in 6 min and baseline resolution was achieved. Concentrations of ~500 µg/mL generated very strong detector responses.

Furfural and furfuryl alcohol and three metabolites (furoic acid, furoylglycine, furanacrylic acid) were well resolved on a C_{18} column and an aminopropyl column ($\lambda = 214$ nm) using either a complex 12-min 7/93 → 25/75 acetonitrile/water (5 mM NaH_2PO_4 buffer at pH 6) gradient or an isocratic 1/99 acetonitrile/water (5 mM NaH_2PO_4 buffer at pH 6) mobile phase, respectively [1252]. Elution was complete in <12 min in both cases and good resolution resulted. Peak shapes on the aminopropyl column were broader than those eluted from the C_{18} column.

A series of nine naphtho[2,3-*b*]furan-4,9-diones (e.g., lapachol, dehydro-α-lapachon, ethyl- and isopropylnaphtho[2,3-*b*]furan-4,9-dione) were extracted from bark and well resolved on a C_{18} column ($\lambda = 254$ nm and 280 nm) using a 20-min 25/20/55 → 45/25/40 acetonitrile/methanol/water (0.1% H_3PO_4) gradient [1253]. An isocratic separation using a 35/25/40 acetonitrile/methanol/water (0.1% H_3PO_4) mobile phase eluted all components in ~10 min but generated poorer resolution for three pair of analytes (as compared with the gradient). Linear working ranges of 0.2–50 mg/L of plant extract were reported.

Stark and Walter [1254] resolved neem oil components (nimbandiol, deacetylsalannin, dimethyldeacetylnimbin, nimbin, 6-acetylnimbandiol, salannin) on a C_8 column ($\lambda = 215$ nm) using a 65-min 28/72 → 95/5 acetonitrile/water gradient. Good separation between compounds of interest and co-extractants was obtained. Peak shapes were excellent. Positive peak identification was made via fraction collection and injection into an MS.

Four products of pyropheophorbide-*a* (hematinic acid, methyl ethyl maleimide, methyl vinyl maleimide, and pyropheophorbide-*a* C-E-ring derivative) were resolved on a polyvinylalcohol C_{18} column ($\lambda = 280$ nm) in 15 min using a 75/25 water (0.1 M ammonium acetate)/acetonitrile mobile phase [1255]. The authors note that a

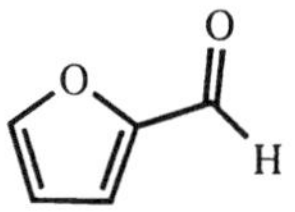

Furfural

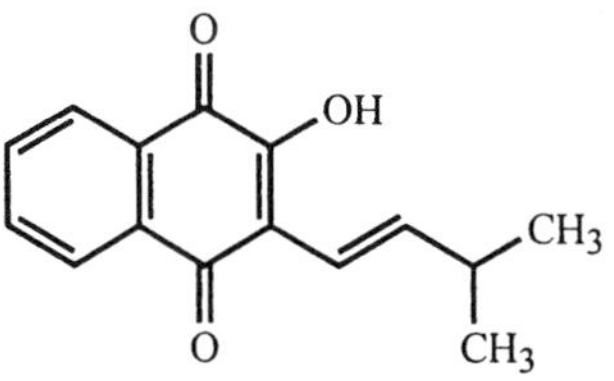

Lapachol

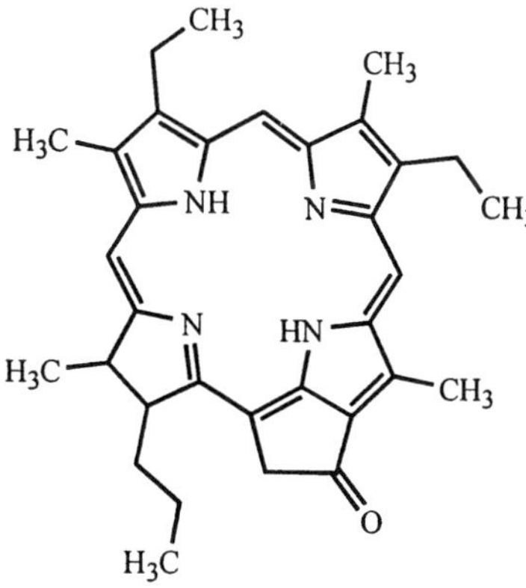

Pyropheophorbide-*a*

generic C_{18} column was ineffective at separating these compounds. A plot of the effect of changing percent acetonitrile on $\log k'$ is presented.

Chlorogenic acid, caffeic acid, echinacoside, and cichoric acid were extracted from *Echinacea* roots and separated on a C_{18} column ($\lambda = 300$ nm) using a $90/10 \rightarrow 75/25$ (at 30 min hold 5 min) water (0.1% H_3PO_4)/acetonitrile gradient [1256]. Good resolution and peak shapes were generated. Sample extracts less than 0.01% by weight were not detectable.

Extracts from the roots of *Polygonum cuspidatum* were analyzed for stilbenes (piceatannol glucoside, resveratroloside, piceid, resveratrol). These compounds were well resolved on a C_{18} column ($\lambda = 254$ nm) using a 15/85/0.15/0.85 acetonitrile/water/triethylamine/formic acid mobile phase [1257]. Elution was complete in 21 min and detected levels of 1–6 mg/g were reported.

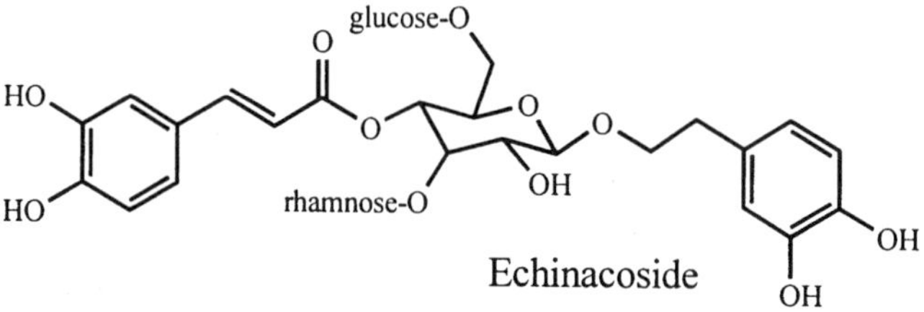

Echinacoside

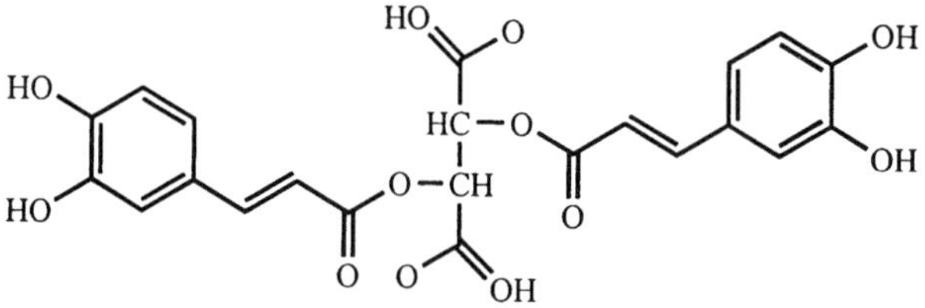

Cichoric acid

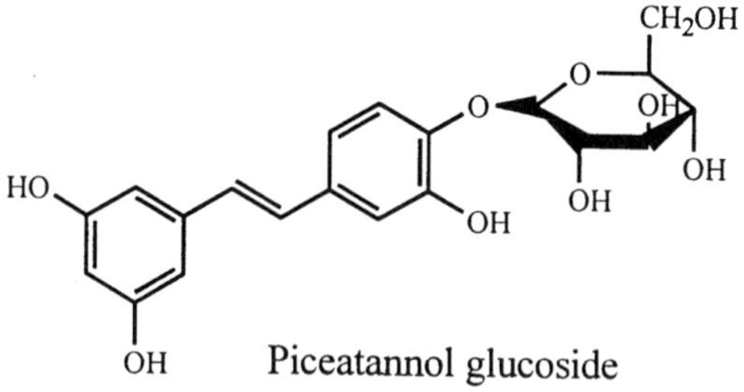

Piceatannol glucoside

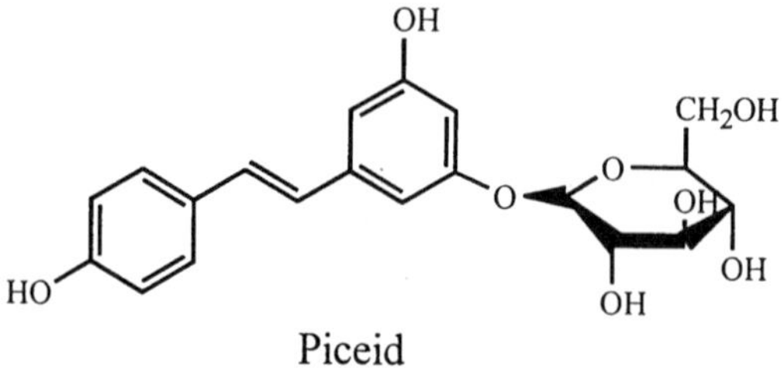

Piceid

Safrole, α- and β-asarone, isosafrole, and anethole extracted from alcoholic beverages and essential oils (e.g., sassafras, nutmeg, and cinnamon) were baseline resolved on a C_{18} column ($\lambda = 290$ nm, ex; 325 nm, em or 310 nm, ex; 355 nm, em) using a 45/55 acetonitrile/water mobile phase [1258]. Elution was complete in <10 min and a quantitation limit of 0.4 ng/μL was reported.

Four cyanogenic glycosides (amygdalin, linamarin, prunasin, mandelonitrile) were extracted from apricots and separated on a C_{18} column ($\lambda = 200$ nm or pulsed amperometric detection, $E_1 = 0.00$ V for 240 ms, $E_2 = 0.60$ V for 60 ms, $E_3 = -0.80$ V for 60 ms). For the amperometric analysis, a 0.5 M NaOH solution was fed into the effluent postcolumn. An isocratic 12 min 82/12 water/acetonitrile run or an 8-min 100/0 → 40/60 water/acetonitrile gradient was used [1259]. Baseline resolution was achieved using the gradient.

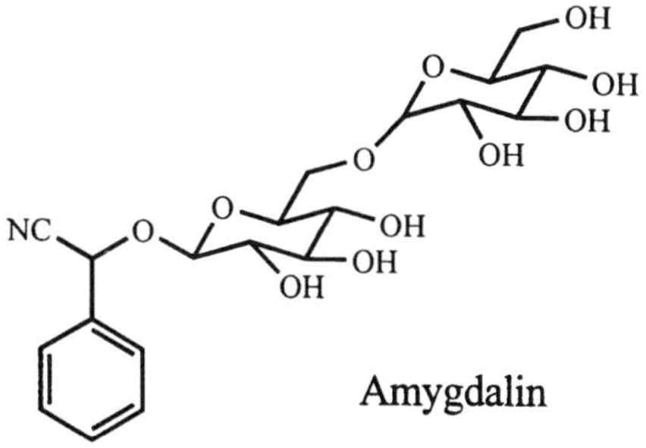

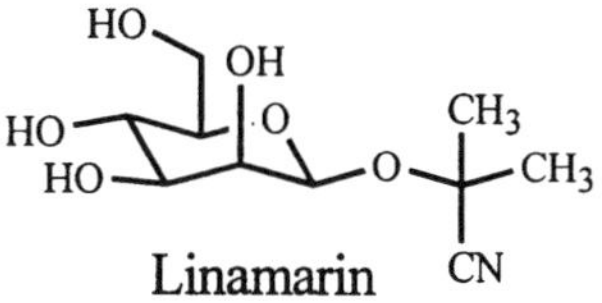

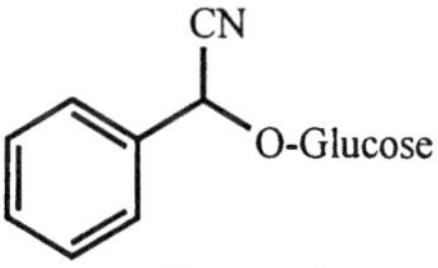

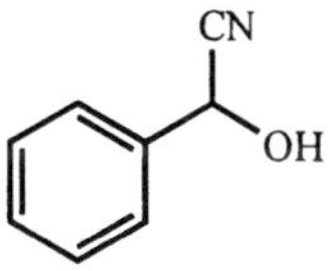

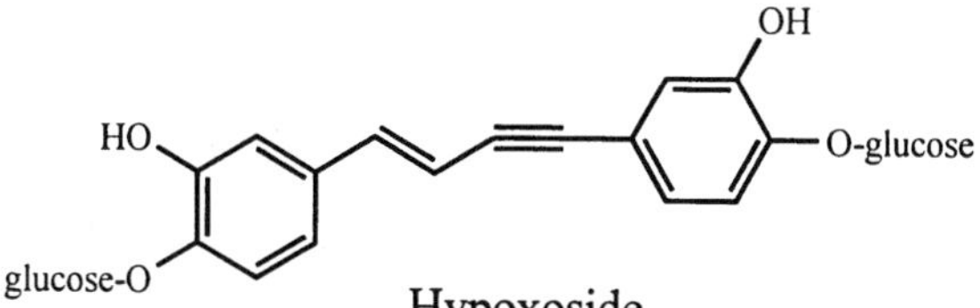

Nine norlignan glucosides (e.g., hypoxoside, nyasol, interjectin, obtuside A and B, and mononyasine A and B) were well resolved on a C_{18} column ($\lambda = 260$ nm)

Safrole

β-Asarone

Amygdalin

Linamarin

Prunasin

Mandelonitrile

Hypoxoside

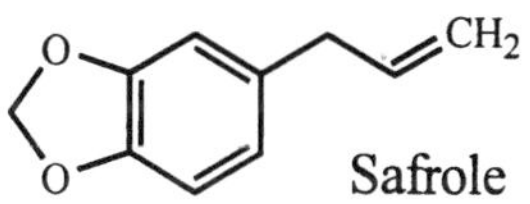

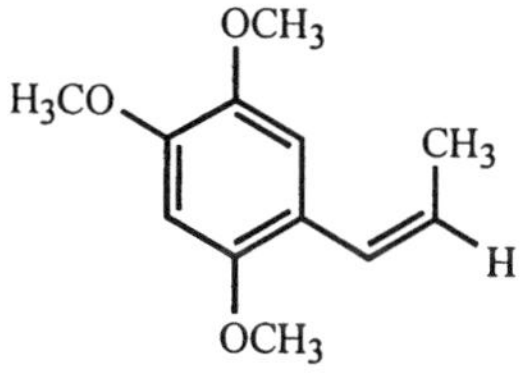

using 20/80 → 70/30 (at 30 min hold 5 min) acetonitrile/water (50 mM phosphate buffer at pH 3.0) gradient [1260]. Working concentration ranges covered the 0.17–15 µg injected levels (analyte dependent). Peak shapes were excellent and elution of the last analyte occurred in 35 min.

Ferulic acid and liguistilide were extacted from *Angelica sinensis* roots and analyzed on a C_{18} column (λ = 303 nm for ferulic acid, 205 nm for liguistilide). A 55/45 methanol/water (1.5% acetic acid) mobile phase generated good peak shapes and elution in 17 min [1261]. Note that the linguistilide work is conducted very near the UV cutoff for methanol. This could lead to reproducibility problems. Linear curves from 10 to 500 µg/mL were reported.

Sixteen shikonin derivatives (red naphthoquinone pigments, e.g., rosmarinic acid, shikonofuran, lithospermic acid, rabdosiin, echinofuran, shikonin, and numerous

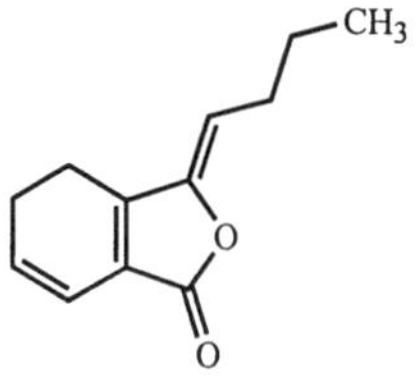

(Z)Ligustilide

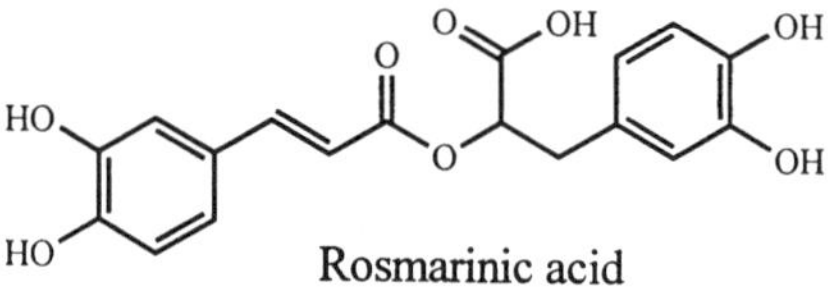

Rosmarinic acid

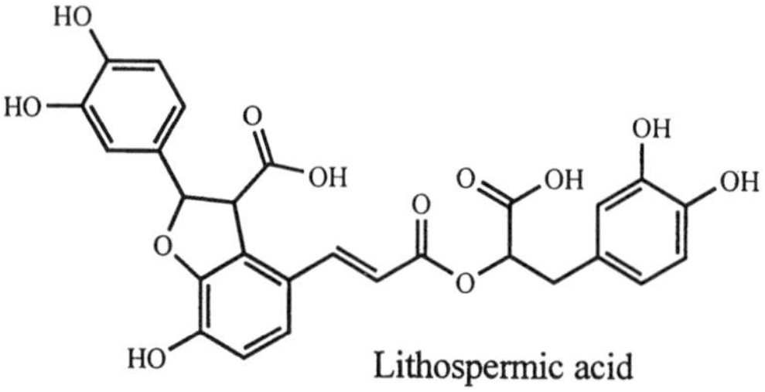

Lithospermic acid

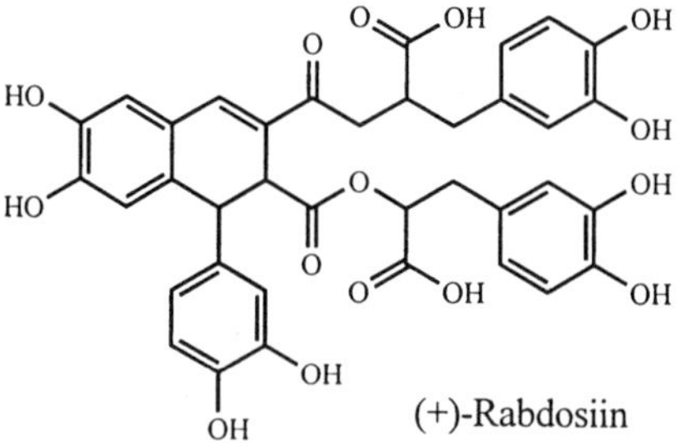

(+)-Rabdosiin

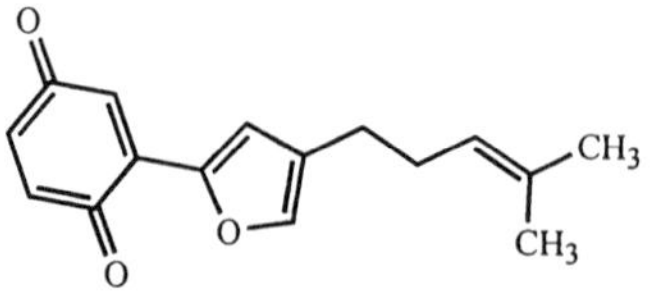

Echinofuran B

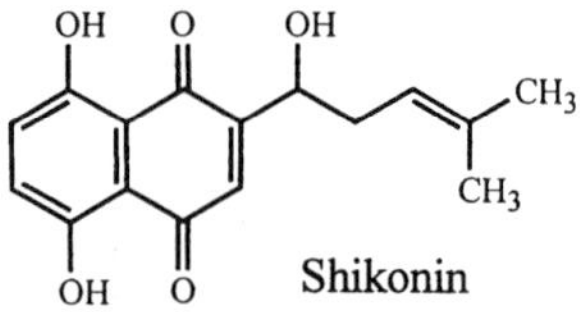

Shikonin

derivatives) were extracted from the root of *Lithospermum erythrorhizon* and separated on a 40°C C_{18} column ($\lambda = 254$ nm) using a water/acetonitrile $40/60 \rightarrow 0/100$ (at 60 min hold 15 min) gradient [1262]. Peak shapes were excellent and separation was outstanding overall. Standards were in the range of 0.3 μg/mL.

Bulk chemical ardacin was analyzed for 10 constituent components (aracidin A, B, B_2, B_3, C_1, C, C_2, D, S, HP_4] on a C_{18} column ($\lambda = 220$ nm) using a 45-min $20/80 \rightarrow 30/70$ acetonitrile/water (10 mM phosphate buffer at pH 6.0) gradient [1263]. This chemical is a fermentation product and the composition varied from lot to lot. Therefore, individual constituent levels must be obtained for each lot. Component levels ranged from 0.5% to 28% (w/w) and were quantitated using 25 μL injections of a 100 μg/mL sample. Excellent resolution and peaks shapes were obtained for these compounds. The method was reported to give linear concentration vs. response results from 50 to 150 μg/mL.

Uro-, heptacarboxyl-, hexacarboxyl (1 and 2)-, pentacarboxyl-, and copropor-phyrin I were resolved on a γ-cyclodextrin column using a 25/75 acetonitrile/water (0.14 M NaH_2PO_4 buffer at pH 6.9) mobile phase (1264). Peaks were quite tailed but separation was achieved nonetheless. Elution was complete in 20 min. A U-shaped plot of k' versus percent acetonitrile was generated in which retention for all compounds decreased for acetonitrile between 20% and 35% and then increased as the acetonitrile level increased to 50%. The authors attribute this result to enhanced cyclodextrin/porphyrin carboxylic acid side chain interactions. Regardless of the retention mechanism, the k' values were generally $\ll 10$ for all mobile phase compositions studied.

A series of α- and β-axial diaquo and aquocyano positional isomer complexes of cobinamide, cobyric acid, and cobinic acids-1, -2 and -3 were separated using a C_{18} column ($\lambda = 365$ nm) and a 30/4/64 acetonitrile/THF/water (80 mM pyridine acetate at pH 3.6) mobile phase [1265]. Retention times varied from 5.4 to 13.6 min and resolution times between isomers was always >0.5 min. The diaquo-cobinamide and diaquocobinic acid-2 solutes generated tailed peaks.

Rontani et al. [1266] studied the photodegradation of chlorophylls using a C_{18} column ($\lambda = 475$ nm) and a 15-min $90/10 \rightarrow 50/50$ acetonitrile/IPA gradient. Fucoxanthin, diadinoxanthin, chlorophyll *a* allomer, chlorophyll *a* and chlorophyll *a* epimer were well resolved. Peak shapes were reasonable and elution was complete in 12 min.

An interesting acetonitrile/methanol/hexane gradient was used to separate eight photosynthesis pigments (β-carotene, lutein, chlorophyll *a* and *b*, violaxanthin, neoxanthin, taraxanthin) extracted from various plant tissues [1267]. A C_{18} column (photodiode array detector, $\lambda = 420$–460 nm) and a 24-min $100/0 \rightarrow 0/100$ (65/15/19/1 acetonitrile/methanol/water/Tris·HCl)/(7/1 methanol/hexane) gradient generated baseline separation.

Hemoglobins (α and β forms of NB C, Hb 0, Hb 1, Hb 2 and Hb (3) were extracted from arctic fish and separated on a C_{18} column ($\lambda = 280$ nm) using a $60/40/0.3$ (hold 5 min) $\rightarrow 40/60/0$ (at 30 min hold 10 min) water/acetonitrile/TFA gradient [1268]. The composition of these hemoglobins was then determined through a tryptic digest on the same type of C_{18} column ($\lambda = 220$ nm) using a two-ramp $100/0$ (hold 5 min) $\rightarrow 50/50$ (at 75 min) $\rightarrow 10/90$ (at 85 min) water (0.1% TFA)/acetonitrile (0.08% TFA) gradient. The ramp time to the $50/50$ varied somewhat (reaching this point anywhere between 75 and 90 min) depending on the complexity of the digest. Peaks and resolutions were very good in all presented separations.

Masala and Manca [1269] characterized and identified globin chains from red blood cells with RP HPLC using a wide pore C_4 column ($\lambda = 220$ nm) and an 80-min $20/80 \rightarrow 60/40$ acetonitrile/water (0.1% TFA) gradient. Blood sample lysates were analyzed. The general elution order was: heme $<$ pre-$\beta < \beta < \delta < \alpha < {}^{G}\gamma < {}^{G}\gamma_1 < {}^{A}\gamma < {}^{A}\gamma_1$. Variants such as ($\beta^x$, $\delta\beta^x$, α^x, and ${}^{A}\gamma^T$) were also separated and identified. The authors noted that a 45 μL lysate sample that was 1% in hemoglobin F (Hb F) contained $\sim$50 μg Hb F, which was enough for one HPLC analysis. Peak shapes were good considering the long elution times for some of the samples.

Four annatto coloring components (*cis*- and *trans*-norbixin and -bixin) were resolved in $<$30 min on a $35°$C C_{18} column ($\lambda = 452$ nm) using a $63/35$ acetonitrile/water (2% acetic acid) mobile phase [1270]. Peak shapes were excellent. Standards were made up in a solution containing 0.2% BHT to prevent oxidation of the analytes. Working concentration standards of 40–400 mg/L generated a linear response for all the analytes.

9.5.9   Summary

As can be seen through the use of acetonitrile with a wide range of compounds (in terms of polarity, functional group composition, and size) it is the most flexible solvent currently used in RP HPLC. Its major limitation to date has been in the few instances where analytes have limited solubility in acetonitrile.

9.6   AMINO ACID, PEPTIDE AND PROTEIN ANALYTES

9.6.1   Amino Acid Analytes

Acetonitrile has found wide use in the separation of amino acids, peptides and proteins. A mainstay separation is that of derivatized amino acids. Classic precolumn derivatization methods include phenyl isothiocyanate (PITC) to give the PTH (phenylthiohydantoin) derivative, dimethylaminonapthalenesulfonyl (dansyl) chloride, *o*-phthalaldehyde (OPA) and 9-fluoromethylchloroformate (FMOC). From there, many variations on a theme have been developed. (The reader is referred to Chapter 4 for some of them.)

Cells and tissue homogenates were extracted to obtain thiols containing amino acids and related compounds (glutatione, glutathione disulfide, cysteine, homocysteine [Hcys], γ-glutylcysteine, cysteinylglycine [CG], *N*-acetylcysteine). These compounds were derivatized with *N*-(1-pyrenyl)maleimide and separated in 15 min on a C_{18} column ($\lambda = 330$ nm, ex; 375 nm, em) with a $65/35/0.1$ acetonitrile/water/H_3PO_4 mobile phase [1271]. Good resolution was achieved for all but the Hcys and CG peaks. A standard curve from 200 fmol to 200 pmol injected was reported. Detection limits of 100 fmol were claimed.

OPA is a reagent that provides very low detection limits. When a 220 nm excitation/450 nm emission fluorescence scheme was used, 50 fmol could be detected [1272]. A major disadvantage to OPA is that it is nonreactive with secondary amino acids (proline, hydroxyproline). Also, even though the derivatization reaction is fast (1–2 min), some compounds do not form stable derivatives (glycine and lysine). A major advantage to OPA is that the reaction is quite clean: no side reactions typically occur. Detection limits of <1 nmol are common. As an example of a modified approach, of which there are many, a very complex acetonitrile/methanol/water (150 mM sodium acetate with 10 mM citric acid and 1 mM EDTA to pH 4.7) gradient on a C_{18} column eluted 25 amino acids and two diamino acids in 24 min with excellent resolution. A linear concentration range of 1–1300 pmol was reported [1273].

In a novel approach, 17 DL-amino acids were generally well resolved on a C_{18} column ($\lambda = 460$ nm, ex; 550 nm, em) using a new derivatizing reagent: $R(-)$-4-(3-isothiocyanatopyrrolidin-1-yl)-7-(N,N-dimethylaminosulfonyl)-2,1,3-benzoxadiazole [1274]. Two 75-min gradient approaches were presented: $70/30/0.1 \to 45/55/0.1$ (at 60 min hold 15 min) water/(7/3 acetonitrile/methanol)/TFA and $75/25/0.1 \to 50/50/0.1$ (at 60 min hold 15 min) water/(9/1 acetonitrile/THF)/TFA. This method was also used in the sequential amino acid profile determination of a 10-amino-acid peptide. Peak shapes were excellent throughout all the separations.

Five biogenic thiols (*l*-cysteine, acetyl-*l*-cysteine, mercaptoacetic acid, γ-glutamylcysteine, glutathione) were isolated from water, derivatized with 7-fluorobenzo-2-oxa-1,3-diazole-4-sulfonate (SBD-F), and analyzed on a C_{18} column ($\lambda = 385$ nm, ex; 515 nm, em). A complex 40-min $90/10 \to 0/100$ water (0.1% TFA)/acetonitrile gradient was used. Standards ranged from 1 to 600 nM with a detection limit of 20 nM reported [1275].

Homocysteine, cysteine, and glutathione were isolated from serum and plasma, derivatized with monobromobimane, and separated on a 35°C C_{18} column ($\lambda = 250$ nm or 300 nm, ex; 470 nm, em). An interesting 25-min gradient based on the volume delivered went from $95/5 \to 55/45$ water (0.1% TFA)/acetonitrile [1276]. A linear range of 0.5–500 nmol/mL and a detection limit (fluorescence) of 5 pmol/mL were reported.

A departure from the acetonitrile/water gradient scheme was presented for 20 amino acids from plant and animal tissue extracts that were analyzed as their FMOC derivatives using a C_{18} column ($\lambda = 264$ nm) and a complex 40-min $100/0/0 \to 50/50/0 \to 0/100/0 \to 0/35/65 \to 0/0/100$ (23/73 acetonitrile/water [15 mM citric acid with 10 mM tetramethylammonium chloride at pH 1.85])/(35/5/60 acetonitrile/THF/water [buffer as above pH 4.5])/(25/62/13 acetonitrile/THF/water [buffer at pH 4.5]) gradient [1277]. An $18/82 \to 99/1$ (9/1 acetonitrile/water)/(15/85 methanol/water [20 mM phosphate buffer pH 6.5]) gradient separated 30 amino acids as their FMOC derivatives on a C_{18} column ($\lambda = 263$ nm, ex; 313 nm, em). The amino acid compositions of four proteins were also tabulated [1278].

Phenylthiocarbamate (PTC) derivatives of cystine and cysteine monomers and dimers have also received attention in their use with MS detectors [1279]. Separation was achieved on a C_{18} column (API MS and photodiode array detector, $\lambda = 200$–350 nm) using a 22-min acetonitrile/methanol/water (50 mM sodium acetate at pH 7.2) gradient. The standards injected contained 1700 pmol of each amino acid. Twenty-six PTC amino acids and galactosamine and glucosamine were resolved on a C_{18} column ($\lambda = 254$ nm) using a complex 70-min 27/75 → 100/0 (60/40 acetonitrile/water)/water (0.14 M sodium acetate with 0.08% TEA at pH 5.5) gradient [1280]. Excellent resolution was obtained.

Proteins (1–5 µg that were recovered from polyacrylamide gels were hydrolyzed and analyzed as their dimethylaminoazobenzene sulfonyl chloride (DAB) and dimethylaminoazobenzene thiohydantoin (DABTH) derivatives. The DAB-amino acids were separated on a C_{18} column ($\lambda = 436$ nm) using a complex 29-min 20/80 → 70/30 (75/25 acetonitrile/IPA)/water (25 mM KH_2PO_4 buffer at pH 6.8) gradient [1281]. Over 27 amino acid peaks were identified and standards of 5–25 pmol were easily detected. The DABTH amino acids were resolved on a C_{18} column ($\lambda = 436$ nm) using a complex 44-min 25/75 → 75/25 acetonitrile/water (25 mM phosphate buffer at pH 6.8) gradient. Standards of 2.5–50 pmol were used. Both methods generated excellent peak shapes and resolution.

α- and β-aspartame in various foods and beverages were separated on a C_{18} column ($\lambda = 210$ nm) using a 20/80 acetonitrile/water (20 mM potassium phosphate buffer at pH 4.0) mobile phase [1282]. The separation was very pH sensitive with retention time, separation, and peak shapes deteriorating below pH 5 and above pH 3. The injection of a 500 ng/mL standard gave detectable peaks. Chromatograms of various powder and soda samples were given. The peaks of interest were completely resolved from other components of the sample. Another sweetener, β-benzyl-N-carbobenzoxy-L-aspartyl-D-alanine and two production contaminants, the aspartic acid and trimethylsilyl ester, were baseline resolved on a C_{18} column ($\lambda = 220$ nm) using a 50/50 acetonitrile/water (0.2 M triethylamine/H_3PO_4 buffer at pH 3) mobile phase [1283]. Elution was complete in <10 min.

9.6.2   Peptide Analytes

In an excellent study, Yoshida et al. [1284], studied the retention characteristics of 100 peptides (ranging from amino acid dimer peptides to peptides comprised of 31 amino acids) on three 40°C supports: amide, diol, and silica ($\lambda = 215$ nm). In each case a 58-min gradient from 90/10/0.2/0.2 → 45/55/0.2/0.2 acetonitrile/water/TFA/triethylamine was run. Retention times ranged from <1 min to almost 40 min. Note that the gradient is *increasing* in water content. This is the opposite of the conventional approach for reversed-phase separations.

This work studied the oxidation of N-terminal seryl and threonyl hexapeptides with periodate and the rate of subsequent cyclization. The first separation baseline resolved the unmodified peptide from the hydrated glyoxylyl peptide from its aldehydic analog. A C_{18} column ($\lambda = 229\,nm$) and a 100/0 (hold 5 min) → 80/20 (at 25 min) water (0.1% w/v TFA)/acetonitrile gradient generated excellent peak shape as well [1285]. The oximation cyclization was monitored through the use of a gradient starting at 100% water (0.1% w/v TFA) held for 5 min then ramped at 1.17% acetonitrile over 40 min ($\lambda = 214\,nm$). Once again, excellent resolution and peak shapes were obtained.

Skin extracts and secretion from the *Rana grylio* frog were isolated, purified, and characterized for peptide content. Eight potential antimicrobial and vasorelaxant peptides (temporins 1Ga, 1Gb, 1Gc, 1Gd; ranatuerins 1Ga, 1Gb; ranalexins 1G, 2G) were found. Initial purification was done using a C_{18} column ($\lambda = 214\,nm$) and a 100/0 → 79/21 (at 10 min) → 51/49 (at 70 min → 30/70 (at 71 min hold 19 min) water (0.1% TFA)/acetonitrile gradient [1286]. Final purification was done using a C_4 column and a 40-min linear gradient from 70/30 to 45/55 water (0.1% TFA)/acetonitrile gradient. Approximately 100 g of skin yielded between 35 and 210 nmol of four of the peptides (480 nmol total).

Galanin (a 29-amino acid peptide) and scyliorhinin I (a 10-amino acid peptide) were isolated from sturgeon and fractionated on a semipreparative C_{18} column ($\lambda = 280\,nm$) using a 100/0 → 79/21 (at 10 min hold 30 min) → 51/49 (at 100 min) water (0.1% TFA)/acetonitrile gradient [1287]. The collected fraction was then subjected to further purification through the sequential injection/fraction collection process on analytical C_4, phenyl, and finally C_{18} columns ($\lambda = 214\,nm$). All used the identical 45-min gradient running from 929/70/1 → 719/280/1 water/acetonitrile/TFA gradient.

The completeness of the deprotection process for solid-phase synthesis of a 9-amino-acid peptide was monitored using a C_{18} column ($\lambda = 220\,nm$) and a 30-min 95/5 → 35/65 water (0.1% TFA)/acetonitrile (0.08% TFA) gradient. Elution of all peaks of interest was achieved in <17 min. Excellent separation between the 9- and three 10-amino-acid peptides was good and peak shapes were excellent [1288].

Three solid-phase synthesized thioredoxin peptides (T37-52, 16 amino acids; T92-107, 16 amino acids; T95-107, 13 amino acids) were purified and analyzed using a C_{18} column ($\lambda = 205\,nm$) and a 40-min 95/5/0.1 → 55/45/0.1 water/acetonitrile/TFA mobile phase [1289]. A noticeable baseline rise was observed in each case due to the low detector wavelength and the presence of TFA. Elution of the peptides occurred between 18 and 22 min. Good separation between impurities and the peptide of interest was obtained in each case.

The peptidic products from the V8-protease hydrolysis of bovine hemoglobin were studied on a C_4 column ($\lambda = 226\,nm$) using a 35-min 100/0 → 33/67 water (0.1% TFA)/(40/60 water [0.1% TFA]/acetonitrile) gradient [1290]. Peptide components from the hydrolysates of 2% w/v hemoglobin samples were readily detected.

A series of peptides, the one of greatest interest being a hypoglycemic octapeptide hormone, were isolated from a crude extract of shrimp sinus glands using a C_{18} column ($\lambda = 206\,nm$) using a 100-min 100/0/0.1 (hold 5 min) → 40/60/0.1 water/acetonitrile/TFA gradient [1291]. The peptide peaks of interest were eluted in the 50–80 min timeframe.

Antimicrobial peptides (ranatuerin-1C, -2Ca and -2Cb, ranalexin-1Ca and -1Cb, 20 to 30 amino acid residue peptides) and five related temporin 13-amino-acid residue peptides were isolated from green frog and partially purified on a C_{18} column ($\lambda = 214$ nm) using a $100/0 \rightarrow 79/21$ (at 10 min) $\rightarrow 37/63$ (at 60 min) water (0.1% TFA)/acetonitrile gradient [1292]. Fractions collected were further purified on a C_4 column using 40-min gradients from $70/30 \rightarrow 45/55$ water (0.1% TFA)/ acetonitrile. A table of amino acid sequences and molecular weights is presented.

An opioid decapeptide, leu-val-val-hemorphin-7, was isolated from broncho-alveolar lavage fluid and purified on a C_{18} column ($\lambda = 214$ nm) using a $100/0 \rightarrow 79/21$ (at 10 min) $\rightarrow 51/49$ (at 70 min) water (0.1% TFA)/acetonitrile gradient. Further purification was achieved using a C_4 column and a 40-min $90/10 \rightarrow 65/35$ water (0.1% TFA)/acetonitrile gradient [1293].

Various tachykinins (substance P-related peptide and neuropeptide γ-related 11- and 21-amino-acid peptides, respectively) were isolated from various fishes (e.g., paddlefish, sturgeon) using a multistep HPLC method. Semiprep cleanup was done on a C_{18} column ($\lambda = 214$ nm) using a $100/0 \rightarrow 79/21$ (at 10 min hold 30 min) $\rightarrow 51/49$ (at 100 min) water (0.1% TFA)/acetonitrile gradient [1294]. Collected fractions were then further sequentially purified on C_4 and phenyl columns ($\lambda = 214$ nm) using a 40-min $86/14 \rightarrow 71/29$ water (0.1% TFA)/acetonitrile gradient.

The production of three peptides, neokyrotropin, VV-hemorphin-4, and brady-kinin-potentiating peptide, was monitored throughout the peptic hydrolysis of bovine hemoglobin using a 65-min $100/0 \rightarrow 33/67$ (at 30 min) $\rightarrow 23/87$ water (0.1% TFA)/(60/40/0.1 acetonitrile/water/TFA) gradient. Peaks of interest were resolved and eluted in 30 min; however, the elution of large peaks was seen until 60 min [1295].

Four lipoheptapeptides, called kurstakins, were extracted from *Bacillus thurin-giensis* and isolated using a semipreparative C_{18} column ($\lambda = 215$ nm) with a 20-min $70/30 \rightarrow 30/70$ water (0.1%TFA)/acetonitrile (0.08% TFA) gradient [1296]. Excellent peak shape and resolution were obtained.

Several *N*-terminally truncated C-type natriuretic peptides were isolated from the venom of the habu snake [1297]. Separation of these components was achieved on a C_{18} column ($\lambda = 214$ nm) using a 30-min $100/0 \rightarrow 40/60$ water (0.1% TFA)/ acetonitrile gradient. Peak shapes were excellent.

A bicyclic 13-amino-acid α-conotoxin chimera was synthesized and purified on a wide-pore C_{18} column ($\lambda = 215$ nm) using a 30-min $95/5 \rightarrow 75/25$ water (0.1% TFA)/(90/10 acetonitrile/water with 0.09% TFA) gradient. The monocyclic peptide and its oxidation product were well resolved from the analyte of interest. Elution was complete in 22 min. A 20 µg injection was readily detected [1298].

Eight defense peptides (e.g., caerulein [10 amino acids], caeridin 1.1 [12 amino acids], caerin 3.1 [22 amino acids]) were isolated from tadpoles and separated on a wide-pore C_{18} column using a $90/10 \rightarrow 30/70$ (at 30 min) water (0.1% TFA)/ acetonitrile gradient [1299]. Very good overall resolution and excellent peak shapes were obtained. The elution order did not follow increasing molecular weight.

A peptidic neurotoxin (37 amino acid residues), huwentoxin II, was extracted from spiders and fractionated using ion-exchange chromatography. The toxin-containing fraction was then analyzed on a C_{18} column ($\lambda = 220$ nm) using a 40-min $100/0 \rightarrow 60/40$ water (0.1% TFA)/acetonitrile gradient [1300]. Huwentoxin II eluted cleanly at 31 min.

Interleukin-6 peptides (from 8 to 21 amino acid residues) were synthesized and subsequently characterized using a wide-pore C_{18} column ($\lambda = 214$ nm) and a 30-min water/acetonitrile/TFA gradient was used. The actual starting and ending compositions were varied according to the peptide analyzed. Initial conditions of 95/5/0.1 to 90/10/0.1 were typical, with final mobile phase compositions ranged from 50/50/0.1 to 20/80/0.1 [1301].

The polypeptide gonadorelin and related analogs goserelin, buserelin, and triptorelin were analyzed on a C_{18} column ($\lambda = 214$ nm). Although each was studied individually, the elution solvent composition for each varied only from 84/16/0.1 to 77/23/0.1 water/acetonitrile/TFA. Elution times were estimated at around 10–15 min. It would therefore be a rather straightforward process to devise a shallow gradient for concomitant analysis of all compounds [1302].

An excellent paper by Sanz-Nebot et al. [1303] focuses on the optimization of separations for therapeutic peptide hormones (e.g., lypressin, oxytocin, bradykinin, triptorelin, buserelin, bovine insulin, salmon calcitonin, met-enkephalin and leu-enkephalin). The optimal mobile phase for a 25°C C_{18} column was 35/65/0.1 acetonitrile/water/TFA. The effects of changing solution pH and percent acetonitrile on both k' and resolution were plotted. The final separation took 28 min with all analytes but calcitonin eluting prior to 8 min.

Leuprolide is a nonapeptide synthetic analog of lutenizing hormone-releasing hormone [1304]. The synthetic mixture and six impurities were separated in 16 min on a C_8 column ($\lambda = 220$ nm) using a 31/69/0.1 acetonitrile/water/TFA mobile phase. Plots of the effects of percent acetonitrile changes on selectivity were presented. Peak shapes were excellent and good resolution was achieved. Solutions were made up at the 1 mg/mL level.

Cardiac troponin I levels in serum are key indicators of myocardial infarction. The identification and quantitation of troponin I were carried out on a C_8 column using positive ion electrospray MS (effluent also monitored at $\lambda = 215$ nm). A 55-min $80/20 \rightarrow 20/80$ water (0.23% TFA)/acetonitrile (0.02% TFA) gradient was used with MS settings: 12 L/min N_2 drying gas at 350°C, nebulizing gas N_2 pressure 25 psi, fragmentor voltage 80–120 V. Elution of the various recombinant troponins occurred between 30 and 40 min [1305].

Funasaki et al. [485] studied the conformational effects of cyclic dipeptides (Xx-L- and D-phenylalanine, where Xx = alanine, valine or leucine; Yy-L- and D-alanine, where Yy = alanine or tryptophan) on retention on a C_{18} column (RI detector) using isocratic 10%, 30%, or 50% acetonitrile in water mobile phases. The elution order for these compounds in 10/90 acetonitrile/water was LL > LD for phenylalanines and LL < DL for tryptophans but was LL < LD and LL < DL for all at 30% and 50% acetonitrile. The authors noted that linear dipeptides always elute LL < LD. Their explanation centers on the position of the phenyl ring with respect to the rest of the cyclic peptide (boat-like conformers).

In an excellent paper, Hearn and co-workers [1306] reviewed 87 papers in which either a C_{18} (78 references), a C_8 (5 references), or a C_4 (4 references) column was used in conjunction with an acetonitrile/water (0.1% TFA) (80 references) or an acetonitrile/IPA/water (0.1% TFA) (7 references) gradient system. The retention contribution for each of 20 amino acid residues was presented graphically for each of 12 different mobile phase/column type combinations (e.g., acetonitrile/IPA/water/TFA and C_{18}; acetonitrile/water/TFA and C_8). Not only are the individual references within this paper valuable, but the initial choice of column type and mobile phase can be attempted if the amino acid composition of the peptide is known.

The vasoconstrictor angiotensin and endothelin peptide groups were extracted from tissue and analyzed using a C_{18} column ($\lambda = 232$ nm). Both groups (angiotensin I, II, II(4-8], III, IV; endothelin 1, 2, 3; and big endothelin) were simultaneously resolved using a $99/1 \rightarrow 78/222$ (at 15 min hold 10 min) (82/18 water/acetonitrile with 4 mM triethylammonium formate [TEAF] and 30 mM formic acid)/(20/80 water/acetonitrile with 4 mM TEAF and 30 mM formic acid) gradient. Excellent resolution and peaks shapes were obtained. A series of chromatograms showing the effect of the removal of all buffer, the presence of TEAF ion only, and the presence of formic acid only led to extremely tailed peaks and very long retention times, poor peak shapes and poor resolution, and good peak shapes but incomplete resolution, respectively. Plots of the effect of changing TEAF concentration and pH were also presented. Detection limits of 3 fmol/g tissue were reported [1307].

Eight cyclic tetrapeptides with the general structure Xx-proline-glycine-Yy, where Xx = Orn, Dab or Dpr and Yy = Glu or Asp, were studied on a C_{18} column ($\lambda = 220$ nm) using a 30-min $0/100/0.25 \rightarrow 70/30/0.025$ acetonitrile/water/TFA gradient [1308]. The spacer groups in the ring system were varied independently from 1 to 4 to give the eight compounds. The retention times for each of these compounds were tabulated and ranged from 12 to 19 min. Interestingly, increased retention time did not strictly correspond to increased carbon content. A series of cyclic tri- to pentapeptides were separated on a C_{18} column ($\lambda = 215$ nm) with a 25-min $1/99 \rightarrow 51/49$ acetonitrile/water (0.1% TFA) gradient [1309]. These peptides included Tyr-Gly-Gly-Phe-Met (3-amino-acid ring), Arg-Gly-Asp-Met (4-amino-acid ring), Tyr-Ala-Ala-Pro-Met (5-amino-acid ring).

Glomosporin, a cyclic depipeptide antifungal, was extracted from a *Glomospora* fermentation broth and analyzed on a C_{18} column (photodiode array) using a complex 20-min $15/85/0.05 \rightarrow 85/15/0.05$ acetonitrile/water/TFA gradient [1310]. Glomosporin eluted at 8 min.

Twenty-three bioactive oligopeptides from 4 to 18 amino acids in size (e.g., leucokinin II, fibrinopeptide, angiotensin II, ACTH 1-10, neurostatin, tyr-somatostatin) were studied on a microbore C_{18} column (electrochemical detector at +0.88 and 0.5 V vs. Ag/AgCl) and a complex 60-min $0/100 \rightarrow 60/40$ acetonitrile/(3/97 IPA/water [0.1% TFA]) gradient [1311]. The authors note that the IPA was present in order to shorten the system re-equilibration time. Various buffers and Cu(II) (at 500 times the analyte concentration) were added to the eluent in a postcolumn step. Calibrations were linear from 0.5 to 100 pmol.

9.6.3   Protein Analytes

Acetonitrile has complex effects on protein retention. Acetonitrile causes denaturation of proteins, as do most organic solvent, which leads to complicated and sometimes unpredictable chromatographic results. This is made even more complex when a gradient elution was added to the variables.

An epidermal growth factor (EGF) was treated with lysyl endopeptidase to produce des-B30 insulin. These compounds were separated on a C_{18} column ($\lambda = 220$ nm) using a 35-min $75/25 \rightarrow 40/60$ water/acetonitrile with 1% TFA gradient [1312]. EGF eluted in roughly 16 min and insulin in 10 min. Peak shapes were excellent.

The metalloenzyme superoxide dimutase was purified from fungal extract by ion-exchange methods and analyzed on a C_{18} column ($\lambda = 214$ nm) using a 60-min $99/1/0.08 \rightarrow 20/80/0.08$ water/acetonitrile/TFA gradient [1313]. Elution of the dismutase was complete after 35 min, so that the gradient could probably be shortened considerably to decrease overall analysis time.

Chicken cholecystokinins 8 and 9 were extracted from intestine and purified by size exclusion and C_{18} reversed-phase HPLC. The final analysis step involved the use of a C_8 column ($\lambda = 214$ nm) and a 45-min $95/5/0.1 \rightarrow 55/450.1$ water/acetonitrile/TFA gradient [1314]. Baseline resolution and excellent peak shapes were obtained.

Insulin was extracted from plasma and analyzed on a C_{18} column ($\lambda = 214$ nm) using a $74/26$ water (0.2 M Na_2SO_4 to pH 2.3 with H_3PO_4)/acetonitrile mobile phase [1315]. Elution was complete in <10 min and peak shape was excellent. A calibration curve from 100 to 800 $\mu IU/mL$ was presented.

Crustacean hyperglycemic hormone (CHH) and vitellogenesis-inhibiting hormone (VIH) were isolated from *Armadillidium vulgare* using a C_{18} column ($\lambda = 214$ nm) and an 80-min $100/0/0.1 \rightarrow 40/60/0.1$ water/acetonitrile/TFA gradient [1316]. Baseline resolution was easily obtained. The tryptic digests of these proteins were characterized using the same setup but a 60-min gradient. The cleavage products from 200 pmol of protein in 100 μL produced readily observable peaks. Peptides of 5 to 42 amino acid residues were separated and further characterized by electrospray MS.

Native and denatured ribonuclease A was characterized using a wide-pore 23°C C_{18} column ($\lambda = 214$ nm) and a 32-min $80/20 \rightarrow 50/50$ water (0.1% TFA)/ (90/10/0.1 acetonitrile/water/TFA) gradient [1317]. The folding and unfolding process was on a timescale such that it could be monitored chromatographically. Ribonuclease A was also reduced with dithiothretol and the native and reduced forms were separated. Samples containing 500 $\mu g/mL$ were used and the resulting products generated easily detected peaks.

Nucleoid proteins were extracted from pea chloroplasts and studied using a C_{18} column ($\lambda = 214\,nm$) and a 54-min $90/10 \rightarrow 55/45$ water (0.1% TFA)/acetonitrile (0.1% TFA) gradient [1318]. Twenty-five separate peaks were identified, but poor overall resolution was achieved. Fractions were collected and tryptic digests were used to further characterize the fractions.

Ovine κ-casein macropeptide was recovered from milk and characterized on a C_{18} column ($\lambda = 214\,nm$). Fractions of aglyco-, nonphosphate-, monophosphate-, and diphosphate- as well as mono- through tetrasaccharide forms of casein were identified. A two-ramp $88.4/21.6 \rightarrow 62.4/37.6$ (at $20\,min$) $\rightarrow 0/100$ (at $40\,min$) water (0.1% TFA)/acetonitrile (0.1% TFA) gradient was used [1319]. Similarly, four casein proteins (β, α_{S2}, α_{S1}, κ) and three other milk proteins (α- and β-lactalbumin, serum albumin) were extracted from ovine milk and analyzed on a 46°C wide-pore C_{18} column ($\lambda = 214\,nm$). A 30-min $67/33/0.1 \rightarrow 51/49/0.1$ water/acetonitrile/ TFA gradient generated excellent separation [1320].

Goat β-lactoglobulin was derived from whey and analyzed on a wide-pore C_4 column ($\lambda = 215\,nm$) using a complex 60-min $100/0/0.1 \rightarrow 0/100/0.1$ water/ acetonitrile/TFA gradient [1321]. β-Lactoglobulin was fully resolved from co-extracted protein serum albumin and α-lactalbumin. Elution was complete in $40\,min$ and no later-eluting peaks were apparent on the chromatogram.

Lysozyme was isolated from chardonnay wines and quantitated on a phenyl column ($\lambda = 225\,nm$) using a 44-min $98.8/1/0.2 \rightarrow 29.8/70/0.2$ water/aceto-nitrile/TFA gradient [1322]. Interestingly, no peaks were visible between $5\,min$ and the elution of lysozyme at $25\,min$, and no later-eluting peaks were noticed. Therefore, an isocratic run should be considered, probably around the $50/50/0.2$ range. The quantitation standard of $75\,mg/L$ generated a $0.06\,AU$ peak.

Armstrong and co-workers [484] considered some of these effects when they separated ribonuclease A, insulin, cytochrome c, lysozyme, bovine serum albumin, myoglobin, and ovalbumin on a C_8 column using a 45-min $10/90 \rightarrow 100/0$ acetonitrile/water (0.1% TFA) gradient. Good resolution and peak shape resulted. However, if these proteins were chromatographed under isocratic conditions, the log k' versus percent acetonitrile plot was strongly U-shaped for all proteins [476]. The common explanation is as follows: At low organic levels in the mobile phase, proteins remain in their native conformation and are eluted as the result of a normal partitioning event. As the organic level increases, retention decreases, as predicted. This indicates that the overall conformation of the protein is only slightly modified. Further increase of the organic level causes large-scale conformational changes and the normally internal hydrophobic amino acid residues now come into contact with the support, so the now partially denatured protein is more retained. As the solvent becomes totally organic in composition, solubility effects can further complicate the retention process. Indeed, split peaks are not atypical when isocratic mobile phase conditions and a denaturing solvent system are used [1323]. Whereas this effect, as noted above, is generally true for all organic mobile phase components, acetonitrile is usually the best-tolerated with respect to not causing protein denaturation [481].

The effect of the level of sample load on the chromatography of proteins has also been considered [1324]. As the sample load of lactalbumin increased from $4\,\mu g$ to $1\,mg$ (C_{18}, $30/70$ acetonitrile/water [$10\,mM$ TFA]) the peak maximum shifted to shorter retention times. It is interesting to note that the 4 μg injection was a broad peak eluting at 9 min. When the chromatographic traces of higher load injections were superimposed, this peak became the "tail" of the larger injection peak. Why? Most likely because there was more than one retention mechanism at work here. Not only does the protein partition onto the hydrophobic octadecyl-bonded moieties, but it also hydrogen bonds with the residual surface silanol groups. The hydrogen bond interactions are very strong (and possibly kinetically controlled) as compared to the partition interactions. However, there are a limited number of these strong adsorption sites. When the number of solute molecules injected begins to exceed the number of strong adsorption sites, the only available retention mechanism is the weaker partitioning mechanism. Hence, the peak maximum will shift to shorter retention times. This does not mean that the hydrogen bond interactions do not take place. These interactions are "seen" as the tail on the protein peaks. When the sample load becomes high enough, this tail, although still present, becomes chromatographically insignificant.

To complicate matters further, proteins are concentrated at the head of the column in a weak mobile phase. If enough protein is injected, a multilayer of protein will form. Since the lower layers cannot desorb until the upper layers desorb, peak tailing can be the result. Therefore, even at very high protein sample loads, a significant tailing problem may persist.

Finally, protein recoveries at low sample loads were usually poor [1325, 1326]. This was most likely due to a combination of "irreversibly" adsorbed proteins, i.e., those proteins that are retained by the strong hydrogen bond interactions, and the loss of area due to the impossibility of accurately integrating a badly tailed peak. It should be noted that irreversible protein loss can occur at metal sites in the HPLC system as well. This fact drove the introduction of "bio-compatible" HPLC components (e.g., titanium, PEEK).

9.6.4 Summary

Acetonitrile (modified with ~0.1% TFA) is the premier organic solvent used in the separation of amino acids, peptides, and proteins. Most separation schemes are developed along the linear of an acetonitrile/water (0.1% TFA) gradient and C_{18}, C_8, or wide-pore C_4 columns. The low viscosity of acetonitrile/water mixtures, the very low UV cutoff, and the comparatively high amount (than of, say, alcohols) of acetonitrile needed to cause protein denaturation make acetonitrile the natural candidate for most separations in this class of compounds.

9.7 PHARMACEUTICAL ANALYTES

9.7.1 Drug Surveys and Screening Procedures

A number of studies have been published that provide systematic approaches to screening and quantitating drug levels. Koves and Wells [1327] generated retention times for 121 basic drug compounds from postmortem blood samples. A C_{18} column (photodiode array detector, $\lambda = 210$–367 nm) was used with a 25/10/5 acetonitrile/water (0.025% H_3PO_4)/water (10 mL/L triethylamine buffer to pH 3.4 with H_3PO_4) mobile phase. Retention times ranged from 3 to 20 min. A phenyl column and a 50/50 acetonitrile/water (0.025% H_3PO_4) mobile phase eluted only half these analytes before the 20-min mark. Representative chromatograms of sets of 12 and 14 well-chosen (so as to show baseline resolution) drug compounds are shown. Working concentration ranges of ~0.1–5 µg/mL (25 µL injected) were tabulated for these compounds. Koves [492] expanded the analyte set to 272 drugs using a C_8 column (photodiode array detector, $\lambda = 210$–367 nm) and a 60/25/15 acetonitrile/water (0.025% H_3PO_4)/water (10 mL/L triethylamine buffer to pH 3.4 with H_3PO_4) mobile phase. Elution was complete in <20 min (~12 compounds did not elute).

In an impressive study, Bogusz and Erkens [1328] tabulated a retention index (based on a 1-nitroalkane scale [1329]) for 383 toxicologically important compounds. These data were generated using a 30-min $0/100 \rightarrow 70/30$ acetonitrile/water (25 mM triethylammonium phosphate buffer at pH 3.0) gradient on a C_{18} column (photodiode array detector). This procedure was used to identify drug components in autopsy blood and liver samples. Absorbance λ_{max} values are given where distinct maxima occur. This work is extremely useful since UV traces for the drug compounds are published as well.

Hill and Kind [1330] expanded the above set to 469 drug compounds. A relative retention index based on 1-nitroalkanes was also used. A 30-min $0/100 \rightarrow 100/0$ acetonitrile/water (80/20 acetonitrile/water (0.15 M H_3PO_4 with 0.05 M triethylamine])/water (0.15 M H_3PO_4 with 0.05 M triethylamine) gradient was used with a C_{18} column (photodiode array detector, $\lambda = 200$–402 nm or 210 nm). Results are tabulated both alphabetically and in an increasing retention index format.

Turcant et al. [1331] generated the same type of database for 350 compounds using a C_{18} column ($\lambda = 210$ nm, 230 nm, 254 nm) and a 20-min $15/85 \rightarrow 80/20$ acetonitrile/water (20 mM H_3PO_4 with 500 µL triethylamine/L) gradient. Retention times were tabulated and ranged from !.5 to 17 min. This method, coupled with photodiode array-generated UV spectra, was used for the identification of drug components in plasma samples. Moving from these excellent general identification studies, the rest of this section will deal with smaller sets of specific pharmaceutical (and related) compounds.

9.7.2   NSAIDs and Analgesic Drugs

Due to side effects of administration of fenbufen and ultimately its metabolite felbinac, with lomefloxacin, these compounds were extracted from plasma and

Fenbufen

Lomefloxacin

quantitated on a strong anion exchange column ($\lambda = 280$ nm) using a 90/10 water (0.1 M phosphate buffer at pH 7.0)/acetonitrile mobile phase [1332]. Excellent peak shapes and resolution were achieved and elution was complete in <8 min. Authors reported calibration standard ranges from 0.2 to 30 µg/mL and detection limits of ~0.05 µg/mL (S/N = 3).

The simultaneous assay of six oxicam NSAIDs (piroxicam, tenoxicam, lornoxicam, cinnoxicam, isoxicam (IS), meloxicam) in pharmaceutical preparations was accomplished on a C_{18} column ($\lambda = 320$ nm for cinnoxicam and 360 nm for all others) using a 715/220/50/15 water (15/85 50 mM Tris/50 mM acetic acid)/acetonitrile/THF/water (8.8 g ammonium phosphate with 25 mM tetrabutylammonium hydroxide to pH 7.5 with H_3PO_4) mobile phase [1333]. Good peak shape, excellent resolution, and elution complete in 25 min were obtained. The calibration curve was reported as 1–11 mg/L with detection limits of 3 ng/injection (S/N = 3) and quantitation limits of 10 ng/injection (S/N = 10).

The analgesic morphine and three metabolites (morphine-3- and 6-glucuronide, hydromorphone) were isolated from plasma and separated on a 35°C C_{18} column (multi-channel electrochemical detector set at 0.2 V, 0.35 V, and 0.4 V and $\lambda = 275$ nm, ex; 345 nm, em for morphine-3-glucosonide only). A 25/75 acetonitrile/water (50 mM phosphate pH 2.1 with 2.5 mM sodium dodecyl sulfate) mobile phase generated baseline resolution in 24 min. Linear ranges of 5–600 ng/mL and detection limits of 0.2 ng/mL were reported [1334].

Mikami et al. [1335] studied three anthranilic acid derivatives (mefenamic acid, flufenamic acid, tolfenamic acid) by extracting them from urine and separating them

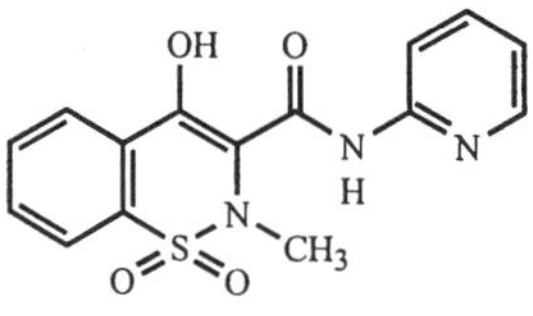

Piroxicam

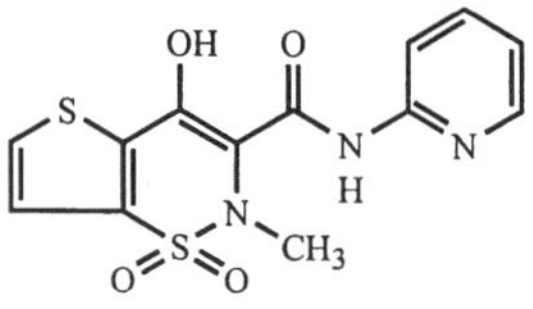

Tenoxicam

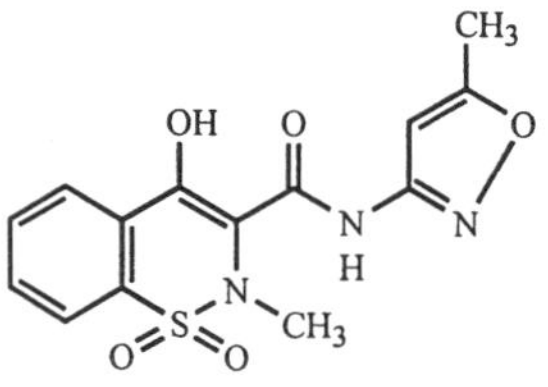

Isoxicam

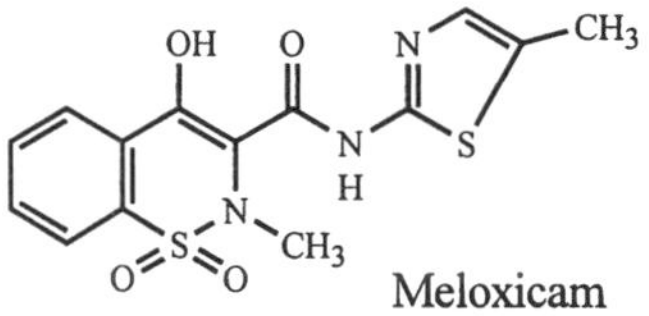

Meloxicam

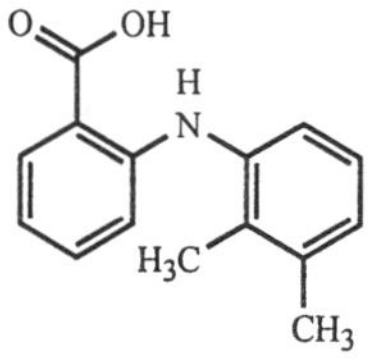

Mefenamic acid

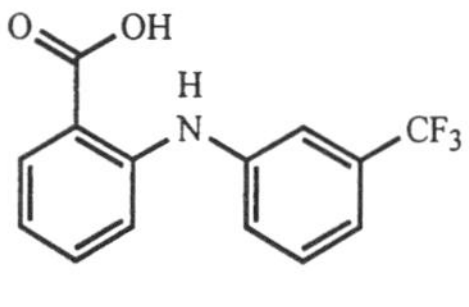

Flufenamic acid

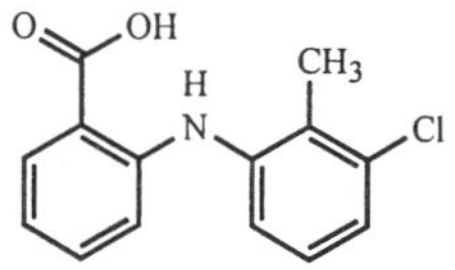

Tolfenamic acid

on a 40°C C_{18} column ($\lambda = 230$ nm and 280 nm). A 45/55 acetonitrile/water (0.14 M sodium acetate with 0.06 M acetic acid) with 1.9 g of tetrapentylammonium bromide (TPAB)/L mobile phase generated baseline resolution and elution in 25 min. A plot of the effect of changing TPAB concentration on k' was presented. A linear range of 2–30 μg/mL and detection limits of ~2 ng injected (S/N = 3) were reported.

Flurbiprofen and its metabolite, 4'-hydroxyflurbiprofen, were isolated from urine and plasma and quantitated on a C_{18} column ($\lambda = 260$ nm, ex; 320 nm, em) using a 40/60 acetonitrile/water (20 mM K_2HPO_4 at pH 3.0) mobile phase [1336]. Peaks were baseline resolved and elution was complete in 18 min. Linear curves of 0.05–15 μg/mL were used and quantitation limits of 0.05–0.5 μg/mL were reported (analyte and matrix dependent).

The NSAID prodrug nabumetone is rapidly converted to the active compound 6-methoxy-2-naphthylacetic acid, which is then isolated from plasma and analyzed on a C_8 column ($\lambda = 280$ nm) using a mobile phase 45/55 acetonitrile/water (20 mM citric acid at pH 2.8) with 300 mg/L heptane sulfonate adjusted to an apparent pH of 3.1 with H_3PO_4 [1337]. Elution was complete in <4 min. A linear range of 0.07–145 μg/mL was used. A quantitation limit of 70 ng/mL (S/N = 8) was reported.

Oxaprozin and nine related compounds and manufacturing precursors and by-products (e.g., benzoic acid, benzaldehyde, benzoin, benzoin hemisuccinate, 4,5-diphenylimidazole-2-propionic acid, diphenylethanedione) were baseline resolved on a C_{18} column ($\lambda = 254$ nm) using a 25/25/50 acetonitrile/methanol/water (10 mM KH_2PO_4 and 5 mM decanesulfonic acid to pH 4.2) mobile phase [1338]. Elution was complete in 25 min. Detection limits of 50 ng/mL were reported. Two acidic compounds, oxaprozin and phenanthro[9,10-d]oxazole-2-propionic acid, were quite tailed. If this was due to incomplete protonation of the analytes, then a slightly lower pH solvent will help correct the tailing problem.

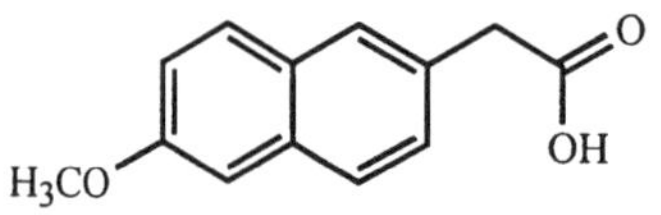

6-Methoxy-2-naphthylacetic acid 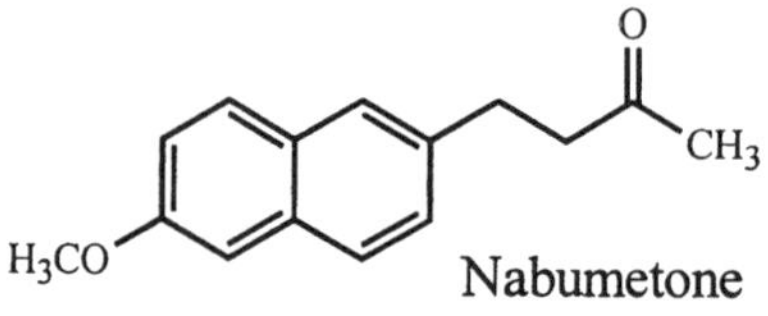   Nabumetone

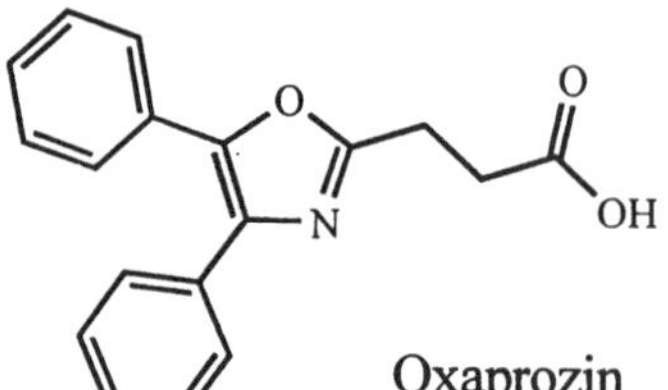

Oxaprozin 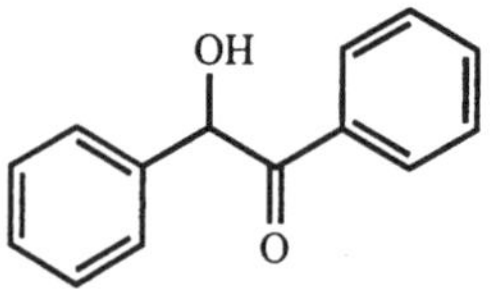   Benzoin

Sulindac and two metabolites (the sulfone and sulfide) were extracted from serum and eluted in 20 min from a C_8 column ($\lambda = 232$ nm, ex; 335 nm, em) using a 55/45 acetonitrile/water (68 mM phosphate buffer at pH 2.5). Peak shapes were excellent. Limits of quantitation of 50 ng/mL and linear concentration ranges of 50–1000 ng/mL were reported [1339].

Ibuprofen and fenbuprofen isomers were extracted from plasma and derivatized with $(R)(+)$-α-phenylethylamine [1340]. The resulting four products were resolved on a C_{18} column ($\lambda = 225$ nm) using a 46.5/53.5/0.1/0.03 acetonitrile/water/acetic acid/TEA mobile phase. Peak shapes were good, as was the resolution. Elution was complete in 25 min. Linear calibration curves over the 0.25–50 µg/mL range were reported.

Indomethacin, hydrocortisone, dexamethasone, phenylbutazone, and oxyphenbutazone were extracted from equine serum and baseline resolved on a C_{18} column ($\lambda = 254$ nm) using a 51/49/0.1 acetonitrile/water/TFA mobile phase [1341]. Excellent peak shape and compete elution were achieved in 18 min. The authors documented the effect of evaporation to dryness and extraction solvent pH on the

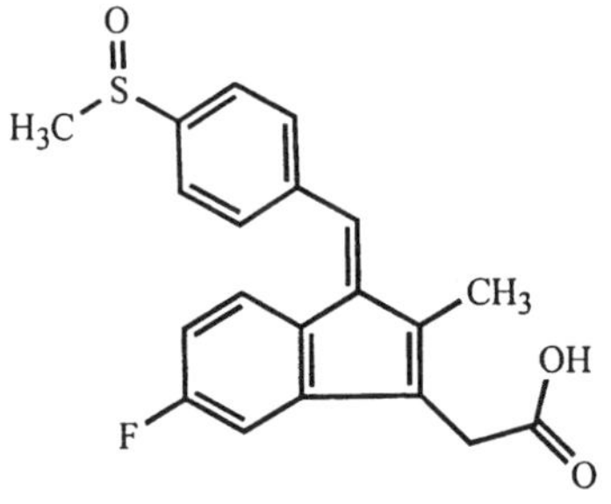

Sulindac

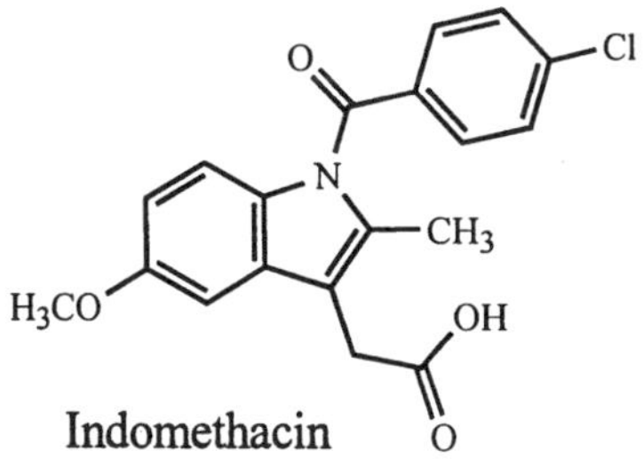

Indomethacin

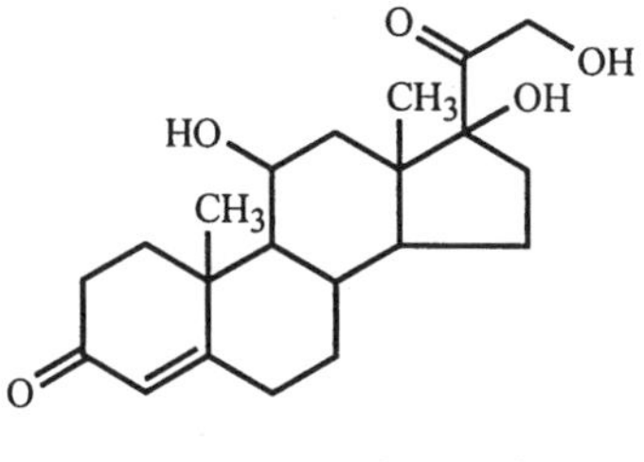

Hydrocortisone

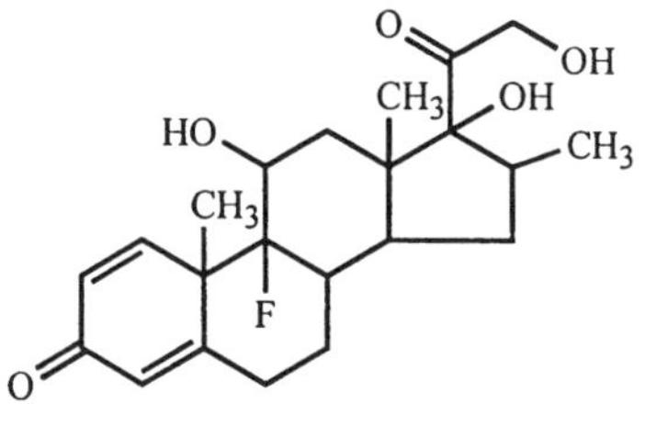

Dexamethasone

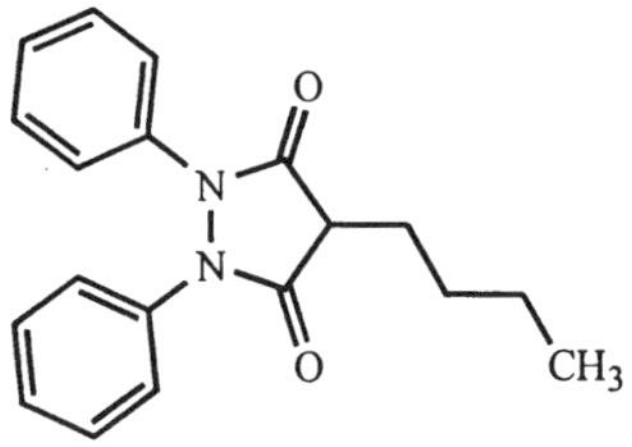

Phenylbutazone

degradation of the samples. Standards ranged from 0.025 to 0.5 mg/mL and detection limits were 0.25–1 μg/mL (S/N = 2, analyte dependent).

Carprofen and the acidic, basic, and photoirradiated breakdown products were separated on a C_{18} column (λ = 270 nm) using a 50/49/1 acetonitrile/water/acetic acid mobile phase. Photodegradation generated at least seven decomposition compounds, whereas acid/base degradation generated two. Separation was adequate and elution was complete in 15 min (<10 min for carprofen alone). A linear range of 2.5–80 μg/mL was reported [1342].

9.7.3   Antibiotic Drugs

Calphostin C is a naturally occurring perylenequinone antibiotic. It was extracted from plasma and then quantitated on a C_{18} column (λ = 479 nm) using a 70/30/0.1/0.1 acetonitrile/water/TFA/triethylamine mobile phase [1343]. The peak eluted interference-free at about 8 min. The authors note that methanol was not effective and that TEA gave critical to good chromatographic performance. A linear range of 0.05–40 μM and a detection limit of 0.02 μM (S/N = 3) was reported.

Cefotaxime, desacetylcefotaxime, ofloxazine, and ciprofloxazine were extracted from ocular aqueous humor and separated in under 15 min. Excellent peak shapes were generated on a C_{18} column (λ = 285 nm) using a 79/15/6 water (10 mM NaH_2PO_4 to pH 3.0 with H_3PO_4)/acetonitrile/N,N-dimethylformamide mobile phase [1344]. Samples were diluted in an HCl-modified buffer. Due to the

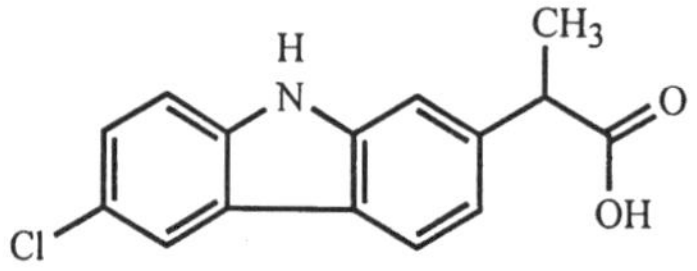

Carprofen

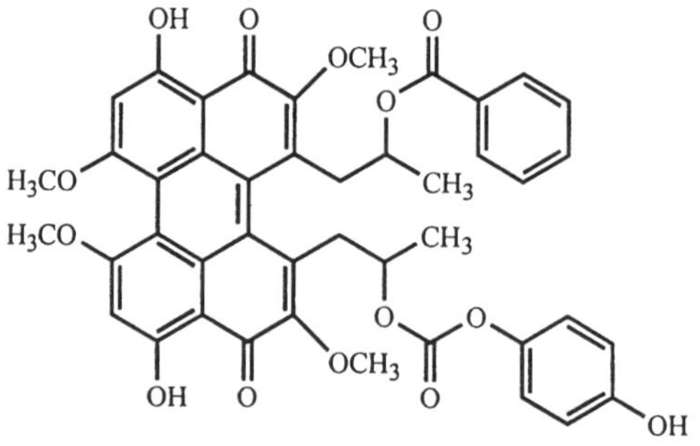

Calphostin C

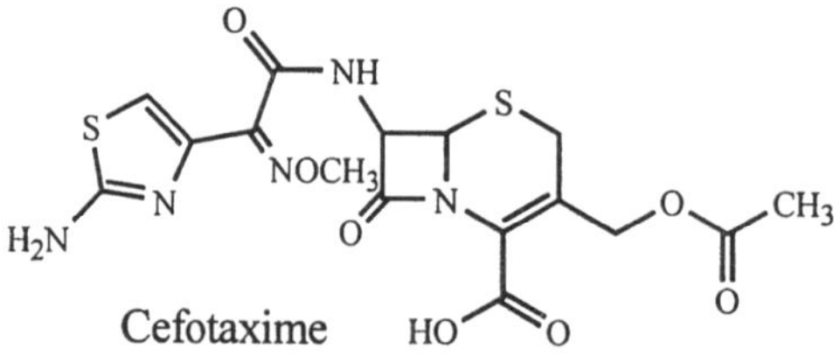

Cefotaxime

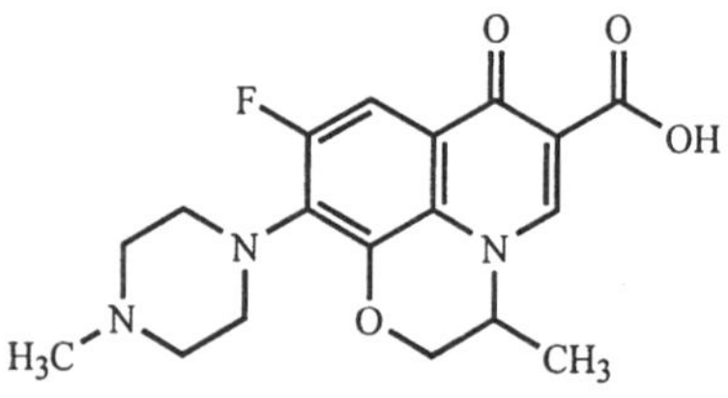

Ofloxazine

aggressiveness of HCl, a better choice of acid is TFA or H_3PO_4. The authors note that peak splitting of desacetylcefotaxime occurs if the sample is injected from methanol. This is a classic example of the ramifications of a solvent mismatch. The mobile phase should be used as the sample diluent whenever possible. A linear range of 0.08–20 μg/mL and a limit of detection of 0.08 μg/mL were reported.

The macrolide antibiotics clarithromycin (and its metabolite 14-hydroxyclari-thromycin) and azithromycin (roxithromycin internal standard) were isolated from plasma and quantitated on a cyanopropyl column (electrochemical detection at +0.85 V). Azithromycin was eluted in 12 min with a 500/600/50 water (50 mM phosphate at pH 6.8)/acetonitrile/methanol mobile phase [1345]. The clarithromy-cins were separated and eluted in 20 min with a 450/300/50 water (50 mM phosphate at pH 7.5)/acetonitrile/methanol mobile phase. The authors noted that phosphate buffers were chosen over ammonium buffers because the background noise was considerably higher with the latter. Linear ranges in the 0.025–5 μg/mL range and detection limits of 0.5–1.5 ng injected (S/N = 3) were reported (analyte dependent).

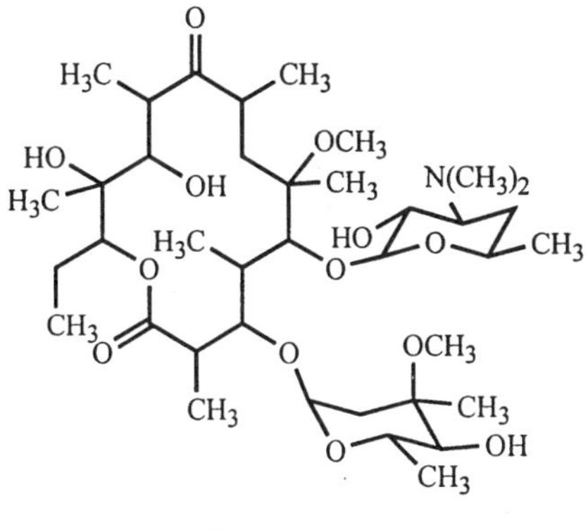

Clarithromycin

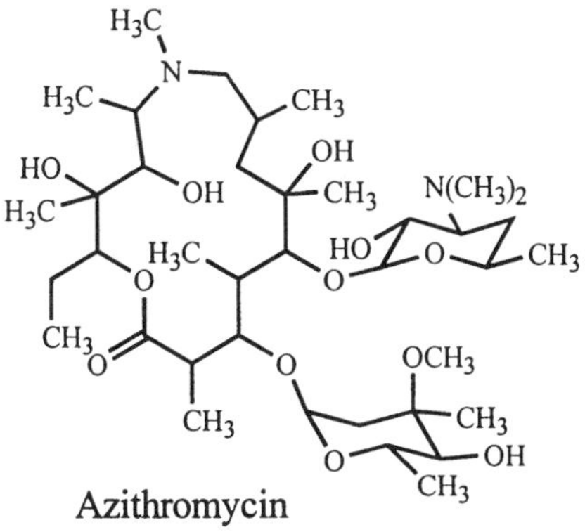

Azithromycin

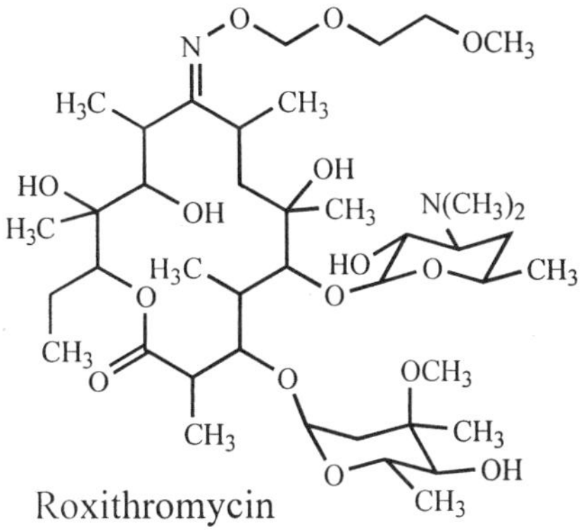

Roxithromycin

Four macrolide antibiotics (azithromycin, erythromycin, roxithromycin, clarithromycin) were extracted from serum, derivatized with FMOC-Cl, and separated on a 50°C base deactivated C_{18} column ($\lambda = 225$ nm, ex; 315 nm, em). A 40/60 water (50 mM KH_2PO_4 with 500 μL triethylamine/L to pH 7.5 with 10% KOH)/ acetonitrile mobile phase generated baseline resolution and complete elution in 22 min [1346]. Calibration curves of 0.1–20 mg/mL and quantitation limits of 1–12 mg/L were reported.

Tanase et al. [1347] isolated four tetracycline antibiotics (tetracycline, demeclocycline, minocycline, oxytetracycline) from powdered teeth. Separation was achieved on a C_{18} column ($\lambda = 354$ nm) using a 74/26 water (50 mM NaH_2PO_4 at pH 1.75)/acetonitrile with 10 mM sodium pentanesulfonate mobile phase. Plots of k' vs. ion pair carbon chain length and ion-pair reagent concentration were presented. Peak shapes and resolution were excellent. Elution was complete in 10 min. Linear ranges of 10 ng/mL to 7.5 μg/mL were reported.

Oxytetracycline and two degradation products (tetracycline and chlorotetracycline) were isolated from sediment and resolved on a C_{18} column ($\lambda = 355$ nm) using a 20-min 70/20/10 → 30/20/50 water (10 mM oxalic acid at pH 2.3–2.55)/methanol/acetonitrile gradient [1348]. Standards of 100 ng injected were readily detected.

Three tetracycline antibiotics (tetracycline, minocycline, demeclocycline) were isolated from serum and separated on a C_8 column ($\lambda = 350$ nm) using a 91/7/2/0.1 water/acetonitrile/methanol/TFA mobile phase [1349]. The authors noted that samples prepared in methanol exhibited severe peak fronting, whereas those made up in the mobile phase produced symmetric peaks. Note that the optimal situation

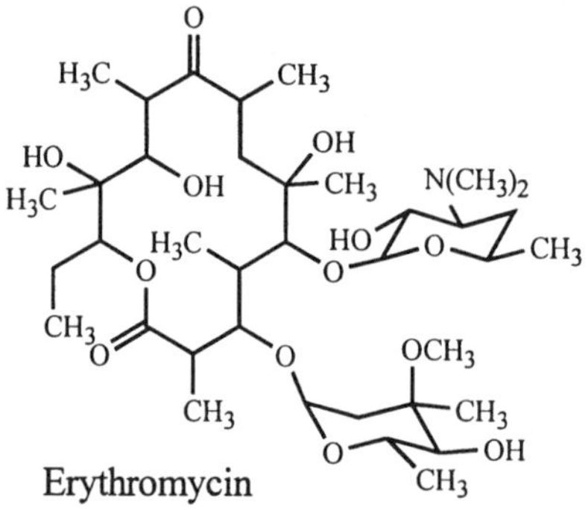

Erythromycin

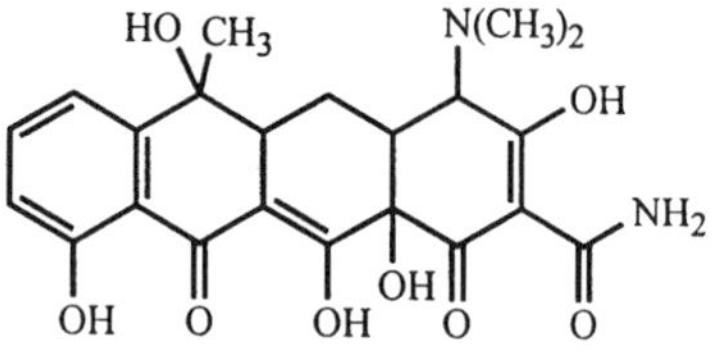

Tetracycline

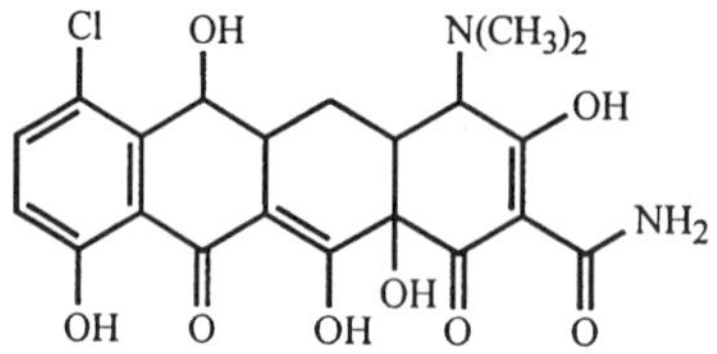

Demeclocycline

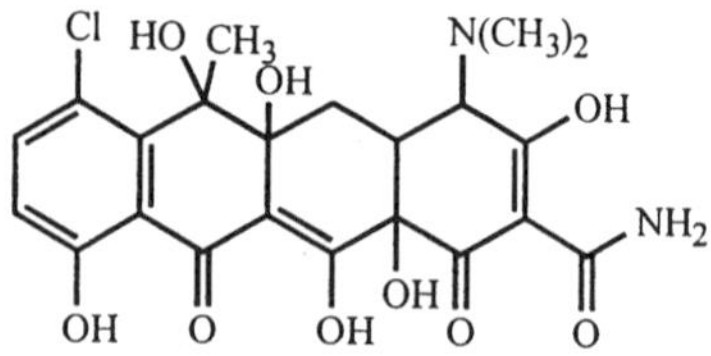

Chlorotetracycline

has a sample prepared in the mobile phase to avoid potential solvent mismatch problems. Elution was complete in 25 min and linearity was achieved from 2 to 255 µg/mL.

The compound 6-pentyl-α-pyrone was extracted from *Trichoderma harzianum* and analyzed on a C_{18} column ($\lambda = 300$ nm). It was separated from other extracted compounds using a $100/0 \rightarrow 80/20$ (at 20 min hold 10 min) water (2.5% acetic acid)/acetonitrile gradient. The pyrone had a retention time of 21 min [1350].

Rapamycin and its C_{39} demethylated metabolite were isolated from pig liver microsomes and separated on a 40°C deactivated base C_{18} column ($\lambda = 276$ nm) using a 35-min $48/52 \rightarrow 100/0$ acetonitrile/water gradient [1351].

Spiramycin is used as a growth promoter and was extracted from chicken eggs and tissues and analyzed on an aminopropyl column ($\lambda = 231$ nm). A mobile phase of 85/15 acetonitrile/water eluted the analyte prior to a large series of co-extracted material. Elution of the analyte occurred at <2 min, which for a 25 cm column operated at 1 mL/min seems to place the peak painfully close to the system void volume. This could lead to quantitation issues. Also, prior clean-up of the sample (e.g., with solid phase extraction) could remove the late-eluting peaks and markedly increase sample throughput and allow for a weaker mobile phase to be used (to move the analyte away from the system void area). Spike levels of 0.1–1.0 ppm were used and a detection limit of 0.1 ppm was reported [1352].

The antibiotic spiramycin consists of multiple components labeled as spiramycin I, II, and III. The ratio of these components varies between manufacturers. These components were baseline resolved in 30 min on a C_{18} column ($\lambda = 232$ nm) using a 20/80 acetonitrile/water (50 mM H_3PO_4 at pH 3.0) mobile phase. Standards of 500 µg/mL were easily detected [1353].

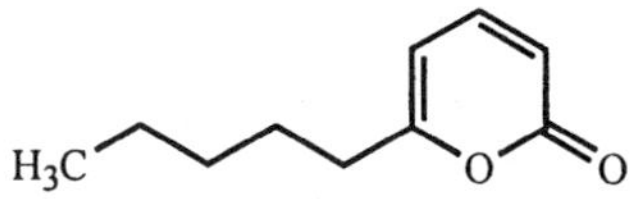

6-Pentylpyrone

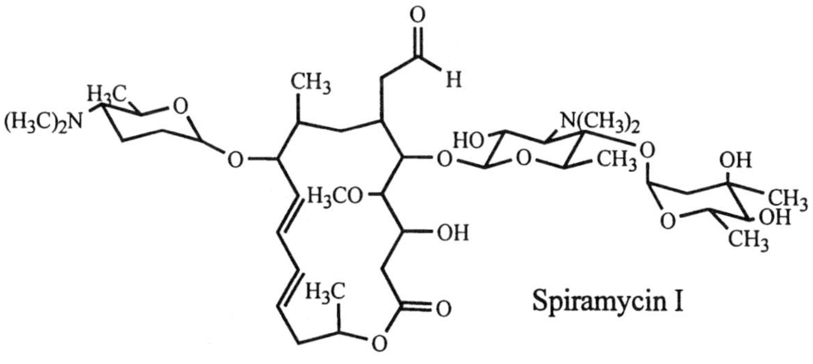

Spiramycin I

A series of six antimycins (A_1, A_2, A_3, A_4, A_7, A_8) were isolated from *Streptomyces* and preparatively separated on a C_{18} column ($\lambda = 210$ nm) using a 30-min $75/25 \rightarrow 90/10$ acetonitrile/water gradient. Good resolution was obtained [1354].

A method by Harang and Westerlund [1355] focuses on the optimization of the separation of erythromycin and eight related compounds: erythromycins A, B, C, and E, erythromycin A enol ether, anhydroerythromycin A, *N*-desmethylerythromycin A, and erythromycin A *N*-oxide. The final separation was run on a 40°C C_{18} column ($\lambda = 215$ nm). A buffer (15.8 g [NH_4]H_2PO_4 with 40.8 g tetrabutylammonium hydrogensulfate/L to pH 7.0 with NH_4OH) was mixed with acetonitrile to generate a 39-min (hold final conditions for 11 min) $71.2/11.0/18.8 \rightarrow 50.4/10.0/39.6$ water/buffer/acetonitrile gradient. Plots of the effect of temperature, ion-pair reagent concentration, and pH were plotted. Good separation for all peaks was generated and the final peak eluted at 47 min.

Dirithromycin purity was determined via separation from erythromycylamine, erythromycin hydrazone, and epidirithromycin on a C_{18} column ($\lambda = 205$ nm) using a 44/19/37 acetonitrile/methanol/water (50 mM potassium phosphate buffer at pH 7.5) mobile phase [1356]. Loading levels of 4–20 μg were easily detected. Peak shape was strongly dependent upon methanol concentration, with tailing decreasing as the methanol level increased. Retention time for all the dirithromycin peaks remained almost invariant over the range of methanol levels examined.

The levels of erythromycins A, B, C, D, E and *N*-desmethylerythromycin A in erythromycin ethylsuccinate were determined by separation on a C_{18} column ($\lambda = 215$ nm) using a 42.5/5/5/47.5 acetonitrile/water (0.2 M tetrabutylammonium hydrogensulfate at pH 6.5)/water (0.2 M phosphate buffer at pH 6.5)/water mobile

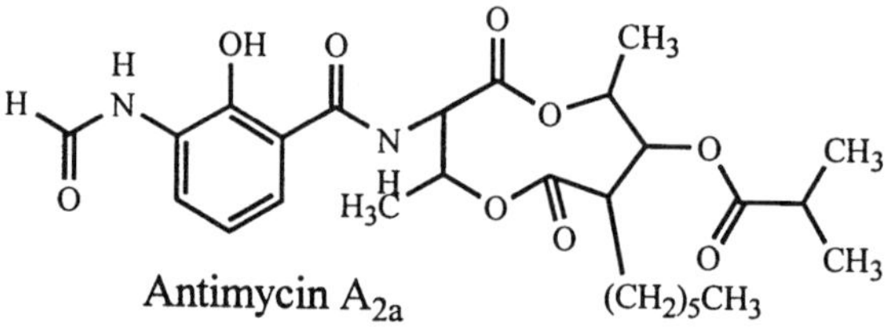

Antimycin A$_{2a}$

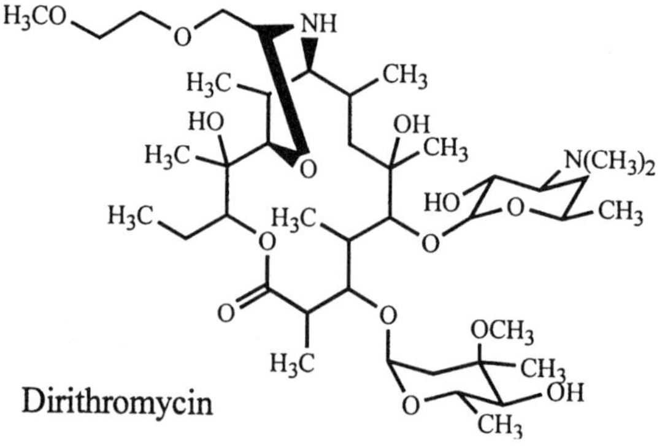

Dirithromycin

phase [1357, 1358]. Erythromycin A eluted at 22 min and the other compounds eluted from 8 to 90 min. The compositions of a number of bulk samples were tabulated with impurity percentages running from 0.1% to 3%. Limits of quantification (300 µg sample injected) ranged from 0.05% to 0.6% (analyte dependent). The efficiency of the method would be improved with a gradient elution scheme.

Five cephalosporins (cephalexin, cefotaxime, cefazolin, cefuroxime, cefotixin) were recovered from urine and separated on a C_{18} column ($\lambda = 254$ nm) using a two-ramp 90/10 → 80/20 (at 2 min, hold 4 min) → 50/50 (at 8 min) water (50 mM NaH_2PO_4 at pH 3)/acetonitrile gradient [1359]. Elution was complete in 6 min with good resolution. Linear ranges of 1.25–500 µg/mL and detection limits of 0.25 µg/mL were reported.

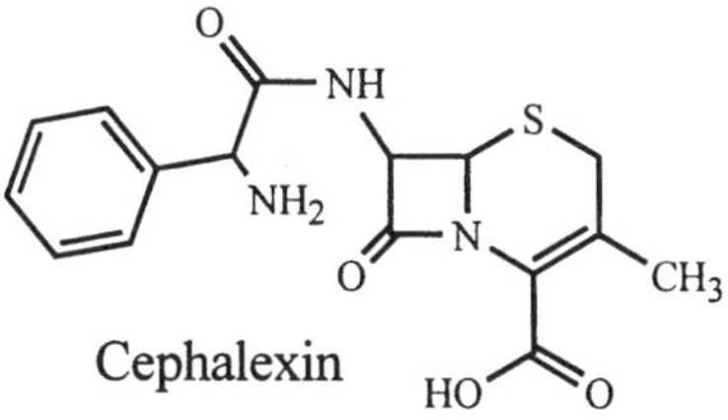

Cephalexin

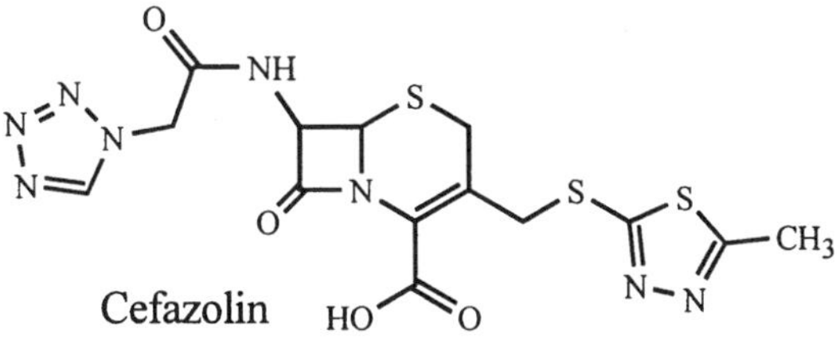

Cefazolin

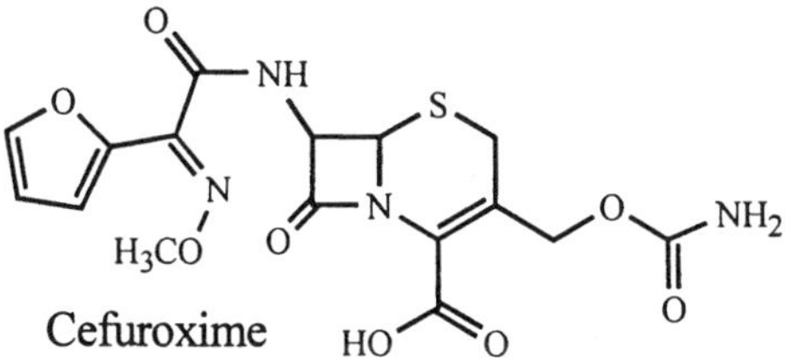

Cefuroxime

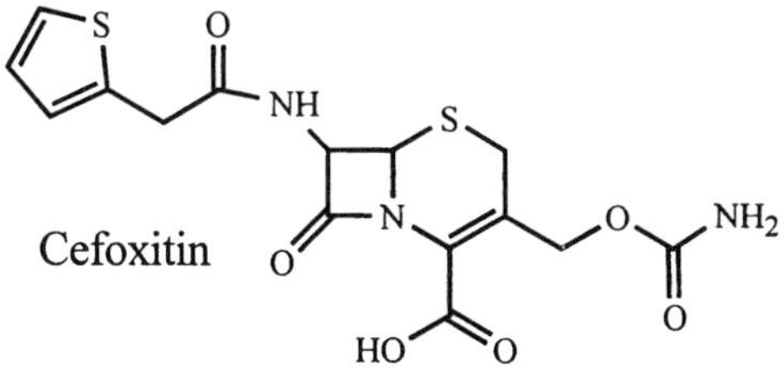

Cefoxitin

Four cephalosporin antibiotics (cefazolin, cefoperazone, cefquinome, ceftiofur) were extracted from bovine milk and analyzed on a phenyl column ($\lambda = 270$ nm). A complex 45-min 86.5/12/1.555/30/15 water (5 mM octanesulfonic acid to pH 2.52 with H_3PO_4)/acetonitrile/methanol gradient [1360] gave excellent peak shapes and resolution. Calibration standards of 50–200 ng/mL were used and detection and quantitation limits of $\sim$10 μg/kg (S/N = 3) and $\sim$15 μg/kg (S/N = 6), respectively, were reported.

Five β-lactam antibiotics (ampicillin, cloxacillin, dicloxacillin, nafcillin, benzyl-penicillin) were isolated from milk and separated on a C_{18} column ($\lambda = 210$ nm) using a 20/10/80 acetonitrile/methanol/water (50 mM KH_2PO_4 to pH 3.5 with H_3PO_4) with 5 mM sodium decanesulfonate. Excellent peak shapes and resolution were obtained with complete elution in 30 min [1361]. Standards were generated from 0.05 to 0.3 μg injected. Detection limits were reported as 0.1–0.35 μg/g (analyte dependent).

Cephapirin (a β-lactam antibiotic) and its metabolite desacetylcephapirin were isolated from milk or beef muscle and separated on a C_{18} column ($\lambda = 295$ nm) using a 40-min $100/0 \rightarrow 40/60$ water (10 mM KH_2PO_4)/acetonitrile gradient

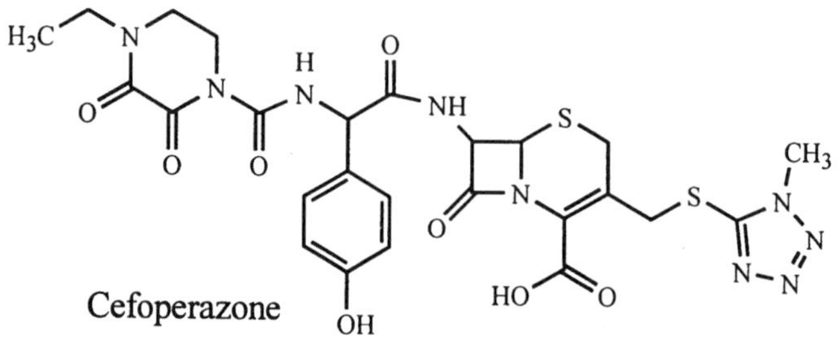

Cefoperazone

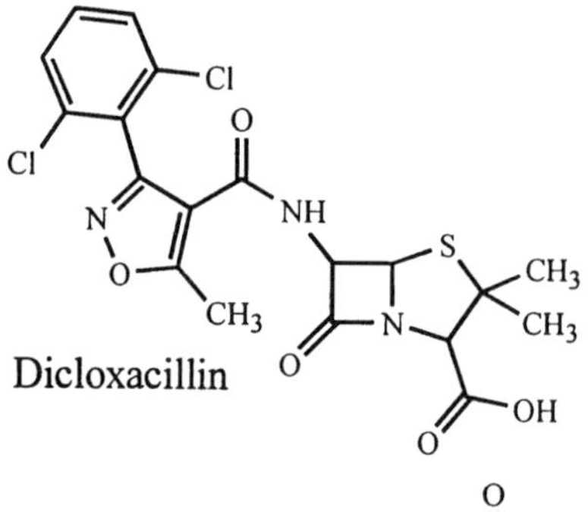

Dicloxacillin

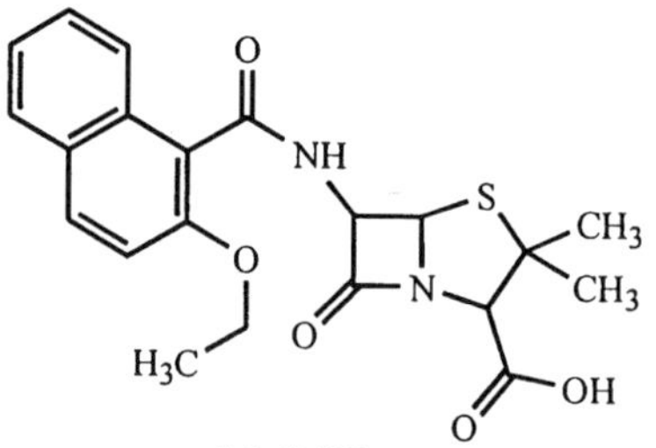

Nafcillin

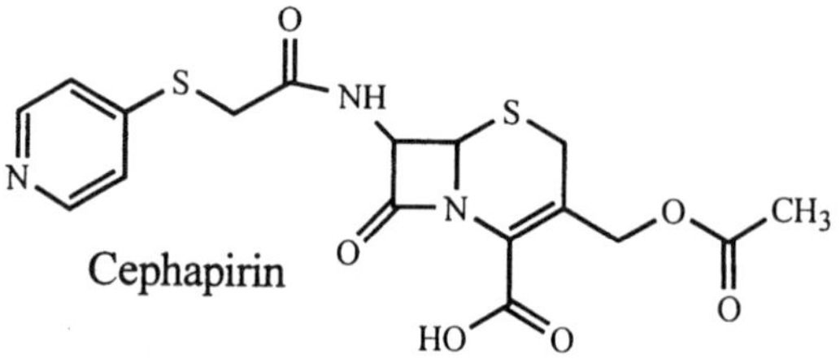

Cephapirin

[1362]. Elution was complete in 22 min and the analytes were well separated from one another, but the metabolite is not cleanly resolved from a co-extracted material in beef muscle samples. Standards of 1–100 µg/mL were utilized.

Kirkland et al. [1363] used a C_8 column and tabulated the mobile phase composition needed to obtain a k' value of ~3: nafcillin, 44/56 acetonitrile/water (0.1% TFA) at 235 nm; cefazolin, 20/80 acetonitrile/water (0.1% TFA) at 272 nm; chloramphenicol, 25/75 acetonitrile/water (0.1% TFA to pH 3 with TEA) at 278 nm; trimethoprim, 16/84 acetonitrile/water (0.1% TFA to pH 3 with TEA) at 220 nm; imipenem, 0.5/99.5 acetonitrile/water (0.1% TFA to pH 7 with TEA) at 296 nm.

Ampicillin, cloxacillin, and their related impurities, such as manufacturing precursors and acid hydrolysis decomposition products (e.g., 6-aminoampicillic acid, phenylglycine, 3-(2-chlorophenyl)-5-methylisoxazole-4-carbonyl chloride), were isolated from capsules and tablets and separated on a C_{18} column ($\lambda = 230$ nm). At least 15 components were found and overall good resolution and peak shape was obtained in 20 min using a 15/85 acetonitrile/water (20 mM phosphate at pH 2.0) made 100 mM in sodium dodecylsulfate [1364]. Parent compound concentration ranges were reported as 50 µg/mL to 300 µg/mL. Detection limits ranged from 20 to 100 ng/mL (analyte dependent).

Penicillin G, procaine, amoxicillin, cloxacillin, ampicillin, cephapirin, and ceftiofur were extracted from milk and baseline resolved on a 40°C phenyl

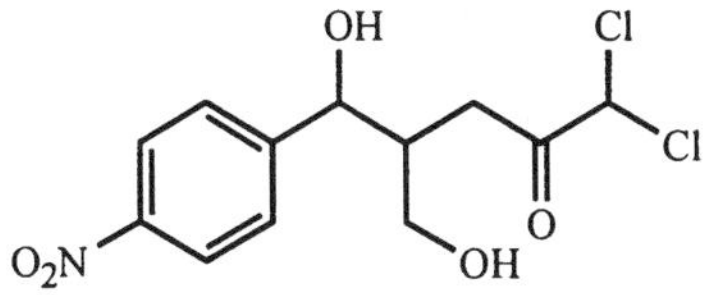

Chloramphenicol

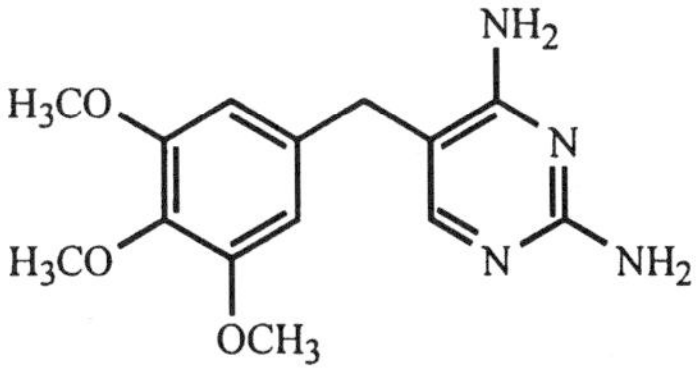

Trimethoprim

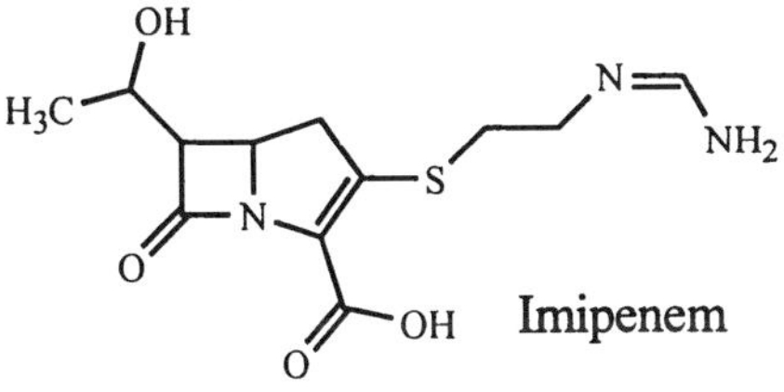

Imipenem

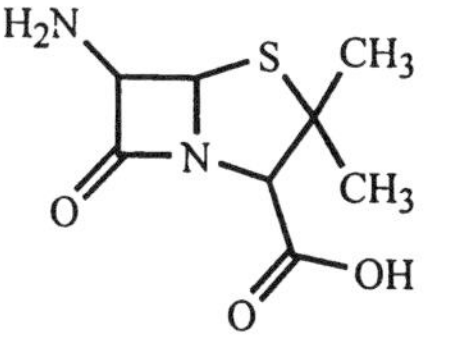

6-Aminoampicillic acid

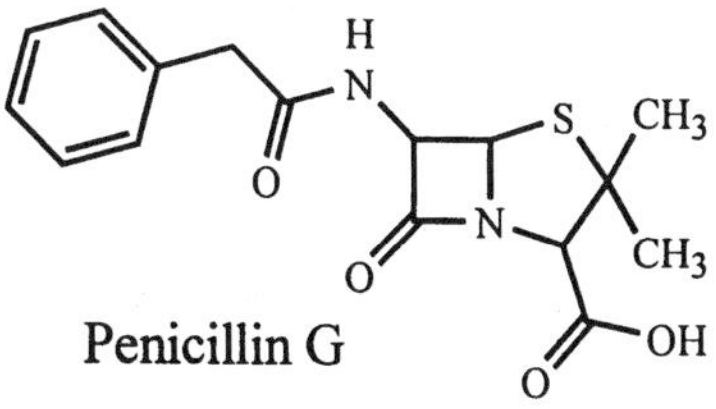

Penicillin G

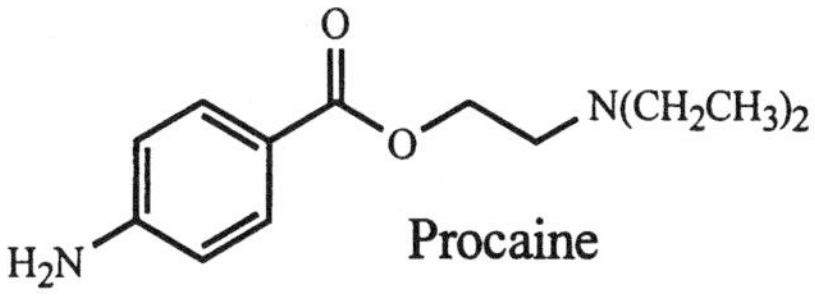

Procaine

column (photodiode array detector, $\lambda = 200–350$ nm or $\lambda = 210$ nm) using an 18/88 acetonitrile/water (0.25% H_3PO_4 with 0.3% TEA, 0.5 mM octanesulfonic acid, and 4.5 mM dodecanesulfonate). Peak shapes were excellent and detection limits of 100 ppb were reported [1365]. Electrospray MS was used to confirm identities and for quantitation down to 100 pg injected. In this case a C_{18} column and a 40/60 acetonitrile/water (1% acetic acid pH 3) mobile phase were used.

Two isomers of paromomycin were extracted from feed and separated on a 40°C C_{18} column (electrochemical detector) using an 80/20 water (3 mM sodium decylsulfonate with 80 mM Na_2SO_4 and 0.1% acetic acid)/acetonitrile mobile phase [1366]. Methanol was not compatible with the system (electrochemically reactive under these conditions). Baseline resolution was achieved with a 10-min elution. The effect of pH on k' was plotted with a dramatic decrease found at pH > 4.5. The authors note that the octyl and dodecyl ion-pair analogs were also tested and the decyl reagent gave the optimal results. Standards ranging from 1.2 to 6.3 mg/L showed a linear relationship. Detection limits of 0.21 mg/L were reported.

Metronidazole and hydroxymetronidazole were extracted from fish fillets and resolved on a C_{18} column ($\lambda = 325$ nm) in 8 min using a 100/0 (hold 7 min) → 0/100 (at 10 min) (88/9/3 water [1.2 mL H_3PO_4 to pH 3.0 with diethylamine]/acetonitrile/methanol)/(20/80/0 water [1.2 mL H_3PO_4 to pH 3.0 with diethylamine]/acetonitrile/methanol) gradient [1367]. The extra run time was used to elute the internal standard. Detection (S/N = 3) and quantitation limits (S/N = 6) were 2 and 4 µg/kg, respectively.

Polymixin B sulfate is a cyclic heptapeptide having a linear tripeptide side chain [1368]. This side chain N-terminal is aceylated and the major by-product components are 6-methyloctanoic acid, isooctanoic acid, and heptanoic acid. These by-products are separated from the parent using a 30°C C_{18} column ($\lambda = 215$ nm) and a 22.25/50/5/22.75 acetonitrile/water (0.7 m/v sodium sulfate)/water (6.8% H_3PO_4)/water mobile phase. Samples were 0.5 mg/mL. Elution was complete in 45 min. The explanation for the long analysis time is that there are another ten unidentified by-products that are resolved from the peaks of interest. Plots of the effects of percent acetonitrile and concentrations of sodium sulfate and H_3PO_4 are presented.

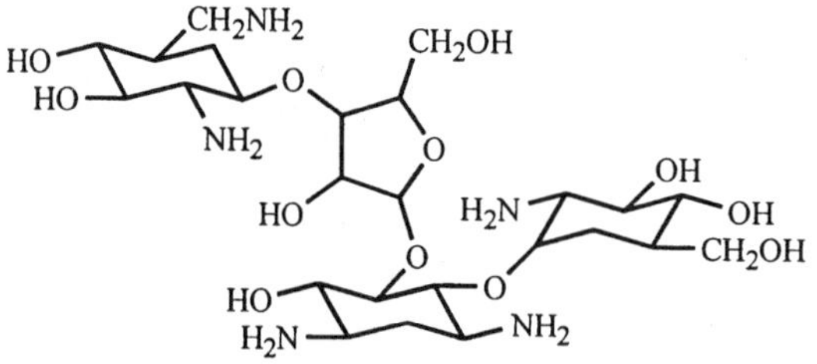

Paromomycin

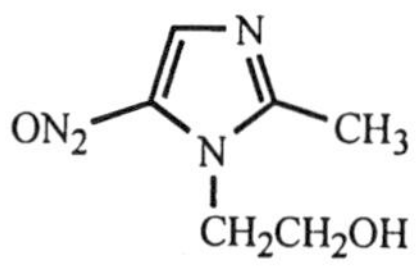

Metronidazole

Minocycline and four production impurities (e.g., 4-epiminocycline, 6-deoxy-6-demethyltetracycline, 7-didemethylminocycline, 7-monodemethylminocycline) were baseline resolved on a C_8 column ($\lambda = 280$ nm) using a 20/55/20/5 DMF/water (0.2 M ammonium oxalate)/water (0.1 M EDTA)/water mobile phase adjusted to pH 7.0 with tetrabutylammonium hydroxide [1369]. The author noted that with DMF levels of more than 25% in the mobile phase, precipitation of the mobile phase EDTA occurred. In these solvents a decrease in the concentration of the EDTA to 0.01 M did not prevent a precipitate from forming. Elution was complete in 40 min.

Injectable streptogramin consists of two components: quinupristin and dalfopristin. Samples were extracted from plasma. The quinupristin and its glutathione and cysteine conjugates were separated on a C_{18} column ($\lambda = 360$ nm, ex; 410 nm, em) using a complex 36-min 70/30 $\rightarrow$ 40/60 water (0.8 mL 70% $HClO_4$/L)/acetonitrile multiple-step gradient [1370]. This did not result in baseline separation of all compounds and it seems likely that a more conventional gradient might be more effective. The dalfopristin and its metabolite pristinamycin were analyzed in the same fashion but detection was done at $\lambda = 235$ nm. Linear ranges ran from 0.008 to 5 mg/L with quantitation limits reported as 0.01–0.025 mg/L (analyte dependent).

Nosiheptide (a gram positive polypeptidic antibiotic) was extracted from meat and eggs and analyzed on a 40°C C_{18} column ($\lambda = 357$ nm, ex; 500 nm, em). A

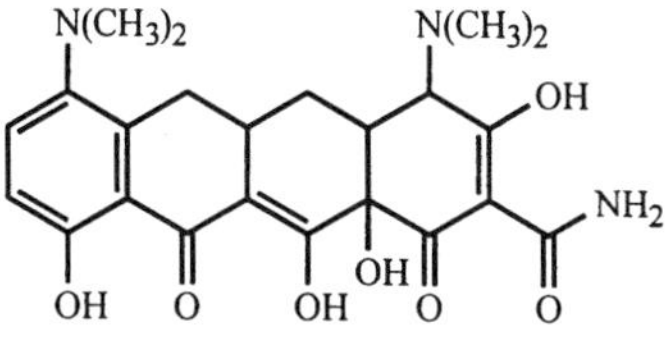

Minocycline

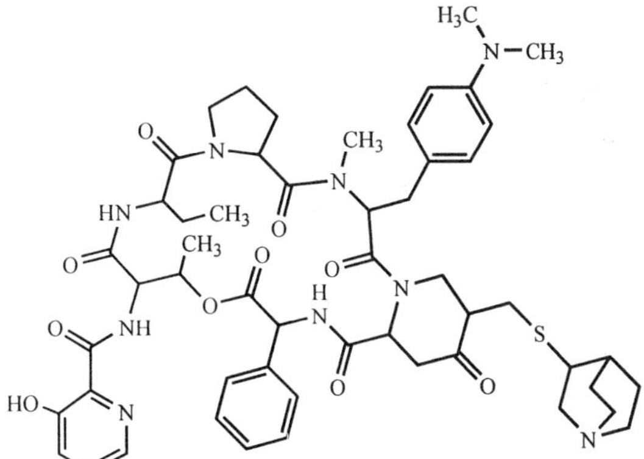

Quinupristin

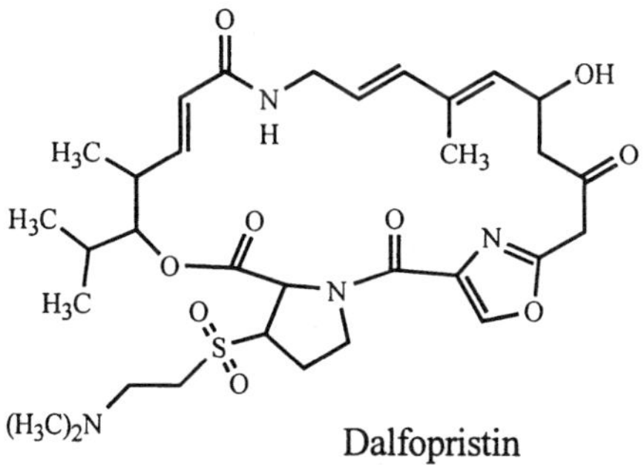

Dalfopristin

50/50 acetonitrile/water (0.025% H_3PO_4) mobile phase generated a 7 min run time. The authors noted that the sample was treated with N,N-diethyldithiocarbamate to remove potential mineral interferences. A linear range of 1–100 ppb (ng/mL) and a detection limit of 10 ppb (S/N = 3) were reported [1371].

A series of six fluoroquinone antibiotics (norfloxacin, sarafloxacin, enrofloxacin, danofloxacin, ciprofloxacin, desethylene ciprofloxacin) were extracted from eggs and analyzed on a $30°C_{18}$ column ($\lambda = 278$ nm, ex; 440 nm, em) using a 15-min $85/15 \rightarrow 65/35$ water (25 mM H_3PO_4 to pH 2.9 with triethylamine)/acetonitrile gradient [1372]. Excellent peak shape and baseline resolution were obtained throughout the separation. Calibration standards of 0.2–200 μg/mL and quantitation limits of 0.3–3 μg/mL were reported (analyte dependent).

The separation of three sulfonamides (pyrimethamine, sulfamethazine, sulfaquinoxaline) from commercial products was achieved on a C_{18} column ($\lambda = 270$ nm) using a 65/35 water (40 mM NaH_2PO_4 at pH 3.0 with 10 mM $NaClO_4$)/acetonitrile mobile phase [1373]. Baseline separation and a 4-min elution were obtained. Plots showing the effect of changing sodium perchlorate and percent acetonitrile on retention are presented. A linear range of 0.2–10 mg/mL with a quantitation limit of 180 μg/mL (S/N = 10) were reported.

Five sulfonamides (sulfaguanidine, sulfadiazine, sulfathiazole, sulfapyridine, sulfamethoxazole) were extracted from honey, milk, and eggs and separated on a C_{18} column ($\lambda = 260$ nm) using a 20-min $3/97 \rightarrow 40/60$ acetonitrile/water gradient [1374]. A plot of k' versus percent acetonitrile (for isocratic mobile phases $0 \rightarrow 40\%$ in acetonitrile) was shown for all analytes. Calibration curves from 0.05 to 5 μg/mL were generated. Sulfadiazine, sulfamethazine, sulfamonomethoxine, sulfadimethoxine, and sulfamethoxazole were extracted from serum, derivatized with fluorescamine, and analyzed on a C_{18} column ($\lambda = 390$ nm, ex; 475 nm, em

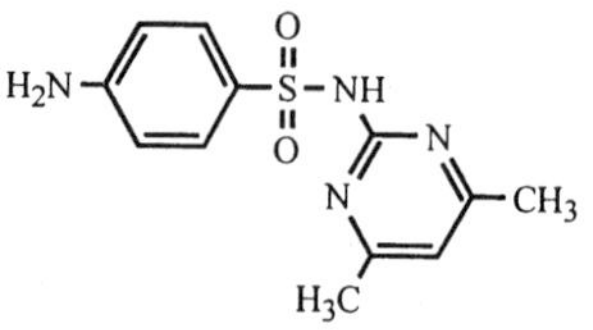

Sulfamethazine

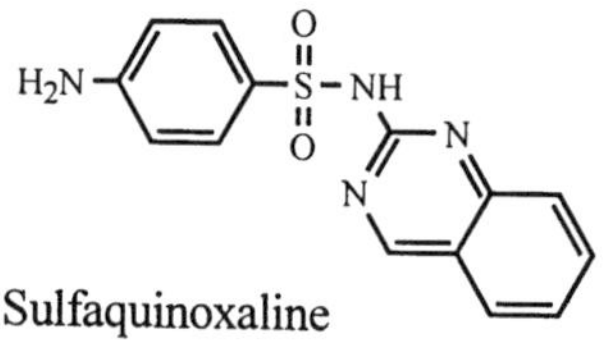

Sulfaquinoxaline

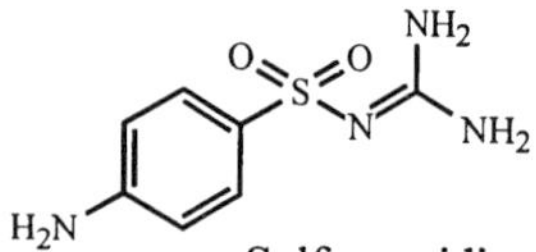

Sulfaguanidine

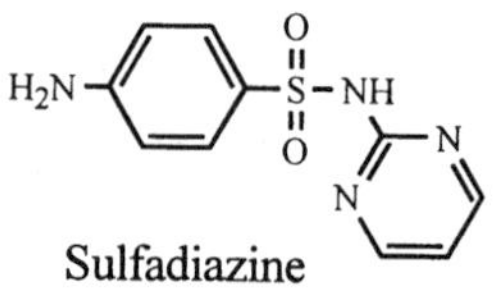

Sulfadiazine

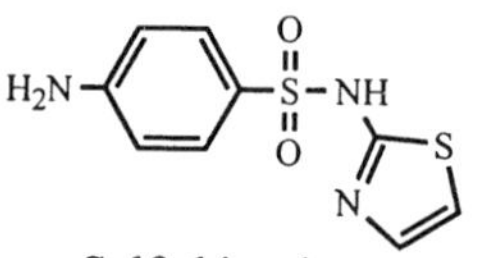

Sulfathiazole

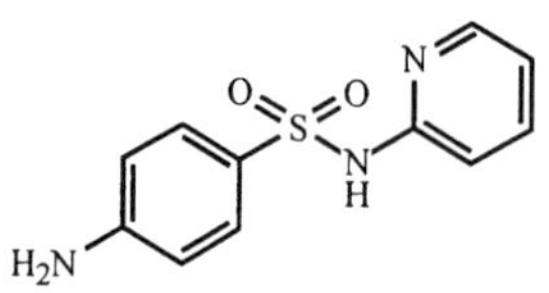

Sulfapyridine

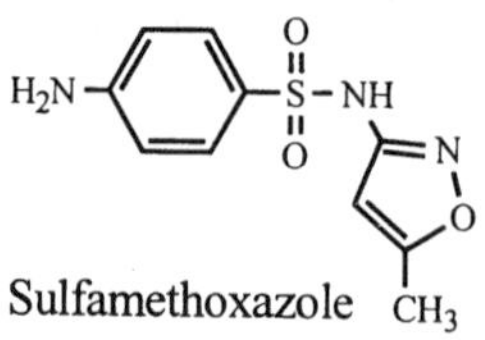

Sulfamethoxazole

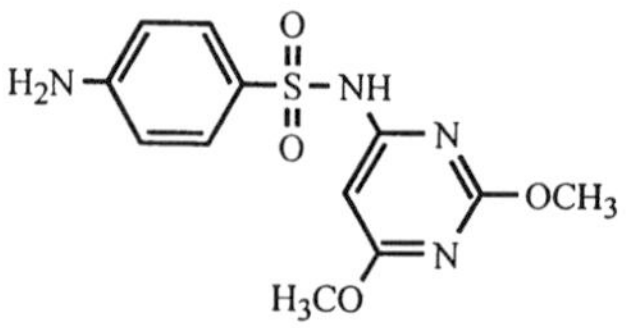

Sulfadimethoxine

using a 30/70 acetonitrile/water (10 mM potassium dihydrogenphosphate) mobile phase [1375]. Detection limits were reported as 0.1 ng/mL and linear ranges were 1–100 ng/mL. Elution was complete in 20 min.

Thirteen veterinary drug residues found in meat were incorporated into this study [1376]. Eight of them (sulfathiazole, sulfamerazine, sulfachloropyridazine, sulfamethazine, sulfamethoxypyridazine, sulfamethoxazole, sulfaquinoxaline [SQX], sulfadimethoxine [SMX]) were resolved on a 35°C C_{18} column ($\lambda = 270$ nm and 365 nm) using a two-ramp 86/14 (hold 5 min) → 78/22 (at 22 min) → 46/54 (at 30 min) water (10 mM ammonium acetate at pH 5.2)/(70/30 acetonitrile/methanol) gradient. Baseline resolution was achieved for all but the SQX and SMX pair. Separately, but using the same column ($\lambda = 290$ nm, 254 nm, 348 nm) and a 56/44 (hold 5 min) → 36/64 gradient (same mobile phases as above), five drug residues (chloramphenicol, thiabendazole, mebendazole, virginiamycin, nicarbazin) were well resolved. Detection limits of 0.8–66 (μg/kg (analyte dependent) were reported.

Tylosin residue was extracted from cow's milk and quantitated on a base deactivated C_{18} column ($\lambda = 287$ nm). Elution was complete in <10 min using a 40/60 water (50 mM H_3PO_4 at pH 2.4)/acetonitrile mobile phase [1377]. The linear range was 0.1–10 μg/mL and the quantitation limit was 0.2 ng injected (S/N = 3).

Doxorubicin, epirubicin, doxorubicinol, epirubicinol, daunorubicin, 7-hydroxy-doxorubicinol, 7-hydroxydoxorubicin aglycone, and 7-deoxydoxorubicin aglycone were baseline resolved on a C_{18} column ($\lambda = 480$ nm, ex; 560 nm, em) in 16 min

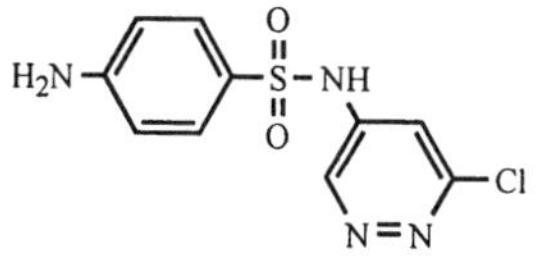

Sulfachloropyridazine

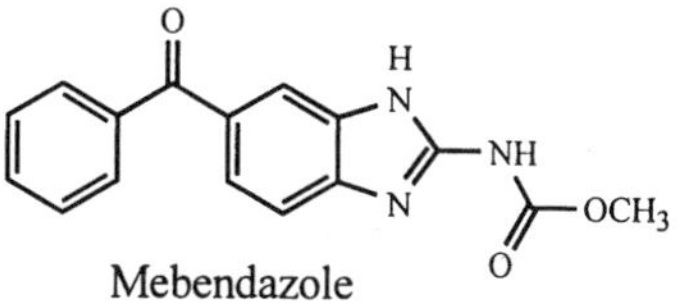

Mebendazole

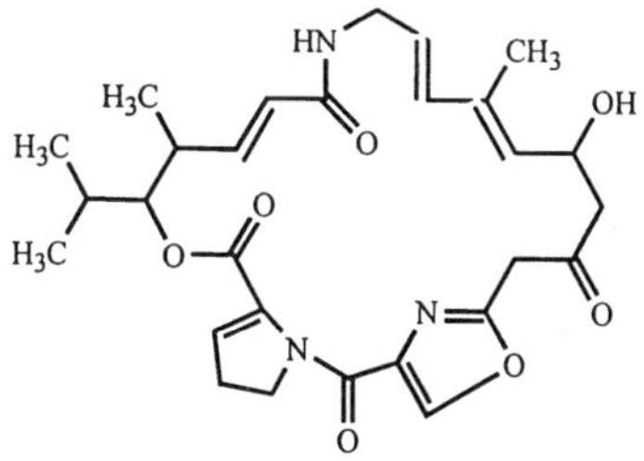

Virginiamycin M_1

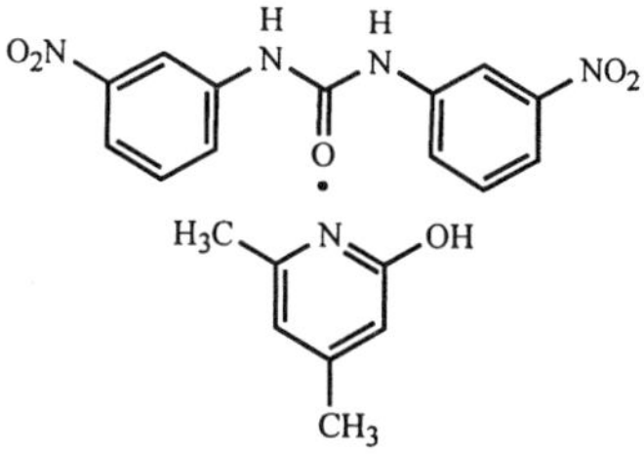

Nicarbazin

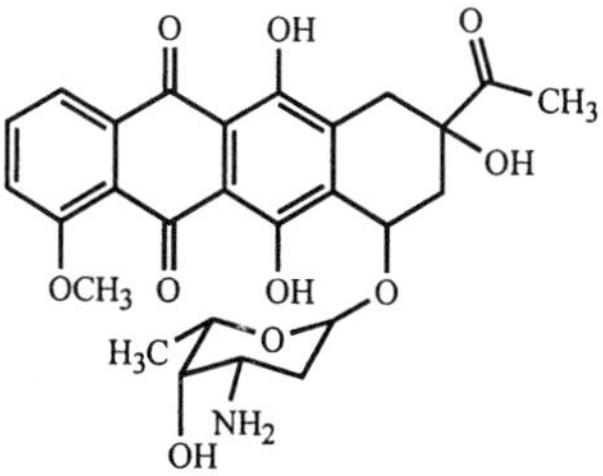

Daunorubicin, daunomycin

using a 35/65 acetonitrile/water (60 mM Na_2HPO_4 and 30 mM citric acid to pH 4.6 with 0.05% triethylamine) mobile phase [1378].

Desacetylcephapirin was extracted from milk samples and quantitated using a PLRP-S column ($\lambda = 290$ nm) using a 16/84 acetonitrile/water (10 mM H_3PO_4 with 10 mM KH_2PO_4 and 10 mM sodium decanesulfonate) mobile phase [1379]. This study also tabulated the conditions for the analysis of amoxicillin, ampicillin, cephapirin, ceftiofur, cloxacillin, and penicillins G and V. Again a C_{18} and an ~30% acetonitrile/water (phosphate buffer) combination were used in each case.

Mitomycin C and 13 metabolites (e.g., albomitomycin C, mitocene, mitosane) were studied through the use of a C_{18} column ($\lambda = 310$ nm) and a $0/100 \rightarrow 15/85$ (at 35 min and hold 7 min) acetonitrile/water (30 mM ammonium acetate) gradient [1380]. The DNA and oligonucleotide adducts of mitomycin C were separated on a C_{18} column using a 60-min $6/94 \rightarrow 18/82$ acetonitrile/water (30 mM potassium phosphate buffer at pH 5.4) gradient [1381]. Interestingly, even though the gradient lasted 60 min, chromatograms of the analytes of interest show that elution was complete in <20 min.

Tobramycin was derivatized with o-phthalaldehyde and analyzed on a C_{18} column ($\lambda = 254$ nm or 340 nm, ex; 450 nm, em) using a 52/48 acetonitrile/water (20 mM phosphate buffer at pH 6.5) mobile phase [1382]. A 25 μL injection of each of a 12.8 μM series of standards was made. The fluorescence-based detection method was more than 40 times more sensitive than the UV method. Precolumn sample derivatization increased the recovery of tobramycin by a factor of 3.

9.7.4 Anticancer Drugs

Doxorubicin and six metabolites (e.g., doxorubicinol, doxorubicine-aglycone and -desoxyaglycone) were isolated from lung tissue and separated on a 6°C (!) C_{18}

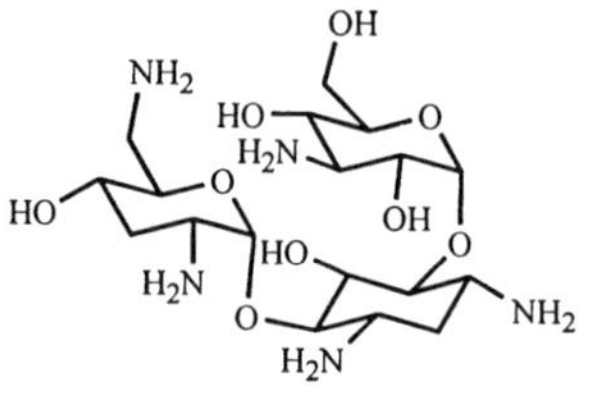

Tobramycin

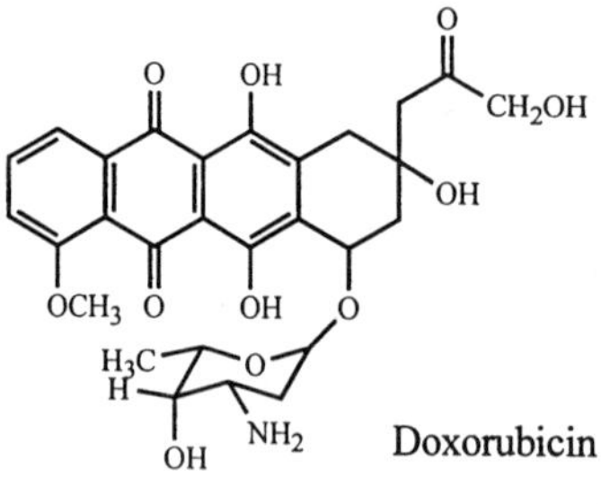

Doxorubicin

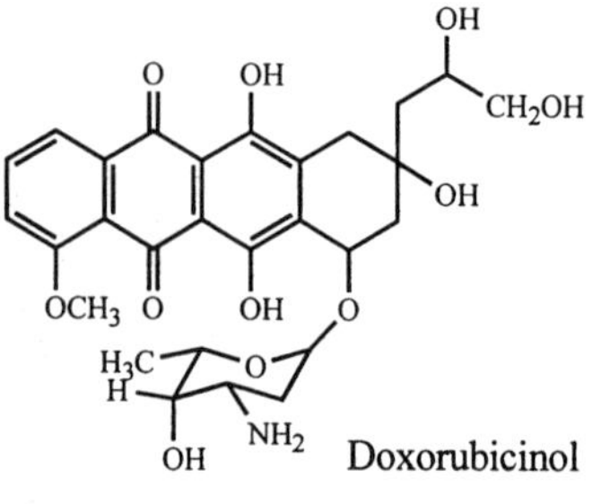

Doxorubicinol

column ($\lambda = 490$ nm, ex; 590 nm, em) using a $100/50/25/5$ water (20 mM citric acid with 0.14% triethylamine at pH 2.4)/acetonitrile/methanol/THF mobile phase [1383]. Elution was complete in 30 min with all peaks being well resolved. The concentration range covered 0.05–$200\,\mu$g/g (analyte dependent).

Epirubicin and doxorubicin and their 13-dihydro metabolites were extracted from plasma and separated on a C_{18} column (electrochemical detector $+400$ mV guard and -300 mV signal). A $71/29$ water (50 mM Na_2HPO_4 and 0.05% triethylamine to pH 4.6 with citric acid)/acetonitrile mobile phase generated baseline resolution in 13 min [1384]. A linear concentration range from 1 to 500 ng/mL was reported.

The prodrug irinotecan (7-ethyl-10-[4-(1-piperidino)-1-piperidine]carbonyloxy-camptothecine) and four metabolites were extracted from hepatic microsomes and separated on a C_{18} column ($\lambda = 355$ nm, ex; 515 nm, em). A 23-min $85/15 \rightarrow 70/30$ water (75 mM ammonium acetate at pH 6.0 and 5 mM tetrabutylammonium phosphate)/acetonitrile gradient baseline resolved all but two components [1385]. In a similar fashion, Escoriaza et al. [1386] recovered irinotecan and its metabolite (7-ethyl-10-hydroxycamptothecine) from plasma and baseline resolved them in <8 min. A $30°C$ C_{18} column ($\lambda = 228$ nm, ex; 450 nm and 543 nm, em) and a $67/33$ water (0.1 M KH_2PO_4 to pH 4.2 with HCl)/acetonitrile phase were used. It should be noted that H_3PO_4 would be a more suitable choice for acidification. It matches the buffer identity and is compatible with stainless steel (HCl is very aggressive toward stainless steel). Calibration standards from 0.5 to 1000 ng/mL and quantitation limits of 1 ng/mL were reported.

Etoposide, a derivative of podophyllotoxin, was isolated from serum and plasma and quantitated on a C_{18} column ($\lambda = 206$ nm or positive ion electrospray MS,

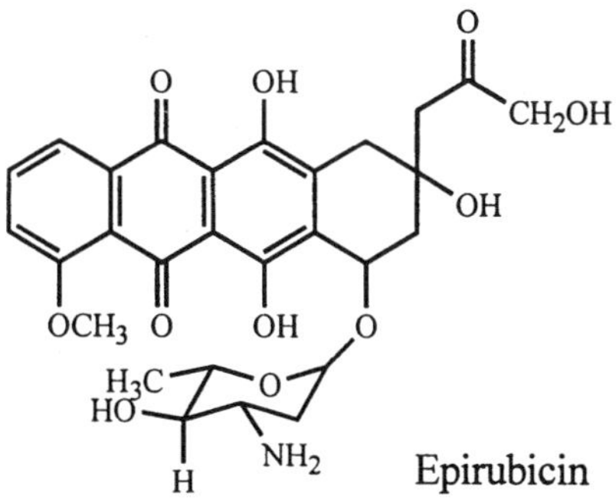

Epirubicin

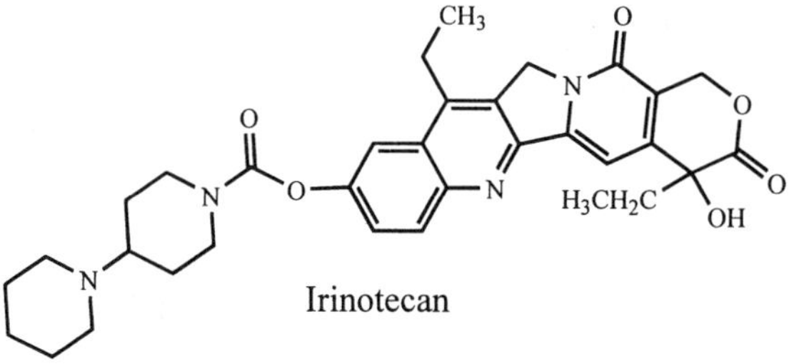

Irinotecan

fragmentor voltage 75 V, drying gas flow 10 L/min and I = 350°C, nebulizer pressure 25 psi, capillary voltage 3.5 kV). A 45/55/0.1 acetonitrile/water/acetic acid mobile phase generated good peak shape and elution in 7 min. A linear range of 0.01–5 μM and detection limit of 5 nM (S/N = 4) were reported [1387].

Finasteride was extracted from plasma and quantitated on a 30°C C_{18} column ($\lambda = 210$ nm). Elution was complete in 18 min using a 60/40 water (15 mM KH_2PO_4)/acetonitrile mobile phase [1388]. Retention was independent of pH in the range 3–6.5. A calibration curve from 4 to 300 ng/mL and a quantitation limit of 4 ng/mL were reported.

Formulations of idarubicin-etoposide (I-E) and idarubicin-ondansetron (I-O) were separated on a C_8 column ($\lambda = 254$ nm). The I-E formulation was fully resolved using a 73/27 water (20 mM NaH_2PO_4)/acetonitrile mobile phase [1389]. Elution was complete in 12 min. Significant tailing of the idarubicin was observed. Linear ranges from 2 to 150 μg/mL were reported. The I-O formulation showed marginal resolution on a C_8 column when a 57/43 water (20 mM KH_2PO_4)/acetonitrile mobile phase was used. Linear ranges of 2–40 μg/mL were reported. Tailing was severe for both compounds. Phosphate buffers are effective acid suppressors but are less effective as silanophilic agents. Since idarulacin and ondansetron have primary and tertiary amines, a TFA or a TFA/triethylamine mobile phase modifier should be considered.

A set of four arylene bis(methylketone) chemotherapeutic analogs (e.g., 3,5-diacetylaniline, 2-amino-4-(3,5-diacetylphenyl)amino-6-methylpyrimidine, and its 3-acetyl analog) were isolated from plasma and separated on a C_8 column ($\lambda = 300$ nm for the pyrimidines and 240 nm for the aniline). A 30-min 97.5/2.5 → 40/60 water (10 mM tetramethylammonium chloride and 10 mM heptane-sulfonic acid to pH 4.2 with H_3PO_4)/acetonitrile gradient generated excellent peak shape and baseline resolution. The authors noted that the extremely weak initial mobile phase requires extensive re-equilibration times between injections. A linear range of 100 ng/mL to 100 μg/mL was reported [1390].

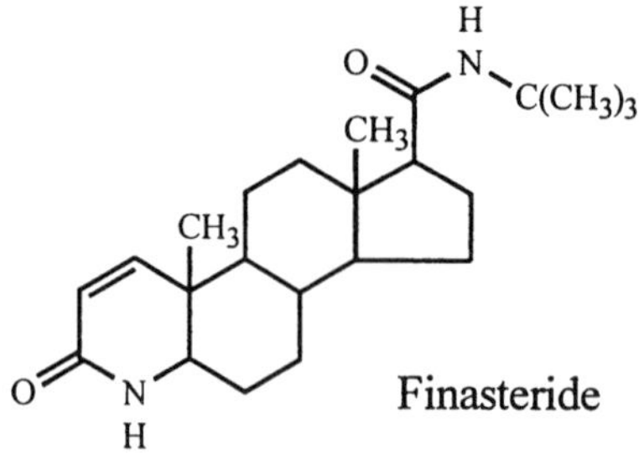

Finasteride

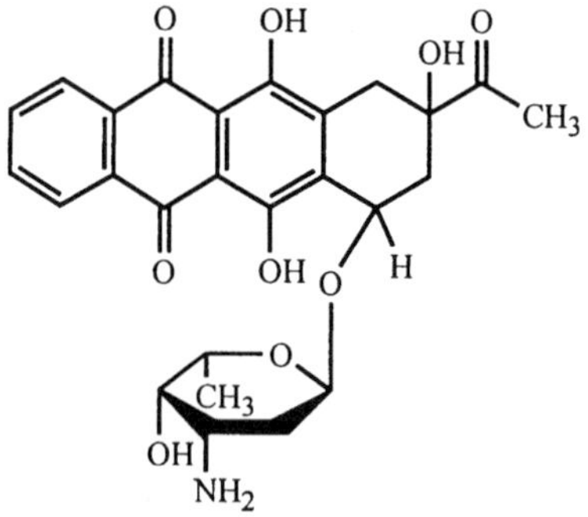

Idarubicin

Xanthenone-4-acetic acid and four metabolites (e.g., 2-hydroxyxanthenone-4-acetic acid) were extracted from urine and resolved on a C_{18} column ($\lambda = 345$ nm, ex; 409 nm, em or UV at $\lambda = 254$ nm) using a 27/75/2 water/acetonitrile/acetic acid mobile phase [1391]. Peak shapes were good and elution was complete in 35 min.

Emara et al. [1392] extracted the antineoplastic anthracycline compounds adriamycin (and its metabolites adriamycinol, adriamycinone) and daunomycin from urine and plasma and resolved them on a C_{18} column ($\lambda = 460$ nm, ex; 555 nm, em) using a 35/65 acetonitrile/water (0.1 M phosphate and 0.3% hepta-fluorobutyric acid to pH 3) mobile phase. The authors noted that the chemical stability of these materials was significantly enhanced when 10 mM γ-cyclodextrin was added to the samples. The analysis was complete in 10 min but the adriamycinol eluted with co-extracted compounds. The reported working concentration range was 2.5–25 nmol/mL. Sodium heptanesufonate at the 0.5% level was found to be an acceptable replacement for the butyric acid material. Useful plots of retention time vs. ion-pair identity and concentration are presented. Detection limits of 0.02 nmol/mL were reported.

Taxol and eight metabolites (e.g., 10-acetyltaxol, 7-epibaccatin taxol, baccatin III, 7-epi-10-deacetyltaxol, 7-epitaxol) were resolved on a pentafluorophenyl column ($\lambda = 227$ nm) using a 50-min 37/63 $\rightarrow$ 60/40 acetonitrile/water (0.1% H_3PO_4) gradient [1393]. A series of 100 μL injections of 0.25–1.5 mg/mL standards were used to generate a calibration curve.

Docetaxel, taxoid anticancer agent was extracted from plasma and analyzed on a C_{18} column ($\lambda = 227$ nm) with paclitaxel as IS [1394]. Good peak shapes and resolution were obtained in 12-min analysis using a 43/57 acetonitrile/water (20 mM ammonium acetate pH 5) mobile phase. Control standards were run from 5 to 1000 ng/mL with a reported quantitation limit of 5 ng/mL.

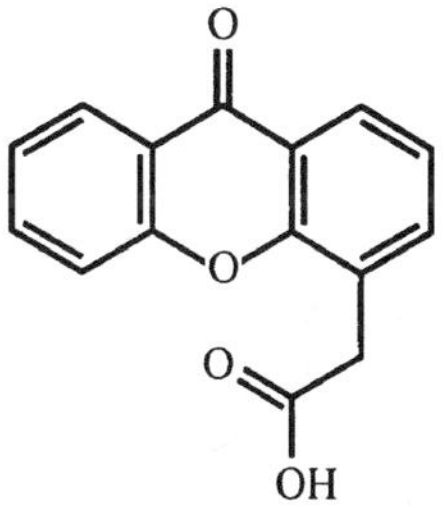

Xanthenone-4-acetic acid

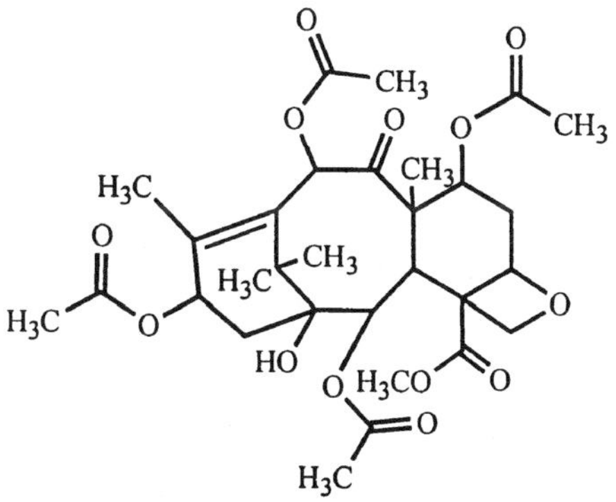

Baccatin III

Two vinca alkaloid anticancer drugs (vincristine and vinblastine) and two precursors (catharanthine and vindoline) were extracted from periwinkle leaves and separated on a C_{18} column ($\lambda = 254$ nm). Good resolution and complete elution were achieved in 15 min using a 62/38/0.3 water (0.1 M phosphate buffer at pH 4.14)/acetonitrile/acetic acid mobile phase [1395a]. Linear response was reported over the range from 0.25–25 µg injected.

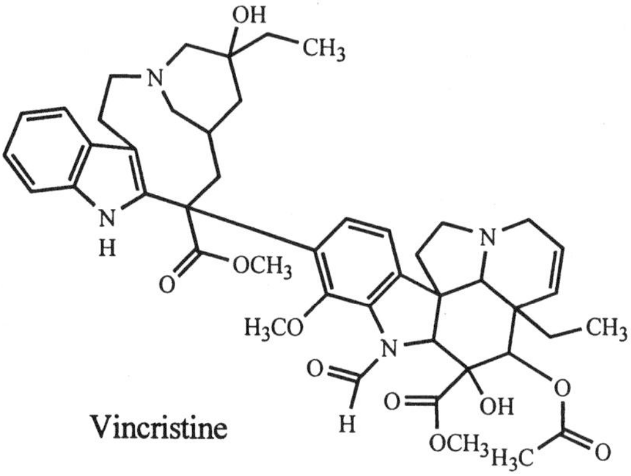

Vincristine

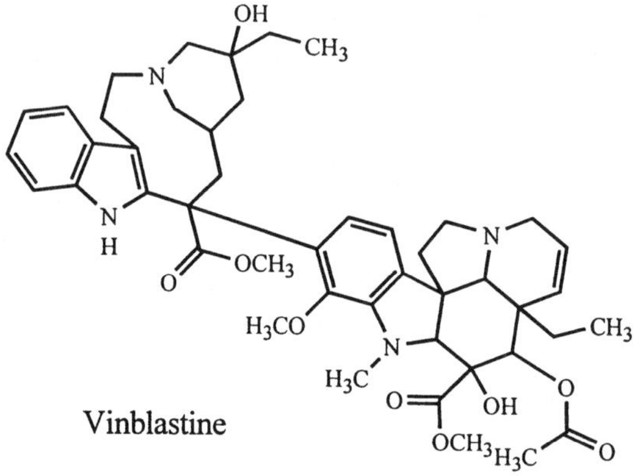

Vinblastine

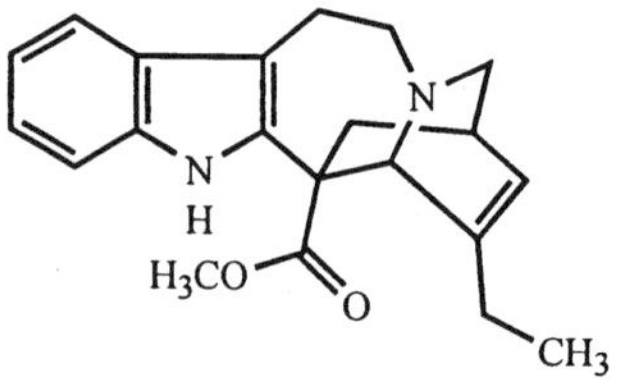

Catharanthine

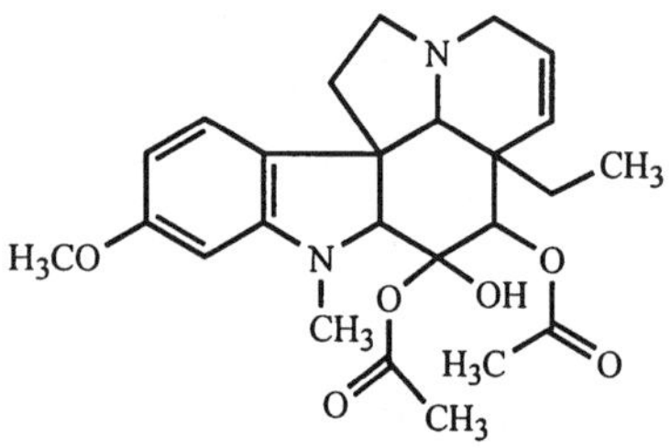

Vindoline

5,6-Dimethoxyxanthenone-4-acetic acid and two liver metabolites (the acyl glucuronide and 6-hydroxymethyl-5-methoxyxanthenone-4-acetic acid) were baseline resolved on a C_{18} column ($\lambda = 345$ nm, ex; 409 nm, em) in <7 min. The mobile phase consisted of 24/76 acetonitrile/water (10 mM ammonium acetate to pH 5.8). Calibration concentration ranges were 0.25–40 µM and detection limits of 0.5 µM (S/N = 10) were reported [1395b].

The chemotherputic cisplatin was extracted from blood, derivatized with bis (salicylaldehyde)tetramethylethylenediamine (synthesis by reaction of 2 : 1 ratio of salicylaldehyde to tetramethylethylenediamine in ethanol) and analyzed on a C_{18} column ($\lambda = 254$ nm) using a 70/20/10 acetonitrile/methanol/water mobile phase [1396]. The reagent reacts with many metals and so the separation included not only Pt(II) from the cisplatin but Fe(II), Ni(II), and Co(II) as well. Analysis was complete in 16 min. The reported linear range was 1.0 µg/mL and the detection limit was 5 ng/injection (S/N = 3).

The analysis of a substituted triazinodiaminopiperidine derivative anticancer drug was achieved on a C_{18} column ($\lambda = 220$ nm) using a 40/60 acetonitrile/water (25 mM phosphate with 0.2% triethylamine to pH 6) mobile phase [1397]. For the investigated compound, $R_1 = R_2 = -CH_2-CH=CH_2$. Separation was complete in 25 min, but the later-eluting peaks were tailed. Increasing the phosphate buffer concentration to 400 mM produced significantly reduced retention times (elution complete in <10 min) and decreased tailing, but the baseline resolution between the early-eluting peaks was lost. The authors noted that triethylamine effectively

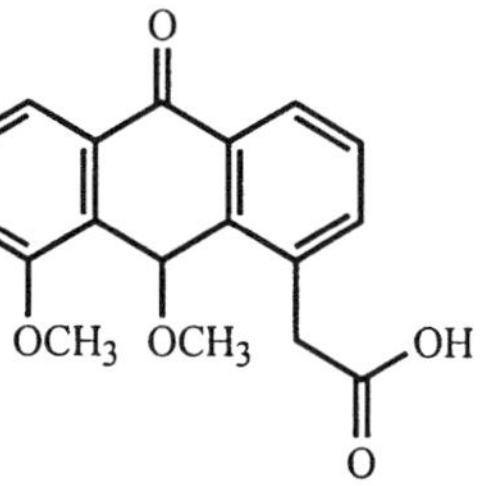

5,6-Dimethoxyxanthenone-4-acetic acid

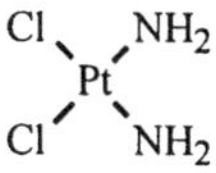

Cisplatin

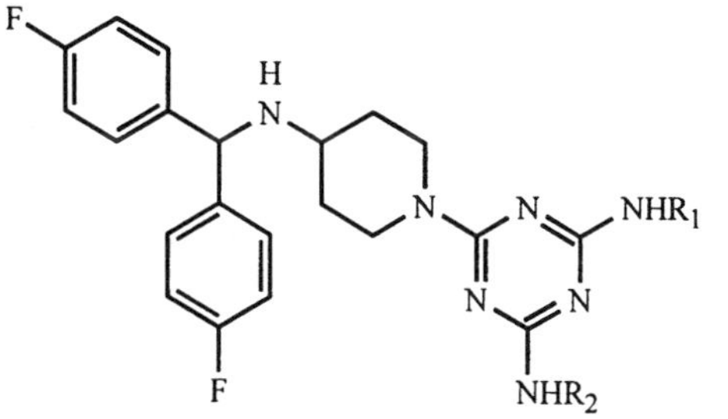

Triazinodiaminopiperidine
derivative anticancer drug

decreased tailing, whereas the addition of THF did not. A table of extraction from cells and assay results for concentrations ranging from 8 ng/mL to 625 µg/mL was presented.

5-Fluorouracil and six anabolites (5-fluoro- and 5-fluoro-2′-deoxyuridine and their monophosphate analogs, and uridine di- and triphosphate) were isolated from excised tumor and separated on a C_{18} column ($\lambda = 254$ nm). Excellent peak shapes and resolution were obtained using a $100/0 \rightarrow 67/33$ (at 25 min) $\rightarrow 10/90$ (at 30 min) $\rightarrow 0/100$ (at 35 min) (99/1 water [1.5 mM ammonium phosphate with 1 mM tetrabutylammonium phosphate pH 3.3]/acetonitrile)/(70/30 water [25 mM ammonium phosphate with 1 mM tetrabutylammonium phosphate pH 3.3]/ acetonitrile) gradient [1398].

9.7.5 Antiepileptic Drugs

Casamenti et al. [1399] developed a method for screening 11 central nervous system drugs (phenobarbital, olanzapine, clozapine, risperidone, loxapine, haloperidol, imipramine, amitriptyline, fluoxetine, chlorpromazine, paroxetine) on a C_{18} column ($\lambda = 230$ nm) using a 20/11.7 water (0.4 g tetramethylammonium perchlorate with 0.2 mL of 7% (m/m) $HClO_4$ to pH 2.8 with ammonia)/acetonitrile mobile phase. Keep in mind that perchlorates, when concentrated with some metals, are hazardous. Elution was complete in 35 min with good resolution for most compounds. Plots of the effects of mobile phase modifier level and percent acetonitrile on overall retention are presented. Linear ranges of 25–5000 ng/mL with detection limits of 10–250 ng/mL (analyte dependent) are reported.

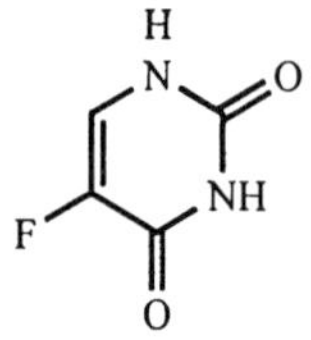

Fluorouracil 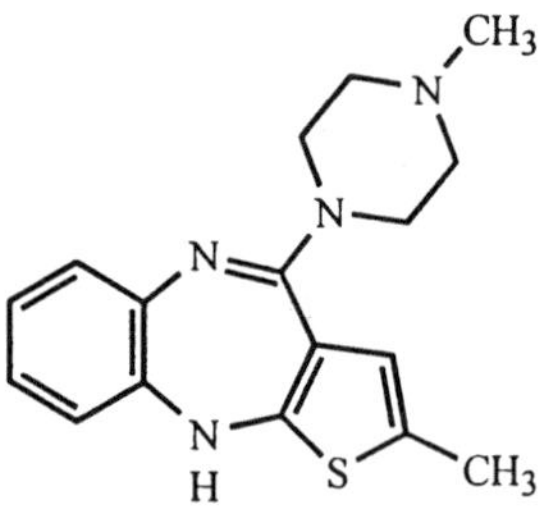 Olanzapine 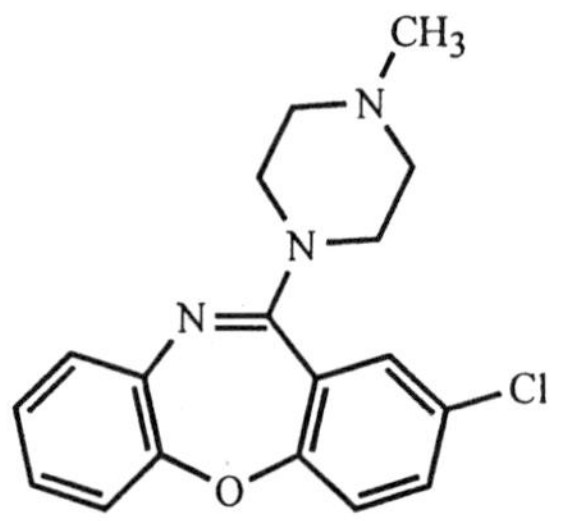 Loxapine 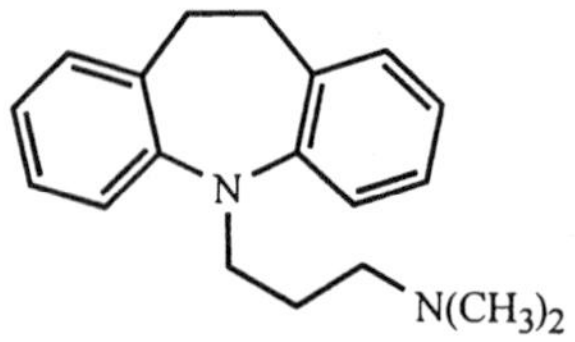 Imipramine 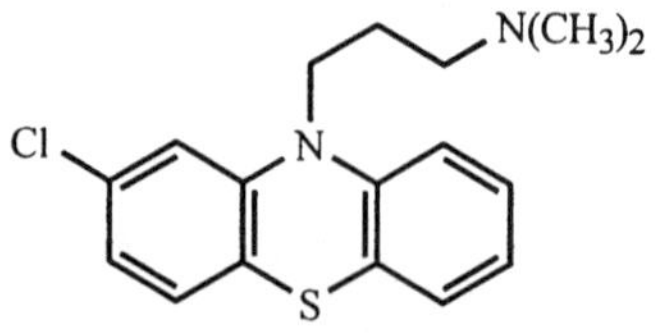 Chlorpromazine

Haloperidol and six metabolites (e.g., 4-(4-chlorophenyl)-4-hydroxypiperidine, haloperidol *N*-oxide, haloperidol-1,2,3,6-tetrahydropyridine) were resolved on a cyanopropyl column ($\lambda = 220$ nm) using a 67/33 acetonitrile/water (10 mM ammonium acetate buffer at pH 5.4) mobile phase [1400]. Peak shapes were good and 10 nmol injections were readily detectable.

Felbamate and three metabolites [2-phenyl-1,3-propanediol, 2-hydroxy-2-phenyl-1,3-propanediol dicarbamate, 2-(4-hydroxyphenyl)-1,3-propanediol dicarbamate) were extracted from heart tissue and baseline resolved on a C_{18} column ($\lambda = 210$ nm) using a 15/5/80 acetonitrile/methanol/water (10 mM potassium phosphate buffer at pH 6.8) mobile phase [1401]. Standard working curves from 0.2–50 µg/mL were generated.

Vigabatrin was extracted from plasma and analyzed on a C_{18} column ($\lambda = 340$ nm) using a 52/35/13 acetonitrile/water (100 µL or 10 mM H_3PO_4)/water (40 mM ammonium acetate buffer at pH 3.5) mobile phase [1402]. Elution was complete in <5 min. A standard curve of 1–50 µg/mL was used and a detection limit of 1 µg/mL was reported.

Remacemide and its desmethylremacemide metabolite were isolated from brain tissue and analyzed on a C_6 column ($\lambda = 218$ nm) using a 29/71 acetonitrile/water (50 mM KH_2PO_4 at pH 3.3) mobile phase [1403]. Excellent peak shapes and resolution were obtained and elution was complete in <10 min. Linear ranges of 0.05–10 mg/mL and detection limits of 30 ng/mL were reported. A list of the retention times for 11 epilepsy drugs is also presented.

Chollet et al. [1404] extracted gabapentin and vigabatrin from serum, derivatized them with *o*-phthalaldehyde, and separated them on a C_{18} column ($\lambda = 235$ nm, ex; 435 nm, em) using a 55/45 acetonitrile/water (22 mM H_3PO_4 at pH 2) mobile phase. The peaks were separated from each other as well as numerous co-extracted materials. Elution was complete in 10 min. Calibration standards from 1 to 40 µg/mL and detection limits of 0.5 µg/mL (S/N = 3) were claimed.

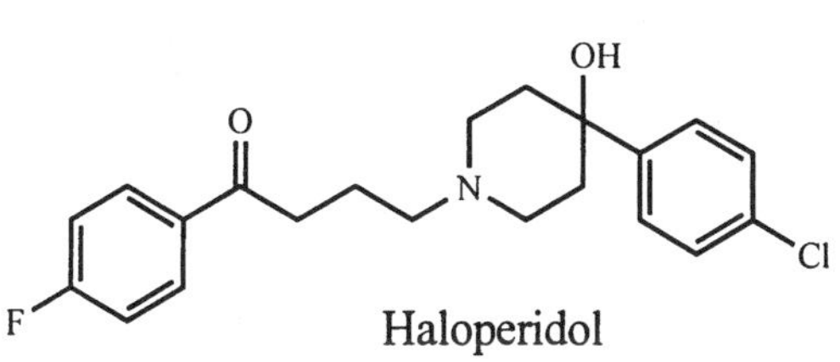

Haloperidol

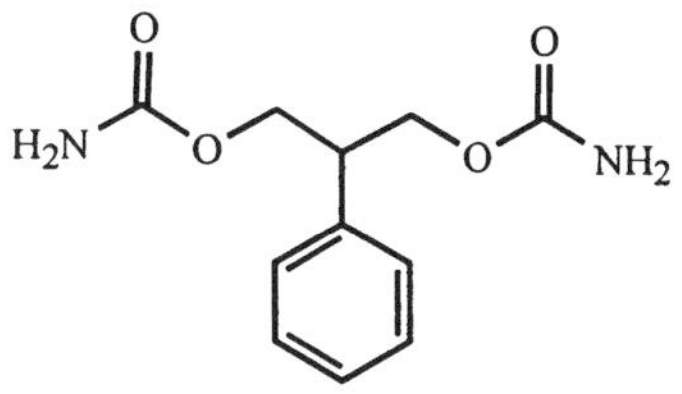

Felbamate

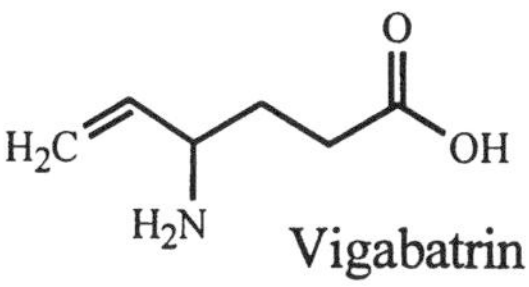

Vigabatrin

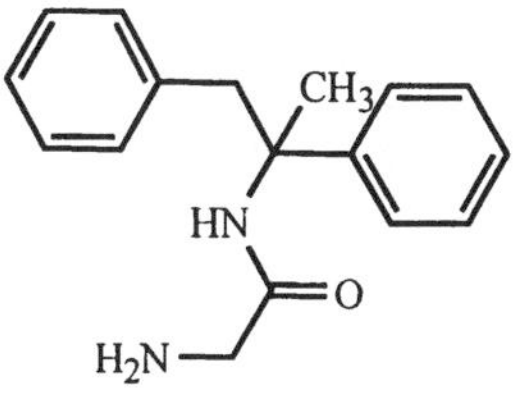

Remacemide

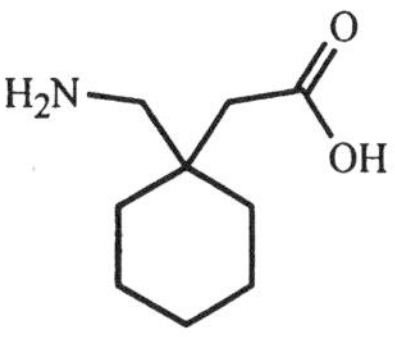

Gabapentin

Primadone and three major metabolites (phenobarbital, p-hydroxyphenobarbital, phenylmethylmalonamide) were extracted from urine and separated on a 40°C C_{18} column ($\lambda = 227$ nm). A 270/30/30 water (10 mM KH_2PO_4 at pH 4.0)/ methanol/acetonitrile mobile phase generated excellent peak shapes and resolution. Elution was complete in 30 min [1405]. Standard curves from 12 to 300 µg/mL (analyte dependent) were used and detection and quantitation limits of 0.5 µg/mL S/N = 3) and 2 µg/mL (S/N = 10), respectively, were reported.

Three metabolites of p-bromophenylacetylurea (N-hydroxy- and N-methyl-p-bromophenylacetylurea, 4-(4-bromophenyl)-3-oxapyrrolidine-2,5-dione) were extracted from urine and separated on a C_{18} column ($\lambda = 240$ nm). A 20-min 75/25 → 50/50 (at 10 min hold 10 min) water (0.1 M ammonium acetate at pH 6.5)/acetonitrile gradient generated excellent peak shapes and resolution [1406].

The decomposition products of 5-phenyltetrahydro-1,2-oxazinane-2,4-dione (a metabolite of felbemate), felbemate acid carbamate, 1-hydroxy-2-phenylpropamide, and 2-phenylacrylamide were separated on a C_{18} column ($\lambda = 214$ nm) in 11 min using a 20/80 acetonitrile/water (1% acetic acid) mobile phase. Peak shapes were good and baseline resolution was obtained [1407]. Fifteen benzodiazepines (e.g., 7-aminoclonazepam, clonazepam, flunitrazepam, 7-aminonitrazepam, nitrazepam,

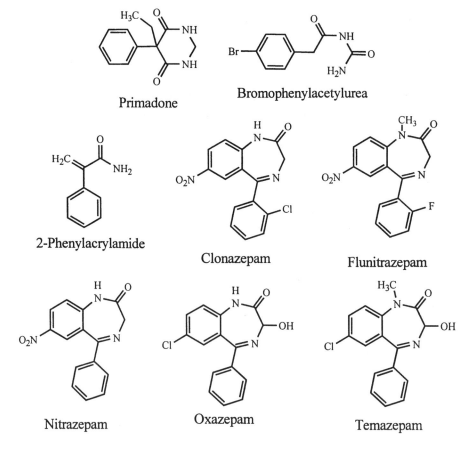

Primadone

Bromophenylacetylurea

2-Phenylacrylamide

Clonazepam

Flunitrazepam

Nitrazepam

Oxazepam

Temazepam

diazepam, oxazepam, nordiazepam, temazepam) were extracted from postmortem blood and separated on a phenyl column ($\lambda = 240$ nm). A complex 60-min $100/0 \rightarrow 10/90$ (15/85 acetonitrile/water [40 mM potassium phosphate buffer at pH 3.8])/(28/72 acetonitrile/water [40 mM potassium phosphate buffer at pH 3.8]) gradient was used for the separation. Linear concentration curves from 0.05 to 2.5 mg/L were reported. Retention times for the benzodiazapines and 14 potential interferents (e.g., salicylic acid, phenytoin, doxepin, desipramine) were also tabulated [1408].

Clobazam and its metabolite, N-desmethylclobazam, were extracted from serum and urine and separated on a deactivated base C_8 column ($\lambda = 228$ nm). Serum sample elution was complete in 8 min using a 440/540/20/0.4 acetonitrile/water/0.5 M KH_2PO_4/H_3PO_4 mobile phase, whereas urine samples were eluted in 16 min using a 360/580/60/0.4 acetonitrile/water/0.5 M KH_2PO_4/H_3PO_4 mobile phase. The reason for the weaker mobile phase in the urine samples was to move the sample peaks away from a significant set of large early-eluting peaks generated by the urine matrix. A table of 11 other benzodiazapine retention times was presented. This was not entirely successful. Linear ranges of 25–3000 ng/mL and detection limits of 1 ng/mL were reported [1409].

Valproic acid was extracted from serum and analyzed on a C_{18} column ($\lambda = 210$ nm) using a 37/63 acetonitrile/water (10 mM sodium phosphate buffer at pH 2.3) mobile phase [1410]. Standard curves from 2.5–200 µg/mL were generated. A detection limit of 2.5 µg/mL was reported. The authors noted that the most serious problem faced in this analysis was the presence of serum proteins. The authors resolved the problem through the precipitation of proteins with acetonitrile prior to analysis.

Phenytoin, 5-(p-hydroxyphenyl)-5-phenylhydantoin, and 5-(m-hydroxyphenyl)-5-phenylhydantoin were extracted from plasma and separated on a C_{18} column ($\lambda = 225$ nm). A 35/65 acetonitrile/water (8 mM sodium phosphate buffer at pH 6) mobile phase generated good peak shapes and elution in 25 min [1411]. The retention times of a number of potential interferent compounds (e.g., ethosuximide, primidone, phenobarbital, diazepam) were also reported. Quantitation limits of 50 ng/mL were obtained.

9.7.6 Anthelmintics

Fenbendazole sulfate is the metabolite of fenbendazole (anthelmintic) that was extracted from pig tissue and analyzed in 6 min on a C_{18} column ($\lambda = 292$ nm) using

Clobazam

Phenytoin

a 32/68 acetonitrile/water (10 mM NaH_2PO_4 at pH 5.5) mobile phase [1412]. Complete conversion of all fenbendazole was accomplished by sample treatment with peracetic acid. Spiked sample concentrations ranged from 20 to 2000 ng/g. A quantitation limit of 20 ng/g was reported. In a similar fashion, Morován, et al. [1413] isolated fenbendazole and praziquantel from plasma and separated them in 12 min on a C_{18} column ($\lambda = 220$ nm) using a 33/76 acetonitrile/water (50 mM KH_2PO_4 at pH 3.0) mobile phase. Whereas praziquantel was well resolved from all other peaks, fenbendazole eluted on the shoulder of a large early-eluting peak. Calibration curves were generated from 15 to 2750 ng/mL with reported detection limits of 5 ng/mL and quantitation limits of 20 ng/mL.

Closantel residue was isolated from milk and quantitated in 8 min using a C_{18} column ($\lambda = 335$ nm, ex; 510 nm, em) and an 85/15 acetonitrile/water (0.05% diethylamine to pH 3.5 with H_3PO_4) mobile phase [1414]. Calibration standards ranged from 50 to 3000 µg/L and a detection limit of 10 µg/L (S/N = 3) was reported.

Triclabendazole and two metabolites (triclabendazole sulfoxide and sulfone) were extracted from milk and separated on a 40°C C_{18} column ($\lambda = 295$ nm). Peaks were well resolved using a 50/50 acetonitrile/water (50 mM ammonium acetate) and elution was complete in 15 min [1415]. The effect of changing percent acetonitrile on k' was plotted. Although the analytes have an absorbance maximum at 220 nm, this wavelength was unsuitable due to the high mobile phase absorbance (i.e., the ammonium acetate). A linear curve from 0.05 to 1.5 µg/mL (1–30 ng injected) was reported as well as detection limits of 5 ng/g (S/N = 3).

Albendazole and two metabolites, albendazole sulfoxide and sulfone, were extracted from plasma and resolved on a C_{18} column ($\lambda = 290$ nm) using a 30/70 acetonitrile/water (0.25 N sodium acetate buffer at pH 5) mobile phase [1416]. Elution was complete in 6 min. Standard concentrations of 0.02–50 µg/mL were used.

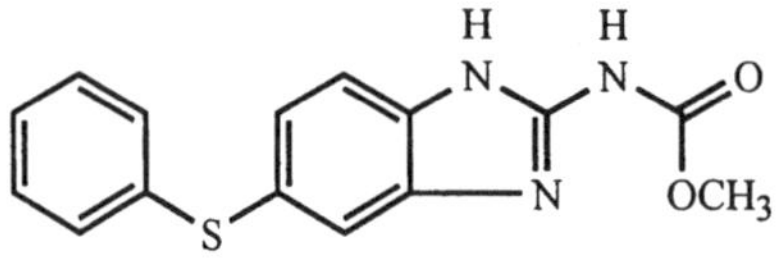

Fenbendazole

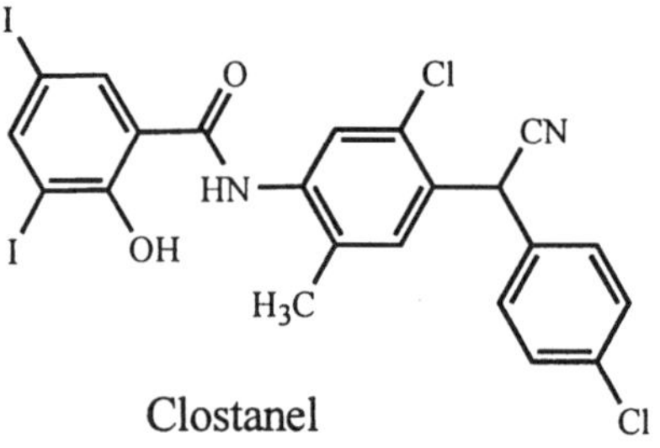

Clostanel

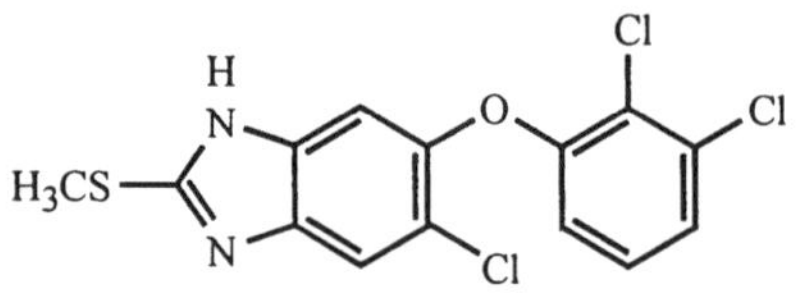

Triclabendazole

Ivermectin, enamectin, abamectin, and milbemectin as well as various metabolites (emanectin B_{1a} and B_{1b} and milbemectin A_3 and A_4) were extracted from crops and analyzed on a C_{18} column ($\lambda = 365$ nm, ex; 470 nm, em). The complex 32-min $80/20 \rightarrow 100/0$ acetonitrile/water gradient used gave excellent separation. Limits of detection of 0.1–0.3 ppt ($\sim$5 ng injected) were reported and chromatograms of concentrations up to 0.1 ppm were shown [1417].

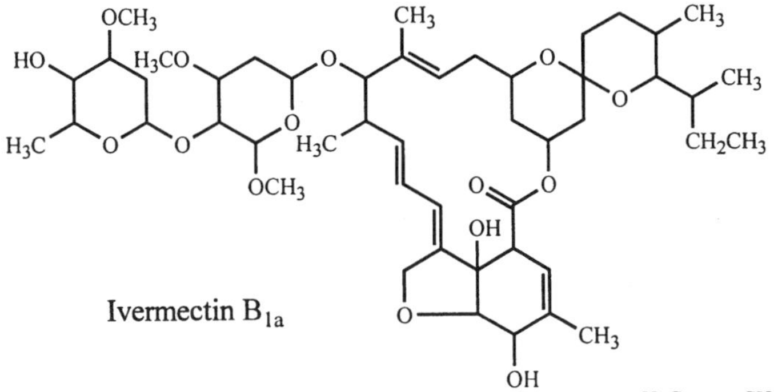

Ivermectin B_{1a}

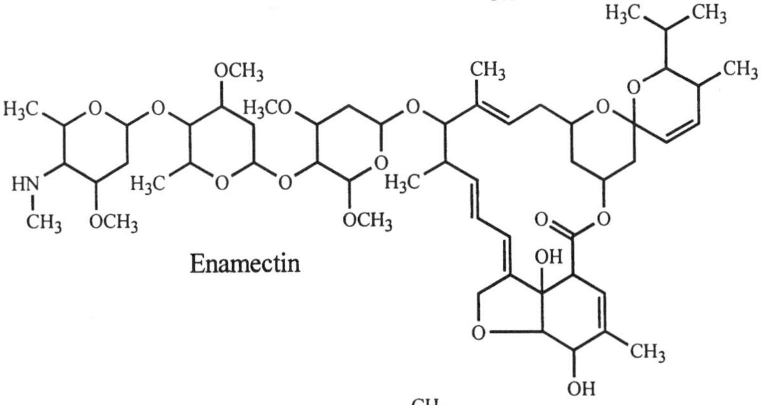

Enamectin

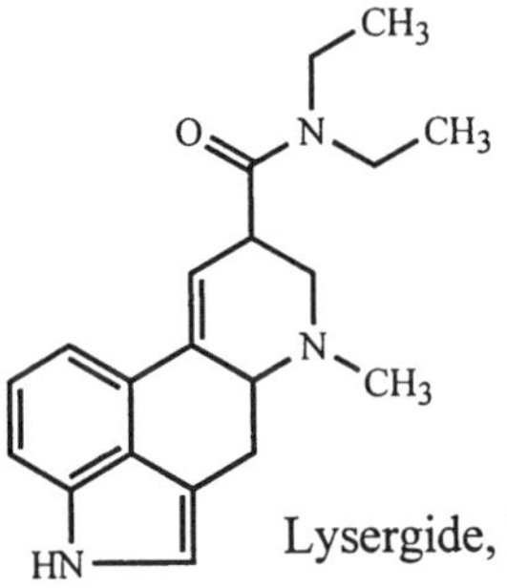

Lysergide, LSD

9.7.7 Illicit and Related Drugs

Lysergic acid diethylamide (lysergide, LSD), methysergide, and lysergic acid methylpropylamide were extracted from urine and separated on a C_{18} column ($\lambda = 330$ nm ex, 420 nm, em). A 30/70 acetonitrile/water (7.7 g ammonium acetate with 2.5 mL triethylamine/L to pH 8 with acetic acid) mobile phase generated baseline separation in 13 min [1418]. Calibration curves from 0.5 to 10 ng/mL were reported. LSD and its liver metabolite, 2-oxo-3-hydroxy-LSD, were extracted from liver hepatocytes and separated on a deactivated base C_{18} column (positive ion MS, collision gas 60 psi, APCI vaporization $T = 425°C$, capillary $T = 325°C$, capillary voltage 4 kV). A 83/17 (hold 3 min) → 68/32 (at 9 min hold 7 min) water (50 mM ammonium acetate with 0.02% triethylamine to pH 8.0)/acetonitrile gradient generated baseline resolution and excellent peak shapes [1419]. Standard concentrations ranged from 1 to 10 µg/mL for LSD and 0.5–8 ng/mL for the metabolite.

Two major impurities [1,3-dimethyl-2-phenylnaphthalene, 1-benzyl-3-methyl-naphthalene) found in illicit synthetic metamphetamine were separated on a C_{18} column ($\lambda = 228$ nm) using a 70/30 acetonitrile/water mobile phase [1420]. Elution was complete in 4 min and peaks were baseline resolved. Fluorescence detection ($\lambda = 228$ nm, ex; 340 nm, em) was approximately 30 times more sensitive (165 pg/mL vs. 3.9 ng/mL, with $S/N = 3$ for both). For fluorescence work the total elution time was 12 min.

Six β-carboline alkaloids (β-carboline, 1-methyl-, 1-methyl-1,2,3,4-tetrahydro-, 1,2,3,4-tetrahydro-, 7-hydroxy-1-methyl-, and 7-methoxy-1-methyl-β-carboline) were procured from crude drug product and assayed on a C_8 column. Two fluorescence detectors in series were used: one programmable ($\lambda = 325$ nm, ex; 417 nm, em from 0 to 7.5 min; 300 nm, ex; 447 nm, em from 7.5 to 10 min; 300 nm, ex; 430 nm, em from 10 to 14 min; 330 nm, ex; 417 nm, em from 14 to 18 min) and one fixed ($\lambda = 275$ nm, ex; 350 nm em). This gave specificity between co-eluting peaks. Baseline separation was achieved in 18 min using an 81.8/18.0/0.2 water/acetonitrile/TFA mobile phase. A linear range of ~0.2–20.0 ng/mL was reported [1421].

Six amphetamines (amphetamine, methamphetamine, ephedrine, methylenedioxyamphetamine, methylenedioxymethamphetamine [ecstasy]) and three adulterants (caffeine, phenylethylamine, paracetamol) were well resolved in 8 min on a C_{18} column ($\lambda = 200$ nm) using a 91/9 water (20 mM NaH_2PO_4 to pH 3.8)/acetonitrile mobile phase [1422]. A plot of the effects of changing the percent acetonitrile on log k' was presented. A linear range of 0.5–20 µg/mL along with detection limits of ~100 ng/mL were reported.

Sadeghipour et al. [1423] baseline resolved a series of six amphetamines (e.g., amphetamine, methamphetamine, ephedrine, and their 3,4-methylenedioxy

1,3-Dimethyl-2-phenylnaphthalene

Amphetamine

Metamphetamine

Ephedrine

Methylenedioxymetamphetamine

analogues) in <8 min using a C_{18} column ($\lambda = 200$ nm) and a 91/9 water (20 mM NaH_2PO_4 to pH 3.8 with HCl)/acetonitrile mobile phase. A linear range of 0.5–20 μg/mL with detection and quantitation limits of 0.1 μg/mL and 0.3 μg/mL, respectively, were reported. Two potential problems arise here. First, HCl is very aggressive to stainless steel in the HPLC system and its use should be avoided; H_3PO_4 seems a more appropriate choice. Second, operation at 200 nm is not recommended due to the variability of the background absorbance generated by the buffer and the lack of specificity (almost everything absorbs at 200 nm). These compounds are all aromatic and so absorbance in the 250–280 nm range should also be acceptable. Regardless of the considerations, good peak shapes and resolution were obtained in this method.

Four methylenedioxyamphetamines [3,4-methylenedioxy-methamphetamine and -ethamphetamine, and N-methyl- and N-ethyl-1-[1,3-benzodioxol-5-yl]-2-butan-amine) were baseline resolved using a C_{18} column ($\lambda = 280$ nm) and a linear 25-min 95/5 → 20/80 water (0.1 M triethylammonium acetate at pH 7.3)/ acetonitrile gradient. Standard concentrations ranged from ~300 μg/mL to 4 mg/mL [1424].

The retention characteristics of sixteen phenylalkylamine derivatives (e.g., amphetamine, phentermine, 5-methoxy-3,4-methylenedioxyamphetamine, mescaline, methoxamine) were monitored on a 40°C C_{18} column (photodiode array detector, $\lambda = 190$–300 nm) using a complex 22-min 5.5/94.5 → 39/61 (90/10 acetonitrile/water [5 mL/L H_3PO_4 and 0.28 mL/L hexylamine])/water (5 mL/L H_3PO_4 and 0.28 mL/L hexylamine) gradient [1425]. Urine and cactus samples were extracted and detection limits of 25–50 ng/mL and 40 ng/g, respectively, were reported.

Nine cannabis products (cannabidiol, cannabidiolic acid, cannabinol, cannabin-diol, (−)-Δ^9-(*trans*)-tetrahydrocannabinol, cannabinolic acid A and B, cannabigerol,

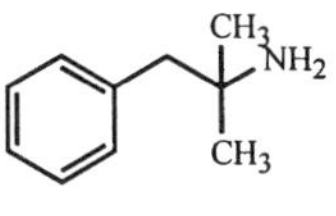

Phentermine

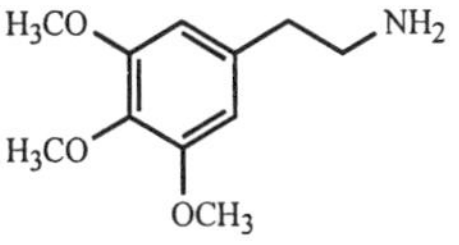

Mescaline

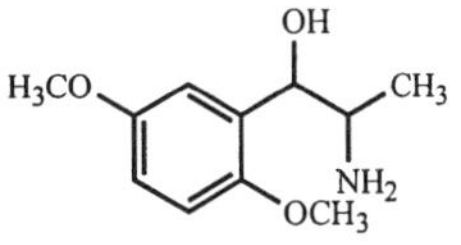

Methoxamine

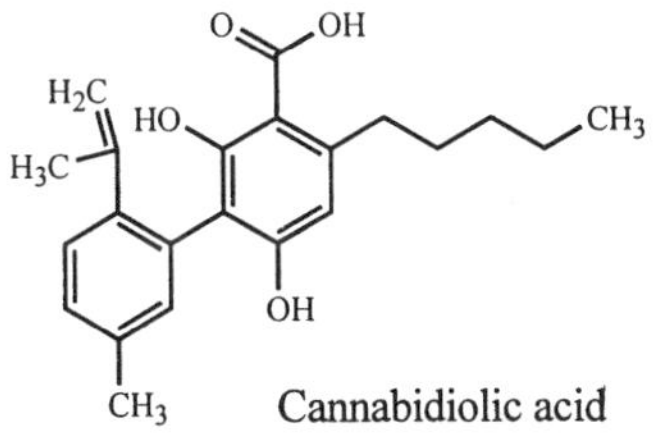

Cannabidiolic acid

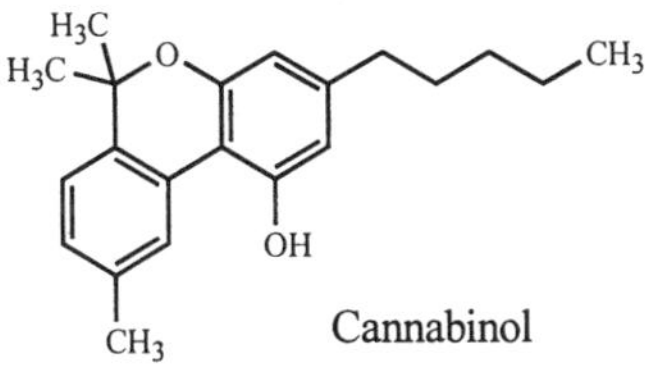

Cannabinol

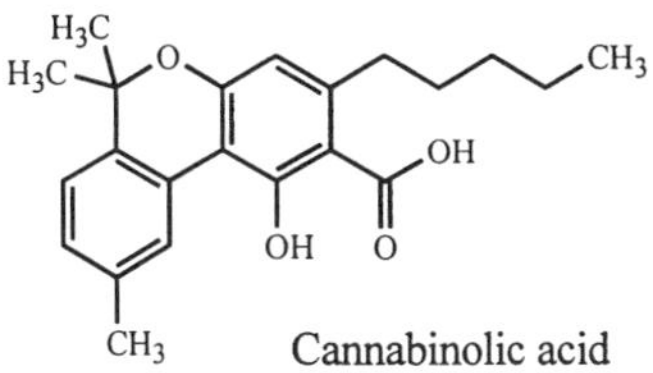

Cannabinolic acid

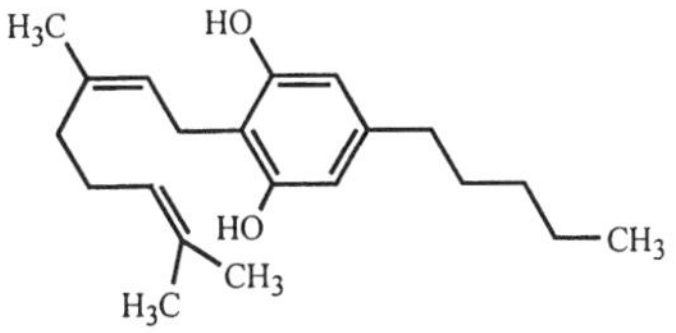

Cannabigerol

and cannabichromene) were well resolved on a C_{18} column ($\lambda = 210$ nm and 224 nm) using a 48-min 47/53 → 70/30 acetonitrile/water (8.64 g/L H_3PO_4) gradient [1426]. Linear concentration curves from 100 to 1000 ng/μL (1 μL injected) were used. Detection limits of 25 ng injected were reported.

In an excellent method, Clerc et al. [1427] separated a number of alkaloid opiates (heroin, codeine, morphine, acetylcodeine, monoacetylmorphine) from other common street "diluents' (procaine, acetaminophen, lidocaine, caffeine, noscapine, papaverine) and baseline resolved them all in <16 min. A C_{18} column ($\lambda = 210$ nm) and a complex 36-min 9/91 → 45/55 (9/1 acetonitrile/water [5 mL/L H_3PO_4 and 0.56 mL hexylamine/L])/water (5 mL/L H_3PO_4 and 0.56 mL hexylamine/L) gradient were used. A table of the working calibration curve concentrations (generally 3–150 μg/mL) was presented. Detection limits from 10 to 100 ng injected were reported. Peak shapes were excellent throughout.

Cocaine, benzoylecgonine, norcocaine, bupivacaine, and cocaethylene were extracted from amniotic fluid and resolved on a C_{18} column (photodiode array detector, $\lambda = 190$–400 nm). Elution was complete in 15 min using a 125/500/12.5 acetonitrile/water (25 mM KH_2PO_4)/butylamine (all to pH 2.9 with H_3PO_4) mobile phase [1428]. Spiked sample recoveries for 0.1–10 μg/mL were tabulated. The minimum quantitation level for cocaine was 100 ng/mL and a minimum detection limit of 30 ng/mL was reported.

9.7.8 Antihistamines

Terfenadine (Seldane[TM], an anthihistamine), terfenadine alcohol, and azacyclonal were isolated from microsomal incubates and separated on a cyanopropyl column (positive ion electrospray MS; source T 100°C probe voltage 3.8 kV, cone voltage 25 V, N_2 nebulizing gas). Elution was complete in 10 min. Linear ranges of

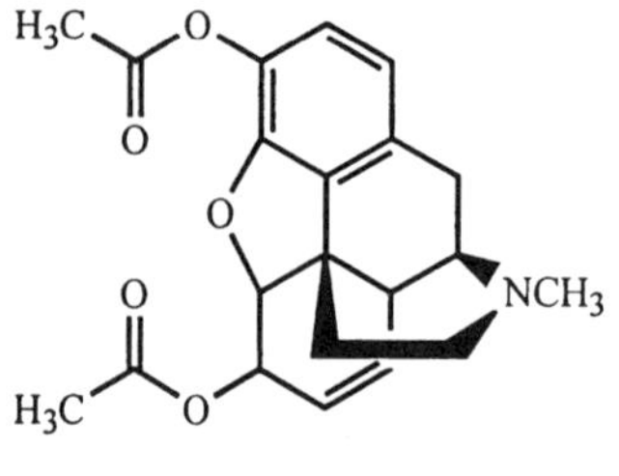

Heroin

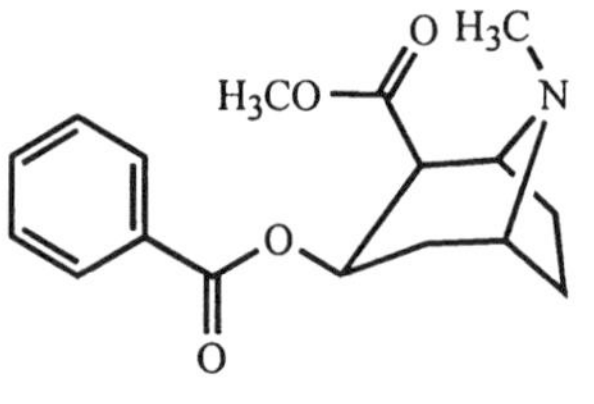

Cocaine

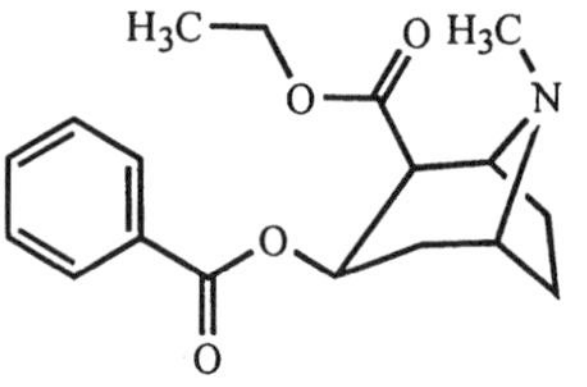

Cocaethylene

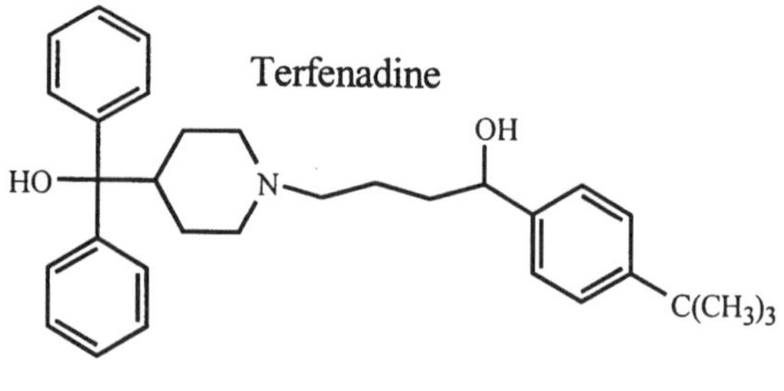

Terfenadine

5–1250 ng/mL incubate and detection limits of 1–5 ng/mL incubate (analyte dependent) were reported [1429].

Histidine, histidinol, histamine, urocanic acid, and imidazole-pyruvic, -acetic, and -lactic acids were recovered from rumen fluid and separated on an aminopropyl column ($\lambda = 220$ nm). Elution was complete in 25 min using a 21/79 water (67 mM K_2HPO_4 at pH 6.45)/(90/10 acetonitrile/water) mobile phase [1430]. Excellent resolution was achieved and peak shapes were excellent except for a very broad imidazolepyruvic acid peak. Decreasing pH increased retention for the analytes. The authors also noted that frequent rinsing of the system with phosphate buffer solution only was needed. A linear range of 1–500 μM along with detection limits of 0.8–5 μM (analyte dependent) were reported.

Chlorpheniramine maleate, methscopolamine nitrate, and phenylephrine hydrochloride levels in capsules were determined using a cyanopropyl column ($\lambda = 262$ nm) and a 70/30 acetonitrile/water (2% acetic acid with 5 mM sodium heptanesulfonate to pH 2.6 with 0.1 N NaOH) mobile phase [1431]. Elution was complete in 12 min and baseline resolution was achieved. Linearity from 2 to 1000 μg/mL and detection limits of 2–10 μg/mL (analyte dependent) were reported.

Four active components [1432] of cough syrup (promethazine, noscapine, ephedrine, bromhexine hydrochloride) were well resolved on a 30°C C_{18} column (photodiode array detector, $\lambda = 200$–350 nm). Separation was achieved using a 100/0 (hold 10 min) → 25/75 (at 20 min) water (5 mM KH_2PO_4 and 5 mM sodium hexanesulfonate to pH 2.8)/(50/50 acetonitrile/water [5 mM KH_2PO_4 with 5 mM sodium hexanesulfonate to pH 2.8]) gradient. Standard concentration ranges were 0.25–0.6 mg/mL (analyte dependent).

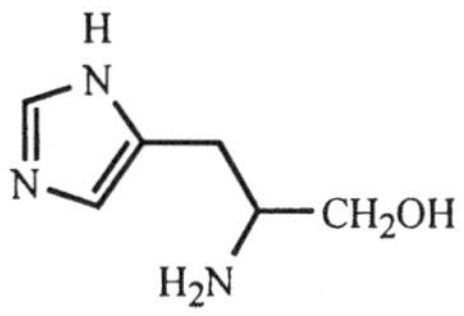

Histidinol

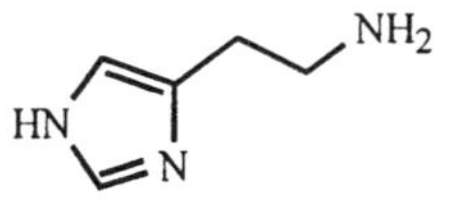

Histamine

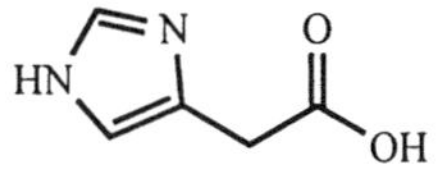

Imidazole-4-acetic acid

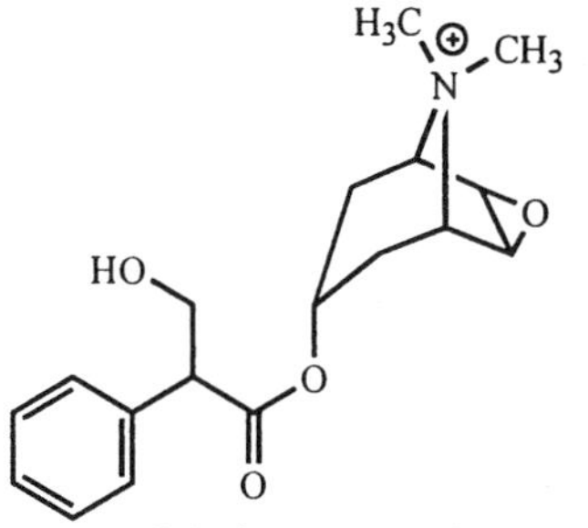

Methscopolamine

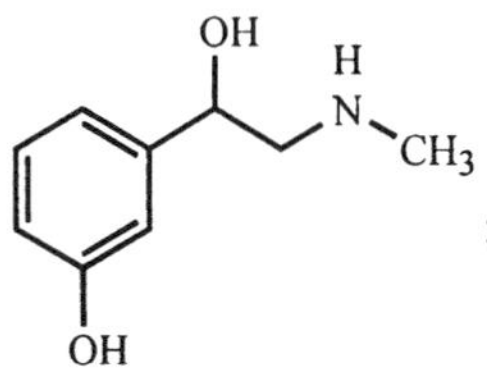

Phenylephrine

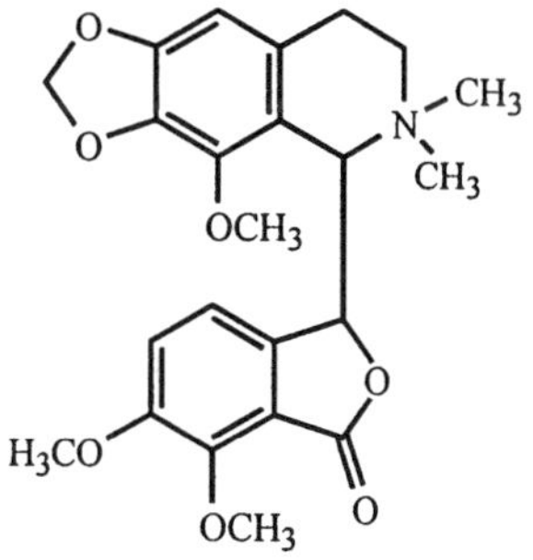

Noscapine

9.7.9   Antiretroviral Drugs

Four HIV protease inhibitors (indinavir, nelfinavir, ritonavir, saquinavir) were extracted from plasma and analyzed on a 40°C C_4 column ($\lambda = 218$ nm and 235 nm for saquinavir). A 52/48 acetonitrile/water (50 mM sodium formate to pH 4.1 with sodium hydroxide) mobile phase generated elution in 14 min. A linear range of 50–20,000 µg/L, detection limits of ~20 µg/L, and quantitation limits of ~40 µg/L (all analyte dependent) were reported [1433].

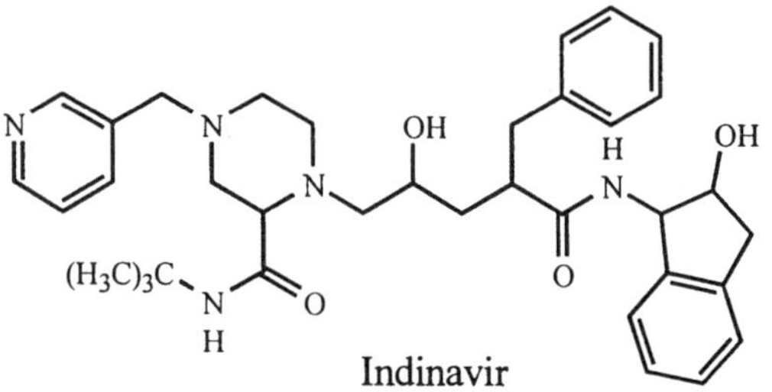

Indinavir

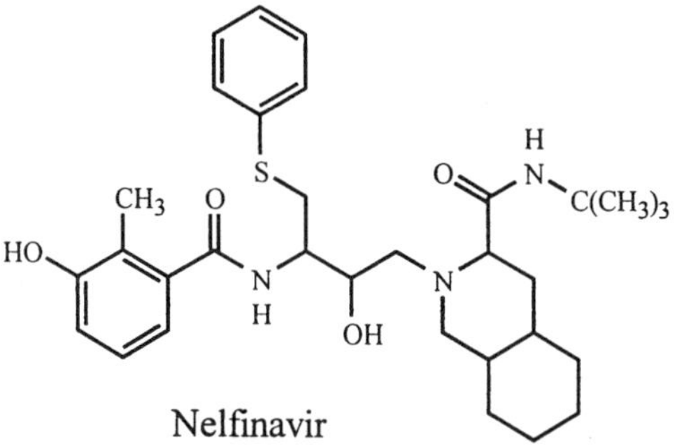

Nelfinavir

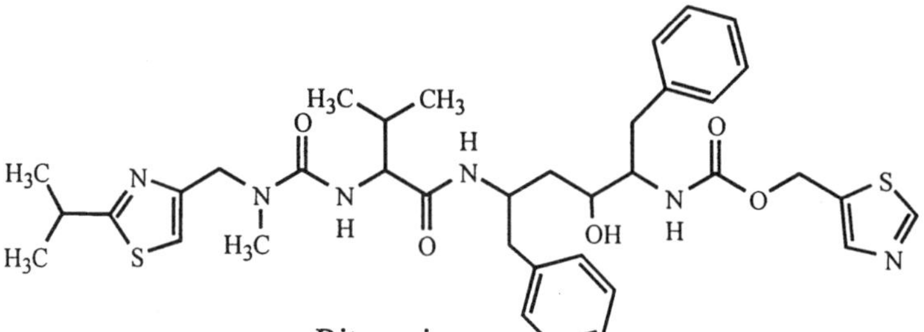

Ritonavir

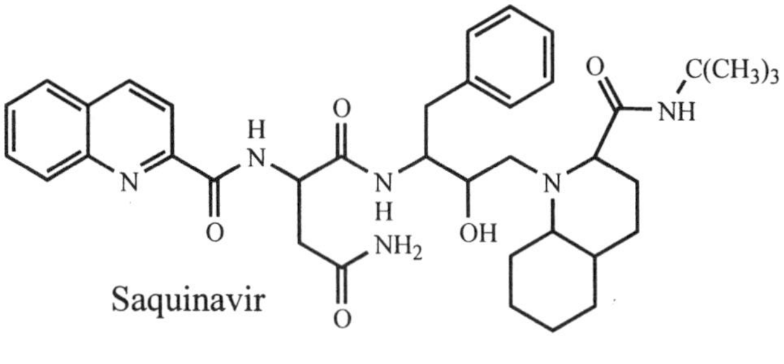

Saquinavir

Abacavir and mycophenolic acid were isolated from plasma and analyzed on a
40°C C_{18} column ($\lambda = 285$ nm for abacivir and 345 nm, ex; 430 nm, em for
mycophenolic acid). A 85/15 water (25 mM Na_2HPO_4 buffer at pH 7.8)/acetonitrile
mobile phase generated elution in <12 min [1434]. Linear curves from 10 to
1000 ng/mL (analyte dependent) and detection limits of <1 ng/mL were reported.

Marzolini et al. [1435] extracted five HIV protease inhibitors (indinavir, amprenavir, saquinavir, ritonavir, nelfinavir) and efavirenz from plasma and separated them
on a C_{18} column ($\lambda = 201$ nm) using a complex 39-min 15/85/0 → 46/54 →
0/10/90 (washout) acetonitrile/water (11.8 mL of 8.5% H_3PO_4 with 0.2 g sodium
heptanesulfonate/L to pH 5.15 with 10 M NaOH)/acetonitrile (0.3% acetic acid)
gradient. Excellent peak shapes and separation was achieved in 32 min. The washout
step generated an extremely large (~100× the analyte peak size) peak. This
indicates that the sample preparation step (here solid phase extraction) should be
modified in order to eliminate this peak. Calibration ranges of 250–10,000 ng/mL
were reported. Limits of detection ranged from 100 to 250 ng/mL (analyte
dependent).

Six antiretroviral agents (ampernavir, nelfinavir, ritonavir, saquinavir, delavirdine,
efavirez) were extracted from plasma and separated on a C_{18} column ($\lambda = 260$ nm).
Use of a 55.2/44.8/0.9/1.0 water (25 mM phosphate at pH 3)/acetonitrile/
diethylamine/THF mobile phase generated a 30-min separation [1436]. Linearity
was reported from 50 to 20,000 ng/mL with quantitation limits of about 75 ng/mL
(analyte dependent).

Abacavir

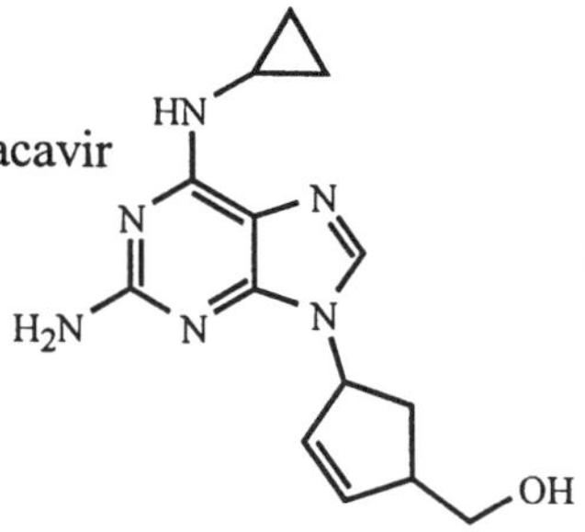

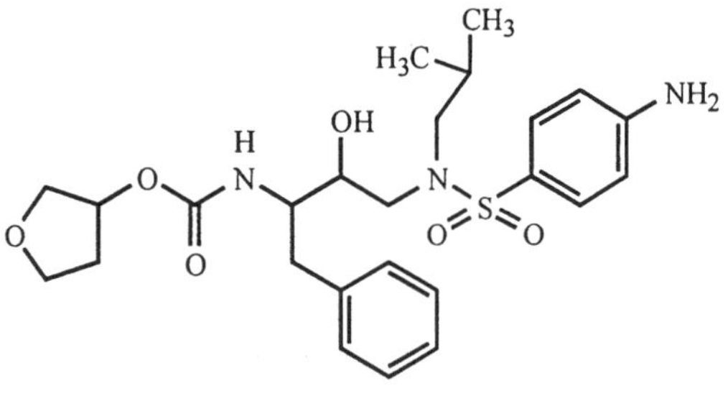

Amprenavir

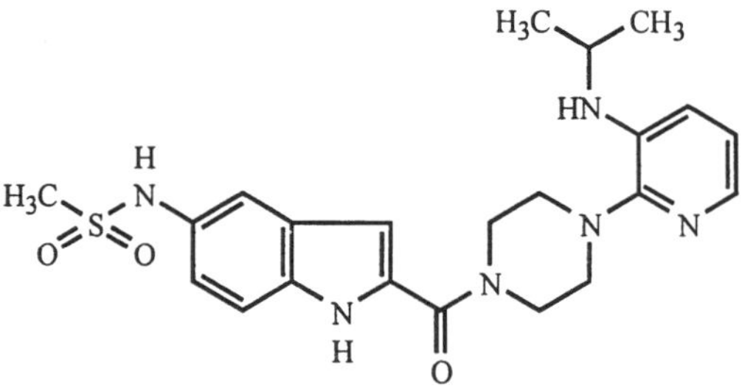

Delavirdine

Famciclovir is an antiviral predrug of penciclovir. Penciclovir and its three metabolites [6-deoxy-, 8-oxo-6-deoxy-, and 8-oxopenciclovir) were baseline resolved in 10 min using a 95/5 water (0.5 mM ammonium acetate at pH 4.65)/acetonitrile mobile phase on a C_{18} column ($\lambda = 280$ nm). Famciclovir and 6-oxo-, desacetyl-, and 6-oxoacetylfamciclovir were separated in 25 min on the same column using an 85/15 water (0.5 mM ammonium acetate pH 4.65)/acetonitrile mobile phase [1437].

Three sets of HIV-1 antiviral compounds were isolated from serum [1438]. Didanosine, lamivudine, and stavudine were separated on a deactivated base C_{18} column ($\lambda = 248$ nm) using a 96/4 water acetonitrile/water with 10 mM phosphate pH 6.9 mobile phase. Elution was complete in 25 min. A calibration curve from 0.02 to 20 mg/mL was reported. Nevirapine and zidovudine (AZT) were separated using a deactivated base C_8 column ($\lambda = 266$ nm) and a 83/17 water/acetonitrile with 15 mM phosphate at pH 7.5 mobile phase. Elution took 13 min and a calibration range from 0.01 to 10 mg/mL was used. Indinavir, delavirdine, nelfinavir, ritonavir, and saquinavir were resolved on a deactivated base C_8 column ($\lambda = 254$ nm) using

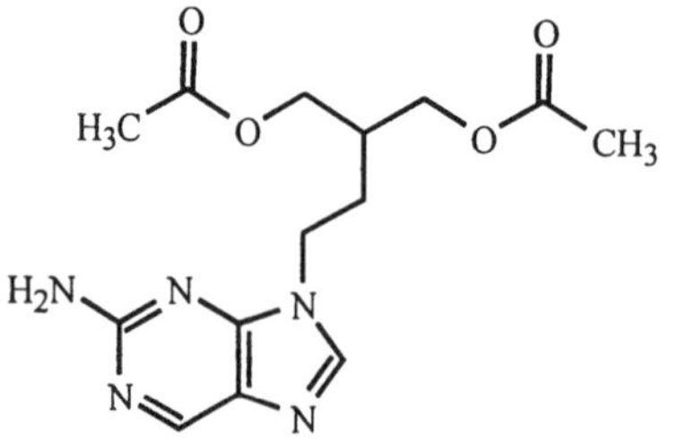

Famciclovir

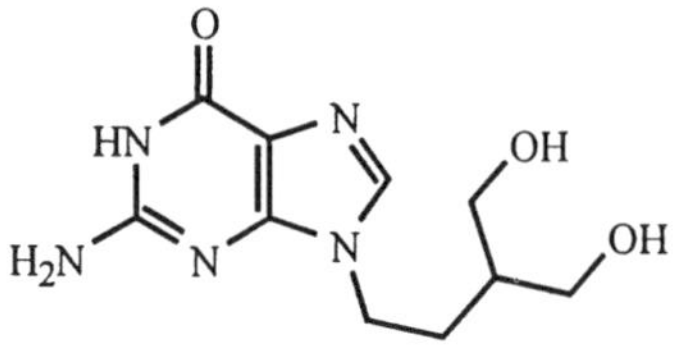

Penciclovir

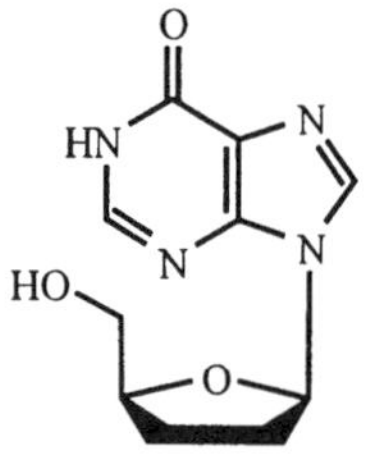

Didanosine

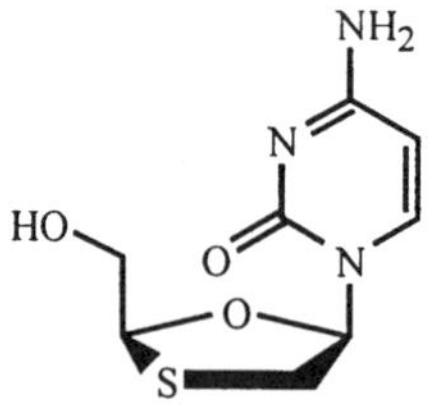

Lamivudine

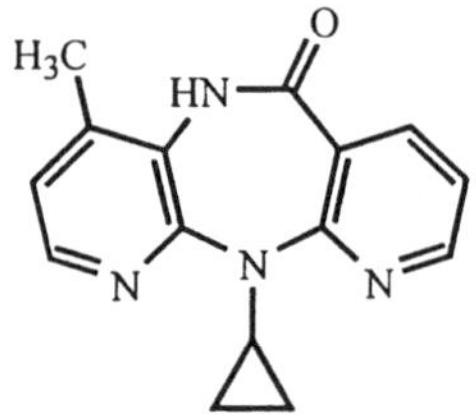

Nevirapine

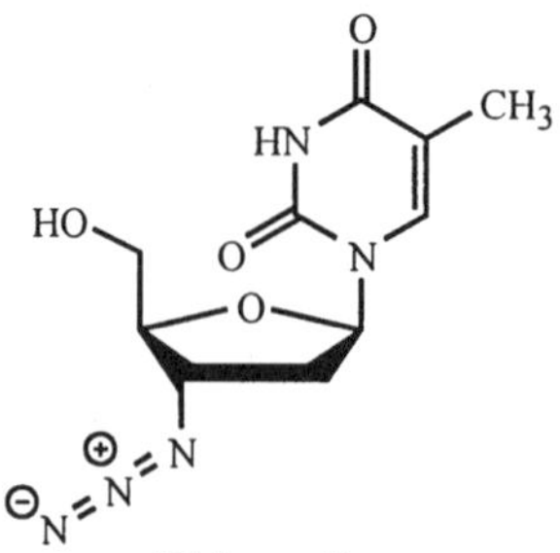

Zidovudine

a 55/45 water/acetonitrile with 15 mM phosphate at pH 7.5 mobile phase. Elution was complete in 25 min and calibration curves from 0.1 to 15 mg/mL were reported.

Amprenavir, an HIV agent, was isolated from plasma analyzed on a 50°C C_{18} column ($\lambda = 270$ nm, ex; 340 nm, em) using a 30/70 acetonitrile/water mobile phase [1439]. Elution was complete in 5 min. A linear range of 1–1000 ng/mL and a quantitation limit of 1 ng/mL were reported.

Atovaquone is a hydroxynaphthoquinone and was isolated from blood and plasma and analyzed on a C_{18} column ($\lambda = 277$ nm) using a 65/35 acetonitrile/water (10 mM KH_2PO_4 at pH 7.0) mobile phase [1440]. The analysis was complete in <4 min with the atovaquone well separated from all other extracted compounds. A linear range of 0.25–15 μM with detection and quantitation limits of 30 nM (S/N = 3) and 150 nM (S/N = 10), respectively, were reported.

Efavirenz was extracted from plasma and analyzed on a C_{18} column ($\lambda = 246$ nm). A mobile phase of 53/47 water (25 mM phosphate to pH 7.5 with KOH)/acetonitrile generated elution in 11 min [1441]. The retention times of seven potential drug interferents were tabulated. Calibration standards ran from 10 to 10,000 ng/mL and a detection limit of 10 ng/mL was reported.

Nevirapine and seven biotransforamtion byproducts (12-hydroxy-, 8-hydroxy-, 2-hydroxy-, and 3-hydroxynevirapine glucuronide, and 4-carboxy-, 12-hydroxy-, and 3-hydroxynevirapine) were extracted from plasma and baseline resolved on a C_{18} column ($\lambda = 240$ nm). A 60-min 100/0 → 65/35 water (50 mM KH_2PO_4 at pH 4.6 with 1% triethylamine)/acetonitrile gradient eluted all compounds in <40 min [1442]. The 2- and 3-hydroxy glucuronide forms were only partially resolved.

The levels of 2′-β-fluoro-2′,3′-dideoxyadenosine 5′-triphosphate and two metabolites were derivatized with chloroacetaldehyde to produce the 1,N^6-ethenoadenine fluorophores. These were then baseline resolved on a C_{18} column (λ= 216 nm, ex; 416 nm, em) using a 60/40 (hold 35 min) → 45/55 (at 60 min) water (60 mM ammonium dihydrogenphosphate with 5 mM tetrabutylammonium phosphate)/(70/30 water [60 mM ammonium dihydrogenphosphate with 5 mM tetrabutylammonium phosphate]/acetonitrile) gradient. Calibration standard levels ranged from 50 to 400 nM [1443].

Two synthetic nucleoside analogs, lamivudine and zidovudine, used in treating HIV were isolated from seminal fluid and separated on a C_{18} column (MS/MS positive ion for lamivudine: source voltage 5 kV, temperature 450°C, orifice voltage 24 V, focusing ring voltage 375 V, entrance voltage −6.5 V; negative ion for

Atoivaquone; Mepron

zidovudine: source voltage -5 kV, temperature 450°C, orifice voltage -34.7 V, focusing voltage -350 V, entrance voltage 8.0 V) using a 15/85 acetonitrile/water mobile phase [1444]. Standard concentrations ranged from 5 to 5000 ng/mL with a detection limit of 5 ng/mL.

9.7.10 Antidepressants

Five tricyclic antidepressants (desipramine, nortriptyline, imipramine, amitriptyline, clomipramine) were extracted from serum and baseline separated on a C_8 column ($\lambda = 254$ nm). A 60/40 acetonitrile/water (10 mM sodium phosphate at buffer pH 3) mobile phase eluted all compounds within 10 min [1445]. The retention times for an additional 11 tricyclic antidepressants using these conditions were tabulated as well. Although not used in the work presented, the authors noted that the addition of the silanol blocking reagent *n*-butylamine enhanced resolution by sharpening the peaks. Detection limits of 10 ng/mL and working concentration ranges from 100 to 500 ng/mL were reported. Similarly, imipramine, amitriptyline, maprotiline, fluoxetine, clomipramine and their metabolites (desipramine, nortriptyline, norfluoxetine, desmethylclomipramine, clovoximine) were extracted from plasma and separated on a C_{18} column ($\lambda = 226$ nm) using a 65/35 water (67 mM KH_2PO_4 to pH 3 with H_3PO_4)/acetonitrile mobile phase [1446]. Elution was complete in <20 min. Imipramine, nortriptyline, and maprolitine were incompletely resolved. A useful table of the retention times (note that the k' values are incorrect) for 32 potential interfering drugs was presented. A linear range of 10–3000 ng/mL and a quantitation limit of 5 ng/mL were reported.

Imipramine and six metabolites (desipramine, iminodibenzyl, imipramine *N*-oxide, 2-hydroxy and 10-hydroxyimipramine, and didesmethylimipramine) were extracted from liver microsomes and well resolved on a cyanopropyl column ($\lambda = 214$ nm). Elution was complete in 14 min when a 40/30/25 acetonitrile/methanol/water (10 mM potassium phosphate buffer at pH 7) mobile phase was used

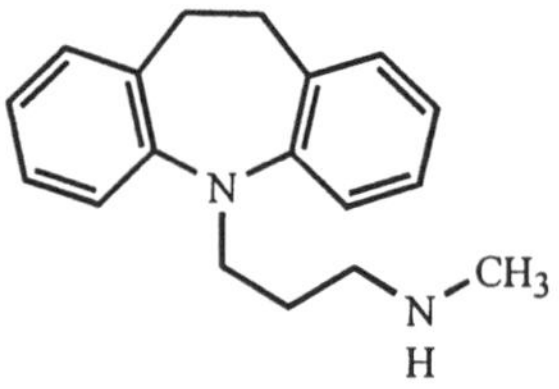

Desipramine

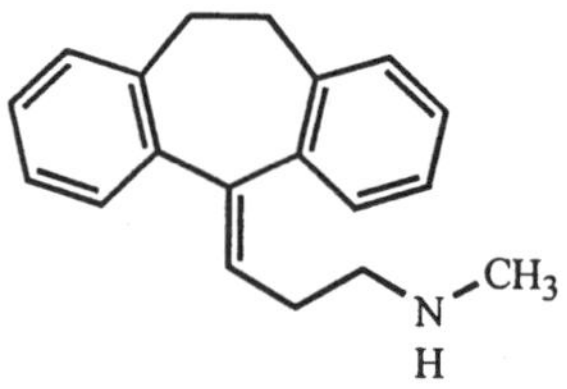

Nortriptyline

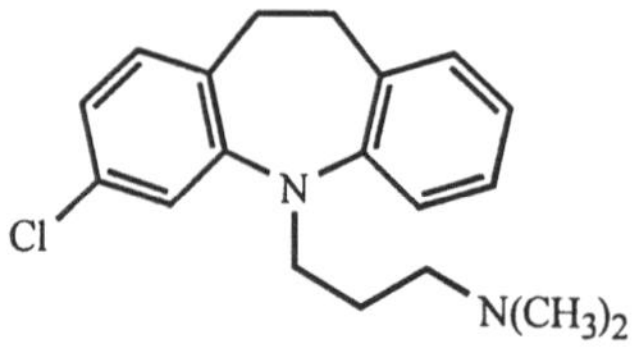

Clomipramine

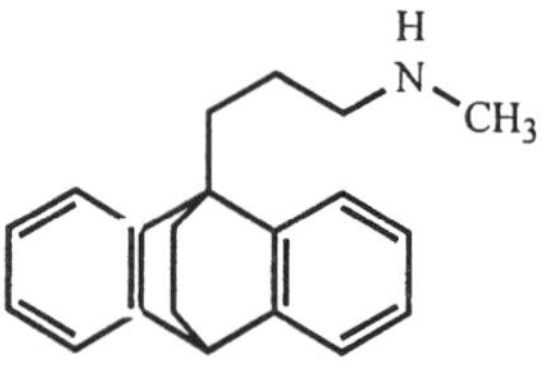

Maprotiline

[1447]. Peak shapes were excellent. A standard curve was generated using 0.02–10.0 nM stock standards.

Tianeptine is a precursor drug to a pentanoic acid side chain analog that is an active antidepressant [1448]. These compounds were extracted from plasma and separated on a C_{18} column ($\lambda = 220$ nm) using a 40/60 acetonitrile/water (1.5 g/L sodium heptansulfonic acid to pH 3 with H_3PO_4) mobile phase. Baseline resolution and good peak shapes were generated in the 13-min separation. Working solutions were prepared from 5 to 500 µg/mL and quantitation limits of 5 µg/mL were reported. A table of 40 potential interferent psychotropic drug retention times was also presented.

Oxcarbazepine, carbamazepine, and six metabolites (e.g., carbamazepine 10,11-epoxide, 10,11-*trans*-dihydroxy-10,11-dihydrocarbamazepine) were extracted from serum and baseline resolved in 28 min [1449]. A C_{18} column ($\lambda = 212$ nm) and a 20/80 acetonitrile/water (0.05% TEA at pH 6.3) mobile phase generated excellent peak shapes. Detection limits of 10 ng/mL were reported.

Kiel et al. [1450], studied the effect of the presence and absence of an amine buffer and solvent pH on the retention and tailing of nortriptyline, desmethylnor-triptyline, and amitriptyline. A C_8 column and a 50/50 acetonitrile/water (25 mM TEA) solvent gave baseline resolution, excellent peak shape, and elution within 4 min. Removal of the TEA led to significant peak tailing and an elution time of nearly 7 min. With no TEA, the mobile phase was buffered to different pH values (2.5–8) with 0.1 M phosphate. A U-shaped plot of k' versus pH resulted. This phenomenon was not because of a protonated-to-deprotonated form of the amine compounds, their pK_a values are >8. Rather, it is due to surface silanol acidity functions and the complex interaction of the solutes with these sites. Therefore, it is important to consider the use of a basic mobile phase modifier when analyzing basic compounds.

Six protriptyline biotransformation products (e.g., *N*-acetoxy-, *N*-acetyl-, *N*-desmethyl-, oxo-*N*-desmethyl-, 2-hydroxy-*N*-acetoxy-protriptyline, and 3-(5-hydro-dibenzo[*b*,*f*][7]annulen-5-yl)propanoic acid) were well resolved on a cyano column ($\lambda = 300$ nm) using a 75/25 water (1.74 g Na_2HPO_4/L to pH 7)/acetonitrile mobile phase [1451]. Peaks were somewhat broad and elution was complete in 25 min.

Six urinary metabolites of doxepin [(*E*)-2-hydroxydoxepin, (*E*)-2-hydroxy-*N*-desmethyldoxepin, (*Z*)- and (*E*)-*N*-desmethyldoxepin and (*Z*)- and (*E*)-doxepin *N*-oxide] were resolved in 35 min on a cyanopropyl column ($\lambda = 250$ nm) using an 84/8/8 acetonitrile/methanol/water (0.1 M ammonium acetate) mobile phase

Tianeptine

Protriptyline

Doxepin

[1452]. The samples were identified by thermospray MS using the same LC system and parameters.

S-Adenosylmethionine (SAMe, AdoMet), adenosine, and S-adenosylhomocysteine were isolated from renal tissue and separated on a 30°C C_{18} column ($\lambda = 254$ nm). A 24-min $97/3/0 \rightarrow 87/3/10$ water (10 mM ammonium dihydrogenphosphate with 0.6 mM sodium heptanesulfonate)/methanol/acetonitrile gradient generated good peak shape and resolution [1453]. A linear range from 0.5 to 100 μM and detection limits of 0.1 μM were reported.

Citalopram and four metabolites (desmethylcitalopram, citralopram N-oxide, citalopram propionic acid, didesmethylcitalopram) were extracted from plasma and baseline resolved on a C_{18} column ($\lambda = 249$ nm, ex; 302 nm, em). A column-switching setup utilized a 1 mM phosphate buffer at pH 3 application solvent and a 30/70 acetonitrile/water/water (20 mM phosphate buffer and 0.1% diethylamine at pH 3) elution solvent [1454]. Peak shapes were excellent and limits of quantitation of 2 ng/mL were reported. Calibration standards from 2 to 150 ng/mL were used.

Centpropazine and its hydroxy metabolite were extracted from serum and separated on a C_{18} column ($\lambda = 250$ nm, ex; 350 nm, em) using a 20-min

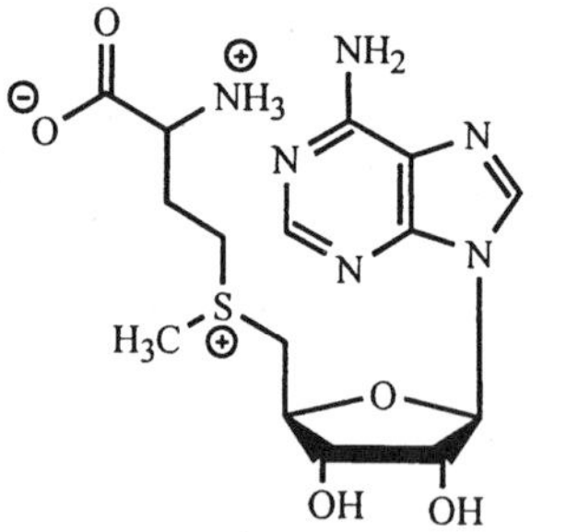

S-Adenosylmethionine; SAMe 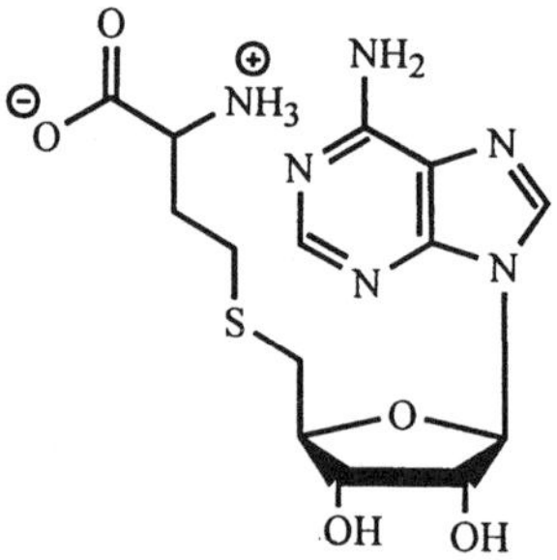   S-Adenosylhomocysteine

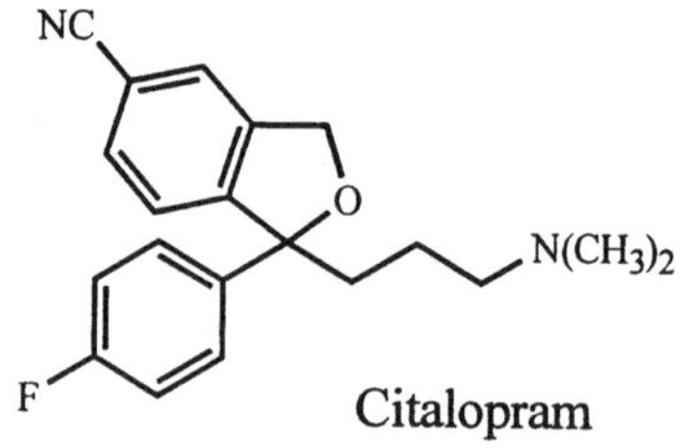

Citalopram

Centpropazine

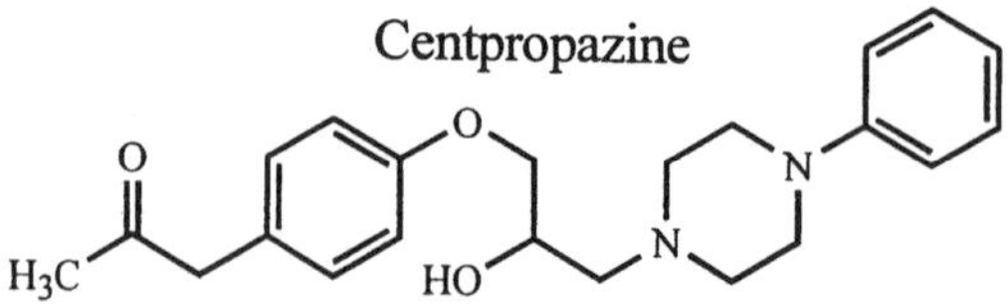

$65/35 \rightarrow 55/45$ water (25 mM phosphate at pH 3)/acetonitrile gradient. Peaks were sharp and baseline resolved. Standards were from 0.625 to 20 ng/mL [1455].

Moclobemide was extracted from blood and urine and analyzed using a 30°C C_8 column ($\lambda = 238$ nm) and a $15/85 \rightarrow 35/65$ (at 6.5 min) $\rightarrow 80/20$ (at 25 min hold 3 min) acetonitrile/water (50 mM NaH_2PO_4 at pH 3.8) gradient. Calibration concentrations ran from 0.2 to 20 μg/mL. The moclobemide eluted at 10 min and the internal standard at 24 min [1456].

9.7.11 Antibacterial Drugs

Rifalazil and three metabolites (25-deacetyl-, 30-hydroxy-, and 32-hydroxyrifalazil) were extracted from urine and separated on a C_{18} column ($\lambda = 230$ nm and 620 nm) using a 50/50 acetonitrile/water (14.05 g $NaClO_4$ with 1.92 g citric acid and 2.94 g trisodium citrate dihydrate/L at pH 3.5–4.5) mobile phase [1457]. Peak shapes were good and all peaks were baseline resolved. The elution was complete in 60 min.

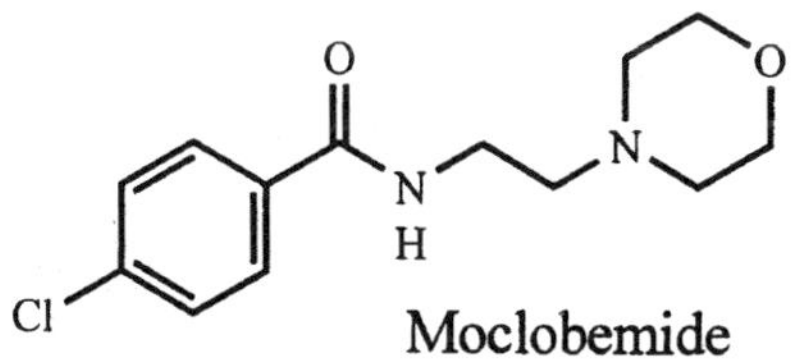

Moclobemide

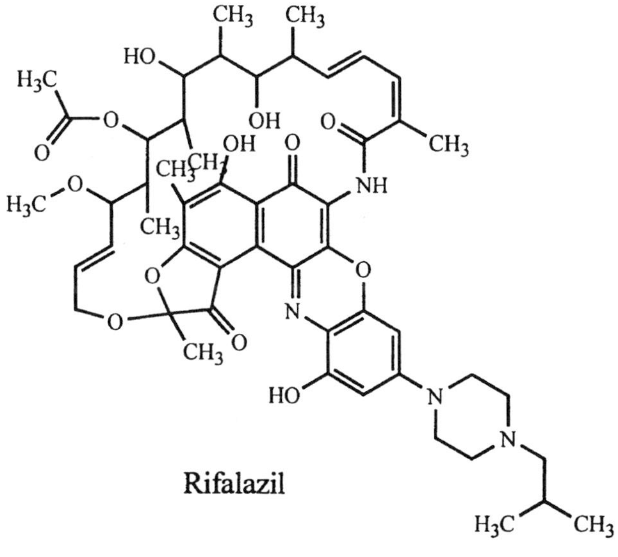

Rifalazil

Horie et al. [1458] extracted five microlide antibacterial agents (josamycin, kitasamycin, mirosamicin, sipamycin, tylosin) from meat and analyzed them on a 35°C C_{18} column ($\lambda = 232$ nm and 287 nm). A 16-min $84/16 \rightarrow 60/40$ water (25 mM NaH_2PO_4 to pH 2.5 with H_3PO_4)/acetonitrile gradient generated baseline resolution, excellent peak shape, and analyte elution in under 15 min. Reported calibration concentrations ranged from 2.5 to 100 ng injected (equivalent to 0.05–2 µg/g) with detection limits of 0.05 µg/g (S/N = 3).

Flumequine and oxolinic acid were extracted from animal tissue and separated on a C_8 column ($\lambda = 328$ nm, ex; 365 nm, em) using a 45/55 acetonitrile/water (10 mM oxalic acid) mobile phase [1459]. Elution was complete in <7 min. The reported linear range was 0.8–500 µg/L with a detection limit of 0.2 µg/L and a quantitation limit of 0.8 µg/L. A plot of k' versus percent acetonitrile is also presented.

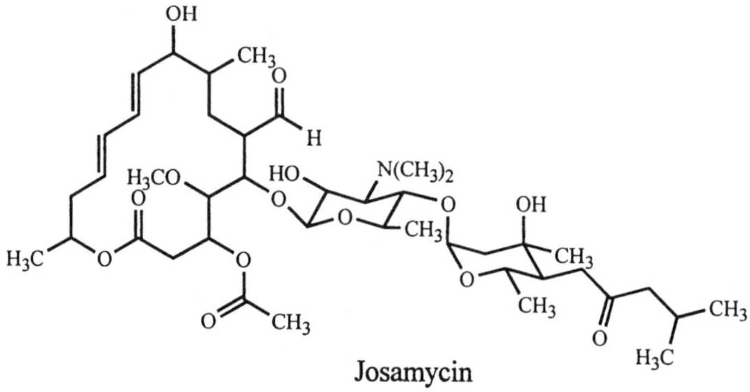

Josamycin

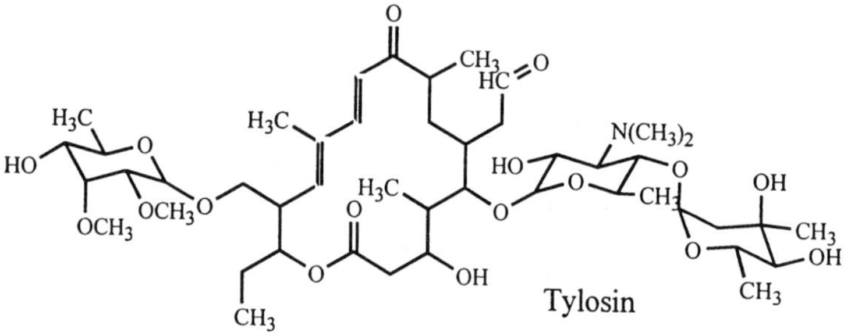

Tylosin

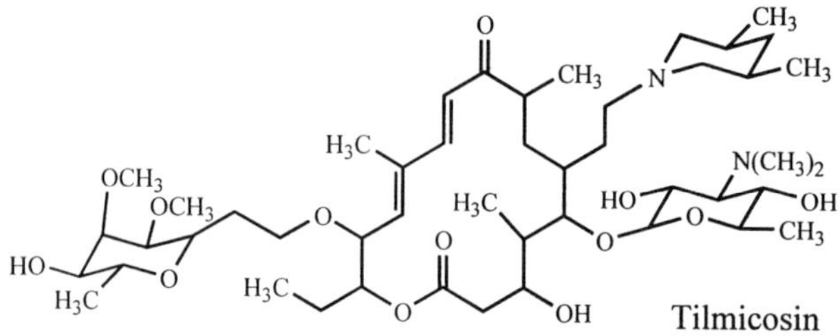

Tilmicosin

The determination of tilmicosin extracted from cow and sheep milk was conducted on a phenyl column ($\lambda = 280$ nm) using a complex 30-min $50/50/0 \to 15/85/0 \to 15/0/85$ acetonitrile/water/water (20 mM dibuytlamine phosphate at pH 2.6) gradient. This successfully separated the analyte from a host of co-extracted peaks [1460]. Standards ranged from 0.01 to 10 µg/mL (sheep milk) and 0.025 to 0.5 µg/mL (cow milk). Detection and quantitation limits were 0.0125 µg/mL and 0.05 µg/mL, respectively.

The levels of norfloxacin and tinidazole in tablets were determined on a C_{18} column ($\lambda = 311$ nm) using an 80/20 water (0.2% triethylamine to pH 2.7 with H_3PO_4)/acetonitrile mobile phase [1461]. The authors noted that when pH > 2.8, peak tailing was unacceptably large. Peaks were somewhat broad but well resolved and elution was complete in 9 min. Linear ranges of 20–300 µg/mL were reported. Detection and quantitation limits were 10 µg/mL (S/N = 3) and 50 µg/mL (S/N = 10), respectively.

Stead and Richards [1462] extracted three aminoglycosides (neomycin, netilmicin, sisomicin) from culture media and plasma, derivatized them with 9-fluorenylmethyl chloroformate (FMOC-Cl), and separated them on a C_{18} column ($\lambda = 260$ nm, ex; 315 nm, em) using a 90/10 acetonitrile/water mobile phase. Good resolution was achieved and elution was complete in 20 min. Linearity was reported over the range 0.1–10 µg/mL but the quantitation limit was designated as 10 ng/mL.

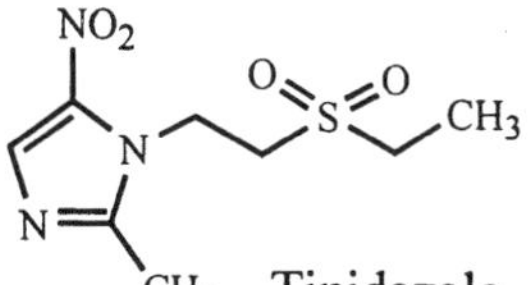

Tinidazole

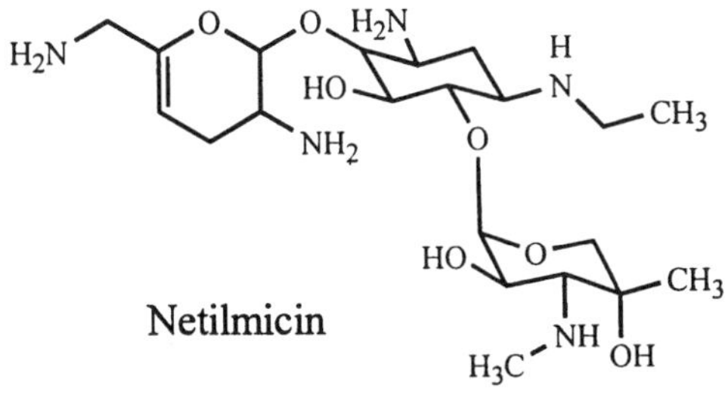

Netilmicin

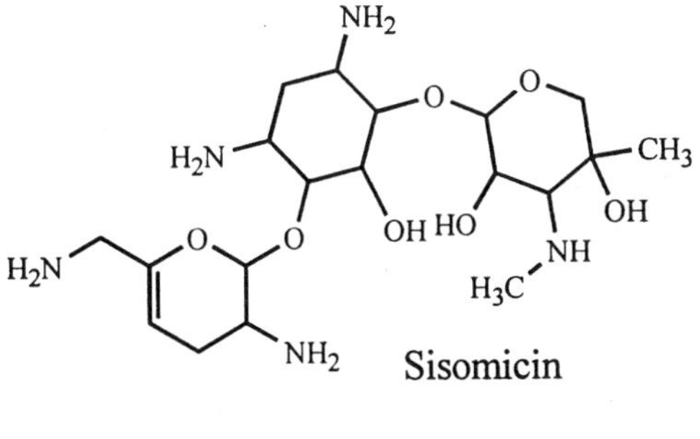

Sisomicin

A series of seven fluoroquinolones (marbofloxacin, danofloxacin, ciprofloxacin, difloxacin, enrofloxacin, norfloxacin, sarafloxacin) were baseline resolved on a C_8 column ($\lambda = 278$ nm) both isocratically, 70 min using 89/11 water/acetonitrile with 10 mM oxalic acid at pH 4.0, and with a 37-min 89/11 (as above) for 13 min then to 75/25 at 33 min gradient [1463]. Peaks were baseline resolved.

The separation of three sulfonamides (pyrimethamine, sulfaquinoxaline, sulfamethazine) was optimized on a C_{18} column ($\lambda = 270$ nm). A 65/35 water (40 mM NaH_2PO_4 at pH 3.0 with 10 mM $NaClO_4$)/acetonitrile mobile phase generated baseline resolution and complete elution in 4 min [1464].

Nalidixic, oxolinic, piromidic, and pipemidic acids and the metabolite 7-hydroxymethylnalidixic acid were baseline resolved on a C_{18} column ($\lambda = 265$ nm) using a 72/28 water (0.1 mM oxalic acid)/acetonitrile mobile phase [1465]. Elution was complete in 8 min. A linear range of 0.25–40 µg/mL was established and detection limits of 3–22 ng/mL were reported (analyte dependent).

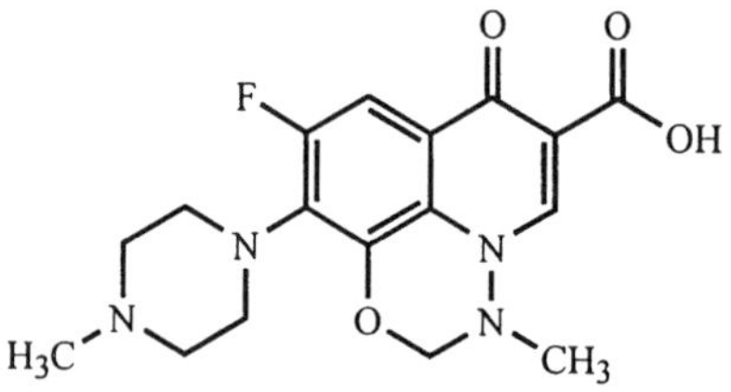

Marbofloxacin

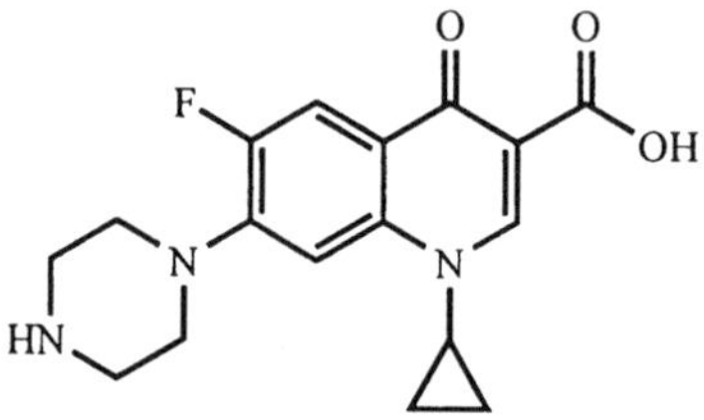

Ciprofloxacin

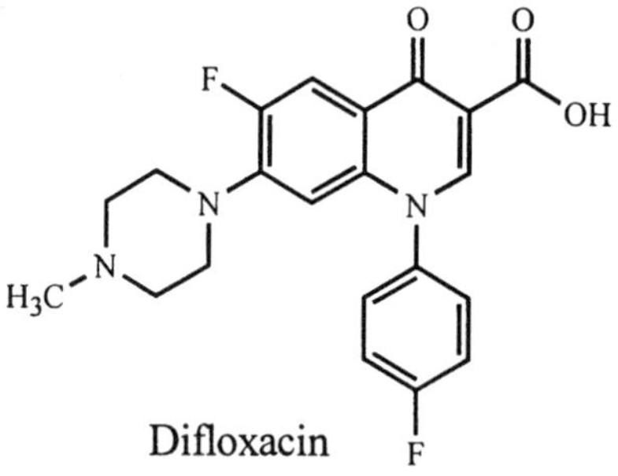

Difloxacin

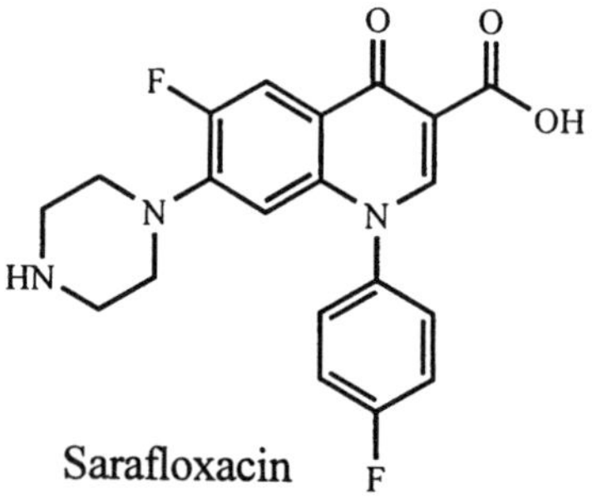

Sarafloxacin

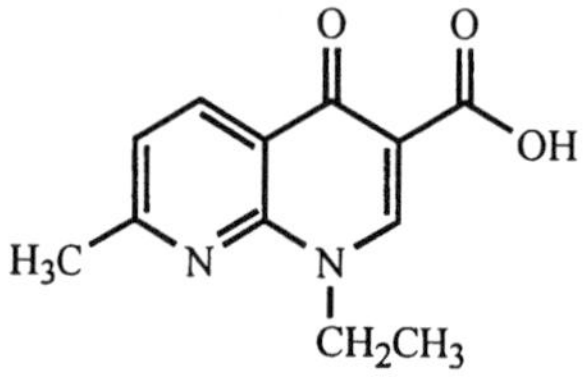

Nalidixic acid

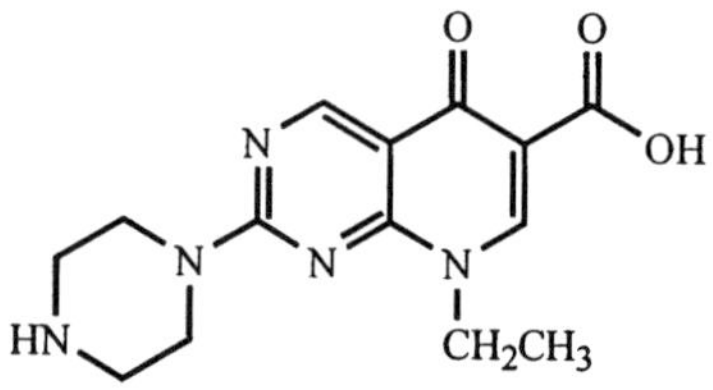

Pipemidic acid

9.7.12 Anesthetics

Ropivacaine and three metabolites (3-hydroxy-, 4-hydroxyropivacaine, 2′,6′-pipecol-oxylidide) were extracted from plasma and analyzed on a 30°C C$_8$ column ($\lambda = 205$ nm). Ropivacaine was analyzed using an 80/20 water (10 mM KH$_2$PO$_4$ at pH 2.1)/acetonitrile mobile phase, whereas the metabolites were baseline resolved using an 85/15 water (5 mM heptanesulfonic acid with 10 mM KH$_2$PO$_4$ to pH 2.1)/acetonitrile mobile phase [1466]. Analysis of ropivacaine took 10 min and the metabolite separation was complete in <20 min. The difference in the mobile phase compositions is such that a simple gradient might prove quite effective in combining these two analyses. A linear range of 1–2000 ng/mL and detection limits of 0.2–200 ng/mL (analyte dependent) were reported. In a similar fashion, ropivacaine, a local anesthetic, and five metabolites (pipecoloxylidide, 3-hydroxy- and isopropyl-pipecoloxylidide, 3- and 4-hydroxyropivacaine) were recovered from urine and eluted in 16 min from a C$_{18}$ column ($\lambda = 210$ nm). A 70/30 water (30 mM phosphate buffer at pH 2.5 with 10 mM 1-octanesulfonic acid)/acetonitrile mobile phase generated good peak shapes and baseline resolution for all peaks. Calibration curves covered the 1–2.4 μM range. Detection limits of 2–18 nM (S/N = 3) were reported [1467].

Ketamine, norketamine, and dehydronorketamine were extracted from plasma and separated on a C$_{18}$ column ($\lambda = 210$ nm) using a 77/23 water (30 mM phosphate at pH 7.2)/acetonitrile mobile phase [1468]. Baseline resolution, excellent peak shapes, and total elution in 18 min were achieved. The linear ranges were reported as 5–5000 μg/L and the quantitation limit as 5 μg/L.

The intravenous anaesthetic propofol (2,6-diisopropylphenol) and five metabolites (2,6,-diisopryopyl-1,4-quinol and 1,4-quinone, the quinol sulfate and glucuronide, and propofol glucuronide) were recovered from urine and separated on a 25°C C$_{18}$ column ($\lambda = 270$ nm or 230 nm) using a 35-min 87/13 → 11/89 water (acetic acid to pH 3.8)/acetonitrile gradient [1469]. Similarly, propofol (2,6-diisopropyl-phenol) and six metabolites (quinol, quinone, quinol-4-sulfate, and propofol-1-,

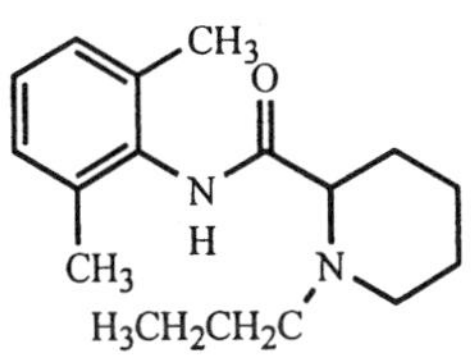

Ropivacaine

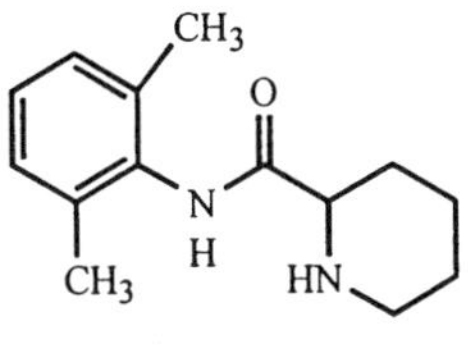

Pipecoloxylidide

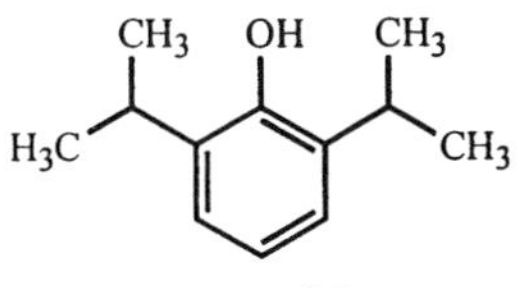

Propofol

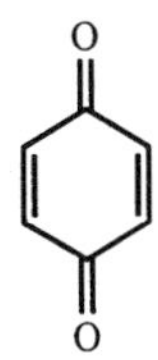

Quinone

quinol-1-, and quinol-4-glucoronide) were recovered from plasma and urine and separated on a C_{18} column ($\lambda = 270$ nm or $\lambda = 270$ nm, ex; 310 nm, em). A 50/40/10 acetonitrile/water/methanol mobile phase resolved all peaks but propofol from quinone. A linear range from 0.1 to 12 μg/mL with a limit of detection of 0.1 μg/mL was reported [1470].

9.7.13   Immunosuppressants

Sirolimus (rapamycin) and three metabolites [29-*O*-acetyl-, 40-*O*-acetyl-, and 28,40-*O*-diacetylsirolimus) were extracted from blood and separated semipreparatively on a 35°C C_8 column ($\lambda = 276$ nm) using a complex 40-min 50/50 → 25/75 water (H_2SO_4 to pH 3)/acetonitrile gradient [1471]. Good resolution and peaks shapes were generated. Calibration curves from 1 to 250 μg/mL were used.

Mycophenolic acid is the metabolite of the mycophenolate mofetil and was extracted from plasma. Quantitation was achieved on a C_{18} column ($\lambda = 254$ nm) using a 45/55 water (50 mM H_3PO_4 at pH 3.2)/acetonitrile mobile phase [1472]. Elution was complete in 10 min. A linear range of 0.1–20 μg/mL and a quantitation limit of 0.1 μg/mL (S/N = 3) were reported.

A semipreparative method for the isolation and recovery of 10 cyclosporin G metabolites was developed on a 70°C C_8 column ($\lambda = 214$ nm) using a 160-min 45/55 → 70/30 (90/10 acetonitrile/methanol)/(90/10 water/methanol) gradient. The metabolites were either hydroxylated or fragments of cyclosporin G and so

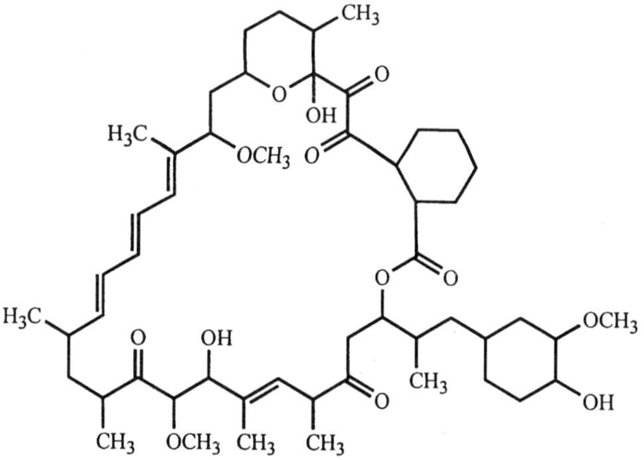

Sirolimus

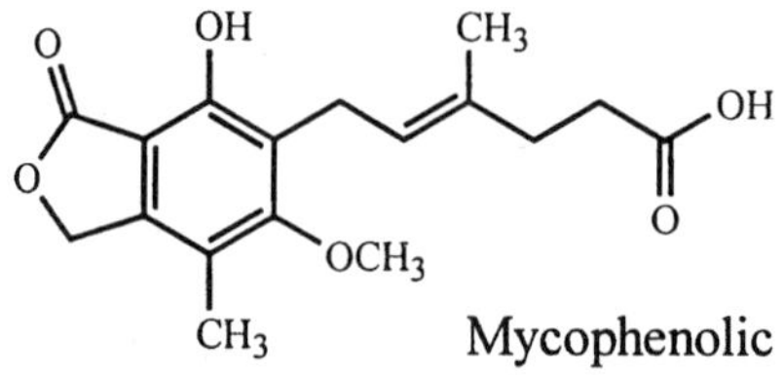

Mycophenolic acid

eluted prior to the parent peak [1473]. It is interesting to note that the peaks exhibited no tailing, even though no buffer was used and cyclosporin is a substituted cyclic polypeptidic compound. Cyclosporins A, B and C were resolved on a 40°C C_8 column ($\lambda = 202$ nm) with an isocratic 50/50 acetonitrile/water (0.1% H_3PO_4) mobile phase [1474]. Temperature (40–70°C), acid concentration (0.001–0.1%), and acetonitrile level (50–70%) were all studied to obtain the optimal result given above. Separation of a 3 µg injection was complete in 15 min. Additionally, 10 metabolites of cyclosporin A were separated on a 70°C C_{18} column ($\lambda = 214$ nm) using a 113-min $43/57 \rightarrow 73/27$ acetonitrile/water gradient [1475]. A table identifying each metabolite was also presented.

9.7.14   Analgesics

Acetaminophen, acetaminophen glucuronide, and acetaminophen sulfate were extracted from serum samples and analyzed on a 30°C C_{18} column ($\lambda = 254$ nm) using a 93/7 water (50 mM sodium sulfate to pH $= 2.2$ with H_3PO_4)/acetonitrile mobile phase [1476]. Elution was complete in 10 min. The glucuronide was not fully resolved from co-extracted materials. Calibration curves from 1 to 500 µg/mL were utilized (analyte dependent). Quantitation limits of 0.15 µg/mL (S/N $= 3$) were reported.

Stability analyses of acetaminophen, salicylamide, and phenyltoloxamine with precursors (p-hydroxyacetphenone, ethylsalicylate) and degradation products (p-aminophenol, p-nitrophenol, salicylic acid) were quantitated on a C_8 column (photodiode array detector, $\lambda = 220$–300 nm). For the assay of the active compounds a 17-min $95/5 \rightarrow 55/45$ water (0.1 M phosphate to pH 2.7 with H_3PO_4)/acetonitrile gradient was used. For separating all the above-mentioned compounds, a complex 35-min $81/15/4 \rightarrow 43/14/43$ water (0.1 M phosphate to pH 2.7 with H_3PO_4)/methanol/acetonitrile gradient was effective [1477]. In this gradient, salicylamide and p-hydroxyacetphenone were incompletely resolved. Linear concentration ranges ran from 8 ng/mL to 300 µg/mL (analyte dependent).

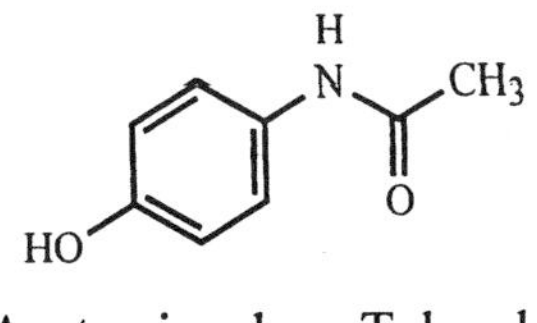

Acetaminophen; Tylenol

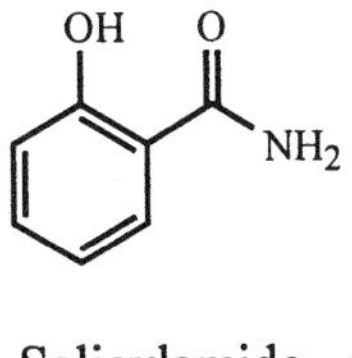

Salicylamide

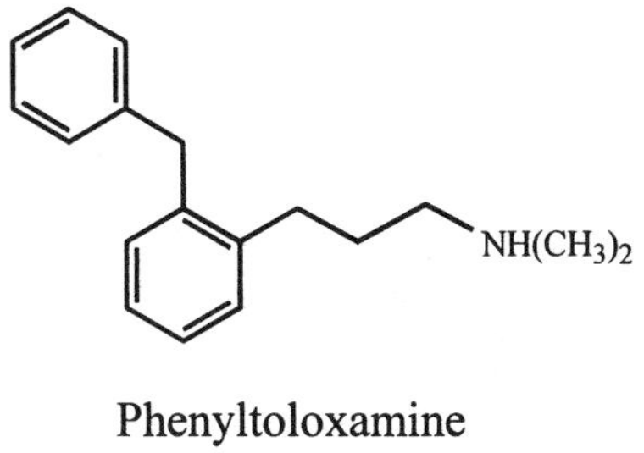

Phenyltoloxamine

Denaverine and its metabolite *N*-monodesmethyldenaverine were extracted from plasma and baseline resolved on C_8 column ($\lambda = 220$ nm) using a 24/17.2/1 acetonitrile/water (0.12 M ammonium phosphate)/THF mobile phase. Peak shapes were good and elution was complete in <7 min. Other columns were tried such as C_{18}, phenyl, and cyanopropyl, but were not as effective. A linear range of 2.5–150 ng/mL with quantitation limits of 2.5 ng/mL was reported [1478].

In an excellent study for cleaning validation work, Nozal et al. [1479] determined acetylsalicylic acid residue levels in 8 min using a 55°C C_{18} column ($\lambda = 226$ nm) and a 79/22/0.1 water/acetonitrile/H_3PO_4 mobile phase. A 69/28/3 water/methanol/acetic acid mobile phase gave the same retention and separation (from excipient peaks), but the sensitivity was only about 1/5 that of the acetonitrile mobile phase. The concentration range of the standard ran from 0.15 to 10 mg/L. Detection limits of 0.04 μg/mL and quantitation limits of 0.14 μg/mL (S/N = 10) were reported.

He et al. [1480] extracted codeine and seven metabolites (e.g., norcodeine, codeine-6-glucuronide, morphine, normorphine) from plasma and urine and separated them in 24 min on a C_8 column ($\lambda = 214$ nm for codeines; electrochemical detector at 350 mV for morphines) using a 77/23 water (5 mM NaH_2PO_4 at pH 2.35 [plasma] or pH 2.90 [urine] with 0.7 mM sodium dodecyl sulfate)/acetonitrile mobile phase. Separation was not outstanding for the UV detection but was excellent for the electrochemical detection. Calibration ranges for UV were 20–10,000 nM and for EC were 3–3500 nM (analyte dependent). Detection limits ranged from 0.02 nM (UV) to 0.003 nM (EC).

9.7.15   Antihypercholesterolemic and Antihyperlipidemic Drugs

Gemfibrozil and four metabolites (the β-D-acylglucuronides of three hydroxy analogs and one carboxyl analog) were extracted from urine, plasma, and tissues

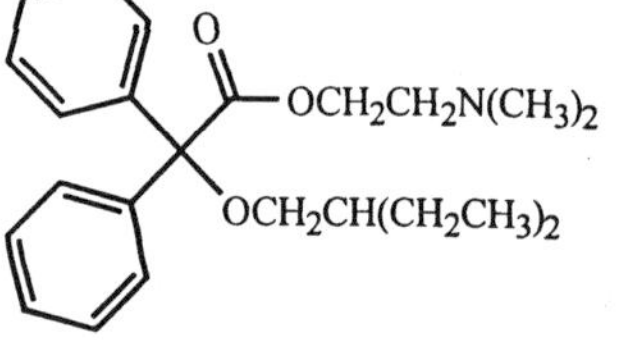

Denaverine

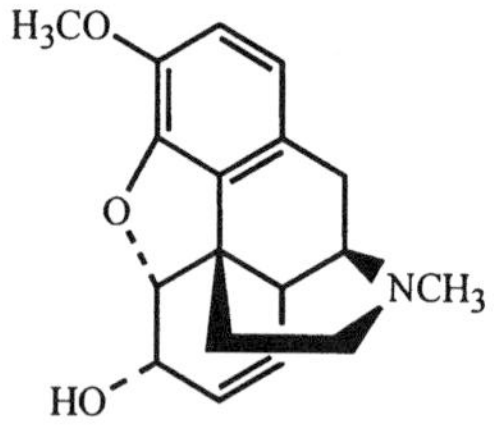

Codeine

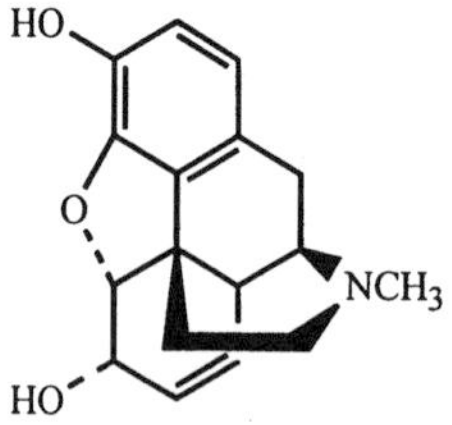

Morphine

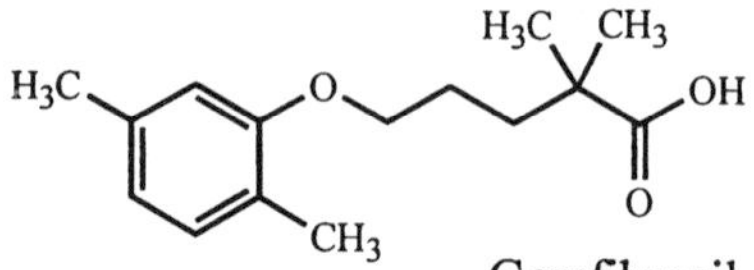

Gemfibrozil

and analyzed on a C_{18} column ($\lambda = 280$–315 nm, ex; 308–450 nm, em; time-programmed). A complex 30-min 100/0 → 50/50 water (10 mM acetate at pH 4.7)/acetonitrile gradient generated excellent resolution and peak shapes [1481]. Concentration ranges varied from 0.1–150 ng/mL to 0.1–50 µg/mL (analyte and matrix dependent).

Four liver metabolites of simvastatin ($6'\beta$-hydroxy-$3'$-hydroxy-, $3',5'$-dihydrodiol-$6'\beta$-exomethylenesimvastatin) were extracted from incubates and analyzed on a C_8 column ($\lambda = 240$ nm) using a 25-min 67/33 → 25/75 water (0.05% phosphoric acid)/acetonitrile gradient [1482]. Peaks were in general well resolved from other extracted compounds. Metabolites from 100 µM incubates were readily detectable.

Lovastatin and six manufacturing impurities (compactin, asterric acid, lovastatin dimer, dehydrolovastatin, dihydrolovastatin, hydroxy acid lovastatin) were fully resolved on a C_8 column ($\lambda = 200$ nm for dihydrolovastatin and then 238 nm). A 14-min 60/40 → 90/10 acetonitrile/water [0.1% H_3PO_4 at pH 2.2) gradient was used to generate the separation [1483]. A quantitation limit of 0.1% was reported. Four metabolites of lovastatin ($6'\beta$-hydroxy-, $3''$-hydroxy-, $6'$-oxomethylene-, and hydroxy acid lovastatin) were extracted from bile and liver tissues and baseline resolved from the parent compound using a C_{18} column ($\lambda = 238$ nm). Elution was complete in 25 min when a 30/70 → 90/10 acetonitrile/water (5 mM formic acid) gradient was used. Peak shapes were excellent [1484].

9.7.16 Antihypertension Drugs

Mibefradil and 10 microsomal metabolites (including 4- and 5-hydroxymibefradil and their glucuronides and N-desmethylmibefradil) were analyzed on a C_6 column

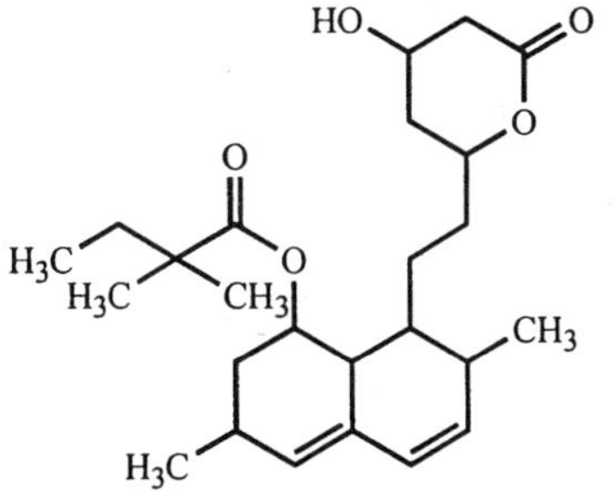

Simvastatin

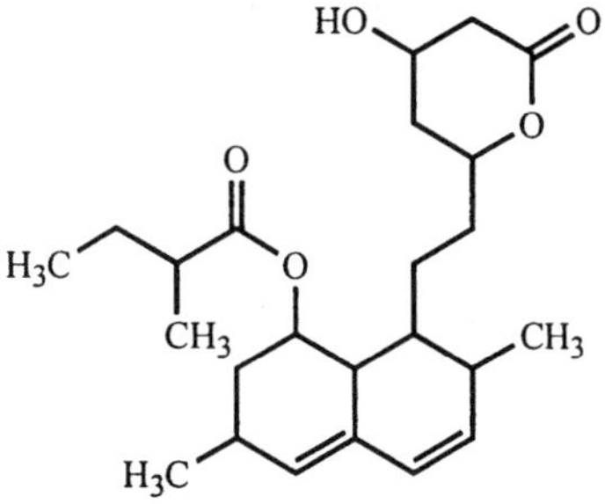

Lovastatin

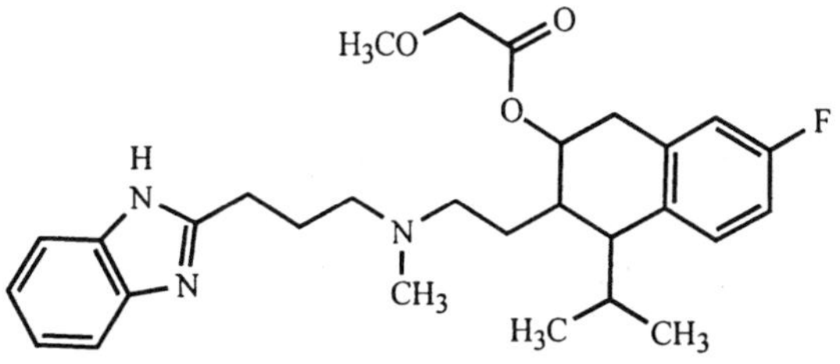

Mibefradil

($\lambda = 270$ nm, ex; 350 nm, em) using a 64-min 75/25 (hold 5 min) → 50/50 (at 40 min hold 24 min) water (0.1 M ammonium acetate at pH 6.9)/acetonitrile gradient [1485a]. Peak shapes were excellent and baseline resolution was achieved.

Nifedipine, a calcium channel antagonist, was recovered from plasma and quantitated in 5 min using a 65/35 acetonitrile/water (13 mM KH_2PO_4/Na_2HPO_4 at pH 7) mobile phase and a C_{18} column ($\lambda = 338$ nm). A linear working concentration range of 2–200 ng/mL and a detection limit of 2 ng/mL were reported [1485b]. The authors note that nifedipine is extremely light sensitive, photodecomposing to dihydronifidepine very rapidly when exposed.

Isradipine and four of its cell-culture metabolites were baseline resolved in 40 min on a C_{18} column ($\lambda = 240$ nm). A 55.4/43.6/1 water (160 mM NaH_2PO_4 with 17 mM tetrabutylammonium hydrogensulfate)/acetonitrile/THF mobile phase gave excellent peak shapes [1486]. A plot of the effect of changing pH on k' was presented. Linear ranges were generated from 1 to 40 μM. Limits of detection and quantitation were given as 0.08–0.23 nmol/mL and 0.2–0.8 nmol/L (analyte dependent), respectively.

Hydrochlorothiazide was extracted from serum and analyzed on a C_{18} column ($\lambda = 270$ nm) using an 80/20 water (1% acetic acid)/acetonitrile mobile phase [1487]. Elution was complete in 6 min but the peak was significantly tailed. The choice of a different mobile phase modifier is recommended (such as TFA). A working curve from 0.5 to 20 μg/mL and a detection limit of 0.2 ng injected were reported.

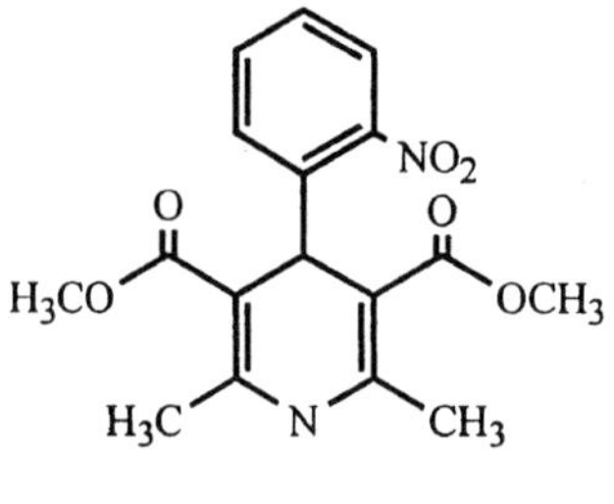

Nifedipine

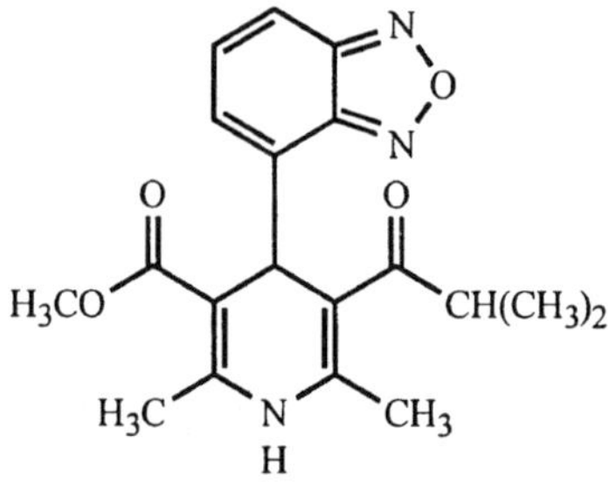

Isradipine

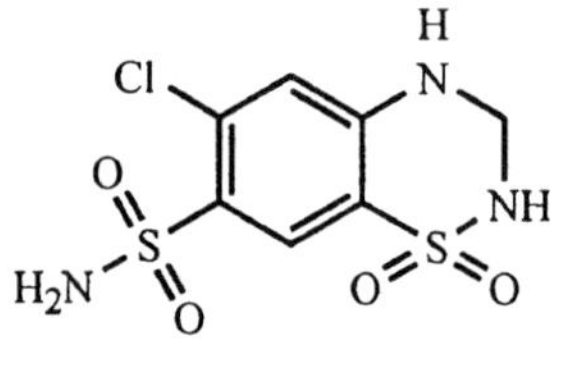

Hydrochlorothiazide

Six calcium channel blocker drugs (atenolol, amlodipine, nifedipine, nitrendipine, nimodipine, felodipine) were separated on a C_{18} column ($\lambda = 250$ nm) using a 50/50 acetonitrile/water (10 mM phosphate at pH 4.5) mobile phase [1488]. Note that the choice of phosphate generates a system with little buffer capacity (halfway between pK_1 and pK_2). Baseline resolution, excellent peak shape and elution time of 13 min were all obtained. Linear ranges from 25 to 3200 ng/mL and detection limits of 15 ng/mL were reported.

The retention characteristics of propranolol, atenolol, metoprolol, verapamil, diltiazem, nifedipine, clonidine, and prazosin were studied on a silica column

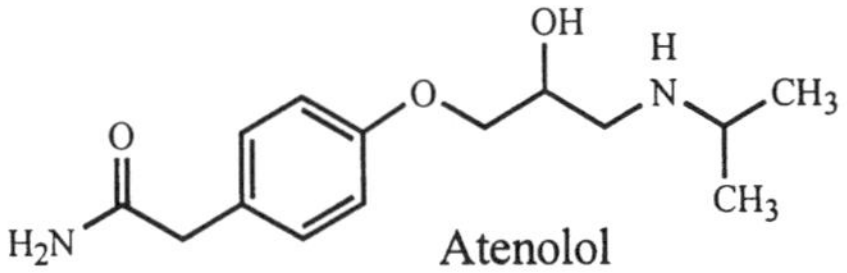

Atenolol

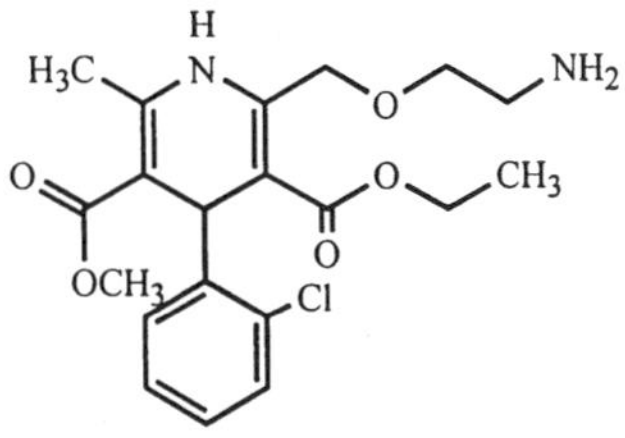

Amlodipine

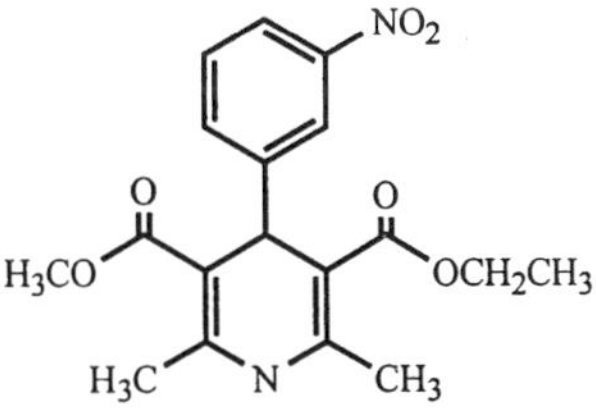

Nitrendipine

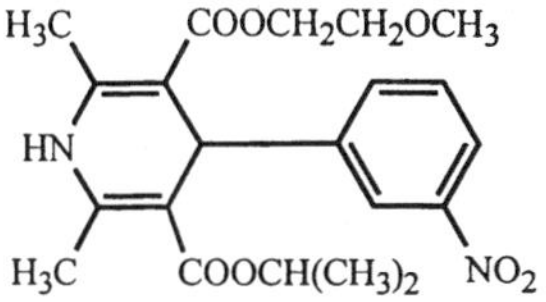

Nimodipine

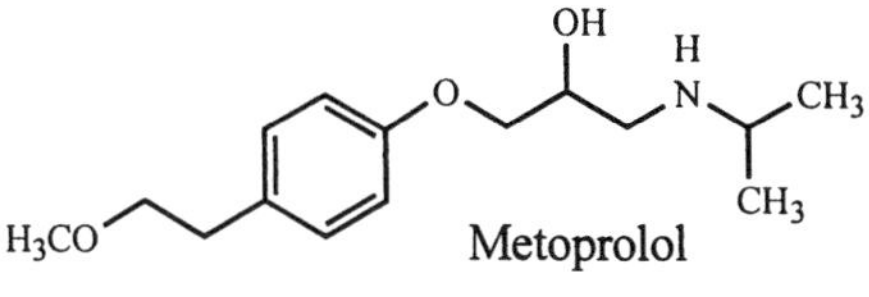

Metoprolol

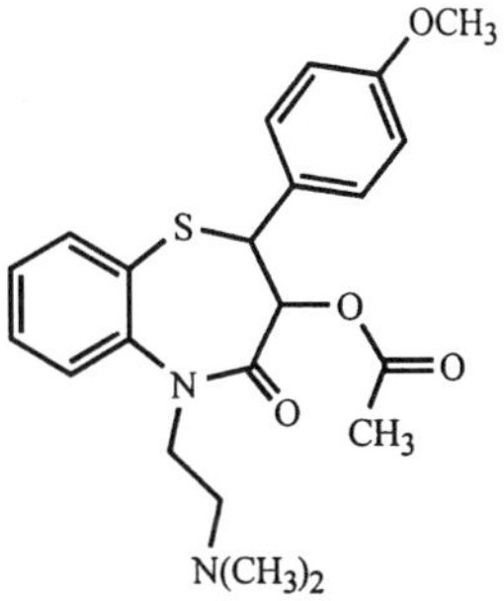

Diltiazem

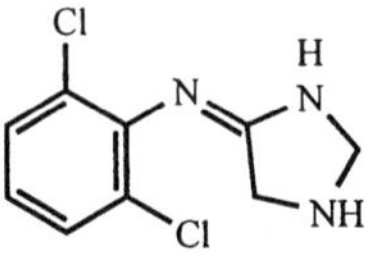

Clonidine

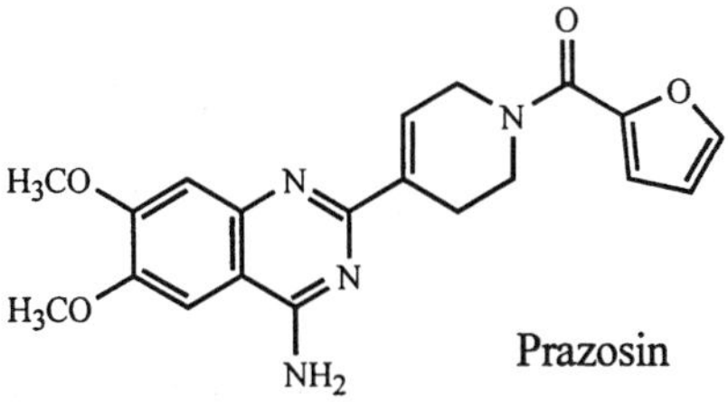

Prazosin

($\lambda = 254$ nm) using 40/60, 50/50, and 60/40 acetonitrile/water (6.25 mM sodium phosphate buffer at pH 3) mobile phases [1489]. The k' values for each analyte and each mobile phase were tabulated. Standards of 25 or 50 µg/mL (50 µL injected) were used. With base silica as the support, the effect of decreasing the concentration of analyte may become extremely important. The diagnostic chromatographic effects are increased peak tailing and longer overall retention times.

Carteolol and its metabolite 8-hydroxycarteolol were extracted from liver microsomes and separated on a C_{18} column ($\lambda = 254$ mn) using a 10.5/89.5/1 acetonitrile/water/acetic acid mobile phase [1490]. Elution was complete in 20 min and baseline resoution was achieved.

Celiprolol (CEL) and oxprenolol (OXP) were extracted from plasma and analyzed on a C_{18} column ($\lambda = 220$ nm for CEL and 232 nm for OXP) using a 16/84 acetonitrile/water (1.2% w/v triethylamine to pH 3 with H_3PO_4) mobile phase [1491]. Elution took 11 min. A linear range of 16–1000 ng/mL was used and quantitation limits of 10 ng/mL were reported.

Verapamil and seven metabolites (e.g, norverapamil, 3,4-dimethoxyphenyl-2-isopropylvaleronitrile, and a series of methoxy $\rightarrow$ hydroxy converted analogs) were isolated from serum samples and analyzed on a C_{18} column ($\lambda = 276$ nm, ex; 310 nm, em) using a 65/35 water (0.3% triethylamine to pH 3.8 with H_3PO_4)/acetonitrile mobile phase [1492]. Elution was complete in 35 min and peak shapes and resolution were very good throughout the separation. Linear curves were generated over the range from 1 to 400 ng/mL with quantitation limits (S/N = 10) of 1–5 ng/mL (analyte dependent) reported.

Ten manufacturing by-products of terazosin were resolved from one another and from terazosin on a C_8 column ($\lambda = 254$ nm) using a 175/50/1775 acetonitrile/IPA/water (50 mM citrate buffer at pH 4.4) mobile phase [1493]. Elution was

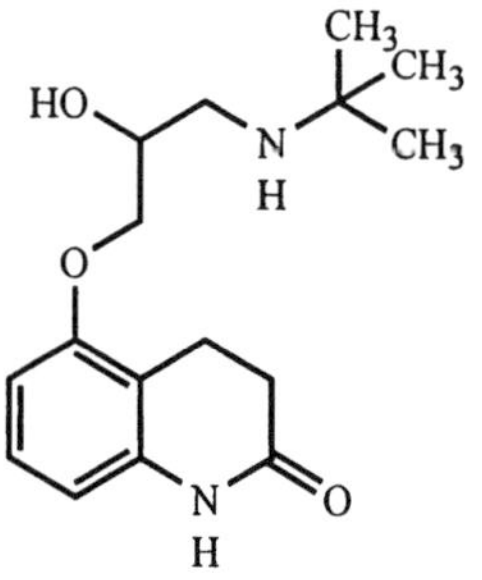

Carteolol

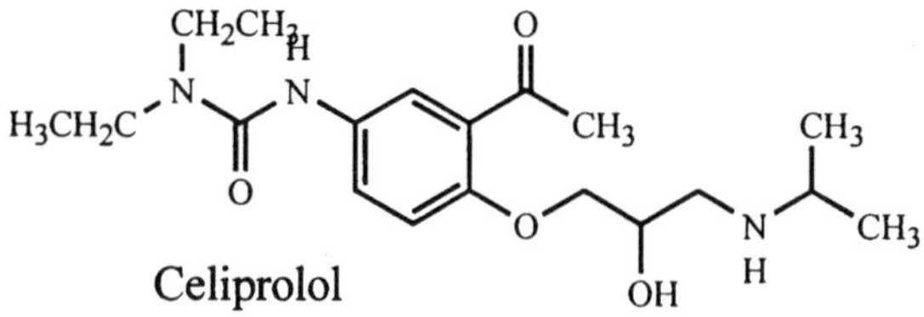

Celiprolol

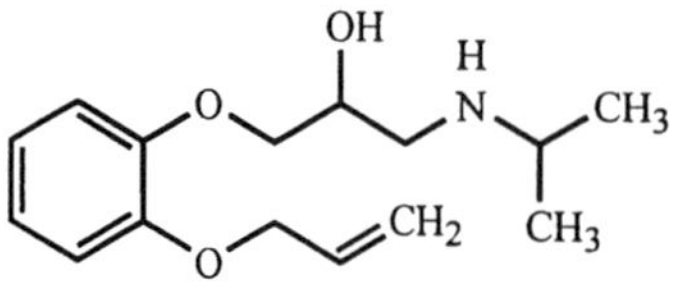

Oxprenolol

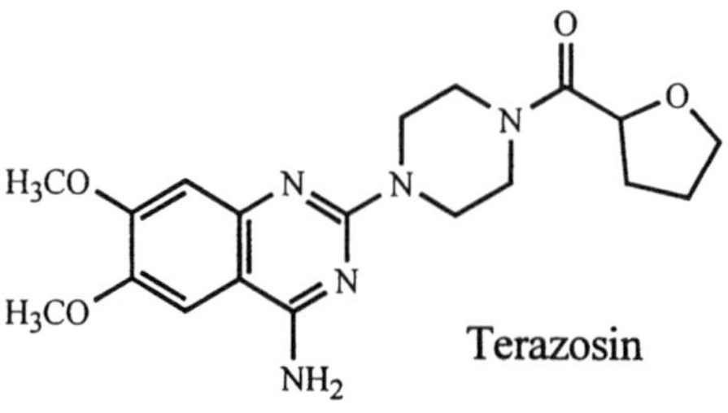

Terazosin

complete in 70 min. Peak shapes were good and impurity levels down to 0.05% were detected.

Armstrong et al. [1494] generated enantiomeric separations of propranolol, metoprolol, timolol, atenolol, carteolol, alprenolol, pindolol, oxprenolol, labetolol, and nadolol on a β-cyclodextrin column ($\lambda = 254$ nm) using a 98/2/0.8/0.6 or 99/1/0.2/0.1 acetonitrile/methanol/acetic acid/TEA mobile phase. Retention times ranged from 15 to 50 min. Labetolol and nadolol have two chiral centers and this method did not baseline resolve both sets. The authors noted that resolution was not achieved with acetonitrile as the only organic component. The addition of methanol played a key role in reducing the retention times of the analytes. A plot of k' and resolution versus percent methanol illustrated the effect of methanol very clearly. TEA and acetic acid were key components in achieving the enantioselectivity of the separation. A plot of k' and resolution vs. modifier level shows this effect very well. This paper is a good reference for general chiral separations of this type.

Diltiazem and eight metabolites (e.g., diltiazem sulfoxide, N-demethyldeacetyl-diltiazem, deacetyldiltiazem, O-demethyldeacetyldiltiazem) were extracted from liver mocrosomes and well resolved on a C_6 column ($\lambda = 240$ nm) using a 53/47 acetonitrile/water (15 mM ammonium phosphate) mobile phase [1495]. Peak shapes were excellent and elution was complete in 16 min.

9.7.17 Antimalarials

Artemisinin and three analog antimalarials (artesunate, α- and β-dihydroartemisinin) were extracted from plasma and separated on a 35°C C_{18} column (electrochemical detector, -1.0 V vs. Ag/AgCl). A 45/55 acetonitrile/water (0.1 M acetic acid at pH

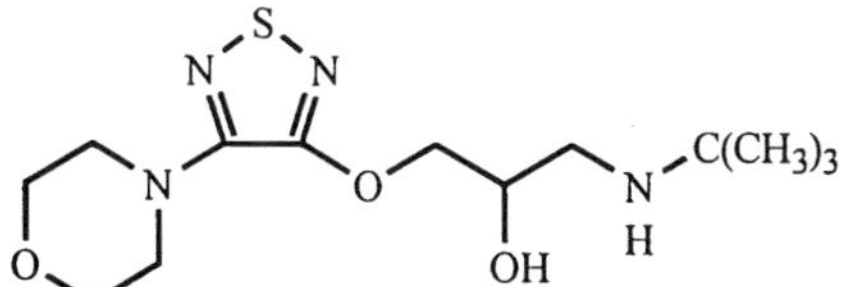

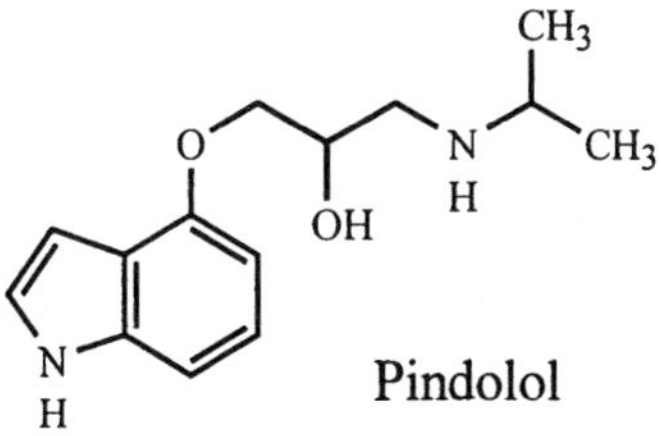

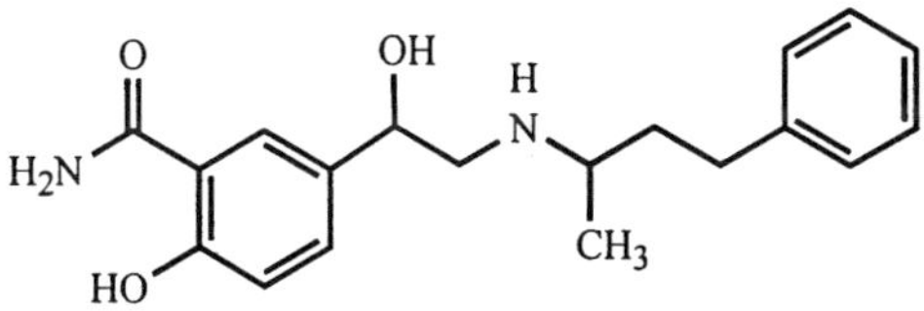

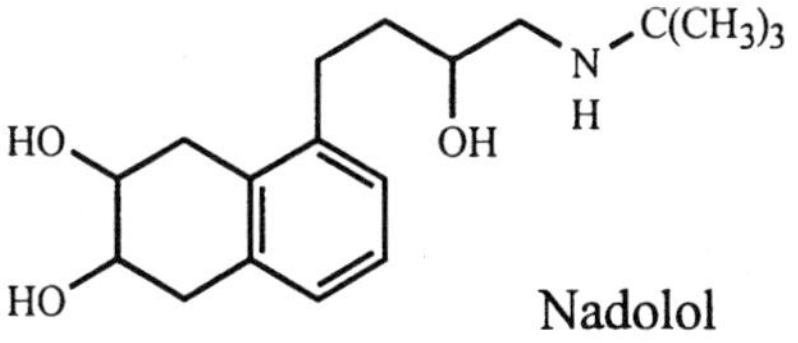

4.8) mobile phase generated good peak shapes and resolution. Elution was complete in <7 min [1496]. Working standard concentrations were 10–800 ng/mL and detection limits were 5 ng/mL (S/N = 3).

A study involving the peroxydisulfate oxidation products of the antimalarial drug primaquine (e.g., 6-methoxy-5,8-di-[4-amino-1'-methylbuylamino]quinoline, N,N-tri-[4-amino-1-methylbutyl]amine) utilized a C_{18} column ($\lambda = 254$ nm) and a 95/30/7/1 water/acetonitrile/methanol/water (1 M perchloric acid) mobile phase [1497]. Primaquine dimers were also studied. The study followed the reaction of 250 µg/mL primaquine. All products were easily detectable.

Pyrimethamine and sulphadoxine were extracted from plasma and separated on a C_{18} column ($\lambda = 240$ nm) using a 65/20/15/0.1 water/acetonitrile/methanol/triethylamine mobile phase [1498]. Peak shapes were excellent and baseline resolution was achieved in 12 min. The reported linear range and detection limit for pyrimethamine was 6.25–2000 ng/mL and 7 ng/mL, and for sulpha-doxine were 0.625 µg/mL and 15 ng/mL.

Chloroquine and three metabolites (desethylchloroquine, bisdesethylchloroquine [BDC], 4-amino-7-chloroquinoline [4AC]) were extracted from plasma and baseline resolved in under 10 min on a C_{18} column ($\lambda = 343$ nm) using a 28/72 acetonitrile/water (20 mM heptanesulfonic acid with 700 µL/L diethylamine to pH 4 with H_3PO_4) mobile phase [1499]. Peak shapes were excellent and detection limits of 2 ng/mL were reported. An interesting modification of the analysis of chloroquine, BDC and 4AC was a separation generated on a silica column ($\lambda = 325$ nm, ex; 380 nm, em). The mobile phase was a 57/40/3 acetonitrile/methanol/ammonia mixture [1500]. Peak shapes were excellent. Detection limits of 4 ng/mL with a linear concentration curve of 25–100 ng/mL were reported. Elution was complete in 15 min.

Quinine, dihydroquinine, 2'-dihydroquininone and 2'-quininone were baseline resolved in 25 min on a C_{18} column ($\lambda = 226$ nm) using a 22/78 aceonitrile/water (0.4% w/v ammonium acetate to pH 3.3 with acetic acid) mobile phase [1501]. Cinchonine, cinchonidine, and their metabolites were separated on the same column under similar conditions.

9.7.18   Steroids

Methylprednisolone suleptanate and its major photodegradation products (methyl-prednisolone, 17-keto- and 11-ketomethoxyprednisolone, and two sulptanate

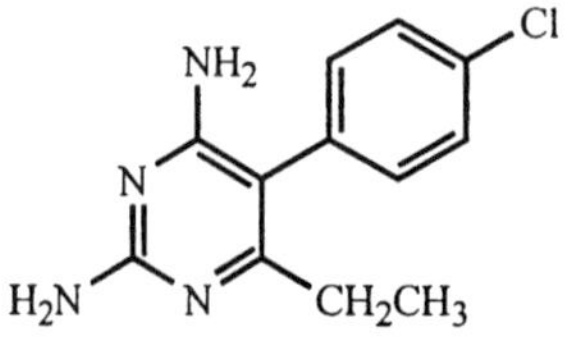

Pyrimethamine 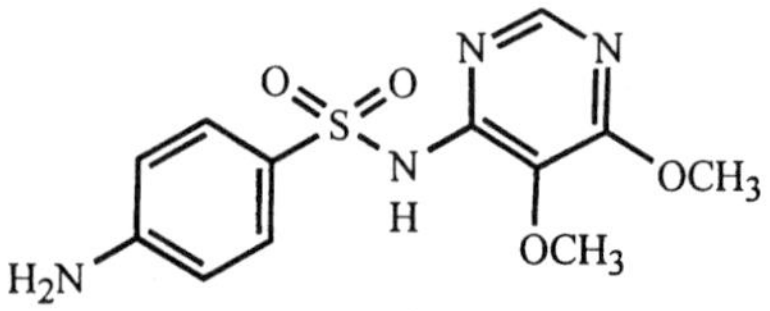   Sulfadoxine

1,11-epoxy analogs) were resolved on a phenyl column ($\lambda = 254$ nm and 288 nm) using a 35-min linear 75/21/4 water (0.2 M acetate at pH 5.8)/ acetonitrile/THF $\rightarrow$ 40/38 water (0.38 M acetate at pH 3.8)/acetonitrile gradient. The initial concentration of the sample (before degradation) was 500 μg/mL but no indication of relative concentrations was given [1502].

Seven 17-ketosteroid sulfates and glucuronides (e.g., androsterone- and epiandrosterone-3-β-sulfate, etiocholanol-17-one-3-β-glucuronide) were extracted from urine and separated on a C_{18} column (negative ion ion trap MS, spray source plate $T = 275°C$ with 1 kV, aperature $T = 170°C$ and 120°C, draft voltage 75 V, focus voltage 35 V). The compounds were eluted isocratically in 20 min with a 75/25 water (100 mM ammonium acetate)/acetonitrile mobile phase. However, a 5-min washout with 75/25 acetonitrile/water (no buffer) and a re-equilibration of 5 min were necessary. Standard concentrations of total ketosteroids from 0.2 to 60 μg/mL were used and quantitation limits of 3–30 ng/mL (analyte dependent) were reported [1503].

Nine corticosteroids (hydrocortisone and its sulfate, cortisone, corticosterone, betamethazone, dexamethazone, prednisone and its methyl analog, triamcinolone acetonide) were extracted from hair and separated on a deactivated base C_{18} column (positive ion electrospray MS, electrospray IS, −5 kV; orifice, −80 V; ring, −344 V). An 80/20 (hold 1 min) $\rightarrow$ 10/90 (at 20 min hold 6 min) gradient of water (126 mg ammonium formate pH 3 with formic acid/L)/acetonitrile was used. Dexamethasone and betamethazone co-eluted. Concentration curves were linear from 0.1 to 200 ng/g hair [1504].

Another study paper focused on the stability of complexes formed between 13 betamethasone-related steriods (e.g., dexamethasone, beclomethasone, and various

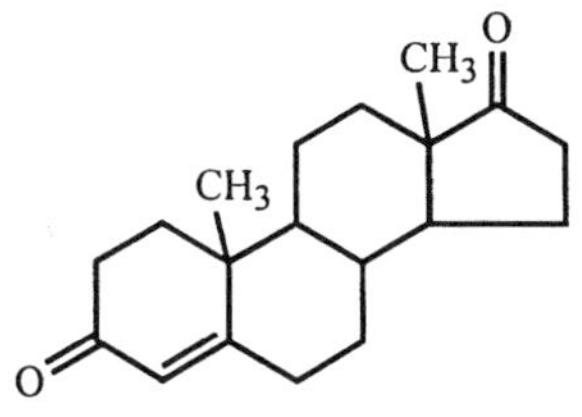

4-Androsterone-3, 17-dione

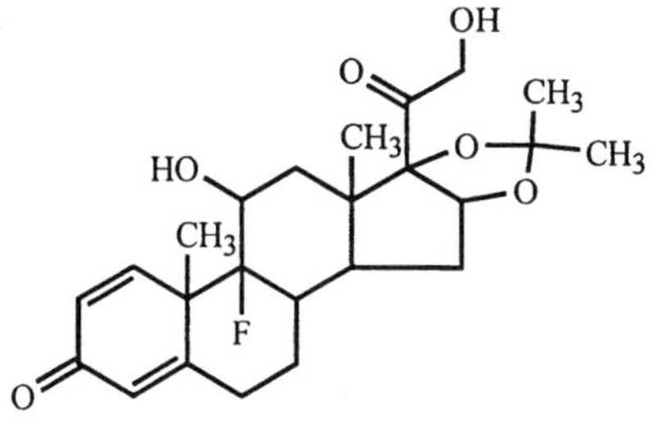

Triamcinolone acetonide

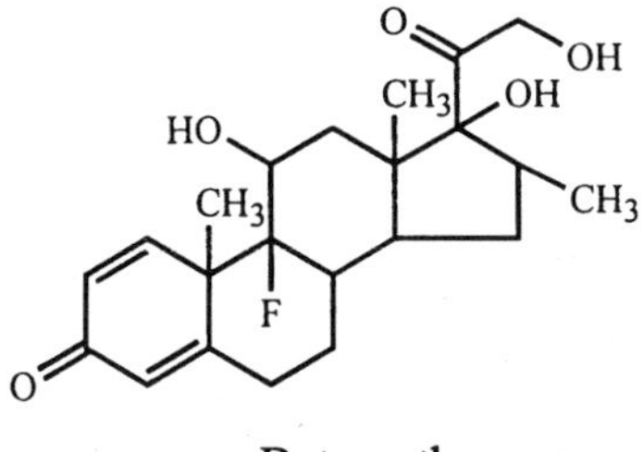

Betamethasone

epoxy, chloro, acetate, and priopionate analogs of betamethasone) and their β- and γ-cyclodextrin complexes. However, chromatographic reasults of the separation are shown for 35/65 acetonitrile/water mobile phases containing various levels of cyclodextrin at temperatures from 10 to 80°C). The best separation was achieved using the 35/65 acetonitrile/water mobile phase containing 7.5 mM γ-cyclodextrin. Elution was complete in ~35 min [1505].

Ten steroid sex hormones (estradiol, norethidrone, ethinyl estradiol, estrone, diethylstilbesterol, levonorgestrel, progesterone, mestranol, ethynodiol diacetate) were separated on a C_{18} column ($\lambda = 225$ nm) using a 40-min 90/10 → 0/100 water/acetonitrile gradient [1506]. Excellent peak shapes were generated. A linear range of 25 ng/mL to 10 µg/mL was used (analyte dependent), whereas detection limits range from 50 to 500 ng/L (S/N = 3).

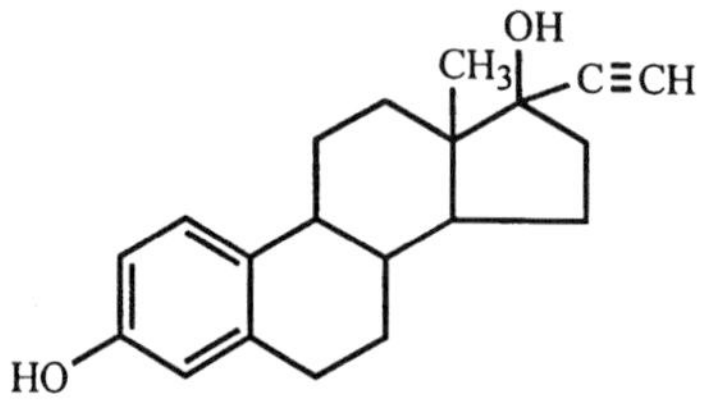

Ethinyl estradiol

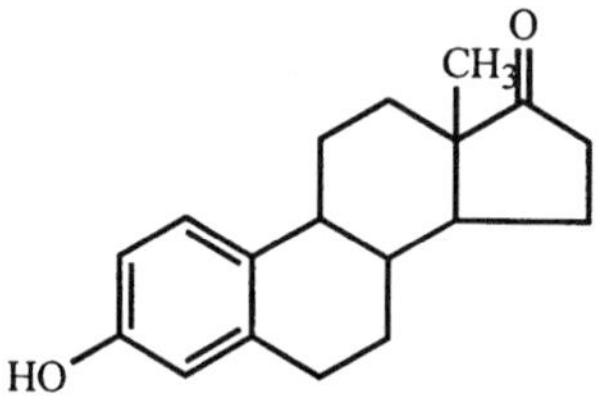

Estrone

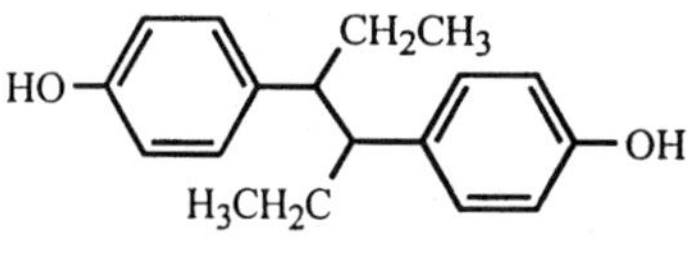

Diethylstilbestrol

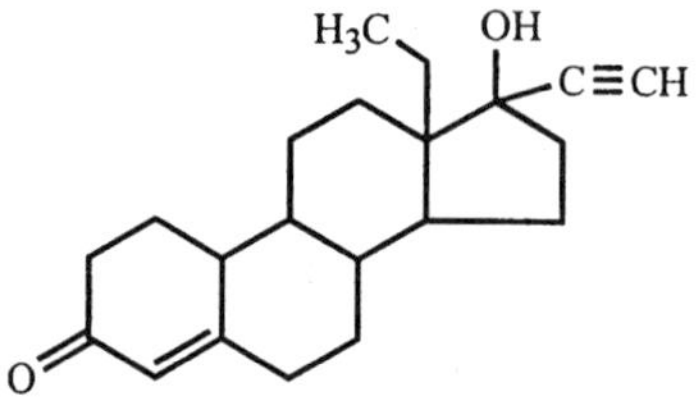

Levonorgestrel

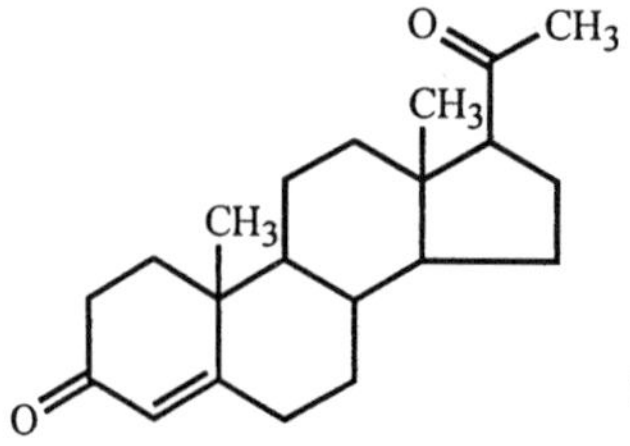

Progesterone

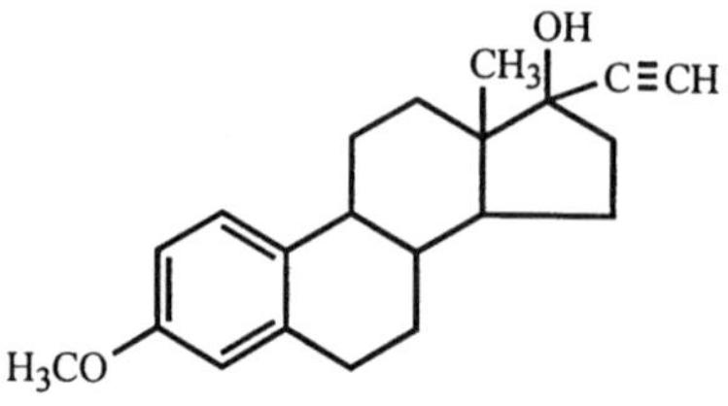

Mestranol

Thirteen urinary anabolics and corticoids (e.g., cortisone, 11-ketotestosterone, hydroxyprogesterone, epitestosterone, androstenolone) and synthetic anabolics (bolderone, bolasterone) were extracted from urine and analyzed on a C_{18} column ($\lambda = 200$ nm) and 245 nm) using a 60/40 water/acetonitrile mobile phase [1507]. Baseline resolution was not achieved for all compounds and elution was complete in 24 min. Calibration data for each compound were generated from 2 to 10 µg/mL. Detection limits were reported as ~0.05 µg/mL. It is interesting to note that the authors used an optimization triangle approach to maximize resolution and minimize analysis time. For 10 of the compounds, a <20 min analysis was achieved using a 56.7/20/13.3/10 water/methanol/acetonitrile/THF mobile phase.

9.7.19 Antipsychotics

The antipsychotic drug CI-1007 and eight metabolites (e.g., the hydroxylated and sulfonated analogs) were extracted from plasma and urine and analyzed on a C_8 column ($\lambda = 245$ nm, ex; 320 nm, em). A 90-min 100/0 → 15/85 (90/10 water [20 mM ammonium acetate at pH 4.0]/acetonitrile)/(25/75 water [20 mM ammonium acetate at pH 4.0]/acetonitrile) gradient produced excellent resolution and peak shapes [1508]. Standards were in the range of 100–1500 ng/mL.

Olanzapine and its metabolite, desmethylolanzapine, were isolated from plasma and analyzed on a C_8 column (amperometric detector, +800 mV vs. Ag/AgCl). The authors note that both pH value and acetonitrile level had a critical role in overall oxidation (i.e., detector response) of the analytes. The optimum response was found

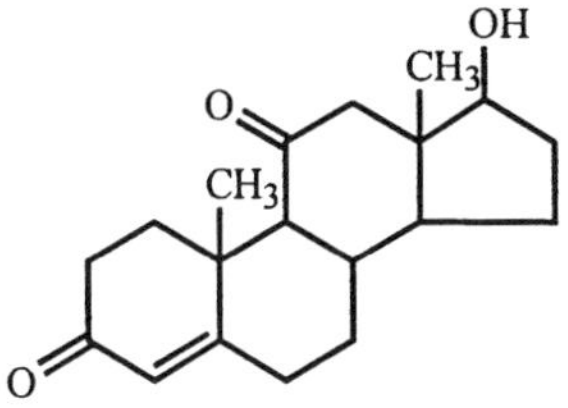

11-Ketotestosterone

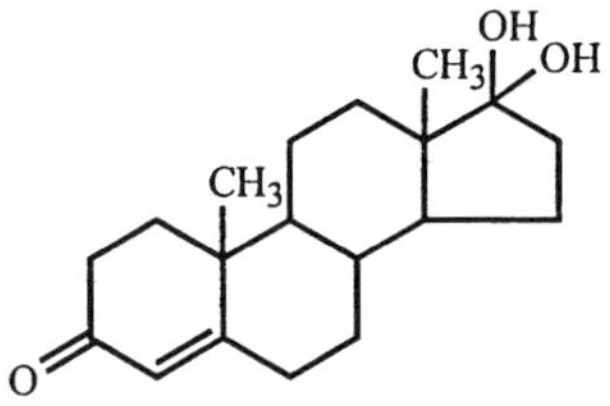

Epistosterone

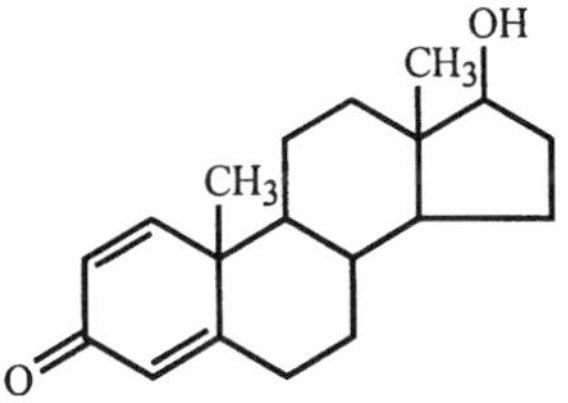

Boldenone

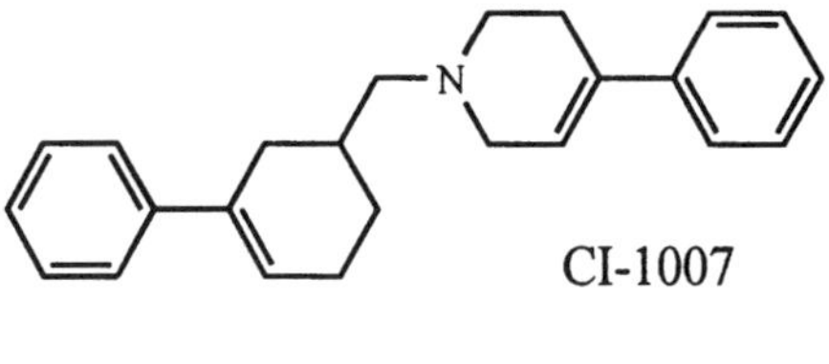

CI-1007

using a 20/80 acetonitrile/water (15 mM phosphate with 19.7 mM triethylamine at pH 3.8) mobile phase. Elution was complete in 15 min. Regardless of the response issue, peaks were visibly tailed but baseline resolution was nonetheless achieved. A linear range of 5–150 ng/mL and detection and quantitation limits of 3 ng/mL and 1 ng/mL, respectively, were reported [1509].

Flupentixol and haloperidol (antipsychotics) were extracted from serum and eluted on a 40°C cyanopropyl column ($\lambda = 254$ nm) using a 92/11/3 acetonitrile/ methanol/water (0.1 M ammonium acetate) with 50 μL of triethylamine added/L mobile phase [1510]. The haloperidol eluted between the *cis*- and *trans*-flupentixol peaks at about 10 min. Calibration curves were constructed from 0.5 to 20 ng/mL. Detection limits were reported as 0.1 ng/m (S/N = 3). Interestingly, in addition to the above analytes, the chromatogram also showed benperdiol, clozapine, levomepromazine, chlorprothixene, olanzapine, respirodone, fluphenazinesulfoxide, melperone, and perazine, all of which are at least partially resolved from one

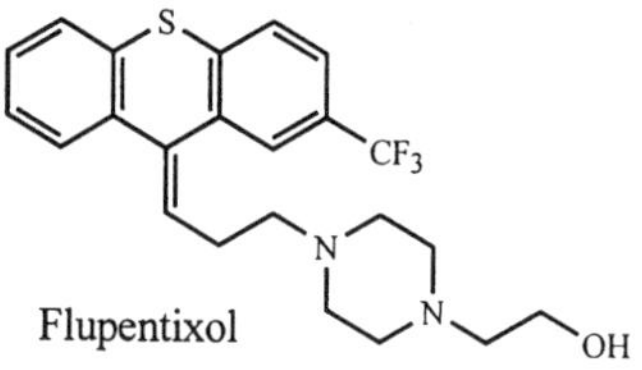

Flupentixol

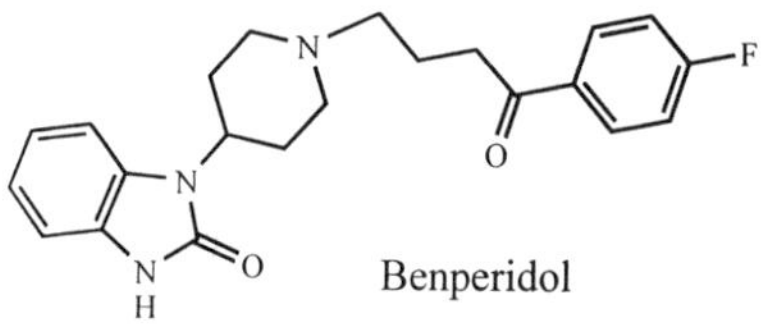

Benperidol

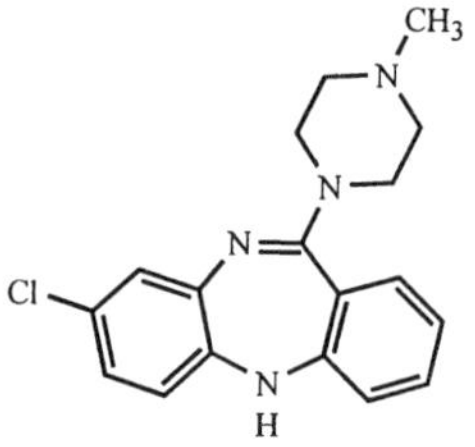

Clozapine

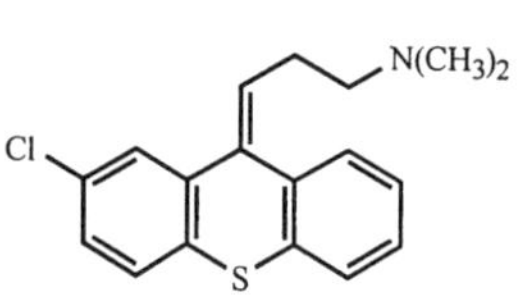

Chlorprothixene

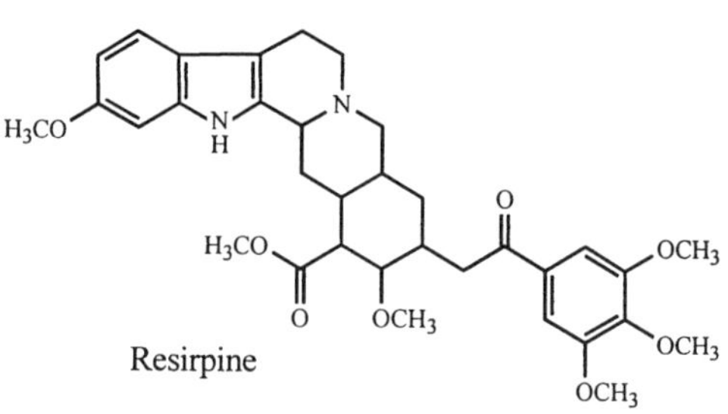

Resirpine

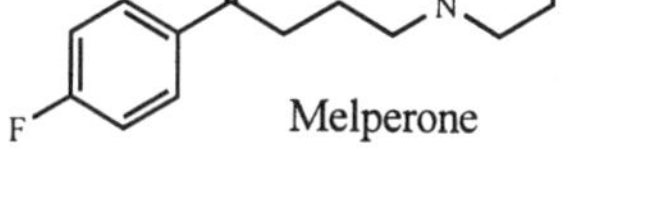

Melperone

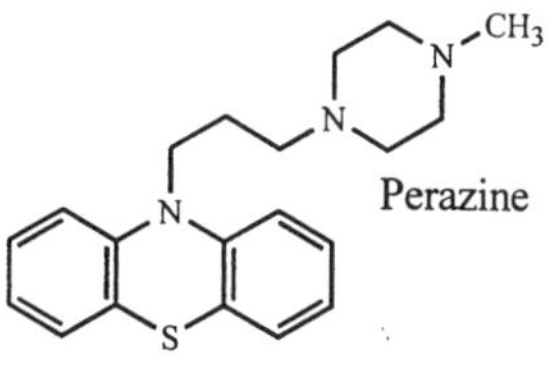

Perazine

another. Elution of all was complete in 30 min. Further, a table of the retention times for 25 antipsychotics (including those above) was presented.

Promazine was extracted from plasma and analyzed on a cyanopropyl column (electrochemical detector at +400 mV and +750 mV) using a 90/10 acetonitrile/water (50 mM ammonium acetate) mobile phase [1511]. Elution was complete in 10 min and the analyte was well resolved from all extracted components. The authors noted that the analyte was light sensitive and so special handling precautions were needed. A linear range of 0.25–25 ng/mL was established and a quantitation limit of 0.25 ng/mL was reported.

Clozapine and the metabolites desmethylclozapine and clozapine N-oxide were isolated from plasma and analyzed on a C_6 column ($\lambda = 254$ nm). Excellent peak shapes and resolution were obtained in <8 min using a 48/52 acetonitrile/water (9 mM heptanesulfonic acid with 60 mM K_2HPO_4 and H_3PO_4 to pH 2.7) mobile phase [1512]. The retention times for 21 potential drug interferences were also tabulated. A linear range of 0.025–2 µg/mL and detection and quantitation limits of 1 ng/mL (S/N = 3) and 3 ng/mL (S/N = 10), respectively, were reported.

Five butyrophenones used as antipsychotics and two analogs (haloperidiol, flouropipamide, moperone, flunarizine, timiperone, bromperidol, pimozide) were

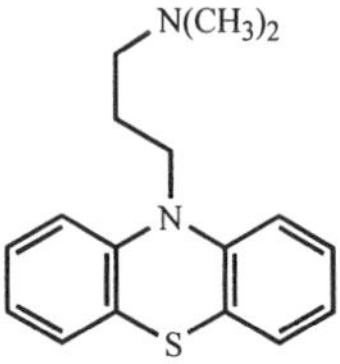

Promazine

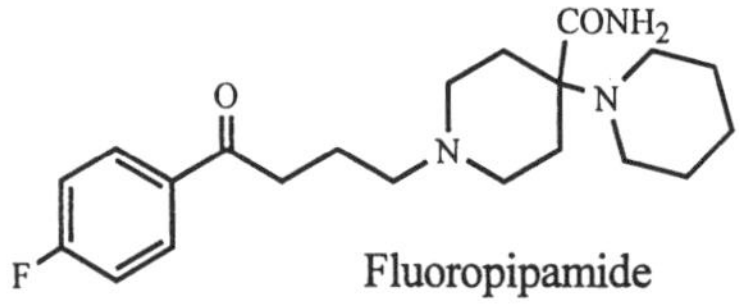

Fluoropipamide

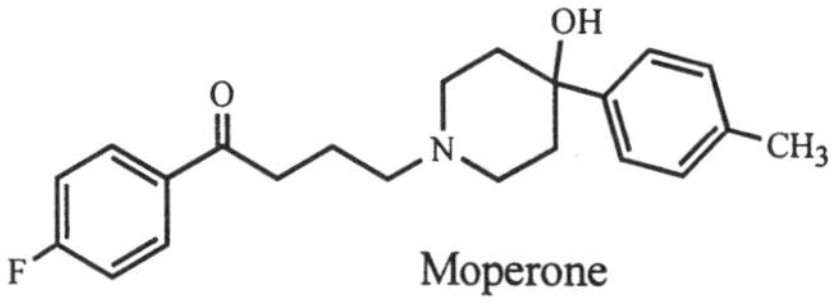

Moperone

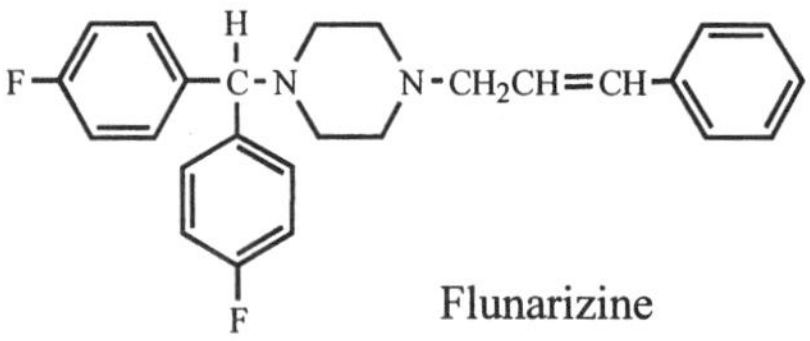

Flunarizine

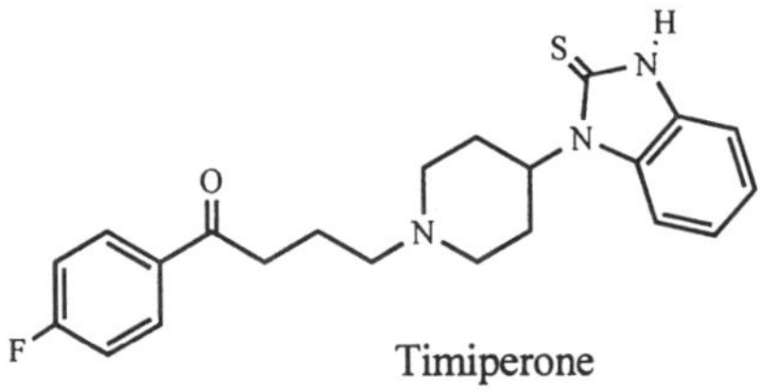

Timiperone

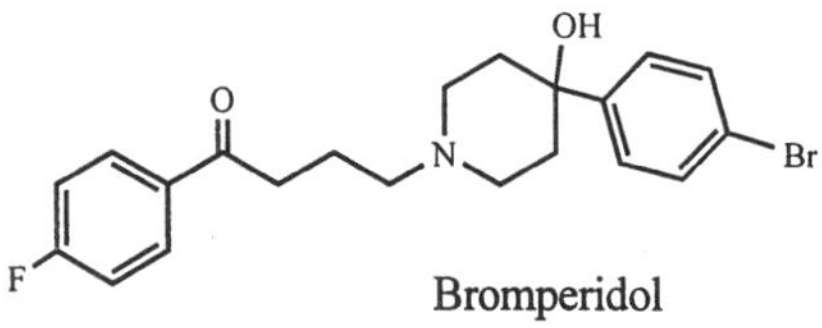

Bromperidol

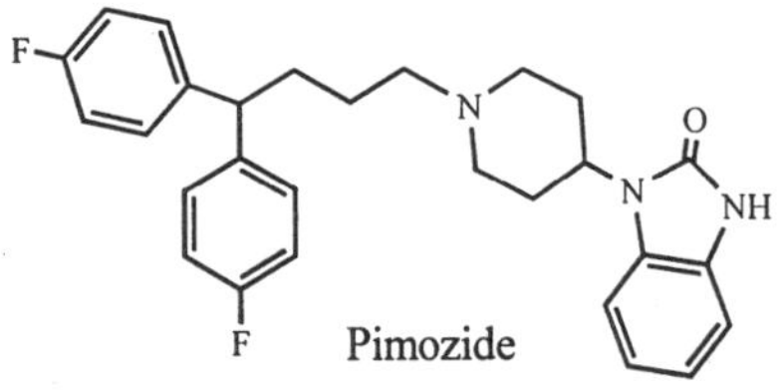

Pimozide

extracted from whole blood and separated on a C_{18} column (positive ion electrospray MS, capillary $T = 230°C$, needle voltage 5.5 kV, sheath gas pressure 80 units, auxiliary gas flow 15 units). An 85/15 (hold 5 min) → 10/90 (at 30 min) water (0.1% formic acid)/acetonitrile gradient was used. Linear ranges of 0.2–0.8 ng/mL and detection limits of 0.1 ng/mL were reported [1513].

Risperidone and its 9-hydroxy metabolite were isolated from plasma and separated on base deactivated C_{18} column ($\lambda = 278$ nm). A 70/30 water (0.1 M KH_2PO_4 to pH 2.2 with H_3PO_4)/acetonitrile mobile phase generated baseline separation and elution in under 5 min. A calibration range of 5–100 ng/mL was reported with detection limits of 1 ng/mL. The retention times for six potential antidepressant drug interferents were tabulated [1514].

9.7.20   Other Drug Analytes

Nefiracetam and six metabolites [3-, 4-, 5-, 3'-, 4'-hydroxynefiracetam and 2-hydroxymethylnefiracetam) were extracted from liver microsomes and separated on a C_{18} column ($\lambda = 210$ nm). Good resolution was obtained for all but the 3- and 4-hydroxynefiracetam peaks using a complex 30-min 80/20 → 45/55 water (25mM phosphate at pH 7)/acetonitrile gradient [1515].

Danofloxacin and its N-desmethyl analog, enrofloxacin, ciprofloxacin, ofloxacin, and norfloxacin were extracted from cattle and chicken tissue and separated on a C_8 column ($\lambda = 280$ nm, ex; 440 nm, em) using an 88/12 water (50 mM phosphate at pH 3.5)/acetonitrile mobile phase [1516]. Orfloxacin and norfloxacin were incompletely resolved, otherwise baseline resolution and excellent peak shapes were generated. Elution was complete in 23 min. A detection limit of 20 pg injected and a working concentration range of 10–500 ng/g were reported.

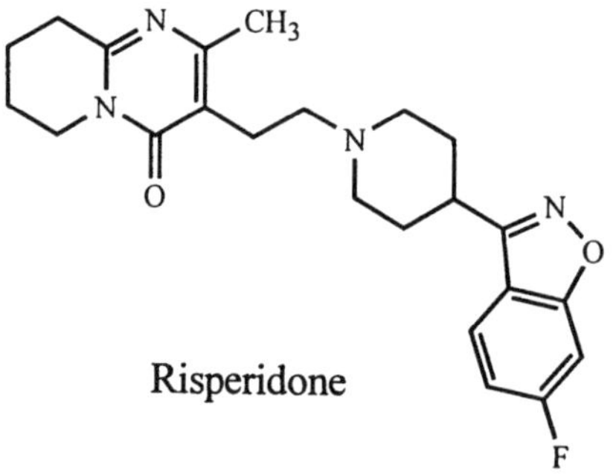

Risperidone

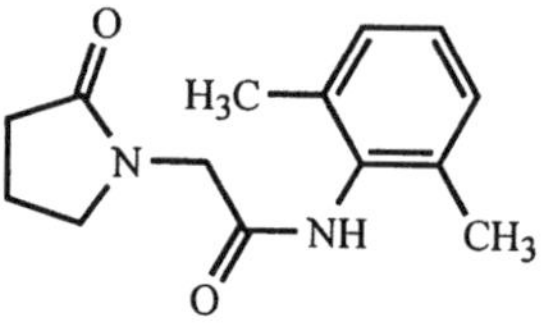

Nefiracetam

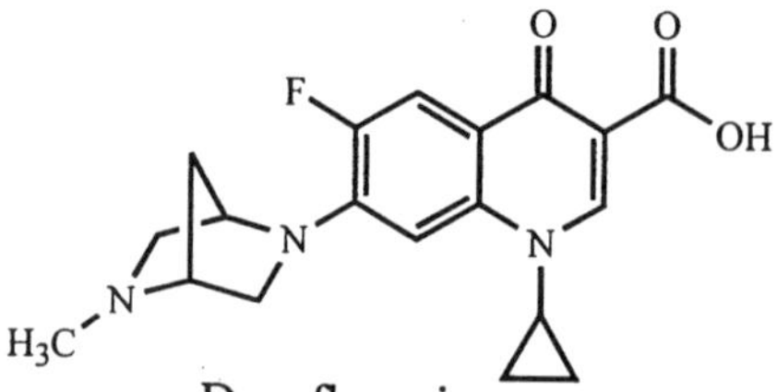

Danofloxacin

Five veterinary drug coccidiostats (alkomide, nitromide, zoalene, ethopabate, dinsed) were extracted from feeds and separated on a C_{18} column ($\lambda = 260$ nm) using a 16-min $20/80 \rightarrow 80/20$ acetonitrile/water gradient [1517]. Peak shapes were good as was the overall resolution. Limits of detection and quantitation were 20–130 ng/g and 70–350 ng/g (analyte dependent), respectively.

Lasalocid was extracted from tissue and eggs and analyzed on a phenylhexyl column ($\lambda = 310$ nm, ex; 430 nm, em) using a 67/10/10/13/0.1 acetonitrile/methanol/THF/water/TFA mobile phase [1518]. Elution was complete in <6 min. A spiked sample range from 0 to 200 ng/g and a detection limit of 5 ng/g were reported.

Voriconizole was extracted from plasma and analyzed on a C_{18} column ($\lambda = 255$ nm) using a 50/50 acetonitrile/water (40 mM ammonium dihydrogenphosphate buffer to pH 6.0 with ammonia) mobile phase [1519]. Linearity was reported over the range of 0.2–10 μg/mL with a detection limit of 0.2 μg/mL. The analyte eluted in <8 min.

A novel imaging agent for breast cancer diagnosis is a technetium-99m N_2S_2-bifunctional (Z)- and (E)-aminotamoxifen conjugate. The (Z)- and (E)-ligands were separated from one another using a C_{18} column ($\lambda = 249$ nm). A $44/56 \rightarrow 53/47$ (at

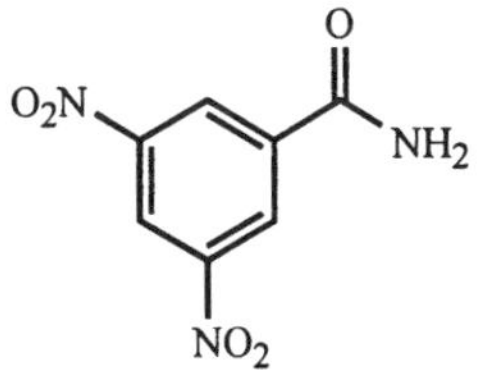

Nitromide

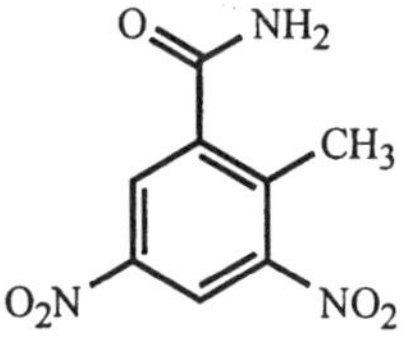

Zoalene

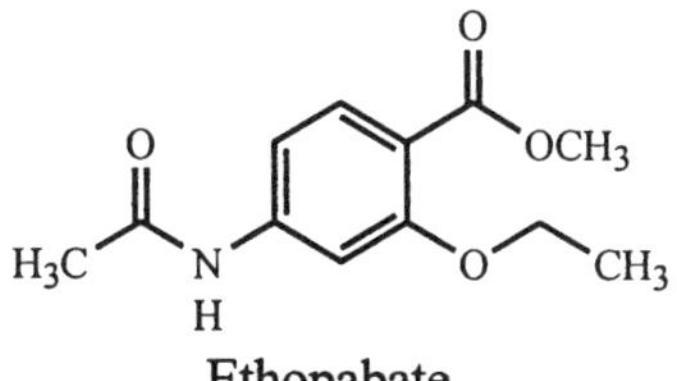

Ethopabate

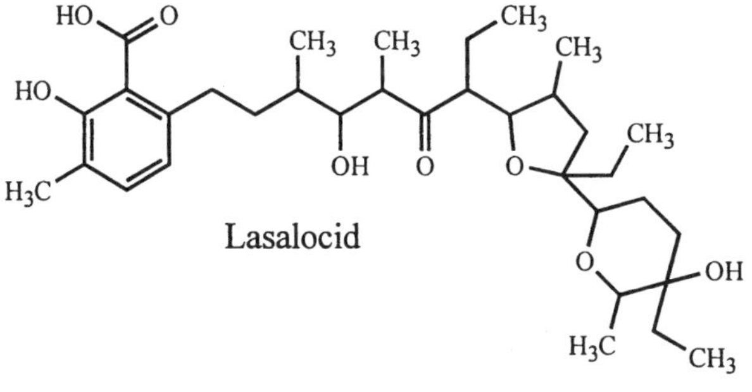

Lasalocid

20 min hold 5 min) acetonitrile/water (10 mM KH_2PO_4) gradient was used. Baseline resolution was achieved. The ^{99m}Tc complex was similarly eluted but baseline resolution between the two forms was incomplete. Elution time was 17 min [1520].

Buspirone and five impurities (e.g., 1-(2'-pyrimidinyl)piperazine, propargyl chloride, 3,3'-tetramethylene glutarimide, propargyl glutarimide) were baseline resolved on a C_{18} column ($\lambda = 210$ nm and 240 nm) using a 27-min 90/10 $\rightarrow$ 65/35 water (10 mM KH_2PO at pH 6.1)/acetonitrile gradient [1521]. All peaks show moderate tailing. Linearity was reported as 1.25–500 µg/mL with detection limits of 0.125 µg/mL and quantitation limits of 1.25 µg/mL.

8-Chlorotheophylline and diphenhydramine are key components of the antiemetic dimenhydrinate. Most pharmacetical formulations also contain caffeine, a stimulant. This product was stress tested and the resulting formulation was analyzed for active levels on a 40°C C_8 column ($\lambda = 229$ nm). A 22/78 acetonitrile/water (10 mM H_3PO_4 with triethylamine to pH 2.8) mobile phase generated excellent peak shape, baseline resolution, and complete elution in 16 min [1522]. The authors note that adjustment of the pH with NaOH or ammonia did not effectively block residual silanol groups and therefore diphenhydramine did not elute before 45 min. They suggest using a highly protected (e.g., end-capped, high carbon load, etc.) column. Triethylamine was effective. Linear ranges were reported as 0.06–0.4 mg/mL (analyte dependent).

The dichloroacetyl quinolinol quinfamide (for intestinal ameobiasis) and its metabolite (1-(dichloroacetyl)-1,2,3,4-tetrahydro-6-quinolinol) were extracted from

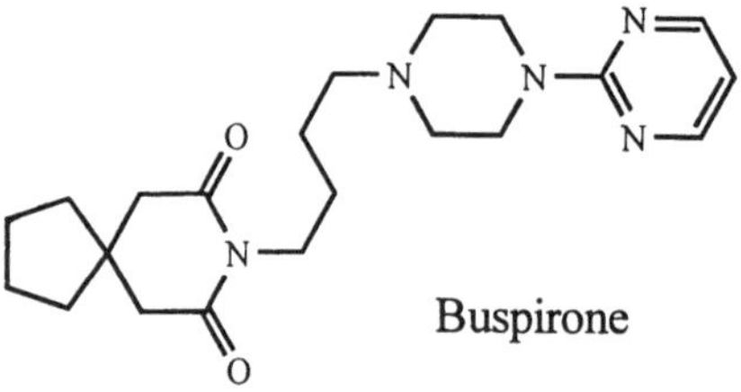

Buspirone

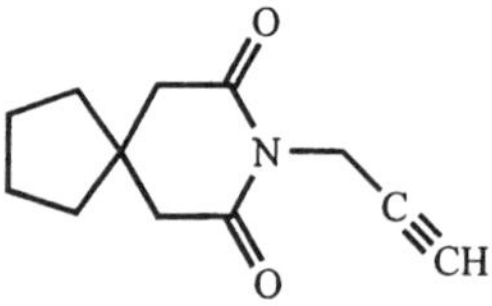

Propargyl glutarimide

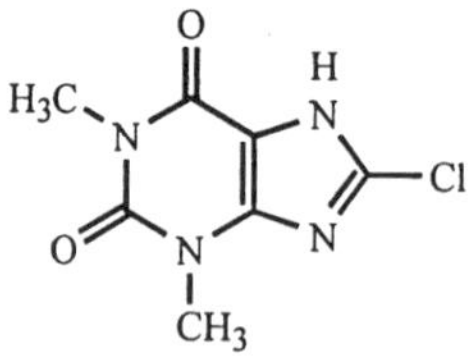

8-Chlorotheophylline

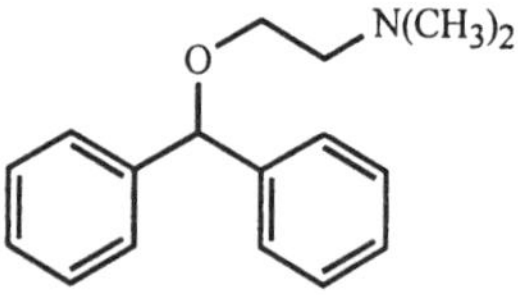

Diphenhydramine

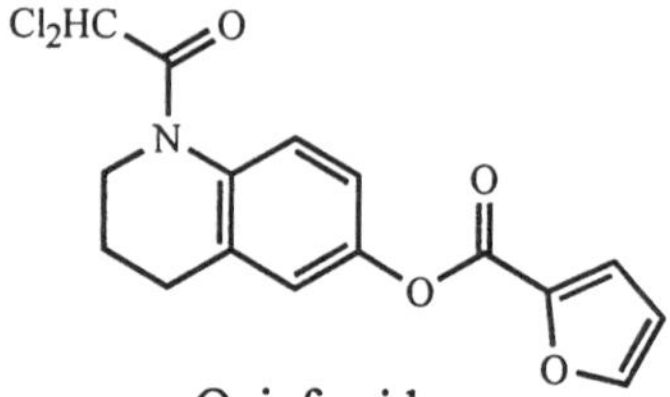

Quinfamide

plasma, urine, or feces and separated on a cyanopropyl column ($\lambda = 269$ nm). A 40/50/10 water/acetonitrile/methanol mobile phase generated excellent peak shape and resolution, and elution in 12 min [1523]. A plot of the effect of changing support material (C_{18}, phenyl, cyanopropyl) concomitantly with a change in the organic mobile phase compositions (acetonitrile, methanol, THF, and water) gave an interesting 3-D resolution response plane. A linear curve from 0.08 to 2 µg/mL and detection (S/N = 3) and quantitation limits (S/N = 10) of 0.05 and 0.08 µg/mL, respectively, were reported.

Tacrine and three metabolites (1-, 2-, and 4-hydroxytacrine) were extracted from plasma and separated on a C_{18} column ($\lambda = 330$ nm, ex; 365 nm, em) using an 83/13 water (0.2 M acetate at pH 4.0)/acetonitrile mobile phase [1524]. Tacrine, the latest-eluting peak was detected at 35 min. Incomplete resolution of 1- and 2-hydroxytacrine occurred under these conditions. Almost 20 min of elution time occurred between the hydroxytacrine and tacrine elution. A gradient system would dramatically decrease analysis time and, if a weaker mobile phase is used initially, resolve the analytes mentioned above. Working standards ran from 6 to 240 nM and quantitation limits of 2–10 nM were reported (analyte dependent).

Ebrotidine (an H2-receptor antagonist) and nine metabolites (e.g., ebrotidine sulfoxide and sulfone, N-[4-bromophenylsulfonyl]acetamide, 2-[4-[[(2-amino-ethyl)thio]methyl]-2-thiazolyl]guanidine and its sulfoxide) were extracted from urine, baseline resolved, and quantitated on a C_{18} column ($\lambda = 235$ nm) using a 30-min 80/20 → 65/35 water (5 mM hexanesulfonic acid to pH 3 with acetic acid)/acetonitrile gradient [1525]. Removal of the ion-pairing reagent in the acetonitrile leads to a significant negative baseline shift over the course of the gradient. Linear ranges of 1–200 µg/mL and detection limits of 25–110 ng/mL (analyte dependent) were reported.

Urine and serum samples were analyzed for olpadronate ([3-dimethylamino-1-hydroxypropylidine] bisphosphonate) level through extraction followed by derivatization with 9-fluorenylmethylchloroformate (FMOC). Separation was accomplished on a C_{18} column ($\lambda = 274$ nm, ex; 307 nm, em) using a 72.5/27.5 water (30 mM phosphate at pH 7 with 5 mM tetrabutylammonium hydroxide and 2 mM etidronate)/acetonitrile mobile phase [1526]. Elution was complete in 6 min. A linear range of

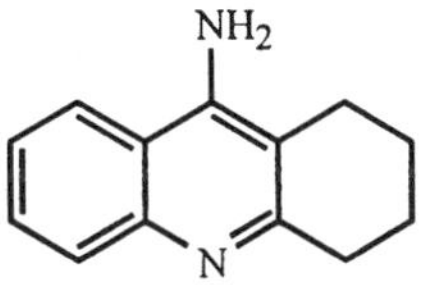

Tacrine

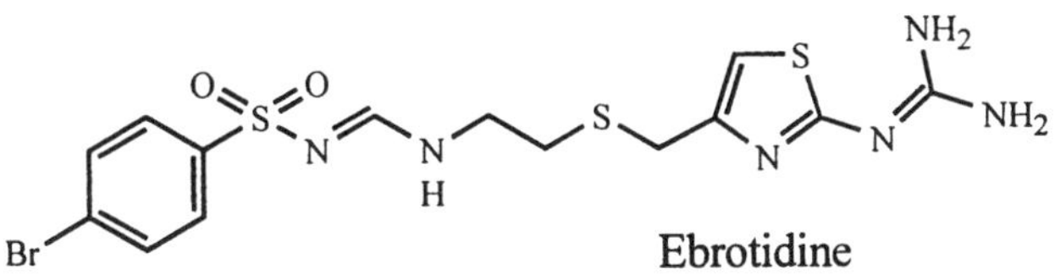

Ebrotidine

10 ng/mL to 10 μg/mL (nonlinear at higher concentration) and a detection limit of 2 ng/mL (S/N = 3) were reported.

Thalidomide levels in raw material and finished products was determined using a C_{18} column ($\lambda = 237$ nm) and an 85/15 water/acetonitrile mobile phase [1527]. Elution was complete in 8 min. The authors note that past mobile phases used a 0.1% H_3PO_4 additive that caused column deterioration much faster than the mobile phase described here. The effect of changing percent acetonitrile on retention time was also reported. A linear curve was established between 10 and 200 μg/mL.

Rofecoxib (a cyclooxygenase inhibitor) and three metabolites (5-hydroxyrofe-coxib and its glucuronide, rofecoxib-3',4'-dihydrodiol) were isolated from urine and resolved on a 40°C C_8 column ($\lambda = 250$ nm, ex; 370 nm, em) using a 45-min 85/15 → 40/60 water (0.1% TFA to pH 3 with ammonium hydroxide)/acetonitrile gradient [1528]. A calibration curve from 1 to 250 ng/mL was used.

Propiverine hydrochloride (for bladder dysfunction) and its N-oxide metabolite were isolated from serum and analyzed on a 35°C C_8 column ($\lambda = 220$ nm) using a 70/30 acetonitrile/water (7 mM Na_2HPO_4 at pH 7.3) mobile phase [1529]. Analytes were well resolved from one another and all co-extracted components. Elution was complete in 17 min. Working curves from 20 to 200 ng/mL and a quantitation limit of 10 ng/mL were reported.

Girisopam (an anxiolytic) and four metabolites (4'-hydroxy-, 7-demethyl-, 4-hydroxymethyl-, demethyl-4-oxogirisopam) were extracted from plasma and base-line resolved on a base deactivated 40°C C_{18} column ($\lambda = 238$ nm). A 98/2 (hold 4.2 min) → 60/40 (at 7 min) → 55/45 (at 10 min) → 50/50 (at 14 min) (20/12/80 acetonitrile/methanol/water [0.98 g ammonium carbonate with 0.90 g ammonium chloride])/(185/110/200 acetonitrile/methanol/water [0.33 g ammonium carbonate with 0.37 g ammonium chloride]) gradient was used [1530]. Linear responses from 10–18,000 ng/mL for girisopam and 50–6000 ng/mL for the metabolites (analyte

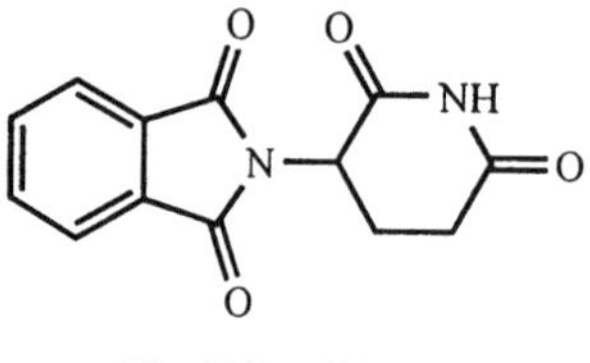

Thalidomide

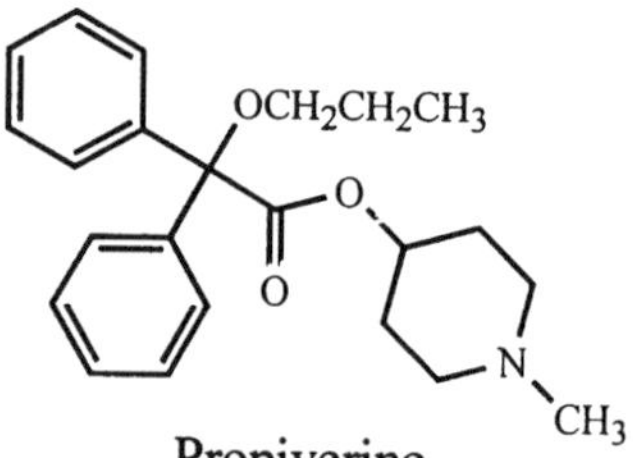

Propiverine

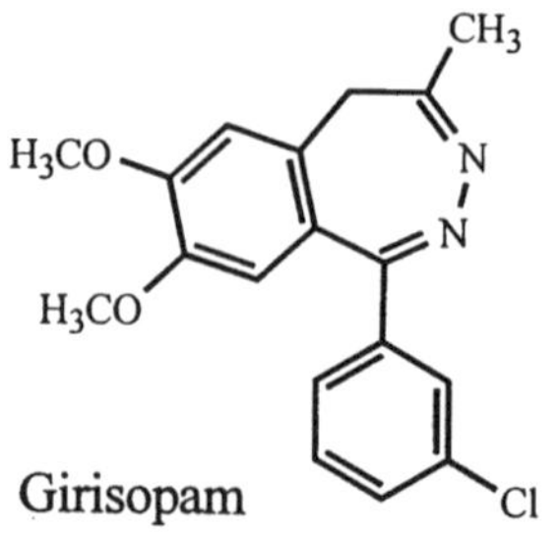

Girisopam

dependent) were obtained. Quantitation limits of 10 ng/mL for girisopam and 50 ng/mL for the metabolites were reported.

2,4,6-Triiodophenol (Bobel-24, anti-inflammatory) was extracted from plasma, synovial fluid, and tissue and analyzed on a C_{18} column ($\lambda = 277$ nm) using a 62/38 water/acetonitrile mobile phase [1531]. Elution was complete in 5 min. Interestingly, the highly substituted phenol should be quite acidic and yet excellent peak shape and acceptable retention time did not require a buffer. Standards of 0.5–20 µg/mL were used and a quantitation limit of 0.1 µg/mL was reported.

The *exo* and *endo* forms of iodixanol (X-ray contrast agent) were isolated from plasma and analyzed in 16 min on a C_{18} column ($\lambda = 244$ nm) using a 91/9 water/acetonitrile mobile phase [1532]. Good resolution between the isomers and co-extracted peaks was obtained. Standards ranged from 0.5 to 500 µg/mL. Detection and quantitation limits of 0.1 µg/mL and 0.3 µg/mL, respectively, were reported.

Rizatriptan benzoate (antimigrane) and two impurities (the benzoate regioisomer and the tryptophol intermediate) were baseline resolved in <10 min on a phenyl column ($\lambda = 280$ nm) using a 16/84 acetonitrile/water (0.1% TFA) mobile phase. Peak shapes were excellent. A standard of 1 mg/mL was used [1533].

Pet shampoos were analyzed for the presence and levels of active ingredients (piperonyl butoxide, dipropylpyridine-2,5-dicarboxylate, *endo*- and *exo*-N-octylbicycloheptanedicarboxamide, and six pyrethrin esters: cinerin I and II, pyrethrin I and

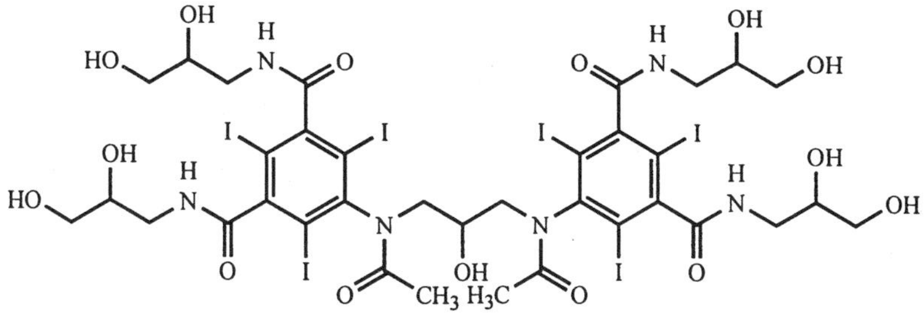

Iodixanol

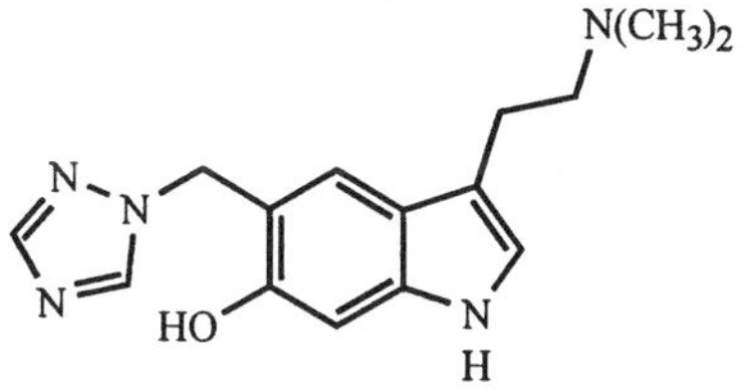

Rizatriptan

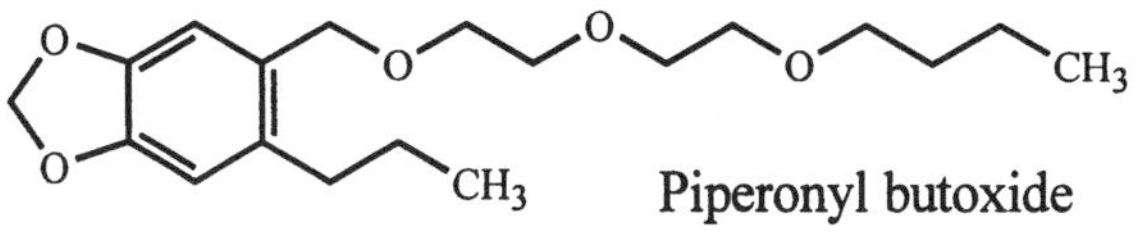

Piperonyl butoxide

II, jasmolin I and II). A C_8 column (photodiode array extracted at $\lambda = 220$ nm, 225 nm, 230 nm, 240 nm) and a complex 35-min $52/38/10 \rightarrow 15/85/0$ water/ acetonitrile/methanol gradient were used to generate good peak shape and resolution [1534].

Three swine stress treatment drugs (azaperone, azaperol, carazolol) were extracted from pig tissue and separated on 50°C C_{18} column (positive ion electro-spray MS; probe $T = 400$°C, source $T = 150$°C, drying gas flow 300 L/h, sheath flow 120 L/h, collision cell pressure 2.4×10^{-3} mbar). An 8-min analysis time was achieved using a 100/0 (hold 1 min) $\rightarrow$ 30/70 (at 7 min) $\rightarrow$ 0/100 (at 8 min) water (0.1 M ammonium acetate)/acetonitrile gradient [1535]. Linear ranges of 5–1000 ng injected were used and limits of detection and quantitation were 0.06–0.14 µg/kg and 0.13–2.6 µg/kg (analyte dependent), respectively.

Diethyldithiocarbamate, a reduction metabolite of disulfiram, is metabolized to a series of compounds: S-(N,N-diethylcarbamoyl)glutathione, S-(N,N-diethylthiocar-bamoyl)glutathione, S-[N-(carboxymethyl)-N-ethylcarbamoyl]glutathione, and S-(N-ethylthiocarbamoyl)glutathione [1536]. These compounds were resolved on a C_{18} column ($\lambda = 214$ nm) using a 40-min $10/90 \rightarrow 50/50$ acetonitrile (0.06% TFA)/water (0.06% TFA) gradient. Peak shapes were good and the last analyte of interest eluted in 25 min.

Theophylline, caffeine and eight metabolites (e.g., theobromine, paraxanthine, 1,7-dimethyluric acid, 1-methylxanthine) were extracted from plasma and well resolved on a 50°C C_{18} column ($\lambda = 270$ nm and 285 nm). A (25/2/83 acetoni-trile/THF/water [10 mM acetate buffer at pH 4])/(0.01/99.99 THF/water [10 mM acetate buffer at pH 4]) gradient starting at 0/100 and changing +2.1%A/min for 22 min was used [1537]. Linear working concentration ranges from 1.5 to 500 µM (20 µL injected) were reported.

Plasma and urine metabolites of isbufylline were extracted from samples and analyzed on a C_{18} column ($\lambda = 280$ nm). Plasma metabolites (1-methyl-7-[2-hydroxy-2-methylpropyl]xanthine, 3-dimethyl-7-[2-hydroxy-2-methylpropyl]-xanthine) were separated in 12 min using a 20/80 acetonitrile/water (0.5% acetic acid) mobile phase [1538]. Three urine metabolites were separated using a 40-min $5/95 \rightarrow 40/60$ acetonitrile/water) 0.1 M ammonium acetate buffer pH 3.7) gradient. Compounds of interest were well resolved and separated from other extracted components for both methods.

Digitoxin, digitoxigenin, digitoxigenin mono- and bisdigitoxoside, and the glucuronides of digitoxin and digitoxigenin monodigitoxoside were extracted from liver microsomes and well resolved on a C_{18} column ($\lambda = 210$ nm) using a 30-min acetonitrile/water (phosphate buffer) mobile phase. Peak shapes were excellent [1539].

Azaperone   Carazolol

Minoxidil and tretinoin were extracted from cosmetic formulations and separated from methyl-, ethyl-, and propylparabens on a cyanopropyl column ($\lambda = 283$ nm and 367 nm). A 30-min $10/90 \rightarrow 70/30$ acetonitrile/water (10 mM $NaClO_4$ to pH 3 with $HClO_4$) gradient was used. Calibration curves were generated for 1–500 μg/mL samples. Detection limits of 5 ng were reported [1540].

Ventura et al. [1541] generated a rapid 10-min screening method for 24 diuretic compounds (e.g., amiloride, triamterene, morazone, benzthiazide, spironolactone) in urine extracts. A C_{18} column (photodiode array detector, $\lambda = 200-400$ nm) and a $10/90 \rightarrow 60/40$ acetonitrile/water (0.1 M ammonium acetate to pH 3 with H_3PO_4) gradient was used. Detection limits of 20–100 ng/mL (analyte dependent) were reported. A chromatogram of 14 select analytes was presented and excellent peak shapes were obtained.

Zopiclone and four potential degradation products were resolved on a C_{18} column ($\lambda = 303$ nm) using an 18/1/81 acetonitrile/THF/water (3.4 g/L sodium hexane-sulfonate with 7.0 g/L KH_2PO_4 at pH 4.55) mobile phase [1542]. Elution was complete in 12 min. The detection limits of impurities were reported as 0.05% (or 1 ng in a 20 μL injection). The authors noted that samples made up in methanol and injected into a 50/50 acetonitrile/water (0.2% TEA) mobile phase were not soluble in this mobile phase. Surprisingly, the addition of 2% THF to the mobile phase increased the retention of zopiclone, but at the 4% level the retention was less than with the mobile phase containing no THF. A 1% THF level was used in the final analysis because of the improved peak symmetry it provided.

Ecabapide and seven metabolites (e.g., 5-acetylamino-2-hydroxybenzamide, 3-aminobenzamide, 5-acetyl-2-hydroxy-N-methylbenzamide) were extracted from urine and separated on a C_{18} column (^{14}C radioactivity detector) using a $0/100 \rightarrow 70/30$ (at 30 min hold for 40 min) (70/30 acetonitrile/water [40 mM phosphate buffer at pH 6.7])/(2/98 acetonitrile/water [40 mM phosphate buffer at pH 6.7]) gradient [1543]. Peak shapes were good.

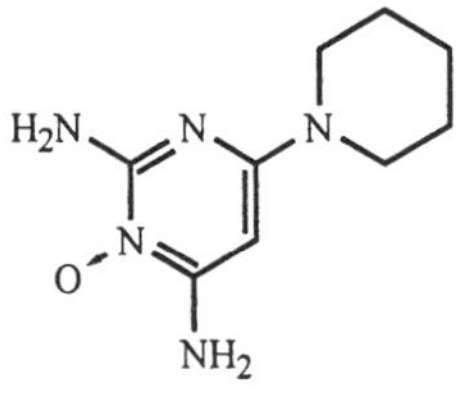

Minoxidil

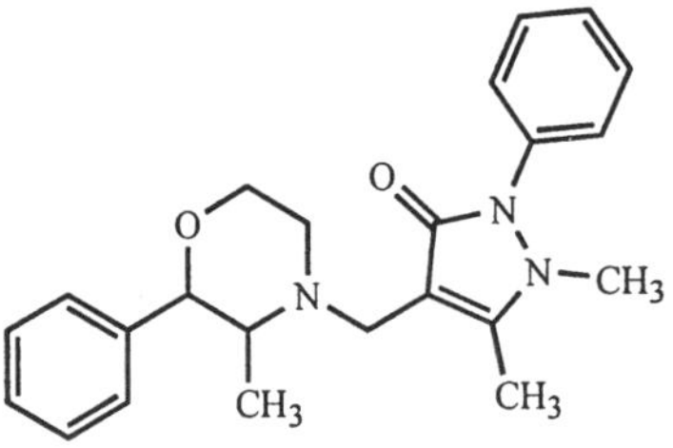

Morazone

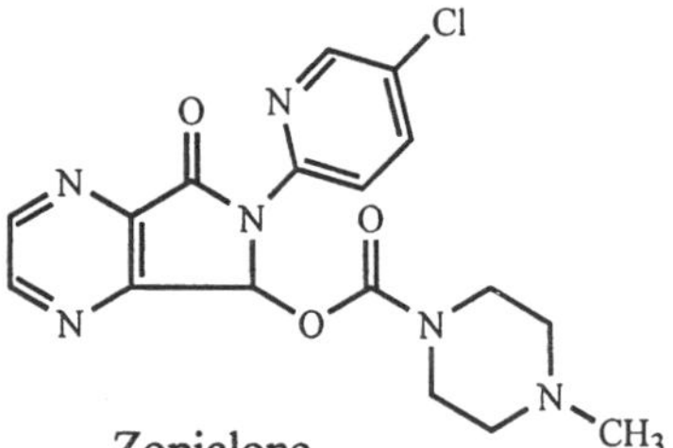

Zopiclone

Panadiplon and eight related compounds were separated on a C_8 column ($\lambda = 229$ nm) using a complex 65-min $5/95 \rightarrow 95/5$ (95/5 acetonitrile/water)/(5/95 acetonitrile/water [20 mg cyclam/L to pH 7 with H_3PO_4] gradient [1544]. Cyclam is a cyclic tetraaza macrocyclic compound that has a strong interaction with residual silanol groups on the silca surface.

Quinidine and four metabolites (3-hydroxyquinidine, O-desmethylquinidine, quinidine-N-oxide, and dihydroquinidine) were extracted from serum and analyzed on a C_{18} column ($\lambda = 235$ nm) using a 4/96 acetonitrile/water (10 mM potassium phosphate buffer at pH 2.4 with 0.37 5 mL/L nonylamine) mobile phase [1545]. Elution was complete in <6 min. Detection limits were reported as 1–5 ng (S/N = 3, analyte dependent). Linear concentration curves from 0.1 to 5.0 µg/mL were used. Peaks were somewhat tailed and complete resolution was not achieved. The use of a different and larger ion-pair reagent may be beneficial here.

The retention characteristics of eight typical active ingredients in cough medications (phenylephrine · HCl, ephedrine · HCl, papaverine · HCl, bromhexine · HCl, chlorpheniramine maleate, diphenhydramine · HCl, codeine phosphate, dextro-methorphan · HBr) were studied on a cyanopropyl column (conductivity detector) [1546]. Retention times were tabulated for these compounds using acetonitrile/ethanol/water (1 mM $HClO_4$) mobile phases at the ratios: 40/2/58, 60/2/38, and 80/2/18. Good separation was achieved with the weakest mobile phase (40/2/58) but elution took over 35 min. Since all these components are not typically monitored simultaneously, it may be possible to use a slightly stronger mobile phase. Amounts of 0.1 µg injected were detectable and standards from 5 to 100 µg/mL were used.

Fluoxetine·HCl and 12 related compounds (e.g., N-methyl-γ-[4-(trifluoromethyl)-phenoxy]benzene propanamine, 1-phenyl-3-N-methylpropanamine, acetophenone) found in manufactured products were analyzed on a cyanopropyl column

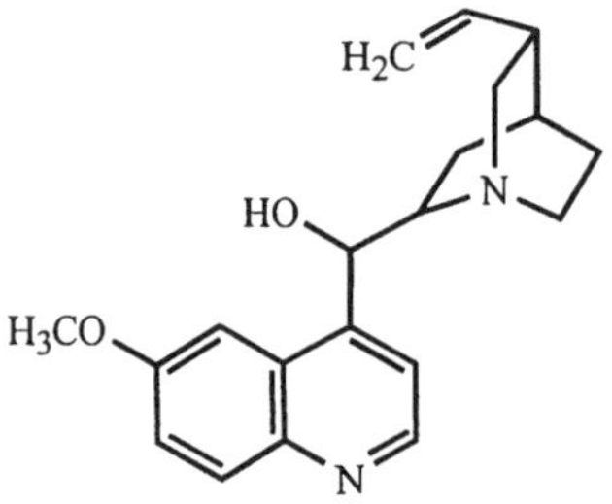

Quinidine

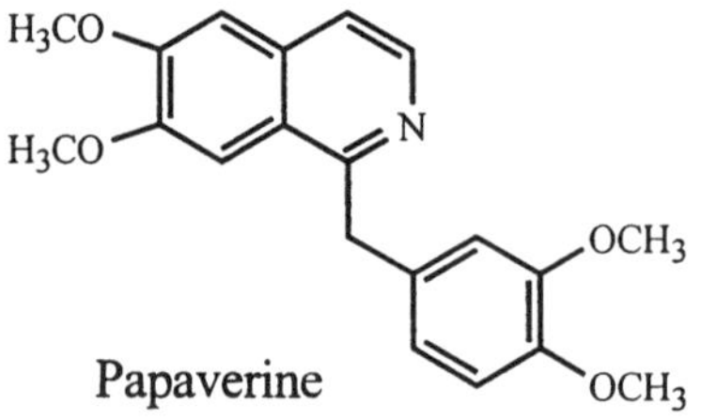

Papaverine

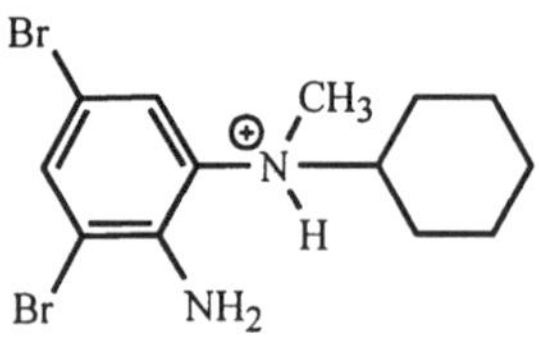

Bromhexine

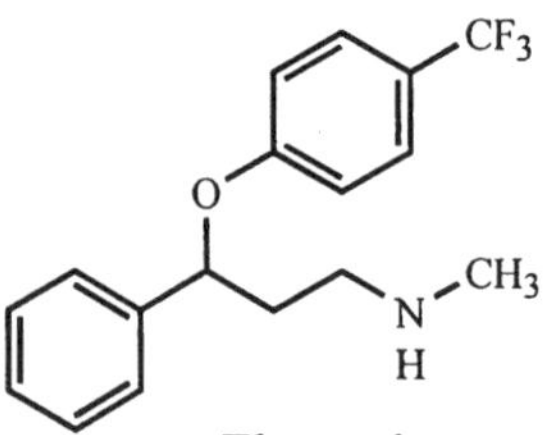

Fluoxetine

($\lambda = 214$ nm) using a 15/5/80 acetonitrile/THF/water (0.25% TFA) mobile phase [1547]. A table of UV maxima and relative retention times (0.2–1.1 vs. fluoxetine·HCl at 15 min) as well as a table of percentages of impurities down to 0.02% were presented. The best reported results had five compounds co-eluting (three in one peak, two in another).

The iron chelator 1,2-diethyl-3-hydroxypyridin-4-one and three metabolites were extracted from urine and analyzed on a porous graphitized carbon column (photodiode array detector, $\lambda = 220$–350nm or UV at $\lambda = 275$ nm). A silica-based column was avoided because of the very strong surface interactions between the analytes and the silanol groups [1548]. A 10/90 acetonitrile/water (phosphate buffer pH 2.9 and 2 mM EDTA) mobile phase was used to generate a 15-min separation. Peak shapes were good.

Tacrine and seven metabolites (e.g., 2-hydroxy-, 4-hydroxy-, and 1-hydroxytacrine) were resolved on a phenyl column (photodiode array detector, $\lambda = 230$–360 nm or UV at $\lambda = 325$ nm) using a 70/30 acetonitrile/water (ammonium formate pH 3.1) mobile phase [1549]. The analysis required 25 min. Preparative LC was conducted with identical mobile phase conditions and a preparative-scale phenyl column.

A number of enantiomers (aminoglutethimide, chlorpheniramine, chlorthalidone, fluoxetine, ibuprofen, ketoprofen, methylphenidate, metoprolol, phensuximide, propranolol, suprofen and mephenytoin) were separated on a β-cyclodextrin column using 40/60 to 20/80 acetonitrile/water (0.1% triethylammonium acetate pH 4.1 or 7.1) mobile phase (analyte dependent) [1550].

Table 9.5 lists some tried and true UPS methods [590].

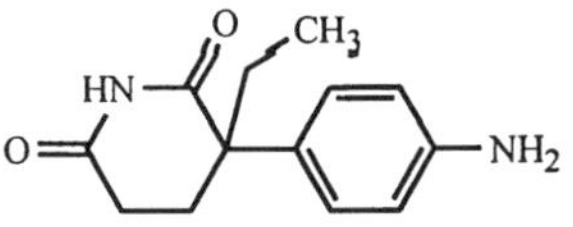

Aminoglutethimide

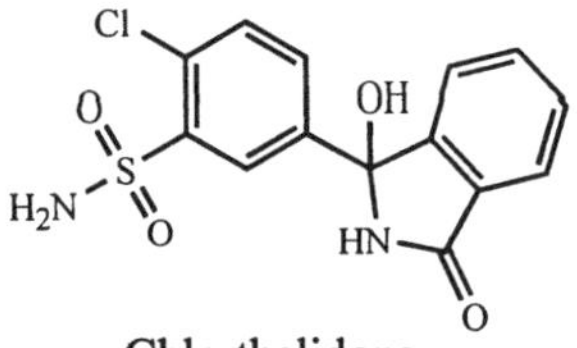

Chlorthalidone

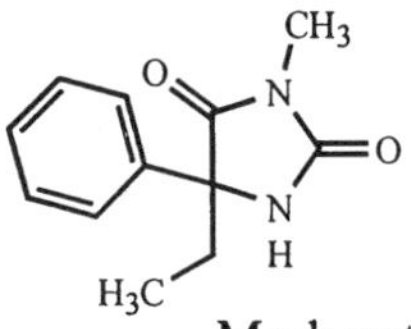

Mephenytoin

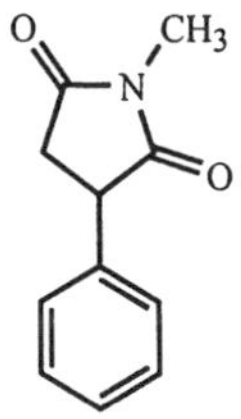

Phensuximide

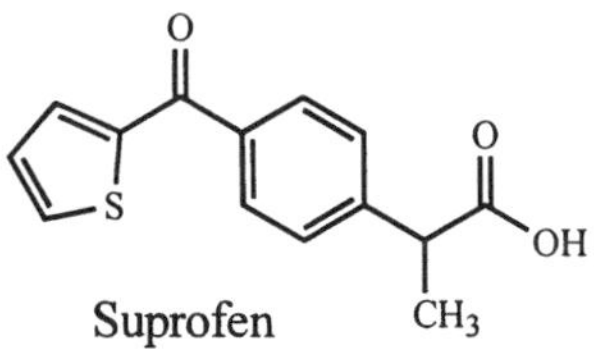

Suprofen

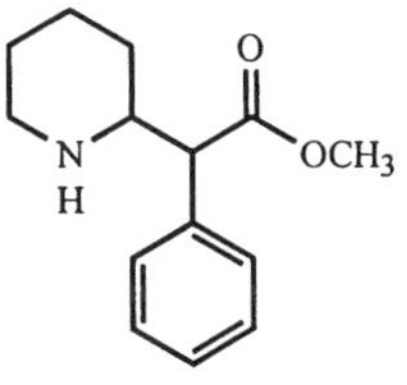

Methylphenidate

9.8 SUMMARY

Acetonitrile is one of the most versatile solvents in HPLC. It is routinely used in both protein and pharmaceutical work. Its chemical and physical properties (low UV cutoff and low-viscosity mixtures with water) give acetonitrile important advantages over the alcohols.

TABLE 9.5 USP Methods[a]

Analyte(s)	Co-analyte (Internal Standard)	Column	Mobile Phase	Wavelength (nm)	USP Page
Antipyrine, benzocaine, and phenylephrine	p-Aminobenzoic acid	Phenyl	12/88 acetonitrile/water (5 mM heptanesulfonate and 1 mL H_3PO_4)	272	154
Benztropine mesylate (injection)		C_8	63/35 acetonitrile/water (5 mM octylamine to pH 3.0 with H_3PO_4)	259	213
Bisacodyl (suppositories)		C_{18}	45/55 acetonitrile/water (74 mM sodium acetate to pH 7.4 with acetic acid)	265	236
Bromocripine mesylate (assay)		C_8	6/4 acetontrile/water (50 mg ammonium carbonate)	300	249
Carboplatin	1,1-cyclobutanedicarboxylic acid	C_{18}	880/100/20 water/ acentonitrile/water (2.1 g tetrabutylammonium HSO_4 and 3.4 mL H_3PO_4 to pH 7.55 with NaOH)	220	311
Cefixime	Cefixime E-isomer	C_{18}	775/225 water (5.3 mL H_3PO_4 to pH 7.0 with NaOH/acetonitrile	254	330
Cyclophosphamide	Ethylparaben	C_{18}	70/30 water/acetonitrile	195	483
Cyclosporin		(70°C) Dimethyl	55/40/5/0.5 acet onitrile/water/ methanol/H_3PO_4	200	487
Dactinomycin		C_{18}	46/25/25 acetonitrile/water (40 mM sodium acetate)/water (70 mM acetic acid)	254	484

Compound	Column	Mobile phase	Detection (nm)	Page
Ergoloid mesylates	C$_{18}$	80/20/2.5 water/acetonitrile/triethylamine	280	654
Dihydro-β-ergocryptine				
Fenprofen calcium	C$_8$	50/49.6/0.4 acetonitrile/water/H$_3$PO$_4$	272	708
Fluorescein sodium and Benoximate hydrochloride	C$_{18}$	60/40 acentonitrile/water with 0.1 g sodium heptanesulfonate, 40 mL acetic acid, 10 mL triethylamine to pH 3 with H$_3$PO$_4$/L	254	731
Gadopentetate dimeglumaine	C$_8$	88/12 water/acetonitrile with 1.37 g tetrabutylammonium perchlorate/L	195	758
Gallamine triethiodide	C$_{18}$	31/69 acetonitrile/water (0.14 M sodium perchlorate to pH 3)	200	760
Glyburide	C$_8$	55/45 acetonitrile/water (2.6 g (NH$_4$)H$_2$PO$_4$ to pH 5.25)	254	779
Hydrochlorothiazide	C$_{18}$	90/10 water (0.1 M NaH$_2$PO$_4$ to pH 3 with H$_3$PO$_4$)/acetonitrile	254	820
Indapamide	C$_{18}$	650/175/175/1 water/acetonitrile/methanol/acetic acid	254	867
Ioxilan (injection)	(30°C) NH$_2$	87/13 acetonitrile/water	254	911
Lactulose concentrate	(40°C) NH$_2$	82/18 acetonitrile/water (1.15 g NaH$_2$PO$_4$/L)	RI	952
Levothyroxine sodium (powder)	CN	65/35 water/acetonitrile with 1 mL H$_2$PO$_4$/L	225	968
Methoxsalen (capsule)	C$_{18}$	65/35 acetonitrile/water	254	1073
Mitoxantrone hydrochloride (assay)	Phenyl	750/250/25 water/acetonitrile/water (22.0 g sodium heptane sulfonate with 32 mL acetic acid/250 mL)	254	1121

(*continued*)

TABLE 9.5 (*continued*)

Analyte(s)	Co-analayte (Internal Standard)	Column	Mobile Phase	Wavelength (nm)	USP Page
Physostigmine	Bezyl alcohol and benzaldehyde	C_{18}	50/50 acetonitrile/water (50 mM ammonium acetate to pH 6)	254	1328
Probucol	Peroxide degradate	C_8	85/15 acetonitrile/water	288	1395
Pyrantel pamoate		Silica	92.8/3/3/1.2 acetonitrile/acetic acid/water/diethylamine	288	1443
Trihexylphenydyl hydrochloride		C_{18}	920/80/0.2 acetonitrile/water/triethylamine	210	1707

[a] From reference [590].

10

WATER, DIMETHYL SULFOXIDE, AND COMMON ACIDIC MODIFIERS

10.1 GENERAL CONSIDERATIONS AND IMPURITIES

Water is the ubiquitous solvent. Its presence is taken for granted as a natural and necessary weak solvent component in RP separations. Its presence at extremely low levels (ppm) in NP separations is critical to determining the success or failure of a separation. Even though water is almost the only component common to all separation mobile phases, either in a necessary or deleterious role, chromatographers often do not worry about the purity of this solvent. Unfortunately, water is often a source of contamination.

As mentioned in Chapter 1, water is available in many different classifications of purity. For chromatographic use, the elimination of metal ions (for use in ion exchange LC), organics (for gradient RP HPLC use), and acids/bases is important.

Laboratories can often prepare LC-quality water through the use of purification and polishing filters and cartridges. The advantage of this is that high-purity water is available on demand; the drawbacks are that the quality of the water varies not only as the cleaning cartridges age but as the source water quality changes. These details are discussed at length in Chapter 1. This requires a constant monitoring of the water quality (not just through the resistivity meter typically found on processing units). Why? Because external sources of contamination are everywhere. These include airborne bacteria, volatile organic compounds, and surface-sorbed chemicals in containers.

It has been found that production of extremely high-purity water (including both the manufacturing and packaging aspects) is quite difficult [1551]. For example, deionization, as expected, leaves low ppb levels of acetone. Interestingly, but not surprisingly, cartridge-purified HPLC-grade water contained ppb levels of chloroform (note that this is a by-product of water treatment process used by most municipal treatment plants.) Finally, bacterial contamination is particularly bad because, if great care is not exercised, bacterial colonies not only quickly establish themselves in the exit tubing of the purification system but also can establish themselves within the filter/cartridge system itself.

This process also takes place in solvent reservoirs that contain water only. In fact, it is a common problem to have the inlet filters clog due to bacteria-related problems. Such filters often are "slippery" to the touch and need to be thoroughly cleaned or replaced. The best ways to prevent clogging from occurring are to clean the inlet filters routinely as a part of a preventive maintenance plan or to make sure that the weak mobile phase always has enough organic (e.g., 10–15% acetonitrile or methanol) or is acidic or basic enough (e.g. pH < 4 or pH > 8) to create a system that will not support bacterial growth.

10.1.2   Solvent Preparation

Although it was discussed in Chapter 1, since water is such a common component of chromatographic mobile phases, this seems a natural place to briefly review the preparation of mobile phases. In most cases, solvent composition is described in terms of volume to volume mixing. Unfortunately, volume, upon mixing, is not an additive process (see Table 1.2). Therefore, the way in which the solvents are mixed is crucial. The preparation technique that generates the most reproducible mobile phase mixture is the one in which individual measurement and transfer of each component into a common container occurs. For example, 1 L of a 50/50 v/v mixture of methanol/water is prepared by carefully measuring 500 mL of methanol and transferring it to the solvent container. Next 500 mL of water is carefully measured and transferred into the container which holds the methanol. A 50/50 v/v methanol/water solvent has now been made. Or has it?

The answer is *no*. Before use, this mobile phase needs to be thoroughly mixed and allowed to come to room temperature. (It will become warm upon mixing (see Table 1.2). In the case of methanol/water mixtures the process is exothermic. Conversely, acetonitrile mixed with water becomes cold; it is an endothermic process.) As is the case when mixing any solutions, make sure to relieve the pressure that may build up from the mixing process through venting the container in a hood (with the opening facing away from you).

As noted above, if the mixing is done in a graduated cylinder, the final volume will be seen not be 1 L. Herein lies the problem. If the solvents are measured *and* mixed concomitantly in a graduated cylider (or volumetric flask) the volume change that occurs upon mixing occurs throughout the entire measurement/mixing process so that, in many cases, the solvent that is added last actually will be present at greater than the 50% level. For example, take 500 mL of water and 500 mL methanol and mix them togther in a 1 L volumetric flask. Once cooled, the total volume will be <1 L. Therefore, an additional amount of solvent needs to be added, but which one? Obviously the addition of either will lead to a solvent that is no longer 50/50 methanol/water and could have disasterous effects on the separation.

Two other critical aspects of solvent preparation are solvent degassing and solvent filtering. Both should be done only when necessary and to the minimal degree needed to prepare the solvent. As discussed in Chapter 1, solvent degassing has become a crucial requirement for use with some HPLC pump systems. It must be remembered that this degassing will remove solvent from the system (i.e., the sparging gas is saturated with solvent). The most volatile component of the mixture will be preferentially removed. If the most volatile component is also a low-level mobile phase constituent (e.g., TFA or triethylamine [TEA]), then the chromatographic ramifications might be observed over the course of a single day! For example, the mobile phase modifier TEA is used to block silanol group interactions with basic solutes. As it is sparged from the mobile phase, a basic solute will begin to exhibit peak tailing and may start to show increased retention. In difficult separations this rapidly ruins chromatographic selectivity. There are three ways to combat this problem:

1. Degas the solvent and then remove the sparge tube from the solvent to a position above the solvent. This blankets the solvent with sparge gas and successfully prevents oxygen from reentering the system.
2. Degas two separate reservoirs that contain the identical mobile phase and then connect them by a bleed tube. Connect reservoir 2 to the pump. Continue to sparge reservoir 1 and let the bleed become the blanket for reservoir 2. This presaturates the sparge gas with mobile phase and helps to maintain a constant mobile phase composition in reservoir 2.
3. Frequently replace the mobile phase with freshly prepared mobile phase.

The second option is unwieldy and not commonly used. The third alternative is typically the most costly, since a higher volume of mobile phase is used when compared with alternatives 1 and 2.

Solvents containing buffers or modifiers prepared from solids must be filtered prior to use. For the best reproducibility, the buffer solutions should be made up separately as individual components of the mobile phase and then filtered separately. For example, a mobile phase having a composition of 50/50 methanol/water (50 mM sodium acetate at pH 4.5 buffer) should have the acetate buffer prepared separately and then combined with the methanol. Avoid adjustments of the pH in nonaqueous solvents, since pH meters are designed to repond to hydrogen ion activity in an aqueous solution only!

Next, when filtering a solvent, remember that it is under vacuum and the most volatile component will be removed to a greater extent. This is important when one component of a buffer is considerably more volatile that the other (e.g., acetic acid versus sodium acetate). Therefore, do not leave the filtered solvent in the filter vessel under vacuum for extended periods after the filtering process is complete.

Note that most high-purity solvents designated for HPLC use do not require filtering before use unless a solid modifier is added to them. In fact, at best filter media are notoriously particulate-laden and often end up adding more particulates to the solvent than were originally present. Furthermore, because of the way filter media are manufactured, there is an ever-present risk that the filtered solvent will actually extract components (this is true on solvent-compatible filter media) that then appear as "inexplicable" peaks in gradient chromatograms. These peaks are almost always incorrectly attributed to impure original solvent. To limit these potential problems (1) make sure that the filter medium chosen is compatible with the solvent that needs filtering, and (2) if filter media are used, then purchase them precleaned if possible.

It should be noted that the pH of water as it comes out of a high-purity processing unit may not be the pH at which it is finally used. This is because water-soluble components of glass will change its pH. Normal high-purity water has a pH close to 7.0. Water bottled in borosilicate glass can have a pH >8.0. While this is not an apparent problem, it should be noted that this level could be significant in low-level ion-pair and ion exchange work. For ACS (American Chemical Society) tested water, glass packaging is most likely unacceptable since the ACS test monograph has a soluble silicate specification. However, from a purely functional point of view, the silicate test really has no impact on the HPLC characteristics of water.

Dimethyl sulfoxide is a very polar, high dipole moment, strong hydrogen bond accepting solvent. It is has a high viscosity nearly equal to that of IPA. Typical impurities from the manufacturing process include dimethyl sulfide (from which DMSO is commonly manufactured) and dimethyl sulfone. Decomposition products include dimethyl sulfide, dimethyl sulfone, methyl mercaptan, and bis(methylthio)-methane [1552].

Dimethyl sulfoxide offers a potentially serious health problem in that it is readily adsorbed through skin. Any compounds dissolved in the DMSO may also penetrate the skin. Therefore, it is always prudent to wear protective gloves when working with DMSO. Note, however, that DMSO also readily passes through many type of gloves, so that the appropriate gloves should be chosen or a contaminated glove should be changed immediately.

Trifluoroacetic acid is a very strong acid with a pK_a value of less than zero (<0). It is very volatile, which makes it ideal for use in separations from which samples are to be collected and reclaimed. Unfortunately, this same volatility makes it unpleasant to work with. It has an sharp odor similar to but more pungent than that of acetic acid. TFA causes severe burns when it comes into contact with skin and appropriate safety precautions need to be taken when it is used. TFA can be purchased in a high-purity form that is suitable for HPLC use.

Tables 10.1–10.4 list some of the important chemical, physical, and chromatographic properties as well as general manufacturing and safety parameters for these

$$CH_3COOH \qquad\qquad CH_3\overset{\displaystyle O}{\overset{\displaystyle \|}{S}}CH_3 \qquad\qquad CF_3COOH$$

Acetic acid   Dimethyl sulfoxide   Trifluoroacetic acid

FIGURE 10.1

TABLE 10.1   Physical Properties of Water, DMSO, and Common Acidic Modifiers[a]

	Water	DMSO	TFA	H_3PO_4[b]	ACH
Molecular weight	18.02	78.13	114.02	98.00	60.05
Density (g/mL)	0.9982	1.100	1.4890	1.88	1.049
Viscosity (cP)	1.00	2.24	0.926		1.314[e]
Solubility in water (%)	—	100	100	100	100
Water solubility in solvent (%)	—	100	100	100	100
Boiling point (°C)	100	189	71.8		117.9
Melting point (°C)	0	18.5	−15.2		16.63
Refractive index (n_D)	1.3330	1.4783			1.3716
Dielectric constant	80.1	46.68	8.55		6.15
Dipole moment (D)	1.87	3.96	2.28		1.74
Surface tension (dyne/cm)	72.8	42.92[c]	13.64[d]		27.59

[a] *Abbreviations:* DMSO, dimethyl sulfoxide, methyl sulfoxide, sulfinylbis[methane]; TFA, trifluoroacetic acid, perfluoroacetic acid; H_3PO_4, phosphoric acid, orthophosphoric acid; ACH, acetic acid (glacial).
[b] >85% phosphoric acid.
[c] At 25°C.
[d] At 24°C.
[e] At 15°C.

solvents [84–92]. Figure 10.1 shows the structure of the solvents listed in Tables 10.1–10.4.

10.2   INDUSTRIAL ANALYTES

Carboxymethylcellulose was characterized by the analysis of completely carboxymethylated samples. The depolymerized sample was then analyzed for 6-mono-*O*-,

TABLE 10.2   Chromatographic Parameters of Water, DMSO, and Common Acidic Modifiers[a]

	Water	DMSO	TFA	H_3PO_4[b]	ACH
Eluotropic strength $\varepsilon°$ on Al_2O_3		0.62			
Eluotropic strength $\varepsilon°$ on SiOH					
Eluotropic strength $\varepsilon°$ on C_{18}					
Solvent strength parameter, P'	10.2	7.2			
Hildebrandt solubility parameter, δ					
Hydrogen bond acidity, α	1.17	0.00			
Hydrogen bond basicity, β	0.18	0.76			
Dipolarity/polarizability, $\pi*$	1.09	1.00	0.50		0.64

[a] *Abbreviations:* DMSO, dimethyl sulfoxide, methyl sulfoxide, sulfinylbis[methane]; TFA, trifluoroacetic acid, perfluoroacetic acid; H_3PO_4, phosphoric acid, orthophosphoric acid; ACH, acetic acid (glacial).

TABLE 10.3   Common Manufacturing Quality Specifications of Water, DMSO, and Common Acidic Modifiers[a]

	Water	DMSO	TFA	H_3PO_4[b]	ACH
UV cutoff (nm)	<190	268	210[b]		
Percent water (maximum)	—	0.04	0.05		
Available as ACS tested[c]	AM[d]	AEJM	n.a.[e]	AJFM	AEJFM
Available as HPLC-grade[c]	ABEFJM	ABEM	BJ	F	JFM
Available through [f]		F	AEFM	E	

[a] *Abbreviations:* DMSO, dimethyl sulfoxide, methyl sulfoxide, sulfinylbis[methane]; TFA, trifluoroacetic acid, perfluoroacetic acid; H_3PO_4, phosphoric acid, orthophosphoric acid; ACH, acetic acid (glacial).
[b] 0.1% solution.
[c] Manufacturer's code: A = Aldrich; B = Burdick & Jackson; E = EM Science; F = Fisher; J = JT Baker; M = Mallinckrodt.
[d] To meet ACS specifications, this product must be stored in plastic bottles.
[e] No ACS test exists for this solvent.
[f] Available as a high-purity solvent but not specifically designated as ACS or HPLC grade. This does *not* mean a lesser quality solvent, just that it is not specifically tested for these applications. If the manufacturers produce either ACS or HPLC solvent, they are not listed under this heading.

2,3-, 2,6-, 3,6-di-*O*-, and 2,3,6-tri-*O*-carboxymethylglucose. Separation was accomplished on a 65°C Aminex 87H resin column (RI detector) using a 0.01 N H_2SO_4 mobile phase. Elution was complete in 25 min. Polymer samples were characterized by the mole fraction of each of the glucose compounds [1553].

Glycerol, propylene glycol, diethylene glycol, IPA, and triethylene glycol were baseline resolved on a 30°C C_{18} column (RI detector) using a water mobile phase [1554]. Elution was complete in <15 min and peak shapes were very good. Detection of a 100 μL injection of 10 mg/L standards was achieved.

Nitric acid, trichloroacetic acid, dichloroacetic acid, chloral, and acetic acid in water were resolved on a C_{18} column ($\lambda = 210$ nm) using an aqueous 150 mM ammonium sulfate mobile phase [1555]. Peak shapes were excellent except for chloral, which was badly tailed. The tailing may be indicative of the instability of this compound. Detection limits of ~10 μg/L (S/N = 4) were reported. Elution was complete in 16 min.

EDTA was analyzed in drinking water samples as its ferric ion complex on a C_{18} column ($\lambda = 254$ nm) using an aqueous 0.3 M acetate buffer containing 20 mL/L of

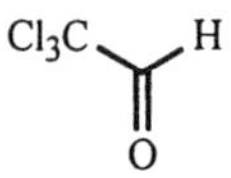

Chloral

TABLE 10.4 Safety Parameters of Water, DMSO, and Common Acidic Modifiers^a

	Water	DMSO	TFA	$H_3PO_4{}^b$	ACH
Flash point^b (TCC) (°C)	None	95	None		40
Vapor pressure (Torr @ 20°C)	17.5	0.6	108^c		
Threshold limit value (ppm)					
CAS number	7732-18-5	67-68-5	76-05-1	7664-38-2	64-19-7
Flammability^d	0	1		0	2
Reactivity^d	0	0		0	0
Health^d	0	1		3	3

^{a} *Abbreviations:* DMSO, dimethyl sulfoxide, methyl sulfoxide, sulfinylbis[methane]; TFA, trifluoroacetic acid, perfluoroacetic acid; H_3PO_4, phosphoric acid, orthophosphoric acid; ACH, acetic acid (glacial).
^{b} TCC = TAG closed cup.
^{c} At 25°C.
^{d} According to National Fire Protection Association ratings [92]:
Fire: 4 = Materials that vaporize at room temperature and pressure and burn readily.
       3 = Liquids or solids that can ignite under room conditions.
       2 = Materials that ignite with elevated temperature or with moderate heat.
       1 = Materials that must be preheated before they ignite.
       0 = Materials that will not burn.
React: 4 = Materials that, by themselves, can deteriorate or explode under room conditions.
       3 = Materials that can detonate or explode but require an initiator (e.g., heat).
       2 = Materials that undergo violent chemical reactions at elevated temperatures or pressures or react with water.
       1 = Materials that are, by themselves, stable but that may become unstable at elevated temperatures and pressures.
       0 = Materials that are stable even under fire conditions and do not react with water.
Health: 4 = Short exposure times to these materials are lethal or cause major residual injury.
       3 = Short exposure times to these materials cause temporary and/or residual injuries.
       2 = Lengthy (but not chronic) exposure to these materials may cause temporary incapacitation and/or minor residual injury.
       1 = Materials that, upon exposure, cause irritation but only minor residual injury.
       0 = Materials that, upon exposure under fire conditions, offer no more hazard than ordinary combustible materials.

20% tetrabutylammonium hydroxide adjusted to pH 4 mobile phase [1556]. The detection limit (S/N = 3) was reported as 0.8 µg/L. Elution was complete in <10 min. Similarly, EDTA, nitrilotriacetic acid, and diethylenetriaminepentaacetic acid (DTPA) were separated and analyzed on a PRP-1 column (amperometric detector, + 1.2 V vs. Ag/AgCl) using a trichloroacetic acid (TCA) in water mobile phase [1557]. The mobile phase pH was adjusted to the proper value through the addition of the TCA. In order to elute DTPA, the mobile phase pH had to be <2.5. Peak shapes were good, as was the resolution. Elution was complete in <15 min. Detection limits of ~2 µM were reported. Finally, the concentrations of the Cd^{2+}, Pb^{2+}, Cu^{2+}, and Zn^{2+} complexes of EDTA were determined in wastewater using a C_{18} column ($\lambda = 258$ nm) and an aqueous 6 mM sodium sulfate with 1 mM HEPES buffer at pH 7 with ammonia. Postcolumn reaction at 40°C with 40 μM ferric nitrate in a 200 mM pH 4.5 sodium formate buffer was used for visualization. Elution was complete in <30 min with excellent peak shape and baseline resolution [1558]. A linear range of 0.1–5 μm with a detection limits of 30–50 ng injected were reported.

4-Fluoro-N-methylaniline and five metabolites (e.g., N-acetyl-N-hydroxy-4-aminobenzene, N-acetyl-4-aminobenzene, 4-fluoroaniline) were extracted from urine and separated on a C_8 column ($\lambda = 300$ nm) using a 50 mM potassium phosphate buffer at pH 7 as the mobile phase [1559]. Elution was complete in 16 min.

Zou et al. [1560] separated 14 phenylamine- and naphthylaminesulfonic acids (e.g., 2-aminonaphthalene-4,6,8-trisulfonic acid, 1,3-diamino-4-sulfonic acid, phenylamine-2-sulfonic acid, 6-chlorophenylamine-3-sulfonic acid, naphthylamine-7-sulfonic acid) in 30 min on a C_{18} column ($\lambda = 254$ nm). Baseline resolution was achieved using an aqueous 10 mM phsophate buffer pH 6.8 mobile phase. The capacity factors for these and 10 additional compounds belonging to these classes of compounds were tabulated.

Five impurities of 8-amino-1-naphthol-3,6-disulfonic acid (e.g., 1-naphthol-4-sulfonic acid, 8-aminonaphthalene-1,3,6-trisulfonic acid, chromotropic acid) were resolved on a C_{18} column ($\lambda = 235$ nm) using an aqueous 0.3 M sodium sulfate mobile phase [1561]. All compounds eluted in <8 min. Detection limits of ~100 μg/mL were reported.

The four isomers of sorbic acid (*trans*-2,*trans*-4; *cis*-2,*trans*-4; *trans*-2,*cis*-4; *cis*-2,*cis*-4) were baseline resolved on a C_{18} column ($\lambda = 254$ nm) using a 90-min 100/0 → 95/5 water (0.1 w% TFA)/methanol gradient. Elution was complete in 55 min. Why the gradient runs an extra 35 min is not explained. Similarly, the ethyl-substituted sorbates were resolved in 75 min using the same column and a gradient from 90/10 → 70/30 (at 60 min hold 15 min) water (0.1 w% TFA)/methanol. Linear response from 5 to 150 μg/mL and detection limits of 1 μg/mL were reported [1562].

10.3 BIOLOGICAL ANALYTES

The disaccharide composition of heparin and heparin sulfate was determined after digestion with heparin lyase, separation on a 50°C C_8 column, and postcolumn derivatization with 2-cyanoacetamide ($\lambda = 346$ nm, ex; 410 nm, em). Eight disaccharides (e.g., 2-acetamido-2-deoxy-4-O- and 2-deoxy-2-sulfamino-4-O-(4-deoxy-α-L-*threo*-hexenepyranosyluronic acid)-D-glucose, and the 6-O-sulfo-D-glucose analogs) were separated in 15 min using a complex 20-min 96/4/0 → 0/4/96 water (1.2 mM tetrabutylammonium hydrogensulfate)/acetonitrile/water (0.1 M cesium chloride) gradient. The calibration range was reported as 1 ng–1 μg injected [1563].

Glycarbylamide (a coccidiostat) was extracted from chicken tissue and analyzed on a 40°C C_{18} column ($\lambda = 260$ nm) using a 50 mM aqueous KH_2PO_4 pH 4.5 mobile phase [1564]. The analyte peak was completely resolved from other extracted compounds and eluted in 18 min. It is interesting to note that although glycarbylamide is sparingly soluble in all but very basic aqueous solutions, the mobile phase was acidic. However, the standard concentrations were well below the published 10 μg/mL solubility limit. A calibration curve from 0.16 to 3 μg/mL was used and a detection limit of 0.05 μg/g tissue was reported.

Four proanthocyanidins (prodelphinidin B3, procyanidin B3, (+)-catechin and (−)-epicatechin) were extracted from beer and baseline resolved on a C_{18} column (electrochemical detector, +350 mV vs. Ag/AgCl). Peak shapes were superb. Elution was complete in 60 min when a 70-min aqueous 0/100 → 100/0 2.5% acetic acid/10% acetic acid gradient was used. Detection limits of 0.1 mg/L were reported [1565].

10.3.1 Carboxylic Acid Analytes

Six carboxylic acids (pyruvic, tartaric, shikimic, lactic, acetic, succinic) were found in Nebbiolo wine and analyzed on a C_{18} column ($\lambda = 230$ nm) using an aqueous 13 mM NaH_2PO_4 with 2.6 mM $(NH_4)_2SO_4$ to pH 2.6 with H_3PO_4 mobile phase [1566]. Standard concentrations were 0.01–2 mg/mL. Good peak shapes and resolution were attained.

All six pyridinedicarboxylic acid isomers were well-resolved using a C_{18} column ($\lambda = 254$ nm). Separation was generated with an aqueous 153.2 mM KH_2PO_4/ K_2HPO_4 buffer at pH 7.3 containing 15 mM tetrabutylammonium phosphate and 2 mM disodium EDTA mobile phase [1567]. Elution was complete in just over 10 min. Peak shapes were excellent. Tables of the effect of buffer and ion-pair reagent concentration on retention are presented. Standards ranging from 200 to 300 μg/mL were easily detected.

Hydroxycitric acid (HCA) was extracted from *Garcinia cambogia* and analyzed on a C_{18} column ($\lambda = 214$ nm) using an aqueous 10 mM sulfuric acid mobile phase [1568]. Elution required 14 min due to co-extracted peaks (HCA was eluted in 4 min). Appropriate concentration ranges for HCA were reported as 2–10 μg/mL.

In an interesting study by Zhou et al. [1569], the separation of six positional isomers of difluorophenylacetic acid (2,3-, 2,4-, 2,5-, 2,6-, 3,4-, 3,5-) were ultimately baseline resolved on a base deactivated 25°C C_{18} column ($\lambda = 210$ nm) using a 95/5 water/acetonitrile with 15 mM sodium phosphate at $pH_{apparent}$ 6.5 mobile phase.

Glycarbylamide

Peak shapes were excellent and elution was complete in 17 min. A plot of the effect of changing percent acetonitrile on $\log k'$ is given. A detection limit of 100 ppm was reported. In this study, a normal-phase separation was also attempted, with the best result obtained on a silica column ($\lambda = 220$ nm) using a 98/1.975/0.025 hexane/ IPA/TFA mobile phase. In this case, the 2,4/3,5-isomers were incompletely resolved but the elution time was only 7 min. A detection limit of 300 ppm was generated for the NP separation.

Five α-hydroxy acids (glycolic, lactic, malic, tartaric, citric) that are commonly used in cosmetic cream and lotion formulations were isolated and separated on a C_{18} column ($\lambda = 210$ nm). An aqueous 5 mM tetrabutylammonium chloride with 10 mM phosphate buffer to pH 2.2 mobile phase generated baseline resolution and excellent peak shapes except for citric acid, which was noticeably tailed. The authors noted that the same mobile phase but at pH 6.5 generated unacceptably long retention times for citric acid. Due to the aggressiveness of chloride toward stainless steel under acidic conditions, a better choice of ion-pair agent would have been the dihydrogenphosphate salt. Elution was complete in 25 min. Stock standard solutions from 7 to 1000 µg/mL (analyte dependent) were used. Detection limits ranged from 8 to 25 ng injected [1570]. A detailed comparison of C_{18} packing materials effect on the retention of these analytes was also presented.

10.3.2   Basic Amine Analytes

Six ephedrines (norephedrine, norpseudoephedrine, ephedrine, pseudoephedrine, N-methylephedrine, ethylephedrine) were recovered from urine and separated on a 40°C C_8 column ($\lambda = 215$ nm). To generate good retention and peak shapes, a base deactivated packing was necessary. An aqueous 50 mM H_3PO_4 with 25 mM triethylamine adjusted to pH 6.5 mobile phase generated baseline resolution and complete elution in 17 min. Plots of the effect of changing triethylamine concentration and pH on k' were presented. A detailed comparison of C_{18} packing materials effect on the retention of these analytes was also presented. A linear range of 1–60 µg/mL and detection limit of 0.5 µg/mL (S/N = 3) were reported [1571].

Four purine metabolites (allantoin, uric acid, xanthine, and hypoxanthine) were isolated from ovine urine and analyzed on $2 \times 250 \times 4.6$ mm C_{18} columns (photodiode array detector, $\lambda = 225$–284 nm). A complex 31-min 100/0 → 95/5 water (2.5 mM ammonium phosphate to pH 3.5 with H_3PO_4)/methanol gradient was used [1572]. Allantoin was poorly retained and minimally resolved from early-eluting co-extracted components. The standard concentration range was 70–1500 µM with quantitation limits of 0.5–2 nmol injected (S/N = 10) reported.

Synephrine and N-methyltyramine were extracted from citrus fruit and separated on a C_{18} column (electrochemical detector, $+1.2$ V vs. Ag/AgCl) using a 70/30 aqueous 20 mM citric acid with 20 mM NaH_2PO_4 at pH 3 mobile phase [1573].

Synephrine

Peaks were slightly tailed but baseline resolution was achieved. Elution was complete in 15 min. Detection limits of 13 ng/mL (S/N = 3) were reported. Linear responses over the concentration range 15–10,000 ng/mL were obtained.

Caffeine, 1,3- and 1,7-dimethylxanthine, and 1,3,7-trimethyluric acid were extracted from liver microsomes and analyzed on a 30°C C_{18} column ($\lambda = 278$ nm). The mobile phase was 96.4/1/1/1.6 water (4 mM acetate buffer at pH 4)/acetonitrile/methanol/THF. Elution was complete in 13 min and good resolution was obtained. Detection limits of 200 nM were reported [1574].

Xanthine, creatine, allantoin, oxonic acid, uric acid, and hypoxanthine were extracted from poultry litter and separated on a C_{18} column ($\lambda = 200$ nm, 215 nm, 235 nm, 290 nm). Elution was complete in 25 min using an aqueous 50 mM monobasic potassium phosphate mobile phase [1575]. Xanthine and hypoxanthine were poorly resolved. Standards of 20 mg/mL with 20 μL injections were easily detected.

Ethanolamine, diethanolamine, and triethanolamine were resolved on a C_{18} column (luminol/hydrogen peroxide postcolumn chemiluminescence suppression detection) using an aqueous 2 mM hexanesulfonic acid mobile phase [1576]. Elution was complete in 15 min. Detection limits of 0.2–1.2 nmol injected (S/N = 2) were reported (analyte dependent).

Levodopa and nine catecholamines (e.g., dopamine, epinephrine, dihydroxyphenylacetic acid, 3-methoxytyramine, norepinephrine) were extracted from brain dialysates and baseline resolved on a C_{18} column (electrochemical detector at +0.75 V vs. Ag/AgCl). An aqueous mobile phase containing 100 mM sodium acetate with 20 mM citric acid and 1 mM dibutylamine adjusted to pH 2.7 with H_3PO_4 generated a 45-min separation [1577]. Detection limits of 30–240 fmol (S/N = 3) were reported (analyte dependent).

10.3.3 Toxins

Patulin and 5-hydroxymethylfurfural (5HMF) were extracted from apple juice and analyzed on a C_{18} column (photodiode array detector, $\lambda = 250–300$ nm). Water alone as a mobile phase did not resolve the peaks, but a 99/1 water/acetonitrile mobile phase resolved and eluted the analytes in <9 min. Concentration ranges of 0.05–10 mg/L for 5HMF and 5–300 μg/L for patulin were reported. Detection limits of 5.4 μg/mL were reported for 5HMF [1578]. Patulin was also isolated from apples and quantitated using a 40°C C_{18} column ($\lambda = 276$ nm) and a 95/5 water/acetonitrile mobile phase [1579]. Patulin is well resolved from other extractants and eluted in 12 min. A linear range of 1–100 μg/mL and detection limits of 2 μg/kg were reported.

Monofluoroacetate, a toxin from the poison leaf plant *Dichapetalum cymosum*, was isolated and analyzed on a specialty organic acid analysis column ($\lambda = 210$ nm) using an aqueous 20 mM H_3PO_4 mobile phase [1580]. This compound was well separated from formic, acetic, and propionic acids. Peak shapes were excellent and elution was complete in 35 min. Linearity was achieved over the range 25–1600 μg/mL, with detection and quantitation limits reported as 12 and 40 μg/mL, respectively.

Oshima [1581] studied a series of 17 saxitoxin analogs as their periodate oxidation products using a C_8 column and fluorescence detection ($\lambda = 330$ nm, ex; 390 nm, em). Five N-sulfocarbamoyl-11-hydroxysulfate toxins were resolved in 15 min using an aqueous 1 mM tetrabutylammonium phosphate buffer adjusted to pH 5.8 with acetic acid mobile phase. Nine gonyautoxins were resolved in 20 min using an aqueous 2 mM sodium heptanesulfonate with 10 mM ammonium phosphate at pH 7.1 mobile phase. Three saxitoxins were resolved in 15 min using an aqueous 2 mM sodium heptanesulfonate and 10 mM ammonium phosphate at pH 7.1 mobile phase. Peak shapes were very good for all separations.

10.3.4 Vitamins

Vitamin B_6 (pyridoxine) and two related compounds (pyridoxamine, pyridoxal; 4-deoxypyridoxine internal standard) were isolated from pork meat and separated on a 30°C C_{18} column ($\lambda = 290$ nm, ex; 395 nm, em) using an aqueous 0.01 M H_2SO_4 mobile phase [1582]. Peak shapes were excellent and elution was complete in 15 min. A table showing the effect of changes in sulfuric acid concentration on retention time is presented. Reported linear ranges of 0.0015–0.8 µg/mL were reported.

Ascorbic acid and dehydroascorbic acid levels were determined in plasma. A C_{18} column (coulometric detector operated at 100 mV) and an aqueous 100 mM Na_2HPO_4 with 2.5 mM EDTA and 2 mM n-dodecyltrimethylammonium chloride (pH 3) mobile phase were used [1583]. Ascorbic acid eluted in <5 min. Dehydroascorbic acid was determined through its reduction to ascorbic acid with diththiothreitol. Subtraction of the original ascorbic acid content from the reduced sample content yielded the dehydroascorbic acid result. A 1 ng injection was easily detected.

Ascorbic acid (AA, vitamin C), dehydroascorbic acid (DAA), isoascorbic acid (IAA), and dehydoisoascorbic acid (DHAA) were extracted from foods and were analyzed on a 20°C C_{18} column ($\lambda = 247$ nm for AA and IAA, postcolumn derivatization with O-phenylenediamine then $\lambda = 350$ nm, ex; 430 nm, em). The system consisted of an aqueous 2.5 mM dodecyltrimethylammonium chloride with 2.5 mM Na_2EDTA and 66 mM phosphate and 20 mM acetate to pH 4.5 mobile phase [1584]. Elution was complete in 9 min and the peaks were well resolved. Concentration ranges were 10–150 µg/mL for AA and 1–50 µg/mL for DHAA.

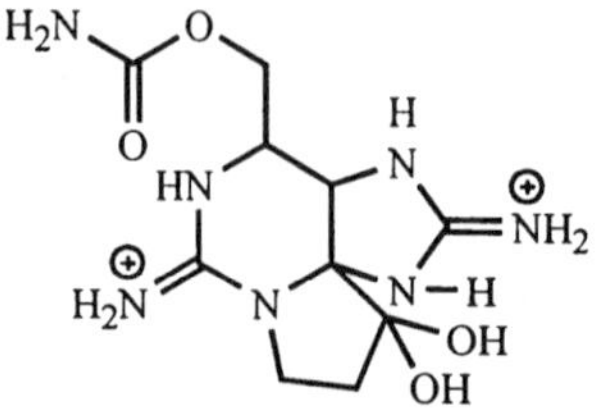

Saxitoxin

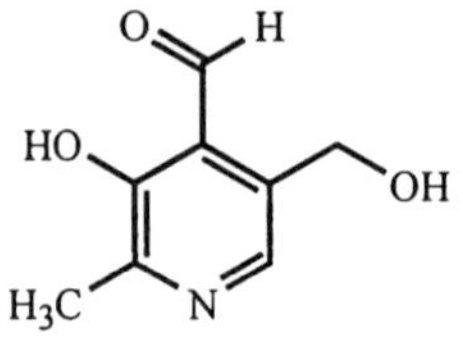

Pyridoxal

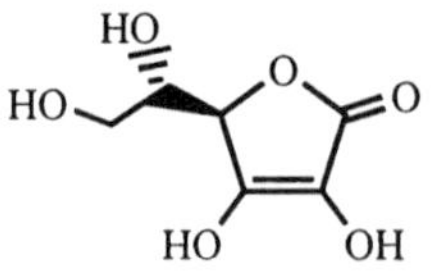

Ascorbic acid

Pyridoxine, pyridoxal, pyridoxamine, pyridoxal phosphate, and pyridoxamine phosphate were extracted from cooked sausage and separated on a 35°C deactivated base C_{18} column ($\lambda = 290$ nm, ex; 395 nm, em) using a 99/1 water (50 mM KH_2PO_4 buffer at pH 3.2)/acetonitrile mobile phase [1585]. Very good peak shapes and resolution were achieved and elution was complete in 4 min. Linear ranges were reported as 10–500 μg injected and detection limits of 0.02–0.5 mg/100 g sample were cited (analyte dependent).

Pyridoxine (vitamin B_6), five vitamers (pyridoxal, pyridoxamine, and the three corresponding 5′-phosphate esters), and the metabolite 4-pyridoxic acid were isolated from baker's yeast, eggs, and milk [1586]. Separation was obtained on a C_{18} column ($\lambda = 290$ nm, ex; 389 nm [long-pass cutoff filter]) using an aqueous 0.15 M NaH_2PO_4 pH 2.5 with 70% $HClO_4$ mobile phase. Very good peak shapes and resolution were obtained and elution was complete in under 20 min. Injected amounts ranged from 15 to 150 pmol.

An excellent method for the separation of orotic acid, inosine-5′-phosphate, uric acid, hypoxanthine, uridine, thymine, oxipurinol, and allopurinol was developed on a C_{18} column (photodiode array detector, $\lambda = 200$–400 nm) using an aqueous 47 mM phosphate buffer at pH 4.65 mobile phase [1587]. Baseline resolution was achieved. Peak shapes were very good and elution was complete in 35 min. The lowest standard concentration levels reported were 3 pmol injected.

Creatinine and ascorbic, uric, and orotic acids were extracted from milk and analyzed on a C_{18} column ($\lambda = 254$ nm). An aqueous 5 mM octylamine adjusted to pH 6.4 with H_3PO_4 mobile phase generated the separation [1588]. Baseline separation was obtained and elution was complete in 25 min.

10.3.5 Nucleosides, Nucleotides, Amino Acids, and Peptides

Homocysteine, cysteinyl-glycine, and cysteine were isolated from plasma and derivatized with SBD-F (ammonium-7-fluorobenzo-2-oxo-1,3-diazole-4-sulfonate) [1589]. These derivatized products were then baseline separated on a base deactivated C_{18} column ($\lambda = 385$ nm, ex; 515 nm, em) using a 99.2/0.8 water (0.1 M KH_2PO_4 at pH 2.1)/acetonitrile mobile phase. Elution was complete in <5 min and peak shapes were excellent. A linear range of 1–60 μM was reported.

Creatinine, allantoin, uric acid, hypoxanthine, and xanthine were extracted from urine and baseline resolved on a C_{18} column ($\lambda = 218$ nm) using an aqueous 10 mM

Orotic acid   Thymine   Creatinine

phosphate buffer at pH 4.0 [1590]. Elution was complete in 10 min and calibration curves from 10 to 400 µg/mL were generated.

Histidine and *cis*- and *trans*-urocanic acid were extracted from skin and analyzed on a C_8 column ($\lambda = 210$ nm) using a 98/2 water (10 mM triethylammonium phosphate to pH 3 with H_3PO_4 and 5 mM octanesulfonic acid)/acetonitrile mobile phase [1591]. The metabolites of histidine, histamine and 1-methylhistamine were also extracted and eluted under these conditions. Overall elution was complete in 20 min and analytes were baseline resolved. The authors also tried a C_{18} and C_4 column but achieved the optimal results with the C_8 column. A linear range of 50 nM to 100 µM and detection limit of 50 nM were claimed.

Thymine, thymidine, and six decomposition compounds (e.g., 5-hydroxy-5,6-dihydrothymine, 5,6-dihydrothymidine, (5S)-5-hydroxy-5,6-dihydrothymidine) were resolved on a C_{18} column (thermospray MS). An aqueous 0.1 M ammonium acetate buffer at pH 6 generated elution in 16 min [1592]. Detection limits of ~0.2 ng injected were reported.

Glutathione and 10 other thiols and disulfides (e.g., homocysteine, cysteine, glutathione disulfide, cysteine-glutathione mixed disulfide) were isolated from plasma and separated on a C_{18} column (electrochemical detector, -1.0 V (disulfide reduction) and $+0.15$ V vs. Ag/AgCl). A 93.25/5/1.75 water (0.1 M chloroacetic acid)/methanol/N,N-dimethylformamide with 2.25 mM heptansulfonic acid to pH 2.8 with NaOH mobile phase generated good resolution and peak shapes. Elution was complete in 18 min [1593]. Detection limits of 8 nM were reported for glutathione.

Nitrated tyrosine was separated from other amino acids (tyrosine, 5-hydroxytryptophan, phenylalanine, tryptophan) on a C_{18} column ($\lambda = 254$ nm) and 95/5 water (50 mM sodium acetate buffer at pH 4.7)/methanol mobile phase [1594]. Excellent peaks shapes and baseline resolution were achieved during the 27 min separation.

In a unique and elegant separation Hu et al. [1595] resolved 13 nucleosides (e.g., adenosine, deoxyadenosine, uracil, guanine, thymidine, uridine, cytosine) on a C_{18} column ($\lambda = 210$ nm) with water as the mobile phase. Varying temperature gradients through the column and initial temperature hold time fine-tuned the separation. A

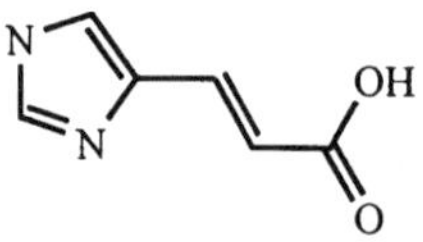

trans-Urocanic acid

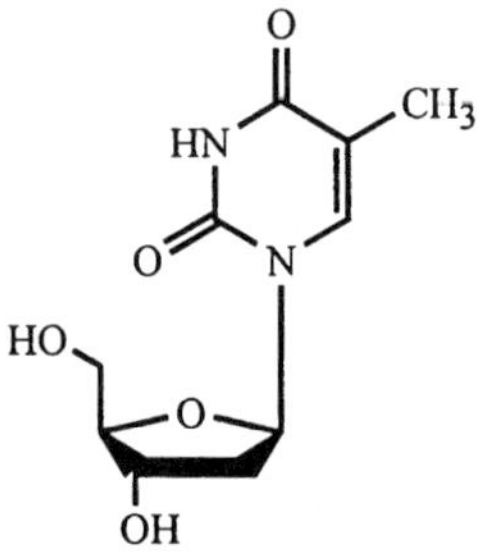

Thymidine

4°C/min temperature gradient yielded a 25-min analysis. Standard concentrations were in the µM range.

10.3.6 Terpenoids, Flavonoids, and Related Analytes

Louche et al. [1596] analyzed various citrus fruits (e.g., lemons, limes, citrons) for their content of phlorin (3,5-dihydroxyphenyl-β-D-glucopyranoside, an orange peel marker compound). After sample preparation, separation was achieved on a base deactivated C_{18} column ($\lambda = 214$ nm) using an aqueous 25 mM KH_2PO_4 (to pH 2 with H_3PO_4) mobile phase. Good separation and peak shapes were obtained and elution was complete in under 11 min. Although no linearity data were cited, phlorin levels from 8 to 235 mg/L were reported.

Five glucosinolates (progoitrin, napoleoferin, gluconapin, 4-hydroxyglucobrassicin, glucobrassicanapin) were isolated from canola seeds and separated on a C_{18} column ($\lambda = 235$ nm). Baseline resolution, good peak shapes, and run times of 25 min were obtained using a 97/3 water (0.1 N ammonium acetate)/methanol mobile phase. A linear range of 100–500 µM was reported [1597].

Seven ephedrine alkaloids (synephrine, norephidrine, ephedrine, pseudoephidrine, norpseudoephidrine, N-methylephedrine, N-methylpseudoephedrine) were extracted from various herbal formulations and baseline resolved on a phenyl column ($\lambda = 255$ nm). The use of a 98/2 water (0.1 M sodium acetate pH 4.8)/acetonitrile mobile phase resulted in a 14-min analysis. Synephrine eluted very close to the system void volume [1598]. A working curve from 4 to 150 µg/mL was used and sample detection limits of 0.05 mg/g were reported.

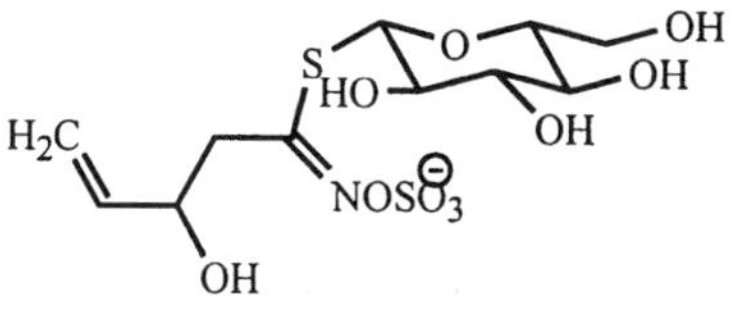

Progoitrin

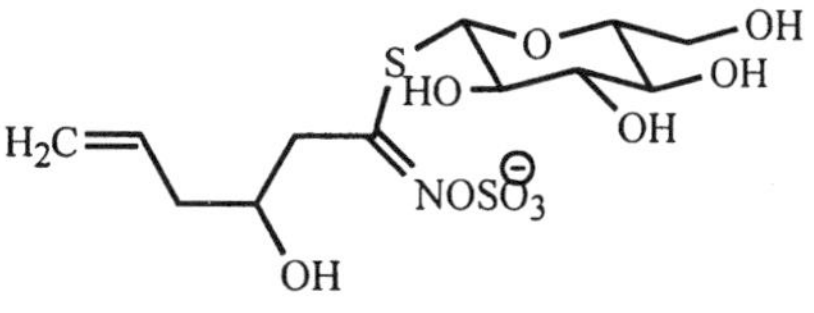

Napoleoferin

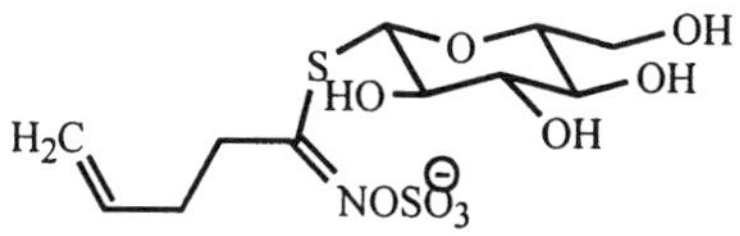

Gluconapin

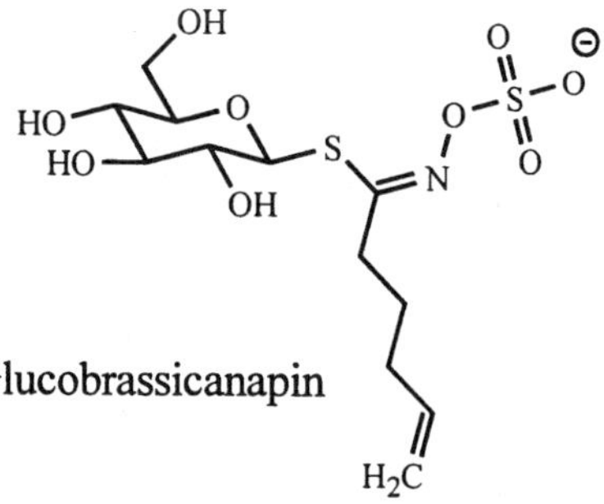

Glucobrassicanapin

10.4   PHARMACEUTICAL ANALYTES

The antibiotic lincomycin was assayed for levels of lincomycin B and 7-epilinco-mycin using a 17-min separation. Good peak shapes and baseline resolution resulted when a 92.5/5/2.25/0.067 water/water (2.72% m/v potassium dihydrogenphosphate with 3.48 m/v dipotassium hydrogenphosphate)/acetonitrile/methanesulfo-methanesulfonic acid mobile phase was used in conjunction with a 50°C C_{18} column ($\lambda = 210$ nm). The authors point out that the column must be base deactivated. The effect of changing the percent acetonitrile and methanesulfonic acid on k' is presented. A linear range of 1.2–2.8 mg/mL was reported [1599].

Zidovudine (AZT, an HIV antiviral), and zidovudine monophosphate were extracted from plasma and analyzed on a C_{18} column ($\lambda = 267$ nm) using a 97.1/2.9 water (0.2 M NaH_2PO_4 with 8 mM tetrabutylammonium hydrogen sulfate pH 7.5) mobile phase. Whereas the phosphate was not well-resolved from plasma extractables, the zidovudine was. Elution was complete in 15 min. Standard concentrations ranged from 0.125 to 5 µg/mL and detection limits of 0.125 µg/mL were reported [1600].

Ganciclovir and acyclovir as internal standard were separated on a 40°C C_{18} column ($\lambda = 254$ nm) using a 98/2 water (50 mM ammonium acetate pH 6.5)/acetonitrile mobile phase [1601]. Elution was complete in <8 min. A linear range of 50–10,000 ng/mL along with a detection limit of 10 ng/mL (S/N = 3) and quantitation limit of 50 ng/mL (S/N = 10) were reported.

Valaciclovir is metabolized to acyclovir in the liver or intestine. Both were isolated from serum and analyzed on a C_8 column ($\lambda = 254$ nm) using a 98/2 water (25 mM $[NH_4]H_2PO_4$ to pH 4.0 with H_3PO_4)/acetonitrile mobile phase [1602]. Elution was complete in 13 min, baseline separation was obtained, and peaks were sharp. Standards were run from 0.5 to 20 µg/mL and detection limits of 50 ng/mL were reported.

The antineoplastic agent carboplatin was isolated by Burns and Embree [1603] from plasma and analyzed directly or directly injected with postcolumn derivatization using a C_{18} column ($\lambda = 230$ nm direct and 290 nm postcolumn). The advantage to the postcolumn method is that no sample extraction step is needed. Elution was <15 min using an aqueous 20 mM NaH_2PO_4 (pH 5.4) mobile phase.

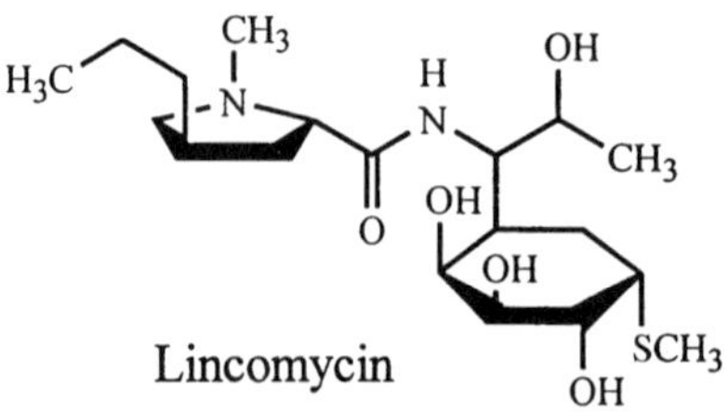

Lincomycin

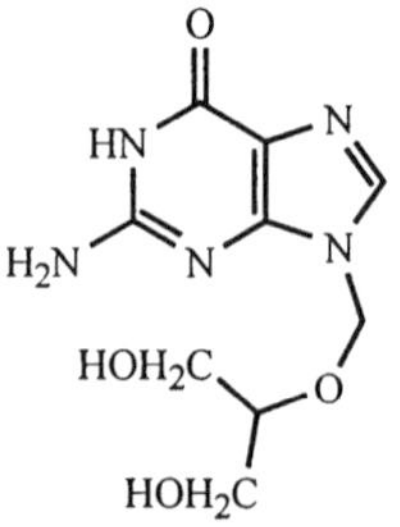

Ganciclovir

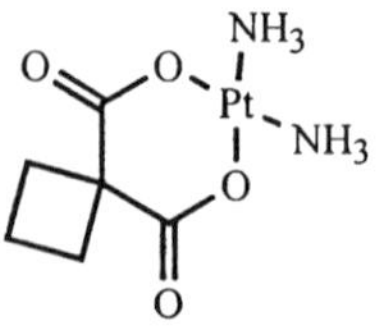

Carboplatin

This buffer was also used with 40 mM sodium bisulfite for the post column reaction solution. The concentration range was 0.05–40 μg/mL with detection limits of 0.025 μg/mL. One potential disadvantage to the direct injection analysis is the long-term buildup of uneluted plasma constituent (water is a very weak solvent).

Muscimol and ibotenic acid are hallucinogens found in *Amanita muscaria*. They were extracted and separated on a C_{18} column ($\lambda = 230$ nm) using an aqueous 5 mM octylammonium ortho-phosphate pH 6.4 mobile phase [1604]. Peaks were some-what tailed but were completely resolved and eluted in 10 min. A linear range of 0.05–1 mg/L was reported with a detection limit of 30 μg/L.

Rosado-Maria et al. [1605] developed an optimal separation of a set of 19 diuretics (althiazide, amiloride, bendro- and hydroflumethiazide, benz- and polythiazide, bumetanide, canrenone, chloralidone, clopamide, dichlophena-mide, ethacrynic acid, furosemide, indapamide, piretanide, spironolactone,

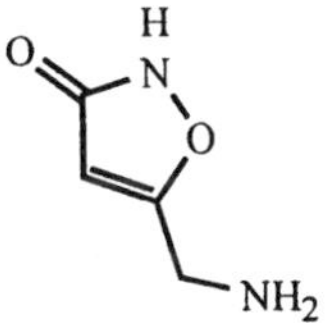

Muscimol

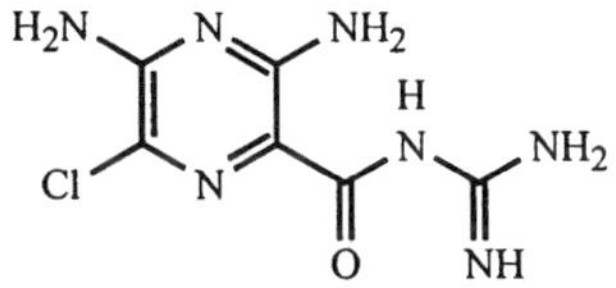

Amiloride

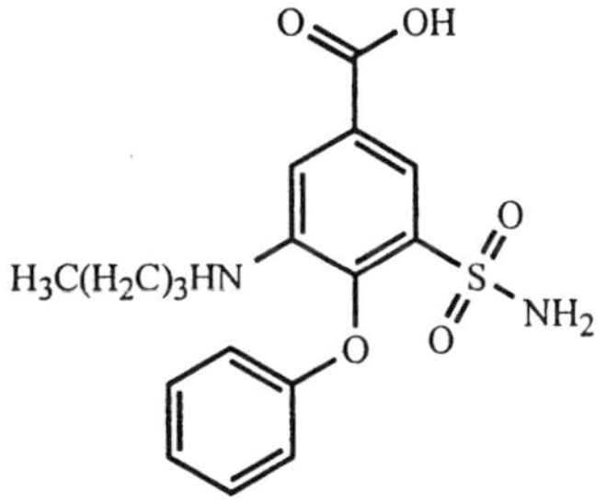

Bumetanide

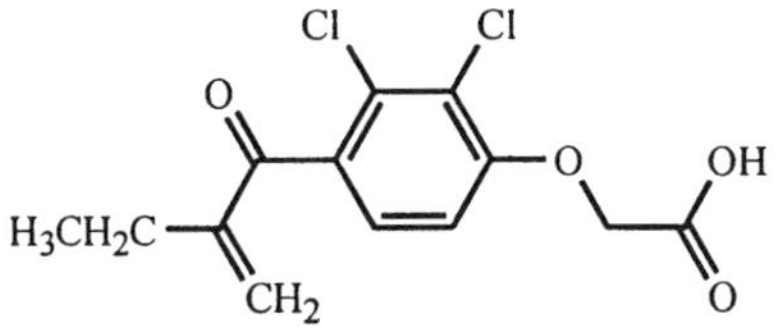

Ethacrynic acid

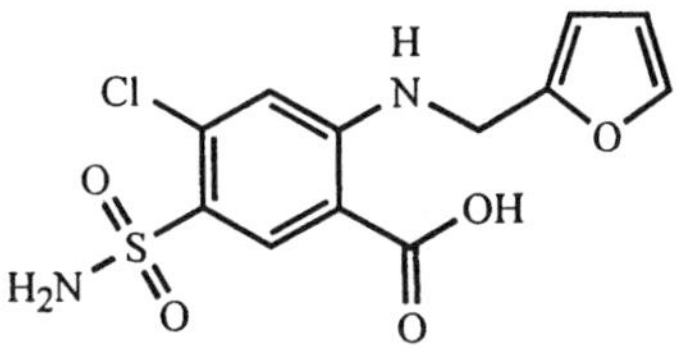

Furosemide

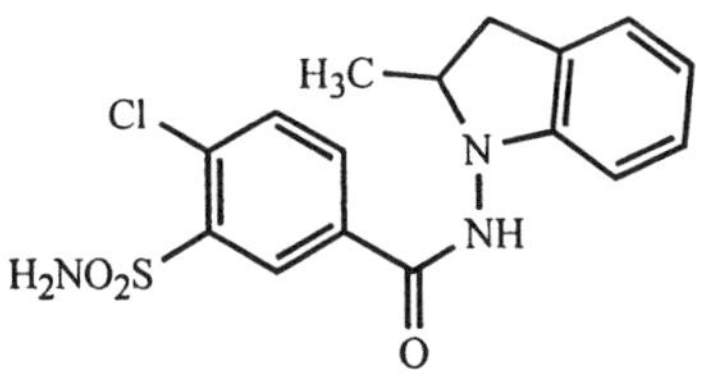

Indapamide

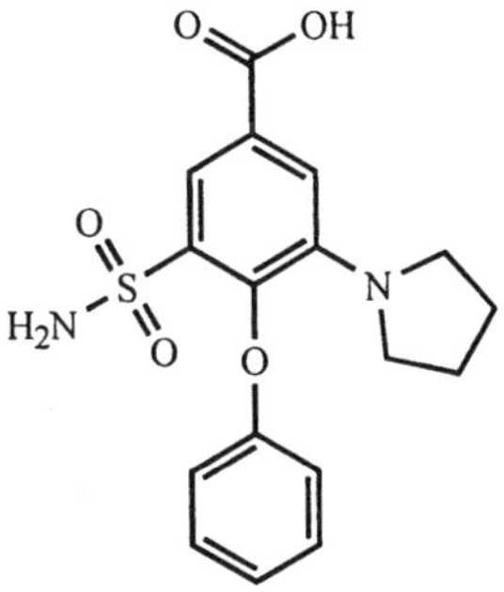

Piretanide

triamterene, trichlormethiazide, xipamide) using a 50°C C_{18} column ($\lambda = 220$ nm). Although complete baseline resolution of all 19 compounds was not achieved, the authors listed the optimal conditions for a number of mobile phases along with retention times for all analytes. The chromatography shown was for a 4/96 THF/water (40 mM sodium dodecyl sulfate [SDS] pH 3.2 with H_3PO_4) mobile phase. Elution was complete in 35 min. Other organic mobile phase modifiers and their optimal levels (the aqueous SDS solvent level adjusted accordingly) were 2% *n*-propyl alcohol, 1.5% *n*-butyl alcohol, 0.3% *n*-pentyl alcohol, and 16% acetonitrile). The effect of changing SDS concentration on k' was plotted. Detection limits were reported as 1–40 ng/mL.

Cisplatin and two hydrolysis products (*cis*-$[Pt(NH_3)_2Cl(H_2O)^+]$, *cis*-$[Pt(NH_3)_2(H_2O)_2{}^{2+}]$) were resolved on a 37°C C_{18} column ($\lambda = 305$ nm) using a 97/3 (0.5 mM sodium dodecyl sulfate to pH 2.5 with triflic acid)/methanol mobile phase [1606]. Elution was complete in 20 min. The reported linear range was 0.05–1.5 mM platinum.

Dihydrocodeine and five metabolites were extracted from blood samples and separated on a C_{18} column ($\lambda = 220$ nm, ex; 340 nm, em). A 96/4 water (25 mM tetraethylammonium phosphate)/acetonitrile mobile phase generated baseline resolution in 35 min [1607]. Concentration ranges were generated from 20 to 2000 ng/mL with detection limits of 15–25 ng/mL (analyte dependent).

Stavudine and its 2′,3′-didehydro-3′-deoxythymidine metabolite were isolated from plasma and urine and analyzed on a C_{18} column ($\lambda = 266$ nm) using a 97/3/1 water (10 mM KH_2PO_4 to pH 2.5 with H_3PO_4)/acetonitrile/triethylamine mobile phase [1608]. Peaks eluted before 10 min and were well resolved. Standards from 0.025 to 25 μg/mL (plasma) and 2 to 150 μg/mL (urine) were utilized. A limit of detection of 12 ng/mL was reported.

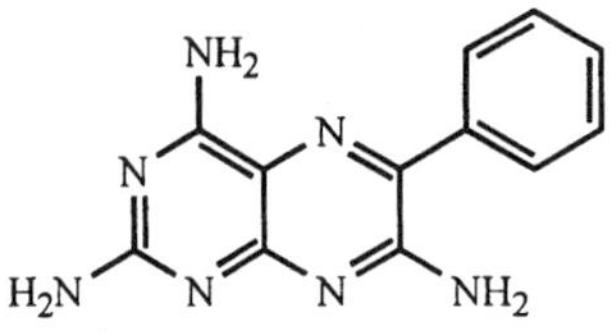

Triamterene

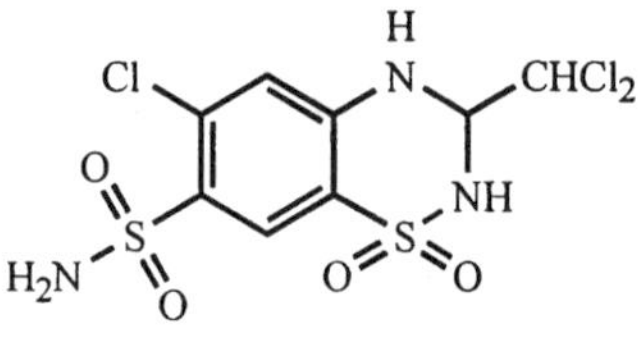

Trichlormethiazide

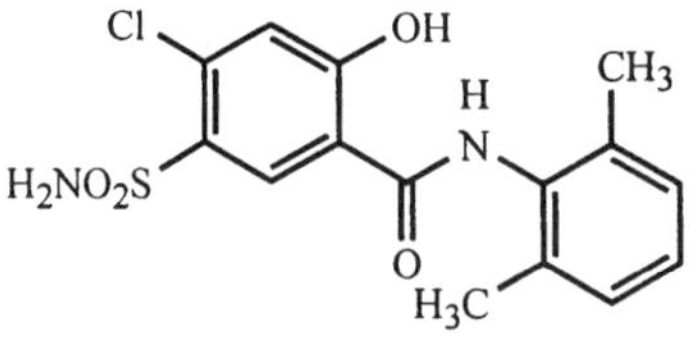

Xipamide

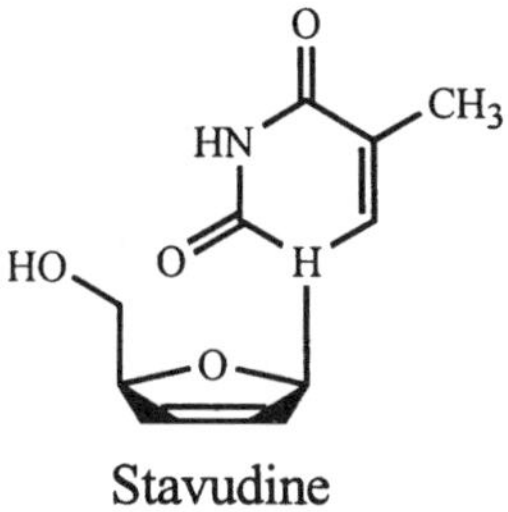

Stavudine

Pharmaceutical formulation mixtures of salicylic acid [SA], caffeine [C], butal-bital [B], and acetaminophen [A] were easily analyzed on a nonporous C_{18} column ($\lambda = 220$ nm). Elution was complete in 3 min and excellent peak shapes with baseline resolution were obtained for both of the following separations. SA, C, and B were separated using a 98/2 water (50 mM KH_2PO_4 at pH 3.0)/acetonitrile mobile phase, whereas A, C, and B were resolved using a 97/3 water (50 mM KH_2PO_4 at pH 2.5)/acetonitrile mobile phase [1609]. Linear ranges of 40–6500 ng/mL were reported along with detection limits of 5–35 ng/mL (analyte dependent).

The antitumor drug 5-fluorouracil was extracted from plasma and separated from uracil and bromouracil (IS) on a 25°C C_{18} column ($\lambda = 260$ nm). An aqueous 10 mM KH_2PO_4 to pH 3.0 with H_3PO_4 mobile phase generated a good separation with elution of 5-bromouracil at 15 min [1610]. The effects of temperature and mobile phase ionic strength on resolution were tabulated. The linear working range was reported to cover 12.5–10,000 µg/L, with detection limits of 3 µg/L.

Table 10.5 lists some promulgated USP methods [590].

10.5 SUMMARY

Water, although rarely used alone, is the single most important LC solvent. Not only does it play a critical role in NP separations, where it is an extremely strong solvent that can readily deactivate the chromatographic surface, but it is a ubiquitous component in almost every RP separation. As seen above, water, with an appropriate buffer and often an ion-pair agent, can also be used effectively in the separation of highly polar compounds. These separations are typically done in the RP mode so that water is an extremely weak solvent. These separations are not typically accomplished using any organic mobile phase modifier, since retention would be too short and all peaks would co-elute or elute in the void volume.

10.6 DIMETHYL SULFOXIDE

The use of DMSO as a mobile phase constituent is not very common. It is an extremely strong solvent and very viscous and has a high UV cutoff. All these factors, coupled with safe use issues argue against its routine use. This said, as discussed previously, DMSO has been used as a low-level mobile phase component in the separation of PAHs [746] and polystyrene isomers [857]. In addition, a small number of other methods have utilized DMSO.

Smets et al. [1611] analyzed chlorotetracycline and four synthesis by-products on a C_8 column using a DMSO/water (1 M $HClO_4$) mobile phase. The level of DMSO used was dependent on the column chosen: 19% for a Zorbax, 40% for a Nucleosil, and 50% for a Spherisorb. Good peak shapes and resolution were obtained. Elution was complete in 20 min. Swedberg et al. [1612] used DMSO to regenerate protein-contaminated C_4 columns. A 20-min 12/88 → 80/20 DMSO/water gradient was used.

TABLE 10.5 USP Methods[a]

Analyte(s)	Co-analyte (Internal Standard)	Column	Mobile Phase	Detector (nm)	USP Page
p-Arsinilic acid	o-Arsinilic acid	C_{18}	985/10 water (4.04 g KH_2PO_4 with 2 mL H_3PO_4)/methanol	242	159
Dacarbazine (limit of 2-azahypoxanthine)		C_{18}	2.2 g docusate sodium with 15 mL acetic acid to 1 L with water	254	493
Hydroxyurea	Uracil	C_{18}	water	214	844
Imipenem		(30°C) C_{18}	135 mg KH_2PO_4 in 1 L water adjusted to pH 6.8	300	863
Ioversol		(35°C) C_8	99.5/0.5 water/acetonitrile	254	909
Isoetharine hydrochloride		C_{18}	0.17N acetic acid in water	278	919
Meprobamate (limit of methyl carbamate)		C_{18}	water	200	1043
Methyldopa and hydrochlorothiazide		C_{18}	95/5 water (11.6 g NaH_2PO_4/L to pH 2.8 with H_3PO_4)/methanol	270	1083
Thiacetarsamide		C_{18}	97/3 water (50 mM KH_2PO_4, 50 mM K_2HPO_4 to pH 7)/acetonitrile	232	1638
Trifluridine	5-(Trifluoromethyl)uracil	C_{18}	0.15% sodium citrate in water to pH 6.8	254	1707

[a] Reference [590].

The levels of the photoisomers of bilirubin were determined in neonatal plasma extracts using a C_{18} column ($\lambda = 455$ nm) and a 32-min 80/20 → 30/70 (60/40 DMSO/water [0.1 M sodium acetate buffer at pH 4.95])/acetonitrile gradient [1613]. Good peak shapes and resolution were obtained. All peaks were identified.

In summary, due to its high UV cutoff and viscosity, DMSO has found very limited use in HPLC. One area where it could be more heavily utilized is in the separation of proteins. DMSO is one of the few solvents that does not denature proteins, even at high concentrations. Unfortunately, the very high boiling point of DMSO also precludes work where sample is to be collected from the eluant after elution.

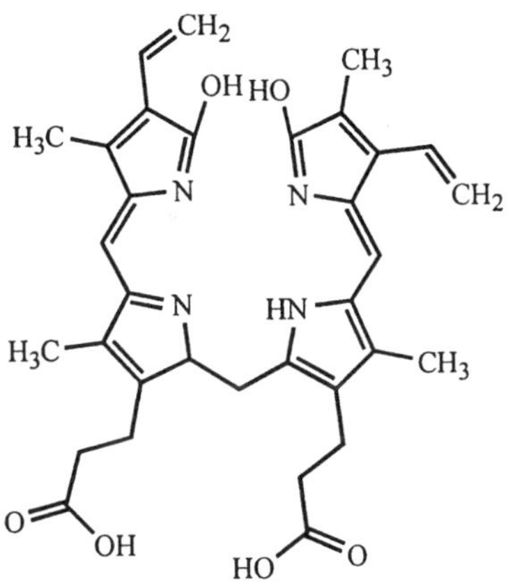

Bilirubin

REFERENCES

1. U.D. Neue, *HPLC Columns: Theory, Technology, and Practice*, Wiley-VCH, New York, 1997.

2. C.A Doyle and J.G. Dorsey, in E. Katz, R. Eksteen, P. Schoenmakers, and N. Miller (eds.), *Handbook of HPLC*, Chromatographic Science Series Volume 78, Marcel Dekker, Inc., New York, 1998, Chapter 8.

3. P.W. Carr D.E. Martire, and L.R. Snyder (eds.), *J. Chromatogr. A* **656** (1993).

4. J.G. Dorsey and W.T. Cooper, *Anal. Chem.* **66**, 857A (1994).

5. L.R. Snyder, J.J. Kirkland, and J.L. Glajch, *Practical Method Development*, 2nd edition, Wiley, New York, 1997, Chapter 6.

6. J. Calvin Giddings, *Unified Separation Science*, Wiley, New York, 1991.

7. W.R. Melander and Cs. Horváth, in Cs. Horváth (ed.), *HPLC: Advances and Perspectives*, Volume 2, Academic Press, New York, 1980, pp. 113–319.

8. T. Fornstedt and D. Westerlund, *J Chromatogr.* **648**, 315 (1993).

9. S. Levin and F. Grushka, *Anal. Chem.* **59**, 1157 (1987).

10. C.W. Hsu and W.T. Cooper, *J. Chromatogr.* **603**, 63 (1992).

11. S. Golshan-Shirazi and G. Guiochon, *Anal. Chem.* **62**, 923 (1990).

12. J.N. Brown, M. Hewins, J.H.M. van der Linden, and R.J. Lynch, *J. Chromatogr.* **204**, 115 (1981).

13. J. Doehl, *J. Chromatogr. Sci.* **2**, 7 (1988).

14. J.W. Dolan and L.R. Snyder, *Troubleshooting LC Systems*, Humana Press, Totowa, NJ, pp. 149–152.

15. R.C. Simpson, in P.R. Brown and R.A. Hartwick (eds.), *High Performance Liquid Chromatography*. Wiley, New York, 1989, p. 384.

16. U.D. Neue, *HPLC Columns: Theory Technology, and Practice*, Wiley-VCH, New York, 1997, p. 30.

17. J.H. Knox, *Chromatogr. Newsletter* **2**, 1 (1973).

18. J.H. Knox, *J. Chromatogr. Sci.* **15**, 352 (1977).

19. C. Carr and J.A. Riddick, *Ind. Eng. Chem.* **43**, 692 (1951).

20. Sj. van der Wal, *Chromatographia* **20**, 274 (1985).

21. W. Hayduk, H. Laudie, and O.H. Smith, *J. Chem. Eng. Data* **18**, 373 (1973).

22. J. Timmermans, *The Physico-Chemical Constants of Binary Systems in Binary Solution*, Volume 4, Wiley, New York, 1960.

23. L.R. Snyder and J.J. Kirkland, *Introduction to Modern Liquid Chromatography*, Wiley, New York, 1979, Chapter 6.

24. R.S. Higgins and S.A. Klinger (eds.), *High Purity Solvent Guide*, 3rd edition, Burdick & Jackson, Muskegon, MI, 1990, p. 197.

25. S.R. Bakalyar, M.P.T. Bradley, and R. Honganen, *J. Chromatogr.* **158**, 277 (1978).

26. P.A. Carson and C.J. Mumford, *The Safe Handling of Chemicals in Industry*, Wiley, New York, 1996.

27. S.G. Luxon (ed.), *Hazards in the Chemical Laboratory*, 5th edition, Royal Society of Chemistry, Cambridge, UK, 1992.

28. I.M. Kolthoff, E.B. Sandell, E.J. Meehan, and S. Bruckenstein, *Quantitative Analytical Analysis*, 4th edition, Macmillan, New York, 1969, p. 854.

29. National Safety Council, Publication Data Sheet I-655, Rev. 87, Chicago, IL, 1987.

30. *Reagent Chemicals*, 9th edition, American Chemical Society, Washington, DC, 1998, p. 296.

31. K. Benedek, S. Dong, and B.L. Karger, *J. Chromatogr.* **317**, 277 (1984).

32. K.A. Cohen, K. Schellenberg, K. Benedek, B.L. Karger, B. Grego, and M.T.W. Hearn, *Anal. Biochem.* **140**, 223 (1984).

33. Gy. Vigh, Z. Varga-Puchony, J. Hlavay, M. Petró-Turcza, and I. Szárföldt-Szalma, *J. Chromatogr.* **193**, 432 (1980).

34. H. Iwase, *J. Chromatogr. A* **881**, 317 (2000).

35. C. Crezcenzi, A. Di Corcia, M.D. Madbouly, and R. Semperi, *Environ. Sci. Technol.* **29**, 2185 (1995).

36. K.D. Nugent, W.G. Burton, T.K. Slattery, B.F. Johnson, and L.R. Snyder, *J. Chromatogr.* **443**, 381 (1988).

37. H.A. Wittcoff and B.G. Reuben, *Industrial Organic Chemicals in Perspective*, Part I, Wiley, New York, 1980, Chapter 2.

38. I.S. Krull, Z. Deyl, and H. Lingeman, *J. Chromatogr. B* **659**, 1 (1994).

39. M. Arninuddin and J.N. Miller, *Talanta* **42**, 775 (1995).

40. H. Lingeman, W.J.M. Underberg, A. Takadate, and A. Hulshoff, *J. Liq. Chromatogr.* **8**, 789 (1985).

41. *Reagent Chemicals*, 9th edition, American Chemical Society, Washington, DC, 1998.

42. B.L. Karger, L.R. Snyder and Cs. Horváth, *An Introduction to Separation Science*, Wiley, New York, 1973, Chapter 13.

43. K.K. Unger and U. Trüdinger, in P.R. Brown and R.A. Hartwick (eds.), *High Performance Liquid Chromatography*, Wiley, New York, 1989, Chapter 3.

44. D.C. Luehrs, D.J. Chesney, and K.A. Godbole, *J. Chromatogr.* **627**, 37 (1992).

45. R.S. Higgins and S.A. Klinger (eds.), *High Purity Solvent Guide*, 3rd edition, Burdick & Jackson, Muskegon, MI, 1990, pp. 182, 185.

46. H. Engelhardt and H. Elgass, in Cs. Horváth (ed.), *HPLC: Advances and Perspectives.* Volume 2, Academic Press, New York, 1980, p. 57.

47. K. Karch, I. Sebestian, I. Halász, and H. Engelhardt, *J. Chromatogr.* **122**, 171 (1976).

48. R.A. Keller and L.R. Snyder, *J. Chromatogr. Sci.* **9**, 346 (1971).

49. L.R. Snyder, *Principles of Adsorption Chromatography*, Marcel Dekker, New York, 1968.

50. M.D. Palamareva and H.E. Palamarev, *J. Chromatogr.* **477**, 235 (1989).

51. L.R. Snyder, *J. Chromatogr.* **15**, 63 (1971).

52. L.R. Snyder, M.A. Quarry, and J.L. Glajch, *Chromatographia* **24**, 33 (1984).

53. L.R. Snyder, *J. Chromatogr.* **92**, 223 (1974).

54. L.R. Snyder, *J. Chromatogr. Sci.* **16**, 223 (1978).

55. D.C. Leuhrs, D.J. Chesney, and K.A. Godbole, *J. Chromatogr. Sci.* **29**, 463 (1991).

56. L.R. Snyder, P.W. Carr, and S.C. Rutan, *J. Chromatogr. A* **656**, 537 (1993).

57. M.J. Kamlet, J.-L.M. Abboud, M.H. Abraham, and R.W. Taft, *J. Org. Chem.* **48**, 2877 (1983).

58. M. Chastrette, M. Rajzmann, M. Chanon, and K.F. Purcell, *J Am. Chem. Soc.* **107**, 1 (1985).

59. P.J. Schoenmakers and T. Blaffert, *J. Chromatogr.* **384**, 117 (1987).

60. P.M.J. Coenegracht., A.K. Smilde, H. Benak, C.H.P. Bruins, H.J. Metting, H. de Vries, and D.A. Doornbos, *J. Chromatogr.* **550**, 397 (1991).

61. H.B. Patel and T.M. Jefferies, *J. Chromatogr.* **389**, 21 (1987).

62. P.J. Schoenmakers, H.A.H. Billiet, and L. de Galan, *J. Chromatogr.* **205**, 13 (1981).

63. P.J. Schoenmakers, A. Bartha, and H.A.H. Billiet, *J. Chromatogr.* **550**, 425 (1991).

64. P.E. Kavanaugh, *J Liq. Chromatogr.* **23**, 1477 (2000).

65. S. Wieliński and A. Olszanowski, *J. Liq. Chromatogr.* **22**, 3115 (1999).

66. S.N. Deming and M.L.H. Turoff, *Anal. Chem.* **50**, 546 (1978).

67. W.P. Price, Jr. and S.N. Deming, *Anal. Chim. Acta* **108**, 227 (1979).

68. John A. Dean (ed.), *Lange's Handbook of Chemistry*, McGraw-Hill. New York, 1985, Table 5-8.

69. P.J. Schoenmakers, H.A.H. Billiet, and L. de Galan, *Chromatographia* **15**, 205 (1982).

70. R. Tijssen, H.A.H. Billiet, and P.J. Schoenmakers, *J. Chromatogr.* **122**, 185 (1976).

71. C. Hansch and A. Leo, *Substituent Constants for Correlation Analysis in Chemistry and Biology*, Wiley, New York, 1979.

72. R. Kaliszan, K. Osrnialowski, S.A. Tomellini, S.-H. Hsu, S.D. Fazio, and R.A. Hartwick, *J. Chromatogr.* **352**, 141 (1986).

73. H. Terada, *Quant. Struct.-Act. Relat.* **5**, 81 (1986).

74. R.E. Koopmans and R.F. Rekker, *J. Chromatogr.* **285**, 271 (1984).

75. R.F. Rekker and H.M. de Kort, *Eur. J. Med Chem.* **6**, 479 (1979).

76. L.R. Snyder, J.J. Kirkland, and J.L. Glajch, *Practical HPLC Method Development*, 2nd edition, Wiley. New York, 1997.

77. V.R. Meyer, *Practical High Performance Liquid Chromatography*, 2nd edition, Wiley, New York, 1994.

78. S. Ahuja, *Selectivity and Detectability in HPLC*, Wiley, New York, 1989.

79. B.A. Bidlingmeyer, *Practical HPLC Methodology and Applications*, Wiley, New York, 1992.

80. M.E. Schwartz and I.S. Krull, *Analytical Method Development and Validation*, Marcel-Dekker, New York, 1997.

81. JR. Taylor, *Introduction to Error Analysis*, Oxford University Press, New York, 1982, pp. 178–184.

82. G. Kateman and L. Buydens, *Quality Control in Analytical Chemistry*, 2nd edition, Wiley, New York, 1994, pp. 118–125.

83. G. Wernimont, in W. Spendley (ed.), *Use of Statistics to Develop and Evaluate Analytical Methods*, AOAC, Arlington, VA, 1975, pp. 78–82.

84. I. Mellan. *Industrial Solvents Handbook*, 2nd edition, Noyes Data Corp., Park Ridge, NJ, 1977.

85. K.L. Hoy, *J. Paint Technol.* **42**, 79 (1970).

86. A.F.M. Barton, *Chem. Rev.* **75**, 731 (1975).

87. M. Windholz (ed.), *The Merck Index*, 11th edition, Merck & Co., Rahway, NJ, 1989.

88. R.S. Higgins and S.A. Klinger (eds.), *High Purity Solvent Guide*, Burdick & Jackson Muskegon, MI; 2nd edition, 1984, & 3rd edition, 1990.

89. R.C. Weast (ed.), *CRC Handbook of Chemistry and Physics*, 61st edition, CRC Press, Boca Raton, FL, 1980.

90. T.J. Bruno, *Chromatographic and Electrophoretic Methods*, Prentice Hall, Englewood Cliffs, NJ, 1991, Chapter 3.

91. D.R. Lide, *Handbook of Organic Solvents*, CRC Press, Boca Raton, FL, 1995.

92. National Fire Protection Agency, *Fire Protection Guide to Hazardous Materials*, 11th edition, Quincy, MA, 1994.

93. Y. Marcus and S. Glikberg, *Pure Appl. Chem.* **57**, 855 (1985).

94. V. Sedivec and J. Flek, *Handbook of Analysis of Organic Solvents*, Ellis Harwood, New York, 1976.

95. V. Yu. Zelvensky, A.S. Lavrenova, S.I. Samolyuk, L.V. Borodai, and G.A. Egorenko, *J. Chromatogr.* **364**, 305 (1986).

96. A. Brega, P. Prandini, C. Amaglio, and E. Pafumi, *J. Chromatogr.* **535**, 311 (1990).

97. V. Coquart and M.-C. Hennion, *J. Chromatogr.* **600**, 195 (1992).

98. O. Busto, J.C. Olucha, and F. Borrull, *Chromatographia* **32**, 423 (1991).

99. O. Busto, J.C. Olucha, and F. Borrull, *Chromatographia* **32**, 566 (1991).

100. A.I. Saiz, G.D. Manrique, and R. Fritz, *J. Agric. Food Chem.* **49**, 98 (2001).

101. P. Chen and M. Zhang, *J. Chromatogr. A* **773**, 365 (1997).

102. K. Othmen and P. Boule, *J. Photochem. Photobiol. A: Chem.* **136**, 79 (2000).

103. B. Li, Y. Yang, Y. Gan, C.D. Eaton, P. He, and A.D. Jones, *J. Chromatogr. A* **873**, 175 (2000).

104. L. Lunar, D. Sicilia, S. Rubio, D. Pérez-Bendito, and U. Nickel. *Water Res.* **34**, 3400 (2000).

105. C. Baiocchi, M.A. Roggero, D. Giacosa, and E. Marengo, *J. Chromatogr. Sci.* **33**, 338 (1995).

106. J.D. Baty and S. Sharp, *J. Chromatogr.* **437**, 13 (1988).

107. E. Bosch, P. Bou, and M. Rosés, *Anal. Chim. Acta* **299**, 219 (1994).

108. J.H. Knox, J. Kriz, and E. Adamcová, *J. Chromatogr.* **447**, 13 (1988).

109. J. Kríz, E. Adamcová, J.H. Knox, and J. Hora, *J. Chromatogr. A* **663**, 151 (1994).

110. D.W. Armstrong, W. DeMond, A. Alak, W.L. Hinze, T.E. Riehl, and K.H. Bui, *Anal. Chem.* **57**, 234 (1985).

111. H. Nohta, F. Sakai, M. Kai, Y. Ohkura, and M. Saito, *Anal. Chim. Acta* **287**, 223 (1994).

112. A. Siwek and J. Sliwiok, *J. Chromatogr.* **506**, 109 (1990).

113. W.T. Kelley, D.L. Coffey, and T.C. Mueller, *J Assoc. Off. Anal. Chem. Int.* **77**, 805 (1994).

114. R. Kimura, N. Ohishi, Y. Kato, S. Yamada, and M. Sato, *Drug Metab. Dispos.* **20**, 161 (1992).

115. S.C. Chua, B.L. Lee, L.S. Liau, and C.N. Ong, *J Anal. Toxicol.* **17**, 129 (1993).

116. A. Opperhuizen, T.L. Sinnige, J.M.D. van der Steen, and O. Hutzinger, *J. Chromatogr.* **388**, 51 (1987).

117. T.H. Dzido and H. Engelhardt, *Chromatographia* **39**, 51 (1994).

118. J. Andrasko, *J. Forensic Sci.* **37**, 1030 (1992).

119. A. Bergens, *Talanta* **42**, 185 (1995).

120. P.J. Schoenmakers and A.C.J.H. Drouen, *Chromatographia* **15**, 688 (1982).

121. B. Markus and C.-H. Kwon, *J. Pharm. Sci.* **83**, 1729 (1994).

122. G. Taibi and M.R. Schiavo, *J. Chromatogr.* **614**, 153 (1993).

123. E. Schenkel, V. Berlaimont, J. Dubois, M. Helson-Cambier, and M. Hanocq, *J. Chromatogr. B* **668**, 189 (1995).

124. D.V. McCalley, *J. Chromatogr. A* **708**, 185 (1995).

125. B.S. Kersten, T. Catalano, and Y. Rozenman, *J. Chromatogr.* **588**, 187 (1991).

126. R.M. Smith and C.M. Burr, *J. Chromatogr.* **550**, 335 (1991).

127. C.H. Collins and MA. Morgano, *J. Chromatogr. A* **846**, 395 (1999).

128. R. Nakajima, K. Shimada, Y. Fujii, A. Yamamoto, and T. Hara, *Bull. Chem. Soc. Jpn.* **64**, 3173 (1991).

129. R. Loos, M.C. Alonso, and D. Barceló, *J. Chromatogr. A* **890**, 225 (2000).

130. Z. Zhang, S.A. Kline, T.A. Kirley, B.D. Goldstein, and G. Witz, *Arch. Toxicol.* **67**, 461 (1993).

131. M.T. Benassi and H.M. Cecchi, *J. Liq. Chromatogr.* **21**, 491 (1998).

132. O. Kirk, T. Damhus, and M.W. Christensen, *J. Chromatogr.* **606**, 49 (1992).

133. M.X. Coutrim, L.A. Nakamura, and C.H. Collins, *Chromatographia* **37**, 185 (1993).

134. A.C. Geng, Z.L. Chen, and G.G. Siu, *Anal. Chim. Acta* **257**, 99 (1992).

135. M. Katayama, Y. Masuda, and H. Taniguchi, *J. Chromatogr.* **585**, 219 (1991).

136. G. Gutnikov and J.R. Streng, *J. Chromatogr.* **587**, 292 (1991).

137. H. König and W. Strobel, *Fresenius Z. Anal. Chem.* **338**, 728 (1990).

138. E.R.J. Wils and A.G. Hulst, *J. Chromatogr.* **454**, 261 (1988).

139. E. Munaf, T. Takeuchi, and T. Miwa, *Anal. Chim. Acta* **418**, 175 (2000).

140. T. Tanimoto, T. Sakaki, T. Iwanaga, and K. Koizumi, *Chem. Pharm. Bull.* **42**, 385 (1994).

141. U. Marquardt, H.J. Möckel, and S.F. Nelsen, *Chromatographia* **27**, 113 (1989).

142. I.W. Wainer, *Trends Anal. Chem.* **6**, 125 (1987).

143. M.R. Hadley, L.A. Damani, A.J. Hutt, HG. Oldham, J. Murphy, and P. Camilleri, *Chromatographia* **37**, 487 (1993).

144. G. Carrea, P. Pasta, S. Colonna, and N. Gaggero, *J. Chromatogr.* **600**, 320 (1992).

145. C. Villani and W.H. Pirkle, *J. Chromatogr. A* **693**, 63 (1995).

146. K. Cammann, M. Robecke, and J. Bettmer, *Fresenius J. Anal. Chem.* **350**, 30 (1994).

147. M. Hempel, H. Hintelmann, and R.-D. Wilken, *Analyst* **117**, 669 (1992).

148. E. Munaf, T. Takeuchi, and T. Miwa, *Anal. Chim. Acta* **418**, 175 (2000).

149. T. Yasui, A. Yuchi, H. Wada, and G. Nakagawa, *J. Chromatogr.* **596**, 73 (1992).

150. S. Dilli and P. Tong, *Anal. Chim. Acta*, **395**, 101 (1999).

151. S. Ichinoki, N. Hongo, and M. Yamazaki, *Anal. Chem.* **60**, 2099 (1988).

152. K. Saitoh, Y. Shibata, and N. Suzuki, *J. Chromatogr.* **542**, 351 (1991).

153. S. Tsukahara, K. Saitoh, and N. Suzuki, *J. Chromatogr.* **634**, 138 (1993).

154. L. Wuping and L. Qiping, *Fresenius J. Anal. Chem.* **350**, 671 (1994).

155. H.-S. Zhang, W.-Y. Mou, and J.-K. Cheng, *Talanta* **41**, 1459 (1994).

156. M.A. Rodriguez Delgado, M.J. Sánchez, V. González, and F. García Montelongo, *Anal. Chim. Acta* **298**, 423 (1994).

157. J.C. Fetzer and W.R. Biggs, *J. Chromatogr.* **386**, 87 (1987).

158. V.C. Anigbogu, A. Muñoz de la Peña, T.T. Ndou, and I.M. Warner, *J. Chromatogr.* **594**, 37 (1992).

159. J.M. Schuette and I.M. Warner, *Talanta* **41**, 647 (1994).

160. G. K.-C. Low, G.E. Batley, and C.I. Brockbank, *J. Chromatogr.* **392**, 199 (1987).

161. G.W. Somsen, L.P.P. van Stee, C. Gooijer, U.A.Th. Brinkman, N.H. Velthorst, and T. Visser, *Anal. Chim. Acta* **290**, 269 (1994).

162. S.-y. Xie, R.-b. Huang, and L.-s. Zheng, *J. Chromatogr. A* **864**, 173 (1999).

163. B.W. Day, Y. Sahali, D.A. Hutchins, M. Wildschütte, R. Pastorelli, T.T. Nguyen, S. Naylor, P.L. Skipper, J.S. Wishnok, and SR. Tannenbaum, *Chem. Res. Toxicol.* **5**, 779 (1992).

164. Y. Sahali, H. Kwon, P.L Skipper, and S.R. Tannenbaum, *Chem. Res. Toxicol.* **5**, 157 (1992).

165. B. Misra, S. Amin, and S.S. Hecht, *Chem. Res. Toxicol.* **5**, 242 (1992).

166. R.V. Nair, A.N. Nettikumara, C. Cortez, R.G. Harvey, and J. DiGiovanni, *Chem. Res. Toxicol.* **5**, 532 (1992).

167. K.L. Platt and M. Schollmeier, *Chem. Res. Toxicol.* **7**, 89 (1994).

168. S. Hegstad, E. Lundanes, J.A. Holme, and J. Alexander, *Xenobiotica* **29**, 1257 (1999).

169. X.R. Michel, C. Beasse, and J.-F. Narbonne, *Arch. Environ. Contam. Toxicol.* **28**, 215 (1995).

170. J.W. Flesher and S.R. Myers, *Drug Metab. Dispos.* **18**, 163 (1990).

171. S.J. Marshman, *Fuel* **70**, 967 (1991).

172. S.W. Tjioe and R.J. Hurtubise, *Talanta* **42**, 59 (1995).

173. E.H.J.M. Jansen, RH. van den Berg, and E. Dinant Kroese, *Anal. Chim. Acta* **290**, 86 (1994).

174. M. Rozbeh and RI. Hurtubise, *J. Liq. Chromatogr.* **18**, 1909 (1995).

175. J.-S. Lai, S.S. Hung, L.E. Unruh, H. Jung, and P.P. Fu, *J. Chromatogr.* **461**, 327 (1989).

176. P.P. Fu, Y. Zhang, Y.-L. Mao, L.S. Von Tungeln, Y. Kim, H. Jung, and M.-J. Jun, *J. Chromatogr.* **642**, 107 (1993).

177. A. Robbat, Jr. and T.-Y. Liu, *J. Chromatogr.* **513**, 117 (1990).

178. W.A. MacCrehan, W.E. May, S.D. Yang, and B.A. Benner, Jr., *Anal. Chem.* **60**, 194 (1988).

179. U. Lewin, L. Wennrich, J. Efer, and W. Engewald, *Chromatographia* **45**, 91 (1997).

180. T.F. Jenkins, M.E. Walsh, P.W. Schumacher, P.H. Miyares, C.F. Bauer, and C.L. Grant, *J. Assoc. Off. Anal. Chem.* **72**, 890 (1989).

181. H. Engelhardt, J. Meister, and P. Kolla, *Chromatographia* **35**, 5 (1993).

182. M.E. Walsh and T.F. Jenkins, *Anal. Chim. Acta* **231**, 313 (1990).

183. J.B.F. Lloyd, *Adv. Chromatogr.* **32**, 173 (1992).

184. T. Watanabe, S. Ishida, M. Kishiji, Y. Takahashi, A. Furuta, T. Kasai, K. Wakabayashi, and T. Hirayama, *J. Chromatogr. A* **839**, 41 (1999).

185. K. Kimata, K. Hosoya, T. Araki, N. Tanaka, E.R. Barnhart, L.R. Alexander, S. Sirimanne, P.C. McClure, J. Grainger, and D.G. Patterson, Jr., *Anal. Chem.* **65**, 2502 (1993).

186. U. Pyell, P. Garrigues, and M.T. Rayez, *J. Chromatogr.* **628**, 3 (1993).

187. M. Sandahl, L. Mathiasson, and J.Å. Jönsson, *J. Chromatogr. A* **893**, 123 (2000).

188. A.P. Cheung, E. Struble, N. Nguyen, P. Liu, *J. Chromatogr. A* **870**, 179 (2000).

189. D.B. Gomis, J.J. Mangas, A. Castaño, and M. Gutiérrez, *Anal. Chem.* **68**, 3867 (1996).

190. R.A. Yokley, L.C. Mayer, R. Rezaaiyan, M.E. Manuli, and M.W. Cheung, *J. Agric. Food Chem.* **48**, 3352 (2000).

191. R. Carabias Martinez, E. Rodriguez Gonzalo, M.J. Amigo Moran, and J. Hernendez Mendez, *J. Chromatogr.* **607**, 37 (1992).

192. J. Liu, J.E. Chambers, and J.R. Coats, *J. Liq. Chromatogr.* **17**, 1995 (1994).

193. C. Molina, M. Honing, and D. Barceló, *Anal. Chem.* **66**, 4444 (1994).

194. B. Szabó-Ravasz, *J. Chromatogr.* **435**, 380 (1988).

195. J.P.G. Wilkins, *Anal. Proc.* **30**, 396 (1993).

196. D. Volmer, A. Preiss, K. Levsen, and G. Wünsch, *J. Chromatogr.* **647**, 235 (1993).

197. S. Sennert, D. Volmer, K. Levsen, and G. Wünsch, *Fresenius J. Anal. Chem.* **351**, 642 (1995).

198. B.D. McGarvey, *J. Chromatogr.* **642**, 89 (1993).

199. M.W. Dong, F.L. Vandemark, W.M. Reuter, and M.V. Pickering, *Am. Environ. Lab.* **6**, 14 (1990).

200. Q.-S. Wang, R.-Y. Gao, and B.-W. Yan, *J. Chromatogr.* **628**, 127 (1993).

201. G. Voos, P.M. Groffman, and M. Pfiel, *J. Agric. Food Chem.* **42**, 2502 (1994).

202. D. Barceló, *Organic Mass Spec.* **24**, 219 (1989).

203. J.V. Sancho-Llopis, F. Hernãndez-Hernãndez, E.A. Hogendoorn, and P. van Zoonen, *Anal. Chim. Acta* **283**, 287 (1993).

204. C. Crescenzi, A. Di Corcia, S. Marchese, and R. Samperi, *Anal. Chem.* **67**, 1968 (1995).

205. C.J. Miles, *J. Chromatogr.* **592**, 283 (1992).

206. N. Font, F. Hernández, E.A. Hogendoorn, R.A. Baumann, and P. van Zoonen, *J. Chromatogr. A* **798**, 179 (1998).

207. A. Laganà, G. Fago, and A. Marino, *J. Chromatogr. A* **796**, 309 (1998).

208. S. Chiron, S. Papilloud, W. Haerdi, and D. Barceló, *Anal. Chem.* **67**, 1637 (1995).

209. J.M. Sanchis-Mallols, S. Sagrado, M.J. Medina-Hernández, R.M. Villanueva Camañas, and E. Bonet-Domingo, *J. Liq. Chromatogr.* **21**, 1871 (1998).

210. J. Fischer and P. Jandera, *J. Chromatogr. A* **684**, 77 (1994).

211. A. Lagana, A. Marino, G. Fago, and B.P. Martinez, *Analusis* **22**, 63 (1994).

212. H. Bagheri, E.R. Brouwer, R.T. Ghijsen, and U.A.Th. Brinkman, *Analusis* **20**, 475 (1992).

213. G. Henze, A. Meyer, and J. Hausen, *Fresenius J. Anal. Chem.* **346**, 761 (1993).

214. M.J. Incorvia Mattina, *J. Chromatogr.* **542**, 385 (1991).

215. H. Bagheri, E.R. Brouwer, R.T. Ghijsen, and U.A.Th. Brinkman, *J. Chromatogr.* **647**, 121 (1993).

216. J. Abián, G. Durand, and D. Barceló, *J. Agric. Food Chem.* **41**, 1264 (1993).

217. C. Mathieu, B. Herbreteau, M. Lafosse, M. Renaud, C. Cardinet, and M. Dreux, *Anal. Chim. Acta* **402**, 87 (1999).

218. H. van Lishaut and W. Schwack, *J Assoc. Off. Anal. Chem.* **83**, 720 (2000).

219. N. Tharsis, J.L. Portillo, F. Broto-Puig, and L. Comellas, *J. Chromatogr. A* **778**, 95 (1997).

220. J.V. Pothuluri, J.P. Freeman, T.M. Heinze, R.D. Beger, and C.E. Cerniglia, *J. Agric. Food. Chem.* **48**, 6138 (2000).

221. K.W. Weissmahr, CL. Houghton, and D.L. Sedlak, *Anal. Chem.* **70**, 4800 (1998).

222. K. Ishizuka, A. Kanayama, H. Satsu, Y. Miyamoto, K. Furihata, and M. Shimizu, *Biosci. Biotechnol. Biochem.* **64**, 1166 (2000).

223. S. Jin and M. Yoshida, *Biosci. Biotechnol. Biochem.* **64**, 1614 (2000).

224. Y. Kondoh, A. Yamada, and S. Takano, *J. Chromatogr.* **541**, 431 (1991).

225. T.M. Schmitt, M.C. Allen, D.K. Brain, K.F. Guin, D.E. Lemmel, and Q.W. Osburn, *J. Am. Oil Chem. Soc.* **67**, 103 (1990).

226. C. Crescenzi, A. Di Corcia, R. Samperi, and A. Marcomini, *Anal. Chem.* **67**, 1797 (1995).

227. C. Elfakir and M. Lafosse, *J. Chromatogr. A* **782**, 191 (1997).

228. T. Tsuda, K. Suga, E. Kaneda, and M. Ohsuga, *J. Chromatogr. B* **746**, 305 (2000).

229. V.M. León, E. González-Mazo, and A. Gómez-Parra, *J. Chromatogr. A* **889**, 211 (2000).

230. A. Di Corcia, M. Marchetti, R. Samperi, and A. Marcomini, *Anal. Chem.* **63**, 1179 (1991).

231. J.L. Jasperse and P.H. Steiger, *J Am. Oil Chem. Soc.* **69**, 621 (1992).

232. A.N. Ageev and Ya.I. Yashin, *Russ. J. Phys. Chem. (Eng. Ed.)* **65**, 1392 (1991).

233. K. Jinno and Y. Yokoyama, *J. Chromatogr.* **550**, 325 (1991).

234. S.C. Matz, *J Chromatogr.* **587**, 205 (1991).

235. A.M. Robertson, D. Farnan, D. Littlejohn, M. Brown, C.J. Dowle, and E. Goodwin, *Anal. Proc.* **30**, 268 (1993).

236. S. Podzimek and L. Hroch, *J Appl. Polym. Sci.* **47**, 2005 (1993).

237. S.B. Ruddy and B.W. Hadzija, *J. Chromatogr. B* **657**, 83 (1994).

238. K. Rissler, U. Fuchslueger and H.J. Grether, *J. Liq. Chromatogr.* **17**, 3109 (1994).

239. B. Trathnigg, A. Gorbunov, and A. Skvortsov, *J. Chromatogr. A* **890**, 195 (2000).

240. L. Bromberg, *J. Appl. Polym. Sci.* **57**, 145 (1995).

241. K. Rissler and U. Fuchslueger, *J. Liq. Chromatogr.* **17**, 2791 (1994).

242. R. Schultz and H. Engelhardt, *Chromatographia* **29**, 205 (1990).

243. G. Glöckner and D. Wolf, *Chromatographia* **34**, 363 (1992).

244. G. Glöckner, D. Wolf, and H. Engelhardt, *Chromatographia* **32**, 107 (1991).

245. G. Glöckner, *Chromatographia* **37**, 7 (1993).

246. G. Glöckner, D. Wolf, and H. Engelhardt, *Chromatographia* **38**, 559 (1994).

247. T.C. Schunk and T.E. Long, *J. Chromatogr. A* **692**, 221 (1995).

248. J. Mijovic, A. Fishbain, and J. Wijaya, *Macromolecules* **25**, 979 (1992).

249. T.T. Chang, *Anal. Chem.* **66**, 3267 (1994).

250. E. Longordo, L.A. Papazian, and T.L. Chang, *J. Liq. Chromatogr.* **14**, 2043 (1991).

251. J.K. Swadesh, C.W. Stewart, Jr., and P.C. Uden, *Analyst* **118**, 1123 (1993).

252. J. Poskrobko, M. Dejnega, and M. Kiedik, *J. Chromatogr. A* **883**, 291 (2000).

253. G. Astarloa-Aierbe, J.M. Echeverría, J.L. Egiburu, M. Ormaetxea, and I. Mondragon, *Polymer* **39**, 3147 (1998).

254. S. Scalia, *J. Chromatogr. A* **870**, 199 (2000).

255. L.-H. Wang, *Chromatographia* **50**, 565 (1999).

256. V. Vanquerp, C. Rodriguez, C. Coiffard, J.M. Coiffard, and Y. De Roeck-Holtzhauer, *J. Chromatogr. A* **832**, 273 (1999).

257. L. Gagliardi, A. Amato, L. Turchetto, G. Cavazzutti, and D. Tonelli, *Anal. Lett.* **23**, 2123 (1990).

258. E. PeI, G. Bordin, and A.R. Rodriguez, *J Liq. Chromatogr.* **21**, 883 (1998).

259. N.-E. Es-Safi, C. Le Guernevé, H. Fulcrand, V. Cheynier, and M. Moutounet, *Int. J. Food Sci. Technol.* **35**, 63 (2000).

260. M.M. Dávila-Jiménez, M.P. Elizalde-González, A. Gutiérrez-González, and A.A. Peláez-Cid, *J. Chromatogr. A* **889**, 253 (2000).

261. P. Novotná, V. Pacáková, Z. Bosáková, and K. Štulik, *J. Chromatogr. A* **863**, 235 (1999).

262. J.J Berzas-Nevado, C. Guiberteau-Cabanillas, and A.M. Contento-Salcedo, *J. Liq. Chromatogr.* **20**, 3073 (1997).

263. M. Chen, D. Moir, F.M. Benoit, and C. Kubwabo, *J. Chromatogr. A* **825**, 37 (1998).

264. J. Yinon and J. Saar, *J. Chromatogr.* **586**, 73 (1991).

265. J.E. McCallum, S.A. Madison, S. Alkan, R.L. DePinto, and R.U.R. Wahl, *Environ. Sci. Technol.* **34**, 5157 (2000).

266. J.J. Harwood and G. Mamantov, *J. Chromatogr. A* **654**, 315 (1993).

267. K. Jinno, T. Uemura, H. Ohta, H. Nagashima, and K. Itoh, *Anal. Chem.* **65**, 2650 (1993).

268. Y. Cui, S.T. Lee, S.V. Olesik, W. Flory, and M. Mearini, *J. Chromatogr.* **625**, 131 (1992).

269. M.P. Gasper and D.W. Armstrong, *J Liq. Chromatogr.* **18**, 1047 (1995).

270. M. Riess and R. van Eldik, *J. Chromatogr. A* **827**, 65 (1998).

271. B.V.M.K. Giuliani and W.J. McGill, *J. Appl. Polym. Sci.* **57**, 1391 (1995).

272. A. Löfgren, B.S. Andrasko, and J. Andrasko, *J Forensic Sci.* **38**, 1151 (1993).

273. C. Sarzanini, M.C. Bruzzoniti, and B. Mentasti, *J. Chromatogr. A* **850**, 197 (1999).

274. R.M. Marcé, M. Calull, J.C. Olucha, F. Borrull, and F.X. Rius, *J. Chromatogr.* **542**, 277 (1991).

275. R.M. Marcé, M. Calull, J.C. Olucha, F. Borrull, and F.X. Rius, *Anal. Chim. Acta* **242**, 25 (1991).

276. V. Rioux, D. Catheline, M. Bouriel, and P. Legrand, *Analusis* **27**, 186 (1999).

277. F.E. Lancaster and J.F. Lawrence, *Food Addit. Contam.* **12**, 9 (1995).

278. V.A. Simon, Y. Hale, and A. Taylor, *LC/GC* **11**, 444 (1993).

279. P.R. Brown, J.M. Beebe, and J. Turcotte, *Crit. Rev. Anal. Chem.* **21**, 193 (1989).

280. C.P. Bicchi, A.E. Binello, G.M. Pellegrino, and AC. Vanni, *J. Agric. Food Chem.* **43**, 1549 (1995).

281. M. Ribotta, N. Rodriguez, and J.M. Marioli, *J. Liq. Chromatogr.* **21**, 1003 (1998).

282. G. Bringmann, D. Feineis, and Ch. Hesselmann, *Anal. Lett.* **25**, 497 (1992).

283. M. Tsimidou, G. Papadopoulos, and D. Boskou, *Food Chem.* **44**, 53 (1992).

284. M. Rajevic and P. Betto, *J. Liq. Chromatogr.* **21**, 2821 (1998).

285. X.X. Zu and G.R. Brown, *Anal. Lett.* **23**, 2011 (1990).

286. T. Herrmann, D. Steinhilber, J. Knospe, and H.J. Roth, *J. Chromatogr.* **428**, 237 (1988).

287. J. Veenstra, H. van de Pol, H. van der Torre, G. Schaafsma, and T. Ockhuizen, *J. Chromatogr.* **431**, 413 (1988).

288. W.R. Mathews, D.M. Guido, B.M. Taylor, and F.F. Sun, *Prostaglandins* **45**, 347 (1993).

289. M. Dawson and C.M. McGee, *J. Chromatogr.* **532**, 379 (1990).

290. B. Wetzka, W. Schafer, M. Scheibel, R. Nusing, and H.P. Zahradnik, *Prostaglandins* **45**, 571 (1993).

291. S. Sakae, A. Harata, T. Kitamori, T. Sawada, A. Okubo, S. Toda, and T. Shimizu, *Microchem. J.* **49**, 355 (1994).

292. M. Amin, *J. Chromatogr.* **404**, 385 (1987).

293. D.W. Armstrong, G.L. Bertrand, K.D. Ward, T.J. Ward, H.V. Secor, and J.I. Seeman, *Anal. Chem.* **62**, 332 (1990).

294. J. Reynolds and S.J. Albazi, *J. Liq. Chromatogr.* **18**, 537 (1995).

295. G.A. Kyerematen, L.H. Taylor, J.D. deBethizy, and ES. Vesell, *J. Chromatogr.* **419**, 191 (1987).

296. S.E. Murphy, D.A. Spina, M.G. Nunes, and D.A. Pullo, *Chem. Res. Toxicol.* **8**, 772 (1995).

297. J. Adachi, Y. Mizoi, T. Naito, K. Yamamoto, S. Fujiwara, and I. Ninomiya, *J. Chromatogr.* **538**, 331 (1991).

298. K. Kotzabasis, M.D. Christakis-Hampsas, and K.A. Roubelakis-Angelakis, *Anal. Biochem.* **214**, 484 (1993).

299. O. Busto, M. Mestres, J. Guasch, and F. Borrull, *Chromatographia* **40**, 404 (1995).

300. C.N. Svendsen, *Analyst* **118**, 123 (1993).

301. M.A. Glomb, D. Rösch, and R.H. Nagaraj, *J. Agric. Food Chem.* **49**, 366 (2001).

302. V.M. Lakshmi, T.V. Zenser, H.D. Goldman, G.G. Spencer, R.C. Gupta, F.F. Hsu, and B.B. Davis, *Chem. Res. Toxicol.* **8**, 711 (1995).

303. K. Zwirner-Baier and H.-G. Neumann, *Arch. Toxicol.* **68**, 8 (1994).

304. K. Yagi, T. Matsumoto, T. Chujo, H. Nojiri, T. Omori, K. Minamisawa, M. Nishiyama, and H. Yamane, *Biosci. Biotechnol. Biochem.* **64**, 1359 (2000).

305. M. Wurst, R. Kysilka, and T. Koza, *J. Chromatogr.* **593**, 201 (1992).

306. Y. Ikarashi and Y. Maruyama, *J. Chromatogr.* **587**, 306 (1991).

307. P.Y. Leung and C.S. Tsao, *J. Chromatogr.* **576**, 245 (1992).

308. B.I. Vázquez, C.A. Fente, C.M. Franco, A. Cepeda, G. Mahuzier, and P. Prognon, *Anal. Commun.* **36**, 5 (1999).

309. D.M. Stresser, G.S. Bailey, and D.E. Williams, *Drug Metab. Dispos.* **22**, 383 (1994).

310. J.F. Gregory III and D.B. Manley, *J Assoc. Off. Anal. Chem.* **65**, 869 (1982).

311. R.M. Beebe, *J. Assoc. Off. Anal. Chem.* **61**, 1347 (1978).

312. M.J. Shepherd, in *Modern Methods in the Analysis and Structural Elucidation of Mycotoxins*, Academic Press, New York, 1986, Chapter 11.

313. P.M. Scott, B.P.-Y. Lau, G.A. Lawrence, and D.A. Lewis, *J. Assoc. Off. Anal.Chem.* **83**, 1313 (2000).

314. M.A. Quilliam, *J. Assoc. Off. Anal. Chem. Int.* **78**, 555 (1995).

315. P. Zöllner, D. Berner, J. Jodlbauer, and W. Lindner, *J. Chromatogr. B* **738**, 233 (2000).

316. M.A. Stander, P.S. Steyn, A. Lübben, A. Miljkovic, P.G. Mantle, and G.J. Marais, *J. Agric. Food Chem.* **48**, 1865 (2000).

317. A. Singh, O.P. Sharma, and S. Ojha, *International Biodeterioration & Biodegradation* **46**, 107 (2000).

318. F. Palmisano, P.G. Zambonin, A. Visconti, and A. Bottalico, *J. Chromatogr.* **465**, 305 (1989).

319. Y. Hua, W. Lu, M.S. Henry, R.H. Pierce, and R.B. Cole, *Anal. Chem.* **67**, 1815 (1995).

320. P. Härmälä, H. Vuorela, P. Lehtonen, and R. Hiltunen, *J. Chromatogr.* **507**, 367 (1990).

321. J.H. Fentem and J.R. Fry, *Xenobiotica* **22**, 357 (1992).

322. T. Chalermchaikit, L.J. Felice, and M.J. Murphy, *J Anal. Toxicol.* **17**, 56 (1993).

323. P.G. Thiel, E.W. Sydenham, G.S. Shephard, and D.J. van Schalkwyk, *J.Assoc. Off. Anal. Chem. Int.* **76**, 361 (1993).

324. T.R. Govindachari, G. Sureshand, and G. Gopalakrishnan, *J. Liq. Chromatogr.* **18**, 3465 (1995).

325. A.K. Jäger, L. Gudiksen, A. Adsersen, and U.W. Smitt, *J. Chromatogr.* **634**, 135 (1993).

326. S. Albalá-Hurtado, M.T. Veciana-Nogués, M. Izquierdo-Pulido, and A. Mariné-Font, *J. Chromatogr. A* **778**, 247 (1997).

327. S.K. Sharma and K. Dakshinamurti, *J. Chromatogr.* **578**, 45 (1992).

328. D.W. Jacobsen, V.J. Gatautis, R. Green, K. Robinson, S.R. Savon, M. Secic, J. Ji, J.M. Otto, and L.M. Taylor, Jr., *Clin. Chem.* **40**, 873 (1994).

329. M.D. Lucock, M. Green, M. Priestnall, I. Daskalakis, M.I. Levene, and R. Hartley, *Food Chem.*, **53**, 329 (1995).

330. H. Chassaingе and R. Łobiński, *Anal. Chim. Acta* **359**, 227 (1998).

331. S.M. El-Gizawy, A.N. Ahmed, and N.A. El-Rabbat, *AnaL Lett.* **24**, 1173 (1991).

332. G.I. Rehner and J. Stein, *Methods Enzymol.* **279**, 286 (1997).

333. E. Kulczykowska and P.M. Iuvone, *J. Chromatogr. Sci.* **36**, 175 (1998).

334. Analytical Methods Committee, *Analyst* **116**, 421 (1991).

335. J. Arnaud, I. Fortis, S. Blachier, D. Kia, and A. Favier, *J. Chromatogr.* **572**, 103 (1991).

336. M. Podda, C. Weber, M.G. Traber, R. Milbradt, and L. Packer, *Methods Enzymol.* **279**, 330 (1997).

337. T. Menke, P. Niklowitz, S. Adam, M. Weber, B. Schlüter, and W. Andler, *Anal. Biochem.* **282**, 209 (2000).

338. S.L. Abidi and T.L. Mounts, *J. Chromatogr. A* **670**, 67 (1994).

339. Y. Yamamoto, N. Maita, A. Fujisawa, J. Takashima, Y. Iishi, and W.C. Dunlap, *J. Nat. Prod.* **62**, 1685 (1999).

340. C.-Y. Tai and B.H. Chen, *J. Agric. Food Chem.* **48**, 5962 (2000).

341. H. Etoh, Y. Utsunorniya, A. Komori, Y. Murakami, S. Oshima, and T. Inakuma, *Biosci. Biotechnol. Biochem.* **64**, 1096 (2000).

342. H.E. Indyk and D.C. Woollard, *J Assoc. Off. Anal. Chem. Int.* **83**, 121 (2000).

343. H. Iwase, *J. Chromatogr. A* **881**, 261 (2000).

344. H.E. Indyk and D.C. Woollard, *Analyst* **122**, 465 (1997).

345. J.M. Conly, *Methods Enzymol.* **282**, 457 (1997).

346. R. Andreoli, M. Careri, P. Manini, G. Mori, and M. Musci, *Chromatographia* **44**, 605 (1997).

347. H. Qian and M. Sheng, *J. Chromatogr. A* **825**, 127 (1998).

348. D. Blanco, M.P. Fernández, and M.D. Gutiérrez, *The Analyst* **125**, 427 (2000).

349. A.B. Barua, *Methods Enzymol.* **189**, 136 (1991).

350. A.B. Barua and J.A. Olson, *J. Chromatogr. B* **707**, 69 (1998).

351. M.E. Cullum and M.H. Zile, *Anal. Biochem.* **153**, 23 (1986).

352. J. Deli, Z. Matus, and J. Szabolcs, *J. Agric. Food Chem.* **40**, 2072 (1992).

353. L.R. Howard, S.T. Talcott, C.H. Brenes, and B. Villalon, *J. Agric. Food Chem.* **48**, 1713 (2000).

354. R.B. van Breemen, *Anal. Chem.* **67**, 2004 (1995).

355. J.-P. Yuan and F. Chen, *J. Agric. Food Chem.* **47**, 3656 (1999).

356. C.M. Bell, L.C. Sander, and S.A. Wise, *J. Chromatogr. A* **757**, 29 (1997).

357. P.P Mouly, E.M. Gaydou, L. Lapierre, and J. Corsetti, *J Agric. Food Sci.* **47**, 4038 (1999).

358. J. Deli, Z. Matus, and G. Tóth, *J. Agric. Food Chem.* **48**, 2793 (2000).

359. H.S. Lee, *J. Agric. Food Chem.* **48**, 1507 (2000).

360. I.M. Lomniczi de Upton, J.R. de la Fuente, J.S. Esteve-Romero, M.C. Garciá-Alvarez-Coque, and S. Carda-Broch, *J. Liq. Chromatogr.* **22**, 909 (1999).

361. M. Ganzera, E. Bedir, and I.A. Khan, *Chromatographia* **52**, 301 (2000).

362. G. Gross, E. Jaccaud, and A.C. Huggett, *Food Chem. Toxicol.* **35**, 547 (1997).

363. C. Laugel, A. Baillet, and D. Ferrier, *J. Liq. Chromatogr.* **21**, 1333 (1998).

364. P. Gómez-Serranillos, O.M. Palomino, A.J. Villarrubia, M.A. Cases, E. Carretero, and A. Villar, *J. Chromatogr. A* **778**, 421 (1997).

365. P. Chen, X.-L. Su, L.-H. Nie, S.-Z. Yao and J.-G. Zeng, *J Chromatogr. Sci.* **36**, 197 (1998).

366. O.P. Sharma, S. Sharma, and R.K. Dawra, *J Chromatogr. A* **786**, 181 (1997).

367. D.S. Weinberg, M.L. Manier, M.D. Richardson, and P.C. Haibach, *J. Agric. Food Chem.* **41**, 42 (1993).

368. I. Martos, F. Ferreres, L. Yao, B. D'Arcy, N. Caffin, and F.A. Tomás-Barberán, *J. Agric. Food Chem.* **48**, 4745 (2000).

369. H.S. Lee, C.B. Jin, H.S. Chong, C.H. Yun, J.S. Park, and D.H. Kim, *Xenobiotica* **24**, 1053 (1994).

370. H. Gagnon, S. Tahara, E. Bliechert, and R.K. Ibrahim, *J. Chromatogr.* **606**, 25 (1992).

371. M. Krause and R. Galensa, *J. Chromatogr.* **588**, 41 (1991).

372. A. Hasler, O. Sticher, and B. Meier, *J. Chromatogr.* **605**, 41 (1992).

373. Y. Miyake, K. Shimoi, S. Kumazawa, K. Yamamoto, N. Kinae, and T. Osawa, *J. Agric. Food Chem.* **48**, 3217 (2000).

374. P.B. Andrade, R. Leitão, R.M. Seabra, M.B. Oliveira, and M.A. Ferreira, *J. Liq. Chromatogr.* **20**, 2023 (1997).

375. K.A. Georga, V.F. Samanidou, and I.N. Papadoyannis, *J. Liq. Chromatogr.* **22**, 2975 (1999).

376. J.F. Lu, X.M. Cao, T. Yi, H.T. Zhuo, and S.S. Ling, *Anal. Lett.* **31**, 613 (1998).

377. D. Blanco, M.P. Fernández, and M.D. Gutiérrez, *The Analyst* **125**, 427 (2000).

378. I. Papadoyannis, V.F. Samanidou, and K.A. Georga, *J. Liq. Chromatogr.* **19**, 2559 (1996).

379. P. Lozano, M.R. Castellar, M.J. Simancas, and J.L. Iborra, *J. Chromatogr. A* **830**, 477 (1999).

380. Y. Amakura, M. Okada, S. Tsuji, and Y. Tonogai, *J. Chromatogr. A* **896**, 87 (2000).

381. M.I. Gil, F.A. Tomás-Barberán, B. Hess-Pierce, D.M. Holcroft, and A.A. Kader, *J. Agric. Food Chem.* **48**, 4581 (2000).

382. M. Rafecas, F. Guardiola, M. Illera, R. Codony, and J. Boatella, *J. Chromatogr. A* **822**, 305 (1998).

383. T. Nurmi and H. Adlercreutz, *Anal. Biochem.* **274**, 110 (1999).

384. N. Li, G. Lin, Y.-W. Kwan, and Z.-D. Min, *J Chromatogr. A* **849**, 349 (1999).

385. E.-S. Ong, S.-O. Woo, and Y.-L. Yong, *J. Chromatogr. A* **904**, 57 (2000).

386. R. Kramell, O. Miersch, G. Schnieder, and C. Westernack, *Chromatographia* **49**, 42 (1999).

387. J. Zhang, Z. Meng, M. Zhang. D. Ma, S. Xu, and H. Kodama, *Clin. Chem. Acta* **289**, 79 (1999).

388. L.V. Bystrykh, J.K. Herrema, W. Kruizinga, and R.M. Kellogg, *Biotechnol. Appl. Biochem.* **26**, 195 (1997).

389. G.E. Rohr, B. Meier, and O. Sticher, *Phytochem. Anal.* **11**, 106 (2000).

390. E. Jagerdeo, F. Passetti, and S.M. Dugar, *J. Assoc. Off. Anal. Chem. Int.* **83**, 237 (2000).

391. F.F. Liu, C.Y.W. Ang, T.M. Heinze, J.D. Rankin, R.D. Beger, J.P. Freeman, and J.O. Lay, Jr., *J. Chromatogr. A* **888**, 85 (2000).

392. G. Piperopoulos, R. Lou, A. Wixforth, T. Schmierer, and K.-P. Zeller, *J. Chromatogr. B* **695**, 309 (1997).

393. F. Yang, T. Zhang, Q. Liu, G. Xu, Y. Zhang, S. Zhang, and Y. Ito, *J. Chromatogr. A* **883**, 67 (2000).

394. S. Schweizer, A.F.W. von Brocke, S.F. Boden, E. Bayer, H.P.T. Ammon, and H. Safayhi, *J Nat Prod.* **63**, 1058 (2000).

395. V.K. Gore and P. Satyamoorthy, *Anal. Lett.* **33**, 337 (2000).

396. P. Cabras, A. Angioni, C. Tuberoso, I. Floris, F. Reniero, C. Guillou, and S. Ghelli, *J. Agric. Food Chem.* **47**, 4064 (1999).

397. D. Deliorman, I. Çaliş, F. Ergun, and U. Tamer, *J. Liq. Chromatogr.* **22**, 3101 (1999).

398. M. Yao, Y. Qi, K. Bi, X. Wang, X. Luo, and C. Che, *J. Chromatogr. Sci.* **38**, 325 (2000).

399. B.L. Lee, D. Koh, H.Y. Ong, and C.N. Ong, *J. Chromatogr. A* **763**, 221 (1997).

400. M. Jiménez and R. Mateo, *J. Chromatogr. A* **778**, 363 (1997).

401. X.-g. He, L.-z. Lian, and L.-z. Lin, *J. Chromatogr. A* **757**, 81 (1997).

402. Y. Shao, K. He, B, Zheng, and Q. Zheng, *J. Chromatogr. A* **825**, 1 (1998).

403. M. Ellnain-Wojtaszek and G. Zgórka, *J. Liq. Chromatogr.* **20**, 1457 (1997).

404. D.M. Goldberg and G.J. Soleas, *Methods Enzymol.* **299**, 122 (1997).

405. A.A. Franke and L.J. Custer, *Clin. Chem.* **42**, 955 (1996).

406. L. Bovanová, E. Brandšteterová, A. Čaniová, K. Argalášová, and A. Lux, *J. Chromatogr. B* **732**, 405 (1999).

407. K. Chervenkova and B, Nikolova-Damyanova, *J. Liq. Chromatogr.* **23**, 741 (2000).

408. F. Yang, T. Zhang, G. Tian, H. Cao, Q. Liu, and Y. Ito, *J. Chromatogr. A* **858**, 103 (1999).

409. B. Bozan, K.H.C. Başer, and S. Kara, *J. Chromatogr. A* **782**, 133 (1997).

410. L.-Z. Lin, X.-G. He, L.-Z. Lian, W. King, and J. Elliott, *J. Chromatogr. A* **810**, 71 (1998).

411. L.G. Arguello, D.K. Sensharma, F. Qiu, A. Nurtaeva, and Z. El Rassi, *J. Assoc. Off. Anal. Chem. Int.* **82**, 1115 (1999).

412. K. Koga, K. Ohmachi, S. Kawashima, K. Takada, and M. Murakami, *J. Chromatogr. B* **738**, 165 (2000).

413. R.K. Verma, K.G. Bhartariya, M.M. Gupta, and S. Kumar, *Phytochem. Anal.* **10**, 191 (1999).

414. M.K. Park, J.H. Park, N.Y Kim, Y.G. Shin, Y.S. Choi, J.G. Lee, K.H. Kim, and S.K. Lee, *Phytochem. Anal.* **9**, 186 (1998).

415. C. Osterhage, M. Schwibbe, G.M. König, and A.D. Wright, *Phytochem. Anal.* **11**, 288 (2000).

416. E.M.J.M. Jansen and P. de Fluiter, *J. Chromatogr.* **580**, 325 (1992).

417. G.P.B. Kraan, K.T. van Wee, B.G. Wolthers, J.C. Vandermolen, G.T. Nagel, N.M. Drayer, and D. van Leusen, *Steroids* **58**, 495 (1993).

418. F. Varin, T.M. Tu, F. Benoît, J.-P. Villeneuve, and Y. Théorêt, *J. Chromatogr.* **574**, 57 (1992).

419. T. Satoh, K.-i. Fujita, H. Munakata, S. Itoh, K. Nakamura, T. Kamataki, S. Itoh, and I. Yoshizawa, *Anal. Biochem.* **286**, 179 (2000).

420. B. Cheng, J. Kowal, and S. Abraham, *J. Lipid Res.* **35**, 1115 (1994).

421. L. Fillion, J.A. Zee, and C. Gosselin, *J. Chromatogr.* **547**, 105 (1991).

422. K. Shimada and S. Nishimura, *J. Liq. Chromatogr.* **18**, 1691 (1995).

423. R. Lafont, N. Kaouadji, E.D. Morgan, and I.D. Wilson, *J. Chromatogr. A* **658**, 55 (1994).

424. M.R. Koupai-Abyazani, J. McCallum, and B.A. Bohm, *J. Chromatogr.* **594**, 117 (1992).

425. S. Héron and A. Tchapla, *Analusis* **22**, 114 (1994).

426. O. Sjovall, A. Kuksis, and H. Kallio, *J. Chromatogr. A* **905**, 119 (2001).

427. T.B. Caligan, K. Peters, J. Ou, E. Wang, J. Saba, and A.H. Merrill, Jr., *Anal. Biochem.* **281**, 36 (2000).

428. A.M. Bernasconi, H.A. Garda, and R.R. Brenner, *Lipids*, **35**, 1335 (2000).

429. T.L. Mounts, S.L. Abidi, and K.A. Rennick, *J. Am. Oil Chem. Soc.* **69**, 438 (1992).

430. S.L. Abidi, T.L. Mounts, and K.A. Rennick, *J. Liq. Chromatogr.* **17**, 3705 (1994).

431. S.L. Melton, *J. Am. Oil Chem. Soc.* **69**, 784 (1992).

432. H.-Y. Kim, T.-C.L. Wang, and Y.-C. Ma, *Anal. Chem.* **66**, 3977 (1994).

433. T.C. Markello, J. Guo, and W.A. Gahl, *Anal. Biochem.* **198**, 368 (1991).

434. W.W. Christie, *J. Lipid Res.* **26**, 507 (1985).

435. W.W. Christie, *J. Chromatogr.* **361**, 396 (1986).

436. M. Nikolova-Karakashian and A.H. Merrill, Jr., *Methods Enzymol.* **311**, 194 (2000).

437. H.M. Liebich, C. Di Stefano, A. Wixforth, and H.R. Schmid, *J. Chromatogr. A* **763**, 193 (1997).

438. A. Klemm, T. Steiner, U. Flötgen, G.A. Cumme, and A. Horn, *Methods Enzymol.* **280**, 171 (1997).

439. V. Micheli and S. Sestini, *Methods Enzymol.* **280**, 211 (1997).

440. G. Xu, C. Di Stefano, H.M. Liebich, Y. Zhang, P. Lu, *J. Chromatogr. B* **732**, 307 (1999).

441. C.K. Kim and T.J. Peters, *J. Chromatogr.* **461**, 259 (1989).

442. A. Fidder, G.W.H. Moes, A.G. Scheffer, G.P. van der Schans, R.A. Baan, L.P.A. de Jong, and H.P. Benschop, *Chem. Res. Toxicol.* **7**, 199 (1994).

443. L. Kronberg, S. Karlsson, and R. Sjöholm, *Chem. Res. Toxicol.* **6**, 495 (1993).

444. M.C. Di Pietro, D. Vannoni, R. Leoncini, G. Liso, R. Guerranti, and F. Marinello, *J. Chromatogr. B* **751**, 87 (2001).

445. K. Valkó, T. Cserháti, and E. Forgács, *J. Chromatogr.* **550**, 667 (1991).

446. J.E. Page, J. Szeliga, S. Amin, S.S. Hecht, and A. Dipple, *Chem. Res. Toxicol.* **8**, 143 (1995).

447. T. Hikita, K. Tadano-Aritomi, N. Iida-Tanaka, H. Toyoda, A. Suzuki, T. Toida, T. Imanari, T. Abe, Y. Yanagawa, and I. Ishizuka, *Anal. Biochem.* **281**, 193 (2000).

448. M. Pescaglini, V. Micheli, H.A. Simmonds, M. Rocchigianil, and. G. Pompucci, *Clin. Chim. Acta* **229**, 15 (1994).

449. L. Almela, J.A. Fernández-López, and M.J. Roca, *J. Chromatogr. A* **870**, 483 (2000).

450. P. Heinmöller, H.-H. Kurth, R. Rabong, W.V. Turner, A. Kettrup, and S. Gäb, *Anal. Chem.* **70**, 1437 (1998).

451. E. Magi, C. Ianni, P. Rivaro, and R. Frache, *J. Chromatogr. A* **905**, 141 (2001).

452. A. Kühnel, U. Gross, and M.O. Doss, *Clin. Biochem.* **33**, 465 (2000).

453. K. Nakayama, A. Takasawa, I. Terai, T. Okui, T. Ohyama, and M. Tamura, *Arch. Biochem. Biophys.* **375**, 240 (2000).

454. W. Buchberger and B. Drda, *Chromatographia* **27**, 101 (1989).

455. K. Takatera and T. Watanabe, *Anal. Chem.* **65**, 759 (1993).

456. B.A. Hill, H.E. Kleiner, E.A. Ryan, D.M. Dulik, T.J. Monks, and S.S. Lau, *Chem. Res. Toxicol.* **6**, 459 (1993).

457. C.B. Eckardt, B.J. Keely, and J.R. Maxwell, *J. Chromatogr.* **557**, 271 (1991).

458. J.L. Allen and J.R. Meinertz, *J. Chromatogr.* **536**, 217 (1991).

459. B. Herbreteau, *Analusis* **20**, 355 (1992).

460. M.C.J. Wilce, M.-I. Aguilar, and M.T.W. Hearn, *Anal. Chem.* **67**, 1210 (1995).

461. M.C.J. Wilce, M.-I. Aguilar, and M.T.W. Hearn, *J. Chromatogr.* **536**, 165 (1991).

462. A.D. Carlson and R.M. Riggin, *Anal. Biochem.* **278**, 150 (2000).

463. G. Noctor and C.H. Foyer, *Anal. Biochem.* **264**, 98 (1998).

464. T. Teerlink, P.A.M. van Leeuwen, and A. Houdijk, *Clin. Chem.* **40**, 245 (1994).

465. K. Fujimura, S. Suzuki, K. Hayashi, and S. Masuda, *Anal. Chem.* **62**, 198 (1990).

466. A. Berthod, S.-C. Chang, and D.W. Armstrong, *Anal. Chem.* **64**, 395 (1992).

467. P.A. Haynes, D. Sheumack, L.G. Greig, J. Kibby, and J.W. Redmond, *J. Chromatogr.* **588**, 107 (1991).

468. S.M. Lunte, T. Mohabbat, O.S. Wong, and T. Kuwana, *Anal. Biochem.* **178**, 202 (1989).

469. M. Dunnett and R.C. Harris, *J. Chromatogr. B* **688**, 47 (1997).

470. Miyazawa, T. Otomatsu, T. Yamada, and S. Kuwata, *Int. J. Peptide Protein Res.* **39**, 229 (1992).

471. C. Griehl and S. Merkel, *Int. J. Peptide Protein Res.* **45**, 217 (1995).

472. S. Abdel-Rahman, Y.M. El-Ayouty and H.A. Kamael, *Int. J. Peptide Protein Res.* **41**, 1 (1993).

473. T.T. Herskovits, B. Gadegbeku, and H. Jaillet, *J. Biol. Chem.* **245**, 2588 (1970).

474. H.B. Bull and K. Breese, *Biopolymers* **17**, 2121 (1978).

475. M.T.W. Hearn and B. Grego, *J. Chromatogr.* **218**, 497 (1981).

476. D.W. Lee and B.Y. Cho, *J. Liq. Chromatogr.* **17**, 2541 (1994).

477. R.J. Simpson, R.L. Moritz, E.C. Nice, and B. Grego, *Eur. J Biochem.* **165**, 21 (1987).

478. F.M. Rabel and D.A. Martin, *J. Liq. Chromatogr.* **6**, 2465 (1983).

479. V.K. Rybin, L.V. Grigorova, and N.V. Makarov, *Russ. J. Phys. Chem.* (*Eng. Ed.*) **65**, 1402 (1991).

480. P. Oroszlan, S. Wicar, G. Teshima, S.-L. Wu, W.S. Hancock, and B.L. Karger, *Anal. Chem.* **64**, 1623 (1992).

481. S. Wicar, M.G. Mulkerrin, G. Bathory, L.H. Khundkar, and B.L. Karger, *Anal. Chem.* **66**, 3908 (1994).

482. J.P. Chang, W.R. Melander, and Cs. Horváth, *J. Chromatogr.* **318**, 11 (1985).

483. A. Fernández and O. Sinanoglu, *Biophys. Chem.* **21**, 163 (1985).

484. R.S. Blanquet, K.H. Bui, and D.W. Armstrong, *J. Liq. Chromatogr.* **9**, 1933 (1986).

485. N. Funasaki, S. Hada, and S. Neya, *Anal. Chem.* **65**, 1861 (1993).

486. G. Teshima and E. Canova-Davis, *J. Chromatogr.* **625**, 207 (1992).

487. M. Kawakatsu, H. Kotaniguchi, R.H. Freiser, and K.M. Gooding, *J. Liq. Chromatogr.* **18**, 633 (1995).

488. A. Tracqui, P. Kintz, and P. Mangin, *J. Forensic Sci.* **40**, 254 (1995).

489. R.M. Smith, J.P. Westlake, R. Gill, and M.D. Osselton, *J. Chromatogr.* **514**, 97 (1990).

490. W.E. Lambert, E. Meyer, and A.P. De Leenheer, *J. Anal. Toxicol.* **19**, 73 (1995).

491. R. Gami-Yilinkou and R. Kaliszan, *J. Chromatogr.* **550**, 573 (1991).

492. E.M. Koves, *J. Chromatogr. A* **692**, 103 (1995).

493. S. Gustafsson, B.-M. Eriksson, and I. Nilsson, *J. Chromatogr.* **506**, 75 (1990).

494. G.B. Cox and R.W. Stout, *J. Chromatogr.* **384**, 315 (1987).

495. M.J, Martín, F. Pablos, and A.G. González, *Talanta* **49**, 453 (1999).

496. O. Corcoran, J.K. Nicholson, F.M. Lenz, F. Abou-Shakra, J. Castro-Perez, A.B. Sage, and I.D. Wilson, *Rapid Commun. Mass Spectrom.* **14**, 2377 (2000).

497. A. Bica, A. Farinha, H. Blume, C.M. Barbosa, *J. Chromatogr. A* **889**, 135 (2000).

498. E.J.G. Portier, K. de Blok, J.J. Butter, and C.J. van Boxtel, *J. Chromatogr. B* **723**, 313 (1999).

499. Geisslinger, K. Dietzel, D. Loew, O. Schuster, G. Rau, C. Lachmann, and K. Brune, *J. Chromatogr.* **491**, 139 (1989).

500. P. Hambleton, W.J. Lough, J. Maltas, and M.J. Mills, *J. Liq. Chromatogr.* **18**, 3205 (1995).

501. M.A. Abuirjeie, M.E. Abdel-Hamid, and E.-S.A. Ibrahim, *Anal. Lett.* **22**, 265 (1989).

502. E.K. Yun, A.J. Prince, J.E. McMillin, and L.E. Welch, *J. Chromatogr. B* **712**, 145 (1998).

503. C.S. Ambekar, B. Cheung, J. Lee, L.C. Chan, R. Liang, and C.R. Kumana, *Eur. J. Clin. Pharmacol.* **56**, 405 (2000).

504. J.P. Cravedi, M. Baradat, L. Dehrauwer, J. Alary, J. Tulliez, and G. Bories, *Drug Metab. Dispos.* **22**, 578 (1994).

505. P.J. Kijak, J. Jackson, and B. Shaikh, *J. Chromatogr. B* **691**, 377 (1997).

506. G. Dusi and V. Gamba, *J. Chromatogr. A* **835**, 243 (1999).

507. S. Croubels, H. Vermeersch, P. De Backer, M.D.F. Santos, J.P. Remon, and C. Van Peteghem, *J. Chromatogr. B* **708**, 145 (1998).

508. F.M. El Anwar, A.H. El Walily, M.H. Abdel Hay, and M. El Swifty, *Anal. Lett.* **24**, 767 (1991).

509. S. Yang and M.G. Khaledi, *J. Chromatogr. A* **692**, 311 (1995).

510. R.F. Straub and R.D. Voyksner, *J. Chromatogr.* **647**, 167 (1993).

511. K.L. Tyczkowska, R.D. Voyksner, and A.L. Aronson, *J. Chromatogr.* **594**, 195 (1992).

512. J. Paesen, E. Roets, and J. Hoogmartens, *Chromatographia* **32**, 162 (1991).

513. E. Forgács and T. Cserháti, *J. Chromatogr. B* **664**, 277 (1995).

514. A. Thaler, T. Hottkowitz, and H. Eibl, *Chem. Phys. Lipids* **107**, 131 (2000).

515. W. Lang, J. Mao, Q. Wang, C. Niu, T.W. Doyle, and B. Almassian, *J. Pharm. Sci.* **89**, 191 (2000).

516. Z.D. Zhang, G. Guetens, G. De Boeck, K. Van Cauwenberghe, R.A.A. Maes, C. Ardiet, A.T. van Oosterom, M. Highley, E.A. de Bruijn, and U.R. Tjaden, *J. Chromatogr. B* **739**, 281 (2000).

517. Y. Su, Y.Y. Hon, Y. Chu, M.E.C. Van de Poll, and M.V. Relling, *J. Chromatogr. B* **732**, 459 (2000).

518. H. Mawatari, Y. Kato, S.-i. Nishimura, N. Sakura, and K. Ueda, *J. Chromatogr. B* **716**, 392 (1998).

519. X. Wang, W. Zhang, A. Zou, and Y. Lou, *J. Chromatogr. B* **746**, 319 (2000).

520. A.P. Cheung, J. He, and Y. Ha, *J. Chromatogr. A* **797**, 283 (1998).

521. J. MacCallum, J. Cummings, J.M. Dixon, and W.R. Miller, *J. Chromatogr. B* **698**, 269 (1997).

522. K. Ogawa, A. Kawasaki, T. Yoshida, H. Nesumi, M. Nakano, Y. Ikoma, and M. Yano, *J. Agric. Food Chem.* **48**, 1763 (2000).

523. S. Bala, G.C. Uniyal, S.K. Chattopdhyay, V. Tripathi, K.V. Sashidhara, M. Kulshrestha, R.P. Sharma, S.P. Jain, A.K. Kukreja, and S. Kumar, *J. Chromatogr. B* **749**, 135 (2000).

524. W.J. Kopycki, H.N. El Sohly, and J.D. McChesney, *J. Liq. Chromatogr.* **17**, 2569 (1994).

525. E. Vidal, C. Pascual, and L. Pou, *J. Chromatogr. B* **736**, 295 (2000).

526. D.M. Reith and G.R. Cannell, *J Liq. Chromatogr.* **22**, 1907 (1999).

527. H. Liu, L.J. Forman, J. Montoya, C. Eggers, C. Barham, and M. Delgado, *J. Chromatogr.* **576**, 163 (1992).

528. I.N. Papadoyannis, *HPLC in Clinical Chemistry*, Marcel Dekker, New York, 1990, Chapter 13.

529. M.R.H. Baltes, J.G. Dubois, and M. Hanocq, *J. Chromatogr. B* **706**, 201 (1998).

530. A.A.M. Stolker, P.L.W.J. Schwillens, L.A. van Ginkel, and U.A.Th. Brinkman, *J. Chromatogr. A* **893**, 55 (2000).

531. A.-H.N. Ahmed, S.M. E1-Gizawy, and N.M. Omar, *Anal. Lett.* **24**, 2207 (1991).

532. G.R. Cannell, R.H. Mortimer, D.J. Maguire, and R.S. Addison, *J. Chromatogr.* **563**, 341 (1991).

533. H. Hirata, T. Kasama, Y. Sawai, and R.R. Fike, *J. Chromatogr. B* **658**, 55 (1994).

534. S. Torres Cartas, M.C. García Alvarez-Coque, and R.M. Villanueva Camañas, *Anal. Chim. Acta* **302**, 163 (1995).

535. Z. Al-Kurdi, T, Al-Jallad, A. Badwan, and A.M.Y. Jaber, *Talanta* **50**, 1089 (1999).

536. M.D. Rose, *The Analyst* **124**, 1023 (1999).

537. M. Sher Ali, T. Sun, G.E. McLeroy, and E.T. Phillippo, *J. Assoc. Off. Anal. Chem. Int.* **83**, 31 (2000).

538. D.A. Barrett, M. Pawula, R.D. Knaggs, and P.N. Shaw, *Chromatographia* **47**, 667 (1998).

539. E.F. O'Connor, S.W.T. Cheng, and W.G. North, *J. Chromatogr.* **491**, 240 (1989).

540. R. Dams, W.E. Lambert, K.W. Clauwaert, and A.P. De Leenheer, *J. Chromatogr. A* **896**, 311 (2000).

541. M.Z. Mesmer, and R.D. Satzger, *J. Forensic Sci.* **43**, 489 (1998).

542. C. Rustichelli, V. Ferioli, F. Vezzalini, M.C. Rossi, and G. Gamberini, *Chromatographia* **43**, 129 (1996).

543. V. Ferioli, C. Rustichelli, C. Pavesi, and G. Gamberini, *Chromatographia* **52**, 39 (1996).

544. B. Yagen and S. Burstein, *J. Chromatogr. B* **740**, 93 (2000).

545. M. Simova and N. Dimov, *Chromatographia* **43**, 436 (1996).

546. D.M. Cross, J.A. Bell, and K. Wilson, *Xenobiotica* **25**, 367 (1995).

547. D.W. Kelly, C.L. Holder, W.A. Korfmacher, and W. Slikker, Jr., *Drug Metab. Dispos.* **18**, 1018 (1990).

548. C.M. Machado, S.M. Thomas, D.F. Hegarty, R.A. Thompson, D.K. Ellison, and J.M. Wyvratt, *J. Liq. Chromatogr.* **21**, 575 (1998).

549. M.S. Benedetti, D. Fraier, U. Painezzola, M.G. Castelli, P. Dostert, and L. Gianni, *Xenobiotica* **23**, 115 (1993).

550. C. Anselmino, L. Voituriez, and J. Cadet, *Chem. Res. Toxicol.* **6**, 858 (1993).

551. A. Khedr and A. Sakr, *J. Chromatogr. Sci.* **37**, 462 (1999).

552. R.M. Azzam, L.J. Notarianni, and H.M. Ali, *J. Chromatogr. B* **708**, 304 (1998).

553. S. Ono, T. Hatanaka, S. Miyazawa, M. Tsutsui, T. Aoyama, F.J. Gonzalez, and T. Satoh, *Xenobiotica* **26**, 1155 (1996).

554. Y. Guillaume and C. Guinchard, *J. Liq. Chromatogr.* **17**, 1443 (1994).

555. C.M. Moore, K. Sato, and Y. Katsumata, *Clin. Chem.* **37**, 804 (1991).

556. I.E. Panderi, H.A. Archontaki, E.E. Gikas, and M. Parissi-Poulou, *J. Liq. Chromatogr.* **21**, 1783 (1998).

557. M. Issar, N.V. Nagarajala, J. Lal, J.K. Paliwal, and R.C. Gupta, *J. Chromatogr. B* **724**, 147 (1999).

558. R. Paroni, B. Comuzzi, C. Arcelloni, S. Brocco, S. de Kreutzenberg, A. Tiengo, A. Ciucci, P. Beck-Peccoz, and S. Genovese, *Clin. Chem.* **46**, 1773 (2000).

559. L. Costa, M. Vega, Y. Díaz, J.L. Marcelo, J.M. Hernández, T. Martino, *J. Chromatogr. A* **907**, 173 (2001).

560. L.M. del Rivero, H. Jung, R. Castillo, and A. Hernández-Campos, *J. Chromatogr. B* **712**, 237 (1998).

561. S.-i. Kuriya, S. Ohmori, M. Hino, C. Senda, K. Sakai, T. Igarashi, and M. Kitada, *J. Chromatogr. B* **744**, 129 (2000).

562. J. Ducharme and R. Farinotti, *J. Chromatogr. B* **698**, 243 (1997).

563. A.S. Gross, A. Nicolay, and A. Eschalier, *J. Chromatogr. B* **728**, 107 (1999).

564. J.D. Goss, *J. Chromatogr. A* **828**, 267 (1998).

565. R. Oertel, K. Richter, and W. Kirch, *J. Chromatogr. A* **846**, 217 (1999).

566. E.M. Clement, J. Odontiadis, and M. Franklin, *J. Chromatogr. B* **705**, 303 (1998).

567. L. Yee, S.H.Y. Wong, and V.A. Skrinska, *J. Anal. Toxicol.* **24**, 651 (2000).

568. S. Singh, T.T. Mariappan, N. Sharda, S. Kumar, and A.K. Chakraborti, *Pharm. Pharmacol. Commun.* **6**, 405 (2000).

569. J.R. Meinertz, G.R. Stehly, T.D. Hubert, and J.A. Bernardy, *J. Chromatogr. A* **855**, 255 (1999).

570. S. Zeng, J. Zhong, L. Pan, and Y. Li, *J. Chromatogr. B* **728**, 151 (1999).

571. H. De Vries and G.M.J. Beijersbergen van Henegouwen, *J. Photochem. Photobiol. B: Biology* **58**, 6 (2000).

572. R.S. Addison, S.L. Duffy, and S.R. Mathers, *J. Chromatogr. Sci.* **37**, 61 (1999).

573. C.A. Lau-Cam and R.W. Roos, *J. Liq. Chromatogr.* **21**, 519 (1998).

574. Q. Wang, Z. Wu, Y. Wang, C. Luo, E. Wu, and X. Gao, *J. Chromatogr. B* **746**, 151 (2000).

575. J.A. López, V. Martínez, R.M. Alonso, and R.M. Jiménez, *J. Chromotogr. A* **870**, 105 (2000).

576. L.A. Shervington, *Anal. Lett.* **30**, 927 (1997).

577. D. Hurtaud-Pessel, B. Delépine, and M. Laurentie, *J. Chromatogr. A* **882**, 89 (2000).

578. L. Grasso, G. Scarapo, E. Imparato, O. Arace, and G. Oliviero, *Food Addit. Contam.* **17**, 749 (2000).

579. M.A. García, C. Soláns, J.J. Aramayona, L.J. Fraile, M.A. Bregante, and J.R. Castillo, *Talanta* **47**, 1245 (1998).

580. P. Overbeck and G. Blaschke, *J. Chromatogr. B* **732**, 185 (1999).

581. S.S. Murthy, C. Mathur, L.T. Kvalo, R.A. Lessor, and J.A. Wilhelm, *Xenobiotica*, **26**, 779 (1996).

582. L.D. Payne, M.B. Hicks, and T.A. Wehner, *J. Agric. Food Chem.* **43**, 1233 (1995).

583. A. Madan and M.D. Faiman, *Drug Metab. Dispos.* **22**, 324 (1994).

584. R.H. Pullen, J.J. Brennan, R. Lammers, and C. Patonay, *Anal. Chem.* **67**, 1903 (1995).

585. B. Law, *Trends Anal. Chem.* **9**, 31 (1990).

586. T. Hamoir, Y. Verlinden, and D.L. Massart, *J. Chromatogr. Sci.* **32**, 14 (1994).

587. F. Li, S.F. Cooper, and M. Côté, *J. Chromatogr. B* **668**, 67 (1995).

588. K.L. Salyers, J. Barr, and I.G. Sipes, *Xenobiotica* **24**, 389 (1994).

589. L.E. Los, S.M. Pitzenberger, H.G. Ramjit, A.B. Coddington, and H.D. Colby, *Drug Metab. Dispos.* **22**, 903 (1994).

590. *United States Pharmacopeia XXIV*, The United States Pharmacopeial Convention, Rockville, MD, 1999.

591. K. Hunchak and I.H. Suffet, *J. Chromatogr.* **392**, 185 (1987).

592. *Hexanes*, JT Baker Bakergram, Lit. #3044, 10/93.

593. K.K. Unger and U. Trüdinger, in R.A. Hartwick and P.R. Brown (eds). *High Performance Liquid Chromatography*, Wiley, New York, 1989, Chapter 3.

594. H. Engelhardt, in Cs. Horváth (ed.), *HPLC: Advances and Perspectives*, Volume 2, Academic Press, New York, 1980, Chapter 2.

595. A.M. Krstulovic (ed.), *Chiral Separations by HPLC*. Wiley, New York, 1989.

596. R.F. Rekker, G. de Vries, and G.J. Bijloo, *J. Chromatogr.* **370**, 355 (1986).

597. J. Kríz, J. Puncochárová, L. Vodicka, and J. Vareka, *J. Chromatogr.* **437**, 177 (1988).

598. G. de Vries and R.F. Rekker, *J. Liq. Chromatogr.* **16**, 383 (1993).

599. S.N. Lanin and Yu.S. Nikitin, *J. Chromatogr.* **520**, 315 (1990).

600. V.R. Meyer, *J. Chromatogr. A* **768**, 315 (1997).

601. P.L. Smith and W.T. Cooper, *J. Chromatogr.* **410**, 249 (1987).

602. Z. Deng, J.Z. Zhang, A.B. Ellis, and S.H. Langer, *J. Chromatogr.* **626**, 159 (1992).

603. S.N. Lanin, and Yu.S. Nikitin, *Talanta* **36**, 573 (1989).

604. H.A. Cooper and R.J. Hurtubise, *J. Chromatogr.* **360**, 327 (1986).

605. L.D. Olsen and R.J. Hurtubise, *J. Chromatogr.* **474**, 347 (1989).

606. A.W. Salotto, E.L. Weiser, K.P. Caffey, R.L. Carty, S.C. Racine, and L.R. Snyder, *J. Chromatogr.* **498**, 55 (1990).

607. M. Waksmundzka-Hajnos, *J. Chromatogr.* **623**, 15 (1992).

608. M. Waksmundzka-Hajnos, T. Wawrzynowicz, and T.H. Dzido, *J. Chromatogr.* **600**, 51 (1992).

609. C.A. Chang and Q. Wu, *J. Liq. Chromatogr.* **10**, 1359 (1987).

610. S. Baj and Z. Kulicki, *J. Chromatogr.* **588**, 33 (1991).

611. A.M. Siouffi, M. Righezza, and G. Guiochon, *J. Chromatogr.* **368**, 189 (1986).

612. M. Lübke, J.-L. Le Quéré, and D. Barron, *J. Chromatogr.* **646**, 307 (1993).

613. B. Walczak, M. Dreux, and J.R. Chretien, *Chromatographia* **31**, 575 (1991).

614. D.P. Nowotnik, P. Nanjappan, W. Zeng, and K. Ramalingam, *J. Liq. Chromatogr.* **18**, 673 (1995).

615. K.-C. Tan, T.S.A. Hor, and H.K. Lee, *J. Liq. Chromatogr.* **17**, 3671 (1994).

616. Y. Yamazaki, N. Morohashi, and K. Hosono, *J. Chromatogr.* **542**, 129 (1991).

617. H. Xu and S. Lesage, *J. Chromatogr.* **607**, 139 (1992).

618. P. Fernandez and J.M. Bayona, *J. Chromatogr.* **625**, 141 (1992).

619. B.E. Il, S.N. Lanin, and Yu.S. Nikitin, *Russ. J. Phys. Chem.* (*Eng. Ed.*) **65**, 1399 (1991).

620. J. Zawadiak, B. Orlińska, and Z. Stec, *Fresenius J. Anal. Chem.* **367**, 502 (2000).

621. Y. Ye, C.C. Duke, and G.M. Holder, *Chem. Res. Toxicol.* **8**, 188 (1995).

622. N. Motohashi, K. Kamata, and R. Meyer, in J.C. Giddings, E. Grushka, and P.R. Brown (eds.), *Advances in Chromatography*, Volume 31, Marcel Dekker, New York, 1992, Chapter 6.

623. H.B. Weems and S.K. Yang, *J. Chromatogr.* **535**, 239 (1990).

624. U.L. Nilsson and A.L. Colmsjö, *Chromatographia* **32**, 335 (1991).

625. S.K. Yang, M. Mushtaq, Z. Bao, H.B. Weems, M. Shou, and X.-L. Lu, *J. Chromatogr.* **461**, 377 (1989).

626. P.J. Tancell, M.M. Rhead, C.J. Trier, M.A. Bell, and D.E. Fussey, *Sci. Total Environ.* **162**, 179 (1995).

627. N. Pasadakis and N. Varotsis, *Energy & Fuels* **14**, 1184 (2000).

628. S.R. Raverkar and N.V. Rama Rao, *Chromatographia* **28**, 412 (1989).

629. B.R. McCord and F.W. Whitehurst, *J. Forensic Sci.* **37**, 1574 (1992).

630. H. Zou, S. Zhou, X. Hu, M. Hong, Y. Zhang, and P. Lu, *Anal. Chim. Acta* **291**, 205 (1994).

631. K. Fukuhara, M. Takei, H. Kageyama, and N. Miyata, *Chem. Res. Toxicol.* **8**, 47 (1995).

632. E. Papadopoulou-Mourkidou, J. Patsias, G. Papaddopoulos, and C. Galanis, *J. Assoc. Off. Anal. Chem. Int.* **79**, 829 (1996).

633. W. Vetter, N.P. Costas, R. Bartha, A.G. Martinez, and B. Luckas, *J. Chromatogr. A* **886**, 123 (2000).

634. V. Tatarkovicová, *Acta Un. Pal. Olom. Fac. Rerum Nat. Chem.* **94**, 203 (1989).

635. B. Køppen, *J. Off. Assoc. Anal. Chem. Int.* **77**, 810 (1994).

636. J.P. Kutter and T.J. Class, *Chromatographia* **33**, 103 (1992).

637. M. Okamoto and H. Nakazawa, *J. Chromatogr.* **588**, 177 (1991).

638. A. Meyer and G. Henze, *Fresenius J. Anal. Chem.* **349**, 650 (1994).

639. W. Yang and P.J. Davis, *Drug Metab. Dispos.* **20**, 38 (1992).

640. N. Márquez, B. Bravo, G. Chávez, F. Ysambertt, and J.L. Salager, *Anal. Chim. Acta* **405**, 267 (2000).

641. W. Miszkiewicz, W. Hreczuch, A. Sobczynska, and J. Szymanowski, *Chromatographia* **51**, 95 (2000).

642. N. Márquez, R.E. Antón, A. Usubillaga, and J.L. Salager, *J. Liq. Chromatogr.* **17**, 1147 (1994).

643. E. Kubeck and C.G. Naylor, *J. Am. Oil Chem. Soc.* **67**, 400 (1990).

644. P. Jandera, S. Urbánek, B. Prokes, and S. Churácek, *J. Chromatogr.* **504**, 297 (1990).

645. S. Brossard, M. Lafosse, and M. Dreux, *J. Chromatogr.* **591**, 149 (1992).

646. N. Martin, *J. Liq. Chromatogr.* **18**, 1173 (1995).

647. C. Zhou, A. Bahr, and G. Schwedt, *Anal. Chim. Acta* **236**, 273 (1990).

648. G. Glöckner, *Chromatographia* **37**, 7 (1993).

649. G. Glöckner, D. Wolf, and H. Engelhardt, *Chromatographia* **38**, 559 (1994).

650. B.L. Neff and H.J. Spinelli, *J. Appl. Polym. Sci.* **42**, 595 (1991).

651. N. Sivaraman, R. Dhamodaran, I. Kaliappan, T.G. Srinivasan, P.R. Vasudeva Rao, and C.K. Mathews, in K.M. Kadish and R.S. Ruoff (eds.), *Recent Advances in the Chemistry and Physics of Fullerenes and Related Materials*, The Electrochemical Society, Pennington, NJ, 1994, pp.156–165.

652. L. Nondek and V. Kuzílek, *Chromatographia* **33**, 344 (1992).

653. H. Ohta, Y. Saito, N. Nagae, J.J. Pesek, M.T. Matyska, and K. Jinno, *J. Chromatogr. A* **883**, 55 (2000).

654. H. Ohta, Y. Saito, K. Jinno, H. Nagashima, and K. Itoh, *Chromatographia* **39**, 453 (1994).

655. Y. Saito, H. Ohta, H. Nagashima, K. Itoh, and K. Jinno, *J. Liq. Chromatogr.* **17**, 2359 (1994).

656. Y. Cui, S.T. Lee, S.V. Olesik, W. Flory, and M. Mearini, *J. Chromatogr.* **625**, 131 (1992).

657. F. Diederich and R.L. Whetten, *Acc. Chem. Res.* **25**, 119 (1992).

658. K. Jinno, T. Uemura, H. Ohta, H. Nagashima, and K. Itoh, *Anal. Chem.* **65**, 2650 (1993).

659. C. Welch, LC205, *J. Chromatogr. Sci.* **32**, 250 (1994).

660. A Herrmann, F. Diederich, C. Thilgen, H,-U. ter Meer, and W.H. Müller, *Helv. Chim. Acta* **77**, 1689 (1994).

661. Y. Itabashi, J.J. Myher, and A. Kuksis, *J. Am. Oil Chem. Soc.* **70**, 1177 (1993).

662. Z. Wu, D.S. Robinson, C. Domoney, and R. Casey, *J. Agric. Food Chem.* **43**, 337 (1995).

663. N. Degousée, C. Triantaphylidès, and J.-L. Montillet, *Plant Physiol.* **104**, 945 (1994).

664. H.C. Gérard, R.A. Moreau, W.F. Fett, and S.F. Osman, *J. Am. Oil Chem. Soc.* **69**, 301 (1992).

665. G.N. Nöll and C. Becker, *J. Chromatogr. A* **881**, 183 (2000).

666. A. Kamal-Eldin, S. Görgen, J. Pettersson, and A.-M. Lampi, *J. Chromatogr. A* **881**, 217 (2000).

667. A.A. Qureshi, H. Mo, L. Packer, and D.M. Peterson, *J. Agric. Food Chem.* **48**, 3130 (2000).

668. T.-S. Shin and J.S. Godber, *J. Am. Oil Chem. Soc.* **70**, 1289 (1993).

669. K. Aitzetmüller, Y. Xin, G. Werner, and M. Grönheim, *J. Chromatogr.* **603**, 165 (1992).

670. G. Panfili, P. Manzi, and L. Pizzoferrato, *Analyst* **119**, 1161 (1994).

671. K. Besler, U. Knecht, and G.N. Nöll, *Fresenius J. Anal. Chem.* **350**, 182 (1994).

672. E. Meyer, W.E. Lambert, and A.P. De Leenheer, *Clin. Chem.* **40**, 48 (1994).

673. A.P. De Leenheer and H.J. Nelis, *Methods Enzymol.* **189**, 50 (1991).

674. F. Khachik, G.R. Beecher, M.B. Goli, W.R. Lusby, and J.C. Smith, *Anal. Chem.* **64**, 2111 (1992).

675. B. Stancher, F. Zonta, and L.G. Favretto, *J. Chromatogr.* **440**, 37 (1988).

676. M. Isaksen and G.W. Francis, *Chromatographia* **27**, 325 (1989).

677. S. Hara, T. Ando, and Y. Nakayama, *J. Liq. Chromatogr.* **12**, 729 (1989).

678. T. Ando, Y. Nakayama, and S. Hara, *J. Liq. Chromatogr.* **12**, 739 (1989).

679. J.C. Herrera, A.J. Rosas Romero, O.E. Crescente, M. Acosta, and S. Pekerar, *J. Chromatogr. A* **740**, 201 (1996).

680. A. Verzera, A. Trozzi, A. Cotroneo, D. Lorenzo, and E. Dellacassa, *J. Agric. Food Chem.* **48**, 2903 (2000).

681. C.W. Huck, C.G. Huber, K.-H. Ongania, and G.K. Bonn, *J. Chromatogr. A* **870**, 453 (2000).

682. L. Piovetti, P. Deffo, R. Valls, and G. Peiffer, *J. Chromatogr.* **588**, 99 (1991).

683. J.-P. Rey, J. Levesque, and J.L. Pousset, *J. Chromatogr.* **605**, 124 (1992).

684. M. Krause and R. Galensa, *J. Chromatogr.* **514**, 147 (1990).

685. M.V. Piretti and P. Doghieri, *J. Chromatogr.* **514**, 334 (1990).

686. N. Okamura, K. Kobayashi, A. Yagi, T. Kitazawa, and K. Shimomura, *J. Chromatogr.* **542**, 317 (1991).

687. K. Kuhajda, J. Kandrac, S. Dobanovic-Slavica, J. Hranisavljevic, and D. Miljkovic, *Lipids* **28**, 863 (1993).

688. L. Steenhorst-Slikkerveer, A. Louter, H.-G. Janssen, and C. Bauer-Plank, *J. Amer. Oil Chem. Soc.* **77**, 837 (2000).

689. M. Mäkinen, A. Kamal-Eldin, A.-M. Lampi, and A. Hopia, *J. Am. Oil Chem. Soc.* **77**, 801 (2000).

690. L. Yang, L.K. Leung, Y. Huang, and Z.-Y. Chen, *J. Agric. Food Chem.* **48**, 3072 (2000).

691. A.R. Brash, *Lipids* **35**, 947 (2000).

692. J.-M. Kuo, D.-B. Yeh, and B.S. Pan, *J Agric. Food Chem.* **47**, 3206 (1999).

693. K. Akasaka, H. Ohrui, and H. Meguro, *J. Chromatogr.* **628**, 31 (1993).

694. A. Bruns, *J. Chromatogr.* **536**, 75 (1991).

695. A. Bruns, D. Berg, and A. Werner-Busse, *J. Chromatogr.* **450**, 111 (1988).

696. P.E. Sonnet, R.L. Dudley, S. Osman, P.E. Pfeffer, and D. Schwartz, *J. Chromatogr.* **586**, 255 (1991).

697. J. Liu, T. Lee, E. Bobik, Jr., M. Guzman-Harty, and C. Hastilow, *J. Am. Oil Chem. Soc.* **70**, 343 (1993).

698. Y. Iwasaki, M. Yasui, T. Ishikawa, R. Irimescu, K. Hata, and T. Yamane, *J. Chromatogr. A* **905**, 111 (2001).

699. T.A. Foglia, K. Petruso, and S.H. Feairheller, *J. Am. Oil Chem. Soc.* **70**, 281 (1993).

700. W.W. Christie, B. Nikolova-Damyanova, P. Laakso, and B. Herslof, *J. Am. Oil Chem. Soc.* **68**, 695 (1991).

701. T. Takagi and T. Suzuki, *Lipids* **28**, 251 (1993).

702. T.A. Foglia and K. Maeda, *Lipids* **27**, 396 (1992).

703. J. Fritsche, S. Fritsche, M.B. Solomon, M.M. Mossoba, M.P. Yurawecz, K. Morehouse, and Y. Ku, *Eur. J. Lipid Sci. Technol.* **102**, 667 (2000).

704. L. Ye, W.O. Landen, and R.R. Eitenmiller, *J. Chromatogr. Sci.* **39**, 1 (2001).

705. H.E. Indyk, *Analyst* **115**, 1525 (1990).

706. P.R. Redden and Y.-S. Huang, *J. Chromatogr.* **567**, 21 (1991)

707. T.L. Mounts and A.M. Nash, *J. Am. Oil Chem. Soc.* **67**, 757 (1990).

708. K.C. Arnoldsson and P. Kaufmann, *Chromatographia* **38**, 317 (1994).

709. B.H. Klein and J.W. Dudenhausen, *J. Liq. Chromatogr.* **17**, 981 (1994).

710. Collaborative data, *Pure Appl. Chem.* **64**, 447 (1992).

711. K. Hyvärinen and P.H. Hynninen, *J. Chromatogr. A* **837**, 107 (1999).

712. F.L. Canjura, R.H. Watkins, and S.J. Schwartz, *J Food Sci.* **64**, 987 (1999).

713. M. Rahmani and A.S. Csallany, *J. Am. Oil Chem. Soc.* **68**, 672 (1991).

714. D. Drexler and K. Ballschmiter, *Fresenius J. Anal. Chem.* **348**, 590 (1994).

715. M. Kobayashi, H. Oh-oka, S. Akutsu, M. Akiyama, K. Tominaga, H. Kise, F. Nishida, T. Watanabe, J. Amesz, M. Koizumi, N. Ishida, and H. Kano, *Photosynthesis Res.* **63**, 269 (2000).

716. S. Andersson, *J. Chromatogr.* **606**, 272 (1992).

717. A.E. Cremesti and A.S. Fischl, *Lipids* **35**, 937 (2000).

718. F.B. Jungalwala, M.R. Natowicz, P. Chaturvedi, and D.S. Newburg, *Methods Enzymol.* **311**, 94 (2000).

719. A.D. Tepper and W.J. Van Blitterswijk, *Methods Enzymol.* **312**, 16 (2000).

720. B. Ramstedt and J.P. Slotte, *Anal. Biochem.* **282**, 245 (2000).

721. E. Neufeld, R. Chayen, and N. Stern, *J. Chromatogr. B* **718**, 273 (1998).

722. L. Siret, A. Tambuté, M. Caude, and R. Rosset, *J. Chromatogr.* **498**, 67 (1990).

723. A. Dobashi, K. Oka, and S. Hara, *J. Am. Chem. Soc.* **102**, 7122 (1980).

724. W.H. Pirkle, T.C. Pochapsky, G.S. Mahler, D.E. Corey, D.S. Reno, and D.M. Alessi, *J. Org. Chem.* **51**, 4991 (1986).

725. N. Ôi, H. Kitahara, and R. Kira, *J. Chromatogr.* **535**, 213 (1990).

726. P.A. Husain, J. Debnath, and S.W. May, *Anal. Chem.* **65**, 1456 (1993).

727. T.A. Shepard, J. Hui, A. Chandrasekaran, R.A. Sams, R.H. Reuning, L.W. Robertson, J.H. Caldwell, and R.L. Donnerberg, *J. Chromatogr.* **380**, 89 (1986).

728. H. Nakashima, K. Tsutsumi, M. Hashiguchi, Y. Kumagai, and A. Ebihara, *J. Chromatogr.* **489**, 425 (1989).

729. A. van Overbeke, W. Baeyens, and C. Dewaele, *J. Liq. Chromatogr.* **18**, 2427 (1995).

730. R. Ferretti, B. Gallinella, F. La Torrej, and C. Villani, *J. Chromatogr. A* **704**, 217 (1995).

731. X.-L. Lu, P. Guengerich, and S.K. Yang, *Drug Metab. Dispos.* **19**, 637 (1991).

732. X.-L. Lu and S.K. Yang, *J. Chromatogr.* **535**, 229 (1990).

732a. J.V. Anderson and K.T. Hansen, *Xenobiotica*, **27**, 901 (1997).

733. T. Cleveland, *J. Liq. Chromatogr.* **18**, 649 (1995).

734. S. Pichini, I. Altieri, A.R. Passa, P. Zuccaro, and R. Pacifici, *J. Liq. Chromatogr.* **18**, 1533 (1995).

735. W.D. Hooper, W.F. Pool, T.F. Woolf, and J. Gal, *Drug Metab. Dispos.* **22**, 719 (1994).

736. V.L. Herring and J.A. Johnson, *J. Chromatogr.* **612**, 215 (1993).

737. A.M. Krstulovic, M.H. Fouchet, J.T. Burke, G. Gillet, and A. Durand, *J. Chromatogr.* **452**, 477 (1988).

738. A.M. Krstulovic, *J. Chromatogr.* **488**, 53 (1989).

739. G.W. Ponder, S.L. Butram, A.G. Adams, C.S. Ramanathan, and J.T. Stewart, *J. Chromatogr. A* **692** 173 (1995).

740. S.A. Jortani and A. Poklis, *J. Anal. Toxicol.* **17**, 374 (1993).

741. B.L. Blake, R.L. Rose, R.B. Mailman, P.E. Levi, and E. Hodgson, *Xenobiotica* **25**, 377 (1995).

742. C. Seaver, J. Przybytek, and N. Roelofs, *LC/GC* **13**, 860 (1995).

743. *Reagent Chemicals*, 9th edition, American Chemical Society, Washington DC, 2000, p. 237.

744. W.J.Th. Brugman, S. Heemstra, and J.C. Kraak, *J. Chromatogr.* **218**, 285 (1981).

745. L. Szepesy, C. Combellas, M. Caude, and R. Rosset, *J. Chromatogr.* **237**, 65 (1982).

746. S.C. Ruckmick and R.J. Hurtubise, *J. Chromatogr.* **360**, 343 (1986).

747. M. Murray and A.M. Butler, *Chem. Res. Toxicol.* **7**, 792 (1994).

748. J. Lehotay, F. Brandsteterová, and D. Oktavec, *J. Liq. Chromatogr.* **15**, 525 (1992).

749. T. Greibrokk, B.E. Berg, S. Hoffmann, H.R. Norli, and Q. Ying, *J. Chromatogr.* **505**, 283 (1990).

750. R.A. Shalliker, P.E. Kavanagh, and I.M. Russell, *J. Chromatogr.* **543**, 157 (1991).

751. R.A. Shalliker, P.E. Kavanagh, I.M. Russell, and D.G. Hawthorne, *Chromatographia* **33**, 427 (1992).

752. H. Sato, K. Ogino, S. Maruo, and M. Sasaki, *J. Polym. Sci., Part B. Polym. Phys.* **29**, 1073 (1991).

753. T.M. Zimina, J.J. Kever, E.Yu. Melenevskaya, and A.F. Fell, *J Chromatogr.* **593**, 233 (1992).

754. S. Mori, *Anal. Chem.* **62**, 1902 (1990).

755. S. Mori and M. Mouri, *Anal. Chem.* **61**, 2171 (1989).

756. S. Mori, *J. Chromatogr.* **541**, 375 (1991).

757. R. Schultz and H. Engelhardt, *Chromatographia* **29**, 325 (1990).

758. S. Mori, *J. Chromatogr.* **503**, 411 (1990).

759. Y. Saito, H. Ohta, H. Nagashima, K. Itoh, K. Jinno, and J.J. Pesek, *J. Microcolumn Sep.* **7**, 41 (1995).

760. R. Ettl, I. Chao, F. Diederich, and R.L. Whetten, *Nature* **353**, 149 (1991).

761. J.F. Anacleto, H. Perreault, R.K. Boyd, S. Pleasance, M.A. Quilliam, P.G. Sim, J.B. Howard, Y. Makarovsky, and A.L. Lafleur, *Rapid Commun. Mass Spectrom.* **6**, 214 (1992).

762. S. Kermasha, S. Kubow, M. Safari, and A. Reid, *J. Am. Oil Chem. Soc.* **70**, 169 (1993).

763. W.W. Christie and G.H.McG. Breckenridge, *J. Chromatogr.* **469**, 261 (1989).

764. G.I. Rehner and J. Stein, *Methods Enzymol.* **279**, 286 (1997).

765. C.A. O'Neil and S.J. Schwartz, *J. Chromatogr.* **624**, 235 (1992).

766. T.M. Chen and B.H. Chen, *Chromatographia* **39**, 346 (1994).

767. M.H. Saleh and B. Tan, *J. Agric. Food Chem.* **39**, 1438 (1991).

768. K.J. Scott, *Food Chem.* **45**, 357 (1992).

769. B. Schoefs, M. Bertrand, and Y. Lemoine, *J. Chromatogr. A* **692**, 239 (1995).

770. C.J. Hogarty, C. Ang, and R.R. Eitenmiller, *J. Food Compos. Anal.* **2**, 200 (1989).

771. A. Guillou, G. Choubert, and J. de la Noüe, *Food Chem.* **46**, 93 (1993).

772. G. Jones, D. J. Hamilton-Trafford, H.L.J. Makin, and B.W. Hollis, in A.P. De Leenheer, W.E. Lambert, and H.J. Nelis (eds.), *Modern Chromatographic Analysis of Vitamins*, 2nd edition, Marcel Dekker, New York, 1992, Chapter 2.

773. K. Gaudin, P. Chaminade, D. Ferrier, and A. Baillet, *J. Liq. Chromatogr.* **23**, 387 (2000).

774. T. Sugawara and T. Miyazawa, *Lipids* **34**, 1231 (1999).

775. W.C. Byrdwell and W.E. Neff, *J. Chromatogr. A* **905**, 85 (2001).

776. P. Laakso and H. Kallio, *J Am. Oil Chem. Soc.* **70**, 1161 (1993).

777. A.J. Palmer and F.J. Palmer, *J. Chromatogr.* **465**, 369 (1989).

778. W.E. Neff, R.O. Adolf, H. Konishi, and D. Weisleder, *J. Am. Oil Chem. Soc.* **70**, 449 (1993).

779. S. Héron, E. Lesellier, and A. Tchapla, *J. Liq. Chromatogr* **18**, 599 (1995).

780. M.N. Vaghela and A. Kilara, *J. Am. Oil Chem. Soc.* **72**, 729 (1995).

781. A. Kawamura, N. Berova, V. Dirsch, A. Mangoni, K. Nakanishi, G. Schwartz, A. Bielawska, Y. Hannun, and I. Kitagawa, *Bioorg. Med. Chem.* **4**, 1035 (1996).

782. T.J. McNabb, A.E. Cremesti, P.R. Brown, and A.S. Fischl, *Anal. Biochem.* **276**, 242 (1999).

783. S.A. Lazarus, G.E. Adamson, J.F. Hammerstone, and H.H. Schmitz, *J. Agric. Food. Chem.* **47**, 3693 (1999).

784. J.P.E. Spenser, F. Chaudry, A.S. Pannala, S.K. Srai, E. Debman, and C. Rice-Evans, *Biochem. Biophys. Res. Commun.* **272**, 236 (2000).

785. M. Baltas, M. Benbakkar, L. Gorrichon, and C. Zedde, *J. Chromatogr.* **600**, 323 (1992).

786. E. Guichard, T.T. Pham, and P. Etievant, *Chromatographia* **37**, 539 (1993).

787. M.I. Selala, A. Musuku, and P.J.C. Schepens, *Anal. Chim. Acta* **244**, 1 (1991).

788. L. Scott Ramos, *J. Chromatogr. Sci.* **32**, 219 (1994).

789. G. Bagur, M. Sánchez, and D. Gázquez, *Analyst* **119**, 1157 (1994).

790. Gy. Szókán, Zs. Majer, E. Kollát, M. Kajtár, M. Hollósi, and M. Peredy-Kajtár, *J. Liq. Chromatogr.* **18**, 941 (1995).

791. T.V. Alfredson, P.W. Bruins, A.H. Maki, and J.-L. Excoffier, *J. Chromatogr. Sci.* **32**, 132 (1994).

792. K.A. Lerro, R. Orlando, H. Zhang, P.N.R. Usherwood, and K. Nakanishi, *Anal. Biochem.* **215**, 38 (1993).

793. E. Reich and A.T. Sneden, *J. Chromatogr. A* **763**, 213 (1997).

794. V.K. Dua, R. Sarin, and A. Prakash, *J. Chromatogr.* **614**, 87 (1993).

795. H.-C. Lin, S.-C. Chang, N.-L. Wang, and L.-R. Chang, *J. Antibiot.* **47**, 675 (1993).

796. W.H. Pirkle and J.A. Burke III, *J. Chromatogr.* **557**, 173 (1991).

797. A.M. El Walily, M.A. Korany, M.M. Bedair, and A. El Gindy, *Anal. Lett.* **23**, 473 (1990).

798. S. Patai (ed.), *The Chemistry of the Ether Linkage*, Wiley, New York, 1967, p. 693.

799. D.D. Perrin and W.L.F. Armarego, *Purification of Laboratory Chemicals*, 3rd edition, Pergamon Press, Oxford, 1988.

800. M. Borremans, J. De Beer, and L. Goeyens, *Chromatographia* **50**, 346 (1999).

801. C.H. Lochmüller, C. Reese, and S.-H. Hsu, *J. Chromatogr. Sci.* **33**, 640 (1995).

802. Machery-Nagel Application Note 145. Printed 1991.

803. L.C. Sander and S.A. Wise, *J. Chromatogr. A* **656**, 335 (1993).

804. L.C. Sander and S.A. Wise, *Anal. Chem.* **61**, 1749 (1989).

805. H. Colin, J.-M. Schmitter, and G. Guiochon, *Anal. Chem.* **53**, 625 (1981).

806. H.J. Issaq, I.Z. Atamna, N.M. Schultz, G.M. Muschik, and J.E. Saavedra, *J. Liq. Chromatogr.* **12**, 771 (1989).

807. T. Cserháti and E. Forgács, *J. Chromatogr.* **643**, 331 (1993).

808. E. Forgács and T. Cserháti, *Analyst* **120**, 1941 (1995).

809. J.A. Stäb, M.J.M. Rozing, B. van Hattum, W.P. Cofino, and U.A.Th. Brinkman, *J. Chromatogr.* **609**, 195 (1992).

810. H. El Mansouri, N. Yagoubi, and D. Ferrier, *Chromatographia* **48**, 491 (1998).

811. H.J.A. Philipsen, H.A. Claessens, P. Jandera, M. Bosman, and B. Klumperman, *Chromatographia* **52**, 325 (2000).

812. G.R. Bear, *J. Chromatogr.* **459**, 91 (1988).

813. M.A. Castles, B.L. Moore, and S.R. Ward, *Anal. Chem.* **61**, 2534 (1989).

814. J.V. Dawkins, T.A. Nicholson, A.J. Handley, E. Meehan, A. Nevin, and P.L. Shaw, *Polymer* **40**, 7331 (1999).

815. G. Glöckner, D. Wolf, and H. Engelhardt, *Chromatographia* **39**, 170 (1994).

816. S. Teramachi, A. Hasegawa, T. Matsumoto, K. Kitahara, Y. Tsukahara, and Y. Yamashita, *Macromolecules* **25**, 4025 (1992).

817. S. Teramachi, A. Hasegawa, and T. Matsumoto, *J. Chromatogr.* **547**, 429 (1991).

818. H.J.A. Philipsen, M. Oestreich, B. Klumperman, and A.L. German, *J. Chromatogr. A* **775**, 157 (1997).

819. S. Podzimek, *Chromatographia* **33**, 377 (1992).

820. D.P. Sheih and D.E. Benton, in W.C. Golton (ed.), *Analysis of Paints and Related Materials: Current Techniques for Solving Coating Problems*, STP 1119, ASTM Publications, Philadelphia, PA, 1992, pp. 41–55.

821. T.C. Schunk, *J. Chromatogr. A* **656**, 591 (1993).

822. J.J. Harwood and G. Mamantov, *J. Chromatogr. A* **654**, 591 (1993).

823. K.L. Calabrese, *J Chromatogr.* **386**, 199 (1987).

824. C.E. Evans and V.L. McGuffin, *Anal. Chem.* **63**, 1393 (1991).

825. R.L. Rouseff, G.R. Dettweiler, R.M. Swaine, M. Naim, and U. Zehavi, *J. Chromatogr. Sci.* **30**, 383 (1992).

826. A. Gachanja and P. Worsfold, *Anal. Chim. Acta* **290**, 226 (1994).

827. K. Warner and T.L. Mounts, *J. Am. Oil Chem. Soc.* **67**, 827 (1990).

828. L.C. Sander, K.E. Sharpless, N.E. Craft, and S.A. Wise, *Anal Chem.* **66**, 1667 (1994).

829. Y.-L. Su, L.K. Leung, Y.-R. Bi, Y. Huang, and Z.-Y. Chen, *J. Am. Oil Chem. Soc.* **77**, 807 (2000).

830. M.-H. Pan, T.-M. Huang, and J.-K. Lin, *Drug Metab. Dispos.* **27**, 486 (1999).

831. N. Kozukue, S. Misoo, T. Yamada, O. Kamijima, and M. Friedman, *J. Agric. Food Chem.* **47**, 4478 (1999).

832. N. Kozukue, H. Tsuchida, and M. Friedman, *J. Agric. Food Chem.* **49**, 92 (2001).

833. H. Ohta, Y. Seto, and N. Tsunoda, *J. Chromatogr. B* **691**, 351 (1997).

834. A. Finger, U.H. Engelhardt, and V. Wray, *J. Sci. Food Agric.* **55**, 313 (1991).

835. A. Finger, S. Kuhr, and U.H. Engelhardt, *J. Chromatogr.* **624**, 293 (1992).

836. T.A. Foglia and K.C. Jones, *J. Liq. Chromatogr.* **20**, 1829 (1997).

837. T. Rezanka and P. Mares, *J. Chromatogr.* **542**, 145 (1991).

838. K. Koba, L.A. Rozee, D.F. Horrobin, and Y.-S. Huang, *Lipids* **29**, 33 (1994).

839. S. Porretta, *J. Chromatogr.* **624**, 211 (1992).

840. C.E. Hendrich, J. Berdecia-Rodriguez, V.T. Wiedmeier, and S.P. Porterfield, *J. Chromatogr.* **577**, 19 (1992).

841. G. Achilli, G.P. Cellerino, and G.M. d'EriI, *J Chromatogr. A* **661**, 201 (1994).

842. R.K. Munns, S.D. Turnipseed, A.P. Pfenning, J.E. Roybal, D.C. Holland, A.R. Long, S.M. Plakas, and J.M. Storey, *J. Assoc. Off. Anal. Chem. Int.* **81**, 825 (1998).

843. A.M. Kaukonen, P. Vuorela, H. Vuorela, and J.-P. Mannermaa, *J. Chromatogr. A* **797**, 271 (1998).

844. B.C. McWhinney, G. Ward, and P.E. Hickman, *Clin. Chem.* **42**, 979 (1996).

845. M.A. Quarry, R.C. Williams, and D.S. Sebastian, *J. Liq. Chromatogr.* **21**, 2841 (1998).

846. P. Parra, A. Limon, S. Ferre, T. Guix, and F. Jane, *J. Chromatogr.* **570**, 185 (1991).

847. A.M. Di Pietra, V. Cavrini, D. Bonazzi, and L. Benfenati, *Chromatographia* **30**, 215 (1990).

848. N. Beaulieu, R.W. Sears, and E.G. Lovering, *J Assoc. Off. Anal. Chem. Int.* **77**, 857 (1994).

849. L. Elrod, Jr., T.G. Golich, and J.A. Morley, *J Chromatogr.* **625**, 362 (1992).

850. G.L. Lensmeyer, C. Onsager, I.H. Carlson, and D.A. Wiebe, *J. Chromatogr. A* **691**, 239 (1995).

851. D.M. McDaniel and B.G. Snider, *J. Chromatogr.* **404**, 123 (1987).

852. C. Wu, A. Akiyama, and J.A. Straub, *J. Chromatogr. A* **684**, 243 (1994).

853. J. Muztar, G. Chari, R. Bhat, S. Ramarao, and D. Vidyasagar, *J Liq. Chromatogr.* **18**, 2635 (1995).

854. R.L. Reeves, in S. Patai (ed.), *The Chemistry of the Carbonyl Group*, Wiley, 1966, pp. 588–591.

855. I. Mellan, *Ketones*, Chemical Publishing, New York, 1968, Chapter 1.

856. L.D. Olsen and R.J. Hurtubise, *J. Chromatogr.* **474**, 347 (1989).

857. S.C. Rastogi and S.S. Johansen, *J. Chromatogr. A* **692**, 53 (1995).

858. H. Pasch, H. Much, and G. Schulz, *J. Appl. Polym. Sci.: Appl. Polym. Symp.* **52**, 79 (1993).

859. D.G. Pereira, A.R.M. Souza-Brito, and N. Durán, *J. Chromatogr. B* **728**, 117 (1999).

860. A. Bortolotti, G. Luchini, M.M. Barzago, F. Stellari, and M. Bonati, *J. Chromatogr.* **617**, 313 (1993).

861. G.M. Landers and J.A. Olson, *J. Chromatogr.* **438**, 383 (1988).

862. M.T.G. Hierro, M.C. Tomas, F. Fernandez-Martin, and G. Santa-María, *J. Chromatogr.* **607**, 329 (1992).

863. K.V. Nurmela and L.T. Satama, *J. Chromatogr.* **435**, 139 (1988).

864. H.M. Ghazali, S. Hamidah, and Y.B. Che Man, *J. Am. Oil Chem. Soc.* **72**, 633 (1995).

865. F. Jahaniaval, Y. Kakuda, and M.F. Marcone, *J. Am. Oil Chem. Soc.* **77**, 847 (2000).

866. Y. Kakuda, F. Jahaniaval, M.F. Marcone, L. Montevirgen, Q. Montevirgen, and J. Umali, *J. Amer. Oil Chem. Soc.* **77**, 991 (2000).

867. J.M. Bland, E.J. Conkerton, and G. Abraham, *J. Am. Oil Chem. Soc.* **68**, 840 (1991).

868. E. Salivaras and A.R. McCurdy, *J. Am. Oil Chem. Soc.* **69**, 935 (1992).

869. R.V. Flor, L. Tiet Hecking, and B.D. Martin, *J Am. Oil Chem. Soc.* **70**, 199 (1993).

870. G. Semporé and J. Bézard, *J. Am. Oil Chem. Soc.* **68**, 702 (1991).

871. D. Firestone, *J. Assoc. Off. Anal. Chem. Int.* **77**, 954 (1994).

872. K.D. Dotson, J.P. Jerrell, M.F. Picciano, and E.G. Perkins, *Lipids* **27**, 933 (1992).

873. Y. Shimada, T. Nagao, Y. Hamasaki, K. Akimoto, A. Sugihara, S. Fujikawa, S. Komemushi, and Y. Tominaga, *Lipids* **77**, 89 (2000).

874. A.H. El-Hamdy and N.K. El-Fizga, *J. Chromatogr. A* **708**, 351 (1995).

875. A. Bruns, *J. Chromatogr.* **536**, 75 (1991).

876. R.B. van Breemen, F.L. Canjura, and S.J. Schwartz, *J. Chromatogr.* **542**, 373 (1991).

877. S. Li and H. Inoue, *Chromatographia* **33**, 567 (1992).

878. S.J. Taylor and I.J. McDowell, *J. Sci. Food Agric.* **57**, 287 (1991).

879. J. Val, E. Monge, and N.R. Baker, *J. Chromatogr. Sci.* **32**, 286 (1994).

880. P. Woitke, C.-D. Martin, S. Nicklisch, and J.-G. Kohl, *Fresenius J. Anal. Chem.* **348**, 762 (1994).

881. M. Amin, K. Harrington, and R. von Wandruska, *Anal. Chem.* **65**, 2346 (1993).

882. C. Vaccher, P. Berthelot, N. Flouquet, and N. Debaert, *J. Chromatogr.* **542**, 502 (1991).

883. J.O. Egekeze, M.C. Danielski, N. Grinberg, G.B. Smith, D.R. Sidler, H.J. Perpall, G.R. Bicker, and P.C. Tway, *Anal. Chem.* **67**, 2292 (1995).

884. J.F. Coetzee and MW. Martin, in J.F. Coetzee (ed.), *Recommended Methods for Purification of Solvents and Tests for Impurities*, Pergamon Press, Oxford, 1982, pp. 10–15.

885. M. Asthana and L.M. Mukherjee, in J.F. Coetzee (ed.), *Recommended Methods for Purification of Solvents and Tests for Impurities*, Pergamon Press, Oxford, 1982, pp. 44–46.

886. J. Juillard, in J.F. Coetzee (ed.), *Recommended Methods for Purification of Solvents and Tests for Impurities*, Pergamon Press, Oxford, 1982, pp. 32–37.

887. N.E. Hoffman, S.-L. Pan, and A.M. Rustum, *J. Chromatogr.* **465**, 189 (1989).

888. D. Vukmanic and M. Chiba, *J. Chromatogr.* **485**, 189 (1989).

889. S. Perlman and J.J. Kirschbaum, *J. Chromatogr.* **357**, 39 (1986).

890. F.M. Pirisi, A. Angioni, P. Cabras, V.L. Garau, M.T. Sanjust di Teulada, M. Kanim dos Santos, and G. Bandino, *J. Chromatogr. A* **768**, 207 (1997).

891. E. Ferrar, A. Alegría, G. Courtois, and R. Farré, *J. Chromatogr. A* **881**, 599 (2000).

892. W. Liu and H.K. Lee, *J. Chromatogr. A* **805**, 109 (1998).

893. M. Buratti, G. Brambilla, S. Fustinoni, O. Pellegrino, S. Pulviremti, and A. Colombi, *J. Chromatogr.* **751**, 305 (2001).

894. C.U. Galdiga and T. Greibrokk, *J. Liq. Chromatogr.* **21**, 855 (1998).

895. T. Hanai, H. Hatano, N. Nimura, and T. Kinoshita, *Analyst* **119**, 1167 (1994).

896. B. Bourguignon, F, Marcenac, H.R. Keller, P.F. de Aguiar, and D.L. Massart, *J. Chromatogr.* **628**, 171 (1993).

897. A. Di Corcia and R. Samperi, *Anal. Chem.* **62**, 1490 (1990).

898. M. Ulgen, J.W. Gorrod, and D. Barlow, *Xenobiotica* **24**, 735 (1994).

899. N.A. Penner and P.N. Nesterenko, *The Analyst* **125**, 1249 (2000).

900. H.M. Pylypiw, Jr. and M.T. Grether, *J. Chromatogr. A* **883**, 299 (2000).

901. Y.-C. Wu and S.-D. Huang, *J. Chromatogr. A* **835**, 127 (1999).

902. D.M. Goldberg, B. Hoffman, J. Yang, and G.J. Soleas, *J. Agric. Food Chem.* **47**, 3978 (1999).

903. M.A. Abuirjeie, M.E. Abdel-Hamid, A.A. Abdel-Aziz, and E.-S.A. Ibrahim, *Anal. Lett.* **23**, 67 (1990).

904. Y. Okamoto, S. Kitamura, M. Takeshita, and S. Ohta, *IUBMB Life* **48**, 543 (1999).

905. M.J.H. Van Haandel, F.C.E. Saraber, M.G. Boersma, C. Laane, Y. Fleming, H. Weenen, and I.M.C.M. Rietjens, *J. Agric. Food Chem.* **48**, 1949 (2000).

906. Y. Itoh, F.H. Ma, H. Hoshi, M. Oka, K. Noda, Y. Ukai, H. Kojima, T. Nagano, and N. Toda, *Anal. Biochem.* **287**, 203 (2000).

907. E. Makatsori, K. Fermani, A. Aletras, N.K. Karamanos, and T. Tsegenidis, *J. Chromatogr. B* **712**, 23 (1998).

908. O. Fiehn, T. Reemtsma, and M. Jekel, *Anal. Chim. Acta* **295**, 297 (1994).

909. C.-l.W. Hsu and T.L. White, *J. Chromatogr. A* **828**, 461 (1998).

910. Y. Miura and A. Kawaoi, *J. Chromatogr. A* **884**, 81 (2000).

911. C. Cerami, X. Zhang, P. Ulrich, M. Bianchi, K.J. Tracey, and B.J. Berger, *J. Chromatogr. B* **675**, 71 (1996).

912. I. Rodriguez, S.F.Y. Li, B.F. Graham, and R.D. Trengove, *J. Liq. Chromatogr.* **20**, 1197 (1997).

913. M. Asakawa, D. Pasini, F.M. Raymo, and J.F. Stoddart, *Anal. Chem.* **68**, 3879 (1996).

914. S. Kitamura, Y. Okamoto, M. Takeshita, and S. Ohta, *Drug Metab. Dispos.* **27**, 767 (1999).

915. W. Pötter, S. Lamotte, H. Engelhardt, and U. Karst, *J. Chromatogr. A* **786**, 47 (1997).

916. M. Possanzini, V. Di Palo, E. Brancaleoni, M. Frattoni, and P. Ciccioli, *J. Chromatogr. A* **883**, 171 (2000).

917. H.-Y. Wu and J.-K. Lin, *Anal. Chem.* **67**, 1603 (1995).

918. B.E. Miller and N.D. Danielson, *Anal. Chem.* **60**, 622 (1988).

919. M.J.M. Wells and C.R. Clark, *Anal. Chem.* **64**, 1660 (1992).

920. H.M.A. Killa and D.L. Rabenstein, *Anal. Chem.* **60**, 2283 (1988).

921. Y. Nagaosa and T. Ishida, *J. Liq. Chromatogr.* **21**, 693 (1998).

922. J. Miura, *Anal. Chem.* **62**, 1424 (1990).

923. J.-F. Jen and S.-M. Yang, *Anal. Chim. Acta* **28**, 997 (1994).

924. C. Li, T.K.L. Neo, H.K. Lee, and T.S.A. Hor, *J. Liq. Chromatogr.* **20**, 1959 (1997).

925. R.A. Parise, D.R. Miles, and M.J. Egorin, *J. Chromatogr. B* **749**, 145 (2000).

926. D.Y. Liu, Z.D. Liu, S.L. Lu, and R.C. Hider, *J. Chromatogr. B* **730**, 135 (1999).

927. P.R. Haddad and N.E. Rochester, *Anal. Chem.* **60**, 536 (1988).

928. R. Falter and H.F Schöler, *Fresenius J. Anal. Chem.* **353**, 34 (1995).

929. G.B. Scarfe, J.C. Lindon, J.K. Nicholson, B. Wright, E. Clayton, and I.D. Wison, *Drug Metab. Dispos.* **27**, 1171 (1999).

930. M.D. Andrés, B. Cañas, R.C. Izquierdo, L.M. Polo, and P. Alarcón, *J. Chromatogr.* **507**, 399 (1990).

931. B. Motamed, J.-L. Böhm, D. Hennequin, H. Texier, R. Mosrati, and D. Barillier, *Analusis* **28**, 592 (2000).

932. A. Di Corcia, S. Marchese, and R. Samperi, *J Chromatogr.* **642**, 175 (1993).

933. V. Coquart and M.-C. Hennion, *Chromatographia* **37**, 392 (1993).

934. S.A. Wise, L.C. Sander, and W.E. May, *J. Chromatogr.* **642**, 329 (1993).

935. S.A. Wise, L.C. Sander, R. Lapouyade, and P. Garrigues, *J. Chromatogr.* **514**, 111 (1990).

936. P.R. Loconto, *J Chromatogr. A* **774**, 223 (1997).

937. C.-T. Kuo and H.-W. Chen, *J. Chromatogr. A* **897**, 393 (2000).

938. M. Wilhelm, G. Matuschek, and A. Kettrup, *J. Chromatogr. A* **878**, 171 (2000).

939. N.K. Karamanos and T. Tsegendis, *J. Liq. Chromatogr.* **19**, 2247 (1996).

940. M.G. Knize, C.P. Salmon, E.C. Hopmans, and J.S. Felton, *J. Chromatogr. A* **763**, 179 (1997).

941. P.A. Guy, E. Gremaud, J. Richoz, and R.J. Turesky, *J. Chromatogr. A* **883**, 89 (2000).

942. Z. Yu, W.L. Hayton, and K.K. Chan, *Drug Metab. Dispos.* **25**, 431 (1997).

943. J. Szeliga, H. Lee, R.G. Harvey, J.E. Page, H.L. Ross, M.N. Routledge, B.D. Hilton, and A. Dipple, *Chem. Res. Toxicol.* **7**, 420 (1994).

944. M. Sayama, M.-A. Mori, Y. Maruyama, M. Inoue, and H. Kozuka, *Xenobiotica* **23**, 123 (1993).

945. F. Ahmad and D.J. Roberts, *J. Chromatogr. A* **693**, 167 (1995).

946. M. Sayama, M. Inoue, M.-A. Mori, Y. Maruyama, and H. Kozuka, *Xenobiotica* **22**, 633 (1992).

947. M. Zheng, C. Fu, and H. Xu, *Analyst* **118**, 269 (1993).

948. J. Barek, P.T. Hai, V. Pacáková, K. Stulík, I. Svagrová, and J. Zima, *Fresenius J. Anal. Chem.* **350**, 678 (1994).

949. A.G. Frenich, J.L. Martínez Vidal, P. Parrilla, and M. Martínez Galera, *J. Chromatogr. A* **778**, 183 (1997).

950. P. Parrilla and J.L. Martinez Vidal, *Anal. Lett.* **30**, 1719 (1997).

951. L. Piedra, A. Tejedor, M.D. Hernado, A. Aguera, D. Barcelo, and A. Frenández-Alba, *Chromatographia* **52**, 631 (2000).

952. M.D. Müller, T. Poiger, and H.-R. Buser, *J. Agric. Food Chem.* **49**, 42 (2001).

953. P. Parilla, M. Martinez Galera, J.L. Martinez Vidal, and A. Garrido Frenich, *Analyst* **119**, 2231 (1994).

954. P. Parilla, J.L. Martinez Vidal, M. Martinez Galera, and A. Garrido Frenich, *Fresenius J. Anal. Chem.* **350**, 633 (1994).

955. J.M. Huen, R. Gillard, A.G. Mayer, B. Baltensperger, and H. Kern, *Fresenius J. Anal Chem.* **348**, 606 (1994).

956. J. Schülein, D. Martens, P. Spitzauer, and A. Kettrup, *Fresenius J. Anal. Chem.* **352**, 565 (1995).

957. B.P. Ioerger and J.S. Smith, *J. Agric. Food Chem.* **41**, 303 (1993).

958. N. Ismail, M. Vairamani, and R.N. Rao, *J. Chromatogr. A* **903**, 255 (2000).

959. E. Corta, A. Bakkali, L.A. Berrueta, B. Gallo, and F. Vicente, *Talanta* **48**, 189 (1999).

960. J.R. Startin, M.D. Sykes, J.C. Taylor, S.J. Hird, K. Jackson, R.J. Fussell, and A.R.C Hill, *J. Assoc. Off. Anal Chem. Int.* **83**, 735 (2000).

961. A.R. Fernández-Alba, A. Tejedor, A. Agüera, M. Contreras, and J. Garrido, *J. Assoc. Off. Anal. Chem. Int.* **83**, 748 (2000).

962. F.J. Schenck and L.H. Lagman, *J. Assoc. Off. Anal. Chem. Int.* **82**, 1340 (1999).

963. A.W. Abu-Qare and M.B. Abou-Donia, *J. Chromatogr. B* **749**, 171 (2000).

964. M.Z. Mesmer and R.A. Flurer, *J Chromatogr. A* **891**, 249 (2000).

965. E. Rodriguez, Z. Gomez de Balugera, M.A. Goicolea, and R.J. Barrio, *J. Liq. Chromatogr.* **21**, 1857 (1998).

966. H.S. Lee, S. Jeong, K. Kim, J.H. Kim, S.K. Lee, B.H. Kang, and J.K. Roh, *Xenobiotica* **27**, 423 (1997).

967. M.C. Bermúdez-Almada, MG. Pérez-Tello, A.I. Valenzuela-Quintanar, and L. Vázquez-Moreno, *J. Food Sci.* **64**, 638 (1999).

968. T.M. Schreier, V.K. Dawson, Y. Choi, N.J. Spanjers, and M.A. Boogaard, *J. Agric. Food Chem.* **48**, 2212 (2000).

969. Y.T. Gebreegzi, G.D. Foster, and S.U. Khan, *J. Agric. Food Chem.* **48**, 5165 (2000).

970. R.S. Tjeerdema and D.G. Crosby, *Xenobiotica* **22**, 681 (1992).

971. G.C. Mattern, G.D. Parker, DL. Green, and G.L. Yeutter, *J. Assoc. Off. Anal. Chem. Int.* **78**, 1286 (1995).

972. W. Blaß, *Fresenius J. Anal. Chem.* **339**, 340 (1991).

973. E. Milchert and W. Pazdzioch, *Analyst* **119**, 1493 (1994).

974. W.H. Newsome, B.P.-Y. Lau, D. Ducharme, and D. Lewis, *J. Assoc. Off. Anal. Chem. Int.* **78**, 1312 (1995).

975. T. Galeano Díaz, M.I. Acedo, A. Muñoz de la Peña, M. Sánchez Peña, and F. Salinas, *Analyst* **119**, 1151 (1994).

976. E.R. Brouwer, E.A. Struys, J.J. Vreuls, and U.A.Th. Brinkman, *Fresenius J. Anal. Chem.* **350**, 487 (1994).

977. S.S. Dehal and D. Kupfer, *Drug Metab. Dispos.* **22**, 937 (1994).

978. E.A.P. Kuijpers, J. den Hartigh, T.J.F. Savelkoul, and F.A. de Wolff, *J Anal. Toxicol*, **19**, 557 (1995).

979. J.R. Meinertz, L.J. Schmidt, G.R. Stehly, and W.H. Gingerich, *J. Assoc. Off. Anal. Chem. Int.* **82**, 1064 (1999).

980. J.-W. Chen, B.-L. Liu, and Y.-M. Tzeng, *J. Chromatogr. A* **830**, 115 (1999).

981. M.A. Kelm, M.G. Nair, and R.A. Schutzki, *Int. J. Pharmacogn.* **35**, 84 (1997).

982. D. Kasaj, A. Reider, L. Krenn, and B. Kopp, *Chromatographia* **50**, 607 (1999).

983. J.P. Alcantara-Licudine, N.L. Bui, Q.X. Li, G.T. McQuate, and S.L. Peck, *J. Assoc. Off. Anal. Chem.* **83**, 563 (2000).

984. S.D. West and L.G. Turner, *J Agric. Food Chem.* **48**, 366 (2000).

985. J.M. Sanchis-Mallols, S. Sagrado, M.J. Medine-Hernández, R.M.V. Camañas, and E. Bonet-Domingo, *J. Liq. Chromatogr.* **21**, 869 (1998).

986. A. Laganà, A. Bacaloni, C. Fago, and A. Marino, *Chromatographia* **52**, 352 (2000).

987. L.J. Marek, W.C. Koskinen, and G.A. Bresnahan, *J. Agric. Food Chem.* **48**, 2797 (2000).

988. M. Ventriglia, P. Restani, P. Morrica, P. David, L. De Angelis, and C.L. Galli, *J. Chromatogr. A* **857**, 327 (1999).

989. L.B. Perkins, R.J. Bushway, and L.E. Katz, *J Assoc. Offic. Anal. Chem. Int.* **82**, 1505 (1999).

990. M.S. Young, *J. Assoc. Off. Anal. Chem.* **81**, 99 (1998).

991. B. Berger, *J. Chromatogr. A* **769**, 338 (1997).

992. P. Cabras, M. Melis, L. Spanedda, and C. Tuberoso, *J. Chromatogr.* **585**, 164 (1991).

993. V. Coquart and M.-C. Hennion. *J. Chromatogr.* **585**, 67 (1991).

994. M. Berg, S.R. Müller, and R.P. Schwarzenbach, *Anal. Chem.* **67**, 1860 (1995).

995. T.R. Steinheimer, *J Agric. Food Chem.* **41**, 588 (1993).

996. J.I. Ademola, L.E. Sedik, R.C. Wester, and H.I. Maibach, *Arch. Toxicol.* **67**, 85 (1993).

997. J.B. Fischer and J.L. Michael, *J. Chromatogr. A* **704**, 131 (1995).

998. T.M. Chichila and S.M. Walters, *J. Assoc. Off. Anal. Chem. Int.*,**74**, 961 (1991).

999. P.J. Lòpez, M. Pujol, and G. Andreu, *Anal. Lett.* **31**, 1351 (1998).

1000. Y. Yamazaki and T. Ninomiya, *J Assoc. Off. Anal. Chem. Int.* **81**, 1252 (1998).

1001. S.M. Delaney, D.V. Mavrodi, R.F. Bonsall, and L.S. Thomashow, *J. Bacteriol.* **183**, 318 (2001).

1002. Y. Yamazaki and T. Ninomiya, *J. Assoc. Off. Anal. Chem. Int.* **82**, 1474 (1999).

1003. L.G. Rushing and E.B. Hansen, Jr., *J. Chromatogr. B* **700**, 223 (1997).

1004. N. Yagoubi, A.E. Baillet, F. Pellerin, and D. Baylocq, *J. Chromatogr.* **522**, 131 (1990).

1005. J.-M.A. Stoll and W. Giger, *Anal Chem.* **69**, 2594 (1997).

1006. I. Ogura, D.L. DuVal, and K. Miyajima, *J. Am. Oil Chem. Soc.* **72**, 827 (1995).

1007. E. Piera, C. Domínguez, P. Clapés, P. Erra, and Ma. R. Infante, *J. Chromatogr. A* **852**, 499 (1999).

1008. M.J. Scarlett, J.A. Fisher, H. Zhang, and M. Ronan, *Water Res.* **10**, 2109 (1994).

1009. G. Cretier, C. Podevin, and J.-L. Rocca, *J. Chromatogr. A* **874**, 305 (2000).

1010. M.Y. Ye, R.G. Walkup, and K.D. Hill, *J. Liq. Chromatogr.* **18**, 2309 (1995).

1011. J.A. Field and T.E. Sawyer, *J. Chromatogr. A* **893**, 253 (2000).

1012. L. Sarrazin, W. Wafo, and P. Rebouillon, *J. Liq. Chromatogr.* **22**, 2511 (1999).

1013. L. Gagliardi, C. Cavazzutti, and D. Tonelli, *Anal. Lett.* **31**, 829 (1998).

1014. L.Y. Zhou, M. Ashraf-Khorassani, and L.T. Taylor, *J. Chromatogr. A* **858**, 209 (1999).

1015. T. Takeuchi and D. Ishii, *J. Chromatogr.* **403**, 324 (1987).

1016. Z. Moldovan, J.L. Martinez, M.V. Delgado Luque, and E.O. Salaverri, *J. Liq. Chromatogr.* **18**, 1633 (1995).

1017. K. Rissler, *J Chromatogr. A* **871**, 243 (2000).

1018. Y. Mengerink, R. Peters, M. Kerkhoff, J. Hellenbrand, H. Omloo, J. Andrien, M. Vestjens, and Sj. van der Wal, *J. Chromatogr. A* **878**, 45 (2000).

1019. P. Paseiro Losada, S. Paz Abuín, L. Vázquez Odériz, J. Simal Lozano, and J. Simal Gándara, *J. Chromatogr.* **585**, 75 (1991).

1020. W.R. Summers, *Anal. Chem.* **62**, 1397 (1990).

1021. G. Baillet, G. Giusti, and R. Guglielmetti, *Bull. Chem. Soc. Japan* **68**, 1220 (1995).

1022. A. Theobald, C. Simoneau, P. Hannaert, P. Roncari, A. Roncari, T. Rudolph, and E. Anklam, *Food Addit. Contam.* **17**, 885 (2000).

1023. H. Shintani, *J Liq. Chromatogr.* **18**, 613 (1995).

1024. A.P. Damant, S.M. Jickells, and L. Castle, *J Assoc. Off. Anal. Chem. Int.* **78**, 711 (1995).

1025. P.J. Gibbs, K.R. Seddon, N.M. Brovenko, Y.A. Petrosyan, and M. Barnard, *Anal. Chem.* **69**, 1965 (1997).

1026. S. Angelino, A.F. Fell, and M.C. Gennaro, *J. Chromatogr. A* **797**, 65 (1998).

1027. M.C. Gennaro, E. Gioannini, S. Angelino, R. Aigotti, and D. Giacosa, *J. Chromatogr. A* **767**, 87 (1997).

1028. M. Pérez-Urquiza, M.D. Prat, and J.L. Beltrán, *J. Chromatogr. A* **871**, 227 (2000).

1029. C. Scarpi, F. Ninci, M. Centini, and C. Anselmi, *J Chromatogr. A* **796**, 319 (1998).

1030. M. Yamada, A. Kawahara, M. Nakamura, and H. Nakazawa, *Food Addit. Contam.* **17**, 665 (2000).

1031. M.D. Peiperl, M.J. Prival, and S.J. Bell, *Educ. Chem. Toxicol.* **33**, 829 (1995).

1032. I.R. Tebbett, C. Chen, M. Fitzgerald, and L. Olson, *J. Forensic Sci.* **37**, 1283 (1992).

1033. P.C. White and T. Catterick, *Analyst* **118**, 791 (1993).

1034. S.-L. Zhao, F.-S. Wei, H-F. Zou, and X.-B. Xu, *J Liq. Chromatogr.* **21**, 717 (1998).

1035. J. Mao, C.R. Pacheco, D.D. Traficante, and W. Rossen, *Fuel* **74**, 880 (1995).

1036. V.R. Dhole and G.K. Ghosal, *J. Liq. Chromatogr.* **18**, 2475 (1995).

1037. K. Gaudin, P. Chaminade, D. Ferrier, A. Baillet, and A. Tchapla, *Chromatographia* **49**, 241 (1999).

1038. Y. Shih and F.-C. Cheng, *J. Chromatogr. A* **876**, 243 (2000).

1039. L. Gagliardi, D. De Orsi, G. Cavazzutti, G. Multari, and D. Tonelli, *Chromatographia* **43**, 76 (1996).

1040. L. Gagliardi, D. De Orsi, G. Multari, G. Cavazzutti, and D. Tonelli, *Analusis* **27**, 163 (1999).

1041. Ph. Dallet, L. Labat, E. Kummer, and J.P. Dubost, *J. Chromatogr. B* **742**, 447 (2000).

1042. R.L. Wolff, W.W. Christie, F. Pédrono, and A.M. Marpeau, *Lipids* **34**, 1083 (1999).

1043. N. Ohashi and M. Yoshikawa, *J. Chromatogr. B* **746**, 17 (2000).

1044. T. Ohya, N. Kudo, E. Suzuki, and Y. Kawashima, *J. Chromatogr. B* **720**, 1 (1998).

1045. G.W. Tindall, R.L. Perry, and A.T. Spaugh, *J. Chromatogr. A* **868**, 41 (2000).

1046. F. Edelkraut and U. Brockmann, *Chromatographia* **30**, 432 (1990).

1047. F.W. Claassen, C. van der Haar, T.A. van Beek, J. Dorado, M.-J. Martínez-Iñigo, and R. Sierra-Alvarez, *Phytochem. Anal.* **11**, 251 (2000).

1048. H.S. Lee, *J. Assoc. Off. Anal. Chem.* **78**, 80 (1995).

1049. H. Pouliquen, F. Armand, and S. Loussouarn, *J. Liq. Chromatogr.* **21**, 591 (1998).

1050. U. Pinkernell, S. Effkemann, and U. Karst, *Anal. Chem.* **69**, 3623 (1997).

1051. E. Makatsori, K. Fermani, A. Aletras, N.K. Karamanos, and T. Tsegenidis, *J. Chromatogr. B* **712**, 23 (1998).

1052. V. Cheynier and M. Moutounet, *J. Agric. Food Chem.* **40**, 2038 (1992).

1053. X.-g. He, M.W. Bernart, G.S. Nolan, L.-z. Lin, and M.P. Lindenmaier, *J. Chromatogr. Sci.* **38**, 169 (2000).

1054. P.G. Pietta, C. Gardana, and P.L, Mauri, *J. Chromatogr. B* **693**, 249 (1997).

1055. T. Sudhakar Johnson, G.A. Ravishankar, and L.V. Venkataraman, *J Agric. Food Chem.* **40**, 2461 (1992).

1056. H. John and W. Schlegel, *J. Chromatogr. B* **698**, 9 (1997).

1057. G. Hofer, C. Bieglmayer, B. Kopp, and H. Janisch, *Prostaglandins* **45**, 413 (1993).

1058. M. Yamane, A. Abe, S. Yamane, and F. Ishikawa, *J. Chromatogr.* **579**, 25 (1992).

1059. H. Miwa, *J. Chromatogr. A* **881**, 365 (2000).

1060. A.B. Kroumova and G.J. Wagner, *Anal. Biochem.* **225**, 270 (1995).

1061. Th.H.M. Roemen and G.J. van der Vusse, *J. Chromatogr.* **570**, 243 (1991).

1062. A.I. Haj-Yehia, P. Assaf, T. Nassar, and J. Katzhendler, *J. Chromatogr. A* **870**, 381 (2000).

1063. M. Pourfarzam and K. Bartlett, *J. Chromatogr.* **570**, 253 (1991).

1064. R. Abushufa, P. Reed, and C. Weinkove, *Clin. Chem.* **40**, 1707 (1994).

1065. R.A. Norton, *Lipids* **30**, 269 (1995).

1066. R. Dekker, R. van der Meer, and C. Olieman, *Chromatographia* **31**, 549 (1991).

1067. C. Dax, M. Vogel, and S. Müllner, *Chromatographia* **40**, 674 (1995).

1068. T. Iida, Y. Yamaguchi, J. Maruyama, M. Nishio, J. Goto, and T. Nambara, *J. Liq. Chromatogr.* **18**, 701 (1995).

1069. M. Prabhakaram and V.V. Mossine, *Preparative Biochem. & Biotechnol.* **28**, 319 (1998).

1070. M. Marcé, D.S. Brown, T. Capell, X. Figueras, and A.F. Tiburcio, *J. Chromatogr. B* **666**, 329 (1995).

1071. S.C. Minocha, R. Minocha, and C.A. Robie, *J. Chromatogr.* **511**, 177 (1990)

1072. H. Yang and C. Thyrion, *J. Liq. Chromatogr.* **21**, 1347 (1998).

1073. F. Xie, P. Wong, K. Yoshioka, R.G. Cooks, and P.T. Kissinger, *J. Liq. Chromatogr.* **21**, 1273 (1998).

1074. J. Ishida, R. Iizuka, and M. Yamaguchi, *Analyst* **118**, 165 (1993).

1075. H.B. He, R.J. Deegan, M. Wood, and A.J.J. Wood, *J. Chromatogr.* **574**, 213 (1992).

1076. J.R. Lee Chin, *J. Chromatogr.* **578**, 17 (1992).

1077. S.R. Aliwell, B.S. Martincigh, and L.F. Salter, *J. Photochem. Photobiol. A: Chem.* **83**, 223 (1994).

1078. F. Berthou, B. Guillois, C. Riche, Y. Dreano, F. Jacqz-Aigrain, and P.H. Beaune, *Xenobiotica* **22**, 671 (1992).

1079. F. Valls, M.T. Sancho, M.A. Fernández-Muiño, and M.A. Checa, *J Agric. Food Chem.* **48**, 3392 (2000).

1080. W. Zwickenpflug, *J. Agric. Food Chem.* **48**, 392 (2000).

1081. B. Sellergren, Å. Zander, T. Renner, and A. Swietlow, *J. Chromatogr. A* **829**, 143 (1998).

1082. J.I. Seeman, H.V. Secor, D.W. Armstrong, K.D. Timmons, and T.J. Ward, *Anal. Chem.* **60**, 2120 (1988).

1083. I. Noyer, B. Fayet, I. Pouliquen-Sonaglia, M. Guerere, and J. Lesgard, *Analusis* **27**, 69 (1999).

1084. P. Kuronen, T. Väänänen, and F. Pehu, *J. Chromatogr. A* **863**, 25 (1999).

1085. A. Sotelo and B. Serrano, *J. Agric. Food Chem.* **48**, 2472 (2000).

1086. M.A. Bacigalupo, A. Ius, R. Longhi, and G. Meroni, *The Analyst* **125**, 1847 (2000).

1087. K.J. Shingfield and N.W. Offer, *J. Chromatogr. B* **706**, 342 (1998).

1088. M. Czauderna and J. Kowalczyk, *J Chromatogr. B* **704**, 89 (1997).

1089. S. Husain, R. Narshima, and R.N. Rao, *J. Chromatogr. A* **863**, 123 (1999).

1090. T. Herraiz, *Food Addit. Contam.* **17**, 859 (2000).

1091. H.-O. Kim, T.D. Durance, C.H. Scaman, and D.D. Kitts, *J. Agric. Food Chem.* **48**, 4187 (2000).

1092. X.-g. He, L.-z. Lin, M.W. Bernart, and L.-z. Lian, *J. Chromatogr. A* **815**, 205 (1998).

1093. G.A. Gross and A. Grüter, *J. Chromatogr.* **592**, 271 (1992).

1094. D. Zhang, R. Shelby, M.A. Savka, Y. Dessaux, and M. Wilson, *J. Chromatogr. A* **813**, 247 (1998).

1095. B. Clement and F. Jung, *Xenobiotica* **25**, 443 (1995).

1096. G. Huhn, J. Mattusch, and H. Schulz, *Fresenius J. Anal. Chem.* **351**, 563 (1995).

1097. J. Gilbert, *Food Addit. Contam.* **10**, 37 (1993).

1098. L.A. Lawton, C. Edwards, and G.A. Codd, *Analyst* **119**, 1526 (1994).

1099. H. Murata, H. Shoji, M. Oshikata, K.-I. Harada, M. Suzuki, F. Kondo, and H. Goto, *J. Chromatogr. A* **693**, 263 (1995).

1100. F. Walker and B. Meier, *J. Assoc. Off. Anal. Chem. Int.* **81**, 741 (1998).

1101. E. Rajakylä, K. Laasasenaho, and P.J.D. Sakkers, *J. Chromatogr.* **384**, 391 (1987).

1102. E. Razzazzi-Fazeli, J. Böhm, and W. Luf, *J. Chromatogr. A* **854**, 45 (1999).

1103. V. Sewram, T.W. Nieuwoudt, W.F.O. Marasas, G.S. Shepard, and A. Ritieni, *J. Chromatogr. A* **858**, 175 (1999).

1104. D. Abramson, E. Usleber, and E. Märtlbauer, *J. Assoc. Off. Anal. Chem. Int.* **82**, 1353 (1999).

1105. K.J. James, A. Furey, I.R. Sherlock, M.A. Stack, M. Twohig, F.B. Caudwell, and O.M. Skulberg, *J. Chromatogr. A* **798**, 147 (1998).

1106. C. Defendenti, E. Bonacina, M. Mauroni, and L. Gelosa, *Forensic Sci. Int.* **92**, 59 (1998).

1107. N. Okamura, A. Yagi, H. Haraguchi, and K. Hashimoto, *J. Chromatogr.* **630**, 418 (1993).

1108. R. Draisci, L. Giannetti, L. Lucentini, C. Marchiafava, K.J. James, A.G. Bishop, B.M. Healy, and S.S. Kelly, *J. Chromatogr. A* **798**, 137 (1998).

1109. T. Suzuki and T. Yasumoto, *J. Chromatogr. A* **874**, 199 (2000).

1110. K.J. James, A.G. Bishop, R. Draisci, L. Palleschi, C. Marchiafava, E. Ferretti, M. Satake, and T. Yasumoto, *J. Chromatogr. A* **844**, 53 (1999).

1111. C.O. Miles, A.L. Wilkins, D.J. Stirling, and A.L. MacKenzie, *J. Agric. Food Chem.* **48**, 1373 (2000).

1112. F. Crespin, M. Calmes, R. Elias, C. Maillard, and G. Balansard, *Chromatographia* **38**, 183 (1994).

1113. M. Friedman and G.M. McDonald, *J. Agric. Food Chem.* **43**, 1501 (1995).

1114. M.W. Dong, J. Lepore, and T. Tarumoto, *J. Chromatogr.* **442**, 81 (1988).

1115. C. Andrés-Lacueva, F. Mattivi, and D. Tonon, *J. Chromatogr. A* **823**, 355 (1998).

1116. C.D. Capo-chichi, J.-L. Guéant, F. Feillet, F. Namour, and M. Vidailhet, *J. Chromatogr. B* **739**, 219 (2000).

1117. P.J. Bagely and J. Selhub, *Methods Enzymol.* **281**, 16 (1997).

1118. E.J.M. Konings, *J. Assos. Off. Anal. Chem. Int.* **82**, 119 (1999).

1119. R.D. Thompson and M. Carlson *J. Assoc. Off. Anal. Chem. Int.* **83**, 847 (2000).

1120. A.F. Hagar. L. Madsen, L. Wales, Jr., and H.B. Bradford, Jr., *J. Assoc. Off. Anal. Chem. Int.* **77**, 1047 (1994).

1121. K. Hayashi, M. Akiyoshi-Shibata, T. Sakaki, and Y. Yabusaki, *Xenobiotica* **28**, 457 (1998).

1122. K. Shimada, K. Mitamura, H. Kaji, and M. Morita, *J. Chromatogr. Sci.* **32**, 107 (1994).

1123. N.K. Andrikopoulos, H. Brueschweiler, H. Felber, and Ch. Taeschler, *J. Am. Oil Chem. Soc.* **68**, 359 (1991).

1124. D.J. Hart and K.J. Scott, *Food Chem.* **54**, 101 (1995).

1125. B. Lyan, V. Azaïs-Braesco, N. Cardinault, V. Tyssandier, P. Borel, M.C. Alexandre-Gouabau, and P. Grolier, *J. Chromatogr. B* **751**, 297 (2001).

1126. F. Khachik, G.R. Beecher, J.T. Vanderslice, and G. Furrow, *Anal. Chem.* **60**, 807 (1988).

1127. M. Kirchmair, R. Pöder, and C.G. Huber, *J. Chromatogr. A* **832**, 247 (1999).

1128. N.-E. Es-Safi, V. Cheynier, and M. Moutounet, *J. Agric. Food Chem.* **48**, 5946 (2000).

1129. B. Günther and H. Wagner, *Phytomedicine* **3**, 59 (1996).

1130. H. Yamamoto, N. Katano, A. Ooi, and K. Inoue, *Phytochemistry* **50**, 417 (1999).

1131. M. Repčák, J. Imrich, and J. Garčár, *Phytochem. Anal.* **10**, 335 (1999).

1132. S. Sturm and H. Stuppner, *Phytochem. Anal.* **11**, 121 (2000).

1133. O. Schaaf, A.P. Jarvis, S.A. van der Esch, G. Giagnocovo, and N.J. Oldham, *J. Chromatogr. A* **886**, 89 (2000).

1134. C. Zidorn, E.P. Ellmerer-Müller, K.-H. Ongania, S. Sturm, and H. Stuppner, *J. Nat. Prod.* **63**, 812 (2000).

1135. J.L. Wolfender, M. Maillard, and K. Hostettmann, *J.Chromatogr.* **647**, 183 (1993).

1136. M.P. Maillard and K. Hostettmann, *J. Chromatogr.* **647**, 137 (1993).

1137. P. Menghinello, L. Cucchiarini, F. Palma, D. Agostini, M. Dachà, and V. Stocchi, *J. Liq. Chromatogr.* **22**, 3007 (1999).

1138. W. Widmer, *J. Assoc. Off. Anal. Chem. Int.* **83**, 1155 (2000).

1139. H. Tsuchiya, *J. Chromatogr. B* **720**, 225 (1998).

1140. M. Esti, L. Cinquanta, and E. La Notte, *J. Agric. Food Chem.* **46**, 32 (1998).

1141. S.A. Ross, D.S. Ziska, K. Zhao, and M.A. ElSohly, *Fitoterapia* **71**, 154 (2000).

1142. J.F. Stevens, A.W. Taylor, and M.L. Deinzer, *J. Chromatogr. A* **832**, 97 (1999).

1143. L.-Z. Lin, X.-G. He, M. Lindenmaier, G. Nolan, J. Yang, M. Cleary, S.-X. Qiu, and G.A. Cordell, *J. Chromatogr. A* **876**, 87 (2000).

1144. H.-L. Lay and C.-C. Chen, *J. Liq. Chromatogr.* **23**, 1439 (2000).

1145. C.T. da Costa, J.J. Dalluge, M.J. Welch, B. Coxon, S.A. Margolis, and D. Horton, *J. Mass Spectrom.* **35**, 540 (2000).

1146. P.P. Mouly, F.M. Gaydou, and C. Arzouyan, *Analusis* **27**, 284 (1999).

1147. P. Mattila, J. Astola, and J. Kumpulainen, *J. Agric. Food. Chem.* **48**, 5834 (2000).

1148. M. Careri, L. Elviri, A. Mangia, and M. Musci, *J. Chromatogr. A* **881**, 449 (2000).

1149. C. Li, M. Homma, and K, Oka, *J. Chromatogr. B* **693**, 191 (1997).

1150. L.S. Hutabarat, H. Greenfield, and M. Mulholland, *J. Chromatogr. A* **886**, 55 (2000).

1151. C. Wang, M. Sherrard, S. Pagadala, R. Wixon, and R.A. Scott, *J. Am. Oil Chem. Soc.* **77**, 483 (2000).

1152. S. Sabatier, M.J. Amiot, M. Tacchini, and S. Aubert, *J Food Sci.* **57**, 773 (1992).

1153. Y.-C. Lee, C.-Y. Huang, K.-C. Wen, and T.-T. Suen, *J. Chromatogr. A* **692**, 137 (1995).

1154. P. Mouly, E.M. Gaydou, and J. Estienne, *J. Chromatogr.* **634**, 129 (1993).

1155. N. Whittle, H. Eldridge, J. Bartley, and G. Organ, *J. Inst. Brewing* **105**, 89 (1999).

1156. P.L. Fernández, M.J. Martín, A.G. González, and F. Pablos, *The Analyst* **125**, 421 (2000).

1157. M.-J. Lee, S. Prabhu, X. Meng, C. Li, and C.S. Yang, *Anal. Biochem.* **279**, 164 (2000).

1158. S.-J. Sheu and H.-R. Chen, *J. Chromatagr. A* **704**, 141 (1995).

1159. M. Murkovic, U. Adam, and W. Pfannhauser, *Fresenius J. Anal. Chem.* **366**, 379 (2000).

1160. B. Berente, D. De la Calle García, M. Reichenbächer, and K. Danzer, *J. Chromatogr. A* **871**, 95 (2000).

1161. J.-P. Goiffon, P.P. Mouly, and E.M. Gaydou, *Anal. Chim. Acta* **382**, 39 (1999).

1162. K.R. Markham, K.S. Gould, C.S. Winefield, K.A. Mitchell, S.J. Bloor, and M.R. Boase, *Phytochemistry* **55**, 327 (2000).

1163. C.L. Price and R.E. Wrolstad, *J. Food Sci.* **60**, 369 (1995).

1164. F. Tomè, M.L. Colombo, and L. Caldiroli, *Phytochem. Anal.* **10**, 264 (1999).

1165. N. Fabre, C. Claparols, S. Richelme, M.-L. Angelin, I. Fourasté, and C. Moulis, *J. Chromatogr. A* **904** 35 (2000).

1166. M. Sellés, J. Bastida, F. Viladomat, and C. Codina, *Analusis* **25**, 156 (1997).

1167. A.S. Moubarak and C.F. Rosenkrans, Jr., *Biochem, Biophys. Res. Commun.* **274**, 746 (2000).

1168. G.M. Ware, G. Price, L. Carter, Jr., and R.R. Eitenmiller, *J. Assoc. Off. Anal. Chem. Int.* **83**, 1395 (2000).

1169. C. Crews, J.R. Startin, and P.A. Clarke, *Food Addit. Contam.* **14**, 419 (1997).

1170. S.-W. Sun, S.-S. Lee, A.-C. Wu, and C.-K. Chen, *J. Chromatogr. A* **799**, 337 (1998).

1171. G. Bringmann, M. Rückert, J. Schlauer, and M. Herderich, *J. Chromatogr.* **810**, 231 (1998).

1172. C.-Q. Niu and L.-Y. He, *J. Chromatogr.* **542**, 193 (1991).

1173. L. Verotta, E. Peterlongo, F. Elisabetsky, T.A. Amador, and D.S. Nunes, *J. Chromatogr. A* **841**, 165 (1999).

1174. A. Rosso and S. Zuccaro, *J. Chromatogr. A* **825**, 96 (1998).

1175. Tsuchiya, H. Hayashi, M. Sato, H. Shimizu, and M. Iinuma, *Phytochem. Anal.* **10**, 247 (1999).

1176. J. Soušek, D. Guédon, T. Adam, H. Bochořáková, E. Táborská, I. Válka, and V. Šimánek, *Phytochem. Anal.* **10**, 6 (1999).

1177. Y. Cui and G. Lin, *J. Chromatogr. A* **903**, 85 (2000).

1178. S.S. Goyal, *Commun. Soil Sci. Plant Anal.* **31**, 1919 (2000).

1179. E. Razzazi-Fazeli, S. Kleineisen, and W. Luf, *J. Chromatogr. A* **896**, 321 (2000).

1180. J. Böswart, P. Kostiuk, J. Vymlátil, T. Schmidt, V. Pacáková, and K. Stulík, *J. Chromatogr.* **571**, 19 (1991).

1181. A. Nuñez and G.J. Piazza, *Lipids* **30**, 129 (1995).

1182. C. Bauer-Plank and L. Steenhorst-Slikkerveer, *J. Am. Oil Chem. Soc.* **77**, 477 (2000).

1183. M. Buchgraber, F. Ulberth, and E. Anklam, *J. Agric. Food Chem.* **48**, 3359 (2000).

1184. S.L. Abidi, *J. Chromatogr.* **587**, 193 (1991).

1185. S. Mawatari and K. Murakami, *Anal. Biochem.* **264**, 118 (1998).

1186. C.S. Ramesha, W.C. Pickett, and D.V.K. Murthy, *J. Chromatogr.* **491**, 37 (1989).

1187. S. Kermasha, S. Kubow, M. Safari, and A. Reid, *J. Am. Oil Chem. Soc.* **70**, 169 (1993).

1188. L. Haalck and F. Spener, *J. Chromatogr.* **498**, 410 (1990).

1189. A. Kuksis, L. Marai, and J.J. Myher, *J. Chromatogr.* **588**, 73 (1991).

1190. J.J. Myher, A. Kuksis, and L. Marai, *J. Am. Oil Chem. Soc.* **70**, 1183 (1993).

1191. L. Marai, A. Kuksis, and J.J. Myher, *J. Chromatogr. A* **672**, 87 (1994).

1192. M.H. Maguire, I. Szabo, P. Slegel, and C.R. King, *J. Chromatogr.* **575**, 243 (1992).

1193. J.-D. Kim, K.H. Row, M.S. So, I.A. Polumina, and A.V. Larin. *J. Liq. Chromatogr.* **18**, 3091 (1995).

1194. C.-H. Tung and S. Stein, *Bioconjugate Chem.* **11**, 606 (2000).

1195. A. Ally and G. Park, *J. Chromatogr.* **575**, 19 (1992).

1196. J. Jankowski, W. Potthoff, M. van der Giet, M. Tepel, W. Zidek, and H. Schlüter, *Anal. Biochem.* **269**, 72 (1999).

1197. J.R. Kalin and D.L. Hill, *J. Chromatogr.* **431**, 184 (1988).

1198. R.E. Holmlin, P.J. Dandliker, and J.K. Barton, *Bioconjugate Chem.* **10**, 1122 (1999).

1199. D. Kanduc, A. Bracalello, G. Troccoli, C. de Benedetto, A. Mosavi, and A. Azzarone, *J. Liq. Chromatogr.* **18**, 505 (1995).

1200. M. Wang, R. Young-Sciame, F.-L. Chung, and S.S. Hecht, *Chem. Res. Toxicol.* **8**, 617 (1995).

1201. G.F.M. Ball, *Food Chem.* **35**, 117 (1990).

1202. L.A.Th. Verhaar and B.F.M. Kuster, *J. Chromatogr.* **234**, 57 (1982).

1203. S. Nojiri, K. Saito, N. Taguchi, M. Oishi, and T. Maki, *J. Assoc. Off. Anal. Chem. Int.* **82**, 134 (1999).

1204. A. Sims, *J. Agric. Food Chem.* **43**, 377 (1995).

1205. S. Nojiri, N. Taguchi, M. Oishi, and S. Suzuki, *J. Chromatogr. A* **893**, 195 (2000).

1206. Y. An, S.P. Young, S.L. Hillman, J.L.K. Van Hove, Y.-T. Chen, and D.S. Millington, *Anal. Biochem.* **287**, 136 (2000).

1207. N. Volpi, *Anal. Biochem.* **277**, 19 (2000).

1208. Y. Guerardel, O. Kol, F. Maes, T. Lefebvre, B. Boilly, M. Davril, and G. Strecker, *Biochem. J.* **352**, 449 (2000).

1209. A. Berthod, S.S.C. Chang, J.P.S. Kullman, and D.W. Armstrong, *Talanta* **47**, 1001 (1998).

1210. P.J. Simms, R.M. Haines, and K.B. Hicks, *J. Chromatogr.* **648**, 131 (1993).

1211. Y. Takahashi, F. Miki, and K. Nagase, *Bull. Chem. Soc. Jpn.* **68**, 1851 (1995).

1212. C.Y. Cheng and Y.-K. Li, *Biotechnol. Appl. Biochem.* **32**, 197 (2000).

1213. S. Vendrell-Pascuas, A.I. Castellote-Bargalló, and M.C. López-Sabater, *J. Chromatogr. A* **881**, 591 (2000).

1214. N.K. Karamanos, P. Vanky, G.N. Tzanakakis, T. Tsegenidis, and A. Hjerpe, *J. Chromatogr. A* **765**, 169 (1997).

1215. H. Kwon and J. Kim, *Anal. Biochem.* **215**, 243 (1993).

1216. M. Kohler and J.A. Leary, *Anal. Chem.* **67**, 3501 (1995).

1217. H. Yamamura, Y. Kawase, M. Kawai, and Y. Butsugan, *Bull Chem. Soc. Jpn.* **66**, 585 (1993).

1218. M. Chiba, X. Xu, J.A. Nichime, S.K. Balani, and J.H. Lin, *Drug Metab. Dispos.* **25**, 1022 (1997).

1219. P. Johnsson, A. Kamal-Eldin, L.N. Lundgren, and P. Åman, *J. Agric. Food Chem.* **48**, 5216 (2000).

1220. B. Matthäus and H. Luftmann, *J. Agric. Food Chem.* **48**, 2234 (2000).

1221. J. Ma, Y. Li, Q. Ye, J. Li, Y. Hua, D. Ju, D. Zhang, R. Cooper, and M. Chang, *J. Agric. Food Chem.* **48**, 5220 (2000).

1222. M. Weissenberg, I. Schaeffler, E. Menagem, M. Barzilai, and A. Levy, *J. Chromatogr. A* **757**, 89 (1999).

1223. S. Zschocke, J.-H. Liu, H. Stuppner, and R. Bauer, *Phytochem. Anal.* **9**, 283 (1998).

1224. L. Krenn, U. Blaeser, and N. Hausknost-Chenicek, *J. Liq. Chromatogr.* **21**, 3149 (1998).

1225. L.P. Meagher, G.R. Beecher, V.P. Flanagan, and B.W. Li, *J. Agric. Food Chem.* **47**, 3173 (1999).

1226. S. Srivastava, R.K. Verma, M.M. Gupta, and S. Kumar, *J. Chromatogr. A* **841**, 123 (1999).

1227. A.M. Hutapea, C. Toskulkao, P. Wilairat, and D. Buddhasukh, *J. Liq. Chromatogr.* **22**, 1161 (1999).

1228. C.-L. Liu, P.-L. Zhu, and M.-C. Liu, *J. Chromatogr. A* **857**, 167 (1999).

1229. V. Van den Bergh, I. Vanhees, R. De Boer, F. Compernolle, and C. Vinckier, *J. Chromatogr. A* **896**, 135 (2000).

1230. S.S. Teng and W. Feldheim, *Chromatographia* **47**, 529 (1998).

1231. D.E. Gray, G.E. Rottinghaus, H.E.G. Garrett, and S.G. Pallardy, *J. Assoc. Off. Anal. Chem. Int.* **83**, 944 (2000).

1232. M. Brolis, B. Gabetta, N. Fuzzati, R. Pace, F. Panzeri, and F. Peterlongo, *J. Chromatogr. A* **825**, 9 (1998).

1233. J.D. Chi and M. Franklin, *J. Chromatogr. B* **735**, 285 (2000).

1234. N. Fuzzati, B. Gabetta, K. Jayakar, R. Pace, G. Ramaschi, and F. Villa, *J. Assoc. Off. Anal. Chem. Int.* **83**, 820 (2000).

1235. W.Y. Kim, J.M. Kim, S.B. Han, S.K. Lee, N.D. Kim, M.K. Park, C.K. Kim, and J.H. Park, *J. Nat. Prod.* **63**, 1702 (2000).

1236. H. Wang, H. Zou, L. Kong, Y. Zhang, H. Pang, C. Su, G. Liu, M. Hui, and L. Fu, *J. Chromatogr. B* **731**, 403 (1999).

1237. X.-g. He, M.W. Bernart, L.-z. Lian, and L.-z. Lin, *J. Chromatogr. A* **796**, 327 (1998).

1238. Y. Sekiwa, K. Kubota, and A. Kobayashi, *J. Agric. Food Chem.* **48**, 373 (2000).

1239. P. Jeandet, A.C. Breuil, M. Adrian, L.A. Weston, S. Debord, P. Meunier, G. Maume, and R. Bessis, *Anal. Chem.* **69**, 5172 (1997).

1240. K. Krizsán, Gy. Szókán, Z.A. Tóth, F. Hollósy, M. Lászlo, and A. Khlafulla, *J. Liq. Chromatogr.* **19**, 2295 (1996).

1241. E.A. Abourashed and I.A. Khan, *J. Assoc. Off. Anal. Chem. Int.* **83**, 789 (2000).

1242. A.R. Bilia, M. Fumarola, S. Gallori, G. Mazzi, and F.F. Vincieri, *J. Agric. Food Chem.* **48**, 4734 (2000).

1243. J. Herraiz, *J. Agric. Food Chem.* **48**, 4900 (2000).

1244. B. Mesrob, C. Nesbitt, R. Misra, and R.C. Pandey, *J. Chromatogr. B* **720**, 189 (1998).

1245. A. Barros, J.A. Rodrigues, P.S. Almeida, and M.T. Oliva-Teles, *J. Liq. Chromatogr.* **22**, 2061 (2000).

1246. N. Dimov, K. Chervenkova, and B. Nikolova-Damyanova, *J. Liq. Chromatogr.* **23**, 935 (2000).

1247. Y.-R. Ku, Y.-T. Lin, K.-C. Wen, J.-H. Lin, and C.-H. Liao, *J. Liq. Chromatogr.* **19**, 3265 (1996).

1248. P. Lehtonen, in L. Nykänen and P. Lehtonen (eds.), *Flavour Research of Alcoholic Beverages*, Alko Symposium, Helsinki, 1984, pp. 121–130.

1249. N.M. Alexander, in W.S. Hancock (ed.), *HPLC for the Separation of Amino Acids, Peptides and Proteins*, Volume I, CRC Press, Boca Raton, FL, 1984, pp. 291–300.

1250. R.D. Hiserodt, D.F.H. Swijter, and C.J. Mussinan, *J. Chromatogr. A* **888**, 103 (2000).

1251. Q. Tian and X. Ding, *J. Chromatogr. A* **874**, 13 (2000).

1252. A.A. Nomeir, D.M. Silveira, M.F. McComish, and M. Chadwick, *Drug Metab. Dispos.* **20**, 198 (1992).

1253. J. Steinert, H. Khalaf, and M. Rimpler, *J. Chromatogr. A* **693**, 281 (1995).

1254. J.D. Stark and J.F. Walter, *J. Agric. Food Chem.* **43**, 507 (1995).

1255. Y. Suzuki, K. Tanabe, and Y. Shioi, *J. Chromatogr. A* **839**, 85 (1999).

1256. C. Hu and D.D. Kitts, *J. Agric. Food Chem.* **48**, 1466 (2000).

1257. B.C. Vastano, Y. Chen, N. Zhu, C.-T. Ho, Z. Zhou, and R.T. Rosen, *J. Agric. Food Chem.* **48**, 253 (2000).

1258. P. Currò, G. Micali, and F. Lanuzza, *J. Chromatogr.* **404**, 273 (1987).

1259. K. Wasserkrug and Z. El Rassi, *J. Liq. Chromatogr.* **20**, 335 (1997).

1260. P. Betto, R. Gabriele, and C. Galeffi, *J. Chromatogr.* **594**, 131 (1992).

1261. H. Wang, L. Kong, H. Zou, J. Ni, and Y. Zhang, *Chromatographia* **50**, 439 (1999).

1262. H. Yamamoto, K. Yazaki, and K. Inoue, *J. Chromatogr. B* **738**, 3 (2000).

1263. H. Abdel-Kader, M.M. Kobylkevich, L.S. Wigman, and G.K. Menon, *J. Assoc. Off. Anal. Chem. Int.* **78**, 289 (1995).

1264. W. Wu and A. Stalcup, *J. Liq. Chromatogr.* **17**, 1111 (1994).

1265. S.H. Ford, A. Nichols, and J.M. Gallery, *J Chromatogr.* **536**, 185 (1991).

1266. J.-F. Rontani, B. Beker, D. Raphel, and G. Baillet, *J. Photochem. Photobiol. A: Chem.* **85**, 137 (1995).

1267. W. Francis, B. Huseby, and Ø.M. Andersen, *Chromatographia* **35**, 189 (1993).

1268. A. Riccio, M. Tamburrini, V. Carratore, and G. di Prisco, *J. Fish Biol.* **57**, 20 (2000).

1269. B. Masala and L. Manca, *Methods Enzymol.* **231**, 21 (1994).

1270. M.J. Scotter, S.A. Thorpe, S.L. Reynolds, L.A. Wilson, and P.R. Strutt, *Food Addit. Contam.* **11**, 301 (1994).

1271. L.A. Ridnour, R.A. Winters, N. Ercal, and D.R. Spitz, *Methods Enzymol.* **299**, 258 (1997).

1272. D.C. Turnell and J.D.H. Cooper, *Clin. Chem.* **28**, 527 (1982).

1273. L. Canevari, R. Vieira, M. Aldegunde and F. Dagani, *Anal. Biochem.* **205**, 137 (1992).

1274. T. Toyo'oka, N. Tomoi, T. Oe, and I. Miyahara, *Anal. Biochem.* **276**, 48 (1999).

1275. D. Tang, L.-S. Wen, and P.H. Santschi, *Anal. Chim. Acta* **408**, 299 (2000).

1276. A.R. Ivanov, I.V. Nazimov, and L.A. Baratova, *J. Chromatogr. A* **870**, 433 (2000).

1277. M.F. Malmer and L.A. Schroeder, *J. Chromatogr.* **514**, 227 (1990).

1278. P.A. Haynes, D. Sheumack, L.G. Greig, J. Kibby, and J.W. Redmond, *J. Chromatogr.* **588**, 107 (1991).

1279. K. Schmeer, M. Khalifa, J. Császár, G. Farkas, E. Bayer, and I. Molnár-Perl, *J. Chromatogr. A* **691**, 285 (1995).

1280. R. Gupta and N. Jentoft, *J. Chromatogr.* **474**, 411 (1989).

1281. V. Stocchi, G. Piccoli, M. Magnani, F. Palma, B. Biagiarelli, and L. Cucchiarini, *Anal. Biochem.* **78**, 107 (1989).

1282. J.F. Lawrence and J.R. Iyengar, *J. Chromatogr.* **404**, 261 (1987).

1283. N. Kim, D.Y. Kwon, H.D. Hong, C. Mok, Y.J. Kim, and Y.J. Nam, *Food Chem.* **47**, 407 (1993).

1284. T. Yoshida, T. Okada, T. Hobo, and R. Chiba, *Chromatographia* **52**, 418 (2000).

1285. K. Rose, J. Chen, M. Dragovic, W. Zeng, D. Jeannerat, P. Kamalaprija, and U. Burger, *Bioconjugate Chem.* **10**, 1038 (1999).

1286. J.B. Kim, T. Halverson, Y.J. Basir, J. Dulka, F.C. Knoop, P.W. Abel, and J.M. Conlon, *Regulatory Peptides* **90**, 53 (2000).

1287. Y. Wang, B.A. Barton, L. Thim, P.F. Nielsen, and J.M. Conlon, *Gen. Comp. Endocrinol.* **113**, 38 (1999).

1288. A. Quesnel and J.P. Briand, *J. Peptide Res.* **52**, 107 (1998).

1289. J.T. Varkey and V.N.R. Pillai, *J. Peptide Res.* **51**, 49 (1998).

1290. D. Vercaigne-Marko, E. Kosciarz, N. Nedjar-Arroume, and D. Guillochon, *Biotechnol. Appl. Biochem.* **31**, 127 (2000).

1291. A. Huberman, M.B. Aguilar, I. Navarro-Quiroga, L. Ramos, I. Fernández, F.M. White, and D.F. Hunt, *Peptides* **21**, 331 (2000).

1292. T. Halverson, Y.J. Basir, F.C. Knoop, and J.M. Conlon, *Peptides* **21**, 469 (2000).

1293. D. Duethman, N. Dewan, and J.M. Conlon, *Peptides* **21**, 137 (2000).

1294. Y. Wang, B.A. Barton, P.F. Nielsen, and J.M. Conlon, *Gen. Comp. Endocrinol.* **113**, 21 (1999).

1295. B. Lignot, R. Froidevaux, N. Nedjar-Arroume, and D. Guillochon, *Biotechnol. Appl. Biochem.* **30**, 201 (1999).

1296. Y. Hathout, Y.-P. Ho, V. Ryzhov, P. Demirev, and C. Fenselau, *J. Nat. Prod.* **63**, 1492 (2000).

1297. G.H. Michel, N. Murayama, T. Sada, M. Nozaki, K. Saguchi, H. Ohi, Y. Fujita, H. Koike, and S. Higuchi, *Peptides* **21**, 609 (2000).

1298. G. Mezó, E. Drakopoulou, V. Paál, É. Rajnavölgyi, C. Vita, and F. Hudecz, *J. Peptide Res.* **55**, 7 (2000).

1299. P.A. Wabnitz, H. Walters, M.J. Tyler, J.C. Wallace, and J.H. Bowie, *J. Peptide Res.* **52**, 477 (1998).

1300. Q. Shu and S.P. Liang, *J. Peptide Res.* **53**, 486 (1999).

1301. S. Bösze, J. Kajtár, R. Szabó, A. Falus, and F. Hudecz, *J. Peptide Res.* **52**, 216 (1998).

1302. M.A. Hoitink, J.H. Beijnen, M.U.S. Boschma, A. Bult, E. Hop, J. Nijholt, C. Versluis, C. Wiese, and W.J.M. Underberg, *Anal. Chem.* **69**, 4972 (1997).

1303. V. Sanz-Nebot, I. Toro, and J. Barbosa, *J. Chromatogr. A* **870**, 335 (2000).

1304. V. Sanz-Nebot, I. Toro, and J. Barbosa, *J. Chromatogr. A* **870**, 315 (2000).

1305. D.M. Bunk, J.J. Dalluge, and M.J. Welch, *Anal. Biochem.* **284**, 191 (2000).

1306. M.C.J. Wilce, M.I. Aguilar, and M.T.W. Hearn, *J. Chromatogr.* **536**, 165 (1991).

1307. G.O.A. Naik and G.W. Moe, *J. Chromatogr. A* **870**, 349 (2000).

1308. M.H. Rao, W. Yang, H. Joshua, J.M. Becker, and F. Naider, *Int. J. Pept. Protein Res.* **45**, 418 (1995).

1309. S.J. Wood and R. Wetzel, *Int. J. Peptide Protein Res.* **39**, 533 (1992).

1310. T. Sato, D. Ishiyama, R. Honda, H. Senda, H. Konno, S. Tokumasu, and S. Kanazawa, *J Antibiotics* **53**, 597 (2000).

1311. J.-G. Chen and S.G. Weber, *Anal. Chem.* **67**, 3596 (1995).

1312. M. Koh, H. Hanagata, S. Ebisu, K. Morihara, and H. Tagaki, *Biosci. Biotechnol. Biochem.* **64**, 1079 (2000).

1313. P. Dolashka, L. Genova, S. Stoeva, B. Stefanow, M. Angelova, R. Hristova, S. Pahova, and W. Voelter, *J. Peptide Res.* **54**, 279 (1999).

1314. L. Jønson, N. Schoeman, H. Saayman, R. Naudé, H. Jensen, and A.H. Johnsen, *Peptides* **21**, 1337 (2000).

1315. G. Khaska, K. Nalini, M. Bhat, and N. Udupa, *Anal. Biochem.* **260**, 92 (1998).

1316. P. Grève, O. Sorokine, T. Berges, C. Lacombe, A. Van Dorsselaer, and G. Martin, *Gen. Comp. Endocrinol.* **115**, 406 (1999).

1317. J.-Y. Chang, *Anal. Biochem.* **268**, 147 (1999).

1318. Yu.P. Oleskina, N.P. Yurina, T.I. Odintsova, Ts.A. Egorov, A. Otto, B. Wittmann-Liebold, and M.S. Odinstova, *Biochem. Mol. Biol. Int.* **47**, 757 (1999).

1319. F.J. Moreno, I. Recio, A. Olano, and R. López-Fandiño, *J. Dairy Res.* **67**, 349 (2000).

1320. A.J. Trujillo, I. Casals, and B. Guamis, *J. Chromatogr. A* **870**, 371 (2000).

1321. F. Sannier, S. Bordenave, and J.-M. Piot, *J. Dairy Res.* **67**, 43 (2000).

1322. R. Marchal, D. Chaboche, L. Marchal-Delahaut, C. Gerland, J.P. Gandon, and P. Jeandet, *J. Agric. Food Chem.* **48**, 3225 (2000).

1323. H. Place, B. Sébille, and C. Vidal-Madjar, *Anal. Chem.* **63**, 1222 (1991).

1324. H. Engelhardt, M. Czok, R. Schultz, and E. Schweinheim, *J. Chromatogr.* **458**, 79 (1988).

1325. K. Büttner, C. Pinilla, J.A. Appel, and R.A. Houghten, *J. Chromatogr.* **625**, 191 (1992).

1326. P.C. Sadek, "Elucidation of the Factors Responsible for Small Solute Retention and Irreversible Protein Binding in Reversed-Phase High Performance Liquid Chromatography," Ph.D. dissertation, University of Minnesota, Minneapolis, 1985.

1327. E.M. Koves and J. Wells, *J. Forensic Sci.* **37**, 42 (1992).

1328. M. Bogusz and M. Erkens, *J. Chromatogr. A* **674**, 97 (1994).

1329. M. Bogusz and M. Wu, *J. Anal. Toxicol.* **15**, 188 (1991).

1330. D.W. Hill and A.J. Kind, *J. Anal. Toxicol.* **18**, 233 (1994).

1331. A. Turcant, A. Premel-Cabic, A. Cailleux and P. Allain, *Clin. Chem.* **37**, 1210 (1991).

1332. G. Carlucci, P. Mazzeo, and G. Palumbo, *Chromatographia* **43**, 261 (1996).

1333. J. Joseph-Charles and M. Bertucat, *J. Liq. Chromatogr.* **22**, 2009 (1999).

1334. Q.C. Meng, M. Soleded Cepeda, T. Kramer, H. Zou, D.J. Matoka, and J. Farrar, *J. Chromatogr. B.* **742**, 115 (2000).

1335. E. Mikami, T. Goto, T. Ohno, H. Matsumoto, K. Inagaki, H. Ishihara, and M. Nishida, *J. Chromatogr. B* **744** 81 (2000).

1336. J.M. Hutzler, R.F. Frye, and T.S. Tracy, *J. Chromatogr. B* **749**, 119 (2000).

1337. A.D. de Jager, H.K.L. Hundt, A.F. Hundt, K.J. Swart, M. Knight, and J. Roberts, *J. Chromatogr. B* **740**, 247 (2000).

1338. F.B. Ibrahim, *J. Liq. Chromatogr.* **18**, 2621 (1995).

1339. M. Siluveru and J.T. Stewart, *J. Chromatogr. B* **673**, 91 (1995).

1340. M.R. Wright, S. Sattari, D.R. Brocks, and F. Jamali, *J. Chromatogr.* **583**, 259 (1992).

1341. E. Grippa, L. Santini, G. Castellano, M.T. Gatto, M.G. Leone, and L. Saso, *J. Chromatogr. B* **738**, 17 (2000).

1342. A.-B. Wu, C.-Y. Chen, S.-D. Chu, Y.-C. Tsai, and F.-A. Chen, *J. Chromatogr. Sci.* **39**, 7 (2001).

1343. C.-L. Chen, H. Chen, D.-M. Zhu, and F.M. Uckun, *J. Chromatogr. B* **724**, 157 (1999).

1344. H.-J. Kraemer, R. Gehrke, A. Breithaupt, and H. Breithaupt, *J. Chromatogr. B* **700**, 147 (1997).

1345. F. Kees, S. Spangler, and M. Wellenhofer, *J. Chromatogr. A* **812**, 287 (1998).

1346. J. Sastre Toraño and H.J. Guchelaar, *J. Chromatogr. B* **720**, 89 (1998).

1347. S. Tanase, H. Tsuchiya, J. Yao, S. Ohmoto, N. Takagi, and S. Yoshida, *J. Chromatogr. B* **706**, 279 (1998).

1348. R. Delépée, D. Maume, B. Le Bizec, and H. Pouliquen, *J. Chromatogr. B* **748**, 369 (2000).

1349. Y.-F. Cheng, D.J. Phillips, and U. Neue, *Chromatographia* **44**, 187 (1997).

1350. R. Pezet, V. Pont, and R. Tabacchi, *Phytochem. Anal.* **10**, 285 (1999).

1351. M.J.M. Nickmilder, D. Latinne, J.-P. De Houx, R.K. Verbeeck, and G.J.J. Lhoëst, *Clin. Chem.* **44**, 532 (1998).

1352. N. Furusawa, *Talanta* **49**, 461 (1999).

1353. H. Oka, K.-i. Harada, M. Suzuki, and Y. Ito, *J. Chromatogr. A* **903**, 93 (2000).

1354. C.J. Barrow, J.J. Oleynek, V. Marinelli, H.H. Sun, P. Kaplita, D.M. Sedlock, A.M. Gillum, C.C. Chadwick, and R. Cooper, *J. Antibiotics* **50**, 729 (1997).

1355. V. Harang and D. Westerlund, *Chromatographia* **50**, 525 (1999).

1356. B.A. Olsen, J.D. Stafford, and D.E. Reed, *J. Chromatogr.* **594**, 203 (1992).

1357. Th. Cachet, P. Lannoo, J. Paesen, G. Janssen, and J. Hoogmartens, *J. Chromatogr.* **600**, 99 (1992).

1358. J. Paesen, P. Claeys, E. Roets, and J. Hoogmartens, *J. Chromatogr.* **630**, 117 (1993).

1359. L.G. Martinez, P. Campíns-Falcó, A. Sevillano-Cabeza, and R. Herráez-Hernández, *J. Liq. Chromatogr.* **21**, 2191 (1998).

1360. L.K. Sørensen and L.K. Snor, *J. Chromatogr. A* **882**, 145 (2000).

1361. K. Takeba, K. Fujinuma, T. Miyazaki, and H. Nakazawa, *J. Chromatogr. A* **812**, 205 (1998).

1362. W.A. Moats, K.L. Anderson, J.E. Rushing, S. Buckley, *J. Agric. Food Chem.* **48**, 498 (2000).

1363. K.M. Kirkland, D.A. McCombs, and J.J. Kirkland, *J. Chromatogr. A* **660**, 327 (1994).

1364. O. Shakoor and R.B. Taylor, *The Analyst* **121**, 1473 (1996).

1365. K.L. Tyczkowska, R.D. Voyksner, R.F. Straub, and A.L. Aronson, *J. Assoc. Off. Anal. Chem. Int.* **77**, 1122 (1994).

1366. P. Pastore, A. Gallina, and F. Magno, *The Analyst* **125**, 1955 (2000).

1367. L.K. Sørenson and H. Hansen, *Food Addit. Contam.* **17**, 197 (2000).

1368. J.A. Orwa, A. Van Gerven, B. Roets, and J. Hoogmartens, *J. Chromatogr. A* **870**, 237 (2000).

1369. W. Naidong, K. Vermeulen, I. Quintens, E. Roets, and J. Hoogmartens, *Chromatographia* **33**, 560 (1992).

1370. A. Le Liboux, O. Pasquier, and G. Montay, *J. Chromatogr. B* **708**, 161 (1998).

1371. S. Horii and N. Oku, *J Assoc. Off. Anal. Chem. Int.* **83**, 17 (2000).

1372. M.J. Schneider and D.J. Donoghue, *J. Assoc. Off. Anal. Chem. Int.* **83**, 1306 (2000).

1373. J.J. Berzas Nevaso, G. Castañeda Peñalvo, and F.J. Guzmán Bernardo, *J. Chromatogr. A* **870**, 169 (2000).

1374. M. Viñas, C. López Erroz, A. Hernández Canals, and M. Hernández Córdoba, *Chromatographia* **40**, 382 (1995).

1375. C.-E. Tsai and F. Kondo, *J Assoc. Off. Anal. Chem. Int.* **78**, 674 (1995).

1376. S. Le Boulaire, J.-C. Bauduret, and F. Andre, *J. Agric. Food Chem.* **45**, 2134 (1997).

1377. E. Dudriková, S. Jozef, and N. Jozef, *J. Assoc. Offic. Anal. Chem. Int.* **82**, 1303 (1999).

1378. G. Nicholls, B.J. Clark and J.E. Brown, *Anal. Proc.* **30**, 51 (1993).

1379. W.A. Moats and R. Harik-Khan, *J. Assoc. Off. Anal. Chem. Int.* **78**, 49 (1995).

1380. M. Sharma and M. Tomasz, *Chem. Res. Toxicol.* **7**, 390 (1994).

1381. M. Sharma, Q.-Y. He, and M. Tomasz, *Chem. Res. Toxicol.* **7**, 401 (1994).

1382. F. Lai and T. Sheehan, *J. Chromatogr.* **609**, 173 (1992).

1383. T.E. Mürdter, B. Sperker, K. Bosslet, P. Fritz, and H.K. Kroemer, *J. Chromatogr. B* **709**, 289 (2000).

1384. R. Ricciarello, S. Pichini, R. Pacifici, I. Altieri, M. Pellegrini, A. Fattorossi, and P. Zuccaro, *J. Chromatogr. B* **707**, 219 (1998).

1385. M.-C. Haaz, C. Riché, L.P. Rivory, and J. Robert, *Drug Metab. Dispos.* **26**, 769 (1998).

1386. J. Escoriaza, A. Aldaz, C. Castellanos, E. Calvo, and J. Giráldez, *J. Chromatogr. B* **740**, 159 (2000).

1387. C.-L. Chen and F.M. Uckun, *J. Chromatogr. B* **744**, 91 (2000).

1388. P. Ptáček, J. Macek, and J. Klíma, *J. Chromatogr. B* **738**, 305 (2000).

1389. H. Zhang, L. Ye, and J.T. Stewart, *J. Liq. Chromatogr.* **21**, 979 (1998).

1390. B.J. Berger, M. Suskin, W.W. Dai, A. Cerami, and P. Ulrich, *J. Chromatogr. B* **691**, 433 (1997).

1391. P. Kestell, G.W. Rewcastle, and B.C. Baguley, *Xenobiotica* **24**, 635 (1994).

1392. S. Emara, I. Morita, K. Tamura, S. Razee, T. Masujima, H.A. Mohamed, S.M. El Gizawy, and N.A. El Rabbat, *J. Liq. Chromatogr.* **21**, 681 (1998).

1393. S.L. Richheimer, D.M. Timmermeier, and D.W. Timmons, *Anal. Chem.* **64**, 2323 (1992).

1394. M.B. Garg and S.P. Ackland, *J. Chromatogr. B* **748**, 383 (2000).

1395a. D.V. Singh, A. Maithy, R.K. Verma, M.M. Gupta, and S. Kumar, *J. Liq. Chromatogr.* **23**, 601 (2000).

1395b. S. Zhou, J.W. Paxton, M.D. Tingle, J. McCall, and P. Kestell, *J. Chromatogr. A* **734**, 129 (1999).

1396. M.Y. Khuhawar, S.N. Lanjwani, and S.A. Memon, *J. Chromatogr. B* **693**, 175 (1997).

1397. J.P. Tassin, J. Dubois, M. Hanocq, and G. Atassi, *Talanta* **42**, 747 (1995).

1398. J.R. Bading, A.H. Shahinian, M.T. Paff, P.B. Yoo, and D.W. Hsia, *Biochem. Pharmacol.* **60**, 963 (2000).

1399. G. Casamenti, R. Mandrioli, C. Sabbioni, F. Bugamelli, V. Volterra, and M.A. Raggi, *J. Liq. Chromatogr.* **23**, 1039 (2000).

1400. J. Fang and J.W. Gorrod, *J. Chromatogr.* **614**, 267 (1993).

1401. A. Jacala, V.E. Adusumalli, N. Kucharczyk, and R.D. Sofia, *J. Chromatogr.* **614**, 285 (1993).

1402. S. George, L. Gill, and R.A. Braithwaite, *Ann. Clin. Biochem.* **37**, 338 (2000).

1403. S. Santangeli, C. McNeill, G.J. Sills, and M.J. Brodie, *J. Chromatogr. B* **746**, 325 (2000).

1404. D.F. Chollet, L. Goumaz, C. Juliano, and G. Anderegg, *J. Chromatogr. B* **746**, 311 (2000).

1405. V. Ferranti, C. Chebenat, S. Ménager, and O. Lafont, *J. Chromatogr. B* **718**, 199 (1998).

1406. J. Xu, J.H. Lamb, R. Jukes, W.M. Purcell, and D.E. Ray, *J. Chromatogr. B* **732**, 349 (1999).

1407. C.D. Thompson, T.A. Miller, M.T. Barthen, C.M. Dieckhaus, R.D. Sofia, and T.M. MacDonald, *Drug Metab. Dispos.* **28**, 434 (2000).

1408. I.M. McIntyre, M.L. Syrjanen, K. Crump, S. Horomidis, A.W. Peace, and O.H. Drummer, *J. Anal. Toxicol.* **17**, 202 (1993).

1409. P.K. Kunicki, *J. Chromatogr. B* **750**, 41 (2001).

1410. L.J. Lovett, G.A. Nygard, G.R. Erdmann, C.Z. Burley, and S.K.W. Khalil, *J. Liq. Chromatogr.* **10**, 877 (1987).

1411. E. Tanaka, N. Sakamoto, N. Inubushi, and S. Misawa, *J. Chromatogr. B* **673**, 147 (1995).

1412. B.P.S. Capece, B. Pérez, E. Castells, M. Arboix, and C. Cristòfol, *J Assoc. Off. Anal. Chem. Int.* **82**, 1007 (1999).

1413. C. Morovján, P. Csokán, L. Makranszki, E.A. Abdellah-Nagy, and K. Tóth, *J. Chromatogr. A* **797**, 237 (1998).

1414. G. Stoev, T. Dakova, and Al. Michailova, *J. Chromatogr. A* **846**, 383 (1999)

1415. K. Takeba, F. Fujinuma, M. Sakamoto, T. Miyazaki, Y. Oka, Y. Itoh, and H. Nakazawa, *J. Chromatogr. A* **882**, 99 (2000).

1416. M.E.C. Valois, O.M. Takayanagui, P.S. Bonato, V.L. Lanchote, and D. Carvalho, *J. Anal. Toxicol.* **18**, 86 (1994).

1417. K. Yoshii, A. Kaihara, Y. Tsumura, S. Ishimitsu, and Y. Tonogai, *J. Chromatogr. A* **896**, 75 (2000).

1418. K.S. Webb, P.B. Baker, N.P. Cassells, J.M. Francis, D.E. Johnston, S.L. Lancaster, P.S. Minty, G.D. Reed, and S.A. White, *J. Forensic Sci.* **41**, 938 (1996).

600 REFERENCES

1419. K.L. Klette, C.J. Anderson, G.K. Poch, A.C. Nimrod, and M.A. ElSohly, *J. Anal. Toxicol.* **24**, 550 (2000).

1420. I.S. Lurie, C.G. Bailey, D.S. Anex, M.J. Bethea, T.D. McKibben, and J.F. Casale, *J. Chromatogr. A* **870**, 53 (2000).

1421. H. Tsuchiya, H. Shimizu, and M. Iinuma, *Chem. Pharm. Bull.* **47**, 440 (1999).

1422. F. Sadeghipour, C. Giroud, L. Rivier, and J.-L. Veuthey, *J. Chromatogr. A* **761**, 71 (1997).

1423. E. Sadeghipour, E. Varesio, C. Giroud, L. Rivier, and J.L. Veuthy, *Forensic Sci. Int.* **86**, 1 (1997).

1424. Ph. Baudot, A. Vicherat, M.-L. Viriot, and M.-C. Carri, *Analusis* **27**, 523 (1999).

1425. H.-J. Helmlin and R. Brenneisen, *J. Chromatogr.* **593**, 87 (1992).

1426. T. Lehmann and R. Brenneisen, *J. Liq. Chromatogr.* **18**, 689 (1995).

1427. K. Grogg-Sulser, H.-J. Helmlin, and J.-T. Clerc, *J. Chromatogr, A* **692**, 121 (1995).

1428. C. Moore, S. Browns, I. Tebbett, A. Negrusz, W. Meyer, and L. Jain, *Forensic Sci. Int.* **56**, 177 (1992).

1429. S. Madani, W.N. Howald, R.F. Lawrence, and D.D. Shen, *J. Chromatogr. B* **741**, 145 (2000).

1430. S. Wadud, R. Onodera, M.M. Or-Rashid, and MR. Amin, *J. Assoc. Off. Anal. Chem. Int.* **83**, 8 (2000).

1431. M.E.-S. Metwally, *Chromatographia* **50**, 113 (1999).

1432. K.-O. Chu and K.-C. Tin, *Anal. Lett.* **31**, 1879 (1998).

1433. R.P. Remmel, S.P. Kawle, D. Weller, and C.V. Fletcher, *Clin. Chem.* **46**, 73 (2000).

1434. R.W. Sparidans, R.M.W. Hoetelmans, and J.H. Beijnen, *J. Chromatogr. B* **750**, 155 (2001).

1435. C. Marzolini, A. Telenti, T. Buclin, J. Biollaz, and L.A. Decosterd, *J. Chromatogr. B* **740**, 43 (2000).

1436. V. Proust, K. Toth, A. Hulin, A.-M. Taburet, F. Gimenez, and E. Singlas, *J. Chromatogr. B* **742**, 453 (2000).

1437. M.R. Rashidi, J.A. Smith, S.E. Clarke, and C. Beedham, *Drug Metab. Dispos.* **25**, 805 (1997).

1438. T.P. Moyer, Z. Temesgen, R. Enger, L. Estes, J. Charlson, L. Oliver, and A. Wright, *Clin. Chem.* **45**, 1465 (1999).

1439. R.W. Sparidans, R.M.W. Hoetelmans, and J.H. Beijnen, *J. Chromatogr. B* **742**, 185 (2000).

1440. N. Lindegårdh and Y. Bergqvist, *J. Chromatogr. B* **744**, 9 (2000).

1441. A.I. Veldlkamp, R.P.G. van Heeswijk, P.L. Meenhorst, J.W. Mulder, J.M.A. Lange, J.H. Beijnen, and R.M.W. Hoetelmans, *J. Chromatogr. B* **734**, 55 (1999).

1442. P. Riska, M. Lamson, T. MacGregor, J. Sabo, S. Hattox, J. Pay, and J. Keirns, *Drug Metab. Dispos.* **27**, 895 (1999).

1443. F. Dai, J.A. Kelley, H. Zhang, N. Malinowski, M.F. Kavlick, J. Lietzau, L. Welles, R. Yarchoan, and H. Ford, Jr., *Anal. Biochem.* **288**, 52 (2001).

1444. A.S. Pereira, K.B. Kenney, M.S. Cohen, J.E. Hall, J.J. Eron, R.R. Tidwell, and J.A. Dunn, *J. Chromatogr. B* **742**, 173 (2000).

1445. M.P. Segatti, G. Nisi, F. Grossi, M. Mangiarotti, and C. Lucarelli, *J. Chromatogr.* **536**, 319 (1991).

1446. G. Aymard, P. Livi, Y.T. Pham, and B. Diquet, *J. Chromatogr. B* **700**, 183 (1997).

1447. D.J. Sequeira and H.W. Strobel, *J. Chromatogr. B* **673**, 251 (1995).

1448. J.-M. Gaulier, P. Marquet, E. Lacassie, R. Desroches, and G. Lachatre, *J. Chromatogr. B* **748**, 407 (2000).

1449. P. Pienmäki, S. Fuchs, J. Isojärvi, and K. Vähäkangas, *J. Chromatogr. B* **673**, 97 (1995).

1450. J.S. Kiel, S.L. Morgan, and R.K. Abramson, *J. Chromatogr.* **320**, 313 (1985).

1451. B.T. Duhart, Jr., D. Zhang, J. Deck, J.P. Freeman, and C.E. Cerniglia, *Xenobiotica*, **29**, 733 (1997).

1452. Y.-Z. Shu, J.W. Hubbard, J.K. Cooper, G. McKay, E.D. Korchinski, R. Kumar, and K.K. Midha, *Drug Metab. Dispos.* **18**, 735 (1990).

1453. G. Luippold, U. Delabar, D. Kloor, and B. Mühlbauser, *J. Chromatogr. B* **724**, 231 (1999).

1454. E. Matsui, M. Hoshino, A. Matsui, and A. Okahira, *J. Chromatogr. B* **668**, 299 (1995).

1455. B.V. Atul and R.C. Gupta, *J. Chromatogr. B* **748**, 331 (2000).

1456. Y. Gaillard and G. Pépin, *Forensic Sci. Int.* **87**, 239 (1997).

1457. T. Mae, E. Konishi, K. Hosoe, and T. Hikada, *Xenobiotica* **29**, 1073 (1999).

1458. M. Horie, K. Saito, R. Ishii, T. Yoshida, Y. Haramaki, and H. Nakazawa, *J. Chromatogr. A* **812**, 295 (1998).

1459. J.A. Hernández-Arteseros, R. Campaño, and M.D. Prat, *Chromatographia* **52**, 58 (2000).

1460. C.M. Stobba-Wiiey and R.S. Readnour. *J. Assoc. Off. Anal. Chem.* **83**, 555 (2000).

1461. A.P. Argekar, S.U. Kapadia, and S.V. Raj, *Anal. Lett.* **29**, 1539 (1996).

1462. D.A. Stead and R.M.E. Richards, *J. Chromatogr. B* **693**, 415 (1997).

1463. J.A. Hernández-Arteseros, I. Boronat, R. Campañó, M.D. Prat, *Chromatographia* **52**, 295 (2000).

1464. J.J. Berzas Nevado, G. Castañeda Peñalvo, and F.J. Guzmán Bernardo, *J. Chromatogr. A* **870**, 169 (2000).

1465. I. Durán Merás, T. Galeano Díaz, M.I. Rodriguez Cáceres, and F. Salinas López, *J. Chromatogr. B* **787**, 119 (1997).

1466. S. Reif, P. Le Corre, G. Dollo, F. Chevanne, and R. Le Verge, *J. Chromatogr. B* **719**, 239 (1998).

1467. J.Å. Jönsson, M. Andersson, C. Melander, J. Norberg, E. Thordarson, and L. Mathiasson, *J. Chromatogr. A* **870**, 151 (2000).

1468. S. Bolze and R. Boulieu, *Clin. Chem.* **44**, 560 (1998).

1469. P. Favetta, J. Guitton, C.S. Degoute, L. Van Daele, and R. Boulieu, *J. Chromatogr. B* **742**, 25 (2000).

1470. T.B. Vree, A.J. Lagerwerf, C.P. Bleeker, and P.M.R.M. de Grood, *J. Chromatogr. B* **721**, 217 (1999).

1471. F. Streit, U. Christians, H.-M. Schiebel, K.L. Napoli, L. Ernst, A. Linck, B.D. Kahan, and K.-F. Sewing, *Clin. Chem.* **42**, 1417 (1996).

1472. K. Na-Bangchang, O. Supasyndh, T. Supaporn, V. Banmairuroi, and I. Karbwang, *J. Chromatogr. B* **738**, 169 (2000).

1473. W.T. Liu, M.B. Tamolang, H. Pang, Y. Ren, and P.Y. Wong, *J. Pharm. Toxicol. Methods* **35**, 121 (1996).

1474. N. George, M. Kuppusamy, and K. Balaraman, *J. Chromatogr.* **604**, 285 (1992).

1475. C.P. Wang, N.R. Hartman, R. Venkataramanan, I. Jardine, F.-T. Lin, J.E. Knapp, T.E. Starzl, and G.J. Burckart, *Drug Metab. Dispos.* **17**, 292 (1989).

1476. L.J. Brunner and S. Bai, *J. Chromatogr. B* **732**, 323 (1999).

1477. J.V. Aukunuru, U.B. Kompella, and G.V. Betageri, *J. Liq. Chromatogr.* **23**, 565 (2000).

1478. A. Staab, S. Scheithauer, H. Fieger-Büschges, E. Mutschler, and H. Blume, *J. Chromatogr. B* **751**, 221 (2001).

1479. M.J. Nozal, J.L. Bernal, L. Toribio, J.J. Jiménez, and M.T. Martin, *J. Chromatogr. A* **870**, 69 (2000).

1480. H. He, S.D. Shay, Y. Caraco, M. Wood, and A.J.J. Wood, *J. Chromatogr. B* **708**, 185 (1998).

1481. A. Hermening, A.-K. Gräfe, G. Baktir, E. Mutschler, and H. Spahn-Langguth, *J. Chromatogr. B* **741**, 129 (2000).

1482. T. Prueksaritanont, L.M. Gorham, B. Ma, L. Liu, X. Yu, J.J. Zhao, D.E. Slaughter, B.H. Arison, and K.P. Vyas, *Drug Metab. Dispos.* **25**, 1191 (1997).

1483. A. Houck, S. Thomas, and D.K. Ellison, *Talanta* **40**, 491 (1993).

1484. K.P. Vyas, P.H. Kari, S.M. Pitzenberger, R.A. Halpin, H.G. Ramjit, B. Arison, J.S. Murphy, W.F. Hoffman, M.S. Schwartz, E.H. Ulm, and D.E. Duggan, *Drug Metab. Dispos.* **18**, 203 (1990).

1485a. H.R. Wiltshire, B.M. Sutton, G. Heeps, A.M. Betty, D.W. Angus, M.J. Madigan, and S.R. Sharp, *Xenobiotica* **27**, 539 (1997).

1485b. M. Yritia, P. Parra, E. Iglesias, and J.M. Barbanoj, *J. Chromatogr. A* **870**, 115 (2000).

1486. S. Bidouil, J. Dubois, and M. Hanocq, *J. Chromatogr. B* **693**, 359 (1997).

1487. I.N. Papadoyannis, V.F. Samanidou, K.A. Georga, and F. Georgarakis, *J. Liq. Chromatogr.* **21**, (1998).

1488. Y.P. Patel, S. Patil, I.C. Bhoir, and M. Sundaresan, *J. Chromatogr. A* **828**, 283 (1998).

1489. B.R. Simmons and J.T. Stewart, *J. Liq. Chromatogr.* **17**, 2675 (1994).

1490. K. Umehara, S. Kudo, and M. Odomi, *Xenobiotica* **27**, 1121 (1997).

1491. A.J. Braza, P. Modamio, and E.L. Mariño, *J. Chromatogr. B* **718**, 267 (1998).

1492. E. Brandšteterová and I.W. Wainer, *J. Chromatogr. B* **732**, 395 (1999).

1493. J.F. Bauer, S.K. Krogh, Z.L. Chang, and C.F. Wong, *J. Chromatogr.* **648**, 175 (1993).

1494. D.W. Armstrong, S. Chen, C. Chang, and S. Chang, *J. Liq. Chromatogr.* **15**, 545 (1992).

1495. L. Pichard, G. Gillet, I. Fabre, I. Dalet-Beluche, C, Bonfils, J.-P. Thenot, and P. Maurel, *Drug Metab. Dispos.* **18**, 711 (1990).

1496. K. Na-Bangchang, K. Congpuong, L.N. Hung, P. Molunto, and J. Karbwang, *J. Chromatogr. B* **708**, 201 (1998).

1497. V.K. Dua, S.N. Sinha, and V.P. Sharma, *J. Chromatogr. B* **708**, 316 (1998).

1498. H. Astier, C. Renard, V. Cheminel, O. Soares, C. Mounier, F. Peyron, and J.F. Chaulet, *J. Chromatogr. B* **698**, 217 (1997).

1499. P. Houzé, A. de Reynies, F.J. Baud, M.F. Benatar, and M. Pays, *J. Chromatogr.* **574**, 305 (1992).

1500. J.F. Chaulet, C. Mounier, O. Soares, and J.L. Brazier, *Anal. Lett.* **24**, 665 (1991).

1501. C. Beedham, Y. Al-Tayib, and J.A. Smith, *Drug Metab. Dispos.* **20**, 889 (1992).

1502. M. Ogata, Y. Noro, M. Yamada, T. Tahara, and T. Nishimura, *J. Pharm. Sci.* **87**, 91 (1998).

1503. Q. Jia, M.-F. Hong, Z.-X. Pan, and S. Orndorff, *J Chromatogr. B* **750**, 81 (2001).

1504. F. Bévalot, Y. Gaillard, M.A. Lhermitte, and G. Pépin, *J. Chromatogr. B* **740**, 227 (2000).

1505. K.G. Flood, E.R. Reynolds, and N.H. Snow, *J. Chromatogr. A* **903**, 49 (2000).

1506. M.J. López de Alda and D. Barceló, *J. Chromatogr. A* **892**, 391 (2000).

1507. R. Gonzalo-Lumbreras and R. Izquierdo-Hornillos, *J. Chromatogr. B* **742**, 47 (2000).

1508. M.R. Feng, J. Atherton, S. Knoll, C.A. Strenkoski, and D.S. Wright, *J. Chromatogr. B.* **693**, 159 (1997).

1509. M.A. Raggi, G. Casamenti, R. Mandrioli, and V. Volterra, *J. Chromatogr. B* **750**, 137 (2001).

1510. S. Walter, S. Bauer, I. Roots, and J. Brockmöller, *J. Chromatogr. B* **720**, 231 (1998).

1511. V. Larsimont, J. Meins, H. Fieger-Büschges, and H. Blume, *J. Chromatogr. B* **719**, 222 (1998).

1512. A. Avenoso, G. Facciolà, G.M. Campo, A. Fazio, and E. Spina, *J. Chromatogr. B* **714**, 299 (1998).

1513. H. Seno, H. Hattori, A. Ishii, T. Kumazawa, K. Watanabe-Suzuki, and O. Suzuki, *J. Chromatogr. B* **746**, 3 (2000).

1514. A. Avenoso, G. Facciolà, M. Salemi, and E. Spina, *J. Chromatogr. B* **746**, 173 (2000).

1515. Y. Fujimaki, H. Hakusui, and Y. Yamazoe, *Xenobiotica* **26**, 821 (1996).

1516. T.J. Strelevitz and M.C. Linhares, *J Chromatogr. B* **675**, 243 (1996).

1517. D.J. Quon, *J. Agric. Food Chem.* **48**, 6421 (2000).

1518. D.M. Matabudul, B. Conway, and I.D. Lumley, *The Analyst* **125**, 2196 (2000).

1519. S. Perea, G.J. Pennick, A. Modak, A.W. Fothergill, D.A. Sutton, D.J. Sheehan, and M.G. Rinaldi, *Antimicrobial Agents and Chemotherapy* **44**, 1209 (2000).

1520. D.H. Hunter and L.G. Luyt, *Bioconjugate Chem.* **11**, 175 (2000).

1521. M. Kartal, A. Khedr, and A. Sakr, *J. Chromatogr. Sci.* **38**, 151 (2000).

1522. C. Barbas, A. García, L. Saavedra, and M. Castro, *J. Chromatogr. A* **870**, 97 (2000).

1523. J.M. Morales, C.H. Jung, A. Alarcón, and A. Barreda, *J. Chromatogr. B* **746**, 133 (2000).

1524. L.L. Hansen, J.T. Larsen, and K. Brøsen, *J. Chromatogr. B* **712**, 183 (1998).

1525. E. Rozman, M.T. Galcerán, and C. Albet, *J. Chromatogr. B* **688**, 107 (1997).

1526. R.W. Sparidans, J. den Hartigh, S. Cremers, J.H. Beijnen, and P. Vermej, *J. Chromatogr. B* **738**, 331 (2000).

1527. J.C. Reepmeyer and D.C. Cox, *J. Assoc. Off. Anal. Chem. Int.* **80**, 767 (1997).

1528. R.A. Halpin, L.A. Geer, K.E. Zhang, T.M. Marks, D.C. Dean, A.N. Jones, D. Melillo, G. Doss, and K.P. Vyas, *Drug Metab. Dispos.* **28**, 1244 (2000).

1529. K. Richter, S. Scheithauer, and D. Thümmler, *J. Chromatogr. B* **708**, 325 (1998).

1530. I. Ürmös, I. Klebovich, and K.B. Nemes, *J. Liq. Chromatogr.* **21**, 803 (1998).

1531. L. García-Capdevila, C. López-Calull, S. Pompermayer, C. Arroyo, A.M. Molins-Pujol, and J. Bonal, *J. Chromatogr. B* **708**, 169 (1998).

1532. P.B. Jacobsen, *J. Chromatogr. B* **749**, 135 (2000).

1533. V. Antonucci, L. Wright, and P. Toma, *J. Liq. Chromatogr.* **21**, 1649 (1998).

1534. I.-H. Wang, R. Moorman, and J. Burleson, *J. Liq. Chromatogr.* **19**, 3293 (1996).

1535. D. Fluchard, S. Kiebooms, M. Dubois, and Ph. Delahaut, *J. Chromatogr. B* **744**, 139 (2000).

1536. L. Jin, M.R. Davis, P. Hu, and T.A. Baillie, *Chem. Res. Toxicol.* **7**, 526 (1994).

1537. T.E.B. Leakey, *J. Chromatogr.* **507**, 199 (1990).

1538. O. Agostini, G. Bonacchi, G. Coppini, E. Toja, A. Triolo, S. Manzini, G. Pieraccini, and G. Moneti, *Drug Metab. Dispos.* **22**, 259 (1994).

1539. M.C. Castle, *Drug Metab. Dispos.* **21**, 1147 (1993).

1540. L. Gagliardi, A. Amato, L. Turchetto, and D. Tonelli, *Anal. Lett.* **24**, 1825 (1991).

1541. R. Ventura, T. Nadal, P. Alcalde, J.A. Pascual, and J. Segura, *J. Chromatogr. A* **655**, 233 (1993).

1542. J.P. Bounine, B. Tardif, P. Beltran, and D.J. Mazzo, *J. Chromatogr. A* **677**, 87 (1994).

1543. Y. Fujimaki, T. Hosokami, and K. Ono, *Xenobiotica* **25**, 501 (1995).

1544. J.P. Scholl, *J. Liq. Chromatogr.* **17**, 3369 (1994).

1545. G.L. Hoyer, D.C. Clawson, L.A. Brookshier, P.E. Nolan, and F.I. Marcus, *J. Chromatogr.* **572**, 159 (1991).

1546. O.-W. Lau and C.-S. Mok, *J. Chromatogr. A* **693**, 45 (1995).

1547. P.M. Lacroix, P.N. Yat, and E.G. Lovering, *J Assoc. Off. Anal. Chem. Int.* **78**, 334 (1995).

1548. S. Singh, R.O. Epemolu, P.S. Dobbin, G.S. Tilbrook, B.L. Ellis, L.A. Damani, and R.C. Hider, *Drug Metab. Dispos.* **20**, 256 (1992).

1549. R.S. Hsu, G.M. Shutske, E.M. Dileo, S.M. Chesson, A.R. Linville, and R.C. Allen, *Drug Metab. Dispos.* **18**, 779 (1990).

1550. A. Berthod, C.-D. Chang, and D.W. Armstrong, *Talanta* **40**, 1367 (1993).

1551. F.L. Cardinali, J.M. McCraw, D.L. Ashley, and M.A. Bonin, *J. Chromatogr. Sci.* **32**, 41 (1994).

1552. C.G. Karakatsanis and T.B. Reddy, in J.F. Coetzee (ed.), *Recommended Methods for the Purification of Solvents and Tests for Impurities*, Pergamon Press, Oxford, 1982, pp. 25–31.

1553. T. Heinze and K. Pfeiffer, *Die Angewandte Makromol. Chemie* **266**, 37 (1999).

1554. L. Nitschke and L. Huber, *Fresenius J. Anal. Chem.* **349**, 451 (1994).

1555. S. Husain, R. Narsimha, S.N. Alvi, and R.N. Rao, *J. Chromatogr.* **600**, 316 (1992).

1556. P.J.M. Bergers and A.C. de Groot, *Water Res.* **28**, 639 (1994).

1557. J. Dai and G.R. Helz, *Ana. Chem.* **60**, 301 (1988).

1558. W.W. Bedsworth and D.L. Sedlak, *J. Chromatogr. A* **905**, 157 (2001).

1559. M.G. Boersma, N.H.P. Cnubben, W.J.H. van Berkel, M. Blom, J. Vervoort, and I.M.C.M. Rietjens, *Drug Metab. Dispos.* **21**, 218 (1993).

1560. H. Zou, Y. Zhang, X. Wen, and P. Lu, *J. Chromatogr.* **523**, 247 (1990).

1561. C.D. Gaitonde and P.V. Pathak, *J. Chromatogr.* **514**, 330 (1990).

1562. I.K. Cigić, J. Plavec, S.S. Možina, and L. Zupančič-Kralj, *J. Chromatogr. A* **905**, 359 (2001).

1563. H. Toyoda, H. Yamamoto, N. Ogino, T. Toida, and T. Imanari, *J. Chromatogr. A* **830**, 197 (1999).

1564. Y. Yamamoto and F. Kondo, *J. Assoc. Off. Anal. Chem. Int.* **82**, 248 (1999).

1565. D. Madigan, I. McMurrough, and M.R. Smyth, *Analyst* **119**, 863 (1994).

1566. E. Marengo, V. Maurino, S. Angelino, R. Aigotti, C. Biaocchi, and M.C. Gennaro, *Anal. Lett.* **32**, 1653 (1999).

1567. F. Pucciarelli, P. Passamonti, and T. Cecchi, *J. Liq. Chromatogr.* **20**, 2233 (1997).

1568. G.K. Jayaprakasha and K.K. Sakariah, *J. Chromatogr. A* **806**, 337 (1998).

1569. L. Zhou, Y. Wu, B.D. Johnson, R. Thompson, and J.M. Wyvratt, *J. Chromatogr. A* **866**, 281 (2000).

1570. I. Nicoletti, C. Corradini, and E. Cogliandro, *Int. J. Cosmet. Sci.* **21**, 265 (1999).

1571. C. Imaz, R. Navajas, D. Carreras, C. Rodríguez, and A.F. Rodríguez, *J. Chromatogr. A* **870**, 23 (2000).

1572. M. Czauderna and I. Kowalczyk, *J. Chromatogr. B* **744**, 129 (2000).

1573. F. Kusu, X.-D. Li and K. Takamura, *Chem. Pharm. Bull* **40**, 3284 (1992).

1574. U. Fuhr, T. Wolff, S. Harder, P. Schymanski, and A.H. Staib, *Drug Metab. Dispos.* **18**, 1005 (1990).

1575. M.A. Eiteman, R.M. Gordillo, and M.L. Cabrera, *Fresenius J. Anal. Chem.* **348**, 680 (1994).

1576. P.J. Worsfold and B. Yan, *Anal. Chim. Acta* **246**, 447 (1991).

1577. S. Sarre, Y. Michotte, P. Herregodts, D, Deleu, N. De Klippel, and G. Ebinger, *J. Chromatogr.* **575**, 207 (1992).

1578. V. Gökmen and J. Acar, *J. Chromatogr. A.* **847**, 69 (1999).

1579. B. Beretta, A.A. Gaiaschi, C.L. Galli, and P. Restani, *Food Addit. Contam.* **17**, 399 (2000).

1580. P.P. Minnaar, G.E. Swan, R.I. McCrindle, W.H.J. de Beer, and T.W. Naudé, *J. Chromatogr. Sci.* **38**, 16 (2000).

1581. Y. Oshima, *J Assoc. Off. Anal. Chem. Int.* **78**, 528 (1995).

1582. M.J. Esteve, R. Farré, A. Frígola, and J.M. García-Cantabella, *J. Chromatogr. A* **795**, 383 (1998).

1583. J. Lykkesfeldt, S. Loft, and H.E. Poulsen, *Anal. Biochem.* **229**, 329 (1995).

1584. M.A. Kall and C. Andersen, *J. Chromatogr. B* **730**, 101 (1999).

1585. F. Valls, M.T. Sancho, M.A. Fernández-Muiño, and M.A. Checa, *J. Agric. Food Chem.* **49**, 38 (2001).

1586. C.J. Argoudelis, *J. Chromatogr. A* **790**, 83 (1997).

1587. R. Kock, B. Delvoux, and H. Greiling, *Eur. J. Chem. Clin. Biochem.* **31**, 303 (1993).

1588. M.C. Gennaro and C. Abrigo, *Fresenius J. Anal. Chem.* **340**, 422 (1991).

1589. R. Accinni, J. Campolo, S. Bartesaghi, G. De Leo, C. Lucarelli, C.F. Cursano, and O. Parodi, *J. Chromatogr. A* **828**, 397 (1998).

1590. J.A. Resines, M.J Arin, and M.T. Díez, *J. Chromatogr.* **607**, 199 (1992).

1591. K. Hermann and D. Abeck, *J. Chromatogr. B* **750**, 71 (2001).

1592. M. Berger, J. Cadet, R. Berube, R. Langlois, and J.E. van Lier, *J. Chromatogr.* **593**, 133 (1992).

1593. W.A. Kleinman and J.P. Richie, Jr., *Biochem. Pharmacol.* **60**, 19 (2000).

1594. R.S. Sodum, S.A. Akerkar, and E.S. Fiala, *Anal. Biochem.* **280**, 278 (2000).

1595. W. Hu, K. Hasebe, and P.R. Haddad, *Anal. Commun.* **34**, 311 (1997).

1596. L.M.-M. Louche, F. Luro, E.M. Gaydou, and J.C. Lesage, *J. Agric. Food Chem.* **48**, 4729 (2000).

1597. A.M. Szmigielska, J.J. Schoenau, and V. Levers, *J. Agric. Food Chem.* **48**, 4487 (2000).

1598. J.A. Hurlbut, J.R. Carr, E.R. Singleton, K.C. Faul, M.R. Madson, J.M. Storey, and T.L. Thomas, *J. Assoc. Off. Anal. Chem. Int.* **81**, 1211 (1998).

1599. J.A. Orwa, F. Bosmans, S. Depuydt, E. Roets, and J. Hoogmartens, *J. Chromatogr. A* **829**, 115 (1998).

1600. X. Tan and F.D. Boudinot, *J. Chromatogr. B* **740**, 281 (2000).

1601. M. Merodio, M.A. Campanero, T. Mirshahi, M. Mirshahi, and J.M. Irache, *J. Chromatogr. A* **870**, 159 (2000).

1602. C. Pham-Huy, F Stathoulopoulou, P. Sandouk, J.-M. Scherrmann, S. Palombo, and C. Girre, *J. Chromatogr. B* **732**, 47 (1999).

1603. R.B. Burns and L. Embree, *J. Chromatogr. B* **744**, 367 (2000).

1604. M.C. Gennaro, D. Giacosa, E. Gioannini, and S. Angelino, *J. Liq. Chromatogr.* **20**, 413 (1997).

1605. A. Rosado-Maria, A.I. Gasco-Lopez, A. Santos-Montes, and R. Izquiero-Hornillos, *J. Chromatogr. B* **748**, 415 (2000).

1606. M. El-Khateeb, T.G. Appleton, B.G. Charles, and L.R. Gahan, *J. Pharm. Sci.* **88**, 319 (1999).

1607. G. Skopp, K. Klinder, F. Pötsch, C. Zimmer, R. Lutz, R. Aderjan, and R. Mattern, *Forensic Sci. Int.* **95**, 99 (1998).

1608. M. Sarasa, N. Riba, L. Zamora and X. Carné, *J. Chromatogr. B* **746**, 183 (1999).

1609. X. Xu and J.T. Stewart, *J. Liq. Chromatogr.* **22**, 769 (1999).

1610. E. Gamelin, M. Boisdron-Celle, A. Turcant, F. Larra, P. Allain, and J. Robert, *J. Chromatogr. B* **695**, 409 (1997).

1611. K. Smets, E. Roets, J.McB. Miller, E. Porqueras, N. Berti, P. Vettori, F. Folliard, and J. Hoogmartens, *Chromatographia* **37**, 640 (1993).

1612. S.A. Swedberg, P.M. Bassett, J.J. Pesek, and A.L. Fink, *J. Chromatogr.* **393**, 317 (1987).

1613. J.-M. Gulian, C. Delmasso, V. Millet, D. Unal, and M. Charrel, *Eur. J Clin. Chem. Clin. Biochem.* **33**, 503 (1995).

SUBJECT INDEX

Absorbance vs. wavelength
  Dissolved O_2 in solvents
  Effects of gradient work, 7,18
  General, 3–9
  Normal-phase solvents
    Ketones, low wavelength use, 15
  Reversed-phase solvents
    UV and non-spectro solvents, 8
  Shifts in curves, 17
  Sparging/volatility effects on, 30
Acetic acid absorbance vs. wavelength curve
  (3% in water), 12
Acetone absorbance vs. wavelength curve
  (3% in water), 12
Acetonitrile
  Absorbance vs. wavelength curve, 4
  Decomposition in HCl, 146
  Decreased sensitivity for gestrinone,
    208
  Ineffective in
    Alkylpyridine seps, 89
    Phenolic acid seps, 127
    Sanfodryl dyes, 121
  Interferences in electrochemical detection
    of pesticides, 107
Alcohols
  General properties, 75–81
  Impurities, 81–2
  Structures, 81
Alkanes
  Decomposition of, 218–20
  General properties, 216–8

Impurities, 218–20
Structures, 223
Antioxidants, added to protect analyte
  Ascorbic acid, 120
  BHT, 144, 147, 238, 276, 278, 386, 388,
    406, 429
  Pyrogallol, 243

Baseline shift issues during gradients, 83,
  149, 193, 194, 202, 205, 211
Buffers and mobile phase additives, 25–9
  Aging effects on absorbance vs. wave-
    length curve, 13

Chlorinated solvents
  Acid formation in, 35
  Decomposition of, 346, 261–2
  General properties, 260–2
  Impurities, 263
  Structures, 264
Cyclodextrins
  Enhancement of aflatoxin fluorescence
    by, 133
  Enhancement of PAH fluorescence by, 97

Decomposition of polymers in unpreserved
  THF, 296
Dichloromethane absorbance vs. wavelength
  curve, 5

2,2-Dimethoxypropane, water removal from silica, 242
Dimethylformamide absorbance vs. wavelength curve, 10
Dimethylsulfoxide (DMSO) absorbance vs. wavelength curve, 10

Eluotropic series, 46–9
Ethanol
    Denaturants used in, 36, 76
    Description of commercially available types, 36
Ethers
    General properties, 32, 287–9
    Impurities
    Tests for peroxides in, 288
    Structures, 292
2-Ethoxyethanol absorbance vs. wavelength curve, 11
Ethyl acetate absorbance vs. wavelength curve, 5
Ethyl ether absorbance vs. wavelength curve, 10

Filtering solvents, 43–3
Fluorescence
    Solvent effects on, 42–3

Gradient absorbance vs. time (baseline) profiles, 7, 18

Hexane absorbance vs. wavelength curve, 5

k' vs. % organic plots, atypical U-shaped
    Caffeine/methanol, 185–6
    Imidazol(in)e/methanol, 186
    Nortryptaline/acetonitrile, 499
    Paraquat-diquat/acetonitrile, 370
    Proteins/acetonitrile, 460
Ketones
    Condensation reactions, 315
    General properties, 314–6
    Impurities, 316
    Structures, 320

Lot-to-lot solvent variability, 16–9
Low-viscosity solvent effects, 270

Manufacturer specifications for solvents
    Individual
    Acids, 538
    Alcohols, 79
    Alkanes, 221
    Chlorinated, 266
    DMSO, 538
    Ethers, 291
    Ketones, 318
    Nitrile/nitrogen-containing, 334
    Water, 538
Mercaptoethanol, precautions in use, 93
Methanol
    Absorbance vs. wavelength curve, 4
    Breakdown of rare earth metals tetraphenylporphyrin complexes in, 95
    Decomposition of aflatoxins in, 134
    Decomposition of cloxacillin in, 192
    UV-background effects at low wavelength, 91, 149, 193, 194
Method optimization techniques
    Brute force, 60–2
    Eluotropic strength, 46–9
    Scouting, 58–60
    Solvent strength, 49–53
    Triangulation, 53–7
    Window diagrams, 60
N-Methylpyrrolidone absorbance vs. wavelength curve, 11
Miscibility
    Chart of, 23
    Definition, 22
    General, 22–5
Mixing effects
    Temperature and volume changes on solvent mixing, 24
Mobile phase modifiers, 9–10

Nitriles and nitrogenous solvents
    General properties, 329–30
    Impurities, 330–1
    Structures, 336

Optimization techniques (See Method optimization techniques)
Oxygen peak in electrochemical detection, 100

Particulates in solvents, 43–4
Peroxide formation
 in ethers, 32–4
 screening for in ethers, 34
pH effect on retention and peak shape, 124,
  126, 140, 164, 165, 173, 198, 206,
  209, 464, 480
pH effect of organic type on, 28
Precolumn use with aggressive solvents,
  97
Protein denaturation by solvents, 181
Pyridine absorbance vs. wavelength curve,
  10

Re-equilibration issues, 480

Safety parameters
 Individual solvent
  Acids, 339
  Alcohols, 80
  Alkanes, 222
  Chlorinated, 267
  DMSO, 539
  Ethers, 292
  Ketones, 319
  Nitrile/nitrogen-containing, 335
  Water, 539
Solubility, 23
 Buffer solubility, 28
 Phosphate in acetonitrile, 28
 Individual solvent in water
  Acids, 539
  Alcohols, 77
  Alkanes, 219
  Chlorinated, 264
  DMSO, 539
  Ethers, 290
  Ketones, 317
  Nitrile/nitrogen-containing, 332
  Water, 539
Solvent contaminants, 41–2
Solvent instability/reactivity, 31–9
 Acetone, 38–9
 Alkanes, 218
 Chlorinated, 34–6
 Ethanol, 36
 Ethers, 32–4
 Water, 36–8
Solute/solvent mismatch effects, 99, 131,
  331, 467, 468

Solvent strength parameters, 49–53
 Individual, 47
  Acids, 537
  Alcohols, 78
  Alkanes, 220
  Chlorinated, 265
  DMSO, 537
  Ethers, 290
  Ketones, 318
  Nitrile/nitrogen-containing, 333
  Water, 537
Solvent variability, 16–9
Solvent volatility, 29–31
System peaks, 14–6, 141, 469
System suitability, 74

Temperature change upon solvent mixing, 24
Tetrabutylammonium hydroxide absorbance
  vs. wavelength curve (50mM in
  water), 14
Tetramethylammonium chloride absorbance
  vs. wavelength curve (50mM in
  water), 14
THF absorbance vs. wavelength curve, 4
THF absorbance vs. wavelength curve
  preserved/unpreserved, 8
THF as mobile phase modifier peak shape
  enhancement, 300, 307, 308, 337, 527
THF causing split peaks, 299
Trifluoroacetic acid (TFA) absorbance vs.
  wavelength curve (3% in water), 12
UV cutoff, 2
 By solvent class, 3
 Definition of, 2
 Effects on detection limit/sensitivity, 104,
   194, 232, 275, 303
 General, 2–3
 Individual solvents
 Acids, 537
 Alcohols, 79
 Alkanes, 221
 Chlorinated, 266
 DMSO, 537
 Ethers, 291
 Ketones, 318
 Nitrile/nitrogen-containing, 334
 Water, 537
  Offset of gradient with additive, 92

"U"-shaped retention vs. % organic plots,
  186

Viscosity
  General effects, 19–22
  Gradient effects, 22
  Individual solvents
    Acids, 537
    Alcohols, 77
    Alkanes, 219
    Chlorinated, 264
    DMSO, 537
    Ethers, 290
    Ketones, 317
    Nitrile/nitrogen-containing, 332
    Water, 537
  Volatility/sparging effects, 320
  Volume change upon mixing solvents, 24

Water, 36–8
Water removal from silica, 242

ANALYTE INDEX*

Abacavir, *495*
Abamectin, 199, 211, *357*, 489
Abietic acid, 162, 384
Acacetin, *410*
Acenaphthene, *97*, 272
Acepromazine, 312
Acesulfame, 256, 309
Acetaldehyde, *91*, 342, 440
Acetanilides, *85*, 355
Acetanilides, alkyl subs, 85
Acetanilides, halogen subs, 85
Acetanilides, hydroxy subs, 85
Acetic acid, 90, 125, 383, 387, 538, 541
Acetic acid, bromo subs, 124
Acetic acid, chloro subs, 124
Acetic anhydride, 383
Acetominophen, 186, 189, 311, 492, *507*, 551
Acetominophen glucuronide, 507
Acetominophen sulfate, 507
Acetonaphthalenes, 268
Acetone, 91, 342, 383, 440
Acetophenone, 224, 528
Acetosyringone, *345*
Acetovanillone, *345*
2-Acetoxybrevifoliol, 195
N-Acetoxyprotriptyline, 499
Acetylacetone metal ion complexes, 94
N-Acetylamino acid butyl esters, 307
5-Actylamino-6-amino-3-methyluracil, 154

N-Acetyl-4-aminobenzene, 540
5-Acetylamino-2-hydroxybenzamide, 527
N-Acetylbenzidine, 132
Acetylcodeine, 200, 492
N-Acetylcysteine, 452
3-Acetyl-4-deoxynivalenol, 398
Acetyldihydrocodeine, 200
2-Acetylfuran, 338
N-Acetylgalactosamine, 434
Acetylgenistein, 415
N-Acetylglucosamine, 434
N-Acetyl-5-hydroxytryptamine, 390
6-Acetylnimbandiol, 447
N-Acetylneuraminic acid, 174
7-O-Acetylokadaic acid, 135
3-N-Acetylproflavin, 349
4α-Acetylpseudoguaian-6β-olide, 148
Acetylsalicylic acid (aspirin) 189, 508
N^1-Acetylspermindine, *88*
10-Acetyltaxol, 481
Acid Orange dyes, *379*
Acid Red dyes, *122*, 379
Acid Violet dye, *122*
Acid Yellow dyes, *379*
Aclonifen, *366*
Aconitine, *303*
Acridine, *228*, 273, 294, 347
Acrolein, *913*, 42
ACTH, 1–10, 458
Actein, *149*

*Structures are presented on italicized page numbers

Acyclovir, 548
N-Acylneuraminic acids, 340, 385
Adenine, 430
Adenosine, 173, 430, 500, 546
Adenosine diphosphate (ADP), 174
Adenosine monophosphate (AMP), 172, *174*
Adenosine triphosphate (ATP), 174, 430
Adenosine, 3':5'-cyclic monophosphate, 173
Adenosylcobalamin, 140
S-Adenosyl-L-homocysteine, 179, *500*
S-Adenosyl-L-methionine, 179, *500*
Adhyperforin, 160, *440*, 441
Adiphenine, *390*
Adlumidiceine, 427
ADP, 174
β-Adrenergic blocking drugs, 212
Adriamycin, 481
Adriamycinol, 481
Adriamycinone, 481
Adrenosterone, 269
Aerosol, 80-I, 374
Aflatoxins, *133*, *134*, 307, 397
Agmatine, *131*
Ajmaline, 206
Alachlor, 107
Albendazole and metabolites, 488
Albomitomycin C, 478
Albumin, serum, 182, 307, 460
Alcohols, ethoxylates, 373
Aldehydes, 342
Aldehydes, aromatic, 228
Aldicarb, *106*, 353, 354, 360, 361
Aldicarb sulfone, 106, 360, 361
Aldicarb sulfoxide, 106, 360, 361
Alditols, 433
Alftalats (polyesters), 376
Aliphatic carboxylic acids (C_1-C_{20}) 300, 383
Alizarin, *121*
Alkamides, 395, 396
Alkaloids, ergot, 422
Alkanals, 342
Alkannins, *164*
Alkyl subs benzenes, 85, 87, 89, 223, 224,
    230, 336, 337
Alkyl subs naphthalenes, 224, 230
Alkyl subs phenols, 228, 270, 336, 337
Alkyl subs pyridine, 89, 337
Alkylacylethanolamonie phospholipids, 429
Alkylamide arginine surfactants, *373*
Alkyl-N-arylsulfaminoyl ester enantiomers,
    252
Alkylbenzenes, 85, 87, 89, 223, 224, 230,
    336, 337

Alkylbenzenesulfonates, 115, *296*
Alkylether sulfates, 296
Alkylhydroxyperoxides, 175
N-Alkyl-N-methylaniline-N-oxides, 93,
    227
Alkylnaphthalenes, 224, 230
Alkylphenolethercarboxylates, 92
Alkylphenylethoxylates, 114, 236, 237
Alkylphenols, ethoxylated, 92, 237
Alkylphosphocholines, 193
Alkylsulfonates, linear, *114*
Allantoin, *382*, 394, 542, 543, 545
Allethrin, *360*
Allocryptopine, *420*, 421
Allooctopine, 396
Allopurinol, 545
Alloxanthin, *327*
Alloxydim, *366*
Aloe-emodin, 164, 414, *439*
Aloenin, 166
Aloesin, *166*
Aloesaponarin II, *158*
Aloins, *166*
Alprenolol enantiomers, 513
Altenuene, 136, *137*
Alternariol, *136*
Altersolanols, *401*
Altertoxins, *136*
Althiazide, 549
Alvaxanthone, 413
α-Amanitin, *401*
Amaranth, 121, *378*
Amentoflavone, 153
Amiloride, 211, 527, *549*
Amino acid enantiomers, N-acylation deriv,
    253, 305
Amino acid enantiomers, O-alkylation deriv,
    253, 305
Amino acids, 178, 179, 180, 253, 452–4
Amino acids, N-succinimidyl-α-
    methoxyphenylacetate deriv, 253
Amino subs benzenes, 227
Amino subs benzoic acids, 269
Amino subs naphthalenes, 227
Amino subs PAHs, 347
Amino subs phenols, 82
Amino subs pyridines, 227
Aminoampicillic acid, *473*
3-Aminobenzamide, 527
2-Aminobenzimidazole, *103*
4-Aminobenzoic acid, 269, 390, 530
2-Amino-5-benzoylbenzimidazole, 198
4-Aminobiphenyl, 380

4-Amino-3-(4-chlorophenyl)butyric acid
  isomers, 328
Aminochrysenes, 347
7-Aminoclonazepam, 486
2-Amino-4-(3,5-diacetylphenyl)amino-6-
  methylpyrimidine, 480
4-Amino-1,5-dichlorobenzene sulfonic acid,
  187
2-Amino-2,5-dichlorobenzophenone, 203
2-Amino-3,8-dimethylimidazole[4,5*f*]
  quinoxaline, *348*, 349
3-Amino-1,4-dimethyl-5*H*-pyrido[4,3*f*]
  indole, 396
Aminodinitrotoluenes, 233
Aminofluorenes, 347
Aminoglutethiamide enantiomers, *529*
Aminoglycosides, 503
Aminoimidazoleazaarenes, 396
2-Amino-6-methyldipyrido[1,2-*a*:3′2′-
  d]imidazole, 396
2-Amino-3-methylimidazole[4,5*f*]
  quinoxaline, *348*, 349
2-Amino-1-methyl-6-phenylimidazo[4,5*b*]
  pyridine, 345, *348*
Aminonaphthalenes, 347
2-Aminonaphthalene-1,5-disulfonate, 90
2-Aminonaphthalene-4,6,8-trisulfonic acid,
  540
8-Amino-1-naphthol-3,6-disulfonic acid and
  impurities, 540
7-Aminonitrazepam, 486
2-Aminonitrophenol, 120
Aminonitrotoluenes, 102, 350
Aminophenols, 82, 120, 225, 338, 507
2-Aminophenoxazine-3-one, *84*
Aminosalicylic acid and impurities, *89*
Amitraz and degrad products, *356*
Amitrole, *235*
Amitryptaline, 211, *212*, 484, 498, 499
Amlodipine, *511*
Amobarbital, *196*
Amoxicillin, *191*, 478
AMP, 172, 174
Amphetamines, 185, *490*, 491
Ampicillin, *191*, 472, 473, 478
Amprenavir, *495*, 497
Amprolium, 214
Amygdalin, *449*
n-Amylbenzene, 224
4-Amylphenol, 270
Anabasine, *130*, 392
Anabolic steroids, 197
Anandamine, *201*

Anatabine, *130*, 392
Anatoxin *a* and degrad, 400
Andalusol, 150
Androstenedione, 196
4-Androsten-17*β*-ol-3-one, 226, *269*
Androstenolone, 517
Androsterone, *226*, 404
Androsterone-3*β*-sulfate, 515
Anemarrhenasaponins, *158*
Anethole, 444, 449
Angelicide, 165
Angelicin, *138*
Angelol A, 437
Angiotensin, 458
Anhydroerythromycin, 470
Anhydrosecoisolariciresinol, 157
Aniline, 84, 225, 227, 380
Aniline, alkyl subs, 369
Aniline, chloro subs, 84, 227, 369
Aniline, methoxy subs, 227
Aniline, methyl subs, 227
Aniline, nitro subs, 227
Anilines, 369
p-Anisaldehyde, 158, 444
Anisic acid, *159i, 345*
o-Anisidine, 380
Anisole, 293
Anka pigments, 440
Ankaflavin, 440
Annatto compounds, *126*, 453
Anox 3114, 115
Anserine, *180*
Antheraxanthins, *146*, 147, 323
Anthocyanins, 419, 420
Anthracene, 97, 337
Anthracenes, nitro subs, 347
Anthralin, 258
Anthranilate, *N*-methyl, *339*
Anthrones, 166
Anthroquinone-2-sulfonic acid, 270
Antimycins, *470*
Antipyrine, 530
Antu, *361*
Apigenin, 152, 410, 413, 414, 416
Aporphine, *421*
Arabinooctose, 433
Arabinose, 280
Aracidins, 451
Arachadonic acid, 338
Arachidonylethanolamine, *201*
Arctigenin, *438*
Argpyrimidine, *132*
Aristolochic acids, *157*

Arkopal surfactants, 238
Aromatic aldehydes, 86
Aromatic sulfonataes, 90
Aromoline, *423*
Arsonilic acid, 552
Artemisinin, 245, *306*, 513
Artesunate, 513
Arvenines, 409
Arylamines, 380
Arylene bis(methyl ketones), 480
Aryloxyphenoxypropionic acids, 108
Asarinin, *438*
Asarones, *449*
Ascorbic acid, 403, *544*, 545
Asiatic acid, 149, *407*
Asiaticoside, 149, *166*, 407
Aspartame, 309
Asperuloside, 164
Asperulosidic acid, *164*, 445
Aspirin (acetylsalicylic acid), 189
Astaxanthin, *146*, 301, 406
Asterric acid, 509
Astragaloside IV, *162*
Atenolol, *511*
Atenolol enantiomers, 513
Atmer, 129, 274
Atovaquone, *497*
ATP, 174, 430
Atranols, *446*
Atrazine, 110, 111, 295, 353, 355, *369*
Atropine, 312
Auroxanthins, 146, *147*
Azaarenes, 228, 347
Azacyclonal, 492
Azadirachtins, 138, 409
4-Azafluorene, *247*
Azaperol, 526
Azaperone, *526*
4-Azapyrene, 347
Azaspirodecanedione, 202
3′-Azido-3′-deoxythymidine (AZT,
    zidovudine) *496*, 497, 548
Azithromycin, *467*, 468
Azo dyes, sulfonated, 122

Baccatin III, *390*, 481
Baccatins, *195*
Baclofen, 214
Bacteriochlorophyll, 663, 251
Bacteriorhodopsin cleavage peptides, 283
Baicalein, *302*, 415, 416
Baicalin, *302*

Balenine, 180
Barban, *107*
Barbitals, *196*
Barbiturates, 186
Basic Green, 121
Basic Violet, 121
Basic Yellow, 122
Baycarb (Fenobucarb), 360
Beauvericin, *400*
Beclomethasone, 215, 515
Bellidifolin-8-*O*-glucoside, 408
Benazolin, *109*
Bendroflumethiazide, 549
Benomyl, *357*, 371
Benoxaprofen enantiomers
Benoximate, 532
Benperidol, *518*
Bentazone, 106, *109*
Benz[*j*]aceanthrylene metabolites, *99*
Benz[*a*]acridine, *228*, 294
Benz[*c*]acridine and subs analogs, *231*
Benzaldehyde, 86, 91, 293, 337, 342, 383,
    464
Benzalkonium chlorides, 310
Benzamides, *342*, 397
Benzamidines, *397*
Benzanilides, *342*
Benz[*a*]anthracene, *97*, 99
Benz[*a*]anthracene and metabolites, 99
Benzene, 85, 88
Benzene, alkyl subs, 85, 87, 89, 223
Benzene, bromo subs, 85
Benzene, chloro subs, 85, 87, 223, 293
Benzene, cyano subs, 88
Benzene, disubs (mixed ligands) 86, 89
Benzene, hydroxy subs (phenols) 85, 83, 84,
    85
Benzene, nitro subs, 85, 101, 232, 293
Benzenedisulfonate, 90
Benzenesulfonic acid, 90, 270
Benzidine, *132*, 380
Benzidine and metabolites, 132
Benzoate, sodium, *338*
Benzocaine, 530
Benzocaine and metabolite, *207*
syn-Benzo[*g*]chrysene-11,12-dihydrodiol-
    13,14-epoxide deoxyadenosine
    adduct, 349
Benzodiazapines, 186, 203, 486
1,3-Benzodioxanes, 438
7,8-Benzoflavone and metabolites, *152*
Benzo[*k*]fluoanthene, *97*
Benzoic acid, 83, 270, 384, 464

Benzoic acid, fluoro subs, 335
Benzoic acid, hydroxy subs, 206
Benzoic acid, nitro subs, 83
Benzoin, *464*
Benzoin hemisuccinate, 464
Benzonitrile, 88, 293
Benzo[*g,h,i*]perylene, 96
Benzophenanthridine, 421
Benzo[*lm*]phenanthro[4,5,6,*abcd*]perylene, 272
Benzophenone sunscreen agents, *119*
Benzo[*a*]pyrene, *97*, 99
Benzo[*a*]pyrene triols, 99, 231
Benzo[*h*]quinoline, *347*
Benzothiazole, *340*
Benzosuberone-4-sulfonic acid, 270
2-Benzothiazole-2,2'-disulfide, 124
5-Benzoyl-2-acetylthiophene, 255
Benzoylecgonine, *312i, 492*
5-Benzoyl-2-ethylthiophene, 255
Benzoylmercury, 93
Benzoylnorecgonine, *312*
4-Benzoyloxyphenol ethers, 291
Benzoyl peroxide, *83*, 377
N-Benzoyl-2,4,6-trimethylaniline, 337
Benzthiazide, 527, 549
Benztropine, 530
Benzyl alcohol, 88
N-Benzylanilide, 202, 226, 331
N-Benzylanilines, 337
S-Benzylcysteine, 282
S-Benzylmercapturic acid, 334
N-Benzylmethylanilines, 337
1-Benzyl-3-methylnaphthalene, 490
Benzylnitrile, 88
N'-Benzylnornicotine, 392
Benzylpenicillin, 472
1-Benzyl-2-phenylpyrrolidine, 392
Berbamine, *423*
Berberine, *337*, 424
Bergapten, *137*
Betamethazone, 211, *212*, 285, *515*
Betaxolol enants, *257*
BHA (butylated hydroxyanisole) *156*, 405
BHEB, 113
BHT (butylated hydroxytoluene) 116, *156*, 405
I3, II8-Biapigenin, 441
Bifenox, 355, *356*, 366
Bifenthrin, *361*
Bifurcadiol, 345
Bile acids, *128*, 346, 388, 389
Bilirubin isomers, *553*

Bilobalide, *150*
Bilobetin, *152*
Biochanin A, *415*
Biotin, *140*, 277
Biphenyl, chloro subs, 87
Biphenyl, nitro subs, 286
Biphenyl-4,4'-dicarboxylic acid alkyl esters, *118*
2,2'-Biquinoline, *294*
α-Bisabolol, *408*
α-Bisabolol oxides, 408
Bisacodyl, 530
Bisbenzylisoquinoline alkaloids, 423
Bis-*N,N'*-(3-chloro-4-methylphenyl)urea, 235
4,4'-Bis(4-chloro-3-sulfonstyryl)biphenyl, 373
Bisdigitoxides, 254
Bis-3-ethylhexylester, 224
2,2'-Bisnalbuphine, *309*
Bisphenol A, *114*, 118
Bisphenol A and F, diglycicyl esters, 376
1,2-Bis(2-pyridyl)ethylene, 272
Bispyrimidinylpiperazinebutane, 202
Bitertanol, *370*
Bixins, *126*, 452
Bolasterone, *237*, 517
Bolderone, *517*
Boswellic acids, *161*
Bovine serum albumin, 182, 307, 460
Bradykinin-potentiating peptide, 457
Brevefoliols, *195*
Brevetoxins, *136*
Brighteners (optical), 115, 373
Brodifacoum, 137, *362*
Bromadiolone, 137, *362*
Bromhexine hydrochloride, 493*, 528*
Bromo subs acetic acid, 124
Bromo subs benzene, 337
Bromo subs phenols, 270, 336, 337
Bromocripine, 530
5-Bromo-2'-deoxyuridine and hydrosylates, *194*
Bromoochratoxin B, 135
Bromophenylacetylurea, *486*
4-(4-Bromophenyl)-3-oxapyrrolidine-2,5-dione, 486
N-(4-Bromophenylsulfonyl)acetamide, 523
Bromopropylate, *103*
Bromothyronines, 176
5-Bromouracil, 194, 551
Bromoxynil, 106, 295, *346*, 355
Bromperidol, *519*

Bronidox, 322
Bronopol, 322
Browning pigments, 303
Bufotenine, *133*
Bufuralol enantiomers, *284*
Bumetanide, 214, *549*
Bupivacaine, *206*, 312, 492
Bupremorphine, 200
Buserelin, 457
Buspirone and degrad prod, *202*
Buspirone and impurities, *522*
2,3-Butadione, 444
Butanal, 342
Butarbital, 551
Butobarbital, *196*
Butocarboxim, *360*
Buturon, *365*
Butylated hydroxyanisole (BHA) *156*, 405
Butylated hydroxytoluene (BHT) 116, *156*, 405
tert-Butylcalix[n]arenes, 341
tert-Butylhydroquinone, 405
4-*t*-Butyl-4′-methoxydibenzoylmethane (Eusolox 9020), 119
N-Butyl-*N*-methylaniline *N*-oxide, 227
Butylparaben, *339*, 382
Butylphenol, 114, 118
Butylphthalide, 165, *437*
N-*t*-Butyl-*N′*-phenylurea, 236
Butyltins, 94
N-Butyltryptophan, 282
Butyraldehyde, 91, 342
Butyric acid, 383, 387
Butyric anhydride, 383
γ-Butyrolactone, *200*

Cadaverine (1,5-diaminopentane), 88, 131, 305, 389, 397
Caeridin, 456
Caerin, 456
Caerulin, 456
Cafestol, 149
Caffeic acid, *86*, 90, 137, 153, 163, 300, 385, 386, 448
Caffeine, *154*, 159, 186, 187, 189, 309, 390, 391, 417, 418, 490, 492, 526, 543, 551
Caffeoylquinic acids, 126
2-*O*-Caffeoyltartaric acid (caftaric acid), *90*
Caftaric acid, 90
Calcifediol, 258
Calcitonin, 457

Californidine, *420*
Calix[n]arenes, *341*
Calphostin C, *466*
Calycanthine, *425*
Calycosin, *412*
Campesterol, *384*
Campholenealdehyde, *440*
Cannabichromene, 200, 492
Cannabidiolic acid, *491*
Cannibigerol, *491*
Cannabidiol, 200, *491*
Cannabinolic acid, *491*
Cannabinols, 200, 491
Canrenone, *213*, 549
Cantaxanthin, *2463*
Capric acid, 387
Caprioc acid, 387
Carpolactam, 376
Caprylic acid, 387
Capsaicin, *105*, 386
Capsanthin, *142*, 147, 301, 323, 436
Capsanthone, *142*
Capsorubin, 145, 147, 323, *436*
Captopril, 214, 333
Capxanthin, 145
Carazolol, *526*
Carbamate pesticides, *105*, 107, 234
Carbamazepine and metabolites, 196, 499
Carbaryl, *359*, 360
Carbazole, 235, 373
Carbendazim, *103*, 112, 357, 370
Carbofuran, 353, 361
Carbolines, 130, 490
Carbon tetrabromide, 89
Carbon tetrachloride, 89
Carbophenathion, *352*, 354
Carboplatin, 530, *548*
Carbonyl metal ion complexes (Mn, Re, Cr, W, Mo), 343
Carboxylic acids, apliphatic (C₆-C₂₀), 300, 383
Carboxylic acids, perfluorinated, 383
Carboxymethyl cellulose, 537
Carboxymethyl glucose, 538
Carminic acid, *121*
Carmosine, 121, 378
Carnosine, *180*
apo-8′-Carotenal, 145
Carotenes, 140, 141, 145, 146, 147, *148*, 250, 278, 301, 324, 406, 451
Carotenoids, 141, 142, 145, 146, 147, 278, 301, 324, 406
Carpofen and degrad products, *466*

Carpofen enantiomers, 255
Carteolol and metab, *512*
Carteolol enantiomers, 513
Caryachine, *420*, 421
Casein, 460
α_5 Casein protease hydrosylates, 183
Cataranthine, *482*
Catechin, 91, 163, *168*, 281, 407, 410, 413, 417, 418
Catechin dimers, *120*
Catechin esters, 413
Catecholamines, 391, 543
Catenanes, 341
Cefachlor, *189*
Cefadroxil, *189*
Cefazolin, 189, *471*, 472, 473
Cefixime, 530
Cefoperazone, *472*
Cefotaxime, 189, *462*, 471
Cefotaxime and metab, 466
Cefoxitin, 189, *471*
Cefquinome, 472
Cefsoludin, 189
Ceftiofur, 472, 478
Cefuroxime, 189, *471*
Celiprolol, *512*
Cellobiose, 434
Centpropazine, *500*
Cephadrin, *189*
Cephalexin, 189, *471*
Cephaloglycin, *189*
Cephalosporins, 471
Cephalothin, *189*
Cephapirin, *191*, 472, 473, 478
Ceramides, 170, 171, 249, 251, 79, 280, 381
Ceramide galactosyl-3'sulfate (sulfatide), *251*
Ceramide monohexoside, 279
Cerebrosides, 280
Chaconines, 303, 393, 397, *402*
Chalcone, 291
4,4'-Chalcones, mono and disubs, *228*
Chamazulene, *408*
Cheleryethrine, *420*
Chelidione, 424
Chelirubine, *421*
Chenodeoxycholic acid, *128*, 246, 389
Chimonanthine, *425*
Chitin, 434
Chitosan digosaccharides, 434
Chloral, *538*
Chloralidone, 549
Chlorambucil, 214
Chloramine-T, *363*

Chloramphenicols, 189, 473, 477
Chloramphenicol and metabolites, 190
Chlorbromuron, 365, *368*
Chlordane, *104*
Chlordiazepoxide, 203
Chlorfenvinphos, *353*, 354
Chloro subs acetic acid, 124
Chloro subs anilines, 87, 227, 336, 346, 358
Chloro subs benzenes, 87, 224, 227, 336, 337
Chloro subs biphenyls, 87, 224
Chloro subs dibenzo-*p*-dioxins, 102
Chloro subs naphthalenes, 87, 224, 227
Chloro subs PAHs, 232
Chloro subs phenols, 270, 336, 337, 345, 346
Chloro subs pyrenes, 98, 231
Chloro subs pyridines, 227
Chloro subs toluenes, 87
8-Chloroadenine, 194
8-Chloroadenosine, 194
Chloroanilines, *84*, 87, 120, 227, 336, 346, 358
Chloroatranonin, 446
Chloroatronal, 446
Chlorobenzene, 293
Chlorobenzilate, *107*
4-Chlorobenzophenone, 224
Chlorobutylazaspirodecanedione, 202
2-Chlorobutyroanilide, 105
α-Chlorocinnamic acid, 202
Chlorocinnamic anilide, 202
Chlorodiamino-*s*-triazine, 111
Chlorogenic acids, 153, 441, 444
8-Chloroinosine, 194
Chloromethoxyanilines, 336, 356
3-Chloro-4-methyl-hydroxycoumarin, 355
4-Chloro-2-methylphenoxyacetic acid (MCPA) 109, 355
4-Chloro-2-methylphenoxybutyric acid (MCPB) 109, 355
4-(4-Chloro-2-methylphenoxy)butyric acid, 109
bis-N,N'-(3-Chloro-4-methylphenyl)urea, 109
6-Chloro-5-(naphthyloxy)-2-methylthiobenzimidazole and metab, 205
4-Chloro-7-nitrobenzo-2-oxa-1,3-diazole (NBD), *171*
Chlorophacinone, 137, 358, *362*
Chlorophenols, 84, 106, 255, 270, 336
Chlorophenoxyacetic acids, 106
6-Chlorophenylamine-3-sulfonic acid, 540

4-(4-Chlorophenyl)-4-hydroxypiperidine, 485
4-Chlorophenylurea, 358
Chlorophyll allomers, 250
Chlorophyllides, *174*, 325
Chlorophylls, 174, *175*, 250, 278, *325*, 451
Chlorophylls *a* and *b* as Mn(III) complexes, 326
Chloropyrenes, 98, 231
Chloroquine, 281
Chloroquine and metab, *205*, 514
2-{4-[7-Chloro-3-quinoxalinyl)oxy]phenoxy} propionic acids and decomp prods, 103
4-Chlororesorcinol, 120
Chlorostilbene oxide, *224*
Chlorotetracycine, *468*, 551
Chlorothalonil, *354*
Chlorotheophylline, *522*
Chlorotriazines, 111
o-Chlorovanillin, 346
Chloroxuron, 105, 110, *365*
Chlorpheniramine, 211, 493, 528
Chlorpheniramine enantiomers, 529
Chlorpromazine, *484*
Chlorpropamide, *204*
Chlorprothixene, *518*
Chlorpyrifos, *353*, 355
Chlorpyrifos-m, 353
Chlorsulfuron, *108*, 367
Chlorthiadone enantiomers, *529*
Chlortoluron, 110, 365, *368*
Chlotiazepam, 203
Cholanoic acids, 246
Cholecalciferol (Vitamin D3) *143*, 144, 244, 300, 323
Cholestanetriol, 167
Cholesterol, 247, 249, *428*
Cholesterol oxides, 167, 248
Cholesteryl arachidonate, 247
Cholesteryl benzoate, 258
Cholesteryl esters, 167, 247, 249, 428
Cholesteryl myristate, 247
Cholic acid, *128*, 388, 389
Chromones, 166
Chromotrope, *378*
Chromotropic acid, 540
Chrysene, 98, 225, 226, 337
Chrysene, methyl subs, 98, 346
Chrysin, 410, 414, 415
Chrysophanol, 164, *439*
Cichoric acid, *448*
Cicloprolol, 257

Cimetidine, 214, *328*, 331, 333
Cimiracemoside A, 149
Cinerins, *364*
Cinnamaldehyde, 416
trans-Cinnamic acid, *416*
Cinnamic acids, 386
N-Cinnamoyl anilide, 202
Cinnoxicam, 463
Cinosulfuron, *108*
Ciprofloxacin, 476, *504*, 520
Ciprofloxazine, 466
Cisapride and metab, *208*
Cisplatin, *483*, 550
Citalopram and metabolites, *500*
Citric acid, 125, 387, 542
Citrinin, 135, *400*
Citroconic anhydride and acid, 334
Clarithromycin, *467*, 468
Clethodim, *366*
Clivoric acid, *427*
Clivorine and metab, *427*
Clobazam and metab, *487*
Clodinafop, *108*
Clomipramine, *498*
Clonazepam, 203, *486*
Clonidine, *511*
Clonixin, *187*
Clopamide, 549
Clostantel, *488*
Clovoximine, 498
Cloxacillin, 191, 472, 473, 478
Cloxacillin and degrad product, *192*
Clozapine, 484, *518*
Clozapine and metab, 519
Cobalamines, 139
Cobinamides, 451
Cobinic acids, 451
Cobyric acids, 451
Cocaethylene, *492*
Cocaine, 185, *312*, 492
Cocaine and metab, 312
Coccine dye, 378
Codeine, 200, 311, 492, *508*, 528
Codeine-6-glucuronide, 508
Coenzyme Q (Ubiquinones), 141, 251
Colchicine, *426*
Colchicosides, 426
Colorflammine, *423*
Columbianedin, 437
Compactin, 509
Conchitriol, 150
Coniferin, *160*, 162
Coniferylaldehyde, 338, 446

Coniferyl ferulate, *437*
α-Conotoxin chimera, 456
Coproporphyrins, 175, *176*, 451
Coptisine, *424*, 427
Coronene, 272, *273*
Coronopilin, *148*
Corticosteroids, 252
Corticosterones, 167, 226, 269, *310*, 515
Cortisol, 197, *308*, 310
Cortisone, 197, *310*, 515, 517
Costunolide, 363
Cotinine, *130*, 392
Coumafuryl, *106*, 137
Coumaphos, 103, *355*
p-Coumarate esters, 388
o-Coumaric acid, 127, 137, 153, *300*, 386
p-Coumaric acid, *86*, 91, 127, 153, 320
Coumarins, 137, 159, 306
Coumarin and metabolites, 137
p-Coumaroylquinic acids, 126
Coumatetralyl, *137*
Coumestrol, *415*
Creatine, 430, 543
Creatinine, *545*
Crepidiasides, *409*
p-Cresol, 270
Cresols (methylphenol) 82, 83, 84, 86, 118,
    225, 227, 270, 293, 346
Crocins, *115*, 156
Crocitin, 154
Crotonaldehyde, 342
Crotonic acid, 334
Crylcoats (polyesters), 376
Cryptocapsin, 147
Cryptotashinone, 245
Cryptoxanthin, 141, 145, 146, 147, *243*, 278,
    301, 323, 406
Cryptoxanthin epoxide, 323
Crystal violet, *177*
Cucumopine, *396*
Cucurbitaxanthin, *142*
Curcubitacins, *408*, 409
Curcumin, *302*
Cyanazine, *111*
Cyanidin-3-glucoside, 156, 419
Cyanidin-3-sambubioside, 419
Cyano complexes of Au, Pd, Pt, 344
Cyanobenzylalcohol, 88
Cyanocobalamin, 139, 140
2-Cyano-3,3-diphenylacrylic acid ethyl ester,
    120
1-Cyanonaphthalene, 269
Cyasorb UV 1084, 116

Cyclodextrin, polytosylated, 435
Cyclodextrins, 98
Cyclohexanone, 383
5,6-Cyclohexenonicotine, 392
Cyclohexylmethylphosphinic acid, *92*
Cyclomaltooctaose, silyl subs, 92
Cyclopenta-1,3-diene-1-carbonitrile, 84
Cyclopenta[c,d]pyrene and metabolites, 98
Cyclopentadienyl metal complexes (Mn, Cr,
    Fe), 229
Cyclophosphamide, 530
Cyclosporins, 507, 530
Cyclosporin G and metab, 506
Cyclothioalaninethioalanine, 282
Cypermethrin, 361
Cypermethrin isomers and enantiomers, *234*
CYREZ resins, 117
Cyromazine, *104*
Cysteine, *139*, 452, 515, 546
Cystein-s-bimane, 139
Cysteineglycine, 452, 545
Cytidine, 430
Cytidine, 5'-monophosphate, 173
Cytochrome c, 182, 307, 460
Cytosine, 270, 546

2,4-D, *106*, 109
Dabsylhydrazine, *342*
Dacarbazine, 552
Dactinomycin, 530
Daidzein, 157, 163, 415
Daidzin, *415*
Dalfopristin, *475*
Danofloxacin, 476, 504, *520*
Daphylloside, *164*, 445
Dapsone, 258, *284*
Daunomycin, 481
Daunorubicin, 477, *478*
Davidigenin, *414*
p,p'-DDT, *295*
10-Deacetylbaccatin III, 195
6-Deacetylnimbin, 409
Decabromodiphenyl oxide, *123*
Decacarbonyl(diphenylphoshpinyl)alkane
    complexes (Mo, Cr, W), 229
Decachloroanthracene, 98
Decachlorobiphenyl, 09
Decacholopyrene, 98
Decacyclene, *272*
Dehydroascorbic acid, 544
7-Decahydrocholesterol, 144
Decahydroisoascorbic acid, 544

Decahydrotomatidine, 394
Decahydrotomatine, 394, 395
Decylbenzene, 85
Deethylatrazine, 111, 369
Deflazacort, *197*
Dehydroepiandrosterone, *404*
Dehydrotanshinone, 245
Delaviridine, *399*, 495, 496
Delphinidin-3-glucoside, *156*, 419
Delphidinin-3-rutinoside-5-glucoside, 419
Demeclocycline, *468*
Demethoxyyangonin, 163
Demethylcotinine, 130
Demeton-*s*-methyl, 104
Demissine, 303
Denaverine and metab, *508*
Deoxophylloerythroetioporphyrin Ni(II) and
    VO(II) complexes, 229
2'-Deoxyadenosine, 173, 546
Deoxyadenosine/chrysene adducts, 174
Deoxycholic acid, *128*, 388, 389
11-Deoxycortisol, 167
Deoxycyclic nucleotides, 172
2'-Deoxycytidine, *173*
2'-Deoxyguanosine, 173
Deoxyloganin, 408
2'-Deoxy-5-methylcytidine, 173
5'-Deoxy-5'-methylthioadenosine, 171
Deoxynivalenol, 162, 398, 399
Deoxypenciclovir, 496
2'-Deoxyribonucleosides, 430
2'-Deoxytaxinine J, 195
Deoxyuridine and derivatives, 174, 194
Dequalinium chloride isomers, *375*
Deschloromethoxuron, 236
Desfenuron, 236
Desipramine, *498*
Desisopropylatrazine
O-Desmethylangeolesin, 157, 163
N-Desmethylclobazam, 487
N-Desmethyldiazepam, 203
Desmethylhumol, 412
Desmethylranitidine, 201
Desmosine, *180*
Destruxins, *363*
Desulfoglucosinates, 436
Detajmium, *206*
Dethiobiotin, 140, 277
Dexamethasone, 197, 308, *465*, 515
Dextromethorphan, 200, 209, 528
Diacetal, 444
Diacetoxyscirpenol, 399
Diacetylaniline, 480

Diacylbenzylglycerides, 248
Diacylchitobiose, 434
Diacylglycerides, 248
Diacylglycerols, 240, 248
Diadenosine phosphates, 431
Diadinoxanthin, 327
3,6-Diaminoacridine, 349
1,4-Diaminobutane (putrecine), *88*, 89, 305,
    389, 397
1,6-Diaminohexane, 89, 305
1,5-Diaminopentane (cadaverine), *88*, 89,
    305, 389, 397
Diaminonitrotoluenes, 102, 233
2,3-Diaminophenazine, *370*
1,3-Diaminopropane, 389
2,6-Diaminopyridine, 120
Diaminotoluene, 120
cis/trans Diarylpropenamines, 322
Diatoxanthin, 327
Diazepam, 203, *255*, 487
Diazepam and metabolites, 203
Diazinon, 359, *361*
Dibenz[*c,h*]acridine, *347*
Dibenz[*a,h*]anthracene, *97*, 98
Dibenzo-*p*-dioxins, chloro subs, 102
Dibenzo[*cd,lm*]perylene, *272*
1,2-Dibromo-2,4-dicyanobutane, 322
4,4'-Dibromobenzophenone, 103
Dibucaine, 214
6¹,6ⁿ-Di-*O*-(t-butyldimethylsilyl)
    cyclomalto-octaose isomers, 92
2,4-Di-*t*-butylphenol, 296
Dibutylphthalate, 270
Di-*n*-butyltin dichloride, 295
Dicamba, 103, 109, *355*
Dicarboxylic acids (C_6-C_{20}), 387
Dichlofluanid, *354*
Dichloran, *352*
Dichloroacetic acid, 538
Dichloroanilines, 84, 110, 346
1,3-Dichlorobenzene, 86
3,3'-Dichlorobenzidine, 132
Dichlorodihydro-PAHs, 231
4,5-Dichloro-4,5-dihydropyrene, 231
Dichloroethenes, 374
Dichlorophenamide, 549
Dichlorophenols, 84
(2,4-Dichlorophenoxy)acetic acid (2,4-D)
    106, 109
S-2,4-Dichlorophenyl cysteine, 86
N-(3,4-Dichlorophenyl)-*N*'-methylurea, 110
Dichlorprop, *106*, 355, 356
Dichlorvos, *353*, 354

Diclofenac and metab, *187*
Diclofop-methyl, 108
Dicloxacillin, *472*
Didanosine, *496*
2,3'-Dideoxyadenosine, 431
2,3'-Dideoxyinosine, 431
Didesmethylimipramine, 498
Didinoxanthin, 451
Didymin, *411*
Dieldrin, *104*
Diesel oil fractionation, 381
Diethanolamine, 543
Diethyldithiocarbamate and metabolites, 211, 526
Diethyldithiocarbamate metal complexes, 94
Diethylene glycol, 538
Diethylenetriaminepentaacetic acid, 539
1,2-Diethyl-3-hydroxypyridin-4-one and metabolites, 529
Diethyllead, 93
Diethylphthalate, *88*, 270
Diethylstilbesterol, *516*
S-(*N,N*-Diethylthiocarbamoyl)glutathione, 526
Difcnacoum, 137, *138*
Difenuroxon, 110
Difethialone, 362
Diflorasone disulfate, 285
Difloxacin, *504*
Diflubenzuron, *105*
Diflubenzuron and metab, 358
2,6-Difluorobenzamide, 358
Difluorobenzoic acids, 334
N'-(2,2-Difluoroethyl)nornicotine, 392
Difluorophenylacetic acids, 541
Digitoxin, *254*, 526
Diglycerides, 249
Digoxigenin, *254*, 526
Digoxigenin mono- and bisdigitoxides, 254, 526
Digoxin, *208*
Dihomo-γ-linolenic aicd, 387
Dihydroanatoxin *a*, 400
Dihydroartemisinins, 513
Dihydrocapsaicin, 105, 386
Dihydrocodeine and metab, 550
Dihydrodaidzein, 157
Dihydrodigoxin, 254
Dihydrogenistein, 157
Dihydrokavain, 163
Dihydrolipoic acid, 387
Dihydromethysticin, 163
Dihydophylloquinone, 143

Dihydropyridines, 209
Dihydroquinine, *284*
Dihydrosphingosine, 280
Dihydrotachysterol, 258
Dihydrothymine, 391
Dihydrouridine, 171
3,4-Dihydroxybenzaldehyde, 86
Dihydroxybenzenes, 227, 338
2,4-Dihydroxybenzophenone (benzophenone 1), *119*
3,4-Dihydroxybenzylamine, 391
4,4'-Dihydroxybiphenyl, 338
1,2-Dihydroxychrysene, 88
3,7-Dihydroxycoumarin, 137
2,2'-Dihydroxy-4,4'-dimethylbenzophenone, 119
Dihydroxyflavones, 410
3,4-Dihydroxymandelic acid, 127
2,2'-Dihydroxy-4-methoxybenzophenone (benzophenone 8), *119*
Dihydroxymethoxyflavones, 244
2,6-Dihydroxymethylphenol, 118
Dihydroxynaphthalenes, 226, 338
3,4-Dihydroxyphenylacetic acid, 127, 391, 543
3,4-Dihydroxyphenylethanol, 334, 411
Dihydroxytoluenes, 226
3,5-Diiodothyronine, 305
3,5-Diiodotyrosine, *176*, 305, 446
Diisobutylketone, 86
Diisopropylbenzenes, 227
2,6-Diisopropylnaphthalene, 230
2,6-Diisopropylphenol (propofol), *505*
Diisopropylphosphoric acid, 92
2,6-Diisopropyl-1,4-quinol, 505
2,6-Diisopropyl-1,4-quinone, 505
Dilaudid (hydromorphone), *199*
Diltiazem, 210, *511*
Diltiazem and metabolites, 513
Dilugistide, 165
Dimethoate, 104, *353*
3,4-Dimethoxybenzoic acid, 270
2,2-Dimethoxypropane, *242*
5,6-Dimethoxyxanthenone-4-acetic acid, *483*
β,β-Dimethylacrylalkannin, 164
Dimethylanilines, 293
3, 4-Dimethylcinnamic acid, 153
Dimethylbenz[*c*]acridine, 228, *349*
9, 10-Dimethyl-1,2-benzanthracene, 347
Dimethylbenzidine, 380
5, 6-Dimethylchrysene, 98
Di(methylethyl)benzene hydroperoxides, 227

Dimethylhydroxybisphenyl, 118
Dimethyllead, 93
Dimethylmalic acid anhydride, 334
2,4-Dimethylphenol, 88
2,4-Dimethylphenylformate, 356
1,3-Dimethyl-2-phenylnaphthalene, *490*
Dimethylphthalate, *115*, 270
Dimethylthiophosphoric acid, 92
5,7-Dimethyltocol, *141*
N,N-Dimethyl-m-toluamide and degrad
    products, 358
N,N-Dimethyl-p-toluidine, 377
Dimethyluric acids, 154, 213, 526
10,10-Dimethylxantan, *118*
Dimethylxanthines, 154, 173, 543
Dimetridazole, *210*
Dinitamine, *368*
Dinitroanilines, 355
1,2-Dinitrobenzene, 225
Dinitrobenzenes, 232
Dinitrobenzoic acid, 100
Dinitrocresols, 100
Dinitrodiphenylamines, 88
Dinitrophenanthrenes, 233
4,6-Dinitro-o-phenol, 106
Dinitrophenols, 100
Dinitrophenylhydrazine, *91*
Dinitropyrenes, 100, 102
2,4-Dinitrotoluene, 87, 273, 350
Dinitrotoluenes, 101, 232
Dinophysistoxins, *135*, 401
Dinoseb, *346*
Dinoterb, *106*
Dinsed, 521
Dioctadecyldisulfide, *321*
Dipalmitins, 247
Dipalmitoylglucerol, 240
Dipeptides, imidazole, 180
Diphacinone, 137, *358*
Diphenamid, *295*
Diphenhydramine, *522*, 528
Diphenyl, 371
Diphenylamine-4-sulfonate, 90
Diphenylborate, 340
1,4-Diphenylbutane, 224
Diphenylethanedione, 464
Diphenylhydramine, 211
4,5-Diphenylimidazole-2-propionic acid,
    464
Diphenyltin, 94
Diphosphatidylglycerol, 250
Dipicrylamine, *101*
Diquat, *370*

Direct dyes, 120, 379
Dirythromycin, *470*
Disaccharides, sulfated, 434
Diselenides, 343
Disperse Blue dyes, 122, *379*
Disperse Brown dye, 122
Disperse Orange dyes, 122
Disperse Red dyes, 122
Disperse Violet dye, *379*
Disperse Yellow dyes, 122
Disterylglycerol, 240
Disulfaton, *107*
Disulfaton sulfone, 107
Disulfiram and metab, *211*
2,2′-(Dithiobis)benzothiazole, *340*
Dithiosalicylic acid, 205
Diuron, 105, 236, 353, 3554, 355, *365*, 368
DNA nucleotides, 173
Docetaxel, 481
Docosatrienoic acid, 387
Dodecanoic acid, 90
Dodeca-2E,4E,8Z,10Z-tetraenioc acid
    isobutylamide, 395, *396*
Dodecylgallate, 156
L-Dopa (levodopa) *132*, 391
Dopamine, 391, 543
Doramectin, 199, *357*
Dowfaxs, *374*
Dowicil-75, 376
Doxepin and metab, *499*
Doxepin-N-oxides, 499
Doxorubicins, 192, 202, 477, *478*
Doxorubicinols, 477, *478*
Doxycycline and epimer, *191*
Drofenine, *390*
Droperidol, 214
Drug screening, 184–6, 461–2
Dyes, 120

Ebrotidine and metab, *523*
Ecabapide and metabolites, 527
Ecdysteroids, 168
Echinacoside, *448*
Echinenone, 146, 301, 406
Echinofuran, *450*
Ecstacy, *490*
EDTA, 538
Efavirenz, 495, 497
Egallic acid, *155*, 156, 339
Eicosapentenoic acid, 387
Eicosatrienoic acid, 388
Elaidic acid, 387

Eleganediol, 245
Elenolic acid, 334, *411*
Elenolic acid glucoside, 411
Emodin, 121, 164, *414*, 418, 439
Enamectins, *489*
α-Endosulfan, *352*
Endosulfans, 353, 354
Endothelin, 458
Endrin, 104
Enrofloxicin, 478, 504, 520
Enterodiol, 157
Enterolactone, *157*
Epanolol, 211, *212*
Ephedrine, 185, *201*, 391, 490, 493, 528, 542, 547
Epiandrosterone-3-β-sulfate, 515
7-Epibaccatin taxol, 481
Epicatechin, *168*, 281, 413, 417, 418
Epicatechin gallate, 413, *417, 418*
Epidermal growth factor, 459
Epigallocatechin, 413, *417, 418*
Epigallocatechin gallate, 413, *417, 418*
Epilinomycin, 548
all-*E*-3'-Epilutein, 243
4-Epiminocycline, 475
Epinephrine, 132, 543
Epiprogoitrin, 436
Epirubicin, 477, *479*
Epirubinicol, 477
7-Epitaxol, 481
Epitestosterone, *517*
Epoxy-amine polymers, 117
Epoxyanatoxin *a*, 400
Epoxycholesterol, 167, 428
Epoxy fatty acids, 241
Epoxy resins, 298
5,6-Epoxyretinoic acid, 145
5,6-Epoxyretinol, 144
Eprinomectin, *199*
Equol, *157*, 163
Ergocalciferol (Vitamin D$_2$) 144, *243*, 285, 300, 323
Ergocornine, *422*
Ergocristine, *422*
Ergocryptines, *422*
Ergoloid mesylates, 531
Ergonovine, *422*
Ergosterol, 144, *384*
Ergotamine and metab, *422*
Eriocitrin, *414*, 416
Eriodictoyl, *153*, 413
Erucic acid, 387
Erythritol, 280, 433

Erythromycin A and related impurities, 192, 470
Erythromycin hydrazone, 470
Erythromycins, *468*, 470
Erythromycylamine, 470
Erythrosine, 121
Eschscholtzine, *420*
Estazolam, 203
Estradiol, *167*, 516
Estriol, 167
Estrone, *516*
Ethacrynic acid, *549*
Ethalfluralin, *368*
Ethanolamines, 543
Ethenoadenosine, 173
Ethenocytidine, 173
Ethinylestradiol, 197, *516*
Ethofumesate, *295*
Ethopabate, *521*
Ethopropazine, 257
Ethosuximide, 487
Ethoxylated alcohol surfactants, *114*, 237, 274
Ethoxylated alkylphenol surfactants, 236, 237, 274
Ethoxylated fatty acids, 237
3-Ethoxyphenol ether, 291
Ethyl, 330 116
4-Ethylbenzaldehyde, 214
Ethylbenzene, 224
Ethylbenzoate, 270
Ethyl/butyl methacrylate co-polymer, 275, 276
Ethylbutyric acid, 387
Ethylenediamine tetraacetic acid (EDTA), 538
Ethyleneglycol dinitrate, *101*
Ethylenethiourea, *273*
Ethylephedrine, 542
Ethylhematommate, 446
2-Ethylhexyl-4-dimethylaminobenzoate, 118
2-Ethylhexyl-*p*-methoxycinnamate, 118
Ethylmercury, 93, 344
Ethylmethacrylate, 276
N-Ethyl-*N*-methylaniline *N*-oxide, 227
Ethylmethylthiophosphinic acid, 92
Ethylmorphine, 200
1-Ethylnaphthalene, 224
Ethylparaben, 215, *339*, 382
Ethyl-parathion, 104
Ethyl-*N*-phenylcarbamate and decomp, *360*
Ethylsalycilate, 507
Ethylvanillin, *159*

Ethynodiolacetate, 516
Etiocholaniol-17-one-3β-glucuronide, 515
Etioporphyrin Ni(II) and VO(II) complexes,
    175
Etoposide, 479, 480
Etrofolan, 360
Euchresraflavones, 413
Eugenol, *159*
Eusolexs, 119, *381*

Falcarnidiol, 437
Falecalcitriol, *405*
Famciclovir, *496*
Famphur, *355*
Famphur oxon, 355
Fargesin, *438*
β-Farnesene, *408*
Fatty acid, hydroxy subs, 241
Fatty acid epoxides, 241
Fatty acid esters, 277
Fatty acids, 91, 125, 299, 383, 387, 422
Fatty alcohol ethoxylates, 325, 485, 486
Fatty alcohols (C_8-C_{20}), 91
Febantel, *198*
Felbamate and metabolites
Felbendazole and metab, 485
Felodipine, 209, 511
Fenamiphos, 104, 107, 354, *360*
Fenbendazole, 198, 487, *488*
Fenbufen, 187, *462*
Fenbuprofen isomers, 465
Fenbutatin oxide, *295*
Fenitrooxon, 104
Fenitrothion, *104*, 359
Fenoprop, *106*
Fenoxaprop, *108*
Fenprofen calcium, 532
Fenpropathrin, 355, *361*
Fentanyl, *188*
Fenthoxon, 355
Fentin hydroxide, *233*
Fenuron, 105, *236*, 365, 368
Fenvalerate, *361*
Ferbam, *112*
Ferruginol, *304*
Ferulic acid, *86*, 91, 153, 163, 300, 320, 386,
    450
Feruloylagmentine, *113*
Feruloylquinic acids, *126*
Fibrinopeptide, 458
Finesteride, *480*
Fisetin, 411

3-Flavanols, 245
Flavanones, 245, 410
Flavin nucleotides, 403
Flavones, 245
3-Flavonols, 245
Flavonones, methoxylated, 244
Fluazifop, *108*
Flubendazole, *204*
Flufenamic acid, *188*, 463
Flumequine, *308*, 502
Flumethasone, *196*
Flumetsulam, *366*
Flunazirine, *519*
Flunitrazepam, *486*
Fluocinolone acetonide, 311
Fluomenturon, *365*
Fluoranthene and metabolites, *97*, 98
Fluorene, *85*
Fluorenes, nitro subs, 347
9-Fluorenylmethylchloroformate (FMOC),
    180
Fluorescein, 532
Fluorescent whitening agents, 373
Fluoroacetate, 543
4-Fluoroacetophenone, 214
4-Fluoroaniline, 540
2-Fluorobenzoic acid, 334
5-Fluoro-2'-deoxyuridine, 484
$2'\beta$-Fluoro-2,3-didoexyadenosine 5'-
    triphosphate, 497
4-Fluoro-*N*-methylaniline and metabolites,
    540
Fluorophenols, 270
Fluoropipamide, *519*
5-Fluorouracil, *484*, 551
5-Fluorouradine, 484
Fluoxetine, 207, 484, *522*
Fluoxetine and enantiomers, 498, 529
Flupentixol and metab, *518*
Fluphenazinesulfoxide, 518
Flupyrazofos and metab, 358
Flupyrazofas oxon, 358
Flurbiprofen, *312*, 464
Flutamide, 214
Fluvalinate, *103*
Fluxinin meglumine, 214
FMOC (9-Fluorenylmethylchloroformate),
    180
Folates, *404*
Folic acid, *139*, 140, 403, *404*
Foliol, *150*
Folpet, 352, 353, 354
Folylmonoglutamates, 139

Formaldehyde, 91, 342, 440
Formononetin, 151, 412, *415*
Formylfolate, 404
N-Formylnorgalanthamine, 421
Fructose, 280, 432, 434
Fucosterol, *245*
2′-Fucosyllactose, 433
Fucoxanthin, *327*, 451
Fullerene, C$_{60}$, 3-methylcyclohexanone
    photoadducts, 239
Fullerenes, 122, 238–40, 276–7, 298–9,
    380–1
Fullerenes, C$_{60}$–C$_{84}$, 238
Fumaric acid, 387
Fumariline, 427
Fumarophycine, 427
Fumartine, 427
Fumonisins A & B, *138*
2-Furaldehyde, 334, *338*
2-Furanalmethyl ketone, 334
Furanacrylic acid, 447
Furans, 334, 338
Furfurals, 334, 407, *447*
Furfuryl alcohol, 487
Furoic acid, 447
Furopyrroleopyridine, 389
Furosemide, *549*
Furosine, 334
Furoylglycine, 447
Fusaproliferin, *400*

Gadopentate dimegumine, 531
Gabapentin, 485
Galactopine, 396
Galactosamine, 434
Galactose, 280, 432, 434
Galangin, *410*, 413, 414, 415
Galanin peptide, 455
Galanthamine, *421*
Gallamine triethiodide, 531
Gallic acid, *90*, 338, 345, 417
Gallocatechin, 168
Gallocatechin dimer, *418*
Galloylglucose, *156*
Ganciclovir, *548*
Ganhuangemin, *302*
Ganhuangenin, *302*
Gasoline adulterants (e.g., kerosene), 381
Gemfibrozil and metab, *508*
Geniposidic acid, 445
Genistein, 157, 163, *415*

Genistin, 415
Gentamicins, *190*
Gentian violet, 371
Gentiobiosilkaempferol, 155
Gentisic acid, 206
6-Geranylnaringenin, 412
Germall-115, 376
Gestrinone, *208*
Gingediols, *442*
Gingedione, 442
Gingerols, *442*
Ginkgetin, 153, *416*
Ginkgolic acid, *385*
Ginkgolides, *150*
Ginkgotoxin, *134*
Ginsenosides, 441, 442
Girisopam and metab, *524*
Glaucine, *421*
Glibenclamide, *204*
Gliclazide, *204*
Glipizide, *204*
Globin chains from blood, 452
γ-Globulin, 183
Glomosporin, 458
Glucitol, 433
Glucoalyssin, 436
Glucobrassicanapin, *547*
Glucobrassin, *165*, 436
Glucobrassin, Desulfo-, 436
Glucoiberin, 165
Gluconapin, *547*
Glucopine, 397
Glucoraphanin, *165*
Glucosamine, 434
Glucose, 280, 432, 433, 434
Glucose oligomers, 433
Glucosinolates, 165
6-*O*-Glucosyl-β-cyclodextran, 92
Glucotropaeolin, *165*, 436
Glucuronic acid, 434
Glutamic acid pyrosylates, 396
Glutamine, *179*
Glutamylcysteine, 179
Glutathione disulfides, 452, 546
Glutathione, 86, *179*, 452, 546
Glutylcysteine, 452
Glyburide, 531
Glycarbylamide, *541*
Glyceraldehyde, 383
Glycerol, *247*, 538
Glycerol diacetate monopropionates, 247
Glyceryl dinitrates, 102
Glycitin, *3415*

Glycodeoxycholic acid, 388
Glycocholic acid, *128*
N-Glycocylneuramininc acid, 174, *340*
Glycol dinitrates, 105
Glycolic acid, 387, 542
Glycyrrhetinic acid, 149, *151*
Glycyrrhizins, 166, 413, 416, 418
Glydant, 376
Glyoxal, 383, *444*
Glyoxylic acid, 383
Gold as cyano complexes, 344
Gonadorelin, 457
Goserelin, 457
Gratioside, *409*
Griseofulvin, 311
Guaiacol, *159*, 346
Guaifenesin, *209*
Guanadrel sulfate, 215
Guanine, 132, 546
Guanosine, 171, 430
Guanylhydrazones, *341*
Guggulsterones, *444*
Guthion, *104*
Gymnodimines, *402*

Haemanthamine, 4221
Haloperidol, 484, 485, 518, 519
Haloperidol and metabolites, 485
Haloperidol-1,2,3,6-tetrahydropyridine, 485
Haloxyfop, 108, 355, *356*
Harmaline, *426*
Harmalol, 426
Harman, *131*, 426
Harmine, *426*
Harmol, 426
α-, β-, γ-Heredin, *402*
Hederagenine, 149, *150*
Hederasaponins, 402
Hematinic acid, 447
Hemoglobins, 451
Heparin, 540
Heparin sulfate, 540
Heptabarbital, 196
Heptachlor, 104
Heptanal, 342
Heptanoic acid, 90, 474
Heptapeptide toxins (cyclic), 181
Heptaporphyrin, 176
Heroin, 200, *492*
Hesperidin, 410, 411, 413, 414, 416
HETEs, *129*, 383
Hexabromobiphenyl, 123

Hexahydro-1,3,5-trinitro-1,3,5-triazine
(RDX), *101*
2,2′,4,4′,6,6′-Hexanitrodiphenylamine,
101
Hexanal, 342
Hexanoic acid, 90
Hexaporphyrin, 176
Hexazinone, *367*, 369
Hexylamine, 131
2-*n*-Hexylnaphthalene, 224
Hexylresorcinol, 120
Hinokinin, *438*
Hippuric acid, 87, 269, 385
Histamine, *181*, 305, 493, 546
Histidine, 180, 493, 546
Histidinol, *493*
HMX, *101*, 102
Hodgkinsine, *425*
Homoanatoxin-*a*, 400
Homoaromaline, 423
Homocarnosine, 180
Homocysteine, *139*, 452, 545, 546
Homocysteine-*S*-bimane, 139
Homoeriodictoyl, 153
Homogentisic acid, *161*
Homovanillic acid, *127*, 269, 391
Honokiol, 413, *416*
Hostanox SE-10, *321*
HT-2 toxin, *399*
Human growth hormone (hGH) variants, 183
Huwentoxin II, 457
Hydrochlorothiazide, *510*, 531
Hydrocinnamoyl anilide, 202
Hydrocinnamoyl chloride and impurities,
202
Hydrocodone, 200
Hydrocortisone, 285, *465*, 515
Hydromorphone (Dilaudid) *199*, 463
Hydroperoxides of unsaturated fatty acids,
175, 395
Hydroperoxyoctadecatrienoic acids, 240
Hydroquinone, *84*, 120, 177, 291, 338
Hydroquinone, glutathione subs, 177
Hydroxy subs benzoic acids, 132, 163
Hydroxy subs naphthalenes, 227
Hydroxy subs pyridines, 227
Hydroxyacetophenone, 507
Hydroxyaloins, 166
2-Hydroxy-3-aminophenazine, 370
4-Hydroxyanilide, 391
Hydroxyatrazine, 111, 369
1-Hydroxybaccatin I, 109
Hydroxybentazones, 109

4-Hydroxybenzaldehyde, 159, 446
4-Hydroxybenzoic acid, 86, 90, 127, 159, 206, 309, 320
Hydroxybenzoic acids, 132, 163, 269, 374
p-Hydroxybenzyl alcohol, 159
N-Hydroxy-p-bromophenylacetylurea, 486
γ-Hydroxybutyrate, 200
3-Hydroxycarbazole, 236
2-Hydroxy-7-chloroquinoxaline, 103
Hydrochlorothiazide, 552
Hydroxycholesterols, 167, 428
1-Hydroxychrysene, 98
Hydroxycinnamic acids, 153, 202, 269, 300
Hydroxycitric acid, 541
6α- and 6β-Hydroxycorticosterone, 167
Hydroxycoumarins, 137
2-Hydroxy-4,6-diamino-s-triazine
Hydroxydoxepines, 499
20-Hydroxyecdysone, 168
Hydroxyeicosatetraenoic acids (HETEs), 129, 383
Hydroxyestradiols, 167
4-Hydroxyfendazole, 197
Hydroxyflavones, 410, 413
4′-Hydroxyflurbiprofen, 464
2′-Hydroxygenistein, 152
Hydroxyglucobrassicin, 547
Hydroxyhippuric acids, 385
3-Hydroxy-N-hydroxymethylcarbazole, 236
Hydroxyimipramines, 498
5-Hydroxyindole and metab, 390
5-Hydroxyindole-3-acetic acid, 269, 391
4-Hydroxyisophthalic acid, 206
Hydroxylovastatins, 509
Hydroxymellein, 167
Hydroxymethoxuron, 109, 236
2-Hydroxy-3-methoxybenzophenone (benzophenone 3), 119
2-Hydroxy-4-methoxybenzophenone, 118, 119, 381
Hydroxymethoxyflavones, 244, 413
4-Hydroxy-3-methoxyphenylacetic acid, 127
4-Hydroxy-3-methoxyphenylethylene glycol, 391
7-Hydroxy-1-methyl-β-carboline, 490
N-Hydroxymethylcarbazole, 236
N-Hydroxy-N-methylbenzylamidine, 397
4-Hydroxy-N-methylcrotonamide, 356
5-(Hydroxymethyl)-2-furfural, 334, 338, 407, 543
2-Hydroxymethylphenol (o-cresol), 118
5-Hydroxy-N-methyltryptophan, 133
5-Hydroxymethyluracil, 391

Hydroxynaphthalenes, 226, 338
6-Hydroxynaphthanlene-2-sulfonic acid, 270
5-Hydroxy-1,4-naphthoquinone, 338
Hydroxynevirapine glucuronides, 497
4-Hydroxy-2-nonenal, 342
Hydroxyochratoxins, 135
Hydroxyoctadecandienoic acid, 240
Hydroxyperoxylinolenic acid, 247
Hydroxyphenazines, 371
p-Hydroxyphenobarbital, 486
Hydroxyphenols, 132, 225
2-(4-Hydroxyphenoxy)propionic acid, 103, 374
o-, p-Hydroxyphenylacetic acid, 137, 269, 374
p-Hydroxyphenylethanol, 334
5-(p-Hydroxyphenyl)-5-phenylhydantoin, 487
1-Hydroxy-2-phenylpropamide, 486
2-Hydroxy-2-phenyl-1,3-propanediol carbamate, 485
Hydroxyprogesterone, 517
1-Hydroxypyrene, 226
3-Hydroxypyridin-4-one Fe complexes, 344
3-Hydroxy-1-(3-pyridyl)-1-butanone, 392
4-Hydroxy-(3-pyridyl)butyric acid, 131
3-Hydroxyquinidine, 528
8-Hydroxyquinoline complexes of V, Ni, Zr, Co, Cu, Al, Fe, Mn, 343
Hydroxyropivacaines, 505
Hydroxystearic acid, 299
15α-Hydroxysteviol, 439
Hydroxytacrines, 257, 523, 529
Hydroxytestosterones, 196
Hydroxythromboxane B_2 anomers, 386
5-Hydroxytryptamine, 390
5-Hydroxytryptophan, 133, 390
5-Hydroxytryptophol, 390
Hydroxytyrosol, 127
Hydroxyurea, 548
Hyocholic acid, 246
Hypaconitine, 303
Hyperglycemic hormone, 459
Hypericin, 159, 160, 440, 441
Hyperforin, 160, 440, 441
Hyperoside, 160, 441
Hypoxanthine, 173, 394, 430, 542, 543, 545
Hypoxoside, 449
Hysterin, 148

Ibotenic acid, 549
Ibuprofen, 187, 255, 465, 529

Ibuprofen enatiomers, 312
Ibuprofen isomers, 529
Ibuprofen and metab, 188
Idarubicin, *480*
Iduronic aicd, 434
Illudins, *407*
Imazosulfuron, *367*
Imidacloprid, 357, *358*
Imidazole carboxylic acids, *493*
Imidazoles, *185*
Imidazol(in)e derivatives, 185
Iminodibenzyl, 498
Imipenem, *473*, 552
Imipramine, *484*, 498
Imipramine-*N*-oxide, 498
IMP, 172, 174, 430
Imperatorin, 137, *306*, 437
Indane, 85
Indane-4-sulfonic acid, 270
1-Indanol, *226*
Indapamide, 531, *549*
Indene, *85*
Indinavir, *494*, 495, 496
Indole, 228
Indole-3-acetaldehyde, 133
Indole-3-acetic acid, *133*
Indole alkaloids, 133
Indomethacin, 188, *465*
Indoprofen, *187*
Inks, priniting, 124, 380
Inosine, *171*, 430
Inosine monophosphate (IMP), 172, 174,
    430, 545
Insulin, 183, 307, 457, 459, 460
Interjectin, 449
Interleukin-6 peptides, 457
Inulin, 434
Iodixanol, *525*
Iododoxorubicin and metabolites, 202
Iodofenphos, 295
Iodohippurate sodium, 215
Iodophenols, 225
Iodothyronines, 176, 305
Iodotyrosines, 176, 305, 446
Ioversol, 552
Ioxilan, 531
Ioxynil, 346, 355, *356*
Iprodione, *352*, 353, 354
Iridoid glucosides, 164, 445
Irgafos, *116*, 274, 296
Irgafos P-EPQ, *372*
Irgafos 168, *372*
Irganox 1010, *115*
Irganox 1076, *372*

Irganox 1330, *372*
Irganox 3114, *372*
Irgarol 1051, 354
Irinotecan and metab, *479*
Isoaloeresin, 166
Isoascorbic acid, 544
Isoaspartic acid, 178, 179
Isbufylline and metab, 526
Isobutylaldehyde, 383
Isobutyric acid, 387
Isocaffeine, 154
Isocaproic acid, 387
Isodesmosine, 180
Isodrin, 104
Isoetharine, 552
Isoflavones, 415
Isoflupredone, 285
Isofoliol, 150
Isoginkgetin, 153
Isoimperatonin, 137, *160*
Isoliquiritigenin, *151*
Isomaltotriose, 433
Isomethyl eugenol, 363
Isomethylnicotinium ion, 130
Isoniazid, *207*, 210
Isonicotinyl hydrozone, 207
Isonox, 116
Isooctanoic acid, 474
Isopimpinellin, 137
Isopropalin, *368*
2-Isopropylnaphthalenes, 230
8-Isoprostaglandin F, 383
Isoproturon, 105, 110, *236*, 365, 368
Isoquercitrin, *159*
Isoquinoline alkaloids, 421, 424
Isorenieratene, 177
Isorhamnetin, 153, 410, 413, 416
Isosafrole, 449
Isosteviol, 439
Isotetrandine, 424
Isotetrinoin, 258
Isovaleric acid, 387
Isovalerylalkannin, 164
Isovanillic acid, 163
Isovanillin, *86*
Isoxanthohumol, 412
Isoxicam, *463*
Isradipine and metab, *510*
Ivermectin, 199, 211, 357

Jacobine, *423*
Jacoline, 423
Jaconine, *423*

Jacozine, *423*
Jasmolins, *364*, 526
Jasmonic acid, *158*
Jatrorrhizine, 377
Jesaconitine, *303*
Josamycin, 502
Juglone, *121*

Kaempferol, 152, 153, 410, 411, 414, 415,
    416
Kaempferol rhamnodiglucosides, 304
Kahweol, 149
Kalopanaxin, 162
Kanamycins, 284
Karanjin, *161*
Kava lactones, *163*
Kavain, *163*
Kerosene in gasoline, 381
Ketamine, *206*, 312, 505
7-Ketocholesterol, *167*, 428
6-Ketolitocholic acid, *246*
17-Ketomethoxyprednisolone, 514
10-Ketonalbuphine, 309
Ketones, 342
Ketooctadecadienoic acid, 240
Ketoprofen, *187*, 255
Ketoprofen enantiomers, 529
17-Ketosteriod glucuronides, 515
17-Ketosteroid sulfates, 515
11-Ketotestosterone, *517*
Kitasamycin, 502
Korupensamines, *424*
Kurstakins, 456

Labetalol enantiomers, *513*
Laccaic acids, *121*
Lacidipine, *209*
Lactalbumins, 182, 460
Lactic acid, 90, 125, 387, 541, 542
Lactoglobulin, 460
γ-Lactones, 306
Lactose, 432
Lactulose, 531
Lagascatriol, *150*
Lamivudine, *496*, 497
Lamotrigine, *195*
Lantadenes, *136, 151*
Lapachols, 121, *447*
Lapachons, 447
Lariciresinol, *438*
Lasalocid, 190, *521*
Lawsone, *121*

Lenacil, *295*
Leucogentian violet, *371*
Leucokinin II, 458
Leucomalachite green, 177, *371*
Leu-enkephalin, 457
Leukotrienes, 128, 129, 386
Leuprolide and impurities, 457
Levistolide, *165*
Levmetamfetamine, 258
Levodopa (L-Dopa), *132*, 391, 543
Levomeprozamine, 518
Levonorgestrel, *516*
Levothyroxine sodiuim, 531
Licoisoflavone A, *152*
Lidocaine, 492
Lignans, 438
Ligstroside dialdehyde, 334
Ligustilide, 165, *450*
Linamarin, *449*
Lincomycin, *548*
Linear alkylsulfonates, *114*
Linearol, *150*
Linoelaidic acid, 387
Linoleic acid isomers and conjugates, *246*,
    248, 384, 388
Linoleic hydroperoxide products, 240
Linolenic acid isomers, 277, 280, 384, 387
Linuron, 105, *110*, 295, 368
Linuron and metabolites, 110, 321
Lipoic acid, 387
Lipoxins, 128
Liquiritigenin, *337*, 414
Liquiritin, 416
Lisianthiuoside, 334, 409
Lithocholic acid, 128, *246*, 388, 389
Lithospermic acid, *450*
Loganin, *408*
Lomefloxacin, *207*, 462
Lorazepam and degrad products, 203, *256*
Lornoxicam, 463
Lovastatin, impurities and metab, *509*
Loxapine, *484*
LSD, *489*
LSD and metab, 489
LTB$_4$ and metabolites, 128, *129*
LTC$_4$, 128
LTD$_4$, 128
LTE$_4$, 128
Lucidin promveroside, *443*
β-Lumicolchicine, *426*
Lupalbigenin, 152
Luteins, *146*, 243, 301, 323, 406, 451
Lutein-5,6-epoxide, 141
Luteolin, *151*, 413, 416

Luteolin 7-glucoside, 411
Luteone, *152*
LXA$_4$, 128
LXB$_4$, 128
Lycopene, 145, *148*, 243, 278, 323, 406
Lypressin, 457
Lysergic acid diethylamide, *489*
Lysine pyroglutamate, 382
Lysolecithin, 250
Lysophosphatidylcholine, 249, 280
Lysophophatidylethoanolamine, 170, 250
Lysozyme, 460

Macluraxanthone, *413*
Madecassic acid, 149, *407*
Madecassoside, 407
MAFBC, 204
Magnolol, 413, *416*
Malachite green, *177*, 371
Malathion, *359*
Maleic anhydride and acid, 334
Malic acid, 90, *125*, 387, 542
Malonaldehyde, 342
Maltilol, 433
Maltose oligomers, 433
Malvidin-3-feroylrutinoside-5-glucoside, 419
Malvidin-3-glucoside, 419
Mancozeb, 111, 112
Mandelic acid, *87*
Mandelonitrile, *449*
Maneb, *112*
Mannitol, 280, 432, 433
Mannopine, *396*
Mannopinic acid, *396*
Mannose, 280, 432, 433, 434
Maprotiline, *498*
Marbofloxacin, *504*
Marrubiin, 245, *306*
Matairesinol, 157, 438
Maytanbutine, 284
Maytansine and maytansinoid homologs, 284
Maysine, 284
Maytanacine, 284
MCPA (4-Chloro-2-methylphenoxyacetic
 acid), 109, 355
MCPB (4-Chloro-2-methylphenoxybutyric
 acid), 109, 355
Mebendazole, *198*, 477
Meclofenoxate, *390*
Mecoprop, *106*
Medazepam, *203*
Medicarpin, *414*

Medroxyprogeserone acetate, 285
Mefenamic acid, *463*
Melamine, 104
Melatonin, *132*, 140, 390
Mellein, 167
Meloxicam, *463*
Melperone, *518*
Menadione, 300, 323
Menaquinones, *143*
Mephenytoin enantiomers, *529*
Meprobamate, 552
2-Mercaptobenzothiazole, *124*, 340
6-Mercaptopurine and metab, *194*
Mercapturic acid, 86
Mersalylic acid, 93
Mesaconitine, 303
Mescaline, *491*
Mesitylate esters, 328
Mesoporphyrin, 176
Mesoporphyrin-II-dipropyl ester Mg(II) and
 Zn(II) chelates, 251
Mestranol, *516*
Metabenzthiazuron, *365*
Metacrolein, 342
Metal carbonyls (Cr, Mo, W, Re, Mn), 343
Metallocenes, tricarbonyl of Cr, Mg, 93
Metallointercalator-DNA conjugates, 431
Metamitron, *355*
Metamphetamine, *490*
Metam-sodium dihydrate, *111*
Metazachlor, 355
Met-enkephalin, 457
Methadone, *200*
Methapyrilene and metabolites, 201
Methiocarb, *353*, 360
Methiophos, 354
Methotrexate, *192*
Methoxamine, *491*
Methoxsalen, 531
Methoxuron, *105*
Methoxy subs flavones 345
4-Methoxybenzaldehyde, 86
3-(4-Methoxybenzilidine)camphor, 381
4–4'-Methoxybis(ethylphenyl)carbamate,
 360
Methoxychlor and metabolites, *362*
4-Methoxycinnamic acid, *153*
Methoxyethylmercury, 93, 344
Methoxyflavones, 247, 413
Methoxymelamine resins, 117
7-Methoxy-1-methyl-β-carboline, 490
(Methoxymethyl)melamine resins, 117
6-Methoxy-2-naphthylacetic acid, *464*

6-Methylocatnoic acid, 474
3-Methoxyphenol ether, 291
Methoxyphenylacetic acid enantiomers, 312
5-Methoxypsoralin, *202*
3-Methoxytyramine, 543
Methscopolamine, *493*
Methyl subs uric acid, 154
Methyl subs xanthine, 154
Methylacrylic anhydride and acid, 334
Methyladenosine, 171, 172
Methylamine, 131
Methylaminedialdehyde, 356
2-Methylamino-7-aminomitosene, 193
2-Methyl-[5-(α-amino-4-fluorobenzyl)
    benzimidazol-2-yl] carbamate, 204
N-Methyl-*p*-aminophenol, *84*
Methylanthrilate, 339
4-Methylbenzaldehyde, 86
N-Methylbenzamide, 397
N-Methylbenzylamidine, 397
N-Methylbenzamidoxime, 397
3-(4-Methylbenzilidine)camphor (Eusolex,
    6300), *119*
2-(6-Methyl-2-benzothiazolylazo)-5-
    diethyl-aminophenol metal ion
    complexes, 95
Methyl/butyl methacrylate co-polymers, 238,
    322
N-Methylcarbazole and metabolites, 236
1-Methyl-*β*-carboline (Harman), *131*, 490
2-Methyl-4-chlorophenoxyacetic acid
    (MCPA), 109, 355
2-Methyl-4-chlorophenoxybutyric acid
    (MCBA), 109, 355
Methylcobalamin, 140
5-Methylcytosine, 270
Methyldibromoglutaronitrile, 322
β-Methyldigoxin, 254
Methyldopa, 552
4,4′-Methylenedianiline, 380
Methylenedioxyamphetamine, 490, 491
Methylenedioxymethamphetamine (Ecstacy),
    490
N-Methylephedrine, 542, 547
Methyl ethyl ketone, 86, 91
2-Methylfuran, 338
Methylfurfural, 334
5-Methyl-2*H*-furanone, 194
Methylgingerol, 442
Methylglyoxal, 444
Methylguanines, 173
Methylguanosine, 172
Methylhematommate

o-Methylhippuric acid, 87
Methylhistidines, 186, 546
8-*O*-Methyl-7-hydroxyaloin, 166
Methylindoles, *86*
Methylinosine, 172
Methylisoricinoleatae stereoisomers, 247
7-Methyljuglone, *437*
Methyllead, 93
Methyllinoleate hydroperoxides, 246, 247
6-Methylmercaptopruine, 194
Methylmercury, 93, 344
Methyl methacrylate, 377
Methyl-*N*-methylanthralinate, *339*
1-Methyl-1-(2-naphthyl)ethylhydroperoxide,
    230
N-Methyl-*N*-nitrosoanilie, 351
Methyl orange, 155
Methyl paraben, 333, *339*, 382
Methyl-parathion, 104
Methylphenidate enantiomers, *529*
Methyl phenols (cresols), 82, 83, 84, 86, 118,
    225, 227, 270, 283
1-Methyl-1-phenylethanol, 227
Methylprednisolone, 285, *308*, 514
Methylpropylamine, 158
N-Methylpseudoephedrine, 547
4′-*O*-Methylpyridoxine (ginkgotoxin), *134*
Methylquinolines, *227*
Methylquinones, 293
Methylscuttellarein, 244
Methysticin, *163*
Methyltestosterone, 197, 327
1-Methyl-1,2,3,4-tetrahydro-*β*-carboline,
    490
5-Methyltetrahydrofolate, 404
N-Methyl-*N*,2,4,6-tetranitroaniline (Tetryl),
    101, 102
2-(Methylthio)benzothiazole, 340
Methylthiotriazines, 107
3-Methyl-4-trifluoromethylaniline and
    metab, 345
N-Methyltyramine, 542
Methyluric acids, *154*, 213
Methyluridines, 172
Methylxanthines, *154*, 173, 213, 526
Metobromuron, 368
Metol, *84*, 120
Metolachlor, *354*
Metomyl, 352, 353, 354
Metoprolol, *213*, 284, 511, 513
Metoprolol enantiomers, 529
Metosulam, *366*
Metoxuron, 110, *365*

Metronidazole, 210, *474*
Metsulfuron-methyl, 108, 355, *367*
Mibefradil and metab, *509*
Michellamines, *424*
Miconazole, 215
Microlides, 502
Midazolam, *188*
Milbemectin, 489
Minocycline 466
Minocycline and impurities, *475*
Minoxidil, *529*
Miromicin, 502
Mitocene, 478
Mitolacol, *193*
Mitomycin C and metabolites, *192*, 478
Mitomycin C, peptide subs, 194
Mitosane, 478
Mitoxantrone, 531
Moclobemide, *501*
Mometosone furoate, 215
Monascidin, 436
Monascin, *440*
Monascorubramine, *436*, 440
Monascorubrin, *436*, 440
Monensin, *190*
Monoacetylmorphine, 200, 492
Monoacylglycerides, 247
Monocolins, *436*
Monocrotophos, *356*
Monolinuron, 368
Mononyasines, 449
Monopalmitin glycorides, 247
Monotropein, 164, *445*
Montelukast and metab, *435*
Monuron, 110, *368*
Moperone, *519*
Morazone, *527*
Morin, 295, 410, 411
Morphine, 185, 186, 200, 492, *508*
Morphine and metab, *199*, 463
Motexafin gadolinium and lutetium, *344*
Moxidectin, *199*, 357
Mucoaldehyde, 90
Mucochloric acid, *173*
Muscimol, *549*
Mutatoxanthines, 146, *147*
Mycocystins, 397, *398*
Mycolic acids, *126*, 281
Mycophenolic acid, 495, *506*
Mycophenoate mofetil, 506
Mycotoxins, paspalitrem, 281
Myoglobin, 307, 460
Myosmine, *392*

Myricetin, 151, 410, 411, 413, 413
Myxoxanthophyll, 327

Nabam, 112
Nabumetone, *464*
NAD, *172*, 430
NADH, 172
NADP, *172*
NADPH, 172
Nadolol enantiomers, *513*
Nafcillin, *472*, 473
Nafronyl, *390*
Nalbuphines and degrad products, *309*
Nalidixic acid, 308, *504*
Nalmefene and metab, *211*
Nalodol, 383
Naloxone, 199, 200
Naltrexone, 199
Naphthalene, 85, 96, 337
Naphthalene, nitro subs, 347
Naphthaleneethanols, 318
Naphthalenemethanols, 318
2-Naphthalenesulfate, 6-hydoxy-5-
    phenylazo-, 380
Naphtho[2,3b]furan-4,9-diones, 447
Naphthoquinone dyes, 121
Naphthaquinones, 164
Naphthols, 227, 318
Naphtholsulfonic acids, 270, 540
Naphthylacetates, 268
1-Naphthylamine, 361
Naphthylaminesulfonic acids, *90*, 540
1-(2-Naphthyl)ethanone, 230
Naphthylisoquinoline alkaloids, 424
Naphthyl-N=N-naphthol dyes, 380
N-Naphthylphthalamic acid, 361
2-(2-Naphthyl)-2-propanol, 230
1-Naphthylthiourea, 361
Napoleoferin, *547*
Napropamide, 355
Naproxen, 187, *188*
Narasin, *190*
Naringenin, *410*, 413, 414
Naringin, 152, *410*, 411, 414, 416
Narirutin, 152, 411, 414, 416
Natapalm, 631
Natriuretic peptides, 456
Natural Yellow, 28 121
NBD chloride, *171*
Neburon, 109, 236, 355, 365
Nefiracetam and metab, *520*
Nelfinavir, *494*, 495, 496

Neoeriocitrin, 410, 414, 416
Neohesperidin, 152, 410, 411, 414, 416
Neokyrotropin, 456
Neomycin, 503
Neoxanthin, 141, 146, *323*, 326, 451
Netilmicin, *503*
Neuraminic acids, *174*
Neurochemicals, *133*
Neurostatin, 458
Nevirapine and biotransformation products,
    496, 497
Niacin (nicotinate), 172, 403
Niacinamide, *403*
Nicarbazin, *477*
Nicardapine, *209*
Niclosamide, *359*
Nicosulfuron, *366*
Nicotinamide, 139, 172, 174, 392
Nicotinamide adenosine dinucleotide (NAD),
    172
Nicotinamide adenosine dinucleotide
    phosphate (NADP), *172*
Nicotinate (niacin), 172
Nicotine, *130*, 392
Nicotine analogs, 392
Nicotine and metabolites, 130
Nicotine-*N*-oxide, *130*, 392
Nicotinic acid, 130, 174, *392*
β-Nicotyrine, *392*
Nifedipine, 209, 210, *510*, 511
Nimbandiol, 447
Nimbin, 138, *409*, 447
Nimodipine, 209, *511*
Nisoldipine, *209*
Nitramine residues, 101
Nitrazepam, 185, 203, *486*
Nitredipine, 210, *511*
Nitric acid, 340, 538
Nitrilotriacetic acid, 539
Nitro subs anilines, 227, 270, 333
Nitro subs benzenes, 101, 227, 232, 293
Nitro subs benzoic acids, 83, 101
Nitro subs biphenyls, 100
Nitro subs fluorenes, *100*, 347
Nitro subs fluorenones, *100*
Nitro subs naphthalenes, 100, 227, 268, 347
Nitro subs PAHs, 100, 233
Nitro subs phenols, 82, 270
Nitro subs pyrenes, 100, 347
Nitro subs pyridines, 227
Nitro subs quinolines, *100*
Nitro subs toluenes, 101, 232
m-Nitroacetophenone, 226

Nitroaminotoluene, 101
o-Nitroaniline, 258, 270, 333
p-Nitroaniline, 155
Nitroanthracene, 100
Nitrobenzene, 101, 102, 224, 225
Nitrobenzoic acids, 83, 100
Nitrobiphenyls, 100
Nitrocellulose, 87
6-Nitrochrysene, 100
Nitrocresols, 100
Nitrodiphenylamines, 88
3-Nitrofluoranthracene, 100
2-Nitrofluorene, *100*
3-Nitrofluorenone, *100*
Nitrofurantoin, 311
Nitroglycerin, *87*, 101
Nitroguanidine, *102*
Nitromersol, *93*
Nitromide, *521*
1-Nitronaphthalene, 227
Nitronaphthalenes, 227
Nitrophenols, 82, 84, 225, 270, 273, 507
4-Nitro-1,2-phenylenediamine, 120
Nitropyrenes, 347
4-(5-Nitro-2-pyridylazo)resorcinol metal ion
    complexes, 95, 320
2-Nitroquinoline, *100*, 347
Nitrosamine metabolites, 131
Nitrosoamines, alkyl subs, 350
N-Nitrosodiphenylamine, *87*
N-Nitrosonipecotic acid conformers, *351*
2-Nitroso-6-nitrotoluene, 350
N-Nitrosonornicotine, *392*
N-Nitrosopipecolinic acid conformers, *351*
N-Nitrososarcosine conformers, *351*
N-Nitrosothiazolidine-4-carboxyllic acid
    conformers, 351
Nitrotoluenes, 350
Nivalenol, 162, *398*
Nobiletin, 244, *398*
Nodularin, *398*
Nomilin glucoside, *446*
Nonanal, 342
2-*trans*-Nonenal, 342
Non-ionic surfactants, 113, *114*
4-Nonylphenol, 114
Nonylphenolethoxylated surfactants, 237,
    373
Nonylphenolglycol ether surfactants, 238
Nopaline, 396
Norbaberine, 423
Norbixins, *126*, 452
Norcocaine, 492

Norcodeine, 200, 508
Nordiazepam, 203, 487
Nordihydroguaiaretic acid, *406*
Norephridrine, 542, 547
Norepinephrine, 391, 543
Norethidrone, 516
Norethisterone acetate, 197
Norfloxacin, 476, 503, 504, 520
Norfluoxetine, 207, 498
Norflurazon, *107*
Norgestrel, *197*
Norharman (β-carboline), *131*, 426
Norketamine, 505
Norleucine, 179
Normannopine, 396
Normaysine, 284
Normethadone, 200
Normorphine, 200, 508
Nornammefene, 211
Norprazepam, 255
Norpseudoephidrine, 542, 547
Nortriobolide, 138
Nortriptyline, *498*, 499
Norvaline, 179
Norverapamil, 512
Noscapine, 200, 492, *493*
Nosiheptide, 475
Notopterol, *160*
Novolac epoxy resins, 116, 297
Nucleosides, 172, 430
Nucleotides, cyclic, 172
Nucleotides, deoxy, 172
Nyasol, 449

Obacunone glucoside, 446
Obamegine, 423
Obtusides, 449
Ochratoxins, 135, *399*
Octabromodiphenyloxide, 123
Octadecanoic acid, 90
Octaethylporphyrin, *175*
Octahydro-1,3,5,7-tetranitro-1,3,5,7-
    tetrazocine (HMX), *101*
Octanal, 342
trans-2-Octenal, 342
Octopamine, *132*
Octopine, 396
Octopinic acid, *396*
exo-N-Octylbicycloheptanedicarboxamide,
    525
Octyldimethyl-p-aminobenzoic acid
    (Eusolex 6007), 119, 381

Octylgallate, 156
Octylglycoside surfactants, *114*
Octylmethoxycinnamate (Eusolex 2292),
    119
4-Octylphenol, 114
Octylphenol polyether alcohol surfactants,
    374
Odapipam and metab, 256
Ofloxacin, 520
Ofloxaicin and enantiomers, 207, *466*
Okadaic acid, *135*, 401
Olanzapine, *484*, 517, 518
Oleanolique acid, 149
Oleic acid, 388
Oleurupein, *411*
Oleurupein aglycone and dialdehyde, 334
Oligosaccharides (maltose, isomaltose,
    inulin, xylose based), 433, 434
Olpadronate, 523
Ondanseton, 480
Ononin, *412*
Opiates, 200
Organomercury compounds, 93, 344
Organophosphorus acids, 92
Organophosphorus compounds 105, 110
Organotin pesticides, 94
Ornithine, *397*
Orotic acid, *545*
Oroxylin, *302*, 415
Oroxylin A, 7-O-glucuronide, 418
Osajaxanthine, 413
Oscillaxanthin, 327
Osthol, *437*
Ostruthol, 137, *306*
Ovalbumin, 307, 460 182
Oxalacetic acid, 383
Oxamyl, *361*
Oxaprozin, *464*
Oxazepam, 203, *486*, 487
Oxcarbazepine and metab, *256*, 499
Oxfendazole and metab, *198*
Oxipurinol, 545
Oxolinic acid, 308, 384, 502, 204
Oxonic acid, 543
8-Oxopenciclovir, 496
4-Oxoretinoic acid, 144, 145
4-Oxoretinol, 144
20-Oxosteroids, 168
Oxprenolol, 284, *512*
Oxprenolol enantiomers, 513
Oxydemeton-methyl, *104*
Oxypeucedanin, 137
Oxyphenbutazone, 465

Oxytetracycline, *359*, 468
Oxytocin, 457

Paclitaxel (Taxol), *193*, 1995
Padimate O, 215
Paeonidin-3-glucoside, 419
PAHs, 85, 96, 230, 232, 271, 272, 293, 346
PAHs, acetylamino subs, 100
PAHs, alkyl subs, 85
PAHs, amino subs, 100, 230
PAHs, dichlorodihyro subs, 231
PAHs, nitro subs, 85, 100, 230, 233
Palmatine, *377*
Palmitic acid, 247, 388
Palmitoleic acid, 388
Panadiplon, 528
Papaverine, 185, 200, 492, *528*
Parabens, *339*, 382, 527
Paracetamol, 490
Paraoxon, 104, *273*, 355
Paraplatin, 192
Paraquat, *370*
Parathion, *273*, 355
Paraxanthine, 154, 309, 391, 526
Parfumine, 427
Parishins, *445*
Paromomycin, *474*
Paroxetine, 484
Parthenin, *148*
Parthenolide, 245, *306*, 363, 443
Paspalicine, *281*
Paspaline, *281*
Paspalinine, *281*
Patulin, *399*, 543
Paxilline, *281*
PCBs, 230, 272
Pectenotoxins, *401*
Penciclovir and metab, *496*
Pendimethalin, *368*
Penicillin G, 191, *473*
Pentachloroanisole, 359
Pentachlorophenol, 106, 359
Pentaerythritol tetranitrate, *101*
Pentahydroxyflavone, 295
Pentanal, 342
Pentaporphyrin, 176
trans-2-Pentenal, 342
6-Pentyl-α-pyrone conjugates, *469*
Peptide oligonucleotides, 430
Peptides, 178, 454–9
Perazine, *518*
Perchloroanthracene, 98

Perchlorobenzene, *98*
Perchlorobiphenyl, *98*
Perchloronaphthalene, 98
PerchloroPAHs, 98
Perchloropyrene, 98
Perfluorocarboxylic acids, 383
Perhalogenated ethylene, 89
Perhelogenated methane, 89
Permethrin, *361*
Permethrin and degrad products, 358
Peroxisomicine A1, *164*
Peroxyacetic acid, 385
Peroxycarboxylic acids (C_8-C_{12}), 91
Perylene, *176*, 225, 272
Perylenes, methyl subs, 346
Petunidin-3-*p*-coumarylrutinoside-5-
    glucoside, 420
Petunidin-3-glucoside, 419
PGB$_2$, 128
PGEs, *129*
PGFs, 129
Phenacetin, 391
Phenanthrene, *225*
Phenazine, 272, *273*
Phenazine-1-carboxylic acid, *371*
Phenethylamine, 131
Phenetole, *224*
Phenobarbital, 196, 484, 486
Phenol, 82, 83, 84, 85, 206, 225, 270, 291
Phenol, alkyl subs, 83, 84, 85, 228
Phenol, amine subs, 82
Phenol, bromo subs, 228
Phenol, chloro subs, 83, 84, 106, 228, 345
Phenol, cyano subs, 87
Phenol, nitro subs, 82, 84, 85, 87, 228, 291,
    345
Phenol, phenyl subs, 371
Phenolic acids, 86
Phenolphthalein, *286*, 382
Phenols, 87, 226, 228, 270
Phenoxyacetic acids, chloro subs, 294
m-Phenoxybenzoic acid, 358
m-Phenoxybenzylalcohol, 358
Phensuximide enantiomers, *529*
Phentermine, *491*
Phenuron, 109
Phenylacetic acid, 374
Phenylacrylamide, *486*
Phenylalanine, *181*, 232, 331, 386, 546
Phenylalcohols, 87
Phenylamide pesticides, 107
2-Phenylamine, 158
Phenylaminesulfonic acids, 540

N-Phenylbenzoylhydroxamic acid complexes
  of vanadium (V), *282*
Phenylborates, 340
4-Phenyl-3-buten-2-ol, 339, 341
4-Phenyl-3-buten-2-one, *339*, 341
4-Phenyl-2-butanone, 339, 341
Phenylbutazone, *465*
Phenylcarbamates, 107
Phenylcarboxylic acids, 87
p-Phenylenediamine, *120*, 338
Phenylephrine, *493*, 528
Phenylesters, 87
Phenylethers, 87
Phenylethylamine, 305, 490
2-Phenylethylbromide, 224
Phenylethyl keyone, 86
Phenylglycine, 282, 473
Phenylglyoxylic acid, *87*
Phenylisobutyl ketone, 86
S-Phenylmercapturic acid, 334
Phenylmercury, 93, 344
Phenylmethylmalonamide, 486
Phenylnitrone, *N*-trimethylphenyl subs, 357
o-Phenylphenol, 270
2-Phenyl-1,3-propanediol, 485
3-Phenylpropanol, 88
Phenylpropionic acid, 374
Phenyl-N=N-naphthol dyes, 380
Phenylpyrones, 166
Phenyl sulfone, *360*
Phenyl sulfoxide, 360
5-Phenyltetrahydro-1,2-oxazine-2,4-dione,
  486
Phenyltins, 94
Phenyltoloxamine, *507*
1-Phenyl-3-trifluoromethyl-5-
  hydroxypyrazole, 358
N-Phenylurea, *109*, 236, 355
Phenylureas, 109, 110, 236, 355, 365, 368
Phenytoin, *487*
Pheophorbides, 174, 177, *326*
Pheophytins, 174, 177, 186, 250, 326
Phlominol, 409
Phlorin, 547
Phloroglucinol (1,3,5-Trihydroxybenzene),
  168, 350
Phloxine B, *364*
Phosmethylan, 105
Phosphatidic acid, *170*, 249, 250
Phosphatidic acid methyl esters (C_6-C_{10}),
  429
Phosphatidylcholine, 169, *170*, 249, 250,
  429

Phosphatidyldimethylethanolamine, 250
Phosphatidylethanolamine, *170*, 249, 250,
  280, 429
Phosphatidylinositol, *170*, 249, 250, 280,
  429
Phosphatidylmonomethylethanolamine, 250
Phosphatidylserine, *170*, 249, 250, 280,
  429
Phosphocreatine, 430
Phospholipids, 170
Phosphorthiolates, 107, 255
o-Phthalaldehyde, *179*
Phthalates, 115
Phthalides, *88*
Phylloquinone (Vitamin K), *142*, 143, 300,
  323
Physcion, 164, *439*
Physostigimine, 532
Δ2,6-Phytadienol chlorophyll *a*, 251
Photalexins, 443
Phytofluene, 146, *147*
Phytol, 170, 249
Phytonadiones, 258
Phytosphingosine, 280
Piceatannol glucoside, *448*
Piceid, *448*
Picenes, 346
2-Picoline, 214
Picric acid (2,4,6-trinitrophenol), *102*
Picricrosin, *155*
Pimobendan and metab, *205*
Pimozide, *519*
Pinacolylmethylphosphonic acid, 92
Pindolol enantiomers, 284, *513*
Pinobanskin, 151, 415
Pinocembrin, *415*
Pinonaldehyde, 440
Pinoresinol, *438*
Pinostrobin, *245*
Pipazethate, *185*
2′,6′-Pipecoloxylidide, *505*
Pipemidic acid, *504*
Piperettine, 393
Piperettyline, 393
Piperine, *393*
Piperolines, 393
Piperonal, *159*
Piperonyl butoxide, *525*
Piperyline, *393*
Piretanide, *549*
Piromidic acid, *308*, 504
Pirprofen, 255
Piroxicam, *463*

Plumbagin, *437*
Polyamide-6, 376
Polyaromatic hydrocarbons (see PAHs)
Polybrominate fire retardants, 123
Polybutyleneglycols, 116
Polybutylmethacrylate, 275
Polydatin, 163
Polyester, 296, 298
Polyethoxylated surfactants, 236
Polyethylene glycol DNA/RNA complexes, 116
Polyethyleneglycol protein complexes, 116
Polyethylene glycols, 116, 375
Polyethylenes, 375, 376
Poly(ethyl methacrylate) polymers, 275
Poly(methyl methacrylate) polymers, 322
Poly(methyl methacrylate)-*graft*-polydimethylsiloxane copolymers, 238
Poly(methyl methacrylate)-*graft*-polystyrene, 297
Polymixin B sulfate, 474
Polypodin B, 168
Polystyrene, 116, 275, 298
Poly(styrene *t*-butyl methacrylate) co-polymer, 275
Poly(styrene methyl methacrylate) co-polymer, 275
Poly(styrene vinyl acetate) block co-polymer, 276
Polysulfides, linear, 93
Polyterahydrofuran polymers, 322
Polythiolates, 341
Poly(vinyl alcohol) polymer, 297
Ponceau 4R, *121*, 378
Ponceau S, 155
Poncirin, 411
Pongamol, *161*
Porfiromycin and metab, *193*
Potassium sorbate, *338*
Practolol, 211, *212*
Prazepam, *255*
Praziquantel, 488
Prazosin, *511*
Prednisolone, 197, 285, *308*, 310
Prednisolone and metabolites, 197
Prednisone, 197, 226, *269*, 285, 308, 310, 515
Pre-ergocalciferol, 258, 285
Pregnenolone, *168, 404*
Prenylflavonoids, 412
Prenylnaringenin, *412*
8-Prenylxanthone, 413
Primaquine, 284, 514

Primicarb and metab, 356
Primidone and metab, *486*
Printing inks, 124
Pristinamycin, 475
Proanthrocyanidins, 541
Proanthocyanins, 168
Probucol, 258, 532
Procaine, *473*, 492
Procymidone, *352*
Procyanidin dimers and trimer, 159
Procyanidin tetramers, 418
Prodelphinidin B3, 541
Prodelphinidin trimers, 418
Proflavine and metab, *349*
Progesterone, 197, 285, 327, *516*
Progoitrin, *165*, 436, 547
Prolintane, *185*
Promazine, *519*
Promethazine, 493
Promethazine enantiomers, *257*
Prometryn, *107*
Pronetholol enantiomers, *284*
Propanal, 342
Propargyl chloride, 522
Propargyl glutarimide, *522*
Propazine, *110*, 369
cic/trans Propenamines (diaryl), 322
Propionaldehyde, 91, 342
Propionaldoxime, 106
Propionic acid, 316, 434
Propionic acid, 2-{4[(7-chloro-2-quinoxalinyl)oxy]phenoxy}, 103
Propionic anhydride, 383
Propineb, *111*
Propiverine and metab, *524*
Propofol and metab, *505*
Propoxur, *107*
Propoxyphene hydrochloride, *189*
Propranolol, 185, *287*, 511, 529
Propranolol enantiomers, 284, 513
Propranolol and metabolites, *257*
Propyldodecylgallate, 278
Propylenediol adipadte polyester plasticizers, 116
Propylene glycol, 538
Propylgallate, *156*
N-Propylneuraminic acid, 174
Propylparaben, *339*, 382
Prostaglandins, 86, *129*, 386
Protamines, 182
Protein retention, 181, 182, 183
Protizinic acid enantiomers, 255
Protocatechualdehyde, 446

Protocatechuic acid, 86, *90*, 163, 320
Protochlorophylls, 278
Protohypericin, *160*
Protopanaxdiols, 441
Protopanaxtriols, *441*
Protopine, 420, *421*, 427
Protopseudohypericin, *160*
Protryptiline and biotransformation products, *499*
Provitamin D glucuronides and sulfates, 405
Prozac, 207
Prunasin, *449*
Prunin, 152
Pseudoehpidrine, 209, 542, 547
Pseudohypericin, 159, 160, 440, 441
Pseudouridine, 171, 172
Psychotridine, *425*
Pterostilbenes, *443*
Pteroylglutamates, 404
Punicalagin, 156
Purines, 173, 430, 542
Purpurin, *121*
Putrescine (1,4-diaminobutane) *88*, 131, 305, 397, 389
Pyrantel, 211, *212*, 532
Pyrazaphos, 255, *356*
Pyrene, 96, 97, 98, 271, 337
Pyrethrins, 364, 525
Pyrethroid pesticides, 234, 321
Pyridine, 89, 228, 337
Pyridine, alkyl subs, 89
Pyridine carboxylic acids, 541
Pyridostigmine and degrad products, *358*
Pyridoxal, 139, 544, 545
Pyridoxal phosphate, 139, 545
Pyridoxamine, 139, *544*, 545
Pyridoxamine phosphate, 139, 545
4-Pyridoxic acid, 139, 545
Pyrimethamine, 284, 476, 504, *513*
1-(2'-Pyrimidinyl)piperazine, 522
Pyridoxine, 139, 140, *403*, 544, 545
3-Pyridylacetic acid, 130
Pyriproxyfen enantiomers, *234*
Pyrocatechol, *338*
Pyrogallol, *350*
L-Pyroglutamic acid, 307
Pyropheoboride, 177, *447*
Pyropheophytins, 177, 250, 326
Pyrones, phenyl, 166
3-(1-Pyrrolin-2-yl)pyridine, 392
Pyrrolizidine alkaloids, 423
Pyruvic acid, 541

Quadrigemines, *425*
Quazepam, 215
Quercetin, 151, 152, 159, 410, 411, 413, 415, *416*
Quercetin rhamnodiglucosides, 304
Quercitrin, 159
Quercitrin-3-rutinoside, 411
Quinacrine, 211, *212*
Quinazoline, *294*
Quinfamide and metab, *522*
Quinidine, 284
Quinidine-*N*-oxide, 528
Quinidine and metabolites, *528*
Quinine, 284, 514
2'-Quininone, 514
Quinols, 505
Quinoline, 228, *294*, 380
8-Quinolinethiol metal ion complexes, *94*, 271
Quinolines, nitro subs, 347
Quinolones, 308
Quinone, 230
Quinones, *505*
Quinoxalines, 294
2-(8-Quinoylazo)-5-(dimethylamino)phenyl metal complexes, 343
Quinupristin, *475*
Quizalofop, *108*

Rabdosiin, *450*
Radiamuls, 142, 274
Ranalexin peptides, 455, 456
Ranatuerin peptides, 455, 456
Ranitidine and metabolites, *201*
Rapamycin and metab, *43–69*, 506
RDX, *101*, 102, 350
Reactive Orange dyes, *122*
Reactive Red dyes, *122*
Remacemide and metab, *485*
Resol, phenolic resin, 118
Resorcinol, *120, 338*
Respirodone, 518
Resveratroloside, 448
Resveratrols, 163, *443*, 448
Retinals, 144, *145*, 243, 278, 300, 323
cis/trans Retinal isomers, 241, 301
Retinal oximes, 322
Retinal palmitate isomers, 241
Retinoic acid, *144*
cis/trans Retinoic acids, 145, 243
Retinoids, 145
Retinol, 144, 145, 300, 322

Terazosin and impurities, *512*
Terbutaline sulfate, 311
Terbuthylazine, *369*
Terfenadine, *492*
Terphenyls, 85
Testosterone, *196*, 214, 258, 327, 404
Testosterone acetate, 327
Testosterone enanthate, 197
Testosterone propionate, 197
Tetrabenzo[*de,hi,op,st*]pentacene, *271*
Tetrabenzo[*a,cdf,lm*]perylene, 96
Tetrabromophthalic anhydride, *123*
Tetracaine, 211, *390*
Tetrachloro-*p*-benzoquinone, 359
Tetrachlorodibenzo-*p*-dioxin isomers, 349
Tetrachloro-*p*-hydroquinone, 359
Tetracycline, *468*
Tetradecanoic acid, 90
Tetradifon, *353*
$^9\Delta$-*trans*-Tetrahydrocannabinol, 491
Tetrahydro-β-carboline, 395
Tetrahydro-β-carboline-3-carboxylic acid,
    395
Tetrahydrocortisol, *252*
Tetrahydrocortisone, *252*
Tetrahydrocurcumin, 302
Tetrahydrodeoxycorticosterone, 252
Tetrahydro-11-deoxycortisol, 252
Tetrahydrofolate, 404
N^2-(3,4,5,6-Tetrahydrofuran-2-yl)
    deoxyguanosine, 432
Tetrahydroharman, 426
N^2-(3,4,5,6-Tetrahydro-2*H*-pyran2-yl)
    deoxyguanosine, 432
Tetrahydronaphthalenes, 85
Tetrahydronaphthols, 318
1,2,3,4-Tetrahydro-1-naphthol, *226*
2,2′,4,4′-Tetrahydroxybenzophenone, 119
3,3′,5,5′-Tetraiodothyronine (T$_4$), 305
Tetramethrin, *361*
Tetramethyl-*O*-scutellarein, 413
Tetramethylthiuram mono and polysulfides,
    111, 124
Tetrandrine, 424
Tetraneurin A, *148*
2,2′6,6′-Tetranitrol-4,4′-azoxytoluene, 350
Tetrapeptides, cyclic
Tetraphenylborates (Na) and degrad products
Tetraphenylporphine rare earth metal ion
    complexes, 95
Tetryl, *101*, 102
Texaphyrin complexes of Gd, Lu, 344
Thalidomide, *524*

Thalrugosine, 423
Thapsigargins, *138*
Thapsitranstigan, 138
Thebacone, *200*
Theobromine, *154*, 309, 391, 526
Theogallin, 417
Theophylline, *154*, 186, 309, 391
Theophylline and metabolites, 526
Thiacetylsemide, 552
Thiabendazole, *112*, 357, 371, 477
Thiamine, *139*, 140, 186, 403
1,3-Thiazolidine carboxylic acids, *127*
Thifensulfuron-methyl, 108, 355, *367*
Thimerosal, *205*
Thiocarbamates, 355
Thiocolchicine, *429*
Thiocolchicoside, 426
Thiocyanate, 341
6-Thioguainine, 194
7α-Thiomethylspirolactone, 213
Thiophanate-methyl and metabolites, *103*,
    357
Thioridazine enantiomers, 259
Thiosalicylic acid, *205*
Thiospirolactones, 213
Thiosulfate, 341
Thiourams, *111*
6-Thioxanthine, *194*
10,10′:Threosennoside B, 418
Thromboxanes, *386*
Thymidine, 430, *546*
Thymine, 270, 391, 545
Thymol, *227*
Thyroxine (T$_4$), *176*
Tianeptine, *499*
Tilmicosin, *502*, 503
Timiperone *519*
Timolol and enantiomers, *513*
Timosaponins, 158
Tinidazole, *503*
Tinuvins, *115*, 274, 296
TNB, 101
TNBA, 101
TNT, 101, 102, 232
Tobacco alkaloids, 130
Tobarrol, *150*
Tobramycin, *478*
α-Tocomonoenol, 141
α-Tocopherol, *140*, 322
Tocopherols (α, β, γ, δ), *141*, *241*, 242, 249,
    278, 300, 323, 406
Tocotrienols, 141, *241*, 242
Tofisopan, *203*

Tolazamide, 258
Tolbutamide, *204*, 258
Tolfenamic acid, *463*
Tolualdehyde, 342
m-Toluamide, 358
Toluene, nitro subs, 232, 350
Toluenediisocyanates, 377
Toluenesulfamide, 309, *363*
m-Toluic acid, 358
o-Toluidine, 380
Tolylmercury, 93
Tomatidine, *393*
α-Tomatine, 303, *393*, 394
Toxaphene, *233*
Toxyloxanthone C, 413
Tramadol and metab, *210*
Trazodone and impurities, *310*
Tretinoin T258, 527
Triacylglycerides, 429
Triacylglycerol hydroperoxy oxidation
    products, 247, 429
Triacylglycerol oxidation products, 246
Triacylglycerols, 168, *247*, 249, 279, 304,
    324, 325, 429, 430
Triamcinolone acetonide, 333, *513*
Triaminotoluenes, 232, 350
Triamterene, 527, *550*
Triaprofenic acid enantiomers and isomers,
    255
Triazines, 110, 355, 369
Triazinodiaminopiperidine, *483*
Triazolam, *203*
Triazophos, 352, *353*
Tribenuron methyl, *367*
2,4,6-Tribromophenol, 123, 270
Tribromothyronine, 176
Tributyltin, 94
Tributylphenolethoxylates, 237
Tricaprin, 324
Tricaproin, 324
Tricaprolyn, 324
Tricetin, 151
Trichlorfon, 104
Trichlormethiazide, *550*
Trichloroacetic acid, 124
Trichloroanilines, 84
Trichloroethene, 374
Trichlorophenols, 84
Trichlorosyringol, *346*
Trichothecenes, 162
Triclabendazole and metab, *488*
Triclosan, *382*
Tricyclic antidepressants, 498

Tridecylbenzene, 85
Triethanolamine, 543
Triethylene glycol, 538
Triethyllead, 93
Trifluralin, 355, *368*
Trifluridine, 552
Trifluorobenzoic acids, 334
Trifluoromethyl-*s*-acetaminobenzoic acid,
    345
Trifluoromethylbenzoic acid, 334
Triglycerides, 169, 324, 406, 429
Trigycerols, 325
Trihexylphenydyl hydrochloride, 532
Trihydroxybenzenes, 84
2,4,6-Triiodophenol, 524
Triiodothyronine (T_3) *176*, 305, 446
3,5,5'-Triiodothyronine (rT_3), 176, 305
1,2,3-Triisopropylbenzene, 85
Trilinolein, 247, 324, 384
Trilobolide, *138*
Trimeprazine enantiomers, *257*
Trimethoprim, *473*
3,4,5-Trimethoxycinnamic acid, 153
3,4,5-Trimethoxyphenylacetic acid, 338
2,4,6-Trimethylaniline, 337
2,2,4-Trimethyl-2*H*-chromen, *118*
1,1,3-Trimethyl-3-(4-hydroxyphenyl)-5-
    indanol, *118*
Trimethylimidazolpyridines, 349
Trimethyluric acids, 391, 543
1,3,7-Trimethylxanthine, 154
Trimipramine enantiomers, *257*
Trimyristin, 324
Trinitrophenanthrenes, 233
2,4,6-Trinitrophenol (picric acid), 100, *102*
1,3,5-Trinitroso-1,3,5-hexahydrotriazine
    isomers, *294*
2,4,6-Trinitrotoluene (TNT), 101, 102, 232,
    273, 350
Trinitrotoluenes, 273
Triolein, 247, 324, 384, 428
Triostin A conformers, 283
Tripalmitin, 247, 324
Tripeptide epimers, 181
Triphenyltin, 146, *147*, 295
Triptorelin, 457
Tristearin, 324
Trollichrome, 146, *147*
Tropinin I (cardiac), 457
Trypsin inhibitor, 182
Tryptamine, 133, *158*, 305, 444
Tryptophan, 133, 305, 331, 391, 444, 541
Tryptophol, 525

Tylosin, 477, *503*
Tyramine, 158, 305
Tyrosine, 391, 546
Tyrosine, nitro subs, 546
Tyrosol, *127*

Ubichromenol, 251
Ubiquinols, 141
Ubiquinone, 141, 251
Ultranox, 296
Umbelliferone, 137, *306*
Undeca-2*E*,4*Z*-diene-8,10-doynoic acid
    isobutylamide, 395, *396*
Undecylenic acid, 196, *208*
Uniblue A dye and degradation comp, *122*
Uracil, 270, 546, 551
Uranine, 364
Urea, 382
Uric acid, 132, 213, *394*, 542, 543, 545
Uric acid, alkyl subs, *546*
Uridine, 172, 430, 545, 546
Urocanic acid, *546*
Uroporphyrins, 451
Ursocholic acid, 128
Ursodeoxycholic acid, *128*, 389
Uvinuls, *119*

Valacyclovir, 548
Valeric acid, 90
Valproic acid, 196, 487
Vanillic acid, 86, *90*, 127, 159, 320
Vanillin, 86, *338*, 446
Vanillylamine, 315, *386*
Vanillylmandelic acid, *127*
Vat Blue dyes, 121, 122
Venlafaxine and metab, *206*
Verapamil, *210*, 511, 512
Veratrole, *224*
Vigabatrin, *485*
Vinblastine, 192, *482*
Vinclozolin, 352, 353
Vinclozolin and biotransformation prods,
    112
Vincristine, *482*
Vindolin, *482*
Viniferin, *443*
Violaxanthin, 146, *323*, 451
Virginiamycin, *477*

Vitamin A, *143*, 144
Vitamin A esters, *143*
Vitamin B₆, 544
pro-Vitamin D₂, 144
Vitamin D₂ (ergocalicferol), 144, *243*, 285,
    300, 323, 404, 405
Vitamin D₃ (cholecalciferol), *143*, 144, 243,
    300, 323, 404, 405
Vitamin D₃ and hydroxy metabolites, 278
Vitamin E, 140, 141, *144*
Vitamin E esters, *143*, 144
Vitamin K₁, *142*, *143*, 144
Vitamin K₂, 142
Vitellogenesis inhibiting hormone, 459
Voriconizole, 521

Warfarin, 106, *137*
Wighteone, *152*
Wogonin, *302*, 415, 418

Xanthenone-4-acetic acid and metabolites,
    481
Xanthine dyes, 364
Xanthines, 154, 173, *394*, 430, 542, 543, 545
Xanthohumols, *415*
Xanthophyll pigments, 326
Xanthophylls, 250, 301
Xanthosine, *172*
Xanthotoxin, 137, *306*
Xipamide, *550*
Xylenes, 86, 224
Xylitol, 280, 433
Xylose, 432
m-Xylylenediamine, 376

Yangonin, *163*

Zearalenols, *135*
Zearalenone, *135*, 399
Zeaxanthin, 145, 146, 177, *243*, 301, 326,
    406
Zibeb, 112
Zidovudine (AZT), *496*, 497, 548
Ziram, *111*, 112
Zoalene, *521*
Zopiclone and degradation compounds, *527*